Matrix Algebra From a Statistician's Perspective

David A. Harville

Matrix Algebra From a Statistician's Perspective

Springer

David A. Harville
IBM T. J. Watson Research Center
Mathematical Sciences Department
Yorktown Heights, NY 10598-0218
USA
harville@us.ibm.com

Updated with an Errata 2012

ISBN 978-0-387-78356-7 e-ISBN 978-0-387-22677-4

Library of Congress Control Number: 2008927514

Printed on acid-free paper.

9 8 7 6 5 4 3 2

springer.com

Preface

Matrix algebra plays a very important role in statistics and in many other disciplines. In many areas of statistics, it has become routine to use matrix algebra in the presentation and the derivation or verification of results. One such area is linear statistical models; another is multivariate analysis. In these areas, a knowledge of matrix algebra is needed in applying important concepts, as well as in studying the underlying theory, and is even needed to use various software packages (if they are to be used with confidence and competence).

On many occasions, I have taught graduate-level courses in linear statistical models. Typically, the prerequisites for such courses include an introductory (undergraduate) course in matrix (or linear) algebra. Also typically, the preparation provided by this prerequisite course is not fully adequate. There are several reasons for this. The level of abstraction or generality in the matrix (or linear) algebra course may have been so high that it did not lead to a "working knowledge" of the subject, or, at the other extreme, the course may have emphasized computations at the expense of fundamental concepts. Further, the content of introductory courses on matrix (or linear) algebra varies widely from institution to institution and from instructor to instructor. Topics such as quadratic forms, partitioned matrices, and generalized inverses that play an important role in the study of linear statistical models may be covered inadequately if at all. An additional difficulty is that several years may have elapsed between the completion of the prerequisite course on matrix (or linear) algebra and the beginning of the course on linear statistical models.

This book* is about matrix algebra. A distinguishing feature is that the content, the ordering of topics, and the level of generality are ones that I consider appropriate for someone with an interest in linear statistical models and perhaps also

* The content of the paperback version is essentially the same as that of the earlier, hardcover version. The paperback version differs from the earlier version in that a number of (mostly minor) corrections and alterations have been incorporated. In addition, the typography has been improved—as a side effect, the content and the numbering of the individual pages differ somewhat from those in the earlier version.

for someone with an interest in another area of statistics or in a related discipline. I have tried to keep the presentation at a level that is suitable for anyone who has had an introductory course in matrix (or linear) algebra. In fact, the book is essentially self-contained, and it is hoped that much, if not all, of the material may be comprehensible to a determined reader with relatively little previous exposure to matrix algebra. To make the material readable for as broad an audience as possible, I have avoided the use of abbreviations and acronyms and have sometimes adopted terminology and notation that may seem more meaningful and familiar to the non-mathematician than those favored by mathematicians. Proofs are provided for essentially all of the results in the book. The book includes a number of results and proofs that are not readily available from standard sources and many others that can be found only in relatively high-level books or in journal articles.

The book can be used as a companion to the textbook in a course on linear statistical models or on a related topic—it can be used to supplement whatever results on matrices may be included in the textbook and as a source of proofs. And, it can be used as a primary or supplementary text in a second course on matrices, including a course designed to enhance the preparation of the students for a course or courses on linear statistical models and/or related topics. Above all, it can serve as a convenient reference book for statisticians and for various other professionals.

While the motivation for the writing of the book came from the statistical applications of matrix algebra, the book itself does not include any appreciable discussion of statistical applications. It is assumed that the book is being read because the reader is aware of the applications (or at least of the potential for applications) or because the material is of intrinsic interest—this assumption is consistent with the uses discussed in the previous paragraph. (In any case, I have found that the discussions of applications that are sometimes interjected into treatises on matrix algebra tend to be meaningful only to those who are already knowledgeable about the applications and can be more of a distraction than a help.)

The book has a number of features that combine to set it apart from the more traditional books on matrix algebra—it also differs in significant respects from those matrix-algebra books that share its (statistical) orientation, such as the books of Searle (1982), Graybill (1983), and Basilevsky (1983). The coverage is restricted to real matrices (i.e., matrices whose elements are real numbers)—complex matrices (i.e., matrices whose elements are complex numbers) are typically not encountered in statistical applications, and their exclusion leads to simplifications in terminology, notation, and results. The coverage includes linear spaces, but only linear spaces whose members are (real) matrices—the inclusion of linear spaces facilitates a deeper understanding of various matrix concepts (e.g., rank) that are very relevant in applications to linear statistical models, while the restriction to linear spaces whose members are matrices makes the presentation more appropriate for the intended audience.

The book features an extensive discussion of generalized inverses and makes heavy use of generalized inverses in the discussion of such standard topics as the solution of linear systems and the rank of a matrix. The discussion of eigenvalues and eigenvectors is deferred until the next-to-last chapter of the book—I have found it unnecessary to use results on eigenvalues and eigenvectors in teaching a first course on linear statistical models and, in any case, find it aesthetically displeasing to use results on eigenvalues and eigenvectors to prove more elementary matrix results. And the discussion of linear transformations is deferred until the very last chapter—in more advanced presentations, matrices are regarded as subservient to linear transformations.

The book provides rather extensive coverage of some nonstandard topics that have important applications in statistics and in many other disciplines. These include matrix differentiation (Chapter 15), the vec and vech operators (Chapter 16), the minimization of a second-degree polynomial (in n variables) subject to linear constraints (Chapter 19), and the ranks, determinants, and ordinary and generalized inverses of partitioned matrices and of sums of matrices (Chapter 18 and parts of Chapters 8, 9, 13 16, 17, and 19). An attempt has been made to write the book in such a way that the presentation is coherent and non-redundant but, at the same time, is conducive to using the various parts of the book selectively.

With the obvious exception of certain of their parts, Chapters 12 through 22 (which comprise approximately three-quarters of the book's pages) can be read in arbitrary order. The ordering of Chapters 1 through 11 (both relative to each other and relative to Chapters 12 through 22) is much more critical. Nevertheless, even Chapters 1 through 11 include sections or subsections that are prerequisites for only a small part of the subsequent material. More often than not, the less essential sections or subsections are deferred until the end of the chapter or section.

The book does not address the computational aspects of matrix algebra in any systematic way, however it does include descriptions and discussion of certain computational strategies and covers a number of results that can be useful in dealing with computational issues. Matrix norms are discussed, but only to a limited extent. In particular, the coverage of matrix norms is restricted to those norms that are defined in terms of inner products.

In writing the book, I was influenced to a considerable extent by Halmos's (1958) book on finite-dimensional vector spaces, by Marsaglia and Styan's (1974) paper on ranks, by Henderson and Searle's (1979, 1981b) papers on the vec and vech operators, by Magnus and Neudecker's (1988) book on matrix differential calculus, and by Rao and Mitra's (1971) book on generalized inverses. And I benefited from conversations with Oscar Kempthorne and from reading some notes (on linear systems, determinants, matrices, and quadratic forms) that he had prepared for a course (on linear statistical models) at Iowa State University. I also benefited from reading the first two chapters (pertaining to linear algebra) of notes prepared by Justus F. Seely for a course (on linear statistical models) at Oregon State University.

The book contains many numbered exercises. The exercises are located at (or near) the ends of the chapters and are grouped by section—some exercises may require the use of results covered in previous sections, chapters, or exercises. Many of the exercises consist of verifying results supplementary to those included in the body of the chapter. By breaking some of the more difficult exercises into parts and/or by providing hints, I have attempted to make all of the exercises appropriate for the intended audience. I have prepared solutions to all of the exercises, and it is my intention to make them available on at least a limited basis.*

The origin and historical development of many of the results covered in the book are difficult (if not impossible) to discern, and I have not made any systematic attempt to do so. However, each of Chapters 15 through 21 ends with a short section entitled Bibliographic and Supplementary Notes. Sources that I have drawn on more-or-less directly for an extensive amount of material are identified in that section. Sources that trace the historical development of various ideas, results, and terminology are also identified. And, for certain of the sections in a chapter, some indication may be given of whether that section is a prerequisite for various other sections (or vice versa).

The book is divided into (22) numbered chapters, the chapters into numbered sections, and (in some cases) the sections into lettered subsections. Sections are identified by two numbers (chapter and section within chapter) separated by a decimal point—thus, the third section of Chapter 9 is referred to as Section 9.3. Within a section, a subsection is referred to by letter alone. A subsection in a different chapter or in a different section of the same chapter is referred to by referring to the section and by appending a letter to the section number—for example, in Section 9.3, Subsection b of Section 9.1 is referred to as Section 9.1b. An exercise in a different chapter is referred to by the number obtained by inserting the chapter number (and a decimal point) in front of the exercise number.

Certain of the displayed "equations" are numbered. An equation number comprises two parts (corresponding to section within chapter and equation within section) separated by a decimal point (and is enclosed in parentheses). An equation in a different chapter is referred to by the "number" obtained by starting with the chapter number and appending a decimal point and the equation number—for example, in Chapter 6, result (2.5) of Chapter 5 is referred to as result (5.2.5). For purposes of numbering (and referring to) equations in the exercises, the exercises in each chapter are to be regarded as forming Section E of that chapter.

Preliminary work on the book dates back to the 1982–1983 academic year, which I spent as a visiting professor in the Department of Mathematics at the University of Texas at Austin (on a faculty improvement leave from my then position as a professor of statistics at Iowa State University). The actual writing began after my return to Iowa State and continued on a sporadic basis (as time permitted) until my departure in December 1995. The work was completed during the first part of my tenure in the Mathematical Sciences Department of the IBM Thomas J. Watson Research Center.

I am indebted to Betty Flehinger, Emmanuel Yashchin, Claude Greengard, and Bill Pulleyblank (all of whom are or were managers at the Research Center) for the time and support they provided for this activity. The most valuable of that support (by far) came in the form of the secretarial help of Peggy Cargiulo, who entered

* The solutions were published by Springer-Verlag in 2001 in a volume entitled *Matrix Algebra: Exercises and Solutions* (ISBN 0-387-95318-3).

the last six chapters of the book in LaTeX and was of immense help in getting the manuscript into final form. I am also indebted to Darlene Wicks (of Iowa State University), who entered Chapters 1 through 16 in LaTeX.

I wish to thank John Kimmel, who has been my editor at Springer-Verlag. He has been everything an author could hope for. I also wish to thank Paul Nikolai (formerly of the Air Force Flight Dynamics Laboratory of Wright-Patterson Air Force Base, Ohio) and Dale Zimmerman (of the Department of Statistics and Actuarial Science of the University of Iowa), whose careful reading and marking of the manuscript led to a number of corrections and improvements. These changes were in addition to ones stimulated by the earlier comments of two anonymous reviewers (and by the comments of the editor).

Contents

1

Matrices

Matrix algebra is a branch of mathematics in which numbers are dealt with collectively (as rectangular arrays of numbers called matrices) rather than individually, as in "ordinary" algebra. The term matrix is formally defined in Section 1.1. Section 1.1 also includes an introduction to various basic terminology used in referring to matrices. And some basic matrix operations (scalar multiplication, matrix addition and subtraction, matrix multiplication, and transposition) are defined and their properties discussed in Section 1.2—not all of the properties of the multiplication of ordinary numbers extend to matrix multiplication.

There are many different types of matrices that are sometimes singled out in the literature for special attention. Some of the most basic of these are introduced in Section 1.3. Various other types of matrices are introduced later in the book (as the need arises). In addition, there are many types of matrices that have received considerable attention in the literature but were regarded as too specialized to be considered here.

1.1 Basic Terminology

A rectangular array of real numbers is called a *matrix*. That is, a matrix is a collection of real numbers, $a_{11}, a_{12}, \ldots, a_{1n}, \ldots, a_{m1}, a_{m2}, \ldots, a_{mn}$, arranged in the general form

$$\begin{pmatrix} a_{11} & a_{12} & \ldots & a_{1n} \\ a_{21} & a_{22} & \ldots & a_{2n} \\ \vdots & \vdots & & \vdots \\ a_{m1} & a_{m2} & \ldots & a_{mn} \end{pmatrix}.$$

For example,

$$\begin{pmatrix} 2 & -3 & 0 \\ 1 & 4 & 2 \end{pmatrix}$$

is a matrix.

Our definition of a matrix is less general than that sometimes adopted. The more general definition applies to any rectangular arrangement of elements, called *scalars*, of a field. (A field is a set, whose elements can be added or multiplied in accordance with rules that have certain properties.) For example, the more general definition covers *complex matrices*, which are rectangular arrays of complex numbers. Here, unless otherwise indicated, the use of the word matrix is confined to *real matrices*, that is, to rectangular arrays of real numbers. Also, scalar is used synonymously with real number.

A matrix having m rows and n columns is referred to as an $m \times n$ matrix, and m and n are called the *dimensions* of the matrix. The scalar located at the intersection of the ith row and the jth column of a matrix is called the ijth *element* or *entry* of the matrix.

Boldface capital letters (e.g., $\mathbf{A}$) are used to represent matrices. The notation $\mathbf{A} = \{a_{ij}\}$ is used in introducing a matrix, the ijth element of which is a_{ij}.

Two matrices $\mathbf{A}$ and $\mathbf{B}$ of the same dimensions are said to be *equal* if each element of $\mathbf{A}$ equals the corresponding element of $\mathbf{B}$, in which case we write $\mathbf{A} = \mathbf{B}$ (and are said to be unequal otherwise, i.e., if some element of $\mathbf{A}$ differs from the corresponding element of $\mathbf{B}$, in which case we write $\mathbf{A} \neq \mathbf{B}$).

1.2 Basic Operations

a. Scalar multiplication

Scalar multiplication is defined for an arbitrary scalar k and an arbitrary $m \times n$ matrix $\mathbf{A} = \{a_{ij}\}$. The *product* of k and $\mathbf{A}$ is written as $k\mathbf{A}$ (or, less commonly, as $\mathbf{A}k$) and is defined to be the $m \times n$ matrix whose ijth element is ka_{ij}. For example,

$$4\begin{pmatrix} 2 & -3 & 0 \\ 1 & 4 & 2 \end{pmatrix} = \begin{pmatrix} 8 & -12 & 0 \\ 4 & 16 & 8 \end{pmatrix}.$$

The matrix $k\mathbf{A}$ is said to be a *scalar multiple* of $\mathbf{A}$.

For any matrix $\mathbf{A}$,

$$1\mathbf{A} = \mathbf{A}.$$

And, for any scalars c and k and any matrix $\mathbf{A}$,

$$c(k\mathbf{A}) = (ck)\mathbf{A} = (kc)\mathbf{A} = k(c\mathbf{A}). \tag{2.1}$$

It is customary to refer to the scalar product $(-1)\mathbf{A}$ of -1 and $\mathbf{A}$ as the *negative* of $\mathbf{A}$ and to abbreviate $(-1)\mathbf{A}$ to $-\mathbf{A}$.

b. Matrix addition and subtraction

Matrix addition and subtraction are defined for any two matrices $\mathbf{A} = \{a_{ij}\}$ and $\mathbf{B} = \{b_{ij}\}$ that have the same number of rows, say m, and the same number of columns, say n. The *sum* of the two $m \times n$ matrices $\mathbf{A}$ and $\mathbf{B}$ is denoted by the symbol $\mathbf{A}+\mathbf{B}$ and is defined to be the $m \times n$ matrix whose ijth element is $a_{ij} + b_{ij}$. For example,

$$\begin{pmatrix} 2 & -3 & 0 \\ 1 & 4 & 2 \end{pmatrix} + \begin{pmatrix} 6 & 1 & -3 \\ -1 & 0 & 4 \end{pmatrix} = \begin{pmatrix} 8 & -2 & -3 \\ 0 & 4 & 6 \end{pmatrix}.$$

Matrix addition is commutative, that is,

$$\mathbf{A} + \mathbf{B} = \mathbf{B} + \mathbf{A}. \tag{2.2}$$

Matrix addition is also associative, that is, taking $\mathbf{C}$ to be a third $m \times n$ matrix,

$$\mathbf{A} + (\mathbf{B} + \mathbf{C}) = (\mathbf{A} + \mathbf{B}) + \mathbf{C}. \tag{2.3}$$

The symbol $\mathbf{A}+\mathbf{B}+\mathbf{C}$ is used to represent the common value of the left and right sides of equality (2.3), and this value is referred to as the *sum* of $\mathbf{A}$, $\mathbf{B}$, and $\mathbf{C}$. This notation and terminology extend in an obvious way to any finite number of $m \times n$ matrices.

For any scalar c, we have that

$$c(\mathbf{A} + \mathbf{B}) = c\mathbf{A} + c\mathbf{B}, \tag{2.4}$$

and, for any scalars c and k, we have that

$$(c + k)\mathbf{A} = c\mathbf{A} + k\mathbf{A}. \tag{2.5}$$

Let us write $\mathbf{A} - \mathbf{B}$ for the sum $\mathbf{A} + (-\mathbf{B})$ or equivalently for the $m \times n$ matrix whose ijth element is $a_{ij} - b_{ij}$ and refer to this matrix as the *difference* between $\mathbf{A}$ and $\mathbf{B}$.

Matrices having the same number of rows and the same number of columns are said to be *conformal* for addition (or subtraction).

c. Matrix multiplication

Let $\mathbf{A} = \{a_{ij}\}$ represent an $m \times n$ matrix, and $\mathbf{B} = \{b_{ij}\}$ a $p \times q$ matrix. When $n = p$, that is, when $\mathbf{A}$ has the same number of columns as $\mathbf{B}$ has rows, the *matrix product* $\mathbf{AB}$ is defined to be the $m \times q$ matrix whose ijth element is

$$\sum_{k=1}^{n} a_{ik}b_{kj} = a_{i1}b_{1j} + a_{i2}b_{2j} + \cdots + a_{in}b_{nj}.$$

For example,

$$\begin{aligned} &\begin{pmatrix} 2 & -3 & 0 \\ 1 & 4 & 2 \end{pmatrix} \begin{pmatrix} 3 & 4 \\ -2 & 0 \\ 1 & 2 \end{pmatrix} \\ &\quad = \begin{pmatrix} 2(3) + (-3)(-2) + 0(1) & 2(4) + (-3)(0) + 0(2) \\ 1(3) + 4(-2) + 2(1) & 1(4) + 4(0) + 2(2) \end{pmatrix} \\ &\quad = \begin{pmatrix} 12 & 8 \\ -3 & 8 \end{pmatrix}. \end{aligned}$$

The formation of the matrix product $\mathbf{AB}$ is referred to as the *premultiplication* of $\mathbf{B}$ by $\mathbf{A}$ or the *postmultiplication* of $\mathbf{A}$ by $\mathbf{B}$. When $n \neq p$, the matrix product $\mathbf{AB}$ is undefined.

Matrix multiplication is associative. Thus, introducing a third matrix $\mathbf{C}$,

$$\mathbf{A}(\mathbf{BC}) = (\mathbf{AB})\mathbf{C}, \tag{2.6}$$

provided that $p = n$ and that $\mathbf{C}$ has q rows so that the matrix multiplications required to form the left and right sides of the equality are defined. The symbol $\mathbf{ABC}$ is used to represent the common value of the left and right sides of equality (2.6), and this value is referred to as the *product* of $\mathbf{A}$, $\mathbf{B}$, and $\mathbf{C}$. This notation and terminology extend in an obvious way to any finite number of matrices.

Matrix multiplication is distributive with respect to addition, that is,

$$\mathbf{A}(\mathbf{B}+\mathbf{C}) = \mathbf{AB} + \mathbf{AC}, \tag{2.7}$$

$$(\mathbf{A}+\mathbf{B})\mathbf{C} = \mathbf{AC} + \mathbf{BC}, \tag{2.8}$$

where, in each equality, it is assumed that the dimensions of $\mathbf{A}$, $\mathbf{B}$, and $\mathbf{C}$ are such that all multiplications and additions are defined. Results (2.7) and (2.8) extend in an obvious way to the postmultiplication or premultiplication of a matrix $\mathbf{A}$ or $\mathbf{C}$ by the sum of any finite number of matrices.

In general, matrix multiplication is not commutative. That is, $\mathbf{AB}$ is not necessarily identical to $\mathbf{BA}$. In fact, when $n = p$ but $m \neq q$ or when $m = q$ but $n \neq p$, one of the matrix products $\mathbf{AB}$ and $\mathbf{BA}$ is defined, while the other is undefined. When $n = p$ and $m = q$, $\mathbf{AB}$ and $\mathbf{BA}$ are both defined, but the dimensions $(m \times m)$ of $\mathbf{AB}$ are the same as those of $\mathbf{BA}$ only if $m = n$. Even when $n = p = m = q$, in which case $\mathbf{A}$ and $\mathbf{B}$ are both $n \times n$ matrices and the two matrix products $\mathbf{AB}$ and $\mathbf{BA}$ are both defined and have the same dimensions, it is not necessarily true that $\mathbf{AB} = \mathbf{BA}$. For example,

$$\begin{pmatrix} 1 & 0 \\ 0 & 2 \end{pmatrix}\begin{pmatrix} 0 & 1 \\ 1 & 0 \end{pmatrix} = \begin{pmatrix} 0 & 1 \\ 2 & 0 \end{pmatrix} \neq \begin{pmatrix} 0 & 2 \\ 1 & 0 \end{pmatrix} = \begin{pmatrix} 0 & 1 \\ 1 & 0 \end{pmatrix}\begin{pmatrix} 1 & 0 \\ 0 & 2 \end{pmatrix}.$$

Two $n \times n$ matrices $\mathbf{A}$ and $\mathbf{B}$ are said to *commute* if $\mathbf{AB} = \mathbf{BA}$. More generally, a collection of $n \times n$ matrices $\mathbf{A}_1, \mathbf{A}_2, \ldots, \mathbf{A}_k$ is said to *commute in pairs* if $\mathbf{A}_i\mathbf{A}_j = \mathbf{A}_j\mathbf{A}_i$ for $j > i = 1, 2, \ldots, k$.

For any scalar c, $m \times n$ matrix $\mathbf{A}$, and $n \times p$ matrix $\mathbf{B}$, it is customary to write $c\mathbf{AB}$ for the product $c(\mathbf{AB})$ of c and the matrix product $\mathbf{AB}$. Note that

$$c\mathbf{AB} = (c\mathbf{A})\mathbf{B} = \mathbf{A}(c\mathbf{B}) \tag{2.9}$$

(as is evident from the very definitions of scalar and matrix multiplication). More generally, for any scalar c and say k matrices $\mathbf{A}_1, \mathbf{A}_2, \ldots, \mathbf{A}_k$ (for which the product $\mathbf{A}_1\mathbf{A}_2\cdots\mathbf{A}_k$ is defined) $c(\mathbf{A}_1\mathbf{A}_2\cdots\mathbf{A}_k)$ is typically abbreviated to $c\mathbf{A}_1\mathbf{A}_2\cdots\mathbf{A}_k$, and we have that

$$\begin{aligned} &c\mathbf{A}_1\mathbf{A}_2\cdots\mathbf{A}_k \\ &\quad = (c\mathbf{A}_1)\mathbf{A}_2\cdots\mathbf{A}_k = \mathbf{A}_1(c\mathbf{A}_2)\mathbf{A}_3\cdots\mathbf{A}_k = \mathbf{A}_1\cdots\mathbf{A}_{k-1}(c\mathbf{A}_k). \end{aligned} \tag{2.10}$$

d. Transposition

The transpose of an $m \times n$ matrix $\mathbf{A}$, to be denoted by the symbol $\mathbf{A}'$, is the $n \times m$ matrix whose ijth element is the jith element of $\mathbf{A}$. For example,

$$\begin{pmatrix} 2 & -3 & 0 \\ 1 & 4 & 2 \end{pmatrix}' = \begin{pmatrix} 2 & 1 \\ -3 & 4 \\ 0 & 2 \end{pmatrix}.$$

Observe that the transpose of a matrix can be formed by rewriting its rows as columns (or its columns as rows).

For any matrix $\mathbf{A}$,

$$(\mathbf{A}')' = \mathbf{A} \tag{2.11}$$

and, for any two matrices $\mathbf{A}$ and $\mathbf{B}$ (that are conformal for addition),

$$(\mathbf{A}+\mathbf{B})' = \mathbf{A}' + \mathbf{B}', \tag{2.12}$$

as is easily verified.

Further, for any two matrices $\mathbf{A}$ and $\mathbf{B}$ (for which the product $\mathbf{AB}$ is defined),

$$(\mathbf{AB})' = \mathbf{B}'\mathbf{A}'. \tag{2.13}$$

To see this, observe that the ijth element of $(\mathbf{AB})'$ equals the jith element of $\mathbf{AB}$, which in turn equals

$$\sum_k a_{jk} b_{ki} = \sum_k b_{ki} a_{jk}. \tag{2.14}$$

Since b_{ki} is the ikth element of $\mathbf{B}'$ and a_{jk} the kjth element of $\mathbf{A}'$, the ijth element of $\mathbf{B}'\mathbf{A}'$, like that of $(\mathbf{AB})'$, equals expression (2.14).

By repeated application of result (2.13), we obtain the more general result that

$$(\mathbf{A}_1\mathbf{A}_2 \cdots \mathbf{A}_k)' = \mathbf{A}_k' \cdots \mathbf{A}_2'\mathbf{A}_1', \tag{2.15}$$

for any k matrices $\mathbf{A}_1, \mathbf{A}_2, \ldots, \mathbf{A}_k$ of appropriate dimensions.

1.3 Some Basic Types of Matrices

a. Square matrices

A matrix having the same number of rows as columns is called a *square matrix*. An $n \times n$ square matrix is said to be of *order n*.

Those n elements $a_{11}, a_{22}, \ldots, a_{nn}$ of an $n \times n$ square matrix

$$\mathbf{A} = \begin{pmatrix} a_{11} & a_{12} & \ldots & a_{1n} \\ a_{21} & a_{22} & \ldots & a_{2n} \\ \vdots & \vdots & \ddots & \vdots \\ a_{n1} & a_{n2} & \ldots & a_{nn} \end{pmatrix}$$

that fall on an imaginary diagonal line extending from the upper left to the lower right corners of the matrix are known as the (first, second, etc.) *diagonal elements*. The diagonal line itself is sometimes referred to simply as the *diagonal*. Those elements of a square matrix other than the diagonal elements (i.e., those elements that lie above and to the right or below and to the left of the diagonal) are called the *off-diagonal elements*.

Note that the product $\mathbf{AA}$ of a matrix $\mathbf{A}$ and itself is defined if and only if $\mathbf{A}$ is square. For a square matrix $\mathbf{A}$, the symbol $\mathbf{A}^2$ is used to represent $\mathbf{AA}$, $\mathbf{A}^3$ to represent $\mathbf{AA}^2 = \mathbf{AAA}$, and, more generally, $\mathbf{A}^k$ to represent $\mathbf{AA}^{k-1} = \mathbf{AAA}^{k-2} = \cdots = \mathbf{AAA} \cdots \mathbf{A}$ $(k = 2, 3, \ldots)$.

b. Symmetric matrices

A matrix $\mathbf{A}$ is said to be *symmetric* if $\mathbf{A}' = \mathbf{A}$. Thus, a symmetric matrix is a square matrix whose ijth element equals its jith element. For example,

$$\begin{pmatrix} 3 & 1 & -4 \\ 1 & 0 & 2 \\ -4 & 2 & -5 \end{pmatrix}$$

is a symmetric matrix.

c. Diagonal matrices

A *diagonal matrix* is a a square matrix whose off-diagonal elements are all equal to 0, that is, a matrix of the general form

$$\begin{pmatrix} d_1 & 0 & \ldots & 0 \\ 0 & d_2 & & 0 \\ \vdots & & \ddots & \\ 0 & 0 & & d_n \end{pmatrix},$$

where $d_1, d_2, \ldots, d_n$ are scalars. The notation $\mathbf{D} = \{d_i\}$ is sometimes used to introduce a diagonal matrix, the ith diagonal element of which is d_i. Also, we sometimes write $\text{diag}(d_1, d_2, \ldots, d_n)$ for such a matrix.

Note that, for any $m \times n$ matrix $\mathbf{A} = \{a_{ij}\}$ and any diagonal matrix $\mathbf{D} = \{d_i\}$ (of order m), the ijth element of the matrix product $\mathbf{DA}$ equals $d_i a_{ij}$. Thus, the effect of premultiplying an $m \times n$ matrix $\mathbf{A}$ by a diagonal matrix $\mathbf{D}$ is simply to multiply each element of the ith row of $\mathbf{A}$ by the ith diagonal element of $\mathbf{D}$ $(i = 1, 2, \ldots, m)$. Similarly, the effect of postmultiplying $\mathbf{A}$ by a diagonal matrix $\mathbf{D}$ (of order n) is to multiply each element of the jth column of $\mathbf{A}$ by the jth diagonal element of $\mathbf{D}$ $(j = 1, 2, \ldots, n)$.

Note also that the effect of multiplying an $m \times n$ matrix $\mathbf{A}$ by a scalar k is the same as that of premultiplying or postmultiplying $\mathbf{A}$ by the $m \times m$ or $n \times n$ matrix $\text{diag}(k, k, \ldots, k)$.

d. Identity matrices

A diagonal matrix

$$\text{diag}(1, 1, \ldots, 1) = \begin{pmatrix} 1 & 0 & \ldots & 0 \\ 0 & 1 & & 0 \\ \vdots & & \ddots & \\ 0 & 0 & & 1 \end{pmatrix}$$

whose diagonal elements are all equal to 1 is called an *identity matrix*. The symbol $\mathbf{I}_n$ is used to represent an identity matrix of order n. In cases where the order is clear from the context, we may simply write $\mathbf{I}$ for an identity matrix. Clearly, for an arbitrary matrix $\mathbf{A}$,

$$\mathbf{IA} = \mathbf{AI} = \mathbf{A}. \tag{3.1}$$

e. Matrices of ones

The symbol $\mathbf{J}_{mn}$ is used to represent an $m \times n$ matrix whose elements are all equal to 1. Or, when the dimensions are clear from the context or are to be left unspecified, we may simply write $\mathbf{J}$ for such a matrix. Thus,

$$\mathbf{J} = \begin{pmatrix} 1 & 1 & \ldots & 1 \\ 1 & 1 & \ldots & 1 \\ \vdots & \vdots & & \vdots \\ 1 & 1 & \ldots & 1 \end{pmatrix}.$$

Also, $\mathbf{J}_n$ is sometimes written for $\mathbf{J}_{nn}$.

Note that

$$\mathbf{J}_{mn}\mathbf{J}_{np} = n\mathbf{J}_{mp}. \tag{3.2}$$

f. Null matrices

A matrix whose elements are all equal to 0 is called a *null matrix*. A null matrix is denoted by the symbol $\mathbf{0}$. Thus,

$$\mathbf{0} = \begin{pmatrix} 0 & 0 & \ldots & 0 \\ 0 & 0 & \ldots & 0 \\ \vdots & \vdots & & \vdots \\ 0 & 0 & \ldots & 0 \end{pmatrix}.$$

Letting $\mathbf{A}$ represent an arbitrary matrix and k an arbitrary scalar, we find that

$$0\mathbf{A} = \mathbf{0}, \qquad k\mathbf{0} = \mathbf{0},$$
$$\mathbf{A} + \mathbf{0} = \mathbf{0} + \mathbf{A} = \mathbf{A},$$
$$\mathbf{A} - \mathbf{A} = \mathbf{0},$$
$$\mathbf{0A} = \mathbf{0}, \qquad \mathbf{A0} = \mathbf{0}$$

(where the dimensions of each null matrix $\mathbf{0}$ can be ascertained from the context).

g. Triangular matrices

If all of the elements of a square matrix that are located below and to the left of the diagonal are 0, the matrix is called an *upper triangular matrix.* Thus, the general form of an upper triangular matrix is

$$\begin{pmatrix} a_{11} & a_{12} & a_{13} & \dots & a_{1n} \\ 0 & a_{22} & a_{23} & \dots & a_{2n} \\ 0 & 0 & a_{33} & \dots & a_{3n} \\ \vdots & \vdots & & \ddots & \vdots \\ 0 & 0 & 0 & & a_{nn} \end{pmatrix}.$$

Similarly, if all of the elements that are located above and to the right of the diagonal are 0, the matrix is called a *lower triangular matrix.* More formally, an $n \times n$ matrix $\mathbf{A} = \{a_{ij}\}$ is upper triangular if $a_{ij} = 0$ for $j < i = 1, \dots, n$, and is lower triangular if $a_{ij} = 0$ for $j > i = 1, \dots, n$. By a *triangular matrix,* we mean a (square) matrix that is upper triangular or lower triangular. Observe that a (square) matrix is a diagonal matrix if and only if it is both upper and lower triangular.

The transpose of an upper triangular matrix is a lower triangular matrix—and vice versa—as is easily verified. Further, the sum of two upper triangular matrices (of the same order) is upper triangular and the sum of two lower triangular matrices is lower triangular, as is also easily verified.

Some basic properties of the product of two upper or two lower triangular matrices are described in the following lemma.

Lemma 1.3.1. Let $\mathbf{A} = \{a_{ij}\}$ and $\mathbf{B} = \{b_{ij}\}$ represent $n \times n$ matrices. If $\mathbf{A}$ and $\mathbf{B}$ are both upper triangular, then their product $\mathbf{AB}$ is upper triangular. Similarly, if $\mathbf{A}$ and $\mathbf{B}$ are both lower triangular, then $\mathbf{AB}$ is lower triangular. Further, if $\mathbf{A}$ and $\mathbf{B}$ are either both upper triangular or both lower triangular, then the ith diagonal element of $\mathbf{AB}$ is the product $a_{ii}b_{ii}$ of the ith diagonal elements a_{ii} and b_{ii} of $\mathbf{A}$ and $\mathbf{B}$ $(i = 1, \dots, n)$.

Proof. By definition, the ijth element of $\mathbf{AB}$ is $\sum_{k=1}^{n} a_{ik}b_{kj}$. Suppose that $\mathbf{A}$ and $\mathbf{B}$ are both upper triangular and hence that $a_{ik} = 0$, for $k < i$, and $b_{kj} = 0$, for $k > j$. If $j \le i$, then clearly $b_{kj} = 0$ for $k > i$. Thus, for $j \le i$,

$$\sum_{k=1}^{n} a_{ik}b_{kj} = \sum_{k=1}^{i-1} (0)b_{kj} + a_{ii}b_{ij} + \sum_{k=i+1}^{n} a_{ik}(0) = a_{ii}b_{ij}.$$

In particular,

$$\sum_{k=1}^{n} a_{ik}b_{ki} = a_{ii}b_{ii}$$

and, for $j < i$,

$$\sum_{k=1}^{n} a_{ik}b_{kj} = a_{ii}(0) = 0.$$

We conclude that the ith diagonal element of $\mathbf{AB}$ is $a_{ii}b_{ii}$ $(i = 1, \ldots, n)$ and that $\mathbf{AB}$ is upper triangular.

Suppose now that $\mathbf{A}$ and $\mathbf{B}$ are both lower triangular, rather than upper triangular. Then, since $\mathbf{B}'$ and $\mathbf{A}'$ are both upper triangular with ith diagonal elements b_{ii} and a_{ii}, respectively, $\mathbf{B}'\mathbf{A}'$ is upper triangular with ith diagonal element $a_{ii}b_{ii}$, implying that $\mathbf{AB} = (\mathbf{B}'\mathbf{A}')'$ is lower triangular with ith diagonal element $a_{ii}b_{ii}$. Q.E.D.

An (upper or lower) triangular matrix is called a *unit (upper or lower) triangular matrix* if all of its diagonal elements equal one. In the special case of unit triangular matrices, Lemma 1.3.1 reduces to the following result.

Corollary 1.3.2. The product of two unit upper triangular matrices (of the same order) is unit upper triangular, and the product of two unit lower triangular matrices is unit lower triangular.

h. Row and column vectors

A matrix that has only one column, that is, a matrix of the form

$$\begin{pmatrix} a_1 \\ a_2 \\ \vdots \\ a_m \end{pmatrix}$$

is called a *column vector*. Similarly, a matrix that has only one row is called a *row vector*. A row or column vector having m elements may be referred to as an m-dimensional row or column vector. Clearly, the transpose of an m-dimensional column vector is an m-dimensional row vector, and vice versa.

Lowercase boldface letters (e.g., $\mathbf{a}$) are customarily used to represent column vectors. This notation is helpful in distinguishing column vectors from matrices that may have more than one column. No further notation is introduced for row vectors. Instead, row vectors are represented as the transposes of column vectors. For example, $\mathbf{a}'$ represents the row vector whose transpose is the column vector $\mathbf{a}$. The notation $\mathbf{a} = \{a_i\}$ or $\mathbf{a}' = \{a_i\}$ is used in introducing a column or row vector whose ith element is a_i.

The symbol $\mathbf{1}_m$ is generally used in place of $\mathbf{J}_{m1}$, that is, $\mathbf{1}_m$ is used to represent an m-dimensional column vector whose elements are all equal to 1. If its dimension is clear from the context or is to be left unspecified, we may simply write $\mathbf{1}$ for such a vector. Note that $\mathbf{J}_{1n}$, or equivalently an n-dimensional row vector whose elements are all equal to 1, may be written as $\mathbf{1}_n'$ or simply as $\mathbf{1}'$.

Observe that

$$\mathbf{1}_m\mathbf{1}_n' = \mathbf{J}_{mn}. \tag{3.3}$$

Null row or column vectors (i.e., null matrices having one row or one column) are sometimes referred to simply as null vectors, and (like null matrices in general) are represented by the symbol $\mathbf{0}$.

i. 1 × 1 matrices

Note that there is a distinction between a scalar, say k, and the 1×1 matrix (k) whose sole element is k. For any column vector $\mathbf{a}$,

$$\mathbf{a}(k) = k\mathbf{a}, \qquad (k)\mathbf{a}' = k\mathbf{a}'.$$

However, in general, the product $k\mathbf{A}$ obtained by multiplying an $m \times n$ matrix $\mathbf{A}$ by the scalar k is not the same as the product $(k)\mathbf{A}$ or $\mathbf{A}(k)$ of the two matrices (k) and $\mathbf{A}$. In fact, $(k)\mathbf{A}$ or $\mathbf{A}(k)$ is undefined unless $m = 1$ or $n = 1$, while $k\mathbf{A}$ is defined for all m and n.

For convenience, we sometimes use the same symbol to represent both a 1×1 matrix and the element of that matrix. Which use is intended is discernible from the context.

Exercises

Section 1.2

1. Show that, for any matrices $\mathbf{A}$, $\mathbf{B}$, and $\mathbf{C}$ (of the same dimensions),

$$(\mathbf{A} + \mathbf{B}) + \mathbf{C} = (\mathbf{C} + \mathbf{A}) + \mathbf{B}.$$

2. Let $\mathbf{A}$ represent an $m \times n$ matrix and $\mathbf{B}$ an $n \times p$ matrix, and let c and k represent arbitrary scalars. Using results (2.9) and (2.1) (or other means), show that

$$(c\mathbf{A})(k\mathbf{B}) = (ck)\mathbf{AB}.$$

3. (a) Verify result (2.6) (on the associativeness of matrix multiplication); that is, show that, for any $m \times n$ matrix $\mathbf{A} = \{a_{ij}\}$, $n \times q$ matrix $\mathbf{B} = \{b_{jk}\}$, and $q \times r$ matrix $\mathbf{C} = \{c_{ks}\}$, $\mathbf{A}(\mathbf{BC}) = (\mathbf{AB})\mathbf{C}$.

   (b) Verify result (2.7) (on the distributiveness with respect to addition of matrix multiplication); that is, show that, for any $m \times n$ matrix $\mathbf{A} = \{a_{ij}\}$ and $n \times q$ matrices $\mathbf{B} = \{b_{jk}\}$ and $\mathbf{C} = \{c_{jk}\}$, $\mathbf{A}(\mathbf{B} + \mathbf{C}) = \mathbf{AB} + \mathbf{AC}$.

4. Let $\mathbf{A} = \{a_{ij}\}$ represent an $m \times n$ matrix and $\mathbf{B} = \{b_{ij}\}$ a $p \times m$ matrix.

   (a) Let $\mathbf{x} = \{x_i\}$ represent an n-dimensional column vector. Show that the ith element of the p-dimensional column vector $\mathbf{BAx}$ is

$$\sum_{j=1}^{m} b_{ij} \sum_{k=1}^{n} a_{jk} x_k \tag{E.1}$$

(b) Let $\mathbf{X} = \{x_{ij}\}$ represent an $n \times q$ matrix. Generalize formula (E.1) by expressing the irth element of the $p \times q$ matrix $\mathbf{BAX}$ in terms of the elements of $\mathbf{A}$, $\mathbf{B}$, and $\mathbf{X}$.

(c) Let $\mathbf{x} = \{x_i\}$ represent an n-dimensional column vector and $\mathbf{C} = \{c_{ij}\}$ a $q \times p$ matrix. Generalize formula (E.1) by expressing the ith element of the q-dimensional column vector $\mathbf{CBAx}$ in terms of the elements of $\mathbf{A}$, $\mathbf{B}$, $\mathbf{C}$, and $\mathbf{x}$.

(d) Let $\mathbf{y} = \{y_i\}$ represent a p-dimensional column vector. Express the ith element of the n-dimensional row vector $\mathbf{y}'\mathbf{BA}$ in terms of the elements of $\mathbf{A}$, $\mathbf{B}$, and $\mathbf{y}$.

Section 1.3

5. Let $\mathbf{A}$ and $\mathbf{B}$ represent $n \times n$ matrices. Show that

$$(\mathbf{A} + \mathbf{B})(\mathbf{A} - \mathbf{B}) = \mathbf{A}^2 - \mathbf{B}^2$$

if and only if $\mathbf{A}$ and $\mathbf{B}$ commute.

6. (a) Show that the product $\mathbf{AB}$ of two $n \times n$ symmetric matrices $\mathbf{A}$ and $\mathbf{B}$ is itself symmetric if and only if $\mathbf{A}$ and $\mathbf{B}$ commute.

   (b) Give an example of two symmetric matrices (of the same order) whose product is not symmetric.

7. Verify (a) that the transpose of an upper triangular matrix is lower triangular and (b) that the sum of two upper triangular matrices (of the same order) is upper triangular.

8. Let $\mathbf{A} = \{a_{ij}\}$ represent an $n \times n$ upper triangular matrix, and suppose that the diagonal elements of $\mathbf{A}$ equal zero (i.e., that $a_{11} = a_{22} = \cdots = a_{nn} = 0$). Further, let p represent an arbitrary positive integer.

   (a) Show that, for $i = 1, \ldots, n$ and $j = 1, \ldots, \min(n, i + p - 1)$, the ijth element of $\mathbf{A}^p$ equals zero.

   (b) Show that, for $i \geq n - p + 1$, the ith row of $\mathbf{A}^p$ is null.

   (c) Show that, for $p \geq n$, $\mathbf{A}^p = \mathbf{0}$.

2

Submatrices and Partitioned Matrices

Two very important (and closely related) concepts are introduced in this chapter: that of a submatrix and that of a partitioned matrix. These concepts arise very naturally in statistics (especially in multivariate analysis and linear models) and in many other disciplines that involve probabilistic ideas. And results on submatrices and partitioned matrices, which can be found in Chapters 8, 9, 13, and 14 (and other of the subsequent chapters), have proved to be very useful. In particular, such results are almost indispensable in work involving the multivariate normal distribution—refer, for example, to Searle (1971, sec. 2.4f).

2.1 Some Terminology and Basic Results

A *submatrix* of a matrix $\mathbf{A}$ is a matrix that can be obtained by striking out rows and/or columns of $\mathbf{A}$. For example, if we strike out the second row of the matrix

$$\begin{pmatrix} 2 & 4 & 3 & 6 \\ 1 & 5 & 7 & 9 \\ -1 & 0 & 2 & 2 \end{pmatrix},$$

we obtain the 2×4 submatrix

$$\begin{pmatrix} 2 & 4 & 3 & 6 \\ -1 & 0 & 2 & 2 \end{pmatrix}.$$

Alternatively, if we strike out the first and third columns, we obtain the 3×2 submatrix

$$\begin{pmatrix} 4 & 6 \\ 5 & 9 \\ 0 & 2 \end{pmatrix};$$

or, if we strike out the second row and the first and third columns, we obtain the 2×2 submatrix

$$\begin{pmatrix} 4 & 6 \\ 0 & 2 \end{pmatrix}.$$

Note that any matrix is a submatrix of itself; it is the submatrix obtained by striking out zero rows and zero columns.

Submatrices of a row or column vector, that is, of a matrix having one row or column, are themselves row or column vectors and are customarily referred to as *subvectors*.

Let $\mathbf{A}_*$ represent an $r \times s$ submatrix of an $m \times n$ matrix $\mathbf{A}$ obtained by striking out the $i_1, \ldots, i_{m-r}$th rows and $j_1, \ldots, j_{n-s}$th columns (of $\mathbf{A}$), and let $\mathbf{B}_*$ represent the $s \times r$ submatrix of $\mathbf{A}'$ obtained by striking out the $j_1, \ldots, j_{n-s}$th rows and $i_1, \ldots, i_{m-r}$th columns (of $\mathbf{A}'$). Then,

$$\mathbf{B}_* = \mathbf{A}'_*, \tag{1.1}$$

as is easily verified.

A submatrix of an $n \times n$ matrix is called a *principal submatrix* if it can be obtained by striking out the same rows as columns (so that the ith row is struck out whenever the ith column is struck out, and vice versa). The $r \times r$ (principal) submatrix of an $n \times n$ matrix obtained by striking out its last $n - r$ rows and columns is referred to as a *leading principal submatrix* ($r = 1, \ldots, n$). A principal submatrix of a symmetric matrix is symmetric, a principal submatrix of a diagonal matrix is diagonal, and a principal submatrix of an upper or lower triangular matrix is respectively upper or lower triangular, as is easily verified.

A matrix can be divided or partitioned into submatrices by drawing horizontal or vertical lines between various of its rows or columns, in which case the matrix is called a *partitioned matrix* and the submatrices are sometimes referred to as *blocks* (as in blocks of elements). For example,

$$\left(\begin{array}{r|rr|r} 2 & 4 & 3 & 6 \\ 1 & 5 & 7 & 9 \\ -1 & 0 & 2 & 2 \end{array}\right), \quad \left(\begin{array}{rrrr} 2 & 4 & 3 & 6 \\ 1 & 5 & 7 & 9 \\ \hline -1 & 0 & 2 & 2 \end{array}\right), \quad \left(\begin{array}{r|rr|r} 2 & 4 & 3 & 6 \\ 1 & 5 & 7 & 9 \\ \hline -1 & 0 & 2 & 2 \end{array}\right)$$

are various partitionings of the same matrix.

In effect, a partitioned $m \times n$ matrix is an $m \times n$ matrix $\mathbf{A} = \{a_{ij}\}$ that has been reexpressed in the general form

$$\begin{pmatrix} \mathbf{A}_{11} & \mathbf{A}_{12} & \ldots & \mathbf{A}_{1c} \\ \mathbf{A}_{21} & \mathbf{A}_{22} & \ldots & \mathbf{A}_{2c} \\ \vdots & \vdots & & \vdots \\ \mathbf{A}_{r1} & \mathbf{A}_{r2} & \ldots & \mathbf{A}_{rc} \end{pmatrix}.$$

Here, $\mathbf{A}_{ij}$ is an $m_i \times n_j$ matrix ($i = 1, \ldots, r$; $j = 1, \ldots, c$), where $m_1, \ldots, m_r$ and $n_1, \ldots, n_c$ are positive integers such that $m_1 + \cdots + m_r = m$ and $n_1 + \cdots + n_c = n$. Or, more explicitly,

$$\mathbf{A}_{ij} = \begin{pmatrix} a_{m_1+\cdots+m_{i-1}+1,\, n_1+\cdots+n_{j-1}+1} & \dots & a_{m_1+\cdots+m_{i-1}+1,\, n_1+\cdots+n_j} \\ \vdots & & \vdots \\ a_{m_1+\cdots+m_i,\, n_1+\cdots+n_{j-1}+1} & \dots & a_{m_1+\cdots+m_i,\, n_1+\cdots+n_j} \end{pmatrix}.$$

(When $i = 1$ or $j = 1$, interpret the degenerate sum $m_1 + \cdots + m_{i-1}$ or $n_1 + \cdots + n_{j-1}$ as zero.) Thus, a partitioned matrix can be regarded as an array or "matrix" of matrices.

Note that a matrix that has been divided by "staggered" lines, for example,

$$\left(\begin{array}{r|cc|c} 2 & 4 & 3 & 6 \\ \cline{4-4} 1 & 5 & 7 & 9 \\ \hline -1 & 0 & 2 & 2 \end{array}\right),$$

does not satisfy our definition of a partitioned matrix. Thus, if a matrix, say

$$\begin{pmatrix} \mathbf{A}_{11} & \mathbf{A}_{12} & \dots & \mathbf{A}_{1c} \\ \mathbf{A}_{21} & \mathbf{A}_{22} & \dots & \mathbf{A}_{2c} \\ \vdots & \vdots & & \vdots \\ \mathbf{A}_{r1} & \mathbf{A}_{r2} & \dots & \mathbf{A}_{rc} \end{pmatrix},$$

is introduced as a partitioned matrix, there is an implicit assumption that each of the submatrices $\mathbf{A}_{i1}, \mathbf{A}_{i2}, \dots, \mathbf{A}_{ic}$ in the ith "row" of submatrices contains the same number of rows ($i = 1, 2, \dots, r$) and similarly that each of the submatrices $\mathbf{A}_{1j}, \mathbf{A}_{2j}, \dots, \mathbf{A}_{rj}$ in the jth "column" of submatrices contains the same number of columns.

It is customary to identify each of the blocks in a partitioned matrix by referring to the row of blocks and the column of blocks in which it appears. Thus, the submatrix $\mathbf{A}_{ij}$ is referred to as the ijth block of the partitioned matrix

$$\begin{pmatrix} \mathbf{A}_{11} & \mathbf{A}_{12} & \dots & \mathbf{A}_{1c} \\ \mathbf{A}_{21} & \mathbf{A}_{22} & \dots & \mathbf{A}_{2c} \\ \vdots & \vdots & & \vdots \\ \mathbf{A}_{r1} & \mathbf{A}_{r2} & \dots & \mathbf{A}_{rc} \end{pmatrix}.$$

In the case of a partitioned $m \times n$ matrix $\mathbf{A}$ of the form

$$\mathbf{A} = \begin{pmatrix} \mathbf{A}_{11} & \mathbf{A}_{12} & \dots & \mathbf{A}_{1r} \\ \mathbf{A}_{21} & \mathbf{A}_{22} & \dots & \mathbf{A}_{2r} \\ \vdots & \vdots & \ddots & \vdots \\ \mathbf{A}_{r1} & \mathbf{A}_{r2} & \dots & \mathbf{A}_{rr} \end{pmatrix} \tag{1.2}$$

(for which the number of rows of blocks equals the number of columns of blocks), the ijth block $\mathbf{A}_{ij}$ of $\mathbf{A}$ is called a *diagonal block* if $j = i$ and an *off-diagonal block* if $j \neq i$. If all of the off-diagonal blocks of $\mathbf{A}$ are null matrices, that is, if

$$\mathbf{A} = \begin{pmatrix} \mathbf{A}_{11} & \mathbf{0} & \ldots & \mathbf{0} \\ \mathbf{0} & \mathbf{A}_{22} & & \mathbf{0} \\ \vdots & & \ddots & \\ \mathbf{0} & \mathbf{0} & & \mathbf{A}_{rr} \end{pmatrix},$$

then $\mathbf{A}$ is called a *block-diagonal matrix*, and sometimes $\mathrm{diag}(\mathbf{A}_{11}, \mathbf{A}_{22}, \ldots, \mathbf{A}_{rr})$ is written for $\mathbf{A}$. If $\mathbf{A}_{ij} = \mathbf{0}$ for $j < i = 1, \ldots, r$, that is, if

$$\mathbf{A} = \begin{pmatrix} \mathbf{A}_{11} & \mathbf{A}_{12} & \ldots & \mathbf{A}_{1r} \\ \mathbf{0} & \mathbf{A}_{22} & \ldots & \mathbf{A}_{2r} \\ \vdots & & \ddots & \vdots \\ \mathbf{0} & \mathbf{0} & & \mathbf{A}_{rr} \end{pmatrix},$$

then $\mathbf{A}$ is called an *upper block-triangular matrix*. Similarly, if $\mathbf{A}_{ij} = \mathbf{0}$ for $j > i = 1, \ldots, r$, that is, if

$$\mathbf{A} = \begin{pmatrix} \mathbf{A}_{11} & \mathbf{0} & \ldots & \mathbf{0} \\ \mathbf{A}_{21} & \mathbf{A}_{22} & & \mathbf{0} \\ \vdots & \vdots & \ddots & \\ \mathbf{A}_{r1} & \mathbf{A}_{r2} & & \mathbf{A}_{rr} \end{pmatrix},$$

then $\mathbf{A}$ is called a *lower block-triangular matrix*. To indicate that $\mathbf{A}$ is either upper or lower block-triangular (without being more specific), $\mathbf{A}$ is referred to simply as *block-triangular*.

Note that a partitioned $m \times n$ matrix $\mathbf{A}$ of the form (1.2) is block-diagonal if and only if it is both upper block-triangular and lower block-triangular. Note also that, if $m = n = r$ (in which case each block of $\mathbf{A}$ consists of a single element), saying that $\mathbf{A}$ is block diagonal or upper or lower block triangular is equivalent to saying that $\mathbf{A}$ is diagonal or upper or lower triangular.

Partitioned matrices having one row or one column are customarily referred to as *partitioned (row or column) vectors*. Thus, a partitioned m-dimensional column vector is an $m \times 1$ vector $\mathbf{a} = \{a_t\}$ that has been reexpressed in the general form

$$\begin{pmatrix} \mathbf{a}_1 \\ \mathbf{a}_2 \\ \vdots \\ \mathbf{a}_r \end{pmatrix}.$$

Here, $\mathbf{a}_i$ is an $m_i \times 1$ vector with elements $a_{m_1+\cdots+m_{i-1}+1}, \ldots, a_{m_1+\cdots+m_{i-1}+m_i}$, respectively ($i = 1, \ldots, r$), where $m_1, \ldots, m_r$ are positive integers such that $m_1 + \cdots + m_r = m$. Similarly, a partitioned m-dimensional row vector is a $1 \times m$ vector $\mathbf{a}' = \{a_t\}$ that has been reexpressed in the general form $(\mathbf{a}_1', \mathbf{a}_2', \ldots, \mathbf{a}_r')$.

2.2 Scalar Multiples, Transposes, Sums, and Products of Partitioned Matrices

Let

$$\mathbf{A} = \begin{pmatrix} \mathbf{A}_{11} & \mathbf{A}_{12} & \dots & \mathbf{A}_{1c} \\ \mathbf{A}_{21} & \mathbf{A}_{22} & \dots & \mathbf{A}_{2c} \\ \vdots & \vdots & & \vdots \\ \mathbf{A}_{r1} & \mathbf{A}_{r2} & \dots & \mathbf{A}_{rc} \end{pmatrix}$$

represent a partitioned $m \times n$ matrix whose ijth block $\mathbf{A}_{ij}$ is of dimensions $m_i \times n_j$. Clearly, for any scalar k,

$$k\mathbf{A} = \begin{pmatrix} k\mathbf{A}_{11} & k\mathbf{A}_{12} & \dots & k\mathbf{A}_{1c} \\ k\mathbf{A}_{21} & k\mathbf{A}_{22} & \dots & k\mathbf{A}_{2c} \\ \vdots & \vdots & & \vdots \\ k\mathbf{A}_{r1} & k\mathbf{A}_{r2} & \dots & k\mathbf{A}_{rc} \end{pmatrix}. \tag{2.1}$$

In particular,

$$-\mathbf{A} = \begin{pmatrix} -\mathbf{A}_{11} & -\mathbf{A}_{12} & \dots & -\mathbf{A}_{1c} \\ -\mathbf{A}_{21} & -\mathbf{A}_{22} & \dots & -\mathbf{A}_{2c} \\ \vdots & \vdots & & \vdots \\ -\mathbf{A}_{r1} & -\mathbf{A}_{r2} & \dots & -\mathbf{A}_{rc} \end{pmatrix}. \tag{2.2}$$

Further, it is a simple exercise to show that

$$\mathbf{A}' = \begin{pmatrix} \mathbf{A}'_{11} & \mathbf{A}'_{21} & \dots & \mathbf{A}'_{r1} \\ \mathbf{A}'_{12} & \mathbf{A}'_{22} & \dots & \mathbf{A}'_{r2} \\ \vdots & \vdots & & \vdots \\ \mathbf{A}'_{1c} & \mathbf{A}'_{2c} & \dots & \mathbf{A}'_{rc} \end{pmatrix}; \tag{2.3}$$

that is, $\mathbf{A}'$ can be expressed as a partitioned matrix, comprising c rows and r columns of blocks, the ijth of which is the transpose $\mathbf{A}'_{ji}$ of the jith block $\mathbf{A}_{ji}$ of $\mathbf{A}$.

Now, let

$$\mathbf{B} = \begin{pmatrix} \mathbf{B}_{11} & \mathbf{B}_{12} & \dots & \mathbf{B}_{1v} \\ \mathbf{B}_{21} & \mathbf{B}_{22} & \dots & \mathbf{B}_{2v} \\ \vdots & \vdots & & \vdots \\ \mathbf{B}_{u1} & \mathbf{B}_{u2} & \dots & \mathbf{B}_{uv} \end{pmatrix}$$

represent a partitioned $p \times q$ matrix whose ijth block $\mathbf{B}_{ij}$ is of dimensions $p_i \times q_j$.

The matrices $\mathbf{A}$ and $\mathbf{B}$ are conformal (for addition) provided that $p = m$ and $q = n$. If $u = r$, $v = c$, $p_i = m_i$ $(i = 1, \dots, r)$, and $q_j = n_j$ $(j = 1, \dots, c)$, that is, if (besides $\mathbf{A}$ and $\mathbf{B}$ being conformal for addition) the rows and columns of $\mathbf{B}$ are partitioned in the same way as those of $\mathbf{A}$, then

$$\mathbf{A}+\mathbf{B} = \begin{pmatrix} \mathbf{A}_{11}+\mathbf{B}_{11} & \mathbf{A}_{12}+\mathbf{B}_{12} & \dots & \mathbf{A}_{1c}+\mathbf{B}_{1c} \\ \mathbf{A}_{21}+\mathbf{B}_{21} & \mathbf{A}_{22}+\mathbf{B}_{22} & \dots & \mathbf{A}_{2c}+\mathbf{B}_{2c} \\ \vdots & \vdots & & \vdots \\ \mathbf{A}_{r1}+\mathbf{B}_{r1} & \mathbf{A}_{r2}+\mathbf{B}_{r2} & \dots & \mathbf{A}_{rc}+\mathbf{B}_{rc} \end{pmatrix}, \tag{2.4}$$

and the partitioning of **A** and **B** is said to be *conformal* (for addition). This result and terminology extend in an obvious way to the addition of any finite number of partitioned matrices.

If **A** and **B** are conformal (for addition), then

$$\mathbf{A}-\mathbf{B} = \begin{pmatrix} \mathbf{A}_{11}-\mathbf{B}_{11} & \mathbf{A}_{12}-\mathbf{B}_{12} & \dots & \mathbf{A}_{1c}-\mathbf{B}_{1c} \\ \mathbf{A}_{21}-\mathbf{B}_{21} & \mathbf{A}_{22}-\mathbf{B}_{22} & \dots & \mathbf{A}_{2c}-\mathbf{B}_{2c} \\ \vdots & \vdots & & \vdots \\ \mathbf{A}_{r1}-\mathbf{B}_{r1} & \mathbf{A}_{r2}-\mathbf{B}_{r2} & \dots & \mathbf{A}_{rc}-\mathbf{B}_{rc} \end{pmatrix}, \tag{2.5}$$

as is evident from results (2.4) and (2.2).

The matrix product **AB** is defined provided that $n = p$. If $c = u$ and $n_k = p_k$ $(k = 1, \dots, c)$ [in which case all of the products $\mathbf{A}_{ik}\mathbf{B}_{kj}$ $(i = 1, \dots, r;\ j = 1, \dots, v;\ k = 1, \dots, c)$, as well as the product **AB**, exist], then

$$\mathbf{AB} = \begin{pmatrix} \mathbf{F}_{11} & \mathbf{F}_{12} & \dots & \mathbf{F}_{1v} \\ \mathbf{F}_{21} & \mathbf{F}_{22} & \dots & \mathbf{F}_{2v} \\ \vdots & \vdots & & \vdots \\ \mathbf{F}_{r1} & \mathbf{F}_{r2} & \dots & \mathbf{F}_{rv} \end{pmatrix}, \tag{2.6}$$

where $\mathbf{F}_{ij} = \sum_{k=1}^{c} \mathbf{A}_{ik}\mathbf{B}_{kj} = \mathbf{A}_{i1}\mathbf{B}_{1j} + \mathbf{A}_{i2}\mathbf{B}_{2j} + \cdots + \mathbf{A}_{ic}\mathbf{B}_{cj}$, and the partitioning of **A** and **B** is said to be *conformal* (for the premultiplication of **B** by **A**).

In the special case where $r = c = u = v = 2$, that is, where

$$\mathbf{A} = \begin{pmatrix} \mathbf{A}_{11} & \mathbf{A}_{12} \\ \mathbf{A}_{21} & \mathbf{A}_{22} \end{pmatrix} \quad \text{and} \quad \mathbf{B} = \begin{pmatrix} \mathbf{B}_{11} & \mathbf{B}_{12} \\ \mathbf{B}_{21} & \mathbf{B}_{22} \end{pmatrix},$$

result (2.6) simplifies to

$$\mathbf{AB} = \begin{pmatrix} \mathbf{A}_{11}\mathbf{B}_{11} + \mathbf{A}_{12}\mathbf{B}_{21} & \mathbf{A}_{11}\mathbf{B}_{12} + \mathbf{A}_{12}\mathbf{B}_{22} \\ \mathbf{A}_{21}\mathbf{B}_{11} + \mathbf{A}_{22}\mathbf{B}_{21} & \mathbf{A}_{21}\mathbf{B}_{12} + \mathbf{A}_{22}\mathbf{B}_{22} \end{pmatrix}. \tag{2.7}$$

If $\mathbf{A} = (\mathbf{A}_1, \mathbf{A}_2, \dots, \mathbf{A}_c)$ is an $m \times n$ matrix that has been partitioned only by columns (for emphasis, we sometimes insert commas between the submatrices of such a partitioned matrix), then

$$\mathbf{A}' = \begin{pmatrix} \mathbf{A}_1' \\ \mathbf{A}_2' \\ \vdots \\ \mathbf{A}_c' \end{pmatrix}, \tag{2.8}$$

and further if

$$\mathbf{B} = \begin{pmatrix} \mathbf{B}_1 \\ \mathbf{B}_2 \\ \vdots \\ \mathbf{B}_c \end{pmatrix}$$

is an $n \times q$ partitioned matrix that has been partitioned only by rows (in a way that is conformal for its premultiplication by $\mathbf{A}$), then

$$\mathbf{AB} = \sum_{k=1}^{c} \mathbf{A}_k \mathbf{B}_k = \mathbf{A}_1\mathbf{B}_1 + \mathbf{A}_2\mathbf{B}_2 + \cdots + \mathbf{A}_c\mathbf{B}_c. \tag{2.9}$$

Similarly, if

$$\mathbf{A} = \begin{pmatrix} \mathbf{A}_1 \\ \mathbf{A}_2 \\ \vdots \\ \mathbf{A}_r \end{pmatrix}$$

is an $m \times n$ matrix that has been partitioned only by rows, then

$$\mathbf{A}' = (\mathbf{A}_1', \mathbf{A}_2', \ldots, \mathbf{A}_r'), \tag{2.10}$$

and further if $\mathbf{B} = (\mathbf{B}_1, \mathbf{B}_2, \ldots, \mathbf{B}_v)$ is an $n \times q$ matrix that has been partitioned only by columns, then

$$\mathbf{AB} = \begin{pmatrix} \mathbf{A}_1\mathbf{B}_1 & \mathbf{A}_1\mathbf{B}_2 & \ldots & \mathbf{A}_1\mathbf{B}_v \\ \mathbf{A}_2\mathbf{B}_1 & \mathbf{A}_2\mathbf{B}_2 & \ldots & \mathbf{A}_2\mathbf{B}_v \\ \vdots & \vdots & & \vdots \\ \mathbf{A}_r\mathbf{B}_1 & \mathbf{A}_r\mathbf{B}_2 & \ldots & \mathbf{A}_r\mathbf{B}_v \end{pmatrix}. \tag{2.11}$$

2.3 Some Results on the Product of a Matrix and a Column Vector

Let $\mathbf{A}$ represent an $m \times n$ matrix and $\mathbf{x}$ an $n \times 1$ vector. Writing $\mathbf{A}$ as $\mathbf{A} = (\mathbf{a}_1, \mathbf{a}_2, \ldots, \mathbf{a}_n)$, where $\mathbf{a}_1, \mathbf{a}_2, \ldots, \mathbf{a}_n$ are the columns of $\mathbf{A}$, and $\mathbf{x}$ as $\mathbf{x} = (x_1, x_2, \ldots, x_n)'$, where $x_1, x_2, \ldots, x_n$ are the elements of $\mathbf{x}$, we find, as a special case of result (2.9), that

$$\mathbf{Ax} = \sum_{k=1}^{n} x_k \mathbf{a}_k = x_1\mathbf{a}_1 + x_2\mathbf{a}_2 + \cdots + x_n\mathbf{a}_n. \tag{3.1}$$

Thus, the effect of postmultiplying a matrix by a column vector is to form a linear combination of the columns of the matrix, where the coefficients in the linear combination are the elements of the column vector. Similarly, the effect of premultiplying a matrix by a row vector is to form a linear combination of the rows of the matrix, where the coefficients in the linear combination are the elements of the row vector.

Representation (3.1) is helpful in establishing the elementary results expressed in the following two lemmas.

Lemma 2.3.1. For any column vector $\mathbf{y}$ and nonnull column vector $\mathbf{x}$, there exists a matrix $\mathbf{A}$ such that $\mathbf{y} = \mathbf{Ax}$.

Proof. Since $\mathbf{x}$ is nonnull, one of its elements, say x_j, is nonzero. Take $\mathbf{A}$ to be the matrix whose jth column is $(1/x_j)\mathbf{y}$ and whose other columns are null. Then, $\mathbf{y} = \mathbf{Ax}$, as is evident from result (3.1). Q.E.D.

Lemma 2.3.2. For any two $m \times n$ matrices $\mathbf{A}$ and $\mathbf{B}$, $\mathbf{A} = \mathbf{B}$ if and only if $\mathbf{Ax} = \mathbf{Bx}$ for every $n \times 1$ vector $\mathbf{x}$.

Proof. It is obvious that, if $\mathbf{A} = \mathbf{B}$, then $\mathbf{Ax} = \mathbf{Bx}$ for every vector $\mathbf{x}$.

To prove the converse, suppose that $\mathbf{Ax} = \mathbf{Bx}$ for every $\mathbf{x}$. Taking $\mathbf{x}$ to be the $n \times 1$ vector whose ith element is 1 and whose other elements are 0, and letting $\mathbf{a}_i$ and $\mathbf{b}_i$ represent the ith columns of $\mathbf{A}$ and $\mathbf{B}$, respectively, it is clear from result (3.1) that

$$\mathbf{a}_i = \mathbf{Ax} = \mathbf{Bx} = \mathbf{b}_i$$

$(i = 1, \ldots, n)$. We conclude that $\mathbf{A} = \mathbf{B}$. Q.E.D.

Note that Lemma 2.3.2 implies, in particular, that $\mathbf{A} = \mathbf{0}$ if and only if $\mathbf{Ax} = \mathbf{0}$ for every $\mathbf{x}$.

2.4 Expansion of a Matrix in Terms of Its Rows, Columns, or Elements

An $m \times n$ matrix $\mathbf{A} = \{a_{ij}\}$ can be expanded in terms of its rows, columns, or elements by making use of formula (2.9). Denote the ith row of $\mathbf{A}$ by $\mathbf{r}_i'$ and the ith column of $\mathbf{I}_m$ by $\mathbf{e}_i$ $(i = 1, 2, \ldots, m)$. Then, writing $\mathbf{I}_m$ as $\mathbf{I}_m = (\mathbf{e}_1, \mathbf{e}_2, \ldots, \mathbf{e}_m)$ and $\mathbf{A}$ as

$$\mathbf{A} = \begin{pmatrix} \mathbf{r}_1' \\ \mathbf{r}_2' \\ \vdots \\ \mathbf{r}_m' \end{pmatrix}$$

and applying formula (2.9) to the product $\mathbf{I}_m\mathbf{A}$, we obtain the expansion

$$\mathbf{A} = \sum_{i=1}^{m} \mathbf{e}_i\mathbf{r}_i' = \mathbf{e}_1\mathbf{r}_1' + \mathbf{e}_2\mathbf{r}_2' + \cdots + \mathbf{e}_m\mathbf{r}_m'. \tag{4.1}$$

Similarly, denote the jth column of $\mathbf{A}$ by $\mathbf{a}_j$ and the jth row of $\mathbf{I}_n$ by $\mathbf{u}_j'$ $(j = 1, 2, \ldots, n)$. Then, writing $\mathbf{A}$ as $\mathbf{A} = (\mathbf{a}_1, \mathbf{a}_2, \ldots, \mathbf{a}_n)$ and $\mathbf{I}_n$ as

$$\mathbf{I}_n = \begin{pmatrix} \mathbf{u}_1' \\ \mathbf{u}_2' \\ \vdots \\ \mathbf{u}_n' \end{pmatrix}$$

and applying formula (2.9) to the product $\mathbf{A}\mathbf{I}_n$, we obtain the alternative expansion

$$\mathbf{A} = \sum_{j=1}^{n} \mathbf{a}_j \mathbf{u}_j' = \mathbf{a}_1 \mathbf{u}_1' + \mathbf{a}_2 \mathbf{u}_2' + \cdots + \mathbf{a}_n \mathbf{u}_n'. \tag{4.2}$$

Moreover, the application of formula (3.1) to the product $\mathbf{I}_m \mathbf{a}_j$ gives the expansion

$$\mathbf{a}_j = \sum_{i=1}^{m} a_{ij} \mathbf{e}_i.$$

Upon substituting this expansion into expansion (4.2), we obtain the further expansion

$$\mathbf{A} = \sum_{i=1}^{m} \sum_{j=1}^{n} a_{ij} \mathbf{U}_{ij}, \tag{4.3}$$

where $\mathbf{U}_{ij} = \mathbf{e}_i \mathbf{u}_j'$ is an $m \times n$ matrix whose ijth element equals 1 and whose remaining $mn - 1$ elements equal 0. In the special case where $n = m$ (i.e., where $\mathbf{A}$ is square), $\mathbf{u}_j = \mathbf{e}_j$ and hence $\mathbf{U}_{ij} = \mathbf{e}_i \mathbf{e}_j'$, and in the further special case where $\mathbf{A} = \mathbf{I}_m$, expansion (4.3) reduces to

$$\mathbf{I}_m = \sum_{i=1}^{m} \mathbf{e}_i \mathbf{e}_i'. \tag{4.4}$$

Note that, as a consequence of result (4.3), we have that

$$\mathbf{e}_i' \mathbf{A} \mathbf{u}_j = \mathbf{e}_i' \left(\sum_{k=1}^{m} \sum_{s=1}^{n} a_{ks} \mathbf{e}_k \mathbf{u}_s' \right) \mathbf{u}_j = \sum_{k=1}^{m} \sum_{s=1}^{n} a_{ks} \mathbf{e}_i' \mathbf{e}_k \mathbf{u}_s' \mathbf{u}_j,$$

which (since $\mathbf{e}_i' \mathbf{e}_k$ equals 1, if $k = i$, and equals 0, if $k \neq i$, and since $\mathbf{u}_s' \mathbf{u}_j$ equals 1, if $s = j$, and equals 0, if $s \neq j$) simplifies to

$$\mathbf{e}_i' \mathbf{A} \mathbf{u}_j = a_{ij}. \tag{4.5}$$

Exercises

Section 2.1

1. Verify result (1.1).

2. Verify (a) that a principal submatrix of a symmetric matrix is symmetric, (b) that a principal submatrix of a diagonal matrix is diagonal, and (c) that a principal submatrix of an upper triangular matrix is upper triangular.

3. Let

$$\mathbf{A} = \begin{pmatrix} \mathbf{A}_{11} & \mathbf{A}_{12} & \dots & \mathbf{A}_{1r} \\ \mathbf{0} & \mathbf{A}_{22} & \dots & \mathbf{A}_{2r} \\ \vdots & & \ddots & \vdots \\ \mathbf{0} & \mathbf{0} & \dots & \mathbf{A}_{rr} \end{pmatrix}$$

represent an $n \times n$ upper block-triangular matrix whose ijth block $\mathbf{A}_{ij}$ is of dimensions $n_i \times n_j$ $(j \geq i = 1, \dots, r)$. Show that $\mathbf{A}$ is upper triangular if and only if each of its diagonal blocks $\mathbf{A}_{11}, \mathbf{A}_{22}, \dots, \mathbf{A}_{rr}$ is upper triangular.

Section 2.2

4. Verify results (2.3) and (2.6).

3

Linear Dependence and Independence

While short in length, the following material on linear dependence and independence is of fundamental importance—so much so that it forms a separate chapter.

3.1 Definitions

Any finite set of row or column vectors, or more generally any finite set of matrices, is either linearly dependent or linearly independent. A nonempty finite set $\{\mathbf{A}_1, \mathbf{A}_2, \ldots, \mathbf{A}_k\}$ of $m \times n$ matrices is said to be *linearly dependent* if there exist scalars $x_1, x_2, \ldots, x_k$, not all zero, such that

$$x_1\mathbf{A}_1 + x_2\mathbf{A}_2 + \cdots + x_k\mathbf{A}_k = \mathbf{0}.$$

If no such scalars exist, the set is called *linearly independent.* The empty set is considered to be linearly independent. Note that if any subset of a finite set of matrices is linearly dependent, then the set itself is linearly dependent.

While technically linear dependence and independence are properties of sets of matrices, it is customary to speak of "a set of linearly dependent (or independent) matrices" or simply of "linearly dependent (or independent) matrices" instead of "a linearly dependent (or independent) set of matrices." In particular, in the case of row or column vectors, it is customary to speak of "linearly dependent (or independent) vectors."

3.2 Some Basic Results

Note that a set consisting of a single matrix is linearly dependent if the matrix is null, and is linearly independent if the matrix is nonnull. An elementary result on the linear dependence or independence of two or more matrices is expressed in the following lemma.

Lemma 3.2.1. A set $\{\mathbf{A}_1, \mathbf{A}_2, \ldots, \mathbf{A}_k\}$ of two or more $m \times n$ matrices is linearly dependent if and only if at least one of the matrices is expressible as a linear combination of the others, that is, if and only if, for some integer j $(1 \le j \le k)$, $\mathbf{A}_j$ is expressible as a linear combination of $\mathbf{A}_1, \ldots, \mathbf{A}_{j-1}, \mathbf{A}_{j+1}, \ldots, \mathbf{A}_k$.

Proof. Suppose that, for some j, $\mathbf{A}_j$ is expressible as a linear combination

$$\mathbf{A}_j = x_1\mathbf{A}_1 + \cdots + x_{j-1}\mathbf{A}_{j-1} + x_{j+1}\mathbf{A}_{j+1} + \cdots + x_k\mathbf{A}_k$$

of the other $k - 1$ $m \times n$ matrices. Then,

$$(-x_1)\mathbf{A}_1 + \cdots + (-x_{j-1})\mathbf{A}_{j-1} + \mathbf{A}_j + (-x_{j+1})\mathbf{A}_{j+1} + \cdots + (-x_k)\mathbf{A}_k = \mathbf{0},$$

implying that $\{\mathbf{A}_1, \mathbf{A}_2, \ldots, \mathbf{A}_k\}$ is a linearly dependent set.

Conversely, suppose that $\{\mathbf{A}_1, \mathbf{A}_2, \ldots, \mathbf{A}_k\}$ is linearly dependent, in which case

$$x_1\mathbf{A}_1 + x_2\mathbf{A}_2 + \cdots + x_k\mathbf{A}_k = \mathbf{0}$$

for some scalars $x_1, x_2, \ldots, x_k$, not all zero. Let j represent any integer for which $x_j \ne 0$. Then,

$$\mathbf{A}_j = (-x_1/x_j)\mathbf{A}_1 + \cdots + (-x_{j-1}/x_j)\mathbf{A}_{j-1} + (-x_{j+1}/x_j)\mathbf{A}_{j+1} + \cdots + (-x_k/x_j)\mathbf{A}_k.$$

Q.E.D.

Another, more profound result on the linear dependence or independence of two or more matrices is expressed in the following lemma.

Lemma 3.2.2. A set $\{\mathbf{A}_1, \mathbf{A}_2, \ldots, \mathbf{A}_k\}$ of two or more $m \times n$ matrices, the first of which is nonnull, is linearly dependent if and only if at least one of the matrices is expressible as a linear combination of the preceding ones, that is, if and only if, for some integer j $(2 \le j \le k)$, $\mathbf{A}_j$ is expressible as a linear combination of $\mathbf{A}_1, \ldots, \mathbf{A}_{j-1}$. (Or, equivalently, $\{\mathbf{A}_1, \mathbf{A}_2, \ldots, \mathbf{A}_k\}$ is linearly independent if and only if none of the matrices is expressible as a linear combination of its predecessors, that is, if and only if, for every integer j $(2 \le j \le k)$, $\mathbf{A}_j$ is not expressible as a linear combination of $\mathbf{A}_1, \ldots, \mathbf{A}_{j-1}$.)

Proof. Suppose that, for some j, $\mathbf{A}_j$ is expressible as a linear combination of $\mathbf{A}_1, \ldots, \mathbf{A}_{j-1}$. Then, it follows immediately from Lemma 3.2.1 that the set $\{\mathbf{A}_1, \ldots, \mathbf{A}_k\}$ is linearly dependent.

Conversely, suppose that $\{\mathbf{A}_1, \ldots, \mathbf{A}_k\}$ is linearly dependent. Define j to be the first integer for which $\{\mathbf{A}_1, \ldots, \mathbf{A}_j\}$ is a linearly dependent set. Then,

$$x_1\mathbf{A}_1 + \cdots + x_j\mathbf{A}_j = \mathbf{0}$$

for some scalars $x_1, \ldots, x_j$, not all zero. Moreover, $x_j \ne 0$, since otherwise $\{\mathbf{A}_1, \ldots, \mathbf{A}_{j-1}\}$ would be a linearly dependent set, contrary to the definition of j. Thus,

$$\mathbf{A}_j = (-x_1/x_j)\mathbf{A}_1 + \cdots + (-x_{j-1}/x_j)\mathbf{A}_{j-1}.$$

Q.E.D.

It is a simple exercise to verify the following corollary of Lemma 3.2.2.

Corollary 3.2.3. Suppose that $\{\mathbf{A}_1, \ldots, \mathbf{A}_k\}$ is a (nonempty) linearly independent set of $m \times n$ matrices and that $\mathbf{A}$ is another $m \times n$ matrix. Then, the set $\{\mathbf{A}_1, \ldots, \mathbf{A}_k, \mathbf{A}\}$ is linearly independent if and only if $\mathbf{A}$ is not expressible as a linear combination of the matrices $\mathbf{A}_1, \ldots, \mathbf{A}_k$.

The following lemma gives a basic result on the linear dependence or independence of linear combinations of matrices.

Lemma 3.2.4. Let $\mathbf{A}_1, \mathbf{A}_2, \ldots, \mathbf{A}_k$ represent $m \times n$ matrices. Further, for $j = 1, \ldots, r$, let

$$\mathbf{C}_j = x_{1j}\mathbf{A}_1 + x_{2j}\mathbf{A}_2 + \cdots + x_{kj}\mathbf{A}_k$$

(where $x_{1j}, x_{2j}, \ldots, x_{kj}$ are scalars), and let $\mathbf{x}_j = (x_{1j}, x_{2j}, \ldots, x_{kj})'$. If $\mathbf{A}_1$, $\mathbf{A}_2, \ldots, \mathbf{A}_k$ are linearly independent and if $\mathbf{x}_1, \mathbf{x}_2, \ldots, \mathbf{x}_r$ are linearly independent, then $\mathbf{C}_1, \mathbf{C}_2, \ldots, \mathbf{C}_r$ are linearly independent. If $\mathbf{x}_1, \mathbf{x}_2, \ldots, \mathbf{x}_r$ are linearly dependent, then $\mathbf{C}_1, \mathbf{C}_2, \ldots, \mathbf{C}_r$ are linearly dependent.

Proof. Observe that, for any scalars $y_1, y_2, \ldots, y_r$,

$$\sum_{j=1}^{r} y_j\mathbf{C}_j = \Big(\sum_{j=1}^{r} y_j x_{1j}\Big)\mathbf{A}_1 + \Big(\sum_{j=1}^{r} y_j x_{2j}\Big)\mathbf{A}_2 + \cdots + \Big(\sum_{j=1}^{r} y_j x_{kj}\Big)\mathbf{A}_k \quad (2.1)$$

and

$$\sum_{j=1}^{r} y_j\mathbf{x}_j = \Big(\sum_{j=1}^{r} y_j x_{1j}, \sum_{j=1}^{r} y_j x_{2j}, \ldots, \sum_{j=1}^{r} y_j x_{kj}\Big)'. \quad (2.2)$$

Now, suppose that $\mathbf{x}_1, \mathbf{x}_2, \ldots, \mathbf{x}_r$ are linearly dependent. Then, there exist scalars $y_1, y_2, \ldots, y_r$, not all of which are zero, such that $\sum_{j=1}^{r} y_j\mathbf{x}_j = \mathbf{0}$ and hence [in light of equalities (2.2) and (2.1)] such that $\sum_{j=1}^{r} y_j\mathbf{C}_j = \mathbf{0}$. Thus, $\mathbf{C}_1, \mathbf{C}_2, \ldots, \mathbf{C}_r$ are linearly dependent.

Alternatively, suppose that $\mathbf{A}_1, \mathbf{A}_2, \ldots, \mathbf{A}_k$ are linearly independent and $\mathbf{x}_1$, $\mathbf{x}_2, \ldots, \mathbf{x}_r$ are linearly independent. And let $y_1, y_2, \ldots, y_r$ represent any scalars such that $\sum_{j=1}^{r} y_j\mathbf{C}_j = \mathbf{0}$. Then, in light of equality (2.1) (and the linear independence of $\mathbf{A}_1, \mathbf{A}_2, \ldots, \mathbf{A}_k$), $\sum_{j=1}^{r} y_j x_{ij} = 0$ (for $i = 1, 2, \ldots, r$) and hence $\sum_{j=1}^{r} y_j\mathbf{x}_j = \mathbf{0}$, implying (in light of the linear independence of $\mathbf{x}_1, \mathbf{x}_2, \ldots$, $\mathbf{x}_r$) that $y_1 = y_2 = \cdots = y_r = 0$. Thus, $\mathbf{C}_1, \mathbf{C}_2, \ldots, \mathbf{C}_r$ are linearly independent.

Q.E.D.

Exercises

Section 3.1

1. For what values of the scalar k are the three row vectors $(k, 1, 0)$, $(1, k, 1)$, and $(0, 1, k)$ linearly dependent, and for what values are they linearly independent? Describe your reasoning.

Section 3.2

2. Let $\mathbf{A}$, $\mathbf{B}$, and $\mathbf{C}$ represent three linearly independent $m \times n$ matrices. Determine whether or not the three pairwise sums $\mathbf{A}+\mathbf{B}$, $\mathbf{A}+\mathbf{C}$, and $\mathbf{B}+\mathbf{C}$ are linearly independent. (*Hint*. Take advantage of Lemma 3.2.4.)

4

Linear Spaces: Row and Column Spaces

Associated with any matrix is a very important characteristic called the rank. The rank of a matrix is the subject of Section 4.4. There are several (consistent) ways of defining the rank. The most fundamental of these is in terms of the dimension of a linear space.

Linear spaces and their dimensions are discussed in Sections 4.1 through 4.3. Any matrix has two characteristics that are even more basic than its rank; these are two linear spaces that are respectively known as the row and column spaces of the matrix—discussion of row and column spaces is included in the coverage of Sections 4.1 through 4.3. It is shown in Section 4.4 that the column space of a matrix is of the same dimension as its row space; the rank of the matrix equals this dimension. The final section of Chapter 4 (Section 4.5) gives some basic results on the ranks and row and column spaces of partitioned matrices and of sums of matrices.

4.1 Some Definitions, Notation, and Basic Relationships and Properties

a. Column spaces

The *column space* of an $m \times n$ matrix $\mathbf{A}$ is the set whose elements consist of all m-dimensional column vectors that are expressible as linear combinations of the n columns of $\mathbf{A}$. Thus, the elements of the column space of $\mathbf{A}$ consist of all m-dimensional column vectors of the general form

$$x_1\mathbf{a}_1 + x_2\mathbf{a}_2 + \cdots + x_n\mathbf{a}_n,$$

where $x_1, x_2, \ldots, x_n$ are scalars and $\mathbf{a}_1, \mathbf{a}_2, \ldots, \mathbf{a}_n$ represent the columns of $\mathbf{A}$, or equivalently of the general form $\mathbf{Ax}$, where $\mathbf{x}$ is an $n \times 1$ column vector. For example, the column space of the 3×4 matrix

$$\begin{pmatrix} 2 & -4 & 0 & 0 \\ -1 & 2 & 0 & 0 \\ 0 & 0 & 1 & 2 \end{pmatrix}$$

includes the column vector

$$\begin{pmatrix} 4 \\ -2 \\ -3 \end{pmatrix} = 2\begin{pmatrix} 2 \\ -1 \\ 0 \end{pmatrix} + 0\begin{pmatrix} -4 \\ 2 \\ 0 \end{pmatrix} - 3\begin{pmatrix} 0 \\ 0 \\ 1 \end{pmatrix} + 0\begin{pmatrix} 0 \\ 0 \\ 2 \end{pmatrix},$$

but not the column vector $(1, 0, 0)'$.

b. Linear spaces

A nonempty set, say $\mathcal{V}$, of matrices (all of which have the same dimensions) is called a *linear space* if: (1) for every matrix $\mathbf{A}$ in $\mathcal{V}$ and every matrix $\mathbf{B}$ in $\mathcal{V}$, the sum $\mathbf{A} + \mathbf{B}$ is in $\mathcal{V}$; and (2) for every matrix $\mathbf{A}$ in $\mathcal{V}$ and every scalar k, the product $k\mathbf{A}$ is in $\mathcal{V}$. [The definition of a linear space can be extended to sets whose elements may not be matrices—refer, for example, to Halmos (1958) for a more general discussion.] If $\mathbf{A}_1, \ldots, \mathbf{A}_k$ are matrices in a linear space $\mathcal{V}$ and $x_1, \ldots, x_k$ are scalars, then it follows from the definition of a linear space that the linear combination $x_1\mathbf{A}_1 + \cdots + x_k\mathbf{A}_k$ is also in $\mathcal{V}$.

Clearly, the column space of any $m \times n$ matrix is a linear space (comprising various $m \times 1$ matrices). And the set consisting of all $m \times n$ matrices is a linear space. Further, since sums and scalar multiples of symmetric matrices are symmetric, the set of all $n \times n$ symmetric matrices is a linear space.

Note that every linear space contains the null matrix $\mathbf{0}$ (of appropriate dimensions), and that the set $\{\mathbf{0}\}$, whose only element is a null matrix, is a linear space. Note also that if a linear space contains a nonnull matrix, then it contains an infinite number of nonnull matrices.

c. Row spaces

The *row space* of an $m \times n$ matrix $\mathbf{A}$ is the set whose elements consist of all n-dimensional row vectors that are expressible as linear combinations of the m rows of $\mathbf{A}$. Thus, the elements of the row space of $\mathbf{A}$ consist of all n-dimensional row vectors of the general form

$$x_1\mathbf{a}_1' + x_2\mathbf{a}_2' + \cdots + x_m\mathbf{a}_m',$$

where $x_1, x_2, \ldots, x_m$ are scalars and $\mathbf{a}_1', \mathbf{a}_2', \ldots, \mathbf{a}_m'$ represent the rows of $\mathbf{A}$, or equivalently of the general form $\mathbf{x}'\mathbf{A}$, where $\mathbf{x}'$ is an m-dimensional row vector. Clearly, a row space, like a column space, is a linear space.

d. Notation

The symbol $\mathcal{C}(\mathbf{A})$ will be used to denote the column space of a matrix $\mathbf{A}$, and $\mathcal{R}(\mathbf{A})$ will be used to denote the row space of $\mathbf{A}$. The symbol $\mathcal{R}^{m\times n}$ will be used to denote the linear space whose elements consist of all $m\times n$ matrices. The symbol $\mathcal{R}^n$ will often be used (in place of $\mathcal{R}^{n\times 1}$ or $\mathcal{R}^{1\times n}$) to represent the set of all n-dimensional column vectors or, depending on the context, the set of all n-dimensional row vectors. Note that $\mathcal{R}(\mathbf{I}_n) = \mathcal{R}^n$ (where $\mathcal{R}^n$ is to be interpreted as the set of all n-dimensional row vectors), and $\mathcal{C}(\mathbf{I}_n) = \mathcal{R}^n$ (where $\mathcal{R}^n$ is to be interpreted as the set of all n-dimensional column vectors).

Sometimes, $x \in S$ will be written as an abbreviation for the statement that x is an element of a set S (or to introduce x as an element of S). Similarly, $x \notin S$ will sometimes be written as an abbreviation for the statement that x is not an element of S.

e. Relationship between row and column spaces

The following lemma relates the column space of a matrix to the row space of its transpose.

Lemma 4.1.1. For any matrix $\mathbf{A}$, $\mathbf{y} \in \mathcal{C}(\mathbf{A})$ if and only if $\mathbf{y}' \in \mathcal{R}(\mathbf{A}')$.

Proof. If $\mathbf{y} \in \mathcal{C}(\mathbf{A})$, then $\mathbf{y} = \mathbf{Ax}$ for some column vector $\mathbf{x}$, implying that $\mathbf{y}' = (\mathbf{Ax})' = \mathbf{x}'\mathbf{A}'$ and hence that $\mathbf{y}' \in \mathcal{R}(\mathbf{A}')$. An analogous argument can be used to show that if $\mathbf{y}' \in \mathcal{R}(\mathbf{A}')$, then $\mathbf{y} \in \mathcal{C}(\mathbf{A})$. Q.E.D.

f. A basic property of linear spaces

The following lemma gives a basic property of linear spaces.

Lemma 4.1.2. Let $\mathbf{B}$ represent an $m \times n$ matrix, and let $\mathcal{V}$ represent a linear space of $m \times n$ matrices. Then, for any matrix $\mathbf{A}$ in $\mathcal{V}$, $\mathbf{A} + \mathbf{B} \in \mathcal{V}$ if and only if $\mathbf{B} \in \mathcal{V}$.

Proof. If $\mathbf{B} \in \mathcal{V}$, then it is apparent from the very definition of a linear space that $\mathbf{A} + \mathbf{B} \in \mathcal{V}$. Conversely, if $\mathbf{A} + \mathbf{B} \in \mathcal{V}$, then, since $\mathbf{B} = (\mathbf{A} + \mathbf{B}) + (-1)\mathbf{A}$, it is clear from the definition of a linear space that $\mathbf{B} \in \mathcal{V}$. Q.E.D.

4.2 Subspaces

A subset $\mathcal{U}$ of a linear space $\mathcal{V}$ is called a *subspace* of $\mathcal{V}$ if $\mathcal{U}$ is itself a linear space. Trivial examples of a subspace of a linear space $\mathcal{V}$ are: (1) the set $\{\mathbf{0}\}$, whose only element is the null matrix, and (2) the entire set $\mathcal{V}$. The column space $\mathcal{C}(\mathbf{A})$ of an $m \times n$ matrix $\mathbf{A}$ is a subspace of $\mathcal{R}^m$ (when $\mathcal{R}^m$ is interpreted as the set of all m-dimensional column vectors), and $\mathcal{R}(\mathbf{A})$ is a subspace of $\mathcal{R}^n$ (when $\mathcal{R}^n$ is interpreted as the set of all n-dimensional row vectors).

Some additional terminology and notation are required in our discussion of subspaces. For any two subsets S and T of a given set (and, in particular, for any two subspaces of $\mathcal{R}^{m\times n}$), S is said to be *contained* in T if every element of S is an element of T. Subsequently, $S \subset T$ (or $T \supset S$) is written as an abbreviation for the statement that S is contained in T. Note that, if $S \subset T$ and $T \subset S$, then $S = T$, that is, the subsets S and T are identical.

Some basic results on row and column spaces are expressed in the following two lemmas.

Lemma 4.2.1. Let $\mathbf{A}$ represent an $m \times n$ matrix. Then, for any subspace $\mathcal{U}$ of $\mathcal{R}^m$, $\mathcal{C}(\mathbf{A}) \subset \mathcal{U}$ if and only if every column of $\mathbf{A}$ belongs to $\mathcal{U}$. Similarly, for any subspace $\mathcal{V}$ of $\mathcal{R}^n$, $\mathcal{R}(\mathbf{A}) \subset \mathcal{V}$ if and only if every row of $\mathbf{A}$ belongs to $\mathcal{V}$.

Proof. Let $\mathbf{a}_1, \ldots, \mathbf{a}_n$ represent the columns of $\mathbf{A}$. If $\mathcal{C}(\mathbf{A}) \subset \mathcal{U}$, then obviously $\mathbf{a}_i \in \mathcal{U}$ $(i = 1, \ldots, n)$.

Conversely, suppose that $\mathbf{a}_i \in \mathcal{U}$ $(i = 1, \ldots, n)$. For $\mathbf{y} \in \mathcal{C}(\mathbf{A})$, there exist scalars $x_1, \ldots, x_n$ such that $\mathbf{y} = x_1\mathbf{a}_1 + \cdots + x_n\mathbf{a}_n$, implying (since $\mathcal{U}$ is a linear space) that $\mathbf{y} \in \mathcal{U}$.

That $\mathcal{R}(\mathbf{A}) \subset \mathcal{V}$ if and only if every row of $\mathbf{A}$ belongs to $\mathcal{V}$ can be proved in a similar fashion. Q.E.D.

Lemma 4.2.2. For any $m \times n$ matrix $\mathbf{A}$ and $m \times p$ matrix $\mathbf{B}$, $\mathcal{C}(\mathbf{B}) \subset \mathcal{C}(\mathbf{A})$ if and only if there exists an $n \times p$ matrix $\mathbf{F}$ such that $\mathbf{B} = \mathbf{AF}$. Similarly, for any $m \times n$ matrix $\mathbf{A}$ and $q \times n$ matrix $\mathbf{C}$, $\mathcal{R}(\mathbf{C}) \subset \mathcal{R}(\mathbf{A})$ if and only if there exists a $q \times m$ matrix $\mathbf{L}$ such that $\mathbf{C} = \mathbf{LA}$.

Proof. Let $\mathbf{b}_1, \ldots, \mathbf{b}_p$ represent the columns of $\mathbf{B}$. Clearly, there exists a matrix $\mathbf{F}$ such that $\mathbf{B} = \mathbf{AF}$ if and only if there exist p column vectors $\mathbf{f}_1, \ldots, \mathbf{f}_p$ (of dimension n) such that $\mathbf{b}_i = \mathbf{Af}_i$ $(i = 1, \ldots, p)$, that is, if and only if (for $i = 1, \ldots, p$) $\mathbf{b}_i \in \mathcal{C}(\mathbf{A})$. We conclude on the basis of Lemma 4.2.1 that $\mathcal{C}(\mathbf{B}) \subset \mathcal{C}(\mathbf{A})$ if and only if there exists a matrix $\mathbf{F}$ such that $\mathbf{B} = \mathbf{AF}$.

That $\mathcal{R}(\mathbf{C}) \subset \mathcal{R}(\mathbf{A})$ if and only if there exists a matrix $\mathbf{L}$ such that $\mathbf{C} = \mathbf{LA}$ follows from an analogous argument. Q.E.D.

The "if" part of Lemma 4.2.2 can be restated as the following corollary.

Corollary 4.2.3. For any $m\times n$ matrix $\mathbf{A}$ and $n\times p$ matrix $\mathbf{F}$, $\mathcal{C}(\mathbf{AF}) \subset \mathcal{C}(\mathbf{A})$. Similarly, for any $m \times n$ matrix $\mathbf{A}$ and $q \times m$ matrix $\mathbf{L}$, $\mathcal{R}(\mathbf{LA}) \subset \mathcal{R}(\mathbf{A})$.

As a further consequence of Lemma 4.2.2, we have the following result.

Corollary 4.2.4. Let $\mathbf{A}$ represent an $m \times n$ matrix, $\mathbf{E}$ an $n \times k$ matrix, $\mathbf{F}$ an $n \times p$ matrix, $\mathbf{L}$ a $q \times m$ matrix, and $\mathbf{T}$ an $s \times m$ matrix. (1) If $\mathcal{C}(\mathbf{E}) \subset \mathcal{C}(\mathbf{F})$, then $\mathcal{C}(\mathbf{AE}) \subset \mathcal{C}(\mathbf{AF})$; and if $\mathcal{C}(\mathbf{E}) = \mathcal{C}(\mathbf{F})$, then $\mathcal{C}(\mathbf{AE}) = \mathcal{C}(\mathbf{AF})$. (2) If $\mathcal{R}(\mathbf{L}) \subset \mathcal{R}(\mathbf{T})$, then $\mathcal{R}(\mathbf{LA}) \subset \mathcal{R}(\mathbf{TA})$; and if $\mathcal{R}(\mathbf{L}) = \mathcal{R}(\mathbf{T})$, then $\mathcal{R}(\mathbf{LA}) = \mathcal{R}(\mathbf{TA})$.

Proof. (1) If $\mathcal{C}(\mathbf{E}) \subset \mathcal{C}(\mathbf{F})$, then (according to Lemma 4.2.2) $\mathbf{E} = \mathbf{FB}$ for some matrix $\mathbf{B}$, implying that $\mathbf{AE} = \mathbf{AFB}$ and hence (according to Lemma 4.2.2) that $\mathcal{C}(\mathbf{AE}) \subset \mathcal{C}(\mathbf{AF})$. Similarly, if $\mathcal{C}(\mathbf{F}) \subset \mathcal{C}(\mathbf{E})$, then $\mathcal{C}(\mathbf{AF}) \subset \mathcal{C}(\mathbf{AE})$. Thus, if $\mathcal{C}(\mathbf{E}) = \mathcal{C}(\mathbf{F})$ [in which case $\mathcal{C}(\mathbf{E}) \subset \mathcal{C}(\mathbf{F})$ and $\mathcal{C}(\mathbf{F}) \subset \mathcal{C}(\mathbf{E})$], then $\mathcal{C}(\mathbf{AE}) \subset \mathcal{C}(\mathbf{AF})$ and $\mathcal{C}(\mathbf{AF}) \subset \mathcal{C}(\mathbf{AE})$ and hence $\mathcal{C}(\mathbf{AE}) = \mathcal{C}(\mathbf{AF})$.

(2) The proof of Part (2) is analogous to that of Part (1). Q.E.D.

Finally, we have the following lemma, which can be easily verified by making use of either Lemma 4.1.1 or Lemma 4.2.2.

Lemma 4.2.5. Let $\mathbf{A}$ represent an $m \times n$ matrix and $\mathbf{B}$ an $m \times p$ matrix. Then, (1) $\mathcal{C}(\mathbf{A}) \subset \mathcal{C}(\mathbf{B})$ if and only if $\mathcal{R}(\mathbf{A}') \subset \mathcal{R}(\mathbf{B}')$, and (2) $\mathcal{C}(\mathbf{A}) = \mathcal{C}(\mathbf{B})$ if and only if $\mathcal{R}(\mathbf{A}') = \mathcal{R}(\mathbf{B}')$.

4.3 Bases

a. Some definitions (and notation)

The *span* of a finite set of matrices (having the same dimensions) is defined as follows: the span of a finite nonempty set $\{\mathbf{A}_1, \ldots, \mathbf{A}_k\}$ is the set consisting of all matrices that are expressible as linear combinations of $\mathbf{A}_1, \ldots, \mathbf{A}_k$, and the span of the empty set is the set $\{\mathbf{0}\}$, whose only element is the null matrix. The span of a finite set S of matrices will be denoted by the symbol $\text{sp}(S)$; $\text{sp}(\{\mathbf{A}_1, \ldots, \mathbf{A}_k\})$, which represents the span of the set $\{\mathbf{A}_1, \ldots, \mathbf{A}_k\}$, will generally be abbreviated to $\text{sp}(\mathbf{A}_1, \ldots, \mathbf{A}_k)$. Clearly, $\text{sp}(S)$ is a linear space.

A finite set S of matrices in a linear space $\mathcal{V}$ is said to *span* $\mathcal{V}$ if $\text{sp}(S) = \mathcal{V}$. Or, equivalently [since $\text{sp}(S) \subset \mathcal{V}$], S spans $\mathcal{V}$ if $\mathcal{V} \subset \text{sp}(S)$.

A *basis* for a linear space $\mathcal{V}$ is a finite set of linearly independent matrices in $\mathcal{V}$ that spans $\mathcal{V}$. Note that the empty set is the (unique) basis for the linear space $\{\mathbf{0}\}$.

Clearly, the column space $\mathcal{C}(\mathbf{A})$ of an $m \times n$ matrix $\mathbf{A}$ is spanned by the set whose members consist of the n columns of $\mathbf{A}$. If this set is linearly independent, then it is a basis for $\mathcal{C}(\mathbf{A})$; otherwise, it is not. Similarly, $\mathcal{R}(\mathbf{A})$ is spanned by the set whose members are the m rows of $\mathbf{A}$, and depending on whether this set is linearly independent or linearly dependent, it may or may not be a basis for $\mathcal{R}(\mathbf{A})$.

b. Natural bases for $\mathcal{R}^{m \times n}$ and for the linear space of all $n \times n$ symmetric matrices

For $i = 1, \ldots, m$ and $j = 1, \ldots, n$, let $\mathbf{U}_{ij}$ represent the $m \times n$ matrix whose ijth element equals 1 and whose remaining $mn - 1$ elements equal 0. The set whose members consist of the mn matrices $\mathbf{U}_{11}, \mathbf{U}_{21}, \ldots, \mathbf{U}_{m1}, \ldots, \mathbf{U}_{1n}, \mathbf{U}_{2n}, \ldots, \mathbf{U}_{mn}$ spans the linear space $\mathcal{R}^{m \times n}$ [as is evident from result (2.4.3)] and is linearly independent (as can be easily verified) and hence is a basis for $\mathcal{R}^{m \times n}$. The set $\{\mathbf{U}_{11}, \mathbf{U}_{21}, \ldots, \mathbf{U}_{m1}, \ldots, \mathbf{U}_{1n}, \mathbf{U}_{2n}, \ldots, \mathbf{U}_{mn}\}$ is sometimes called the *natural basis* for $\mathcal{R}^{m \times n}$. When $m = 1$ or $n = 1$ (in which case the members of $\mathcal{R}^{m \times n}$ are n-dimensional row vectors or m-dimensional column vectors), the natural basis for $\mathcal{R}^{m \times n}$ is the set whose members consist of the n rows $(1, 0, \ldots, 0), \ldots, (0, \ldots, 0, 1)$ of $\mathbf{I}_n$ or the m columns of $\mathbf{I}_m$.

For $i = 1, \ldots, n$, let $\mathbf{U}_{ii}^*$ represent the $n \times n$ matrix whose ith diagonal element equals 1 and whose remaining $n^2 - 1$ elements equal 0. And, for $j < i = 1, \ldots, n$, let $\mathbf{U}_{ij}^*$ represent the $n \times n$ matrix whose ijth and jith elements equal 1 and whose remaining n^2-2 elements equal 0. Then, for any $n \times n$ symmetric matrix $\mathbf{A} = \{a_{ij}\}$,

$$\mathbf{A} = \sum_{i=1}^{n} \sum_{j=1}^{i} a_{ij} \mathbf{U}_{ij}^* \tag{3.1}$$

[as is readily apparent from result (2.4.3) or upon comparing the corresponding elements of the left and right sides of equality (3.1)]. And it follows that the set whose members consist of the $n(n+1)/2$ matrices $\mathbf{U}_{11}^*, \mathbf{U}_{21}^*, \ldots, \mathbf{U}_{n1}^*, \ldots, \mathbf{U}_{ii}^*, \mathbf{U}_{i+1,i}^*, \ldots, \mathbf{U}_{ni}^*, \ldots, \mathbf{U}_{nn}^*$ spans the linear space of all $n \times n$ symmetric matrices. Moreover, this set is linearly independent (as can be easily verified). Thus, the set $\{\mathbf{U}_{11}^*, \mathbf{U}_{21}^*, \ldots, \mathbf{U}_{n1}^*, \ldots, \mathbf{U}_{ii}^*, \mathbf{U}_{i+1,i}^*, \ldots, \mathbf{U}_{ni}^*, \ldots, \mathbf{U}_{nn}^*\}$ is a basis for the linear space of all $n \times n$ symmetric matrices.

c. Existence of a basis

It was determined in Subsection b that the linear space $\mathcal{R}^{m \times n}$ has a basis and that the linear space of all $n \times n$ symmetric matrices has a basis. Does every linear space (of $m \times n$ matrices) have a basis? This question is answered (in the affirmative) as the culmination of a series of basic results—presented in the form of a lemma and two theorems—on linearly independent sets and spanning sets.

Lemma 4.3.1. Let $\mathbf{A}_1, \ldots, \mathbf{A}_p$ and $\mathbf{B}_1, \ldots, \mathbf{B}_q$ represent matrices in a linear space $\mathcal{V}$. Then, if the set $\{\mathbf{A}_1, \ldots, \mathbf{A}_p\}$ spans $\mathcal{V}$, so does the set $\{\mathbf{A}_1, \ldots, \mathbf{A}_p, \mathbf{B}_1, \ldots, \mathbf{B}_q\}$. Moreover, if the set $\{\mathbf{A}_1, \ldots, \mathbf{A}_p, \mathbf{B}_1, \ldots, \mathbf{B}_q\}$ spans $\mathcal{V}$ and if $\mathbf{B}_1, \ldots, \mathbf{B}_q$ are expressible as linear combinations of $\mathbf{A}_1, \ldots, \mathbf{A}_p$, then the set $\{\mathbf{A}_1, \ldots, \mathbf{A}_p\}$ spans $\mathcal{V}$.

Lemma 4.3.1 can be proved rather simply by showing that if $\mathbf{B}_1, \ldots, \mathbf{B}_q$ are expressible as linear combinations of $\mathbf{A}_1, \ldots, \mathbf{A}_p$, then any linear combination of the matrices $\mathbf{A}_1, \ldots, \mathbf{A}_p, \mathbf{B}_1, \ldots, \mathbf{B}_q$ is expressible as a linear combination of $\mathbf{A}_1, \ldots, \mathbf{A}_p$ and vice versa. The details are left as Exercise 6.

Theorem 4.3.2. Suppose that $\mathcal{V}$ is a linear space that is spanned by a set of r matrices, and let S represent any set of k linearly independent matrices in $\mathcal{V}$. Then, $k \leq r$, and if $k = r$, S is a basis for $\mathcal{V}$.

Proof. Let us restrict attention to the case $r > 0$. (The proof for the case $r = 0$ is trivial: if $r = 0$, then $\mathcal{V} = \{\mathbf{0}\}$, and the only linearly independent set of matrices in $\mathcal{V}$ is the empty set, which contains 0 members and is the basis for $\mathcal{V}$.) Denote by $\mathbf{A}_1, \ldots, \mathbf{A}_k$ the k matrices in the linearly independent set S, and let $\{\mathbf{B}_1, \ldots, \mathbf{B}_r\}$ represent a set of r matrices that spans $\mathcal{V}$.

To prove that $k \leq r$, assume—with the intent of establishing a contradiction—the contrary, that is, assume that $k > r$.

Consider the set $\{\mathbf{A}_1, \mathbf{B}_1, \ldots, \mathbf{B}_r\}$, obtained by inserting the matrix $\mathbf{A}_1$ at the beginning of the set $\{\mathbf{B}_1, \ldots, \mathbf{B}_r\}$. According to Lemma 4.3.1, this set spans $\mathcal{V}$. Moreover, since $\mathbf{A}_1$ is expressible as a linear combination of $\mathbf{B}_1, \ldots, \mathbf{B}_r$, the set $\{\mathbf{A}_1, \mathbf{B}_1, \dots, \mathbf{B}_r\}$ is (according to Lemma 3.2.1) linearly dependent, implying (in light of Lemma 3.2.2) that, among the matrices $\mathbf{B}_1, \dots, \mathbf{B}_r$, there exists a matrix, say $\mathbf{B}_p$, that is expressible as a linear combination of the matrices that precede it in the sequence $\mathbf{A}_1, \mathbf{B}_1, \dots, \mathbf{B}_r$. Thus, it follows from Lemma 4.3.1 that the set $\{\mathbf{A}_1, \mathbf{B}_1, \dots, \mathbf{B}_{p-1}, \mathbf{B}_{p+1}, \dots, \mathbf{B}_r\}$, obtained by deleting $\mathbf{B}_p$ from the set $\{\mathbf{A}_1, \mathbf{B}_1, \dots, \mathbf{B}_r\}$, spans $\mathcal{V}$.

Consider now the set $\{\mathbf{A}_1, \mathbf{A}_2, \mathbf{B}_1, \dots, \mathbf{B}_{p-1}, \mathbf{B}_{p+1}, \dots, \mathbf{B}_r\}$ obtained by inserting the matrix $\mathbf{A}_2$ after $\mathbf{A}_1$ in the set $\{\mathbf{A}_1, \mathbf{B}_1, \dots, \mathbf{B}_{p-1}, \mathbf{B}_{p+1}, \dots, \mathbf{B}_r\}$. This set spans $\mathcal{V}$. Moreover, this set is linearly dependent, so that, among the matrices $\mathbf{B}_1, \dots, \mathbf{B}_{p-1}, \mathbf{B}_{p+1}, \dots, \mathbf{B}_r$, there exists a matrix that is expressible as a linear combination of the matrices that precede it in the sequence $\mathbf{A}_1, \mathbf{A}_2, \mathbf{B}_1, \dots, \mathbf{B}_{p-1}, \mathbf{B}_{p+1}, \dots, \mathbf{B}_r$. Deleting this matrix from the set $\{\mathbf{A}_1, \mathbf{A}_2, \mathbf{B}_1, \dots, \mathbf{B}_{p-1}, \mathbf{B}_{p+1}, \dots, \mathbf{B}_r\}$, we obtain a subset of r matrices that spans $\mathcal{V}$.

Continuing this process of inserting "**A** matrices" and deleting "**B** matrices," we eventually (after a total of r steps) find that the set $\{\mathbf{A}_1, \dots, \mathbf{A}_r\}$ spans $\mathcal{V}$. It follows that the matrices $\mathbf{A}_{r+1}, \dots, \mathbf{A}_k$ are expressible as linear combinations of $\mathbf{A}_1, \dots, \mathbf{A}_r$, which, since S is a linearly independent set, results in a contradiction. We conclude that $k \le r$.

If $k = r$, then by inserting "**A** matrices" and deleting "**B** matrices" via the same algorithm employed in proving that $k \le r$, we find that the set S spans $\mathcal{V}$ and hence, since S is linearly independent, that S is a basis for $\mathcal{V}$. Q.E.D.

As discussed in Subsection b, the linear space $\mathcal{R}^{m \times n}$ (of all $m \times n$ matrices) is spanned by mn matrices, and the linear space of all symmetric matrices of order n is spanned by $n(n+1)/2$ matrices. Thus, we obtain, as a corollary of Theorem 4.3.2, the following result.

Corollary 4.3.3. The number of matrices in a linearly independent set of $m \times n$ matrices cannot exceed mn, and the number of matrices in a linearly independent set of $n \times n$ symmetric matrices cannot exceed $n(n+1)/2$.

Theorem 4.3.4. Every linear space (of $m \times n$ matrices) has a basis.

Proof. Let $\mathcal{V}$ represent an arbitrary linear space of $m \times n$ matrices. The proof of Theorem 4.3.4 consists of a description of a stepwise procedure for constructing a (finite) set B of linearly independent matrices that spans $\mathcal{V}$, that is, for constructing a basis B for $\mathcal{V}$.

If $\mathcal{V} = \{\mathbf{0}\}$, that is, if there are no nonnull matrices in $\mathcal{V}$, then terminate the procedure immediately and take B to be the empty set. Otherwise, locate a nonnull matrix $\mathbf{A}_1$ in $\mathcal{V}$. If the linearly independent set $\{\mathbf{A}_1\}$, whose only member is $\mathbf{A}_1$, spans $\mathcal{V}$, then terminate the procedure and take $B = \{\mathbf{A}_1\}$. If this set does not span $\mathcal{V}$, then locate a matrix $\mathbf{A}_2$ that cannot be expressed as a scalar multiple of $\mathbf{A}_1$. Necessarily (according to Corollary 3.2.3), the set $\{\mathbf{A}_1, \mathbf{A}_2\}$ is linearly independent. If the set $\{\mathbf{A}_1, \mathbf{A}_2\}$ spans $\mathcal{V}$, then terminate the procedure, taking $B = \{\mathbf{A}_1, \mathbf{A}_2\}$. If $\{\mathbf{A}_1, \mathbf{A}_2\}$ does not span $\mathcal{V}$, then locate a matrix $\mathbf{A}_3$ in $\mathcal{V}$ that is not expressible as a linear combination of $\mathbf{A}_1$ and $\mathbf{A}_2$. Necessarily, the set $\{\mathbf{A}_1, \mathbf{A}_2, \mathbf{A}_3\}$ is linearly independent.

Continuing, on the kth step, either we terminate the procedure, taking $B = \{\mathbf{A}_1, \ldots, \mathbf{A}_k\}$, or else we locate a matrix $\mathbf{A}_{k+1}$ in $\mathcal{V}$ such that the set $\{\mathbf{A}_1, \ldots, \mathbf{A}_{k+1}\}$ is linearly independent. The stepwise procedure terminates after no more than mn steps, since otherwise there would exist a set of more thn mn linearly independent matrices, contradicting Corollary 4.3.3. Q.E.D.

d. Uniqueness of representation

Any particular basis for a linear space has the following property.

Lemma 4.3.5. A matrix $\mathbf{A}$ in a linear space $\mathcal{V}$ has a unique representation in terms of any particular basis $\{\mathbf{A}_1, \ldots, \mathbf{A}_k\}$; that is, the coefficients $x_1, \ldots, x_k$ in the linear combination

$$\mathbf{A} = x_1\mathbf{A}_1 + \cdots + x_k\mathbf{A}_k$$

are uniquely determined.

Proof. Let $x_1, \ldots, x_k$ and $y_1, \ldots, y_k$ represent any scalars such that $\mathbf{A} = \sum_{i=1}^k x_i\mathbf{A}_i$ and $\mathbf{A} = \sum_{i=1}^k y_i\mathbf{A}_i$. Then,

$$\sum_{i=1}^k (y_i - x_i)\mathbf{A}_i = \sum_{i=1}^k y_i\mathbf{A}_i - \sum_{i=1}^k x_i\mathbf{A}_i = \mathbf{A} - \mathbf{A} = \mathbf{0}.$$

Since $\mathbf{A}_1, \ldots, \mathbf{A}_k$ are linearly independent matrices, we conclude that $y_i - x_i = 0$ or equivalently that $y_i = x_i$ $(i = 1, \ldots, k)$. Q.E.D.

e. Dimension

With the exception of the linear space $\{\mathbf{0}\}$, for which the only basis is the empty set, every linear space of $m \times n$ matrices has an infinite number of bases. However, it follows from Theorem 4.3.2 that if two sets, one containing k_1 matrices and the other containing k_2 matrices, are both bases for the same linear space, then $k_1 \leq k_2$ and $k_2 \leq k_1$. Thus, we have the following result.

Theorem 4.3.6. Any two bases for the same linear space contain the same number of matrices.

The number of matrices in a basis for a linear space $\mathcal{V}$ is called the *dimension* of $\mathcal{V}$ and is denoted by the symbol dim $\mathcal{V}$ or dim($\mathcal{V}$). (Note that the term dimension is used, not only in reference to the number of matrices in a basis, but also in reference to the number of rows or columns in a matrix—which usage is intended can be determined from the context.)

As an essentially immediate consequence of Theorem 4.3.2, we obtain the results expressed in the following three theorems.

Theorem 4.3.7. If a linear space $\mathcal{V}$ is spanned by a set of r matrices, then dim $\mathcal{V} \leq r$, and if there is a set of k linearly independent matrices in $\mathcal{V}$, then dim $\mathcal{V} \geq k$.

Theorem 4.3.8. If $\mathcal{U}$ is a subspace of a linear space $\mathcal{V}$, then $\dim \mathcal{U} \leq \dim \mathcal{V}$.

Theorem 4.3.9. Any set of r linearly independent matrices in an r-dimensional linear space $\mathcal{V}$ is a basis for $\mathcal{V}$.

A further, related result is expressed in the following theorem.

Theorem 4.3.10. Let $\mathcal{U}$ and $\mathcal{V}$ represent linear spaces of $m \times n$ matrices. If $\mathcal{U} \subset \mathcal{V}$ (that is, if $\mathcal{U}$ is a subspace of $\mathcal{V}$) and if in addition $\dim \mathcal{U} = \dim \mathcal{V}$, then $\mathcal{U} = \mathcal{V}$.

Proof. Let $r = \dim \mathcal{U}$, and let S represent any set of r linearly independent matrices that is a basis for $\mathcal{U}$. Suppose that $\mathcal{U} \subset \mathcal{V}$, in which case all r of the matrices in S are in $\mathcal{V}$. If $\dim \mathcal{U} = \dim \mathcal{V}$, then (according to Theorem 4.3.9) S is a basis for $\mathcal{V}$, implying that $\mathcal{V} = \text{sp}(S) = \mathcal{U}$. Q.E.D.

It follows from the discussion of Subsection b that

$$\dim(\mathcal{R}^{m \times n}) = mn. \tag{3.2}$$

This result, together with Theorem 4.3.8, implies that

$$\dim \mathcal{U} \leq mn \tag{3.3}$$

for any linear space $\mathcal{U}$ of $m \times n$ matrices (as is also deducible from Corollary 4.3.3).

f. Extracting a basis from a spanning set

A basis for a linear space can be extracted from any spanning set, as indicated by the following theorem.

Theorem 4.3.11. Any set S that spans a linear space $\mathcal{V}$ (of $m \times n$ matrices) contains a subset that is a basis for $\mathcal{V}$.

Proof. Assume that S contains at least one nonnull matrix. (The proof for the case where S is empty or consists entirely of null matrices is trivial—in that case, $\mathcal{V} = \{\mathbf{0}\}$, and the empty set is a basis for $\mathcal{V}$.)

Let $\mathbf{A}_1, \ldots, \mathbf{A}_k$ represent the matrices in S. Define S^* to be the subset of S obtained by successively applying to each of the matrices $\mathbf{A}_1, \ldots, \mathbf{A}_k$ the following algorithm: include the matrix as a member of S^* if it is nonnull and if it is not expressible as a linear combination of any matrices already included in S^*. It is clear from Lemma 3.2.2 that the subset S^* formed in this way is linearly independent. Moreover, every matrix in S either belongs to S^* or is expressible as a linear combination of matrices that belong to S^*, implying (according to Lemma 4.3.1) that S^* spans $\mathcal{V}$. We conclude that S^* is a basis for $\mathcal{V}$. Q.E.D.

Given any particular set that spans a linear space $\mathcal{V}$, the algorithm described in the proof of Theorem 4.3.11 can be used to construct a basis for $\mathcal{V}$ and, in the process, to determine the dimension of $\mathcal{V}$.

g. Inclusion of any particular linearly independent set in a basis

Any set of linearly independent matrices in a linear space can be augmented to form a basis, as indicated by the following theorem, which generalizes Theorem 4.3.4.

Theorem 4.3.12. For any set S of r linearly independent matrices in a k-dimensional linear space $\mathcal{V}$, there exists a basis for $\mathcal{V}$ that includes all r of the matrices in S (and $k - r$ additional matrices).

Proof. Assume that $r > 0$. (In the special case $r = 0$, the theorem essentially reduces to Theorem 4.3.4.)

Denote by $\mathbf{A}_1, \ldots, \mathbf{A}_r$ the matrices in S, and let $\{\mathbf{B}_1, \ldots, \mathbf{B}_k\}$ represent any basis for $\mathcal{V}$. An alternative basis for $\mathcal{V}$ can be extracted from the set $\{\mathbf{A}_1, \ldots, \mathbf{A}_r, \mathbf{B}_1, \ldots, \mathbf{B}_k\}$, which (according to Lemma 4.3.1) spans $\mathcal{V}$, by applying the algorithm described in the proof of Theorem 4.3.11. It follows from Lemma 3.2.2 (and from the method of extraction) that the alternative basis includes all r of the matrices $\mathbf{A}_1, \ldots, \mathbf{A}_r$. Q.E.D.

Given any particular set that spans a k-dimensional linear space $\mathcal{V}$, the procedure described in the proof of Theorem 4.3.12 can be used to construct a basis that includes as a subset any particular set $\{\mathbf{A}_1, \ldots, \mathbf{A}_r\}$ of linearly independent matrices. An alternative approach, not requiring knowledge of a spanning set, is to adopt the procedure described in the proof of Theorem 4.3.4. In the alternative approach, a basis is constructed in $k - r$ steps, the ith of which consists of locating a matrix $\mathbf{A}_{r+i}$ in $\mathcal{V}$ that is not expressible as a linear combination of $\mathbf{A}_1, \ldots, \mathbf{A}_{r+i-1}$. Clearly, the set $\{\mathbf{A}_1, \ldots, \mathbf{A}_k\}$ is a basis for $\mathcal{V}$.

4.4 Rank of a Matrix

a. Some definitions and basic results

The *row rank* of a matrix $\mathbf{A}$ is defined to be the dimension of the row space of $\mathbf{A}$, and the *column rank* of $\mathbf{A}$ is defined to be the dimension of the column space of $\mathbf{A}$. A fundamental result on the row and column spaces of a matrix is expressed in the following theorem.

Theorem 4.4.1. The row rank of any matrix $\mathbf{A}$ equals the column rank of $\mathbf{A}$.

Preliminary to proving Theorem 4.4.1, it is convenient to prove the following result, which is of some interest in its own right.

Theorem 4.4.2. Let $\mathbf{A}$ represent an $m \times n$ nonnull matrix of row rank r and column rank c. Then, there exists an $m \times c$ matrix $\mathbf{B}$ and a $c \times n$ matrix $\mathbf{L}$ such that $\mathbf{A} = \mathbf{BL}$. Similarly, there exists an $m \times r$ matrix $\mathbf{K}$ and an $r \times n$ matrix $\mathbf{T}$ such that $\mathbf{A} = \mathbf{KT}$.

Proof. According to Theorem 4.3.4, there exists a set $\{\mathbf{b}_1, \ldots, \mathbf{b}_c\}$ of c vectors that is a basis for $\mathcal{C}(\mathbf{A})$. Take $\mathbf{B}$ to be the $m \times c$ matrix whose columns are $\mathbf{b}_1, \ldots, \mathbf{b}_c$. Then, $\mathcal{C}(\mathbf{B}) = \text{sp}(\mathbf{b}_1, \ldots, \mathbf{b}_c) = \mathcal{C}(\mathbf{A})$, and it follows from Lemma 4.2.2 that there exists a $c \times n$ matrix $\mathbf{L}$ such that $\mathbf{A} = \mathbf{BL}$. The existence of an $m \times r$ matrix $\mathbf{K}$ and an $r \times n$ matrix $\mathbf{T}$ such that $\mathbf{A} = \mathbf{KT}$ can be established via an analogous argument. Q.E.D.

Proof (of Theorem 4.4.1). Let r represent the row rank and c the column rank of a matrix $\mathbf{A}$. Assume that $\mathbf{A}$ is nonnull. (The proof for the case $\mathbf{A} = \mathbf{0}$ is trivial: If $\mathbf{A} = \mathbf{0}$, then $r = 0 = c$.) According to Theorem 4.4.2, there exists an $m \times c$ matrix $\mathbf{B}$ and a $c \times n$ matrix $\mathbf{L}$ such that $\mathbf{A} = \mathbf{BL}$, and similarly there exists an $m \times r$ matrix $\mathbf{K}$ and an $r \times n$ matrix $\mathbf{T}$ such that $\mathbf{A} = \mathbf{KT}$.

It follows from Lemma 4.2.2 that $\mathcal{R}(\mathbf{A}) \subset \mathcal{R}(\mathbf{L})$ and $\mathcal{C}(\mathbf{A}) \subset \mathcal{C}(\mathbf{K})$. Thus, making use of Theorems 4.3.8 and 4.3.7 [and observing that $\mathcal{R}(\mathbf{L})$ is spanned by the c rows of $\mathbf{L}$ and $\mathcal{C}(\mathbf{K})$ by the r columns of $\mathbf{K}$], we find that

$$r \leq \dim[\mathcal{R}(\mathbf{L})] \leq c$$

and similarly that

$$c \leq \dim[\mathcal{C}(\mathbf{K})] \leq r.$$

We conclude that $r = c$. Q.E.D.

In light of Theorem 4.4.1, it is not necessary to distinguish between the row and column ranks of a matrix $\mathbf{A}$. Their common value is called the *rank* of $\mathbf{A}$ and is denoted by the symbol rank $\mathbf{A}$ or $\text{rank}(\mathbf{A})$.

b. Restatement of results on dimensions

Our results on the dimensions of linear spaces and subspaces can be specialized to row and column spaces and restated in terms of ranks. Result (3.3) implies that the row rank of a matrix cannot exceed the number of columns and that its column rank cannot exceed the number of rows, leading to the following conclusion.

Lemma 4.4.3. For any $m \times n$ matrix $\mathbf{A}$, $\text{rank}(\mathbf{A}) \leq m$ and $\text{rank}(\mathbf{A}) \leq n$.

Theorem 4.3.8 has the following implication.

Theorem 4.4.4. Let $\mathbf{A}$ represent an $m \times n$ matrix, $\mathbf{B}$ an $m \times p$ matrix, and $\mathbf{C}$ a $q \times n$ matrix. If $\mathcal{C}(\mathbf{B}) \subset \mathcal{C}(\mathbf{A})$, then $\text{rank}(\mathbf{B}) \leq \text{rank}(\mathbf{A})$. Similarly, if $\mathcal{R}(\mathbf{C}) \subset \mathcal{R}(\mathbf{A})$, then $\text{rank}(\mathbf{C}) \leq \text{rank}(\mathbf{A})$.

In light of Corollary 4.2.3, we have the following corollary of Theorem 4.4.4.

Corollary 4.4.5. For any $m \times n$ matrix $\mathbf{A}$ and $n \times p$ matrix $\mathbf{F}$, $\text{rank}(\mathbf{AF}) \leq \text{rank}(\mathbf{A})$ and $\text{rank}(\mathbf{AF}) \leq \text{rank}(\mathbf{F})$.

Theorem 4.3.10 has the following implications.

Theorem 4.4.6. Let $\mathbf{A}$ represent an $m \times n$ matrix, $\mathbf{B}$ an $m \times p$ matrix, and $\mathbf{C}$ a $q \times n$ matrix. If $\mathcal{C}(\mathbf{B}) \subset \mathcal{C}(\mathbf{A})$ and $\text{rank}(\mathbf{B}) = \text{rank}(\mathbf{A})$, then $\mathcal{C}(\mathbf{B}) = \mathcal{C}(\mathbf{A})$. Similarly, if $\mathcal{R}(\mathbf{C}) \subset \mathcal{R}(\mathbf{A})$ and $\text{rank}(\mathbf{C}) = \text{rank}(\mathbf{A})$, then $\mathcal{R}(\mathbf{C}) = \mathcal{R}(\mathbf{A})$.

Corollary 4.4.7. Let $\mathbf{A}$ represent an $m \times n$ matrix and $\mathbf{F}$ an $n \times p$ matrix. If $\text{rank}(\mathbf{AF}) = \text{rank}(\mathbf{A})$, then $\mathcal{C}(\mathbf{AF}) = \mathcal{C}(\mathbf{A})$. Similarly, if $\text{rank}(\mathbf{AF}) = \text{rank}(\mathbf{F})$, then $\mathcal{R}(\mathbf{AF}) = \mathcal{R}(\mathbf{F})$.

c. Nonsingular matrices and matrices of full row or column rank

Among $m \times n$ matrices, the maximum rank is $\min(m, n)$, as indicated by Lemma 4.4.3. The minimum rank is 0, which is achieved by the $m \times n$ null matrix $\mathbf{0}$ (and by no other $m \times n$ matrix).

An $m \times n$ matrix $\mathbf{A}$ is said to have *full row rank* if $\text{rank}(\mathbf{A}) = m$, that is, if its rank equals the number of rows, and to have *full column rank* if $\text{rank}(\mathbf{A}) = n$. Clearly, an $m \times n$ matrix can have full row rank only if $m \leq n$, that is, only if the number of rows does not exceed the number of columns, and can have full column rank only if $n \leq m$.

A matrix is said to be *nonsingular* if it has both full row rank and full column rank. Clearly, any nonsingular matrix is square. By definition, an $n \times n$ matrix $\mathbf{A}$ is nonsingular if and only if $\text{rank}(\mathbf{A}) = n$. An $n \times n$ matrix of rank less than n is said to be *singular*.

An $m \times n$ matrix can be expressed as the product of a matrix having full column rank and a matrix having full row rank as indicated by the following theorem, which is a restatement and extension of Theorem 4.4.2.

Theorem 4.4.8. Let $\mathbf{A}$ represent an $m \times n$ nonnull matrix of rank r. Then, there exist an $m \times r$ matrix $\mathbf{B}$ and an $r \times n$ matrix $\mathbf{T}$ such that $\mathbf{A} = \mathbf{BT}$. Moreover, for any $m \times r$ matrix $\mathbf{B}$ and $r \times n$ matrix $\mathbf{T}$ such that $\mathbf{A} = \mathbf{BT}$, $\text{rank}(\mathbf{B}) = \text{rank}(\mathbf{T}) = r$, that is, $\mathbf{B}$ has full column rank and $\mathbf{T}$ has full row rank.

Proof. The existence of an $m \times r$ matrix $\mathbf{B}$ and an $r \times n$ matrix $\mathbf{T}$ such that $\mathbf{A} = \mathbf{BT}$ follows from Theorem 4.4.2. Moreover, for any such $m \times r$ matrix $\mathbf{B}$ and $r \times n$ matrix $\mathbf{T}$, it follows from Lemma 4.4.3 that $\text{rank}(\mathbf{B}) \leq r$ and $\text{rank}(\mathbf{T}) \leq r$ and from Corollary 4.4.5 that $\text{rank}(\mathbf{B}) \geq r$ and $\text{rank}(\mathbf{T}) \geq r$, leading to the conclusion that $\text{rank}(\mathbf{B}) = r$ and $\text{rank}(\mathbf{T}) = r$. Q.E.D.

d. A decomposition of a rectangular matrix

The decomposition of a rectangular matrix given in Theorem 4.4.8 can be expressed in an alternative form, as described in the following theorem.

Theorem 4.4.9. Let $\mathbf{A}$ represent an $m \times n$ nonnull matrix of rank r. Then, there exist an $m \times m$ nonsingular matrix $\mathbf{B}$ and an $n \times n$ nonsingular matrix $\mathbf{K}$ such that

$$\mathbf{A} = \mathbf{B}\begin{pmatrix} \mathbf{I}_r & \mathbf{0} \\ \mathbf{0} & \mathbf{0} \end{pmatrix}\mathbf{K}.$$

Proof. According to Theorem 4.4.8, there exist an $m \times r$ matrix $\mathbf{B}_1$ of full column rank and an $r \times n$ matrix $\mathbf{K}_1$ of full row rank such that $\mathbf{A} = \mathbf{B}_1\mathbf{K}_1$. Let $\mathbf{b}_1, \ldots, \mathbf{b}_r$ represent the first, ..., rth columns, respectively, of $\mathbf{B}_1$ and $\mathbf{k}_1', \ldots, \mathbf{k}_r'$ the first, ..., rth rows, respectively, of $\mathbf{K}_1$. Clearly, $\mathbf{b}_1, \ldots, \mathbf{b}_r$ are linearly independent, and $\mathbf{k}_1', \ldots, \mathbf{k}_r'$ are linearly independent. Then, according to Theorem 4.3.12, there exist $m - r$ vectors $\mathbf{b}_{r+1}, \ldots, \mathbf{b}_m$ such that $\mathbf{b}_1, \ldots, \mathbf{b}_r, \mathbf{b}_{r+1}, \ldots, \mathbf{b}_m$ form a basis for $\mathcal{R}^m$, and, similarly, there exist $n - r$ vectors $\mathbf{k}_{r+1}', \ldots, \mathbf{k}_n'$ such that $\mathbf{k}_1', \ldots, \mathbf{k}_r', \mathbf{k}_{r+1}', \ldots, \mathbf{k}_n'$ form a basis for $\mathcal{R}^n$.

Take $\mathbf{B} = (\mathbf{B}_1, \mathbf{B}_2)$ and $\mathbf{K} = \begin{pmatrix} \mathbf{K}_1 \\ \mathbf{K}_2 \end{pmatrix}$, where $\mathbf{B}_2 = (\mathbf{b}_{r+1}, \ldots, \mathbf{b}_m)$ and $\mathbf{K}_2 = \begin{pmatrix} \mathbf{k}'_{r+1} \\ \vdots \\ \mathbf{k}'_n \end{pmatrix}$. Then, clearly, $\mathbf{B}$ and $\mathbf{K}$ are nonsingular, and

$$\mathbf{A} = \mathbf{B}_1\mathbf{K}_1 = (\mathbf{B}_1, \mathbf{B}_2)\begin{pmatrix} \mathbf{I}_r & \mathbf{0} \\ \mathbf{0} & \mathbf{0} \end{pmatrix}\begin{pmatrix} \mathbf{K}_1 \\ \mathbf{K}_2 \end{pmatrix} = \mathbf{B}\begin{pmatrix} \mathbf{I}_r & \mathbf{0} \\ \mathbf{0} & \mathbf{0} \end{pmatrix}\mathbf{K}.$$

Q.E.D.

e. Some equivalent definitions

The following theorem can be used to characterize the rank of a matrix in terms of the ranks of its submatrices.

Theorem 4.4.10. Let $\mathbf{A}$ represent any $m \times n$ matrix of rank r. Then, $\mathbf{A}$ contains r linearly independent rows and r linearly independent columns, and, for any r linearly independent rows and r linearly independent columns of $\mathbf{A}$, the $r \times r$ submatrix, obtained by striking out the other $m - r$ rows and $n - r$ columns, is nonsingular. Moreover, any set of more than r rows or more than r columns (of $\mathbf{A}$) is linearly dependent, and there exists no submatrix of $\mathbf{A}$ whose rank exceeds r.

Proof. Since the m rows and n columns of $\mathbf{A}$ span $\mathcal{R}(\mathbf{A})$ and $\mathcal{C}(\mathbf{A})$, respectively, it follows from Theorem 4.3.11 that $\mathbf{A}$ contains r linearly independent rows and r linearly independent columns. Moreover, it follows from Theorem 4.3.7 that any set of more than r rows or more than r columns is linearly dependent.

Consider now the $r \times r$ submatrix $\mathbf{B}$ of $\mathbf{A}$ obtained by striking out all of the rows and columns of $\mathbf{A}$ other than r linearly independent rows, say rows $i_1, \ldots, i_r$, respectively, and r linearly independent columns, say columns $j_1, \ldots, j_r$, respectively. Letting a_{ij} represent the ijth element of $\mathbf{A}$, the tsth element of $\mathbf{B}$ is by definition

$$b_{ts} = a_{i_t j_s}.$$

Let $\mathbf{U}$ represent the $m \times r$ matrix whose columns are columns $j_1, \ldots, j_r$ of $\mathbf{A}$. According to Theorem 4.3.9, rows $i_1, \ldots, i_r$ of $\mathbf{A}$ form a basis for $\mathcal{R}(\mathbf{A})$, implying that every row of $\mathbf{A}$ is expressible as a linear combination of rows $i_1, \ldots, i_r$ or, equivalently, that

$$a_{tj} = \sum_{k=1}^{r} f_{tk} a_{i_k j} \qquad (j = 1, \ldots, n)$$

for some scalars $f_{t1}, \ldots, f_{tr}$ $(t = 1, \ldots, m)$. Thus, the tsth element of $\mathbf{U}$ is

$$u_{ts} = a_{tj_s} = \sum_k f_{tk} a_{i_k j_s} = \sum_k f_{tk} b_{ks}$$

$(t = 1, \ldots, m;\ s = 1, \ldots, r)$, so that

$$\mathbf{U} = \mathbf{FB},$$

where $\mathbf{F}$ is the $m \times r$ matrix whose tkth element is f_{tk}. Since (according to Lemma 4.4.3 and Corollary 4.4.5) $\text{rank}(\mathbf{B}) \leq r$ and $\text{rank}(\mathbf{B}) \geq \text{rank}(\mathbf{U}) = r$, we conclude that $\text{rank}(\mathbf{B}) = r$ or, equivalently, that $\mathbf{B}$ is nonsingular.

It remains to show that there exists no submatrix of $\mathbf{A}$ whose rank exceeds r. Let $\mathbf{H}$ represent a matrix obtained by striking out all of the rows and columns of $\mathbf{A}$ except rows $t_1, \ldots, t_k$ and columns $s_1, \ldots, s_p$. Define $\mathbf{W}$ to be the $m \times p$ matrix whose columns are columns $s_1, \ldots, s_p$ of $\mathbf{A}$. It is clear that $\mathcal{C}(\mathbf{W}) \subset \mathcal{C}(\mathbf{A})$ and (since the rows of $\mathbf{H}$ are rows $t_1, \ldots, t_k$ of $\mathbf{W}$) that $\mathcal{R}(\mathbf{H}) \subset \mathcal{R}(\mathbf{W})$. Thus, applying Theorem 4.4.4,

$$\text{rank}(\mathbf{H}) \leq \text{rank}(\mathbf{W}) \leq r.$$

Q.E.D.

As applied to symmetric matrices, Theorem 4.4.10 has the following implication.

Corollary 4.4.11. Any symmetric matrix of rank r contains an $r \times r$ nonsingular principal submatrix.

The rank of a matrix was defined in terms of the dimension of the row and column spaces. Theorem 4.4.10 suggests some equivalent definitions. The rank of a matrix $\mathbf{A}$ is interpretable as the size of the largest linearly independent set that can be formed from the rows of $\mathbf{A}$. Similarly, it is interpretable as the size of the largest linearly independent set that can be formed from the columns of $\mathbf{A}$. The rank of $\mathbf{A}$ is also interpretable as the size (number of rows or columns) of the largest nonsingular (square) submatrix of $\mathbf{A}$.

Clearly, an $m \times n$ matrix has full row rank if and only if all m of its rows are linearly independent and has full column rank if and only if all n of its columns are linearly independent. An $n \times n$ matrix is nonsingular if and only if all of its rows are linearly independent; similarly, it is nonsingular if and only if all of its columns are linearly independent.

f. Some elementary equalities

Let $\mathbf{A}$ represent an arbitrary matrix. Then, it is a simple exercise to show that

$$\text{rank}(\mathbf{A}') = \text{rank}(\mathbf{A}) \tag{4.1}$$

and that, for any *nonzero* scalar k,

$$\text{rank}(k\mathbf{A}) = \text{rank}(\mathbf{A}). \tag{4.2}$$

As a special case of result (4.2), we have that

$$\text{rank}(-\mathbf{A}) = \text{rank}(\mathbf{A}). \tag{4.3}$$

4.5 Some Basic Results on Partitioned Matrices and on Sums of Matrices

In what follows, the symbol $\mathcal{C}(\mathbf{A}, \mathbf{B})$ is used as an abbreviation for $\mathcal{C}[(\mathbf{A}, \mathbf{B})]$ and $\mathcal{R}\begin{pmatrix}\mathbf{A}\\ \mathbf{C}\end{pmatrix}$ as an abbreviation for $\mathcal{R}\left[\begin{pmatrix}\mathbf{A}\\ \mathbf{C}\end{pmatrix}\right]$. Thus, $\mathcal{C}(\mathbf{A}, \mathbf{B})$ represents the column space of a partitioned matrix $(\mathbf{A}, \mathbf{B})$ comprising two blocks $\mathbf{A}$ and $\mathbf{B}$ arranged in a row, and $\mathcal{R}\begin{pmatrix}\mathbf{A}\\ \mathbf{C}\end{pmatrix}$ represents the row space of a partitioned matrix $\begin{pmatrix}\mathbf{A}\\ \mathbf{C}\end{pmatrix}$ comprising two blocks $\mathbf{A}$ and $\mathbf{C}$ arranged in a column.

The following five lemmas (and two corollaries) give some basic results on the ranks and row and column spaces of matrices that have been partitioned into two blocks.

Lemma 4.5.1. Let $\mathbf{A}$ represent an $m \times n$ matrix, $\mathbf{B}$ an $m \times p$ matrix, and $\mathbf{C}$ a $q \times n$ matrix. Then,

$$\mathcal{C}(\mathbf{A}) \subset \mathcal{C}(\mathbf{A}, \mathbf{B}), \qquad \mathcal{C}(\mathbf{B}) \subset \mathcal{C}(\mathbf{A}, \mathbf{B}),$$

$$\mathcal{R}(\mathbf{A}) \subset \mathcal{R}\begin{pmatrix}\mathbf{A}\\ \mathbf{C}\end{pmatrix}, \qquad \mathcal{R}(\mathbf{C}) \subset \mathcal{R}\begin{pmatrix}\mathbf{A}\\ \mathbf{C}\end{pmatrix}.$$

Moreover,

$$\mathcal{C}(\mathbf{A}) = \mathcal{C}(\mathbf{A}, \mathbf{B}) \Leftrightarrow \mathcal{C}(\mathbf{B}) \subset \mathcal{C}(\mathbf{A}), \qquad \mathcal{C}(\mathbf{B}) = \mathcal{C}(\mathbf{A}, \mathbf{B}) \Leftrightarrow \mathcal{C}(\mathbf{A}) \subset \mathcal{C}(\mathbf{B}),$$

$$\mathcal{R}(\mathbf{A}) = \mathcal{R}\begin{pmatrix}\mathbf{A}\\ \mathbf{C}\end{pmatrix} \Leftrightarrow \mathcal{R}(\mathbf{C}) \subset \mathcal{R}(\mathbf{A}), \qquad \mathcal{R}(\mathbf{C}) = \mathcal{R}\begin{pmatrix}\mathbf{A}\\ \mathbf{C}\end{pmatrix} \Leftrightarrow \mathcal{R}(\mathbf{A}) \subset \mathcal{R}(\mathbf{C}).$$

Proof. That $\mathcal{C}(\mathbf{A}) \subset \mathcal{C}(\mathbf{A}, \mathbf{B})$, $\mathcal{C}(\mathbf{B}) \subset \mathcal{C}(\mathbf{A}, \mathbf{B})$, $\mathcal{R}(\mathbf{A}) \subset \mathcal{R}\begin{pmatrix}\mathbf{A}\\ \mathbf{C}\end{pmatrix}$, and $\mathcal{R}(\mathbf{C}) \subset \mathcal{R}\begin{pmatrix}\mathbf{A}\\ \mathbf{C}\end{pmatrix}$ follows from Lemma 4.2.2 upon observing that

$$\mathbf{A} = (\mathbf{A}, \mathbf{B}) \begin{pmatrix}\mathbf{I}\\ \mathbf{0}\end{pmatrix}, \qquad \mathbf{B} = (\mathbf{A}, \mathbf{B}) \begin{pmatrix}\mathbf{0}\\ \mathbf{I}\end{pmatrix},$$

$$\mathbf{A} = (\mathbf{I}, \mathbf{0}) \begin{pmatrix}\mathbf{A}\\ \mathbf{C}\end{pmatrix}, \qquad \mathbf{C} = (\mathbf{0}, \mathbf{I}) \begin{pmatrix}\mathbf{A}\\ \mathbf{C}\end{pmatrix}.$$

Suppose now that $\mathcal{C}(\mathbf{B}) \subset \mathcal{C}(\mathbf{A})$. Then, according to Lemma 4.2.2, there exists a matrix $\mathbf{F}$ such that $\mathbf{B} = \mathbf{AF}$ and hence such that $(\mathbf{A}, \mathbf{B}) = \mathbf{A}(\mathbf{I}, \mathbf{F})$. Thus, $\mathcal{C}(\mathbf{A}, \mathbf{B}) \subset \mathcal{C}(\mathbf{A})$, implying [since $\mathcal{C}(\mathbf{A}) \subset \mathcal{C}(\mathbf{A}, \mathbf{B})$] that $\mathcal{C}(\mathbf{A}) = \mathcal{C}(\mathbf{A}, \mathbf{B})$.

Conversely, suppose that $\mathcal{C}(\mathbf{A}) = \mathcal{C}(\mathbf{A}, \mathbf{B})$. Then, since $\mathcal{C}(\mathbf{B}) \subset \mathcal{C}(\mathbf{A}, \mathbf{B})$, $\mathcal{C}(\mathbf{B}) \subset \mathcal{C}(\mathbf{A})$. Thus, we have established that $\mathcal{C}(\mathbf{A}) = \mathcal{C}(\mathbf{A}, \mathbf{B}) \Leftrightarrow \mathcal{C}(\mathbf{B}) \subset \mathcal{C}(\mathbf{A})$.

That $\mathcal{C}(\mathbf{B}) = \mathcal{C}(\mathbf{A}, \mathbf{B}) \Leftrightarrow \mathcal{C}(\mathbf{A}) \subset \mathcal{C}(\mathbf{B})$, that $\mathcal{R}(\mathbf{A}) = \mathcal{R}\begin{pmatrix}\mathbf{A}\\ \mathbf{C}\end{pmatrix} \Leftrightarrow \mathcal{R}(\mathbf{C}) \subset \mathcal{R}(\mathbf{A})$, and that $\mathcal{R}(\mathbf{C}) = \mathcal{R}\begin{pmatrix}\mathbf{A}\\ \mathbf{C}\end{pmatrix} \Leftrightarrow \mathcal{R}(\mathbf{A}) \subset \mathcal{R}(\mathbf{C})$ can be established via similar arguments. Q.E.D.

Corollary 4.5.2. Let $\mathbf{A}$ represent an $m \times n$ matrix, $\mathbf{B}$ an $m \times p$ matrix, and $\mathbf{C}$ a $q \times n$ matrix. Then,

(1) $\operatorname{rank}(\mathbf{A}) \le \operatorname{rank}(\mathbf{A}, \mathbf{B})$, with equality holding if and only if $\mathcal{C}(\mathbf{B}) \subset \mathcal{C}(\mathbf{A})$;
(2) $\operatorname{rank}(\mathbf{B}) \le \operatorname{rank}(\mathbf{A}, \mathbf{B})$, with equality holding if and only if $\mathcal{C}(\mathbf{A}) \subset \mathcal{C}(\mathbf{B})$;
(3) $\operatorname{rank}(\mathbf{A}) \le \operatorname{rank}\begin{pmatrix}\mathbf{A}\\ \mathbf{C}\end{pmatrix}$, with equality holding if and only if $\mathcal{R}(\mathbf{C}) \subset \mathcal{R}(\mathbf{A})$;
(4) $\operatorname{rank}(\mathbf{C}) \le \operatorname{rank}\begin{pmatrix}\mathbf{A}\\ \mathbf{C}\end{pmatrix}$, with equality holding if and only if $\mathcal{R}(\mathbf{A}) \subset \mathcal{R}(\mathbf{C})$.

Proof. Since (according to Lemma 4.5.1) $\mathcal{C}(\mathbf{A}) \subset \mathcal{C}(\mathbf{A}, \mathbf{B})$, it follows from Theorem 4.4.4 that $\operatorname{rank}(\mathbf{A}) \le \operatorname{rank}(\mathbf{A}, \mathbf{B})$. Moreover, if $\mathcal{C}(\mathbf{B}) \subset \mathcal{C}(\mathbf{A})$, then (according to Lemma 4.5.1) $\mathcal{C}(\mathbf{A}) = \mathcal{C}(\mathbf{A}, \mathbf{B})$ and, consequently, $\operatorname{rank}(\mathbf{A}) = \operatorname{rank}(\mathbf{A}, \mathbf{B})$. And, conversely, if $\operatorname{rank}(\mathbf{A}) = \operatorname{rank}(\mathbf{A}, \mathbf{B})$, then, since $\mathcal{C}(\mathbf{A}) \subset \mathcal{C}(\mathbf{A}, \mathbf{B})$, it follows from Theorem 4.4.6 that $\mathcal{C}(\mathbf{A}) = \mathcal{C}(\mathbf{A}, \mathbf{B})$ and hence (in light of Lemma 4.5.1) that $\mathcal{C}(\mathbf{B}) \subset \mathcal{C}(\mathbf{A})$. Thus, Part (1) of the corollary is valid. The validity of Parts (2), (3), and (4) can be established in similar fashion. Q.E.D.

Lemma 4.5.3. For any $m \times n$ matrix $\mathbf{A}$, $m \times p$ matrix $\mathbf{B}$, and $q \times n$ matrix $\mathbf{C}$,

$$\mathcal{C}(\mathbf{B}, \mathbf{A}) = \mathcal{C}(\mathbf{A}, \mathbf{B}), \qquad \operatorname{rank}(\mathbf{B}, \mathbf{A}) = \operatorname{rank}(\mathbf{A}, \mathbf{B}),$$
$$\mathcal{R}\begin{pmatrix}\mathbf{C}\\ \mathbf{A}\end{pmatrix} = \mathcal{R}\begin{pmatrix}\mathbf{A}\\ \mathbf{C}\end{pmatrix}, \qquad \operatorname{rank}\begin{pmatrix}\mathbf{C}\\ \mathbf{A}\end{pmatrix} = \operatorname{rank}\begin{pmatrix}\mathbf{A}\\ \mathbf{C}\end{pmatrix}.$$

Proof. Since

$$(\mathbf{B}, \mathbf{A}) = (\mathbf{A}, \mathbf{B})\begin{pmatrix}\mathbf{0} & \mathbf{I}_n\\ \mathbf{I}_p & \mathbf{0}\end{pmatrix}, \qquad (\mathbf{A}, \mathbf{B}) = (\mathbf{B}, \mathbf{A})\begin{pmatrix}\mathbf{0} & \mathbf{I}_p\\ \mathbf{I}_n & \mathbf{0}\end{pmatrix},$$

we have (as a consequence of Lemma 4.2.2) that $\mathcal{C}(\mathbf{B}, \mathbf{A}) \subset \mathcal{C}(\mathbf{A}, \mathbf{B})$ and $\mathcal{C}(\mathbf{A}, \mathbf{B}) \subset \mathcal{C}(\mathbf{B}, \mathbf{A})$ and hence that $\mathcal{C}(\mathbf{B}, \mathbf{A}) = \mathcal{C}(\mathbf{A}, \mathbf{B})$ [which implies that $\operatorname{rank}(\mathbf{B}, \mathbf{A}) = \operatorname{rank}(\mathbf{A}, \mathbf{B})$]. That $\mathcal{R}\begin{pmatrix}\mathbf{C}\\ \mathbf{A}\end{pmatrix} = \mathcal{R}\begin{pmatrix}\mathbf{A}\\ \mathbf{C}\end{pmatrix}$ [and that $\operatorname{rank}\begin{pmatrix}\mathbf{C}\\ \mathbf{A}\end{pmatrix} = \operatorname{rank}\begin{pmatrix}\mathbf{A}\\ \mathbf{C}\end{pmatrix}$] follows from an analogous argument. Q.E.D.

Lemma 4.5.4. For any $m \times n$ matrix $\mathbf{A}$, $m \times p$ matrix $\mathbf{B}$, and $n \times p$ matrix $\mathbf{L}$,

$$\mathcal{C}(\mathbf{A}, \mathbf{B}) = \mathcal{C}(\mathbf{A}, \mathbf{B} - \mathbf{AL}), \qquad \operatorname{rank}(\mathbf{A}, \mathbf{B}) = \operatorname{rank}(\mathbf{A}, \mathbf{B} - \mathbf{AL}).$$

Similarly, for any $m \times n$ matrix $\mathbf{A}$, $q \times n$ matrix $\mathbf{B}$, and $q \times m$ matrix $\mathbf{L}$,

$$\mathcal{R}\begin{pmatrix}\mathbf{A}\\ \mathbf{B}\end{pmatrix} = \mathcal{R}\begin{pmatrix}\mathbf{A}\\ \mathbf{B} - \mathbf{LA}\end{pmatrix}, \qquad \operatorname{rank}\begin{pmatrix}\mathbf{A}\\ \mathbf{B}\end{pmatrix} = \operatorname{rank}\begin{pmatrix}\mathbf{A}\\ \mathbf{B} - \mathbf{LA}\end{pmatrix}.$$

Proof. To establish the first part of the lemma, observe that

$$(\mathbf{A}, \mathbf{B} - \mathbf{AL}) = (\mathbf{A}, \mathbf{B})\begin{pmatrix}\mathbf{I} & -\mathbf{L}\\ \mathbf{0} & \mathbf{I}\end{pmatrix}, \qquad (\mathbf{A}, \mathbf{B}) = (\mathbf{A}, \mathbf{B} - \mathbf{AL})\begin{pmatrix}\mathbf{I} & \mathbf{L}\\ \mathbf{0} & \mathbf{I}\end{pmatrix},$$

implying (in light of Lemma 4.2.2) that $\mathcal{C}(\mathbf{A}, \mathbf{B}-\mathbf{AL}) \subset \mathcal{C}(\mathbf{A}, \mathbf{B})$ and $\mathcal{C}(\mathbf{A}, \mathbf{B}) \subset \mathcal{C}(\mathbf{A}, \mathbf{B}-\mathbf{AL})$ and, consequently, that $\mathcal{C}(\mathbf{A}, \mathbf{B}) = \mathcal{C}(\mathbf{A}, \mathbf{B}-\mathbf{AL})$ and $\text{rank}(\mathbf{A}, \mathbf{B}) = \text{rank}(\mathbf{A}, \mathbf{B} - \mathbf{AL})$. The second part of the lemma can be established in similar fashion. Q.E.D.

Lemma 4.5.5. (1) If $\mathbf{A}$ is an $m \times n$ matrix and $\mathbf{E}$ an $m \times q$ matrix such that $\mathcal{C}(\mathbf{E}) \subset \mathcal{C}(\mathbf{A})$ and if $\mathbf{B}$ is an $m \times p$ matrix and $\mathbf{F}$ an $m \times r$ matrix such that $\mathcal{C}(\mathbf{F}) \subset \mathcal{C}(\mathbf{B})$, then

$$\mathcal{C}(\mathbf{E}, \mathbf{F}) \subset \mathcal{C}(\mathbf{A}, \mathbf{B}), \qquad \text{rank}(\mathbf{E}, \mathbf{F}) \leq \text{rank}(\mathbf{A}, \mathbf{B}).$$

(2) Similarly, if $\mathbf{A}$ is an $m \times n$ matrix and $\mathbf{E}$ a $p \times n$ matrix such that $\mathcal{R}(\mathbf{E}) \subset \mathcal{R}(\mathbf{A})$ and if $\mathbf{B}$ is a $q \times n$ matrix and $\mathbf{F}$ an $r \times n$ matrix such that $\mathcal{R}(\mathbf{F}) \subset \mathcal{R}(\mathbf{B})$, then

$$\mathcal{R}\begin{pmatrix} \mathbf{E} \\ \mathbf{F} \end{pmatrix} \subset \mathcal{R}\begin{pmatrix} \mathbf{A} \\ \mathbf{B} \end{pmatrix}, \qquad \text{rank}\begin{pmatrix} \mathbf{E} \\ \mathbf{F} \end{pmatrix} \leq \text{rank}\begin{pmatrix} \mathbf{A} \\ \mathbf{B} \end{pmatrix}.$$

Proof. (1) Suppose that $\mathcal{C}(\mathbf{E}) \subset \mathcal{C}(\mathbf{A})$ and $\mathcal{C}(\mathbf{F}) \subset \mathcal{C}(\mathbf{B})$. Then, according to Lemma 4.2.2, there exist matrices $\mathbf{K}$ and $\mathbf{L}$ such that $\mathbf{E} = \mathbf{AK}$ and $\mathbf{F} = \mathbf{BL}$. Thus, $(\mathbf{E}, \mathbf{F}) = (\mathbf{A}, \mathbf{B})\,\text{diag}(\mathbf{K}, \mathbf{L})$, implying that $\mathcal{C}(\mathbf{E}, \mathbf{F}) \subset \mathcal{C}(\mathbf{A}, \mathbf{B})$ and further (in light of Theorem 4.4.4) that $\text{rank}(\mathbf{E}, \mathbf{F}) \leq \text{rank}(\mathbf{A}, \mathbf{B})$.

(2) The proof of Part (2) is analogous to that of Part (1). Q.E.D.

Corollary 4.5.6. (1) If $\mathbf{A}$ is an $m \times n$ matrix and $\mathbf{E}$ an $m \times q$ matrix such that $\mathcal{C}(\mathbf{E}) = \mathcal{C}(\mathbf{A})$, and if $\mathbf{B}$ is an $m \times p$ matrix and $\mathbf{F}$ an $m \times r$ matrix such that $\mathcal{C}(\mathbf{F}) = \mathcal{C}(\mathbf{B})$, then

$$\mathcal{C}(\mathbf{E}, \mathbf{F}) = \mathcal{C}(\mathbf{A}, \mathbf{B}), \qquad \text{rank}(\mathbf{E}, \mathbf{F}) = \text{rank}(\mathbf{A}, \mathbf{B}).$$

(2) Similarly, if $\mathbf{A}$ is an $m \times n$ matrix and $\mathbf{E}$ a $p \times n$ matrix such that $\mathcal{R}(\mathbf{E}) = \mathcal{R}(\mathbf{A})$, and if $\mathbf{B}$ is a $q \times n$ matrix and $\mathbf{F}$ an $r \times n$ matrix such that $\mathcal{R}(\mathbf{F}) = \mathcal{R}(\mathbf{B})$, then

$$\mathcal{R}\begin{pmatrix} \mathbf{E} \\ \mathbf{F} \end{pmatrix} = \mathcal{R}\begin{pmatrix} \mathbf{A} \\ \mathbf{B} \end{pmatrix}, \qquad \text{rank}\begin{pmatrix} \mathbf{E} \\ \mathbf{F} \end{pmatrix} = \text{rank}\begin{pmatrix} \mathbf{A} \\ \mathbf{B} \end{pmatrix}.$$

Proof. (1) Suppose that $\mathcal{C}(\mathbf{E}) = \mathcal{C}(\mathbf{A})$ and $\mathcal{C}(\mathbf{F}) = \mathcal{C}(\mathbf{B})$. Then, as a consequence of Lemma 4.5.5, we have that $\mathcal{C}(\mathbf{E}, \mathbf{F}) \subset \mathcal{C}(\mathbf{A}, \mathbf{B})$ and $\mathcal{C}(\mathbf{A}, \mathbf{B}) \subset \mathcal{C}(\mathbf{E}, \mathbf{F})$. Thus, $\mathcal{C}(\mathbf{E}, \mathbf{F}) = \mathcal{C}(\mathbf{A}, \mathbf{B})$, and further $\text{rank}(\mathbf{E}, \mathbf{F}) = \text{rank}(\mathbf{A}, \mathbf{B})$.

(2) The proof of Part (2) is analogous to that of Part (1). Q.E.D.

Lemma 4.5.7. For any $m \times n$ matrix $\mathbf{A}$, $m \times p$ matrix $\mathbf{B}$, and $q \times n$ matrix $\mathbf{C}$,

$$\text{rank}(\mathbf{A}, \mathbf{B}) \leq \text{rank}(\mathbf{A}) + \text{rank}(\mathbf{B}), \tag{5.1}$$

$$\text{rank}\begin{pmatrix} \mathbf{A} \\ \mathbf{C} \end{pmatrix} \leq \text{rank}(\mathbf{A}) + \text{rank}(\mathbf{C}). \tag{5.2}$$

Proof. For purposes of establishing inequality (5.1), assume that both $\mathbf{A}$ and $\mathbf{B}$ are nonnull—if either $\mathbf{A}$ or $\mathbf{B}$ is null, then, clearly, inequality (5.1) holds as an equality. Let $\mathbf{A}^*$ represent an $m \times r$ matrix whose columns form a basis for $\mathcal{C}(\mathbf{A})$

and $\mathbf{B}^*$ an $m \times s$ matrix whose columns form a basis for $\mathcal{C}(\mathbf{B})$. Then, clearly, $\mathcal{C}(\mathbf{A}^*) = \mathcal{C}(\mathbf{A})$ and $\mathcal{C}(\mathbf{B}^*) = \mathcal{C}(\mathbf{B})$. Thus, in light of Corollary 4.5.6 and Lemma 4.4.3, we have that

$$\text{rank}(\mathbf{A}, \mathbf{B}) = \text{rank}(\mathbf{A}^*, \mathbf{B}^*) \leq r + s = \text{rank}(\mathbf{A}) + \text{rank}(\mathbf{B}),$$

which establishes inequality (5.1). Inequality (5.2) can be established via an analogous argument. Q.E.D.

By repeated application of result (5.1), we obtain the more general result that, for any matrices $\mathbf{A}_1, \mathbf{A}_2, \ldots, \mathbf{A}_k$ having m rows,

$$\text{rank}(\mathbf{A}_1, \mathbf{A}_2, \ldots, \mathbf{A}_k) \leq \text{rank}(\mathbf{A}_1) + \text{rank}(\mathbf{A}_2) + \cdots + \text{rank}(\mathbf{A}_k). \tag{5.3}$$

Similarly, by repeated application of result (5.2), we find that, for any matrices $\mathbf{A}_1, \mathbf{A}_2, \ldots, \mathbf{A}_k$ having n columns,

$$\text{rank}\begin{pmatrix} \mathbf{A}_1 \\ \mathbf{A}_2 \\ \vdots \\ \mathbf{A}_k \end{pmatrix} \leq \text{rank}(\mathbf{A}_1) + \text{rank}(\mathbf{A}_2) + \cdots + \text{rank}(\mathbf{A}_k). \tag{5.4}$$

The following lemma establishes a basic relationship between the rank or the row or column space of a sum of two matrices and the rank or the row or column space of a certain partitioned matrix.

Lemma 4.5.8. For any $m \times n$ matrices $\mathbf{A}$ and $\mathbf{B}$,

$$\mathcal{C}(\mathbf{A} + \mathbf{B}) \subset \mathcal{C}(\mathbf{A}, \mathbf{B}), \qquad \text{rank}(\mathbf{A} + \mathbf{B}) \leq \text{rank}(\mathbf{A}, \mathbf{B}), \tag{5.5}$$

$$\mathcal{R}(\mathbf{A} + \mathbf{B}) \subset \mathcal{R}\begin{pmatrix} \mathbf{A} \\ \mathbf{B} \end{pmatrix}, \qquad \text{rank}(\mathbf{A} + \mathbf{B}) \leq \text{rank}\begin{pmatrix} \mathbf{A} \\ \mathbf{B} \end{pmatrix}. \tag{5.6}$$

Proof. We have that

$$\mathbf{A} + \mathbf{B} = (\mathbf{A}, \mathbf{B})\begin{pmatrix} \mathbf{I} \\ \mathbf{I} \end{pmatrix} = (\mathbf{I}, \mathbf{I})\begin{pmatrix} \mathbf{A} \\ \mathbf{B} \end{pmatrix},$$

implying (in light of Lemma 4.2.2) that $\mathcal{C}(\mathbf{A} + \mathbf{B}) \subset \mathcal{C}(\mathbf{A}, \mathbf{B})$ and $\mathcal{R}(\mathbf{A} + \mathbf{B}) \subset \mathcal{R}\begin{pmatrix} \mathbf{A} \\ \mathbf{B} \end{pmatrix}$ and further (in light of Theorem 4.4.4) that $\text{rank}(\mathbf{A} + \mathbf{B}) \leq \text{rank}(\mathbf{A}, \mathbf{B})$ and $\text{rank}(\mathbf{A} + \mathbf{B}) \leq \text{rank}\begin{pmatrix} \mathbf{A} \\ \mathbf{B} \end{pmatrix}$. Q.E.D.

In light of Lemma 4.5.7, we have the following corollary of Lemma 4.5.8.

Corollary 4.5.9. For any $m \times n$ matrices $\mathbf{A}$ and $\mathbf{B}$,

$$\text{rank}(\mathbf{A} + \mathbf{B}) \leq \text{rank}(\mathbf{A}) + \text{rank}(\mathbf{B}). \tag{5.7}$$

By repeated application of results (5.5) through (5.7), we obtain the more general results that, for any $m \times n$ matrices $\mathbf{A}_1, \mathbf{A}_2, \ldots, \mathbf{A}_k$,

$$\mathcal{C}(\mathbf{A}_1 + \mathbf{A}_2 + \cdots + \mathbf{A}_k) \subset \mathcal{C}(\mathbf{A}_1, \mathbf{A}_2, \ldots, \mathbf{A}_k), \tag{5.8a}$$

$$\text{rank}(\mathbf{A}_1 + \mathbf{A}_2 + \cdots + \mathbf{A}_k) \leq \text{rank}(\mathbf{A}_1, \mathbf{A}_2, \ldots, \mathbf{A}_k), \tag{5.8b}$$

$$\mathcal{R}(\mathbf{A}_1 + \mathbf{A}_2 + \cdots + \mathbf{A}_k) \subset \mathcal{R}\begin{pmatrix} \mathbf{A}_1 \\ \mathbf{A}_2 \\ \vdots \\ \mathbf{A}_k \end{pmatrix}, \tag{5.9a}$$

$$\text{rank}(\mathbf{A}_1 + \mathbf{A}_2 + \cdots + \mathbf{A}_k) \leq \text{rank}\begin{pmatrix} \mathbf{A}_1 \\ \mathbf{A}_2 \\ \vdots \\ \mathbf{A}_k \end{pmatrix}, \tag{5.9b}$$

$$\text{rank}(\mathbf{A}_1 + \mathbf{A}_2 + \cdots + \mathbf{A}_k) \leq \text{rank}(\mathbf{A}_1) + \text{rank}(\mathbf{A}_2) + \cdots + \text{rank}(\mathbf{A}_k). \tag{5.10}$$

A further result on the row and column spaces of a sum of matrices is given by the following lemma.

Lemma 4.5.10. Let $\mathbf{A}$ and $\mathbf{B}$ represent $m \times n$ matrices. Then, for any $m \times p$ matrix $\mathbf{E}$ such that $\mathcal{C}(\mathbf{A}) \subset \mathcal{C}(\mathbf{E})$,

$$\mathcal{C}(\mathbf{A} + \mathbf{B}) \subset \mathcal{C}(\mathbf{E}) \quad \Leftrightarrow \quad \mathcal{C}(\mathbf{B}) \subset \mathcal{C}(\mathbf{E}).$$

And, similarly, for any $q \times n$ matrix $\mathbf{F}$ such that $\mathcal{R}(\mathbf{A}) \subset \mathcal{R}(\mathbf{F})$,

$$\mathcal{R}(\mathbf{A} + \mathbf{B}) \subset \mathcal{R}(\mathbf{F}) \quad \Leftrightarrow \quad \mathcal{R}(\mathbf{B}) \subset \mathcal{R}(\mathbf{F}).$$

Proof. According to Lemma 4.2.2, $\mathbf{A} = \mathbf{ER}$ for some matrix $\mathbf{R}$. Further, if $\mathcal{C}(\mathbf{B}) \subset \mathcal{C}(\mathbf{E})$, $\mathbf{B} = \mathbf{ES}$ for some matrix $\mathbf{S}$, implying that $\mathbf{A} + \mathbf{B} = \mathbf{E}(\mathbf{R} + \mathbf{S})$ and hence (in light of Lemma 4.2.2) that $\mathcal{C}(\mathbf{A} + \mathbf{B}) \subset \mathcal{C}(\mathbf{E})$. Conversely, if $\mathcal{C}(\mathbf{A} + \mathbf{B}) \subset \mathcal{C}(\mathbf{E})$, then $\mathbf{A} + \mathbf{B} = \mathbf{ET}$ for some matrix $\mathbf{T}$, implying that $\mathbf{B} = (\mathbf{A} + \mathbf{B}) - \mathbf{A} = \mathbf{E}(\mathbf{T} - \mathbf{R})$ and hence that $\mathcal{C}(\mathbf{B}) \subset \mathcal{C}(\mathbf{E})$.

It can be established in similar fashion that $\mathcal{R}(\mathbf{A} + \mathbf{B}) \subset \mathcal{R}(\mathbf{F})$ if and only if $\mathcal{R}(\mathbf{B}) \subset \mathcal{R}(\mathbf{F})$. Q.E.D.

The following lemma gives the rank of a block-diagonal matrix in terms of the ranks of the diagonal blocks.

Lemma 4.5.11. For any matrices $\mathbf{A}$ and $\mathbf{B}$,

$$\text{rank}\begin{pmatrix} \mathbf{A} & \mathbf{0} \\ \mathbf{0} & \mathbf{B} \end{pmatrix} = \text{rank}(\mathbf{A}) + \text{rank}(\mathbf{B}). \tag{5.11}$$

Proof. Suppose that both $\mathbf{A}$ and $\mathbf{B}$ are nonnull—if $\mathbf{A}$ or $\mathbf{B}$ is null, then equality (5.11) is clearly valid. Let $\mathbf{A}^*$ represent an $m \times r$ matrix whose columns form a basis for $\mathcal{C}(\mathbf{A})$ and $\mathbf{B}^*$ an $n \times s$ matrix whose columns form a basis for $\mathcal{C}(\mathbf{B})$.

Then, $\mathcal{C}(\mathbf{A}^*) = \mathcal{C}(\mathbf{A})$, and $\mathcal{C}(\mathbf{B}^*) = \mathcal{C}(\mathbf{B})$, implying (in light of Lemma 4.2.2) that $\mathbf{A}^* = \mathbf{AK}$ and $\mathbf{B}^* = \mathbf{BL}$ for some matrices $\mathbf{K}$ and $\mathbf{L}$ and hence that

$$\begin{pmatrix} \mathbf{A}^* & \mathbf{0} \\ \mathbf{0} & \mathbf{B}^* \end{pmatrix} = \begin{pmatrix} \mathbf{A} & \mathbf{0} \\ \mathbf{0} & \mathbf{B} \end{pmatrix} \begin{pmatrix} \mathbf{K} & \mathbf{0} \\ \mathbf{0} & \mathbf{L} \end{pmatrix}.$$

Moreover, for any $r \times 1$ vector $\mathbf{c}$ and $s \times 1$ vector $\mathbf{d}$ such that

$$\begin{pmatrix} \mathbf{A}^* & \mathbf{0} \\ \mathbf{0} & \mathbf{B}^* \end{pmatrix} \begin{pmatrix} \mathbf{c} \\ \mathbf{d} \end{pmatrix} = \mathbf{0},$$

we find that $\mathbf{A}^*\mathbf{c} = \mathbf{0}$ and $\mathbf{B}^*\mathbf{d} = \mathbf{0}$, implying (since the columns of $\mathbf{A}^*$ are linearly independent) that $\mathbf{c} = \mathbf{0}$ and, similarly, that $\mathbf{d} = \mathbf{0}$. It follows that the columns of $\text{diag}(\mathbf{A}^*, \mathbf{B}^*)$ are linearly independent. Thus,

$$\text{rank}\begin{pmatrix} \mathbf{A} & \mathbf{0} \\ \mathbf{0} & \mathbf{B} \end{pmatrix} \geq \text{rank}\begin{pmatrix} \mathbf{A}^* & \mathbf{0} \\ \mathbf{0} & \mathbf{B}^* \end{pmatrix} = r + s = \text{rank}(\mathbf{A}) + \text{rank}(\mathbf{B}). \tag{5.12}$$

Further, it follows from Lemma 4.5.7 that

$$\text{rank}\begin{pmatrix} \mathbf{A} & \mathbf{0} \\ \mathbf{0} & \mathbf{B} \end{pmatrix} \leq \text{rank}\begin{pmatrix} \mathbf{A} \\ \mathbf{0} \end{pmatrix} + \text{rank}\begin{pmatrix} \mathbf{0} \\ \mathbf{B} \end{pmatrix} \leq \text{rank}(\mathbf{A}) + \text{rank}(\mathbf{B}). \tag{5.13}$$

Together, inequalities (5.12) and (5.13) imply equality (5.11). Q.E.D.

Repeated application of result (5.11) gives the following formula for the rank of an arbitrary block-diagonal matrix:

$$\text{rank}[\text{diag}(\mathbf{A}_1, \mathbf{A}_2, \ldots, \mathbf{A}_k)] = \text{rank}(\mathbf{A}_1) + \text{rank}(\mathbf{A}_2) + \cdots + \text{rank}(\mathbf{A}_k). \tag{5.14}$$

Note that result (5.14) implies in particular that the rank of a diagonal matrix $\mathbf{D}$ equals the number of nonzero diagonal elements in $\mathbf{D}$.

Exercises

Section 4.1

1. Which of the following two sets are linear spaces: (a) the set of all $n \times n$ upper triangular matrices; (b) the set of all $n \times n$ nonsymmetric matrices?

Section 4.2

2. Verify Lemma 4.2.5.

3. Let $\mathcal{U}$ and $\mathcal{W}$ represent subspaces of a linear space $\mathcal{V}$. Show that if every matrix in $\mathcal{V}$ belongs to $\mathcal{U}$ or $\mathcal{W}$, then $\mathcal{U} = \mathcal{V}$ or $\mathcal{W} = \mathcal{V}$.

Section 4.3

4. Let $\mathbf{A}$, $\mathbf{B}$, and $\mathbf{C}$ represent three matrices (having the same dimensions) such that $\mathbf{A} + \mathbf{B} + \mathbf{C} = \mathbf{0}$. Show that $\mathrm{sp}(\mathbf{A}, \mathbf{B}) = \mathrm{sp}(\mathbf{A}, \mathbf{C})$.*

5. Let $\mathbf{A}_1, \ldots, \mathbf{A}_k$ represent any matrices in a linear space $\mathcal{V}$. Show that $\mathrm{sp}(\mathbf{A}_1, \ldots, \mathbf{A}_k)$ is a subspace of $\mathcal{V}$ and that, among all subspaces of $\mathcal{V}$ that contain $\mathbf{A}_1, \ldots \mathbf{A}_k$, it is the smallest [in the sense that, for any subspace $\mathcal{U}$ (of $\mathcal{V}$) that contains $\mathbf{A}_1, \ldots, \mathbf{A}_k$, $\mathrm{sp}(\mathbf{A}_1, \ldots, \mathbf{A}_k) \subset \mathcal{U}$].

6. Prove Lemma 4.3.1.

7. Suppose that $\{\mathbf{A}_1, \ldots, \mathbf{A}_k\}$ is a set of matrices that spans a linear space $\mathcal{V}$ but is not a basis for $\mathcal{V}$. Show that, for any matrix $\mathbf{A}$ in $\mathcal{V}$, the representation of $\mathbf{A}$ in terms of $\mathbf{A}_1, \ldots, \mathbf{A}_k$ is nonunique.

Section 4.4

8. Let

$$\mathbf{A} = \begin{pmatrix} 0 & 1 & 0 & -3 & 2 \\ 0 & -2 & 0 & 6 & 2 \\ 0 & 2 & 2 & 5 & 2 \\ 0 & -4 & -2 & 1 & 0 \end{pmatrix}.$$

(a) Show that each of the two column vectors $(2, -1, 3, -4)'$ and $(0, 9, -3, 12)'$ is expressible as a linear combination of the columns of $\mathbf{A}$ [and hence is in $\mathcal{C}(\mathbf{A})$].

(b) Find a basis for $\mathcal{C}(\mathbf{A})$ by applying the algorithm described in the proof of Theorem 4.3.11. (In applying the algorithm, take the spanning set to be the set consisting of the columns of $\mathbf{A}$.)

(c) What is the value of $\mathrm{rank}(\mathbf{A})$? Explain your reasoning.

(d) Find a basis for $\mathcal{C}(\mathbf{A})$ that includes the two column vectors from Part (a). Do so by applying the algorithm described in the proof of Theorem 4.3.12. (In applying the algorithm, take the spanning set to be the set consisting of the columns of $\mathbf{A}$.)

9. Let $\mathbf{A}$ represent a $q \times p$ matrix, $\mathbf{B}$ a $p \times n$ matrix, and $\mathbf{C}$ an $m \times q$ matrix. Show that (a) if $\mathrm{rank}(\mathbf{CAB}) = \mathrm{rank}(\mathbf{C})$, then $\mathrm{rank}(\mathbf{CA}) = \mathrm{rank}(\mathbf{C})$ and (b) if $\mathrm{rank}(\mathbf{CAB}) = \mathrm{rank}(\mathbf{B})$, then $\mathrm{rank}(\mathbf{AB}) = \mathrm{rank}(\mathbf{B})$.

---

* Following the first printing, this exercise was relocated (to better reflect its content) and was renumbered accordingly. In the first printing, it was located under Section 4.2 and ahead of what is now Exercise 3 (formerly Exercise 4).

10. Let $\mathbf{A}$ represent an $m \times n$ matrix of rank r. Show that $\mathbf{A}$ can be expressed as the sum of r matrices of rank 1.

Section 4.5

11. Let $\mathbf{A}$ represent an $m \times n$ matrix and $\mathbf{C}$ a $q \times n$ matrix.

    (a) Add to the proof of Lemma 4.5.1 by confirming that

$$\mathcal{R}(\mathbf{C}) = \mathcal{R}\begin{pmatrix} \mathbf{A} \\ \mathbf{C} \end{pmatrix} \Leftrightarrow \mathcal{R}(\mathbf{A}) \subset \mathcal{R}(\mathbf{C}).$$

    (b) Add to the proof of Corollary 4.5.2 by confirming that $\operatorname{rank}(\mathbf{C}) \leq \operatorname{rank}\begin{pmatrix} \mathbf{A} \\ \mathbf{C} \end{pmatrix}$, with equality holding if and only if $\mathcal{R}(\mathbf{A}) \subset \mathcal{R}(\mathbf{C})$.

5

Trace of a (Square) Matrix

The primary subject of this relatively short chapter is a basic characteristic of a square matrix called the trace. There are many areas of statistics in which the trace of a matrix is encountered. For example, in the design of experiments, one of the criteria (A-optimality) that is used for comparing designs involves the trace of a matrix (Fedorov 1972).

The trace of a matrix is defined and its basic properties described in Section 5.1. Some very useful results on the trace of a product of matrices are covered in Section 5.2. One of these results gives rise to some very useful mathematical equivalences, which are presented in Section 5.3. That the results on the trace of a matrix and the related equivalences are placed in a separate chapter is indicative not only of their importance but of the fact that they don't fit particularly well into any of the other chapters.

5.1 Definition and Basic Properties

The *trace* of a square matrix $\mathbf{A} = \{a_{ij}\}$ of order n is defined to be the sum of the n diagonal elements of $\mathbf{A}$ and is to be denoted by the symbol $\text{tr}(\mathbf{A})$. Thus,

$$\text{tr}(\mathbf{A}) = a_{11} + a_{22} + \cdots + a_{nn}.$$

In particular,

$$\text{tr}(\mathbf{I}_n) = n, \qquad \text{tr}(\mathbf{J}_n) = n, \tag{1.1}$$

and, in the case of a 1×1 matrix (k), whose only element is the scalar k,

$$\text{tr}[(k)] = k. \tag{1.2}$$

Clearly, for any scalar k and any $n \times n$ matrices $\mathbf{A}$ and $\mathbf{B}$,

$$\text{tr}(k\mathbf{A}) = k\,\text{tr}(\mathbf{A}), \tag{1.3}$$

$$\text{tr}(\mathbf{A} + \mathbf{B}) = \text{tr}(\mathbf{A}) + \text{tr}(\mathbf{B}), \tag{1.4}$$

$$\text{tr}(\mathbf{A}') = \text{tr}(\mathbf{A}). \tag{1.5}$$

Further, for any r scalars $k_1, k_2, \ldots, k_r$ and for any r matrices $\mathbf{A}_1, \mathbf{A}_2, \ldots, \mathbf{A}_r$ of dimensions $n \times n$,

$$\mathrm{tr}\Big(\sum_{i=1}^{r} k_i \mathbf{A}_i\Big) = \sum_{i=1}^{r} k_i \,\mathrm{tr}(\mathbf{A}_i), \tag{1.6}$$

as can be easily verified by, for example, the repeated application of results (1.4) and (1.3).

If the diagonal blocks $\mathbf{A}_{11}, \mathbf{A}_{22}, \ldots, \mathbf{A}_{kk}$ of the partitioned matrix

$$\mathbf{A} = \begin{pmatrix} \mathbf{A}_{11} & \mathbf{A}_{12} & \ldots & \mathbf{A}_{1k} \\ \mathbf{A}_{21} & \mathbf{A}_{22} & \ldots & \mathbf{A}_{2k} \\ \vdots & \vdots & \ddots & \vdots \\ \mathbf{A}_{k1} & \mathbf{A}_{k2} & \ldots & \mathbf{A}_{kk} \end{pmatrix}$$

are square (in which case $\mathbf{A}$ itself is square), then clearly

$$\mathrm{tr}(\mathbf{A}) = \mathrm{tr}(\mathbf{A}_{11}) + \mathrm{tr}(\mathbf{A}_{22}) + \cdots + \mathrm{tr}(\mathbf{A}_{kk}). \tag{1.7}$$

5.2 Trace of a Product

Let $\mathbf{A} = \{a_{ij}\}$ represent an $m \times n$ matrix and $\mathbf{B} = \{b_{ji}\}$ an $n \times m$ matrix. Then,

$$\mathrm{tr}(\mathbf{AB}) = \sum_{i=1}^{m} \sum_{j=1}^{n} a_{ij} b_{ji}, \tag{2.1}$$

as is evident upon observing that the ith diagonal element of $\mathbf{AB}$ is $\sum_{j=1}^{n} a_{ij} b_{ji}$. Thus, since the jith element of $\mathbf{B}$ is the ijth element of $\mathbf{B}'$, the trace of the matrix product $\mathbf{AB}$ can be formed by multiplying the ijth element of $\mathbf{A}$ by the corresponding (ijth) element of $\mathbf{B}'$ and by then summing (over i and j).

It follows from results (1.5) and (1.2.13) that, for any $m \times n$ matrix $\mathbf{A}$ and $n \times m$ matrix $\mathbf{B}$,

$$\mathrm{tr}(\mathbf{AB}) = \mathrm{tr}(\mathbf{B}'\mathbf{A}'). \tag{2.2}$$

A further, very basic result on the product of two matrices is expressed in the following lemma.

Lemma 5.2.1. For any $m \times n$ matrix $\mathbf{A}$ and $n \times m$ matrix $\mathbf{B}$,

$$\mathrm{tr}(\mathbf{AB}) = \mathrm{tr}(\mathbf{BA}). \tag{2.3}$$

Proof. Let a_{ij} represent the ijth element of $\mathbf{A}$ and b_{ji} the jith element of $\mathbf{B}$, and observe that the jth diagonal element of $\mathbf{BA}$ is $\sum_{i=1}^{m} b_{ji} a_{ij}$. Thus, making use of result (2.1), we find that

$$\mathrm{tr}(\mathbf{AB}) = \sum_{i=1}^{m} \sum_{j=1}^{n} a_{ij} b_{ji} = \sum_{j=1}^{n} \sum_{i=1}^{m} a_{ij} b_{ji} = \sum_{j=1}^{n} \sum_{i=1}^{m} b_{ji} a_{ij} = \mathrm{tr}(\mathbf{BA}).$$

Q.E.D.

Now, let $\mathbf{A} = \{a_{ij}\}$ and $\mathbf{B} = \{b_{ij}\}$ represent $m \times n$ matrices, and let $\mathbf{a}_j = (a_{1j}, \ldots, a_{mj})'$ and $\mathbf{b}_j = (b_{1j}, \ldots, b_{mj})'$ represent the jth columns of $\mathbf{A}$ and $\mathbf{B}$. Then, since the jith element of $\mathbf{B}'$ is the ijth element of $\mathbf{B}$ and since $\mathbf{b}_j'\mathbf{a}_j = \mathbf{a}_j'\mathbf{b}_j = \sum_{i=1}^m a_{ij}b_{ij}$, it follows from results (2.3) and (2.1) that

$$\operatorname{tr}(\mathbf{AB}') = \operatorname{tr}(\mathbf{B}'\mathbf{A}) = \sum_{j=1}^n \mathbf{b}_j'\mathbf{a}_j = \sum_{j=1}^n \mathbf{a}_j'\mathbf{b}_j = \sum_{i=1}^m \sum_{j=1}^n a_{ij}b_{ij}. \tag{2.4}$$

In particular,

$$\operatorname{tr}(\mathbf{AA}') = \operatorname{tr}(\mathbf{A}'\mathbf{A}) = \sum_{j=1}^n \mathbf{a}_j'\mathbf{a}_j = \sum_{i=1}^m \sum_{j=1}^n a_{ij}^2 \geq 0. \tag{2.5}$$

Thus, both $\operatorname{tr}(\mathbf{AA}')$ and $\operatorname{tr}(\mathbf{A}'\mathbf{A})$ equal the sum of squares of the elements of $\mathbf{A}$ (and both are nonnegative).

As special cases of results (2.4) and (2.5), we have that, for any m-dimensional column vectors $\mathbf{a} = \{a_i\}$ and $\mathbf{b} = \{b_i\}$,

$$\operatorname{tr}(\mathbf{ab}') = \mathbf{b}'\mathbf{a} = \mathbf{a}'\mathbf{b} = \sum_{i=1}^m a_i b_i, \tag{2.6}$$

$$\operatorname{tr}(\mathbf{aa}') = \mathbf{a}'\mathbf{a} = \sum_{i=1}^m a_i^2 \geq 0. \tag{2.7}$$

Note that, together, results (2.2) and (2.3) imply that for any $m \times n$ matrix $\mathbf{A}$ and $n \times m$ matrix $\mathbf{B}$,

$$\operatorname{tr}(\mathbf{AB}) = \operatorname{tr}(\mathbf{B}'\mathbf{A}') = \operatorname{tr}(\mathbf{BA}) = \operatorname{tr}(\mathbf{A}'\mathbf{B}'). \tag{2.8}$$

Thus, in evaluating the trace of the product of two matrices, the matrices can be permuted or replaced by their transposes or both.

Consider now the trace of the product $\mathbf{ABC}$ of an $m \times n$ matrix $\mathbf{A}$, an $n \times p$ matrix $\mathbf{B}$, and a $p \times m$ matrix $\mathbf{C}$. Since $\mathbf{ABC}$ can be regarded as the product of the two matrices $\mathbf{AB}$ and $\mathbf{C}$ or, alternatively, of $\mathbf{A}$ and $\mathbf{BC}$, it follows from Lemma 5.2.1 that

$$\operatorname{tr}(\mathbf{ABC}) = \operatorname{tr}(\mathbf{CAB}) = \operatorname{tr}(\mathbf{BCA}). \tag{2.9}$$

More generally, for any k matrices $\mathbf{A}_1, \mathbf{A}_2, \ldots, \mathbf{A}_k$ for which the matrix product $\mathbf{A}_1\mathbf{A}_2\cdots\mathbf{A}_k$ is defined,

$$\operatorname{tr}(\mathbf{A}_1\mathbf{A}_2\cdots\mathbf{A}_k) = \operatorname{tr}(\mathbf{A}_{j+1}\cdots\mathbf{A}_k\mathbf{A}_1\cdots\mathbf{A}_j) \tag{2.10}$$

$(j = 1, \ldots, k-1)$.

Results (2.9) and (2.10) indicate that, for purposes of evaluating the trace of a product of matrices, the matrices can be permuted. However, aside from special cases, they can be permuted only in a certain way. For example, even if the three matrices $\mathbf{A}$, $\mathbf{B}$, and $\mathbf{C}$ are square and of the same order so that $\mathbf{BAC}$, like $\mathbf{ABC}$, is defined and square, $\operatorname{tr}(\mathbf{BAC})$ is not necessarily equal to $\operatorname{tr}(\mathbf{ABC})$.

5.3 Some Equivalent Conditions

The following lemma gives a basic equivalence.

Lemma 5.3.1. For any $m \times n$ matrix $\mathbf{A} = \{a_{ij}\}$, $\mathbf{A} = \mathbf{0}$ if and only if $\text{tr}(\mathbf{A}'\mathbf{A}) = 0$.

Proof. If $\mathbf{A} = \mathbf{0}$, then obviously $\text{tr}(\mathbf{A}'\mathbf{A}) = 0$. Conversely, if $\text{tr}(\mathbf{A}'\mathbf{A}) = 0$, then it follows from result (2.5) that $a_{ij}^2 = 0$ and hence that $a_{ij} = 0$ $(i = 1, \ldots, m;$ $j = 1, \ldots, n)$ or equivalently that $\mathbf{A} = \mathbf{0}$. Q.E.D.

As an essentially immediate consequence of Lemma 5.3.1, we have the following corollary.

Corollary 5.3.2. For any $m \times n$ matrix $\mathbf{A}$, $\mathbf{A} = \mathbf{0}$ if and only if $\mathbf{A}'\mathbf{A} = \mathbf{0}$.

The following corollary generalizes Corollary 5.3.2.

Corollary 5.3.3. (1) For any $m \times n$ matrix $\mathbf{A}$ and $n \times p$ matrices $\mathbf{B}$ and $\mathbf{C}$, $\mathbf{AB} = \mathbf{AC}$ if and only if $\mathbf{A}'\mathbf{AB} = \mathbf{A}'\mathbf{AC}$. (2) Similarly, for any $m \times n$ matrix $\mathbf{A}$ and $p \times n$ matrices $\mathbf{B}$ and $\mathbf{C}$, $\mathbf{BA}' = \mathbf{CA}'$ if and only if $\mathbf{BA}'\mathbf{A} = \mathbf{CA}'\mathbf{A}$.

Proof (of Corollary 5.3.3). (1) If $\mathbf{AB} = \mathbf{AC}$, then obviously $\mathbf{A}'\mathbf{AB} = \mathbf{A}'\mathbf{AC}$. Conversely, if $\mathbf{A}'\mathbf{AB} = \mathbf{A}'\mathbf{AC}$, then

$$(\mathbf{AB} - \mathbf{AC})'(\mathbf{AB} - \mathbf{AC}) = (\mathbf{B}' - \mathbf{C}')(\mathbf{A}'\mathbf{AB} - \mathbf{A}'\mathbf{AC}) = \mathbf{0},$$

and it follows from Corollary 5.3.2 that $\mathbf{AB} - \mathbf{AC} = \mathbf{0}$ or, equivalently, that $\mathbf{AB} = \mathbf{AC}$.

(2) To establish Part (2), simply take the transpose of each side of the two equivalent equalities $\mathbf{AB}' = \mathbf{AC}'$ and $\mathbf{A}'\mathbf{AB}' = \mathbf{A}'\mathbf{AC}'$. [The equivalence of these two equalities follows from part (1).] Q.E.D.

Note that, as a special case of Part (1) of Corollary 5.3.3 (the special case where $\mathbf{C} = \mathbf{0}$), we have that $\mathbf{AB} = \mathbf{0}$ if and only if $\mathbf{A}'\mathbf{AB} = \mathbf{0}$, and, as a special case of Part (2), we have that $\mathbf{BA}' = \mathbf{0}$ if and only if $\mathbf{BA}'\mathbf{A} = \mathbf{0}$.

Exercises

Section 5.2

1. Show that for any $m \times n$ matrix $\mathbf{A}$, $n \times p$ matrix $\mathbf{B}$, and $p \times m$ matrix $\mathbf{C}$,

$$\text{tr}(\mathbf{ABC}) = \text{tr}(\mathbf{B}'\mathbf{A}'\mathbf{C}') = \text{tr}(\mathbf{A}'\mathbf{C}'\mathbf{B}').$$

2. Let $\mathbf{A}$, $\mathbf{B}$, and $\mathbf{C}$ represent $n \times n$ matrices.

   (a) Using the result of Exercise 1 (or otherwise), show that if $\mathbf{A}$, $\mathbf{B}$, and $\mathbf{C}$ are symmetric, then $\text{tr}(\mathbf{ABC}) = \text{tr}(\mathbf{BAC})$.

   (b) Show that [aside from special cases like that considered in Part (a)] $\text{tr}(\mathbf{BAC})$ is not necessarily equal to $\text{tr}(\mathbf{ABC})$.

Section 5.3

3. Let $\mathbf{A}$ represent an $n \times n$ matrix such that $\mathbf{A}'\mathbf{A} = \mathbf{A}^2$.

   (a) Show that $\text{tr}[(\mathbf{A} - \mathbf{A}')'(\mathbf{A} - \mathbf{A}')] = 0$.

   (b) Show that $\mathbf{A}$ is symmetric.

6

Geometrical Considerations

This chapter provides an introduction (at a somewhat general level) to some rather fundamental geometrical ideas and results. Some knowledge of this material is an important prerequisite for assimilating ideas in several areas of statistics, including linear statistical models and multivariate analysis.

Among those who teach and write about the more theoretical aspects of linear statistical models, there is a considerable difference of opinion about the extent to which geometrical ideas should be emphasized (relative to "algebraic" ideas). Those who prefer a "geometrical approach" (e.g., Scheffé 1959, Christensen 1996) argue that it is more general, more elegant, and (perhaps most important) more intuitive. Those who prefer a more algebraic approach (e.g., Searle 1971) find it to be more rigorous, more concrete, more palatable to those with a limited mathematical background, and (perhaps most important) more suggestive of computational approaches.

6.1 Definitions: Norm, Distance, Angle, Inner Product, and Orthogonality

Certain definitions from plane and solid geometry can be extended in a natural way to an arbitrary linear space and can be otherwise generalized, as is now discussed. Let us begin by reviewing the usual definitions in the familiar settings of $\mathbb{R}^2$ and $\mathbb{R}^3$.

a. Usual definitions in $\mathbb{R}^2$ and $\mathbb{R}^3$

Each point in the plane can be identified by its coordinates (relative to a horizontal line and a vertical line—known as the coordinate axes—labeled so that their point of intersection—known as the origin—has coordinates of zero). A vector $\mathbf{x} = (x_1, x_2)'$ in $\mathbb{R}^2$ can be regarded as representing a point in the plane; namely, the point whose coordinates are x_1 and x_2. Similarly, each point in 3-dimensional space can be identified by its coordinates (relative to 3 mutually perpendicular axes whose intersection, called the origin, has coordinates of zero); and a vector $\mathbf{x} = (x_1, x_2, x_3)'$ in $\mathcal{R}^3$ can be regarded as representing the point in space whose coordinates are x_1, x_2, and x_3. (For definiteness, it is supposed that $\mathcal{R}^2 = \mathcal{R}^{2\times 1}$ and $\mathcal{R}^3 = \mathcal{R}^{3\times 1}$, so that the members of $\mathcal{R}^2$ and $\mathcal{R}^3$ are column vectors rather than row vectors.)

In addition to representing a point, a vector in $\mathcal{R}^2$ or $\mathcal{R}^3$ is also used to represent the directed line segment that starts at the origin and ends at that point. The intended usage can be ascertained from the context.

The usual definition of the *norm* $\|\mathbf{x}\|$ (also known as the *length* or *magnitude*) of a vector $\mathbf{x} = (x_1, x_2)'$ in $\mathcal{R}^2$ is

$$\|\mathbf{x}\| = (x_1^2 + x_2^2)^{1/2}.$$

The *distance* $\delta(\mathbf{x}, \mathbf{y})$ between two vectors $\mathbf{x} = (x_1, x_2)'$ and $\mathbf{y} = (y_1, y_2)'$ in $\mathcal{R}^2$ is usually taken to be

$$\delta(\mathbf{x}, \mathbf{y}) = \|\mathbf{x} - \mathbf{y}\| = [(x_1 - y_1)^2 + (x_2 - y_2)^2]^{1/2}.$$

Similarly, the norm $\|\mathbf{x}\|$ of a vector $\mathbf{x} = (x_1, x_2, x_3)'$ in $\mathcal{R}^3$ is usually defined to be

$$\|\mathbf{x}\| = (x_1^2 + x_2^2 + x_3^2)^{1/2},$$

and the distance $\delta(\mathbf{x}, \mathbf{y})$ between two vectors $\mathbf{x} = (x_1, x_2, x_3)'$ and $\mathbf{y} = (y_1, y_2, y_3)'$ (in $\mathcal{R}^3$) to be

$$\delta(\mathbf{x}, \mathbf{y}) = \|\mathbf{x} - \mathbf{y}\| = [(x_1 - y_1)^2 + (x_2 - y_2)^2 + (x_3 - y_3)^2]^{1/2}.$$

Clearly, for any vector $\mathbf{x}$ in $\mathcal{R}^2$ or $\mathcal{R}^3$,

$$\|\mathbf{x}\| = (\mathbf{x}'\mathbf{x})^{1/2},$$

and $\|\mathbf{x}\| = \delta(\mathbf{x}, \mathbf{0})$, that is, the norm of $\mathbf{x}$ is the distance between $\mathbf{x}$ and $\mathbf{0}$.

Consider now the angle that two nonnull vectors $\mathbf{x}$ and $\mathbf{y}$ in $\mathcal{R}^2$ or $\mathcal{R}^3$ form with each other. There are actually two such angles, as depicted in Figure 6.1. The first angle θ_1 is that determined by rotating the directed line segment $\mathbf{x}$ in a counterclockwise direction (about the origin and within the plane containing $\mathbf{x}$ and $\mathbf{y}$) until the directed line segment $\mathbf{y}$ is encountered. The second angle $\theta_2 = 2\pi - \theta_1$ is that determined by rotating $\mathbf{y}$ in a counterclockwise direction until $\mathbf{x}$ is encountered. (Unless otherwise indicated, angles are to be expressed in radians.) Let us arbitrarily define the angle θ between $\mathbf{x}$ and $\mathbf{y}$ to be the smaller of the two angles θ_1 and θ_2 (which is the usual convention). Thus,

$$\begin{aligned} \theta &= \theta_1, && \text{if } 0 \le \theta_1 \le \pi, \\ &= \theta_2 = 2\pi - \theta_1, && \text{if } \pi < \theta_1 < 2\pi. \end{aligned}$$

Clearly, $0 \le \theta \le \pi$.

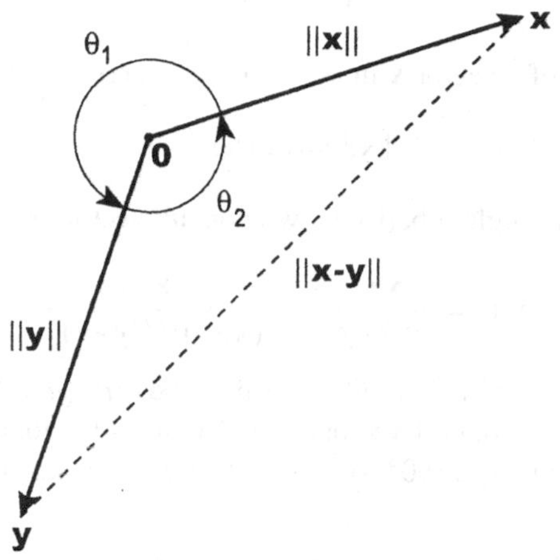

FIGURE 6.1: The angles formed by two vectors in $\mathcal{R}^2$ or $\mathcal{R}^3$ with each other.

Let us now consider the cosine of the angle θ between two nonnull vectors $\mathbf{x}$ and $\mathbf{y}$ in $\mathcal{R}^2$ or $\mathcal{R}^3$. Applying a well-known trigonometric identity, called the law of cosines, to the triangle whose vertices are $\mathbf{0}$, $\mathbf{x}$ and $\mathbf{y}$, we obtain the equality

$$\|\mathbf{x}-\mathbf{y}\|^2 = \|\mathbf{x}\|^2 + \|\mathbf{y}\|^2 - 2\|\mathbf{x}\|\,\|\mathbf{y}\|\cos\theta. \tag{1.1}$$

Then, substituting the expressions

$$\|\mathbf{x}-\mathbf{y}\|^2 = (\mathbf{x}-\mathbf{y})'(\mathbf{x}-\mathbf{y}) = \mathbf{x}'\mathbf{x} - 2\mathbf{x}'\mathbf{y} + \mathbf{y}'\mathbf{y},$$
$$\|\mathbf{x}\|^2 = \mathbf{x}'\mathbf{x}, \qquad \|\mathbf{y}\|^2 = \mathbf{y}'\mathbf{y}$$

in equality (1.1) and canceling terms common to both sides, we find that

$$-2\mathbf{x}'\mathbf{y} = -2(\mathbf{x}'\mathbf{x})^{1/2}(\mathbf{y}'\mathbf{y})^{1/2}\cos\theta$$

and hence that

$$\cos\theta = \frac{\mathbf{x}'\mathbf{y}}{(\mathbf{x}'\mathbf{x})^{1/2}(\mathbf{y}'\mathbf{y})^{1/2}}.$$

The usual definition of the *inner* (or *dot*) *product* $\mathbf{x}\bullet\mathbf{y}$ of two vectors $\mathbf{x} = (x_1, x_2)'$ and $\mathbf{y} = (y_1, y_2)'$ in $\mathcal{R}^2$ is

$$\mathbf{x}\bullet\mathbf{y} = x_1y_1 + x_2y_2 \tag{1.2}$$

and that of two vectors $\mathbf{x} = (x_1, x_2, x_3)'$ and $\mathbf{y} = (y_1, y_2, y_3)'$ in $\mathcal{R}^3$ is

$$\mathbf{x}\bullet\mathbf{y} = x_1y_1 + x_2y_2 + x_3y_3. \tag{1.3}$$

Clearly, for any two vectors $\mathbf{x}$ and $\mathbf{y}$ in $\mathcal{R}^2$ or $\mathcal{R}^3$,

$$\mathbf{x} \bullet \mathbf{y} = \mathbf{x}'\mathbf{y}.$$

Note that the norm of a vector $\mathbf{x}$ in $\mathcal{R}^2$ or $\mathcal{R}^3$ is expressible as

$$\|\mathbf{x}\| = (\mathbf{x} \bullet \mathbf{x})^{1/2},$$

and the cosine of the angle θ between two nonnull vectors $\mathbf{x}$ and $\mathbf{y}$ in $\mathcal{R}^2$ or $\mathcal{R}^3$ is expressible as

$$\cos\theta = \frac{\mathbf{x} \bullet \mathbf{y}}{\|\mathbf{x}\|\,\|\mathbf{y}\|} = \frac{\mathbf{x} \bullet \mathbf{y}}{(\mathbf{x} \bullet \mathbf{x})^{1/2}(\mathbf{y} \bullet \mathbf{y})^{1/2}}.$$

Two vectors $\mathbf{x}$ and $\mathbf{y}$ in $\mathcal{R}^2$ or $\mathcal{R}^3$ are said to be *orthogonal* (or *perpendicular*) if $\mathbf{x} \bullet \mathbf{y} = 0$. Clearly, nonnull vectors $\mathbf{x}$ and $\mathbf{y}$ are orthogonal if and only if the angle between them is $\pi/2$ (90°) or, equivalently, the cosine of that angle is 0.

b. General definitions

The usual definitions of norm, distance, angle, inner product, and orthogonality can be extended, in a natural way, from $\mathcal{R}^2$ and $\mathcal{R}^3$ to an arbitrary linear space and can be otherwise generalized, though—except for linear spaces of 2- or 3-dimensional row or column vectors—the extended definitions have no obvious geometrical interpretations. To do so, it suffices to extend and generalize the definition of inner product. The other definitions can then be extended and generalized by supposing that the concepts of norm, distance, angle, and orthogonality are related to each other and to the concept of inner product in the same way as (under the usual definitions) in $\mathcal{R}^2$ and $\mathcal{R}^3$.

The usual definition of the *inner* (or *dot*) *product* $\mathbf{x} \bullet \mathbf{y}$ of two n-dimensional (column) vectors $\mathbf{x} = (x_1, x_2, \ldots, x_n)'$ and $\mathbf{y} = (y_1, y_2, \ldots, y_n)'$ in $\mathcal{R}^n$, or, more generally, in a subspace of $\mathcal{R}^n$, is

$$\mathbf{x} \bullet \mathbf{y} = \mathbf{x}'\mathbf{y} = x_1 y_1 + x_2 y_2 + \ldots + x_n y_n. \tag{1.4}$$

Formula (1.4) is the natural extension of formulas (1.2) and (1.3) for the usual inner product of two vectors in $\mathcal{R}^2$ or $\mathcal{R}^3$.

More generally, the usual definition of the *inner* (or *dot*) *product* $\mathbf{A} \bullet \mathbf{B}$ of two $m \times n$ matrices $\mathbf{A}$ and $\mathbf{B}$ in a linear space $\mathcal{V}$ of $m \times n$ matrices is

$$\mathbf{A} \bullet \mathbf{B} = \text{tr}(\mathbf{A}'\mathbf{B}) \tag{1.5}$$

or, equivalently [in light of result (5.2.8)],

$$\mathbf{A} \bullet \mathbf{B} = \text{tr}(\mathbf{A}\mathbf{B}'). \tag{1.6}$$

In the special case of two n-dimensional row vectors $\mathbf{x}' = (x_1, x_2, \ldots, x_n)$ and $\mathbf{y}' = (y_1, y_2, \ldots, y_n)$, formula (1.6) for the usual inner product simplifies to

$$\mathbf{x}' \bullet \mathbf{y}' = \mathbf{x}'\mathbf{y} = x_1 y_1 + x_2 y_2 + \cdots + x_n y_n,$$

which is the same as formula (1.4) for the usual inner product of the corresponding column vectors $\mathbf{x}$ and $\mathbf{y}$.

The usual inner product $\mathbf{A} \bullet \mathbf{B}$ of two matrices $\mathbf{A}$ and $\mathbf{B}$ in a linear space $\mathcal{V}$ of $m \times n$ matrices can be regarded as the value assigned to $\mathbf{A}$ and $\mathbf{B}$ by a function whose domain consists of all (ordered) pairs of matrices in $\mathcal{V}$. As is evident from results (5.1.3), (5.1.4), (5.2.3), and (5.2.5) and from Lemma 5.3.1, this function has the following four basic properties:

(1) $\mathbf{A} \bullet \mathbf{B} = \mathbf{B} \bullet \mathbf{A}$;

(2) $\mathbf{A} \bullet \mathbf{A} \geq 0$, with equality holding if and only if $\mathbf{A} = \mathbf{0}$;

(3) $(k\mathbf{A}) \bullet \mathbf{B} = k(\mathbf{A} \bullet \mathbf{B})$;

(4) $(\mathbf{A} + \mathbf{B}) \bullet \mathbf{C} = (\mathbf{A} \bullet \mathbf{C}) + (\mathbf{B} \bullet \mathbf{C})$

(where $\mathbf{A}$, $\mathbf{B}$, and $\mathbf{C}$ represent arbitrary matrices in $\mathcal{V}$ and k represents an arbitrary scalar).

It is sometimes useful to adopt a different definition for the inner product. Subsequently, we define the *inner product* of two matrices $\mathbf{A}$ and $\mathbf{B}$ in a linear space $\mathcal{V}$ (of $m \times n$ matrices) to be the value $\mathbf{A} \bullet \mathbf{B}$ assigned to $\mathbf{A}$ and $\mathbf{B}$ by any designated function f (whose domain consists of all pairs of matrices in $\mathcal{V}$) having Properties (1)–(4). The term inner product is applied to the function f, as well as to its values. When f is chosen to be the function whose values are given by formula (1.5) or (1.6), we obtain the usual inner product.

Other properties of the inner product can be derived from the four basic properties. We find, in particular, that (for any $\mathbf{A}$ in $\mathcal{V}$)

$$\mathbf{0} \bullet \mathbf{A} = 0, \tag{1.7}$$

as is evident from Property (3) upon observing that

$$\mathbf{0} \bullet \mathbf{A} = (0\,\mathbf{A}) \bullet \mathbf{A} = 0\,(\mathbf{A} \bullet \mathbf{A}) = 0.$$

Further (for any matrices $\mathbf{A}_1, \ldots, \mathbf{A}_k$, and $\mathbf{B}$ in $\mathcal{V}$ and any scalars $x_1, \ldots, x_k$),

$$(x_1\mathbf{A}_1 + \cdots + x_k\mathbf{A}_k) \bullet \mathbf{B} = x_1(\mathbf{A}_1 \bullet \mathbf{B}) + \cdots + x_k(\mathbf{A}_k \bullet \mathbf{B}), \tag{1.8}$$

as can be readily verified by the repeated application of Properties (3) and (4).

Note that if a function f qualifies as an inner product for a linear space $\mathcal{V}$ of $m \times n$ matrices, then f also qualifies as an inner product for any subspace of $\mathcal{V}$—or, more precisely, the function obtained by restricting the domain of f (which consists of pairs of matrices in $\mathcal{V}$) to pairs of matrices in the subspace qualifies as an inner product. In fact, if f is the designated inner product for a linear space $\mathcal{V}$, then (unless otherwise indicated) it is implicitly assumed that the same function f is the designated inner product for any subspace of $\mathcal{V}$.

The *norm* $\|\mathbf{A}\|$ of a matrix $\mathbf{A} = \{a_{ij}\}$ in a linear space $\mathcal{V}$ of $m \times n$ matrices is

$$\|\mathbf{A}\| = (\mathbf{A} \bullet \mathbf{A})^{1/2},$$

which, in the case of the usual inner product, gives

$$\|\mathbf{A}\| = \left[\operatorname{tr}(\mathbf{A}'\mathbf{A})\right]^{1/2} = \Big(\sum_{i=1}^{m}\sum_{j=1}^{n} a_{ij}^2\Big)^{1/2}. \tag{1.9}$$

If $\mathcal{V}$ is a linear space of m-dimensional column vectors, then, letting $\mathbf{x} = \{x_i\}$ represent any vector in $\mathcal{V}$, formula (1.9) can be restated as

$$\|\mathbf{x}\| = (\mathbf{x}'\mathbf{x})^{1/2} = (x_1^2 + \cdots + x_m^2)^{1/2}. \tag{1.10}$$

The norm defined by formula (1.10) or more generally by formula (1.9) is called the *usual norm*.

Letting $\mathbf{A}$ represent an arbitrary matrix in a linear space $\mathcal{V}$ and k an arbitrary scalar, it is clear that

$$\|\mathbf{A}\| > 0, \quad \text{if } \mathbf{A} \neq \mathbf{0}, \tag{1.11a}$$
$$= 0, \quad \text{if } \mathbf{A} = \mathbf{0}, \tag{1.11b}$$

and that

$$\|k\mathbf{A}\| = |k|\|\mathbf{A}\|. \tag{1.12}$$

As a special case of equality (1.12) (that where $k = -1$), we have that

$$\|-\mathbf{A}\| = \|\mathbf{A}\|. \tag{1.13}$$

The *distance* $\delta(\mathbf{A}, \mathbf{B})$ between two matrices $\mathbf{A}$ and $\mathbf{B}$ in a linear space $\mathcal{V}$ is

$$\delta(\mathbf{A}, \mathbf{B}) = \|\mathbf{A} - \mathbf{B}\|. \tag{1.14}$$

If $\mathcal{V}$ is a linear space of n-dimensional column vectors, then, letting $\mathbf{x} = \{x_i\}$ represent any vector in $\mathcal{V}$, formula (1.14) simplifies, in the case of the usual norm, to

$$\delta(\mathbf{x}, \mathbf{y}) = [(\mathbf{x} - \mathbf{y})'(\mathbf{x} - \mathbf{y})]^{1/2} = [(x_1 - y_1)^2 + \cdots + (x_n - y_n)^2]^{1/2}. \tag{1.15}$$

For two nonnull vectors $\mathbf{x} = \{x_i\}$ and $\mathbf{y} = \{y_i\}$ in a linear space $\mathcal{V}$ of n-dimensional column vectors, the *angle* θ $(0 \leq \theta \leq \pi)$ between $\mathbf{x}$ and $\mathbf{y}$ is defined (indirectly in terms of its cosine) by

$$\cos\theta = \frac{\mathbf{x} \bullet \mathbf{y}}{\|\mathbf{x}\|\,\|\mathbf{y}\|}$$

or equivalently, in the case of the usual inner product and norm, by

$$\cos\theta = \frac{x_1 y_1 + \cdots + x_n y_n}{(x_1^2 + \cdots + x_n^2)^{1/2}(y_1^2 + \cdots + y_n^2)^{1/2}}.$$

More generally, the angle θ $(0 \leq \theta \leq \pi)$ between two nonnull matrices $\mathbf{A}$ and $\mathbf{B}$ in a linear space $\mathcal{V}$ of $m \times n$ matrices is defined by

$$\cos\theta = \frac{\mathbf{A} \bullet \mathbf{B}}{\|\mathbf{A}\| \, \|\mathbf{B}\|}.$$

The Schwarz inequality, which is to be introduced in Section 6.3, implies that

$$-1 \leq \frac{\mathbf{A} \bullet \mathbf{B}}{\|\mathbf{A}\| \, \|\mathbf{B}\|} \leq 1 \tag{1.16}$$

and hence insures that this angle is well-defined.

Two vectors $\mathbf{x}$ and $\mathbf{y}$ in a linear space $\mathcal{V}$ of n-dimensional column vectors are said to be *orthogonal* if $\mathbf{x} \bullet \mathbf{y} = 0$. More generally, two matrices $\mathbf{A}$ and $\mathbf{B}$ in a linear space $\mathcal{V}$ are said to be orthogonal if $\mathbf{A} \bullet \mathbf{B} = 0$. The statement that two matrices $\mathbf{A}$ and $\mathbf{B}$ are orthogonal is sometimes abbreviated to $\mathbf{A} \perp \mathbf{B}$.

Clearly, two nonnull matrices are orthogonal if and only if the angle between them is $\pi/2$ or, equivalently, the cosine of that angle is 0. Whether two matrices are orthogonal depends on the choice of inner product; that is, two matrices that are orthogonal with respect to one inner product (one choice for the function f) may not be orthogonal with respect to another.

6.2 Orthogonal and Orthonormal Sets

A finite set of matrices in a linear space $\mathcal{V}$ is said to be *orthogonal* if every pair of matrices in the set arc orthogonal. Thus, the empty set and any set containing only one matrix are orthogonal sets, and a finite set $\{\mathbf{A}_1, \ldots, \mathbf{A}_k\}$ of two or more matrices in $\mathcal{V}$ is an orthogonal set if $\mathbf{A}_i \bullet \mathbf{A}_j = 0$ for $j \neq i = 1, \ldots, k$. A finite set of matrices in $\mathcal{V}$ is said to be *orthonormal* if the set is orthogonal and if the norm of every matrix in the set equals 1. In the special case of a set of (row or column) vectors, the expression "set of orthogonal (or orthonormal) vectors," or simply "orthogonal (or orthonormal) vectors," is often used in lieu of the technically more correct expression "orthogonal (or orthonormal) set of vectors."

Note that if $\mathbf{A}$ is any nonnull matrix in a linear space $\mathcal{V}$, then the norm of the matrix

$$\mathbf{B} = \|\mathbf{A}\|^{-1}\mathbf{A} \tag{2.1}$$

equals 1. The expression "normalizing a (nonnull) matrix" is often used in referring to the conversion—via formula (2.1)—of the matrix into a matrix whose norm is 1.

If $\{\mathbf{A}_1, \ldots, \mathbf{A}_k\}$ is an orthogonal set of nonnull matrices in a linear space $\mathcal{V}$, then the matrices

$$\mathbf{B}_1 = \|\mathbf{A}_1\|^{-1}\mathbf{A}_1, \ldots, \mathbf{B}_k = \|\mathbf{A}_k\|^{-1}\mathbf{A}_k$$

form an orthonormal set. Thus, any orthogonal set of nonnull matrices can be converted into an orthonormal set by normalizing each of the matrices.

There is a relationship between orthogonality and linear independence, which is described in the following lemma.

Lemma 6.2.1. An orthogonal set of nonnull matrices is linearly independent.

Proof. If the orthogonal set is the empty set, then the result is clearly true (since, by convention, the empty set is linearly independent). Suppose then that $\{\mathbf{A}_1, \ldots, \mathbf{A}_k\}$ is any nonempty orthogonal set of nonnull matrices. Let $x_1, \ldots, x_k$ represent arbitrary scalars such that

$$x_1\mathbf{A}_1 + \cdots + x_k\mathbf{A}_k = \mathbf{0}.$$

For $i = 1, \ldots, k$, we find that

$$\begin{aligned} 0 = \mathbf{0} \bullet \mathbf{A}_i &= (x_1\mathbf{A}_1 + \cdots + x_k\mathbf{A}_k) \bullet \mathbf{A}_i \\ &= x_1(\mathbf{A}_1 \bullet \mathbf{A}_i) + \cdots + x_k(\mathbf{A}_k \bullet \mathbf{A}_i) \\ &= x_i(\mathbf{A}_i \bullet \mathbf{A}_i), \end{aligned}$$

implying (since $\mathbf{A}_i$ is nonnull) that $x_i = 0$. We conclude that the set $\{\mathbf{A}_1, \ldots, \mathbf{A}_k\}$ is linearly independent. Q.E.D.

Note that Lemma 6.2.1 implies in particular that any orthonormal set of matrices is linearly independent. Note also that the converse of Lemma 6.2.1 is not necessarily true; that is, a linearly independent set is not necessarily orthogonal. For example, the set consisting of the two 2-dimensional row vectors $(1, 0)$ and $(1, 1)$ is linearly independent but is not orthogonal (with respect to the usual inner product).

6.3 Schwarz Inequality

The following theorem gives a famous inequality, known as the Schwarz inequality.*

Theorem 6.3.1 (Schwarz inequality). For any two matrices $\mathbf{A}$ and $\mathbf{B}$ in a linear space $\mathcal{V}$,

$$|\mathbf{A} \bullet \mathbf{B}| \leq \|\mathbf{A}\| \, \|\mathbf{B}\|, \tag{3.1}$$

with equality holding if and only if $\mathbf{B} = \mathbf{0}$ or $\mathbf{A} = k\mathbf{B}$ for some scalar k.

Proof. Suppose first that $\|\mathbf{B}\| = 0$ or equivalently that $\mathbf{B} = \mathbf{0}$. Then, $|\mathbf{A} \bullet \mathbf{B}| = 0 = \|\mathbf{A}\| \, \|\mathbf{B}\|$, and inequality (3.1) holds as an equality.

Now, suppose that $\|\mathbf{B}\| > 0$. Then (repeatedly making use of the basic properties of inner products), we find that, for any scalars x and y,

$$\begin{aligned} 0 &\leq (x\mathbf{A} - y\mathbf{B}) \bullet (x\mathbf{A} - y\mathbf{B}) \\ &= x^2\|\mathbf{A}\|^2 - 2xy(\mathbf{A} \bullet \mathbf{B}) + y^2\|\mathbf{B}\|^2. \end{aligned}$$

Choosing $x = \|\mathbf{B}\|$ and $y = (\mathbf{A} \bullet \mathbf{B})/\|\mathbf{B}\|$, we obtain the inequality

* This inequality is also known as the Cauchy-Schwarz (or Cauchy-Bunyakovsky-Schwarz) inequality.

$$0 \leq \|\mathbf{B}\|^2\|\mathbf{A}\|^2 - (\mathbf{A}\bullet\mathbf{B})^2,$$

which is equivalent to the inequality

$$(\mathbf{A}\bullet\mathbf{B})^2 \leq \|\mathbf{A}\|^2\|\mathbf{B}\|^2$$

and hence to inequality (3.1).

If $\mathbf{A} = k\mathbf{B}$, then $\|\mathbf{A}\| = |k|\|\mathbf{B}\|$, so that

$$|\mathbf{A}\bullet\mathbf{B}| = |k|(\mathbf{B}\bullet\mathbf{B}) = |k|\|\mathbf{B}\|^2 = \|\mathbf{A}\|\ \|\mathbf{B}\|.$$

Conversely, suppose that $|\mathbf{A}\bullet\mathbf{B}| = \|\mathbf{A}\|\ \|\mathbf{B}\|$. Then, letting $\mathbf{C} = \|\mathbf{B}\|^2\mathbf{A} - (\mathbf{A}\bullet\mathbf{B})\mathbf{B}$, we find that $\mathbf{C}\bullet\mathbf{C} = 0$, implying that $\mathbf{C} = \mathbf{0}$, that is, that

$$\|\mathbf{B}\|^2\mathbf{A} = (\mathbf{A}\bullet\mathbf{B})\mathbf{B}.$$

We conclude that either $\mathbf{B} = \mathbf{0}$ or

$$\mathbf{A} = [(\mathbf{A}\bullet\mathbf{B})/\|\mathbf{B}\|^2]\mathbf{B}.$$

Q.E.D.

As a special case of Theorem 6.3.1, we have that, for any two n-dimensional column vectors $\mathbf{x}$ and $\mathbf{y}$,

$$|\mathbf{x}'\mathbf{y}| \leq (\mathbf{x}'\mathbf{x})^{1/2}(\mathbf{y}'\mathbf{y})^{1/2}, \tag{3.2}$$

with equality holding if and only if $\mathbf{y} = \mathbf{0}$ or $\mathbf{x} = k\mathbf{y}$ for some scalar k.

Note that the result of Theorem 6.3.1 can be reexpressed as follows.

Corollary 6.3.2. Let $\mathcal{V}$ represent a linear space of $m \times n$ matrices. Then, for any matrix $\mathbf{B}$ in $\mathcal{V}$,

$$\max_{\{\mathbf{A}\in\mathcal{V}\,:\,\mathbf{A}\neq\mathbf{0}\}} \frac{|\mathbf{A}\bullet\mathbf{B}|}{\|\mathbf{A}\|} = \|\mathbf{B}\|.$$

Moreover, if $\mathbf{B} = \mathbf{0}$, the maximum is attained at every nonnull $\mathbf{A} \in \mathcal{V}$; and if $\mathbf{B} \neq \mathbf{0}$, the maximum is attained at every nonnull $\mathbf{A} \in \mathcal{V}$ that is proportional to $\mathbf{B}$.

6.4 Orthonormal Bases

An *orthonormal basis* for a linear space $\mathcal{V}$ is an orthonormal set of matrices in $\mathcal{V}$ that is a basis for $\mathcal{V}$. The technique of Gram-Schmidt orthogonalization, which is described in Subsection a below, is used in Subsection b to establish the existence of an orthonormal basis.

a. Gram-Schmidt orthogonalization

A nonempty linearly independent set of matrices can be transformed into an orthogonal or orthonormal set of matrices by applying the results of the following theorem and corollary.

Theorem 6.4.1. Let $\{\mathbf{A}_1, \ldots, \mathbf{A}_k\}$ represent a nonempty linearly independent set of matrices in a linear space $\mathcal{V}$. Then, there exist unique scalars x_{ij} $(i < j = 1, \ldots, k)$ such that the set comprising the k matrices

$$\begin{aligned}
\mathbf{B}_1 &= \mathbf{A}_1, \\
\mathbf{B}_2 &= \mathbf{A}_2 - x_{12}\mathbf{B}_1, \\
&\vdots \\
\mathbf{B}_j &= \mathbf{A}_j - x_{j-1,j}\mathbf{B}_{j-1} - \cdots - x_{1j}\mathbf{B}_1, \\
&\vdots \\
\mathbf{B}_k &= \mathbf{A}_k - x_{k-1,k}\mathbf{B}_{k-1} - \cdots - x_{1k}\mathbf{B}_1
\end{aligned}$$

is orthogonal. Further, $\mathbf{B}_1, \mathbf{B}_2, \ldots, \mathbf{B}_k$ are nonnull, and (for $i < j = 1, \ldots, k$)

$$x_{ij} = \frac{\mathbf{A}_j \bullet \mathbf{B}_i}{\mathbf{B}_i \bullet \mathbf{B}_i}. \tag{4.1}$$

Proof. The proof is by mathematical induction. The theorem is obviously true for $k = 1$. Suppose now that the theorem is true for a set of $k - 1$ linearly independent matrices. Then, there exist unique scalars x_{ij} $(i < j = 1, \ldots, k-1)$ such that the set comprising the $k-1$ matrices $\mathbf{B}_1, \ldots, \mathbf{B}_{k-1}$ is orthogonal. Further, $\mathbf{B}_1, \ldots, \mathbf{B}_{k-1}$ are nonnull, and (for $i < j = 1, \ldots, k - 1$)

$$x_{ij} = \frac{\mathbf{A}_j \bullet \mathbf{B}_i}{\mathbf{B}_i \bullet \mathbf{B}_i}.$$

And, for $i = 1, \ldots, k - 1$, we find that

$$\begin{aligned}
\mathbf{B}_k \bullet \mathbf{B}_i &= \mathbf{A}_k \bullet \mathbf{B}_i - x_{k-1,k}(\mathbf{B}_{k-1} \bullet \mathbf{B}_i) - \cdots - x_{1k}(\mathbf{B}_1 \bullet \mathbf{B}_i) \\
&= \mathbf{A}_k \bullet \mathbf{B}_i - x_{ik}(\mathbf{B}_i \bullet \mathbf{B}_i)
\end{aligned}$$

and thus that $\mathbf{B}_k \bullet \mathbf{B}_i = 0$ if and only if

$$\mathbf{A}_k \bullet \mathbf{B}_i - x_{ik}(\mathbf{B}_i \bullet \mathbf{B}_i) = 0$$

or, equivalently (since $\mathbf{B}_i$ is nonnull), if and only if

$$x_{ik} = \frac{\mathbf{A}_k \bullet \mathbf{B}_i}{\mathbf{B}_i \bullet \mathbf{B}_i}.$$

We conclude that there exist unique scalars x_{ij} $(i < j = 1, \ldots, k)$ such that the set comprising the k matrices $\mathbf{B}_1, \ldots, \mathbf{B}_k$ is orthogonal. We further conclude that $\mathbf{B}_1, \ldots, \mathbf{B}_k$ are nonnull and that (for $i < j = 1, \ldots, k$) x_{ij} is given by formula (4.1). (The matrix $\mathbf{B}_k = \mathbf{A}_k - x_{k-1,k}\mathbf{B}_{k-1} - \cdots - x_{1k}\mathbf{B}_1$ is nonnull since, by repeated substitution, it can be rewritten as a nontrivial linear combination of the linearly independent matrices $\mathbf{A}_1, \ldots, \mathbf{A}_k$.) Q.E.D.

Corollary 6.4.2. Let $\{\mathbf{A}_1, \ldots, \mathbf{A}_k\}$ represent a nonempty linearly independent set of matrices in a linear space $\mathcal{V}$. Then, the set comprising the k matrices

$$\mathbf{C}_1 = \|\mathbf{B}_1\|^{-1}\mathbf{B}_1, \ldots, \mathbf{C}_k = \|\mathbf{B}_k\|^{-1}\mathbf{B}_k,$$

obtained by normalizing the k matrices $\mathbf{B}_1, \ldots, \mathbf{B}_k$ defined in Theorem 6.4.1, is orthonormal.

Corollary 6.4.2 is an essentially immediate consequence of Theorem 6.4.1. The process of constructing an orthonormal set $\{\mathbf{C}_1, \ldots, \mathbf{C}_k\}$ from a linearly independent set $\{\mathbf{A}_1, \ldots, \mathbf{A}_k\}$ in accordance with the formulas of Theorem 6.4.1 and Corollary 6.4.2 is known as *Gram-Schmidt orthogonalization*.

b. Existence of orthonormal bases and a basic property of orthonormal bases

According to Theorem 4.3.4, every linear space (of $m \times n$ matrices) has a basis. The basis of the linear space $\{\mathbf{0}\}$, whose only member is the $m \times n$ null matrix $\mathbf{0}$, is the empty set, which is an orthonormal set. The basis of any other linear space $\mathcal{V}$ is a nonempty (linearly independent) set, say $\{\mathbf{A}_1, \ldots, \mathbf{A}_r\}$. By applying Gram-Schmidt orthogonalization to the set $\{\mathbf{A}_1, \ldots, \mathbf{A}_r\}$, we can construct an orthonormal set $\{\mathbf{C}_1, \ldots, \mathbf{C}_r\}$ of r matrices in $\mathcal{V}$. It follows from Lemma 6.2.1 that this set is linearly independent and hence—in light of Theorem 4.3.9— that it is a basis for $\mathcal{V}$. Thus, we have the following theorem.

Theorem 6.4.3. Every linear space (of $m \times n$ matrices) has an orthonormal basis.

The following theorem describes a basic property of orthonormal bases.

Theorem 6.4.4. Let $\mathbf{A}_1, \mathbf{A}_2, \ldots, \mathbf{A}_r$ represent matrices that form an orthonormal basis for a linear space $\mathcal{V}$. Then, for any matrix $\mathbf{A}$ in $\mathcal{V}$,

$$\mathbf{A} = (\mathbf{A} \bullet \mathbf{A}_1)\mathbf{A}_1 + (\mathbf{A} \bullet \mathbf{A}_2)\mathbf{A}_2 + \cdots + (\mathbf{A} \bullet \mathbf{A}_r)\mathbf{A}_r.$$

Proof. There exist (unique) scalars $x_1, x_2, \ldots, x_r$ such that $\mathbf{A} = \sum_{i=1}^{r} x_i \mathbf{A}_i$. And, for $j = 1, \ldots, r$,

$$\mathbf{A} \bullet \mathbf{A}_j = \Bigl(\sum_{i=1}^{r} x_i \mathbf{A}_i\Bigr) \bullet \mathbf{A}_j = \sum_{i=1}^{r} x_i (\mathbf{A}_i \bullet \mathbf{A}_j) = x_j.$$

Thus, $\mathbf{A} = \sum_{i=1}^{r} (\mathbf{A} \bullet \mathbf{A}_i)\mathbf{A}_i$. Q.E.D.

c. Inclusion of any particular orthonormal set in an orthonormal basis

Suppose that $\{\mathbf{A}_1, \ldots, \mathbf{A}_r\}$ is an orthonormal set of matrices in a k-dimensional linear space $\mathcal{V}$. Then, in light of Lemma 6.2.1, the matrices $\mathbf{A}_1, \ldots, \mathbf{A}_r$ are linearly independent, and it follows from Theorem 4.3.12 that there exists a basis for

$\mathcal{V}$ that includes $\mathbf{A}_1, \ldots, \mathbf{A}_r$; that is, there exist $k - r$ additional matrices, say $\mathbf{A}_{r+1}, \ldots, \mathbf{A}_k$, such that the set $\{\mathbf{A}_1, \ldots, \mathbf{A}_r, \mathbf{A}_{r+1}, \ldots, \mathbf{A}_k\}$ is a basis for $\mathcal{V}$.

Does there exist an orthonormal basis for $\mathcal{V}$ that includes $\mathbf{A}_1, \ldots, \mathbf{A}_r$? This question can be answered (in the affirmative) by observing that an orthonormal set $\{\mathbf{C}_1, \ldots, \mathbf{C}_r, \mathbf{C}_{r+1}, \ldots, \mathbf{C}_k\}$ of k matrices in $\mathcal{V}$ can be constructed by applying Gram-Schmidt orthogonalization to the set $\{\mathbf{A}_1, \ldots, \mathbf{A}_r, \mathbf{A}_{r+1}, \ldots, \mathbf{A}_k\}$. Since the subset comprising the first r of the matrices in the set $\{\mathbf{A}_1, \ldots, \mathbf{A}_r, \mathbf{A}_{r+1}, \ldots, \mathbf{A}_k\}$ is orthonormal, the matrices in this subset are not altered by the Gram-Schmidt orthogonalization; that is,

$$\mathbf{C}_i = \mathbf{A}_i \quad (i = 1, \ldots, r).$$

Moreover, it follows from Lemma 6.2.1 and Theorem 4.3.9 that the orthonormal set $\{\mathbf{C}_1, \ldots, \mathbf{C}_r, \mathbf{C}_{r+1}, \ldots, \mathbf{C}_k\}$ is a basis for $\mathcal{V}$. Thus, we have the following theorem.

Theorem 6.4.5. For any orthonormal set S of r matrices in a k-dimensional linear space $\mathcal{V}$, there exists an orthonormal basis for $\mathcal{V}$ that includes all r of the matrices in S (and $k - r$ additional matrices).

d. QR decomposition

Let $\mathbf{A}$ represent an $m \times k$ matrix of full column rank, that is, of rank k. And denote the first, ..., kth columns of $\mathbf{A}$ by $\mathbf{a}_1, \ldots, \mathbf{a}_k$, respectively. Then, according to Theorem 6.4.1, there exist unique scalars x_{ij} $(i < j = 1, \ldots, k)$ such that the k column vectors $\mathbf{b}_1, \ldots, \mathbf{b}_k$ defined recursively by the equalities

$$\begin{aligned}
\mathbf{b}_1 &= \mathbf{a}_1, \\
\mathbf{b}_2 &= \mathbf{a}_2 - x_{12}\mathbf{b}_1, \\
&\vdots \\
\mathbf{b}_j &= \mathbf{a}_j - x_{j-1,j}\mathbf{b}_{j-1} - \cdots - x_{1j}\mathbf{b}_1, \\
&\vdots \\
\mathbf{b}_k &= \mathbf{a}_k - x_{k-1,k}\mathbf{b}_{k-1} - \cdots - x_{1k}\mathbf{b}_1
\end{aligned}$$

or, equivalently, by the equalities

$$\begin{aligned}
\mathbf{a}_1 &= \mathbf{b}_1, \\
\mathbf{a}_2 &= \mathbf{b}_2 + x_{12}\mathbf{b}_1, \\
&\vdots \\
\mathbf{a}_j &= \mathbf{b}_j + x_{j-1,j}\mathbf{b}_{j-1} + \cdots + x_{1j}\mathbf{b}_1, \\
&\vdots \\
\mathbf{a}_k &= \mathbf{b}_k + x_{k-1,k}\mathbf{b}_{k-1} + \cdots + x_{1k}\mathbf{b}_1
\end{aligned}$$

form an orthogonal set. Further, $\mathbf{b}_1, \ldots, \mathbf{b}_k$ are nonnull, and (for $i < j = 1, \ldots, k$)

$$x_{ij} = \frac{\mathbf{a}_j \cdot \mathbf{b}_i}{\mathbf{b}_i \cdot \mathbf{b}_i}.$$

Now, let $\mathbf{B}$ represent the $m \times k$ matrix whose first, ..., kth columns are $\mathbf{b}_1, \ldots, \mathbf{b}_k$, respectively, and let $\mathbf{X}$ represent the $k \times k$ unit upper triangular matrix whose ijth element is (for $i < j = 1, \ldots, k$) x_{ij}. Then,

$$\mathbf{A} = \mathbf{BX}, \tag{4.2}$$

as is evident upon observing that the first column of $\mathbf{BX}$ is $\mathbf{b}_1$ and that (for $j = 2, \ldots, k$) the jth column of $\mathbf{BX}$ is $\mathbf{b}_j + x_{j-1,j}\mathbf{b}_{j-1} + \cdots + x_{1j}\mathbf{b}_1$.

Thus, $\mathbf{A}$ can be expressed as the product of an $m \times k$ matrix with (nonnull) orthogonal columns and of a $k \times k$ unit upper triangular matrix. Moreover, this decomposition or factorization of $\mathbf{A}$ is unique; that is, if $\mathbf{B}^*$ is an $m \times k$ matrix with orthogonal columns and $\mathbf{X}^*$ is a $k \times k$ unit upper triangular matrix such that $\mathbf{A} = \mathbf{B}^*\mathbf{X}^*$, then $\mathbf{B}^* = \mathbf{B}$ and $\mathbf{X}^* = \mathbf{X}$. To see this, denote the first, ..., kth columns of $\mathbf{B}^*$ by $\mathbf{b}_1^*, \ldots, \mathbf{b}_k^*$, respectively, denote the ijth element of $\mathbf{X}^*$ by x_{ij}^*, and observe that (by definition) $\mathbf{a}_1 = \mathbf{b}_1^*$ and (for $j = 2, \ldots, k$)

$$\mathbf{a}_j = \mathbf{b}_j^* + x_{j-1,j}^*\mathbf{b}_{j-1}^* + \cdots + x_{1j}^*\mathbf{b}_1^*,$$

or, equivalently, that $\mathbf{b}_1^* = \mathbf{a}_1$ and (for $j = 2, \ldots, k$)

$$\mathbf{b}_j^* = \mathbf{a}_j - x_{j-1,j}^*\mathbf{b}_{j-1}^* - \cdots - x_{1j}^*\mathbf{b}_1^*.$$

Then, it follows from Theorem 6.4.1 that (for $i < j = 1, \ldots, k$) $x_{ij}^* = x_{ij}$ and, as a consequence, that (for all j) $\mathbf{b}_j^* = \mathbf{b}_j$, so that $\mathbf{X}^* = \mathbf{X}$ and $\mathbf{B}^* = \mathbf{B}$.

A variation on decomposition (4.2) can be obtained by defining $\mathbf{Q} = \mathbf{BD}$, where $\mathbf{D} = \text{diag}(\|\mathbf{b}_1\|^{-1}, \ldots, \|\mathbf{b}_k\|^{-1})$, and $\mathbf{R} = \mathbf{EX}$, where $\mathbf{E} = \text{diag}(\|\mathbf{b}_1\|, \ldots, \|\mathbf{b}_k\|)$; or equivalently by defining $\mathbf{Q}$ to be the $m \times k$ matrix with jth column $\mathbf{q}_j = \|\mathbf{b}_j\|^{-1}\mathbf{b}_j$ and $\mathbf{R} = \{r_{ij}\}$ to be the $k \times k$ matrix with

$$r_{ij} = \begin{cases} \|\mathbf{b}_i\| x_{ij}, & \text{for } j > i, \\ \|\mathbf{b}_i\|, & \text{for } j = i, \\ 0, & \text{for } j < i. \end{cases}$$

Then, clearly, the columns of $\mathbf{Q}$ are orthonormal (they are the orthonormal vectors obtained by applying Gram-Schmidt orthogonalization to the columns of $\mathbf{A}$), $\mathbf{R}$ is an upper triangular matrix with positive diagonal elements, and

$$\mathbf{A} = \mathbf{QR}. \tag{4.3}$$

Moreover, decomposition (4.3) is unique; that is, if $\mathbf{Q}^*$ is an $m \times k$ matrix with orthonormal columns $\mathbf{q}_1^*, \ldots, \mathbf{q}_k^*$, respectively, and if $\mathbf{R}^* = \{r_{ij}^*\}$ is a $k \times k$ upper triangular matrix (with positive diagonal elements) such that $\mathbf{A} = \mathbf{Q}^*\mathbf{R}^*$, then $\mathbf{Q}^* = \mathbf{Q}$ and $\mathbf{R}^* = \mathbf{R}$. To see this, let $\mathbf{B}^* = \mathbf{Q}^*\mathbf{E}^*$, where $\mathbf{E}^* = \text{diag}(r_{11}^*, \ldots, r_{kk}^*)$, and let $\mathbf{X}^* = \mathbf{D}^*\mathbf{R}^*$, where $\mathbf{D}^* = \text{diag}(r_{11}^{*-1}, \ldots, r_{kk}^{*-1})$;

or, equivalently, let $\mathbf{B}^*$ represent the $m \times k$ matrix with jth column $\mathbf{b}_j^* = r_{jj}^* \mathbf{q}_j^*$, and let $\mathbf{X}^*$ represent the $k \times k$ matrix with ijth element $x_{ij}^* = r_{ii}^{*-1} r_{ij}^*$. Then, the columns of $\mathbf{B}^*$ are orthogonal, $\mathbf{X}^*$ is unit upper triangular, and $\mathbf{A} = \mathbf{B}^*\mathbf{X}^*$. Thus, in light of the uniqueness of decomposition (4.2), $\mathbf{B}^* = \mathbf{B}$ and $\mathbf{X}^* = \mathbf{X}$, so that $r_{jj}^* \mathbf{q}_j^* = \mathbf{b}_j^* = \mathbf{b}_j = \|\mathbf{b}_j\| \mathbf{q}_j$ (for all j) and $r_{ii}^{*-1} r_{ij}^* = x_{ij}^* = x_{ij} = \|\mathbf{b}_i\|^{-1} r_{ij}$ (for all j and $i < j$), implying that

$$r_{jj}^* = \left(r_{jj}^{*2} \mathbf{q}_j^{*\prime} \mathbf{q}_j^*\right)^{1/2} = \left(\|\mathbf{b}_j\|^2 \mathbf{q}_j' \mathbf{q}_j\right)^{1/2} = \|\mathbf{b}_j\| \ (= r_{jj})$$

(for all j), and hence that $\mathbf{q}_j = \mathbf{q}_j^*$ (for all j) and $r_{ij}^* = r_{ij}$ (for all j and $i \leq j$), or, equivalently, that $\mathbf{Q}^* = \mathbf{Q}$ and $\mathbf{R}^* = \mathbf{R}$.

Note that decompositions (4.2) and (4.3) depend on the choice of inner product (with respect to which the columns of $\mathbf{B}$ and the columns of $\mathbf{Q}$ form orthogonal or orthonormal sets). Let us refer to the version of decomposition (4.3) corresponding to the usual inner product as the *QR decomposition*—for that version, $\mathbf{Q}'\mathbf{Q} = \mathbf{I}_k$. This usage of the term QR decomposition conforms to that of Stewart (1973, p. 214) and Lancaster and Tismenetsky (1985, p. 111), but differs somewhat from that of Golub and Van Loan (1989, sec. 5.2), who refer to decomposiiton (4.3) as the "skinny" QR decomposition.

In principle, the QR decomposition of the matrix $\mathbf{A}$ could be computed by Gram-Schmidt orthogonalization. However, as a computational algorithm, Gram-Schmidt orthogonalization is numerically unstable. Numerically more stable algorithms for computing the QR decomposition are discussed by Golub and Van Loan (1989, sec. 5.2)—one such algorithm is a variant on Gram-Schmidt orthogonalization called *modified Gram-Schmidt orthogonalization.*

Exercises

Section 6.3

1. Use the Schwarz inequality to show that, for any two matrices $\mathbf{A}$ and $\mathbf{B}$ in a linear space $\mathcal{V}$,

$$\|\mathbf{A} + \mathbf{B}\| \leq \|\mathbf{A}\| + \|\mathbf{B}\|,$$

with equality holding if and only if $\mathbf{B} = \mathbf{0}$ or $\mathbf{A} = k\mathbf{B}$ for some nonnegative scalar k. (This inequality is known as the triangle inequality.)

2. Letting $\mathbf{A}$, $\mathbf{B}$, and $\mathbf{C}$ represent arbitrary matrices in a linear space $\mathcal{V}$, show that

(a) $\delta(\mathbf{B}, \mathbf{A}) = \delta(\mathbf{A}, \mathbf{B})$, that is, the distance between $\mathbf{B}$ and $\mathbf{A}$ is the same as that between $\mathbf{A}$ and $\mathbf{B}$;

(b)

$$\begin{aligned} \delta(\mathbf{A}, \mathbf{B}) &> 0, \quad \text{if } \mathbf{A} \neq \mathbf{B}, \\ &= 0, \quad \text{if } \mathbf{A} = \mathbf{B}, \end{aligned}$$

that is, the distance between any two matrices is greater than zero, unless the two matrices are identical, in which case the distance between them is zero;

(c) $\delta(\mathbf{A}, \mathbf{B}) \leq \delta(\mathbf{A}, \mathbf{C}) + \delta(\mathbf{C}, \mathbf{B})$, that is, the distance between **A** and **B** is less than or equal to the sum of the distances between **A** and **C** and between **C** and **B**;

(d) $\delta(\mathbf{A}, \mathbf{B}) = \delta(\mathbf{A}+\mathbf{C},\ \mathbf{B}+\mathbf{C})$, that is, distance is unaffected by a translation of "axes."

[For Part (c), use the result of Exercise 1, i.e., the triangle inequality.]

Section 6.4

3. Let $\mathbf{w}_1', \mathbf{w}_2'$, and $\mathbf{w}_3'$ represent the three linearly independent 4-dimensional row vectors $(6, 0, -2, 3)$, $(-2, 4, 4, 2)$, and $(0, 5, -1, 2)$, respectively, in the linear space $\mathcal{R}^4$, and adopt the usual definition of inner product.

   (a) Use Gram-Schmidt orthogonalization to find an orthonormal basis for the linear space $\text{sp}(\mathbf{w}_1', \mathbf{w}_2', \mathbf{w}_3')$.

   (b) Find an orthonormal basis for $\mathcal{R}^4$ that includes the three orthonormal vectors from Part (a). Do so by extending the results of the Gram-Schmidt orthogonalization [from Part (a)] to a fourth linearly independent row vector such as $(0, 1, 0, 0)$.

4. Let $\{\mathbf{A}_1, \ldots, \mathbf{A}_k\}$ represent a nonempty (possibly linearly dependent) set of matrices in a linear space $\mathcal{V}$.

   (a) Generalize Theorem 6.4.1 by showing (1) that there exist scalars x_{ij} $(i < j = 1, \ldots, k)$ such that the set comprising the k matrices

$$
\begin{aligned}
\mathbf{B}_1 &= \mathbf{A}_1, \\
\mathbf{B}_2 &= \mathbf{A}_2 - x_{12}\mathbf{B}_1, \\
&\vdots \\
\mathbf{B}_j &= \mathbf{A}_j - x_{j-1,j}\mathbf{B}_{j-1} - \cdots - x_{1j}\mathbf{B}_1, \\
&\vdots \\
\mathbf{B}_k &= \mathbf{A}_k - x_{k-1,k}\mathbf{B}_{k-1} - \cdots - x_{1k}\mathbf{B}_1
\end{aligned}
$$

   is orthogonal; (2) that, for $j = 1, \ldots, k$ and for those $i < j$ such that $\mathbf{B}_i$ is nonnull, x_{ij} is given uniquely by

$$x_{ij} = \frac{\mathbf{A}_j \bullet \mathbf{B}_i}{\mathbf{B}_i \bullet \mathbf{B}_i};$$

   and (3) that the number of nonnull matrices among $\mathbf{B}_1, \ldots, \mathbf{B}_k$ equals $\dim[\text{sp}(\mathbf{A}_1, \ldots, \mathbf{A}_k)]$.

   (b) Describe a procedure for constructing an orthonormal basis for $\text{sp}(\mathbf{A}_1, \ldots, \mathbf{A}_k)$.

5. Let $\mathbf{A}$ represent an $m \times k$ matrix of rank r (where r is possibly less than k). Generalize result (4.3) by using the results of Exercise 4 to obtain a decomposition of the form $\mathbf{A} = \mathbf{Q}\mathbf{R}_1$, where $\mathbf{Q}$ is an $m \times r$ matrix with orthonormal columns and $\mathbf{R}_1$ is an $r \times k$ submatrix whose rows are the r nonnull rows of a $k \times k$ upper triangular matrix $\mathbf{R}$ having r positive diagonal elements and $k - r$ null rows.

7

Linear Systems: Consistency and Compatibility

In many instances, the solution to a problem encountered in statistics (or in some other discipline) can be reduced to a problem of solving a system of linear equations. For example, the problem of obtaining the least squares estimates of the parameters in a linear statistical model can be reduced to the problem of solving a system of linear equations called the normal equations or (if the parameters are subject to linear constraints) a system of linear equations sometimes called the constrained normal equations (e.g., Searle 1971).

The focus in this chapter is on questions about the existence of (one or more) solutions to a system of linear equations—questions about the solutions themselves (when solutions exist) are deferred until Chapter 11. Such questions of existence arise in the theory of linear statistical models in determining which parametric functions are estimable (i.e., which parametric functions can be estimated from the data) and in the design of experiments (e.g., Searle 1971). The results of Sections 7.2 and 7.3 are general (i.e., applicable to any system of linear equations), while those of Section 7.4 are specific to systems of linear equations of the form of the normal equations or constrained normal equations—some of the terminology employed in discussing systems of linear equations is introduced in Section 7.1.

7.1 Some Basic Terminology

Consider a set of m equations of the general form

$$\begin{aligned} a_{11}x_1 + a_{12}x_2 + \cdots + a_{1n}x_n &= b_1 \\ &\vdots \\ a_{m1}x_1 + a_{m2}x_2 + \cdots + a_{mn}x_n &= b_m, \end{aligned}$$

where $a_{11}, a_{12}, \ldots, a_{1n}, \ldots, a_{m1}, a_{m2}, \ldots, a_{mn}$ and $b_1, \ldots, b_m$ represent fixed scalars, and $x_1, x_2, \ldots, x_n$ are scalar-valued unknowns or variables. Each of these equations is linear in the unknowns $x_1, x_2, \ldots, x_n$. Collectively, these equations are called a *system of linear equations* (in unknowns $x_1, x_2, \ldots, x_n$) or simply a *linear system* (in $x_1, x_2, \ldots, x_n$).

The linear system can be rewritten in matrix form as

$$\mathbf{Ax} = \mathbf{b}, \tag{1.1}$$

where $\mathbf{A}$ is the $m \times n$ matrix whose ijth element is a_{ij} ($i = 1, \ldots, m$; $j = 1, \ldots, n$), $\mathbf{b} = (b_1, \ldots, b_m)'$, and $\mathbf{x} = (x_1, x_2, \ldots, x_n)'$. The matrix $\mathbf{A}$ is called the *coefficient matrix* of the linear system, and $\mathbf{b}$ is called the *right side*. Any value of the vector $\mathbf{x}$ of unknowns that satisfies $\mathbf{Ax} = \mathbf{b}$ is called a *solution* to the linear system, and the process of finding a solution (when one exists) is called solving the linear system.

There may be occasion to solve the linear system for more than one right side, that is, to solve each of p linear systems

$$\mathbf{Ax}_k = \mathbf{b}_k \quad (k = 1, \ldots, p) \tag{1.2}$$

(in vectors $\mathbf{x}_1, \ldots, \mathbf{x}_p$, respectively, of unknowns) that have the same coefficient matrix $\mathbf{A}$ but right sides $\mathbf{b}_1, \ldots, \mathbf{b}_p$, respectively, that may differ. By forming an $n \times p$ matrix $\mathbf{X}$ whose first, ..., pth columns are $\mathbf{x}_1, \ldots, \mathbf{x}_p$, respectively, and an $m \times p$ matrix $\mathbf{B}$ whose columns are $\mathbf{b}_1, \ldots, \mathbf{b}_p$, respectively, the p linear systems (1.2) can be rewritten collectively as

$$\mathbf{AX} = \mathbf{B}. \tag{1.3}$$

As in the special case (1.1) where $p = 1$, $\mathbf{AX} = \mathbf{B}$ is called a *linear system* (in $\mathbf{X}$), $\mathbf{A}$ and $\mathbf{B}$ are called the *coefficient matrix* and the *right side*, respectively, and any value of $\mathbf{X}$ that satisfies $\mathbf{AX} = \mathbf{B}$ is called a *solution*.

In the special case

$$\mathbf{AX} = \mathbf{0},$$

where the right side $\mathbf{B}$ of linear system (1.3) is a null matrix, the linear system is said to be *homogeneous*. If $\mathbf{B}$ is nonnull, linear system (1.3) is said to be *nonhomogeneous*.

7.2 Consistency

A linear system is said to be *consistent* if it has one or more solutions. If no solution exists, the linear system is said to be *inconsistent*.

Every homogeneous linear system is consistent—one solution to a homogeneous linear system is the null matrix (of appropriate dimensions). A nonhomogeneous linear system may be either consistent or inconsistent. Some necessary and sufficient conditions for a linear system to be consistent are given by the following theorem.

Theorem 7.2.1. *Each* of the following conditions is necessary and sufficient for a linear system $\mathbf{AX} = \mathbf{B}$ (in $\mathbf{X}$) to be consistent:

(1) $\mathcal{C}(\mathbf{B}) \subset \mathcal{C}(\mathbf{A})$;
(2) every column of **B** belongs to $\mathcal{C}(\mathbf{A})$;
(3) $\mathcal{C}(\mathbf{A}, \mathbf{B}) = \mathcal{C}(\mathbf{A})$;
(4) $\text{rank}(\mathbf{A}, \mathbf{B}) = \text{rank}(\mathbf{A})$.

Proof. That Condition (1) is necessary and sufficient for the consistency of $\mathbf{AX} = \mathbf{B}$ is an immediate consequence of Lemma 4.2.2. Further, it follows from Lemma 4.2.1 that Condition (2) is equivalent to Condition (1), from Lemma 4.5.1 that Condition (3) is equivalent to Condition (1), and from Corollary 4.5.2 that Condition (4) is equivalent to Condition (1). Thus, each of Conditions (2) through (4) is also necessary and sufficient for the consistency of $\mathbf{AX} = \mathbf{B}$. Q.E.D.

A sufficient (but in general not a necessary) condition for the consistency of a linear system is given by the following theorem.

Theorem 7.2.2. If the coefficient matrix **A** of a linear system $\mathbf{AX} = \mathbf{B}$ (in **X**) has full row rank, then $\mathbf{AX} = \mathbf{B}$ is consistent.

Proof. Let m represent the number of rows in **A**, and suppose that **A** has full row rank or, equivalently, that $\text{rank}(\mathbf{A}) = m$. Then, making use of Lemma 4.4.3, we find that

$$\text{rank}(\mathbf{A}, \mathbf{B}) \leq m = \text{rank}(\mathbf{A}).$$

Moreover, since $\mathcal{C}(\mathbf{A}) \subset \mathcal{C}(\mathbf{A}, \mathbf{B})$, it follows from Theorem 4.4.4 that $\text{rank}(\mathbf{A}, \mathbf{B}) \geq \text{rank}(\mathbf{A})$. Thus, $\text{rank}(\mathbf{A}, \mathbf{B}) = \text{rank}(\mathbf{A})$, and the consistency of $\mathbf{AX} = \mathbf{B}$ follows from Theorem 7.2.1. Q.E.D.

7.3 Compatibility

A linear system $\mathbf{AX} = \mathbf{B}$ (in **X**) is said to be *compatible* if every linear relationship that exists among the rows of **A** also exists among the rows of **B**. That is, $\mathbf{AX} = \mathbf{B}$ is compatible if $\mathbf{k}'\mathbf{B} = \mathbf{0}$ for every row vector $\mathbf{k}'$ (of appropriate dimension) such that $\mathbf{k}'\mathbf{A} = \mathbf{0}$.

Compatibility is equivalent to consistency as indicated by the following theorem.

Theorem 7.3.1. A linear system $\mathbf{AX} = \mathbf{B}$ (in **X**) is consistent if and only if it is compatible.

Proof. Suppose that the linear system $\mathbf{AX} = \mathbf{B}$ is consistent. Then, there exists a matrix $\mathbf{X}^*$ such that $\mathbf{AX}^* = \mathbf{B}$, and, for every row vector $\mathbf{k}'$ such that $\mathbf{k}'\mathbf{A} = \mathbf{0}$, we have that

$$\mathbf{k}'\mathbf{B} = \mathbf{k}'\mathbf{AX}^* = \mathbf{0}.$$

Thus, $\mathbf{AX} = \mathbf{B}$ is compatible.

To prove the converse, let $\mathbf{a}'_1, \ldots, \mathbf{a}'_m$ and $\mathbf{b}'_1, \ldots, \mathbf{b}'_m$ represent the rows of **A** and **B**, respectively, and observe that if a value of **X** is a solution to all m of the equations

$$\mathbf{a}'_i\mathbf{X} = \mathbf{b}'_i \quad (i = 1, \ldots, m),$$

then it is a solution to $\mathbf{AX} = \mathbf{B}$.

According to Theorem 4.3.11, there exists a subset $S = \{k_1, \ldots, k_r\}$ of the first m positive integers such that the set $\{\mathbf{a}'_{k_1}, \ldots, \mathbf{a}'_{k_r}\}$ is a basis for $\mathcal{R}(\mathbf{A})$. Taking $\mathbf{A}_1$ to be a matrix whose rows are $\mathbf{a}'_{k_1}, \ldots, \mathbf{a}'_{k_r}$ and $\mathbf{B}_1$ to be a matrix whose rows are $\mathbf{b}'_{k_1}, \ldots, \mathbf{b}'_{k_r}$, it follows from Theorem 7.2.2 that the linear system $\mathbf{A}_1\mathbf{X} = \mathbf{B}_1$ (in $\mathbf{X}$) is consistent. Thus, there exists a matrix $\mathbf{X}^*$ such that $\mathbf{A}_1\mathbf{X}^* = \mathbf{B}_1$ or, equivalently, such that

$$\mathbf{a}'_{k_j}\mathbf{X}^* = \mathbf{b}'_{k_j} \quad (j = 1, \ldots, r).$$

To complete the proof, it suffices to show that if $\mathbf{AX} = \mathbf{B}$ is compatible, then $\mathbf{a}'_i\mathbf{X}^* = \mathbf{b}'_i$ for $i \notin S$. For each $i \notin S$, there exist scalars $\lambda_{i1}, \ldots, \lambda_{ir}$ such that

$$\mathbf{a}'_i = \sum_{j=1}^{r} \lambda_{ij}\mathbf{a}'_{k_j}$$

or, equivalently, such that

$$\mathbf{a}'_i - \sum_{j=1}^{r} \lambda_{ij}\mathbf{a}'_{k_j} = \mathbf{0}.$$

If $\mathbf{AX} = \mathbf{B}$ is compatible, then, for each $i \notin S$,

$$\mathbf{b}'_i - \sum_{j=1}^{r} \lambda_{ij}\mathbf{b}'_{k_j} = \mathbf{0}$$

or, equivalently,

$$\mathbf{b}'_i = \sum_{j=1}^{r} \lambda_{ij}\mathbf{b}'_{k_j},$$

in which case

$$\mathbf{a}'_i\mathbf{X}^* = \sum_{j=1}^{r} \lambda_{ij}\mathbf{a}'_{k_j}\mathbf{X}^* = \sum_{j=1}^{r} \lambda_{ij}\mathbf{b}'_{k_j} = \mathbf{b}'_i.$$

Q.E.D.

7.4 Linear Systems of the Form A′AX = A′B

a. A basic result

Theorem 7.3.1 can be used to establish the following result.

Theorem 7.4.1. For any $m \times n$ matrix $\mathbf{A}$ and $m \times p$ matrix $\mathbf{B}$, the linear system $\mathbf{A}'\mathbf{AX} = \mathbf{A}'\mathbf{B}$ (in $\mathbf{X}$) is consistent.

Proof. To prove Theorem 7.4.1, it suffices (in light of Theorem 7.3.1) to show that the linear system $\mathbf{A'AX} = \mathbf{A'B}$ is compatible.

Let $\mathbf{k}'$ represent any row vector such that $\mathbf{k'A'A} = \mathbf{0}$. Then, according to Corollary 5.3.3, we have that $\mathbf{k'A'} = \mathbf{0}$ and hence that $\mathbf{k'A'B} = \mathbf{0}$. Thus, the linear system $\mathbf{A'AX} = \mathbf{A'B}$ is compatible. Q.E.D.

In the special case where $\mathbf{B} = \mathbf{I}$, Theorem 7.4.1 reduces to the following result.

Corollary 7.4.2. For any matrix $\mathbf{A}$, the linear system $\mathbf{A'AX} = \mathbf{A'}$ (in $\mathbf{X}$) is consistent.

b. Some results on row and column spaces and on ranks

Corollary 7.4.2 can be used to establish the following result.

Theorem 7.4.3. For any $m \times n$ matrix $\mathbf{A}$ and $n \times s$ matrix $\mathbf{T}$, $\mathcal{C}(\mathbf{T'A'A}) = \mathcal{C}(\mathbf{T'A'})$ and $\mathcal{R}(\mathbf{A'AT}) = \mathcal{R}(\mathbf{AT})$.

Proof. According to Corollary 7.4.2, there exists a matrix $\mathbf{X}$ such that $\mathbf{A'AX} = \mathbf{A'}$. Thus, making use of Corollary 4.2.3, we find that $\mathcal{C}(\mathbf{T'A'}) = \mathcal{C}(\mathbf{T'A'AX}) \subset \mathcal{C}(\mathbf{T'A'A})$ and also that $\mathcal{C}(\mathbf{T'A'A}) \subset \mathcal{C}(\mathbf{T'A'})$, implying that $\mathcal{C}(\mathbf{T'A'A}) = \mathcal{C}(\mathbf{T'A'})$ and further (in light of Lemma 4.2.5) that $\mathcal{R}(\mathbf{A'AT}) = \mathcal{R}(\mathbf{AT})$. Q.E.D.

One implication of Theorem 7.4.3 is given by the following corollary.

Corollary 7.4.4. For any $m \times n$ matrix $\mathbf{A}$ and $n \times s$ matrix $\mathbf{T}$, $\text{rank}(\mathbf{T'A'A}) = \text{rank}(\mathbf{T'A'})$ and $\text{rank}(\mathbf{A'AT}) = \text{rank}(\mathbf{AT})$.

In the special case where $\mathbf{T} = \mathbf{I}$, the results of Theorem 7.4.3 and Corollary 7.4.4 reduce to the results given by the following corollary.

Corollary 7.4.5. For any matrix $\mathbf{A}$, $\mathcal{C}(\mathbf{A'A}) = \mathcal{C}(\mathbf{A'})$, $\mathcal{R}(\mathbf{A'A}) = \mathcal{R}(\mathbf{A})$, and $\text{rank}(\mathbf{A'A}) = \text{rank}(\mathbf{A})$.

One implication of Corollary 7.4.5 is described in the following, additional corollary.

Corollary 7.4.6. For any matrix $\mathbf{A}$ of full column rank, $\mathbf{A'A}$ is nonsingular.

In light of Corollary 4.5.6, the following theorem is an immediate consequence of Theorem 7.4.3 (and Corollary 7.4.4).

Theorem 7.4.7. Let $\mathbf{A}$ represent an $m \times n$ matrix, $\mathbf{T}$ an $n \times s$ matrix, and $\mathbf{K}$ a $q \times s$ matrix. Then,

$$\mathcal{C}(\mathbf{T'A'A}, \mathbf{K'}) = \mathcal{C}(\mathbf{T'A'}, \mathbf{K'}), \qquad \text{rank}(\mathbf{T'A'A}, \mathbf{K'}) = \text{rank}(\mathbf{T'A'}, \mathbf{K'}),$$

$$\mathcal{R}\begin{pmatrix} \mathbf{A'AT} \\ \mathbf{K} \end{pmatrix} = \mathcal{R}\begin{pmatrix} \mathbf{AT} \\ \mathbf{K} \end{pmatrix}, \qquad \text{rank}\begin{pmatrix} \mathbf{A'AT} \\ \mathbf{K} \end{pmatrix} = \text{rank}\begin{pmatrix} \mathbf{AT} \\ \mathbf{K} \end{pmatrix}.$$

c. Extensions

The following theorem extends Theorem 7.4.1.

Theorem 7.4.8. For any $m \times n$ matrix $\mathbf{A}$, $m \times p$ matrix $\mathbf{B}$, and $n \times q$ matrix $\mathbf{L}$, and for any $q \times p$ matrix $\mathbf{C}$ such that $\mathcal{C}(\mathbf{C}) \subset \mathcal{C}(\mathbf{L}')$, the linear system

$$\begin{pmatrix} \mathbf{A}'\mathbf{A} & \mathbf{L} \\ \mathbf{L}' & \mathbf{0} \end{pmatrix} \begin{pmatrix} \mathbf{X} \\ \mathbf{Y} \end{pmatrix} = \begin{pmatrix} \mathbf{A}'\mathbf{B} \\ \mathbf{C} \end{pmatrix} \tag{4.1}$$

(in $\mathbf{X}$ and $\mathbf{Y}$) is consistent.

Proof. It suffices (in light of Theorem 7.3.1) to show that linear system (4.1) is compatible.

Let $\mathbf{k}_1'$ and $\mathbf{k}_2'$ represent any $n \times 1$ and $q \times 1$ vectors such that

$$(\mathbf{k}_1', \mathbf{k}_2') \begin{pmatrix} \mathbf{A}'\mathbf{A} & \mathbf{L} \\ \mathbf{L}' & \mathbf{0} \end{pmatrix} = \mathbf{0}.$$

Then,

$$\mathbf{k}_1'\mathbf{A}'\mathbf{A} = -\mathbf{k}_2'\mathbf{L}', \qquad \mathbf{L}'\mathbf{k}_1 = \mathbf{0}, \tag{4.2}$$

implying that

$$(\mathbf{A}\mathbf{k}_1)'\mathbf{A}\mathbf{k}_1 = \mathbf{k}_1'\mathbf{A}'\mathbf{A}\mathbf{k}_1 = -\mathbf{k}_2'\mathbf{L}'\mathbf{k}_1 = 0$$

and hence (in light of Corollary 5.3.2) that

$$\mathbf{A}\mathbf{k}_1 = \mathbf{0}. \tag{4.3}$$

Moreover, there exists a matrix $\mathbf{F}$ such that $\mathbf{C} = \mathbf{L}'\mathbf{F}$, implying—in light of results (4.2) and (4.3)—that

$$\mathbf{k}_2'\mathbf{C} = \mathbf{k}_2'\mathbf{L}'\mathbf{F} = -\mathbf{k}_1'\mathbf{A}'\mathbf{A}\mathbf{F} = -(\mathbf{A}\mathbf{k}_1)'\mathbf{A}\mathbf{F} = \mathbf{0}. \tag{4.4}$$

Together, results (4.3) and (4.4) imply that

$$(\mathbf{k}_1', \mathbf{k}_2') \begin{pmatrix} \mathbf{A}'\mathbf{B} \\ \mathbf{C} \end{pmatrix} = (\mathbf{A}\mathbf{k}_1)'\mathbf{B} + \mathbf{k}_2'\mathbf{C} = \mathbf{0}.$$

Thus, linear system (4.1) is compatible. Q.E.D.

Theorem 7.4.8 is generalized in the following corollary.

Corollary 7.4.9. For any $m \times n$ matrix $\mathbf{A}$, $m \times p$ matrix $\mathbf{B}$, $n \times q$ matrix $\mathbf{L}$, and $q \times p$ matrix $\mathbf{R}$, and for any $q \times p$ matrix $\mathbf{C}$ such that $\mathcal{C}(\mathbf{C}) \subset \mathcal{C}(\mathbf{L}')$, the linear system

$$\begin{pmatrix} \mathbf{A}'\mathbf{A} & \mathbf{L} \\ \mathbf{L}' & \mathbf{0} \end{pmatrix} \begin{pmatrix} \mathbf{X} \\ \mathbf{Y} \end{pmatrix} = \begin{pmatrix} \mathbf{A}'\mathbf{B} + \mathbf{L}\mathbf{R} \\ \mathbf{C} \end{pmatrix} \tag{4.5}$$

(in $\mathbf{X}$ and $\mathbf{Y}$) is consistent.

Proof. According to Theorem 7.4.8, the linear system

$$\begin{pmatrix} \mathbf{A}'\mathbf{A} & \mathbf{L} \\ \mathbf{L}' & \mathbf{0} \end{pmatrix} \begin{pmatrix} \mathbf{X} \\ \mathbf{Y} \end{pmatrix} = \begin{pmatrix} \mathbf{A}'\mathbf{B} \\ \mathbf{C} \end{pmatrix}$$

(in $\mathbf{X}$ and $\mathbf{Y}$) has a solution, say $\begin{pmatrix} \mathbf{X}_* \\ \mathbf{Y}_* \end{pmatrix}$. Then, clearly, the matrix $\begin{pmatrix} \mathbf{X}_* \\ \mathbf{Y}_* + \mathbf{R} \end{pmatrix}$ is a solution to linear system (4.5). Thus, linear system (4.5) is consistent. Q.E.D.

Exercise

Section 7.4

1. (a) Let $\mathbf{A}$ represent an $m \times n$ matrix, $\mathbf{C}$ an $n \times q$ matrix, and $\mathbf{B}$ a $q \times p$ matrix. Show that if $\text{rank}(\mathbf{AC}) = \text{rank}(\mathbf{C})$, then

$$\mathcal{R}(\mathbf{ACB}) = \mathcal{R}(\mathbf{CB}) \quad \text{and} \quad \text{rank}(\mathbf{ACB}) = \text{rank}(\mathbf{CB})$$

and that if $\text{rank}(\mathbf{CB}) = \text{rank}(\mathbf{C})$, then

$$\mathcal{C}(\mathbf{ACB}) = \mathcal{C}(\mathbf{AC}) \quad \text{and} \quad \text{rank}(\mathbf{ACB}) = \text{rank}(\mathbf{AC}),$$

thereby extending the results of Corollary 4.4.7, Theorem 7.4.3, and Corollary 7.4.4.

(b) Let $\mathbf{A}$ and $\mathbf{B}$ represent $m \times n$ matrices. (1) Show that if $\mathbf{C}$ is an $r \times q$ matrix and $\mathbf{D}$ a $q \times m$ matrix such that $\text{rank}(\mathbf{CD}) = \text{rank}(\mathbf{D})$, then $\mathbf{CDA} = \mathbf{CDB}$ implies $\mathbf{DA} = \mathbf{DB}$, thereby extending the result of Part (1) of Corollary 5.3.3. {*Hint*. To show that $\mathbf{DA} = \mathbf{DB}$, it suffices to show that $\text{rank}[\mathbf{D}(\mathbf{A} - \mathbf{B})] = 0$.} (2) Similarly, show that if $\mathbf{C}$ is an $n \times q$ matrix and $\mathbf{D}$ a $q \times p$ matrix such that $\text{rank}(\mathbf{CD}) = \text{rank}(\mathbf{C})$, then $\mathbf{ACD} = \mathbf{BCD}$ implies $\mathbf{AC} = \mathbf{BC}$, thereby extending the result of Part (2) of Corollary 5.3.3.

8

Inverse Matrices

Corresponding to any nonsingular matrix is another nonsingular matrix that is referred to as the inverse of the first matrix. Inverse matrices are introduced and defined in Section 8.1 and are discussed in the subsequent sections of this chapter. Inverse matrices are of interest because there is a close relationship between the solution to a linear system having a nonsingular coefficient matrix and the inverse of the coefficient matrix—this relationship is described in Section 8.2a. Moreover, there are many situations in statistics and in other disciplines that give rise to inverse matrices that are of interest in their own right.

8.1 Some Definitions and Basic Results

A *right inverse* of an $m \times n$ matrix $\mathbf{A}$ is an $n \times m$ matrix $\mathbf{R}$ such that

$$\mathbf{AR} = \mathbf{I}_m.$$

Similarly, a *left inverse* of an $m \times n$ matrix $\mathbf{A}$ is an $n \times m$ matrix $\mathbf{L}$ such that

$$\mathbf{LA} = \mathbf{I}_n$$

or, equivalently, such that

$$\mathbf{A}'\mathbf{L}' = \mathbf{I}_n.$$

A matrix may or may not have a right or left inverse, as indicated by the following lemma.

Lemma 8.1.1. An $m \times n$ matrix $\mathbf{A}$ has a right inverse if and only if $\text{rank}(\mathbf{A}) = m$ (i.e., if and only if $\mathbf{A}$ has full row rank) and has a left inverse if and only if $\text{rank}(\mathbf{A}) = n$ (i.e., if and only if $\mathbf{A}$ has full column rank).

Proof. If $\text{rank}(\mathbf{A}) = m$, then it follows from Theorem 7.2.2 that there exists a matrix $\mathbf{R}$ such that $\mathbf{AR} = \mathbf{I}_m$, that is, that $\mathbf{A}$ has a right inverse. Conversely, if there exists a matrix $\mathbf{R}$ such that $\mathbf{AR} = \mathbf{I}_m$, then

$$\text{rank}(\mathbf{A}) \geq \text{rank}(\mathbf{AR}) = \text{rank}(\mathbf{I}_m) = m,$$

implying [since, according to Lemma 4.4.3, $\text{rank}(\mathbf{A}) \leq m$] that $\text{rank}(\mathbf{A}) = m$. Thus, $\mathbf{A}$ has a right inverse if and only if $\text{rank}(\mathbf{A}) = m$. That $\mathbf{A}$ has a left inverse if and only if $\text{rank}(\mathbf{A}) = n$ is evident upon observing that $\mathbf{A}$ has a left inverse if and only if $\mathbf{A}'$ has a right inverse [and that $\text{rank}(\mathbf{A}') = \text{rank}(\mathbf{A})$]. Q.E.D.

As an almost immediate consequence of Lemma 8.1.1, we have the following corollary.

Corollary 8.1.2. A matrix $\mathbf{A}$ has both a right inverse and a left inverse if and only if $\mathbf{A}$ is a (square) nonsingular matrix.

If there exists a matrix $\mathbf{B}$ that is both a right and left inverse of a matrix $\mathbf{A}$ (so that $\mathbf{AB} = \mathbf{I}$ and $\mathbf{BA} = \mathbf{I}$), then $\mathbf{A}$ is said to be *invertible* and $\mathbf{B}$ is referred to as an *inverse* of $\mathbf{A}$. Only a (square) nonsingular matrix can be invertible, as is evident from Corollary 8.1.2.

The following lemma and theorem include some basic results on the existence and uniqueness of inverse matrices.

Lemma 8.1.3. If a square matrix $\mathbf{A}$ has a right or left inverse $\mathbf{B}$, then $\mathbf{A}$ is nonsingular and $\mathbf{B}$ is an inverse of $\mathbf{A}$.

Proof. Suppose that $\mathbf{A}$ has a right inverse $\mathbf{R}$. Then, it follows from Lemma 8.1.1 that $\mathbf{A}$ is nonsingular and further that $\mathbf{A}$ has a left inverse $\mathbf{L}$. Observing that

$$\mathbf{L} = \mathbf{LI} = \mathbf{LAR} = \mathbf{IR} = \mathbf{R}$$

and hence that $\mathbf{RA} = \mathbf{I}$, we conclude that $\mathbf{R}$ is an inverse of $\mathbf{A}$.

A similar argument can be used to show that if $\mathbf{A}$ has a left inverse $\mathbf{L}$, then $\mathbf{A}$ is nonsingular and $\mathbf{L}$ is an inverse of $\mathbf{A}$. Q.E.D.

Theorem 8.1.4. A matrix is invertible if and only if it is a (square) nonsingular matrix. Further, any nonsingular matrix has a unique inverse $\mathbf{B}$ and has no right or left inverse other than $\mathbf{B}$.

Proof. Suppose that $\mathbf{A}$ is a nonsingular matrix. Then, it follows from Lemma 8.1.1 that $\mathbf{A}$ has a right inverse $\mathbf{B}$ and from Lemma 8.1.3 that $\mathbf{B}$ is an inverse of $\mathbf{A}$. Thus, $\mathbf{A}$ is invertible. Moreover, for any inverse $\mathbf{C}$ of $\mathbf{A}$, we find that

$$\mathbf{C} = \mathbf{CI} = \mathbf{CAB} = \mathbf{IB} = \mathbf{B},$$

implying that $\mathbf{A}$ has a unique inverse and further—in light of Lemma 8.1.3—that $\mathbf{A}$ has no right or left inverse other than $\mathbf{B}$.

That any invertible matrix is nonsingular is (as noted earlier) evident from Corollary 8.1.2. Q.E.D.

The symbol $\mathbf{A}^{-1}$ is used to denote the inverse of a nonsingular matrix $\mathbf{A}$. By definition,

$$\mathbf{AA}^{-1} = \mathbf{A}^{-1}\mathbf{A} = \mathbf{I}.$$

For an $n \times n$ nonsingular diagonal matrix $\mathbf{D} = \{d_i\}$,

$$\mathbf{D}^{-1} = \text{diag}(1/d_1, 1/d_2, \ldots, 1/d_n). \tag{1.1}$$

(Clearly, the diagonal elements of a nonsingular diagonal matrix are nonzero.)

For any 2×2 nonsingular matrix $\mathbf{A} = \begin{pmatrix} a_{11} & a_{12} \\ a_{21} & a_{22} \end{pmatrix}$,

$$\mathbf{A}^{-1} = (1/k) \begin{pmatrix} a_{22} & -a_{12} \\ -a_{21} & a_{11} \end{pmatrix}, \tag{1.2}$$

where $k = a_{11}a_{22} - a_{12}a_{21}$, as is easily verified. [Necessarily, $k \neq 0$. To see this, assume—for purposes of establishing a contradiction—the contrary (i.e., assume that $k = 0$). Then, if $a_{11} = 0$, we find that $a_{12}a_{21} = 0$, or, equivalently, that $a_{12} = 0$ or $a_{21} = 0$, and hence that $\mathbf{A}$ has a null row or column. If $a_{11} \neq 0$, we find that $a_{22} = a_{12}a_{21}/a_{11}$ and hence that $(a_{21}, a_{22}) = (a_{21}/a_{11})(a_{11}, a_{12})$, so that the second row of $\mathbf{A}$ is a scalar multiple of the first. Thus, in either case ($a_{11} = 0$ or $a_{11} \neq 0$), the assumption that $k = 0$ leads to the conclusion that $\mathbf{A}$ is singular, thereby establishing the sought-after contradiction.]

8.2 Properties of Inverse Matrices

a. Relationship of inverse matrices to the solution of linear systems

There is an intimate relationship between the inverse $\mathbf{A}^{-1}$ of a nonsingular matrix $\mathbf{A}$ and the solution of linear systems whose coefficient matrix is $\mathbf{A}$. This relationship is described in the following theorem.

Theorem 8.2.1. Let $\mathbf{A}$ represent any $n \times n$ nonsingular matrix, $\mathbf{G}$ any $n \times n$ matrix, and p any positive integer. Then, $\mathbf{GB}$ is a solution to a linear system $\mathbf{AX} = \mathbf{B}$ (in $\mathbf{X}$) for every $n \times p$ matrix $\mathbf{B}$ if and only if $\mathbf{G} = \mathbf{A}^{-1}$.

Proof. If $\mathbf{G} = \mathbf{A}^{-1}$, then, for every $n \times p$ matrix $\mathbf{B}$,

$$\mathbf{A}(\mathbf{GB}) = \mathbf{AA}^{-1}\mathbf{B} = \mathbf{IB} = \mathbf{B}.$$

Conversely, suppose that $\mathbf{GB}$ is a solution to $\mathbf{AX} = \mathbf{B}$ (i.e., that $\mathbf{AGB} = \mathbf{B}$) for every $n \times p$ matrix $\mathbf{B}$ and, in particular, for every $n \times p$ matrix of the form

$$\mathbf{B} = (\mathbf{0}, \ldots, \mathbf{0}, \mathbf{b}, \mathbf{0}, \ldots, \mathbf{0}).$$

Then, $\mathbf{AGb} = \mathbf{b} = \mathbf{Ib}$ for every $n \times 1$ vector $\mathbf{b}$. We conclude (on the basis of Lemma 2.3.2) that $\mathbf{AG} = \mathbf{I}$ and hence (in light of Lemma 8.1.3) that $\mathbf{G} = \mathbf{A}^{-1}$. Q.E.D.

b. Some elementary properties

For any nonsingular matrix $\mathbf{A}$ and any nonzero scalar k, $k\mathbf{A}$ is nonsingular and

$$(k\mathbf{A})^{-1} = (1/k)\mathbf{A}^{-1}, \tag{2.1}$$

as is easily verified. In the special case $k = -1$, equality (2.1) reduces to

$$(-\mathbf{A})^{-1} = -\mathbf{A}^{-1}. \tag{2.2}$$

It is easy to show that, for any nonsingular matrix $\mathbf{A}$, $\mathbf{A}'$ is nonsingular, and

$$(\mathbf{A}')^{-1} = (\mathbf{A}^{-1})'. \tag{2.3}$$

In the special case of a symmetric matrix $\mathbf{A}$, equality (2.3) reduces to

$$\mathbf{A}^{-1} = (\mathbf{A}^{-1})'. \tag{2.4}$$

Thus, the inverse of any nonsingular symmetric matrix is symmetric.

The inverse $\mathbf{A}^{-1}$ of an $n \times n$ nonsingular matrix $\mathbf{A}$ is invertible, or, equivalently (in light of Theorem 8.1.4),

$$\text{rank}(\mathbf{A}^{-1}) = n, \tag{2.5}$$

and

$$(\mathbf{A}^{-1})^{-1} = \mathbf{A}, \tag{2.6}$$

that is, the inverse of $\mathbf{A}^{-1}$ is $\mathbf{A}$ (as is evident from the definition of $\mathbf{A}^{-1}$).

For any two $n \times n$ nonsingular matrices $\mathbf{A}$ and $\mathbf{B}$,

$$\text{rank}(\mathbf{AB}) = n, \tag{2.7}$$

that is, $\mathbf{AB}$ is nonsingular, and

$$(\mathbf{AB})^{-1} = \mathbf{B}^{-1}\mathbf{A}^{-1}. \tag{2.8}$$

Results (2.7) and (2.8) can be easily verified by observing that $\mathbf{ABB}^{-1}\mathbf{A}^{-1} = \mathbf{I}$ (so that $\mathbf{B}^{-1}\mathbf{A}^{-1}$ is a right inverse of $\mathbf{AB}$) and applying Lemma 8.1.3. (If either or both of two $n \times n$ matrices $\mathbf{A}$ and $\mathbf{B}$ are singular, then their product $\mathbf{AB}$ is singular, as is evident from Corollary 4.4.5.) Repeated application of results (2.7) and (2.8) leads to the conclusion that, for any k $n \times n$ nonsingular matrices $\mathbf{A}_1, \mathbf{A}_2, \ldots, \mathbf{A}_k$,

$$\text{rank}(\mathbf{A}_1\mathbf{A}_2 \cdots \mathbf{A}_k) = n \tag{2.9}$$

and

$$(\mathbf{A}_1\mathbf{A}_2 \cdots \mathbf{A}_k)^{-1} = \mathbf{A}_k^{-1} \cdots \mathbf{A}_2^{-1}\mathbf{A}_1^{-1}. \tag{2.10}$$

8.3 Premultiplication or Postmultiplication by a Matrix of Full Column or Row Rank

Lemma 8.1.1 is useful in investigating the effects of premultiplication or postmultiplication by a matrix having full column or row rank. In particular, it can be used to establish the following two lemmas.

Lemma 8.3.1. Let $\mathbf{A}$ and $\mathbf{B}$ represent $m \times n$ matrices. Then, for any $r \times m$ matrix $\mathbf{C}$ of full column rank (i.e., of rank m) and any $n \times p$ matrix $\mathbf{D}$ of full row rank (i.e., of rank n), (1) $\mathbf{CA} = \mathbf{CB}$ implies $\mathbf{A} = \mathbf{B}$, (2) $\mathbf{AD} = \mathbf{BD}$ implies $\mathbf{A} = \mathbf{B}$, and (3) $\mathbf{CAD} = \mathbf{CBD}$ implies $\mathbf{A} = \mathbf{B}$.

Proof. Parts (1) and (2) are special cases of Part (3) (those where $\mathbf{D} = \mathbf{I}$ and $\mathbf{C} = \mathbf{I}$, respectively). Thus, it suffices to prove Part (3).

According to Lemma 8.1.1, $\mathbf{C}$ has a left inverse $\mathbf{L}$ and $\mathbf{D}$ a right inverse $\mathbf{R}$. Consequently, if $\mathbf{CAD} = \mathbf{CBD}$, then

$$\mathbf{A} = \mathbf{IAI} = \mathbf{LCADR} = \mathbf{LCBDR} = \mathbf{IBI} = \mathbf{B}.$$

Q.E.D.

Lemma 8.3.2. Let $\mathbf{A}$ represent an $m \times n$ matrix and $\mathbf{B}$ an $n \times p$ matrix. If $\mathbf{A}$ has full column rank, then

$$\mathcal{R}(\mathbf{AB}) = \mathcal{R}(\mathbf{B}) \quad \text{and} \quad \text{rank}(\mathbf{AB}) = \text{rank}(\mathbf{B}).$$

Similarly, if $\mathbf{B}$ has full row rank, then

$$\mathcal{C}(\mathbf{AB}) = \mathcal{C}(\mathbf{A}) \quad \text{and} \quad \text{rank}(\mathbf{AB}) = \text{rank}(\mathbf{A}).$$

Proof. It is clear from Corollary 4.2.3 that $\mathcal{R}(\mathbf{AB}) \subset \mathcal{R}(\mathbf{B})$ and $\mathcal{C}(\mathbf{AB}) \subset \mathcal{C}(\mathbf{A})$. If $\mathbf{A}$ has full column rank, then (according to Lemma 8.1.1) it has a left inverse $\mathbf{L}$, implying that

$$\mathcal{R}(\mathbf{B}) = \mathcal{R}(\mathbf{IB}) = \mathcal{R}(\mathbf{LAB}) \subset \mathcal{R}(\mathbf{AB})$$

and hence that $\mathcal{R}(\mathbf{AB}) = \mathcal{R}(\mathbf{B})$ [which implies, in turn, that $\text{rank}(\mathbf{AB}) = \text{rank}(\mathbf{B})$]. Similarly, if $\mathbf{B}$ has full row rank, then it has a right inverse $\mathbf{R}$, implying that $\mathcal{C}(\mathbf{A}) = \mathcal{C}(\mathbf{ABR}) \subset \mathcal{C}(\mathbf{AB})$ and hence that $\mathcal{C}(\mathbf{AB}) = \mathcal{C}(\mathbf{A})$ [and $\text{rank}(\mathbf{AB}) = \text{rank}(\mathbf{A})$]. Q.E.D.

As an immediate consequence of Lemma 8.3.2, we have the following corollary.

Corollary 8.3.3. If $\mathbf{A}$ is an $n \times n$ nonsingular matrix, then, for any $n \times p$ matrix $\mathbf{B}$,

$$\mathcal{R}(\mathbf{AB}) = \mathcal{R}(\mathbf{B}) \quad \text{and} \quad \text{rank}(\mathbf{AB}) = \text{rank}(\mathbf{B}).$$

Similarly, if $\mathbf{B}$ is an $n \times n$ nonsingular matrix, then, for any $m \times n$ matrix $\mathbf{A}$,

$$\mathcal{C}(\mathbf{AB}) = \mathcal{C}(\mathbf{A}) \quad \text{and} \quad \text{rank}(\mathbf{AB}) = \text{rank}(\mathbf{A}).$$

A further consequence of Lemma 8.3.2 is described in the following additional corollary.

Corollary 8.3.4. Let $\mathbf{A}$ represent an $m \times n$ nonnull matrix, and let $\mathbf{B}$ represent an $m \times r$ matrix of full column rank r and $\mathbf{T}$ an $r \times n$ matrix of full row rank r such that $\mathbf{A} = \mathbf{BT}$. Then, $r = \text{rank}(\mathbf{A})$.

Proof (of Corollary 8.3.4). In light of Lemma 8.3.2, we have that

$$\text{rank}(\mathbf{A}) = \text{rank}(\mathbf{BT}) = \text{rank}(\mathbf{T}) = r.$$

Q.E.D.

8.4 Orthogonal Matrices

a. Definition and basic properties and results

A (square) matrix $\mathbf{A}$ is said to be an *orthogonal matrix* if

$$\mathbf{A}'\mathbf{A} = \mathbf{A}\mathbf{A}' = \mathbf{I}$$

or, equivalently, if $\mathbf{A}$ is nonsingular and $\mathbf{A}^{-1} = \mathbf{A}'$. To show that a (square) matrix $\mathbf{A}$ is orthogonal, it suffices (in light of Lemma 8.1.3) to demonstrate that $\mathbf{A}'\mathbf{A} = \mathbf{I}$ or, alternatively, that $\mathbf{A}\mathbf{A}' = \mathbf{I}$.

The 2×2 matrix

$$(1/\sqrt{2})\begin{pmatrix} 1 & 1 \\ -1 & 1 \end{pmatrix}$$

is an example of an orthogonal matrix. More generally, for any angle θ, the 2×2 matrix

$$\begin{pmatrix} \cos\theta & \sin\theta \\ -\sin\theta & \cos\theta \end{pmatrix}$$

is an orthogonal matrix. Further, the $n \times n$ identity matrix $\mathbf{I}_n$ is an orthogonal matrix.

Clearly, for an $n \times n$ matrix $\mathbf{A}$, $\mathbf{A}'\mathbf{A} = \mathbf{I}$ if and only if the columns $\mathbf{a}_1, \ldots, \mathbf{a}_n$ of $\mathbf{A}$ are such that

$$\begin{aligned} \mathbf{a}_i'\mathbf{a}_j &= 1, \quad \text{for } j = i = 1, \ldots, n, \\ &= 0, \quad \text{for } j \neq i = 1, \ldots, n. \end{aligned}$$

Thus, a square matrix is orthogonal if and only if its columns form an orthonormal (with respect to the usual inner product) set of vectors. Similarly, a square matrix is orthogonal if and only if its rows form an orthonormal set of vectors.

Note that if $\mathbf{A}$ is an orthogonal matrix, then its transpose $\mathbf{A}'$ is also orthogonal.

If $\mathbf{P}$ and $\mathbf{Q}$ are both $n \times n$ orthogonal matrices, then it follows from results (2.7) and (2.8) that $\mathbf{PQ}$ is nonsingular and that

$$(\mathbf{PQ})^{-1} = \mathbf{Q}^{-1}\mathbf{P}^{-1} = \mathbf{Q}'\mathbf{P}' = (\mathbf{PQ})'. \tag{4.1}$$

Thus, the product of two $n \times n$ orthogonal matrices is an $(n \times n)$ orthogonal matrix. Repeated application of this result leads to the following extension.

Lemma 8.4.1. If each of the matrices $\mathbf{Q}_1, \mathbf{Q}_2, \ldots, \mathbf{Q}_k$ is an $n \times n$ orthogonal matrix, then their product $\mathbf{Q}_1\mathbf{Q}_2 \cdots \mathbf{Q}_k$ is an $(n \times n)$ orthogonal matrix.

A further result on orthogonal matrices is given by the following lemma.

Lemma 8.4.2. Let $\mathbf{A}$ represent an $m \times n$ matrix. Then, when the norm is taken to be the usual norm,

$$\|\mathbf{PAQ}\| = \|\mathbf{A}\| \tag{4.2}$$

for any $m \times m$ orthogonal matrix $\mathbf{P}$ and any $n \times n$ orthogonal matrix $\mathbf{Q}$.

Proof. Making use of Lemma 5.2.1, we find that

$$\begin{aligned}\|\mathbf{PAQ}\|^2 &= \text{tr}[(\mathbf{PAQ})'\mathbf{PAQ}] \\ &= \text{tr}(\mathbf{Q}'\mathbf{A}'\mathbf{P}'\mathbf{PAQ}) \\ &= \text{tr}(\mathbf{Q}'\mathbf{A}'\mathbf{AQ}) = \text{tr}(\mathbf{A}'\mathbf{AQQ}') = \text{tr}(\mathbf{A}'\mathbf{A}) = \|\mathbf{A}\|^2.\end{aligned}$$

Q.E.D.

Note that, as special cases of result (4.2) (those obtained by setting $\mathbf{Q} = \mathbf{I}_n$ or $\mathbf{P} = \mathbf{I}_m$), we have that (for any $m \times m$ orthogonal matrix $\mathbf{P}$) $\|\mathbf{PA}\| = \|\mathbf{A}\|$ and (for any $n \times n$ orthogonal matrix $\mathbf{Q}$) $\|\mathbf{AQ}\| = \|\mathbf{A}\|$.

b. Helmert matrix

Let $\mathbf{a}' = (a_1, a_2, \ldots, a_n)$ represent any row vector such that $a_i \neq 0$ $(i = 1, 2, \ldots, n)$, that is, any row vector whose elements are all nonzero. Suppose that we require an $n \times n$ orthogonal matrix, one row of which is proportional to $\mathbf{a}'$. In what follows, one such matrix $\mathbf{P}$ is derived.

Let $\mathbf{p}_1', \ldots, \mathbf{p}_n'$ represent the rows of $\mathbf{P}$, and take the first row $\mathbf{p}_1'$ to be the row of $\mathbf{P}$ that is proportional to $\mathbf{a}'$. Take the second row $\mathbf{p}_2'$ to be proportional to the n-dimensional row vector

$$(a_1, -a_1^2/a_2, 0, 0, \ldots, 0),$$

the third row $\mathbf{p}_3'$ proportional to

$$[a_1, a_2, -(a_1^2 + a_2^2)/a_3, 0, 0, \ldots, 0],$$

and more generally the second through nth rows $\mathbf{p}_2', \ldots, \mathbf{p}_n'$ proportional to

$$(a_1, a_2, \ldots, a_{k-1}, -\textstyle\sum_{i=1}^{k-1} a_i^2/a_k, 0, 0, \ldots, 0) \quad (k = 2, \ldots, n), \tag{4.3}$$

respectively.

It is easy to confirm that the $n - 1$ vectors (4.3) are orthogonal to each other and to the vector $\mathbf{a}'$. To obtain explicit expressions for $\mathbf{p}_1', \ldots, \mathbf{p}_n'$, it remains to normalize $\mathbf{a}'$ and the vectors (4.3). Observing that the (usual) norm of the kth of the vectors (4.3) is

$$\left[\textstyle\sum_{i=1}^{k-1} a_i^2 + \left(\sum_{i=1}^{k-1} a_i^2\right)^2/a_k^2\right]^{1/2} = \left[\left(\sum_{i=1}^{k-1} a_i^2\right)\left(\sum_{i=1}^{k} a_i^2\right)/a_k^2\right]^{1/2},$$

we find that

$$\mathbf{p}_1' = \left(\textstyle\sum_{i=1}^{n} a_i^2\right)^{-1/2}(a_1, a_2, \ldots, a_n), \tag{4.4}$$

and that, for $k = 2, \ldots, n$,

$$\mathbf{p}_k' = \left[\frac{a_k^2}{(\sum_{i=1}^{k-1} a_i^2)(\sum_{i=1}^{k} a_i^2)}\right]^{1/2} \times (a_1, a_2, \ldots, a_{k-1}, -\textstyle\sum_{i=1}^{k-1} a_i^2/a_k, 0, 0, \ldots, 0). \tag{4.5}$$

When $\mathbf{a}' = (1, 1, \ldots, 1)$, formulas (4.4) and (4.5) simplify to

$$\mathbf{p}_1' = n^{-1/2}(1, 1, \ldots, 1),$$
$$\mathbf{p}_k' = [k(k-1)]^{-1/2}(1, 1, \ldots, 1,\ 1-k,\ 0, 0, \ldots, 0)$$

$(k = 2, \ldots, n)$, and $\mathbf{P}$ reduces to a matrix known as the *Helmert matrix* (of order n). (In some presentations, it is the transpose of this matrix that is called the Helmert matrix.) For example, the Helmert matrix of order 4 is

$$\begin{pmatrix} 1/2 & 1/2 & 1/2 & 1/2 \\ 1/\sqrt{2} & -1/\sqrt{2} & 0 & 0 \\ 1/\sqrt{6} & 1/\sqrt{6} & -2/\sqrt{6} & 0 \\ 1/\sqrt{12} & 1/\sqrt{12} & 1/\sqrt{12} & -3/\sqrt{12} \end{pmatrix}.$$

c. Permutation matrices

A *permutation matrix* is a square matrix whose columns can be obtained by permuting (rearranging) the columns of an identity matrix. Thus, letting $\mathbf{u}_1, \mathbf{u}_2, \ldots, \mathbf{u}_n$ represent the first, second, ..., nth columns, respectively, of $\mathbf{I}_n$, an $n \times n$ permutation matrix is a matrix of the general form

$$(\mathbf{u}_{k_1}, \mathbf{u}_{k_2}, \ldots, \mathbf{u}_{k_n}),$$

where $k_1, k_2, \ldots, k_n$ represents any permutation of the first n positive integers $1, 2, \ldots, n$. For example, one permutation matrix of order $n = 3$ is the 3×3 matrix

$$(\mathbf{u}_3, \mathbf{u}_1, \mathbf{u}_2) = \begin{pmatrix} 0 & 1 & 0 \\ 0 & 0 & 1 \\ 1 & 0 & 0 \end{pmatrix},$$

whose columns are the third, first, and second columns, respectively, of $\mathbf{I}_3$.

It is clear that the columns of any permutation matrix form an orthonormal (with respect to the usual inner product) set and hence that any permutation matrix is an orthogonal matrix. Thus, if $\mathbf{P}$ is a permutation matrix, then $\mathbf{P}$ is nonsingular and $\mathbf{P}^{-1} = \mathbf{P}'$.

Clearly, the jth element of the k_jth row of the $n \times n$ permutation matrix $(\mathbf{u}_{k_1}, \mathbf{u}_{k_2}, \ldots, \mathbf{u}_{k_n})$ is 1 and its other $n-1$ elements are 0. That is, the jth row $\mathbf{u}_j'$ of $\mathbf{I}_n$ is the k_jth row of $(\mathbf{u}_{k_1}, \mathbf{u}_{k_2}, \ldots, \mathbf{u}_{k_n})$ or, equivalently, the jth column $\mathbf{u}_j$ of $\mathbf{I}_n$ is the k_jth column of $(\mathbf{u}_{k_1}, \mathbf{u}_{k_2}, \ldots, \mathbf{u}_{k_n})'$. Thus, the transpose of any permutation matrix is itself a permutation matrix. Further, the rows of any permutation matrix are a permutation of the rows of an identity matrix, and conversely any matrix whose rows can be obtained by permuting the rows of an identity matrix is a permutation matrix.

The effect of postmultiplying an $m \times n$ matrix $\mathbf{A}$ by an $n \times n$ permutation matrix $\mathbf{P}$ is to permute the columns of $\mathbf{A}$ in the same way that the columns of $\mathbf{I}_n$ were permuted in forming $\mathbf{P}$. Thus, if $\mathbf{a}_1, \mathbf{a}_2, \ldots, \mathbf{a}_n$ are the first, second, ..., nth columns of $\mathbf{A}$, the first, second, ..., nth columns of the product $\mathbf{A}(\mathbf{u}_{k_1}, \mathbf{u}_{k_2}, \ldots, \mathbf{u}_{k_n})$ of $\mathbf{A}$ and the $n \times n$ permutation matrix $(\mathbf{u}_{k_1}, \mathbf{u}_{k_2}, \ldots, \mathbf{u}_{k_n})$ are $(\mathbf{a}_{k_1}, \mathbf{a}_{k_2}, \ldots, \mathbf{a}_{k_n})$, respectively. When $n = 3$, we have, for example, that

$$\mathbf{A}(\mathbf{u}_3, \mathbf{u}_1, \mathbf{u}_2) = (\mathbf{a}_1, \mathbf{a}_2, \mathbf{a}_3) \begin{pmatrix} 0 & 1 & 0 \\ 0 & 0 & 1 \\ 1 & 0 & 0 \end{pmatrix} = (\mathbf{a}_3, \mathbf{a}_1, \mathbf{a}_2).$$

Further, the first, second ..., nth columns $\mathbf{a}_1, \mathbf{a}_2, \ldots, \mathbf{a}_n$ of $\mathbf{A}$ are the $k_1, k_2, \ldots, k_n$th columns, respectively, of the product $\mathbf{A}(\mathbf{u}_{k_1}, \mathbf{u}_{k_2}, \ldots, \mathbf{u}_{k_n})'$ of $\mathbf{A}$ and the permutation matrix $(\mathbf{u}_{k_1}, \mathbf{u}_{k_2}, \ldots, \mathbf{u}_{k_n})'$. When $n = 3$, we have, for example, that

$$\mathbf{A}(\mathbf{u}_3, \mathbf{u}_1, \mathbf{u}_2)' = (\mathbf{a}_1, \mathbf{a}_2, \mathbf{a}_3) \begin{pmatrix} 0 & 0 & 1 \\ 1 & 0 & 0 \\ 0 & 1 & 0 \end{pmatrix} = (\mathbf{a}_2, \mathbf{a}_3, \mathbf{a}_1).$$

Similarly, the effect of premultiplying an $n \times m$ matrix $\mathbf{A}$ by an $n \times n$ permutation matrix is to permute the rows of $\mathbf{A}$. If the first, second, ..., nth rows of $\mathbf{A}$ are $\mathbf{a}'_1, \mathbf{a}'_2, \ldots, \mathbf{a}'_n$, respectively, then the first, second, ..., nth rows of the product $(\mathbf{u}_{k_1}, \mathbf{u}_{k_2}, \ldots, \mathbf{u}_{k_n})'\mathbf{A}$ of the permutation matrix $(\mathbf{u}_{k_1}, \mathbf{u}_{k_2}, \ldots, \mathbf{u}_{k_n})'$ and $\mathbf{A}$ are $(\mathbf{a}'_{k_1}, \mathbf{a}'_{k_2}, \ldots, \mathbf{a}'_{k_n})$, respectively; and $\mathbf{a}'_1, \mathbf{a}'_2, \ldots, \mathbf{a}'_n$ are the $k_1, k_2, \ldots, k_n$th rows, respectively, of $(\mathbf{u}_{k_1}, \mathbf{u}_{k_2}, \ldots, \mathbf{u}_{k_n})\mathbf{A}$. When $k = 3$, we have, for example, that

$$(\mathbf{u}_3, \mathbf{u}_1, \mathbf{u}_2)'\mathbf{A} = \begin{pmatrix} 0 & 0 & 1 \\ 1 & 0 & 0 \\ 0 & 1 & 0 \end{pmatrix} \begin{pmatrix} \mathbf{a}'_1 \\ \mathbf{a}'_2 \\ \mathbf{a}'_3 \end{pmatrix} = \begin{pmatrix} \mathbf{a}'_3 \\ \mathbf{a}'_1 \\ \mathbf{a}'_2 \end{pmatrix},$$

$$(\mathbf{u}_3, \mathbf{u}_1, \mathbf{u}_2)\mathbf{A} = \begin{pmatrix} 0 & 1 & 0 \\ 0 & 0 & 1 \\ 1 & 0 & 0 \end{pmatrix} \begin{pmatrix} \mathbf{a}'_1 \\ \mathbf{a}'_2 \\ \mathbf{a}'_3 \end{pmatrix} = \begin{pmatrix} \mathbf{a}'_2 \\ \mathbf{a}'_3 \\ \mathbf{a}'_1 \end{pmatrix}.$$

Letting $\mathbf{e}_1, \mathbf{e}_2, \ldots, \mathbf{e}_m$ represent the first, second, ..., mth columns of $\mathbf{I}_m$ and letting $r_1, r_2, \ldots, r_m$ represent any permutation of the first m positive integers $1, 2, \ldots, m$, the ijth element of the matrix

$$(\mathbf{e}_{r_1}, \mathbf{e}_{r_2}, \ldots, \mathbf{e}_{r_m})'\mathbf{A}(\mathbf{u}_{k_1}, \mathbf{u}_{k_2}, \ldots, \mathbf{u}_{k_n}),$$

obtained by premultiplying and postmultiplying an $m \times n$ matrix $\mathbf{A}$ by the permutation matrices $(\mathbf{e}_{r_1}, \mathbf{e}_{r_2}, \ldots, \mathbf{e}_{r_m})'$ and $(\mathbf{u}_{k_1}, \mathbf{u}_{k_2}, \ldots, \mathbf{u}_{k_n})$, respectively, is the $r_i k_j$th element of $\mathbf{A}$. And the ijth element of $\mathbf{A}$ is the $r_i k_j$th element of the matrix

$$(\mathbf{e}_{r_1}, \mathbf{e}_{r_2}, \ldots, \mathbf{e}_{r_m})\mathbf{A}(\mathbf{u}_{k_1}, \mathbf{u}_{k_2}, \ldots, \mathbf{u}_{k_n})',$$

obtained by premultiplying and postmultiplying $\mathbf{A}$ by the permutation matrices $(\mathbf{e}_{r_1}, \mathbf{e}_{r_2}, \ldots, \mathbf{e}_{r_m})$ and $(\mathbf{u}_{k_1}, \mathbf{u}_{k_2}, \ldots, \mathbf{u}_{k_n})'$, respectively.

Note that the product of two or more $n \times n$ permutation matrices is another $n \times n$ permutation matrix.

8.5 Some Basic Results on the Ranks and Inverses of Partitioned Matrices

a. A lemma

The following lemma is useful in deriving results on the ranks of partitioned matrices.

Lemma 8.5.1. Let $\mathbf{T}$ represent an $m \times p$ matrix, $\mathbf{U}$ an $m \times q$ matrix, $\mathbf{V}$ an $n \times p$ matrix, and $\mathbf{W}$ an $n \times q$ matrix. Then,

$$\text{rank}\begin{pmatrix} \mathbf{T} & \mathbf{U} \\ \mathbf{V} & \mathbf{W} \end{pmatrix} = \text{rank}\begin{pmatrix} \mathbf{U} & \mathbf{T} \\ \mathbf{W} & \mathbf{V} \end{pmatrix} = \text{rank}\begin{pmatrix} \mathbf{V} & \mathbf{W} \\ \mathbf{T} & \mathbf{U} \end{pmatrix} = \text{rank}\begin{pmatrix} \mathbf{W} & \mathbf{V} \\ \mathbf{U} & \mathbf{T} \end{pmatrix}. \tag{5.1}$$

Proof. Lemma 8.5.1 is obtained by successively applying Lemma 4.5.3 to $\begin{pmatrix} \mathbf{U} & \mathbf{T} \\ \mathbf{W} & \mathbf{V} \end{pmatrix}$ [taking $\mathbf{A} = \begin{pmatrix} \mathbf{U} \\ \mathbf{W} \end{pmatrix}$ and $\mathbf{B} = \begin{pmatrix} \mathbf{T} \\ \mathbf{V} \end{pmatrix}$], to $\begin{pmatrix} \mathbf{V} & \mathbf{W} \\ \mathbf{T} & \mathbf{U} \end{pmatrix}$ [taking $\mathbf{A} = (\mathbf{V}, \mathbf{W})$ and $\mathbf{C} = (\mathbf{T}, \mathbf{U})$], and to $\begin{pmatrix} \mathbf{W} & \mathbf{V} \\ \mathbf{U} & \mathbf{T} \end{pmatrix}$ [taking $\mathbf{A} = \begin{pmatrix} \mathbf{W} \\ \mathbf{U} \end{pmatrix}$ and $\mathbf{B} = \begin{pmatrix} \mathbf{V} \\ \mathbf{T} \end{pmatrix}$ or taking $\mathbf{A} = (\mathbf{W}, \mathbf{V})$ and $\mathbf{C} = (\mathbf{U}, \mathbf{T})$]. Q.E.D.

b. Block-diagonal matrices

Let $\mathbf{T}$ represent an $m \times m$ matrix and $\mathbf{W}$ an $n \times n$ matrix. Then, the $(m+n) \times (m+n)$ block-diagonal matrix $\begin{pmatrix} \mathbf{T} & \mathbf{0} \\ \mathbf{0} & \mathbf{W} \end{pmatrix}$ is nonsingular if and only if both $\mathbf{T}$ and $\mathbf{W}$ are nonsingular, as is evident from Lemma 4.5.11. Moreover, if $\mathbf{T}$ and $\mathbf{W}$ are nonsingular, then

$$\begin{pmatrix} \mathbf{T} & \mathbf{0} \\ \mathbf{0} & \mathbf{W} \end{pmatrix}^{-1} = \begin{pmatrix} \mathbf{T}^{-1} & \mathbf{0} \\ \mathbf{0} & \mathbf{W}^{-1} \end{pmatrix}, \tag{5.2}$$

as is easily verified.

More generally, for any square matrices $\mathbf{A}_1, \mathbf{A}_2, \ldots, \mathbf{A}_k$, the block-diagonal matrix $\text{diag}(\mathbf{A}_1, \mathbf{A}_2, \ldots, \mathbf{A}_k)$ is nonsingular if and only if $\mathbf{A}_1, \mathbf{A}_2, \ldots, \mathbf{A}_k$ are all nonsingular [as is evident from result (4.5.14)], in which case

$$[\text{diag}(\mathbf{A}_1, \mathbf{A}_2, \ldots, \mathbf{A}_k)]^{-1} = \text{diag}(\mathbf{A}_1^{-1}, \mathbf{A}_2^{-1}, \ldots, \mathbf{A}_k^{-1}). \tag{5.3}$$

c. Block-triangular matrices

Consider now block-triangular matrices of the form $\begin{pmatrix} \mathbf{I}_m & \mathbf{0} \\ \mathbf{V} & \mathbf{I}_n \end{pmatrix}$ or $\begin{pmatrix} \mathbf{I}_n & \mathbf{V} \\ \mathbf{0} & \mathbf{I}_m \end{pmatrix}$, where $\mathbf{V}$ is an $n \times m$ matrix. Note that for any $m \times p$ matrix $\mathbf{A}$ and $n \times p$ matrix $\mathbf{B}$,

$$\begin{pmatrix} \mathbf{I}_m & \mathbf{0} \\ \mathbf{V} & \mathbf{I}_n \end{pmatrix} \begin{pmatrix} \mathbf{A} \\ \mathbf{B} \end{pmatrix} = \begin{pmatrix} \mathbf{A} \\ \mathbf{B} + \mathbf{VA} \end{pmatrix}, \tag{5.4a}$$

$$\begin{pmatrix} \mathbf{I}_n & \mathbf{V} \\ \mathbf{0} & \mathbf{I}_m \end{pmatrix} \begin{pmatrix} \mathbf{B} \\ \mathbf{A} \end{pmatrix} = \begin{pmatrix} \mathbf{B} + \mathbf{VA} \\ \mathbf{A} \end{pmatrix} \tag{5.4b}$$

and similarly that, for any $p \times m$ matrix $\mathbf{A}$ and $p \times n$ matrix $\mathbf{B}$,

$$(\mathbf{A}, \mathbf{B}) \begin{pmatrix} \mathbf{I}_m & \mathbf{0} \\ \mathbf{V} & \mathbf{I}_n \end{pmatrix} = (\mathbf{A} + \mathbf{BV}, \mathbf{B}), \tag{5.5a}$$

$$(\mathbf{B}, \mathbf{A}) \begin{pmatrix} \mathbf{I}_n & \mathbf{V} \\ \mathbf{0} & \mathbf{I}_m \end{pmatrix} = (\mathbf{B}, \mathbf{A} + \mathbf{BV}). \tag{5.5b}$$

Further, recalling (from Theorem 8.1.4) that an invertible matrix is nonsingular, and observing that

$$\begin{pmatrix} \mathbf{I} & \mathbf{0} \\ -\mathbf{V} & \mathbf{I} \end{pmatrix} \begin{pmatrix} \mathbf{I} & \mathbf{0} \\ \mathbf{V} & \mathbf{I} \end{pmatrix} = \begin{pmatrix} \mathbf{I} & \mathbf{0} \\ \mathbf{0} & \mathbf{I} \end{pmatrix},$$

$$\begin{pmatrix} \mathbf{I} & -\mathbf{V} \\ \mathbf{0} & \mathbf{I} \end{pmatrix} \begin{pmatrix} \mathbf{I} & \mathbf{V} \\ \mathbf{0} & \mathbf{I} \end{pmatrix} = \begin{pmatrix} \mathbf{I} & \mathbf{0} \\ \mathbf{0} & \mathbf{I} \end{pmatrix},$$

we obtain the following result.

Lemma 8.5.2. For any $n \times m$ matrix $\mathbf{V}$, the $(m+n) \times (m+n)$ partitioned matrices $\begin{pmatrix} \mathbf{I}_m & \mathbf{0} \\ \mathbf{V} & \mathbf{I}_n \end{pmatrix}$ and $\begin{pmatrix} \mathbf{I}_n & \mathbf{V} \\ \mathbf{0} & \mathbf{I}_m \end{pmatrix}$ are nonsingular, and

$$\begin{pmatrix} \mathbf{I}_m & \mathbf{0} \\ \mathbf{V} & \mathbf{I}_n \end{pmatrix}^{-1} = \begin{pmatrix} \mathbf{I}_m & \mathbf{0} \\ -\mathbf{V} & \mathbf{I}_n \end{pmatrix}, \tag{5.6a}$$

$$\begin{pmatrix} \mathbf{I}_n & \mathbf{V} \\ \mathbf{0} & \mathbf{I}_m \end{pmatrix}^{-1} = \begin{pmatrix} \mathbf{I}_n & -\mathbf{V} \\ \mathbf{0} & \mathbf{I}_m \end{pmatrix}. \tag{5.6b}$$

Formula (4.5.11) for the rank of a block-diagonal matrix can be extended to certain block-triangular matrices, as indicated by the following lemma.

Lemma 8.5.3. Let $\mathbf{T}$ represent an $m \times p$ matrix, $\mathbf{V}$ an $n \times p$ matrix, and $\mathbf{W}$ an $n \times q$ matrix. If $\mathbf{T}$ has full column rank or $\mathbf{W}$ has full row rank, that is, if $\text{rank}(\mathbf{T}) = p$ or $\text{rank}(\mathbf{W}) = n$, then

$$\text{rank}\begin{pmatrix} \mathbf{T} & \mathbf{0} \\ \mathbf{V} & \mathbf{W} \end{pmatrix} = \text{rank}\begin{pmatrix} \mathbf{W} & \mathbf{V} \\ \mathbf{0} & \mathbf{T} \end{pmatrix} = \text{rank}(\mathbf{T}) + \text{rank}(\mathbf{W}). \tag{5.7}$$

Proof. Suppose that $\text{rank}(\mathbf{T}) = p$. Then, according to Lemma 8.1.1, there exists a matrix $\mathbf{L}$ that is a left inverse of $\mathbf{T}$, in which case

$$\begin{pmatrix} \mathbf{I} & -\mathbf{VL} \\ \mathbf{0} & \mathbf{I} \end{pmatrix} \begin{pmatrix} \mathbf{W} & \mathbf{V} \\ \mathbf{0} & \mathbf{T} \end{pmatrix} = \begin{pmatrix} \mathbf{W} & \mathbf{0} \\ \mathbf{0} & \mathbf{T} \end{pmatrix}.$$

Since (according to Lemma 8.5.2) $\begin{pmatrix} \mathbf{I} & -\mathbf{VL} \\ \mathbf{0} & \mathbf{I} \end{pmatrix}$ is nonsingular, we conclude (on the basis of Corollary 8.3.3) that

$$\text{rank}\begin{pmatrix} \mathbf{W} & \mathbf{V} \\ \mathbf{0} & \mathbf{T} \end{pmatrix} = \text{rank}\begin{pmatrix} \mathbf{W} & \mathbf{0} \\ \mathbf{0} & \mathbf{T} \end{pmatrix}$$

and hence (in light of Lemmas 8.5.1 and 4.5.11) that

$$\text{rank}\begin{pmatrix} \mathbf{T} & \mathbf{0} \\ \mathbf{V} & \mathbf{W} \end{pmatrix} = \text{rank}\begin{pmatrix} \mathbf{W} & \mathbf{V} \\ \mathbf{0} & \mathbf{T} \end{pmatrix} = \text{rank}(\mathbf{T}) + \text{rank}(\mathbf{W}).$$

That result (5.7) holds if $\text{rank}(\mathbf{W}) = n$ can be established via an analogous argument. Q.E.D.

The results of Lemma 8.5.2 can be extended to additional block-triangular matrices, as detailed in the following lemma.

Lemma 8.5.4. Let $\mathbf{T}$ represent an $m \times m$ matrix, $\mathbf{V}$ an $n \times m$ matrix, and $\mathbf{W}$ an $n \times n$ matrix. Then, the $(m+n) \times (m+n)$ partitioned matrix $\begin{pmatrix} \mathbf{T} & \mathbf{0} \\ \mathbf{V} & \mathbf{W} \end{pmatrix}$, or equivalently $\begin{pmatrix} \mathbf{W} & \mathbf{V} \\ \mathbf{0} & \mathbf{T} \end{pmatrix}$, is nonsingular if and only if both $\mathbf{T}$ and $\mathbf{W}$ are nonsingular, in which case

$$\begin{pmatrix} \mathbf{T} & \mathbf{0} \\ \mathbf{V} & \mathbf{W} \end{pmatrix}^{-1} = \begin{pmatrix} \mathbf{T}^{-1} & \mathbf{0} \\ -\mathbf{W}^{-1}\mathbf{V}\mathbf{T}^{-1} & \mathbf{W}^{-1} \end{pmatrix}, \tag{5.8a}$$

$$\begin{pmatrix} \mathbf{W} & \mathbf{V} \\ \mathbf{0} & \mathbf{T} \end{pmatrix}^{-1} = \begin{pmatrix} \mathbf{W}^{-1} & -\mathbf{W}^{-1}\mathbf{V}\mathbf{T}^{-1} \\ \mathbf{0} & \mathbf{T}^{-1} \end{pmatrix}. \tag{5.8b}$$

Proof. If $\begin{pmatrix} \mathbf{T} & \mathbf{0} \\ \mathbf{V} & \mathbf{W} \end{pmatrix}$ is nonsingular, then, making use of Lemmas 4.5.7 and 4.4.3, we find that

$$\begin{aligned} m + n = \text{rank}\begin{pmatrix} \mathbf{T} & \mathbf{0} \\ \mathbf{V} & \mathbf{W} \end{pmatrix} &\\ &\le \text{rank}(\mathbf{T}, \mathbf{0}) + \text{rank}(\mathbf{V}, \mathbf{W}) \\ &\le \text{rank}(\mathbf{T}) + \text{rank}(\mathbf{0}) + \text{rank}(\mathbf{V}, \mathbf{W}) \\ &\le \text{rank}(\mathbf{T}) + n \le m + n, \end{aligned}$$

implying that $m \le \text{rank}(\mathbf{T}) \le m$ and hence that $\text{rank}(\mathbf{T}) = m$ or, equivalently, that $\mathbf{T}$ is nonsingular. It can likewise be shown that if $\begin{pmatrix} \mathbf{T} & \mathbf{0} \\ \mathbf{V} & \mathbf{W} \end{pmatrix}$ is nonsingular, then $\mathbf{W}$ is nonsingular.

Suppose now that both $\mathbf{T}$ and $\mathbf{W}$ are nonsingular. Then,

$$\begin{pmatrix} \mathbf{T} & \mathbf{0} \\ \mathbf{V} & \mathbf{W} \end{pmatrix} = \begin{pmatrix} \mathbf{I}_m & \mathbf{0} \\ \mathbf{0} & \mathbf{W} \end{pmatrix} \begin{pmatrix} \mathbf{I}_m & \mathbf{0} \\ \mathbf{W}^{-1}\mathbf{V}\mathbf{T}^{-1} & \mathbf{I}_n \end{pmatrix} \begin{pmatrix} \mathbf{T} & \mathbf{0} \\ \mathbf{0} & \mathbf{I}_n \end{pmatrix},$$

as is easily verified, and (in light of Lemma 8.5.3) $\begin{pmatrix} \mathbf{I} & \mathbf{0} \\ \mathbf{0} & \mathbf{W} \end{pmatrix}$, $\begin{pmatrix} \mathbf{I} & \mathbf{0} \\ \mathbf{W}^{-1}\mathbf{V}\mathbf{T}^{-1} & \mathbf{I} \end{pmatrix}$, and $\begin{pmatrix} \mathbf{T} & \mathbf{0} \\ \mathbf{0} & \mathbf{I} \end{pmatrix}$ are nonsingular. Thus, it follows from result (2.9) that $\begin{pmatrix} \mathbf{T} & \mathbf{0} \\ \mathbf{V} & \mathbf{W} \end{pmatrix}$ is nonsingular and from results (2.10), (5.2), and (5.6) that

$$\begin{aligned} \begin{pmatrix} \mathbf{T} & \mathbf{0} \\ \mathbf{V} & \mathbf{W} \end{pmatrix}^{-1} &= \begin{pmatrix} \mathbf{T} & \mathbf{0} \\ \mathbf{0} & \mathbf{I} \end{pmatrix}^{-1} \begin{pmatrix} \mathbf{I} & \mathbf{0} \\ \mathbf{W}^{-1}\mathbf{V}\mathbf{T}^{-1} & \mathbf{I} \end{pmatrix}^{-1} \begin{pmatrix} \mathbf{I} & \mathbf{0} \\ \mathbf{0} & \mathbf{W} \end{pmatrix}^{-1} \\ &= \begin{pmatrix} \mathbf{T}^{-1} & \mathbf{0} \\ \mathbf{0} & \mathbf{I} \end{pmatrix} \begin{pmatrix} \mathbf{I} & \mathbf{0} \\ -\mathbf{W}^{-1}\mathbf{V}\mathbf{T}^{-1} & \mathbf{I} \end{pmatrix} \begin{pmatrix} \mathbf{I} & \mathbf{0} \\ \mathbf{0} & \mathbf{W}^{-1} \end{pmatrix} \\ &= \begin{pmatrix} \mathbf{T}^{-1} & \mathbf{0} \\ -\mathbf{W}^{-1}\mathbf{V}\mathbf{T}^{-1} & \mathbf{W}^{-1} \end{pmatrix}. \end{aligned}$$

The second part of result (5.8) can be derived in similar fashion. Q.E.D.

The following theorem extends the first part of Lemma 8.5.4.

Theorem 8.5.5. Let

$$\mathbf{A} = \begin{pmatrix} \mathbf{A}_{11} & \mathbf{A}_{12} & \dots & \mathbf{A}_{1r} \\ \mathbf{0} & \mathbf{A}_{22} & \dots & \mathbf{A}_{2r} \\ \vdots & & \ddots & \vdots \\ \mathbf{0} & \mathbf{0} & & \mathbf{A}_{rr} \end{pmatrix} \quad \text{and} \quad \mathbf{B} = \begin{pmatrix} \mathbf{B}_{11} & \mathbf{0} & \dots & \mathbf{0} \\ \mathbf{B}_{21} & \mathbf{B}_{22} & & \mathbf{0} \\ \vdots & \vdots & \ddots & \\ \mathbf{B}_{r1} & \mathbf{B}_{r2} & \dots & \mathbf{B}_{rr} \end{pmatrix}$$

represent, respectively, an $n \times n$ upper block-triangular matrix whose ijth block $\mathbf{A}_{ij}$ is of dimensions $n_i \times n_j$ ($j \geq i = 1, 2, \dots, r$) and an $n \times n$ lower block-triangular matrix whose ijth block $\mathbf{B}_{ij}$ is of dimensions $n_i \times n_j$ ($j \leq i = 1, 2, \dots, r$). Then, $\mathbf{A}$ is nonsingular if and only if its diagonal blocks $\mathbf{A}_{11}, \mathbf{A}_{22}, \dots, \mathbf{A}_{rr}$ are all nonsingular. Likewise, $\mathbf{B}$ is nonsingular if and only if its diagonal blocks $\mathbf{B}_{11}, \mathbf{B}_{22}, \dots, \mathbf{B}_{rr}$ are all nonsingular.

Proof. Observe that the transpose $\mathbf{A}'$ of the upper block-triangular matrix $\mathbf{A}$ is lower block-triangular with diagonal blocks $\mathbf{A}'_{11}, \mathbf{A}'_{22}, \dots, \mathbf{A}'_{rr}$, that $\text{rank}(\mathbf{A}') = \text{rank}(\mathbf{A})$ and $\text{rank}(\mathbf{A}'_{ii}) = \text{rank}(\mathbf{A}_{ii})$ ($i = 1, 2, \dots, r$), and that, as a consequence, it suffices to prove the part of the theorem pertaining to the lower block-triangular matrix $\mathbf{B}$.

The proof is by mathematical induction. Lemma 8.5.4 implies that the theorem is valid for $r = 2$. Suppose then that the theorem is valid for $r = k - 1$. Defining

$$\mathbf{T} = \begin{pmatrix} \mathbf{B}_{11} & \mathbf{0} & \dots & \mathbf{0} \\ \mathbf{B}_{21} & \mathbf{B}_{22} & & \mathbf{0} \\ \vdots & \vdots & \ddots & \\ \mathbf{B}_{k-1,1} & \mathbf{B}_{k-1,2} & \dots & \mathbf{B}_{k-1,k-1} \end{pmatrix} \quad \text{and} \quad \mathbf{V} = (\mathbf{B}_{k1}, \mathbf{B}_{k2}, \dots, \mathbf{B}_{k,k-1}),$$

we need to show that $\begin{pmatrix} \mathbf{T} & \mathbf{0} \\ \mathbf{V} & \mathbf{B}_{kk} \end{pmatrix}$ is nonsingular if and only if the matrices $\mathbf{B}_{11}, \mathbf{B}_{22}, \dots, \mathbf{B}_{kk}$ are all nonsingular.

If $\mathbf{B}_{11}, \mathbf{B}_{22}, \ldots, \mathbf{B}_{kk}$ are nonsingular, then, by supposition, $\mathbf{T}$ is nonsingular and, based on Lemma 8.5.4, we conclude that $\begin{pmatrix} \mathbf{T} & \mathbf{0} \\ \mathbf{V} & \mathbf{B}_{kk} \end{pmatrix}$ is nonsingular.

Conversely, if $\begin{pmatrix} \mathbf{T} & \mathbf{0} \\ \mathbf{V} & \mathbf{B}_{kk} \end{pmatrix}$ is nonsingular, then, according to Lemma 8.5.4, $\mathbf{T}$ and $\mathbf{B}_{kk}$ are both nonsingular. Moreover, if $\mathbf{T}$ is nonsingular, then, by supposition, $\mathbf{B}_{11}, \ldots, \mathbf{B}_{k-1,k-1}$ are nonsingular. Q.E.D.

In the special case of a triangular matrix (i.e., the special case where $r = n$), Theorem 8.5.5 can be restated as follows.

Corollary 8.5.6. A triangular matrix is nonsingular if and only if its diagonal elements are all nonzero.

Certain of the results given in Lemma 8.5.4 on the inverse of a block-triangular matrix with two rows and columns of blocks are extended in the following theorem to a block-triangular matrix with an arbitrary number r of rows and columns of blocks.

Theorem 8.5.7. Let

$$\mathbf{A} = \begin{pmatrix} \mathbf{A}_{11} & \mathbf{A}_{12} & \ldots & \mathbf{A}_{1r} \\ \mathbf{A}_{21} & \mathbf{A}_{22} & \ldots & \mathbf{A}_{2r} \\ \vdots & \vdots & \ddots & \vdots \\ \mathbf{A}_{r1} & \mathbf{A}_{r2} & \ldots & \mathbf{A}_{rr} \end{pmatrix}$$

represent a nonsingular partitioned matrix whose ijth block $\mathbf{A}_{ij}$ is of dimensions $n_i \times n_j$ $(i, j = 1, \ldots, r)$. Partition $\mathbf{A}^{-1}$ as

$$\mathbf{F} = \begin{pmatrix} \mathbf{F}_{11} & \mathbf{F}_{12} & \ldots & \mathbf{F}_{1r} \\ \mathbf{F}_{21} & \mathbf{F}_{22} & \ldots & \mathbf{F}_{2r} \\ \vdots & \vdots & \ddots & \vdots \\ \mathbf{F}_{r1} & \mathbf{F}_{r2} & \ldots & \mathbf{F}_{rr} \end{pmatrix},$$

where $\mathbf{F}_{ij}$ is of the same dimensions as $\mathbf{A}_{ij}$ $(i, j = 1, \ldots, r)$. If $\mathbf{A}$ is upper block-triangular, then $\mathbf{A}^{-1}$ is upper block-triangular; that is, if $\mathbf{A}_{ij} = \mathbf{0}$ for $j < i = 1, \ldots, r$, then $\mathbf{F}_{ij} = \mathbf{0}$ for $j < i = 1, \ldots, r$. Similarly, if $\mathbf{A}$ is lower block-triangular, then $\mathbf{A}^{-1}$ is lower block-triangular. Further, if $\mathbf{A}$ is (lower or upper) block-triangular, then $\mathbf{F}_{ii} = \mathbf{A}_{ii}^{-1}$; that is, the ith diagonal block of $\mathbf{A}^{-1}$ equals the inverse of the ith diagonal block of $\mathbf{A}$ $(i = 1, \ldots, r)$.

Proof. The proof is by mathematical induction. Lemma 8.5.4 implies that the theorem is valid for $r = 2$. Suppose then that the theorem is valid for $r = k - 1$. Define

$$\mathbf{T} = \begin{pmatrix} \mathbf{A}_{11} & \mathbf{A}_{12} & \ldots & \mathbf{A}_{1,k-1} \\ \mathbf{A}_{21} & \mathbf{A}_{22} & \ldots & \mathbf{A}_{2,k-1} \\ \vdots & \vdots & \ddots & \vdots \\ \mathbf{A}_{k-1,1} & \mathbf{A}_{k-1,2} & \ldots & \mathbf{A}_{k-1,k-1} \end{pmatrix}, \qquad \mathbf{U} = \begin{pmatrix} \mathbf{A}_{1k} \\ \mathbf{A}_{2k} \\ \vdots \\ \mathbf{A}_{k-1,k} \end{pmatrix},$$

and $\mathbf{V} = (\mathbf{A}_{k1}, \mathbf{A}_{k2}, \ldots, \mathbf{A}_{k,k-1})$, so that

$$\begin{pmatrix} \mathbf{T} & \mathbf{U} \\ \mathbf{V} & \mathbf{A}_{kk} \end{pmatrix} = \begin{pmatrix} \mathbf{A}_{11} & \mathbf{A}_{12} & \ldots & \mathbf{A}_{1k} \\ \mathbf{A}_{21} & \mathbf{A}_{22} & \ldots & \mathbf{A}_{2k} \\ \vdots & \vdots & \ddots & \vdots \\ \mathbf{A}_{k1} & \mathbf{A}_{k2} & \ldots & \mathbf{A}_{kk} \end{pmatrix},$$

and partition $\begin{pmatrix} \mathbf{T} & \mathbf{U} \\ \mathbf{V} & \mathbf{A}_{kk} \end{pmatrix}^{-1}$ as

$$\begin{pmatrix} \mathbf{T} & \mathbf{U} \\ \mathbf{V} & \mathbf{A}_{kk} \end{pmatrix}^{-1} = \begin{pmatrix} \mathbf{B}_{11} & \mathbf{B}_{12} & \ldots & \mathbf{B}_{1k} \\ \mathbf{B}_{21} & \mathbf{B}_{22} & \ldots & \mathbf{B}_{2k} \\ \vdots & \vdots & \ddots & \vdots \\ \mathbf{B}_{k1} & \mathbf{B}_{k2} & \ldots & \mathbf{B}_{kk} \end{pmatrix},$$

where $\mathbf{B}_{ij}$ is of dimensions $n_i \times n_j$ $(i, j = 1, \ldots, k)$. It suffices to show that if $\mathbf{A}_{ij} = \mathbf{0}$ for $j < i = 1, \ldots, k$, then $\mathbf{B}_{ij} = \mathbf{0}$ for $j < i = 1, \ldots, k$; if $\mathbf{A}_{ij} = \mathbf{0}$ for $i < j = 1, \ldots, k$, then $\mathbf{B}_{ij} = \mathbf{0}$ for $i < j = 1, \ldots, k$; and if either $\mathbf{A}_{ij} = \mathbf{0}$ for $j < i = 1, \ldots, k$ or $\mathbf{A}_{ij} = \mathbf{0}$ for $j > i = 1, \ldots, k$, then $\mathbf{B}_{ii} = \mathbf{A}_{ii}^{-1}$ $(i = 1, \ldots, k)$.

If $\mathbf{A}_{ij} = \mathbf{0}$ for $j < i = 1, \ldots, k$, then $\mathbf{V} = \mathbf{0}$, so that (according to Lemma 8.5.4) $\mathbf{B}_{kj} = \mathbf{0}$ for $j = 1, \ldots, k-1$, $\mathbf{B}_{kk} = \mathbf{A}_{kk}^{-1}$, and

$$\begin{pmatrix} \mathbf{B}_{11} & \mathbf{B}_{12} & \ldots & \mathbf{B}_{1,k-1} \\ \mathbf{B}_{21} & \mathbf{B}_{22} & \ldots & \mathbf{B}_{2,k-1} \\ \vdots & \vdots & \ddots & \vdots \\ \mathbf{B}_{k-1,1} & \mathbf{B}_{k-1,2} & \ldots & \mathbf{B}_{k-1,k-1} \end{pmatrix} = \mathbf{T}^{-1};$$

and $\mathbf{T} = \begin{pmatrix} \mathbf{A}_{11} & \mathbf{A}_{12} & \ldots & \mathbf{A}_{1,k-1} \\ \mathbf{0} & \mathbf{A}_{22} & \ldots & \mathbf{A}_{2,k-1} \\ \vdots & & \ddots & \vdots \\ \mathbf{0} & \mathbf{0} & & \mathbf{A}_{k-1,k-1} \end{pmatrix}$ is upper block-triangular, so that (by supposition) $\mathbf{B}_{ij} = \mathbf{0}$ for $j < i = 1, \ldots, k-1$ and $\mathbf{B}_{ii} = \mathbf{A}_{ii}^{-1}$ $(i = 1, \ldots, k-1)$.

Similarly, if $\mathbf{A}_{ij} = \mathbf{0}$ for $i < j = 1, \ldots, k$, then $\mathbf{U} = \mathbf{0}$, so that (according to Lemma 8.5.4) $\mathbf{B}_{ik} = \mathbf{0}$ for $i = 1, \ldots, k-1$, $\mathbf{B}_{kk} = \mathbf{A}_{kk}^{-1}$, and

$$\begin{pmatrix} \mathbf{B}_{11} & \mathbf{B}_{12} & \ldots & \mathbf{B}_{1,k-1} \\ \mathbf{B}_{21} & \mathbf{B}_{22} & \ldots & \mathbf{B}_{2,k-1} \\ \vdots & \vdots & \ddots & \vdots \\ \mathbf{B}_{k-1,1} & \mathbf{B}_{k-1,2} & \ldots & \mathbf{B}_{k-1,k-1} \end{pmatrix} = \mathbf{T}^{-1};$$

and $\mathbf{T} = \begin{pmatrix} \mathbf{A}_{11} & \mathbf{0} & \ldots & \mathbf{0} \\ \mathbf{A}_{21} & \mathbf{A}_{22} & & \mathbf{0} \\ \vdots & \vdots & \ddots & \\ \mathbf{A}_{k-1,1} & \mathbf{A}_{k-1,2} & \ldots & \mathbf{A}_{k-1,k-1} \end{pmatrix}$ is lower block-triangular, so that (by supposition) $\mathbf{B}_{ij} = \mathbf{0}$ for $i < j = 1, \ldots, k-1$ and $\mathbf{B}_{ii} = \mathbf{A}_{ii}^{-1}$ ($i = 1, \ldots, k-1$). Q.E.D.

In the special case of a triangular matrix (i.e., the special case where $r = n$), Theorem 8.5.7 can be restated as follows.

Corollary 8.5.8. Let $\mathbf{A} = \{a_{ij}\}$ represent an $n \times n$ nonsingular matrix. If $\mathbf{A}$ is upper triangular, then $\mathbf{A}^{-1}$ is also upper triangular. Similarly, if $\mathbf{A}$ is lower triangular, then $\mathbf{A}^{-1}$ is lower triangular. Further, if $\mathbf{A}$ is (lower or upper) triangular, then the ith diagonal element of $\mathbf{A}^{-1}$ is the reciprocal $1/a_{ii}$ of the ith diagonal element a_{ii} of $\mathbf{A}$ $(i = 1, \ldots, n)$.

In the special case of a unit triangular matrix, Corollary 8.5.8 reduces to the following result.

Corollary 8.5.9. The inverse of a unit upper triangular matrix is unit upper triangular. Similarly, the inverse of a unit lower triangular matrix is unit lower triangular.

d. A recursive algorithm for finding the inverse of a (nonsingular) triangular or block-triangular matrix

Let

$$\mathbf{A} = \begin{pmatrix} \mathbf{A}_{11} & \mathbf{A}_{12} & \ldots & \mathbf{A}_{1r} \\ \mathbf{0} & \mathbf{A}_{22} & \ldots & \mathbf{A}_{2r} \\ \vdots & & \ddots & \vdots \\ \mathbf{0} & \mathbf{0} & & \mathbf{A}_{rr} \end{pmatrix}$$

represent an $n \times n$ upper block-triangular matrix whose ijth block $\mathbf{A}_{ij}$ is of dimensions $n_i \times n_j$ $(j \geq i = 1, 2, \ldots, r)$. Suppose that the diagonal blocks $\mathbf{A}_{11}, \mathbf{A}_{22}, \ldots, \mathbf{A}_{rr}$ of $\mathbf{A}$ are all nonsingular, or, equivalently (according to Theorem 8.5.5), that $\mathbf{A}$ itself is nonsingular.

We now describe a procedure, derived by repeated application of result (5.8), for building up the inverse of $\mathbf{A}$ one "row" of blocks at a time, or in the special case of an upper triangular matrix, one row (of elements) at a time. According to result (5.8), the inverse of the submatrix

$$\begin{pmatrix} \mathbf{A}_{r-1,r-1} & \mathbf{A}_{r-1,r} \\ \mathbf{0} & \mathbf{A}_{rr} \end{pmatrix}$$

is expressible as

$$\begin{pmatrix} \mathbf{F}_{r-1,r-1} & \mathbf{F}_{r-1,r} \\ \mathbf{0} & \mathbf{F}_{rr} \end{pmatrix},$$

where $\mathbf{F}_{rr} = \mathbf{A}_{rr}^{-1}$ and

$$\mathbf{F}_{r-1,r-1} = \mathbf{A}_{r-1,r-1}^{-1}, \qquad \mathbf{F}_{r-1,r} = -\mathbf{A}_{r-1,r-1}^{-1}\mathbf{A}_{r-1,r}\mathbf{F}_{rr}.$$

By using this representation in an obvious way, the inverse of the submatrix

$\begin{pmatrix} \mathbf{A}_{r-1,r-1} & \mathbf{A}_{r-1,r} \\ \mathbf{0} & \mathbf{A}_{rr} \end{pmatrix}$ can be generated from that of the sub-submatrix $\mathbf{A}_{rr}$.

More generally, result (5.8) indicates that the inverse of the submatrix

$$\begin{pmatrix} \mathbf{A}_{ii} & \mathbf{A}_{i,i+1} & \dots & \mathbf{A}_{ir} \\ \mathbf{0} & \mathbf{A}_{i+1,i+1} & \dots & \mathbf{A}_{i+1,r} \\ \vdots & & \ddots & \vdots \\ \mathbf{0} & \mathbf{0} & & \mathbf{A}_{rr} \end{pmatrix}$$

can be generated by bordering the inverse matrix

$$\begin{pmatrix} \mathbf{F}_{i+1,i+1} & \mathbf{F}_{i+1,i+2} & \dots & \mathbf{F}_{i+1,r} \\ \mathbf{0} & \mathbf{F}_{i+2,i+2} & \dots & \mathbf{F}_{i+2,r} \\ \vdots & & \ddots & \vdots \\ \mathbf{0} & \mathbf{0} & & \mathbf{F}_{rr} \end{pmatrix} \tag{5.9a}$$

$$= \begin{pmatrix} \mathbf{A}_{i+1,i+1} & \mathbf{A}_{i+1,i+2} & \dots & \mathbf{A}_{i+1,r} \\ \mathbf{0} & \mathbf{A}_{i+2,i+2} & \dots & \mathbf{A}_{i+2,r} \\ \vdots & & \ddots & \vdots \\ \mathbf{0} & \mathbf{0} & & \mathbf{A}_{rr} \end{pmatrix}^{-1} \tag{5.9b}$$

with the matrices

$$\mathbf{F}_{ii} = \mathbf{A}_{ii}^{-1}, \qquad \mathbf{F}_{ij} = -\mathbf{A}_{ii}^{-1} \sum_{k=i+1}^{j} \mathbf{A}_{ik}\mathbf{F}_{kj} \quad (j = i+1, \dots, r), \tag{5.10}$$

to form the matrix

$$\begin{pmatrix} \mathbf{F}_{ii} & \mathbf{F}_{i,i+1} & \dots & \mathbf{F}_{ir} \\ \mathbf{0} & \mathbf{F}_{i+1,i+1} & \dots & \mathbf{F}_{i+1,r} \\ \vdots & & \ddots & \vdots \\ \mathbf{0} & \mathbf{0} & & \mathbf{F}_{rr} \end{pmatrix} \tag{5.11}$$

$(i = r-1, r-2, \dots, 1)$. Note that formulas (5.10) are in terms of the entries of the inverse matrix (5.9) and of the submatrices $\mathbf{A}_{ii}, \mathbf{A}_{i,i+1}, \dots, \mathbf{A}_{ir}$, belonging to the ith "row" of $\mathbf{A}$. Note also that, for $i = 1$, matrix (5.11) equals $\mathbf{A}^{-1}$.

These observations suggest an algorithm for computing $\mathbf{A}^{-1}$ in r steps. The first step is to compute the matrix $\mathbf{F}_{rr} = \mathbf{A}_{rr}^{-1}$. The $(r-i+1)$th step is to compute the matrices $\mathbf{F}_{ii}, \mathbf{F}_{i,i+1}, \dots, \mathbf{F}_{ir}$ from formulas (5.10) $(i = r-1, r-2, \dots, 1)$. Note that if the matrices $\mathbf{F}_{i,i+1}, \dots, \mathbf{F}_{ir}$ are computed in reverse order, that is, in the order $\mathbf{F}_{ir}, \dots, \mathbf{F}_{i,i+1}$, then, once the matrix $\mathbf{F}_{ij}$ $(j \geq i)$ has been computed, the submatrix $\mathbf{A}_{ij}$ is no longer needed (at least not for purposes of computing $\mathbf{A}^{-1}$), a feature that allows us to save storage space in implementing the algorithm on a computer. The savings is achieved by storing the elements of $\mathbf{F}_{ij}$ in the locations previously occupied by those of $\mathbf{A}_{ij}$. Then, at the completion of the algorithm, the locations that were previously occupied by the elements of $\mathbf{A}$ are occupied by the corresponding elements of $\mathbf{A}^{-1}$.

A possible variation on this algorithm is to use formulas (5.10) to build-up $\mathbf{A}^{-1}$ one "column" at a time, rather than one "row" at a time. As a first step, we could compute the diagonal blocks $\mathbf{F}_{rr}, \mathbf{F}_{r-1,r-1}, \ldots, \mathbf{F}_{11}$. As the $(r-j+2)$th step, we could compute in succession the submatrices $\mathbf{F}_{j-1,j}, \mathbf{F}_{j-2,j}, \ldots, \mathbf{F}_{1j}$ $(j = r, r-1, \ldots, 2)$. Note that this alternative scheme, like the original, is such that the elements of $\mathbf{F}_{ij}$ can share storage locations with those of $\mathbf{A}_{ij}$.

For purposes of illustration, we apply the algorithm to the 3×3 upper triangular matrix

$$\mathbf{A} = \begin{pmatrix} 8 & 1 & 6 \\ 0 & 5 & 4 \\ 0 & 0 & 2 \end{pmatrix},$$

which has an inverse of the form

$$\mathbf{F} = \begin{pmatrix} f_{11} & f_{12} & f_{13} \\ 0 & f_{22} & f_{23} \\ 0 & 0 & f_{33} \end{pmatrix}.$$

On the first step of the algorithm, we compute

$$f_{33} = 1/2 = 0.5;$$

on the second step,

$$f_{22} = 1/5 = 0.2, \qquad f_{23} = -(0.2)(4)(0.5) = -0.4;$$

and, on the third (and final) step,

$$f_{11} = 1/8 = 0.125, \qquad f_{13} = -(0.125)[(1)(-0.4) + (6)(0.5)] = -0.325,$$
$$f_{12} = -(0.125)(1)(0.2) = -0.025;$$

giving

$$\mathbf{A}^{-1} = \mathbf{F} = \begin{pmatrix} 0.125 & -0.025 & -0.325 \\ 0 & 0.2 & -0.4 \\ 0 & 0 & 0.5 \end{pmatrix}.$$

Let us now switch our attention from the upper block-triangular matrix $\mathbf{A}$ to a lower block-triangular matrix

$$\mathbf{B} = \begin{pmatrix} \mathbf{B}_{11} & \mathbf{0} & \ldots & \mathbf{0} \\ \mathbf{B}_{21} & \mathbf{B}_{22} & & \mathbf{0} \\ \vdots & \vdots & \ddots & \\ \mathbf{B}_{r1} & \mathbf{B}_{r2} & \ldots & \mathbf{B}_{rr} \end{pmatrix},$$

whose ijth block $\mathbf{B}_{ij}$ is of dimensions $n_i \times n_j$ $(j \le i = 1, \ldots, r)$. Suppose that the diagonal blocks $\mathbf{B}_{11}, \mathbf{B}_{22}, \ldots, \mathbf{B}_{rr}$ of $\mathbf{B}$ are all nonsingular, or, equivalently, that $\mathbf{B}$ itself is nonsingular.

According to result (5.8), the inverse of the submatrix

$$\begin{pmatrix} \mathbf{B}_{11} & \mathbf{0} & \dots & \mathbf{0} \\ \mathbf{B}_{21} & \mathbf{B}_{22} & & \mathbf{0} \\ \vdots & \vdots & \ddots & \\ \mathbf{B}_{i1} & \mathbf{B}_{i2} & \dots & \mathbf{B}_{ii} \end{pmatrix}$$

can be generated by bordering the inverse matrix

$$\begin{pmatrix} \mathbf{G}_{11} & \mathbf{0} & \dots & \mathbf{0} \\ \mathbf{G}_{21} & \mathbf{G}_{22} & & \mathbf{0} \\ \vdots & \vdots & \ddots & \\ \mathbf{G}_{i-1,1} & \mathbf{G}_{i-2,2} & \dots & \mathbf{G}_{i-1,i-1} \end{pmatrix} \tag{5.12a}$$

$$= \begin{pmatrix} \mathbf{B}_{11} & \mathbf{0} & \dots & \mathbf{0} \\ \mathbf{B}_{21} & \mathbf{B}_{22} & & \mathbf{0} \\ \vdots & \vdots & \ddots & \\ \mathbf{B}_{i-1,1} & \mathbf{B}_{i-1,2} & \dots & \mathbf{B}_{i-1,i-1} \end{pmatrix}^{-1} \tag{5.12b}$$

with the matrices

$$\mathbf{G}_{ii} = \mathbf{B}_{ii}^{-1}, \qquad \mathbf{G}_{ij} = -\mathbf{B}_{ii}^{-1} \sum_{k=j}^{i-1} \mathbf{B}_{ik}\mathbf{G}_{kj} \quad (j = 1, \dots, i-1), \tag{5.13}$$

to form the matrix

$$\begin{pmatrix} \mathbf{G}_{11} & \mathbf{0} & \dots & \mathbf{0} \\ \mathbf{G}_{21} & \mathbf{G}_{22} & & \mathbf{0} \\ \vdots & \vdots & \ddots & \\ \mathbf{G}_{i1} & \mathbf{G}_{i2} & \dots & \mathbf{G}_{ii} \end{pmatrix}. \tag{5.14}$$

We see that formulas (5.13) are in terms of the entries of the inverse matrix (5.12) and of the submatrices $\mathbf{B}_{i1}, \mathbf{B}_{i2}, \dots, \mathbf{B}_{ii}$, belonging to the ith "row" of $\mathbf{B}$. We see also that, for $i = r$, matrix (5.14) equals $\mathbf{B}^{-1}$.

These results suggest an algorithm for computing $\mathbf{B}^{-1}$ in r steps. The first step is to compute $\mathbf{G}_{11} = \mathbf{B}_{11}^{-1}$. The ith step is to compute the matrices $\mathbf{G}_{i1}, \mathbf{G}_{i2}, \dots,$ $\mathbf{G}_{ii}$ from formulas (5.13) ($i = 2, \dots, r$). Note that if $\mathbf{G}_{i1}, \mathbf{G}_{i2}, \dots, \mathbf{G}_{i,i-1}$ are computed in the order listed, then, once the matrix $\mathbf{G}_{ij}$ ($j \le i$) has been computed, the submatrix $\mathbf{B}_{ij}$ is no longer required and hence, in implementing the algorithm on a computer, storage can be saved by "overwriting" the matrix $\mathbf{B}_{ij}$ with the matrix $\mathbf{G}_{ij}$.

e. Matrices partitioned into two rows and two columns of submatrices

The ranks and inverses of block-diagonal and block-triangular matrices were considered in Subsections a through d. In this subsection, we consider the ranks and inverses of partitioned matrices that are not necessarily block-diagonal or block-triangular.

The following theorem can (when applicable) be used to express the rank of a partitioned matrix in terms of the rank of a matrix of smaller dimensions.

Theorem 8.5.10. Let $\mathbf{T}$ represent an $m \times m$ matrix, $\mathbf{U}$ an $m \times q$ matrix, $\mathbf{V}$ an $n \times m$ matrix, and $\mathbf{W}$ an $n \times q$ matrix. If $\text{rank}(\mathbf{T}) = m$, that is, if $\mathbf{T}$ is nonsingular, then

$$\begin{aligned} \text{rank}\begin{pmatrix} \mathbf{T} & \mathbf{U} \\ \mathbf{V} & \mathbf{W} \end{pmatrix} = \text{rank}\begin{pmatrix} \mathbf{U} & \mathbf{T} \\ \mathbf{W} & \mathbf{V} \end{pmatrix} &= \text{rank}\begin{pmatrix} \mathbf{V} & \mathbf{W} \\ \mathbf{T} & \mathbf{U} \end{pmatrix} \\ &= \text{rank}\begin{pmatrix} \mathbf{W} & \mathbf{V} \\ \mathbf{U} & \mathbf{T} \end{pmatrix} \\ &= m + \text{rank}(\mathbf{W} - \mathbf{V}\mathbf{T}^{-1}\mathbf{U}). \qquad (5.15) \end{aligned}$$

Proof. Suppose that $\text{rank}(\mathbf{T}) = m$. Then, in light of Lemma 8.5.1, it suffices to show that

$$\text{rank}\begin{pmatrix} \mathbf{T} & \mathbf{U} \\ \mathbf{V} & \mathbf{W} \end{pmatrix} = m + \text{rank}(\mathbf{W} - \mathbf{V}\mathbf{T}^{-1}\mathbf{U}).$$

It is easy to verify that

$$\begin{pmatrix} \mathbf{I}_m & \mathbf{0} \\ -\mathbf{V}\mathbf{T}^{-1} & \mathbf{I}_n \end{pmatrix} \begin{pmatrix} \mathbf{T} & \mathbf{U} \\ \mathbf{V} & \mathbf{W} \end{pmatrix} = \begin{pmatrix} \mathbf{T} & \mathbf{U} \\ \mathbf{0} & \mathbf{W}-\mathbf{V}\mathbf{T}^{-1}\mathbf{U} \end{pmatrix}.$$

Thus, since (according to Lemma 8.5.2) $\begin{pmatrix} \mathbf{I} & \mathbf{0} \\ -\mathbf{V}\mathbf{T}^{-1} & \mathbf{I} \end{pmatrix}$ is nonsingular, it follows from Corollary 8.3.3 and Lemma 8.5.3 that

$$\begin{aligned} \text{rank}\begin{pmatrix} \mathbf{T} & \mathbf{U} \\ \mathbf{V} & \mathbf{W} \end{pmatrix} &= \text{rank}\begin{pmatrix} \mathbf{T} & \mathbf{U} \\ \mathbf{0} & \mathbf{W}-\mathbf{V}\mathbf{T}^{-1}\mathbf{U} \end{pmatrix} \\ &= \text{rank}(\mathbf{T}) + \text{rank}(\mathbf{W} - \mathbf{V}\mathbf{T}^{-1}\mathbf{U}) \\ &= m + \text{rank}(\mathbf{W} - \mathbf{V}\mathbf{T}^{-1}\mathbf{U}). \end{aligned}$$

Q.E.D.

Result (5.8) on the inverse of a block-triangular matrix can be extended to other partitioned matrices, as detailed in the following theorem.

Theorem 8.5.11. Let $\mathbf{T}$ represent an $m \times m$ matrix, $\mathbf{U}$ an $m \times n$ matrix, $\mathbf{V}$ an $n \times m$ matrix, and $\mathbf{W}$ an $n \times n$ matrix. Suppose that $\mathbf{T}$ is nonsingular. Then, $\begin{pmatrix} \mathbf{T} & \mathbf{U} \\ \mathbf{V} & \mathbf{W} \end{pmatrix}$, or equivalently $\begin{pmatrix} \mathbf{W} & \mathbf{V} \\ \mathbf{U} & \mathbf{T} \end{pmatrix}$, is nonsingular if and only if the $n \times n$ matrix

$$\mathbf{Q} = \mathbf{W} - \mathbf{V}\mathbf{T}^{-1}\mathbf{U}$$

is nonsingular, in which case

$$\begin{aligned} \begin{pmatrix} \mathbf{T} & \mathbf{U} \\ \mathbf{V} & \mathbf{W} \end{pmatrix}^{-1} &= \begin{pmatrix} \mathbf{T}^{-1} + \mathbf{T}^{-1}\mathbf{U}\mathbf{Q}^{-1}\mathbf{V}\mathbf{T}^{-1} & -\mathbf{T}^{-1}\mathbf{U}\mathbf{Q}^{-1} \\ -\mathbf{Q}^{-1}\mathbf{V}\mathbf{T}^{-1} & \mathbf{Q}^{-1} \end{pmatrix} \qquad (5.16\text{a}) \\ &= \begin{pmatrix} \mathbf{T}^{-1} & \mathbf{0} \\ \mathbf{0} & \mathbf{0} \end{pmatrix} + \begin{pmatrix} -\mathbf{T}^{-1}\mathbf{U} \\ \mathbf{I}_n \end{pmatrix} \mathbf{Q}^{-1}(-\mathbf{V}\mathbf{T}^{-1},\ \mathbf{I}_n), \qquad (5.16\text{b}) \end{aligned}$$

$$\begin{pmatrix} \mathbf{W} & \mathbf{V} \\ \mathbf{U} & \mathbf{T} \end{pmatrix}^{-1} = \begin{pmatrix} \mathbf{Q}^{-1} & -\mathbf{Q}^{-1}\mathbf{V}\mathbf{T}^{-1} \\ -\mathbf{T}^{-1}\mathbf{U}\mathbf{Q}^{-1} & \mathbf{T}^{-1}+\mathbf{T}^{-1}\mathbf{U}\mathbf{Q}^{-1}\mathbf{V}\mathbf{T}^{-1} \end{pmatrix} \tag{5.17a}$$

$$= \begin{pmatrix} \mathbf{0} & \mathbf{0} \\ \mathbf{0} & \mathbf{T}^{-1} \end{pmatrix} + \begin{pmatrix} \mathbf{I}_n \\ -\mathbf{T}^{-1}\mathbf{U} \end{pmatrix} \mathbf{Q}^{-1}(\mathbf{I}_n, -\mathbf{V}\mathbf{T}^{-1}). \tag{5.17b}$$

Proof. That $\begin{pmatrix} \mathbf{T} & \mathbf{U} \\ \mathbf{V} & \mathbf{W} \end{pmatrix}$ is nonsingular if and only if $\mathbf{Q}$ is nonsingular is an immediate consequence of Theorem 8.5.10.

Suppose now that $\mathbf{Q}$ is nonsingular, and observe that

$$\begin{pmatrix} \mathbf{T} & \mathbf{0} \\ \mathbf{V} & \mathbf{Q} \end{pmatrix} = \begin{pmatrix} \mathbf{T} & \mathbf{U} \\ \mathbf{V} & \mathbf{W} \end{pmatrix} \begin{pmatrix} \mathbf{I} & -\mathbf{T}^{-1}\mathbf{U} \\ \mathbf{0} & \mathbf{I} \end{pmatrix}. \tag{5.18}$$

Then (in light of Lemma 8.5.4), $\begin{pmatrix} \mathbf{T} & \mathbf{0} \\ \mathbf{V} & \mathbf{Q} \end{pmatrix}$, as well as $\begin{pmatrix} \mathbf{T} & \mathbf{U} \\ \mathbf{V} & \mathbf{W} \end{pmatrix}$, is nonsingular. Premultiplying both sides of equality (5.18) by $\begin{pmatrix} \mathbf{T} & \mathbf{U} \\ \mathbf{V} & \mathbf{W} \end{pmatrix}^{-1}$ and postmultiplying both sides by $\begin{pmatrix} \mathbf{T} & \mathbf{0} \\ \mathbf{V} & \mathbf{Q} \end{pmatrix}^{-1}$ and making further use of Lemma 8.5.4, we find that

$$\begin{aligned} \begin{pmatrix} \mathbf{T} & \mathbf{U} \\ \mathbf{V} & \mathbf{W} \end{pmatrix}^{-1} &= \begin{pmatrix} \mathbf{I} & -\mathbf{T}^{-1}\mathbf{U} \\ \mathbf{0} & \mathbf{I} \end{pmatrix} \begin{pmatrix} \mathbf{T} & \mathbf{0} \\ \mathbf{V} & \mathbf{Q} \end{pmatrix}^{-1} \\ &= \begin{pmatrix} \mathbf{I} & -\mathbf{T}^{-1}\mathbf{U} \\ \mathbf{0} & \mathbf{I} \end{pmatrix} \begin{pmatrix} \mathbf{T}^{-1} & \mathbf{0} \\ -\mathbf{Q}^{-1}\mathbf{V}\mathbf{T}^{-1} & \mathbf{Q}^{-1} \end{pmatrix} \\ &= \begin{pmatrix} \mathbf{T}^{-1}+\mathbf{T}^{-1}\mathbf{U}\mathbf{Q}^{-1}\mathbf{V}\mathbf{T}^{-1} & -\mathbf{T}^{-1}\mathbf{U}\mathbf{Q}^{-1} \\ -\mathbf{Q}^{-1}\mathbf{V}\mathbf{T}^{-1} & \mathbf{Q}^{-1} \end{pmatrix}, \end{aligned}$$

which establishes formula (5.16). Formula (5.17) can be derived in an analogous fashion. (As an alternative to deriving these formulas from results on the inverse of a block-triangular matrix, which is the approach taken here, their validity could be established simply by verifying that the product of expression (5.16) or (5.17) and $\begin{pmatrix} \mathbf{T} & \mathbf{U} \\ \mathbf{V} & \mathbf{W} \end{pmatrix}$ or $\begin{pmatrix} \mathbf{W} & \mathbf{V} \\ \mathbf{U} & \mathbf{T} \end{pmatrix}$, respectively, equals $\mathbf{I}_{m+n}$.) Q.E.D.

When $\mathbf{T}$ is nonsingular, the matrix $\mathbf{Q} = \mathbf{W} - \mathbf{V}\mathbf{T}^{-1}\mathbf{U}$, which appears in formulas (5.16) and (5.17) for the inverse of a partitioned matrix and also in formula (5.15) for the rank of a partitioned matrix, is called the *Schur complement* of $\mathbf{T}$ in $\begin{pmatrix} \mathbf{T} & \mathbf{U} \\ \mathbf{V} & \mathbf{W} \end{pmatrix}$ or $\begin{pmatrix} \mathbf{W} & \mathbf{V} \\ \mathbf{U} & \mathbf{T} \end{pmatrix}$, respectively. Moreover, when the context is clear, it is sometimes referred to simply as the Schur complement of $\mathbf{T}$ or even more simply as the Schur complement. Note that if $\begin{pmatrix} \mathbf{T} & \mathbf{U} \\ \mathbf{V} & \mathbf{W} \end{pmatrix}$ or $\begin{pmatrix} \mathbf{W} & \mathbf{V} \\ \mathbf{U} & \mathbf{T} \end{pmatrix}$ is symmetric (in which case $\mathbf{T}' = \mathbf{T}$, $\mathbf{W}' = \mathbf{W}$, and $\mathbf{V} = \mathbf{U}'$), then the Schur complement of $\mathbf{T}$ is symmetric.

Among other things, the following corollary (of Theorem 8.5.11) expresses the inverse of a diagonal block of a partitioned matrix in terms of the inverse of the partitioned matrix.

Corollary 8.5.12. Let $\mathbf{T}$ represent an $m \times m$ matrix, $\mathbf{U}$ an $m \times n$ matrix, $\mathbf{V}$ an $n \times m$ matrix, and $\mathbf{W}$ an $n \times n$ matrix. Suppose that the partitioned matrix $\begin{pmatrix} \mathbf{T} & \mathbf{U} \\ \mathbf{V} & \mathbf{W} \end{pmatrix}$ is nonsingular, define $\mathbf{B} = \begin{pmatrix} \mathbf{T} & \mathbf{U} \\ \mathbf{V} & \mathbf{W} \end{pmatrix}^{-1}$, and partition $\mathbf{B}$ as $\mathbf{B} = \begin{pmatrix} \mathbf{B}_{11} & \mathbf{B}_{12} \\ \mathbf{B}_{21} & \mathbf{B}_{22} \end{pmatrix}$, where the dimensions of $\mathbf{B}_{11}, \mathbf{B}_{12}, \mathbf{B}_{21}$, and $\mathbf{B}_{22}$ are the same as those of $\mathbf{T}$, $\mathbf{U}$, $\mathbf{V}$, and $\mathbf{W}$, respectively. If $\mathbf{T}$ is nonsingular, then $\mathbf{B}_{22}$ is nonsingular,

$$\mathbf{T}^{-1} = \mathbf{B}_{11} - \mathbf{B}_{12}\mathbf{B}_{22}^{-1}\mathbf{B}_{21} \tag{5.19}$$

(i.e., $\mathbf{T}^{-1}$ equals the Schur complement of $\mathbf{B}_{22}$),

$$\mathbf{T}^{-1}\mathbf{U} = -\mathbf{B}_{12}\mathbf{B}_{22}^{-1}, \qquad \mathbf{V}\mathbf{T}^{-1} = -\mathbf{B}_{22}^{-1}\mathbf{B}_{21}, \tag{5.20}$$

and

$$\mathbf{W} - \mathbf{V}\mathbf{T}^{-1}\mathbf{U} = \mathbf{B}_{22}^{-1} \tag{5.21}$$

(i.e., the Schur complement of $\mathbf{T}$ equals $\mathbf{B}_{22}^{-1}$). Similarly, if $\mathbf{W}$ is nonsingular, then $\mathbf{B}_{11}$ is nonsingular,

$$\mathbf{W}^{-1} = \mathbf{B}_{22} - \mathbf{B}_{21}\mathbf{B}_{11}^{-1}\mathbf{B}_{12} \tag{5.22}$$

(i.e., $\mathbf{W}^{-1}$ equals the Schur complement of $\mathbf{B}_{11}$),

$$\mathbf{W}^{-1}\mathbf{V} = -\mathbf{B}_{21}\mathbf{B}_{11}^{-1}, \qquad \mathbf{U}\mathbf{W}^{-1} = -\mathbf{B}_{11}^{-1}\mathbf{B}_{12}, \tag{5.23}$$

and

$$\mathbf{T} - \mathbf{U}\mathbf{W}^{-1}\mathbf{V} = \mathbf{B}_{11}^{-1} \tag{5.24}$$

(i.e., the Schur complement of $\mathbf{W}$ equals $\mathbf{B}_{11}^{-1}$).

Proof. Suppose that $\mathbf{T}$ is nonsingular, and let $\mathbf{Q} = \mathbf{W} - \mathbf{V}\mathbf{T}^{-1}\mathbf{U}$. Then, it follows from Theorem 8.5.11 that $\mathbf{Q}$ is nonsingular, that $\mathbf{B}_{22} = \mathbf{Q}^{-1}$, implying [in light of results (2.5) and (2.6)] that $\mathbf{B}_{22}$ is nonsingular and $\mathbf{B}_{22}^{-1} = \mathbf{Q}$, and that

$$-\mathbf{B}_{12}\mathbf{B}_{22}^{-1} = -(-\mathbf{T}^{-1}\mathbf{U}\mathbf{Q}^{-1})\mathbf{Q} = \mathbf{T}^{-1}\mathbf{U},$$
$$-\mathbf{B}_{22}^{-1}\mathbf{B}_{21} = -\mathbf{Q}(-\mathbf{Q}^{-1}\mathbf{V}\mathbf{T}^{-1}) = \mathbf{V}\mathbf{T}^{-1},$$

and

$$\mathbf{B}_{11} - \mathbf{B}_{12}\mathbf{B}_{22}^{-1}\mathbf{B}_{21} = \mathbf{T}^{-1} + \mathbf{T}^{-1}\mathbf{U}\mathbf{Q}^{-1}\mathbf{V}\mathbf{T}^{-1} + \mathbf{T}^{-1}\mathbf{U}(-\mathbf{Q}^{-1}\mathbf{V}\mathbf{T}^{-1}) = \mathbf{T}^{-1}.$$

The second part of the corollary can be established via an analogous argument. Q.E.D.

Suppose that we wish to find the rank of a matrix $\mathbf{A}$; or if $\mathbf{A}$ is nonsingular, that we wish to find $\mathbf{A}^{-1}$ or a submatrix of $\mathbf{A}^{-1}$, or that we wish to find the inverse of a nonsingular submatrix of $\mathbf{A}$. Then, depending on the nature of $\mathbf{A}$, Theorems 8.5.10 and 8.5.11 and Corollary 8.5.12 can be helpful, as will now be discussed.

Let $\mathbf{A} = \begin{pmatrix} \mathbf{A}_{11} & \mathbf{A}_{12} \\ \mathbf{A}_{21} & \mathbf{A}_{22} \end{pmatrix}$ represent an arbitrary partitioning of $\mathbf{A}$ into two rows and columns of submatrices. If $\mathbf{A}_{11}$ or $\mathbf{A}_{22}$ is nonsingular, then, by making use of Theorem 8.5.10, the problem of finding the rank of $\mathbf{A}$ can be replaced by that of (1) inverting $\mathbf{A}_{11}$ or $\mathbf{A}_{22}$, (2) forming the Schur complement $\mathbf{A}_{22} - \mathbf{A}_{21}\mathbf{A}_{11}^{-1}\mathbf{A}_{12}$ or $\mathbf{A}_{11} - \mathbf{A}_{12}\mathbf{A}_{22}^{-1}\mathbf{A}_{21}$, and (3) finding the rank of the Schur complement. This can be advantageous if the inverse of $\mathbf{A}_{11}$ or $\mathbf{A}_{22}$ is known—as would be the case, for example, if $\mathbf{A}_{11}$ or $\mathbf{A}_{22}$ were a (nonsingular) diagonal matrix—and if $\mathbf{A}_{11}$ or $\mathbf{A}_{22}$, respectively, is of relatively large order. Thus, whether Theorem 8.5.10 is helpful in finding the rank of $\mathbf{A}$ depends on whether the partitioning of $\mathbf{A}$ can be carried out in such a way that $\mathbf{A}_{11}$ or $\mathbf{A}_{22}$ is a relatively large order, nonsingular matrix with a known inverse.

Now, suppose that $\mathbf{A}$ is nonsingular, let $\mathbf{B} = \mathbf{A}^{-1}$, and partition $\mathbf{B}$ as $\mathbf{B} = \begin{pmatrix} \mathbf{B}_{11} & \mathbf{B}_{12} \\ \mathbf{B}_{21} & \mathbf{B}_{22} \end{pmatrix}$, where the dimensions of $\mathbf{B}_{11}, \mathbf{B}_{12}, \mathbf{B}_{21}$, and $\mathbf{B}_{22}$ are the same as those of $\mathbf{A}_{11}, \mathbf{A}_{12}, \mathbf{A}_{21}$, and $\mathbf{A}_{22}$, respectively. If $\mathbf{A}_{11}$ or $\mathbf{A}_{22}$ is nonsingular, then, by making use of Theorem 8.5.11, the problem of constructing $\mathbf{A}^{-1}$ can be reformulated as one of (1) finding $\mathbf{A}_{11}^{-1}$ or $\mathbf{A}_{22}^{-1}$, (2) forming and inverting the Schur complement $\mathbf{A}_{22} - \mathbf{A}_{21}\mathbf{A}_{11}^{-1}\mathbf{A}_{12}$ or $\mathbf{A}_{11} - \mathbf{A}_{12}\mathbf{A}_{22}^{-1}\mathbf{A}_{21}$, and (3) determining, based on formula (5.16) or (5.17), respectively, $\mathbf{B}_{11}, \mathbf{B}_{12}, \mathbf{B}_{21}$, and $\mathbf{B}_{22}$. This can be advantageous if $\mathbf{A}_{11}$ or $\mathbf{A}_{22}$ has a known inverse and is of relatively large order and/or if only the lower-right or upper-left part ($\mathbf{B}_{22}$ or $\mathbf{B}_{11}$) of $\mathbf{A}^{-1}$ is of interest. Thus, whether Theorem 8.5.11 is helpful in constructing $\mathbf{A}^{-1}$ depends on whether $\mathbf{A}$ can be partitioned in such a way that $\mathbf{A}_{11}$ or $\mathbf{A}_{22}$ is a relatively large order matrix with a known inverse and possibly on whether only part of $\mathbf{A}^{-1}$ is of interest.

Finally, suppose that the upper-left or lower-right part ($\mathbf{A}_{11}$ or $\mathbf{A}_{22}$) of $\mathbf{A}$ is nonsingular and that we wish to construct $\mathbf{A}_{11}^{-1}$ or $\mathbf{A}_{22}^{-1}$. If $\mathbf{A}$ itself is nonsingular, then, by making use of Corollary 8.5.12, we can compute $\mathbf{A}_{11}^{-1}$ or $\mathbf{A}_{22}^{-1}$ from $\mathbf{A}^{-1}$, which can be advantageous if $\mathbf{A}^{-1}$ is known and if $\mathbf{A}_{11}$ or $\mathbf{A}_{22}$, respectively, is of relatively large order.

f. Use of permutation matrices to extend results on partitioned matrices

One important use of permutation matrices is to extend results on partitioned matrices. Consider, for example, result (5.16) on the inverse of a partitioned matrix. This result gives the inverse of an $(m+n)\times(m+n)$ matrix, say $\mathbf{A}$, in terms of a submatrix $\mathbf{T}$ formed from the first m rows and columns of $\mathbf{A}$, a submatrix $\mathbf{U}$ formed from the first m rows and last n columns of $\mathbf{A}$, a submatrix $\mathbf{V}$ formed from the last n rows and first m columns of $\mathbf{A}$, and a submatrix $\mathbf{W}$ formed from the last n rows and columns of $\mathbf{A}$. Taking $i_1, i_2, \ldots, i_{m+n}$ and $j_1, j_2, \ldots, j_{m+n}$ to be any two (not necessarily different) permutations of the first $m+n$ positive integers $1, 2, \ldots, m+n$, result (5.16) can be extended to the case where the submatrix $\mathbf{T}$ is formed from rows

$i_1, \dots, i_m$, respectively, and columns $j_1, \dots, j_m$, respectively (of **A**), **U** is formed from rows $i_1, \dots, i_m$, respectively, and columns $j_{m+1}, \dots, j_{m+n}$, respectively, **V** from rows $i_{m+1}, \dots, i_{m+n}$, respectively, and columns $j_1, \dots, j_m$, respectively, and **W** from rows $i_{m+1}, \dots, i_{m+n}$, respectively, and columns $j_{m+1}, \dots, j_{m+n}$, respectively. In fact, result (5.17) can be viewed as one particularly simple extension of this type.

To present explicitly a version of result (5.16) that would be applicable to arbitrary permutations $i_1, i_2, \dots, i_{m+n}$ and $j_1, j_2, \dots, j_{m+n}$ would require very elaborate and cumbersome notation. However, by explicitly or implicitly making use of appropriately chosen permutation matrices, result (5.16), which is for the special case where both $i_1, i_2, \dots, i_{m+n}$ and $j_1, j_2, \dots, j_{m+n}$ are the sequence $1, 2, \dots, m+n$, can be applied to any other choice for $i_1, i_2, \dots, i_{m+n}$ and $j_1, j_2, \dots, j_{m+n}$.

Letting $\mathbf{u}_1, \mathbf{u}_2, \dots, \mathbf{u}_{m+n}$ represent the first, second, ..., $(m+n)$th columns of $\mathbf{I}_{m+n}$, define

$$\mathbf{B} = \mathbf{PAQ},$$

where

$$\mathbf{P} = (\mathbf{u}_{i_1}, \mathbf{u}_{i_2}, \dots, \mathbf{u}_{i_{m+n}})' \quad \text{and} \quad \mathbf{Q} = (\mathbf{u}_{j_1}, \mathbf{u}_{j_2}, \dots, \mathbf{u}_{j_{m+n}}).$$

Partition **B** as

$$\mathbf{B} = \begin{pmatrix} \mathbf{T} & \mathbf{U} \\ \mathbf{V} & \mathbf{W} \end{pmatrix},$$

where **T** is $m \times m$, **U** is $m \times n$, **V** is $n \times m$, and **W** is $n \times n$. Note that **T** is the matrix formed from rows $i_1, \dots, i_m$, respectively, and columns $j_1, \dots, j_m$, respectively, of **A**; the submatrices **U**, **V**, and **W** have analogous interpretations.

Suppose that we invert **B** by, for example, making use of result (5.16) or (5.17) (assuming that **A**, and hence **B**, is invertible). Then, to obtain $\mathbf{A}^{-1}$ from $\mathbf{B}^{-1}$, we observe that

$$\mathbf{A} = \mathbf{P}'\mathbf{B}\mathbf{Q}'$$

and, consequently, that

$$\mathbf{A}^{-1} = \mathbf{Q}\mathbf{B}^{-1}\mathbf{P}. \tag{5.25}$$

Thus, to form $\mathbf{A}^{-1}$ from $\mathbf{B}^{-1}$, it suffices to permute the rows of $\mathbf{B}^{-1}$ so that its first row becomes the j_1th row, its second row becomes the j_2th row, and, in general, its rth row becomes the j_rth row, and to then permute the columns so that the first column becomes the i_1th column, the second column becomes the i_2th column, and, in general, the sth column becomes the i_sth column. Or, combining the row and column operations, we can form $\mathbf{A}^{-1}$ from $\mathbf{B}^{-1}$ simply by rearranging the elements of $\mathbf{B}^{-1}$ so that the rsth element becomes the $j_r i_s$th element.

Consider, for example, the matrix

$$\mathbf{A} = \begin{pmatrix} 0 & 1 & 0 & 1 \\ -6 & 2 & 4 & 8 \\ 0 & 5 & 1 & 0 \\ 1 & 3 & 0 & 0 \end{pmatrix}.$$

Observe that the 3×3 matrix formed from the fourth, third, and first rows, respectively, and the first, third, and fourth columns, respectively, of $\mathbf{A}$ is the 3×3 identity matrix, whose inverse is known. Suppose then that we take

$$\mathbf{B} = (\mathbf{u}_4, \mathbf{u}_3, \mathbf{u}_1, \mathbf{u}_2)'\mathbf{A}(\mathbf{u}_1, \mathbf{u}_3, \mathbf{u}_4, \mathbf{u}_2) = \begin{pmatrix} 1 & 0 & 0 & 3 \\ 0 & 1 & 0 & 5 \\ 0 & 0 & 1 & 1 \\ -6 & 4 & 8 & 2 \end{pmatrix}.$$

Applying result (5.16) with $\mathbf{T} = \mathbf{I}_3$, $\mathbf{U}' = (3, 5, 1)$, $\mathbf{V} = (-6, 4, 8)$, and $\mathbf{W} = (2)$, we find easily that

$$\mathbf{B}^{-1} = \begin{pmatrix} 3.25 & -1.5 & -3 & 0.375 \\ 3.75 & -1.5 & -5 & 0.625 \\ 0.75 & -0.5 & 0 & 0.125 \\ -0.75 & 0.5 & 1 & -0.125 \end{pmatrix}.$$

Rearranging the elements of $\mathbf{B}^{-1}$ in accordance with formula (5.25), we obtain

$$\mathbf{A}^{-1} = \begin{pmatrix} -3 & 0.375 & -1.5 & 3.25 \\ 1 & -0.125 & 0.5 & -0.75 \\ -5 & 0.625 & -1.5 & 3.75 \\ 0 & 0.125 & -0.5 & 0.75 \end{pmatrix}.$$

Exercises

Section 8.1

1. Let $\mathbf{A}$ represent an $m \times n$ matrix. Show that (a) if $\mathbf{A}$ has a right inverse, then $n \geq m$ and (b) if $\mathbf{A}$ has a left inverse, then $m \geq n$.

2. An $n \times n$ matrix $\mathbf{A}$ is said to be *involutory* if $\mathbf{A}^2 = \mathbf{I}$, that is, if $\mathbf{A}$ is invertible and is its own inverse.

   (a) Show that an $n \times n$ matrix $\mathbf{A}$ is involutory if and only if $(\mathbf{I}-\mathbf{A})(\mathbf{I}+\mathbf{A}) = \mathbf{0}$.

   (b) Show that a 2×2 matrix $\mathbf{A} = \begin{pmatrix} a & b \\ c & d \end{pmatrix}$ is involutory if and only if (1) $a^2 + bc = 1$ and $d = -a$ or (2) $b = c = 0$ and $d = a = \pm 1$.

Section 8.3

3. Let $\mathbf{A}$ represent an $n \times n$ nonnull symmetric matrix, and let $\mathbf{B}$ represent an $n \times r$ matrix of full column rank r and $\mathbf{T}$ an $r \times n$ matrix of full row rank r such that $\mathbf{A} = \mathbf{BT}$. Show that the $r \times r$ matrix $\mathbf{TB}$ is nonsingular. (*Hint.* Observe that $\mathbf{A}'\mathbf{A} = \mathbf{A}^2 = \mathbf{BTBT}$.)

Section 8.4

4. Let $\mathbf{A}$ represent an $n \times n$ matrix, and partition $\mathbf{A}$ as $\mathbf{A} = (\mathbf{A}_1, \mathbf{A}_2)$.

   (a) Show that if $\mathbf{A}$ is invertible and $\mathbf{A}^{-1}$ is partitioned as $\mathbf{A}^{-1} = \begin{pmatrix} \mathbf{B}_1 \\ \mathbf{B}_2 \end{pmatrix}$ (where $\mathbf{B}_1$ has the same number of rows as $\mathbf{A}_1$ has columns), then

$$\mathbf{B}_1\mathbf{A}_1 = \mathbf{I}, \quad \mathbf{B}_1\mathbf{A}_2 = \mathbf{0}, \quad \mathbf{B}_2\mathbf{A}_1 = \mathbf{0}, \quad \mathbf{B}_2\mathbf{A}_2 = \mathbf{I}, \tag{E.1}$$

$$\mathbf{A}_1\mathbf{B}_1 = \mathbf{I} - \mathbf{A}_2\mathbf{B}_2, \qquad \mathbf{A}_2\mathbf{B}_2 = \mathbf{I} - \mathbf{A}_1\mathbf{B}_1. \tag{E.2}$$

   (b) Show that if $\mathbf{A}$ is orthogonal, then

$$\mathbf{A}_1'\mathbf{A}_1 = \mathbf{I}, \quad \mathbf{A}_1'\mathbf{A}_2 = \mathbf{0}, \quad \mathbf{A}_2'\mathbf{A}_1 = \mathbf{0}, \quad \mathbf{A}_2'\mathbf{A}_2 = \mathbf{I}, \tag{E.3}$$

$$\mathbf{A}_1\mathbf{A}_1' = \mathbf{I} - \mathbf{A}_2\mathbf{A}_2', \qquad \mathbf{A}_2\mathbf{A}_2' = \mathbf{I} - \mathbf{A}_1\mathbf{A}_1'. \tag{E.4}$$

5. Let $\mathbf{A}$ represent an $m \times n$ nonnull matrix of rank r. Show that there exists an $m \times m$ orthogonal matrix whose first r columns span $\mathcal{C}(\mathbf{A})$.

Section 8.5

6. Let $\mathbf{T}$ represent an $n \times n$ triangular matrix. Show that $\text{rank}(\mathbf{T})$ is greater than or equal to the number of nonzero diagonal elements in $\mathbf{T}$.

7. Let

$$\mathbf{A} = \begin{pmatrix} \mathbf{A}_{11} & \mathbf{A}_{12} & \dots & \mathbf{A}_{1r} \\ \mathbf{0} & \mathbf{A}_{22} & \dots & \mathbf{A}_{2r} \\ \vdots & & \ddots & \vdots \\ \mathbf{0} & \mathbf{0} & & \mathbf{A}_{rr} \end{pmatrix}, \qquad \mathbf{B} = \begin{pmatrix} \mathbf{B}_{11} & \mathbf{0} & \dots & \mathbf{0} \\ \mathbf{B}_{21} & \mathbf{B}_{22} & & \mathbf{0} \\ \vdots & \vdots & \ddots & \\ \mathbf{B}_{r1} & \mathbf{B}_{r2} & \dots & \mathbf{B}_{rr} \end{pmatrix}$$

   represent, respectively, an $n \times n$ upper block-triangular matrix whose ijth block $\mathbf{A}_{ij}$ is of dimensions $n_i \times n_j$ ($j \geq i = 1, \dots, r$) and an $n \times n$ lower block-triangular matrix whose ijth block $\mathbf{B}_{ij}$ is of dimensions $n_i \times n_j$ ($j \leq i = 1, \dots, r$).

   (a) Assuming that $\mathbf{A}$ and $\mathbf{B}$ are invertible, show that

$$\mathbf{A}^{-1} = \begin{pmatrix} \mathbf{F}_{11} & \mathbf{F}_{12} & \dots & \mathbf{F}_{1r} \\ \mathbf{0} & \mathbf{F}_{22} & \dots & \mathbf{F}_{2r} \\ \vdots & & \ddots & \vdots \\ \mathbf{0} & \mathbf{0} & & \mathbf{F}_{rr} \end{pmatrix}, \qquad \mathbf{B}^{-1} = \begin{pmatrix} \mathbf{G}_{11} & \mathbf{0} & \dots & \mathbf{0} \\ \mathbf{G}_{21} & \mathbf{G}_{22} & & \mathbf{0} \\ \vdots & \vdots & \ddots & \\ \mathbf{G}_{r1} & \mathbf{G}_{r2} & \dots & \mathbf{G}_{rr} \end{pmatrix},$$

   where

$$\mathbf{F}_{jj} = \mathbf{A}_{jj}^{-1}, \quad \mathbf{F}_{ij} = -\Big(\sum_{k=i}^{j-1} \mathbf{F}_{ik}\mathbf{A}_{kj}\Big)\mathbf{A}_{jj}^{-1} \quad (i < j = 1, \ldots r), \quad \text{(E.5)}$$

$$\mathbf{G}_{jj} = \mathbf{B}_{jj}^{-1}, \quad \mathbf{G}_{ij} = -\Big(\sum_{k=j+1}^{i} \mathbf{G}_{ik}\mathbf{B}_{kj}\Big)\mathbf{B}_{jj}^{-1} \quad (i > j = 1, \ldots r), \quad \text{(E.6)}$$

Do so by applying the results of Section 8.5d to $\mathbf{A}'$ and $\mathbf{B}'$ [as opposed to mimicking the derivations of formulas (5.10) and (5.13)].

(b) Describe how the formulas in Part (a) can be used to devise r-step algorithms for computing $\mathbf{A}^{-1}$ and $\mathbf{B}^{-1}$, and indicate how these algorithms differ from those described in Section 8.5d.

9

Generalized Inverses

In statistics (and in many other disciplines), it is common to encounter linear systems whose coefficient matrices are square but not nonsingular (i.e., they are singular) and hence are not invertible (i.e., they lack inverses). For a linear system that has a nonsingular coefficient matrix, there is an intimate relationship—refer to Section 8.2a—between the solution of a linear system and the inverse of the coefficient matrix. What about a linear system whose coefficient matrix is not nonsingular? Is there a matrix comparable to the inverse (i.e., that relates to the solution of the linear system in a similar way)? The answer is yes! In fact, there are an infinite number of such matrices. These matrices, which are called generalized inverses, are the subject of this chapter. Over time, the use of generalized inverses in statistical discourse (especially that related to linear statistical models and multivariate analysis) has become increasingly routine.

There is a particular generalized inverse, known as the Moore-Penrose inverse, that is sometimes singled out for special attention. Discussion of the Moore-Penrose inverse is deferred to Chapter 20. For many purposes, one generalized inverse is as good as another.

9.1 Definition, Existence, and a Connection to the Solution of Linear Systems

A *generalized inverse* of an $m \times n$ matrix $\mathbf{A}$ is any $n \times m$ matrix $\mathbf{G}$ such that

$$\mathbf{AGA} = \mathbf{A}.$$

(The use of the term generalized inverse to refer to any such matrix is widespread, but not universal—among the alternative names found in the literature are pseudoinverse and conditional inverse.) For example, each of the two 3×2 matrices

$$\begin{pmatrix} 1 & 0 \\ 0 & 0 \\ 0 & 0 \end{pmatrix} \quad \text{and} \quad \begin{pmatrix} -42 & -1 \\ 5 & 3 \\ 2 & 2 \end{pmatrix}$$

is a generalized inverse of the 2×3 matrix

$$\begin{pmatrix} 1 & 3 & 2 \\ 2 & 6 & 4 \end{pmatrix}.$$

For a nonsingular matrix $\mathbf{A}$, it is clear that $\mathbf{A}\mathbf{A}^{-1}\mathbf{A} = \mathbf{A}$ and that if $\mathbf{G}$ is a generalized inverse of $\mathbf{A}$, then $\mathbf{G} = \mathbf{A}^{-1}\mathbf{A}\mathbf{G}\mathbf{A}\mathbf{A}^{-1} = \mathbf{A}^{-1}\mathbf{A}\mathbf{A}^{-1} = \mathbf{A}^{-1}$. Thus, we have the following lemma.

Lemma 9.1.1. The inverse $\mathbf{A}^{-1}$ of a nonsingular matrix $\mathbf{A}$ is a generalized inverse of $\mathbf{A}$, and a nonsingular matrix has no generalized inverse other than its inverse.

There is an intimate relationship between any generalized inverse of an $m \times n$ matrix $\mathbf{A}$ and the solution of linear systems whose coefficient matrix is $\mathbf{A}$. This relationship is described in the following theorem, which generalizes Theorem 8.2.1.

Theorem 9.1.2. Let $\mathbf{A}$ represent any $m \times n$ matrix, $\mathbf{G}$ any $n \times m$ matrix, and p any positive integer. Then, $\mathbf{GB}$ is a solution to a linear system $\mathbf{AX} = \mathbf{B}$ (in $\mathbf{X}$) for every $m \times p$ matrix $\mathbf{B}$ for which the linear system is consistent if and only if $\mathbf{G}$ is a generalized inverse of $\mathbf{A}$.

Proof. Suppose that $\mathbf{G}$ is a generalized inverse of $\mathbf{A}$. Let $\mathbf{B}$ represent any $m \times p$ matrix for which $\mathbf{AX} = \mathbf{B}$ is consistent, and take $\mathbf{X}^*$ to be any solution to $\mathbf{AX} = \mathbf{B}$. Then,

$$\mathbf{A}(\mathbf{GB}) = (\mathbf{AG})\mathbf{B} = \mathbf{AGAX}^* = \mathbf{AX}^* = \mathbf{B}.$$

Conversely, suppose that $\mathbf{GB}$ is a solution to $\mathbf{AX} = \mathbf{B}$ (i.e., that $\mathbf{AGB} = \mathbf{B}$) for every $m \times p$ matrix $\mathbf{B}$ for which $\mathbf{AX} = \mathbf{B}$ is consistent. Letting $\mathbf{a}_i$ represent the ith column of $\mathbf{A}$, observe that $\mathbf{AX} = \mathbf{B}$ is consistent in particular for

$$\mathbf{B} = (\mathbf{a}_i, \mathbf{0}, \ldots, \mathbf{0})$$

—for this $\mathbf{B}$, one solution to $\mathbf{AX} = \mathbf{B}$ is the matrix $(\mathbf{u}_i, \mathbf{0}, \ldots, \mathbf{0})$, where $\mathbf{u}_i$ is the ith column of $\mathbf{I}_n$ $(i = 1, \ldots, n)$. It follows that

$$\mathbf{AG}(\mathbf{a}_i, \mathbf{0}, \ldots, \mathbf{0}) = (\mathbf{a}_i, \mathbf{0}, \ldots, \mathbf{0})$$

and hence that $\mathbf{AGa}_i = \mathbf{a}_i$ $(i = 1, \ldots, n)$, implying that $\mathbf{AGA} = \mathbf{A}$. Q.E.D.

Let us now consider the existence of generalized inverses. Does every matrix have at least one generalized inverse? The answer to this question is yes, as can be shown by making use of the following theorem, which is of interest in its own right.

Theorem 9.1.3. Let $\mathbf{B}$ represent an $m \times r$ matrix of full column rank and $\mathbf{T}$ an $r \times n$ matrix of full row rank. Then, $\mathbf{B}$ has a left inverse, say $\mathbf{L}$, and $\mathbf{T}$ has a right inverse, say $\mathbf{R}$; and $\mathbf{RL}$ is a generalized inverse of $\mathbf{BT}$.

Proof. That $\mathbf{B}$ has a left inverse $\mathbf{L}$ and $\mathbf{T}$ a right inverse $\mathbf{R}$ is an immediate consequence of Lemma 8.1.1. Moreover,

$$\mathbf{BT}(\mathbf{RL})\mathbf{BT} = \mathbf{B}(\mathbf{TR})(\mathbf{LB})\mathbf{T} = \mathbf{BIIT} = \mathbf{BT},$$

that is, $\mathbf{RL}$ is a generalized inverse of $\mathbf{BT}$. Q.E.D.

Now, consider an arbitrary $m \times n$ matrix $\mathbf{A}$. If $\mathbf{A} = \mathbf{0}$, then clearly any $n \times m$ matrix is a generalized inverse of $\mathbf{A}$. If $\mathbf{A} \neq \mathbf{0}$, then (according to Theorem 4.4.8) there exist a matrix $\mathbf{B}$ of full column rank and a matrix $\mathbf{T}$ of full row rank such that $\mathbf{A} = \mathbf{BT}$, and hence, according to Theorem 9.1.3, $\mathbf{A}$ has a generalized inverse. Thus, we arrive at the following conclusion.

Corollary 9.1.4. Every matrix has at least one generalized inverse.

The symbol $\mathbf{A}^-$ is used to denote an arbitrary generalized inverse of an $m \times n$ matrix $\mathbf{A}$. By definition,

$$\mathbf{AA}^-\mathbf{A} = \mathbf{A}.$$

9.2 Some Alternative Characterizations

a. A generalized inverse in terms of the inverse of a nonsingular submatrix

For an $m \times n$ matrix $\mathbf{A}$ of the general form

$$\mathbf{A} = \begin{pmatrix} \mathbf{A}_{11} & \mathbf{0} \\ \mathbf{0} & \mathbf{0} \end{pmatrix},$$

where $\mathbf{A}_{11}$ is an $r \times r$ nonsingular matrix, and an $n \times m$ matrix

$$\mathbf{G} = \begin{pmatrix} \mathbf{G}_{11} & \mathbf{G}_{12} \\ \mathbf{G}_{21} & \mathbf{G}_{22} \end{pmatrix}$$

(where $\mathbf{G}_{11}$ is of dimensions $r \times r$), we find that

$$\mathbf{AGA} = \begin{pmatrix} \mathbf{A}_{11}\mathbf{G}_{11}\mathbf{A}_{11} & \mathbf{0} \\ \mathbf{0} & \mathbf{0} \end{pmatrix},$$

implying that $\mathbf{G}$ is a generalized inverse of $\mathbf{A}$ if and only if

$$\mathbf{A}_{11}\mathbf{G}_{11}\mathbf{A}_{11} = \mathbf{A}_{11},$$

or equivalently (in light of Lemma 9.1.1) if and only if $\mathbf{G}_{11} = \mathbf{A}_{11}^{-1}$, and hence that $\mathbf{G}$ is a generalized inverse of $\mathbf{A}$ if and only if

$$\mathbf{G} = \begin{pmatrix} \mathbf{A}_{11}^{-1} & \mathbf{X} \\ \mathbf{Y} & \mathbf{Z} \end{pmatrix}$$

for some matrices $\mathbf{X}$, $\mathbf{Y}$, and $\mathbf{Z}$ (of appropriate dimensions). This result can be generalized as follows:

Theorem 9.2.1. Let $\mathbf{A}$ represent an $m \times n$ matrix of rank r, and partition $\mathbf{A}$ as

$$\mathbf{A} = \begin{pmatrix} \mathbf{A}_{11} & \mathbf{A}_{12} \\ \mathbf{A}_{21} & \mathbf{A}_{22} \end{pmatrix},$$

where $\mathbf{A}_{11}$ is of dimensions $r \times r$. Suppose that $\mathbf{A}_{11}$ is nonsingular. Then, an $n \times m$ matrix $\mathbf{G}$ is a generalized inverse of $\mathbf{A}$ if and only if

$$\mathbf{G} = \begin{pmatrix} \mathbf{A}_{11}^{-1} - \mathbf{X}\mathbf{A}_{21}\mathbf{A}_{11}^{-1} - \mathbf{A}_{11}^{-1}\mathbf{A}_{12}\mathbf{Y} - \mathbf{A}_{11}^{-1}\mathbf{A}_{12}\mathbf{Z}\mathbf{A}_{21}\mathbf{A}_{11}^{-1} & \mathbf{X} \\ \mathbf{Y} & \mathbf{Z} \end{pmatrix} \tag{2.1}$$

for some matrices $\mathbf{X}$, $\mathbf{Y}$, and $\mathbf{Z}$ (of appropriate dimensions).

To prove Theorem 9.2.1, we require the following lemma, which is of interest in its own right.

Lemma 9.2.2. Let

$$\mathbf{A} = \begin{pmatrix} \mathbf{A}_{11} & \mathbf{A}_{12} \\ \mathbf{A}_{21} & \mathbf{A}_{22} \end{pmatrix}$$

represent an $m \times n$ partitioned matrix of rank r, where $\mathbf{A}_{11}$ is of dimensions $r \times r$. If $\operatorname{rank}(\mathbf{A}_{11}) = r$, then

$$\mathbf{A}_{22} = \mathbf{A}_{21}\mathbf{A}_{11}^{-1}\mathbf{A}_{12}.$$

Proof (of Lemma 9.2.2). Suppose that $\operatorname{rank}(\mathbf{A}_{11}) = r$. Then, applying Theorem 8.5.10 (with $\mathbf{T} = \mathbf{A}_{11}$, $\mathbf{U} = \mathbf{A}_{12}$, $\mathbf{V} = \mathbf{A}_{21}$, and $\mathbf{W} = \mathbf{A}_{22}$), we find that

$$r = \operatorname{rank}(\mathbf{A}) = r + \operatorname{rank}(\mathbf{A}_{22} - \mathbf{A}_{21}\mathbf{A}_{11}^{-1}\mathbf{A}_{12}),$$

implying that

$$\operatorname{rank}(\mathbf{A}_{22} - \mathbf{A}_{21}\mathbf{A}_{11}^{-1}\mathbf{A}_{12}) = 0$$

and hence that

$$\mathbf{A}_{22} - \mathbf{A}_{21}\mathbf{A}_{11}^{-1}\mathbf{A}_{12} = \mathbf{0}$$

or equivalently that

$$\mathbf{A}_{22} = \mathbf{A}_{21}\mathbf{A}_{11}^{-1}\mathbf{A}_{12}.$$

Q.E.D.

Proof (of Theorem 9.2.1). It follows from Lemma 9.2.2 that

$$\mathbf{A} = \begin{pmatrix} \mathbf{A}_{11} \\ \mathbf{A}_{21} \end{pmatrix} \mathbf{A}_{11}^{-1}(\mathbf{A}_{11}, \mathbf{A}_{12}).$$

By definition, $\mathbf{G}$ is a generalized inverse of $\mathbf{A}$ if and only if $\mathbf{AGA} = \mathbf{A}$, or, equivalently, if and only if

$$\begin{pmatrix} \mathbf{A}_{11} \\ \mathbf{A}_{21} \end{pmatrix} \mathbf{A}_{11}^{-1}(\mathbf{A}_{11}, \mathbf{A}_{12})\mathbf{G} \begin{pmatrix} \mathbf{A}_{11} \\ \mathbf{A}_{21} \end{pmatrix} \mathbf{A}_{11}^{-1}(\mathbf{A}_{11}, \mathbf{A}_{12}) = \begin{pmatrix} \mathbf{A}_{11} \\ \mathbf{A}_{21} \end{pmatrix} \mathbf{A}_{11}^{-1}(\mathbf{A}_{11}, \mathbf{A}_{12}). \tag{2.2}$$

Partition $\mathbf{G}$ as

$$\mathbf{G} = \begin{pmatrix} \mathbf{G}_{11} & \mathbf{G}_{12} \\ \mathbf{G}_{21} & \mathbf{G}_{22} \end{pmatrix}$$

(where $\mathbf{G}_{11}$ is of dimensions $r \times r$). It is clear that

$$\text{rank}\begin{pmatrix} \mathbf{A}_{11} \\ \mathbf{A}_{21} \end{pmatrix} = r \qquad \text{and} \qquad \text{rank}(\mathbf{A}_{11}, \mathbf{A}_{12}) = r$$

and hence (in light of Lemma 8.3.1) that $\mathbf{G}$ satisfies condition (2.2) if and only if

$$\mathbf{A}_{11}^{-1}(\mathbf{A}_{11}, \mathbf{A}_{12})\mathbf{G}\begin{pmatrix} \mathbf{A}_{11} \\ \mathbf{A}_{21} \end{pmatrix}\mathbf{A}_{11}^{-1} = \mathbf{A}_{11}^{-1},$$

or, equivalently [since

$$\mathbf{A}_{11}^{-1}(\mathbf{A}_{11}, \mathbf{A}_{12})\mathbf{G}\begin{pmatrix} \mathbf{A}_{11} \\ \mathbf{A}_{21} \end{pmatrix}\mathbf{A}_{11}^{-1}$$
$$= \mathbf{G}_{11} + \mathbf{A}_{11}^{-1}\mathbf{A}_{12}\mathbf{G}_{21} + \mathbf{G}_{12}\mathbf{A}_{21}\mathbf{A}_{11}^{-1} + \mathbf{A}_{11}^{-1}\mathbf{A}_{12}\mathbf{G}_{22}\mathbf{A}_{21}\mathbf{A}_{11}^{-1}],$$

if and only if

$$\mathbf{G}_{11} = \mathbf{A}_{11}^{-1} - \mathbf{A}_{11}^{-1}\mathbf{A}_{12}\mathbf{G}_{21} - \mathbf{G}_{12}\mathbf{A}_{21}\mathbf{A}_{11}^{-1} - \mathbf{A}_{11}^{-1}\mathbf{A}_{12}\mathbf{G}_{22}\mathbf{A}_{21}\mathbf{A}_{11}^{-1}.$$

Q.E.D.

Any $m \times n$ matrix $\mathbf{A}$ of rank r can be partitioned as

$$\mathbf{A} = \begin{pmatrix} \mathbf{A}_{11} & \mathbf{A}_{12} \\ \mathbf{A}_{21} & \mathbf{A}_{22} \end{pmatrix},$$

where $\mathbf{A}_{11}$ is of dimensions $r \times r$. [If $m = r$ or $n = r$, then $\mathbf{A}_{22}$ and $\mathbf{A}_{21}$ or $\mathbf{A}_{12}$, respectively, are regarded as degenerate, that is

$$\mathbf{A} = (\mathbf{A}_{11}, \mathbf{A}_{12}) \qquad \text{or} \qquad \mathbf{A} = \begin{pmatrix} \mathbf{A}_{11} \\ \mathbf{A}_{21} \end{pmatrix},$$

respectively.] Theorem 9.2.1 describes (for the special case where $\mathbf{A}_{11}$ is nonsingular) the general form of a generalized inverse of $\mathbf{A}$ (in terms of the submatrices $\mathbf{A}_{11}$, $\mathbf{A}_{12}$, and $\mathbf{A}_{21}$). [If $m = r$ or $n = r$, then, in Theorem 9.2.1, $\mathbf{Z}$ and $\mathbf{X}$ or $\mathbf{Y}$, respectively, are regarded as degenerate, and—assuming $\text{rank}(\mathbf{A}_{11}) = r$—we have respectively that $\mathbf{G}$ is a generalized inverse of $\mathbf{A} = (\mathbf{A}_{11}, \mathbf{A}_{12})$ if and only if

$$\mathbf{G} = \begin{pmatrix} \mathbf{A}_{11}^{-1} - \mathbf{A}_{11}^{-1}\mathbf{A}_{12}\mathbf{Y} \\ \mathbf{Y} \end{pmatrix}$$

for some $\mathbf{Y}$, or that $\mathbf{G}$ is a genealized inverse of

$$\mathbf{A} = \begin{pmatrix} \mathbf{A}_{11} \\ \mathbf{A}_{21} \end{pmatrix}$$

if and only if

$$\mathbf{G} = (\mathbf{A}_{11}^{-1} - \mathbf{X}\mathbf{A}_{21}\mathbf{A}_{11}^{-1},\ \mathbf{X})$$

for some $\mathbf{X}$.] Note that Theorem 9.2.1 implies, in particular, that, if $\mathbf{A}_{11}$ is nonsingular, then one generalized inverse of $\mathbf{A}$ is the $n \times m$ matrix

$$\mathbf{G} = \begin{pmatrix} \mathbf{A}_{11}^{-1} & \mathbf{0} \\ \mathbf{0} & \mathbf{0} \end{pmatrix}.$$

The $r \times r$ submatrix $\mathbf{A}_{11}$ may be singular. For example, if

$$\mathbf{A} = \begin{pmatrix} 1 & 0 & 5 \\ 0 & 0 & 0 \\ 5 & 0 & 2 \end{pmatrix},$$

then $r = 2$ and

$$\mathbf{A}_{11} = \begin{pmatrix} 1 & 0 \\ 0 & 0 \end{pmatrix}.$$

If $\mathbf{A}_{11}$ is singular or equivalently (in light of Theorem 4.4.10) if the first r rows or first r columns of $\mathbf{A}$ are linearly dependent, then the expression given by Theorem 9.2.1 for the general form of a generalized inverse is not applicable to $\mathbf{A}$. The following theorem, which generalizes Theorem 9.2.1, gives an expression for the general form of a generalized inverse of an $m \times n$ matrix $\mathbf{A}$ that is applicable even if the first r rows or columns of $\mathbf{A}$ are linearly dependent.

Theorem 9.2.3. Let $\mathbf{A}$ represent an $m \times n$ matrix of rank r. Let $i_1, \ldots, i_r$ represent integers, chosen from the first m positive integers $1, \ldots, m$ in such a way that the $i_1, \ldots, i_r$th rows of $\mathbf{A}$ are linearly independent, and let $j_1, \ldots, j_r$ represent integers, chosen from the first n positive integers $1, \ldots, n$ in such a way that the $j_1, \ldots, j_r$th columns of $\mathbf{A}$ are linearly independent. Take $\mathbf{P}$ to be any $m \times m$ permutation matrix that has as its first r rows the $i_1, \ldots, i_r$th rows of the $m \times m$ identity matrix, and $\mathbf{Q}$ to be any $n \times n$ permutation matrix that has as its first r columns the $j_1, \ldots, j_r$th columns of the $n \times n$ identity matrix. Let $\mathbf{B} = \mathbf{PAQ}$, and partition $\mathbf{B}$ as

$$\mathbf{B} = \begin{pmatrix} \mathbf{B}_{11} & \mathbf{B}_{12} \\ \mathbf{B}_{21} & \mathbf{B}_{22} \end{pmatrix},$$

where $\mathbf{B}_{11}$ is of dimensions $r \times r$. Then, an $n \times m$ matrix $\mathbf{G}$ is a generalized inverse of $\mathbf{A}$ if and only if

$$\mathbf{G} = \mathbf{Q} \begin{pmatrix} \mathbf{B}_{11}^{-1} - \mathbf{X}\mathbf{B}_{21}\mathbf{B}_{11}^{-1} - \mathbf{B}_{11}^{-1}\mathbf{B}_{12}\mathbf{Y} - \mathbf{B}_{11}^{-1}\mathbf{B}_{12}\mathbf{Z}\mathbf{B}_{21}\mathbf{B}_{11}^{-1} & \mathbf{X} \\ \mathbf{Y} & \mathbf{Z} \end{pmatrix} \mathbf{P} \tag{2.3}$$

for some matrices $\mathbf{X}$, $\mathbf{Y}$, and $\mathbf{Z}$ (of appropriate dimensions).

With regard to Theorem 9.2.3, it follows from Theorem 4.4.10 that $\mathbf{A}$ contains r linearly independent rows and r linearly independent columns. Furthermore, if $i_1 < \cdots < i_r$ and $j_1 < \cdots < j_r$, then $\mathbf{B}_{11}$ is a submatrix of $\mathbf{A}$, namely, the

$r \times r$ submatrix obtained by striking out all of the rows and columns of $\mathbf{A}$ except the $i_1, \ldots, i_r$th rows and $j_1, \ldots, j_r$th columns—more generally, $\mathbf{B}_{11}$ is an $r \times r$ matrix obtained from that submatrix by permuting its rows and columns. That $\mathbf{B}_{11}$ is invertible is a consequence of Theorem 4.4.10. If the first r rows and first r columns of $\mathbf{A}$ are linearly independent, then $i_1, \ldots, i_r$ and $j_1, \ldots, j_r$ can be chosen to be the first r positive integers $1, \ldots, r$, and $\mathbf{P}$ and $\mathbf{Q}$ to be identity matrices, in which case Theorem 9.2.3 reduces to Theorem 9.2.1.

It is convenient, in proving Theorem 9.2.3, to make use of the following lemma, which generalizes result (8.2.8) and is of interest in its own right.

Lemma 9.2.4. Let $\mathbf{B}$ represent an $m \times n$ matrix and $\mathbf{G}$ an $n \times m$ matrix. Then, for any $m \times m$ nonsingular matrix $\mathbf{A}$ and $n \times n$ nonsingular matrix $\mathbf{C}$, (1) $\mathbf{G}$ is a generalized inverse of $\mathbf{AB}$ if and only if $\mathbf{G} = \mathbf{HA}^{-1}$ for some generalized inverse $\mathbf{H}$ of $\mathbf{B}$, (2) $\mathbf{G}$ is a generalized inverse of $\mathbf{BC}$ if and only if $\mathbf{G} = \mathbf{C}^{-1}\mathbf{H}$ for some generalized inverse $\mathbf{H}$ of $\mathbf{B}$, and (3) $\mathbf{G}$ is a generalized inverse of $\mathbf{ABC}$ if and only if $\mathbf{G} = \mathbf{C}^{-1}\mathbf{HA}^{-1}$ for some generalized inverse $\mathbf{H}$ of $\mathbf{B}$.

Proof (of Lemma 9.2.4). Parts (1) and (2) are special cases of Part (3) (those where $\mathbf{C} = \mathbf{I}$ and $\mathbf{A} = \mathbf{I}$, respectively). Thus, it suffices to prove Part (3).

By definition, $\mathbf{G}$ is a generalized inverse of $\mathbf{ABC}$ if and only if

$$\mathbf{ABCGABC} = \mathbf{ABC}$$

or, equivalently (in light of Lemma 8.3.1), if and only if

$$\mathbf{BCGAB} = \mathbf{B}.$$

Thus, $\mathbf{G}$ is a generalized inverse of $\mathbf{ABC}$ if and only if $\mathbf{CGA} = \mathbf{H}$ for some generalized inverse $\mathbf{H}$ of $\mathbf{B}$ or, equivalently, if and only if $\mathbf{G} = \mathbf{C}^{-1}\mathbf{HA}^{-1}$ for some generalized inverse $\mathbf{H}$ of $\mathbf{B}$. Q.E.D.

Proof (of Theorem 9.2.3). Since $\mathbf{P}$ and $\mathbf{Q}$ are permutation matrices, they are orthogonal. Thus,

$$\mathbf{A} = \mathbf{P}'\mathbf{PAQQ}' = \mathbf{P}'\mathbf{BQ}',$$

and it follows from Lemma 9.2.4 that $\mathbf{G}$ is a generalized inverse of $\mathbf{A}$ if and only if $\mathbf{G} = \mathbf{QHP}$ for some generalized inverse $\mathbf{H}$ of $\mathbf{B}$. Moreover, it follows from Theorem 9.2.1 that an $n \times m$ matrix $\mathbf{H}$ is a generalized inverse of $\mathbf{B}$ if and only if

$$\mathbf{H} = \begin{pmatrix} \mathbf{B}_{11}^{-1} - \mathbf{XB}_{21}\mathbf{B}_{11}^{-1} - \mathbf{B}_{11}^{-1}\mathbf{B}_{12}\mathbf{Y} - \mathbf{B}_{11}^{-1}\mathbf{B}_{12}\mathbf{ZB}_{21}\mathbf{B}_{11}^{-1} & \mathbf{X} \\ \mathbf{Y} & \mathbf{Z} \end{pmatrix}$$

for some matrices $\mathbf{X}$, $\mathbf{Y}$, and $\mathbf{Z}$. Q.E.D.

By setting the matrices $\mathbf{X}, \mathbf{Y}$, and $\mathbf{Z}$ in expression (2.3) equal to null matrices, we find that one generalized inverse of an $m \times n$ matrix $\mathbf{A}$ of rank r is

$$\mathbf{G} = \mathbf{Q}\begin{pmatrix} \mathbf{B}_{11}^{-1} & \mathbf{0} \\ \mathbf{0} & \mathbf{0} \end{pmatrix}\mathbf{P} \tag{2.4}$$

(where $\mathbf{B}_{11}$, $\mathbf{P}$, and $\mathbf{Q}$—and $i_1, \ldots, i_r$ and $j_1, \ldots, j_r$—are as defined in Theorem 9.2.3). Note that this generalized inverse can be formed by carrying out the following 5-step procedure:

(1) Find the rank r of $\mathbf{A}$ (by, e.g., determining the size of the largest linearly independent set that can be formed from the rows or columns of $\mathbf{A}$).

(2) Determine values for $i_1, \ldots, i_r$ and $j_1, \ldots, j_r$ (such that $i_1 < \cdots < i_r$ and $j_1 < \cdots < j_r$); that is, locate r linearly independent rows of $\mathbf{A}$ and r linearly independent columns.

(3) Form a submatrix of $\mathbf{A}$ by striking out all of the rows and columns of $\mathbf{A}$ except the $i_1, \ldots, i_r$th rows and $j_1, \ldots, j_r$th columns—this matrix is the matrix $\mathbf{B}_{11}$.

(4) Form the inverse $\mathbf{B}_{11}^{-1}$ of $\mathbf{B}_{11}$.

(5) Write out the generalized inverse $\mathbf{G}$ of $\mathbf{A}$, taking the $j_s i_t$th element of $\mathbf{G}$ to be the stth element of $\mathbf{B}_{11}^{-1}$ (for $s = 1, \ldots, r$ and $t = 1, \ldots, r$) and taking the other elements of $\mathbf{G}$ to be zero.

Suppose, for example, that

$$\mathbf{A} = \begin{pmatrix} -6 & 2 & -2 & -3 \\ 3 & -1 & 5 & 2 \\ -3 & 1 & 3 & -1 \end{pmatrix}.$$

The first and second rows of $\mathbf{A}$ are linearly independent (as is easily verified), and the third row is the sum of the first two, so that $r = 2$. Let us choose $i_1 = 1$, $i_2 = 3$, $j_1 = 2$, and $j_2 = 4$—clearly, the first and third rows and the second and fourth columns of $\mathbf{A}$ are linearly independent. Then,

$$\mathbf{B}_{11} = \begin{pmatrix} 2 & -3 \\ 1 & -1 \end{pmatrix}.$$

Applying formula (8.1.2) for the inverse of a 2×2 nonsingular matrix, we find that

$$\mathbf{B}_{11}^{-1} = \begin{pmatrix} -1 & 3 \\ -1 & 2 \end{pmatrix}.$$

Thus, one generalized inverse of $\mathbf{A}$ is

$$\mathbf{G} = \begin{pmatrix} 0 & 0 & 0 \\ -1 & 0 & 3 \\ 0 & 0 & 0 \\ -1 & 0 & 2 \end{pmatrix}.$$

How many generalized inverses does an $m \times n$ matrix $\mathbf{A}$ of rank r possess? If $\mathbf{A}$ is nonsingular, then (according to Theorem 8.1.4) it has a unique generalized inverse, namely $\mathbf{A}^{-1}$. Suppose that $\mathbf{A}$ is not nonsingular (i.e., that $\mathbf{A}$ is either not square or is square but singular). Then, in expression (2.3) for the general form of a generalized inverse, at least one of the three matrices $\mathbf{X}$, $\mathbf{Y}$, and $\mathbf{Z}$ is nondegenerate, and it follows from Lemma 8.3.1 that distinct choices for $\mathbf{X}$, $\mathbf{Y}$, and $\mathbf{Z}$ produce distinct values of $\mathbf{G}$, that is, distinct generalized inverses. Since there are an infinite number of choices for these matrices, we conclude that $\mathbf{A}$ has an infinite number of generalized inverses.

b. Generalized inverses of symmetric matrices

The expression given by Theorem 9.2.3 for the general form of a generalized inverse of an $m \times n$ matrix $\mathbf{A}$ of rank r is, of course, applicable to a symmetric matrix. If the $i_1, \ldots, i_r$th rows of a symmetric matrix $\mathbf{A}$ are linearly independent, then clearly the corresponding ($i_1, \ldots, i_r$th) columns of $\mathbf{A}$ are likewise linearly independent. Thus, for symmetric matrices, we have the following special case of Theorem 9.2.3.

Theorem 9.2.5. Let $\mathbf{A}$ represent an $n \times n$ symmetric matrix of rank r. Let $i_1, \ldots, i_r$ represent integers chosen from the first n positive integers $1, \ldots, n$ in such a way that the $i_1, \ldots, i_r$th rows of $\mathbf{A}$ are linearly independent, and take $\mathbf{P}$ to be any $n \times n$ permutation matrix that has as its first r rows the $i_1, \ldots, i_r$th rows of the $n \times n$ identity matrix. Let $\mathbf{B} = \mathbf{PAP}'$, and partition $\mathbf{B}$ as

$$\mathbf{B} = \begin{pmatrix} \mathbf{B}_{11} & \mathbf{B}_{21} \\ \mathbf{B}_{21} & \mathbf{B}_{22} \end{pmatrix},$$

where $\mathbf{B}_{11}$ is of dimensions $r \times r$. Then, an $n \times n$ matrix $\mathbf{G}$ is a generalized inverse of $\mathbf{A}$ if and only if

$$\mathbf{G} = \mathbf{P}' \begin{pmatrix} \mathbf{B}_{11}^{-1} - \mathbf{X}\mathbf{B}_{21}\mathbf{B}_{11}^{-1} - \mathbf{B}_{11}^{-1}\mathbf{B}_{12}\mathbf{Y} - \mathbf{B}_{11}^{-1}\mathbf{B}_{12}\mathbf{Z}\mathbf{B}_{21}\mathbf{B}_{11}^{-1} & \mathbf{X} \\ \mathbf{Y} & \mathbf{Z} \end{pmatrix} \mathbf{P} \tag{2.5}$$

for some matrices $\mathbf{X}$, $\mathbf{Y}$, and $\mathbf{Z}$ (of appropriate dimensions).

As indicated in Section 8.2b, the inverse of any nonsingular symmetric matrix is symmetric. Is a generalized inverse of a singular symmetric matrix necessarily symmetric? To answer this question, observe that the generalized inverses of an $n \times n$ symmetric matrix $\mathbf{A}$ of rank r consist of all matrices of the form $\mathbf{G} = \mathbf{P}'\mathbf{C}\mathbf{P}$, where

$$\mathbf{C} = \begin{pmatrix} \mathbf{B}_{11}^{-1} - \mathbf{X}\mathbf{B}_{21}\mathbf{B}_{11}^{-1} - \mathbf{B}_{11}^{-1}\mathbf{B}_{12}\mathbf{Y} - \mathbf{B}_{11}^{-1}\mathbf{B}_{12}\mathbf{Z}\mathbf{B}_{21}\mathbf{B}_{11}^{-1} & \mathbf{X} \\ \mathbf{Y} & \mathbf{Z} \end{pmatrix}$$

and the matrices $\mathbf{X}$, $\mathbf{Y}$, and $\mathbf{Z}$ are arbitrary. (Here, $\mathbf{P}$ and $\mathbf{B}$—and the submatrices $\mathbf{B}_{11}, \mathbf{B}_{12}, \mathbf{B}_{21}$, and $\mathbf{B}_{22}$ of $\mathbf{B}$—are as defined in Theorem 9.2.5.) Since $\mathbf{G}' = \mathbf{P}'\mathbf{C}'\mathbf{P}$, it is clear from Lemma 8.3.1 that $\mathbf{G}' = \mathbf{G}$ (i.e., that $\mathbf{G}$ is symmetric) if and only if $\mathbf{C}' = \mathbf{C}$.

Observing that $\mathbf{B}' = \mathbf{PA}'\mathbf{P}' = \mathbf{PAP}' = \mathbf{B}$, that is, $\mathbf{B}$ is symmetric (so that $\mathbf{B}_{11}$ is symmetric—and consequently $\mathbf{B}_{11}^{-1}$ is symmetric—and $\mathbf{B}_{21}' = \mathbf{B}_{12}$), we find that

$$\mathbf{C}' = \begin{pmatrix} \mathbf{B}_{11}^{-1} - \mathbf{B}_{11}^{-1}\mathbf{B}_{12}\mathbf{X}' - \mathbf{Y}'\mathbf{B}_{21}\mathbf{B}_{11}^{-1} - \mathbf{B}_{11}^{-1}\mathbf{B}_{12}\mathbf{Z}'\mathbf{B}_{21}\mathbf{B}_{11}^{-1} & \mathbf{Y}' \\ \mathbf{X}' & \mathbf{Z}' \end{pmatrix}.$$

It follows that $\mathbf{C}' = \mathbf{C}$ if and only if $\mathbf{Y}' = \mathbf{X}$ and $\mathbf{Z}' = \mathbf{Z}$. We conclude that, if $n > r > 0$ (in which case $\mathbf{A}$ is singular and the matrices $\mathbf{X}$, $\mathbf{Y}$, and $\mathbf{Z}$ are nondegenerate), then $\mathbf{A}$ has both symmetric and nonsymmetric generalized inverses. For example, if we take $\mathbf{X}$, $\mathbf{Y}$, and $\mathbf{Z}$ all to be null, then $\mathbf{G}$ is symmetric; however if we take $\mathbf{X}$ to be null but $\mathbf{Y}$ to be nonnull, then $\mathbf{G}$ is nonsymmetric. In conclusion, we have the following lemma.

Lemma 9.2.6. Every singular symmetric matrix (of order two or more) has both symmetric and nonsymmetric generalized inverses.

c. Every generalized inverse in terms of any particular generalized inverse

The general form of a generalized inverse of an $m \times n$ matrix $\mathbf{A}$ can be expressed in terms of any particular generalized inverse of $\mathbf{A}$, as described in the following theorem.

Theorem 9.2.7. Let $\mathbf{A}$ represent an $m \times n$ matrix, and $\mathbf{G}$ any particular generalized inverse of $\mathbf{A}$. Then, an $n \times m$ matrix $\mathbf{G}^*$ is a generalized inverse of $\mathbf{A}$ if and only if

$$\mathbf{G}^* = \mathbf{G} + \mathbf{Z} - \mathbf{GAZAG} \tag{2.6}$$

for some $n \times m$ matrix $\mathbf{Z}$. Also, $\mathbf{G}^*$ is a generalized inverse of $\mathbf{A}$ if and only if

$$\mathbf{G}^* = \mathbf{G} + (\mathbf{I} - \mathbf{GA})\mathbf{T} + \mathbf{S}(\mathbf{I} - \mathbf{AG}) \tag{2.7}$$

for some $n \times m$ matrices $\mathbf{T}$ and $\mathbf{S}$.

Proof. It is a simple exercise to verify that any matrix $\mathbf{G}^*$ that is expressible in the form (2.6) or the form (2.7) is a generalized inverse of $\mathbf{A}$. Conversely, if $\mathbf{G}^*$ is any generalized inverse of $\mathbf{A}$, then

$$\begin{aligned}\mathbf{G}^* &= \mathbf{G} + (\mathbf{G}^* - \mathbf{G}) - \mathbf{GA}(\mathbf{G}^* - \mathbf{G})\mathbf{AG} \\ &= \mathbf{G} + \mathbf{Z} - \mathbf{GAZAG},\end{aligned}$$

where $\mathbf{Z} = \mathbf{G}^* - \mathbf{G}$, and

$$\begin{aligned}\mathbf{G}^* &= \mathbf{G} + (\mathbf{I} - \mathbf{GA})\mathbf{G}^*\mathbf{AG} + (\mathbf{G}^* - \mathbf{G})(\mathbf{I} - \mathbf{AG}) \\ &= \mathbf{G} + (\mathbf{I} - \mathbf{GA})\mathbf{T} + \mathbf{S}(\mathbf{I} - \mathbf{AG}),\end{aligned}$$

where $\mathbf{T} = \mathbf{G}^*\mathbf{AG}$ and $\mathbf{S} = \mathbf{G}^* - \mathbf{G}$. Q.E.D.

All generalized inverses of the $m \times n$ matrix $\mathbf{A}$ can be generated from expression (2.6) by letting $\mathbf{Z}$ range over all $n \times m$ matrices. Alternatively, all generalized inverses of $\mathbf{A}$ can be generated from expression (2.7) by letting both $\mathbf{T}$ and $\mathbf{S}$ range over all $n \times m$ matrices. Note that distinct choices for $\mathbf{Z}$ or for $\mathbf{T}$ or $\mathbf{S}$ may not result in distinct generalized inverses.

d. Generalized inverses of matrices having full column or row rank

The following lemma extends the results of Lemma 8.1.3.

Lemma 9.2.8. Let $\mathbf{A}$ represent a matrix of full column rank and $\mathbf{B}$ a matrix of full row rank. Then, (1) a matrix $\mathbf{G}$ is a generalized inverse of $\mathbf{A}$ if and only if $\mathbf{G}$ is a left inverse of $\mathbf{A}$. And (2) a matrix $\mathbf{G}$ is a generalized inverse of $\mathbf{B}$ if and only if $\mathbf{G}$ is a right inverse of $\mathbf{B}$.

Proof. Let $\mathbf{L}$ represent any left inverse of $\mathbf{A}$— that $\mathbf{A}$ has a left inverse is guaranteed by Lemma 8.1.1. Then, $\mathbf{ALA} = \mathbf{AI} = \mathbf{A}$, so that $\mathbf{L}$ is a generalized inverse of $\mathbf{A}$. And the proof of Part (1) of Lemma 9.2.8 is complete upon observing that

$$\mathbf{A}^-\mathbf{A} = \mathbf{IA}^-\mathbf{A} = \mathbf{LAA}^-\mathbf{A} = \mathbf{LA} = \mathbf{I}$$

and hence that $\mathbf{A}^-$ is a left inverse of $\mathbf{A}$. The validity of Part (2) can be established via an analogous argument. Q.E.D.

If a matrix $\mathbf{A}$ has full column rank, then (according to Corollary 7.4.6) $\mathbf{A}'\mathbf{A}$ is nonsingular, and similarly if $\mathbf{A}$ has full row rank, then $\mathbf{AA}'$ is nonsingular. Further, if $\mathbf{A}$ has full column or row rank, then, by making use of the following (easily verifiable) lemma, a generalized inverse of $\mathbf{A}$ can (in light of Lemma 9.2.8) be obtained from the ordinary inverse of $\mathbf{A}'\mathbf{A}$ or $\mathbf{AA}'$.

Lemma 9.2.9. If a matrix $\mathbf{A}$ has full column rank, then the matrix $(\mathbf{A}'\mathbf{A})^{-1}\mathbf{A}'$ is a left inverse of $\mathbf{A}$. Similarly, if $\mathbf{A}$ has full row rank, then $\mathbf{A}'(\mathbf{AA}')^{-1}$ is a right inverse of $\mathbf{A}$.

9.3 Some Elementary Properties

Let us begin by considering the extension (to generalized inverses) of various elementary properties of inverse matrices. It is easy to verify the following lemma, which is a generalization of result (8.2.1).

Lemma 9.3.1. For any matrix $\mathbf{A}$ and any nonzero scalar k, $(1/k)\mathbf{A}^-$ is a generalized inverse of $k\mathbf{A}$.

Setting $k = -1$ in Lemma 9.3.1, we obtain the following corollary, which generalizes result (8.2.2).

Corollary 9.3.2. For any matrix $\mathbf{A}$, $-\mathbf{A}^-$ is a generalized inverse of $-\mathbf{A}$.

For any matrix $\mathbf{A}$, we find that

$$\mathbf{A}'(\mathbf{A}^-)'\mathbf{A}' = (\mathbf{AA}^-\mathbf{A})' = \mathbf{A}'.$$

Thus, we have the following lemma, which generalizes result (8.2.3).

Lemma 9.3.3. For any matrix $\mathbf{A}$, $(\mathbf{A}^-)'$ is a generalized inverse of $\mathbf{A}'$.

While (as discussed in Section 9.2b) the transpose $(\mathbf{A}^-)'$ of a generalized inverse $\mathbf{A}^-$ of a singular symmetric matrix $\mathbf{A}$ is not necessarily the same as $\mathbf{A}^-$, Lemma 9.3.3 implies that $(\mathbf{A}^-)'$ has the following weaker property.

Corollary 9.3.4. For any symmetric matrix $\mathbf{A}$, $(\mathbf{A}^-)'$ is a generalized inverse of $\mathbf{A}$.

For a nonsingular matrix $\mathbf{A}$, we have (by definition) that $\mathbf{A}\mathbf{A}^{-1} = \mathbf{A}^{-1}\mathbf{A} = \mathbf{I}$. The following two lemmas describe, for an arbitrary matrix $\mathbf{A}$, some properties of $\mathbf{A}\mathbf{A}^{-}$ and $\mathbf{A}^{-}\mathbf{A}$.

Lemma 9.3.5. Let $\mathbf{A}$ represent an $m \times n$ matrix. Then, for any $m \times p$ matrix $\mathbf{B}$, $\mathcal{C}(\mathbf{B}) \subset \mathcal{C}(\mathbf{A})$ if and only if $\mathbf{B} = \mathbf{A}\mathbf{A}^{-}\mathbf{B}$, or, equivalently, if and only if $(\mathbf{I} - \mathbf{A}\mathbf{A}^{-})\mathbf{B} = \mathbf{0}$. And, for any $q \times n$ matrix $\mathbf{C}$, $\mathcal{R}(\mathbf{C}) \subset \mathcal{R}(\mathbf{A})$ if and only if $\mathbf{C} = \mathbf{C}\mathbf{A}^{-}\mathbf{A}$, or, equivalently, if and only if $\mathbf{C}(\mathbf{I} - \mathbf{A}^{-}\mathbf{A}) = \mathbf{0}$.

Proof. If $\mathbf{B} = \mathbf{A}\mathbf{A}^{-}\mathbf{B}$, then it follows immediately from Lemma 4.2.2 that $\mathcal{C}(\mathbf{B}) \subset \mathcal{C}(\mathbf{A})$. Conversely, if $\mathcal{C}(\mathbf{B}) \subset \mathcal{C}(\mathbf{A})$, then, according to the same lemma, there exists a matrix $\mathbf{F}$ such that $\mathbf{B} = \mathbf{A}\mathbf{F}$, implying that

$$\mathbf{B} = \mathbf{A}\mathbf{A}^{-}\mathbf{A}\mathbf{F} = \mathbf{A}\mathbf{A}^{-}\mathbf{B}.$$

Thus, $\mathcal{C}(\mathbf{B}) \subset \mathcal{C}(\mathbf{A})$ if and only if $\mathbf{B} = \mathbf{A}\mathbf{A}^{-}\mathbf{B}$. That $\mathcal{R}(\mathbf{C}) \subset \mathcal{R}(\mathbf{A})$ if and only if $\mathbf{C} = \mathbf{C}\mathbf{A}^{-}\mathbf{A}$ follows from an analogous argument. Q.E.D.

In the special case where $p = 1$ and $q = 1$, Lemma 9.3.5 reduces to the following corollary.

Corollary 9.3.6. Let $\mathbf{A}$ represent an $m \times n$ matrix. Then, for any m-dimensional column vector $\mathbf{x}$, $\mathbf{x} \in \mathcal{C}(\mathbf{A})$ if and only if $\mathbf{x} = \mathbf{A}\mathbf{A}^{-}\mathbf{x}$, and, for any n-dimensional row vector $\mathbf{y}'$, $\mathbf{y}' \in \mathcal{R}(\mathbf{A})$ if and only if $\mathbf{y}' = \mathbf{y}'\mathbf{A}^{-}\mathbf{A}$.

Lemma 9.3.7. For any matrix $\mathbf{A}$,

$$\mathcal{R}(\mathbf{A}^{-}\mathbf{A}) = \mathcal{R}(\mathbf{A}) \qquad \text{and} \qquad \mathcal{C}(\mathbf{A}\mathbf{A}^{-}) = \mathcal{C}(\mathbf{A}).$$

Proof. It follows from Corollary 4.2.3 that $\mathcal{R}(\mathbf{A}^{-}\mathbf{A}) \subset \mathcal{R}(\mathbf{A})$ and also, since $\mathbf{A} = \mathbf{A}(\mathbf{A}^{-}\mathbf{A})$, that $\mathcal{R}(\mathbf{A}) \subset \mathcal{R}(\mathbf{A}^{-}\mathbf{A})$. Thus, $\mathcal{R}(\mathbf{A}^{-}\mathbf{A}) = \mathcal{R}(\mathbf{A})$. That $\mathcal{C}(\mathbf{A}\mathbf{A}^{-}) = \mathcal{C}(\mathbf{A})$ follows from an analogous argument. Q.E.D.

As an immediate consequence of Lemma 9.3.7, we have the following corollary.

Corollary 9.3.8. For any matrix $\mathbf{A}$,

$$\operatorname{rank}(\mathbf{A}^{-}\mathbf{A}) = \operatorname{rank}(\mathbf{A}\mathbf{A}^{-}) = \operatorname{rank}(\mathbf{A}).$$

It follows from this corollary, together with Corollary 4.4.5, that, for any matrix $\mathbf{A}$,

$$\operatorname{rank}(\mathbf{A}^{-}) \geq \operatorname{rank}(\mathbf{A}^{-}\mathbf{A}) = \operatorname{rank}(\mathbf{A}).$$

Thus, we have the following lemma, which extends result (8.2.5).

Lemma 9.3.9. For any matrix $\mathbf{A}$,

$$\operatorname{rank}(\mathbf{A}^{-}) \geq \operatorname{rank}(\mathbf{A}).$$

The following two lemmas give some elementary results on generalized inverses of partitioned matrices.

Lemma 9.3.10. Let $\mathbf{A}$ and $\mathbf{B}$ represent $m \times n$ partitioned matrices of the form $\mathbf{A} = (\mathbf{T}, \mathbf{0})$ and $\mathbf{B} = \begin{pmatrix} \mathbf{W} \\ \mathbf{0} \end{pmatrix}$. Then, an $n \times m$ partitioned matrix $\mathbf{G} = \begin{pmatrix} \mathbf{G}_1 \\ \mathbf{G}_2 \end{pmatrix}$ (where the number of rows in $\mathbf{G}_1$ equals the number of columns in $\mathbf{T}$) is a generalized inverse of $\mathbf{A}$ if and only if $\mathbf{G}_1$ is a generalized inverse of $\mathbf{T}$. Similarly, an $n \times m$ partitioned matrix $\mathbf{H} = (\mathbf{H}_1, \mathbf{H}_2)$ (where the number of columns in $\mathbf{H}_1$ equals the number of rows in $\mathbf{W}$) is a generalized inverse of $\mathbf{B}$ if and only if $\mathbf{H}_1$ is a generalized inverse of $\mathbf{W}$.

Proof. The lemma becomes obvious upon observing that $\mathbf{AGA} = (\mathbf{TG}_1\mathbf{T}, \mathbf{0})$ and $\mathbf{BHB} = \begin{pmatrix} \mathbf{WH}_1\mathbf{W} \\ \mathbf{0} \end{pmatrix}$. Q.E.D.

Lemma 9.3.11. Let $\mathbf{A}$ represent an $m \times n$ matrix, $\mathbf{B}$ an $m \times p$ matrix, and $\mathbf{C}$ a $q \times n$ matrix. Then, the $(n + p) \times m$ matrix $\begin{pmatrix} \mathbf{A}^- \\ \mathbf{B}^- \end{pmatrix}$ is a generalized inverse of the partitioned matrix $(\mathbf{A}, \mathbf{B})$ if and only if $\mathbf{AA}^-\mathbf{B} = \mathbf{0}$ and $\mathbf{BB}^-\mathbf{A} = \mathbf{0}$. Similarly, the $n \times (m+q)$ matrix $(\mathbf{A}^-, \mathbf{C}^-)$ is a generalized inverse of the partitioned matrix $\begin{pmatrix} \mathbf{A} \\ \mathbf{C} \end{pmatrix}$ if and only if $\mathbf{CA}^-\mathbf{A} = \mathbf{0}$ and $\mathbf{AC}^-\mathbf{C} = \mathbf{0}$.

Proof. The proof becomes obvious upon observing that

$$\begin{aligned}(\mathbf{A}, \mathbf{B})\begin{pmatrix} \mathbf{A}^- \\ \mathbf{B}^- \end{pmatrix}(\mathbf{A}, \mathbf{B}) &= (\mathbf{AA}^-\mathbf{A} + \mathbf{BB}^-\mathbf{A},\ \mathbf{AA}^-\mathbf{B} + \mathbf{BB}^-\mathbf{B}) \\ &= (\mathbf{A} + \mathbf{BB}^-\mathbf{A},\ \mathbf{AA}^-\mathbf{B} + \mathbf{B})\end{aligned}$$

and that

$$\begin{pmatrix} \mathbf{A} \\ \mathbf{C} \end{pmatrix}(\mathbf{A}^-, \mathbf{C}^-)\begin{pmatrix} \mathbf{A} \\ \mathbf{C} \end{pmatrix} = \begin{pmatrix} \mathbf{AA}^-\mathbf{A} + \mathbf{AC}^-\mathbf{C} \\ \mathbf{CA}^-\mathbf{A} + \mathbf{CC}^-\mathbf{C} \end{pmatrix} = \begin{pmatrix} \mathbf{A} + \mathbf{AC}^-\mathbf{C} \\ \mathbf{CA}^-\mathbf{A} + \mathbf{C} \end{pmatrix}.$$

Q.E.D.

9.4 Invariance to the Choice of a Generalized Inverse

Suppose that a generalized inverse $\mathbf{A}^-$ of a matrix $\mathbf{A}$ is premultiplied by a matrix $\mathbf{B}$ and postmultiplied by a matrix $\mathbf{C}$. Under certain conditions, the product $\mathbf{BA}^-\mathbf{C}$ is invariant to the choice of $\mathbf{A}^-$. These conditions are given by the following theorem.

Theorem 9.4.1. Let $\mathbf{A}$ represent an $m \times n$ matrix, $\mathbf{B}$ a $p \times n$ matrix, and $\mathbf{C}$ an $m \times q$ matrix. If $\mathcal{R}(\mathbf{B}) \subset \mathcal{R}(\mathbf{A})$ and $\mathcal{C}(\mathbf{C}) \subset \mathcal{C}(\mathbf{A})$, then $\mathbf{BA}^-\mathbf{C}$ is invariant to the choice of the generalized inverse $\mathbf{A}^-$. Conversely, if $\mathbf{BA}^-\mathbf{C}$ is invariant to the choice of $\mathbf{A}^-$ and if in addition $\mathbf{C}$ is nonnull, then $\mathcal{R}(\mathbf{B}) \subset \mathcal{R}(\mathbf{A})$; and, if $\mathbf{BA}^-\mathbf{C}$ is invariant to the choice of $\mathbf{A}^-$ and if in addition $\mathbf{B}$ is nonnull, then $\mathcal{C}(\mathbf{C}) \subset \mathcal{C}(\mathbf{A})$.

Proof. Suppose that $\mathcal{R}(\mathbf{B}) \subset \mathcal{R}(\mathbf{A})$ and $\mathcal{C}(\mathbf{C}) \subset \mathcal{C}(\mathbf{A})$. Then, there exist matrices $\mathbf{L}$ and $\mathbf{R}$ such that $\mathbf{B} = \mathbf{LA}$ and $\mathbf{C} = \mathbf{AR}$, in which case

$$\mathbf{BA^-C} = \mathbf{LAA^-AR} = \mathbf{LAR}.$$

Thus, $\mathbf{BA^-C}$ is invariant to the choice of $\mathbf{A}^-$.

Conversely, suppose that $\mathbf{BA^-C}$ is invariant to the choice of $\mathbf{A}^-$. Then, letting $\mathbf{G}$ represent any particular choice for $\mathbf{A}^-$, it follows from Theorem 9.2.7 that

$$\mathbf{BGC} = \mathbf{B[G + (I - GA)T + S(I - AG)]C}$$

for all $n \times m$ matrices $\mathbf{T}$ and $\mathbf{S}$, or equivalently that

$$\mathbf{B(I - GA)TC + BS(I - AG)C} = \mathbf{0}$$

for all $\mathbf{T}$ and $\mathbf{S}$, and hence that

$$\mathbf{B(I - GA)TC} = \mathbf{0}$$

for all $\mathbf{T}$ and

$$\mathbf{BS(I - AG)C} = \mathbf{0}$$

for all $\mathbf{S}$.

Suppose further that $\mathbf{C}$ is nonnull. Then, some column, say the kth column $\mathbf{c}_k$, of $\mathbf{C}$ is nonnull. And, letting $\mathbf{u}_i$ represent the ith column of $\mathbf{I}_n$, there exists an $n \times m$ matrix $\mathbf{T}_i$ such that $\mathbf{T}_i\mathbf{c}_k = \mathbf{u}_i$ $(i = 1, \ldots, n)$.

We have (for $i = 1, \ldots, n$) that $\mathbf{B(I-GA)T}_i\mathbf{C} = \mathbf{0}$. Moreover, the kth column of $\mathbf{B(I - GA)T}_i\mathbf{C}$ is $\mathbf{B(I - GA)T}_i\mathbf{c}_k = \mathbf{B(I - GA)u}_i$, which is the ith column of $\mathbf{B(I - GA)}$. Thus, the ith column of $\mathbf{B(I - GA)}$ is null $(i = 1, \ldots, n)$. We conclude that $\mathbf{B(I - GA)} = \mathbf{0}$ and (in light of Lemma 9.3.5) that $\mathcal{R}(\mathbf{B}) \subset \mathcal{R}(\mathbf{A})$.

That $\mathcal{C}(\mathbf{C}) \subset \mathcal{C}(\mathbf{A})$ if $\mathbf{B}$ is nonnull (and $\mathbf{BA^-C}$ is invariant to the choice of $\mathbf{A}^-$) can be established in similar fashion. Q.E.D.

Theorem 9.4.1 has the following implication.

Corollary 9.4.2. Let $\mathbf{A}$ represent a $q \times p$ matrix, $\mathbf{B}$ a $p \times n$ matrix, and $\mathbf{C}$ an $m \times q$ matrix. If $\text{rank}(\mathbf{CAB}) = \text{rank}(\mathbf{C}) = \text{rank}(\mathbf{B})$, then $\mathbf{B(CAB)^-C}$ is invariant to the choice of the generalized inverse $(\mathbf{CAB})^-$.

Proof. Suppose that $\text{rank}(\mathbf{CAB}) = \text{rank}(\mathbf{C}) = \text{rank}(\mathbf{B})$. Then, it follows from Corollary 4.4.7 that $\mathcal{R}(\mathbf{B}) = \mathcal{R}(\mathbf{CAB})$ and $\mathcal{C}(\mathbf{C}) = \mathcal{C}(\mathbf{CAB})$. Based on Theorem 9.4.1, we conclude that $\mathbf{B(CAB)^-C}$ is invariant to the choice of $(\mathbf{CAB})^-$. Q.E.D.

9.5 A Necessary and Sufficient Condition for the Consistency of a Linear System

According to Theorem 7.2.1, a linear system $\mathbf{AX} = \mathbf{B}$ (in $\mathbf{X}$) is consistent if and only if $\mathcal{C}(\mathbf{B}) \subset \mathcal{C}(\mathbf{A})$. In light of Lemma 9.3.5, this result can be restated as follows:

Lemma 9.5.1. A linear system $\mathbf{AX} = \mathbf{B}$ (in $\mathbf{X}$) is consistent if and only if

$$\mathbf{AA^-B} = \mathbf{B}$$

or, equivalently, if and only if

$$(\mathbf{I} - \mathbf{AA^-})\mathbf{B} = \mathbf{0}.$$

According to Lemma 9.5.1, either of the two matrices $\mathbf{AA^-}$ or $\mathbf{I} - \mathbf{AA^-}$ can be used to determine whether any linear system having $\mathbf{A}$ as a coefficient matrix is consistent or inconsistent. If the right side of the linear system is unaffected by premultiplication by $\mathbf{AA^-}$, then the linear system is consistent; otherwise, it is inconsistent. Similarly, if the premultiplication of the right side by $\mathbf{I}-\mathbf{AA^-}$ produces a null matrix, then the linear system is consistent; otherwise, it is inconsistent.

Consider, for example, the linear system $\mathbf{Ax} = \mathbf{b}$ (in $\mathbf{x}$), where

$$\mathbf{A} = \begin{pmatrix} -6 & 2 & -2 & -3 \\ 3 & -1 & 5 & 2 \\ -3 & 1 & 3 & -1 \end{pmatrix}.$$

As determined in Section 9.2a, one generalized inverse of $\mathbf{A}$ is

$$\mathbf{G} = \begin{pmatrix} 0 & 0 & 0 \\ -1 & 0 & 3 \\ 0 & 0 & 0 \\ -1 & 0 & 2 \end{pmatrix}.$$

Clearly,

$$\mathbf{AG} = \begin{pmatrix} 1 & 0 & 0 \\ -1 & 0 & 1 \\ 0 & 0 & 1 \end{pmatrix}.$$

If $\mathbf{b} = (3, 2, 5)'$, then

$$\mathbf{AGb} = (3, 2, 5)' = \mathbf{b},$$

in which case the linear system $\mathbf{Ax} = \mathbf{b}$ is consistent. However, if $\mathbf{b} = (1, 2, 1)'$, then

$$\mathbf{AGb} = (1, 0, 1)' \neq \mathbf{b},$$

in which case $\mathbf{Ax} = \mathbf{b}$ is inconsistent.

9.6 Some Results on the Ranks and Generalized Inverses of Partitioned Matrices

a. Basic results

Consider the partitioned matrix $\mathbf{A} = \begin{pmatrix} \mathbf{T} & \mathbf{U} \\ \mathbf{V} & \mathbf{W} \end{pmatrix}$. In the special case where $\mathbf{T}$ is nonsingular, Theorem 8.5.10 gives rank($\mathbf{A}$) in terms of the Schur complement $\mathbf{Q} = \mathbf{W} - \mathbf{V}\mathbf{T}^{-1}\mathbf{U}$, which is of smaller dimensions than $\mathbf{A}$. Moreover, in the special case where $\mathbf{A}$, as well as $\mathbf{T}$, is nonsingular, Theorem 8.5.11 gives $\mathbf{A}^{-1}$ in terms of $\mathbf{T}^{-1}$ and $\mathbf{Q}^{-1}$ (and $\mathbf{U}$ and $\mathbf{V}$). The following theorem extends the results of Theorems 8.5.10 and 8.5.11 to a larger class of partitioned matrices.

Theorem 9.6.1. Let $\mathbf{T}$ represent an $m \times p$ matrix, $\mathbf{U}$ an $m \times q$ matrix, $\mathbf{V}$ an $n \times p$ matrix, and $\mathbf{W}$ an $n \times q$ matrix, and define $\mathbf{Q} = \mathbf{W} - \mathbf{V}\mathbf{T}^{-}\mathbf{U}$. Suppose that $\mathcal{C}(\mathbf{U}) \subset \mathcal{C}(\mathbf{T})$ and $\mathcal{R}(\mathbf{V}) \subset \mathcal{R}(\mathbf{T})$. Then,

$$\operatorname{rank}\begin{pmatrix} \mathbf{T} & \mathbf{U} \\ \mathbf{V} & \mathbf{W} \end{pmatrix} = \operatorname{rank}\begin{pmatrix} \mathbf{W} & \mathbf{V} \\ \mathbf{U} & \mathbf{T} \end{pmatrix} = \operatorname{rank}(\mathbf{T}) + \operatorname{rank}(\mathbf{Q}). \tag{6.1}$$

Further, the partitioned matrices

$$\begin{pmatrix} \mathbf{T}^{-} + \mathbf{T}^{-}\mathbf{U}\mathbf{Q}^{-}\mathbf{V}\mathbf{T}^{-} & -\mathbf{T}^{-}\mathbf{U}\mathbf{Q}^{-} \\ -\mathbf{Q}^{-}\mathbf{V}\mathbf{T}^{-} & \mathbf{Q}^{-} \end{pmatrix} \tag{6.2a}$$

$$= \begin{pmatrix} \mathbf{T}^{-} & \mathbf{0} \\ \mathbf{0} & \mathbf{0} \end{pmatrix} + \begin{pmatrix} -\mathbf{T}^{-}\mathbf{U} \\ \mathbf{I}_q \end{pmatrix} \mathbf{Q}^{-}(-\mathbf{V}\mathbf{T}^{-}, \mathbf{I}_n) \tag{6.2b}$$

and

$$\begin{pmatrix} \mathbf{Q}^{-} & -\mathbf{Q}^{-}\mathbf{V}\mathbf{T}^{-} \\ -\mathbf{T}^{-}\mathbf{U}\mathbf{Q}^{-} & \mathbf{T}^{-} + \mathbf{T}^{-}\mathbf{U}\mathbf{Q}^{-}\mathbf{V}\mathbf{T}^{-} \end{pmatrix} \tag{6.3a}$$

$$= \begin{pmatrix} \mathbf{0} & \mathbf{0} \\ \mathbf{0} & \mathbf{T}^{-} \end{pmatrix} + \begin{pmatrix} \mathbf{I}_q \\ -\mathbf{T}^{-}\mathbf{U} \end{pmatrix} \mathbf{Q}^{-}(\mathbf{I}_n, -\mathbf{V}\mathbf{T}^{-}) \tag{6.3b}$$

are generalized inverses of $\begin{pmatrix} \mathbf{T} & \mathbf{U} \\ \mathbf{V} & \mathbf{W} \end{pmatrix}$ and $\begin{pmatrix} \mathbf{W} & \mathbf{V} \\ \mathbf{U} & \mathbf{T} \end{pmatrix}$, respectively.

In the special case of a block-triangular matrix, Theorem 9.6.1 can be restated as the following result, which generalizes results (8.5.7) and (8.5.8).

Corollary 9.6.2. Let $\mathbf{T}$ represent an $m \times p$ matrix, $\mathbf{V}$ an $n \times p$ matrix, and $\mathbf{W}$ an $n \times q$ matrix. If $\mathcal{R}(\mathbf{V}) \subset \mathcal{R}(\mathbf{T})$ or $\mathcal{C}(\mathbf{V}) \subset \mathcal{C}(\mathbf{W})$, then

$$\operatorname{rank}\begin{pmatrix} \mathbf{T} & \mathbf{0} \\ \mathbf{V} & \mathbf{W} \end{pmatrix} = \operatorname{rank}\begin{pmatrix} \mathbf{W} & \mathbf{V} \\ \mathbf{0} & \mathbf{T} \end{pmatrix} = \operatorname{rank}(\mathbf{T}) + \operatorname{rank}(\mathbf{W}),$$

and the block-triangular matrices

$$\begin{pmatrix} \mathbf{T}^{-} & \mathbf{0} \\ -\mathbf{W}^{-}\mathbf{V}\mathbf{T}^{-} & \mathbf{W}^{-} \end{pmatrix} \quad \text{and} \quad \begin{pmatrix} \mathbf{W}^{-} & -\mathbf{W}^{-}\mathbf{V}\mathbf{T}^{-} \\ \mathbf{0} & \mathbf{T}^{-} \end{pmatrix}$$

are generalized inverses of $\begin{pmatrix} \mathbf{T} & \mathbf{0} \\ \mathbf{V} & \mathbf{W} \end{pmatrix}$ and $\begin{pmatrix} \mathbf{W} & \mathbf{V} \\ \mathbf{0} & \mathbf{T} \end{pmatrix}$, respectively.

As a special case of Corollary 9.6.2, we obtain the following generalization of result (8.5.2).

Corollary 9.6.3. For any matrices $\mathbf{T}$ and $\mathbf{W}$, the partitioned matrix $\begin{pmatrix} \mathbf{T}^- & \mathbf{0} \\ \mathbf{0} & \mathbf{W}^- \end{pmatrix}$ is a generalized inverse of the block-diagonal matrix $\begin{pmatrix} \mathbf{T} & \mathbf{0} \\ \mathbf{0} & \mathbf{W} \end{pmatrix}$.

To prove Theorem 9.6.1, let us first prove the following result, which is of some interest in its own right.

Theorem 9.6.4. Let $\mathbf{T}$ represent an $m \times p$ matrix, $\mathbf{R}$ a $p \times q$ matrix, $\mathbf{L}$ an $n \times m$ matrix, and $\mathbf{W}$ an $n \times q$ matrix, and define $\mathbf{Q} = \mathbf{W} - \mathbf{LTR}$. Then,

$$\operatorname{rank}\begin{pmatrix} \mathbf{T} & \mathbf{TR} \\ \mathbf{LT} & \mathbf{W} \end{pmatrix} = \operatorname{rank}\begin{pmatrix} \mathbf{W} & \mathbf{LT} \\ \mathbf{TR} & \mathbf{T} \end{pmatrix} = \operatorname{rank}(\mathbf{T}) + \operatorname{rank}(\mathbf{Q}). \tag{6.4}$$

Furthermore, the partitioned matrices

$$\begin{pmatrix} \mathbf{T}^- + \mathbf{RQ}^-\mathbf{L} & -\mathbf{RQ}^- \\ -\mathbf{Q}^-\mathbf{L} & \mathbf{Q}^- \end{pmatrix} = \begin{pmatrix} \mathbf{T}^- & \mathbf{0} \\ \mathbf{0} & \mathbf{0} \end{pmatrix} + \begin{pmatrix} -\mathbf{R} \\ \mathbf{I}_q \end{pmatrix} \mathbf{Q}^-(-\mathbf{L}, \mathbf{I}_n) \tag{6.5}$$

and

$$\begin{pmatrix} \mathbf{Q}^- & -\mathbf{Q}^-\mathbf{L} \\ -\mathbf{RQ}^- & \mathbf{T}^- + \mathbf{RQ}^-\mathbf{L} \end{pmatrix} = \begin{pmatrix} \mathbf{0} & \mathbf{0} \\ \mathbf{0} & \mathbf{T}^- \end{pmatrix} + \begin{pmatrix} \mathbf{I}_q \\ -\mathbf{R} \end{pmatrix} \mathbf{Q}^-(\mathbf{I}_n, -\mathbf{L}) \tag{6.6}$$

are generalized inverses of $\begin{pmatrix} \mathbf{T} & \mathbf{TR} \\ \mathbf{LT} & \mathbf{W} \end{pmatrix}$ and $\begin{pmatrix} \mathbf{W} & \mathbf{LT} \\ \mathbf{TR} & \mathbf{T} \end{pmatrix}$, respectively.

Proof (of Theorem 9.6.4). Observe that

$$\begin{pmatrix} \mathbf{I} & \mathbf{0} \\ -\mathbf{L} & \mathbf{I} \end{pmatrix} \begin{pmatrix} \mathbf{T} & \mathbf{TR} \\ \mathbf{LT} & \mathbf{W} \end{pmatrix} \begin{pmatrix} \mathbf{I} & -\mathbf{R} \\ \mathbf{0} & \mathbf{I} \end{pmatrix} = \begin{pmatrix} \mathbf{T} & \mathbf{0} \\ \mathbf{0} & \mathbf{Q} \end{pmatrix} \tag{6.7}$$

and that (according to Lemma 8.5.2) the matrices $\begin{pmatrix} \mathbf{I} & \mathbf{0} \\ -\mathbf{L} & \mathbf{I} \end{pmatrix}$ and $\begin{pmatrix} \mathbf{I} & -\mathbf{R} \\ \mathbf{0} & \mathbf{I} \end{pmatrix}$ are nonsingular. Thus, making use of Corollary 8.3.3 and Lemma 4.5.11, we find that

$$\operatorname{rank}\begin{pmatrix} \mathbf{T} & \mathbf{TR} \\ \mathbf{LT} & \mathbf{W} \end{pmatrix} = \operatorname{rank}\begin{pmatrix} \mathbf{T} & \mathbf{0} \\ \mathbf{0} & \mathbf{Q} \end{pmatrix} = \operatorname{rank}(\mathbf{T}) + \operatorname{rank}(\mathbf{Q}),$$

which (in light of Lemma 8.5.1) establishes result (6.4).

Equality (6.7) further implies that

$$\begin{pmatrix} \mathbf{T} & \mathbf{TR} \\ \mathbf{LT} & \mathbf{W} \end{pmatrix} = \begin{pmatrix} \mathbf{I} & \mathbf{0} \\ -\mathbf{L} & \mathbf{I} \end{pmatrix}^{-1} \begin{pmatrix} \mathbf{T} & \mathbf{0} \\ \mathbf{0} & \mathbf{Q} \end{pmatrix} \begin{pmatrix} \mathbf{I} & -\mathbf{R} \\ \mathbf{0} & \mathbf{I} \end{pmatrix}^{-1}.$$

Observing that

$$\begin{pmatrix} \mathbf{T} & \mathbf{0} \\ \mathbf{0} & \mathbf{Q} \end{pmatrix} \begin{pmatrix} \mathbf{T}^- & \mathbf{0} \\ \mathbf{0} & \mathbf{Q}^- \end{pmatrix} \begin{pmatrix} \mathbf{T} & \mathbf{0} \\ \mathbf{0} & \mathbf{Q} \end{pmatrix} = \begin{pmatrix} \mathbf{T} & \mathbf{0} \\ \mathbf{0} & \mathbf{Q} \end{pmatrix}$$

and hence that $\begin{pmatrix} \mathbf{T}^- & \mathbf{0} \\ \mathbf{0} & \mathbf{Q}^- \end{pmatrix}$ is a generalized inverse of $\begin{pmatrix} \mathbf{T} & \mathbf{0} \\ \mathbf{0} & \mathbf{Q} \end{pmatrix}$, we conclude, on the basis of Lemma 9.2.4, that the matrix

$$\begin{pmatrix} \mathbf{I} & -\mathbf{R} \\ \mathbf{0} & \mathbf{I} \end{pmatrix} \begin{pmatrix} \mathbf{T}^- & \mathbf{0} \\ \mathbf{0} & \mathbf{Q}^- \end{pmatrix} \begin{pmatrix} \mathbf{I} & \mathbf{0} \\ -\mathbf{L} & \mathbf{I} \end{pmatrix} = \begin{pmatrix} \mathbf{T}^- + \mathbf{R}\mathbf{Q}^-\mathbf{L} & -\mathbf{R}\mathbf{Q}^- \\ -\mathbf{Q}^-\mathbf{L} & \mathbf{Q}^- \end{pmatrix}$$

is a generalized inverse of $\begin{pmatrix} \mathbf{T} & \mathbf{TR} \\ \mathbf{LT} & \mathbf{W} \end{pmatrix}$. That matrix (6.6) is a generalized inverse of $\begin{pmatrix} \mathbf{W} & \mathbf{LT} \\ \mathbf{TR} & \mathbf{T} \end{pmatrix}$ follows from a similar argument. (An alternative way of establishing that matrices (6.5) and (6.6) are generalized inverses of $\begin{pmatrix} \mathbf{T} & \mathbf{TR} \\ \mathbf{LT} & \mathbf{W} \end{pmatrix}$ and $\begin{pmatrix} \mathbf{W} & \mathbf{LT} \\ \mathbf{TR} & \mathbf{T} \end{pmatrix}$, respectively, is by direct verification; that is, by premultiplying and postmultiplying matrix (6.5) by $\begin{pmatrix} \mathbf{T} & \mathbf{TR} \\ \mathbf{LT} & \mathbf{W} \end{pmatrix}$ and premultiplying and postmultiplying matrix (6.6) by $\begin{pmatrix} \mathbf{W} & \mathbf{LT} \\ \mathbf{TR} & \mathbf{T} \end{pmatrix}$ and by verifying that the resultant products equal $\begin{pmatrix} \mathbf{T} & \mathbf{TR} \\ \mathbf{LT} & \mathbf{W} \end{pmatrix}$ and $\begin{pmatrix} \mathbf{W} & \mathbf{LT} \\ \mathbf{TR} & \mathbf{T} \end{pmatrix}$, respectively.) Q.E.D.

Proof (of Theorem 9.6.1). According to Lemma 9.3.5, $\mathbf{U} = \mathbf{TT}^-\mathbf{U}$ and $\mathbf{V} = \mathbf{VT}^-\mathbf{T}$. It follows that

$$\begin{pmatrix} \mathbf{T} & \mathbf{U} \\ \mathbf{V} & \mathbf{W} \end{pmatrix} = \begin{pmatrix} \mathbf{T} & \mathbf{TT}^-\mathbf{U} \\ \mathbf{VT}^-\mathbf{T} & \mathbf{W} \end{pmatrix} \quad \text{and} \quad \begin{pmatrix} \mathbf{W} & \mathbf{V} \\ \mathbf{U} & \mathbf{T} \end{pmatrix} = \begin{pmatrix} \mathbf{W} & \mathbf{VT}^-\mathbf{T} \\ \mathbf{TT}^-\mathbf{U} & \mathbf{T} \end{pmatrix}.$$

Thus, Theorem 9.6.1 can be obtained from Theorem 9.6.4 by setting $\mathbf{R} = \mathbf{T}^-\mathbf{U}$ and $\mathbf{L} = \mathbf{VT}^-$ and observing that

$$\mathbf{LTR} = \mathbf{VT}^-\mathbf{TT}^-\mathbf{U} = \mathbf{VT}^-\mathbf{U}.$$

Q.E.D.

The matrix $\mathbf{Q} = \mathbf{W} - \mathbf{VT}^-\mathbf{U}$, which appears in formulas (6.1)–(6.3) for the rank and generalized inverse of a partitioned matrix, is called the *Schur complement* of $\mathbf{T}$ in $\begin{pmatrix} \mathbf{T} & \mathbf{U} \\ \mathbf{V} & \mathbf{W} \end{pmatrix}$ or $\begin{pmatrix} \mathbf{W} & \mathbf{V} \\ \mathbf{U} & \mathbf{T} \end{pmatrix}$ relative to $\mathbf{T}^-$. Moreover, when the context is clear, $\mathbf{Q}$ is sometimes referred to simply as the Schur complement or the Schur complement of $\mathbf{T}$. In general, the Schur complement varies with the choice of the generalized inverse $\mathbf{T}^-$. However, in the special case where the conditions of Theorem 9.6.1 are satisfied [i.e., where $\mathcal{C}(\mathbf{U}) \subset \mathcal{C}(\mathbf{T})$ and $\mathcal{R}(\mathbf{V}) \subset \mathcal{R}(\mathbf{T})$], it follows from Theorem 9.4.1 that $\mathbf{VT}^-\mathbf{U}$ is invariant to the choice of the generalized inverse $\mathbf{T}^-$ and hence that the Schur complement of $\mathbf{T}$ is invariant to the choice of $\mathbf{T}^-$. Moreover, if $\begin{pmatrix} \mathbf{T} & \mathbf{U} \\ \mathbf{V} & \mathbf{W} \end{pmatrix}$ or $\begin{pmatrix} \mathbf{W} & \mathbf{V} \\ \mathbf{U} & \mathbf{T} \end{pmatrix}$ is symmetric (or, equivalently, if $\mathbf{T}' = \mathbf{T}$, $\mathbf{W}' = \mathbf{W}$, and $\mathbf{V} = \mathbf{U}'$), then, in the special case where $\mathcal{C}(\mathbf{U}) \subset \mathcal{C}(\mathbf{T})$, the Schur complement of $\mathbf{T}$ is symmetric (as well as invariant to the choice of

$\mathbf{T}^-$), as is evident upon observing that (in this special case) there exists a matrix $\mathbf{R}$ such that $\mathbf{U} = \mathbf{TR}$ and hence such that

$$\mathbf{Q} = \mathbf{W} - \mathbf{R}'\mathbf{T}\mathbf{T}^-\mathbf{T}\mathbf{R} = \mathbf{W} - \mathbf{R}'\mathbf{T}\mathbf{R}.$$

b. A further result

According to Theorem 9.6.1, there exists a generalized inverse of a partitioned matrix $\mathbf{A} = \begin{pmatrix} \mathbf{T} & \mathbf{U} \\ \mathbf{V} & \mathbf{W} \end{pmatrix}$ [where $\mathcal{C}(\mathbf{U}) \subset \mathcal{C}(\mathbf{T})$ and $\mathcal{R}(\mathbf{V}) \subset \mathcal{R}(\mathbf{W})$] whose lower right corner is a generalized inverse of the Schur complement $\mathbf{Q} = \mathbf{W} - \mathbf{V}\mathbf{T}^-\mathbf{U}$. The following theorem indicates that every generalized inverse of $\mathbf{A}$ has this property.

Theorem 9.6.5. Let $\mathbf{T}$ represent an $m \times p$ matrix, $\mathbf{U}$ an $m \times q$ matrix, $\mathbf{V}$ an $n \times p$ matrix, and $\mathbf{W}$ an $n \times q$ matrix, and define $\mathbf{Q} = \mathbf{W} - \mathbf{V}\mathbf{T}^-\mathbf{U}$. Suppose that $\mathcal{C}(\mathbf{U}) \subset \mathcal{C}(\mathbf{T})$ and $\mathcal{R}(\mathbf{V}) \subset \mathcal{R}(\mathbf{T})$. Then, for any generalized inverse $\mathbf{B} = \begin{pmatrix} \mathbf{B}_{11} & \mathbf{B}_{12} \\ \mathbf{B}_{21} & \mathbf{B}_{22} \end{pmatrix}$ of the partitioned matrix $\begin{pmatrix} \mathbf{T} & \mathbf{U} \\ \mathbf{V} & \mathbf{W} \end{pmatrix}$, the $(q \times n)$ submatrix $\mathbf{B}_{22}$ is a generalized inverse of $\mathbf{Q}$. Similarly, for any generalized inverse $\mathbf{C} = \begin{pmatrix} \mathbf{C}_{11} & \mathbf{C}_{12} \\ \mathbf{C}_{21} & \mathbf{C}_{22} \end{pmatrix}$ of the matrix $\begin{pmatrix} \mathbf{W} & \mathbf{V} \\ \mathbf{U} & \mathbf{T} \end{pmatrix}$, the $(q \times n)$ submatrix $\mathbf{C}_{11}$ is a generalized inverse of $\mathbf{Q}$.

Proof. Let $\mathbf{A} = \begin{pmatrix} \mathbf{T} & \mathbf{U} \\ \mathbf{V} & \mathbf{W} \end{pmatrix}$, and take $\mathbf{G}$ to be the generalized inverse of $\mathbf{A}$ given by expression (6.2). Then, according to Theorem 9.2.7,

$$\mathbf{B} = \mathbf{G} + \mathbf{Z} - \mathbf{GAZAG}$$

for some $(p+q) \times (m+n)$ matrix $\mathbf{Z}$. Thus, partitioning $\mathbf{Z}$ as

$$\mathbf{Z} = \begin{pmatrix} \mathbf{Z}_{11} & \mathbf{Z}_{12} \\ \mathbf{Z}_{21} & \mathbf{Z}_{22} \end{pmatrix}$$

(where the dimensions of $\mathbf{Z}_{22}$ are $q \times n$) and observing (in light of Lemma 9.3.5) that $\mathbf{U} = \mathbf{T}\mathbf{T}^-\mathbf{U}$ and $\mathbf{V} = \mathbf{V}\mathbf{T}^-\mathbf{T}$, we find that

$$\begin{aligned} \mathbf{B}_{22} &= \mathbf{Q}^- + \mathbf{Z}_{22} - (-\mathbf{Q}^-\mathbf{V}\mathbf{T}^-, \ \mathbf{Q}^-)\mathbf{AZA}\begin{pmatrix} -\mathbf{T}^-\mathbf{U}\mathbf{Q}^- \\ \mathbf{Q}^- \end{pmatrix} \\ &= \mathbf{Q}^- + \mathbf{Z}_{22} - (\mathbf{0}, \ \mathbf{Q}^-\mathbf{Q})\mathbf{Z}\begin{pmatrix} \mathbf{0} \\ \mathbf{Q}\mathbf{Q}^- \end{pmatrix} \\ &= \mathbf{Q}^- + \mathbf{Z}_{22} - \mathbf{Q}^-\mathbf{Q}\mathbf{Z}_{22}\mathbf{Q}\mathbf{Q}^- \end{aligned}$$

and hence that

$$\mathbf{Q}\mathbf{B}_{22}\mathbf{Q} = \mathbf{Q}\mathbf{Q}^-\mathbf{Q} + \mathbf{Q}\mathbf{Z}_{22}\mathbf{Q} - \mathbf{Q}\mathbf{Q}^-\mathbf{Q}\mathbf{Z}_{22}\mathbf{Q}\mathbf{Q}^-\mathbf{Q} = \mathbf{Q}.$$

We conclude that $\mathbf{B}_{22}$ is a generalized inverse of $\mathbf{Q}$. That $\mathbf{C}_{11}$ is a generalized inverse of $\mathbf{Q}$ can be established via an analogous argument. Q.E.D.

9.7 Extension of Some Results on Systems of the Form $\mathbf{AX} = \mathbf{B}$ to Systems of the Form $\mathbf{AXC} = \mathbf{B}$

It was established in Section 9.5 that a linear system $\mathbf{AX} = \mathbf{B}$ (with an $m \times n$ coefficient matrix $\mathbf{A}$, $m \times p$ right side $\mathbf{B}$, and $n \times p$ matrix of unknowns $\mathbf{X}$) is consistent if and only if $\mathbf{AA^-B} = \mathbf{B}$, in which case $\mathbf{A^-B}$ is a solution to $\mathbf{AX} = \mathbf{B}$. Consider now the generalization of this result to a system of the form $\mathbf{AXC} = \mathbf{B}$, where $\mathbf{A}$ is an $m \times n$ matrix, $\mathbf{C}$ a $p \times q$ matrix, $\mathbf{B}$ an $m \times q$ matrix, and $\mathbf{X}$ an $n \times p$ matrix of unknowns. A system of this form (like one of the form $\mathbf{AX} = \mathbf{B}$) is said to be *consistent* if it has one or more solutions.

The desired generalization is given by the following theorem.

Theorem 9.7.1. The system $\mathbf{AXC} = \mathbf{B}$ (in $\mathbf{X}$) is consistent if and only if $\mathbf{AA^-BC^-C} = \mathbf{B}$, or equivalently if and only if $\mathbf{AA^-B} = \mathbf{B}$ and $\mathbf{BC^-C} = \mathbf{B}$, in which case $\mathbf{A^-BC^-}$ is a solution to $\mathbf{AXC} = \mathbf{B}$.

Proof. If $\mathbf{AA^-BC^-C} = \mathbf{B}$, then clearly $\mathbf{A^-BC^-}$ is a solution to the system $\mathbf{AXC} = \mathbf{B}$, and this system is consistent.

Conversely, suppose that the system $\mathbf{AXC} = \mathbf{B}$ is consistent. Then, by definition, this system has a solution, say $\mathbf{X}_*$. Thus,

$$\mathbf{AA^-BC^-C} = \mathbf{AA^-AX_*CC^-C} = \mathbf{AX_*C} = \mathbf{B}.$$

It remains to show that $\mathbf{AA^-BC^-C} = \mathbf{B}$ if and only if $\mathbf{AA^-B} = \mathbf{B}$ and $\mathbf{BC^-C} = \mathbf{B}$. If $\mathbf{AA^-B} = \mathbf{B}$ and $\mathbf{BC^-C} = \mathbf{B}$, then clearly $\mathbf{AA^-BC^-C} = \mathbf{BC^-C} = \mathbf{B}$. Conversely, if $\mathbf{AA^-BC^-C} = \mathbf{B}$, then clearly

$$\mathbf{AA^-B} = \mathbf{AA^-AA^-BC^-C} = \mathbf{AA^-BC^-C} = \mathbf{B}$$

and similarly

$$\mathbf{BC^-C} = \mathbf{AA^-BC^-CC^-C} = \mathbf{AA^-BC^-C} = \mathbf{B}.$$

Q.E.D.

In light of Lemma 9.3.5, we have the following corollary, which generalizes Part (1) of Theorem 7.2.1.

Corollary 9.7.2. The system $\mathbf{AXC} = \mathbf{B}$ (in $\mathbf{X}$) is consistent if and only if $\mathcal{C}(\mathbf{B}) \subset \mathcal{C}(\mathbf{A})$ and $\mathcal{R}(\mathbf{B}) \subset \mathcal{R}(\mathbf{C})$.

Exercises

Section 9.1

1. Let $\mathbf{A}$ represent any $m \times n$ matrix and $\mathbf{B}$ any $m \times p$ matrix. Show that if $\mathbf{AHB} = \mathbf{B}$ for some $n \times m$ matrix $\mathbf{H}$, then $\mathbf{AGB} = \mathbf{B}$ for every generalized inverse $\mathbf{G}$ of $\mathbf{A}$.

2. (a) Let $\mathbf{A}$ represent an $m \times n$ matrix. Show that any $n \times m$ matrix $\mathbf{X}$ such that $\mathbf{A}'\mathbf{A}\mathbf{X} = \mathbf{A}'$ is a generalized inverse of $\mathbf{A}$ and similarly that any $n \times m$ matrix $\mathbf{Y}$ such that $\mathbf{A}\mathbf{A}'\mathbf{Y}' = \mathbf{A}$ is a generalized inverse of $\mathbf{A}$.
   (b) Use Part (a), together with Corollary 7.4.2, to arrive at the same conclusion as Corollary 9.1.4.

3. Let $\mathbf{A}$ represent an $m \times n$ nonnull matrix, let $\mathbf{B}$ represent a matrix of full column rank and $\mathbf{T}$ a matrix of full row rank such that $\mathbf{A} = \mathbf{B}\mathbf{T}$, and let $\mathbf{L}$ represent a left inverse of $\mathbf{B}$ and $\mathbf{R}$ a right inverse of $\mathbf{T}$.

   (a) Show that the matrix $\mathbf{R}(\mathbf{B}'\mathbf{B})^{-1}\mathbf{R}'$ is a generalized inverse of the matrix $\mathbf{A}'\mathbf{A}$ and that the matrix $\mathbf{L}'(\mathbf{T}\mathbf{T}')^{-1}\mathbf{L}$ is a generalized inverse of the matrix $\mathbf{A}\mathbf{A}'$.
   (b) Show that if $\mathbf{A}$ is symmetric, then the matrix $\mathbf{R}(\mathbf{T}\mathbf{B})^{-1}\mathbf{L}$ is a generalized inverse of the matrix $\mathbf{A}^2$. (If $\mathbf{A}$ is symmetric, then it follows from the result of Exercise 8.3 that $\mathbf{T}\mathbf{B}$ is nonsingular.)

Section 9.2

4. Use formula (2.4) to find a generalized inverse of the matrix

$$\mathbf{A} = \begin{pmatrix} 0 & 0 & 0 \\ 0 & 4 & 2 \\ 0 & -2 & -1 \\ 0 & 3 & 3 \\ 0 & 8 & 4 \end{pmatrix}.$$

5. Let $\mathbf{A}$ represent an $m \times n$ nonnull matrix of rank r. Take $\mathbf{B}$ and $\mathbf{K}$ to be nonsingular matrices (of orders m and n, respectively) such that

$$\mathbf{A} = \mathbf{B}\begin{pmatrix} \mathbf{I}_r & \mathbf{0} \\ \mathbf{0} & \mathbf{0} \end{pmatrix}\mathbf{K}$$

(the existence of which is guaranteed by Theorem 4.4.9). Show (a) that an $n \times m$ matrix $\mathbf{G}$ is a generalized inverse of $\mathbf{A}$ if and only if $\mathbf{G}$ is expressible in the form

$$\mathbf{G} = \mathbf{K}^{-1}\begin{pmatrix} \mathbf{I}_r & \mathbf{U} \\ \mathbf{V} & \mathbf{W} \end{pmatrix}\mathbf{B}^{-1} \tag{E.1}$$

for some $r \times (m-r)$ matrix $\mathbf{U}$, $(n-r) \times r$ matrix $\mathbf{V}$, and $(n-r) \times (m-r)$ matrix $\mathbf{W}$, and (b) that distinct choices for $\mathbf{U}$, $\mathbf{V}$, and/or $\mathbf{W}$ lead to distinct generalized inverses.

Section 9.3

6. Let $\mathbf{A}$ represent an $m \times n$ matrix, $\mathbf{B}$ an $m \times p$ matrix, and $\mathbf{C}$ a $q \times n$ matrix, and let k represent a nonzero scalar. Generalize Lemma 9.3.1 by showing (a) that,

for any generalized inverse $\begin{pmatrix} \mathbf{G}_1 \\ \mathbf{G}_2 \end{pmatrix}$ of the partitioned matrix $(\mathbf{A}, \mathbf{B})$ (where $\mathbf{G}_1$ is of dimensions $n \times m$), $\begin{pmatrix} \mathbf{G}_1 \\ k^{-1}\mathbf{G}_2 \end{pmatrix}$ is a generalized inverse of $(\mathbf{A}, k\mathbf{B})$ and (b) that, for any generalized inverse $(\mathbf{H}_1, \mathbf{H}_2)$ of the partitioned matrix $\begin{pmatrix} \mathbf{A} \\ \mathbf{C} \end{pmatrix}$ (where $\mathbf{H}_1$ is of dimensions $n \times m$), $(\mathbf{H}_1, k^{-1}\mathbf{H}_2)$ is a generalized inverse of $\begin{pmatrix} \mathbf{A} \\ k\mathbf{C} \end{pmatrix}$.

Section 9.6

7. Let $\mathbf{T}$ represent an $m \times p$ matrix and $\mathbf{W}$ an $n \times q$ matrix.

   (a) Show that, unless $\mathbf{T}$ and $\mathbf{W}$ are both nonsingular, there exist generalized inverses of $\begin{pmatrix} \mathbf{T} & \mathbf{0} \\ \mathbf{0} & \mathbf{W} \end{pmatrix}$ that are not of the form $\begin{pmatrix} \mathbf{T}^- & \mathbf{0} \\ \mathbf{0} & \mathbf{W}^- \end{pmatrix}$. [*Hint*. Make use of result (2.7).]

   (b) Take $\mathbf{U}$ to be an $m \times q$ matrix and $\mathbf{V}$ an $n \times p$ matrix such that $\mathcal{C}(\mathbf{U}) \subset \mathcal{C}(\mathbf{T})$ and $\mathcal{R}(\mathbf{V}) \subset \mathcal{R}(\mathbf{T})$. Generalize the result of Part (a) by showing that, unless $\mathbf{T}$ and the matrix $\mathbf{Q} = \mathbf{W} - \mathbf{V}\mathbf{T}^-\mathbf{U}$ are both nonsingular, there exist generalized inverses of $\begin{pmatrix} \mathbf{T} & \mathbf{U} \\ \mathbf{V} & \mathbf{W} \end{pmatrix}$ that are not of the form (6.2). [*Hint*. Use Part (a), together with Lemma 9.2.4.]

8. Let $\mathbf{T}$ represent an $m \times p$ matrix, $\mathbf{U}$ an $m \times q$ matrix, $\mathbf{V}$ an $n \times p$ matrix, and $\mathbf{W}$ an $n \times q$ matrix, take $\mathbf{A} = \begin{pmatrix} \mathbf{T} & \mathbf{U} \\ \mathbf{V} & \mathbf{W} \end{pmatrix}$, and define $\mathbf{Q} = \mathbf{W} - \mathbf{V}\mathbf{T}^-\mathbf{U}$.

   (a) Show that the matrix

   $$\mathbf{G} = \begin{pmatrix} \mathbf{T}^- + \mathbf{T}^-\mathbf{U}\mathbf{Q}^-\mathbf{V}\mathbf{T}^- & -\mathbf{T}^-\mathbf{U}\mathbf{Q}^- \\ -\mathbf{Q}^-\mathbf{V}\mathbf{T}^- & \mathbf{Q}^- \end{pmatrix}$$

   is a generalized inverse of the matrix $\mathbf{A}$ if and only if

   (1) $(\mathbf{I} - \mathbf{T}\mathbf{T}^-)\mathbf{U}(\mathbf{I} - \mathbf{Q}^-\mathbf{Q}) = \mathbf{0}$,

   (2) $(\mathbf{I} - \mathbf{Q}\mathbf{Q}^-)\mathbf{V}(\mathbf{I} - \mathbf{T}^-\mathbf{T}) = \mathbf{0}$, and

   (3) $(\mathbf{I} - \mathbf{T}\mathbf{T}^-)\mathbf{U}\mathbf{Q}^-\mathbf{V}(\mathbf{I} - \mathbf{T}^-\mathbf{T}) = \mathbf{0}$.

   (b) Verify that the conditions $[\mathcal{C}(\mathbf{U}) \subset \mathcal{C}(\mathbf{T})$ and $\mathcal{R}(\mathbf{V}) \subset \mathcal{R}(\mathbf{T})]$ of Theorem 9.6.1 imply Conditions (1)–(3) of Part (a).

   (c) Exhibit matrices $\mathbf{T}$, $\mathbf{U}$, $\mathbf{V}$, and $\mathbf{W}$ that (regardless of how the generalized inverses $\mathbf{T}^-$ and $\mathbf{Q}^-$ are chosen) satisfy Conditions (1)–(3) of Part (a) but do not satisfy (both of) the conditions of Theorem 9.6.1.

9. (a) Suppose that a matrix $\mathbf{A}$ is partitioned as

$$\mathbf{A} = \begin{pmatrix} \mathbf{A}_{11} & \mathbf{A}_{12} & \mathbf{A}_{13} \\ \mathbf{A}_{21} & \mathbf{A}_{22} & \mathbf{A}_{23} \\ \mathbf{A}_{31} & \mathbf{A}_{32} & \mathbf{A}_{33} \end{pmatrix}$$

and that $\mathcal{C}(\mathbf{A}_{12}) \subset \mathcal{C}(\mathbf{A}_{11})$ and $\mathcal{R}(\mathbf{A}_{21}) \subset \mathcal{R}(\mathbf{A}_{11})$. Take $\mathbf{Q}$ to be the Schur complement of $\mathbf{A}_{11}$ in $\mathbf{A}$ relative to $\mathbf{A}_{11}^-$, and partition $\mathbf{Q}$ as

$$\mathbf{Q} = \begin{pmatrix} \mathbf{Q}_{11} & \mathbf{Q}_{12} \\ \mathbf{Q}_{21} & \mathbf{Q}_{22} \end{pmatrix}$$

(where $\mathbf{Q}_{11}$, $\mathbf{Q}_{12}$, $\mathbf{Q}_{21}$, and $\mathbf{Q}_{22}$ are of the same dimensions as $\mathbf{A}_{22}$, $\mathbf{A}_{23}$, $\mathbf{A}_{32}$, and $\mathbf{A}_{33}$, respectively), so that $\mathbf{Q}_{11} = \mathbf{A}_{22} - \mathbf{A}_{21}\mathbf{A}_{11}^-\mathbf{A}_{12}$, $\mathbf{Q}_{12} = \mathbf{A}_{23} - \mathbf{A}_{21}\mathbf{A}_{11}^-\mathbf{A}_{13}$, $\mathbf{Q}_{21} = \mathbf{A}_{32} - \mathbf{A}_{31}\mathbf{A}_{11}^-\mathbf{A}_{12}$, and $\mathbf{Q}_{22} = \mathbf{A}_{33} - \mathbf{A}_{31}\mathbf{A}_{11}^-\mathbf{A}_{13}$. Let

$$\mathbf{G} = \begin{pmatrix} \mathbf{A}_{11}^- + \mathbf{A}_{11}^-\mathbf{A}_{12}\mathbf{Q}_{11}^-\mathbf{A}_{21}\mathbf{A}_{11}^- & -\mathbf{A}_{11}^-\mathbf{A}_{12}\mathbf{Q}_{11}^- \\ -\mathbf{Q}_{11}^-\mathbf{A}_{21}\mathbf{A}_{11}^- & \mathbf{Q}_{11}^- \end{pmatrix}.$$

Define $\mathbf{T} = \begin{pmatrix} \mathbf{A}_{11} & \mathbf{A}_{12} \\ \mathbf{A}_{21} & \mathbf{A}_{22} \end{pmatrix}$, $\mathbf{U} = \begin{pmatrix} \mathbf{A}_{13} \\ \mathbf{A}_{23} \end{pmatrix}$, and $\mathbf{V} = (\mathbf{A}_{31}, \mathbf{A}_{32})$, or equivalently define $\mathbf{T}$, $\mathbf{U}$, and $\mathbf{V}$ to satisfy

$$\mathbf{A} = \begin{pmatrix} \mathbf{T} & \mathbf{U} \\ \mathbf{V} & \mathbf{A}_{33} \end{pmatrix}.$$

Show that:

(1) $\mathbf{G}$ is a generalized inverse of $\mathbf{T}$;

(2) the Schur complement $\mathbf{Q}_{22} - \mathbf{Q}_{21}\mathbf{Q}_{11}^-\mathbf{Q}_{12}$ of $\mathbf{Q}_{11}$ in $\mathbf{Q}$ relative to $\mathbf{Q}_{11}^-$ equals the Schur complement $\mathbf{A}_{33} - \mathbf{VGU}$ of $\mathbf{T}$ in $\mathbf{A}$ relative to $\mathbf{G}$;

(3) $\mathbf{GU} = \begin{pmatrix} \mathbf{A}_{11}^-\mathbf{A}_{13} - \mathbf{A}_{11}^-\mathbf{A}_{12}\mathbf{Q}_{11}^-\mathbf{Q}_{12} \\ \mathbf{Q}_{11}^-\mathbf{Q}_{12} \end{pmatrix}$, and

$\mathbf{VG} = (\mathbf{A}_{31}\mathbf{A}_{11}^- - \mathbf{Q}_{21}\mathbf{Q}_{11}^-\mathbf{A}_{21}\mathbf{A}_{11}^-, \;\; \mathbf{Q}_{21}\mathbf{Q}_{11}^-)$.

(b) Let $\mathbf{A}$ represent an $n \times n$ matrix (where $n \geq 2$), let $n_1, \ldots, n_k$ represent positive integers such that $n_1 + \cdots + n_k = n$ (where $k \geq 2$), and (for $i = 1, \ldots, k$) let $n_i^* = n_1 + \cdots + n_i$. Define (for $i = 1, \ldots, k$) $\mathbf{A}_i$ to be the leading principal submatrix of $\mathbf{A}$ of order n_i^* and define (for $i = 1, \ldots, k-1$) $\mathbf{U}_i$ to be the $n_i^* \times (n - n_i^*)$ matrix obtained by striking out all of the rows and columns of $\mathbf{A}$ except the first n_i^* rows and the last $n - n_i^*$ columns, $\mathbf{V}_i$ to be the $(n - n_i^*) \times n_i^*$ matrix obtained by striking out all of the rows and columns of $\mathbf{A}$ except the last $n - n_i^*$ rows and first n_i^* columns, and $\mathbf{W}_i$ to be the $(n - n_i^*) \times (n - n_i^*)$ submatrix obtained by striking out all of the rows and columns of $\mathbf{A}$ except the last $n - n_i^*$ rows and columns, so that (for $i = 1, \ldots, k-1$)

$$\mathbf{A} = \begin{pmatrix} \mathbf{A}_i & \mathbf{U}_i \\ \mathbf{V}_i & \mathbf{W}_i \end{pmatrix}.$$

Suppose that (for $i = 1, \ldots, k-1$) $\mathcal{C}(\mathbf{U}_i) \subset \mathcal{C}(\mathbf{A}_i)$ and $\mathcal{R}(\mathbf{V}_i) \subset \mathcal{R}(\mathbf{A}_i)$. Let

$$\mathbf{B}^{(i)} = \begin{pmatrix} \mathbf{B}_{11}^{(i)} & \mathbf{B}_{12}^{(i)} \\ \mathbf{B}_{21}^{(i)} & \mathbf{B}_{22}^{(i)} \end{pmatrix}$$

($i = 1, \ldots, k-1$) and $\mathbf{B}^{(k)} = \mathbf{B}_{11}^{(k)}$, where $\mathbf{B}_{11}^{(1)} = \mathbf{A}_1^-$, $\mathbf{B}_{12}^{(1)} = \mathbf{A}_1^-\mathbf{U}_1$, $\mathbf{B}_{21}^{(1)} = \mathbf{V}_1\mathbf{A}_1^-$, and $\mathbf{B}_{22}^{(1)} = \mathbf{W}_1 - \mathbf{V}_1\mathbf{A}_1^-\mathbf{U}_1$, and where (for $i \geq 2$) $\mathbf{B}_{11}^{(i)}$, $\mathbf{B}_{12}^{(i)}$, $\mathbf{B}_{21}^{(i)}$, and $\mathbf{B}_{22}^{(i)}$ are defined recursively by partitioning $\mathbf{B}_{12}^{(i-1)}$, $\mathbf{B}_{21}^{(i-1)}$, and $\mathbf{B}_{22}^{(i-1)}$ as

$$\mathbf{B}_{12}^{(i-1)} = (\mathbf{X}_1^{(i-1)}, \mathbf{X}_2^{(i-1)}), \qquad \mathbf{B}_{21}^{(i-1)} = \begin{pmatrix} \mathbf{Y}_1^{(i-1)} \\ \mathbf{Y}_2^{(i-1)} \end{pmatrix},$$

and

$$\mathbf{B}_{22}^{(i-1)} = \begin{pmatrix} \mathbf{Q}_{11}^{(i-1)} & \mathbf{Q}_{12}^{(i-1)} \\ \mathbf{Q}_{21}^{(i-1)} & \mathbf{Q}_{22}^{(i-1)} \end{pmatrix}$$

(in such a way that $\mathbf{X}_1^{(i-1)}$ has n_i columns, $\mathbf{Y}_1^{(i-1)}$ has n_i rows, and $\mathbf{Q}_{11}^{(i-1)}$ is of dimensions $n_i \times n_i$) and (using $\mathbf{Q}_{11}^{-(i-1)}$ to represent a generalized inverse of $\mathbf{Q}_{11}^{(i-1)}$) by taking

$$\mathbf{B}_{11}^{(i)} = \begin{pmatrix} \mathbf{B}_{11}^{(i-1)} + \mathbf{X}_1^{(i-1)}\mathbf{Q}_{11}^{-(i-1)}\mathbf{Y}_1^{(i-1)} & -\mathbf{X}_1^{(i-1)}\mathbf{Q}_{11}^{-(i-1)} \\ -\mathbf{Q}_{11}^{-(i-1)}\mathbf{Y}_1^{(i-1)} & \mathbf{Q}_{11}^{-(i-1)} \end{pmatrix},$$

$$\mathbf{B}_{12}^{(i)} = \begin{pmatrix} \mathbf{X}_2^{(i-1)} - \mathbf{X}_1^{(i-1)}\mathbf{Q}_{11}^{-(i-1)}\mathbf{Q}_{12}^{(i-1)} \\ \mathbf{Q}_{11}^{-(i-1)}\mathbf{Q}_{12}^{(i-1)} \end{pmatrix},$$

$$\mathbf{B}_{21}^{(i)} = (\mathbf{Y}_2^{(i-1)} - \mathbf{Q}_{21}^{(i-1)}\mathbf{Q}_{11}^{-(i-1)}\mathbf{Y}_1^{(i-1)}, \; \mathbf{Q}_{21}^{(i-1)}\mathbf{Q}_{11}^{-(i-1)}),$$

and

$$\mathbf{B}_{22}^{(i)} = \mathbf{Q}_{22}^{(i-1)} - \mathbf{Q}_{21}^{(i-1)}\mathbf{Q}_{11}^{-(i-1)}\mathbf{Q}_{12}^{(i-1)}.$$

Show that:

(1) $\mathbf{B}_{11}^{(i)}$ is a generalized inverse of $\mathbf{A}_i$ ($i = 1, \ldots, k$);

(2) $\mathbf{B}_{22}^{(i)}$ is the Schur complement of $\mathbf{A}_i$ in $\mathbf{A}$ relative to $\mathbf{B}_{11}^{(i)}$ ($i = 1, \ldots, k-1$);

(3) $\mathbf{B}_{12}^{(i)} = \mathbf{B}_{11}^{(i)}\mathbf{U}_i$, and $\mathbf{B}_{21}^{(i)} = \mathbf{V}_i\mathbf{B}_{11}^{(i)}$ ($i = 1, \ldots, k-1$).

[*Note*. The recursive formulas given in Part (b) for the sequence of matrices $\mathbf{B}^{(1)}, \ldots, \mathbf{B}^{(k-1)}, \mathbf{B}^{(k)}$ can be used to generate $\mathbf{B}^{(k-1)}$ in $k-1$ steps or to generate $\mathbf{B}^{(k)}$ in k steps—the formula for generating $\mathbf{B}^{(i)}$ from $\mathbf{B}^{(i-1)}$ involves a generalized inverse of the $n_i \times n_i$ matrix $\mathbf{Q}_{11}^{(i-1)}$. The various parts of $\mathbf{B}^{(k-1)}$ consist of a generalized inverse $\mathbf{B}_{11}^{(k-1)}$ of $\mathbf{A}_{k-1}$, the Schur complement $\mathbf{B}_{22}^{(k-1)}$ of $\mathbf{A}_{k-1}$ in $\mathbf{A}$ relative to $\mathbf{B}_{11}^{(k-1)}$, a solution $\mathbf{B}_{12}^{(k-1)}$ of the linear system $\mathbf{A}_{k-1}\mathbf{X} = \mathbf{U}_{k-1}$ (in $\mathbf{X}$), and a solution $\mathbf{B}_{21}^{(k-1)}$ of the linear system $\mathbf{Y}\mathbf{A}_{k-1} = \mathbf{V}_{k-1}$ (in $\mathbf{Y}$). The matrix $\mathbf{B}^{(k)}$ is a generalized inverse of $\mathbf{A}$. In the special case where $n_i = 1$, the process of generating the elements of the $n \times n$ matrix $\mathbf{B}^{(i)}$ from those of the $n \times n$ matrix $\mathbf{B}^{(i-1)}$ is called a *sweep operation*—see, e.g., Goodnight (1979).]

10. Let $\mathbf{T}$ represent an $m \times p$ matrix and $\mathbf{W}$ an $n \times q$ matrix, and let $\mathbf{G} = \begin{pmatrix} \mathbf{G}_{11} & \mathbf{G}_{12} \\ \mathbf{G}_{21} & \mathbf{G}_{22} \end{pmatrix}$ (where $\mathbf{G}_{11}$ is of dimensions $p \times m$) represent an arbitrary generalized inverse of the $(m + n) \times (p + q)$ block-diagonal matrix $\mathbf{A} = \begin{pmatrix} \mathbf{T} & \mathbf{0} \\ \mathbf{0} & \mathbf{W} \end{pmatrix}$. Show that $\mathbf{G}_{11}$ is a generalized inverse of $\mathbf{T}$ and $\mathbf{G}_{22}$ a generalized inverse of $\mathbf{W}$. Show also that $\mathbf{T}\mathbf{G}_{12}\mathbf{W} = \mathbf{0}$ and $\mathbf{W}\mathbf{G}_{21}\mathbf{T} = \mathbf{0}$.

11. Let $\mathbf{T}$ represent an $m \times p$ matrix, $\mathbf{U}$ an $m \times q$ matrix, $\mathbf{V}$ an $n \times p$ matrix, and $\mathbf{W}$ an $n \times q$ matrix, and define $\mathbf{Q} = \mathbf{W} - \mathbf{V}\mathbf{T}^{-}\mathbf{U}$. Suppose that $\mathcal{C}(\mathbf{U}) \subset \mathcal{C}(\mathbf{T})$ and $\mathcal{R}(\mathbf{V}) \subset \mathcal{R}(\mathbf{T})$. Devise an alternative proof of the first part of Theorem 9.6.5 by showing that

$$\begin{pmatrix} \mathbf{I} & \mathbf{0} \\ -\mathbf{V}\mathbf{T}^{-} & \mathbf{I} \end{pmatrix} \begin{pmatrix} \mathbf{T} & \mathbf{U} \\ \mathbf{V} & \mathbf{W} \end{pmatrix} \begin{pmatrix} \mathbf{I} & -\mathbf{T}^{-}\mathbf{U} \\ \mathbf{0} & \mathbf{I} \end{pmatrix} = \begin{pmatrix} \mathbf{T} & \mathbf{0} \\ \mathbf{0} & \mathbf{Q} \end{pmatrix}$$

and by then using Lemma 9.2.4 and the result of Exercise 10 to show that, for any generalized inverse $\mathbf{G} = \begin{pmatrix} \mathbf{G}_{11} & \mathbf{G}_{12} \\ \mathbf{G}_{21} & \mathbf{G}_{22} \end{pmatrix}$ of the partitioned matrix $\begin{pmatrix} \mathbf{T} & \mathbf{U} \\ \mathbf{V} & \mathbf{W} \end{pmatrix}$, the $(q \times n)$ submatrix $\mathbf{G}_{22}$ is a generalized inverse of $\mathbf{Q}$.

12. Let $\mathbf{T}$ represent an $m \times p$ matrix, $\mathbf{U}$ an $m \times q$ matrix, $\mathbf{V}$ an $n \times p$ matrix, and $\mathbf{W}$ an $n \times q$ matrix, and take $\mathbf{A} = \begin{pmatrix} \mathbf{T} & \mathbf{U} \\ \mathbf{V} & \mathbf{W} \end{pmatrix}$.

    (a) Define $\mathbf{Q} = \mathbf{W} - \mathbf{V}\mathbf{T}^{-}\mathbf{U}$, and let $\mathbf{G} = \begin{pmatrix} \mathbf{G}_{11} & \mathbf{G}_{12} \\ \mathbf{G}_{21} & \mathbf{G}_{22} \end{pmatrix}$, where $\mathbf{G}_{11} = \mathbf{T}^{-} + \mathbf{T}^{-}\mathbf{U}\mathbf{Q}^{-}\mathbf{V}\mathbf{T}^{-}$, $\mathbf{G}_{12} = -\mathbf{T}^{-}\mathbf{U}\mathbf{Q}^{-}$, $\mathbf{G}_{21} = -\mathbf{Q}^{-}\mathbf{V}\mathbf{T}^{-}$, and $\mathbf{G}_{22} = \mathbf{Q}^{-}$. Show that the matrix

    $$\mathbf{G}_{11} - \mathbf{G}_{12}\mathbf{G}_{22}^{-}\mathbf{G}_{21}$$

    is a generalized inverse of $\mathbf{T}$. {*Note.* If the conditions [$\mathcal{C}(\mathbf{U}) \subset \mathcal{C}(\mathbf{T})$ and $\mathcal{R}(\mathbf{V}) \subset \mathcal{R}(\mathbf{T})$] of Theorem 9.6.1 are satisfied or, more generally, if Conditions (1)–(3) of Part (a) of Exercise 8 are satisfied, then $\mathbf{G}$ is a generalized inverse of $\mathbf{A}$.}

    (b) Show by example that, for some values of $\mathbf{T}$, $\mathbf{U}$, $\mathbf{V}$, and $\mathbf{W}$, there exists a generalized inverse $\mathbf{G} = \begin{pmatrix} \mathbf{G}_{11} & \mathbf{G}_{12} \\ \mathbf{G}_{21} & \mathbf{G}_{22} \end{pmatrix}$ of $\mathbf{A}$ (where $\mathbf{G}_{11}$ is of dimensions

$p \times m$, $\mathbf{G}_{12}$ of dimensions $p \times n$, $\mathbf{G}_{21}$ of dimensions $q \times m$, and $\mathbf{G}_{22}$ of dimensions $q \times n$) such that the matrix

$$\mathbf{G}_{11} - \mathbf{G}_{12}\mathbf{G}_{22}^{-}\mathbf{G}_{21}$$

is *not* a generalized inverse of $\mathbf{T}$.

10

Idempotent Matrices

This chapter is devoted to a very important class of matrices called idempotent matrices. It provides coverage of some basic properties of idempotent matrices and also of some basic results pertaining to idempotent matrices. Idempotent matrices play an important role in the theory of linear statistical models (especially in connection with the theory of least squares and the analysis of variance) and (not coincidentally) appear prominently in several of the ensuing chapters of this book (including Chapters 12 and 17). Making idempotent matrices the subject of a separate chapter (even though this results in a very short chapter) is convenient and serves to emphasize their importance.

10.1 Definition and Some Basic Properties

A square matrix $\mathbf{A}$ is said to be *idempotent* if

$$\mathbf{A}^2 = \mathbf{A}.$$

Examples of $n \times n$ idempotent matrices are the identity matrix $\mathbf{I}_n$, the $n \times n$ null matrix $\mathbf{0}$, and the matrix $(1/n)\mathbf{J}_n$, each element of which equals $1/n$.

As indicated by the following lemma, $n \times n$ idempotent matrices are, with one exception, singular.

Lemma 10.1.1. The only $n \times n$ idempotent matrix of rank n is the $n \times n$ identity matrix $\mathbf{I}_n$.

Proof. Suppose that $\mathbf{A}$ is an $n \times n$ idempotent matrix of rank n. Then,

$$\mathbf{A} = \mathbf{I}_n\mathbf{A} = \mathbf{A}^{-1}\mathbf{A}\mathbf{A} = \mathbf{A}^{-1}\mathbf{A} = \mathbf{I}_n.$$

Q.E.D.

If a square matrix $\mathbf{A}$ is idempotent, then

$$(\mathbf{A}')^2 = (\mathbf{A}\mathbf{A})' = \mathbf{A}'$$

and

$$(\mathbf{I} - \mathbf{A})^2 = \mathbf{I} - 2\mathbf{A} + \mathbf{A}^2 = \mathbf{I} - 2\mathbf{A} + \mathbf{A} = \mathbf{I} - \mathbf{A}.$$

Thus, upon observing that $\mathbf{A} = (\mathbf{A}')'$ and $\mathbf{A} = \mathbf{I} - (\mathbf{I} - \mathbf{A})$, we have the following lemma.

Lemma 10.1.2. Let $\mathbf{A}$ represent a square matrix. Then, (1) $\mathbf{A}'$ is idempotent if and only if $\mathbf{A}$ is idempotent, and (2) $\mathbf{I} - \mathbf{A}$ is idempotent if and only if $\mathbf{A}$ is idempotent.

For any idempotent matrix $\mathbf{A}$, we have that $\mathbf{AAA} = (\mathbf{AA})\mathbf{A} = \mathbf{AA} = \mathbf{A}$, giving rise to the following lemma.

Lemma 10.1.3. If a (square) matrix $\mathbf{A}$ is idempotent, then $\mathbf{A}$ is a generalized inverse of itself (i.e., of $\mathbf{A}$).

10.2 Some Basic Results

Suppose that $\mathbf{A}$ is a square matrix such that

$$\mathbf{A}^2 = k\mathbf{A}$$

for some nonzero scalar k. Or, equivalently, suppose that

$$[(1/k)\mathbf{A}]^2 = (1/k)\mathbf{A},$$

that is, $(1/k)\mathbf{A}$ is an idempotent matrix (so that, depending on whether $k = 1$ or $k \neq 1$, $\mathbf{A}$ is either an idempotent matrix or a scalar multiple of an idempotent matrix). Then, in light of the following theorem,

$$\text{rank}(\mathbf{A}) = (1/k)\,\text{tr}(\mathbf{A}),$$

and consequently the rank of $\mathbf{A}$ is determinable from the trace of $\mathbf{A}$.

Theorem 10.2.1. For any square matrix $\mathbf{A}$ such that $\mathbf{A}^2 = k\mathbf{A}$ for some scalar k,

$$\text{tr}(\mathbf{A}) = k\,\text{rank}(\mathbf{A}).$$

Proof. Let us restrict attention to the case where $\mathbf{A}$ is nonnull. [The case where $\mathbf{A} = \mathbf{0}$ is trivial—if $\mathbf{A} = \mathbf{0}$, then $\text{tr}(\mathbf{A}) = 0 = k\,\text{rank}(\mathbf{A})$.]

Let n denote the order of $\mathbf{A}$, and let $r = \text{rank}(\mathbf{A})$. Then, according to Theorem 4.4.8, there exists an $n \times r$ matrix $\mathbf{B}$ and an $r \times n$ matrix $\mathbf{L}$ such that $\mathbf{A} = \mathbf{BL}$. Moreover, $\text{rank}(\mathbf{B}) = \text{rank}(\mathbf{L}) = r$. We have that

$$\mathbf{BLBL} = \mathbf{A}^2 = k\mathbf{A} = k\mathbf{BL} = \mathbf{B}(k\mathbf{I})\mathbf{L},$$

implying (in light of Lemma 8.3.1) that

$$\mathbf{LB} = k\mathbf{I}.$$

Thus, making use of Lemma 5.2.1, we find that

$$\text{tr}(\mathbf{A}) = \text{tr}(\mathbf{BL}) = \text{tr}(\mathbf{LB}) = \text{tr}(k\mathbf{I}) = k\,\text{tr}(\mathbf{I}_r) = kr.$$

Q.E.D.

In the special case where $k = 1$, Theorem 10.2.1 can be restated as follows.

Corollary 10.2.2. For any idempotent matrix $\mathbf{A}$,

$$\text{rank}(\mathbf{A}) = \text{tr}(\mathbf{A}).$$

Upon observing that only a null matrix can be of rank zero, we obtain the following, additional corollary (of Theorem 10.2.1).

Corollary 10.2.3. If the trace of an idempotent matrix $\mathbf{A}$ equals 0, then $\mathbf{A} = \mathbf{0}$.

Making use of Lemma 10.1.2 and Corollary 10.2.2, we find that, for any $n \times n$ idempotent matrix $\mathbf{A}$,

$$\text{rank}(\mathbf{I} - \mathbf{A}) = \text{tr}(\mathbf{I} - \mathbf{A}) = \text{tr}(\mathbf{I}_n) - \text{tr}(\mathbf{A}) = n - \text{rank}(\mathbf{A}),$$

thereby establishing the following lemma.

Lemma 10.2.4. For any $n \times n$ idempotent matrix $\mathbf{A}$,

$$\text{rank}(\mathbf{I} - \mathbf{A}) = \text{tr}(\mathbf{I} - \mathbf{A}) = n - \text{rank}(\mathbf{A}).$$

For any matrix $\mathbf{A}$, $\mathbf{A}^-\mathbf{A}\mathbf{A}^-\mathbf{A} = \mathbf{A}^-\mathbf{A}$ and $\mathbf{A}\mathbf{A}^-\mathbf{A}\mathbf{A}^- = \mathbf{A}\mathbf{A}^-$. Thus, we have the following lemma.

Lemma 10.2.5. Let $\mathbf{A}$ represent an $m \times n$ matrix. Then, the $n \times n$ matrix $\mathbf{A}^-\mathbf{A}$ and the $m \times m$ matrix $\mathbf{A}\mathbf{A}^-$ are both idempotent.

Note that this lemma, together with Corollary 9.3.8 and Corollary 10.2.2, implies that, for any matrix $\mathbf{A}$,

$$\text{rank}(\mathbf{A}) = \text{rank}(\mathbf{A}^-\mathbf{A}) = \text{rank}(\mathbf{A}\mathbf{A}^-) = \text{tr}(\mathbf{A}^-\mathbf{A}) = \text{tr}(\mathbf{A}\mathbf{A}^-). \tag{2.1}$$

In light of Lemma 10.2.4 [and of result (2.1)], a further implication of Lemma 10.2.5 is as follows.

Lemma 10.2.6. Let $\mathbf{A}$ represent an $m \times n$ matrix. Then,

$$\text{rank}(\mathbf{I} - \mathbf{A}^-\mathbf{A}) = \text{tr}(\mathbf{I} - \mathbf{A}^-\mathbf{A}) = n - \text{rank}(\mathbf{A}), \tag{2.2}$$

$$\text{rank}(\mathbf{I} - \mathbf{A}\mathbf{A}^-) = \text{tr}(\mathbf{I} - \mathbf{A}\mathbf{A}^-) = m - \text{rank}(\mathbf{A}). \tag{2.3}$$

The following theorem expands on the results of Lemma 10.2.5 and of Corollary 9.3.8.

Theorem 10.2.7. Let $\mathbf{A}$ represent an $m \times n$ matrix and $\mathbf{B}$ an $n \times m$ matrix. Then, $\mathbf{B}$ is a generalized inverse of $\mathbf{A}$ if and only if $\mathbf{BA}$ is idempotent and $\text{rank}(\mathbf{BA}) = \text{rank}(\mathbf{A})$. Similarly, $\mathbf{B}$ is a generalized inverse of $\mathbf{A}$ if and only if $\mathbf{AB}$ is idempotent and $\text{rank}(\mathbf{AB}) = \text{rank}(\mathbf{A})$.

Proof. In light of Lemma 10.2.5 and Corollary 9.3.8, it suffices to show that $\mathbf{B}$ is a generalized inverse of $\mathbf{A}$ if $\mathbf{BA}$ is idempotent and $\text{rank}(\mathbf{BA}) = \text{rank}(\mathbf{A})$ or if $\mathbf{AB}$ is idempotent and $\text{rank}(\mathbf{AB}) = \text{rank}(\mathbf{A})$.

Suppose that $\mathbf{BA}$ is idempotent and $\text{rank}(\mathbf{BA}) = \text{rank}(\mathbf{A})$. Then, according to Corollary 4.4.7, $\mathcal{R}(\mathbf{A}) = \mathcal{R}(\mathbf{BA})$, and it follows from Lemma 4.2.2 that $\mathbf{A} = \mathbf{SBA}$ for some matrix $\mathbf{S}$. Thus,

$$\mathbf{ABA} = \mathbf{SBABA} = \mathbf{SBA} = \mathbf{A}.$$

Alternatively, suppose that $\mathbf{AB}$ is idempotent and $\text{rank}(\mathbf{AB}) = \text{rank}(\mathbf{A})$. Then, according to Corollary 4.4.7, $\mathcal{C}(\mathbf{A}) = \mathcal{C}(\mathbf{AB})$, and consequently $\mathbf{A} = \mathbf{ABT}$ for some matrix $\mathbf{T}$. Thus,

$$\mathbf{ABA} = \mathbf{ABABT} = \mathbf{ABT} = \mathbf{A}.$$

Q.E.D.

Exercises

Section 10.1

1. Show that if an $n \times n$ matrix $\mathbf{A}$ is idempotent, then

   (a) for any $n \times n$ nonsingular matrix $\mathbf{B}$, $\mathbf{B}^{-1}\mathbf{AB}$ is idempotent;

   (b) for any integer k greater than or equal to 2, $\mathbf{A}^k = \mathbf{A}$.

2. Let $\mathbf{P}$ represent an $m \times n$ matrix (where $m \geq n$) such that $\mathbf{P}'\mathbf{P} = \mathbf{I}_n$, or equivalently an $m \times n$ matrix whose columns are orthonormal (with respect to the usual inner product). Show that the $m \times m$ symmetric matrix $\mathbf{PP}'$ is idempotent.

3. Show that, for any symmetric idempotent matrix $\mathbf{A}$, the matrix $\mathbf{I} - 2\mathbf{A}$ is orthogonal.

4. Let $\mathbf{A}$ represent an $m \times n$ matrix. Show that if $\mathbf{A}'\mathbf{A}$ is idempotent, then $\mathbf{AA}'$ is idempotent.

5. Let $\mathbf{A}$ represent a symmetric matrix and k an integer greater than or equal to 1. Show that if $\mathbf{A}^{k+1} = \mathbf{A}^k$, then $\mathbf{A}$ is idempotent.

6. Let $\mathbf{A}$ represent an $n \times n$ matrix. Show that $(1/2)(\mathbf{I} + \mathbf{A})$ is idempotent if and only if $\mathbf{A}$ is involutory (where involutory is as defined in Exercise 8.2).

7. Let $\mathbf{A}$ and $\mathbf{B}$ represent $n \times n$ symmetric idempotent matrices. Show that if $\mathcal{C}(\mathbf{A}) = \mathcal{C}(\mathbf{B})$, then $\mathbf{A} = \mathbf{B}$.

8. Let $\mathbf{A}$ represent an $r \times m$ matrix and $\mathbf{B}$ an $m \times n$ matrix.

   (a) Show that $\mathbf{B}^-\mathbf{A}^-$ is a generalized inverse of $\mathbf{AB}$ if and only if $\mathbf{A}^-\mathbf{ABB}^-$ is idempotent.

   (b) Show that if $\mathbf{A}$ has full column rank or $\mathbf{B}$ has full row rank, then $\mathbf{B}^-\mathbf{A}^-$ is a generalized inverse of $\mathbf{AB}$.

Section 10.2

9. Let $\mathbf{T}$ represent an $m \times p$ matrix, $\mathbf{U}$ an $m \times q$ matrix, $\mathbf{V}$ an $n \times p$ matrix, and $\mathbf{W}$ an $n \times q$ matrix, and define $\mathbf{Q} = \mathbf{W} - \mathbf{VT}^-\mathbf{U}$. Using Part (a) of Exercise 9.8, together with result (2.1), show that if

   (1) $(\mathbf{I} - \mathbf{TT}^-)\mathbf{U}(\mathbf{I} - \mathbf{Q}^-\mathbf{Q}) = \mathbf{0}$,
   (2) $(\mathbf{I} - \mathbf{QQ}^-)\mathbf{V}(\mathbf{I} - \mathbf{T}^-\mathbf{T}) = \mathbf{0}$, and
   (3) $(\mathbf{I} - \mathbf{TT}^-)\mathbf{UQ}^-\mathbf{V}(\mathbf{I} - \mathbf{T}^-\mathbf{T}) = \mathbf{0}$,

   then
   $$\operatorname{rank}\begin{pmatrix} \mathbf{T} & \mathbf{U} \\ \mathbf{V} & \mathbf{W} \end{pmatrix} = \operatorname{rank}(\mathbf{T}) + \operatorname{rank}(\mathbf{Q}).$$

10. Let $\mathbf{T}$ represent an $m \times p$ matrix, $\mathbf{U}$ an $m \times q$ matrix, $\mathbf{V}$ an $n \times p$ matrix, and $\mathbf{W}$ an $n \times q$ matrix, take $\mathbf{A} = \begin{pmatrix} \mathbf{T} & \mathbf{U} \\ \mathbf{V} & \mathbf{W} \end{pmatrix}$, and define $\mathbf{Q} = \mathbf{W} - \mathbf{VT}^-\mathbf{U}$. Further, let

    $$\mathbf{E}_T = \mathbf{I} - \mathbf{TT}^-,\ \mathbf{F}_T = \mathbf{I} - \mathbf{T}^-\mathbf{T},\ \mathbf{X} = \mathbf{E}_T\mathbf{U},\ \mathbf{Y} = \mathbf{VF}_T,\ \mathbf{E}_Y = \mathbf{I} - \mathbf{YY}^-,$$
    $$\mathbf{F}_X = \mathbf{I} - \mathbf{X}^-\mathbf{X},\ \mathbf{Z} = \mathbf{E}_Y\mathbf{QF}_X,\ \text{and}\ \mathbf{Q}^* = \mathbf{F}_X\mathbf{Z}^-\mathbf{E}_Y.$$

    (a) (Meyer 1973, Theorem 3.1) Show that the matrix
    $$\mathbf{G} = \mathbf{G}_1 + \mathbf{G}_2, \tag{E.1}$$
    where
    $$\mathbf{G}_1 = \left(\begin{array}{c|c} \begin{matrix} \mathbf{T}^- - \mathbf{T}^-\mathbf{U}(\mathbf{I} - \mathbf{Q}^*\mathbf{Q})\mathbf{X}^-\mathbf{E}_T \\ -\mathbf{F}_T\mathbf{Y}^-(\mathbf{I} - \mathbf{QQ}^*)\mathbf{VT}^- \\ -\mathbf{F}_T\mathbf{Y}^-(\mathbf{I} - \mathbf{QQ}^*)\mathbf{QX}^-\mathbf{E}_T \end{matrix} & \mathbf{F}_T\mathbf{Y}^-(\mathbf{I} - \mathbf{QQ}^*) \\ \hline (\mathbf{I} - \mathbf{Q}^*\mathbf{Q})\mathbf{X}^-\mathbf{E}_T & \mathbf{0} \end{array}\right)$$
    and
    $$\mathbf{G}_2 = \begin{pmatrix} -\mathbf{T}^-\mathbf{U} \\ \mathbf{I}_q \end{pmatrix} \mathbf{Q}^*(-\mathbf{VT}^-, \mathbf{I}_n),$$
    is a generalized inverse of $\mathbf{A}$.

    (b) (Meyer 1973, Theorem 4.1) Show that
    $$\operatorname{rank}(\mathbf{A}) = \operatorname{rank}(\mathbf{T}) + \operatorname{rank}(\mathbf{X}) + \operatorname{rank}(\mathbf{Y}) + \operatorname{rank}(\mathbf{Z}). \tag{E.2}$$
    [*Hint.* Use Part (a) together with result (2.1).]

(c) Show that if $\mathcal{C}(\mathbf{U}) \subset \mathcal{C}(\mathbf{T})$ and $\mathcal{R}(\mathbf{V}) \subset \mathcal{R}(\mathbf{T})$, then formula (E.2) for $\text{rank}(\mathbf{A})$ reduces to formula (9.6.1), and formula (9.6.2) for a generalized inverse of $\mathbf{A}$ can be obtained as a special case of formula (E.1).

11

Linear Systems: Solutions

The subject of this chapter is linear systems. This subject was considered previously in Chapter 7. The emphasis in Chapter 7 was on whether the linear system is consistent, that is, whether it has solutions, whereas the emphasis in the present chapter is on the solutions themselves. The presentation makes heavy use of the results of Chapters 9 and 10 (on generalized inverses and idempotent matrices).

11.1 Some Terminology, Notation, and Basic Results

The collection of all solutions to a linear system $\mathbf{AX} = \mathbf{B}$ (in $\mathbf{X}$) is called the *solution set* of the linear system. Clearly, a linear system is consistent if and only if its solution set is nonempty.

Is the solution set of a linear system $\mathbf{AX} = \mathbf{B}$ (in $\mathbf{X}$) a linear space? The answer depends on whether the linear system is homogeneous or nonhomogeneous, that is, on whether the right side $\mathbf{B}$ is null or nonnull.

Consider first the solution set of a homogeneous linear system $\mathbf{AX} = \mathbf{0}$. Since a homogeneous linear system is consistent, the solution set of $\mathbf{AX} = \mathbf{0}$ is nonempty. Furthermore, if $\mathbf{X}_1$ and $\mathbf{X}_2$ are solutions to $\mathbf{AX} = \mathbf{0}$ and k is a scalar, then $\mathbf{A}(\mathbf{X}_1 + \mathbf{X}_2) = \mathbf{AX}_1 + \mathbf{AX}_2 = \mathbf{0}$, and $\mathbf{A}(k\mathbf{X}_1) = k(\mathbf{AX}_1) = \mathbf{0}$, so that $\mathbf{X}_1 + \mathbf{X}_2$ and $k\mathbf{X}_1$ are also solutions to $\mathbf{AX} = \mathbf{0}$. Thus, the solution set of a homogeneous linear system is a linear space. Accordingly, the solution set of a homogeneous linear system $\mathbf{AX} = \mathbf{0}$ may be called the *solution space* of $\mathbf{AX} = \mathbf{0}$.

The solution space of a homogeneous linear system $\mathbf{Ax} = \mathbf{0}$ (in a column vector $\mathbf{x}$) is called the *null space* of the matrix $\mathbf{A}$ and is denoted by the symbol $\mathcal{N}(\mathbf{A})$. Thus, for any $m \times n$ matrix $\mathbf{A}$,

$$\mathcal{N}(\mathbf{A}) = \{\mathbf{x} \in \mathcal{R}^{n\times 1} \; : \; \mathbf{Ax} = \mathbf{0}\}.$$

Note that, for any $n \times n$ matrix $\mathbf{A}$,

$$\mathcal{N}(\mathbf{A}) = \{\mathbf{x} \in \mathcal{R}^{n\times 1} \; : \; (\mathbf{I} - \mathbf{A})\mathbf{x} = \mathbf{x}\} \subset \mathcal{C}(\mathbf{I} - \mathbf{A}), \tag{1.1}$$

$$\mathcal{N}(\mathbf{I} - \mathbf{A}) = \{\mathbf{x} \in \mathcal{R}^{n\times 1} \; : \; \mathbf{Ax} = \mathbf{x}\} \subset \mathcal{C}(\mathbf{A}). \tag{1.2}$$

The solution set of a nonhomogeneous linear system is not a linear space (as can be easily seen by, e.g., observing that the solution set does not contain the null matrix).

11.2 General Form of a Solution

a. Homogeneous linear systems

The following theorem gives an expression for the general form of a solution to a homogeneous linear system in terms of any particular generalized inverse of the coefficient matrix.

Theorem 11.2.1. A matrix $\mathbf{X}^*$ is a solution to a homogeneous linear system $\mathbf{AX} = \mathbf{0}$ (in $\mathbf{X}$) if and only if

$$\mathbf{X}^* = (\mathbf{I} - \mathbf{A}^-\mathbf{A})\mathbf{Y}$$

for some matrix $\mathbf{Y}$.

Proof. If $\mathbf{X}^* = (\mathbf{I} - \mathbf{A}^-\mathbf{A})\mathbf{Y}$ for some matrix $\mathbf{Y}$, then

$$\mathbf{AX}^* = (\mathbf{A} - \mathbf{AA}^-\mathbf{A})\mathbf{Y} = (\mathbf{A} - \mathbf{A})\mathbf{Y} = \mathbf{0},$$

so that $\mathbf{X}^*$ is a solution to $\mathbf{AX} = \mathbf{0}$. Conversely, if $\mathbf{X}^*$ is a solution to $\mathbf{AX} = \mathbf{0}$, then

$$\mathbf{X}^* = \mathbf{X}^* - \mathbf{A}^-(\mathbf{AX}^*) = (\mathbf{I} - \mathbf{A}^-\mathbf{A})\mathbf{X}^*,$$

so that $\mathbf{X}^* = (\mathbf{I} - \mathbf{A}^-\mathbf{A})\mathbf{Y}$ for $\mathbf{Y} = \mathbf{X}^*$. Q.E.D.

According to Theorem 11.2.1, all solutions to a homogeneous linear system $\mathbf{AX} = \mathbf{0}$ can be generated by setting

$$\mathbf{X} = (\mathbf{I} - \mathbf{A}^-\mathbf{A})\mathbf{Y}$$

and allowing $\mathbf{Y}$ to range over all matrices (of the appropriate dimensions).

As a special case of Theorem 11.2.1, we have that a column vector $\mathbf{x}^*$ is a solution to a homogeneous linear system $\mathbf{Ax} = \mathbf{0}$ (in a column vector $\mathbf{x}$) if and only if

$$\mathbf{x}^* = (\mathbf{I} - \mathbf{A}^-\mathbf{A})\mathbf{y}$$

for some column vector $\mathbf{y}$. Thus, we have the following corollary of Theorem 11.2.1.

Corollary 11.2.2. For any matrix $\mathbf{A}$,

$$\mathcal{N}(\mathbf{A}) = \mathcal{C}(\mathbf{I} - \mathbf{A}^-\mathbf{A}).$$

b. Nonhomogeneous linear systems

The following theorem relates the solutions of an arbitrary (consistent) linear system $\mathbf{AX} = \mathbf{B}$ to those of the linear homogeneous system $\mathbf{AZ} = \mathbf{0}$ (in $\mathbf{Z}$) having the same coefficient matrix. (To avoid confusion, the matrix of unknowns in the homogeneous linear system is now denoted by $\mathbf{Z}$, rather than $\mathbf{X}$.)

Theorem 11.2.3. Let $\mathbf{X}_0$ represent any particular solution to a consistent linear system $\mathbf{AX} = \mathbf{B}$ (in $\mathbf{X}$). Then, a matrix $\mathbf{X}^*$ is a solution to $\mathbf{AX} = \mathbf{B}$ if and only if

$$\mathbf{X}^* = \mathbf{X}_0 + \mathbf{Z}^*$$

for some solution $\mathbf{Z}^*$ to the homogeneous linear system $\mathbf{AZ} = \mathbf{0}$ (in $\mathbf{Z}$).

Proof. If $\mathbf{X}^* = \mathbf{X}_0 + \mathbf{Z}^*$ for some solution $\mathbf{Z}^*$ to $\mathbf{AZ} = \mathbf{0}$, then

$$\mathbf{AX}^* = \mathbf{AX}_0 + \mathbf{AZ}^* = \mathbf{B} + \mathbf{0} = \mathbf{B},$$

so that $\mathbf{X}^*$ is a solution to $\mathbf{AX} = \mathbf{B}$. Conversely, if $\mathbf{X}^*$ is a solution to $\mathbf{AX} = \mathbf{B}$, then, defining $\mathbf{Z}^* = \mathbf{X}^* - \mathbf{X}_0$, we find that

$$\mathbf{X}^* = \mathbf{X}_0 + \mathbf{Z}^*$$

and, since

$$\mathbf{AZ}^* = \mathbf{AX}^* - \mathbf{AX}_0 = \mathbf{B} - \mathbf{B} = \mathbf{0},$$

that $\mathbf{Z}^*$ is a solution to $\mathbf{AZ} = \mathbf{0}$. Q.E.D.

The upshot of Theorem 11.2.3 is that all of the matrices in the solution set of a consistent linear system $\mathbf{AX} = \mathbf{B}$ can be generated from any particular solution $\mathbf{X}_0$ by setting

$$\mathbf{X} = \mathbf{X}_0 + \mathbf{Z}$$

and allowing $\mathbf{Z}$ to range over all the matrices in the solution space of the linear homogeneous system $\mathbf{AZ} = \mathbf{0}$.

It follows from Theorem 9.1.2 that one solution to a consistent linear system $\mathbf{AX} = \mathbf{B}$ is the matrix $\mathbf{A}^-\mathbf{B}$. Thus, in light of Theorem 11.2.3, we have the following extension of Theorem 11.2.1.

Theorem 11.2.4. A matrix $\mathbf{X}^*$ is a solution to a consistent linear system $\mathbf{AX} = \mathbf{B}$ (in $\mathbf{X}$) if and only if

$$\mathbf{X}^* = \mathbf{A}^-\mathbf{B} + (\mathbf{I} - \mathbf{A}^-\mathbf{A})\mathbf{Y}$$

for some matrix $\mathbf{Y}$.

As a special case of Theorem 11.2.4, we have that a column vector $\mathbf{x}^*$ is a solution to a consistent linear system $\mathbf{Ax} = \mathbf{b}$ (in a column vector $\mathbf{x}$) if and only if

$$\mathbf{x}^* = \mathbf{A}^-\mathbf{b} + (\mathbf{I} - \mathbf{A}^-\mathbf{A})\mathbf{y} \tag{2.1}$$

for some column vector $\mathbf{y}$.

Consider, for example, expression (2.1) as applied to the linear system

$$\begin{pmatrix} -6 & 2 & -2 & -3 \\ 3 & -1 & 5 & 2 \\ -3 & 1 & 3 & -1 \end{pmatrix} \mathbf{x} = \begin{pmatrix} 3 \\ 2 \\ 5 \end{pmatrix}, \tag{2.2}$$

whose consistency was established in Section 9.5. Taking $\mathbf{A}$ and $\mathbf{b}$ to be the coefficient matrix and right side of linear system (2.2), choosing

$$\mathbf{A}^{-} = \begin{pmatrix} 0 & 0 & 0 \\ -1 & 0 & 3 \\ 0 & 0 & 0 \\ -1 & 0 & 2 \end{pmatrix},$$

and denoting the elements of $\mathbf{y}$ by y_1, y_2, y_3, and y_4, respectively, we find that

$$\mathbf{A}^{-}\mathbf{b} + (\mathbf{I} - \mathbf{A}^{-}\mathbf{A})\mathbf{y} = \begin{pmatrix} y_1 \\ 12 + 3y_1 - 11y_3 \\ y_3 \\ 7 - 8y_3 \end{pmatrix}. \tag{2.3}$$

Thus, the members of the solution set of linear system (2.2) consist of all vectors of the general form (2.3).

11.3 Number of Solutions

a. Homogeneous linear systems

What is the dimension of the solution space of a homogeneous linear system $\mathbf{AX} = \mathbf{0}$ (in $\mathbf{X}$)? The following lemma answers this question in the special case of a homogeneous linear system $\mathbf{Ax} = \mathbf{0}$ (in a column vector $\mathbf{x}$).

Lemma 11.3.1. Let $\mathbf{A}$ represent an $m \times n$ matrix. Then,

$$\dim[\mathcal{N}(\mathbf{A})] = n - \text{rank}(\mathbf{A}).$$

That is, the dimension of the solution space of the homogeneous linear system $\mathbf{Ax} = \mathbf{0}$ (in an n-dimensional column vector $\mathbf{x}$) equals $n - \text{rank}(\mathbf{A})$.

Proof. Recalling (from Corollary 11.2.2) that $\mathcal{N}(\mathbf{A}) = \mathcal{C}(\mathbf{I} - \mathbf{A}^{-}\mathbf{A})$ and making use of Lemma 10.2.6, we find that

$$\dim[\mathcal{N}(\mathbf{A})] = \dim[\mathcal{C}(\mathbf{I} - \mathbf{A}^{-}\mathbf{A})] = \text{rank}(\mathbf{I} - \mathbf{A}^{-}\mathbf{A}) = n - \text{rank}(\mathbf{A}).$$

Q.E.D.

The solution space of a homogeneous linear system $\mathbf{Ax} = \mathbf{0}$ (in an n-dimensional column vector $\mathbf{x}$) is a subspace of the linear space $\mathcal{R}^n$ of all n-dimensional column vectors. According to Lemma 11.3.1, the dimension of this subspace equals $n - \text{rank}(\mathbf{A})$. Thus, if $\text{rank}(\mathbf{A}) = n$, that is, if $\mathbf{A}$ has full column rank, then the homogeneous linear system $\mathbf{Ax} = \mathbf{0}$ has a unique solution, namely, the null vector $\mathbf{0}$. And, if $\text{rank}(\mathbf{A}) < n$, then $\mathbf{Ax} = \mathbf{0}$ has an infinite number of solutions.

Consider now the dimension of the solution space of the homogeneous linear system $\mathbf{AX} = \mathbf{0}$ (in an $n \times p$ matrix $\mathbf{X}$). Clearly, a matrix $\mathbf{X}^*$ is a solution to $\mathbf{AX} = \mathbf{0}$ if and only if each of its columns is a solution to the homogeneous linear system $\mathbf{Ax} = \mathbf{0}$ (in the n-dimensional column vector $\mathbf{x}$).

Suppose that $\text{rank}(\mathbf{A}) = n$. Then, the only solution to $\mathbf{Ax} = \mathbf{0}$ is the n-dimensional null vector, and hence the only solution to $\mathbf{AX} = \mathbf{0}$ is the $n \times p$ null matrix.

Alternatively, suppose that $\text{rank}(\mathbf{A}) < n$. Let $s = n - \text{rank}(\mathbf{A})$, and take $\mathbf{x}_1, \ldots, \mathbf{x}_s$ to be any s linearly independent solutions to $\mathbf{Ax} = \mathbf{0}$. (According to Lemma 11.3.1, the solution space of $\mathbf{Ax} = \mathbf{0}$ is s-dimensional.) Then, it is a straightforward exercise to show that the ps $n \times p$ matrices $(\mathbf{x}_1, \mathbf{0}, \ldots, \mathbf{0})$, ..., $(\mathbf{x}_s, \mathbf{0}, \ldots, \mathbf{0})$, ..., $(\mathbf{0}, \ldots, \mathbf{0}, \mathbf{x}_1)$, ..., $(\mathbf{0}, \ldots, \mathbf{0}, \mathbf{x}_s)$ form a basis for the solution space of $\mathbf{AX} = \mathbf{0}$ and hence that the dimension of the solution space of $\mathbf{AX} = \mathbf{0}$ equals ps. Thus, we have the following generalization of Lemma 11.3.1.

Lemma 11.3.2. The dimension of the solution space of a homogeneous linear system $\mathbf{AX} = \mathbf{0}$ (in an $n \times p$ matrix $\mathbf{X}$) equals $p[n - \text{rank}(\mathbf{A})]$.

The solution space of a homogeneous linear system $\mathbf{AX} = \mathbf{0}$ (in an $n \times p$ matrix $\mathbf{X}$) is a subspace of the linear space $\mathcal{R}^{n \times p}$ of all $n \times p$ matrices. According to Lemma 11.3.2, the dimension of this subspace equals $p[n - \text{rank}(\mathbf{A})]$. Thus, if $\text{rank}(\mathbf{A}) = n$, then the homogeneous linear system $\mathbf{AX} = \mathbf{0}$ has a unique solution, namely, the null matrix $\mathbf{0}$. And, if $\text{rank}(\mathbf{A}) < n$, then $\mathbf{AX} = \mathbf{0}$ has an infinite number of solutions.

b. Nonhomogeneous linear systems

Consider the size of the solution set of a possibly nonhomogeneous linear system $\mathbf{AX} = \mathbf{B}$ (in an $n \times p$ matrix $\mathbf{X}$). If $\mathbf{AX} = \mathbf{B}$ is inconsistent, then, by definition, it has zero solutions.

If $\mathbf{AX} = \mathbf{B}$ is consistent, then it follows from Theorem 11.2.3 that there is a one-to-one correspondence between solutions to $\mathbf{AX} = \mathbf{B}$ and solutions to the corresponding homogeneous linear system $\mathbf{AZ} = \mathbf{0}$. Thus, in light of the discussion of Subsection a, we conclude (1) that if the linear system $\mathbf{AX} = \mathbf{B}$ (in the $n \times p$ matrix $\mathbf{X}$) is consistent and if the coefficient matrix $\mathbf{A}$ has full column rank (i.e., rank n), then $\mathbf{AX} = \mathbf{B}$ has a unique solution, and (2) that if $\mathbf{AX} = \mathbf{B}$ is consistent but $\mathbf{A}$ is of less than full column rank, then $\mathbf{AX} = \mathbf{B}$ has an infinite number of solutions.

Moreover, on the basis of Theorem 7.2.2, we reach the following, additional conclusion.

Theorem 11.3.3. If the coefficient matrix $\mathbf{A}$ of the linear system $\mathbf{AX} = \mathbf{B}$ (in $\mathbf{X}$) is nonsingular, then $\mathbf{AX} = \mathbf{B}$ has a solution and that solution is unique.

11.4 A Basic Result on Null Spaces

The following lemma gives a basic result on null spaces.

Lemma 11.4.1. Let $\mathbf{A}$ represent an $m \times n$ matrix and $\mathbf{X}$ an $n \times p$ matrix. (1) If $\mathbf{AX} = \mathbf{0}$, then $\mathcal{C}(\mathbf{X}) \subset \mathcal{N}(\mathbf{A})$. (2) If $\mathbf{AX} = \mathbf{0}$ and $\text{rank}(\mathbf{X}) = n - \text{rank}(\mathbf{A})$, then $\mathcal{C}(\mathbf{X}) = \mathcal{N}(\mathbf{A})$.

Proof. (1) Suppose that $\mathbf{AX} = \mathbf{0}$. Then, for any $n \times 1$ vector $\mathbf{x}$ in $\mathcal{C}(\mathbf{X})$, $\mathbf{x} = \mathbf{Xr}$ for some $p \times 1$ vector $\mathbf{r}$, and consequently $\mathbf{Ax} = \mathbf{AXr} = \mathbf{0}$, so that $\mathbf{x} \in \mathcal{N}(\mathbf{A})$. Thus, $\mathcal{C}(\mathbf{X}) \subset \mathcal{N}(\mathbf{A})$.

(2) Suppose that $\mathbf{AX} = \mathbf{0}$ and that $\text{rank}(\mathbf{X}) = n - \text{rank}(\mathbf{A})$. Then, in light of Part (1) and Lemma 11.3.1, we have that $\mathcal{C}(\mathbf{X}) \subset \mathcal{N}(\mathbf{A})$ and {since $\dim[\mathcal{C}(\mathbf{X})] = \text{rank}(\mathbf{X})$} that $\dim[\mathcal{C}(\mathbf{X})] = \dim[\mathcal{N}(\mathbf{A})]$. Thus, it follows from Theorem 4.3.10 that $\mathcal{C}(\mathbf{X}) = \mathcal{N}(\mathbf{A})$. Q.E.D.

11.5 An Alternative Expression for the General Form of a Solution

It follows from Theorem 9.1.2 that one solution to a consistent linear system $\mathbf{AX} = \mathbf{B}$ (in $\mathbf{X}$) is $\mathbf{A}^{-}\mathbf{B}$. If $\mathbf{X}^*$ is a solution to $\mathbf{AX} = \mathbf{B}$, is it necessarily the case that $\mathbf{X}^* = \mathbf{GB}$ for some generalized inverse $\mathbf{G}$ of $\mathbf{A}$? That is, can all solutions to $\mathbf{AX} = \mathbf{B}$ be generated by setting $\mathbf{X} = \mathbf{GB}$ and letting $\mathbf{G}$ range over all generalized inverses of $\mathbf{A}$? The answer to this question depends on $\mathbf{A}$ and $\mathbf{B}$.

Consider, for example, the special case of a consistent linear system $\mathbf{Ax} = \mathbf{b}$ (in an $n \times 1$ column vector $\mathbf{x}$). If $\text{rank}(\mathbf{A}) < n$ and $\mathbf{b} = \mathbf{0}$, then (as discussed in Section 11.3a) $\mathbf{Ax} = \mathbf{b}$ has an infinite number of solutions; however, only one of them is expressible in the form $\mathbf{Gb}$ for some generalized inverse $\mathbf{G}$ of $\mathbf{A}$, namely, the $n \times 1$ null vector $\mathbf{0}$.

The following theorem gives conditions on $\mathbf{A}$ and $\mathbf{B}$ under which every solution to $\mathbf{AX} = \mathbf{B}$ is expressible in the form $\mathbf{GB}$ for some generalized inverse $\mathbf{G}$ of $\mathbf{A}$.

Theorem 11.5.1. Suppose that $\mathbf{AX} = \mathbf{B}$ is a consistent linear system (in an $n \times p$ matrix $\mathbf{X}$) such that $\text{rank}(\mathbf{A}) = n$ or $\text{rank}(\mathbf{B}) = p$. Then, a matrix $\mathbf{X}^*$ is a solution to $\mathbf{AX} = \mathbf{B}$ if and only if

$$\mathbf{X}^* = \mathbf{GB}$$

for some generalized inverse $\mathbf{G}$ of $\mathbf{A}$.

In the special case where $p = 1$, Theorem 11.5.1 reduces to the following result.

Corollary 11.5.2. Suppose that $\mathbf{Ax} = \mathbf{b}$ is a consistent linear system (in an n-dimensional column vector $\mathbf{x}$) such that $\text{rank}(\mathbf{A}) = n$ or $\mathbf{b} \neq \mathbf{0}$. Then, $\mathbf{x}^*$ is a solution to $\mathbf{Ax} = \mathbf{b}$ if and only if

$$\mathbf{x}^* = \mathbf{Gb}$$

for some generalized inverse $\mathbf{G}$ of $\mathbf{A}$.

Proof (of Theorem 11.5.1). If $\mathbf{X}^* = \mathbf{GB}$ for some generalized inverse $\mathbf{G}$ of $\mathbf{A}$, then $\mathbf{X}^*$ is a solution to $\mathbf{AX} = \mathbf{B}$, as is evident from Theorem 9.1.2.

Conversely, suppose that $\mathbf{X}^*$ is a solution to $\mathbf{AX} = \mathbf{B}$. By assumption, $\text{rank}(\mathbf{A}) = n$ or $\text{rank}(\mathbf{B}) = p$.

Assume that $\text{rank}(\mathbf{A}) = n$. Then, $\mathbf{AX} = \mathbf{B}$ has a unique solution (as discussed in Section 11.3), and hence (since—according to Theorem 9.1.2—$\mathbf{A}^-\mathbf{B}$ is a solution to $\mathbf{AX} = \mathbf{B}$) $\mathbf{X}^* = \mathbf{A}^-\mathbf{B}$; that is, $\mathbf{X}^* = \mathbf{GB}$ for any generalized inverse $\mathbf{G}$ of $\mathbf{A}$.

Alternatively, assume that $\text{rank}(\mathbf{B}) = p$. According to Theorem 11.2.4, there exists a matrix $\mathbf{Y}$ such that

$$\mathbf{X}^* = \mathbf{A}^-\mathbf{B} + (\mathbf{I} - \mathbf{A}^-\mathbf{A})\mathbf{Y},$$

and, since the matrix $\mathbf{B}'$ is of full row rank (i.e., rank p), there exists (in light of Theorem 7.2.2) a matrix $\mathbf{K}$ such that

$$\mathbf{B}'\mathbf{K} = \mathbf{Y}'.$$

Furthermore, according to Lemma 9.5.1,

$$\mathbf{AA}^-\mathbf{B} = \mathbf{B}.$$

Thus,

$$\begin{aligned}\mathbf{X}^* = \mathbf{A}^-\mathbf{B} + (\mathbf{I} - \mathbf{A}^-\mathbf{A})(\mathbf{B}'\mathbf{K})' &= \mathbf{A}^-\mathbf{B} + \mathbf{K}'\mathbf{B} - \mathbf{A}^-\mathbf{AK}'\mathbf{B} \\ &= (\mathbf{A}^- + \mathbf{K}' - \mathbf{A}^-\mathbf{AK}'\mathbf{AA}^-)\mathbf{B},\end{aligned}$$

and hence, in light of Theorem 9.2.7, $\mathbf{X}^* = \mathbf{GB}$ for some generalized inverse $\mathbf{G}$ of $\mathbf{A}$. Q.E.D.

Theorem 11.5.1 gives an expression—alternative to that given by Theorem 11.2.4—for the general form of a solution to a consistent linear system $\mathbf{AX} = \mathbf{B}$ (in an $n \times p$ matrix $\mathbf{X}$). However, Theorem 11.5.1 is applicable only if $\text{rank}(\mathbf{A}) = n$ or $\text{rank}(\mathbf{B}) = p$, whereas Theorem 11.2.4 is applicable to any consistent linear system.

11.6 Equivalent Linear Systems

If two linear systems have the same solution set, they are said to be *equivalent*.

The linear system $\mathbf{CAX} = \mathbf{CB}$ obtained by premultiplying both sides of a linear system $\mathbf{AX} = \mathbf{B}$ (in $\mathbf{X}$) by a matrix $\mathbf{C}$ may or may not be equivalent to $\mathbf{AX} = \mathbf{B}$. The following two lemmas, which are immediate consequences of Lemma 8.3.1 and Corollary 5.3.3, respectively, give sufficient conditions for the two linear systems to be equivalent.

Lemma 11.6.1. Let $\mathbf{A}$ represent an $m \times n$ matrix and $\mathbf{B}$ an $m \times p$ matrix. If $\mathbf{C}$ is an $r \times m$ matrix of full column rank (i.e., of rank m), then the linear system $\mathbf{CAX} = \mathbf{CB}$ is equivalent to the linear system $\mathbf{AX} = \mathbf{B}$ (in $\mathbf{X}$).

Lemma 11.6.2. For any $m \times n$ matrix $\mathbf{A}$ and $n \times p$ matrix $\mathbf{F}$, the linear system $\mathbf{A}'\mathbf{A}\mathbf{X} = \mathbf{A}'\mathbf{A}\mathbf{F}$ is equivalent to the linear system $\mathbf{A}\mathbf{X} = \mathbf{A}\mathbf{F}$ (in $\mathbf{X}$).

Note that, as a special case of Lemma 11.6.2 (the special case where $\mathbf{F} = \mathbf{0}$), we have that the linear system $\mathbf{A}'\mathbf{A}\mathbf{X} = \mathbf{0}$ is equivalent to the linear system $\mathbf{A}\mathbf{X} = \mathbf{0}$.

11.7 Null and Column Spaces of Idempotent Matrices

It was noted earlier [in results (1.1) and (1.2)] that, for any square matrix $\mathbf{A}$, $\mathcal{N}(\mathbf{A}) \subset \mathcal{C}(\mathbf{I} - \mathbf{A})$ and $\mathcal{N}(\mathbf{I} - \mathbf{A}) \subset \mathcal{C}(\mathbf{A})$. Is it possible for $\mathcal{N}(\mathbf{A})$ to equal $\mathcal{C}(\mathbf{I} - \mathbf{A})$ and/or for $\mathcal{N}(\mathbf{I} - \mathbf{A})$ to equal $\mathcal{C}(\mathbf{A})$? And, if so, under what condition(s)? These questions are answered in the following theorem and corollary.

Theorem 11.7.1. Let $\mathbf{A}$ represent an $n \times n$ matrix. Then,

$$\mathcal{N}(\mathbf{A}) = \mathcal{C}(\mathbf{I} - \mathbf{A})$$

if and only if $\mathbf{A}$ is idempotent.

Proof. Suppose that $\mathbf{A}$ is idempotent. Then, since (according to Lemma 10.1.3) $\mathbf{A}$ is a generalized inverse of itself, it follows from Corollary 11.2.2 that

$$\mathcal{N}(\mathbf{A}) = \mathcal{C}(\mathbf{I} - \mathbf{A}\mathbf{A}) = \mathcal{C}(\mathbf{I} - \mathbf{A}).$$

Conversely, suppose that $\mathcal{N}(\mathbf{A}) = \mathcal{C}(\mathbf{I} - \mathbf{A})$. Then, for every $n \times 1$ vector $\mathbf{x}$, $(\mathbf{I} - \mathbf{A})\mathbf{x} \in \mathcal{N}(\mathbf{A})$ or equivalently $\mathbf{A}(\mathbf{I} - \mathbf{A})\mathbf{x} = \mathbf{0}$. Thus, in light of Lemma 2.3.2, we have that $\mathbf{A} - \mathbf{A}^2 = \mathbf{A}(\mathbf{I} - \mathbf{A}) = \mathbf{0}$ and hence that $\mathbf{A}^2 = \mathbf{A}$. Q.E.D.

Corollary 11.7.2. Let $\mathbf{A}$ represent an $n \times n$ matrix. Then,

$$\mathcal{C}(\mathbf{A}) = \mathcal{N}(\mathbf{I} - \mathbf{A})$$

if and only if $\mathbf{A}$ is idempotent.

Proof. According to Theorem 11.7.1,

$$\mathcal{N}(\mathbf{I} - \mathbf{A}) = \mathcal{C}[\mathbf{I} - (\mathbf{I} - \mathbf{A})]$$

if and only if $\mathbf{I} - \mathbf{A}$ is idempotent. The proof is complete upon observing that $\mathbf{I} - (\mathbf{I} - \mathbf{A}) = \mathbf{A}$ and further (in light of Lemma 10.1.2) that $\mathbf{I} - \mathbf{A}$ is idempotent if and only if $\mathbf{A}$ is idempotent. Q.E.D.

11.8 Linear Systems With Nonsingular Triangular or Block-Triangular Coefficient Matrices

Let

$$\mathbf{A} = \begin{pmatrix} \mathbf{A}_{11} & \mathbf{A}_{12} & \dots & \mathbf{A}_{1r} \\ \mathbf{0} & \mathbf{A}_{22} & \dots & \mathbf{A}_{2r} \\ \vdots & & \ddots & \vdots \\ \mathbf{0} & \mathbf{0} & & \mathbf{A}_{rr} \end{pmatrix}$$

represent an $n \times n$ upper block-triangular matrix whose ijth block $\mathbf{A}_{ij}$ is of dimensions $n_i \times n_j$ ($j \geq i = 1, 2, \ldots, r$). Suppose that the diagonal blocks $\mathbf{A}_{11}, \mathbf{A}_{22}, \ldots, \mathbf{A}_{rr}$ of $\mathbf{A}$ are all nonsingular, or, equivalently (in light of Theorem 8.5.5), that $\mathbf{A}$ itself is nonsingular.

Let $\mathbf{B}$ represent an $n \times p$ matrix, and consider the linear system $\mathbf{AX} = \mathbf{B}$ (in the $n \times p$ matrix $\mathbf{X}$). Partition $\mathbf{X}$ and $\mathbf{B}$ as

$$\mathbf{X} = \begin{pmatrix} \mathbf{X}_1 \\ \mathbf{X}_2 \\ \vdots \\ \mathbf{X}_r \end{pmatrix} \quad \text{and} \quad \mathbf{B} = \begin{pmatrix} \mathbf{B}_1 \\ \mathbf{B}_2 \\ \vdots \\ \mathbf{B}_r \end{pmatrix},$$

where $\mathbf{X}_i$ and $\mathbf{B}_i$ are of dimensions $n_i \times p$ ($i = 1, \ldots, r$), and rewrite the linear system as

$$\begin{pmatrix} \mathbf{A}_{11} & \mathbf{A}_{12} & \ldots & \mathbf{A}_{1r} \\ \mathbf{0} & \mathbf{A}_{22} & \ldots & \mathbf{A}_{2r} \\ \vdots & & \ddots & \vdots \\ \mathbf{0} & \mathbf{0} & & \mathbf{A}_{rr} \end{pmatrix} \begin{pmatrix} \mathbf{X}_1 \\ \mathbf{X}_2 \\ \vdots \\ \mathbf{X}_r \end{pmatrix} = \begin{pmatrix} \mathbf{B}_1 \\ \mathbf{B}_2 \\ \vdots \\ \mathbf{B}_r \end{pmatrix} \tag{8.1}$$

or, equivalently, as

$$\mathbf{A}_{ii}\mathbf{X}_i + \sum_{j=i+1}^{r} \mathbf{A}_{ij}\mathbf{X}_j = \mathbf{B}_i \qquad (i = 1, \ldots, r-1),$$

$$\mathbf{A}_{rr}\mathbf{X}_r = \mathbf{B}_r.$$

In solving linear system (8.1), the block-triangularity of the coefficient matrix can be exploited by proceeding in r steps. The first step is to solve the linear system $\mathbf{A}_{rr}\mathbf{X}_r = \mathbf{B}_r$ for $\mathbf{X}_r$—one representation for the solution is $\mathbf{X}_r = \mathbf{A}_{rr}^{-1}\mathbf{B}_r$. The second step is to solve the linear system

$$\mathbf{A}_{r-1,r-1}\mathbf{X}_{r-1} + \mathbf{A}_{r-1,r}\mathbf{X}_r = \mathbf{B}_{r-1}$$

for $\mathbf{X}_{r-1}$ in terms of $\mathbf{X}_r$—one representation for this solution is

$$\mathbf{X}_{r-1} = \mathbf{A}_{r-1,r-1}^{-1}(\mathbf{B}_{r-1} - \mathbf{A}_{r-1,r}\mathbf{X}_r).$$

More generally, the $(r - i + 1)$th step is to solve the linear system

$$\mathbf{A}_{ii}\mathbf{X}_i + \sum_{j=i+1}^{r} \mathbf{A}_{ij}\mathbf{X}_j = \mathbf{B}_i$$

for $\mathbf{X}_i$ in terms of the quantities $\mathbf{X}_r, \mathbf{X}_{r-1}, \ldots, \mathbf{X}_{i+1}$ determined during the first, second, ..., $(r - i)$th steps, respectively—one representation for this solution is

$$\mathbf{X}_i = \mathbf{A}_{ii}^{-1}\Big(\mathbf{B}_i - \sum_{j=i+1}^{r} \mathbf{A}_{ij}\mathbf{X}_j\Big) \tag{8.2}$$

$(i = r-1, r-2, \ldots, 1)$. This approach to the solution of linear system (8.1) is called *back substitution*.

Let us now specialize by taking $\mathbf{U} = \{u_{ij}\}$ to be an $n \times n$ nonsingular upper triangular matrix and $\mathbf{b} = \{b_i\}$ to be an $n \times 1$ vector, and by considering the linear system $\mathbf{Ux} = \mathbf{b}$ in the $n \times 1$ vector $\mathbf{x} = \{x_i\}$. As applied to this linear system, back substitution consists of determining $x_n, x_{n-1}, \ldots, x_1$, recursively, from the n formulas

$$x_n = b_n/u_{nn}, \tag{8.3a}$$

$$x_i = \Bigl(b_i - \sum_{j=i+1}^{n} u_{ij}x_j\Bigr)/u_{ii} \quad (i = n-1, \ldots, 1). \tag{8.3b}$$

For example, in the case of the linear system

$$\begin{pmatrix} 8 & 1 & 6 \\ 0 & 5 & 4 \\ 0 & 0 & 2 \end{pmatrix} \begin{pmatrix} x_1 \\ x_2 \\ x_3 \end{pmatrix} = \begin{pmatrix} -9 \\ 3 \\ -6 \end{pmatrix},$$

these formulas give

$$\begin{aligned} x_3 &= -6/2 = -3, \\ x_2 &= [3 - 4(-3)]/5 = 3, \\ x_1 &= [-9 - 1(3) - 6(-3)]/8 = 0.75. \end{aligned}$$

Suppose now that, instead of being upper block-triangular, the coefficient matrix of the linear system $\mathbf{AX} = \mathbf{B}$ is a lower block-triangular matrix

$$\mathbf{A} = \begin{pmatrix} \mathbf{A}_{11} & \mathbf{0} & \ldots & \mathbf{0} \\ \mathbf{A}_{21} & \mathbf{A}_{22} & & \mathbf{0} \\ \vdots & \vdots & \ddots & \\ \mathbf{A}_{r1} & \mathbf{A}_{r2} & \ldots & \mathbf{A}_{rr} \end{pmatrix}$$

(with nonsingular diagonal blocks $\mathbf{A}_{11}, \mathbf{A}_{22}, \ldots, \mathbf{A}_{rr}$), in which case the linear system is expressible as

$$\begin{pmatrix} \mathbf{A}_{11} & \mathbf{0} & \ldots & \mathbf{0} \\ \mathbf{A}_{21} & \mathbf{A}_{22} & & \mathbf{0} \\ \vdots & \vdots & \ddots & \\ \mathbf{A}_{r1} & \mathbf{A}_{r2} & \ldots & \mathbf{A}_{rr} \end{pmatrix} \begin{pmatrix} \mathbf{X}_1 \\ \mathbf{X}_2 \\ \vdots \\ \mathbf{X}_r \end{pmatrix} = \begin{pmatrix} \mathbf{B}_1 \\ \mathbf{B}_2 \\ \vdots \\ \mathbf{B}_r \end{pmatrix} \tag{8.4}$$

or, equivalently, as

$$\begin{aligned} \mathbf{A}_{11}\mathbf{X}_1 &= \mathbf{B}_1, \\ \mathbf{A}_{ii}\mathbf{X}_i + \sum_{j=1}^{i-1} \mathbf{A}_{ij}\mathbf{X}_j &= \mathbf{B}_i \quad (i = 2, 3, \ldots, r). \end{aligned}$$

In solving linear system (8.4), the block-triangularity of the coefficient matrix can (as in the case of an upper block-triangular coefficient matrix) be exploited by proceeding in r steps. The first step is to solve the linear system $\mathbf{A}_{11}\mathbf{X}_1 = \mathbf{B}_1$ for $\mathbf{X}_1$—one representation for this solution is $\mathbf{X}_1 = \mathbf{A}_{11}^{-1}\mathbf{B}_1$. The ith step is to solve the linear system

$$\mathbf{A}_{ii}\mathbf{X}_i + \sum_{j=1}^{i-1} \mathbf{A}_{ij}\mathbf{X}_j = \mathbf{B}_i$$

for $\mathbf{X}_i$ in terms of the quantities $\mathbf{X}_1, \ldots, \mathbf{X}_{i-1}$ determined during the first, second, $\ldots$, $(i-1)$th steps, respectively—one representation for this solution is

$$\mathbf{X}_i = \mathbf{A}_{ii}^{-1}\Big(\mathbf{B}_i - \sum_{j=1}^{i-1} \mathbf{A}_{ij}\mathbf{X}_j\Big) \tag{8.5}$$

$(i = 2, 3, \ldots, r)$. This approach to the solution of linear system (8.4) is called *forward elimination*.

Let us now specialize by taking $\mathbf{L} = \{\ell_{ij}\}$ to be an $n \times n$ nonsingular lower triangular matrix and $\mathbf{b} = \{b_i\}$ to be an $n \times 1$ vector and by considering the linear system $\mathbf{Lx} = \mathbf{b}$ in the $n \times 1$ vector $\mathbf{x} = \{x_i\}$. As applied to this linear system, forward elimination consists of determining $x_1, x_2, \ldots, x_n$, recursively, from the n formulas

$$x_1 = b_1/\ell_{11}, \tag{8.6a}$$

$$x_i = \Big(b_i - \sum_{j=1}^{i-1} \ell_{ij}x_j\Big)/\ell_{ii} \quad (i = 2, 3, \ldots, n). \tag{8.6b}$$

11.9 A Computational Approach

Consider the problem of computing a solution to a consistent linear system $\mathbf{AX} = \mathbf{B}$ (in $\mathbf{X}$). Depending on the nature of the coefficient matrix $\mathbf{A}$, this problem can be relatively easy or relatively difficult. One case where it is realtively easy is that where $\mathbf{A}$ is an orthogonal matrix or more generally where the columns of $\mathbf{A}$ are orthonormal (with respect to the usual inner product). In that special case, the (unique) solution to $\mathbf{AX} = \mathbf{B}$ is $\mathbf{X} = \mathbf{A}'\mathbf{B}$. Another case where the computation of a solution to $\mathbf{AX} = \mathbf{B}$ is relatively easy is that where $\mathbf{A}$ is a triangular matrix. If $\mathbf{A}$ is upper or lower triangular, then a solution to the linear system can be computed by back substitution or forward elimination—refer to Section 11.8—or (in the event $\mathbf{A}$ is singular) by a variant of back substitution or forward elimination.

More generally, the computation of a solution to $\mathbf{AX} = \mathbf{B}$ can be approached by (implicitly or explicitly) decomposing $\mathbf{A}$ into the product $\mathbf{A} = \mathbf{KT}$ of two matrices $\mathbf{K}$ and $\mathbf{T}$, where $\mathbf{T}$ is of full row rank, and by proceeding in two steps as follows: (1) compute a solution $\mathbf{Y}_*$ to the linear system $\mathbf{KY} = \mathbf{B}$ (in a matrix $\mathbf{Y}$) and then (2) compute a solution $\mathbf{X}_*$ to the linear system $\mathbf{TX} = \mathbf{Y}_*$ (in $\mathbf{X}$).

[Upon observing that $\mathcal{C}(\mathbf{B}) \subset \mathcal{C}(\mathbf{A}) \subset \mathcal{C}(\mathbf{K})$, it follows from Theorem 7.2.1 that the linear system $\mathbf{KY} = \mathbf{B}$ is consistent; and it follows from Theorem 7.2.2 that the linear system $\mathbf{TX} = \mathbf{Y}_*$ is consistent.] Clearly,

$$\mathbf{AX}_* = \mathbf{KTX}_* = \mathbf{KY}_* = \mathbf{B},$$

so that the matrix $\mathbf{X}_*$ computed in Step (2) is a solution to $\mathbf{AX} = \mathbf{B}$. Here, the decomposition $\mathbf{A} = \mathbf{KT}$ should be such that the solutions of the linear systems $\mathbf{KY} = \mathbf{B}$ and $\mathbf{TX} = \mathbf{Y}_*$ are relatively easy to compute. One choice for this decomposition (in the special case where $\mathbf{A}$ is of full column rank) is the QR decomposition, considered in Section 6.4d. Other decompositions that could (depending on the nature of $\mathbf{A}$) be suitable choices will be considered subsequently (in Chapters 14 and 21).

11.10 Linear Combinations of the Unknowns

Suppose that $\mathbf{Ax} = \mathbf{b}$ is a consistent linear system (in an n-dimensional column vector $\mathbf{x}$). Often, $\mathbf{k}'\mathbf{x}$, where $\mathbf{k}$ is an n-dimensional column vector, or more generally $\mathbf{K}'\mathbf{x}$, where $\mathbf{K}$ is an $n \times q$ matrix, may be of interest rather than $\mathbf{x}$ itself—note that $\mathbf{k}'\mathbf{x}$ is a linear combination of the unknowns (the elements of $\mathbf{x}$) and that $\mathbf{K}'\mathbf{x}$ is a q-dimensional column vector, each element of which is a linear combination of the unknowns. For instance, in the case of the so-called normal equations (which arise from fitting a linear statistical model by least squares), interest may be confined to those linear combinations of the unknowns that correspond to estimable functions—refer, for example, to Searle (1971, sec. 5.4).

More generally, suppose that $\mathbf{AX} = \mathbf{B}$ is a consistent linear system (in an $n \times p$ matrix $\mathbf{X}$). Then, $\mathbf{K}'\mathbf{X}$ may be of interest.

a. Invariance to choice of solution

Under what circumstances is the value of $\mathbf{K}'\mathbf{X}$ invariant to the choice of solution to $\mathbf{AX} = \mathbf{B}$? The answer to this question is provided by the following theorem.

Theorem 11.10.1. Suppose that $\mathbf{AX} = \mathbf{B}$ is a consistent linear system (in an $n \times p$ matrix $\mathbf{X}$), and let $\mathbf{K}$ represent an $n \times q$ matrix. Then, the value of $\mathbf{K}'\mathbf{X}$ is the same for every solution to $\mathbf{AX} = \mathbf{B}$ if and only if $\mathcal{R}(\mathbf{K}') \subset \mathcal{R}(\mathbf{A})$ or equivalently if and only if every row of $\mathbf{K}'$ belongs to $\mathcal{R}(\mathbf{A})$.

Proof. Let $\mathbf{X}_0$ represent any particular solution to $\mathbf{AX} = \mathbf{B}$.

Suppose that $\mathcal{R}(\mathbf{K}') \subset \mathcal{R}(\mathbf{A})$. Then, according to Lemma 4.2.2, there exists a matrix $\mathbf{L}$ such that $\mathbf{K}' = \mathbf{LA}$. Letting $\mathbf{X}^*$ represent any solution to $\mathbf{AX} = \mathbf{B}$, we find that

$$\mathbf{K}'\mathbf{X}^* = \mathbf{LAX}^* = \mathbf{LB} = \mathbf{LAX}_0 = \mathbf{K}'\mathbf{X}_0.$$

Thus, the value of $\mathbf{K}'\mathbf{X}$ is the same for every solution to $\mathbf{AX} = \mathbf{B}$.

Conversely, suppose that the value of $\mathbf{K}'\mathbf{X}$ is the same for every solution to $\mathbf{AX} = \mathbf{B}$. Then, it follows from Theorem 11.2.4 that the value of

$$\mathbf{K}'[\mathbf{A}^-\mathbf{B} + (\mathbf{I} - \mathbf{A}^-\mathbf{A})\mathbf{Y}]$$

is the same for every matrix $\mathbf{Y}$ (of appropriate dimensions) or equivalently that $\mathbf{K}'(\mathbf{I} - \mathbf{A}^-\mathbf{A})\mathbf{Y} = \mathbf{0}$ for every matrix $\mathbf{Y}$. In particular, $\mathbf{K}'(\mathbf{I} - \mathbf{A}^-\mathbf{A})\mathbf{Y} = \mathbf{0}$ for every matrix $\mathbf{Y}$ of the general form $(\mathbf{y}, \mathbf{0}, \ldots, \mathbf{0})$. Thus, $\mathbf{K}'(\mathbf{I} - \mathbf{A}^-\mathbf{A})\mathbf{y} = \mathbf{0}$ for every $n \times 1$ column vector $\mathbf{y}$, or equivalently $\mathbf{K}'\mathbf{y} = \mathbf{K}'\mathbf{A}^-\mathbf{A}\mathbf{y}$ for every $\mathbf{y}$, implying (in light of Lemma 2.3.2) that $\mathbf{K}' = \mathbf{K}'\mathbf{A}^-\mathbf{A}$. We conclude, on the basis of Lemma 9.3.5, that $\mathcal{R}(\mathbf{K}') \subset \mathcal{R}(\mathbf{A})$. Q.E.D.

As a special case of Theorem 11.10.1, we have the following corollary.

Corollary 11.10.2. Suppose that $\mathbf{Ax} = \mathbf{b}$ is a consistent linear system (in an n-dimensional column vector $\mathbf{x}$), and let $\mathbf{k}'$ represent an n-dimensional row vector. Then, the value of $\mathbf{k}'\mathbf{x}$ is the same for every solution to $\mathbf{Ax} = \mathbf{b}$ if and only if $\mathbf{k}' \in \mathcal{R}(\mathbf{A})$.

b. An alternative representation

As described in the following theorem, the solution set of the linear system $\mathbf{AX} = \mathbf{B}$ is related to that of a linear system $\mathbf{A}'\mathbf{Y} = \mathbf{K}$ (in $\mathbf{Y}$).

Theorem 11.10.3. Suppose that $\mathbf{AX} = \mathbf{B}$ and $\mathbf{A}'\mathbf{Y} = \mathbf{K}$ are consistent linear systems (in $\mathbf{X}$ and $\mathbf{Y}$, respectively). Then,

$$\mathbf{K}'\mathbf{X}_0 = \mathbf{Y}_0'\mathbf{B} \tag{10.1}$$

for any solutions $\mathbf{X}_0$ and $\mathbf{Y}_0$ to $\mathbf{AX} = \mathbf{B}$ and $\mathbf{A}'\mathbf{Y} = \mathbf{K}$, respectively.

Proof. We have that

$$\mathbf{K}'\mathbf{X}_0 = (\mathbf{A}'\mathbf{Y}_0)'\mathbf{X}_0 = \mathbf{Y}_0'\mathbf{A}\mathbf{X}_0 = \mathbf{Y}_0'\mathbf{B}.$$

Q.E.D.

Suppose that $\mathbf{AX} = \mathbf{B}$ is a consistent linear system (in $\mathbf{X}$) and that, for some $n \times q$ matrix $\mathbf{K}$ such that $\mathcal{R}(\mathbf{K}') \subset \mathcal{R}(\mathbf{A})$, the value of $\mathbf{K}'\mathbf{X}$ is of interest. (It follows from Theorem 11.10.1 that the value of $\mathbf{K}'\mathbf{X}$ is the same for every solution to $\mathbf{AX} = \mathbf{B}$.) An obvious way to determine the value of $\mathbf{K}'\mathbf{X}$ is to find a solution $\mathbf{X}_0$ to $\mathbf{AX} = \mathbf{B}$ and to then form the product $\mathbf{K}'\mathbf{X}_0$. According to Theorem 11.10.3, there is an alternative approach—the value of $\mathbf{K}'\mathbf{X}$ can be determined by finding a solution $\mathbf{Y}_0$ to $\mathbf{A}'\mathbf{Y} = \mathbf{K}$ and then forming the product $\mathbf{Y}_0'\mathbf{B}$.

c. Two augmented linear systems

It is easy to verify the following two theorems.

Theorem 11.10.4. Let $\mathbf{A}$ represent an $m \times n$ matrix, $\mathbf{B}$ an $m \times p$ matrix, and $\mathbf{K}$ an $n \times q$ matrix. If $\mathbf{X}^*$ and $\mathbf{L}^*$ are the first and second parts, respectively, of any solution to the linear system

$$\begin{pmatrix} \mathbf{A} & \mathbf{0} \\ -\mathbf{K}' & \mathbf{I} \end{pmatrix} \begin{pmatrix} \mathbf{X} \\ \mathbf{L} \end{pmatrix} = \begin{pmatrix} \mathbf{B} \\ \mathbf{0} \end{pmatrix} \tag{10.2}$$

(in **X** and **L**), then $\mathbf{X}^*$ is a solution to the linear system $\mathbf{AX} = \mathbf{B}$ (in **X**), and $\mathbf{L}^* = \mathbf{K}'\mathbf{X}^*$. Conversely, if $\mathbf{X}^*$ is any solution to $\mathbf{AX} = \mathbf{B}$, then $\mathbf{X}^*$ and $\mathbf{K}'\mathbf{X}^*$ are the first and second parts, respectively, of some solution to linear system (10.2).

Theorem 11.10.5. Let **A** represent an $n \times n$ matrix, **B** an $n \times p$ matrix, and **K** an $n \times q$ matrix. If $\mathbf{X}^*$ and $\mathbf{L}^*$ are the first and second parts, respectively, of any solution to the linear system

$$\begin{pmatrix} \mathbf{A} + \mathbf{KK}' & -\mathbf{K} \\ -\mathbf{K}' & \mathbf{I} \end{pmatrix} \begin{pmatrix} \mathbf{X} \\ \mathbf{L} \end{pmatrix} = \begin{pmatrix} \mathbf{B} \\ \mathbf{0} \end{pmatrix} \tag{10.3}$$

(in **X** and **L**), then $\mathbf{X}^*$ is a solution to the linear system $\mathbf{AX} = \mathbf{B}$ (in **X**) and $\mathbf{L}^* = \mathbf{K}'\mathbf{X}^*$. Conversely, if $\mathbf{X}^*$ is any solution to $\mathbf{AX} = \mathbf{B}$, then $\mathbf{X}^*$ and $\mathbf{K}'\mathbf{X}^*$ are the first and second parts, respectively, of some solution to linear system (10.3).

Note that Theorem 11.10.4 implies that linear system (10.2) is consistent if and only if the linear system $\mathbf{AX} = \mathbf{B}$ is consistent. Similarly, Theorem 11.10.5 implies that linear system (10.3) is consistent if and only if the linear system $\mathbf{AX} = \mathbf{B}$ is consistent.

Theorems 11.10.4 and 11.10.5 indicate that a solution $\mathbf{X}^*$ to a consistent linear system $\mathbf{AX} = \mathbf{B}$ (in **X**) and the corresponding value $\mathbf{K}'\mathbf{X}^*$ of $\mathbf{K}'\mathbf{X}$ can both be obtained as parts of the solution to a single linear system, namely linear system (10.2) or (in the special case where **A** is square) linear system (10.3).

11.11 Absorption

Consider the problem of determining whether a linear system

$$a_{11}x_1 + a_{12}x_2 = b_1, \tag{11.1a}$$
$$a_{21}x_1 + a_{22}x_2 = b_2 \tag{11.1b}$$

of two equations in two unknowns (x_1 and x_2) is consistent and, if it is consistent, of solving it. Suppose that $a_{11} \neq 0$. Then, one approach to the problem is to solve the first equation for x_1 in terms of x_2, obtaining

$$x_1 = a_{11}^{-1}b_1 - a_{11}^{-1}a_{12}x_2, \tag{11.2}$$

and to eliminate x_1 from the second equation by replacing x_1 with expression (11.2), which gives

$$(a_{22} - a_{21}a_{11}^{-1}a_{12})x_2 = b_2 - a_{21}a_{11}^{-1}b_1. \tag{11.3}$$

Clearly, linear system (11.1) is consistent if and only if equation (11.3) has a solution (for x_2), say x_2^*, in which case a solution to linear system (11.1) is x_1^* and x_2^*, where

$$x_1^* = a_{11}^{-1}b_1 - a_{11}^{-1}a_{12}x_2^*$$

is the value of x_1 obtained by setting $x_2 = x_2^*$ in expression (11.2). Moreover, if $a_{22} \neq a_{21}a_{11}^{-1}a_{12}$, equation (11.3) has a unique solution, namely,

$$x_2^* = (a_{22} - a_{21}a_{11}^{-1}a_{12})^{-1}(b_2 - a_{21}a_{11}^{-1}b_1).$$

If $a_{22} = a_{21}a_{11}^{-1}a_{12}$, then, depending on whether or not $b_2 = a_{21}a_{11}^{-1}b_1$, either any value of x_2 can serve as a solution to equation (11.3), or there is no solution to equation (11.3).

Let us now consider the extension of this approach to the problem of solving linear system (11.1) to that of solving a linear system comprising an arbitrary number of equations in an arbitrary number of unknowns. The basis for such an extension is provided by the following theorem.

Theorem 11.11.1. Let $\mathbf{A}$ represent an $m \times n$ matrix and $\mathbf{B}$ an $m \times p$ matrix, and consider the linear system $\mathbf{AX} = \mathbf{B}$ (in an $n \times p$ matrix $\mathbf{X}$). Partition $\mathbf{A}$, $\mathbf{B}$, and $\mathbf{X}$ as

$$\mathbf{A} = \begin{pmatrix} \mathbf{A}_{11} & \mathbf{A}_{12} \\ \mathbf{A}_{21} & \mathbf{A}_{22} \end{pmatrix}, \quad \mathbf{B} = \begin{pmatrix} \mathbf{B}_1 \\ \mathbf{B}_2 \end{pmatrix}, \quad \text{and} \quad \mathbf{X} = \begin{pmatrix} \mathbf{X}_1 \\ \mathbf{X}_2 \end{pmatrix},$$

where $\mathbf{A}_{11}, \mathbf{A}_{12}$, and $\mathbf{B}_1$ have m_1 rows, $\mathbf{A}_{21}, \mathbf{A}_{22}$, and $\mathbf{B}_2$ have m_2 rows, $\mathbf{A}_{11}$ and $\mathbf{A}_{21}$ have n_1 columns, $\mathbf{A}_{12}$ and $\mathbf{A}_{22}$ have n_2 columns, and $\mathbf{X}_1$ and $\mathbf{X}_2$ have n_1 rows and n_2 rows, respectively. Suppose that

$$\mathcal{C}(\mathbf{A}_{12}) \subset \mathcal{C}(\mathbf{A}_{11}), \quad \mathcal{C}(\mathbf{B}_1) \subset \mathcal{C}(\mathbf{A}_{11}), \quad \text{and} \quad \mathcal{R}(\mathbf{A}_{21}) \subset \mathcal{R}(\mathbf{A}_{11}),$$

and let $\mathbf{C} = \mathbf{A}_{11}^{-}\mathbf{B}_1$ and $\mathbf{K} = \mathbf{A}_{11}^{-}\mathbf{A}_{12}$ or, more generally, let $\mathbf{C}$ represent any matrix such that $\mathbf{A}_{11}\mathbf{C} = \mathbf{B}_1$ and $\mathbf{K}$ any matrix such that $\mathbf{A}_{11}\mathbf{K} = \mathbf{A}_{12}$. Then, (1) the matrix $\mathbf{X}^* = \begin{pmatrix} \mathbf{X}_1^* \\ \mathbf{X}_2^* \end{pmatrix}$ (where $\mathbf{X}_1^*$ has n_1 rows) is a solution to the linear system $\mathbf{AX} = \mathbf{B}$ if and only if $\mathbf{X}_2^*$ is a solution to the linear system

$$(\mathbf{A}_{22} - \mathbf{A}_{21}\mathbf{K})\mathbf{X}_2 = \mathbf{B}_2 - \mathbf{A}_{21}\mathbf{C} \quad (\text{in } \mathbf{X}_2) \tag{11.4}$$

and $\mathbf{X}_1^*$ and $\mathbf{X}_2^*$ form a solution to the linear system

$$\mathbf{A}_{11}\mathbf{X}_1 + \mathbf{A}_{12}\mathbf{X}_2 = \mathbf{B}_1 \quad (\text{in } \mathbf{X}_1 \text{ and } \mathbf{X}_2); \tag{11.5}$$

(2) the linear system $\mathbf{AX} = \mathbf{B}$ is consistent if and only if linear system (11.4) is consistent; (3) for any particular value of $\mathbf{X}_2$, $\begin{pmatrix} \mathbf{C} - \mathbf{KX}_2 \\ \mathbf{X}_2 \end{pmatrix}$ is a solution to linear system (11.5), that is

$$\mathbf{A}_{11}(\mathbf{C} - \mathbf{KX}_2) + \mathbf{A}_{12}\mathbf{X}_2 = \mathbf{B}_1.$$

In connection with Theorem 11.11.1, observe that the linear system $\mathbf{AX} = \mathbf{B}$ can be reexpressed in two parts as follows:

$$\mathbf{A}_{11}\mathbf{X}_1 + \mathbf{A}_{12}\mathbf{X}_2 = \mathbf{B}_1,$$
$$\mathbf{A}_{21}\mathbf{X}_1 + \mathbf{A}_{22}\mathbf{X}_2 = \mathbf{B}_2.$$

Observe also (in light of Lemma 4.2.2) that the existence of matrices $\mathbf{C}$ and $\mathbf{K}$ such that $\mathbf{A}_{11}\mathbf{C} = \mathbf{B}_1$ and $\mathbf{A}_{11}\mathbf{K} = \mathbf{A}_{12}$ is insured by the assumptions that $\mathcal{C}(\mathbf{B}_1) \subset \mathcal{C}(\mathbf{A}_{11})$ and $\mathcal{C}(\mathbf{A}_{12}) \subset \mathcal{C}(\mathbf{A}_{11})$.

Proof (of Theorem 11.11.1). It is clear that, for any particular value of $\mathbf{X}_2$,

$$\mathbf{A}_{11}(\mathbf{C} - \mathbf{K}\mathbf{X}_2) + \mathbf{A}_{12}\mathbf{X}_2 = \mathbf{B}_1 - \mathbf{A}_{12}\mathbf{X}_2 + \mathbf{A}_{12}\mathbf{X}_2 = \mathbf{B}_1,$$

which establishes Part (3) of the theorem.

For purposes of proving Part (1), observe (in light of Lemma 4.2.2) that, since $\mathcal{R}(\mathbf{A}_{21}) \subset \mathcal{R}(\mathbf{A}_{11})$, there exists a matrix $\mathbf{L}$ such that

$$\mathbf{A}_{21} = \mathbf{L}\mathbf{A}_{11}.$$

Thus, it follows from Part (3) that, for any solution $\mathbf{X}_1^*$ and $\mathbf{X}_2^*$ to linear system (11.5),

$$\begin{aligned}\mathbf{A}_{21}\mathbf{X}_1^* = \mathbf{L}\mathbf{A}_{11}\mathbf{X}_1^* = \mathbf{L}(\mathbf{B}_1 - \mathbf{A}_{12}\mathbf{X}_2^*) &= \mathbf{L}\mathbf{A}_{11}(\mathbf{C} - \mathbf{K}\mathbf{X}_2^*) \\ &= \mathbf{A}_{21}(\mathbf{C} - \mathbf{K}\mathbf{X}_2^*).\end{aligned} \tag{11.6}$$

Suppose now that $\mathbf{X}^* = \begin{pmatrix}\mathbf{X}_1^* \\ \mathbf{X}_2^*\end{pmatrix}$ is a solution to $\mathbf{A}\mathbf{X} = \mathbf{B}$. Then, it is clear that $\mathbf{X}_1^*$ and $\mathbf{X}_2^*$ form a solution to linear system (11.5) and, in light of result (11.6), that

$$\mathbf{A}_{21}(\mathbf{C} - \mathbf{K}\mathbf{X}_2^*) + \mathbf{A}_{22}\mathbf{X}_2^* = \mathbf{B}_2,$$

or equivalently that

$$(\mathbf{A}_{22} - \mathbf{A}_{21}\mathbf{K})\mathbf{X}_2^* = \mathbf{B}_2 - \mathbf{A}_{21}\mathbf{C},$$

and hence that $\mathbf{X}_2^*$ is a solution to linear system (11.4).

Conversely, suppose that $\mathbf{X}_1^*$ and $\mathbf{X}_2^*$ form a solution to linear system (11.5) and additionally that $\mathbf{X}_2^*$ is a solution to linear system (11.4). Then,

$$\mathbf{A}_{21}(\mathbf{C} - \mathbf{K}\mathbf{X}_2^*) + \mathbf{A}_{22}\mathbf{X}_2^* = \mathbf{B}_2,$$

implying [in light of result (11.6)] that

$$\mathbf{A}_{21}\mathbf{X}_1^* + \mathbf{A}_{22}\mathbf{X}_2^* = \mathbf{B}_2$$

and hence that $\begin{pmatrix}\mathbf{X}_1^* \\ \mathbf{X}_2^*\end{pmatrix}$ is a solution to $\mathbf{A}\mathbf{X} = \mathbf{B}$, which completes the proof of Part (1) of the theorem.

Finally, consider Part (2). That the consistency of $\mathbf{A}\mathbf{X} = \mathbf{B}$ implies the consistency of linear system (11.4) is an obvious consequence of Part (1). Conversely, if linear system (11.4) is consistent, then by definition this linear system has a solution, say $\mathbf{X}_2^*$, and, letting $\mathbf{X}_1^* = \mathbf{C} - \mathbf{K}\mathbf{X}_2^*$, it follows from Part (3) that $\mathbf{X}_1^*$ and $\mathbf{X}_2^*$ form a solution to linear system (11.5) and hence, in light of Part (1), that $\begin{pmatrix} \mathbf{X}_1^* \\ \mathbf{X}_2^* \end{pmatrix}$ is a solution to $\mathbf{AX} = \mathbf{B}$, implying that $\mathbf{AX} = \mathbf{B}$ is consistent. Q.E.D.

Theorem 11.11.1 can be translated into the following four-step procedure for solving the linear system $\mathbf{AX} = \mathbf{B}$:

(1) find matrices $\mathbf{C}$ and $\mathbf{K}$ such that $\mathbf{A}_{11}\mathbf{C} = \mathbf{B}_1$ and $\mathbf{A}_{11}\mathbf{K} = \mathbf{A}_{12}$;

(2) form the coefficient matrix $\mathbf{A}_{22} - \mathbf{A}_{21}\mathbf{K}$ and the right side $\mathbf{B}_2 - \mathbf{A}_{21}\mathbf{C}$ of linear system (11.4);

(3) obtain a solution $\mathbf{X}_2^*$ to linear system (11.4); and

(4) form the matrix $\mathbf{X}_1^* = \mathbf{C} - \mathbf{K}\mathbf{X}_2^*$, and take $\begin{pmatrix} \mathbf{X}_1^* \\ \mathbf{X}_2^* \end{pmatrix}$ to be the solution to $\mathbf{AX} = \mathbf{B}$.

In intuitive terms, this procedure consists of:

(1) solving the first m_1 equations of the linear system $\mathbf{AX} = \mathbf{B}$ for $\mathbf{X}_1$ in terms of $\mathbf{X}_2$;

(2) "absorbing" the first m_1 equations into the last m_2 equations by substituting this solution for $\mathbf{X}_1$ into the latter equations, thereby eliminating $\mathbf{X}_1$ from these equations and in effect "reducing" them to linear system (11.4);

(3) solving linear system (11.4); and

(4) "back-solving" for $\mathbf{X}_1$, using the expression for $\mathbf{X}_1$ obtained in solving the first m_1 equations for $\mathbf{X}_1$ in terms of $\mathbf{X}_2$.

Accordingly, in the context of this procedure, linear system (11.4) is sometimes called the *reduced linear system*, and the procedure itself is sometimes referred to as *absorption*.

The circumstances under which the use of absorption is advantageous are similar to those under which the use of formula (8.5.16) for the inverse of a partitioned matrix is advantageous, or more generally under which the use of formula (9.6.2) for the generalized inverse of a partioned matrix is advantageous. More specifically, the use of absorption can be advantageous if matrices $\mathbf{C}$ and $\mathbf{K}$ such that $\mathbf{A}_{11}\mathbf{C} = \mathbf{B}_1$ and $\mathbf{A}_{11}\mathbf{K} = \mathbf{A}_{12}$ are known and if the dimensions of $\mathbf{A}_{11}$ are relatively large. Under these circumstances, the effect of absorption is to replace the problem of solving the linear system $\mathbf{AX} = \mathbf{B}$, which comprises m equations in n unknowns, with the problem of solving linear system (11.4), which comprises "only" m_2 equations in "only" n_2 unknowns.

In describing absorption, it has been supposed that the first n_1 rows of $\mathbf{X}$ are to be eliminated by absorbing the first m_1 equations of the linear system into the last m_2 equations. By modifying the results of Theorem 11.11.1 in a straightforward way, we could, in principle, obtain formulas for eliminating an arbitrary set of n_1 rows of $\mathbf{X}$ by absorbing an arbitrary set of m_1 equations of the linear system into the remaining equations. Another means to the same end is to rewrite any particular linear system in such a way that the unknowns to be eliminated are those in the first n_1 rows and the equations to be absorbed are the first m_1 equations, which can be accomplished by permuting the rows and columns of the coefficient matrix and also the rows of the right side and of the matrix of unknowns.

Consider, for example, the linear system

$$\begin{pmatrix} 0 & 1 & 0 & 1 \\ -6 & 2 & 4 & 8 \\ 0 & 5 & 1 & 0 \\ 1 & 3 & 0 & 0 \end{pmatrix} \begin{pmatrix} x_1 \\ x_2 \\ x_3 \\ x_4 \end{pmatrix} = \begin{pmatrix} -2 \\ 6 \\ -1 \\ 5 \end{pmatrix}. \tag{11.7}$$

By permuting the rows and columns of the coefficient matrix, the elements of the right side, and the unknowns, this linear system can be rewritten as

$$\begin{pmatrix} 1 & 0 & 0 & 3 \\ 0 & 1 & 0 & 5 \\ 0 & 0 & 1 & 1 \\ -6 & 4 & 8 & 2 \end{pmatrix} \begin{pmatrix} x_1 \\ x_3 \\ x_4 \\ x_2 \end{pmatrix} = \begin{pmatrix} 5 \\ -1 \\ -2 \\ 6 \end{pmatrix}.$$

Now, let us apply Theorem 11.11.1, taking

$$\mathbf{A}_{11} = \begin{pmatrix} 1 & 0 & 0 \\ 0 & 1 & 0 \\ 0 & 0 & 1 \end{pmatrix}, \quad \mathbf{A}_{12} = \begin{pmatrix} 3 \\ 5 \\ 1 \end{pmatrix}, \quad \mathbf{B}_1 = \begin{pmatrix} 5 \\ -1 \\ -2 \end{pmatrix}, \quad \mathbf{X}_1 = \begin{pmatrix} x_1 \\ x_3 \\ x_4 \end{pmatrix},$$
$$\mathbf{A}_{21} = (-6,\ 4,\ 8), \quad \mathbf{A}_{22} = (2), \quad \mathbf{B}_2 = (6), \quad \text{and} \quad \mathbf{X}_2 = (x_2).$$

It is immediately apparent that the unique choices for $\mathbf{C}$ and $\mathbf{K}$ (such that $\mathbf{A}_{11}\mathbf{C} = \mathbf{B}_1$ and $\mathbf{A}_{11}\mathbf{K} = \mathbf{A}_{12}$) are

$$\mathbf{C} = \begin{pmatrix} 5 \\ -1 \\ -2 \end{pmatrix} \quad \text{and} \quad \mathbf{K} = \begin{pmatrix} 3 \\ 5 \\ 1 \end{pmatrix}.$$

Thus,

$$\mathbf{A}_{22} - \mathbf{A}_{21}\mathbf{K} = -8 \quad \text{and} \quad \mathbf{B}_2 - \mathbf{A}_{21}\mathbf{C} = 56,$$

and the unique solution to linear system (11.4) is $x_2 = 56/(-8) = -7$. Moreover, for $x_2 = -7$,

$$\mathbf{C} - \mathbf{K}\mathbf{X}_2 = \begin{pmatrix} 26 \\ 34 \\ 5 \end{pmatrix}.$$

We conclude that the unique solution to linear system (11.7) is $x_1 = 26$, $x_2 = -7$, $x_3 = 34$, and $x_4 = 5$.

In solving some linear systems, it may be helpful to make repeated use of absorption. That is, the linear system (11.4) obtained by absorbing the first m_1 equations of the linear system $\mathbf{AX} = \mathbf{B}$ into the last m_2 equations could itself possibly be solved by absorption.

For some linear statistical models, the use of absorption can greatly facilitate the solution of the so-called normal equations (which arise in fitting the model by least squares). One linear statistical model for which this is the case is the additive version of the two-way crossed-classification model—refer to Searle (1971, sec. 7.1).

11.12 Extensions to Systems of the Form $\mathbf{AXC} = \mathbf{B}$

Let $\mathbf{X}_0 = \mathbf{A}^-\mathbf{B}$ or, more generally, let $\mathbf{X}_0$ represent any particular solution to a consistent linear system $\mathbf{AX} = \mathbf{B}$ (with an $m \times n$ coefficient matrix $\mathbf{A}$, $m \times p$ right side $\mathbf{B}$, and $n \times p$ matrix of unknowns $\mathbf{X}$). It was established in Section 11.2 that a matrix $\mathbf{X}_*$ is a solution to $\mathbf{AX} = \mathbf{B}$ if and only if $\mathbf{X}_* = \mathbf{X}_0 + \mathbf{Z}_*$ for some solution $\mathbf{Z}_*$ to the homogeneous linear system $\mathbf{AZ} = \mathbf{0}$ (in $\mathbf{Z}$). Also, in light of Theorem 11.2.1, $\mathbf{X}_*$ is a solution to $\mathbf{AX} = \mathbf{B}$ if and only if $\mathbf{X}_* = \mathbf{X}_0 + (\mathbf{I} - \mathbf{A}^-\mathbf{A})\mathbf{Y}$ for some matrix $\mathbf{Y}$.

The following theorem generalizes these results to any consistent system of the form $\mathbf{AXC} = \mathbf{B}$ (where $\mathbf{A}$ is an $m \times n$ matrix, $\mathbf{C}$ a $p \times q$ matrix, and $\mathbf{B}$ an $m \times q$ matrix, and where $\mathbf{X}$ is an $n \times p$ matrix of unknowns).

Theorem 11.12.1. Let $\mathbf{X}_0 = \mathbf{A}^-\mathbf{B}\mathbf{C}^-$ or, more generally, let $\mathbf{X}_0$ represent any particular solution to a consistent system $\mathbf{AXC} = \mathbf{B}$ (in $\mathbf{X}$). Then, (1) a matrix $\mathbf{X}_*$ is a solution to $\mathbf{AXC} = \mathbf{B}$ if and only if

$$\mathbf{X}_* = \mathbf{X}_0 + \mathbf{Z}_*$$

for some solution $\mathbf{Z}_*$ to the system $\mathbf{AZC} = \mathbf{0}$ (in $\mathbf{Z}$); (2) $\mathbf{X}_*$ is a solution to $\mathbf{AXC} = \mathbf{B}$ if and only if

$$\mathbf{X}_* = \mathbf{X}_0 + \mathbf{Y} - \mathbf{A}^-\mathbf{A}\mathbf{Y}\mathbf{C}\mathbf{C}^-$$

for some matrix $\mathbf{Y}$ (of suitable dimensions); and (3) $\mathbf{X}_*$ is a solution to $\mathbf{AXC} = \mathbf{B}$ if and only if

$$\mathbf{X}_* = \mathbf{X}_0 + \mathbf{A}^-\mathbf{A}\mathbf{R}(\mathbf{I} - \mathbf{C}\mathbf{C}^-) + (\mathbf{I} - \mathbf{A}^-\mathbf{A})\mathbf{S}\mathbf{C}\mathbf{C}^- + (\mathbf{I} - \mathbf{A}^-\mathbf{A})\mathbf{T}(\mathbf{I} - \mathbf{C}\mathbf{C}^-)$$

for some matrices $\mathbf{R}$, $\mathbf{S}$, and $\mathbf{T}$ (of suitable dimensions).

Proof. Note that it follows from Theorem 9.7.1 that $\mathbf{A}^-\mathbf{B}\mathbf{C}^-$ is a solution to $\mathbf{AXC} = \mathbf{B}$.

(1) If $\mathbf{X}_* = \mathbf{X}_0 + \mathbf{Z}_*$ for some solution $\mathbf{Z}_*$ to $\mathbf{AZC} = \mathbf{0}$, then

$$\mathbf{A}\mathbf{X}_*\mathbf{C} = \mathbf{A}\mathbf{X}_0\mathbf{C} + \mathbf{A}\mathbf{Z}_*\mathbf{C} = \mathbf{B} + \mathbf{0} = \mathbf{B},$$

so that $\mathbf{X}_*$ is a solution to $\mathbf{AXC} = \mathbf{B}$. Conversely, suppose that $\mathbf{X}_*$ is a solution to $\mathbf{AXC} = \mathbf{B}$. Then, $\mathbf{X}_* = \mathbf{X}_0 + \mathbf{Z}_*$, where $\mathbf{Z}_* = \mathbf{X}_* - \mathbf{X}_0$. And, since

$$\mathbf{A}\mathbf{Z}_*\mathbf{C} = \mathbf{A}\mathbf{X}_*\mathbf{C} - \mathbf{A}\mathbf{X}_0\mathbf{C} = \mathbf{B} - \mathbf{B} = \mathbf{0},$$

$\mathbf{Z}_*$ is a solution to $\mathbf{AZC} = \mathbf{0}$.

(2) If $\mathbf{X}_* = \mathbf{X}_0 + \mathbf{Y} - \mathbf{A}^-\mathbf{AYCC}^-$ for some matrix $\mathbf{Y}$, then

$$\mathbf{AX}_*\mathbf{C} = \mathbf{AX}_0\mathbf{C} + \mathbf{AYC} - \mathbf{AA}^-\mathbf{AYCC}^-\mathbf{C} = \mathbf{B} + \mathbf{AYC} - \mathbf{AYC} = \mathbf{B},$$

so that $\mathbf{X}_*$ is a solution to $\mathbf{AXC} = \mathbf{B}$. Conversely, if $\mathbf{X}_*$ is a solution to $\mathbf{AXC} = \mathbf{B}$, then [as a consequence of Part (1)] $\mathbf{X}_* = \mathbf{X}_0 + \mathbf{Z}_*$ for some solution $\mathbf{Z}_*$ to $\mathbf{AZC} = \mathbf{0}$ and hence

$$\mathbf{X}_* = \mathbf{X}_0 + \mathbf{Z}_* - \mathbf{A}^-\mathbf{0}\mathbf{C}^- = \mathbf{X}_0 + \mathbf{Z}_* - \mathbf{A}^-\mathbf{AZ}_*\mathbf{CC}^-,$$

so that $\mathbf{X}_* = \mathbf{X}_0 + \mathbf{Y} - \mathbf{A}^-\mathbf{AYCC}^-$ for $\mathbf{Y} = \mathbf{Z}_*$.

(3) If

$$\mathbf{X}_* = \mathbf{X}_0 + \mathbf{A}^-\mathbf{AR}(\mathbf{I} - \mathbf{CC}^-) + (\mathbf{I} - \mathbf{A}^-\mathbf{A})\mathbf{SCC}^- + (\mathbf{I} - \mathbf{A}^-\mathbf{A})\mathbf{T}(\mathbf{I} - \mathbf{CC}^-)$$

for some matrices $\mathbf{R}$, $\mathbf{S}$, and $\mathbf{T}$, then

$$\begin{aligned}\mathbf{AX}_*\mathbf{C} &= \mathbf{AX}_0\mathbf{C} + \mathbf{AA}^-\mathbf{AR}(\mathbf{C} - \mathbf{CC}^-\mathbf{C}) \\ &\qquad + (\mathbf{A} - \mathbf{AA}^-\mathbf{A})\mathbf{SCC}^-\mathbf{C} + (\mathbf{A} - \mathbf{AA}^-\mathbf{A})\mathbf{T}(\mathbf{C} - \mathbf{CC}^-\mathbf{C}) \\ &= \mathbf{B} + \mathbf{0} + \mathbf{0} + \mathbf{0} = \mathbf{B},\end{aligned}$$

so that $\mathbf{X}_*$ is a solution to $\mathbf{AXC} = \mathbf{B}$. Conversely, if $\mathbf{X}_*$ is a solution to $\mathbf{AXC} = \mathbf{B}$, then [as a consequence of Part (2)]

$$\mathbf{X}_* = \mathbf{X}_0 + \mathbf{Y} - \mathbf{A}^-\mathbf{AYCC}^-$$

for some matrix $\mathbf{Y}$, implying {since

$$\begin{aligned}\mathbf{Y} &= [\mathbf{A}^-\mathbf{A} + (\mathbf{I} - \mathbf{A}^-\mathbf{A})]\mathbf{Y}[\mathbf{CC}^- + (\mathbf{I} - \mathbf{CC}^-)] \\ &= \mathbf{A}^-\mathbf{AYCC}^- + \mathbf{A}^-\mathbf{AY}(\mathbf{I} - \mathbf{CC}^-) \\ &\qquad + (\mathbf{I} - \mathbf{A}^-\mathbf{A})\mathbf{YCC}^- + (\mathbf{I} - \mathbf{A}^-\mathbf{A})\mathbf{Y}(\mathbf{I} - \mathbf{CC}^-)\end{aligned}$$

and hence since

$$\begin{aligned}\mathbf{Y} - \mathbf{A}^-\mathbf{AYCC}^- &= \mathbf{A}^-\mathbf{AY}(\mathbf{I} - \mathbf{CC}^-) + (\mathbf{I} - \mathbf{A}^-\mathbf{A})\mathbf{YCC}^- \\ &\qquad + (\mathbf{I} - \mathbf{A}^-\mathbf{A})\mathbf{Y}(\mathbf{I} - \mathbf{CC}^-)\}\end{aligned}$$

that

$$\mathbf{X}_* = \mathbf{X}_0 + \mathbf{A}^-\mathbf{AR}(\mathbf{I} - \mathbf{CC}^-) + (\mathbf{I} - \mathbf{A}^-\mathbf{A})\mathbf{SCC}^- + (\mathbf{I} - \mathbf{A}^-\mathbf{A})\mathbf{T}(\mathbf{I} - \mathbf{CC}^-)$$

for $\mathbf{R} = \mathbf{S} = \mathbf{T} = \mathbf{Y}$. Q.E.D.

Exercises

Section 11.2

1. Show that, for any matrix $\mathbf{A}$,

$$\mathcal{C}(\mathbf{A}) = \mathcal{N}(\mathbf{I} - \mathbf{A}\mathbf{A}^-).$$

2. Show that if $\mathbf{X}_1, \ldots, \mathbf{X}_k$ are solutions to a linear system $\mathbf{AX} = \mathbf{B}$ (in $\mathbf{X}$) and $c_1, \ldots, c_k$ are scalars such that $\sum_{i=1}^k c_i = 1$, then the matrix $\sum_{i=1}^k c_i \mathbf{X}_i$ is a solution to $\mathbf{AX} = \mathbf{B}$.

Section 11.3

3. Let $\mathbf{A}$ and $\mathbf{Z}$ represent $n \times n$ matrices. Suppose that $\text{rank}(\mathbf{A}) = n - 1$, and let $\mathbf{x}$ and $\mathbf{y}$ represent nonnull n-dimensional column vectors such that $\mathbf{Ax} = \mathbf{0}$ and $\mathbf{A}'\mathbf{y} = \mathbf{0}$.

   (a) Show that $\mathbf{AZ} = \mathbf{0}$ if and only if $\mathbf{Z} = \mathbf{x}\mathbf{k}'$ for some n-dimensional row vector $\mathbf{k}'$.

   (b) Show that $\mathbf{AZ} = \mathbf{ZA} = \mathbf{0}$ if and only if $\mathbf{Z} = c\mathbf{x}\mathbf{y}'$ for some scalar c.

4. Suppose that $\mathbf{AX} = \mathbf{B}$ is a nonhomogeneous linear system (in an $n \times p$ matrix $\mathbf{X}$). Let $s = p[n - \text{rank}(\mathbf{A})]$, and take $\mathbf{Z}_1, \ldots, \mathbf{Z}_s$ to be any s $n \times p$ matrices that form a basis for the solution space of the homogeneous linear system $\mathbf{AZ} = \mathbf{0}$ (in an $n \times p$ matrix $\mathbf{Z}$). Define $\mathbf{X}_0$ to be any particular solution to $\mathbf{AX} = \mathbf{B}$, and let $\mathbf{X}_i = \mathbf{X}_0 + \mathbf{Z}_i$ $(i = 1, \ldots, s)$.

   (a) Show that the $s + 1$ matrices $\mathbf{X}_0, \mathbf{X}_1, \ldots, \mathbf{X}_s$ are linearly independent solutions to $\mathbf{AX} = \mathbf{B}$.

   (b) Show that every solution to $\mathbf{AX} = \mathbf{B}$ is expressible as a linear combination of $\mathbf{X}_0, \mathbf{X}_1, \ldots, \mathbf{X}_s$.

   (c) Show that a linear combination $\sum_{i=0}^s k_i \mathbf{X}_i$ of $\mathbf{X}_0, \mathbf{X}_1, \ldots, \mathbf{X}_s$ is a solution to $\mathbf{AX} = \mathbf{B}$ if and only if the scalars $k_0, k_1, \ldots, k_s$ are such that $\sum_{i=0}^s k_i = 1$.

   (d) Show that the solution set of $\mathbf{AX} = \mathbf{B}$ is a proper subset of the linear space $\text{sp}(\mathbf{X}_0, \mathbf{X}_1, \ldots, \mathbf{X}_s)$.

Section 11.5

5. Suppose that $\mathbf{AX} = \mathbf{B}$ is a consistent linear system (in an $n \times p$ matrix $\mathbf{X}$). Show that if $\text{rank}(\mathbf{A}) < n$ and $\text{rank}(\mathbf{B}) < p$, then there exists a solution $\mathbf{X}^*$ to $\mathbf{AX} = \mathbf{B}$ that is not expressible as $\mathbf{X}^* = \mathbf{GB}$ for any generalized inverse $\mathbf{G}$ of $\mathbf{A}$.

Section 11.6

6. Use the result of Part (b) of Exercise 7.1 to generalize Lemma 11.6.1.

7. Let $\mathbf{A}$ represent an $m \times n$ matrix, $\mathbf{B}$ an $m \times p$ matrix, and $\mathbf{C}$ a $q \times m$ matrix, and suppose that $\mathbf{AX} = \mathbf{B}$ and $\mathbf{CAX} = \mathbf{CB}$ are linear systems (in $\mathbf{X}$).

   (a) Show that if $\text{rank}[\mathbf{C}(\mathbf{A}, \mathbf{B})] = \text{rank}(\mathbf{A}, \mathbf{B})$, then $\mathbf{CAX} = \mathbf{CB}$ is equivalent to $\mathbf{AX} = \mathbf{B}$, thereby generalizing Lemmas 11.6.1 and 11.6.2.

   (b) Show that if $\text{rank}[\mathbf{C}(\mathbf{A}, \mathbf{B})] < \text{rank}(\mathbf{A}, \mathbf{B})$ and if $\mathbf{CAX} = \mathbf{CB}$ is consistent, then the solution set of $\mathbf{AX} = \mathbf{B}$ is a proper subset of that of $\mathbf{CAX} = \mathbf{CB}$ (i.e., there exists a solution to $\mathbf{CAX} = \mathbf{CB}$ that is not a solution to $\mathbf{AX} = \mathbf{B}$).

   (c) Show, by example, that if $\text{rank}[\mathbf{C}(\mathbf{A}, \mathbf{B})] < \text{rank}(\mathbf{A}, \mathbf{B})$ and if $\mathbf{AX} = \mathbf{B}$ is inconsistent, then $\mathbf{CAX} = \mathbf{CB}$ can be either consistent or inconsistent.

Section 11.10

8. Let $\mathbf{A}$ represent a $q \times n$ matrix, $\mathbf{B}$ an $m \times p$ matrix, and $\mathbf{C}$ an $m \times q$ matrix; and suppose that the linear system $\mathbf{CAX} = \mathbf{B}$ (in an $n \times p$ matrix $\mathbf{X}$) is consistent. Show that the value of $\mathbf{AX}$ is the same for every solution to $\mathbf{CAX} = \mathbf{B}$ if and only if $\text{rank}(\mathbf{CA}) = \text{rank}(\mathbf{A})$.

9. Verify (1) Theorem 11.10.4 and (2) Theorem 11.10.5.

12

Projections and Projection Matrices

Projections and projection matrices, which are introduced and discussed in this chapter, are frequently encountered in discourse on linear statistical models related to the estimation of parameters and to the analysis of variance. Their appearance in such discourse can be attributed to their connection to thc so-called least squares problem—one long-standing approach to the estimation of the parameters of a linear statistical model is based on "fitting" the model by least squares. Their connection to the least squares problem is described and discussed in Section 12.4.

12.1 Some General Results, Terminology, and Notation

If a matrix $\mathbf{Y}$ in a linear space of matrices is orthogonal to every matrix in a subspace $\mathcal{U}$, $\mathbf{Y}$ is said to be *orthogonal* to $\mathcal{U}$. The statement that $\mathbf{Y}$ is orthogonal to $\mathcal{U}$ is sometimes abbreviated to $\mathbf{Y} \perp \mathcal{U}$. Similarly, to indicate that every matrix in a subspace $\mathcal{U}$ is orthogonal to every matrix in a subspace $\mathcal{W}$, one says that $\mathcal{U}$ is *orthogonal* to $\mathcal{W}$ or writes $\mathcal{U} \perp \mathcal{W}$.

The following lemma is easy to verify.

Lemma 12.1.1. Let $\mathbf{Y}$ represent a matrix in a linear space $\mathcal{V}$, let $\mathcal{U}$ and $\mathcal{W}$ represent subspaces of $\mathcal{V}$, and take $\{\mathbf{X}_1, \ldots, \mathbf{X}_s\}$ to be a set of matrices that spans $\mathcal{U}$ and $\{\mathbf{Z}_1, \ldots, \mathbf{Z}_t\}$ to be a set that spans $\mathcal{W}$. Then, $\mathbf{Y} \perp \mathcal{U}$ if and only if $\mathbf{Y} \bullet \mathbf{X}_i = 0$ for $i = 1, \ldots, s$; that is, $\mathbf{Y}$ is orthogonal to $\mathcal{U}$ if and only if $\mathbf{Y}$ is orthogonal to each of the matrices $\mathbf{X}_1, \ldots, \mathbf{X}_s$. And, similarly, $\mathcal{U} \perp \mathcal{W}$ if and only if $\mathbf{X}_i \bullet \mathbf{Z}_j = 0$ for $i = 1, \ldots, s$ and $j = 1, \ldots, t$; that is, $\mathcal{U}$ is orthogonal to $\mathcal{W}$ if and only if each of the matrices $\mathbf{X}_1, \ldots, \mathbf{X}_s$ is orthogonal to each of the matrices $\mathbf{Z}_1, \ldots, \mathbf{Z}_t$.

By applying Lemma 12.1.1 in the special case where $\mathcal{V} = \mathcal{R}^{m \times 1}$ and $\mathcal{U}$ and $\mathcal{W}$ are the column spaces of two matrices (each of which has m rows), we obtain the following corollary.

Corollary 12.1.2. Let $\mathbf{y}$ represent an m-dimensional column vector, and let $\mathbf{X}$ represent an $m \times n$ matrix and $\mathbf{Z}$ an $m \times p$ matrix. Then, $\mathbf{y}$ is orthogonal to $\mathcal{C}(\mathbf{X})$ (with respect to the usual inner product) if and only if $\mathbf{X}'\mathbf{y} = \mathbf{0}$ (or equivalently if and only if $\mathbf{y}'\mathbf{X} = \mathbf{0}$). Similarly, $\mathcal{C}(\mathbf{X})$ is orthogonal to $\mathcal{C}(\mathbf{Z})$ (with respect to the usual inner product) if and only if $\mathbf{X}'\mathbf{Z} = \mathbf{0}$ (or equivalently if and only if $\mathbf{Z}'\mathbf{X} = \mathbf{0}$).

Consider the following theorem.

Theorem 12.1.3. Let $\mathbf{Y}$ represent a matrix in a linear space $\mathcal{V}$, and let $\mathcal{U}$ represent an r-dimensional subspace of $\mathcal{V}$. Then, there exists a unique matrix $\mathbf{Z}$ in $\mathcal{U}$ such that $(\mathbf{Y} - \mathbf{Z}) \perp \mathcal{U}$, that is, such that the difference between $\mathbf{Y}$ and $\mathbf{Z}$ is orthogonal to every matrix in $\mathcal{U}$. If $r = 0$, then $\mathbf{Z} = \mathbf{0}$, and, if $r > 0$, $\mathbf{Z}$ is expressible as

$$\mathbf{Z} = c_1\mathbf{X}_1 + \cdots + c_r\mathbf{X}_r, \tag{1.1}$$

where $\{\mathbf{X}_1, \ldots, \mathbf{X}_r\}$ is any orthonormal basis for $\mathcal{U}$ and $c_j = \mathbf{Y} \bullet \mathbf{X}_j$ $(j = 1, \ldots, r)$. Moreover, $\mathbf{Z} = \mathbf{Y}$ if and only if $\mathbf{Y} \in \mathcal{U}$.

Proof. Consider first the case where $r = 0$. In this case, the only matrix in $\mathcal{U}$ is the null matrix $\mathbf{0}$. Clearly, $\mathbf{Y} - \mathbf{0}$ is orthogonal to $\mathbf{0}$. Thus, there exists a unique matrix $\mathbf{Z}$ in $\mathcal{U}$ such that $(\mathbf{Y} - \mathbf{Z}) \perp \mathcal{U}$, namely, $\mathbf{Z} = \mathbf{0}$. Moreover, it is clear that $\mathbf{Y} = \mathbf{Z}$ if and only if $\mathbf{Y} \in \mathcal{U}$.

Consider now the case where $r > 0$. Take $\{\mathbf{X}_1, \ldots, \mathbf{X}_r\}$ to be any orthonormal basis for $\mathcal{U}$, and define $c_j = \mathbf{Y} \bullet \mathbf{X}_j$ $(j = 1, \ldots, r)$. Clearly, $\sum_j c_j\mathbf{X}_j \in \mathcal{U}$, and

$$\Big(\mathbf{Y} - \sum_j c_j\mathbf{X}_j\Big) \bullet \mathbf{X}_i = (\mathbf{Y} \bullet \mathbf{X}_i) - c_i = 0$$

for $i = 1, \ldots, r$, implying (in light of Lemma 12.1.1) that $(\mathbf{Y} - \sum_j c_j\mathbf{X}_j) \perp \mathcal{U}$. Moreover, for any matrix $\mathbf{X}$ such that $\mathbf{X} \in \mathcal{U}$ and $(\mathbf{Y} - \mathbf{X}) \perp \mathcal{U}$, we find that $(\mathbf{X} - \sum_j c_j\mathbf{X}_j) \in \mathcal{U}$ and hence that

$$\begin{aligned}\Big(\mathbf{X} - \sum_j c_j\mathbf{X}_j\Big) \bullet \Big(\mathbf{X} - \sum_j c_j\mathbf{X}_j\Big) &= \Big(\mathbf{Y} - \sum_j c_j\mathbf{X}_j\Big) \bullet \Big(\mathbf{X} - \sum_j c_j\mathbf{X}_j\Big) - (\mathbf{Y} - \mathbf{X}) \bullet \Big(\mathbf{X} - \sum_j c_j\mathbf{X}_j\Big) \\ &= 0 - 0 = 0,\end{aligned}$$

so that $\mathbf{X} - \sum_j c_j\mathbf{X}_j = \mathbf{0}$ or equivalently $\mathbf{X} = \sum_j c_j\mathbf{X}_j$. We conclude that there exists a unique matrix $\mathbf{Z}$ in $\mathcal{U}$ such that $(\mathbf{Y} - \mathbf{Z}) \perp \mathcal{U}$, namely, $\mathbf{Z} = \sum_j c_j\mathbf{X}_j$. To complete the proof, observe that, if $\mathbf{Y} = \mathbf{Z}$, then obviously $\mathbf{Y} \in \mathcal{U}$, and conversely, if $\mathbf{Y} \in \mathcal{U}$, then, since $\mathbf{Y} - \mathbf{Y} = \mathbf{0}$ is orthogonal to $\mathcal{U}$, $\mathbf{Y} = \mathbf{Z}$. Q.E.D.

Suppose that $\mathcal{U}$ is a subspace of a linear space $\mathcal{V}$. Then, it follows from Theorem 12.1.3 that, corresponding to each matrix $\mathbf{Y}$ in $\mathcal{V}$, there exists a unique matrix $\mathbf{Z}$ in $\mathcal{U}$ such that $\mathbf{Y} - \mathbf{Z}$ is orthogonal to $\mathcal{U}$ or, equivalently (if $\mathbf{Y} \notin \mathcal{U}$), such that $\mathbf{Y} - \mathbf{Z}$ forms an angle of $\pi/2$ with every nonnull matrix in $\mathcal{U}$. The matrix

$\mathbf{Z}$ is called the *orthogonal projection of* $\mathbf{Y}$ *on* $\mathcal{U}$ or simply the *projection of* $\mathbf{Y}$ *on* $\mathcal{U}$.

Note that, for any matrix $\mathbf{Y}$ in $\mathcal{U}$, $\mathbf{Y}$ itself is the projection of $\mathbf{Y}$ on $\mathcal{U}$.

The following theorem relates the projection of a linear combination of p matrices $\mathbf{Y}_1, \ldots, \mathbf{Y}_p$ (on a subspace $\mathcal{U}$ of a linear space $\mathcal{V}$) to the projections of $\mathbf{Y}_1, \ldots, \mathbf{Y}_p$ (on $\mathcal{U}$).

Theorem 12.1.4. Let $\mathbf{Y}_1, \ldots, \mathbf{Y}_p$ represent matrices in a linear space $\mathcal{V}$, let $\mathcal{U}$ represent a subspace of $\mathcal{V}$, and let $\mathbf{Z}_1, \ldots, \mathbf{Z}_p$ represent the projections of $\mathbf{Y}_1, \ldots, \mathbf{Y}_p$, respectively, on $\mathcal{U}$. Then, for any scalars $k_1, \ldots, k_p$, the projection of the linear combination $k_1\mathbf{Y}_1 + \cdots + k_p\mathbf{Y}_p$ (on $\mathcal{U}$) is the corresponding linear combination $k_1\mathbf{Z}_1 + \cdots + k_p\mathbf{Z}_p$ of $\mathbf{Z}_1, \ldots, \mathbf{Z}_p$.

Proof. By definition, $\mathbf{Z}_i \in \mathcal{U}$ and $(\mathbf{Y}_i - \mathbf{Z}_i) \perp \mathcal{U}$ $(i = 1, \ldots, p)$. Thus, $(k_1\mathbf{Z}_1 + \cdots + k_p\mathbf{Z}_p) \in \mathcal{U}$. Moreover, for every matrix $\mathbf{X}$ in $\mathcal{U}$,

$$\begin{aligned}
[k_1\mathbf{Y}_1 + \cdots + k_p\mathbf{Y}_p - (k_1\mathbf{Z}_1 + \cdots + k_p\mathbf{Z}_p)] \bullet \mathbf{X}& \\
&= [k_1(\mathbf{Y}_1 - \mathbf{Z}_1) + \cdots + k_p(\mathbf{Y}_p - \mathbf{Z}_p)] \bullet \mathbf{X} \\
&= k_1[(\mathbf{Y}_1 - \mathbf{Z}_1) \bullet \mathbf{X}] + \cdots + k_p[(\mathbf{Y}_p - \mathbf{Z}_p) \bullet \mathbf{X}] \\
&= k_1 0 + \cdots + k_p 0 = 0,
\end{aligned}$$

so that $[k_1\mathbf{Y}_1 + \cdots + k_p\mathbf{Y}_p - (k_1\mathbf{Z}_1 + \cdots + k_p\mathbf{Z}_p)] \perp \mathcal{U}$. We conclude that $k_1\mathbf{Z}_1 + \cdots + k_p\mathbf{Z}_p$ is the projection of $k_1\mathbf{Y}_1 + \cdots + k_p\mathbf{Y}_p$ on $\mathcal{U}$. Q.E.D.

12.2 Projection of a Column Vector

a. Main results

The following theorem gives an algebraic expression for the projection (with respect to the usual inner product) of an n-dimensional column vector on a subspace of $\mathcal{R}^n$.

Theorem 12.2.1. Let $\mathbf{z}$ represent the projection (with respect to the usual inner product) of an n-dimensional column vector $\mathbf{y}$ on a subspace $\mathcal{U}$ of $\mathcal{R}^n$, and let $\mathbf{X}$ represent any $n \times p$ matrix whose columns span $\mathcal{U}$. Then,

$$\mathbf{z} = \mathbf{X}\mathbf{b}^* \tag{2.1}$$

for any solution $\mathbf{b}^*$ to the (consistent) linear system

$$\mathbf{X}'\mathbf{X}\mathbf{b} = \mathbf{X}'\mathbf{y} \text{ (in } \mathbf{b}). \tag{2.2}$$

Proof. Suppose that $\mathbf{b}^*$ is a solution to the (consistent) linear system $\mathbf{X}'\mathbf{X}\mathbf{b} = \mathbf{X}'\mathbf{y}$. (The consistency of this linear system is evident from Theorem 7.4.1.) Then, $\mathbf{X}'(\mathbf{y} - \mathbf{X}\mathbf{b}^*) = \mathbf{0}$, implying (in light of Corollary 12.1.2) that $\mathbf{y} - \mathbf{X}\mathbf{b}^*$ is orthogonal to $\mathcal{C}(\mathbf{X})$ and hence [since $\mathcal{C}(\mathbf{X}) = \mathcal{U}$] to $\mathcal{U}$. Since $\mathbf{X}\mathbf{b}^* \in \mathcal{U}$, we conclude that $\mathbf{X}\mathbf{b}^*$ is the projection of $\mathbf{y}$ on $\mathcal{U}$ and hence, by definition, that $\mathbf{z} = \mathbf{X}\mathbf{b}^*$. Q.E.D.

Linear system (2.2) comprises p equations. These p equations are known as the *normal equations*. When the columns, say $\mathbf{x}_1, \ldots, \mathbf{x}_p$, of the matrix $\mathbf{X}$ are taken to be an orthonormal basis for $\mathcal{U}$, we find that $\mathbf{X}'\mathbf{X} = \mathbf{I}$ and hence that the normal equations (2.2) have a unique solution $\mathbf{b}^* = (\mathbf{x}_1'\mathbf{y}, \ldots, \mathbf{x}_p'\mathbf{y})'$, in which case expression (2.1) for the projection $\mathbf{z}$ of $\mathbf{y}$ on $\mathcal{U}$ reduces to a special case of expression (1.1).

Since the vector $(\mathbf{X}'\mathbf{X})^-\mathbf{X}'\mathbf{y}$ is one solution to the normal equations (2.2), we have the following corollary of Theorem 12.2.1.

Corollary 12.2.2. Let $\mathbf{z}$ represent the projection (with respect to the usual inner product) of an n-dimensional column vector $\mathbf{y}$ on a subspace $\mathcal{U}$ of $\mathcal{R}^n$, and let $\mathbf{X}$ represent any $n \times p$ matrix whose columns span $\mathcal{U}$. Then,

$$\mathbf{z} = \mathbf{X}(\mathbf{X}'\mathbf{X})^-\mathbf{X}'\mathbf{y}. \tag{2.3}$$

A further implication of Theorem 12.2.1 is as follows:

Corollary 12.2.3. Let $\mathbf{y}$ represent an n-dimensional column vector, $\mathbf{X}$ an $n \times p$ matrix, and $\mathbf{W}$ any $n \times q$ matrix such that $\mathcal{C}(\mathbf{W}) = \mathcal{C}(\mathbf{X})$. Then,

$$\mathbf{W}\mathbf{a}^* = \mathbf{X}\mathbf{b}^*$$

for any solution $\mathbf{a}^*$ to the linear system $\mathbf{W}'\mathbf{W}\mathbf{a} = \mathbf{W}'\mathbf{y}$ (in $\mathbf{a}$) and any solution $\mathbf{b}^*$ to the linear system $\mathbf{X}'\mathbf{X}\mathbf{b} = \mathbf{X}'\mathbf{y}$ (in $\mathbf{b}$).

As a special case of Corollary 12.2.3, we have the following corollary.

Corollary 12.2.4. Let $\mathbf{y}$ represent an n-dimensional column vector, and $\mathbf{X}$ an $n \times p$ matrix. Then, $\mathbf{X}\mathbf{b}_1 = \mathbf{X}\mathbf{b}_2$ for any two solutions $\mathbf{b}_1$ and $\mathbf{b}_2$ to the linear system $\mathbf{X}'\mathbf{X}\mathbf{b} = \mathbf{X}'\mathbf{y}$ (in $\mathbf{b}$).

The following theorem gives a converse of Theorem 12.2.1.

Theorem 12.2.5. Let $\mathbf{z}$ represent the projection (with respect to the usual inner product) of an n-dimensional column vector $\mathbf{y}$ on a subspace $\mathcal{U}$ of $\mathcal{R}^n$, and let $\mathbf{X}$ represent any $n \times p$ matrix whose columns span $\mathcal{U}$. Then, any $p \times 1$ vector $\mathbf{b}^*$ such that $\mathbf{z} = \mathbf{X}\mathbf{b}^*$ is a solution to the linear system $\mathbf{X}'\mathbf{X}\mathbf{b} = \mathbf{X}'\mathbf{y}$ (in $\mathbf{b}$).

Proof. In light of Theorem 12.2.1, $\mathbf{X}'\mathbf{X}\mathbf{b} = \mathbf{X}'\mathbf{y}$ has a solution, say $\mathbf{a}$, and $\mathbf{z} = \mathbf{X}\mathbf{a}$. Thus,

$$\mathbf{X}'\mathbf{X}\mathbf{b}^* = \mathbf{X}'\mathbf{z} = \mathbf{X}'\mathbf{X}\mathbf{a} = \mathbf{X}'\mathbf{y}.$$

Q.E.D.

b. Two-dimensional example

Let us find the projection (with respect to the usual inner product) of an n-dimensional column vector $\mathbf{y}$ on a subspace $\mathcal{U}$ of $\mathcal{R}^n$ in the special case where $n = 2$, $\mathbf{y} = (4, 8)'$, and $\mathcal{U} = \text{sp}(\mathbf{x})$, with $\mathbf{x} = (3, 1)'$.

Upon taking $\mathbf{X}$ to be the 2×1 matrix whose only column is $\mathbf{x}$, the linear system $\mathbf{X}'\mathbf{X}\mathbf{b} = \mathbf{X}'\mathbf{y}$ becomes $(10)\mathbf{b} = (20)$, which has the unique solution $\mathbf{b} = (2)$. Thus, the projection of $\mathbf{y}$ on $\mathcal{U}$ is

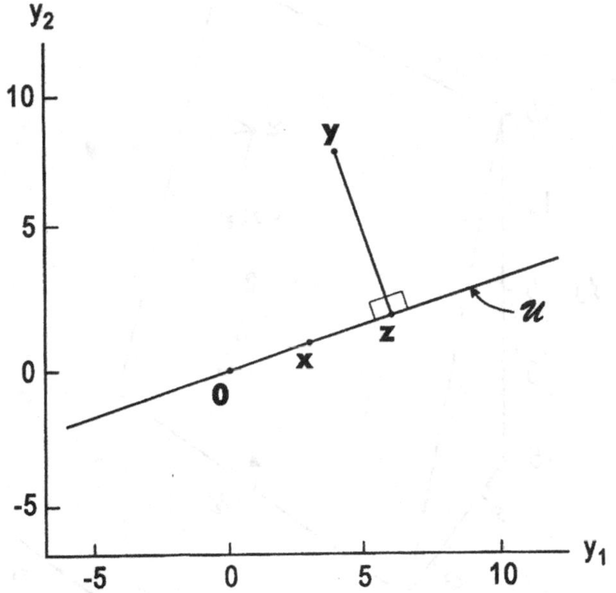

FIGURE 12.1: The projection **z** of a two-dimensional column vector **y** on a one-dimensional subspace $\mathcal{U}$ of $\mathcal{R}^2$.

$$\mathbf{z} = \begin{pmatrix} 3 \\ 1 \end{pmatrix} (2) = \begin{pmatrix} 6 \\ 2 \end{pmatrix},$$

as depicted in Figure 12.1.

c. Three-dimensional example

Let us find the projection (with respect to the usual inner product) of an n-dimensional column vector $\mathbf{y}$ on a subspace $\mathcal{U}$ of $\mathcal{R}^n$ in the special case where $n = 3$, $\mathbf{y} = (3, -38/5, 74/5)'$, and $\mathcal{U} = \text{sp}(\mathbf{x}_1, \mathbf{x}_2, \mathbf{x}_3)$, with

$$\mathbf{x}_1 = \begin{pmatrix} 0 \\ 3 \\ 6 \end{pmatrix}, \quad \mathbf{x}_2 = \begin{pmatrix} -2 \\ 2 \\ 4 \end{pmatrix}, \quad \mathbf{x}_3 = \begin{pmatrix} -2 \\ 1 \\ 2 \end{pmatrix}. \tag{2.4}$$

Clearly, $\mathbf{x}_1$ and $\mathbf{x}_2$ are linearly independent, and $\mathbf{x}_3 = \mathbf{x}_2 - (1/3)\mathbf{x}_1$. Thus, $\dim(\mathcal{U}) = 2$.

Upon taking $\mathbf{X}$ to be the 3×3 matrix whose columns are $\mathbf{x}_1, \mathbf{x}_2$ and $\mathbf{x}_3$, respectively, the normal equations $\mathbf{X}'\mathbf{X}\mathbf{b} = \mathbf{X}'\mathbf{y}$ become

$$\begin{pmatrix} 45 & 30 & 15 \\ 30 & 24 & 14 \\ 15 & 14 & 9 \end{pmatrix} \mathbf{b} = \begin{pmatrix} 66 \\ 38 \\ 16 \end{pmatrix}.$$

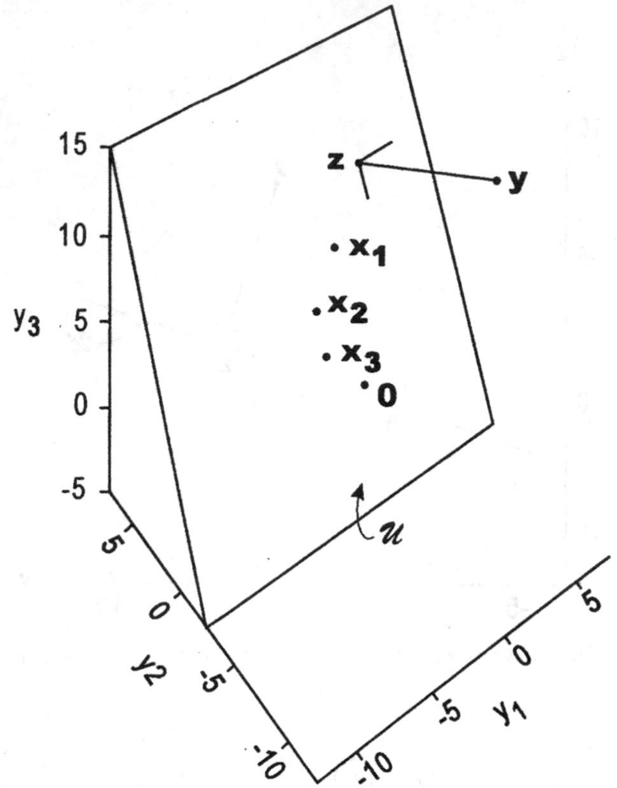

FIGURE 12.2: The projection **z** of a three-dimensional column vector **y** on a two-dimensional subspace $\mathcal{U}$ of $\mathcal{R}^3$.

One solution to these equations is the vector $(32/15, -1/2, -1)'$. Thus, the projection of **y** on $\mathcal{U}$ is

$$\mathbf{z} = \begin{pmatrix} 0 & -2 & -2 \\ 3 & 2 & 1 \\ 6 & 4 & 2 \end{pmatrix} \begin{pmatrix} 32/15 \\ -1/2 \\ -1 \end{pmatrix} = \begin{pmatrix} 3 \\ 22/5 \\ 44/5 \end{pmatrix},$$

as depicted in Figure 12.2.

12.3 Projection Matrices

Subsequently, for any matrix **X**, the symbol $\mathbf{P_X}$ is used to represent the square matrix $\mathbf{X}(\mathbf{X}'\mathbf{X})^-\mathbf{X}'$. Note that $\mathbf{P_X}$ is invariant to the choice of the generalized inverse $(\mathbf{X}'\mathbf{X})^-$, as (in light of Corollary 7.4.5) is evident from Theorem 9.4.1 and as is implicit in the following theorem.

Theorem 12.3.1. Let $\mathcal{U}$ represent any subspace of the linear space $\mathcal{R}^n$ of all n-dimensional column vectors. Then, there exists a unique matrix $\mathbf{A}$ (of dimensions $n \times n$) such that $\mathbf{Ay}$ is the projection (with respect to the usual inner product) of $\mathbf{y}$ on $\mathcal{U}$ for every column vector $\mathbf{y}$ in $\mathcal{R}^n$. Moreover, $\mathbf{A} = \mathbf{P_X}$ for any matrix $\mathbf{X}$ such that $\mathcal{C}(\mathbf{X}) = \mathcal{U}$.

Proof. Let $\mathbf{X}$ represent any matrix such that $\mathcal{C}(\mathbf{X}) = \mathcal{U}$. Then, it follows from Corollary 12.2.2 that, for every $\mathbf{y} \in \mathcal{R}^n$, $\mathbf{P_X y}$ is the projection of $\mathbf{y}$ on $\mathcal{U}$. Moreover, if a matrix $\mathbf{A}$ is such that $\mathbf{Ay}$ is the projection of $\mathbf{y}$ on $\mathcal{U}$ for every $\mathbf{y} \in \mathcal{R}^n$, then $\mathbf{Ay} = \mathbf{P_X y}$ for every $\mathbf{y}$, implying (in light of Lemma 2.3.2) that $\mathbf{A} = \mathbf{P_X}$. We conclude that there exists a unique matrix $\mathbf{A}$ such that $\mathbf{Ay}$ is the projection of $\mathbf{y}$ on $\mathcal{U}$ for every $\mathbf{y} \in \mathcal{R}^n$ and that $\mathbf{A} = \mathbf{P_X}$. Q.E.D.

Suppose that $\mathcal{U}$ is a subspace of the linear space $\mathcal{R}^n$ of all n-dimensional column vectors. Then, according to Theorem 12.3.1, there exists a unique matrix $\mathbf{A}$ (of dimensions $n \times n$) such that $\mathbf{Ay}$ is the projection (with respect to the usual inner product) of $\mathbf{y}$ on $\mathcal{U}$ for every $\mathbf{y} \in \mathcal{R}^n$. This matrix is called the *orthogonal projection matrix for* $\mathcal{U}$ or simply the *projection matrix for* $\mathcal{U}$. The meaning of a statement that an $n \times n$ matrix is a projection matrix is that there exists some subspace of $\mathcal{R}^n$ for which that matrix is the projection matrix.

The last part of Theorem 12.3.1 can now be restated as follows:

Corollary 12.3.2. Let $\mathbf{A}$ represent the projection matrix for a subspace $\mathcal{U}$ of the linear space $\mathcal{R}^n$ of all n-dimensional column vectors. Then, $\mathbf{A} = \mathbf{P_X}$ for any matrix $\mathbf{X}$ such that $\mathcal{C}(\mathbf{X}) = \mathcal{U}$.

For any matrix $\mathbf{X}$, it is clear that $\mathbf{P_X}$ is the projection matrix for $\mathcal{C}(\mathbf{X})$. Thus, we have the following additional corollary of Theorem 12.3.1.

Corollary 12.3.3. A matrix $\mathbf{A}$ is a projection matrix if and only if $\mathbf{A} = \mathbf{P_X}$ for some matrix $\mathbf{X}$.

Certain basic properties of projection matrices are listed in the following two theorems.

Theorem 12.3.4. Let $\mathbf{X}$ represent any $n \times p$ matrix. Then,

(1) $\mathbf{P_X X} = \mathbf{X}$; that is, $\mathbf{X(X'X)^-X'X} = \mathbf{X}$; that is, $\mathbf{(X'X)^-X'}$ is a generalized inverse of $\mathbf{X}$;

(2) $\mathbf{P_X} = \mathbf{XB^*}$ for any solution $\mathbf{B^*}$ to the (consistent) linear system $\mathbf{X'XB} = \mathbf{X'}$ (in $\mathbf{B}$);

(3) $\mathbf{P_X'} = \mathbf{P_X}$; that is, $\mathbf{P_X}$ is symmetric; that is, $\mathbf{X[(X'X)^-]'X'} = \mathbf{X(X'X)^-X'}$;

(4) $\mathbf{X[(X'X)^-]'X'X} = \mathbf{X}$; that is, $\mathbf{[(X'X)^-]'X'}$ is a generalized inverse of $\mathbf{X}$;

(5) $\mathbf{X'P_X} = \mathbf{X'P_X'} = \mathbf{X'}$; that is, $\mathbf{X'X(X'X)^-X'} = \mathbf{X'X[(X'X)^-]'X'} = \mathbf{X'}$; that is, $\mathbf{X(X'X)^-}$ and $\mathbf{X[(X'X)^-]'}$ are generalized inverses of $\mathbf{X'}$;

(6) $\mathbf{P_X^2} = \mathbf{P_X}$; that is, $\mathbf{P_X}$ is idempotent;

(7) $\mathcal{C}(\mathbf{P_X}) = \mathcal{C}(\mathbf{X})$, and $\mathcal{R}(\mathbf{P_X}) = \mathcal{R}(\mathbf{X'})$;

(8) $\text{rank}(\mathbf{P_X}) = \text{rank}(\mathbf{X})$;

(9) $(\mathbf{I} - \mathbf{P_X})^2 = \mathbf{I} - \mathbf{P_X} = (\mathbf{I} - \mathbf{P_X})'$; that is, $\mathbf{I} - \mathbf{P_X}$ is symmetric and idempotent;

(10) $\text{rank}(\mathbf{I} - \mathbf{P_X}) = n - \text{rank}(\mathbf{X})$.

Proof. (1) It follows from the very definition of a generalized inverse that $\mathbf{X}'\mathbf{X}(\mathbf{X}'\mathbf{X})^{-}\mathbf{X}'\mathbf{X} = \mathbf{X}'\mathbf{X} = \mathbf{X}'\mathbf{X}\mathbf{I}$, and, upon applying Corollary 5.3.3, we conclude that $\mathbf{X}(\mathbf{X}'\mathbf{X})^{-}\mathbf{X}'\mathbf{X} = \mathbf{X}\mathbf{I} = \mathbf{X}$.

[An alternative proof of (1) is as follows: Let $\mathbf{x}_1, \dots, \mathbf{x}_p$ represent the columns of $\mathbf{X}$. Then, according to Theorem 12.3.1, $\mathbf{P_X}\mathbf{x}_i$ is the projection of $\mathbf{x}_i$ on $\mathcal{C}(\mathbf{X})$, and, since $\mathbf{x}_i \in \mathcal{C}(\mathbf{X})$, it follows from Theorem 12.1.3 that the projection of $\mathbf{x}_i$ on $\mathcal{C}(\mathbf{X})$ is $\mathbf{x}_i$. Thus, $\mathbf{P_X}\mathbf{X} = (\mathbf{P_X}\mathbf{x}_1, \dots, \mathbf{P_X}\mathbf{x}_p) = (\mathbf{x}_1, \dots, \mathbf{x}_p) = \mathbf{X}$.]

(2) If $\mathbf{B}^*$ is a solution to the linear system $\mathbf{X}'\mathbf{X}\mathbf{B} = \mathbf{X}'$ (whose consistency was established in Corollary 7.4.2), then (according to Theorem 11.2.4)

$$\mathbf{B}^* = (\mathbf{X}'\mathbf{X})^{-}\mathbf{X}' + [\mathbf{I} - (\mathbf{X}'\mathbf{X})^{-}\mathbf{X}'\mathbf{X}]\mathbf{Y}$$

for some matrix $\mathbf{Y}$, implying [in light of Part (1)] that

$$\mathbf{X}\mathbf{B}^* = \mathbf{P_X} + (\mathbf{X} - \mathbf{P_X}\mathbf{X})\mathbf{Y} = \mathbf{P_X}.$$

(3) According to Corollary 9.3.4, $[(\mathbf{X}'\mathbf{X})^{-}]'$ is a generalized inverse of $\mathbf{X}'\mathbf{X}$, so that $[(\mathbf{X}'\mathbf{X})^{-}]'\mathbf{X}'$ is a solution to the (consistent) linear system $\mathbf{X}'\mathbf{X}\mathbf{B} = \mathbf{X}'$ (in $\mathbf{B}$). Thus, applying Part (2), we find that

$$\mathbf{X}(\mathbf{X}'\mathbf{X})^{-}\mathbf{X}' = \mathbf{P_X} = \mathbf{X}[(\mathbf{X}'\mathbf{X})^{-}]'\mathbf{X}'.$$

(4) Making use of Parts (3) and (1), we find that

$$\mathbf{X}[(\mathbf{X}'\mathbf{X})^{-}]'\mathbf{X}'\mathbf{X} = \mathbf{P}_{\mathbf{X}}'\mathbf{X} = \mathbf{P_X}\mathbf{X} = \mathbf{X}.$$

(5) Making use of Parts (3) and (1), we find that

$$\mathbf{X}'\mathbf{P_X} = \mathbf{X}'\mathbf{P}_{\mathbf{X}}' = (\mathbf{P_X}\mathbf{X})' = \mathbf{X}'.$$

(6) Making use of Part (1), we find that

$$\mathbf{P}_{\mathbf{X}}^2 = \mathbf{P_X}\mathbf{X}(\mathbf{X}'\mathbf{X})^{-}\mathbf{X}' = \mathbf{X}(\mathbf{X}'\mathbf{X})^{-}\mathbf{X}' = \mathbf{P_X}.$$

(7) Recalling Corollary 4.2.3, it is clear from the definition of $\mathbf{P_X}$ that $\mathcal{C}(\mathbf{P_X}) \subset \mathcal{C}(\mathbf{X})$ and from Part (1) that $\mathcal{C}(\mathbf{X}) \subset \mathcal{C}(\mathbf{P_X})$. Thus, $\mathcal{C}(\mathbf{P_X}) = \mathcal{C}(\mathbf{X})$, and (in light of Lemma 4.2.5) $\mathcal{R}(\mathbf{P}_{\mathbf{X}}') = \mathcal{R}(\mathbf{X}')$ or, equivalently [in light of Part (3)], $\mathcal{R}(\mathbf{P_X}) = \mathcal{R}(\mathbf{X}')$.

(8) That $\text{rank}(\mathbf{P_X}) = \text{rank}(\mathbf{X})$ is an immediate consequence of Part (7).

(9) It follows from Part (3) that the matrix $\mathbf{I} - \mathbf{P_X}$ is symmetric. And, in light of Lemma 10.1.2, it follows from Part (6) that $\mathbf{I} - \mathbf{P_X}$ is idempotent.

(10) Making use of Lemma 10.2.4 and Parts (6) and (8), we find that

$$\text{rank}(\mathbf{I} - \mathbf{P_X}) = n - \text{rank}(\mathbf{P_X}) = n - \text{rank}(\mathbf{X}).$$

Q.E.D.

Theorem 12.3.5. Let $\mathbf{X}$ represent an $n \times p$ matrix, and let $\mathbf{W}$ represent any $n \times q$ matrix such that $\mathcal{C}(\mathbf{W}) \subset \mathcal{C}(\mathbf{X})$. Then,

(1) $\mathbf{P_X W} = \mathbf{W}$, and $\mathbf{W'P_X} = \mathbf{W'}$;

(2) $\mathbf{P_X P_W} = \mathbf{P_W P_X} = \mathbf{P_W}$.

Proof. (1) According to Lemma 4.2.2, there exists a matrix $\mathbf{F}$ such that $\mathbf{W} = \mathbf{XF}$. Thus, making use of Parts (1) and (3) of Theorem 12.3.4, we find that

$$\mathbf{P_X W} = \mathbf{P_X XF} = \mathbf{XF} = \mathbf{W}$$

and

$$\mathbf{W'} = (\mathbf{P_X W})' = \mathbf{W'P_X'} = \mathbf{W'P_X}.$$

(2) Making use of Part (1), we find that

$$\mathbf{P_X P_W} = \mathbf{P_X W(W'W)^- W'} = \mathbf{W(W'W)^- W'} = \mathbf{P_W}$$

and similarly that

$$\mathbf{P_W P_X} = \mathbf{W(W'W)^- W'P_X} = \mathbf{W(W'W)^- W'} = \mathbf{P_W}.$$

Q.E.D.

In the special case where $\mathbf{W} = \mathbf{X}$, the results of Theorem 12.3.5 reduce to results included in Parts (1), (5), and (6) of Theorem 12.3.4.

Note that if $\mathbf{X}$ and $\mathbf{W}$ are any two matrices such that $\mathcal{C}(\mathbf{W}) = \mathcal{C}(\mathbf{X})$, then it follows from Part (2) of Theorem 12.3.5 that

$$\mathbf{P_W} = \mathbf{P_X P_W} = \mathbf{P_X}.$$

(That $\mathbf{P_W} = \mathbf{P_X}$ can also be deduced from Theorem 12.3.1.) Thus, we have the following corollary of Theorem 12.3.5.

Corollary 12.3.6. For any two matrices $\mathbf{X}$ and $\mathbf{W}$ such that $\mathcal{C}(\mathbf{W}) = \mathcal{C}(\mathbf{X})$, $\mathbf{P_W} = \mathbf{P_X}$.

Another result that can be deduced from Part (2) of Theorem 12.3.5 is given by the following corollary.

Corollary 12.3.7. Let $\mathbf{X}$ represent an $n \times p$ matrix. Then, for any matrix $\mathbf{A}$ that is the projection matrix for some subspace $\mathcal{U}$ of $\mathcal{C}(\mathbf{X})$,

$$\mathbf{P_X A} = \mathbf{A P_X} = \mathbf{A}.$$

Proof. Let $\mathbf{W}$ represent any matrix whose columns span $\mathcal{U}$. Then, according to Corollary 12.3.2, $\mathbf{A} = \mathbf{P_W}$. Thus, since $\mathcal{C}(\mathbf{W}) \subset \mathcal{C}(\mathbf{X})$, it follows from Part (2) of Theorem 12.3.5 that

$$\mathbf{P_X A} = \mathbf{A P_X} = \mathbf{A}.$$

Q.E.D.

Suppose that a matrix $\mathbf{A}$ is the projection matrix for a subspace $\mathcal{U}$ of the linear space $\mathcal{R}^n$ of all n-dimensional column vectors. Let $\mathbf{X}$ represent any matrix whose columns span $\mathcal{U}$. Then, according to Corollary 12.3.2, $\mathbf{A} = \mathbf{P_X}$. Thus, making use of Part (7) of Theorem 12.3.4, we find that

$$\mathcal{U} = \mathcal{C}(\mathbf{X}) = \mathcal{C}(\mathbf{P_X}) = \mathcal{C}(\mathbf{A})$$

[in which case $\dim(\mathcal{U}) = \text{rank}(\mathbf{A})$], thereby establishing the following theorem.

Theorem 12.3.8. If a matrix $\mathbf{A}$ is the projection matrix for a subspace $\mathcal{U}$ of the linear space $\mathcal{R}^n$ of all n-dimensional column vectors, then $\mathcal{U} = \mathcal{C}(\mathbf{A})$, and $\dim(\mathcal{U}) = \text{rank}(\mathbf{A})$.

Projection matrices can be characterized as follows:

Theorem 12.3.9. A matrix is a projection matrix if and only if it is symmetric and idempotent.

Proof. Together, Corollary 12.3.3 and Parts (3) and (6) of Theorem 12.3.4 imply that every projection matrix is symmetric and idempotent. Thus, to prove Theorem 12.3.9, it suffices to show that every symmetric idempotent matrix is a projection matrix.

Suppose that $\mathbf{A}$ is a symmetric idempotent matrix. Then, $\mathbf{A}'\mathbf{A} = \mathbf{A}^2 = \mathbf{A}$, so that any generalized inverse of $\mathbf{A}'\mathbf{A}$ is a generalized inverse of $\mathbf{A}$. Thus,

$$\mathbf{P_A} = \mathbf{A}(\mathbf{A}'\mathbf{A})^-\mathbf{A}' = \mathbf{A}(\mathbf{A}'\mathbf{A})^-\mathbf{A} = \mathbf{A}.$$

We conclude, on the basis of Corollary 12.3.3, that $\mathbf{A}$ is a projection matrix. Q.E.D.

Note that, together, Theorems 12.3.9 and 12.3.8 imply that if a matrix $\mathbf{A}$ is symmetric and idempotent, then it is the projection matrix for $\mathcal{C}(\mathbf{A})$.

12.4 Least Squares Problem

Suppose that $\mathbf{Y}$ is a matrix in a linear space $\mathcal{V}$ and that $\mathcal{U}$ is a subspace of $\mathcal{V}$. Then, as indicated by Theorem 12.1.3, there exists a unique matrix $\mathbf{Z}$ in $\mathcal{U}$, called the projection of $\mathbf{Y}$ on $\mathcal{U}$, such that the difference $\mathbf{Y} - \mathbf{Z}$ between $\mathbf{Y}$ and $\mathbf{Z}$ is orthogonal to $\mathcal{U}$. A fundamental property of the projection of $\mathbf{Y}$ on $\mathcal{U}$ is described in the following theorem.

Theorem 12.4.1. Let $\mathbf{Y}$ represent a matrix in a linear space $\mathcal{V}$, and let $\mathcal{U}$ represent a subspace of $\mathcal{V}$. Then, for $\mathbf{W} \in \mathcal{U}$, the distance $\|\mathbf{Y} - \mathbf{W}\|$ between $\mathbf{Y}$ and $\mathbf{W}$ is minimized uniquely by taking $\mathbf{W}$ to be the projection $\mathbf{Z}$ of $\mathbf{Y}$ on $\mathcal{U}$. Moreover,

$$\|\mathbf{Y} - \mathbf{Z}\|^2 = \mathbf{Y} \bullet (\mathbf{Y} - \mathbf{Z}).$$

Proof. For any matrix $\mathbf{W}$ in $\mathcal{U}$,

$$\begin{aligned}\|\mathbf{Y} - \mathbf{W}\|^2 &= \|(\mathbf{Y} - \mathbf{Z}) - (\mathbf{W} - \mathbf{Z})\|^2 \\ &= (\mathbf{Y} - \mathbf{Z}) \bullet (\mathbf{Y} - \mathbf{Z}) - 2(\mathbf{Y} - \mathbf{Z}) \bullet (\mathbf{W} - \mathbf{Z}) \\ &\qquad + (\mathbf{W} - \mathbf{Z}) \bullet (\mathbf{W} - \mathbf{Z}).\end{aligned}$$

Further, $\mathbf{W} - \mathbf{Z}$ is in $\mathcal{U}$ and, by definition, $\mathbf{Y} - \mathbf{Z}$ is orthogonal to every matrix in $\mathcal{U}$. Thus, $(\mathbf{Y} - \mathbf{Z}) \bullet (\mathbf{W} - \mathbf{Z}) = 0$, and hence

$$\begin{aligned}\|\mathbf{Y} - \mathbf{W}\|^2 &= (\mathbf{Y} - \mathbf{Z}) \bullet (\mathbf{Y} - \mathbf{Z}) + (\mathbf{W} - \mathbf{Z}) \bullet (\mathbf{W} - \mathbf{Z}) \\ &= \|\mathbf{Y} - \mathbf{Z}\|^2 + \|\mathbf{W} - \mathbf{Z}\|^2 \\ &\geq \|\mathbf{Y} - \mathbf{Z}\|^2,\end{aligned}$$

with equality holding if and only if $\mathbf{W} = \mathbf{Z}$. It follows that, for $\mathbf{W} \in \mathcal{U}$, $\|\mathbf{Y} - \mathbf{W}\|^2$ and, consequently, $\|\mathbf{Y} - \mathbf{W}\|$ are minimized uniquely by taking $\mathbf{W} = \mathbf{Z}$.

That $\|\mathbf{Y} - \mathbf{Z}\|^2 = \mathbf{Y} \bullet (\mathbf{Y} - \mathbf{Z})$ is clear upon observing that $\mathbf{Z} \in \mathcal{U}$ and hence that $\mathbf{Z} \bullet (\mathbf{Y} - \mathbf{Z}) = 0$. Q.E.D.

Suppose that $\mathbf{y} = \{y_i\}$ is an n-dimensional column vector and that $\mathcal{U}$ is a subspace of the linear space $\mathcal{R}^n$ of all n-dimensional column vectors. Consider the problem of minimizing, for $\mathbf{w} = \{w_i\}$ in $\mathcal{U}$, the sum of squares $(\mathbf{y} - \mathbf{w})'(\mathbf{y} - \mathbf{w}) = \sum_{i=1}^n (y_i - w_i)^2$ or, equivalently, the distance $[(\mathbf{y} - \mathbf{w})'(\mathbf{y} - \mathbf{w})]^{1/2}$ (with respect to the usual norm) between $\mathbf{y}$ and $\mathbf{w}$. This minimization problem is known as the *least squares problem.*

The solution of the least squares problem can be obtained as a special case of Theorem 12.4.1. Theorem 12.4.1 implies that, for $\mathbf{w} \in \mathcal{U}$, the distance between $\mathbf{y}$ and $\mathbf{w}$ is minimized uniquely by taking $\mathbf{w}$ to be the projection of $\mathbf{y}$ on $\mathcal{U}$. Thus, in light of Theorem 12.2.1, we reach the following conclusion.

Theorem 12.4.2. Let $\mathcal{U}$ represent a subspace of the linear space $\mathcal{R}^n$ of all n-dimensional column vectors, take $\mathbf{X}$ to be an $n \times p$ matrix such that $\mathcal{C}(\mathbf{X}) = \mathcal{U}$, and let $\mathbf{y}$ represent a vector in $\mathcal{R}^n$. Then, for $\mathbf{w} \in \mathcal{U}$, the sum of squares $(\mathbf{y} - \mathbf{w})'(\mathbf{y} - \mathbf{w})$ of the elements of the difference $\mathbf{y} - \mathbf{w}$ between $\mathbf{y}$ and $\mathbf{w}$ is minimized uniquely by taking $\mathbf{w} = \mathbf{X}\mathbf{b}^*$, where $\mathbf{b}^*$ is any solution to the normal equations $\mathbf{X}'\mathbf{X}\mathbf{b} = \mathbf{X}'\mathbf{y}$, or, equivalently, by taking $\mathbf{w} = \mathbf{P}_{\mathbf{X}}\mathbf{y}$. Further, the minimum value of the sum of squares is expressible as

$$(\mathbf{y} - \mathbf{X}\mathbf{b}^*)'(\mathbf{y} - \mathbf{X}\mathbf{b}^*) = \mathbf{y}'(\mathbf{y} - \mathbf{X}\mathbf{b}^*) = \mathbf{y}'(\mathbf{I} - \mathbf{P}_{\mathbf{X}})\mathbf{y}.$$

In connection with Theorem 12.4.2, observe that $\mathcal{U}$ comprises every $n \times 1$ vector $\mathbf{w}$ that is expressible as $\mathbf{w} = \mathbf{X}\mathbf{b}$ for some vector $\mathbf{b}$. Thus, in light of Theorem 12.2.5, we have the following variant of Theorem 12.4.2.

Theorem 12.4.3. Let $\mathbf{X} = \{x_{ij}\}$ represent an $n \times p$ matrix, and $\mathbf{y} = \{y_i\}$ an n-dimensional column vector. Then, for $\mathbf{b} = \{b_i\}$ in $\mathcal{R}^p$, the sum of squares $(\mathbf{y} - \mathbf{X}\mathbf{b})'(\mathbf{y} - \mathbf{X}\mathbf{b}) = \sum_{i=1}^n (y_i - \sum_{j=1}^p x_{ij} b_j)^2$ has a minimum at a point $\mathbf{b}^*$ if and only if $\mathbf{b}^*$ is a solution to the normal equations $\mathbf{X}'\mathbf{X}\mathbf{b} = \mathbf{X}'\mathbf{y}$, in which case $\mathbf{X}\mathbf{b}^* = \mathbf{P}_{\mathbf{X}}\mathbf{y}$ and

$$(\mathbf{y} - \mathbf{X}\mathbf{b}^*)'(\mathbf{y} - \mathbf{X}\mathbf{b}^*) = \mathbf{y}'(\mathbf{y} - \mathbf{X}\mathbf{b}^*) = \mathbf{y}'(\mathbf{I} - \mathbf{P}_{\mathbf{X}})\mathbf{y}.$$

12.5 Orthogonal Complements

a. Some general results, terminology, and notation

The set of all matrices in a linear space $\mathcal{V}$ that are orthogonal to a subspace $\mathcal{U}$ of $\mathcal{V}$ is called the *orthogonal complement of* $\mathcal{U}$ *relative to* $\mathcal{V}$ or simply the *orthogonal complement of* $\mathcal{U}$. The symbol $\mathcal{U}^{\perp}$ is used to represent the orthogonal complement of $\mathcal{U}$.

Note that, by definition, the orthogonal complement of a subspace $\mathcal{U}$ of a linear space $\mathcal{V}$ depends on $\mathcal{V}$ as well as $\mathcal{U}$ (and also on the choice of inner product). Since this dependence is not evident in the symbol $\mathcal{U}^{\perp}$, the use of this symbol is restricted to settings where the identity of the linear space $\mathcal{V}$, relative to which the orthogonal complement of $\mathcal{U}$ is defined, is clear from the context.

If $\mathcal{U}$ is a subspace of a linear space $\mathcal{V}$, then $\mathcal{U}^{\perp}$ is also a subspace of $\mathcal{V}$, as is easily verified.

As an immediate consequence of Lemma 12.1.1, we have the following lemma.

Lemma 12.5.1. Let $\mathcal{U}$ represent a subspace of a linear space $\mathcal{V}$, and let $\{\mathbf{B}_1, \ldots, \mathbf{B}_k\}$ represent any set of matrices that spans $\mathcal{U}$. Then, a matrix $\mathbf{A}$ in $\mathcal{V}$ belongs to $\mathcal{U}^{\perp}$ if and only if $\mathbf{A}$ is orthogonal to each of the matrices $\mathbf{B}_1, \ldots, \mathbf{B}_k$.

For any $n \times p$ matrix $\mathbf{X}$, the symbol $\mathcal{C}^{\perp}(\mathbf{X})$ is used to represent the orthogonal complement of $\mathcal{C}(\mathbf{X})$ relative to the linear space $\mathcal{R}^n$ of all n-dimensional column vectors (taking the inner product to be the usual inner product for $\mathcal{R}^n$). Similarly, the symbol $\mathcal{R}^{\perp}(\mathbf{X})$ is used to represent the orthogonal complement of $\mathcal{R}(\mathbf{X})$ relative to the linear space $\mathcal{R}^p$ of all p-dimensional row vectors.

It follows from Corollary 12.1.2 that $\mathcal{C}^{\perp}(\mathbf{X})$ is the set of all solutions to the homogeneous linear system $\mathbf{X}'\mathbf{z} = \mathbf{0}$ (in $\mathbf{z}$). Thus, recalling (from Theorem 12.3.4) that $\mathbf{X}(\mathbf{X}'\mathbf{X})^{-}$ is a generalized inverse of $\mathbf{X}'$, we have (in light of Corollary 11.2.2) the following lemma.

Lemma 12.5.2. For any $n \times p$ matrix $\mathbf{X}$,

$$\mathcal{C}^{\perp}(\mathbf{X}) = \mathcal{N}(\mathbf{X}') = \mathcal{C}(\mathbf{I} - \mathbf{P_X}).$$

As a consequence of Lemma 11.3.1 or, alternatively, Theorem 12.3.4, we have the following corollary of Lemma 12.5.2.

Corollary 12.5.3. For any $n \times p$ matrix $\mathbf{X}$,

$$\dim[\mathcal{C}^{\perp}(\mathbf{X})] = n - \operatorname{rank}(\mathbf{X}) = n - \dim[\mathcal{C}(\mathbf{X})]. \tag{5.1}$$

The following theorem describes a fundamental property of the orthogonal complement of a subspace $\mathcal{U}$ of a linear space $\mathcal{V}$.

Theorem 12.5.4. A matrix $\mathbf{A}$ in a linear space $\mathcal{V}$ belongs to a subspace $\mathcal{U}$ of $\mathcal{V}$ if and only if $\mathbf{A}$ is orthogonal to every matrix in $\mathcal{U}^{\perp}$ [i.e., if and only if $\mathbf{A}$ belongs to $(\mathcal{U}^{\perp})^{\perp}$]. Or, equivalently, $\mathcal{U} = (\mathcal{U}^{\perp})^{\perp}$.

Proof. Suppose that $\mathbf{A} \in \mathcal{U}$. Then, it follows immediately from the definition of $\mathcal{U}^{\perp}$ that $\mathbf{A}$ is orthogonal to every matrix in $\mathcal{U}^{\perp}$.

Conversely, suppose that $\mathbf{A}$ is orthogonal to every matrix in $\mathcal{U}^{\perp}$. Let $\mathbf{B}$ represent the projection of $\mathbf{A}$ on $\mathcal{U}$. Then, by definition, $(\mathbf{A} - \mathbf{B}) \perp \mathcal{U}$ or, equivalantly, $(\mathbf{A} - \mathbf{B}) \in \mathcal{U}^{\perp}$. Thus,

$$(\mathbf{A} - \mathbf{B}) \bullet (\mathbf{A} - \mathbf{B}) = \mathbf{A} \bullet (\mathbf{A} - \mathbf{B}) - \mathbf{B} \bullet (\mathbf{A} - \mathbf{B}) = 0 - 0 = 0,$$

implying that $\mathbf{A} - \mathbf{B} = \mathbf{0}$, or equivalently that $\mathbf{A} = \mathbf{B}$, and hence that $\mathbf{A} \in \mathcal{U}$. Q.E.D.

Various implications of Theorem 12.5.4 are set forth in the following three corollaries.

Corollary 12.5.5. A subspace $\mathcal{W}$ of a linear space $\mathcal{V}$ is contained in a subspace $\mathcal{U}$ of $\mathcal{V}$ if and only if $\mathcal{W} \perp \mathcal{U}^{\perp}$ (i.e., if and only if $\mathcal{W}$ is orthogonal to $\mathcal{U}^{\perp}$).

Proof. Corollary 12.5.5 follows from Theorem 12.5.4 upon observing that $\mathcal{W} \perp \mathcal{U}^{\perp} \Leftrightarrow \mathcal{W} \subset (\mathcal{U}^{\perp})^{\perp}$. Q.E.D.

Corollary 12.5.6. Let $\mathcal{U}$ and $\mathcal{W}$ represent subspaces of a linear space $\mathcal{V}$. Then, (1) $\mathcal{W} \subset \mathcal{U}$ if and only if $\mathcal{U}^{\perp} \subset \mathcal{W}^{\perp}$, and (2) $\mathcal{W} = \mathcal{U}$ if and only if $\mathcal{W}^{\perp} = \mathcal{U}^{\perp}$.

Proof. (1) Making use of Corollary 12.5.5 and Theorem 12.5.4, we find that

$$\mathcal{W} \subset \mathcal{U} \Leftrightarrow \mathcal{W} \perp \mathcal{U}^{\perp} \Leftrightarrow (\mathcal{W}^{\perp})^{\perp} \perp \mathcal{U}^{\perp} \Leftrightarrow \mathcal{U}^{\perp} \perp (\mathcal{W}^{\perp})^{\perp} \Leftrightarrow \mathcal{U}^{\perp} \subset \mathcal{W}^{\perp}.$$

(2) If $\mathcal{W}^{\perp} = \mathcal{U}^{\perp}$, then $\mathcal{W}^{\perp} \subset \mathcal{U}^{\perp}$ and $\mathcal{U}^{\perp} \subset \mathcal{W}^{\perp}$, implying [in light of Part (1)] that $\mathcal{U} \subset \mathcal{W}$ and $\mathcal{W} \subset \mathcal{U}$ and hence that $\mathcal{W} = \mathcal{U}$. Conversely, if $\mathcal{W} = \mathcal{U}$, then $\mathcal{W} \subset \mathcal{U}$ and $\mathcal{U} \subset \mathcal{W}$, implying that $\mathcal{U}^{\perp} \subset \mathcal{W}^{\perp}$ and $\mathcal{W}^{\perp} \subset \mathcal{U}^{\perp}$ and hence that $\mathcal{W}^{\perp} = \mathcal{U}^{\perp}$. Q.E.D.

Corollary 12.5.7. Let $\mathbf{X}$ represent an $n \times p$ matrix, and let $\mathbf{Z}$ represent any $n \times s$ matrix whose columns span $\mathcal{C}^{\perp}(\mathbf{X})$ or, equivalently, span $\mathcal{N}(\mathbf{X}')$. Then, for any $n \times q$ matrix $\mathbf{Y}$, $\mathcal{C}(\mathbf{Y}) \subset \mathcal{C}(\mathbf{X})$ if and only if $\mathbf{Z}'\mathbf{Y} = \mathbf{0}$ or, equivalently, if and only if $\mathbf{Y}'\mathbf{Z} = \mathbf{0}$.

Proof. In light of Corollary 12.1.2 and Corollary 12.5.5, we have that

$$\begin{aligned}
\mathbf{Z}'\mathbf{Y} = \mathbf{0} &\Leftrightarrow \mathcal{C}(\mathbf{Y}) \perp \mathcal{C}(\mathbf{Z}) \\
&\Leftrightarrow \mathcal{C}(\mathbf{Y}) \perp \mathcal{C}^{\perp}(\mathbf{X}) \text{ [since, by definition, } \mathcal{C}(\mathbf{Z}) = \mathcal{C}^{\perp}(\mathbf{X})] \\
&\Leftrightarrow \mathcal{C}(\mathbf{Y}) \subset \mathcal{C}(\mathbf{X}).
\end{aligned}$$

Q.E.D.

Note that, for any matrix $\mathbf{Y}$ in the orthogonal complement $\mathcal{U}^{\perp}$ of a subspace $\mathcal{U}$ of a linear space $\mathcal{V}$ of $m \times n$ matrices, the $m \times n$ null matrix $\mathbf{0}$ is the projection of $\mathbf{Y}$ on $\mathcal{U}$.

b. Projection on an orthogonal complement

For any matrix $\mathbf{Y}$ in a linear space $\mathcal{V}$, there is a simple relationship between the projection of $\mathbf{Y}$ on a subspace $\mathcal{U}$ and the projection of $\mathbf{Y}$ on the orthogonal complement $\mathcal{U}^{\perp}$ of $\mathcal{U}$. This relationship is described in the following theorem.

Theorem 12.5.8. Let $\mathbf{Y}$ represent a matrix in a linear space $\mathcal{V}$, and let $\mathcal{U}$ represent a subspace of $\mathcal{V}$. Then, the projection of $\mathbf{Y}$ on $\mathcal{U}^{\perp}$ equals $\mathbf{Y} - \mathbf{Z}$, where $\mathbf{Z}$ is the projection of $\mathbf{Y}$ on $\mathcal{U}$.

Proof. By definition, $\mathbf{Y} - \mathbf{Z}$ is orthogonal to $\mathcal{U}$, and hence $(\mathbf{Y} - \mathbf{Z}) \in \mathcal{U}^{\perp}$. Thus, to show that $\mathbf{Y} - \mathbf{Z}$ is the projection of $\mathbf{Y}$ on $\mathcal{U}^{\perp}$, it suffices to show that $\mathbf{Y} - (\mathbf{Y} - \mathbf{Z})$ is orthogonal to $\mathcal{U}^{\perp}$. That is, it suffices to show that $\mathbf{Z}$ is orthogonal to every matrix in $\mathcal{U}^{\perp}$ or, equivalently (in light of Theorem 12.5.4), that $\mathbf{Z} \in \mathcal{U}$. That $\mathbf{Z} \in \mathcal{U}$ is clear from the definition of $\mathbf{Z}$. Q.E.D.

In light of Theorem 12.2.1, we have the following two corollaries of Theorem 12.5.8.

Corollary 12.5.9. Let $\mathbf{X}$ represent an $n \times p$ matrix, and let $\mathbf{z}$ represent the projection (with respect to the usual inner product) of a column vector $\mathbf{y}$ on $\mathcal{C}^{\perp}(\mathbf{X})$. Then,

$$\mathbf{z} = \mathbf{y} - \mathbf{X}\mathbf{b}^*$$

for any solution $\mathbf{b}^*$ to the normal equations $\mathbf{X}'\mathbf{X}\mathbf{b} = \mathbf{X}'\mathbf{y}$ (in $\mathbf{b}$). In particular,

$$\mathbf{z} = \mathbf{y} - \mathbf{P_X}\mathbf{y} = (\mathbf{I} - \mathbf{P_X})\mathbf{y}.$$

Corollary 12.5.10. Let $\mathbf{X}$ represent an $n \times p$ matrix. Then, the projection matrix for $\mathcal{C}^{\perp}(\mathbf{X})$ is $\mathbf{I} - \mathbf{P_X}$.

Various of the results included in Theorems 12.1.3 and 12.5.8 can be combined and restated as follows:

Theorem 12.5.11. Let $\mathbf{Y}$ represent a matrix in a linear space $\mathcal{V}$, and let $\mathcal{U}$ represent a subspace of $\mathcal{V}$. Then, $\mathcal{U}$ contains a unique matrix $\mathbf{Z}$ and $\mathcal{U}^{\perp}$ a unique matrix $\mathbf{W}$ such that $\mathbf{Y} = \mathbf{Z} + \mathbf{W}$. Moreover, $\mathbf{Z}$ is the projection of $\mathbf{Y}$ on $\mathcal{U}$, and $\mathbf{W}$ is the projection of $\mathbf{Y}$ on $\mathcal{U}^{\perp}$.

c. Two-dimensional example revisited

Let us find, for the two-dimensional example considered previously in Section 12.2b, the projection (with respect to the usual inner product) of an n-dimensional column vector $\mathbf{y}$ on the orthogonal complement $\mathcal{U}^{\perp}$ of a subspace $\mathcal{U}$ of $\mathcal{R}^n$. Recall that $\mathbf{y} = (4, 8)'$, $\mathcal{U} = \text{sp}(\mathbf{x}) = \mathcal{C}(\mathbf{X})$, where $\mathbf{x} = (3, 1)'$ and $\mathbf{X}$ is the 2×1 matrix whose only column is $\mathbf{x}$, and that the projection of $\mathbf{y}$ on $\mathcal{U}$ is $\mathbf{z} = (6, 2)'$.

It follows from Corollary 12.5.3 and Lemma 12.5.2 that

$$\dim(\mathcal{U}^{\perp}) = \dim[\mathcal{C}^{\perp}(\mathbf{X})] = n - \text{rank}(\mathbf{X}) = 2 - 1 = 1$$

and that $\mathcal{U}^{\perp} = \mathcal{C}^{\perp}(\mathbf{X}) = \text{sp}(\mathbf{v}^*)$, where $\mathbf{v}^*$ is any nonnull solution to the honogeneous linear system $\mathbf{X}'\mathbf{v} = \mathbf{0}$ (in $\mathbf{v}$), for example, $\mathbf{v}^* = (1, -3)'$. Furthermore, the projection of $\mathbf{y}$ on $\mathcal{U}^{\perp}$ is $\mathbf{w} = \mathbf{y} - \mathbf{z} = (-2, 6)'$, as depicted in Figure 12.3.

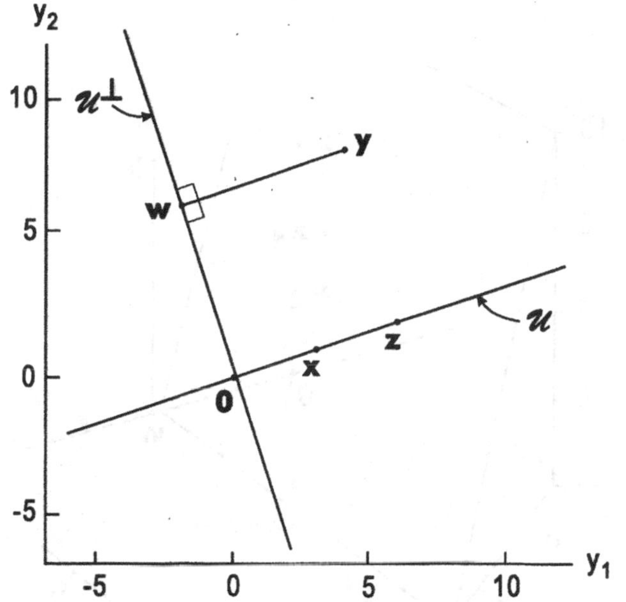

FIGURE 12.3: The projection $\mathbf{w}$ of a two-dimensional column vector $\mathbf{y}$ on the orthogonal complement $\mathcal{U}^{\perp}$ of a one-dimensional subspace $\mathcal{U}$ of $\mathcal{R}^2$.

d. Three-dimensional example revisited

Let us find, for the three-dimensional example considered previously in Section 12.2c, the projection (with respect to the usual inner product) of an n-dimensional column vector $\mathbf{y}$ on the orthogonal complement $\mathcal{U}^{\perp}$ of a subspace $\mathcal{U}$ of $\mathcal{R}^n$. Recall that $\mathbf{y} = (3, -38/5, 74/5)'$, $\mathcal{U} = \text{sp}(\mathbf{x}_1, \mathbf{x}_2, \mathbf{x}_3) = \mathcal{C}(\mathbf{X})$, where $\mathbf{x}_1$, $\mathbf{x}_2$, and $\mathbf{x}_3$ are given by expressions (2.4) and $\mathbf{X}$ is the 3×3 matrix with columns $\mathbf{x}_1$, $\mathbf{x}_2$, and $\mathbf{x}_3$, and the projection of $\mathbf{y}$ on $\mathcal{U}$ is $\mathbf{z} = (3, 22/5, 44/5)'$.

We find that $\dim(\mathcal{U}^{\perp}) = n - \text{rank}(\mathbf{X}) = 3 - 2 = 1$ and that $\mathcal{U}^{\perp} = \mathcal{C}^{\perp}(\mathbf{X}) = \text{sp}(\mathbf{v}^*)$, where $\mathbf{v}^*$ is any nonnull solution to the homogeneous linear system $\mathbf{X}'\mathbf{v} = \mathbf{0}$, for example, $\mathbf{v}^* = (0, 2, -1)'$. Furthermore, the projection of $\mathbf{y}$ on $\mathcal{U}^{\perp}$ is $\mathbf{w} = \mathbf{y} - \mathbf{z} = (0, -12, 6)'$, as depicted in Figure 12.4.

e. Dimension of an orthogonal complement

Formula (5.1) for the dimension of the orthogonal complement $\mathcal{C}^{\perp}(\mathbf{X})$ of the column space $\mathcal{C}(\mathbf{X})$ of an $n \times p$ matrix $\mathbf{X}$ can be extended to the orthogonal complement of any subspace of any linear space. To do so, suppose that $\mathcal{U}$ is an r-dimensional subspace of an n-dimensional linear space $\mathcal{V}$, define $k = \dim(\mathcal{U}^{\perp})$, and let $M = \{\mathbf{A}_1, \ldots, \mathbf{A}_r\}$ and $T = \{\mathbf{B}_1, \ldots, \mathbf{B}_k\}$ represent orthonormal bases for $\mathcal{U}$ and $\mathcal{U}^{\perp}$, respectively. In addition, take $S = \{\mathbf{A}_1, \ldots, \mathbf{A}_r, \mathbf{B}_1, \ldots, \mathbf{B}_k\}$ to be

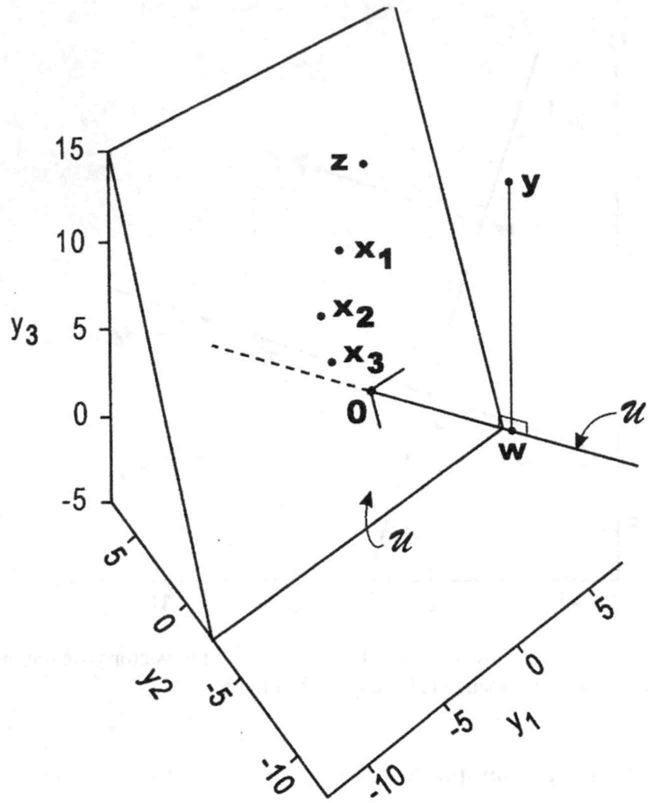

FIGURE 12.4: The projection $\mathbf{w}$ of a three-dimensional column vector $\mathbf{y}$ on the orthogonal complement $\mathcal{U}^\perp$ of a two-dimensional subspace $\mathcal{U}$ of $\mathcal{R}^3$.

the set whose members consist of the r matrices in M together with the k matrices in T.

Theorem 12.5.11 implies that every matrix $\mathbf{Y}$ in $\mathcal{V}$ is expressible as $\mathbf{Y} = \mathbf{Z}+\mathbf{W}$ for some matrix $\mathbf{Z}$ in $\mathcal{U}$ and some matrix $\mathbf{W}$ in $\mathcal{U}^\perp$ and hence that the set S spans $\mathcal{V}$. Moreover, since $\mathbf{A}_i \in \mathcal{U}$ and $\mathbf{B}_j \in \mathcal{U}^\perp$, $\mathbf{A}_i$ and $\mathbf{B}_j$ are orthogonal ($i = 1, \ldots, r$; $j = 1, \ldots, k$), implying that S is orthonormal and consequently linearly independent. We conclude that S is a basis for $\mathcal{V}$ and hence that $n = r+k$, thereby establishing the following generalization of Corollary 12.5.3.

Theorem 12.5.12. Let $\mathcal{U}$ represent a subspace of a linear space $\mathcal{V}$. Then,

$$\dim(\mathcal{V}) = \dim(\mathcal{U}) + \dim(\mathcal{U}^\perp)$$

or, equivalently, $\dim(\mathcal{U}^\perp) = \dim(\mathcal{V}) - \dim(\mathcal{U})$.

Exercises

Section 12.1

1. Verify Lemma 12.1.1.

2. Let $\mathcal{U}$ and $\mathcal{V}$ represent subspaces of $\mathcal{R}^{m\times n}$. Show that if $\dim(\mathcal{V}) > \dim(\mathcal{U})$, then $\mathcal{V}$ contains a nonnull matrix that is orthogonal to $\mathcal{U}$.

Section 12.2

3. Let $\mathcal{U}$ represent a subspace of the linear space $\mathcal{R}^m$ of all m-dimensional column vectors. Take $\mathcal{M}$ to be the subspace of $\mathcal{R}^{m\times n}$ defined by $\mathbf{W} \in \mathcal{M}$ if and only if $\mathbf{W} = (\mathbf{w}_1, \ldots, \mathbf{w}_n)$ for some vectors $\mathbf{w}_1, \ldots, \mathbf{w}_n$ in $\mathcal{U}$. Let $\mathbf{Z}$ represent the projection (with respect to the usual inner product) of an $m \times n$ matrix $\mathbf{Y}$ on $\mathcal{M}$, and let $\mathbf{X}$ represent any $m \times p$ matrix whose columns span $\mathcal{U}$. Show that $\mathbf{Z} = \mathbf{X}\mathbf{B}^*$ for any solution $\mathbf{B}^*$ to the linear system

$$\mathbf{X}'\mathbf{X}\mathbf{B} = \mathbf{X}'\mathbf{Y} \text{ (in } \mathbf{B}).$$

4. For the example of Section 12.2c, recompute the projection of $\mathbf{y}$ on $\mathcal{U}$ by taking $\mathbf{X}$ to be the 3×2 matrix

$$\begin{pmatrix} 0 & -2 \\ 3 & 2 \\ 6 & 4 \end{pmatrix}$$

and carrying out the following two steps:

(1) compute the solution to the normal equations $\mathbf{X}'\mathbf{X}\mathbf{b} = \mathbf{X}'\mathbf{y}$;

(2) postmultiply $\mathbf{X}$ by the solution you computed in Step (1).

Section 12.3

5. Let $\mathbf{X}$ represent any $n \times p$ matrix. It follows from Parts (1)–(3) of Theorem 12.3.4 that if a $p \times n$ matrix $\mathbf{B}^*$ is a solution to the linear system $\mathbf{X}'\mathbf{X}\mathbf{B} = \mathbf{X}'$ (in $\mathbf{B}$), then $\mathbf{B}^*$ is a generalized inverse of $\mathbf{X}$, and $\mathbf{X}\mathbf{B}^*$ is symmetric. Show that, conversely, if a $p \times n$ matrix $\mathbf{G}$ is a generalized inverse of $\mathbf{X}$, and $\mathbf{X}\mathbf{G}$ is symmetric, then $\mathbf{X}'\mathbf{X}\mathbf{G} = \mathbf{X}'$ (i.e., $\mathbf{G}$ is a solution to $\mathbf{X}'\mathbf{X}\mathbf{B} = \mathbf{X}'$).

6. Using the result of Part (b) of Exercise 9.3 (or otherwise), show that, for any nonnull symmetric matrix $\mathbf{A}$,

$$\mathbf{P}_{\mathbf{A}} = \mathbf{B}(\mathbf{T}\mathbf{B})^{-1}\mathbf{T},$$

where $\mathbf{B}$ is any matrix of full column rank and $\mathbf{T}$ any matrix of full row rank such that $\mathbf{A} = \mathbf{B}\mathbf{T}$. (That $\mathbf{T}\mathbf{B}$ is nonsingular follows from the result of Exercise 8.3.)

7. Let $\mathcal{V}$ represent a k-dimensional subspace of the linear space $\mathcal{R}^n$ of all n-dimensional column vectors. Take $\mathbf{X}$ to be any $n \times p$ matrix whose columns span $\mathcal{V}$, let $\mathcal{U}$ represent a subspace of $\mathcal{V}$, and define $\mathbf{A}$ to be the projection matrix for $\mathcal{U}$. Show (1) that a matrix $\mathbf{B}$ (of dimensions $n \times n$) is such that $\mathbf{By}$ is the projection of $\mathbf{y}$ on $\mathcal{U}$ for every $\mathbf{y} \in \mathcal{V}$ if and only if $\mathbf{B} = \mathbf{A} + \mathbf{Z}_*'$ for some solution $\mathbf{Z}_*$ to the honogeneous linear system $\mathbf{X}'\mathbf{Z} = \mathbf{0}$ (in an $n \times n$ matrix $\mathbf{Z}$), and (2) that, unless $k = n$, there is more than one matrix $\mathbf{B}$ such that $\mathbf{By}$ is the projection of $\mathbf{y}$ on $\mathcal{U}$ for every $\mathbf{y} \in \mathcal{V}$.

Section 12.5

8. Let $\{\mathbf{A}_1, \ldots, \mathbf{A}_k\}$ represent a nonempty linearly independent set of matrices in a linear space $\mathcal{V}$, and define $\mathbf{B}_1, \ldots, \mathbf{B}_k$ as in Theorem 6.4.1. Show that $\mathbf{B}_j$ is the (orthogonal) projection of $\mathbf{A}_j$ on some subspace $\mathcal{U}_j$ (of $\mathcal{V}$) and describe $\mathcal{U}_j$ $(j = 2, \ldots, k)$.

13

Determinants

Determinants are encountered with considerable frequency in the statistics literature (and in the literature of various other disciplines that involve the notion of randomness). A determinant appears in the "normalizing constant" of the probability density function of the all-important multivariate normal distribution (e.g., Searle 1971, sec. 2.4f). And the definition of the generalized variance (or generalized dispersion) of a random vector involves a determinant—in the design of experiments, D-optimal designs are obtained by minimizing a generalized variance (e.g., Fedorov 1972).

13.1 Some Definitions, Notation, and Special Cases

a. Negative versus positive pairs

Associated with any square matrix is a scalar that is known as the determinant of the matrix. As a preliminary to defining the determinant, it is convenient to introduce a convention for classifying various pairs of matrix elements as either positive or negative.

Let $\mathbf{A} = \{a_{ij}\}$ represent an arbitrary $n \times n$ matrix. Consider any pair of elements of $\mathbf{A}$ that do not lie either in the same row or the same column, say a_{ij} and $a_{i'j'}$ (where $i' \neq i$ and $j' \neq j$). The pair is said to be a *negative pair* if one of the elements is located above and to the right of the other, or, equivalently, if either $i' > i$ and $j' < j$ or $i' < i$ and $j' > j$. Otherwise (if one of the elements is located above and to the left of the other, or, equivalently, if either $i' > i$ and $j' > j$ or $i' < i$ and $j' < j$), the pair is said to be a *positive pair*. Thus, the pair a_{ij} and $a_{i'j'}$ is classified as positive or negative in accordance with the following two-way table:

	$i' > i$	$i' < i$
$j' > j$	$+$	$-$
$j' < j$	$-$	$+$

For example (supposing that $n \geq 4$), the pair a_{34} and a_{22} is positive, whereas the pair a_{34} and a_{41} is negative.

Note that whether the pair a_{ij} and $a_{i'j'}$ is positive or negative is completely determined by the relative locations of a_{ij} and $a_{i'j'}$ and has nothing to do with whether a_{ij} and $a_{i'j'}$ are positive or negative numbers.

Now, consider n elements of $\mathbf{A}$, no two of which lie either in the same row or the same column, say the $i_1 j_1, \ldots, i_n j_n$th elements (where both $i_1, \ldots, i_n$ and $j_1, \ldots, j_n$ are permutations of the first n positive integers). A total of $\binom{n}{2}$ pairs can be formed from these n elements. The symbol $\sigma_n(i_1, j_1; \ldots; i_n, j_n)$ is to be used to represent the number of these $\binom{n}{2}$ pairs that are negative pairs.

Observe that $\sigma_n(i_1, j_1; \ldots; i_n, j_n)$ has the following two properties:

(1) the value of $\sigma_n(i_1, j_1; \ldots; i_n, j_n)$ is not affected by permuting its n pairs of arguments; in particular, it is not affected if the n pairs are permuted so that they are ordered by row number or by column number [e.g., $\sigma_3(2,3;1,2;3,1) = \sigma_3(1,2;2,3;3,1) = \sigma_3(3,1;1,2;2,3)$];

(2) $$\sigma_n(i_1, j_1; \ldots; i_n, j_n) = \sigma_n(j_1, i_1; \ldots; j_n, i_n)$$
[e.g., $\sigma_3(2,3;1,2;3,1) = \sigma_3(3,2;2,1;1,3)$].

For any sequence of n distinct integers $i_1, \ldots, i_n$, define

$$\phi_n(i_1, \ldots, i_n) = p_1 + \cdots + p_{n-1},$$

where p_k represents the number of integers in the subsequence $i_{k+1}, \ldots, i_n$ that are smaller than i_k $(k = 1, \ldots, n-1)$. For example,

$$\phi_5(3,7,2,1,4) = 2 + 3 + 1 + 0 = 6.$$

Then, clearly, for any permutation $i_1, \ldots, i_n$ of the first n positive integers,

$$\sigma_n(1, i_1; \ldots; n, i_n) = \sigma_n(i_1, 1; \ldots; i_n, n) = \phi_n(i_1, \ldots, i_n). \tag{1.1}$$

b. Definition of determinant

The *determinant* of an $n \times n$ matrix $\mathbf{A} = \{a_{ij}\}$, to be denoted by $|\mathbf{A}|$ or (to avoid confusion with the absolute value of a scalar) by $\det \mathbf{A}$ or $\det(\mathbf{A})$, is defined by

$$|\mathbf{A}| = \sum (-1)^{\sigma_n(1, j_1; \ldots; n, j_n)} a_{1j_1} \cdots a_{nj_n} \tag{1.2a}$$

or equivalently by

$$|\mathbf{A}| = \sum (-1)^{\phi_n(j_1, \ldots, j_n)} a_{1j_1} \cdots a_{nj_n}, \tag{1.2b}$$

where $j_1, \ldots, j_n$ is a permutation of the first n positive integers and the summation is over all such permutations.

Thus, the determinant of an $n \times n$ matrix $\mathbf{A}$ can (at least in principle) be obtained via the following process:

(1) Form all possible products, each of n factors, that can be obtained by picking one and only one element from each row and column of $\mathbf{A}$.

(2) In each product, count the number of negative pairs among the $\binom{n}{2}$ pairs of elements that can be generated from the n elements that contribute to this particular product. If the number of negative pairs is an even number, attach a plus sign to the product; if it is an odd number, attach a minus sign.

(3) Sum the signed products.

In particular, the determinant of a 1×1 matrix $\mathbf{A} = (a_{11})$ is

$$|\mathbf{A}| = a_{11}; \tag{1.3}$$

the determinant of a 2×2 matrix $\mathbf{A} = \{a_{ij}\}$ is

$$\begin{aligned} |\mathbf{A}| &= (-1)^0 a_{11}a_{22} + (-1)^1 a_{12}a_{21} \\ &= a_{11}a_{22} - a_{12}a_{21}; \end{aligned} \tag{1.4}$$

and the determinant of a 3×3 matrix $\mathbf{A} = \{a_{ij}\}$ is

$$\begin{aligned} |\mathbf{A}| &= (-1)^0 a_{11}a_{22}a_{33} + (-1)^1 a_{11}a_{23}a_{32} + (-1)^1 a_{12}a_{21}a_{33} \\ &\quad + (-1)^2 a_{12}a_{23}a_{31} + (-1)^2 a_{13}a_{21}a_{32} + (-1)^3 a_{13}a_{22}a_{31} \\ &= a_{11}a_{22}a_{33} + a_{12}a_{23}a_{31} + a_{13}a_{21}a_{32} \\ &\quad - a_{11}a_{23}a_{32} - a_{12}a_{21}a_{33} - a_{13}a_{22}a_{31}. \end{aligned} \tag{1.5}$$

In the case of a partitioned matrix, say $\begin{pmatrix} \mathbf{A}_{11} & \mathbf{A}_{12} \\ \mathbf{A}_{21} & \mathbf{A}_{22} \end{pmatrix}$, it is customary to write $\left|\begin{pmatrix} \mathbf{A}_{11} & \mathbf{A}_{12} \\ \mathbf{A}_{21} & \mathbf{A}_{22} \end{pmatrix}\right|$ in the abbreviated form $\begin{vmatrix} \mathbf{A}_{11} & \mathbf{A}_{12} \\ \mathbf{A}_{21} & \mathbf{A}_{22} \end{vmatrix}$.

An alternative definition of the determinant of an $n \times n$ matrix $\mathbf{A} = \{a_{ij}\}$ is

$$|\mathbf{A}| = \sum (-1)^{\sigma_n(i_1,1;\ldots;i_n,n)} a_{i_1 1} \cdots a_{i_n n} \tag{1.6a}$$

$$= \sum (-1)^{\phi_n(i_1,\ldots,i_n)} a_{i_1 1} \cdots a_{i_n n}, \tag{1.6b}$$

where $i_1, \ldots, i_n$ is a permutation of the first n positive integers and the summation is over all such permutations.

Definition (1.6) is equivalent to definition (1.2). To see this, observe that the product $a_{1j_1} \cdots a_{nj_n}$, which appears in definition (1.2), can be reexpressed by permuting the n factors $a_{1j_1}, \ldots, a_{nj_n}$ so that they are ordered by column number, giving

$$a_{1j_1} \cdots a_{nj_n} = a_{i_1 1} \cdots a_{i_n n},$$

where $i_1, \ldots, i_n$ is a permutation of the first n positive integers that is defined uniquely by

$$j_{i_1} = 1, \ \ldots, \ j_{i_n} = n.$$

Further,

$$\sigma_n(1, j_1; \ldots; n, j_n) = \sigma_n(i_1, j_{i_1}; \ldots; i_n, j_{i_n}) = \sigma_n(i_1, 1; \ldots; i_n, n),$$

so that

$$(-1)^{\sigma_n(1,j_1;\ldots;n,j_n)} a_{1j_1} \cdots a_{nj_n} = (-1)^{\sigma_n(i_1,1;\ldots;i_n,n)} a_{i_1 1} \cdots a_{i_n n}.$$

Thus, we can establish a one-to-one correspondence between the terms of the sum (1.6) and the terms of the sum (1.2) such that the corresponding terms are equal. We conclude that the two sums are themselves equal.

The number of terms in the sum (1.2) or (1.6) equals $n!$. As n increases, the number of terms grows rapidly. As a consequence, the computations required for the direct numerical evaluation of this sum can be extensive and (for even moderately large values of n) prohibitive. Representations that are better suited for computational purposes can be devised by making use of various results to be presented subsequently.

c. Diagonal, triangular, and permutation matrices

There is a simple expression for the determinant of a triangular matrix, which is given by the following lemma.

Lemma 13.1.1. If an $n \times n$ matrix $\mathbf{A} = \{a_{ij}\}$ is (upper or lower) triangular, then

$$|\mathbf{A}| = a_{11}a_{22} \cdots a_{nn}, \tag{1.7}$$

that is, the determinant of a triangular matrix equals the product of its diagonal elements.

Proof. Consider a lower triangular matrix

$$\mathbf{A} = \begin{pmatrix} a_{11} & 0 & \ldots & 0 \\ a_{21} & a_{22} & & 0 \\ \vdots & \vdots & \ddots & \\ a_{n1} & a_{n2} & \ldots & a_{nn} \end{pmatrix}.$$

That $|\mathbf{A}| = a_{11}a_{22} \cdots a_{nn}$ follows immediately upon observing that the only term in the sum (1.2) that can be nonzero is that corresponding to the permutation $j_1 = 1, \ldots, j_n = n$ [and that $\phi_n(1, 2, \ldots, n) = 0$]. (To verify formally that only this one term can be nonzero, let $j_1, \ldots, j_n$ represent an arbitrary permutation of the first n positive integers, and suppose that $a_{1j_1} \cdots a_{nj_n} \neq 0$ or equivalently that $a_{ij_i} \neq 0$ for $i = 1, \ldots, n$. Then, it is clear that $j_1 = 1$ and that, if $j_1 = 1, \ldots, j_{i-1} = i - 1$, then $j_i = i$. We conclude, on the basis of mathematical induction, that $j_1 = 1, \ldots, j_n = n$).

The validity of formula (1.7) as applied to an upper triangular matrix follows from a similar argument. Q.E.D.

Corollary 13.1.2. The determinant of a diagonal matrix equals the product of its diagonal elements.

As obvious special cases of Corollary 13.1.2, we have that

$$|\mathbf{0}| = 0, \tag{1.8}$$

$$|\mathbf{I}| = 1. \tag{1.9}$$

The following lemma gives a simple expression for the determinant of a permutation matrix.

Lemma 13.1.3. For an $n \times n$ permutation matrix $\mathbf{P}$ whose first, $\ldots$, nth rows or columns are respectively the $k_1, \ldots, k_n$th rows or columns of the $n \times n$ identity matrix,

$$|\mathbf{P}| = (-1)^{\phi_n(k_1,\ldots,k_n)}.$$

Proof. Putting $\mathbf{A} = \mathbf{P}$, we find that all of the terms in sum (1.6) are zero except the term corresponding to the permutation $i_1 = k_1, \ldots, i_n = k_n$, which equals $(-1)^{\phi_n(k_1,\ldots,k_n)}$. Q.E.D.

13.2 Some Basic Properties of Determinants

The following lemma relates the determinant of a matrix to the determinant of its transpose.

Lemma 13.2.1. For any $n \times n$ matrix $\mathbf{A}$,

$$|\mathbf{A}'| = |\mathbf{A}|. \tag{2.1}$$

Proof. Let a_{ij} and b_{ij} represent the ijth elements of $\mathbf{A}$ and $\mathbf{A}'$, respectively ($i, j = 1, \ldots, n$). Then, in light of the equivalence of definitions (1.6) and (1.2),

$$\begin{aligned} |\mathbf{A}'| &= \sum (-1)^{\phi_n(j_1,\ldots,j_n)} b_{1j_1} \cdots b_{nj_n} \\ &= \sum (-1)^{\phi_n(j_1,\ldots,j_n)} a_{j_1 1} \cdots a_{j_n n} \\ &= |\mathbf{A}|, \end{aligned}$$

where $j_1, \ldots, j_n$ is a permutation of the first n positive integers and the summations are over all such permutations. Q.E.D.

As an immediate consequence of the definition of a determinant, we have the following lemma.

Lemma 13.2.2. If an $n \times n$ matrix $\mathbf{B}$ is formed from an $n \times n$ matrix $\mathbf{A}$ by multiplying all of the elements of one row or one column of $\mathbf{A}$ by the same scalar k (and leaving the elements of the other $n - 1$ rows or columns unchanged), then

$$|\mathbf{B}| = k|\mathbf{A}|.$$

As a corollary of Lemma 13.2.2, we obtain the following generalization of result (1.8).

Corollary 13.2.3. If one or more rows (or columns) of an $n \times n$ matrix $\mathbf{A}$ are null, then

$$|\mathbf{A}| = 0.$$

Proof. Suppose that the ith row of $\mathbf{A}$ is null, and let $\mathbf{B}$ represent an $n \times n$ matrix formed from $\mathbf{A}$ by multiplying all of the elements of the ith row of $\mathbf{A}$ by zero. Then, $\mathbf{A} = \mathbf{B}$, and we find that

$$|\mathbf{A}| = |\mathbf{B}| = 0|\mathbf{A}| = 0.$$

Q.E.D.

The following corollary (of Lemma 13.2.2) relates the determinant of a scalar multiple of a matrix $\mathbf{A}$ to that of $\mathbf{A}$ itself.

Corollary 13.2.4. For any $n \times n$ matrix $\mathbf{A}$ and any scalar k,

$$|k\mathbf{A}| = k^n |\mathbf{A}|. \tag{2.2}$$

Proof. This result follows from Lemma 13.2.2 upon observing that $k\mathbf{A}$ can be formed from $\mathbf{A}$ by successively multiplying each of the n rows of $\mathbf{A}$ by k. Q.E.D.

As a special case of Corollary 13.2.4, we have the following, additional corollary.

Corollary 13.2.5. For any $n \times n$ matrix $\mathbf{A}$,

$$|-\mathbf{A}| = (-1)^n |\mathbf{A}|. \tag{2.3}$$

The following two theorems describe how the determinant of a matrix is affected by permuting its rows or columns in certain ways.

Theorem 13.2.6. If an $n \times n$ matrix $\mathbf{B} = \{b_{ij}\}$ is formed from an $n \times n$ matrix $\mathbf{A} = \{a_{ij}\}$ by interchanging two rows or two columns of $\mathbf{A}$, then

$$|\mathbf{B}| = -|\mathbf{A}|.$$

Proof. Consider first the case where $\mathbf{B}$ is formed from $\mathbf{A}$ by interchanging two adjacent rows, say the ith and $(i+1)$th rows. Then,

$$\begin{aligned}
|\mathbf{B}| &= \sum (-1)^{\phi_n(j_1,\ldots,j_n)} b_{1j_1} \cdots b_{nj_n} \\
&= \sum (-1)^{\phi_n(j_1,\ldots,j_n)} a_{1j_1} \cdots a_{i-1,j_{i-1}} a_{i+1,j_i} a_{i,j_{i+1}} a_{i+2,j_{i+2}} \cdots a_{nj_n} \\
&= -\sum (-1)^{\phi_n(j_1,\ldots,j_{i-1},j_{i+1},j_i,j_{i+2},\ldots,j_n)} \\
&\qquad\qquad \times a_{1j_1} \cdots a_{i-1,j_{i-1}} a_{i,j_{i+1}} a_{i+1,j_i} a_{i+2,j_{i+2}} \cdots a_{nj_n} \\
&\qquad [\text{since } \phi_n(j_1, \ldots, j_{i-1}, j_{i+1}, j_i, j_{i+2} \ldots, j_n) \\
&\qquad\qquad = \phi_n(j_1, \ldots, j_n) + 1, \quad \text{if } j_{i+1} > j_i, \\
&\qquad\qquad = \phi_n(j_1, \ldots, j_n) - 1, \quad \text{if } j_{i+1} < j_i] \\
&= -|\mathbf{A}|,
\end{aligned}$$

where $j_1, \ldots, j_n$ (and hence $j_1, \ldots, j_{i-1}, j_{i+1}, j_i, j_{i+2}, \ldots, j_n$) is a permutation of the first n positive integers and the summation is over all such permutations.

Consider now the case where $\mathbf{B}$ is formed from $\mathbf{A}$ by interchanging two not-necessarily-adjacent rows, say the ith and kth rows where $k > i$. Suppose that we successively interchange the kth row of $\mathbf{A}$ with the $k - i$ rows immediately preceding it, putting the n rows of $\mathbf{A}$ in the order $1, \ldots, i-1, k, i, i+1, \ldots, k-1, k+1, \ldots, n$. Suppose that we then further reorder the rows of $\mathbf{A}$ by successively interchanging what was originally the ith row with the $k-i-1$ rows immediately succeeding it, putting the n rows in the order $1, \ldots, i-1, k, i+1, \ldots, k-1, i, k+1, \ldots, n$. Thus, by executing $2(k-i)-1$ successive interchanges of adjacent rows, we have in effect interchanged the ith and kth rows of $\mathbf{A}$. Since each interchange of adjacent rows changes the sign of the determinant, we conclude that

$$|\mathbf{B}| = (-1)^{2(k-i)-1}|\mathbf{A}| = -|\mathbf{A}|.$$

By employing an analogous argument, we find that the interchange of any two columns of $\mathbf{A}$ likewise changes the sign of the determinant. Q.E.D.

Theorem 13.2.7. If $\mathbf{B}$ is an $n \times p$ matrix (where $p < n$) and $\mathbf{C}$ an $n \times q$ matrix (where $q = n - p$), then

$$|\mathbf{B},\ \mathbf{C}| = (-1)^{pq}|\mathbf{C},\ \mathbf{B}|. \tag{2.4}$$

Similarly, if $\mathbf{B}$ is a $p \times n$ matrix and $\mathbf{C}$ a $q \times n$ matrix, then

$$\begin{vmatrix} \mathbf{B} \\ \mathbf{C} \end{vmatrix} = (-1)^{pq} \begin{vmatrix} \mathbf{C} \\ \mathbf{B} \end{vmatrix}. \tag{2.5}$$

Proof. Suppose that the dimensions of $\mathbf{B}$ are $n \times p$ and those of $\mathbf{C}$ are $n \times q$. Let $\mathbf{b}_1, \ldots, \mathbf{b}_p$ represent the columns of $\mathbf{B}$ and $\mathbf{c}_1, \ldots, \mathbf{c}_q$ the columns of $\mathbf{C}$. Then, $(\mathbf{C},\ \mathbf{B}) = (\mathbf{c}_1, \ldots, \mathbf{c}_q, \mathbf{b}_1, \ldots, \mathbf{b}_p)$.

Suppose that, in the matrix $(\mathbf{C},\ \mathbf{B})$, we successively interchange column $\mathbf{b}_1$ with the columns $\mathbf{c}_q, \ldots, \mathbf{c}_1$, producing the matrix $(\mathbf{b}_1, \mathbf{c}_1, \ldots, \mathbf{c}_q, \mathbf{b}_2, \ldots, \mathbf{b}_p)$. Suppose then that, in the latter matrix, we successively interchange column $\mathbf{b}_2$ with the columns $\mathbf{c}_q, \ldots, \mathbf{c}_1$, producing the matrix $(\mathbf{b}_1, \mathbf{b}_2, \mathbf{c}_1, \ldots, \mathbf{c}_q, \mathbf{b}_3, \ldots, \mathbf{b}_p)$. Continuing in this fashion, we produce (after p steps) the matrix $(\mathbf{b}_1, \ldots, \mathbf{b}_p, \mathbf{c}_1, \ldots, \mathbf{c}_q) = (\mathbf{B},\ \mathbf{C})$.

It is now clear that we can obtain the matrix $(\mathbf{B},\ \mathbf{C})$ from the matrix $(\mathbf{C},\ \mathbf{B})$ via a total of pq successive interchanges of columns. Thus, it follows from Theorem 13.2.6 that

$$|\mathbf{B},\ \mathbf{C}| = (-1)^{pq}|\mathbf{C},\ \mathbf{B}|.$$

Result (2.5) can be derived via an analogous approach. Q.E.D.

A (square) matrix that has one or more null rows or columns has (according to Corollary 13.2.3) a zero determinant. Other matrices whose determinants are zero are described in the following two lemmas.

Lemma 13.2.8. If two rows or two columns of an $n \times n$ matrix $\mathbf{A}$ are identical, then $|\mathbf{A}| = 0$.

Proof. Suppose that two rows of $\mathbf{A}$ are identical, say the ith and kth rows, and let $\mathbf{B}$ represent a matrix formed from $\mathbf{A}$ by interchanging its ith and kth rows. Obviously, $\mathbf{B} = \mathbf{A}$ and hence $|\mathbf{B}| = |\mathbf{A}|$. Moreover, according to Theorem 13.2.6, $|\mathbf{B}| = -|\mathbf{A}|$. Thus, $|\mathbf{A}| = |\mathbf{B}| = -|\mathbf{A}|$, implying that $|\mathbf{A}| = 0$.

That the determinant of a (square) matrix having two identical columns equals zero can be proved via an analogous argument. Q.E.D.

Lemma 13.2.9. If a row or column of an $n \times n$ matrix $\mathbf{A}$ is a scalar multiple of another row or column, then $|\mathbf{A}| = 0$.

Proof. Let $\mathbf{a}_1', \ldots, \mathbf{a}_n'$ represent the rows of $\mathbf{A}$. Suppose that one row is a scalar multiple of another, that is, that $\mathbf{a}_s' = k\mathbf{a}_i'$ for some s and i (with $s \neq i$) and some scalar k. Let $\mathbf{B}$ represent a matrix formed from $\mathbf{A}$ by multiplying the ith row of $\mathbf{A}$ by the scalar k. Then, according to Lemmas 13.2.2 and 13.2.8,

$$k|\mathbf{A}| = |\mathbf{B}| = 0. \tag{2.6}$$

If $k \neq 0$, then it follows from equality (2.6) that $|\mathbf{A}| = 0$. If $k = 0$, then $\mathbf{a}_s' = \mathbf{0}$, and it follows from Corollary 13.2.3 that $|\mathbf{A}| = 0$. Thus, in either case, $|\mathbf{A}| = 0$.

An analogous argument shows that if one column of a (square) matrix is a scalar multiple of another, then again the determinant of the matrix equals zero. Q.E.D.

The transposition of a (square) matrix does not (according to Lemma 13.2.1) affect its determinant. Other operations that do not affect the determinant of a matrix are described in the following two theorems.

Theorem 13.2.10. Let $\mathbf{B}$ represent a matrix formed from an $n \times n$ matrix $\mathbf{A}$ by adding, to any one row or column of $\mathbf{A}$, scalar multiples of one or more other rows or columns. Then,

$$|\mathbf{B}| = |\mathbf{A}|.$$

Proof. Let $\mathbf{a}_i' = (a_{i1}, \ldots, a_{in})$ and $\mathbf{b}_i' = (b_{i1}, \ldots, b_{in})$ represent the ith rows of $\mathbf{A}$ and $\mathbf{B}$, respectively ($i = 1, \ldots, n$). For some integer s ($1 \leq s \leq n$) and some scalars $k_1, \ldots, k_{s-1}, k_{s+1}, \ldots, k_n$,

$$\mathbf{b}_s' = \mathbf{a}_s' + \sum_{i \neq s} k_i \mathbf{a}_i' \quad \text{and} \quad \mathbf{b}_i' = \mathbf{a}_i' \; (i \neq s).$$

Thus,

$$\begin{aligned}
|\mathbf{B}| &= \sum (-1)^{\phi_n(j_1,\ldots,j_n)} b_{1j_1} \cdots b_{nj_n} \\
&= \sum (-1)^{\phi_n(j_1,\ldots,j_n)} \\
&\qquad \times a_{1j_1} \cdots a_{s-1,j_{s-1}} \Big(a_{sj_s} + \sum_{i \neq s} k_i a_{ij_s}\Big) a_{s+1,j_{s+1}} \cdots a_{nj_n} \\
&= |\mathbf{A}| + \sum_{i \neq s} \sum (-1)^{\phi_n(j_1,\ldots,j_n)} \\
&\qquad \times a_{1j_1} \cdots a_{s-1,j_{s-1}} (k_i a_{ij_s}) a_{s+1,j_{s+1}} \cdots a_{nj_n} \\
&= |\mathbf{A}| + \sum_{i \neq s} |\mathbf{B}_i|,
\end{aligned}$$

where $\mathbf{B}_i$ is a matrix formed from $\mathbf{A}$ by replacing the sth row of $\mathbf{A}$ with $k_i\mathbf{a}_i'$ and where $j_1, \ldots, j_n$ is a permutation of the first n positive integers and the (unlabeled) summations are over all such permutations.

Since (according to Lemma 13.2.9)

$$|\mathbf{B}_i| = 0 \quad (i \neq s),$$

we conclude that $|\mathbf{B}| = |\mathbf{A}|$.

An analogous argument shows that $|\mathbf{B}| = |\mathbf{A}|$ when $\mathbf{B}$ is formed from $\mathbf{A}$ by adding, to a column of $\mathbf{A}$, scalar multiples of other columns. Q.E.D.

Theorem 13.2.11. For any $n \times n$ matrix $\mathbf{A}$ and any (upper or lower) unit triangular matrix $\mathbf{T}$,

$$|\mathbf{AT}| = |\mathbf{TA}| = |\mathbf{A}|. \tag{2.7}$$

Proof. Consider the case where $\mathbf{A}$ is postmultiplied by $\mathbf{T}$ and $\mathbf{T}$ is unit lower triangular. Define $\mathbf{T}_i$ to be a matrix formed from $\mathbf{I}_n$ by replacing the ith column of $\mathbf{I}_n$ with the ith column of $\mathbf{T}$ $(i = 1, \ldots, n)$. It is easy to verify that $\mathbf{T} = \mathbf{T}_1\mathbf{T}_2 \cdots \mathbf{T}_n$ and hence that

$$\mathbf{AT} = \mathbf{AT}_1\mathbf{T}_2 \cdots \mathbf{T}_n.$$

Define $\mathbf{B}_0 = \mathbf{A}$, and $\mathbf{B}_i = \mathbf{AT}_1\mathbf{T}_2 \cdots \mathbf{T}_i$ $(i = 1, \ldots, n-1)$. Clearly, to show that $|\mathbf{AT}| = |\mathbf{A}|$, it suffices to show that, for $i = 1, \ldots, n$, the postmultiplication of $\mathbf{B}_{i-1}$ by $\mathbf{T}_i$ does not alter the determinant of $\mathbf{B}_{i-1}$. Observe that the columns of $\mathbf{B}_{i-1}\mathbf{T}_i$ are the same as those of $\mathbf{B}_{i-1}$, except for the ith column of $\mathbf{B}_{i-1}\mathbf{T}_i$, which consists of the ith column of $\mathbf{B}_{i-1}$ plus scalar multiples of the $(i+1), \ldots,$ nth columns of $\mathbf{B}_{i-1}$. Thus, it follows from Theorem 13.2.10 that

$$|\mathbf{B}_{i-1}\mathbf{T}_i| = |\mathbf{B}_{i-1}|.$$

We conclude that $|\mathbf{AT}| = |\mathbf{A}|$.

That result (2.7) is valid for the cases of postmultiplication by a unit upper triangular matrix, and premultiplication by a unit upper or lower triangular matrix can be established via similar arguments. Q.E.D.

13.3 Partitioned Matrices, Products of Matrices, and Inverse Matrices

Formula (1.7) for the determinant of a triangular matrix can be extended to a block-triangular matrix, as indicated by the following theorem.

Theorem 13.3.1. Let $\mathbf{T}$ represent an $m \times m$ matrix, $\mathbf{V}$ an $n \times m$ matrix, and $\mathbf{W}$ an $n \times n$ matrix. Then,

$$\begin{vmatrix} \mathbf{T} & \mathbf{0} \\ \mathbf{V} & \mathbf{W} \end{vmatrix} = \begin{vmatrix} \mathbf{W} & \mathbf{V} \\ \mathbf{0} & \mathbf{T} \end{vmatrix} = |\mathbf{T}||\mathbf{W}|. \tag{3.1}$$

Proof. Let

$$\mathbf{A} = \begin{pmatrix} \mathbf{T} & \mathbf{0} \\ \mathbf{V} & \mathbf{W} \end{pmatrix},$$

and let a_{ij} represent the ijth element of $\mathbf{A}$ $(i, j = 1, \ldots, m+n)$. By definition,

$$|\mathbf{A}| = \sum (-1)^{\phi_{m+n}(j_1,\ldots,j_{m+n})} a_{1j_1} \cdots a_{m+n,j_{m+n}}, \tag{3.2}$$

where $j_1, \ldots, j_{m+n}$ is a permutation of the first $m+n$ positive integers and the summation is over all such permutations.

Clearly, the only terms of the sum (3.2) that can be nonzero are those for which $j_1, \ldots, j_m$ constitutes a permutation of the first m positive integers and thus for which $j_{m+1}, \ldots, j_{m+n}$ constitutes a permutation of the integers $m+1, \ldots, m+n$. For any such permutation, we have that

$$\begin{aligned} a_{1j_1} \cdots a_{m+n,j_{m+n}} &= t_{1j_1} \cdots t_{mj_m} w_{1,j_{m+1}-m} \cdots w_{n,j_{m+n}-m} \\ &= t_{1j_1} \cdots t_{mj_m} w_{1k_1} \cdots w_{nk_n}, \end{aligned}$$

where t_{ij} represents the ijth element of $\mathbf{T}$ $(i, j = 1, \ldots, m)$ and w_{ij} the ijth element of $\mathbf{W}$ $(i, j = 1, \ldots, n)$ and where $k_1 = j_{m+1} - m, \ \ldots, \ k_n = j_{m+n} - m$, and that

$$\begin{aligned} \phi_{m+n}(j_1, \ldots, j_{m+n}) &= \phi_m(j_1, \ldots, j_m) + \phi_n(j_{m+1}, \ldots, j_{m+n}) \\ &= \phi_m(j_1, \ldots, j_m) + \phi_n(j_{m+1} - m, \ldots, j_{m+n} - m) \\ &= \phi_m(j_1, \ldots, j_m) + \phi_n(k_1, \ldots, k_n). \end{aligned}$$

Thus,

$$\begin{aligned} |\mathbf{A}| &= \sum \sum (-1)^{\phi_m(j_1,\ldots,j_m)+\phi_n(k_1,\ldots,k_n)} t_{1j_1} \cdots t_{mj_m} w_{1k_1} \cdots w_{nk_n} \\ &= \sum (-1)^{\phi_m(j_1,\ldots,j_m)} t_{1j_1} \cdots t_{mj_m} \sum (-1)^{\phi_n(k_1,\ldots,k_n)} w_{1k_1} \cdots w_{nk_n} \\ &= |\mathbf{T}||\mathbf{W}|, \end{aligned}$$

where $j_1, \ldots, j_m$ is a permutation of the first m positive integers and $k_1, \ldots, k_n$ a permutation of the first n positive integers, and where the summations are over all such permutations.

That $\begin{vmatrix} \mathbf{W} & \mathbf{V} \\ \mathbf{0} & \mathbf{T} \end{vmatrix} = |\mathbf{T}||\mathbf{W}|$ can be established via a similar argument. Q.E.D.

The repeated application of Theorem 13.3.1 leads to the following formulas for the determinant of an arbitrary (square) upper or lower block-triangular matrix (with square diagonal blocks):

$$\begin{vmatrix} \mathbf{A}_{11} & \mathbf{A}_{12} & \ldots & \mathbf{A}_{1r} \\ \mathbf{0} & \mathbf{A}_{22} & \ldots & \mathbf{A}_{2r} \\ \vdots & & \ddots & \vdots \\ \mathbf{0} & \mathbf{0} & & \mathbf{A}_{rr} \end{vmatrix} = |\mathbf{A}_{11}||\mathbf{A}_{22}| \cdots |\mathbf{A}_{rr}|; \tag{3.3}$$

$$\begin{vmatrix} \mathbf{B}_{11} & \mathbf{0} & \dots & \mathbf{0} \\ \mathbf{B}_{21} & \mathbf{B}_{22} & & \mathbf{0} \\ \vdots & \vdots & \ddots & \\ \mathbf{B}_{r1} & \mathbf{B}_{r2} & \dots & \mathbf{B}_{rr} \end{vmatrix} = |\mathbf{B}_{11}||\mathbf{B}_{22}| \cdots |\mathbf{B}_{rr}|. \tag{3.4}$$

In the special case of a block-diagonal matrix, formula (3.3) becomes

$$\text{diag}(\mathbf{A}_{11}, \mathbf{A}_{22}, \dots, \mathbf{A}_{rr}) = |\mathbf{A}_{11}||\mathbf{A}_{22}| \cdots |\mathbf{A}_{rr}|. \tag{3.5}$$

By making use of Theorem 13.2.7, we obtain the following corollary of Theorem 13.3.1.

Corollary 13.3.2. Let $\mathbf{T}$ represent an $m \times m$ matrix, $\mathbf{V}$ an $n \times m$ matrix, and $\mathbf{W}$ an $n \times n$ matrix. Then,

$$\begin{vmatrix} \mathbf{0} & \mathbf{T} \\ \mathbf{W} & \mathbf{V} \end{vmatrix} = \begin{vmatrix} \mathbf{V} & \mathbf{W} \\ \mathbf{T} & \mathbf{0} \end{vmatrix} = (-1)^{mn} |\mathbf{T}||\mathbf{W}|. \tag{3.6}$$

The following corollary gives a simplified version of formula (3.6) for the special case where $m = n$ and $\mathbf{T} = -\mathbf{I}_n$.

Corollary 13.3.3. For $n \times n$ matrices $\mathbf{W}$ and $\mathbf{V}$,

$$\begin{vmatrix} \mathbf{0} & -\mathbf{I}_n \\ \mathbf{W} & \mathbf{V} \end{vmatrix} = \begin{vmatrix} \mathbf{V} & \mathbf{W} \\ -\mathbf{I}_n & \mathbf{0} \end{vmatrix} = |\mathbf{W}|. \tag{3.7}$$

Proof (of Corollary 13.3.3). Corollary 13.3.3 can be derived from the special case of Corollary 13.3.2 where $m = n$ and $\mathbf{T} = -\mathbf{I}_n$ by observing that

$$(-1)^{nn} |-\mathbf{I}_n||\mathbf{W}| = (-1)^{nn}(-1)^n |\mathbf{W}| = (-1)^{n(n+1)} |\mathbf{W}|$$

and that either n or $n + 1$ is an even number and consequently $n(n + 1)$ is an even number. Q.E.D.

By using Theorem 13.3.1 (and Corollary 13.3.3), together with Theorem 13.2.11, we find that, for $n \times n$ matrices $\mathbf{A}$ and $\mathbf{B}$,

$$|\mathbf{A}||\mathbf{B}| = \begin{vmatrix} \mathbf{A} & \mathbf{0} \\ -\mathbf{I} & \mathbf{B} \end{vmatrix} = \left| \begin{pmatrix} \mathbf{A} & \mathbf{0} \\ -\mathbf{I} & \mathbf{B} \end{pmatrix} \begin{pmatrix} \mathbf{I} & \mathbf{B} \\ \mathbf{0} & \mathbf{I} \end{pmatrix} \right| = \begin{vmatrix} \mathbf{A} & \mathbf{AB} \\ -\mathbf{I} & \mathbf{0} \end{vmatrix} = |\mathbf{AB}|,$$

thereby establishing the following, very important result.

Theorem 13.3.4. For $n \times n$ matrices $\mathbf{A}$ and $\mathbf{B}$,

$$|\mathbf{AB}| = |\mathbf{A}||\mathbf{B}|. \tag{3.8}$$

The repeated application of Theorem 13.3.4 leads to the following formula for the determinant of the product of an arbitrary number of $n \times n$ matrices $\mathbf{A}_1, \mathbf{A}_2, \dots,$ $\mathbf{A}_k$:

$$|\mathbf{A}_1\mathbf{A}_2 \cdots \mathbf{A}_k| = |\mathbf{A}_1||\mathbf{A}_2| \cdots |\mathbf{A}_k|. \tag{3.9}$$

As a special case of this formula, we obtain the following formula for the determinant of the kth power of an $n \times n$ matrix $\mathbf{A}$:

$$|\mathbf{A}^k| = |\mathbf{A}|^k \tag{3.10}$$

($k = 1, 2, \ldots$).

In light of Lemma 13.2.1, we have the following corollary of Theorem 13.3.4.

Corollary 13.3.5. For any $n \times n$ matrix $\mathbf{A}$,

$$|\mathbf{A}'\mathbf{A}| = |\mathbf{A}|^2. \tag{3.11}$$

According to Lemma 13.1.3, a permutation matrix has a determinant whose absolute value is one. The following corollary indicates that every orthogonal matrix has this property.

Corollary 13.3.6. For an orthogonal matrix $\mathbf{P}$,

$$|\mathbf{P}| = \pm 1.$$

Proof (of Corollary 13.3.6). Using Corollary 13.3.5, together with result (1.9), we find that

$$|\mathbf{P}|^2 = |\mathbf{P}'\mathbf{P}| = |\mathbf{I}| = 1.$$

Q.E.D.

Corollary 13.2.3 and Lemmas 13.2.8 and 13.2.9 identify certain matrices that have zero determinants. These results are generalized as part of the following theorem.

Theorem 13.3.7. Let $\mathbf{A}$ represent an $n \times n$ matrix. Then, $\mathbf{A}$ is nonsingular (or, equivalently, $\mathbf{A}$ is invertible) if and only if $|\mathbf{A}| \neq 0$, in which case

$$|\mathbf{A}^{-1}| = 1/|\mathbf{A}|. \tag{3.12}$$

Proof. It suffices to show that, if $\mathbf{A}$ is nonsingular, then $|\mathbf{A}| \neq 0$ and $|\mathbf{A}^{-1}| = 1/|\mathbf{A}|$ and that, if $\mathbf{A}$ is singular, then $|\mathbf{A}| = 0$.

Suppose that $\mathbf{A}$ is nonsingular. Then, according to Theorem 13.3.4 and result (1.9),

$$|\mathbf{A}^{-1}||\mathbf{A}| = |\mathbf{A}^{-1}\mathbf{A}| = |\mathbf{I}| = 1,$$

implying that $|\mathbf{A}| \neq 0$ and further that $|\mathbf{A}^{-1}| = 1/|\mathbf{A}|$.

Suppose now that $\mathbf{A}$ is singular. Then, some column of $\mathbf{A}$, say the sth column $\mathbf{a}_s$, can be expressed as a linear combination of the other $n-1$ columns $\mathbf{a}_1, \ldots, \mathbf{a}_{s-1}, \mathbf{a}_{s+1}, \ldots, \mathbf{a}_n$; that is,

$$\mathbf{a}_s = \sum_{i \neq s} k_i \mathbf{a}_i$$

for some scalars $k_1, \ldots, k_{s-1}, k_{s+1}, \ldots, k_n$. Now, let $\mathbf{B}$ represent a matrix formed from $\mathbf{A}$ by adding the vector $-\sum_{i \neq s} k_i \mathbf{a}_i$ to the sth column of $\mathbf{A}$. Clearly, the sth column of $\mathbf{B}$ is null, and it follows from Corollary 13.2.3 that $|\mathbf{B}| = 0$. Moreover, it follows from Theorem 13.2.10 that $|\mathbf{A}| = |\mathbf{B}|$. Thus, $|\mathbf{A}| = 0$. Q.E.D.

The following theorem gives formulas for the determinant of a partitioned matrix that are analogous to formulas (8.5.16) and (8.5.17) for the inverse of a partitioned matrix.

Theorem 13.3.8. Let $\mathbf{T}$ represent an $m \times m$ matrix, $\mathbf{U}$ an $m \times n$ matrix, $\mathbf{V}$ an $n \times m$ matrix, and $\mathbf{W}$ an $n \times n$ matrix. If $\mathbf{T}$ is nonsingular, then

$$\begin{vmatrix} \mathbf{T} & \mathbf{U} \\ \mathbf{V} & \mathbf{W} \end{vmatrix} = \begin{vmatrix} \mathbf{W} & \mathbf{V} \\ \mathbf{U} & \mathbf{T} \end{vmatrix} = |\mathbf{T}||\mathbf{W} - \mathbf{V}\mathbf{T}^{-1}\mathbf{U}|. \tag{3.13}$$

Proof. Suppose that $\mathbf{T}$ is nonsingular. Then,

$$\begin{pmatrix} \mathbf{T} & \mathbf{U} \\ \mathbf{V} & \mathbf{W} \end{pmatrix} = \begin{pmatrix} \mathbf{I} & \mathbf{0} \\ \mathbf{V}\mathbf{T}^{-1} & \mathbf{W} - \mathbf{V}\mathbf{T}^{-1}\mathbf{U} \end{pmatrix} \begin{pmatrix} \mathbf{T} & \mathbf{U} \\ \mathbf{0} & \mathbf{I} \end{pmatrix}.$$

Applying Theorems 13.3.4 and 13.3.1, we find that

$$\begin{vmatrix} \mathbf{T} & \mathbf{U} \\ \mathbf{V} & \mathbf{W} \end{vmatrix} = |\mathbf{T}||\mathbf{W} - \mathbf{V}\mathbf{T}^{-1}\mathbf{U}|.$$

That $\begin{vmatrix} \mathbf{W} & \mathbf{V} \\ \mathbf{U} & \mathbf{T} \end{vmatrix} = |\mathbf{T}||\mathbf{W} - \mathbf{V}\mathbf{T}^{-1}\mathbf{U}|$ can be proved in similar fashion. Q.E.D.

13.4 A Computational Approach

Formula (3.8) and, more generally, formula (3.9) (for the determinant of a product of matrices) can be very useful from a computational standpoint. Suppose that $\mathbf{A}$ is an $n \times n$ matrix whose determinant $|\mathbf{A}|$ is of interest and that $\mathbf{A}$ can be decomposed into the product of some number of $n \times n$ matrices, each of whose determinants is easy to "evaluate." Then, formula (3.9) can be used to evaluate $|\mathbf{A}|$.

Suppose, for instance, that we have obtained the QR decomposition $\mathbf{A} = \mathbf{QR}$ (where $\mathbf{Q}$ is orthogonal and $\mathbf{R}$ triangular) of an $n \times n$ matrix $\mathbf{A}$. Then, in light of Corollary 13.3.6, $|\mathbf{A}| = \pm|\mathbf{R}|$; and, in light of Lemma 13.1.1, $|\mathbf{A}|$ can be determined (up to its absolute value) by forming the product of the diagonal elements of $\mathbf{R}$.

13.5 Cofactors

Let $\mathbf{A} = \{a_{ij}\}$ represent an $n \times n$ matrix. Let $\mathbf{A}_{ij}$ represent the $(n-1) \times (n-1)$ submatrix of $\mathbf{A}$ obtained by striking out the row and the column that contain the element a_{ij}, that is, the ith row and the jth column. The determinant $|\mathbf{A}_{ij}|$ of this submatrix is called the *minor* of the element a_{ij}. The "signed" minor $(-1)^{i+j}|\mathbf{A}_{ij}|$ is called the *cofactor* of a_{ij}.

The determinant of **A** can be expanded in terms of the cofactors of the n elements of any particular row or column of **A**, as described in the following theorem.

Theorem 13.5.1. Let a_{ij} represent the ijth element of an $n \times n$ matrix **A**, and let α_{ij} represent the cofactor of a_{ij} $(i, j = 1, \ldots, n)$. Then, for $i = 1, \ldots, n$,

$$|\mathbf{A}| = \sum_{j=1}^{n} a_{ij}\alpha_{ij} = a_{i1}\alpha_{i1} + \cdots + a_{in}\alpha_{in}, \tag{5.1}$$

and, for $j = 1, \ldots, n$,

$$|\mathbf{A}| = \sum_{i=1}^{n} a_{ij}\alpha_{ij} = a_{1j}\alpha_{1j} + \cdots + a_{nj}\alpha_{nj}. \tag{5.2}$$

Proof. Let $\mathbf{A}_{ij}$ represent the matrix obtained by striking out the ith row and the jth column of **A** $(i, j = 1, \ldots, n)$. Consider first result (5.1) for the case $i = 1$. Denote by $a_{ts}^{(j)}$ the tsth element of the matrix $\mathbf{A}_{1j}$ $(t, s = 1, \ldots, n-1;$ $j = 1, \ldots, n)$. We find that

$$|\mathbf{A}| = \sum (-1)^{\phi_n(k_1,\ldots,k_n)} a_{1k_1} \cdots a_{nk_n}$$

(where $k_1, \ldots, k_n$ is a permutation of the first n positive integers and the summation is over all such permutations)

$$= a_{11} \sum (-1)^{\phi_{n-1}(k_2,\ldots,k_n)} a_{2k_2} \cdots a_{nk_n} + \cdots$$
$$+ a_{1j} \sum (-1)^{j-1+\phi_{n-1}(k_2,\ldots,k_n)} a_{2k_2} \cdots a_{nk_n} + \cdots$$
$$+ a_{1n} \sum (-1)^{n-1+\phi_{n-1}(k_2,\ldots,k_n)} a_{2k_2} \cdots a_{nk_n}$$

(where, in the jth of the n sums, $k_2, \ldots, k_n$ is a permutation of the $n-1$ integers $1, \ldots, j-1, j+1, \ldots, n$ and the summation is over all such permutations)

$$= a_{11} \sum (-1)^{\phi_{n-1}(s_1,\ldots,s_{n-1})} a_{1s_1}^{(1)} \cdots a_{n-1,s_{n-1}}^{(1)} + \cdots$$
$$+ a_{1j}(-1)^{j-1} \sum (-1)^{\phi_{n-1}(s_1,\ldots,s_{n-1})} a_{1s_1}^{(j)} \cdots a_{n-1,s_{n-1}}^{(j)} + \cdots$$
$$+ a_{nj}(-1)^{n-1} \sum (-1)^{\phi_{n-1}(s_1,\ldots,s_{n-1})} a_{1s_1}^{(n)} \cdots a_{n-1,s_{n-1}}^{(n)}$$

(where, in each of the n sums, $s_1, \ldots, s_{n-1}$ is a permutation of the first $n-1$ positive integers and the summation is over all such permutations)

$$= \sum_{j=1}^{n} a_{1j}(-1)^{j-1}|\mathbf{A}_{1j}|$$
$$= \sum_{j=1}^{n} a_{1j}(-1)^{j+1}|\mathbf{A}_{1j}|$$

$$= \sum_{j=1}^{n} a_{1j}\alpha_{1j}.$$

Consider now result (5.1) for the case $i > 1$. Let $\mathbf{B}$ represent the $n \times n$ matrix whose first row is the ith row of $\mathbf{A}$, whose second, ..., ith rows are the first, ..., $(i-1)$th rows, respectively, of $\mathbf{A}$, and whose $(i+1), \ldots, n$th rows are the same as those of $\mathbf{A}$. Observe that $\mathbf{A}$ can be obtained from $\mathbf{B}$ by successively interchanging the first row of $\mathbf{B}$ with its second, ..., ith rows, so that, according to Theorem 13.2.6,

$$|\mathbf{A}| = (-1)^{i-1}|\mathbf{B}|.$$

Let $\mathbf{B}_{1j}$ represent the matrix obtained by striking out the first row and the jth column of $\mathbf{B}$, and let b_{1j} represent the jth element of the first row of $\mathbf{B}$ $(j = 1, \ldots, n)$. Then, $\mathbf{B}_{1j} = \mathbf{A}_{ij}$ $(j = 1, \ldots, n)$, and we find that

$$\begin{aligned} |\mathbf{A}| = (-1)^{i-1}|\mathbf{B}| &= (-1)^{i-1} \sum_{j=1}^{n} b_{1j}(-1)^{1+j}|\mathbf{B}_{1j}| \\ &= (-1)^{i-1} \sum_{j=1}^{n} a_{ij}(-1)^{1+j}|\mathbf{A}_{ij}| \\ &= \sum_{j=1}^{n} a_{ij}\alpha_{ij}. \end{aligned}$$

Finally, consider result (5.2). Observe that the matrix obtained by striking out the jth row and the ith column of $\mathbf{A}'$ is $\mathbf{A}'_{ij}$ and hence that the cofactor of the jith element of $\mathbf{A}'$ is

$$(-1)^{j+i}|\mathbf{A}'_{ij}| = (-1)^{i+j}|\mathbf{A}_{ij}| = \alpha_{ij}$$

$(i, j = 1, \ldots, n)$. Thus, since the jith element of $\mathbf{A}'$ is the ijth element of $\mathbf{A}$ $(i, j = 1, \ldots, n)$, it follows from result (5.1) (and from Lemma 13.2.1) that

$$|\mathbf{A}| = |\mathbf{A}'| = \sum_{i=1}^{n} a_{ij}\alpha_{ij}$$

$(j = 1, \ldots, n)$. Q.E.D.

Note that if the jth element a_{ij} of the ith row of $\mathbf{A}$ equals zero, then the jth term in formula (5.1) also equals zero. Similarly, if $a_{ij} = 0$, then the ith term in formula (5.2) equals zero. Thus, if the $n \times n$ matrix $\mathbf{A}$ contains a row or column that includes many zeroes, then formula (5.1) or (5.2) can be used to reexpress $|\mathbf{A}|$ in terms of the determinants of a relatively small number of $(n-1) \times (n-1)$ submatrices of $\mathbf{A}$.

The following theorem can be regarded as a companion to Theorem 13.5.1.

Theorem 13.5.2. Let a_{ij} represent the ijth element of an $n \times n$ matrix $\mathbf{A}$, and let α_{ij} represent the cofactor of a_{ij} $(i, j = 1, \ldots, n)$. Then, for $i' \neq i = 1, \ldots, n$,

$$\sum_{j=1}^{n} a_{ij}\alpha_{i'j} = a_{i1}\alpha_{i'1} + \cdots + a_{in}\alpha_{i'n} = 0, \tag{5.3}$$

and, for $j' \neq j = 1, \ldots, n$,

$$\sum_{i=1}^{n} a_{ij}\alpha_{ij'} = a_{1j}\alpha_{1j'} + \cdots + a_{nj}\alpha_{nj'} = 0. \tag{5.4}$$

Proof. Consider result (5.3). Let **B** represent a matrix whose i'th row equals the ith row of **A** and whose first, ..., $(i'-1)$th, $(i'+1)$th, ..., nth rows are identical to those of **A** (where $i' \neq i$). Observe that the i'th row of **B** is a duplicate of its ith row and hence, according to Lemma 13.2.8, that $|\mathbf{B}| = 0$.

Let b_{kj} represent the kjth element of **B** $(k, j = 1, \ldots, n)$. Clearly, the cofactor of $b_{i'j}$ is the same as that of a_{ij} $(j = 1, \ldots, n)$. Thus, making use of Theorem 13.5.1, we find that

$$\sum_{j=1}^{n} a_{ij}\alpha_{i'j} = \sum_{j=1}^{n} b_{i'j}\alpha_{i'j} = |\mathbf{B}| = 0,$$

which establishes result (5.3). Result (5.4) can be proved via an analogous argument. Q.E.D.

For any $n \times n$ matrix $\mathbf{A} = \{a_{ij}\}$, the $n \times n$ matrix whose ijth element is the cofactor α_{ij} of a_{ij} is called the *matrix of cofactors* (or *cofactor matrix*) of **A**. The transpose of this matrix is called the *adjoint matrix* of **A** and is denoted by the symbol adj **A** or adj(**A**). Thus,

$$\text{adj}\,\mathbf{A} = \begin{pmatrix} \alpha_{11} & \ldots & \alpha_{1n} \\ \vdots & \ddots & \vdots \\ \alpha_{n1} & \ldots & \alpha_{nn} \end{pmatrix}' = \begin{pmatrix} \alpha_{11} & \ldots & \alpha_{n1} \\ \vdots & \ddots & \vdots \\ \alpha_{1n} & \ldots & \alpha_{nn} \end{pmatrix}.$$

If **A** is symmetric, then the matrix of cofactors is also symmetric (or equivalently the adjoint matrix equals the matrix of cofactors), as is easily verified.

There is a close relationship between the adjoint of a nonsingular matrix **A** and the inverse of **A**, as is evident from the following theorem and is made explicit in the corollary of this theorem.

Theorem 13.5.3. For an $n \times n$ matrix **A**,

$$\mathbf{A}\,\text{adj}(\mathbf{A}) = (\text{adj}\,\mathbf{A})\mathbf{A} = |\mathbf{A}|\mathbf{I}_n. \tag{5.5}$$

Proof. Let a_{ij} represent the ijth element of **A**, and let α_{ij} represent the cofactor of a_{ij} $(i, j = 1, \ldots, n)$. Then, the ii'th element of the matrix product **A** adj(**A**) is $\sum_{j=1}^{n} a_{ij}\alpha_{i'j}$ $(i, i' = 1, \ldots, n)$. According to Theorems 13.5.1 and 13.5.2,

$$\sum_{j=1}^{n} a_{ij}\alpha_{i'j} = |\mathbf{A}|, \qquad \text{if } i' = i,$$

$$= 0, \qquad \text{if } i' \neq i.$$

Thus, $\mathbf{A}\,\text{adj}(\mathbf{A}) = |\mathbf{A}|\mathbf{I}$.

That $(\text{adj}\,\mathbf{A})\mathbf{A} = |\mathbf{A}|\mathbf{I}$ can be established via a similar argument. Q.E.D.

Corollary 13.5.4. If $\mathbf{A}$ is an $n \times n$ nonsingular matrix, then

$$\text{adj}(\mathbf{A}) = |\mathbf{A}|\mathbf{A}^{-1} \tag{5.6}$$

or equivalently

$$\mathbf{A}^{-1} = (1/|\mathbf{A}|)\,\text{adj}(\mathbf{A}). \tag{5.7}$$

13.6 Vandermonde Matrices

Consider the matrix

$$\mathbf{V} = \begin{pmatrix} 1 & x_1 & x_1^2 & \dots & x_1^{n-1} \\ 1 & x_2 & x_2^2 & \dots & x_2^{n-1} \\ \vdots & \vdots & \vdots & & \vdots \\ 1 & x_n & x_n^2 & \dots & x_n^{n-1} \end{pmatrix},$$

where $x_1, x_2, \dots, x_n$ are arbitrary scalars. A matrix of this general form is called a Vandermonde matrix, after Alexandre Theophile Vandermonde (1735–1796). Vandermonde matrices are encountered in statistics in fitting (by least squares) a polynomial (in an explanatory variable x) to the observed values of a "dependent" variable y. Let us find the determinant of $\mathbf{V}$.

Observe that

$$\mathbf{VT} = \begin{pmatrix} \mathbf{1}_{n-1} & \mathbf{A} \\ 1 & \mathbf{0} \end{pmatrix}, \tag{6.1}$$

where

$$\mathbf{T} = \begin{pmatrix} 1 & -x_n & 0 & \dots & 0 & 0 \\ 0 & 1 & -x_n & & 0 & 0 \\ 0 & 0 & 1 & & & \\ \vdots & & \vdots & & & \\ 0 & 0 & 0 & & 1 & -x_n \\ 0 & 0 & 0 & & 0 & 1 \end{pmatrix}$$

and

$$\mathbf{A} = \begin{pmatrix} x_1 - x_n & x_1(x_1 - x_n) & \dots & x_1^{n-2}(x_1 - x_n) \\ x_2 - x_n & x_2(x_2 - x_n) & \dots & x_2^{n-2}(x_2 - x_n) \\ \vdots & \vdots & \ddots & \vdots \\ x_{n-1} - x_n & x_{n-1}(x_{n-1} - x_n) & \dots & x_{n-1}^{n-2}(x_{n-1} - x_n) \end{pmatrix}.$$

Note that the effect of postmultiplying $\mathbf{V}$ by $\mathbf{T}$ is to add, to the jth column of $\mathbf{V}$, a scalar multiple of the preceding column ($j = 2, \ldots, n$), thereby creating a matrix whose last row is $(1, 0, 0, \ldots, 0)$.

Observe also that $\mathbf{A}$ is expressible as

$$\mathbf{A} = \mathbf{DW}, \tag{6.2}$$

where $\mathbf{D} = \operatorname{diag}(x_1 - x_n, x_2 - x_n, \ldots, x_{n-1} - x_n)$ and

$$\mathbf{W} = \begin{pmatrix} 1 & x_1 & x_1^2 & \ldots & x_1^{n-2} \\ 1 & x_2 & x_2^2 & \ldots & x_2^{n-2} \\ \vdots & \vdots & \vdots & & \vdots \\ 1 & x_{n-1} & x_{n-1}^2 & \ldots & x_{n-1}^{n-2} \end{pmatrix}.$$

Thus, $\mathbf{A}$ is expressible as a product of a diagonal matrix and of the $(n-1) \times (n-1)$ submatrix of $\mathbf{V}$ obtained by deleting the last row and column of $\mathbf{V}$. Note that this submatrix (i.e., the matrix $\mathbf{W}$) is an $(n-1) \times (n-1)$ Vandermonde matrix.

Making use of results (6.1) and (6.2) and of various properties of determinants (described in Theorems 13.2.11 and 13.3.4 and Corollaries 13.3.2, 13.2.5, and 13.1.2), we find that

$$\begin{aligned} |\mathbf{V}| = |\mathbf{VT}| &= \begin{vmatrix} \mathbf{1}_{n-1} & \mathbf{A} \\ 1 & \mathbf{0} \end{vmatrix} = (-1)^{n-1}|\mathbf{A}| \\ &= (-1)^{n-1}|\mathbf{D}||\mathbf{W}| = |-\mathbf{D}||\mathbf{W}| \\ &= (x_n - x_1)(x_n - x_2)\cdots(x_n - x_{n-1})|\mathbf{W}|. \end{aligned} \tag{6.3}$$

Formula (6.3) serves to relate the determinant of an $n \times n$ Vandermonde matrix to that of an $(n-1) \times (n-1)$ Vandermonde matrix, and its repeated application allows us to evaluate the determinant of any Vandermonde matrix.

Clearly, when $n = 2$,

$$|\mathbf{V}| = x_2 - x_1;$$

when $n = 3$,

$$|\mathbf{V}| = (x_3 - x_1)(x_3 - x_2)(x_2 - x_1);$$

and, in general,

$$\begin{aligned} |\mathbf{V}| &= \prod_{\substack{i,j \\ (j<i)}} (x_i - x_j) \\ &= (x_n - x_1)(x_n - x_2)\cdots(x_n - x_{n-1}) \\ &\quad \times (x_{n-1} - x_1)(x_{n-1} - x_2)\cdots(x_{n-1} - x_{n-2})\cdots(x_2 - x_1), \end{aligned} \tag{6.4}$$

as can be formally verified by a simple mathematical induction argument based on relationship (6.3).

It is evident from formula (6.4) that $|\mathbf{V}| \neq 0$ if and only if $x_j \neq x_i$ for $j < i = 1, \ldots, n$. Thus, $\mathbf{V}$ is nonsingular if and only if the n scalars $x_1, x_2, \ldots, x_n$ are distinct.

In fact,

$$\text{rank}(\mathbf{V}) = r, \tag{6.5}$$

where r is the number of distinct values represented among $x_1, \ldots, x_n$. To see this, let $i_1, i_2, \ldots, i_r$ ($i_1 < i_2 < \cdots < i_r$) represent r integers such that $x_{i_1}, x_{i_2}, \ldots, x_{i_r}$ are distinct. Then, the $i_1, \ldots, i_r$th rows of $\mathbf{V}$ span $\mathcal{R}(\mathbf{V})$ (since each of the remaining $n - r$ rows of $\mathbf{V}$ is a duplicate of one of these rows), implying that

$$\text{rank}(\mathbf{V}) = \dim[\mathcal{R}(\mathbf{V})] \leq r. \tag{6.6}$$

Moreover, the $r \times r$ submatrix of $\mathbf{V}$ formed from the first r columns of $\mathbf{V}$ and from the $i_1, \ldots, i_r$th rows, that is, the submatrix

$$\begin{pmatrix} 1 & x_{i_1} & x_{i_1}^2 & \ldots & x_{i_1}^{r-1} \\ 1 & x_{i_2} & x_{i_2}^2 & \ldots & x_{i_2}^{r-1} \\ \vdots & \vdots & \vdots & & \vdots \\ 1 & x_{i_r} & x_{i_r}^2 & \ldots & x_{i_r}^{r-1} \end{pmatrix},$$

is a Vandermonde matrix, which (since $x_{i_1}, \ldots, x_{i_r}$ are distinct) is nonsingular. Thus, according to Theorem 4.4.10, $\text{rank}(\mathbf{V}) \geq r$, which together with inequality (6.6) implies that $\text{rank}(\mathbf{V}) = r$.

13.7 Some Results on the Determinant of the Sum of Two Matrices

Consider the determinant of the sum of two $n \times n$ matrices $\mathbf{A}$ and $\mathbf{B}$. Except in special cases, $|\mathbf{A} + \mathbf{B}| \neq |\mathbf{A}| + |\mathbf{B}|$. However, as described in the following theorem, $|\mathbf{A} + \mathbf{B}|$ can, for any particular integer k ($1 \leq k \leq n$), be expressed as the sum of the determinants of 2^k $n \times n$ matrices; the ith row of each of these 2^k matrices is identical to the ith row of $\mathbf{A}$, $\mathbf{B}$, or $\mathbf{A} + \mathbf{B}$ ($i = 1, \ldots, n$).

Theorem 13.7.1. For any two $n \times n$ matrices $\mathbf{A}$ and $\mathbf{B}$,

$$|\mathbf{A} + \mathbf{B}| = \sum_{\{i_1,\ldots,i_r\}} |\mathbf{C}_k^{\{i_1,\ldots,i_r\}}|, \tag{7.1}$$

where k is any particular one of the first n positive integers $1, \ldots, n$, and $\{i_1, \ldots, i_r\}$ is a subset of the first k positive integers $1, \ldots, k$ (and the summation is over all 2^k such subsets), and where $\mathbf{C}_k^{\{i_1,\ldots,i_r\}}$ is an $n \times n$ matrix whose last $n - k$ rows are identical to the last $n - k$ rows of $\mathbf{A} + \mathbf{B}$, whose $i_1, \ldots, i_r$th rows are identical to the $i_1, \ldots, i_r$th rows of $\mathbf{A}$, and whose remaining rows, say the $i_{r+1}, \ldots, i_k$th rows, are identical to the $i_{r+1}, \ldots, i_k$th rows of $\mathbf{B}$.

Preliminary to proving Theorem 13.7.1, it is convenient to prove the following result, which is of some interest in its own right.

Theorem 13.7.2. Let **A**, **B**, and **C** represent $n \times n$ matrices, and denote by $\mathbf{a}_i'$, $\mathbf{b}_i'$, and $\mathbf{c}_i'$ the ith rows of **A**, **B**, and **C**, respectively ($i = 1, \ldots, n$). If, for some k,

$$\mathbf{c}_k' = \mathbf{a}_k' + \mathbf{b}_k'$$

and

$$\mathbf{c}_i' = \mathbf{a}_i' = \mathbf{b}_i' \quad (i = 1, \ldots, k-1, k+1, \ldots, n),$$

then

$$|\mathbf{C}| = |\mathbf{A}| + |\mathbf{B}|.$$

Proof (of Theorem 13.7.2). Denote by a_{kj}, b_{kj}, and c_{kj} the kjth elements of **A**, **B**, and **C**, respectively, and let α_{kj} represent the cofactor of the kjth element of **A** ($j = 1, \ldots, n$). Clearly, the cofactor of the kjth element of **B** and the cofactor of the kjth element of **C** are the same as the cofactor of the kjth element of **A** ($j = 1, \ldots, n$). Thus, making use of Theorem 13.5.1, we find that

$$\begin{aligned} |\mathbf{C}| = \sum_{j=1}^{n} c_{kj}\alpha_{kj} &= \sum_{j=1}^{n} (a_{kj} + b_{kj})\alpha_{kj} \\ &= \sum_{j=1}^{n} a_{kj}\alpha_{kj} + \sum_{j=1}^{n} b_{kj}\alpha_{kj} \\ &= |\mathbf{A}| + |\mathbf{B}|. \end{aligned}$$

Q.E.D.

Proof (of Theorem 13.7.1). The proof is by mathematical induction. It follows from Theorem 13.7.2 that

$$|\mathbf{A} + \mathbf{B}| = |\mathbf{C}_1^{\{1\}}| + |\mathbf{C}_1^{\emptyset}|$$

(where $\emptyset$ denotes the empty set), so that result (7.1) is valid for $k = 1$.

Suppose that result (7.1) is valid for $k = k^* - 1$, that is,

$$|\mathbf{A} + \mathbf{B}| = \sum_{\{i_1,\ldots,i_r\}} |\mathbf{C}_{k^*-1}^{\{i_1,\ldots,i_r\}}|,$$

where $\{i_1, \ldots, i_r\}$ is a subset of the first $k^* - 1$ positive integers (and the summation is over all 2^{k^*-1} such subsets). We must show that result (7.1) is valid for $k = k^*$.

Using Theorem 13.7.2, we find that

$$|\mathbf{C}_{k^*-1}^{\{i_1,\ldots,i_r\}}| = |\mathbf{C}_{k^*}^{\{i_1,\ldots,i_r,k^*\}}| + |\mathbf{C}_{k^*}^{\{i_1,\ldots,i_r\}}|,$$

for any subset $\{i_1, \ldots, i_r\}$ of the first $k^* - 1$ positive integers. Moreover, the 2^{k^*} possible subsets of the first k^* positive integers consist of the 2^{k^*-1} possible subsets of the first $k^* - 1$ positive integers plus the 2^{k^*-1} subsets obtained by augmenting each of the latter subsets with the integer k^*. We conclude that

$$|\mathbf{A} + \mathbf{B}| = \sum_{\{i_1,\ldots,i_r\}} |\mathbf{C}_{k^*}^{\{i_1,\ldots,i_r\}}|,$$

where $\{i_1, \ldots, i_r\}$ is a subset of the first k^* positive integers (and the summation is over all 2^{k^*} such subsets). Q.E.D.

The following theorem gives an expansion for the determinant of the sum of two $n \times n$ matrices in the special case where one of the matrices is diagonal.

Theorem 13.7.3. Let $\mathbf{B}$ represent an $n \times n$ matrix, and let $\mathbf{D}$ represent an $n \times n$ diagonal matrix whose diagonal elements are $d_1, \ldots, d_n$. Then,

$$|\mathbf{D} + \mathbf{B}| = \sum_{\{i_1,\ldots,i_r\}} d_{i_1} \cdots d_{i_r} |\mathbf{B}^{\{i_1,\ldots,i_r\}}|, \tag{7.2}$$

where $\{i_1, \ldots, i_r\}$ is a subset of the first n positive integers $1, \ldots, n$ (and the summation is over all 2^n such subsets) and where $\mathbf{B}^{\{i_1,\ldots,i_r\}}$ is the $(n-r) \times (n-r)$ principal submatrix of $\mathbf{B}$ obtained by striking out the $i_1, \ldots, i_r$th rows and columns. [The term in sum (7.2) corresponding to the empty set is to be interpreted as $|\mathbf{B}|$, and the term corresponding to the set $\{1, 2, \ldots, n\}$ is to be interpreted as $d_1 d_2 \cdots d_n$.]

Proof. Applying result (7.1) with $\mathbf{A} = \mathbf{D}$ and $k = n$, we find that

$$|\mathbf{D} + \mathbf{B}| = \sum_{\{i_1,\ldots,i_r\}} |\mathbf{C}^{\{i_1,\ldots,i_r\}}|,$$

where $\mathbf{C}^{\{i_1,\ldots,i_r\}}$ is an $n \times n$ matrix whose $i_1, \ldots, i_r$th rows are identical to the $i_1, \ldots, i_r$th rows of $\mathbf{D}$ and whose remaining rows, say the $i_{r+1}, \ldots, i_n$th rows $(i_{r+1} < \cdots < i_n)$, are identical to the $i_{r+1}, \ldots, i_n$th rows of $\mathbf{B}$. Thus, to verify result (7.2), it suffices to show that

$$|\mathbf{C}^{\{i_1,\ldots,i_r\}}| = d_{i_1} \cdots d_{i_r} |\mathbf{B}^{\{i_1,\ldots,i_r\}}|.$$

Let $\mathbf{P}$ represent the $n \times n$ (permutation) matrix whose first, ..., rth, $(r+1)$th, ..., nth columns are the $i_1, \ldots, i_r, i_{r+1}, \ldots, i_n$th columns, respectively, of $\mathbf{I}_n$. Then, clearly,

$$\mathbf{P}'\mathbf{C}^{\{i_1,\ldots,i_r\}}\mathbf{P} = \begin{pmatrix} \text{diag}(d_{i_1}, \ldots, d_{i_r}) & \mathbf{0} \\ \mathbf{F}^{\{i_1,\ldots,i_r\}} & \mathbf{B}^{\{i_1,\ldots,i_r\}} \end{pmatrix},$$

where $\mathbf{F}^{\{i_1,\ldots,i_r\}}$ is an $(n-r) \times r$ matrix whose stth element is the $i_{r+s}i_t$th element of $\mathbf{B}$.

Thus, using Lemmas 13.2.1 and 13.1.3, Theorems 13.3.4 and 13.3.1, and Corollary 13.1.2, we find that

$$\begin{aligned}|\mathbf{C}^{\{i_1,\ldots,i_r\}}| &= |\mathbf{P}'||\mathbf{C}^{\{i_1,\ldots,i_r\}}||\mathbf{P}| \\ &= |\mathbf{P}'\mathbf{C}^{\{i_1,\ldots,i_r\}}\mathbf{P}| \\ &= |\mathrm{diag}(d_{i_1},\ldots,d_{i_r})||\mathbf{B}^{\{i_1,\ldots,i_r\}}| \\ &= d_{i_1}\cdots d_{i_r}|\mathbf{B}^{\{i_1,\ldots,i_r\}}|.\end{aligned}$$

Q.E.D.

In the special case where $d_1 = \cdots = d_n$, result (7.2) simplifies as follows:

Corollary 13.7.4. For any $n \times n$ matrix $\mathbf{B}$ and any scalar x,

$$|\mathbf{B} + x\mathbf{I}_n| = \sum_{r=0}^{n} x^r \sum_{\{i_1,\ldots,i_r\}} |\mathbf{B}^{\{i_1,\ldots,i_r\}}|, \tag{7.3}$$

where $\{i_1,\ldots,i_r\}$ is an r-dimensional subset of the first n positive integers $1,\ldots,n$ (and the second summation is over all $\binom{n}{r}$ such subsets) and where $\mathbf{B}^{\{i_1,\ldots,i_r\}}$ is the $(n-r)\times(n-r)$ principal submatrix of $\mathbf{B}$ obtained by striking out the $i_1,\ldots,i_r$th rows and columns. (For $r = n$, the sum $\sum_{\{i_1,\ldots,i_r\}} |\mathbf{B}^{\{i_1,\ldots,i_r\}}|$ is to be interpreted as 1.)

Note that expression (7.3) is a polynomial in x and that the coefficient of x^0 (i.e., the constant term of the polynomial) equals $|\mathbf{B}|$ and the coefficient of x^{n-1} equals $\mathrm{tr}(\mathbf{B})$.

13.8 Laplace's Theorem and the Binet-Cauchy Formula

The following theorem, which is due to Pierre Simon Laplace (1749–1827), is a generalization of Theorem 13.5.1.

Theorem 13.8.1 (Laplace's theorem). Let $\mathbf{A}$ represent an $n \times n$ matrix, let r represent any particular one of the first $n-1$ positive integers $1,\ldots,n-1$, and let $\{i_1,\ldots,i_r\}$ and $\{j_1,\ldots,j_r\}$ (where $i_1 < \cdots < i_r$ and $j_1 < \cdots < j_r$) represent arbitrary r-dimensional subsets of the first n positive integers $1,\ldots,n$. Further, let $\mathbf{A}_{i_1,\ldots,i_r}^{j_1,\ldots,j_r}$ represent the $r \times r$ submatrix of $\mathbf{A}$ obtained by striking out all of the rows and columns except the $i_1,\ldots,i_r$th rows and $j_1,\ldots,j_r$th columns, and let $\bar{\mathbf{A}}_{i_1,\ldots,i_r}^{j_1,\ldots,j_r}$ represent the $(n-r)\times(n-r)$ submatrix of $\mathbf{A}$ obtained by striking out the $i_1,\ldots,i_r$th rows and $j_1,\ldots,j_r$th columns. Then, for fixed $i_1,\ldots,i_r$,

$$|\mathbf{A}| = \sum_{\{j_1,\ldots,j_r\}} (-1)^{i_1+\cdots+i_r+j_1+\cdots+j_r} |\mathbf{A}_{i_1,\ldots,i_r}^{j_1,\ldots,j_r}||\bar{\mathbf{A}}_{i_1,\ldots,i_r}^{j_1,\ldots,j_r}|, \tag{8.1}$$

and, for fixed $j_1,\ldots,j_r$,

$$|\mathbf{A}| = \sum_{\{i_1,\ldots,i_r\}} (-1)^{i_1+\cdots+i_r+j_1+\cdots+j_r} |\mathbf{A}_{i_1,\ldots,i_r}^{j_1,\ldots,j_r}||\bar{\mathbf{A}}_{i_1,\ldots,i_r}^{j_1,\ldots,j_r}|. \tag{8.2}$$

Proof. Consider first result (8.1) for the case $i_1 = 1, \ldots, i_r = r$. The $n!$ permutations of the first n positive integers can be divided into $\binom{n}{r}$ classes—one class for each subset $\{j_1, \ldots, j_r\}$. The class corresponding to any particular subset $\{j_1, \ldots, j_r\}$ consists of the $r!(n-r)!$ permutations of the general form $k_1, \ldots, k_r, k_{r+1}, \ldots, k_n$, where $k_1, \ldots, k_r$ is a permutation of the integers $j_1, \ldots, j_r$ and $k_{r+1}, \ldots, k_n$ is a permutation of the other $n-r$ of the first n positive integers. Note that, for each permutation $k_1, \ldots, k_n$ corresponding to any particular subset $\{j_1, \ldots, j_r\}$,

$$\phi_n(k_1, \ldots, k_n) = \phi_r(k_1, \ldots, k_r) + \phi_{n-r}(k_{r+1}, \ldots, k_n) + (j_1 - 1) + (j_2 - 2) + \cdots + (j_r - r)$$

and consequently

$$\begin{aligned}(-1)^{\phi_n(k_1,\ldots,k_n)} &= (-1)^{\phi_n(k_1,\ldots,k_n)+2(1+2+\cdots+r)} \\ &= (-1)^{\phi_r(k_1,\ldots,k_r)+\phi_{n-r}(k_{r+1},\ldots,k_n)+j_1+j_2+\cdots+j_r+1+2+\cdots+r}.\end{aligned}$$

Denote by $a_{ts}^{\{j_1,\ldots,j_r\}}$ the tsth element of $\mathbf{A}_{1,\ldots,r}^{j_1,\ldots,j_r}$ $(t, s = 1, \ldots, r)$, and by $\bar{a}_{ts}^{\{j_1,\ldots,j_r\}}$ the tsth element of $\bar{\mathbf{A}}_{1,\ldots,r}^{j_1,\ldots,j_r}$ $(t, s = 1, \ldots, n-r)$. We find that

$$\begin{aligned}|\mathbf{A}| &= \sum (-1)^{\phi_n(k_1,\ldots,k_n)} a_{1k_1} \cdots a_{nk_n} \\ &\qquad \text{(where } k_1, \ldots, k_n \text{ is a permutation of the first } n \text{ positive integers} \\ &\qquad \text{and the summation is over all such permutations)} \\ &= \sum_{\{j_1,\ldots,j_r\}} \sum_{k_1,\ldots,k_r} \sum_{k_{r+1},\ldots,k_n} (-1)^{\phi_n(k_1,\ldots,k_n)} a_{1k_1} \cdots a_{nk_n} \\ &\qquad \text{[where } k_1, \ldots, k_r \text{ is a permutation of the integers } j_1, \ldots, j_r \\ &\qquad \text{(and the second summation is over all such permutations) and} \\ &\qquad \text{where } k_{r+1}, \ldots, k_n \text{ is a permutation of those } n-r \text{ of the first } n \\ &\qquad \text{positive integers not contained in the set } \{j_1, \ldots, j_r\} \\ &\qquad \text{(and the third summation is over all such permutations)]} \\ &= \sum_{\{j_1,\ldots,j_r\}} (-1)^{1+2+\cdots+r+j_1+j_2+\cdots+j_r} \\ &\qquad \times \sum_{k_1,\ldots,k_r} (-1)^{\phi_r(k_1,\ldots,k_r)} a_{1k_1} \cdots a_{rk_r} \\ &\qquad \times \sum_{k_{r+1},\ldots,k_n} (-1)^{\phi_{n-r}(k_{r+1},\ldots,k_n)} a_{r+1,k_{r+1}} \cdots a_{nk_n} \\ &= \sum_{\{j_1,\ldots,j_r\}} (-1)^{1+2+\cdots+r+j_1+j_2+\cdots+j_r} \\ &\qquad \times \sum_{t_1,\ldots,t_r} (-1)^{\phi_r(t_1,\ldots,t_r)} a_{1t_1}^{\{j_1,\ldots,j_r\}} \cdots a_{rt_r}^{\{j_1,\ldots,j_r\}} \\ &\qquad \times \sum_{t_{r+1},\ldots,t_n} (-1)^{\phi_{n-r}(t_{r+1},\ldots,t_n)} \bar{a}_{1t_{r+1}}^{\{j_1,\ldots,j_r\}} \cdots \bar{a}_{n-r,t_n}^{\{j_1,\ldots,j_r\}}\end{aligned}$$

[where $t_1, \ldots, t_r$ is a permutation of the first r positive integers (and the second summation is over all such permutations) and where $t_{r+1}, \ldots, t_n$ is a permutation of the first $n-r$ positive integers (and the third summation is over all such permutations)]

$$= \sum_{\{j_1,\ldots,j_r\}} (-1)^{1+2+\cdots+r+j_1+j_2+\cdots+j_r} |\mathbf{A}_{1,\ldots,r}^{j_1,\ldots,j_r}||\bar{\mathbf{A}}_{1,\ldots,r}^{j_1,\ldots,j_r}|.$$

Consider now result (8.1) for the case of an arbitary subset $\{i_1, \ldots, i_r\}$. Let $i_{r+1}, \ldots, i_n$ (where $i_{r+1} < \cdots < i_n$) represent those $n-r$ of the first n positive integers not contained in $\{i_1, \ldots, i_r\}$, let $\mathbf{a}'_1, \ldots, \mathbf{a}'_n$ represent the first, ..., nth rows of $\mathbf{A}$, and let

$$\mathbf{B} = \begin{pmatrix} \mathbf{a}'_{i_1} \\ \vdots \\ \mathbf{a}'_{i_r} \\ \mathbf{a}'_{i_{r+1}} \\ \vdots \\ \mathbf{a}'_{i_n} \end{pmatrix}.$$

The matrix $\mathbf{A}$ can be obtained from $\mathbf{B}$ by a successive interchange of rows. Specifically, suppose that we successively interchange the rth row of $\mathbf{B}$ with the next $i_r - r$ rows, thereby producing a matrix, say $\mathbf{B}^*$, whose i_rth row is the same as that of $\mathbf{A}$. Next, suppose that we successively interchange the $(r-1)$th row of $\mathbf{B}^*$ with the next $i_{r-1} - (r-1)$ rows, thereby producing a matrix whose i_{r-1} and i_rth rows are the same as those of $\mathbf{A}$. Continuing in this fashion, we produce, after r steps involving a total of $(i_r - r) + [i_{r-1} - (r-1)] + \cdots + (i_1 - 1) = i_1 + \cdots + i_r - (1 + \cdots + r)$ interchanges, the matrix $\mathbf{A}$. We conclude, on the basis of Theorem 13.2.6, that

$$|\mathbf{A}| = (-1)^{i_1+\cdots+i_r-(1+\cdots+r)} |\mathbf{B}|.$$

Let $\mathbf{B}_{1,\ldots,r}^{j_1,\ldots,j_r}$ represent the $r \times r$ submatrix of $\mathbf{B}$ obtained by striking out all of the rows and columns except the first, ..., rth rows and $j_1, \ldots, j_r$th columns, and let $\bar{\mathbf{B}}_{1,\ldots,r}^{j_1,\ldots,j_r}$ represent the $(n-r) \times (n-r)$ submatrix of $\mathbf{B}$ obtained by striking out the first, ..., rth rows and $j_1, \ldots, j_r$th columns. Then, $\mathbf{B}_{1,\ldots,r}^{j_1,\ldots,j_r} = \mathbf{A}_{i_1,\ldots,i_r}^{j_1,\ldots,j_r}$ and $\bar{\mathbf{B}}_{1,\ldots,r}^{j_1,\ldots,j_r} = \bar{\mathbf{A}}_{i_1,\ldots,i_r}^{j_1,\ldots,j_r}$, and we find that

$$\begin{aligned} |\mathbf{A}| &= (-1)^{i_1+\cdots+i_r-(1+\cdots+r)} |\mathbf{B}| \\ &= (-1)^{i_1+\cdots+i_r-(1+\cdots+r)} \\ &\quad \times \sum_{\{j_1,\ldots,j_r\}} (-1)^{1+\cdots+r+j_1+\cdots+j_r} |\mathbf{B}_{1,\ldots,r}^{j_1,\ldots,j_r}||\bar{\mathbf{B}}_{1,\ldots,r}^{j_1,\ldots,j_r}| \\ &= \sum_{\{j_1,\ldots,j_r\}} (-1)^{i_1+\cdots+i_r+j_1+\cdots+j_r} |\mathbf{A}_{i_1,\ldots,i_r}^{j_1,\ldots,j_r}||\bar{\mathbf{A}}_{i_1,\ldots,i_r}^{j_1,\ldots,j_r}|. \end{aligned}$$

Finally, consider result (8.2). Observe that the matrix obtained by striking out all of the rows and columns of $\mathbf{A}'$ except the $j_1, \ldots, j_r$th rows and $i_1, \ldots, i_r$th columns is $(\mathbf{A}_{i_1,\ldots,i_r}^{j_1,\ldots,j_r})'$ and that the matrix obtained by striking out the $j_1, \ldots, j_r$th rows and $i_1, \ldots, i_r$th columns of $\mathbf{A}'$ is $(\bar{\mathbf{A}}_{i_1,\ldots,i_r}^{j_1,\ldots,j_r})'$. Thus, it follows from result (8.1) [and result (2.1)] that

$$\begin{aligned}|\mathbf{A}| = |\mathbf{A}'| &= \sum_{\{i_1,\ldots,i_r\}} (-1)^{j_1+\cdots+j_r+i_1+\cdots+i_r} |(\mathbf{A}_{i_1,\ldots,i_r}^{j_1,\ldots,j_r})'||(\bar{\mathbf{A}}_{i_1,\ldots,i_r}^{j_1,\ldots,j_r})'| \\ &= \sum_{\{i_1,\ldots,i_r\}} (-1)^{i_1+\cdots+i_r+j_1+\cdots+j_r} |\mathbf{A}_{i_1,\ldots,i_r}^{j_1,\ldots,j_r}||\bar{\mathbf{A}}_{i_1,\ldots,i_r}^{j_1,\ldots,j_r}|.\end{aligned}$$

Q.E.D.

Note that, in the special case where $r = 1$, expressions (8.1) and (8.2) are essentially the same as expressions (5.1) and (5.2).

The following theorem is a generalization of Theorem 13.3.4.

Theorem 13.8.2. Let $\mathbf{A}$ represent an $m \times n$ matrix and $\mathbf{B}$ an $n \times m$ matrix, where $m \leq n$. Then,

$$|\mathbf{AB}| = \sum_{\{i_1,\ldots,i_m\}} |\mathbf{A}^{\{i_1,\ldots,i_m\}}||\mathbf{B}_{\{i_1,\ldots,i_m\}}|, \tag{8.3}$$

where $\{i_1, \ldots, i_m\}$ is an m-dimensional subset of the first n positive integers $1, \ldots, n$ [and the summation is over all $\binom{n}{m}$ such subsets], where $\mathbf{A}^{\{i_1,\ldots,i_m\}}$ is the $m \times m$ submatrix of $\mathbf{A}$ obtained by striking out all of the columns of $\mathbf{A}$ except the $i_1, \ldots, i_m$th columns, and where $\mathbf{B}_{\{i_1,\ldots,i_m\}}$ is the $m \times m$ submatrix of $\mathbf{B}$ obtained by striking out all of the rows of $\mathbf{B}$ except the $i_1, \ldots, i_m$th rows.

Proof. To prove Theorem 13.8.2, let us first express $|\mathbf{AB}|$ in terms of the determinant of the $(m+n) \times (m+n)$ partitioned matrix

$$\mathbf{C} = \begin{pmatrix} \mathbf{A} & \mathbf{0} \\ -\mathbf{I}_n & \mathbf{B} \end{pmatrix},$$

and then use Laplace's theorem to evaluate $|\mathbf{C}|$.

It follows from Corollary 13.3.2 (and from Corollary 13.1.2) that

$$\begin{vmatrix} \mathbf{A} & \mathbf{AB} \\ -\mathbf{I}_n & \mathbf{0} \end{vmatrix} = (-1)^{nm}|-\mathbf{I}_n||\mathbf{AB}| = (-1)^{nm}(-1)^n|\mathbf{AB}| = (-1)^{n(m+1)}|\mathbf{AB}|.$$

Thus, making use of Theorem 13.2.11, we find that

$$\begin{aligned}|\mathbf{AB}| = (-1)^{-n(m+1)} \begin{vmatrix} \mathbf{A} & \mathbf{AB} \\ -\mathbf{I}_n & \mathbf{0} \end{vmatrix} &= (-1)^{-n(m+1)} \left| \begin{pmatrix} \mathbf{A} & \mathbf{0} \\ -\mathbf{I}_n & \mathbf{B} \end{pmatrix} \begin{pmatrix} \mathbf{I}_n & \mathbf{B} \\ \mathbf{0} & \mathbf{I}_m \end{pmatrix} \right| \\ &= (-1)^{-n(m+1)} |\mathbf{C}|. \end{aligned} \tag{8.4}$$

Moreover, it follows from result (8.1) that

$$|\mathbf{C}| = \sum_{\{j_1,\ldots,j_m\}} (-1)^{1+\cdots+m+j_1+\cdots+j_m} |\mathbf{C}_1^{\{j_1,\ldots,j_m\}}||\bar{\mathbf{C}}_2^{\{j_1,\ldots,j_m\}}|,$$

where $\{j_1, \ldots, j_m\}$ is an m-dimensional subset of the first $m+n$ positive integers $1, \ldots, m+n$ [and the sum is over all $\binom{m+n}{m}$ such subsets], where $\mathbf{C}_1^{\{j_1,\ldots,j_m\}}$ is the $m \times m$ submatrix of the $m \times (m+n)$ matrix $(\mathbf{A},\ \mathbf{0})$ obtained by striking out all of the columns except the $j_1, \ldots, j_m$th columns, and where $\bar{\mathbf{C}}_2^{\{j_1,\ldots,j_m\}}$ is the $n \times n$ submatrix of the $n \times (m+n)$ matrix $(-\mathbf{I}_n,\ \mathbf{B})$ obtained by striking out the $j_1, \ldots, j_m$th columns. Clearly, if $j_s \le n$ for $s = 1, \ldots, m$, then $\mathbf{C}_1^{\{j_1,\ldots,j_m\}} = \mathbf{A}^{\{j_1,\ldots,j_m\}}$; and, if $j_s > n$ for some integer s $(1 \le s \le m)$, then at least one column of $\mathbf{C}_1^{\{j_1,\ldots,j_m\}}$ is null. Therefore, in light of Corollary 13.2.3, we have that

$$|\mathbf{C}| = \sum_{\{i_1,\ldots,i_m\}} (-1)^{1+\cdots+m+i_1+\cdots+i_m} |\mathbf{A}^{\{i_1,\ldots,i_m\}}||-\bar{\mathbf{I}}^{\{i_1,\ldots,i_m\}},\ \mathbf{B}|, \tag{8.5}$$

where $\bar{\mathbf{I}}^{\{i_1,\ldots,i_m\}}$ is the $n \times (n-m)$ matrix obtained from $\mathbf{I}_n$ by striking out the i_1, $\ldots$, i_mth columns (and where $\{i_1, \ldots, i_m\}$ is an m-dimensional subset of the first n positive integers).

Now, applying result (8.2), we find that

$$|-\bar{\mathbf{I}}^{\{i_1,\ldots,i_m\}},\ \mathbf{B}| = \sum_{\{k_1,\ldots,k_m\}} (-1)^{k_1+\cdots+k_m+(n-m+1)+\cdots+(n-1)+n} \times |\mathbf{B}_{\{k_1,\ldots,k_m\}}||-\bar{\mathbf{I}}_{k_1,\ldots,k_m}^{i_1,\ldots,i_m}|,$$

where $\{k_1, \ldots, k_m\}$ is an m-dimensional subset of the first n positive integers $1, \ldots, n$ [and the summation is over all $\binom{n}{m}$ such subsets] and where $\bar{\mathbf{I}}_{k_1,\ldots,k_m}^{i_1,\ldots,i_m}$ is the $(n-m) \times (n-m)$ submatrix of $\mathbf{I}_n$ obtained by striking out the $k_1, \ldots, k_m$th rows and $i_1, \ldots, i_m$th columns. Moreover, unless $\{k_1, \ldots, k_m\} = \{i_1, \ldots, i_m\}$, at least one row of $\bar{\mathbf{I}}_{k_1,\ldots,k_m}^{i_1,\ldots,i_m}$ is null, so that

$$\begin{aligned} |-\bar{\mathbf{I}}_{k_1,\ldots,k_m}^{i_1,\ldots,i_m}| &= |-\mathbf{I}_{n-m}|, && \text{if } \{k_1, \ldots, k_m\} = \{i_1, \ldots, i_m\}, \\ &= 0, && \text{otherwise.} \end{aligned}$$

Therefore,

$$\begin{aligned} &|-\bar{\mathbf{I}}^{\{i_1,\ldots,i_m\}},\ \mathbf{B}| \\ &\quad = (-1)^{i_1+\cdots+i_m+(n-m+1)+\cdots+(n-1)+n} |\mathbf{B}_{\{i_1,\ldots,i_m\}}||-\mathbf{I}_{n-m}| \\ &\quad = (-1)^{i_1+\cdots+i_m+(n-m+1)+\cdots+(n-1)+n} |\mathbf{B}_{\{i_1,\ldots,i_m\}}|(-1)^{n-m} \\ &\quad = (-1)^{i_1+\cdots+i_m+(n-m)+\cdots+(n-1)+n} |\mathbf{B}_{\{i_1,\ldots,i_m\}}|. \end{aligned} \tag{8.6}$$

Finally, substituting from result (8.6) in expression (8.5), we obtain

$$\begin{aligned} |\mathbf{C}| &= \sum_{\{i_1,\ldots,i_m\}} (-1)^{1+\cdots+m+2(i_1+\cdots+i_m)+(n-m)+\cdots+(n-1)+n} \times |\mathbf{A}^{\{i_1,\ldots,i_m\}}||\mathbf{B}_{\{i_1,\ldots,i_m\}}| \\ &= \sum_{\{i_1,\ldots,i_m\}} (-1)^{2(i_1+\cdots+i_m)+n(m+1)} |\mathbf{A}^{i_1,\ldots,i_m\}}||\mathbf{B}_{\{i_1,\ldots,i_m\}}|, \end{aligned}$$

which, together with result (8.4), implies that

$$\begin{aligned} |\mathbf{AB}| &= \sum_{\{i_1,\ldots,i_m\}} (-1)^{2(i_1+\cdots+i_m)} |\mathbf{A}^{\{i_1,\ldots,i_m\}}||\mathbf{B}_{\{i_1,\ldots,i_m\}}| \\ &= \sum_{\{i_1,\ldots,i_m\}} |\mathbf{A}^{\{i_1,\ldots,i_m\}}||\mathbf{B}_{\{i_1,\ldots,i_m\}}|. \end{aligned}$$

Q.E.D.

Formula (8.3) is sometimes called the *Binet-Cauchy formula*. In the special case where $m = n$, it reduces to the formula $|\mathbf{AB}| = |\mathbf{A}||\mathbf{B}|$, given by Theorem 13.3.4. Note (in light of Lemma 4.4.3 and Corollary 4.4.5) that the rank of the product $\mathbf{AB}$ of an $m \times n$ matrix $\mathbf{A}$ and an $n \times m$ matrix $\mathbf{B}$ is at most n, implying (in light of Theorem 13.3.7) that $|\mathbf{AB}| = 0$ if $m > n$.

Exercises

Section 13.1

1. Let

$$\mathbf{A} = \begin{pmatrix} a_{11} & a_{12} & a_{13} & \boxed{a_{14}} \\ \boxed{a_{21}} & a_{22} & a_{23} & a_{24} \\ a_{31} & a_{32} & \boxed{a_{33}} & a_{34} \\ a_{41} & \boxed{a_{42}} & a_{43} & a_{44} \end{pmatrix}.$$

   (a) Write out all of the pairs that can be formed from the four boxed elements of $\mathbf{A}$.

   (b) Indicate which of the pairs from Part (a) are positive and which are negative.

   (c) Use formula (1.1) to compute the number of pairs from Part (a) that are negative, and check that the result of this computation is consistent with your answer to Part (b).

Section 13.2

2. Consider the $n \times n$ matrix

$$\mathbf{A} = \begin{pmatrix} x+\lambda & x & \ldots & x \\ x & x+\lambda & & x \\ \vdots & & \ddots & \\ x & x & & x+\lambda \end{pmatrix}.$$

   Use Theorem 13.2.10 to show that

$$|\mathbf{A}| = \lambda^{n-1}(nx+\lambda).$$

(*Hint*. Add the last $n-1$ columns of $\mathbf{A}$ to the first column, and then subtract the first row of the resultant matrix from each of the last $n-1$ rows.)

Section 13.3

3. Let $\mathbf{A}$ represent an $n \times n$ nonsingular matrix. Show that if the elements of $\mathbf{A}$ and $\mathbf{A}^{-1}$ are all integers, then $|\mathbf{A}| = \pm 1$.

4. Let $\mathbf{T}$ represent an $m \times m$ matrix, $\mathbf{U}$ an $m \times n$ matrix, $\mathbf{V}$ an $n \times m$ matrix, and $\mathbf{W}$ an $n \times n$ matrix. Show that if $\mathbf{T}$ is nonsingular, then

$$\begin{vmatrix} \mathbf{V} & \mathbf{W} \\ \mathbf{T} & \mathbf{U} \end{vmatrix} = \begin{vmatrix} \mathbf{U} & \mathbf{T} \\ \mathbf{W} & \mathbf{V} \end{vmatrix} = (-1)^{mn} |\mathbf{T}| |\mathbf{W} - \mathbf{V}\mathbf{T}^{-1}\mathbf{U}|.$$

Section 13.5

5. Compute the determinant of the 4×4 matrix

$$\mathbf{A} = \begin{pmatrix} 0 & 4 & 0 & 5 \\ 1 & 0 & -1 & 2 \\ 0 & 3 & 0 & -2 \\ 0 & 0 & -6 & 0 \end{pmatrix}$$

in each of the following two ways:

(a) by finding and summing the nonzero terms in expression (1.2) or (1.6);

(b) by repeated expansion in terms of cofactors [use formula (5.1) or (5.2) to expand $|\mathbf{A}|$ in terms of the determinants of 3×3 matrices, to expand the determinants of the 3×3 matrices in terms of the determinants of 2×2 matrices, and finally to expand the determinants of the 2×2 matrices in terms of the determinants of 1×1 matrices].

6. Let $\mathbf{A} = \{a_{ij}\}$ represent an $n \times n$ matrix. Verify that if $\mathbf{A}$ is symmetric, then the matrix of cofactors (of $\mathbf{A}$) is also symmetric.

7. Let $\mathbf{A}$ represent an $n \times n$ matrix.

(a) Show that if $\mathbf{A}$ is singular, then $\text{adj}(\mathbf{A})$ is singular.

(b) Show that $\det[\text{adj}(\mathbf{A})] = [\det(\mathbf{A})]^{n-1}$.

8. Use formula (5.7) to verify formula (8.1.2) for the inverse of a 2×2 nonsingular matrix.

9. Let

$$\mathbf{A} = \begin{pmatrix} 2 & 0 & -1 \\ -1 & 3 & 1 \\ 0 & -4 & 5 \end{pmatrix}.$$

(a) Compute the cofactor of each element of $\mathbf{A}$.

(b) Compute $|\mathbf{A}|$ by expanding $|\mathbf{A}|$ in terms of the cofactors of the elements of the second row of $\mathbf{A}$, and then check your answer by expanding $|\mathbf{A}|$ in terms of the cofactors of the elements of the second column of $\mathbf{A}$.

(c) Use formula (5.7) to compute $\mathbf{A}^{-1}$.

10. Let $\mathbf{A} = \{a_{ij}\}$ represent an $n \times n$ matrix (where $n \geq 2$), and let α_{ij} represent the cofactor of a_{ij}.

(a) Show [by, for instance, making use of the result of Part (b) of Exercise 11.3] that if $\text{rank}(\mathbf{A}) = n - 1$, then there exists a scalar c such that $\text{adj}(\mathbf{A}) = c\mathbf{x}\mathbf{y}'$, where $\mathbf{x} = \{x_j\}$ and $\mathbf{y} = \{y_i\}$ are any nonnull n-dimensional column vectors such that $\mathbf{A}\mathbf{x} = \mathbf{0}$ and $\mathbf{A}'\mathbf{y} = \mathbf{0}$. Show also that c is nonzero and is expressible as $c = \alpha_{ij}/(y_i x_j)$ for any i and j such that $y_i \neq 0$ and $x_j \neq 0$.

(b) Show that if $\text{rank}(\mathbf{A}) \leq n - 2$, then $\text{adj}(\mathbf{A}) = \mathbf{0}$.

11. Let $\mathbf{A}$ represent an $n \times n$ nonsingular matrix and $\mathbf{b}$ an $n \times 1$ vector. Show that the solution to the linear system $\mathbf{A}\mathbf{x} = \mathbf{b}$ (in $\mathbf{x}$) is the $n \times 1$ vector whose jth component is

$$|\mathbf{A}_j|/|\mathbf{A}|,$$

where $\mathbf{A}_j$ is a matrix formed from $\mathbf{A}$ by substituting $\mathbf{b}$ for the jth column of $\mathbf{A}$ $(j = 1, \ldots, n)$. [This result is called Cramer's rule, after Gabriel Cramer (1704–1752).]

12. Let c represent a scalar, let $\mathbf{x}$ and $\mathbf{y}$ represent $n \times 1$ vectors, and let $\mathbf{A}$ represent an $n \times n$ matrix.

(a) Show that

$$\begin{vmatrix} \mathbf{A} & \mathbf{y} \\ \mathbf{x}' & c \end{vmatrix} = c|\mathbf{A}| - \mathbf{x}' \,\text{adj}(\mathbf{A})\mathbf{y}. \tag{E.1}$$

(b) Show that, in the special case where $\mathbf{A}$ is nonsingular, result (E.1) can be reexpressed as

$$\begin{vmatrix} \mathbf{A} & \mathbf{y} \\ \mathbf{x}' & c \end{vmatrix} = |\mathbf{A}|(c - \mathbf{x}'\mathbf{A}^{-1}\mathbf{y}),$$

in agreement with result (3.13).

Section 13.6

13. Let $\mathbf{V}_k$ represent the $(n - 1) \times (n - 1)$ submatrix of the $n \times n$ Vandermonde matrix

$$\mathbf{V} = \begin{pmatrix} 1 & x_1 & x_1^2 & \dots & x_1^{n-1} \\ 1 & x_2 & x_2^2 & \dots & x_2^{n-1} \\ \vdots & \vdots & \vdots & & \vdots \\ 1 & x_n & x_n^2 & \dots & x_n^{n-1} \end{pmatrix}$$

(where $x_1, x_2, \dots, x_n$ are arbitrary scalars) obtained by striking out the kth row and the nth (last) column (of $\mathbf{V}$). Show that

$$|\mathbf{V}| = |\mathbf{V}_k|(-1)^{n-k} \prod_{i \neq k} (x_k - x_i).$$

Section 13.8

14. Show that, for $n \times n$ matrices $\mathbf{A}$ and $\mathbf{B}$,

$$\text{adj}(\mathbf{AB}) = \text{adj}(\mathbf{B})\,\text{adj}(\mathbf{A}).$$

[*Hint.* Use the Binet-Cauchy formula to establish that the ijth elements of $\text{adj}(\mathbf{AB})$ and $\text{adj}(\mathbf{B})\text{adj}(\mathbf{A})$ are equal.]

14

Linear, Bilinear, and Quadratic Forms

This chapter is devoted to certain functions (of an n-dimensional vector) that are defined in Section 14.1 and that are known as linear, bilinear, and quadratic forms. Those bilinear forms that have certain properties (known as symmetry and positive definiteness) are of interest (at least in part) because they qualify as inner products for $\mathcal{R}^n$ (as is discussed in Section 14.10). Quadratic (and bilinear) forms are of considerable relevance in statistics; the "sums of squares (or products)" in an analysis of variance (or covariance) table are quadratic (or bilinear) forms.

14.1 Some Terminology and Basic Results

Let $\mathbf{a} = (a_1, \ldots, a_n)'$ represent an arbitrary n-dimensional column vector, and consider the function that assigns to each vector $\mathbf{x} = (x_1, \ldots, x_n)'$ in $\mathcal{R}^n$ the value $\mathbf{a}'\mathbf{x} = \sum_i a_i x_i$. A function of $\mathbf{x}$ (with domain $\mathcal{R}^n$) that is expressible as $\mathbf{a}'\mathbf{x}$ for some vector $\mathbf{a}$ is called a *linear form* (in $\mathbf{x}$). Thus, a linear form is a homogeneous polynomial of degree one. Note that the linear form $\mathbf{a}'\mathbf{x}$ can be reexpressed as $\mathbf{x}'\mathbf{a}$. It is customary to refer to $a_1, \ldots, a_n$ as the *coefficients of the linear form* and to the row or column vector $\mathbf{a}'$ or $\mathbf{a}$ as the *coefficient vector.*

A function, say $f(\mathbf{x})$, of $\mathbf{x}$ (with domain $\mathcal{R}^n$) is said to be *linear* if (1) $f(\mathbf{x}_1 + \mathbf{x}_2) = f(\mathbf{x}_1) + f(\mathbf{x}_2)$ for all vectors $\mathbf{x}_1$ and $\mathbf{x}_2$ in $\mathcal{R}^n$ and (2) $f(c\mathbf{x}) = cf(\mathbf{x})$ for every scalar c and every vector $\mathbf{x}$ in $\mathcal{R}^n$. Or, equivalently, $f(\mathbf{x})$ is linear if

$$f(c_1\mathbf{x}_1 + c_2\mathbf{x}_2 + \cdots + c_k\mathbf{x}_k) = c_1 f(\mathbf{x}_1) + c_2 f(\mathbf{x}_2) + \cdots + c_k f(\mathbf{x}_k)$$

for all scalars $c_1, c_2, \ldots, c_k$ and all (n-dimensional) vectors $\mathbf{x}_1, \mathbf{x}_2, \ldots, \mathbf{x}_k$ (where k is an integer greater than or equal to 2).

A linear form $\mathbf{a}'\mathbf{x} = \sum_i a_i x_i$ is a linear function of $\mathbf{x} = (x_1, \ldots, x_n)'$, as is easily verified. The converse is also true; that is, a linear function $f(\mathbf{x})$ of $\mathbf{x}$ is expressible as a linear form (in $\mathbf{x}$). To see this, take $a_i = f(\mathbf{e}_i)$, where $\mathbf{e}_i$ is the ith column of $\mathbf{I}_n$ ($i = 1, \ldots, n$), and observe that

$$\begin{aligned} f(\mathbf{x}) &= f(x_1\mathbf{e}_1 + x_2\mathbf{e}_2 + \cdots + x_n\mathbf{e}_n) \\ &= x_1 f(\mathbf{e}_1) + x_2 f(\mathbf{e}_2) + \cdots + x_n f(\mathbf{e}_n) \\ &= a_1x_1 + a_2x_2 + \cdots + a_nx_n = \mathbf{a}'\mathbf{x}. \end{aligned}$$

If the coefficient vectors $\mathbf{a} = (a_1, \ldots, a_n)'$ and $\mathbf{b} = (b_1, \ldots, b_n)'$ of two linear forms $\mathbf{a}'\mathbf{x}$ and $\mathbf{b}'\mathbf{x}$ (in $\mathbf{x}$) are equal, then clearly the two linear forms are identically equal (i.e., equal for all $\mathbf{x}$). Conversely, if the two linear forms $\mathbf{a}'\mathbf{x}$ and $\mathbf{b}'\mathbf{x}$ are identically equal, then $\mathbf{a} = \mathbf{b}$, as is evident from Lemma 2.3.2.

Now, let $\mathbf{A} = \{a_{ij}\}$ represent an arbitrary $m \times n$ matrix, and consider the function that assigns to each pair of vectors $\mathbf{x} = (x_1, \ldots, x_m)'$ and $\mathbf{y} = (y_1, \ldots, y_n)'$ (in $\mathcal{R}^m$ and $\mathcal{R}^n$, respectively) the value $\mathbf{x}'\mathbf{A}\mathbf{y} = \sum_{i,j} a_{ij}x_iy_j$. A function of $\mathbf{x}$ and $\mathbf{y}$ that is expressible in the form $\mathbf{x}'\mathbf{A}\mathbf{y}$ is called a *bilinear form* (in $\mathbf{x}$ and $\mathbf{y}$). Thus, a bilinear form is a special kind of second-degree homogeneous polynomial (in the $m+n$ variables $x_1, \ldots, x_m, y_1, \ldots, y_n$)—one which, for fixed $\mathbf{x}$, is homogeneous of first degree in $y_1, \ldots, y_n$ and which, for fixed $\mathbf{y}$, is homogeneous of first degree in $x_1, \ldots, x_m$. It is customary to refer to $\mathbf{A}$ as the *matrix of the bilinear form* $\mathbf{x}'\mathbf{A}\mathbf{y}$. Note that the bilinear form $\mathbf{x}'\mathbf{A}\mathbf{y}$ can be reexpressed as $\mathbf{y}'\mathbf{A}'\mathbf{x}$—a bilinear form in $\mathbf{y}$ and $\mathbf{x}$ whose matrix is $\mathbf{A}'$.

A function, say $f(\mathbf{x}, \mathbf{y})$, of $\mathbf{x}$ and $\mathbf{y}$ is said to be *bilinear* if, for every fixed $\mathbf{x}$, it is a linear function of $\mathbf{y}$ and if, for every fixed $\mathbf{y}$, it is a linear function of $\mathbf{x}$. For every fixed $\mathbf{y}$, the bilinear form $\mathbf{x}'\mathbf{A}\mathbf{y}$ is a linear form in $\mathbf{x}$—one with coefficient vector $\mathbf{A}\mathbf{y}$. (And, for every fixed $\mathbf{x}$, $\mathbf{x}'\mathbf{A}\mathbf{y}$ is a linear form in $\mathbf{y}$.) Since (as noted earlier) a linear form is a linear function, a bilinear form is a bilinear function. Conversely, any bilinear function is expressible as a bilinear form. To see this, let $f(\mathbf{x}, \mathbf{y})$ represent an arbitrary bilinear function in $\mathbf{x} = (x_1, \ldots, x_m)'$ and $\mathbf{y} = (y_1, \ldots, y_n)'$, and take $\mathbf{A}$ to be the $m \times n$ matrix with ijth element $a_{ij} = f(\mathbf{e}_i, \mathbf{u}_j)$, where $\mathbf{e}_i$ denotes the ith column of $\mathbf{I}_m$, and $\mathbf{u}_j$ the jth column of $\mathbf{I}_n$. Then,

$$\begin{aligned} f(\mathbf{x}, \mathbf{y}) = f\Big(\sum_i x_i\mathbf{e}_i, \mathbf{y}\Big) &= \sum_i x_i f(\mathbf{e}_i, \mathbf{y}) \\ &= \sum_i x_i f\Big(\mathbf{e}_i, \sum_j y_j\mathbf{u}_j\Big) \\ &= \sum_i x_i \sum_j y_j f(\mathbf{e}_i, \mathbf{u}_j) = \sum_{i,j} a_{ij}x_iy_j = \mathbf{x}'\mathbf{A}\mathbf{y}. \end{aligned}$$

If the matrices $\mathbf{A} = \{a_{ij}\}$ and $\mathbf{B} = \{b_{ij}\}$ of two bilinear forms $\mathbf{x}'\mathbf{A}\mathbf{y}$ and $\mathbf{x}'\mathbf{B}\mathbf{y}$ (in $\mathbf{x}$ and $\mathbf{y}$) are equal, then clearly the two bilinear forms are identically equal (i.e., equal for all $\mathbf{x}$ and $\mathbf{y}$). Conversely, if the two bilinear forms are identically equal, then $\mathbf{A} = \mathbf{B}$, as is evident upon observing that, for $\mathbf{x} = \mathbf{e}_i$ (the ith column of $\mathbf{I}_m$) and $\mathbf{y} = \mathbf{u}_j$ (the jth column of $\mathbf{I}_n$), $a_{ij} = \mathbf{x}'\mathbf{A}\mathbf{y} = \mathbf{x}'\mathbf{B}\mathbf{y} = b_{ij}$ ($i = 1, \ldots, m$; $j = 1, \ldots, n$).

A bilinear form $\mathbf{x}'\mathbf{A}\mathbf{y}$ in vectors $\mathbf{x} = (x_1, \ldots, x_n)'$ and $\mathbf{y} = (y_1, \ldots, y_n)'$ (of the same dimensions) is said to be *symmetric* if $\mathbf{x}'\mathbf{A}\mathbf{y} = \mathbf{y}'\mathbf{A}\mathbf{x}$ for all $\mathbf{x}$ and $\mathbf{y}$.

Since $\mathbf{y}'\mathbf{A}\mathbf{x} = (\mathbf{y}'\mathbf{A}\mathbf{x})' = \mathbf{x}'\mathbf{A}'\mathbf{y}$, the bilinear form $\mathbf{x}'\mathbf{A}\mathbf{y}$ is symmetric if and only if $\mathbf{x}'\mathbf{A}\mathbf{y} = \mathbf{x}'\mathbf{A}'\mathbf{y}$ for all $\mathbf{x}$ and $\mathbf{y}$ and hence if and only if $\mathbf{A} = \mathbf{A}'$. Thus, the bilinear form is symmetric if and only if the matrix of the bilinear form is symmetric.

Finally, let $\mathbf{A} = \{a_{ij}\}$ represent an arbitrary $n \times n$ matrix, and consider the function that assigns to each vector $\mathbf{x} = (x_1, \ldots, x_n)'$ (in $\mathcal{R}^n$) the value

$$\mathbf{x}'\mathbf{A}\mathbf{x} = \sum_{i,j} a_{ij}x_i x_j = \sum_i a_{ii}x_i^2 + \sum_{i,j\neq i} a_{ij}x_i x_j .$$

A function of $\mathbf{x}$ that is expressible in the form $\mathbf{x}'\mathbf{A}\mathbf{x}$ is called a *quadratic form* (in $\mathbf{x}$). Thus, a quadratic form is a homogeneous polynomial of degree two. It is customary to refer to $\mathbf{A}$ as the *matrix of the quadratic form* $\mathbf{x}'\mathbf{A}\mathbf{x}$. Clearly, the quadratic form $\mathbf{x}'\mathbf{A}\mathbf{x}$ can be obtained from the bilinear form $\mathbf{x}'\mathbf{A}\mathbf{y}$ (in $\mathbf{x}$ and $\mathbf{y}$) simply by setting $\mathbf{y} = \mathbf{x}$.

Let $\mathbf{B} = \{b_{ij}\}$ represent a second $n \times n$ matrix. Under what circumstances are the two quadratic forms $\mathbf{x}'\mathbf{A}\mathbf{x}$ and $\mathbf{x}'\mathbf{B}\mathbf{x}$ identically equal? Clearly, a sufficient condition for $\mathbf{x}'\mathbf{A}\mathbf{x} \equiv \mathbf{x}'\mathbf{B}\mathbf{x}$ is that $\mathbf{A} = \mathbf{B}$. However, except in the special case $n = 1$, $\mathbf{A} = \mathbf{B}$ is not a necessary condition.

For the purpose of establishing a necessary condition, suppose that $\mathbf{x}'\mathbf{A}\mathbf{x} \equiv \mathbf{x}'\mathbf{B}\mathbf{x}$. Setting $\mathbf{x}$ equal to the ith column of $\mathbf{I}_n$, we find that

$$a_{ii} = \mathbf{x}'\mathbf{A}\mathbf{x} = \mathbf{x}'\mathbf{B}\mathbf{x} = b_{ii} \quad (i = 1, \ldots, n). \tag{1.1}$$

That is, the diagonal elements of $\mathbf{A}$ are the same as those of $\mathbf{B}$. Consider now the off-diagonal elements of $\mathbf{A}$ and $\mathbf{B}$. Setting $\mathbf{x}$ equal to the n-dimensional column vector whose ith and jth elements equal one and whose remaining elements equal zero, we find that

$$\begin{aligned} &a_{ii} + a_{ij} + a_{ji} + a_{jj} \\ &\quad = \mathbf{x}'\mathbf{A}\mathbf{x} = \mathbf{x}'\mathbf{B}\mathbf{x} = b_{ii} + b_{ij} + b_{ji} + b_{jj} \quad (j \neq i = 1, \ldots, n). \end{aligned} \tag{1.2}$$

Together, results (1.1) and (1.2) imply that

$$a_{ii} = b_{ii}, \quad a_{ij} + a_{ji} = b_{ij} + b_{ji} \quad (j \neq i = 1, \ldots, n),$$

or equivalently that

$$\mathbf{A} + \mathbf{A}' = \mathbf{B} + \mathbf{B}'. \tag{1.3}$$

Thus, condition (1.3) is a necessary condition for $\mathbf{x}'\mathbf{A}\mathbf{x}$ and $\mathbf{x}'\mathbf{B}\mathbf{x}$ to be identically equal. It is also a sufficient condition. To see this, observe that (since a 1×1 matrix is symmetric) condition (1.3) implies that

$$\begin{aligned} \mathbf{x}'\mathbf{A}\mathbf{x} = (1/2)[\mathbf{x}'\mathbf{A}\mathbf{x} + (\mathbf{x}'\mathbf{A}\mathbf{x})'] &= (1/2)(\mathbf{x}'\mathbf{A}\mathbf{x} + \mathbf{x}'\mathbf{A}'\mathbf{x}) \\ &= (1/2)\mathbf{x}'(\mathbf{A} + \mathbf{A}')\mathbf{x} \\ &= (1/2)\mathbf{x}'(\mathbf{B} + \mathbf{B}')\mathbf{x} \\ &= (1/2)[\mathbf{x}'\mathbf{B}\mathbf{x} + (\mathbf{x}'\mathbf{B}\mathbf{x})'] = \mathbf{x}'\mathbf{B}\mathbf{x}. \end{aligned}$$

In summary, we have the following lemma.

Lemma 14.1.1. Let $\mathbf{A} = \{a_{ij}\}$ and $\mathbf{B} = \{b_{ij}\}$ represent arbitrary $n \times n$ matrices. The two quadratic forms $\mathbf{x}'\mathbf{A}\mathbf{x}$ and $\mathbf{x}'\mathbf{B}\mathbf{x}$ (in $\mathbf{x}$) are identically equal if and only if, for $j \neq i = 1, \ldots, n$, $a_{ii} = b_{ii}$ and $a_{ij} + a_{ji} = b_{ij} + b_{ji}$ or, equivalently, if and only if $\mathbf{A} + \mathbf{A}' = \mathbf{B} + \mathbf{B}'$.

Note that Lemma 14.1.1 implies in particular that $\mathbf{x}'\mathbf{A}'\mathbf{x} = \mathbf{x}'\mathbf{A}\mathbf{x}$ for all $\mathbf{x}$.

When $\mathbf{B}$ is symmetric, the condition $\mathbf{A} + \mathbf{A}' = \mathbf{B} + \mathbf{B}'$ is equivalent to the condition $\mathbf{B} = (1/2)(\mathbf{A} + \mathbf{A}')$, and, when both $\mathbf{A}$ and $\mathbf{B}$ are symmetric, the condition $\mathbf{A} + \mathbf{A}' = \mathbf{B} + \mathbf{B}'$ is equivalent to the condition $\mathbf{A} = \mathbf{B}$. Thus, we have the following two corollaries of Lemma 14.1.1.

Corollary 14.1.2. Corresponding to any quadratic form $\mathbf{x}'\mathbf{A}\mathbf{x}$, there is a unique symmetric matrix $\mathbf{B}$ such that $\mathbf{x}'\mathbf{B}\mathbf{x} = \mathbf{x}'\mathbf{A}\mathbf{x}$ for all $\mathbf{x}$, namely, the matrix $\mathbf{B} = (1/2)(\mathbf{A} + \mathbf{A}')$.

Corollary 14.1.3. For any pair of $n \times n$ symmetric matrices $\mathbf{A}$ and $\mathbf{B}$, the two quadratic forms $\mathbf{x}'\mathbf{A}\mathbf{x}$ and $\mathbf{x}'\mathbf{B}\mathbf{x}$ (in $\mathbf{x}$) are identically equal (i.e., $\mathbf{x}'\mathbf{A}\mathbf{x} = \mathbf{x}'\mathbf{B}\mathbf{x}$ for all $\mathbf{x}$) if and only if $\mathbf{A} = \mathbf{B}$.

As a special case of Corollary 14.1.3 (that where $\mathbf{B} = \mathbf{0}$), we have the following, additional corollary.

Corollary 14.1.4. Let $\mathbf{A}$ represent an $n \times n$ symmetric matrix. If $\mathbf{x}'\mathbf{A}\mathbf{x} = 0$ for every $(n \times 1)$ vector $\mathbf{x}$, then $\mathbf{A} = \mathbf{0}$.

14.2 Nonnegative Definite Quadratic Forms and Matrices

a. Definitions

Let $\mathbf{x}'\mathbf{A}\mathbf{x}$ represent an arbitrary quadratic form [in an n-dimensional vector $\mathbf{x} = (x_1, \ldots, x_n)'$]. The quadratic form $\mathbf{x}'\mathbf{A}\mathbf{x}$ is said to be *nonnegative definite* if $\mathbf{x}'\mathbf{A}\mathbf{x} \geq 0$ for every $\mathbf{x}$ in $\mathcal{R}^n$.

Note that $\mathbf{x}'\mathbf{A}\mathbf{x} = 0$ for at least one value of $\mathbf{x}$, namely, $\mathbf{x} = \mathbf{0}$. If $\mathbf{x}'\mathbf{A}\mathbf{x}$ is nonnegative definite and if, in addition, the null vector $\mathbf{0}$ is the only value of $\mathbf{x}$ for which $\mathbf{x}'\mathbf{A}\mathbf{x} = 0$, then $\mathbf{x}'\mathbf{A}\mathbf{x}$ is said to be *positive definite*. That is, $\mathbf{x}'\mathbf{A}\mathbf{x}$ is positive definite if $\mathbf{x}'\mathbf{A}\mathbf{x} > 0$ for every $\mathbf{x}$ except $\mathbf{x} = \mathbf{0}$. A quadratic form that is nonegative definite, but not positive definite, is called *positive semidefinite*. Thus, $\mathbf{x}'\mathbf{A}\mathbf{x}$ is positive semidefinite if $\mathbf{x}'\mathbf{A}\mathbf{x} \geq 0$ for every $\mathbf{x} \in \mathcal{R}^n$ and $\mathbf{x}'\mathbf{A}\mathbf{x} = 0$ for some nonnull $\mathbf{x}$.

Consider, for example, the two quadratic forms $\mathbf{x}'\mathbf{I}\mathbf{x} = x_1^2 + x_2^2 + \cdots + x_n^2$ and $\mathbf{x}'\mathbf{J}\mathbf{x} = \mathbf{x}'\mathbf{1}\mathbf{1}'\mathbf{x} = (\mathbf{1}'\mathbf{x})'\mathbf{1}'\mathbf{x} = (x_1 + x_2 + \cdots + x_n)^2$. Clearly, $\mathbf{x}'\mathbf{I}\mathbf{x}$ and $\mathbf{x}'\mathbf{J}\mathbf{x}$ are both nonnegative definite. Moreover, $\mathbf{x}'\mathbf{I}\mathbf{x} > 0$ for all nonnull $\mathbf{x}$, while (assuming that $n \geq 2$) $\mathbf{x}'\mathbf{J}\mathbf{x} = 0$ for the nonnull vector $\mathbf{x} = (1-n, 1, 1, \ldots, 1)'$. Thus, $\mathbf{x}'\mathbf{I}\mathbf{x}$ is positive definite, while $\mathbf{x}'\mathbf{J}\mathbf{x}$ is positive semidefinite.

The quadratic form $\mathbf{x}'\mathbf{A}\mathbf{x}$ is said to be *nonpositive definite*, *negative definite*, or *negative semidefinite* if $-\mathbf{x}'\mathbf{A}\mathbf{x}$ is nonnegative definite, positive definite, or positive semidefinite, respectively. Thus, $\mathbf{x}'\mathbf{A}\mathbf{x}$ is nonpositive definite if $\mathbf{x}'\mathbf{A}\mathbf{x} \leq 0$ for every

$\mathbf{x}$ in $\mathcal{R}^n$, is negative definite if $\mathbf{x}'\mathbf{A}\mathbf{x} < 0$ for every nonnull $\mathbf{x}$ in $\mathcal{R}^n$, and is negative semidefinite if $\mathbf{x}'\mathbf{A}\mathbf{x} \leq 0$ for every $\mathbf{x}$ in $\mathcal{R}^n$ and $\mathbf{x}'\mathbf{A}\mathbf{x} = 0$ for some nonnull $\mathbf{x}$.

A quadratic form that is neither nonnegative definite nor nonpositive definite is said to be *indefinite*. Thus, $\mathbf{x}'\mathbf{A}\mathbf{x}$ is indefinite if $\mathbf{x}'\mathbf{A}\mathbf{x} < 0$ for some $\mathbf{x}$ and $\mathbf{x}'\mathbf{A}\mathbf{x} > 0$ for some (other) $\mathbf{x}$.

The terms nonnegative definite, positive definite, positive semidefinite, nonpositive definite, negative definite, negative semidefinite, and indefinite are applied to matrices as well as to quadratic forms. An $n \times n$ matrix $\mathbf{A}$ is said to be *nonnegative definite, positive definite*, or *positive semidefinite* if the quadratic form $\mathbf{x}'\mathbf{A}\mathbf{x}$ (in $\mathbf{x}$) is nonnegative definite, positive definite, or positive semidefinite, respectively. Similarly, $\mathbf{A}$ is said to be *nonpositive definite, negative definite*, or *negative semidefinite* if the quadratic form $\mathbf{x}'\mathbf{A}\mathbf{x}$ is nonpositive definite, negative definite, or negative semidefinite, respectively, or equivalently if $-\mathbf{A}$ is nonnegative definite, positive definite, or positive semidefinite. Further, $\mathbf{A}$ is said to be indefinite if the quadratic form $\mathbf{x}'\mathbf{A}\mathbf{x}$ is indefinite or equivalently if $\mathbf{A}$ is neither nonnegative definite nor nonpositive definite.

(Symmetric) nonnegative definite matrices are encountered with considerable frequency in linear statistical models and in other areas of statistics. In particular, the variance-covariance matrix of any random vector is inherently nonnegative definite (and symmetric).

Our usage of the terms nonnegative definite, positive definite, positive semidefinite, nonpositive definite, negative definite, negative semidefinite, and indefinite differs somewhat from that employed in various other presentations. In particular, we apply these terms to both symmetric and nonsymmetric matrices, whereas in many other presentations their application to matrices is confined to symmetric matrices. Moreover, our usage of the terms positive semidefinite and negative semidefinite is not completely standard. In some presentations, these terms are used in the same way that nonnegative definite and nonpositive definite are being used here.

It is instructive to consider the following lemma, which characterizes the concepts of nonnegative definiteness, positive definiteness, and positive semidefiniteness as applied to diagonal matrices and which is easy to verify.

Lemma 14.2.1. Let $\mathbf{D} = \{d_i\}$ represent an $n \times n$ diagonal matrix. Then, (1) $\mathbf{D}$ is nonnegative definite if and only if $d_1, \ldots, d_n$ are nonnegative; (2) $\mathbf{D}$ is positive definite if and only if $d_1, \ldots, d_n$ are positive; and (3) $\mathbf{D}$ is positive semidefinite if and only if $d_i \geq 0$ for $i = 1, \ldots, n$ with equality holding for one or more values of i.

Suppose that an $n \times n$ symmetric matrix $\mathbf{A}$ is both nonnegative definite and nonpositive definite. Then, for every $n \times 1$ vector $\mathbf{x}$, $\mathbf{x}'\mathbf{A}\mathbf{x} \geq 0$ and $-\mathbf{x}'\mathbf{A}\mathbf{x} \geq 0$ (or equivalently $\mathbf{x}'\mathbf{A}\mathbf{x} \leq 0$). We conclude that $\mathbf{x}'\mathbf{A}\mathbf{x} = 0$ for every $\mathbf{x}$ and hence (in light of Corollary 14.1.4) that $\mathbf{A} = \mathbf{0}$. Thus, we have the following lemma.

Lemma 14.2.2. The only $n \times n$ symmetric matrix that is both nonnegative definite and nonpositive definite is the $n \times n$ null matrix.

b. Some basic properties of nonnegative definite matrices

We now consider some of the more elementary properties of nonnegative definite matrices. We do not explicitly consider the properties of nonpositive definite matrices. However, since $\mathbf{A}$ is nonpositive definite, negative definite, or negative semidefinite if and only if $-\mathbf{A}$ is nonnegative definite, positive definite, or positive semidefinite, respectively, the properties of nonpositive definite matrices can be readily deduced from those of nonnegative definite matrices.

The following two lemmas give some basic results on scalar multiples and sums of nonnegative definite matrices.

Lemma 14.2.3. Let $k > 0$ represent a (positive) scalar, and $\mathbf{A}$ an $n \times n$ matrix. If $\mathbf{A}$ is positive definite, then $k\mathbf{A}$ is also positive definite; if $\mathbf{A}$ is positive semidefinite, then $k\mathbf{A}$ is also positive semidefinite.

Proof. Consider the quadratic forms $\mathbf{x}'\mathbf{A}\mathbf{x}$ and $\mathbf{x}'(k\mathbf{A})\mathbf{x}$ (in $\mathbf{x}$). Clearly, $\mathbf{x}'(k\mathbf{A})\mathbf{x} = k\mathbf{x}'\mathbf{A}\mathbf{x}$. Thus, if $\mathbf{x}'\mathbf{A}\mathbf{x}$ is positive definite, then $\mathbf{x}'(k\mathbf{A})\mathbf{x}$ is positive definite; similarly, if $\mathbf{x}'\mathbf{A}\mathbf{x}$ is positive semidefinite, then $\mathbf{x}'(k\mathbf{A})\mathbf{x}$ is positive semidefinite. Or, equivalently, if $\mathbf{A}$ is positive definite, then $k\mathbf{A}$ is positive definite; and if $\mathbf{A}$ is positive semidefinite, then $k\mathbf{A}$ is positive semidefinite. Q.E.D.

Lemma 14.2.4. Let $\mathbf{A}$ and $\mathbf{B}$ represent $n \times n$ matrices. If $\mathbf{A}$ and $\mathbf{B}$ are both nonnegative definite, then $\mathbf{A} + \mathbf{B}$ is nonnegative definite. Moreover, if either $\mathbf{A}$ or $\mathbf{B}$ is positive definite and the other is nonnegative definite (i.e., positive definite or positive semidefinite), then $\mathbf{A} + \mathbf{B}$ is positive definite.

Proof. Suppose that one of the two matrices, say $\mathbf{A}$, is positive definite and that the other ($\mathbf{B}$) is nonnegative definite. Then, for every nonnull vector $\mathbf{x}$ in $\mathcal{R}^n$, $\mathbf{x}'\mathbf{A}\mathbf{x} > 0$ and $\mathbf{x}'\mathbf{B}\mathbf{x} \geq 0$, and hence

$$\mathbf{x}'(\mathbf{A} + \mathbf{B})\mathbf{x} = \mathbf{x}'\mathbf{A}\mathbf{x} + \mathbf{x}'\mathbf{B}\mathbf{x} > 0.$$

Thus, $\mathbf{A} + \mathbf{B}$ is positive definite.

A similar argument shows that if $\mathbf{A}$ and $\mathbf{B}$ are both nonnegative definite, then $\mathbf{A} + \mathbf{B}$ is nonnegative definite. Q.E.D.

The repeated application of Lemma 14.2.4 leads to the following generalization.

Corollary 14.2.5. Let $\mathbf{A}_1, \ldots, \mathbf{A}_k$ represent $n \times n$ matrices. If $\mathbf{A}_1, \ldots, \mathbf{A}_k$ are all nonnegative definite, then their sum $\mathbf{A}_1 + \cdots + \mathbf{A}_k$ is also nonnegative definite. Moreover, if one or more of the matrices $\mathbf{A}_1, \ldots, \mathbf{A}_k$ are positive definite and the others are nonnegative definite, then $\mathbf{A}_1 + \cdots + \mathbf{A}_k$ is positive definite.

As an essentially immediate consequence of Lemma 14.1.1, we have the following lemma and corollary.

Lemma 14.2.6. Let $\mathbf{A}$ represent an $n \times n$ matrix, and take $\mathbf{B}$ to be any $n \times n$ matrix such that $\mathbf{B} + \mathbf{B}' = \mathbf{A} + \mathbf{A}'$. Then, $\mathbf{A}$ is positive definite if and only if $\mathbf{B}$ is positive definite, is positive semidefinite if and only if $\mathbf{B}$ is positive semidefinite, and is nonnegative definite if and only if $\mathbf{B}$ is nonnegative definite.

Corollary 14.2.7. An $n \times n$ matrix $\mathbf{A}$ is positive definite if and only if $\mathbf{A}'$ is positive definite and if and only if $(1/2)(\mathbf{A} + \mathbf{A}')$ is positive definite; is positive semidefinite if and only if $\mathbf{A}'$ is positive semidefinite and if and only if $(1/2)(\mathbf{A} + \mathbf{A}')$ is positive semidefinite; and is nonnegative definite if and only if $\mathbf{A}'$ is nonnegative definite and if and only if $(1/2)(\mathbf{A} + \mathbf{A}')$ is nonnegative definite.

If $\mathbf{A}$ is a symmetric matrix, then $\mathbf{A}' = \mathbf{A}$ and $(1/2)(\mathbf{A} + \mathbf{A}') = \mathbf{A}$. Thus, it is only in the case of a nonsymmetric positive definite, positive semidefinite, or nonnegative definite matrix that Corollary 14.2.7 is meaningful.

A basic property of positive definite matrices is described in the following lemma.

Lemma 14.2.8. Any positive definite matrix is nonsingular.

Proof. Let $\mathbf{A}$ represent an $n \times n$ positive definite matrix. For purposes of establishing a contradiction, suppose that $\mathbf{A}$ is singular or, equivalently, that $\text{rank}(\mathbf{A}) < n$. Then, the columns of $\mathbf{A}$ are linearly dependent, and hence there exists a nonnull vector $\mathbf{x}_*$ such that $\mathbf{A}\mathbf{x}_* = \mathbf{0}$. We find that

$$\mathbf{x}_*'\mathbf{A}\mathbf{x}_* = \mathbf{x}_*'(\mathbf{A}\mathbf{x}_*) = \mathbf{x}_*'\mathbf{0} = 0,$$

which (since $\mathbf{A}$ is positive definite) establishes the desired contradication. Q.E.D.

Lemma 14.2.8 implies that singular nonnegative definite matrices are positive semidefinite. However, the converse is not necessarily true. That is, there exist nonsingular positive semidefinite matrices as well as singular positive semidefinite matrices.

Additional basic properties of nonnegative definite matrices are described in the following theorem and corollaries.

Theorem 14.2.9. Let $\mathbf{A}$ represent an $n \times n$ matrix, and $\mathbf{P}$ an $n \times m$ matrix. (1) If $\mathbf{A}$ is nonnegative definite, then $\mathbf{P}'\mathbf{A}\mathbf{P}$ is nonnegative definite. (2) If $\mathbf{A}$ is nonnegative definite and $\text{rank}(\mathbf{P}) < m$, then $\mathbf{P}'\mathbf{A}\mathbf{P}$ is positive semidefinite. (3) If $\mathbf{A}$ is positive definite and $\text{rank}(\mathbf{P}) = m$, then $\mathbf{P}'\mathbf{A}\mathbf{P}$ is positive definite.

Proof. Suppose that $\mathbf{A}$ is nonnegative definite (either positive definite or positive semidefinite). Then, $\mathbf{y}'\mathbf{A}\mathbf{y} \geq 0$ for every $\mathbf{y}$ in $\mathcal{R}^n$ and in particular for every $\mathbf{y}$ that is expressible in the form $\mathbf{P}\mathbf{x}$. Thus, for every m-dimensional vector $\mathbf{x}$,

$$\mathbf{x}'(\mathbf{P}'\mathbf{A}\mathbf{P})\mathbf{x} = (\mathbf{P}\mathbf{x})'\mathbf{A}\mathbf{P}\mathbf{x} \geq 0, \tag{2.1}$$

which establishes that $\mathbf{P}'\mathbf{A}\mathbf{P}$ is nonnegative definite, thereby completing the proof of Part (1).

If $\text{rank}(\mathbf{P}) < m$, then

$$\text{rank}(\mathbf{P}'\mathbf{A}\mathbf{P}) \leq \text{rank}(\mathbf{P}) < m,$$

which (in light of Lemma 14.2.8) establishes that $\mathbf{P}'\mathbf{A}\mathbf{P}$ is not positive definite and hence (since $\mathbf{P}'\mathbf{A}\mathbf{P}$ is nonnegative definite) that $\mathbf{P}'\mathbf{A}\mathbf{P}$ is positive semidefinite, thereby completing the proof of Part (2).

If $\mathbf{A}$ is positive definite, then equality is attained in inequality (2.1) only when $\mathbf{P}\mathbf{x} = \mathbf{0}$. Moreover, if $\text{rank}(\mathbf{P}) = m$, then (in light of Lemma 11.3.1) $\mathbf{P}\mathbf{x} = \mathbf{0}$

implies $\mathbf{x} = \mathbf{0}$. Thus, if $\mathbf{A}$ is positive definite and $\text{rank}(\mathbf{P}) = m$, then equality is attained in inequality (2.1) only when $\mathbf{x} = \mathbf{0}$, implying (since $\mathbf{P}'\mathbf{A}\mathbf{P}$ is nonnegative definite) that $\mathbf{P}'\mathbf{A}\mathbf{P}$ is positive definite. Q.E.D.

Corollary 14.2.10. Let $\mathbf{A}$ represent an $n \times n$ matrix and $\mathbf{P}$ an $n \times n$ nonsingular matrix. (1) If $\mathbf{A}$ is positive definite, then $\mathbf{P}'\mathbf{A}\mathbf{P}$ is positive definite. (2) If $\mathbf{A}$ is positive semidefinite, then $\mathbf{P}'\mathbf{A}\mathbf{P}$ is positive semidefinite.

Proof. (1) That $\mathbf{P}'\mathbf{A}\mathbf{P}$ is positive definite if $\mathbf{A}$ is positive definite is a direct consequence of Part (3) of Theorem 14.2.9.

(2) Suppose that $\mathbf{A}$ is positive semidefinite. Then, according to Part (1) of Theorem 14.2.9, $\mathbf{P}'\mathbf{A}\mathbf{P}$ is nonnegative definite. Further, there exists a nonnull vector $\mathbf{y}$ such that $\mathbf{y}'\mathbf{A}\mathbf{y} = 0$. Let $\mathbf{x} = \mathbf{P}^{-1}\mathbf{y}$. Then, $\mathbf{y} = \mathbf{P}\mathbf{x}$, and we find that $\mathbf{x} \neq \mathbf{0}$ (since, otherwise, we would have that $\mathbf{y} = \mathbf{0}$) and that

$$\mathbf{x}'(\mathbf{P}'\mathbf{A}\mathbf{P})\mathbf{x} = (\mathbf{P}\mathbf{x})'\mathbf{A}\mathbf{P}\mathbf{x} = \mathbf{y}'\mathbf{A}\mathbf{y} = 0.$$

We conclude that $\mathbf{P}'\mathbf{A}\mathbf{P}$ is positive semidefinite. Q.E.D.

Corollary 14.2.11. (1) A positive definite matrix is invertible, and its inverse is positive definite. (2) If a positive semidefinite matrix is nonsingular, then it is invertible and its inverse is positive semidefinite.

Proof. (1) Let $\mathbf{A}$ represent a positive definite matrix. Then, according to Lemma 14.2.8, $\mathbf{A}$ is nonsingular and hence (according to Theorem 8.1.4) invertible. Since $(\mathbf{A}^{-1})' = (\mathbf{A}^{-1})'\mathbf{A}\mathbf{A}^{-1}$, it follows from Part (1) of Corollary 14.2.10 [together with result (8.2.5)] that $(\mathbf{A}^{-1})'$ is positive definite and hence (in light of Corollary 14.2.7) that $\mathbf{A}^{-1}$ is positive definite.

(2) By employing similar reasoning, it can be shown that Part (2) of Corollary 14.2.11 follows from Part (2) of Corollary 14.2.10. Q.E.D.

Corollary 14.2.12. Any principal submatrix of a positive definite matrix is positive definite; any principal submatrix of a positive semidefinite matrix is nonnegative definite.

Proof. Let $\mathbf{A}$ represent an $n \times n$ matrix, and consider the principal submatrix of $\mathbf{A}$ obtained by striking out all of its rows and columns except its $i_1, i_2, \ldots, i_m$th rows and columns (where $i_1 < i_2 < \cdots < i_m$). This submatrix is expressible as $\mathbf{P}'\mathbf{A}\mathbf{P}$, where $\mathbf{P}$ is the $n \times m$ matrix whose columns are the $i_1, i_2, \ldots, i_m$th columns of $\mathbf{I}_n$. Since $\text{rank}(\mathbf{P}) = m$, it follows from Part (3) of Theorem 14.2.9 that $\mathbf{P}'\mathbf{A}\mathbf{P}$ is positive definite if $\mathbf{A}$ is positive definite. Further, it follows from Part (1) of Theorem 14.2.9 that $\mathbf{P}'\mathbf{A}\mathbf{P}$ is nonnegative definite if $\mathbf{A}$ is nonnegative definite (and, in particular, if $\mathbf{A}$ is positive semidefinite). Q.E.D.

Corollary 14.2.13. The diagonal elements of a positive definite matrix are positive; the diagonal elements of a positive semidefinite matrix are nonnegative.

Proof. This corollary follows immediately from Corollary 14.2.12 upon observing (1) that the ith diagonal element of a (square) matrix $\mathbf{A}$ is the element of a 1×1 principal submatrix (that obtained by striking out all of the rows and columns of $\mathbf{A}$ except the ith row and column) and (2) that the element of a 1×1 positive definite matrix is positive and the element of a 1×1 nonnegative definite matrix is nonnegative. Q.E.D.

Corollary 14.2.14. Let $\mathbf{P}$ represent an arbitrary $n \times m$ matrix. The $m \times m$ matrix $\mathbf{P}'\mathbf{P}$ is nonnegative definite. If $\text{rank}(\mathbf{P}) = m$, $\mathbf{P}'\mathbf{P}$ is positive definite; otherwise [if $\text{rank}(\mathbf{P}) < m$], $\mathbf{P}'\mathbf{P}$ is positive semidefinite.

Proof. This corollary follows from Theorem 14.2.9 upon observing that $\mathbf{P}'\mathbf{P} = \mathbf{P}'\mathbf{I}\mathbf{P}$ and that (as demonstrated in Subsection a) $\mathbf{I}$ is positive definite. Q.E.D.

Corollary 14.2.15. Let $\mathbf{P}$ represent an $n \times n$ nonsingular matrix and $\mathbf{D} = \{d_i\}$ an $n \times n$ diagonal matrix. Then, (1) $\mathbf{P}'\mathbf{D}\mathbf{P}$ is nonnegative definite if and only if $d_1, \ldots, d_n$ are nonnegative; (2) $\mathbf{P}'\mathbf{D}\mathbf{P}$ is positive definite if and only if $d_1, \ldots, d_n$ are positive; and (3) $\mathbf{P}'\mathbf{D}\mathbf{P}$ is positive semidefinite if and only if $d_i \geq 0$ for $i = 1, \ldots, n$ with equality holding for one or more values of i.

Proof. Let $\mathbf{A} = \mathbf{P}'\mathbf{D}\mathbf{P}$. Then, clearly, $\mathbf{D} = (\mathbf{P}^{-1})'\mathbf{A}\mathbf{P}^{-1}$. Thus, it follows from Part (1) of Theorem 14.2.9 that $\mathbf{A}$ is nonnegative definite if and only if $\mathbf{D}$ is nonnegative definite; and it follows, respectively, from Parts (1) and (2) of Corollary 14.2.10 that $\mathbf{A}$ is positive definite if and only if $\mathbf{D}$ is positive definite and that $\mathbf{A}$ is positive semidefinite if and only if $\mathbf{D}$ is positive semidefinite. In light of Lemma 14.2.1, the proof is complete. Q.E.D.

Corollary 14.2.16. Let $\mathbf{A}$ represent an $n \times n$ symmetric matrix and $\mathbf{D} = \{d_i\}$ an $n \times n$ diagonal matrix such that $\mathbf{P}'\mathbf{A}\mathbf{P} = \mathbf{D}$ for some $n \times n$ nonsingular matrix $\mathbf{P}$. Then, (1) $\mathbf{A}$ is nonnegative definite if and only if $d_1, \ldots, d_n$ are nonnegative; (2) $\mathbf{A}$ is positive definite if and only if $d_1, \ldots, d_n$ are positive; and (3) $\mathbf{A}$ is positive semidefinite if and only if $d_i \geq 0$ for $i = 1, \ldots, n$ with equality holding for one or more values of i.

Proof. Corollary 14.2.16 follows from Corollary 14.2.15 upon observing that $\mathbf{A} = (\mathbf{P}^{-1})'\mathbf{D}\mathbf{P}^{-1}$. Q.E.D.

Note that Corollaries 14.2.15 and 14.2.16 can be regarded as generalizations of Lemma 14.2.1.

c. The nonnegative definiteness of symmetric idempotent matrices

Suppose that $\mathbf{A}$ is a symmetric idempotent matrix. Then, $\mathbf{A} = \mathbf{A}\mathbf{A} = \mathbf{A}'\mathbf{A}$, and it follows from Corollary 14.2.14 that $\mathbf{A}$ is nonnegative definite. Thus, we have the following lemma.

Lemma 14.2.17. Every symmetric idempotent matrix is nonnegative definite.

Note (in light of Lemmas 14.2.8 and 10.1.1) that the $n \times n$ identity matrix $\mathbf{I}_n$ is the only $n \times n$ idempotent matrix (symmetric or otherwise) that is positive definite.

14.3 Decomposition of Symmetric and Symmetric Nonnegative Definite Matrices

According to Corollary 14.2.14, every matrix $\mathbf{A}$ that is expressible in the form $\mathbf{A} = \mathbf{P}'\mathbf{P}$ is a (symmetric) nonnegative definite matrix. Is the converse true? That is, is every symmetric nonnegative definite matrix $\mathbf{A}$ expressible in the form $\mathbf{A} = \mathbf{P}'\mathbf{P}$? In Subsection b, it is established that the answer to this question is yes.

As a preliminary, consideration is given to a closely related question, which is of some interest in its own right. Is a symmetric matrix $\mathbf{A}$ expressible in the form $\mathbf{A} = \mathbf{P}'\mathbf{D}\mathbf{P}$, where $\mathbf{P}$ is a nonsingular matrix and $\mathbf{D}$ a diagonal matrix? In Subsection a, it is established that the answer to this question is also yes.

In what follows, it will be convenient to have at our disposal the following lemma.

Lemma 14.3.1. Let $\mathbf{A}$ represent an $n \times n$ matrix and $\mathbf{D}$ an $n \times n$ diagonal matrix such that $\mathbf{A} = \mathbf{PDQ}$ for some $n \times n$ nonsingular matrices $\mathbf{P}$ and $\mathbf{Q}$. Then, $\text{rank}(\mathbf{A})$ equals the number of nonzero diagonal elements in $\mathbf{D}$.

Proof. Making use of Corollary 8.3.3, we find that

$$\text{rank}(\mathbf{A}) = \text{rank}(\mathbf{PDQ}) = \text{rank}(\mathbf{DQ}) = \text{rank}(\mathbf{D}).$$

Moreover, $\text{rank}(\mathbf{D})$ equals the number of nonzero diagonal elements in $\mathbf{D}$. Q.E.D.

a. Decomposition of symmetric matrices

Let us begin by establishing the following two lemmas.

Lemma 14.3.2. Let $\mathbf{A} = \{a_{ij}\}$ represent an $n \times n$ symmetric matrix (where $n \geq 2$). Define $\mathbf{B} = \{b_{ij}\} = \mathbf{L}'\mathbf{A}\mathbf{L}$, where $\mathbf{L}$ is an $n \times n$ unit lower triangular matrix of the form $\mathbf{L} = \begin{pmatrix} 1 & \mathbf{0} \\ \boldsymbol{\ell} & \mathbf{I}_{n-1} \end{pmatrix}$. Partition $\mathbf{A}$ as $\mathbf{A} = \begin{pmatrix} a_{11} & \mathbf{a}' \\ \mathbf{a} & \mathbf{A}_{22} \end{pmatrix}$. Suppose that $a_{11} = 0$, but that $\mathbf{a} \neq \mathbf{0}$ or, equivalently, that $a_{1j} \neq 0$ for some j (greater than 1), say $j = k$. Then, the $(n-1)$-dimensional vector $\boldsymbol{\ell}$ can be chosen so that $b_{11} \neq 0$; this can be done by taking the $(k-1)$th element of $\boldsymbol{\ell}$ to be any nonzero scalar c such that $ca_{kk} \neq -2a_{1k}$ and by taking the other $n-2$ elements of $\boldsymbol{\ell}$ to be zero.

Proof. Since $a_{1k} \neq 0$, there exists (whether or not $a_{kk} = 0$) a nonzero scalar c such that $ca_{kk} \neq -2a_{1k}$. Taking the $(k-1)$th element of $\boldsymbol{\ell}$ to be any such nonzero scalar c and the remaining elements of $\boldsymbol{\ell}$ to be zero, we find that

$$\begin{aligned} b_{11} &= a_{11} + \boldsymbol{\ell}'\mathbf{a} + \mathbf{a}'\boldsymbol{\ell} + \boldsymbol{\ell}'\mathbf{A}_{22}\boldsymbol{\ell} \\ &= 0 + ca_{k1} + ca_{1k} + c^2 a_{kk} \\ &= c(ca_{kk} + 2a_{1k}) \neq 0. \end{aligned}$$

Q.E.D.

Lemma 14.3.3. Let $\mathbf{A} = \{a_{ij}\}$ represent an $n \times n$ symmetric matrix (where $n \geq 2$). Define $\mathbf{B} = \{b_{ij}\} = \mathbf{U}'\mathbf{A}\mathbf{U}$, where $\mathbf{U}$ is an $n \times n$ unit upper triangular matrix of the form $\mathbf{U} = \begin{pmatrix} 1 & \mathbf{u}' \\ \mathbf{0} & \mathbf{I}_{n-1} \end{pmatrix}$. Partition $\mathbf{B}$ and $\mathbf{A}$ as $\mathbf{B} = \begin{pmatrix} b_{11} & \mathbf{b}' \\ \mathbf{b} & \mathbf{B}_{22} \end{pmatrix}$ and $\mathbf{A} = \begin{pmatrix} a_{11} & \mathbf{a}' \\ \mathbf{a} & \mathbf{A}_{22} \end{pmatrix}$. Suppose that $a_{11} \neq 0$. Then, the $(n-1)$-dimensional vector $\mathbf{u}$ can be chosen so that $\mathbf{b} = \mathbf{0}$; this can be done by taking $\mathbf{u} = -a_{11}^{-1}\mathbf{a}$.

Proof. We find that

$$\mathbf{b} = (\mathbf{u},\ \mathbf{I})\mathbf{A}\begin{pmatrix} 1 \\ \mathbf{0} \end{pmatrix} = a_{11}\mathbf{u} + \mathbf{a},$$

which, for $\mathbf{u} = -a_{11}^{-1}\mathbf{a}$, gives $\mathbf{b} = \mathbf{a} - \mathbf{a} = \mathbf{0}$. Q.E.D.

We are now in a position to establish the following theorem.

Theorem 14.3.4. Corresponding to any $n \times n$ symmetric matrix $\mathbf{A}$, there exists a nonsingular matrix $\mathbf{Q}$ such that $\mathbf{Q}'\mathbf{A}\mathbf{Q}$ is a diagonal matrix.

Proof. The proof is by mathematical induction. The theorem is clearly true for any 1×1 matrix. Suppose now that it is true for any $(n-1) \times (n-1)$ symmetric matrix, and consider an arbitrary $n \times n$ symmetric matrix $\mathbf{A} = \{a_{ij}\}$. For purposes of establishing the existence of a nonsingular matrix $\mathbf{Q}$ such that $\mathbf{Q}'\mathbf{A}\mathbf{Q}$ is diagonal, it is convenient to partition $\mathbf{A}$ as $\mathbf{A} = \begin{pmatrix} a_{11} & \mathbf{a}' \\ \mathbf{a} & \mathbf{A}_{22} \end{pmatrix}$ and to proceed case-by-case.

Case (1): $\mathbf{a} = \mathbf{0}$. The submatrix $\mathbf{A}_{22}$ is an $(n-1) \times (n-1)$ symmetric matrix. Thus, by supposition, there exists a nonsingular matrix $\mathbf{Q}_*$ such that $\mathbf{Q}_*'\mathbf{A}_{22}\mathbf{Q}_*$ is a diagonal matrix. Take $\mathbf{Q} = \mathrm{diag}(1,\ \mathbf{Q}_*)$. Then, $\mathbf{Q}$ is nonsingular, and $\mathbf{Q}'\mathbf{A}\mathbf{Q} = \mathrm{diag}(a_{11},\ \mathbf{Q}_*'\mathbf{A}_{22}\mathbf{Q}_*)$, which (like $\mathbf{Q}_*'\mathbf{A}_{22}\mathbf{Q}_*$) is a diagonal matrix.

Case (2): $\mathbf{a} \neq \mathbf{0}$ *and* $a_{11} \neq 0$. According to Lemma 14.3.3, there exists a unit upper triangular matrix $\mathbf{U}$ such that $\mathbf{U}'\mathbf{A}\mathbf{U} = \mathrm{diag}(b_{11},\ \mathbf{B}_{22})$ for some scalar b_{11} and some $(n-1) \times (n-1)$ matrix $\mathbf{B}_{22}$. Moreover, by supposition, there exists a nonsingular matrix $\mathbf{Q}_*$ such that $\mathbf{Q}_*'\mathbf{B}_{22}\mathbf{Q}_*$ is a diagonal matrix. Take $\mathbf{Q} = \mathbf{U}\,\mathrm{diag}(1,\ \mathbf{Q}_*)$. Then, $\mathbf{Q}$ is nonsingular (since $\mathbf{Q}$ is the product of two nonsingular matrices), and

$$\mathbf{Q}'\mathbf{A}\mathbf{Q} = \mathrm{diag}(1,\ \mathbf{Q}_*')\,\mathrm{diag}(b_{11},\ \mathbf{B}_{22})\,\mathrm{diag}(1,\ \mathbf{Q}_*) = \mathrm{diag}(b_{11},\ \mathbf{Q}_*'\mathbf{B}_{22}\mathbf{Q}_*),$$

which (like $\mathbf{Q}_*'\mathbf{B}_{22}\mathbf{Q}_*$) is a diagonal matrix.

Case (3): $\mathbf{a} \neq \mathbf{0}$, *but* $a_{11} = 0$. Let $\mathbf{B} = \{b_{ij}\} = \mathbf{L}'\mathbf{A}\mathbf{L}$, where $\mathbf{L}$ is a unit lower triangular matrix chosen so that $b_{11} \neq 0$—the existence of such a choice is guaranteed by Lemma 14.3.2. Then, according to Lemma 14.3.3, there exists a unit upper triangular matrix $\mathbf{U}$ such that $\mathbf{U}'\mathbf{B}\mathbf{U} = \mathrm{diag}(c_{11},\ \mathbf{C}_{22})$ for some scalar c_{11} and some $(n-1) \times (n-1)$ matrix $\mathbf{C}_{22}$. Moreover, by supposition, there exists a nonsingular matrix $\mathbf{Q}_*$ such that $\mathbf{Q}_*'\mathbf{C}_{22}\mathbf{Q}_*$ is a diagonal matrix. Take $\mathbf{Q} = \mathbf{L}\mathbf{U}\,\mathrm{diag}(1,\ \mathbf{Q}_*)$. Then, $\mathbf{Q}$ is nonsingular (since $\mathbf{Q}$ is the product of three nonsingular matrices), and

$$\begin{aligned}\mathbf{Q}'\mathbf{A}\mathbf{Q} &= \text{diag}(1,\ \mathbf{Q}_*')\mathbf{U}'\mathbf{B}\mathbf{U}\ \text{diag}(1,\ \mathbf{Q}_*)\\ &= \text{diag}(1,\ \mathbf{Q}_*')\ \text{diag}(c_{11},\ \mathbf{C}_{22})\ \text{diag}(1,\ \mathbf{Q}_*) = \text{diag}(c_{11},\ \mathbf{Q}_*'\mathbf{C}_{22}\mathbf{Q}_*),\end{aligned}$$

which (like $\mathbf{Q}_*'\mathbf{C}_{22}\mathbf{Q}_*$) is a diagonal matrix. Q.E.D.

Note that if $\mathbf{Q}$ is a nonsingular matrix such that $\mathbf{Q}'\mathbf{A}\mathbf{Q} = \mathbf{D}$ for some diagonal matrix $\mathbf{D}$, then

$$\mathbf{A} = (\mathbf{Q}^{-1})'\mathbf{Q}'\mathbf{A}\mathbf{Q}\mathbf{Q}^{-1} = (\mathbf{Q}^{-1})'\mathbf{D}\mathbf{Q}^{-1}.$$

Thus, we have the following corollary of Theorem 14.3.4.

Corollary 14.3.5. Corresponding to any $n \times n$ symmetric matrix $\mathbf{A}$, there exist a nonsingular matrix $\mathbf{P}$ and a diagonal matrix $\mathbf{D}$ such that $\mathbf{A} = \mathbf{P}'\mathbf{D}\mathbf{P}$.

Corollary 14.3.5 leads to the following result on quadratic forms.

Corollary 14.3.6. Let $\mathbf{A}$ represent an $n \times n$ matrix. Then, there exist a nonsingular matrix $\mathbf{P}$ and n scalars $d_1, \ldots, d_n$ such that the quadratic form $\mathbf{x}'\mathbf{A}\mathbf{x}$ (in an n-dimensional vector $\mathbf{x}$) is expressible as a linear combination $\sum_{i=1}^{n} d_i y_i^2$ of the squares of the elements $y_1, \ldots, y_n$ of the transformed vector $\mathbf{y} = \mathbf{P}\mathbf{x}$.

Proof. According to Corollary 14.1.2, there is a unique symmetric matrix $\mathbf{B}$ such that $\mathbf{x}'\mathbf{A}\mathbf{x} = \mathbf{x}'\mathbf{B}\mathbf{x}$ for all $\mathbf{x}$, namely, the matrix $\mathbf{B} = (1/2)(\mathbf{A} + \mathbf{A}')$. And, according to Corollary 14.3.5, there exist a nonsingular matrix $\mathbf{P}$ and a diagonal matrix $\mathbf{D}$ such that $\mathbf{B} = \mathbf{P}'\mathbf{D}\mathbf{P}$. Thus, letting $d_1, \ldots, d_n$ represent the diagonal elements of $\mathbf{D}$ and $y_1, \ldots, y_n$ the elements of the vector $\mathbf{y} = \mathbf{P}\mathbf{x}$, we find that

$$\mathbf{x}'\mathbf{A}\mathbf{x} = \mathbf{x}'\mathbf{P}'\mathbf{D}\mathbf{P}\mathbf{x} = (\mathbf{P}\mathbf{x})'\mathbf{D}\mathbf{P}\mathbf{x} = \sum_i d_i y_i^2.$$

Q.E.D.

b. Decomposition of symmetric nonnegative definite matrices

Let us now specialize to symmetric nonnegative definite matrices, beginning with the following theorem.

Theorem 14.3.7. An $n \times n$ (nonnull) matrix $\mathbf{A}$ is a symmetric nonnegative definite matrix of rank r if and only if there exists an $r \times n$ matrix $\mathbf{P}$ of rank r such that $\mathbf{A} = \mathbf{P}'\mathbf{P}$.

Proof. Suppose that $\mathbf{A}$ is a symmetric nonnegative definite matrix of rank r. Then, according to Corollary 14.3.5, there exist a nonsingular matrix $\mathbf{T}$ and a diagonal matrix $\mathbf{D}$ such that $\mathbf{A} = \mathbf{T}'\mathbf{D}\mathbf{T}$.

Let $d_1, \ldots, d_n$ represent the first, …, nth diagonal elements of $\mathbf{D}$, and $\mathbf{t}_1', \ldots, \mathbf{t}_n'$ the first, …, nth rows of $\mathbf{T}$. It follows from Lemma 14.3.1 that r of the diagonal elements of $\mathbf{D}$, say the $k_1, \ldots, k_r$th diagonal elements, are nonzero (and hence, in light of Corollary 14.2.15, are positive). Now, take $\mathbf{P}$ to be the $r \times n$ matrix whose ith row is $\sqrt{d_{k_i}}\,\mathbf{t}_{k_i}'$. Then,

$$\mathbf{A} = \sum_{k=1}^{n} d_k \mathbf{t}_k \mathbf{t}_k' = \sum_{i=1}^{r} d_{k_i} \mathbf{t}_{k_i} \mathbf{t}_{k_i}' = \sum_{i=1}^{r} \sqrt{d_{k_i}}\,\mathbf{t}_{k_i} \left(\sqrt{d_{k_i}}\,\mathbf{t}_{k_i}'\right) = \mathbf{P}'\mathbf{P},$$

and $\text{rank}(\mathbf{P}) = \text{rank}(\mathbf{P}'\mathbf{P}) = \text{rank}(\mathbf{A}) = r$.

Conversely, if there exists an $r \times n$ matrix $\mathbf{P}$ of rank r such that $\mathbf{A} = \mathbf{P}'\mathbf{P}$, then

$$\text{rank}(\mathbf{A}) = \text{rank}(\mathbf{P}'\mathbf{P}) = \text{rank}(\mathbf{P}) = r,$$

$\mathbf{A}$ is symmetric, and (according to Corollary 14.2.14) $\mathbf{A}$ is nonnegative definite. Q.E.D.

In light of Corollary 14.2.14 (and upon observing tht $\mathbf{0} = \mathbf{0}'\mathbf{0}$), we have the following corollary of Theorem 14.3.7.

Corollary 14.3.8. An $n \times n$ matrix $\mathbf{A}$ is a symmetric nonnegative definite matrix if and only if there exists a matrix $\mathbf{P}$ (having n columns) such that $\mathbf{A} = \mathbf{P}'\mathbf{P}$.

As a further corollary of Theorem 14.3.7, we have the following result on quadratic forms.

Corollary 14.3.9. Let $\mathbf{A}$ represent an $n \times n$ matrix, and let $r = \text{rank}(\mathbf{A} + \mathbf{A}')$. Then, assuming that $r > 0$, the quadratic form $\mathbf{x}'\mathbf{A}\mathbf{x}$ (in an n-dimensional vector $\mathbf{x}$) is nonnegative definite if and only if, for some $r \times n$ matrix $\mathbf{P}$ of rank r, $\mathbf{x}'\mathbf{A}\mathbf{x} = (\mathbf{P}\mathbf{x})'\mathbf{P}\mathbf{x}$ for all $\mathbf{x}$ (i.e., $\mathbf{x}'\mathbf{A}\mathbf{x}$ is expressible as the sum of squares of the elements of the vector $\mathbf{P}\mathbf{x}$).

Proof. According to Corollary 14.1.2, there is a unique symmetric matrix $\mathbf{B}$ such that $\mathbf{x}'\mathbf{A}\mathbf{x} = \mathbf{x}'\mathbf{B}\mathbf{x}$ for all $\mathbf{x}$, namely, the matrix $\mathbf{B} = (1/2)(\mathbf{A} + \mathbf{A}')$. Suppose that the quadratic form $\mathbf{x}'\mathbf{A}\mathbf{x}$, or equivalently the quadratic form $\mathbf{x}'\mathbf{B}\mathbf{x}$, is nonnegative definite, in which case $\mathbf{B}$ is a nonnegative definite matrix. Then, it follows from Theorem 14.3.7 that there exists an $r \times n$ matrix $\mathbf{P}$ of rank r such that $\mathbf{B} = \mathbf{P}'\mathbf{P}$ and hence such that $\mathbf{x}'\mathbf{B}\mathbf{x} = (\mathbf{P}\mathbf{x})'\mathbf{P}\mathbf{x}$ for all $\mathbf{x}$ or, equivalently, such that $\mathbf{x}'\mathbf{A}\mathbf{x} = (\mathbf{P}\mathbf{x})'\mathbf{P}\mathbf{x}$ for all $\mathbf{x}$.

Conversely, suppose that, for some $r \times n$ matrix $\mathbf{P}$ of rank r, $\mathbf{x}'\mathbf{A}\mathbf{x} = (\mathbf{P}\mathbf{x})'\mathbf{P}\mathbf{x}$ for all $\mathbf{x}$ or, equivalently, $\mathbf{x}'\mathbf{A}\mathbf{x} = \mathbf{x}'\mathbf{P}'\mathbf{P}\mathbf{x}$ for all $\mathbf{x}$. Since (according to Theorem 14.3.7 or, alternatively, according to Corollary 14.2.14) $\mathbf{P}'\mathbf{P}$ is a nonnegative definite matrix, $\mathbf{x}'\mathbf{P}'\mathbf{P}\mathbf{x}$, or equivalently $\mathbf{x}'\mathbf{A}\mathbf{x}$, is a nonnegative definite quadratic form. Q.E.D.

Further implications of Theorem 14.3.7 are given in Corollaries 14.3.10 to 14.3.12.

Corollary 14.3.10. Let $\mathbf{A}$ represent an $n \times n$ matrix, let $r = \text{rank}(\mathbf{A})$, and take m to be any positive integer greater than or equal to r. If $\mathbf{A}$ is symmetric and nonnegative definite, then there exists an $m \times n$ matrix $\mathbf{P}$ such that $\mathbf{A} = \mathbf{P}'\mathbf{P}$.

Proof. Suppose that $\mathbf{A}$ is symmetric and nonnegative definite. And assume that $r > 0$. (When $r = 0$, $\mathbf{A} = \mathbf{0}'\mathbf{0}$.) According to Theorem 14.3.7, there exists an $r \times n$ matrix $\mathbf{P}_1$ such that $\mathbf{A} = \mathbf{P}_1'\mathbf{P}_1$. Take $\mathbf{P} = \begin{pmatrix} \mathbf{P}_1 \\ \mathbf{0} \end{pmatrix}$. Then, clearly, $\mathbf{A} = \mathbf{P}'\mathbf{P}$. Q.E.D.

Corollary 14.3.11. For any $n \times m$ matrix $\mathbf{X}$ and any $n \times n$ symmetric nonnegative definite matrix $\mathbf{A}$, $\mathbf{A}\mathbf{X} = \mathbf{0}$ if and only if $\mathbf{X}'\mathbf{A}\mathbf{X} = \mathbf{0}$.

Proof. According to Corollary 14.3.8, there exists a matrix $\mathbf{P}$ such that $\mathbf{A} = \mathbf{P}'\mathbf{P}$ and hence such that $\mathbf{X}'\mathbf{A}\mathbf{X} = (\mathbf{P}\mathbf{X})'\mathbf{P}\mathbf{X}$. Thus, if $\mathbf{X}'\mathbf{A}\mathbf{X} = \mathbf{0}$, then (in light of Corollary 5.3.2) $\mathbf{P}\mathbf{X} = \mathbf{0}$, implying that $\mathbf{A}\mathbf{X} = \mathbf{P}'(\mathbf{P}\mathbf{X}) = \mathbf{0}$.

That $\mathbf{X}'\mathbf{A}\mathbf{X} = \mathbf{0}$ if $\mathbf{A}\mathbf{X} = \mathbf{0}$ is obvious. Q.E.D.

Corollary 14.3.12. A symmetric nonnegative definite matrix is positive definite if and only if it is nonsingular (or, equivalently, is positive semidefinite if and only if it is singular).

Proof. Let $\mathbf{A}$ represent an $n \times n$ symmetric nonnegative definite matrix. If $\mathbf{A}$ is positive definite, then we have, as an immediate consequence of Lemma 14.2.8, that $\mathbf{A}$ is nonsingular.

Suppose now that the symmetric nonnegative definite matrix $\mathbf{A}$ is nonsingular, and consider the quadratic form $\mathbf{x}'\mathbf{A}\mathbf{x}$ (in $\mathbf{x}$). If $\mathbf{x}'\mathbf{A}\mathbf{x} = 0$, then, according to Corollary 14.3.11, $\mathbf{A}\mathbf{x} = \mathbf{0}$, and consequently $\mathbf{x} = \mathbf{A}^{-1}\mathbf{A}\mathbf{x} = \mathbf{0}$. Thus, the quadratic form $\mathbf{x}'\mathbf{A}\mathbf{x}$ is positive definite and hence the matrix $\mathbf{A}$ is positive definite. Q.E.D.

As a special case of Theorem 14.3.7 (that where $r = n$), we have (in light of Corollary 14.3.12) the following corollary.

Corollary 14.3.13. An $n \times n$ matrix $\mathbf{A}$ is a symmetric positive definite matrix if and only if there exists a nonsingular matrix $\mathbf{P}$ such that $\mathbf{A} = \mathbf{P}'\mathbf{P}$.

Corollary 14.3.13 is a variant of Theorem 14.3.7 that is specific to symmetric positive definite matrices. Similarly, the following corollary is a variant of Corollary 14.3.9 that is specific to positive definite quadratic forms.

Corollary 14.3.14. A quadratic form $\mathbf{x}'\mathbf{A}\mathbf{x}$ (in an n-dimensional vector $\mathbf{x}$) is positive definite if and only if, for some $n \times n$ nonsingular matrix $\mathbf{P}$, $\mathbf{x}'\mathbf{A}\mathbf{x} = (\mathbf{P}\mathbf{x})'\mathbf{P}\mathbf{x}$ for all $\mathbf{x}$ (i.e., $\mathbf{x}'\mathbf{A}\mathbf{x}$ is expressible as the sum of squares of the elements of the vector $\mathbf{P}\mathbf{x}$).

Proof. If $\mathbf{x}'\mathbf{A}\mathbf{x}$ is positive definite, then (according to Corollary 14.2.7) $(1/2)(\mathbf{A} + \mathbf{A}')$ is positive definite and hence (in light of Lemma 14.2.8) is of rank n, so that it follows from Corollary 14.3.9 that, for some $n \times n$ nonsingular matrix $\mathbf{P}$, $\mathbf{x}'\mathbf{A}\mathbf{x} = (\mathbf{P}\mathbf{x})'\mathbf{P}\mathbf{x}$ for all $\mathbf{x}$. Conversely, if for some $n \times n$ nonsingular matrix $\mathbf{P}$, $\mathbf{x}'\mathbf{A}\mathbf{x} = (\mathbf{P}\mathbf{x})'\mathbf{P}\mathbf{x}$ for all $\mathbf{x}$, then, since (according to Corollary 14.2.14 or Corollary 14.3.13) $\mathbf{P}'\mathbf{P}$ is positive definite, $\mathbf{x}'\mathbf{A}\mathbf{x}$ (which equals $\mathbf{x}'\mathbf{P}'\mathbf{P}\mathbf{x}$) is positive definite. Q.E.D.

14.4 Generalized Inverses of Symmetric Nonnegative Definite Matrices

A symmetric positive definite matrix is invertible, and its inverse is positive definite and symmetric [Corollary 14.2.11 and result (8.2.4)]. A symmetric positive semidefinite matrix is not invertible (Corollary 14.3.12). What is the nature of a generalized inverse of a symmetric positive semidefinite matrix? In particular, are some or all of the generalized inverses of a symmetric positive semidefinite matrix nonnegative definite?

Let **A** represent an $n \times n$ matrix of rank r, and suppose that

$$\mathbf{A} = \mathbf{PDQ}, \tag{4.1}$$

where **P** and **Q** are $(n \times n)$ nonsingular matrices and $\mathbf{D} = \{d_i\}$ is an $n \times n$ diagonal matrix. According to Lemma 14.3.1, r of the diagonal elements of **D**, say the $i_1, \ldots, i_r$th diagonal elements, are nonzero and the other $n - r$ diagonal elements of **D** equal zero. Let $S = \{i_1, i_2, \ldots, i_r\}$ and denote by $\bar{S}$ the set whose elements consist of those $n - r$ of the first n positive integers $1, 2, \ldots, n$ not contained in S.

Let $\mathbf{D}^*$ represent an $n \times n$ matrix of the general form

$$\mathbf{D}^* = \text{diag}(d_1^*, d_2^*, \ldots, d_n^*),$$

where, for $i \in S$, $d_i^* = 1/d_i$ and, for $i \in \bar{S}$, d_i^* is an arbitrary scalar. Note that $\mathbf{D}^*$ is a generalized inverse of **D**.

Define $\mathbf{G} = \mathbf{Q}^{-1}\mathbf{D}^-\mathbf{P}^{-1}$ and $\mathbf{G}^* = \mathbf{Q}^{-1}\mathbf{D}^*\mathbf{P}^{-1}$. According to Lemma 9.2.4, **G** is a generalized inverse of **A**. In particular, $\mathbf{G}^*$ is a generalized inverse of **A**.

Suppose now that **A** is symmetric and that decomposition (4.1) is such that $\mathbf{P} = \mathbf{Q}'$, so that decomposition (4.1) is of the form

$$\mathbf{A} = \mathbf{Q}'\mathbf{DQ}. \tag{4.2}$$

Then, in light of result (8.2.3),

$$\mathbf{G}^* = \mathbf{Q}^{-1}\mathbf{D}^*(\mathbf{Q}^{-1})'. \tag{4.3}$$

Thus, $\mathbf{G}^*$ is a symmetric generalized inverse of **A**. Moreover, if d_i^* is chosen to be nonzero for every i in $\bar{S}$, then $\mathbf{G}^*$ is a (symmetric) nonsingular generalized inverse of **A**.

Next, suppose that **A** is nonnegative definite, as well as symmetric—and continue to suppose that $\mathbf{P} = \mathbf{Q}'$ and hence that decomposition (4.1) is of the form (4.2). Then, it follows from Corollary 14.2.15 that $d_i^* > 0$ for every i in S. Thus, if d_i^* is chosen to be greater than zero for every i in $\bar{S}$, then (in light of Corollary 14.2.15) $\mathbf{G}^*$ is a symmetric positive definite generalized inverse of **A**. (If d_i is chosen to be greater than or equal to zero for all i in $\bar{S}$ and equal to zero for at least one i in $\bar{S}$, then $\mathbf{G}^*$ is positive semidefinite; if d_i is chosen to be less than zero for at least one i in $\bar{S}$, then—assuming that **A** is nonnull—$\mathbf{G}^*$ is indefinite.) Since (according to Corollary 14.3.5) every symmetric matrix has a decomposition of the form (4.2), we have the following lemma.

Lemma 14.4.1. Every symmetric nonnegative definite matrix has a symmetric positive definite generalized inverse.

14.5 LDU, U′DU, and Cholesky Decompositions

It was established in Section 14.3 that every symmetric nonnegative definite matrix **A** has a decomposition of the form $\mathbf{A} = \mathbf{P}'\mathbf{P}$. Stronger versions of this result are possible. In fact, it is shown in the present section that every symmetric nonnegative definite matrix $\mathbf{A}$ has a decomposition of the form $\mathbf{A} = \mathbf{T}'\mathbf{T}$, where $\mathbf{T}$ is an upper triangular matrix. [And, in a subsequent chapter (Section 21.9), it is shown that $\mathbf{A}$ has a (unique) decomposition of the form $\mathbf{A} = \mathbf{R}^2$, where $\mathbf{R}$ is a symmetric nonnegative definite matrix.]

It is convenient to begin by considering some general questions about the decomposition of (not necessarily nonnegative definite) matrices. Is a square matrix $\mathbf{A}$ expressible in the form $\mathbf{A} = \mathbf{LDU}$, where $\mathbf{L}$ is a unit lower triangular matrix, $\mathbf{D}$ is a diagonal matrix, and $\mathbf{U}$ is a unit upper triangular matrix? And, more specifically, is a symmetric matrix $\mathbf{A}$ expressible in the form $\mathbf{A} = \mathbf{U}'\mathbf{DU}$? Subsequently, the term *LDU decomposition* is used for any decomposition of the form $\mathbf{A} = \mathbf{LDU}$ and the term *U′DU decomposition* for any decomposition of the form $\mathbf{A} = \mathbf{U}'\mathbf{DU}$.

a. Some preliminary results

The following lemma relates the LDU decomposition of an $n \times n$ matrix $\mathbf{A}$ to that of the leading $(n-1) \times (n-1)$ principal submatrix of $\mathbf{A}$.

Lemma 14.5.1. Let $\mathbf{A}$ represent an $n \times n$ matrix, $\mathbf{L}$ an $n \times n$ unit lower triangular matrix, $\mathbf{U}$ an $n \times n$ unit upper triangular matrix, and $\mathbf{D}$ an $n \times n$ diagonal matrix (where $n \geq 2$). Partition $\mathbf{A}$, $\mathbf{L}$, $\mathbf{U}$, and $\mathbf{D}$ as

$$\mathbf{A} = \begin{pmatrix} \mathbf{A}_* & \mathbf{a} \\ \mathbf{b}' & c \end{pmatrix}, \quad \mathbf{L} = \begin{pmatrix} \mathbf{L}_* & \mathbf{0} \\ \boldsymbol{\ell}' & 1 \end{pmatrix}, \quad \mathbf{U} = \begin{pmatrix} \mathbf{U}_* & \mathbf{u} \\ \mathbf{0} & 1 \end{pmatrix}, \text{ and } \mathbf{D} = \begin{pmatrix} \mathbf{D}_* & \mathbf{0} \\ \mathbf{0} & k \end{pmatrix},$$

where $\mathbf{A}_*$, $\mathbf{L}_*$, $\mathbf{U}_*$, and $\mathbf{D}_*$ are of dimensions $(n-1) \times (n-1)$. Then, $\mathbf{A} = \mathbf{LDU}$ if and only if

$$\mathbf{L}_*\mathbf{D}_*\mathbf{U}_* = \mathbf{A}_*, \tag{5.1a}$$

$$\mathbf{L}_*\mathbf{D}_*\mathbf{u} = \mathbf{a}, \tag{5.1b}$$

$$\mathbf{U}_*'\mathbf{D}_*\boldsymbol{\ell} = \mathbf{b}, \tag{5.1c}$$

and

$$k = c - \boldsymbol{\ell}'\mathbf{D}_*\mathbf{u}. \tag{5.1d}$$

Lemma 14.5.1 can be verified simply by equating the four submatrices $\mathbf{A}_*$, $\mathbf{a}$, $\mathbf{b}'$, and c to the corresponding submatrices of the matrix product $\mathbf{LDU}$. With regard to condition (5.1a), note that $\mathbf{L}_*$ is a principal submatrix of a unit lower triangular matrix and hence is itself unit lower triangular, and similarly that $\mathbf{U}_*$ is unit upper triangular and $\mathbf{D}_*$ is diagonal. Thus, condition (5.1a) states that $\mathbf{L}_*\mathbf{D}_*\mathbf{U}_*$ is an LDU decomposition of $\mathbf{A}_*$.

The following theorem relates the existence and construction of an LDU or U′DU decomposition of an $n \times n$ matrix to the existence and construction of an LDU or U′DU decomposition of the leading $(n-1) \times (n-1)$ principal submatrix of $\mathbf{A}$.

Theorem 14.5.2. Let $\mathbf{A}$ represent an $n \times n$ matrix (where $n \geq 2$), and partition $\mathbf{A}$ as

$$\mathbf{A} = \begin{pmatrix} \mathbf{A}_* & \mathbf{a} \\ \mathbf{b}' & c \end{pmatrix},$$

where $\mathbf{A}_*$ is of dimensions $(n-1) \times (n-1)$.

(1) If $\mathbf{A}_*$ has an LDU decomposition, say $\mathbf{A}_* = \mathbf{L}_*\mathbf{D}_*\mathbf{U}_*$, and if $\mathbf{a} \in \mathcal{C}(\mathbf{A}_*)$ and $\mathbf{b}' \in \mathcal{R}(\mathbf{A}_*)$, then there exist vectors $\boldsymbol{\ell}$ and $\mathbf{u}$ such that $\mathbf{U}_*'\mathbf{D}_*\boldsymbol{\ell} = \mathbf{b}$ and $\mathbf{L}_*\mathbf{D}_*\mathbf{u} = \mathbf{a}$, and, taking $\boldsymbol{\ell}$ and $\mathbf{u}$ to be any such vectors and taking $k = c - \boldsymbol{\ell}'\mathbf{D}_*\mathbf{u}$, an LDU decomposition of $\mathbf{A}$ is

$$\mathbf{A} = \begin{pmatrix} \mathbf{L}_* & \mathbf{0} \\ \boldsymbol{\ell}' & 1 \end{pmatrix} \begin{pmatrix} \mathbf{D}_* & \mathbf{0} \\ \mathbf{0} & k \end{pmatrix} \begin{pmatrix} \mathbf{U}_* & \mathbf{u} \\ \mathbf{0} & 1 \end{pmatrix}. \tag{5.2}$$

(1′) If $\mathbf{A}$ is symmetric (in which case $\mathbf{A}_*' = \mathbf{A}_*$ and $\mathbf{b} = \mathbf{a}$), if $\mathbf{A}_*$ has a U′DU decomposition, say $\mathbf{A}_* = \mathbf{U}_*'\mathbf{D}_*\mathbf{U}_*$, and if $\mathbf{a} \in \mathcal{C}(\mathbf{A}_*)$, then there exists a vector $\mathbf{u}$ such that $\mathbf{U}_*'\mathbf{D}_*\mathbf{u} = \mathbf{a}$, and, taking $\mathbf{u}$ to be any such vector and taking $k = c - \mathbf{u}'\mathbf{D}_*\mathbf{u}$, a U′DU decomposition of $\mathbf{A}$ is

$$\mathbf{A} = \begin{pmatrix} \mathbf{U}_* & \mathbf{u} \\ \mathbf{0} & 1 \end{pmatrix}' \begin{pmatrix} \mathbf{D}_* & \mathbf{0} \\ \mathbf{0} & k \end{pmatrix} \begin{pmatrix} \mathbf{U}_* & \mathbf{u} \\ \mathbf{0} & 1 \end{pmatrix}. \tag{5.3}$$

(2) The matrix $\mathbf{A}$ has an LDU decomposition only if $\mathbf{A}_*$ has an LDU decomposition, $\mathbf{a} \in \mathcal{C}(\mathbf{A}_*)$, and $\mathbf{b}' \in \mathcal{R}(\mathbf{A}_*)$.

(2′) The matrix $\mathbf{A}$ has a U′DU decomposition only if $\mathbf{A}$ is symmetric, $\mathbf{A}_*$ has a U′DU decomposition, and $\mathbf{a} \in \mathcal{C}(\mathbf{A}_*)$.

Proof. (1) Suppose that $\mathbf{A}_*$ has an LDU decomposition, say $\mathbf{A}_* = \mathbf{L}_*\mathbf{D}_*\mathbf{U}_*$, and that $\mathbf{a} \in \mathcal{C}(\mathbf{A}_*)$ and $\mathbf{b}' \in \mathcal{R}(\mathbf{A}_*)$. Then, there exist vectors $\mathbf{r}$ and $\mathbf{s}$ such that $\mathbf{a} = \mathbf{A}_*\mathbf{r}$ and $\mathbf{b}' = \mathbf{s}'\mathbf{A}_*$. Since $\mathbf{U}_*'\mathbf{D}_*(\mathbf{L}_*'\mathbf{s}) = \mathbf{A}_*'\mathbf{s} = \mathbf{b}$, and $\mathbf{L}_*\mathbf{D}_*(\mathbf{U}_*\mathbf{r}) = \mathbf{A}_*\mathbf{r} = \mathbf{a}$, there exist vectors $\boldsymbol{\ell}$ and $\mathbf{u}$ such that $\mathbf{U}_*'\mathbf{D}_*\boldsymbol{\ell} = \mathbf{b}$ and $\mathbf{L}_*\mathbf{D}_*\mathbf{u} = \mathbf{a}$. Moreover, equality (5.2) is an immediate consequence of Lemma 14.5.1.

(1′) If $\mathbf{A}_*' = \mathbf{A}_*$, then $\mathbf{a} \in \mathcal{C}(\mathbf{A}_*)$ implies $\mathbf{a}' \in \mathcal{R}(\mathbf{A}_*)$. Thus, Part (1′) can be obtained as a special case of Part (1) by putting $\mathbf{b} = \mathbf{a}$ and $\mathbf{L}_* = \mathbf{U}_*'$ and setting $\boldsymbol{\ell} = \mathbf{u}$.

(2) and (2′). Suppose that $\mathbf{A}$ has an LDU decomposition, say $\mathbf{A} = \mathbf{LDU}$. Partition $\mathbf{L}$, $\mathbf{U}$, and $\mathbf{D}$ as

$$\mathbf{L} = \begin{pmatrix} \mathbf{L}_* & \mathbf{0} \\ \boldsymbol{\ell}' & 1 \end{pmatrix}, \quad \mathbf{U} = \begin{pmatrix} \mathbf{U}_* & \mathbf{u} \\ \mathbf{0} & 1 \end{pmatrix}, \quad \text{and } \mathbf{D} = \begin{pmatrix} \mathbf{D}_* & \mathbf{0} \\ \mathbf{0} & k \end{pmatrix},$$

where $\mathbf{L}_*$, $\mathbf{U}_*$, and $\mathbf{D}_*$ are of dimensions $(n-1) \times (n-1)$. Then, it follows from Lemma 14.5.1 that $\mathbf{A}_* = \mathbf{L}_*\mathbf{D}_*\mathbf{U}_*$ and hence that $\mathbf{A}_*$ has an LDU decomposition. As a further consequence, we have that $\mathbf{a} = \mathbf{L}_*\mathbf{D}_*\mathbf{u} = \mathbf{A}_*\mathbf{U}_*^{-1}\mathbf{u} \in \mathcal{C}(\mathbf{A}_*)$ and $\mathbf{b}' = \boldsymbol{\ell}'\mathbf{D}_*\mathbf{U}_* = \boldsymbol{\ell}'\mathbf{L}_*^{-1}\mathbf{A}_* \in \mathcal{R}(\mathbf{A}_*)$.

In the special case where $\mathbf{L} = \mathbf{U}'$ (i.e., where $\mathbf{A} = \mathbf{LDU}$ is a U′DU decomposition), we have that $\mathbf{L}_* = \mathbf{U}_*'$ and hence that $\mathbf{A}_* = \mathbf{U}_*'\mathbf{D}_*\mathbf{U}_*$. Thus, if $\mathbf{A}$ has a U′DU decomposition, then $\mathbf{A}_*$ also has a U′DU decomposition. Moreover, if $\mathbf{A}$ has a U′DU decomposition, then clearly $\mathbf{A}$ is symmetric. Q.E.D.

Note that, together, Parts (1) and (2) of Theorem 14.5.2 imply, in particular, that $\mathbf{A}$ has an LDU decomposition if and only if $\mathbf{A}_*$ has an LDU decomposition, $\mathbf{a} \in \mathcal{C}(\mathbf{A}_*)$, and $\mathbf{b}' \in \mathcal{R}(\mathbf{A}_*)$. Similarly, Parts (1′) and (2′) imply that $\mathbf{A}$ has a U′DU decomposition if and only if $\mathbf{A}$ is symmetric, $\mathbf{A}_*$ has a U′DU decomposition, and $\mathbf{a} \in \mathcal{C}(\mathbf{A}_*)$.

The following theorem relates the LDU decomposition of *any* leading principal submatrix of an $n \times n$ matrix $\mathbf{A}$ to that of $\mathbf{A}$ itself.

Theorem 14.5.3. Let $\mathbf{A}$ represent an $n \times n$ matrix (where $n \geq 2$), and let $\mathbf{A}_{11}$ represent the $k \times k$ leading principal submatrix of $\mathbf{A}$ (where $1 \leq k \leq n-1$). Suppose that $\mathbf{A}$ has an LDU decomposition, say $\mathbf{A} = \mathbf{LDU}$, and partition $\mathbf{L}$, $\mathbf{U}$, and $\mathbf{D}$ as

$$\mathbf{L} = \begin{pmatrix} \mathbf{L}_{11} & \mathbf{0} \\ \mathbf{L}_{21} & \mathbf{L}_{22} \end{pmatrix}, \quad \mathbf{U} = \begin{pmatrix} \mathbf{U}_{11} & \mathbf{U}_{12} \\ \mathbf{0} & \mathbf{U}_{22} \end{pmatrix}, \quad \text{and } \mathbf{D} = \begin{pmatrix} \mathbf{D}_1 & \mathbf{0} \\ \mathbf{0} & \mathbf{D}_2 \end{pmatrix},$$

where the dimensions of $\mathbf{L}_{11}$, $\mathbf{U}_{11}$, and $\mathbf{D}_1$ are $k \times k$. Then, an LDU decomposition of $\mathbf{A}_{11}$ is $\mathbf{A}_{11} = \mathbf{L}_{11}\mathbf{D}_1\mathbf{U}_{11}$.

Theorem 14.5.3 can be verified simply by equating $\mathbf{A}_{11}$ to the $k \times k$ leading principal submatrix of the matrix product $\mathbf{LDU}$. Note that, in the special case of Theorem 14.5.3 where $\mathbf{A} = \mathbf{LDU}$ is a U′DU decomposition (i.e., where $\mathbf{L} = \mathbf{U}'$), $\mathbf{L}_{11} = \mathbf{U}'_{11}$ and hence $\mathbf{A}_{11} = \mathbf{L}_{11}\mathbf{D}_1\mathbf{U}_{11}$ is a U′DU decomposition of $\mathbf{A}_{11}$. Theorem 14.5.3 implies in particular that if $\mathbf{A}$ has an LDU or U′DU decomposition, then each of its leading principal submatrices has, respectively, an LDU or U′DU decomposition.

b. Existence, recursive construction, and uniqueness of an LDU or U′DU decomposition

It is easy to see that a 1×1 matrix $\mathbf{A}$ has a (unique) LDU (and U′DU) decomposition, namely, $\mathbf{A} = (1)\mathbf{A}(1)$. The following theorem gives results on the existence and recursive construction of an LDU or U′DU decomposition of a (square) matrix of order two or more.

Theorem 14.5.4. Let $\mathbf{A} = \{a_{ij}\}$ represent an $n \times n$ matrix (where $n \geq 2$). Denote by $\mathbf{A}_i$ the leading principal submatrix of $\mathbf{A}$ of order i $(i = 1, \ldots, n)$. For $i = 2, \ldots, n$, take $\mathbf{a}_i = (a_{1i}, \ldots, a_{i-1,i})'$ and $\mathbf{b}'_i = (a_{i1}, \ldots, a_{i,i-1})$, or equivalently define $\mathbf{a}_i$ and $\mathbf{b}'_i$ by

$$\mathbf{A}_i = \begin{pmatrix} \mathbf{A}_{i-1} & \mathbf{a}_i \\ \mathbf{b}'_i & a_{ii} \end{pmatrix}.$$

(1) Let $\mathbf{L}_1 = (1)$, $\mathbf{U}_1 = (1)$, and $\mathbf{D}_1 = (a_{11})$, and suppose that, for $i = 2, \ldots, n$, $\mathbf{a}_i \in \mathcal{C}(\mathbf{A}_{i-1})$ and $\mathbf{b}'_i \in \mathcal{R}(\mathbf{A}_{i-1})$. Then, for $i = 2, \ldots, n$, there exist a unit

lower triangular matrix $\mathbf{L}_i$, a unit upper triangular matrix $\mathbf{U}_i$, and a diagonal matrix $\mathbf{D}_i$ such that

$$\mathbf{L}_i = \begin{pmatrix} \mathbf{L}_{i-1} & \mathbf{0} \\ \boldsymbol{\ell}_i' & 1 \end{pmatrix}, \quad \mathbf{U}_i = \begin{pmatrix} \mathbf{U}_{i-1} & \mathbf{u}_i \\ \mathbf{0} & 1 \end{pmatrix}, \quad \mathbf{D}_i = \begin{pmatrix} \mathbf{D}_{i-1} & \mathbf{0} \\ \mathbf{0} & d_i \end{pmatrix}, \tag{5.4}$$

where $\mathbf{U}_{i-1}'\mathbf{D}_{i-1}\boldsymbol{\ell}_i = \mathbf{b}_i$, $\mathbf{L}_{i-1}\mathbf{D}_{i-1}\mathbf{u}_i = \mathbf{a}_i$, and $d_i = a_{ii} - \boldsymbol{\ell}_i'\mathbf{D}_{i-1}\mathbf{u}_i$; and, taking $\mathbf{L}_i$, $\mathbf{U}_i$, and $\mathbf{D}_i$ to be any such matrices, an LDU decomposition of $\mathbf{A}_i$ is $\mathbf{A}_i = \mathbf{L}_i\mathbf{D}_i\mathbf{U}_i$. In particular, an LDU decomposition of $\mathbf{A}$ is $\mathbf{A} = \mathbf{L}_n\mathbf{D}_n\mathbf{U}_n$.

(1′) Let $\mathbf{U}_1 = (1)$ and $\mathbf{D}_1 = (a_{11})$, and suppose that $\mathbf{A}$ is symmetric (in which case $\mathbf{A}_{i-1}' = \mathbf{A}_{i-1}$ and $\mathbf{b}_i = \mathbf{a}_i$ for $i = 2, \ldots, n$) and that, for $i = 2, \ldots, n$, $\mathbf{a}_i \in \mathcal{C}(\mathbf{A}_{i-1})$. Then, for $i = 2, \ldots, n$, there exist a unit upper triangular matrix $\mathbf{U}_i$ and a diagonal matrix $\mathbf{D}_i$ such that

$$\mathbf{U}_i = \begin{pmatrix} \mathbf{U}_{i-1} & \mathbf{u}_i \\ \mathbf{0} & 1 \end{pmatrix}, \qquad \mathbf{D}_i = \begin{pmatrix} \mathbf{D}_{i-1} & \mathbf{0} \\ \mathbf{0} & d_i \end{pmatrix}, \tag{5.5}$$

where $\mathbf{U}_{i-1}'\mathbf{D}_{i-1}\mathbf{u}_i = \mathbf{a}_i$ and $d_i = a_{ii} - \mathbf{u}_i'\mathbf{D}_{i-1}\mathbf{u}_i$; and taking $\mathbf{U}_i$ and $\mathbf{D}_i$ to be any such matrices, a U′DU decomposition of $\mathbf{A}_i$ is $\mathbf{A}_i = \mathbf{U}_i'\mathbf{D}_i\mathbf{U}_i$. In particular, a U′DU decomposition of $\mathbf{A}$ is $\mathbf{A} = \mathbf{U}_n'\mathbf{D}_n\mathbf{U}_n$.

(2) The matrix $\mathbf{A}$ has an LDU decomposition only if, for $i = 2, \ldots, n$, $\mathbf{a}_i \in \mathcal{C}(\mathbf{A}_{i-1})$ and $\mathbf{b}_i' \in \mathcal{R}(\mathbf{A}_{i-1})$.

(2′) The matrix $\mathbf{A}$ has a U′DU decomposition only if $\mathbf{A}$ is symmetric and, for $i = 2, \ldots, n$, $\mathbf{a}_i \in \mathcal{C}(\mathbf{A}_{i-1})$.

Proof. (1) For purposes of proving Part (1), suppose that, for $i = 2, \ldots, n$, $\mathbf{a}_i \in \mathcal{C}(\mathbf{A}_{i-1})$ and $\mathbf{b}_i' \in \mathcal{R}(\mathbf{A}_{i-1})$. The proof is by mathematical induction. It follows from Theorem 14.5.2 that, for $i = 2$, there exist a unit lower triangular matrix $\mathbf{L}_i$, a unit upper triangular matrix $\mathbf{U}_i$, and a diagonal matrix $\mathbf{D}_i$ of the form (5.4) and that $\mathbf{A}_2 = \mathbf{L}_2\mathbf{D}_2\mathbf{U}_2$ for any such matrices $\mathbf{L}_2$, $\mathbf{U}_2$, and $\mathbf{D}_2$. Suppose now that, for $i = k - 1$ (where $3 \le k \le n$), there exist a unit lower triangular matrix $\mathbf{L}_i$, a unit upper triangular matrix $\mathbf{U}_i$, and a diagonal matrix $\mathbf{D}_i$ of the form (5.4) and, taking $\mathbf{L}_{k-1}$, $\mathbf{U}_{k-1}$, and $\mathbf{D}_{k-1}$ to be any such matrices, that $\mathbf{A}_{k-1} = \mathbf{L}_{k-1}\mathbf{D}_{k-1}\mathbf{U}_{k-1}$. The proof is complete upon observing that, for $i = k$, there exist (as a consequence of Theorem 14.5.2) a unit lower triangular matrix $\mathbf{L}_i$, a unit upper triangular matrix $\mathbf{U}_i$, and a diagonal matrix $\mathbf{D}_i$ of the form (5.4) and that $\mathbf{A}_k = \mathbf{L}_k\mathbf{D}_k\mathbf{U}_k$ for any such matrices $\mathbf{L}_k$, $\mathbf{U}_k$, and $\mathbf{D}_k$.

(1′) A proof of Part (1′), analogous to that of Part (1), can be constructed by making use of Part (1′) of Theorem 14.5.2.

(2) Suppose that $\mathbf{A}$ has an LDU decomposition. Then, it follows from Theorem 14.5.3 that the leading principal submatrices $\mathbf{A}_2, \mathbf{A}_3, \ldots, \mathbf{A}_{n-1}$ of orders two through $n - 1$ (as well as the leading principal submatrix $\mathbf{A}_n = \mathbf{A}$ of order n) have LDU decompositions. Based on Part (2) of Theorem 14.5.2, we conclude that, for $i = 2, \ldots, n$, $\mathbf{a}_i \in \mathcal{C}(\mathbf{A}_{i-1})$ and $\mathbf{b}_i' \in \mathcal{R}(\mathbf{A}_{i-1})$.

(2′) A proof of Part (2′), analogous to that of Part (2), can be constructed by making use of Theorem 14.5.3 and Part (2′) of Theorem 14.5.2. Q.E.D.

Note that, together, Parts (1) and (2) of Theorem 14.5.4 imply in particular that $\mathbf{A}$ has an LDU decomposition if and only if, for $i = 2, \ldots, n$, $\mathbf{a}_i \in \mathcal{C}(\mathbf{A}_{i-1})$ and $\mathbf{b}'_i \in \mathcal{R}(\mathbf{A}_{i-1})$. Similarly, Parts (1′) and (2′) imply that $\mathbf{A}$ has a U′DU decomposition if and only if $\mathbf{A}$ is symmetric and, for $i = 2, \ldots, n$, $\mathbf{a}_i \in \mathcal{C}(\mathbf{A}_{i-1})$.

For purposes of illustration, consider a 2×2 matrix $\mathbf{A} = \{a_{ij}\}$ and suppose that $a_{11} = 0$. Then, $\mathbf{A}$ has an LDU decomposition if and only if $a_{21} = a_{12} = 0$, that is, if and only if $\mathbf{A}$ is of the form

$$\mathbf{A} = \begin{pmatrix} 0 & 0 \\ 0 & a_{22} \end{pmatrix}.$$

When $\mathbf{A}$ is of this form, an LDU decomposition of $\mathbf{A}$ is

$$\mathbf{A} = \begin{pmatrix} 1 & 0 \\ \ell & 1 \end{pmatrix} \begin{pmatrix} 0 & 0 \\ 0 & a_{22} \end{pmatrix} \begin{pmatrix} 1 & u \\ 0 & 1 \end{pmatrix},$$

where ℓ and u are arbitrary scalars.

To what extent is the LDU decomposition of an $n \times n$ matrix unique? This question is addressed in the following theorem.

Theorem 14.5.5. Let $\mathbf{A}$ represent an $n \times n$ matrix of rank r that has an LDU decomposition. Let $\mathbf{L} = \{\ell_{ij}\}$ represent an $n \times n$ unit lower triangular matrix, $\mathbf{U} = \{u_{ij}\}$ an $n \times n$ unit upper triangular matrix, and $\mathbf{D} = \{d_i\}$ an $n \times n$ diagonal matrix such that $\mathbf{A} = \mathbf{LDU}$; that is, such that $\mathbf{A} = \mathbf{LDU}$ is an LDU decomposition of $\mathbf{A}$. (1) The diagonal matrix $\mathbf{D}$ is unique. (2) Suppose that the $i_1, \ldots, i_r$th diagonal elements of $\mathbf{D}$ (where $i_1 < \cdots < i_r$) are nonzero (and the remaining diagonal elements of $\mathbf{D}$ are zero). Then, for $i \in \{i_1, \ldots, i_r\}$ and $j > i$, u_{ij} and ℓ_{ji} are unique, and, for $i \notin \{i_1, \ldots, i_r\}$ and $j > i$, u_{ij} and ℓ_{ji} are completely arbitrary. That is, of the elements of $\mathbf{U}$ above the diagonal, those in the $i_1, \ldots, i_r$th rows are unique and those in the remaining rows are completely arbitrary; and, similarly, of the elements of $\mathbf{L}$ below the diagonal, those in the $i_1, \ldots, i_r$th columns are unique and those in the remaining columns are completely arbitrary.

Proof. Let $\mathbf{L}^* = \{\ell^*_{ij}\}$ represent a unit lower triangular matrix, $\mathbf{U}^* = \{u^*_{ij}\}$ a unit upper triangular matrix, and $\mathbf{D}^* = \{d^*_i\}$ a diagonal matrix such that $\mathbf{A} = \mathbf{L}^*\mathbf{D}^*\mathbf{U}^*$, so that $\mathbf{A} = \mathbf{L}^*\mathbf{D}^*\mathbf{U}^*$, like $\mathbf{A} = \mathbf{LDU}$, is an LDU decomposition of $\mathbf{A}$. Then,

$$\mathbf{L}^{*-1}\mathbf{LD} = \mathbf{L}^{*-1}(\mathbf{LDU})\mathbf{U}^{-1} = \mathbf{L}^{*-1}(\mathbf{L}^*\mathbf{D}^*\mathbf{U}^*)\mathbf{U}^{-1} = \mathbf{D}^*\mathbf{U}^*\mathbf{U}^{-1}. \quad (5.6)$$

Since (according to Corollary 8.5.9) $\mathbf{L}^{*-1}$ is unit lower triangular and $\mathbf{U}^{-1}$ is unit upper triangular, it follows from Lemma 1.3.1 that the diagonal elements of $\mathbf{L}^{*-1}\mathbf{LD}$ are $d_1, \ldots, d_n$, respectively, and that the diagonal elements of $\mathbf{D}^*\mathbf{U}^*\mathbf{U}^{-1}$ are $d^*_1, \ldots, d^*_n$, respectively, implying—in light of equality (5.6)—that $d^*_i = d_i$ $(i = 1, \ldots, n)$ or equivalently that $\mathbf{D}^* = \mathbf{D}$, which establishes Part (1) of Theorem 14.5.5.

For purposes of proving Part (2), take $\mathbf{A}_{11}$ to be the $r \times r$ submatrix obtained by striking out all of the rows and columns of $\mathbf{A}$ except the $i_1, \ldots, i_r$th rows and columns. Let $\mathbf{D}_1 = \text{diag}(d_{i_1}, \ldots, d_{i_r})$. Denote by $\mathbf{L}_1$ and $\mathbf{L}_1^*$ the $r \times n$ submatrices obtained by striking out all of the rows of $\mathbf{L}$ and $\mathbf{L}^*$, respectively, except the i_1, ..., i_rth rows, and by $\mathbf{L}_{11}$ and $\mathbf{L}_{11}^*$ the submatrices obtained by striking out all of the rows and columns of $\mathbf{L}$ and $\mathbf{L}^*$, respectively, except the $i_1, \ldots, i_r$th rows and columns. Similarly, denote by $\mathbf{U}_1$ and $\mathbf{U}_1^*$ the $n \times r$ submatrices obtained by striking out all of the columns of $\mathbf{U}$ and $\mathbf{U}^*$, respectively, except the $i_1, \ldots,$ i_rth columns, and by $\mathbf{U}_{11}$ and $\mathbf{U}_{11}^*$ the $r \times r$ submatrices obtained by striking out all of the rows and columns of $\mathbf{U}$ and $\mathbf{U}^*$, respectively, except the $i_1, \ldots, i_r$th rows and columns. Then, $\mathbf{A}_{11} = \mathbf{L}_1\mathbf{D}\mathbf{U}_1 = \mathbf{L}_{11}\mathbf{D}_1\mathbf{U}_{11}$ and (since $\mathbf{D}^* = \mathbf{D}$) $\mathbf{A}_{11} = \mathbf{L}_1^*\mathbf{D}^*\mathbf{U}_1^* = \mathbf{L}_1^*\mathbf{D}\mathbf{U}_1^* = \mathbf{L}_{11}^*\mathbf{D}_1\mathbf{U}_{11}^*$, implying that

$$\mathbf{L}_{11}\mathbf{D}_1\mathbf{U}_{11} = \mathbf{L}_{11}^*\mathbf{D}_1\mathbf{U}_{11}^*. \tag{5.7}$$

The matrices $\mathbf{L}_{11}$ and $\mathbf{L}_{11}^*$ are principal submatrices of unit lower triangular matrices and hence are themselves unit lower triangular. Similarly, $\mathbf{U}_{11}$ and $\mathbf{U}_{11}^*$ are unit upper triangular. Premultiplying both sides of equality (5.7) by $\mathbf{L}_{11}^{*-1}$ and postmultiplying both sides by $\mathbf{U}_{11}^{-1}\mathbf{D}_1^{-1}$, we find that

$$\mathbf{L}_{11}^{*-1}\mathbf{L}_{11} = \mathbf{D}_1\mathbf{U}_{11}^*\mathbf{U}_{11}^{-1}\mathbf{D}_1^{-1}. \tag{5.8}$$

According to Corollary 8.5.9, $\mathbf{L}_{11}^{*-1}$ is unit lower triangular and $\mathbf{U}_{11}^{-1}$ is unit upper triangular. Observe that (according to Lemma 1.3.1) $\mathbf{L}_{11}^{*-1}\mathbf{L}_{11}$ is unit lower triangular. Since $\mathbf{D}_1\mathbf{U}_{11}^*\mathbf{U}_{11}^{-1}\mathbf{D}_1^{-1}$ is upper triangular, it follows from equality (5.8) that $\mathbf{L}_{11}^{*-1}\mathbf{L}_{11} = \mathbf{I}$, or, equivalently, that

$$\mathbf{L}_{11}^* = \mathbf{L}_{11},$$

and—in light of equality (5.7) and the nonsingularity of $\mathbf{L}_{11}$ and $\mathbf{D}_1$—that

$$\mathbf{U}_{11}^* = \mathbf{U}_{11}.$$

Now, take $\mathbf{A}_1$ to be the $n \times r$ submatrix obtained by striking out all of the columns of $\mathbf{A}$ except the $i_1, \ldots, i_r$th columns, and take $\mathbf{A}_2$ to be the $r \times n$ submatrix obtained by striking out all of the rows of $\mathbf{A}$ except the $i_1, \ldots, i_r$th rows. Denote by $\mathbf{L}_2$ and $\mathbf{L}_2^*$ the $n \times r$ submatrices obtained by striking out all of the columns of $\mathbf{L}$ and $\mathbf{L}^*$, respectively, except the $i_1, \ldots, i_r$th columns. Similarly, denote by $\mathbf{U}_2$ and $\mathbf{U}_2^*$ the $r \times n$ submatrices obtained by striking out all of the rows of $\mathbf{U}$ and $\mathbf{U}^*$, respectively, except the $i_1, \ldots, i_r$th rows. Then, $\mathbf{A}_1 = \mathbf{L}\mathbf{D}\mathbf{U}_1 = \mathbf{L}_2\mathbf{D}_1\mathbf{U}_{11}$ and (since $\mathbf{D}^* = \mathbf{D}$ and $\mathbf{U}_{11}^* = \mathbf{U}_{11}$) $\mathbf{A}_1 = \mathbf{L}^*\mathbf{D}^*\mathbf{U}_1^* = \mathbf{L}^*\mathbf{D}\mathbf{U}_1^* = \mathbf{L}_2^*\mathbf{D}_1\mathbf{U}_{11}^* = \mathbf{L}_2^*\mathbf{D}_1\mathbf{U}_{11}$, implying that $\mathbf{L}_2^*\mathbf{D}_1\mathbf{U}_{11} = \mathbf{L}_2\mathbf{D}_1\mathbf{U}_{11}$, and hence (since $\mathbf{D}_1$ and $\mathbf{U}_{11}$ are nonsingular) that

$$\mathbf{L}_2^* = \mathbf{L}_2. \tag{5.9}$$

Similarly, $\mathbf{A}_2 = \mathbf{L}_1\mathbf{D}\mathbf{U} = \mathbf{L}_{11}\mathbf{D}_1\mathbf{U}_2$ and $\mathbf{A}_2 = \mathbf{L}_1^*\mathbf{D}^*\mathbf{U}^* = \mathbf{L}_1^*\mathbf{D}\mathbf{U}^* = \mathbf{L}_{11}^*\mathbf{D}_1\mathbf{U}_2^* = \mathbf{L}_{11}\mathbf{D}_1\mathbf{U}_2^*$, implying that $\mathbf{L}_{11}\mathbf{D}_1\mathbf{U}_2^* = \mathbf{L}_{11}\mathbf{D}_1\mathbf{U}_2$, and hence that

$$\mathbf{U}_2^* = \mathbf{U}_2. \tag{5.10}$$

The kjth elements of $\mathbf{U}_2$ and $\mathbf{U}_2^*$ are $u_{i_k j}$ and $u_{i_k j}^*$, respectively, and the jkth elements of $\mathbf{L}_2$ and $\mathbf{L}_2^*$ are ℓ_{ji_k} and $\ell_{ji_k}^*$, respectively ($k = 1, \ldots, r; j = 1, \ldots, n$). Thus, it follows from equalities (5.9) and (5.10) that, for $i \in \{i_1, \ldots, i_r\}$ and for $j = 1, \ldots, n$ (and in particular for $j = i+1, \ldots, n$) that $u_{ij}^* = u_{ij}$ and $\ell_{ji}^* = \ell_{ji}$.

To complete the proof of Part (2) of Theorem 14.5.5, observe that $\mathbf{A} = \mathbf{LDU} = \mathbf{L}_2\mathbf{D}_1\mathbf{U}_2$ and hence that $\mathbf{A} = \mathbf{LDU}$ even if those elements of $\mathbf{L}$ (below the diagonal) that are not elements of $\mathbf{L}_2$ and/or those elements of $\mathbf{U}$ (above the diagonal) that are not elements of $\mathbf{U}_2$ are changed arbitrarily. Q.E.D.

Under what circumstances does an $n \times n$ matrix $\mathbf{A}$ have a unique LDU decomposition? This question is addressed in the following two corollaries.

Corollary 14.5.6. If an $n \times n$ matrix $\mathbf{A}$ has an LDU decomposition and if the leading principal submatrix (of $\mathbf{A}$) of order $n - 1$ is nonsingular, then the LDU decomposition of $\mathbf{A}$ is unique.

Proof. Let $\mathbf{A}_{11}$ represent the $(n-1) \times (n-1)$ leading principal submatrix of $\mathbf{A}$. Suppose that $\mathbf{A}$ has an LDU decomposition, say $\mathbf{A} = \mathbf{LDU}$, and that $\mathbf{A}_{11}$ is nonsingular. Partition $\mathbf{L}$, $\mathbf{U}$, and $\mathbf{D}$ as

$$\mathbf{L} = \begin{pmatrix} \mathbf{L}_{11} & \mathbf{0} \\ \boldsymbol{\ell}' & 1 \end{pmatrix}, \quad \mathbf{U} = \begin{pmatrix} \mathbf{U}_{11} & \mathbf{u} \\ \mathbf{0} & 1 \end{pmatrix}, \quad \text{and } \mathbf{D} = \begin{pmatrix} \mathbf{D}_1 & \mathbf{0} \\ \mathbf{0} & d \end{pmatrix}$$

[where $\mathbf{L}_{11}$, $\mathbf{U}_{11}$, and $\mathbf{D}_1$ are of dimensions $(n-1) \times (n-1)$]. According to Lemma 14.5.1, an LDU decomposition of $\mathbf{A}_{11}$ is $\mathbf{A}_{11} = \mathbf{L}_{11}\mathbf{D}_1\mathbf{U}_{11}$. Thus (since $\mathbf{A}_{11}$ is nonsingular), it follows from Lemma 14.3.1 that all $n-1$ diagonal elements of $\mathbf{D}_1$ are nonzero or, equivalently, that the first $n - 1$ diagonal elements of $\mathbf{D}$ are nonzero. Based on Theorem 14.5.5, we conclude that the LDU decomposition of $\mathbf{A}$ is unique. Q.E.D.

Corollary 14.5.7. Let $\mathbf{A}$ represent an $n \times n$ matrix (where $n \geq 2$), and, for $i = 1, 2, \ldots, n-1$, let $\mathbf{A}_i$ represent the leading principal submatrix (of $\mathbf{A}$) of order i. If $\mathbf{A}_1, \mathbf{A}_2, \ldots, \mathbf{A}_{n-1}$ are nonsingular, then $\mathbf{A}$ has a unique LDU decomposition.

Proof. Let a_{ij} represent the ijth element of $\mathbf{A}$ ($i, j = 1, \ldots, n$); and, for $i = 2, \ldots, n$, define $\mathbf{a}_i = (a_{1i}, \ldots, a_{i-1,i})'$ and $\mathbf{b}_i' = (a_{i1}, \ldots, a_{i,i-1})$. Suppose that $\mathbf{A}_1, \mathbf{A}_2, \ldots, \mathbf{A}_{n-1}$ are nonsingular. Then, for $i = 2, \ldots, n$, $\mathbf{a}_i \in \mathcal{C}(\mathbf{A}_{i-1})$ and $\mathbf{b}_i' \in \mathcal{R}(\mathbf{A}_{i-1})$, and it follows from Part (1) of Theorem 14.5.4 that $\mathbf{A}$ has an LDU decomposition. That this decomposition is unique is (since $\mathbf{A}_{n-1}$ is nonsingular) an immediate consequence of Corollary 14.5.6. Q.E.D.

For a symmetric matrix, the uniqueness of an LDU decomposition has the following implication.

Lemma 14.5.8. If a symmetric matrix $\mathbf{A}$ has a unique LDU decomposition, say $\mathbf{A} = \mathbf{LDU}$, then $\mathbf{L} = \mathbf{U}'$, that is, the unique LDU decomposition is a U′DU decomposition.

Proof. Suppose that the symmetric matrix $\mathbf{A}$ has the unique LDU decomposition $\mathbf{A} = \mathbf{LDU}$. Then, $\mathbf{A} = \mathbf{A}' = \mathbf{U}'\mathbf{D}\mathbf{L}'$, implying (since $\mathbf{U}'$ is unit lower triangular and $\mathbf{L}'$ is unit upper triangular) that $\mathbf{A} = \mathbf{U}'\mathbf{D}\mathbf{L}'$ is an LDU decomposition of $\mathbf{A}$ and hence—in light of the uniqueness of the LDU decomposition—that $\mathbf{L} = \mathbf{U}'$ (or equivalently $\mathbf{U} = \mathbf{L}'$). Q.E.D.

Note that if a (square) matrix $\mathbf{A}$ has an LDU decomposition, say $\mathbf{A} = \mathbf{LDU}$, then, letting $\mathbf{L}^* = \mathbf{LD}$ and $\mathbf{U}^* = \mathbf{DU}$, it can also be decomposed as $\mathbf{A} = \mathbf{L}^*\mathbf{U}$, which is the product of a lower triangular matrix and a unit upper triangular matrix, or as $\mathbf{A} = \mathbf{LU}^*$, which is the product of a unit lower triangular matrix and an upper triangular matrix.

The term *LU decomposition* is sometimes used in referring to any decomposition of a (square) matrix $\mathbf{A}$ of the general form $\mathbf{A} = \mathbf{LU}$, where $\mathbf{L}$ is a lower triangular matrix and $\mathbf{U}$ an upper triangular matrix—see, e.g., Stewart (1973). In the special case where $\mathbf{L}$ is unit lower triangular, Stewart refers to such a decomposition as a *Doolittle decomposition*, and in the special case where $\mathbf{U}$ is unit upper triangular, he refers to such a decomposition as a *Crout decomposition*. Golub and Van Loan (1989) restrict their use of the term LU decomposition to the special case where $\mathbf{L}$ is unit lower triangular, that is, to what Stewart calls a Doolittle decomposition.

c. Decomposition of positive definite matrices

Let us now consider the implications of the results of Subsection b as applied to positive definite matrices. Does a positive definite matrix necessarily have an LDU decomposition, and is the LDU decomposition of a positive definite matrix unique? These questions are answered (in the affirmative) by the following theorem, which also provides some information about the nature of the LDU decomposition of a positive definite matrix.

Theorem 14.5.9. An $n \times n$ positive definite matrix $\mathbf{A}$ has a unique LDU decomposition, say $\mathbf{A} = \mathbf{LDU}$, and the diagonal elements of the diagonal matrix $\mathbf{D}$ are positive.

Proof. According to Corollary 14.2.12, any principal submatrix of a positive definite matrix is positive definite and hence (according to Lemma 14.2.8) is nonsingular. Thus, it follows from Corollary 14.5.7 that $\mathbf{A}$ has a unique LDU decomposition, say $\mathbf{A} = \mathbf{LDU}$.

It remains to show that the diagonal elements, say $d_1, \ldots, d_n$, of the diagonal matrix $\mathbf{D}$ are positive. For this purpose, consider the matrix $\mathbf{B} = \mathbf{DU}(\mathbf{L}^{-1})'$. Since (in light of Corollary 8.5.9) $(\mathbf{L}^{-1})'$ (like $\mathbf{U}$) is unit upper triangular, it follows from Lemma 1.3.1 that the diagonal elements of $\mathbf{B}$ are the same as the diagonal elements $d_1, \ldots, d_n$ of $\mathbf{D}$. Moreover,

$$\mathbf{B} = \mathbf{L}^{-1}(\mathbf{LDU})(\mathbf{L}^{-1})' = \mathbf{L}^{-1}\mathbf{A}(\mathbf{L}^{-1})',$$

implying (in light of Corollary 14.2.10) that $\mathbf{B}$ is positive definite. We conclude (on the basis of Corollary 14.2.13) that $d_1, \ldots, d_n$ are positive. Q.E.D.

In the special case of a symmetric positive definite matrix, the conclusions of Theorem 14.5.9 can (as a consequence of Lemma 14.5.8) be made more specific, as described in the following corollary.

Corollary 14.5.10. An $n \times n$ symmetric positive definite matrix $\mathbf{A}$ has a unique U′DU decomposition, say $\mathbf{A} = \mathbf{U}'\mathbf{D}\mathbf{U}$, and the diagonal elements of the diagonal matrix $\mathbf{D}$ are positive.

Note that, as an immediate consequence of Corollary 14.2.15, we have the following converse to Corollary 14.5.10: If an $n \times n$ matrix $\mathbf{A}$ has a U′DU decomposition, say $\mathbf{A} = \mathbf{U}'\mathbf{D}\mathbf{U}$, and if the diagonal elements of the diagonal matrix $\mathbf{D}$ are positive, then $\mathbf{A}$ is positive definite (and symmetric).

An alternative decomposition of a symmetric positive definite matrix is described in the following theorem.

Theorem 14.5.11. For any $n \times n$ symmetric positive definite matrix $\mathbf{A}$, there exists a unique upper triangular matrix $\mathbf{T}$ with positive diagonal elements such that

$$\mathbf{A} = \mathbf{T}'\mathbf{T}. \tag{5.11}$$

Moreover, taking $\mathbf{U}$ to be the unique unit upper triangular matrix and $\mathbf{D} = \{d_i\}$ to be the unique diagonal matrix such that $\mathbf{A} = \mathbf{U}'\mathbf{D}\mathbf{U}$,

$$\mathbf{T} = \mathbf{D}^{1/2}\mathbf{U},$$

where $\mathbf{D}^{1/2} = \mathrm{diag}(\sqrt{d_1}, \sqrt{d_2}, \ldots, \sqrt{d_n})$.

Proof. Note that Corollary 14.5.10 guarantees the existence and uniqueness of the unit upper triangular matrix $\mathbf{U}$ and the diagonal matrix $\mathbf{D}$ and also guarantees that the diagonal elements of $\mathbf{D}$ are positive. Note further that $\mathbf{D}^{1/2}\mathbf{U}$ is upper triangular, that the diagonal elements of $\mathbf{D}^{1/2}\mathbf{U}$ are $\sqrt{d_1}, \sqrt{d_2}, \ldots, \sqrt{d_n}$, and that $\mathbf{A} = (\mathbf{D}^{1/2}\mathbf{U})'\mathbf{D}^{1/2}\mathbf{U}$. Thus, there exists an upper triangular matrix $\mathbf{T}$ with positive diagonal elements such that $\mathbf{A} = \mathbf{T}'\mathbf{T}$. It remains to show that there is only one such upper triangular matrix with positive diagonal elements.

Suppose that $\mathbf{T} = \{t_{ij}\}$ is an $n \times n$ upper triangular matrix with positive diagonal elements such that $\mathbf{A} = \mathbf{T}'\mathbf{T}$. Define $\mathbf{D}_* = \mathrm{diag}(t_{11}^2, t_{22}^2, \ldots, t_{nn}^2)$ and

$$\mathbf{U}_* = [\mathrm{diag}(t_{11}, t_{22}, \ldots, t_{nn})]^{-1}\mathbf{T} = \mathrm{diag}(1/t_{11}, 1/t_{22}, \ldots, 1/t_{nn})\mathbf{T}.$$

Then, $\mathbf{U}_*$ is unit upper triangular, and $\mathbf{A} = \mathbf{U}_*'\mathbf{D}_*\mathbf{U}_*$, so that $\mathbf{A} = \mathbf{U}_*'\mathbf{D}_*\mathbf{U}_*$ is a U′DU decomposition of $\mathbf{A}$. Thus, it follows from Corollary 14.5.10 that $\mathbf{D}_* = \mathbf{D}$ or, equivalently (since t_{ii} is positive), that $t_{ii} = \sqrt{d_i}$ $(i = 1, \ldots, n)$, and that $\mathbf{U}_* = \mathbf{U}$. Consequently, $\mathbf{T} = \mathbf{D}^{1/2}\mathbf{U}_* = \mathbf{D}^{1/2}\mathbf{U}$. We conclude that the only upper triangular matrix $\mathbf{T}$ with positive diagonal elements such that $\mathbf{A} = \mathbf{T}'\mathbf{T}$ is $\mathbf{T} = \mathbf{D}^{1/2}\mathbf{U}$. Q.E.D.

Decomposition (5.11) of the symmetric positive definite matrix $\mathbf{A}$ is known as the *Cholesky decomposition*.

d. Decomposition of symmetric nonnegative definite matrices

Let us now extend the results of Corollary 14.5.10 and Theorem 14.5.11 (which apply to symmetric positive definite matrices) to symmetric nonnegative definite matrices that may not be positive definite. The following theorem provides the basis for these extensions.

Theorem 14.5.12. Corresponding to any $n \times n$ symmetric nonnegative definite matrix $\mathbf{A}$, there exists a unit upper triangular matrix $\mathbf{U}$ such that $\mathbf{U}'\mathbf{A}\mathbf{U}$ is a diagonal matrix.

For purposes of proving this theorem, it is helpful to establish the following lemma.

Lemma 14.5.13. Let $\mathbf{A} = \{a_{ij}\}$ represent an $n \times n$ nonnegative definite matrix. If $a_{ii} = 0$, then, for $j = 1, \ldots, n$, $a_{ij} = -a_{ji}$; that is, if the ith diagonal element of $\mathbf{A}$ equals 0, then $(a_{i1}, \ldots, a_{in})$, which is the ith row of $\mathbf{A}$, equals $-(a_{1i}, \ldots, a_{ni})$, which is -1 times the transpose of the ith column of $\mathbf{A}$ $(i = 1, \ldots, n)$.

Proof (of Lemma 14.5.13). Suppose that $a_{ii} = 0$, and take $\mathbf{x} = \{x_k\}$ to be an n-dimensional column vector such that $x_i < -a_{jj}$, $x_j = a_{ij} + a_{ji}$, and $x_k = 0$ for k other than $k = i$ and $k = j$ (where $j \neq i$). Then,

$$\begin{aligned} \mathbf{x}'\mathbf{A}\mathbf{x} &= a_{ii}x_i^2 + (a_{ij} + a_{ji})x_i x_j + a_{jj}x_j^2 \\ &= (a_{ij} + a_{ji})^2(x_i + a_{jj}) \\ &\leq 0, \end{aligned} \tag{5.12}$$

with equality holding only if $a_{ij} + a_{ji} = 0$ or, equivalently, only if $a_{ij} = -a_{ji}$. Moreover, since $\mathbf{A}$ is nonnegative definite, $\mathbf{x}'\mathbf{A}\mathbf{x} \geq 0$, which—together with inequality (5.12)—implies that $\mathbf{x}'\mathbf{A}\mathbf{x} = 0$. We conclude that $a_{ij} = -a_{ji}$. Q.E.D.

If the nonnegative definite matrix $\mathbf{A}$ in Lemma 14.5.13 is symmetric, then $a_{ij} = -a_{ji} \Leftrightarrow 2a_{ij} = 0 \Leftrightarrow a_{ij} = 0$. Thus, we have the following corollary.

Corollary 14.5.14. Let $\mathbf{A} = \{a_{ij}\}$ represent an $n \times n$ symmetric nonnegative definite matrix. If $a_{ii} = 0$, then, for $j = 1, \ldots, n$, $a_{ji} = a_{ij} = 0$; that is, if the ith diagonal element of $\mathbf{A}$ equals zero, then the ith column $(a_{1i}, \ldots, a_{ni})'$ of $\mathbf{A}$ and the ith row $(a_{i1}, \ldots, a_{in})$ of $\mathbf{A}$ are null. Moreover, if all n diagonal elements $a_{11}, a_{22}, \ldots, a_{nn}$ of $\mathbf{A}$ equal zero, then $\mathbf{A} = \mathbf{0}$.

Proof (of Theorem 14.5.12). The proof is by mathematical induction and is similar to the proof of Theorem 14.3.4. The theorem is clearly true for any 1×1 matrix. Suppose now that it is true for any $(n-1) \times (n-1)$ symmetric nonnegative definite matrix, and consider an arbitrary $n \times n$ symmetric nonnegative definite matrix $\mathbf{A} = \{a_{ij}\}$. For purposes of establishing the existence of a unit upper triangular matrix $\mathbf{U}$ such that $\mathbf{U}'\mathbf{A}\mathbf{U}$ is diagonal, it is convenient to consider separately the case $a_{11} = 0$ and the case $a_{11} \neq 0$.

Case (1): $a_{11} = 0$. It follows from Corollary 14.5.14 that the first row and column of $\mathbf{A}$ are null or, equivalently, that $\mathbf{A} = \text{diag}(0, \mathbf{A}_{22})$ for some $(n-1) \times (n-1)$ matrix $\mathbf{A}_{22}$. Moreover, since $\mathbf{A}_{22}$ is a principal submatrix of $\mathbf{A}$, $\mathbf{A}_{22}$ is symmetric and nonnegative definite, and hence, by supposition, there exists a unit upper triangular matrix $\mathbf{U}_*$ such that $\mathbf{U}_*'\mathbf{A}_{22}\mathbf{U}_*$ is a diagonal matrix. Take $\mathbf{U} = \text{diag}(1, \mathbf{U}_*)$. Then, $\mathbf{U}$ is unit upper triangular, and $\mathbf{U}'\mathbf{A}\mathbf{U} = \text{diag}(0, \mathbf{U}_*'\mathbf{A}_{22}\mathbf{U}_*)$, which (like $\mathbf{U}_*'\mathbf{A}_{22}\mathbf{U}_*$) is a diagonal matrix.

Case (2): $a_{11} \neq 0$. According to Lemma 14.3.3, there exists a unit upper triangular matrix $\mathbf{U}_1$ such that $\mathbf{U}_1'\mathbf{A}\mathbf{U}_1 = \text{diag}(b_{11}, \mathbf{B}_{22})$ for some scalar b_{11} and some $(n-1) \times (n-1)$ matrix $\mathbf{B}_{22}$. Moreover, $\mathbf{U}_1'\mathbf{A}\mathbf{U}_1$ is (in light of Theorem

14.2.9) symmetric and nonnegative definite, and, since $\mathbf{B}_{22}$ is a principal submatrix of $\mathbf{U}_1'\mathbf{A}\mathbf{U}_1$, $\mathbf{B}_{22}$ is symmetric and nonnegative definite. Thus, by supposition, there exists a unit upper triangular matrix $\mathbf{U}_2$ such that $\mathbf{U}_2'\mathbf{B}_{22}\mathbf{U}_2$ is a diagonal matrix. Take $\mathbf{U} = \mathbf{U}_1 \operatorname{diag}(1, \mathbf{U}_2)$. Then, $\mathbf{U}$ is a unit upper triangular matrix [since $\operatorname{diag}(1, \mathbf{U}_2)$ is unit upper triangular and the product of two unit upper triangular matrices is unit upper triangular], and

$$\mathbf{U}'\mathbf{A}\mathbf{U} = \operatorname{diag}(1, \mathbf{U}_2') \operatorname{diag}(b_{11}, \mathbf{B}_{22}) \operatorname{diag}(1, \mathbf{U}_2) = \operatorname{diag}(b_{11}, \mathbf{U}_2'\mathbf{B}_{22}\mathbf{U}_2),$$

which (like $\mathbf{U}_2'\mathbf{B}_{22}\mathbf{U}_2$) is a diagonal matrix. Q.E.D.

Corollary 14.5.10 indicates that, in the special case of a symmetric positive definite matrix $\mathbf{A}$, the nonsingular matrix $\mathbf{P}$ in Corollary 14.3.5 can be chosen to be a unit upper triangular matrix, in which case the decomposition $\mathbf{A} = \mathbf{P}'\mathbf{D}\mathbf{P}$ is a U′DU decomposition. The following corollary (of Theorem 14.5.12) indicates that $\mathbf{P}$ can be chosen to be unit upper triangular even if $\mathbf{A}$ is only (symmetric) positive semidefinite.

Corollary 14.5.15. An $n \times n$ symmetric nonnegative definite matrix $\mathbf{A}$ has a U′DU decomposition, and, for any such decomposition $\mathbf{A} = \mathbf{U}'\mathbf{D}\mathbf{U}$, the diagonal elements of the diagonal matrix $\mathbf{D}$ are nonnegative.

Proof. In light of Corollary 14.2.15, it suffices to show that $\mathbf{A}$ has a U′DU decomposition. According to Theorem 14.5.12, there exists a unit upper triangular matrix $\mathbf{T}$ such that $\mathbf{T}'\mathbf{A}\mathbf{T} = \mathbf{D}$ for some diagonal matrix $\mathbf{D}$. Take $\mathbf{U} = \mathbf{T}^{-1}$. Then, $\mathbf{A} = \mathbf{U}'\mathbf{D}\mathbf{U}$, and (according to Corollary 8.5.9) $\mathbf{U}$ is unit upper triangular. Thus, $\mathbf{A} = \mathbf{U}'\mathbf{D}\mathbf{U}$ is a U′DU decomposition of $\mathbf{A}$. Q.E.D.

Note that, as an immediate consequence of Corollary 14.2.15, we have the following converse to Corollary 14.5.15: If an $n \times n$ matrix $\mathbf{A}$ has a U′DU decomposition, say $\mathbf{A} = \mathbf{U}'\mathbf{D}\mathbf{U}$, and if the diagonal elements of the diagonal matrix $\mathbf{D}$ are nonnegative, then $\mathbf{A}$ is nonnegative definite (and symmetric).

Theorem 14.5.11 establishes that any symmetric positive definite matrix $\mathbf{A}$ has a unique decomposition of the form (5.11) and relates this decomposition to the U′DU decomposition of $\mathbf{A}$. The following theorem extends these results to any symmetric nonnegative definite matrix.

Theorem 14.5.16. Let $\mathbf{A}$ represent an $n \times n$ symmetric nonnegative definite matrix, and let $r = \operatorname{rank}(\mathbf{A})$. Then, there exists a unique upper triangular matrix $\mathbf{T}$ with r positive diagonal elements and with $n - r$ null rows such that

$$\mathbf{A} = \mathbf{T}'\mathbf{T}. \tag{5.13}$$

Moreover, taking $\mathbf{U}$ to be a unit upper triangular matrix and $\mathbf{D} = \{d_i\}$ to be the unique diagonal matrix such that $\mathbf{A} = \mathbf{U}'\mathbf{D}\mathbf{U}$,

$$\mathbf{T} = \mathbf{D}^{1/2}\mathbf{U}, \tag{5.14}$$

where $\mathbf{D}^{1/2} = \operatorname{diag}(\sqrt{d_1}, \sqrt{d_2}, \ldots, \sqrt{d_n})$.

Proof. Note that Corollary 14.5.15 and Theorem 14.5.5 guarantee the existence of the unit upper triangular matrix $\mathbf{U}$ and the existence and uniqueness of the diagonal matrix $\mathbf{D}$ and also guarantee that the diagonal elements of $\mathbf{D}$ are nonnegative. Note further (on the basis of Lemma 14.3.1) that r of the diagonal elements of $\mathbf{D}$, say the $i_1, \ldots, i_r$th diagonal elements (where $i_1 < \cdots < i_r$), are nonzero and hence positive, and that the remaining $n - r$ diagonal elements of $\mathbf{D}$ equal zero. Moreover, $\mathbf{D}^{1/2}\mathbf{U}$ is upper triangular, the diagonal elements of $\mathbf{D}^{1/2}\mathbf{U}$ are $\sqrt{d_1}, \sqrt{d_2}, \ldots, \sqrt{d_n}$ (r of which are positive), $n - r$ rows of $\mathbf{D}^{1/2}\mathbf{U}$ are null (those rows other than the $i_1, \ldots, i_r$th rows), and $\mathbf{A} = (\mathbf{D}^{1/2}\mathbf{U})'\mathbf{D}^{1/2}\mathbf{U}$. Thus, there exists an upper triangular matrix $\mathbf{T}$ with r positive diagonal elements and with $n - r$ null rows such that $\mathbf{A} = \mathbf{T}'\mathbf{T}$. It remains to show that there is only one such upper triangular matrix with r positive diagonal elements and with $n - r$ null rows.

Suppose that $\mathbf{T} = \{t_{ij}\}$ is an $n \times n$ upper triangular matrix with r positive diagonal elements—say the $k_1, \ldots, k_r$th diagonal elements (where $k_1 < \cdots < k_r$)—and with $n - r$ null rows such that $\mathbf{A} = \mathbf{T}'\mathbf{T}$. Take $\mathbf{D}_*$ to be the $n \times n$ diagonal matrix whose $k_1, \ldots, k_r$th diagonal elements are $t_{k_1k_1}^2, \ldots, t_{k_rk_r}^2$, respectively, and whose other $n - r$ diagonal elements equal zero. Further, let $\mathbf{t}'_{k_1}, \ldots, \mathbf{t}'_{k_r}$ represent the $k_1, \ldots, k_r$th rows of $\mathbf{T}$, and take $\mathbf{U}_*$ to be any unit upper triangular matrix whose $k_1, \ldots, k_r$th rows are $(1/t_{k_1k_1})\mathbf{t}'_{k_1}, \ldots, (1/t_{k_rk_r})\mathbf{t}'_{k_r}$, respectively—clearly, such a unit upper triangular matrix exists. Then,

$$\mathbf{A} = \mathbf{t}_{k_1}\mathbf{t}'_{k_1} + \cdots + \mathbf{t}_{k_r}\mathbf{t}'_{k_r} = \mathbf{U}'_*\mathbf{D}_*\mathbf{U}_*,$$

so that $\mathbf{A} = \mathbf{U}'_*\mathbf{D}_*\mathbf{U}_*$ is a U$'$DU decomposition of $\mathbf{A}$. Thus, it follows from Theorem 14.5.5 that $\mathbf{D}_* = \mathbf{D}$, in which case $k_1 = i_1, \ldots, k_r = i_r$ and $t_{i_1i_1} = \sqrt{d_{i_1}}, \ldots, t_{i_ri_r} = \sqrt{d_{i_r}}$, and that the $i_1, \ldots, i_r$th rows of $\mathbf{U}_*$ are respectively equal to the $i_1, \ldots, i_r$th rows of $\mathbf{U}$. Consequently,

$$\mathbf{T} = \mathbf{D}^{1/2}\mathbf{U}_* = \mathbf{D}^{1/2}\mathbf{U}.$$

We conclude that the only upper triangular matrix $\mathbf{T}$ with r positive diagonal elements and with $n - r$ null rows such that $\mathbf{A} = \mathbf{T}'\mathbf{T}$ is $\mathbf{T} = \mathbf{D}^{1/2}\mathbf{U}$. Q.E.D.

In the special case where the symmetric nonnegative definite matrix $\mathbf{A}$ is positive definite, decomposition (5.13) simplifies to decomposition (5.11). And, as in that special case, the decomposition is called the *Cholesky decomposition.*

e. Recursive formulas for LDU and Cholesky decompositions

Let $\mathbf{A} = \{a_{ij}\}$ represent an $n \times n$ matrix. Consider the problem of finding a unit upper triangular matrix $\mathbf{U} = \{u_{ij}\}$, a unit lower triangular matrix $\mathbf{L} = \{\ell_{ji}\}$, and a diagonal matrix $\mathbf{D} = \{d_i\}$ such that $\mathbf{A} = \mathbf{LDU}$ (when they exist). The formulas given in Part (1) of Theorem 14.5.4 can be used to construct $\mathbf{L}$, $\mathbf{D}$, and $\mathbf{U}$ in n steps. At each step, an additional column of $\mathbf{U}$, an additional row of $\mathbf{L}$, and an additional diagonal element of $\mathbf{D}$ are obtained by solving linear systems with triangular coefficient matrices. By making use of the results of Section 11.8 (on the solution of linear systems with nonsingular triangular or block-triangular coefficient matrices), the formulas given in Part (1) of Theorem 14.5.4 can be reexpressed as follows:

$$d_1 = a_{11},$$

$$d_1 u_{1j} = a_{1j},$$

$$d_i u_{ij} = a_{ij} - \sum_{k=1}^{i-1} \ell_{ik} d_k u_{kj} \quad (i = 2, 3, \ldots, j-1),$$

$$d_1 \ell_{j1} = a_{j1},$$

$$d_i \ell_{ji} = a_{ji} - \sum_{k=1}^{i-1} u_{ki} d_k \ell_{jk} \quad (i = 2, 3, \ldots, j-1),$$

$$d_j = a_{jj} - \sum_{k=1}^{j-1} u_{kj} d_k \ell_{jk} \tag{5.15}$$

$(j = 2, 3, \ldots, n)$.

The first step in the n-step procedure for constructing **U**, **L**, and **D** consists of setting $d_1 = a_{11}$. The jth step $(2 \leq j \leq n)$ consists of successively determining u_{ij} and ℓ_{ji} for $i = 1, 2, \ldots, j-1$, and of then determining [from formula (5.15)] d_j. If $d_i = 0$, then u_{ij} and ℓ_{ji} can be chosen arbitrarily; for example, choose $u_{ij} = \ell_{ji} = 0$. If $d_i \neq 0$, then, depending on whether $i = 1$ or $i > 1$, take $u_{ij} = a_{ij}/d_i$ and $\ell_{ji} = a_{ji}/d_i$ or

$$u_{ij} = \bigl(a_{ij} - \textstyle\sum_{k=1}^{i-1} \ell_{ik} d_k u_{kj}\bigr)/d_i \text{ and } \ell_{ji} = \bigl(a_{ji} - \textstyle\sum_{k=1}^{j-1} u_{ki} d_k \ell_{jk}\bigr)/d_i,$$

respectively.

There is an alternative n-step procedure for constructing **U**, **L**, and **D** in which **U** is formed row-by-row, rather than column-by-column, and **L** is formed column-by-column, rather than row-by-row. The first step consists of setting $d_1 = a_{11}$ and, depending on whether $d_1 = 0$ or $d_1 \neq 0$, choosing $u_{12}, \ldots, u_{1n}$ and $\ell_{21}, \ldots, \ell_{n1}$ arbitrarily or (for $j = 2, \ldots, n$) setting $u_{1j} = a_{1j}/d_1$ and $\ell_{j1} = a_{j1}/d_1$. The ith step $(2 \leq i \leq n-1)$ consists of determining [from formula (5.15)] d_i and, depending on whether $d_i = 0$ or $d_i \neq 0$, choosing $u_{i,i+1}, \ldots, u_{in}$ and $\ell_{i+1,i}, \ldots, \ell_{ni}$ arbitrarily or (for $j = i+1, \ldots, n$) setting

$$u_{ij} = \bigl(a_{ij} - \textstyle\sum_{k=1}^{i-1} \ell_{ik} d_k u_{kj}\bigr)/d_i \text{ and } \ell_{ji} = \bigl(a_{ji} - \textstyle\sum_{k=1}^{i-1} u_{ki} d_k \ell_{jk}\bigr)/d_i.$$

The final (nth) step consists of setting

$$d_n = a_{nn} - \sum_{k=1}^{n-1} u_{kn} d_k \ell_{nk}.$$

Note that (with either of these two n-step procedures), once d_j has been determined, a_{jj} is no longer needed (at least not for purposes of constructing an LDU decomposition). Similarly, once u_{ij} has been determined, a_{ij} is no longer required, and, once ℓ_{ji} has been determined, a_{ji} is no longer required. Thus, in implementing either n-step procedure on a computer, storage can be saved if, as d_j, u_{ij}, and ℓ_{ji} are determined, they are placed in the locations previously occupied by a_{jj}, a_{ij}, and a_{ji}, respectively.

So far, it has been presumed that **A** has an LDU decomposition. Suppose that there is uncertainty about the existence of an LDU decomposition. The nonexistence of an LDU decomposition can (at least in principle) be determined during the course of the n-step procedures by incorporating certain tests. If $d_1 = 0$, then—prior to assigning values to u_{1j} and ℓ_{j1}—we should test whether $a_{1j} = 0$ and $a_{j1} = 0$; similarly, if $d_i = 0$ $(2 \leq i \leq n-1)$, then—prior to assigning values to u_{ij} and ℓ_{ji}—we should test whether

$$a_{ij} - \sum_{k=1}^{i-1} \ell_{ik} d_k u_{kj} = 0 \quad \text{and} \quad a_{ji} - \sum_{k=1}^{i-1} u_{ki} d_k \ell_{jk} = 0$$

or, equivalently, whether

$$a_{ij} = \sum_{k=1}^{i-1} \ell_{ik} d_k u_{kj} \quad \text{and} \quad a_{ji} = \sum_{k=1}^{i-1} u_{ki} d_k \ell_{jk}.$$

If the result of any test is negative, we terminate the n-step procedure and conclude (on the basis of Theorem 14.5.4) that **A** does not have an LDU decomposition. If the result of every test is positive, then the n-step procedure is carried to completion and produces an LDU decomposition.

Suppose now that **A** is symmetric, and consider the problem of determining whether **A** has a U′DU decomposition and of constructing a U′DU decomposition. This problem can be "solved" by employing a simplified version of the procedure for checking for the existence of an LDU decomposition and for constructing an LDU decomposition. The simplification comes from setting $\ell_{ji} = u_{ij}$, for $j = 2, \ldots, n$ and $i = 1, \ldots, j-1$, and more specifically from the resulting reduction in computational and storage requirements.

Finally, suppose that **A** is a symmetric nonnegative definite matrix, and consider the problem of finding the Cholesky decomposition of **A**. That is, consider the problem of finding the unique upper triangular matrix $\mathbf{T} = \{t_{ij}\}$ with r positive diagonal elements and $n - r$ null rows [where $r = \text{rank}(\mathbf{A})$] such that $\mathbf{A} = \mathbf{T}'\mathbf{T}$. One approach is to find the U′DU decomposition of **A** and to then determine **T** from formula (5.14). However, there is a more direct method, which is sometimes called the *square root method.* The square root method is based on the following formulas, which can be obtained from the formulas of Part (1′) of Theorem 14.5.4 by applying the results of Section 11.8 and making use of relationship (5.14):

$$t_{11} = \sqrt{a_{11}}, \tag{5.16}$$

$$t_{1j} = a_{1j}/t_{11}, \quad \text{if } t_{11} > 0, \tag{5.17a}$$

$$= 0, \quad \text{if } t_{11} = 0, \tag{5.17b}$$

$$t_{ij} = \left(a_{ij} - \textstyle\sum_{k=1}^{i-1} t_{ki}t_{kj}\right)/t_{ii}, \qquad \text{if } t_{ii} > 0, \tag{5.18a}$$
$$= 0, \qquad \text{if } t_{ii} = 0 \tag{5.18b}$$

$(i = 2, 3, \ldots, j-1)$,

$$t_{jj} = \left(a_{jj} - \textstyle\sum_{k=1}^{j-1} t_{kj}^2\right)^{1/2} \tag{5.19}$$

$(j = 2, 3, \ldots, n)$.

In the square root method, **T** is constructed row-by-row or column-by-column in n steps, one row or column per step. For example, the first step of the row-by-row version consists of determining [from formulas (5.16) and (5.17)] $t_{11}, t_{12}, \ldots, t_{1n}$; the ith step $(2 \le i \le n-1)$ consists of determining [from formulas (5.19) and (5.18)] $t_{ii}, t_{i,i+1}, \ldots, t_{in}$; and the final ($n$th) step consists of determining [from formula (5.19)] t_{nn}.

If there is uncertainty about whether **A** is nonnegative definite, then, in applying the square root method to **A**, various modifications should be incorporated. We should (prior to determining t_{11}) test whether $a_{11} \ge 0$, and if $a_{11} = 0$, we should (prior to setting $t_{1j} = 0$) test whether $a_{1j} = 0$. Similarly, for $2 \le i \le n-1$, we should (prior to determining t_{ii}) test whether $a_{ii} - \sum_{k=1}^{i-1} t_{ki}^2 \ge 0$, and if $a_{ii} - \sum_{k=1}^{i-1} t_{ki}^2 = 0$, we should (prior to setting $t_{ij} = 0$) test whether $a_{ij} - \sum_{k=1}^{i-1} t_{ki}t_{kj} = 0$. Finally, we should (prior to determining t_{nn}) test whether $a_{nn} - \sum_{k=1}^{n-1} t_{kn}^2 \ge 0$. If the result of any test is negative, it follows from Theorems 14.5.4 and 14.5.16 that either **A** does not have a U′DU decomposition or, for a U′DU decomposition (of **A**), say $\mathbf{A} = \mathbf{U}'\mathbf{D}\mathbf{U}$, one or more of the diagonal elements of the diagonal matrix **D** are negative. Thus, if the result of any test is negative, we terminate the square root procedure and conclude (on the basis of Corollary 14.5.15) that **A** is not nonnegative definite. If the result of every test is positive, then **A** is nonnegative definite, and the procedure is carried to completion and produces the Cholesky decomposition.

Suppose, for example, that

$$\mathbf{A} = \begin{pmatrix} 4 & 2 & -2 \\ 2 & 1 & -1 \\ -2 & -1 & 10 \end{pmatrix}.$$

Let us apply the modified square root method (the row-by-row version). Observing that $a_{11} > 0$, we find that $t_{11} = \sqrt{4} = 2$, $t_{12} = 2/2 = 1$, and $t_{13} = -2/2 = -1$. Further, upon observing that $a_{22} - t_{12}^2 = 0$, we find that $t_{22} = 0$, and, after observing that $a_{23} - t_{12}t_{13} = 0$, we set $t_{23} = 0$. Finally, upon observing that $a_{33} - \sum_{k=1}^{2} t_{k3}^2 = 9$, we find that $t_{33} = \sqrt{9} = 3$. We conclude that **A** is nonnegative definite—since $t_{22} = 0$, it is positive semidefinite—and that the Cholesky decomposition of **A** is

$$\mathbf{A} = \begin{pmatrix} 2 & 1 & -1 \\ 0 & 0 & 0 \\ 0 & 0 & 3 \end{pmatrix}' \begin{pmatrix} 2 & 1 & -1 \\ 0 & 0 & 0 \\ 0 & 0 & 3 \end{pmatrix}.$$

Note that if $a_{23} = a_{32}$ had equaled any number other than -1, it would have been the case that $a_{23} - t_{12}t_{13} \neq 0$, and $\mathbf{A}$ would not have been nonnegative definite. Or, if the value of a_{22} had been smaller than one, we would have had that $a_{22} - t_{12}^2 < 0$, and $\mathbf{A}$ would not have been nonnegative definite.

f. Use of an LDU, U′DU, or Cholesky decomposition in obtaining a generalized inverse

Suppose that $\mathbf{A}$ is a (square) matrix that has an LDU decomposition, say $\mathbf{A} = \mathbf{LDU}$. Then, it follows from the discussion in Section 14.4 that the matrix

$$\mathbf{G} = \mathbf{U}^{-1}\mathbf{D}^{-}\mathbf{L}^{-1} \tag{5.20}$$

is a generalized inverse of $\mathbf{A}$ and that one choice for $\mathbf{D}^{-}$ is the diagonal matrix obtained by replacing the nonzero diagonal elements of the (diagonal) matrix $\mathbf{D}$ with their reciprocals.

Thus, one way to obtain a generalized inverse of $\mathbf{A}$ is to obtain the LDU decomposition of $\mathbf{A}$ and to then apply formula (5.20). In this regard, note that the recursive procedure described in Section 8.5d can be used to invert $\mathbf{U}$ and $\mathbf{L}$. Note also that, in the special case where $\mathbf{L} = \mathbf{U}'$, formula (5.20) simplifies to

$$\mathbf{G} = \mathbf{U}^{-1}\mathbf{D}^{-}(\mathbf{U}^{-1})'. \tag{5.21}$$

Suppose now that $\mathbf{A}$ is an $n \times n$ symmetric nonnegative definite matrix of rank r. Then, we can obtain a generalized inverse of $\mathbf{A}$ in terms of its U′DU decomposition [using formula (5.21)]. We can also obtain a generalized inverse of $\mathbf{A}$ directly in terms of its Cholesky decomposition. To see this, let $\mathbf{A} = \mathbf{T}'\mathbf{T}$ represent the Cholesky decomposition of $\mathbf{A}$, so that, by definition, $\mathbf{T}$ is an upper triangular matrix with r positive diagonal elements, say its $i_1, i_2, \ldots, i_r$th diagonal elements (where $i_1 < i_2 < \cdots < i_r$), and with $n - r$ null rows.

Take $\mathbf{T}_1$ to be the $r \times n$ matrix obtained by striking out the null rows of $\mathbf{T}$. Then, clearly, $\mathbf{A} = \mathbf{T}_1'\mathbf{T}_1$, and $\text{rank}(\mathbf{T}_1) = \text{rank}(\mathbf{A}) = r$. It follows from Lemma 8.1.1 that $\mathbf{T}_1$ has a right inverse, say $\mathbf{R}$. And, since $\mathbf{ARR}'\mathbf{A} = \mathbf{T}_1'\mathbf{T}_1\mathbf{R}(\mathbf{T}_1\mathbf{R})'\mathbf{T}_1 = \mathbf{T}_1'\mathbf{T}_1 = \mathbf{A}$, $\mathbf{RR}'$ is a generalized inverse of $\mathbf{A}$.

A right inverse of $\mathbf{T}_1$ can be obtained by making use of the procedure described in Section 8.5d for inverting a nonsingular triangular matrix. To see this, let $\mathbf{T}_{11}$ represent the $r \times r$ submatrix obtained by striking out all of the columns of $\mathbf{T}_1$ except the $i_1, i_2, \ldots, i_r$th columns. Note that $\mathbf{T}_{11}$ is a principal submatrix of $\mathbf{T}$ and hence (like $\mathbf{T}$) is upper triangular and (since the diagonal elements of $\mathbf{T}_{11}$ are nonzero) that $\mathbf{T}_{11}$ is nonsingular. Take $\mathbf{R}$ to be the $n \times r$ matrix whose $i_1, i_2, \ldots, i_r$th rows are the first, second, ..., rth rows of $\mathbf{T}_{11}^{-1}$ and whose other $n - r$ rows are null. Then, clearly, $\mathbf{T}_1\mathbf{R} = \mathbf{T}_{11}\mathbf{T}_{11}^{-1} = \mathbf{I}$, so that $\mathbf{R}$ is a right inverse of $\mathbf{T}_1$.

14.6 Skew-Symmetric Matrices

An $n \times n$ matrix $\mathbf{A} = \{a_{ij}\}$ is said to be *skew-symmetric* if $\mathbf{A}' = -\mathbf{A}$ or, equivalently (since $a_{ii} = -a_{ii} \Leftrightarrow 2a_{ii} = 0 \Leftrightarrow a_{ii} = 0$), if $a_{ii} = 0$ for $i = 1, \ldots, n$ and

$a_{ji} = -a_{ij}$ for $j \neq i = 1, \ldots, n$. For example, the 2×2 matrix $\begin{pmatrix} 0 & 1 \\ -1 & 0 \end{pmatrix}$ is skew-symmeric.

A principal submatrix of a skew-symmetric matrix is skew-symmetric, as is easily verified. Other basic properties of skew-symmetric matrices are described in the following two lemmas.

Lemma 14.6.1. The only $n \times n$ matrix that is both symmetric and skew-symmetric is the $n \times n$ null matrix.

Proof. If an $n \times n$ matrix $\mathbf{A}$ is both symmetric and skew-symmetric, then $\mathbf{A} = \mathbf{A}' = -\mathbf{A}$, implying that $2\mathbf{A} = \mathbf{0}$ and hence that $\mathbf{A} = \mathbf{0}$. Q.E.D.

Lemma 14.6.2. Let $\mathbf{A}$ represent an $n \times n$ matrix and $\mathbf{P}$ an $n \times m$ matrix. If $\mathbf{A}$ is skew-symmetric, then $\mathbf{P}'\mathbf{A}\mathbf{P}$ is skew-symmetric.

Proof. If $\mathbf{A}$ is skew-symmetric, then

$$(\mathbf{P}'\mathbf{A}\mathbf{P})' = \mathbf{P}'\mathbf{A}'\mathbf{P} = \mathbf{P}'(-\mathbf{A})\mathbf{P} = -\mathbf{P}'\mathbf{A}\mathbf{P}.$$

Q.E.D.

The following lemma, which is an immediate consequence of Lemma 14.1.1, characterizes the skew-symmetry of a (square) matrix $\mathbf{A}$ in terms of the quadratic form whose matrix is $\mathbf{A}$.

Lemma 14.6.3. An $n \times n$ matrix $\mathbf{A}$ is skew-symmetric if and only if, for every n-dimensional column vector $\mathbf{x}$, $\mathbf{x}'\mathbf{A}\mathbf{x} = 0$.

There is a relationship between skew-symmetry and nonnegative definiteness, which is described in the following lemma.

Lemma 14.6.4. An $n \times n$ matrix $\mathbf{A} = \{a_{ij}\}$ is skew-symmetric if and only if it is nonnegative definite and its diagonal elements $a_{11}, a_{22}, \ldots, a_{nn}$ equal zero.

Proof. Suppose that $\mathbf{A}$ is skew-symmetric. Then, by definition, $a_{ii} = 0$ $(i = 1, 2, \ldots, n)$. Moreover, it follows from Lemma 14.6.3 that $\mathbf{A}$ is nonnegative definite.

Conversely, suppose that $\mathbf{A}$ is nonnegative definite and that its diagonal elements equal zero. Then, it follows from Lemma 14.5.13 that $a_{ji} = -a_{ij}$ for $i, j = 1, 2, \ldots, n$. Thus, $\mathbf{A}$ is skew-symmetric. Q.E.D.

14.7 Trace of a Nonnegative Definite Matrix

a. Basic results

The following result is an immediate consequence of Corollary 14.2.13.

Theorem 14.7.1. For any positive definite matrix $\mathbf{A}$, $\mathrm{tr}(\mathbf{A}) > 0$.

The following theorem extends the result of Theorem 14.7.1 to nonnegative definite matrices that are not positive definite, that is, to positive semidefinite matrices.

Theorem 14.7.2. Let $\mathbf{A} = \{a_{ij}\}$ represent an $n \times n$ nonnegative definite matrix. Then, $\text{tr}(\mathbf{A}) \geq 0$, with equality holding if and only if the diagonal elements $a_{11}, a_{22}, \ldots, a_{nn}$ of $\mathbf{A}$ equal zero or, equivalently, if and only if $\mathbf{A}$ is skew-symmetric.

Proof. According to Corollary 14.2.13, $a_{ii} \geq 0 \; (i = 1, 2, \ldots, n)$. Thus,

$$\text{tr}(\mathbf{A}) = a_{11} + a_{22} + \cdots + a_{nn} \geq 0,$$

with $\text{tr}(\mathbf{A}) = 0$ if and only if $a_{11}, a_{22}, \ldots, a_{nn}$ equal zero. That the nonnegative definite matrix $\mathbf{A}$ being skew-symmetric is equivalent to the diagonal elements of $\mathbf{A}$ being zero is an immediate consequence of Lemma 14.6.4. Q.E.D.

In the special case where $\mathbf{A}$ is a symmetric nonnegative definite matrix, Theorem 14.7.2 can (in light of Lemma 14.6.1) be restated as follows.

Corollary 14.7.3. Let $\mathbf{A} = \{a_{ij}\}$ represent an $n \times n$ symmetric nonnegative definite matrix. Then, $\text{tr}(\mathbf{A}) \geq 0$, with equality holding if and only if the diagonal elements $a_{11}, a_{22}, \ldots, a_{nn}$ of $\mathbf{A}$ equal zero or, equivalently, if and only if $\mathbf{A} = \mathbf{0}$.

b. Extensions to products of nonnegative definite matrices

The following theorem extends Theorem 14.7.1.

Theorem 14.7.4. For any $n \times n$ positive definite matrix $\mathbf{A}$ and any $n \times n$ nonnull symmetric nonnegative definite matrix $\mathbf{B}$, $\text{tr}(\mathbf{AB}) > 0$.

Proof. Let $r = \text{rank}(\mathbf{B})$. Then, according to Theorem 14.3.7, $\mathbf{B} = \mathbf{Q}'\mathbf{Q}$ for some $r \times n$ matrix $\mathbf{Q}$ of rank r. Since (according to Theorem 14.2.9) $\mathbf{QAQ}'$ is positive definite, we find (using Lemma 5.2.1 and Theorem 14.7.1) that

$$\text{tr}(\mathbf{AB}) = \text{tr}(\mathbf{AQ}'\mathbf{Q}) = \text{tr}(\mathbf{QAQ}') > 0.$$

Q.E.D.

As an essentially immediate consequence of Theorem 14.7.4, we have the following corollary, which extends Corollary 14.7.3.

Corollary 14.7.5. Let $\mathbf{A}$ represent an $n \times n$ positive definite matrix and $\mathbf{B}$ an $n \times n$ symmetric nonnegative definite matrix. Then, $\text{tr}(\mathbf{AB}) \geq 0$, with equality holding if and only if $\mathbf{B} = \mathbf{0}$.

In the special case where $\mathbf{B} = \mathbf{I}$, Theorem 14.7.4 reduces to Theorem 14.7.1, and, in the special case where $\mathbf{A} = \mathbf{I}$, Corollary 14.7.5 reduces to Corollary 14.7.3.

The following theorem extends Theorem 14.7.2.

Theorem 14.7.6. Let $\mathbf{A}$ represent an $n \times n$ nonnegative definite matrix and $\mathbf{B}$ an $n \times n$ symmetric nonnegative definite matrix. Then, $\text{tr}(\mathbf{AB}) \geq 0$, with equality holding if and only if $\mathbf{BAB}$ is skew-symmetric.

Proof. The theorem is true in the special case where $\mathbf{B} = \mathbf{0}$, since in that special case $\text{tr}(\mathbf{AB}) = 0$ and $\mathbf{BAB} = \mathbf{0}$. Thus, in proving the theorem, it suffices to restrict attention to the case where $\mathbf{B} \neq \mathbf{0}$.

Suppose that $\mathbf{B} \neq \mathbf{0}$, and let $r = \text{rank}(\mathbf{B})$. Then, according to Theorem 14.3.7, $\mathbf{B} = \mathbf{Q}'\mathbf{Q}$ for some $r \times n$ matrix $\mathbf{Q}$ of rank r. Observe [in light of result (5.2.3)] that

$$\text{tr}(\mathbf{AB}) = \text{tr}(\mathbf{AQ}'\mathbf{Q}) = \text{tr}(\mathbf{QAQ}'). \tag{7.1}$$

According to Theorem 14.2.9, $\mathbf{QAQ}'$ is nonnegative definite. Thus, it follows from Theorem 14.7.2 that $\text{tr}(\mathbf{QAQ}') \geq 0$, or, equivalently, that $\text{tr}(\mathbf{AB}) \geq 0$, with equality holding if and only if $\mathbf{QAQ}'$ is skew-symmetric or, equivalently, if and only if

$$\mathbf{QA}'\mathbf{Q}' = -\mathbf{QAQ}'. \tag{7.2}$$

Since $\mathbf{Q}$ is of full row rank and $\mathbf{Q}'$ of full column rank, we have (as a consequence of Lemma 8.3.1) that condition (7.2) is equivalent to the condition

$$\mathbf{Q}'\mathbf{QA}'\mathbf{Q}'\mathbf{Q} = -\mathbf{Q}'\mathbf{QAQ}'\mathbf{Q}. \tag{7.3}$$

Moreover, condition (7.3) can be reexpressed as

$$(\mathbf{BAB})' = -\mathbf{BAB}.$$

Thus, condition (7.3) is satisfied if and only if $\mathbf{BAB}$ is skew-symmetric. Q.E.D.

In the special case where $\mathbf{B} = \mathbf{I}$, Theorem 14.7.6 reduces to Theorem 14.7.2. If $\mathbf{A}$ is symmetric (as well as nonnegative definite), then the results of Theorem 14.7.6 can be sharpened, as described in the following corollary, which generalizes (in a different way than Corollary 14.7.5) Corollary 14.7.3.

Corollary 14.7.7. Let $\mathbf{A}$ and $\mathbf{B}$ represent $n \times n$ symmetric nonnegative definite matrices. Then, $\text{tr}(\mathbf{AB}) \geq 0$, with equality holding if and only if $\mathbf{AB} = \mathbf{0}$.

Proof. Clearly, $(\mathbf{BAB})' = \mathbf{B}'\mathbf{A}'\mathbf{B}' = \mathbf{BAB}$; that is, $\mathbf{BAB}$ is symmetric, implying (in light of Lemma 14.6.1) that $\mathbf{BAB}$ is skew-symmetric if and only if $\mathbf{BAB} = \mathbf{0}$. Thus, it follows from Theorem 14.7.6 that $\text{tr}(\mathbf{AB}) \geq 0$, with equality holding if and only if

$$\mathbf{BAB} = \mathbf{0}. \tag{7.4}$$

To complete the proof, it suffices to show that condition (7.4) is equivalent to the condition $\mathbf{AB} = \mathbf{0}$. In the special case where $\mathbf{A} = \mathbf{0}$, these two conditions are clearly equivalent. Thus, it suffices to restrict attention to the case where $\mathbf{A} \neq \mathbf{0}$.

Suppose that $\mathbf{A} \neq \mathbf{0}$, and let $r = \text{rank}(\mathbf{A})$. According to Theorem 14.3.7, $\mathbf{A} = \mathbf{Q}'\mathbf{Q}$ for some $r \times n$ matrix $\mathbf{Q}$ of rank r. Thus, condition (7.4) can be reexpressed as $(\mathbf{QB})'\mathbf{QB} = \mathbf{0}$, which (according to Corollary 5.3.2) is equivalent to the condition

$$\mathbf{QB} = \mathbf{0}. \tag{7.5}$$

Moreover, since $\mathbf{Q}'$ is of full column rank, it follows from Lemma 8.3.1 that condition (7.5) is equivalent to the condition

$$\mathbf{Q}'\mathbf{QB} = \mathbf{Q}'\mathbf{0},$$

which is reexpressible as $\mathbf{AB} = \mathbf{0}$. We conclude that condition (7.4) is equivalent to the condition $\mathbf{AB} = \mathbf{0}$. Q.E.D.

In the special case where $\mathbf{A} = \mathbf{I}$ or $\mathbf{B} = \mathbf{I}$, Corollary 14.7.7 reduces to Corollary 14.7.3.

14.8 Partitioned Nonnegative Definite Matrices

a. Applicability of formulas for ranks, determinants, inverses, and generalized inverses

Consider a (square) partitioned matrix $\mathbf{A}$ of the form

$$\mathbf{A} = \begin{pmatrix} \mathbf{T} & \mathbf{U} \\ \mathbf{V} & \mathbf{W} \end{pmatrix},$$

where $\mathbf{T}$ is an $m \times m$ (square) matrix, $\mathbf{W}$ is an $n \times n$ (square) matrix, $\mathbf{U}$ is of dimensions $m \times n$, and $\mathbf{V}$ is of dimensions $n \times m$, or consider a partitioned matrix $\mathbf{B}$ of the form

$$\mathbf{B} = \begin{pmatrix} \mathbf{W} & \mathbf{V} \\ \mathbf{U} & \mathbf{T} \end{pmatrix}.$$

Suppose that $\mathbf{T}$ is nonsingular. Then, formula (13.3.13) can be used to express $|\mathbf{A}|$ or $|\mathbf{B}|$ in terms of the determinant of $\mathbf{T}$ and the determinant of the Schur complement $\mathbf{W} - \mathbf{V}\mathbf{T}^{-1}\mathbf{U}$ of $\mathbf{T}$, and formula (8.5.15) can be used to express $\text{rank}(\mathbf{A})$ or $\text{rank}(\mathbf{B})$ in terms of the order m of $\mathbf{T}$ and the rank of the Schur complement of $\mathbf{T}$. Suppose further that $\mathbf{A}$ or $\mathbf{B}$ is nonsingular. Then, formula (8.5.16) or (8.5.17) can be used to express $\mathbf{A}^{-1}$ or $\mathbf{B}^{-1}$ in terms of the inverse of $\mathbf{T}$ and the inverse of the Schur complement of $\mathbf{T}$.

Is $\mathbf{A}$ or $\mathbf{B}$ nonsingular—and is $\mathbf{T}$ nonsingular—if $\mathbf{A}$ or $\mathbf{B}$ is nonnegative definite? If $\mathbf{A}$ or $\mathbf{B}$ is positive definite, then it is clear from Lemma 14.2.8 that $\mathbf{A}$ or $\mathbf{B}$, respectively, is nonsingular and from Corollary 14.2.12 that $\mathbf{T}$ is nonsingular. If $\mathbf{A}$ or $\mathbf{B}$ is a symmetric positive semidefinite matrix, then it is clear from Corollary 14.3.12 that $\mathbf{A}$ or $\mathbf{B}$, respectively, is singular. If $\mathbf{A}$ or $\mathbf{B}$ is a nonsymmetric positive semidefinite matrix, then $\mathbf{A}$ or $\mathbf{B}$, respectively, may or may not be nonsingular. Moreover, even if $\mathbf{A}$ or $\mathbf{B}$ is a nonsingular positive semidefinite matrix, $\mathbf{T}$ may be singular.

Suppose now that $\mathbf{T}$ is possibly singular. If $\mathcal{C}(\mathbf{U}) \subset \mathcal{C}(\mathbf{T})$ and $\mathcal{R}(\mathbf{V}) \subset \mathcal{R}(\mathbf{T})$, then formula (9.6.2) or (9.6.3) gives a generalized inverse of $\mathbf{A}$ or $\mathbf{B}$, respectively, in terms of an arbitrary generalized inverse $\mathbf{T}^-$ of $\mathbf{T}$ and an arbitrary generalized inverse of the Schur complement $\mathbf{W} - \mathbf{V}\mathbf{T}^-\mathbf{U}$ of $\mathbf{T}$, and formula (9.6.1) can be used to express $\text{rank}(\mathbf{A})$ or $\text{rank}(\mathbf{B})$ in terms of the rank of $\mathbf{T}$ and the rank of the Schur complement of $\mathbf{T}$.

Are the conditions $\mathcal{C}(\mathbf{U}) \subset \mathcal{C}(\mathbf{T})$ and $\mathcal{R}(\mathbf{V}) \subset \mathcal{R}(\mathbf{T})$ satisfied if $\mathbf{A}$ or $\mathbf{B}$ is positive semidefinite? If $\mathbf{A}$ or $\mathbf{B}$ is symmetric as well as positive semidefinite, the answer to this question is yes, as is evident from the corollary of the following lemma.

Lemma 14.8.1. Let $\mathbf{A}$ represent a symmetric nonnegative definite matrix that has been partitioned as

$$\mathbf{A} = \begin{pmatrix} \mathbf{T} & \mathbf{U} \\ \mathbf{V} & \mathbf{W} \end{pmatrix},$$

where $\mathbf{T}$ (and hence $\mathbf{W}$) is square. Then, there exist matrices $\mathbf{R}$ and $\mathbf{S}$ such that

$$\mathbf{T} = \mathbf{R}'\mathbf{R}, \quad \mathbf{U} = \mathbf{R}'\mathbf{S}, \quad \mathbf{V} = \mathbf{S}'\mathbf{R}, \quad \mathbf{W} = \mathbf{S}'\mathbf{S}.$$

Proof. According to Corollary 14.3.8, there exists a matrix $\mathbf{X}$ such that $\mathbf{A} = \mathbf{X}'\mathbf{X}$. Partition $\mathbf{X}$ as $\mathbf{X} = (\mathbf{R}, \ \mathbf{S})$ (where $\mathbf{R}$ has the same number of columns as $\mathbf{T}$). Then,

$$\mathbf{A} = \mathbf{X}'\mathbf{X} = \begin{pmatrix} \mathbf{R}' \\ \mathbf{S}' \end{pmatrix} (\mathbf{R}, \ \mathbf{S}) = \begin{pmatrix} \mathbf{R}'\mathbf{R} & \mathbf{R}'\mathbf{S} \\ \mathbf{S}'\mathbf{R} & \mathbf{S}'\mathbf{S} \end{pmatrix}.$$

Since $\mathbf{R}'\mathbf{R}$, $\mathbf{R}'\mathbf{S}$, $\mathbf{S}'\mathbf{R}$, and $\mathbf{S}'\mathbf{S}$ are of the same dimensions as $\mathbf{T}$, $\mathbf{U}$, $\mathbf{V}$, and $\mathbf{W}$, respectively, $\mathbf{T} = \mathbf{R}'\mathbf{R}$, $\mathbf{U} = \mathbf{R}'\mathbf{S}$, $\mathbf{V} = \mathbf{S}'\mathbf{R}$, and $\mathbf{W} = \mathbf{S}'\mathbf{S}$. Q.E.D.

Corollary 14.8.2. Let $\mathbf{A}$ represent a symmetric nonnegative definite matrix that has been partitioned as

$$\mathbf{A} = \begin{pmatrix} \mathbf{T} & \mathbf{U} \\ \mathbf{V} & \mathbf{W} \end{pmatrix},$$

where $\mathbf{T}$ (and hence $\mathbf{W}$) is square. Then, $\mathcal{C}(\mathbf{U}) \subset \mathcal{C}(\mathbf{T})$ and $\mathcal{R}(\mathbf{V}) \subset \mathcal{R}(\mathbf{T})$. Similarly, $\mathcal{C}(\mathbf{V}) \subset \mathcal{C}(\mathbf{W})$ and $\mathcal{R}(\mathbf{U}) \subset \mathcal{R}(\mathbf{W})$.

Proof. According to Lemma 14.8.1, there exist matrices $\mathbf{R}$ and $\mathbf{S}$ such that

$$\mathbf{T} = \mathbf{R}'\mathbf{R}, \quad \mathbf{U} = \mathbf{R}'\mathbf{S}, \quad \mathbf{V} = \mathbf{S}'\mathbf{R}, \quad \mathbf{W} = \mathbf{S}'\mathbf{S}.$$

Thus, making use of Corollaries 4.2.3 and 7.4.5, we find that

$$\mathcal{C}(\mathbf{U}) \subset \mathcal{C}(\mathbf{R}') = \mathcal{C}(\mathbf{T})$$

and

$$\mathcal{R}(\mathbf{V}) \subset \mathcal{R}(\mathbf{R}) = \mathcal{R}(\mathbf{T}).$$

It can be shown in similar fashion that $\mathcal{C}(\mathbf{V}) \subset \mathcal{C}(\mathbf{W})$ and $\mathcal{R}(\mathbf{U}) \subset \mathcal{R}(\mathbf{W})$. Q.E.D.

b. Block-diagonal matrices

Lemma 14.2.1 characterizes the concepts of nonnegative definiteness, positive definiteness, and positive semidefiniteness as applied to diagonal matrices. The following result extends the results of Lemma 14.2.1 to block-diagonal matrices.

Lemma 14.8.3. Let $\mathbf{A}_i$ represent an $n_i \times n_i$ matrix ($i = 1, 2, \ldots, k$), let $n = n_1 + n_2 + \cdots + n_k$, and define $\mathbf{A}$ to be the $n \times n$ block-diagonal matrix

$$\mathbf{A} = \text{diag}(\mathbf{A}_1, \mathbf{A}_2, \ldots, \mathbf{A}_k).$$

Then, (1) $\mathbf{A}$ is nonnegative definite if and only if $\mathbf{A}_1, \mathbf{A}_2, \ldots, \mathbf{A}_k$ are nonnegative definite; (2) $\mathbf{A}$ is positive definite if and only if $\mathbf{A}_1, \mathbf{A}_2, \ldots, \mathbf{A}_k$ are positive definite; and (3) $\mathbf{A}$ is positive semidefinite if and only if the diagonal blocks $\mathbf{A}_1, \mathbf{A}_2, \ldots, \mathbf{A}_k$ are nonnegative definite with at least one of the diagonal blocks being positive semidefinite.

Proof. Consider the quadratic form $\mathbf{x}'\mathbf{A}\mathbf{x}$ (in an n-dimensional column vector $\mathbf{x}$) whose matrix is $\mathbf{A}$. Partition $\mathbf{x}'$ as $\mathbf{x}' = (\mathbf{x}_1', \mathbf{x}_2', \ldots, \mathbf{x}_k')$, where (for $i = 1, 2, \ldots, k$) $\mathbf{x}_i$ is an n_i-dimensional (column) vector. Then, clearly,

$$\mathbf{x}'\mathbf{A}\mathbf{x} = \mathbf{x}_1'\mathbf{A}_1\mathbf{x}_1 + \mathbf{x}_2'\mathbf{A}_2\mathbf{x}_2 + \cdots + \mathbf{x}_k'\mathbf{A}_k\mathbf{x}_k.$$

(1) If $\mathbf{A}_1, \mathbf{A}_2, \ldots, \mathbf{A}_k$ are nonnegative definite, then, by definition, $\mathbf{x}_i'\mathbf{A}_i\mathbf{x}_i \geq 0$ for every $\mathbf{x}_i$ ($i = 1, 2, \ldots, k$), implying that $\mathbf{x}'\mathbf{A}\mathbf{x} \geq 0$ for every $\mathbf{x}$ and hence that $\mathbf{A}$ is nonnegative definite. Conversely, if $\mathbf{A}$ is nonnegative definite, then it follows from Corollary 14.2.12 that $\mathbf{A}_1, \mathbf{A}_2, \ldots, \mathbf{A}_k$ are nonnegative definite.

(2) If $\mathbf{A}_1, \mathbf{A}_2, \ldots, \mathbf{A}_k$ are positive definite, then, by definition, $\mathbf{x}_i'\mathbf{A}_i\mathbf{x}_i > 0$ for every nonnull $\mathbf{x}_i$ ($i = 1, 2, \ldots, k$), implying that $\mathbf{x}'\mathbf{A}\mathbf{x} > 0$ for every nonnull $\mathbf{x}$ and hence that $\mathbf{A}$ is positive definite. Conversely, if $\mathbf{A}$ is positive definite, then it follows from Corollary 14.2.12 that $\mathbf{A}_1, \mathbf{A}_2, \ldots, \mathbf{A}_k$ are positive definite.

(3) Suppose that $\mathbf{A}_1, \mathbf{A}_2, \ldots, \mathbf{A}_k$ are nonnegative definite and that, for some i, say $i = i^*$, $\mathbf{A}_i$ is positive semidefinite. Then, by definition, $\mathbf{x}_i'\mathbf{A}_i\mathbf{x}_i \geq 0$ for every $\mathbf{x}_i$ ($i = 1, 2, \ldots, k$), and there exists some nonnull value of $\mathbf{x}_{i^*}$, say $\mathbf{x}_{i^*} = \tilde{\mathbf{x}}_{i^*}$, for which $\mathbf{x}_{i^*}'\mathbf{A}_{i^*}\mathbf{x}_{i^*} = 0$. It follows that $\mathbf{x}'\mathbf{A}\mathbf{x} \geq 0$ for every $\mathbf{x}$, with equality holding for $\mathbf{x}' = (\mathbf{0}, \ldots, \mathbf{0}, \tilde{\mathbf{x}}_{i^*}', \mathbf{0}, \ldots, \mathbf{0})$. Thus, $\mathbf{A}$ is positive semidefinite.

Conversely, suppose that $\mathbf{A}$ is positive semidefinite. Then, it follows from Part (1) that $\mathbf{A}_1, \mathbf{A}_2, \ldots, \mathbf{A}_k$ are nonnegative definite. Moreover, it follows from Part (2) that not all of the matrices $\mathbf{A}_1, \mathbf{A}_2, \ldots, \mathbf{A}_k$ are positive definite and hence (since they are nonnegative definite) that at least one of them is positive semidefinite. Q.E.D.

c. Nonnegative definiteness of a Schur complement

Sufficient conditions for the positive definiteness or positive semidefiniteness of a Schur complement are given by the following theorem.

Theorem 14.8.4. Let

$$\mathbf{A} = \begin{pmatrix} \mathbf{T} & \mathbf{U} \\ \mathbf{V} & \mathbf{W} \end{pmatrix},$$

where $\mathbf{T}$ is of dimensions $m \times m$, $\mathbf{U}$ of dimensions $m \times n$, $\mathbf{V}$ of dimensions $n \times m$, and $\mathbf{W}$ of dimensions $n \times n$.

(1) If $\mathbf{A}$ is positive definite, then the Schur complement $\mathbf{W} - \mathbf{V}\mathbf{T}^{-1}\mathbf{U}$ of $\mathbf{T}$ and the Schur complement $\mathbf{T} - \mathbf{U}\mathbf{W}^{-1}\mathbf{V}$ of $\mathbf{W}$ are positive definite.

(2) If $\mathbf{A}$ is symmetric positive semidefinite, then the Schur complement $\mathbf{W} - \mathbf{V}\mathbf{T}^{-}\mathbf{U}$ of $\mathbf{T}$ and the Schur complement $\mathbf{T} - \mathbf{U}\mathbf{W}^{-}\mathbf{V}$ of $\mathbf{W}$ are nonnegative definite.

(3) If $\mathbf{A}$ is positive semidefinite (but not necessarily symmetric) and if $\mathcal{C}(\mathbf{U}) \subset \mathcal{C}(\mathbf{T})$ or $\mathcal{R}(\mathbf{V}) \subset \mathcal{R}(\mathbf{T})$, then the Schur complement $\mathbf{W} - \mathbf{V}\mathbf{T}^{-}\mathbf{U}$ of $\mathbf{T}$ (relative to $\mathbf{T}^{-}$) is nonnegative definite.

(3′) If $\mathbf{A}$ is positive semidefinite and if $\mathcal{C}(\mathbf{V}) \subset \mathcal{C}(\mathbf{W})$ or $\mathcal{R}(\mathbf{U}) \subset \mathcal{R}(\mathbf{W})$, then the Schur complement $\mathbf{T} - \mathbf{U}\mathbf{W}^{-}\mathbf{V}$ of $\mathbf{W}$ (relative to $\mathbf{W}^{-}$) is nonnegative definite.

Proof. Let us confine our attention to Result (3) and to the parts of Results (1) and (2) that pertain to the Schur complement of $\mathbf{T}$. Result (3′) and the parts of Results (1) and (2) that pertain to the Schur complement of $\mathbf{W}$ can be established via analogous arguments.

Define

$$\mathbf{P} = \begin{pmatrix} \mathbf{I}_m & -\mathbf{T}^-\mathbf{U} \\ \mathbf{0} & \mathbf{I}_n \end{pmatrix}.$$

According to Lemma 8.5.2, $\mathbf{P}$ is nonsingular. Further,

$$\mathbf{P}'\mathbf{A}\mathbf{P} = \begin{pmatrix} \mathbf{T} & (\mathbf{I} - \mathbf{T}\mathbf{T}^-)\mathbf{U} \\ \mathbf{V} - (\mathbf{T}^-\mathbf{U})'\mathbf{T} & \mathbf{Q} \end{pmatrix},$$

where $\mathbf{Q} = \mathbf{W} - \mathbf{V}\mathbf{T}^-\mathbf{U} - (\mathbf{T}^-\mathbf{U})'(\mathbf{I} - \mathbf{T}\mathbf{T}^-)\mathbf{U}$.

Suppose that $\mathbf{A}$ is positive definite. Then, according to Corollary 14.2.10, $\mathbf{P}'\mathbf{A}\mathbf{P}$ is positive definite. Thus, it follows from Corollary 14.2.12 that $\mathbf{Q}$ is positive definite. A further implication of Corollary 14.2.12 is that $\mathbf{T}$ is positive definite and hence nonsingular, so that $\mathbf{Q} = \mathbf{W} - \mathbf{V}\mathbf{T}^{-1}\mathbf{U}$. We conclude that if $\mathbf{A}$ is positive definite, then the Schur complement of $\mathbf{T}$ is positive definite.

Suppose now that $\mathbf{A}$ is positive semidefinite. Then, according to Corollary 14.2.10, $\mathbf{P}'\mathbf{A}\mathbf{P}$ is positive semidefinite. Thus, it follows from Corollary 14.2.12 that $\mathbf{Q}$ is nonnegative definite. If $\mathcal{C}(\mathbf{U}) \subset \mathcal{C}(\mathbf{T})$—which (in light of Corollary 14.8.2) would be the case if $\mathbf{A}$ were symmetric—then (according to Lemma 9.3.5) $(\mathbf{I} - \mathbf{T}\mathbf{T}^-)\mathbf{U} = \mathbf{0}$ and hence $\mathbf{Q} = \mathbf{W} - \mathbf{V}\mathbf{T}^-\mathbf{U}$. We conclude that if $\mathbf{A}$ is symmetric positive semidefinite or, more generally, if $\mathbf{A}$ is positive semidefinite and $\mathcal{C}(\mathbf{U}) \subset \mathcal{C}(\mathbf{T})$, then the Schur complement of $\mathbf{T}$ is nonnegative definite.

That the Schur complement of $\mathbf{T}$ is nonnegative definite if $\mathbf{A}$ is positive semidefinite and $\mathcal{R}(\mathbf{V}) \subset \mathcal{R}(\mathbf{T})$ can be established via a similar argument. Q.E.D.

As a partial "converse" to Theorem 14.8.4, we have the following result.

Theorem 14.8.5. Let

$$\mathbf{A} = \begin{pmatrix} \mathbf{T} & \mathbf{U} \\ \mathbf{U}' & \mathbf{W} \end{pmatrix},$$

where $\mathbf{T}$ is of dimensions $m \times m$, $\mathbf{U}$ of dimensions $m \times n$, and $\mathbf{W}$ of dimensions $n \times n$.

(1) If $\mathbf{T}$ is symmetric positive definite and the Schur complement $\mathbf{W} - \mathbf{U}'\mathbf{T}^{-1}\mathbf{U}$ of $\mathbf{T}$ is positive definite, or if $\mathbf{W}$ is symmetric positive definite and the Schur complement $\mathbf{T} - \mathbf{U}\mathbf{W}^{-1}\mathbf{U}'$ of $\mathbf{W}$ is positive definite, then $\mathbf{A}$ is positive definite.

(2) Suppose that $\mathbf{T}$ is symmetric nonnegative definite and that the Schur complement $\mathbf{W} - \mathbf{U}'\mathbf{T}^-\mathbf{U}$ of $\mathbf{T}$ is nonnegative definite. Suppose further that $\mathcal{C}(\mathbf{U}) \subset \mathcal{C}(\mathbf{T})$. Then, $\mathbf{A}$ is nonnegative definite.

(2′) Suppose that $\mathbf{W}$ is symmetric nonnegative definite and that the Schur complement $\mathbf{T} - \mathbf{U}\mathbf{W}^-\mathbf{U}'$ of $\mathbf{W}$ is nonnegative definite. Suppose further that $\mathcal{R}(\mathbf{U}) \subset \mathcal{R}(\mathbf{W})$. Then, $\mathbf{A}$ is nonnegative definite.

Proof. Suppose that $\mathbf{T}$ is symmetric positive definite (in which case $\mathbf{T}$ is nonsingular) or more generally that $\mathbf{T}$ is symmetric nonnegative definite and $\mathcal{C}(\mathbf{U}) \subset \mathcal{C}(\mathbf{T})$. Define

$$\mathbf{P} = \begin{pmatrix} \mathbf{I}_m & -\mathbf{T}^-\mathbf{U} \\ \mathbf{0} & \mathbf{I}_n \end{pmatrix}.$$

According to Lemma 8.5.2, $\mathbf{P}$ is nonsingular. Moreover,

$$\mathbf{P}'\mathbf{A}\mathbf{P} = \text{diag}(\mathbf{T},\ \mathbf{W} - \mathbf{U}'\mathbf{T}^-\mathbf{U}),$$

as can be easily verified by making use of Lemma 9.3.5, or equivalently

$$\mathbf{A} = (\mathbf{P}^{-1})'\,\text{diag}(\mathbf{T},\ \mathbf{W} - \mathbf{U}'\mathbf{T}^-\mathbf{U})\mathbf{P}^{-1}. \tag{8.1}$$

If $\mathbf{W} - \mathbf{U}'\mathbf{T}^-\mathbf{U}$ is nonnegative definite, then it follows from Lemma 14.8.3 that $\text{diag}(\mathbf{T},\ \mathbf{W} - \mathbf{U}'\mathbf{T}^-\mathbf{U})$ is nonnegative definite, implying [in light of equality (8.1) and Corollary 14.2.10] that $\mathbf{A}$ is nonnegative definite. Similarly, if $\mathbf{T}$ is positive definite (and hence invertible) and $\mathbf{W} - \mathbf{U}'\mathbf{T}^{-1}\mathbf{U}$ is positive definite, then $\text{diag}(\mathbf{T},\ \mathbf{W} - \mathbf{U}'\mathbf{T}^{-1}\mathbf{U})$ is positive definite, implying that $\mathbf{A}$ is positive definite.

We conclude that if $\mathbf{T}$ is symmetric nonnegative definite, $\mathbf{W} - \mathbf{U}'\mathbf{T}^-\mathbf{U}$ is nonnegative definite, and $\mathcal{C}(\mathbf{U}) \subset \mathcal{C}(\mathbf{T})$, then $\mathbf{A}$ is nonnegative definite, and that if $\mathbf{T}$ is symmetric positive definite and $\mathbf{W} - \mathbf{U}'\mathbf{T}^{-1}\mathbf{U}$ is positive definite, then $\mathbf{A}$ is positive definite. It can be shown, in similar fashion, that if $\mathbf{W}$ is symmetric nonnegative definite, $\mathbf{T} - \mathbf{U}\mathbf{W}^-\mathbf{U}'$ is nonnegative definite, and $\mathcal{R}(\mathbf{U}) \subset \mathcal{R}(\mathbf{W})$, then $\mathbf{A}$ is nonnegative definite, and that if $\mathbf{W}$ is symmetric positive definite and $\mathbf{T} - \mathbf{U}\mathbf{W}^{-1}\mathbf{U}'$ is positive definite, then $\mathbf{A}$ is positive definite. Q.E.D.

In the special case of a symmetric partitioned matrix, we have (in light of Theorem 14.8.4 and Corollary 14.2.12) the following corollary of Theorem 14.8.5.

Corollary 14.8.6. Suppose that a symmetric matrix $\mathbf{A}$ is partitioned as

$$\mathbf{A} = \begin{pmatrix} \mathbf{T} & \mathbf{U} \\ \mathbf{U}' & \mathbf{W} \end{pmatrix}$$

(where $\mathbf{T}$ and $\mathbf{W}$ are square). Then, $\mathbf{A}$ is positive definite if and only if $\mathbf{T}$ and the Schur complement $\mathbf{W} - \mathbf{U}'\mathbf{T}^{-1}\mathbf{U}$ of $\mathbf{T}$ are both positive definite. Similarly, $\mathbf{A}$ is positive definite if and only if $\mathbf{W}$ and the Schur complement $\mathbf{T} - \mathbf{U}\mathbf{W}^{-1}\mathbf{U}'$ of $\mathbf{W}$ are both positive definite.

14.9 Some Results on Determinants

a. Determinant of a symmetric nonnegative definite matrix

Let $\mathbf{A}$ represent an $n \times n$ symmetric positive definite matrix. Then, according to Corollary 14.3.13, there exists a nonsingular matrix $\mathbf{P}$ such that $\mathbf{A} = \mathbf{P}'\mathbf{P}$. Since (in light of Theorem 13.3.7) $|\mathbf{P}| \neq 0$, it follows that

$$|\mathbf{A}| = |\mathbf{P}'||\mathbf{P}| = |\mathbf{P}|^2 > 0.$$

Thus, recalling Corollary 14.3.12, we have the following theorem.

Lemma 14.9.1. The determinant of a symmetric positive definite matrix is (strictly) positive. The determinant of a symmetric positive semidefinite matrix equals zero.

b. Determinant of a (not necessarily symmetric) nonnegative definite matrix

Those nonsymmetric nonnegative definite matrices that are nonsingular have nonzero determinants—refer to Theorem 13.3.7. Can anything further be said about their determinants? In particular, is the determinant of a nonsingular nonsymmetric nonnegative definite matrix, like that of a symmetric positive definite matrix, necessarily positive? In what follows, the answer to this question is shown to be yes.

Preliminary to considering the determinant of a possibly nonsymmetric nonnegative definite matrix, it is convenient to establish the following result on the determinants of skew-symmetric matrices—recall (from Lemma 14.6.4) that a matrix is skew-symmetric if and only if it is nonnegative definite and its diagonal elements equal zero.

Theorem 14.9.2. Let $\mathbf{A}$ represent an $n \times n$ skew-symmetric matrix. (1) If n is an odd number, then $\det(\mathbf{A}) = 0$. (2) If n is an even number, then $\det(\mathbf{A}) \geq 0$.

Proof. (1) By definition, $\mathbf{A}' = -\mathbf{A}$, so that, according to Lemma 13.2.1 and Corollary 13.2.5,

$$|\mathbf{A}| = |\mathbf{A}'| = |-\mathbf{A}| = (-1)^n|\mathbf{A}|.$$

Consequently, if n is an odd number, $|\mathbf{A}| = -|\mathbf{A}|$, implying that $2|\mathbf{A}| = 0$ and hence that $|\mathbf{A}| = 0$.

(2) The proof that $\det(\mathbf{A}) \geq 0$ when n is an even number is by mathematical induction. When $n = 2$,

$$\mathbf{A} = \begin{pmatrix} 0 & c \\ -c & 0 \end{pmatrix}$$

for some scalar c, in which case

$$|\mathbf{A}| = c^2 \geq 0.$$

Suppose now that the determinant of every $(n-2) \times (n-2)$ skew-symmetric matrix is greater than or equal to zero, where n is an arbitrary even number (greater than or equal to 4). To complete the induction argument, we must show that the determinant of an arbitrary $n \times n$ skew-symmetric matrix $\mathbf{A} = \{a_{ij}\}$ is greater than or equal to zero.

For this purpose, it suffices to restrict attention to the case where at least one of the last $n-1$ elements of the first row of $\mathbf{A}$ is nonzero. (The case where all $n-1$ of these elements are zero is trivial, since in that case the first row of $\mathbf{A}$ is null and hence it follows from Corollary 13.2.3 that $|\mathbf{A}| = 0$.) Assume then that $a_{1m} \neq 0$ for some integer m $(2 \leq m \leq n)$.

Let $\mathbf{B} = \mathbf{P}'\mathbf{A}\mathbf{P}$, where $\mathbf{P}$ is the permutation matrix obtained by interchanging the second and mth columns of $\mathbf{I}_n$. (If $m = 2$, take $\mathbf{P} = \mathbf{I}$.) Observe (in light of Theorem 13.3.4 and Corollary 13.3.6) that

$$|\mathbf{B}| = |\mathbf{P}|^2|\mathbf{A}| = |\mathbf{A}|, \tag{9.1}$$

and partition $\mathbf{B}$ as $\mathbf{B} = \begin{pmatrix} \mathbf{B}_{11} & \mathbf{B}_{12} \\ \mathbf{B}_{21} & \mathbf{B}_{22} \end{pmatrix}$, where the dimensions of $\mathbf{B}_{11}$ are 2×2.

According to Lemma 14.6.2, $\mathbf{B}$ is skew-symmetric, implying that $\mathbf{B}_{11} = \begin{pmatrix} 0 & a_{1m} \\ -a_{1m} & 0 \end{pmatrix}$, $\mathbf{B}'_{12} = -\mathbf{B}_{21}$, and $\mathbf{B}'_{22} = -\mathbf{B}_{22}$. Since $a_{1m} \neq 0$, we have that

$$|\mathbf{B}_{11}| = a_{1m}^2 > 0,$$

and it follows from Theorem 13.3.8 that

$$|\mathbf{B}| = a_{1m}^2|\mathbf{B}_{22} - \mathbf{B}_{21}\mathbf{B}_{11}^{-1}\mathbf{B}_{12}|. \tag{9.2}$$

Moreover,

$$\begin{aligned}(\mathbf{B}_{22} - \mathbf{B}_{21}\mathbf{B}_{11}^{-1}\mathbf{B}_{12})' &= \mathbf{B}'_{22} - \mathbf{B}'_{12}(\mathbf{B}'_{11})^{-1}\mathbf{B}'_{21} \\ &= -\mathbf{B}_{22} - (-\mathbf{B}_{21})(-\mathbf{B}_{11}^{-1})(-\mathbf{B}_{12}) \\ &= -(\mathbf{B}_{22} - \mathbf{B}_{21}\mathbf{B}_{11}^{-1}\mathbf{B}_{12}),\end{aligned}$$

so that $\mathbf{B}_{22} - \mathbf{B}_{21}\mathbf{B}_{11}^{-1}\mathbf{B}_{12}$ [which is of dimensions $(n-2) \times (n-2)$] is skew-symmetric. Thus, by supposition,

$$|\mathbf{B}_{22} - \mathbf{B}_{21}\mathbf{B}_{11}^{-1}\mathbf{B}_{12}| \geq 0. \tag{9.3}$$

Together, results (9.2) and (9.3) imply that $|\mathbf{B}| \geq 0$ and hence [in light of result (9.1)] that $|\mathbf{A}| \geq 0$. Q.E.D.

We are now in a position to establish the following result on the determinant of a nonnegative definite matrix.

Theorem 14.9.3. For any nonnegative definite matrix $\mathbf{A}$, $\det(\mathbf{A}) \geq 0$.

Proof. The proof is by mathematical induction. Clearly, the determinant of any 1×1 nonnegative definite matrix is nonnegative.

Suppose now that the determinant of every $(n-1) \times (n-1)$ nonnegative definite matrix is nonnegative, and consider the determinant of an $n \times n$ nonnegative definite matrix $\mathbf{A} = \{a_{ij}\}$. To complete the induction argument, we must show that $\det(\mathbf{A}) \geq 0$.

In light of Theorem 14.9.2, it suffices to restrict attention to the case where at least one of the diagonal elements $a_{11}, a_{22}, \ldots, a_{nn}$ of $\mathbf{A}$ is nonzero and hence positive. [If all n of the diagonal elements of $\mathbf{A}$ are zero, then (according to Lemma

14.6.4) $\mathbf{A}$ is skew-symmetric, and we have, as an immediate consequence of Theorem 14.9.2, that $\det(\mathbf{A}) \geq 0$.] Assume then that $a_{mm} > 0$ for some integer m $(1 \leq m \leq n)$.

Let $\mathbf{B} = \mathbf{P}'\mathbf{A}\mathbf{P}$, where $\mathbf{P}$ is the permutation matrix obtained by interchanging the first and mth columns of $\mathbf{I}_n$. (If $m = 1$, take $\mathbf{P} = \mathbf{I}$.) Observe (in light of Theorem 14.2.9) that $\mathbf{B}$ is nonnegative definite and that

$$|\mathbf{B}| = |\mathbf{P}|^2|\mathbf{A}| = |\mathbf{A}|. \tag{9.4}$$

Partition $\mathbf{B}$ as $\mathbf{B} = \begin{pmatrix} \mathbf{B}_{11} & \mathbf{B}_{12} \\ \mathbf{B}_{21} & \mathbf{B}_{22} \end{pmatrix}$, where the dimensions of $\mathbf{B}_{11}$ are 1×1.

Clearly, $\mathbf{B}_{11} = (a_{mm})$, and consequently (since $a_{mm} \neq 0$) it follows from Theorem 13.3.8 that

$$|\mathbf{B}| = a_{mm}|\mathbf{B}_{22} - \mathbf{B}_{21}\mathbf{B}_{11}^{-1}\mathbf{B}_{12}|. \tag{9.5}$$

Moreover, according to Parts (1) and (3) of Theorem 14.8.4, $\mathbf{B}_{22} - \mathbf{B}_{21}\mathbf{B}_{11}^{-1}\mathbf{B}_{12}$ [which is of dimensions $(n-1) \times (n-1)$] is nonnegative definite, so that, by supposition,

$$|\mathbf{B}_{22} - \mathbf{B}_{21}\mathbf{B}_{11}^{-1}\mathbf{B}_{12}| \geq 0. \tag{9.6}$$

Since $a_{mm} > 0$, we conclude [on the basis of results (9.5) and (9.6)] that $|\mathbf{B}| \geq 0$ and hence [in light of result (9.4)] that $|\mathbf{A}| \geq 0$. Q.E.D.

In light of Lemma 14.2.8 and Corollary 14.3.12, we have, as a corollary of Theorem 14.9.3, the following result, which generalizes Lemma 14.9.1.

Corollary 14.9.4. The determinant of a positive definite matrix or of a nonsingular nonsymmetric positive semidefinite matrix is (strictly) positive. The determinant of a symmetric positive semidefinite matrix or of a singular nonsymmetric positive semidefinite matrix equals 0.

c. A necessary and sufficient condition for the positive definiteness of a symmetric matrix

Whether a symmetric matrix is positive definite can be ascertained from the determinants of its leading principal submatrices. The following theorem provides the basis for doing so.

Theorem 14.9.5. Let $\mathbf{A} = \{a_{ij}\}$ represent an $n \times n$ symmetric matrix, and, for $k = 1, 2, \ldots, n$, let

$$\mathbf{A}_k = \begin{pmatrix} a_{11} & a_{12} & \ldots & a_{1k} \\ a_{21} & a_{22} & \ldots & a_{2k} \\ \vdots & \vdots & \ddots & \vdots \\ a_{k1} & a_{k2} & \ldots & a_{kk} \end{pmatrix}$$

represent the leading principal submatrix of $\mathbf{A}$ of order k. Then, $\mathbf{A}$ is positive definite if and only if, for $k = 1, 2, \ldots, n$, $\det(\mathbf{A}_k) > 0$, that is, if and only if the determinants of all n of the leading principal submatrices $\mathbf{A}_1, \mathbf{A}_2, \ldots, \mathbf{A}_n$ of $\mathbf{A}$ are positive.

In proving Theorem 14.9.5, it is convenient to make use of the following result, which is of some interest in its own right.

Lemma 14.9.6. Let $\mathbf{A}$ represent an $n \times n$ symmetric matrix (where $n \geq 2$), and partition $\mathbf{A}$ as

$$\mathbf{A} = \begin{pmatrix} \mathbf{A}_* & \mathbf{a} \\ \mathbf{a}' & c \end{pmatrix},$$

where the dimensions of $\mathbf{A}_*$ are $(n-1) \times (n-1)$. Then, $\mathbf{A}$ is positive definite if and only if $\mathbf{A}_*$ is positive definite and $|\mathbf{A}| > 0$.

Proof (of Lemma 14.9.6). If $\mathbf{A}$ is positive definite, then it is clear from Corollary 14.2.12 that $\mathbf{A}_*$ is positive definite and from Lemma 14.9.1 that $|\mathbf{A}| > 0$.

Conversely, suppose that $\mathbf{A}_*$ is positive definite (and hence nonsingular) and that $|\mathbf{A}| > 0$. Then, according to Theorem 13.3.8,

$$|\mathbf{A}| = |\mathbf{A}_*|(c - \mathbf{a}'\mathbf{A}_*^{-1}\mathbf{a}).$$

Since (according to Lemma 14.9.1) $|\mathbf{A}_*| > 0$ (and since $|\mathbf{A}| > 0$), we conclude that the Schur complement $c - \mathbf{a}'\mathbf{A}_*^{-1}\mathbf{a}$ of $\mathbf{A}_*$ (like $\mathbf{A}_*$ itself) is positive definite and hence [in light of Part (1) of Theorem 14.8.5] that $\mathbf{A}$ is positive definite. Q.E.D.

Proof (of Theorem 14.9.5). That the determinants of $\mathbf{A}_1, \mathbf{A}_2, \ldots, \mathbf{A}_n$ are positive if $\mathbf{A}$ is positive definite is an immediate consequence of Corollary 14.2.12 and Lemma 14.9.1.

For purposes of proving the converse, suppose that the determinants of $\mathbf{A}_1$, $\mathbf{A}_2, \ldots, \mathbf{A}_n$ are positive. The proof consists of establishing, via a mathematical induction argument, that $\mathbf{A}_1, \mathbf{A}_2, \ldots, \mathbf{A}_n$ are positive definite, which (since $\mathbf{A} = \mathbf{A}_n$) implies in particular that $\mathbf{A}$ is positive definite.

Clearly, $\mathbf{A}_1$ is positive definite. Suppose now that $\mathbf{A}_{k-1}$ is positive definite (where $2 \leq k \leq n$), and partition $\mathbf{A}_k$ as

$$\mathbf{A}_k = \begin{pmatrix} \mathbf{A}_{k-1} & \mathbf{a}_k \\ \mathbf{a}_k' & a_{kk} \end{pmatrix},$$

where $\mathbf{a}_k = (a_{1k}, a_{2k}, \ldots, a_{k-1,k})'$. Since $\mathbf{A}_{k-1}$ is (by supposition) positive definite (and since $|\mathbf{A}_k| > 0$), it follows from Lemma 14.9.6 that $\mathbf{A}_k$ is positive definite.

We conclude, on the basis of the induction argument, that $\mathbf{A}_1, \mathbf{A}_2, \ldots, \mathbf{A}_n$ are positive definite and that $\mathbf{A}$ in particular is positive definite. Q.E.D.

d. A necessary and sufficient condition for the nonnegative definiteness of a symmetric matrix

Is it the case that an $n \times n$ symmetric matrix is nonnegative definite if and only if the determinants of all n of its leading principal submatrices are nonnegative (as might be conjectured on the basis of Theorem 14.9.5)? It is clear (from Corollary 14.2.12 and Lemma 14.9.1) that the nonnegativity of the determinants of the n leading principal submatrices is a necessary condition (for nonnegative definiteness). It can be shown, however, that this condition is not sufficient.

In this section, we establish a necessary and sufficient condition for the nonnegative definiteness of a symmetric matrix. This condition is analogous to the necessary and sufficient condition for positive definiteness given by the following theorem (which differs from the necessary and sufficient condition for positive definiteness given by Theorem 14.9.5).

Theorem 14.9.7. Let $\mathbf{A}$ represent an $n \times n$ symmetric matrix. Then, $\mathbf{A}$ is positive definite if and only if (1) $\mathbf{A}$ is nonsingular and (2) the determinant of every principal submatrix of $\mathbf{A}$ is nonnegative.

To prove Theorem 14.9.7, we require the following result, which is of some interest in its own right.

Theorem 14.9.8. Let $\mathbf{A}$ represent an $n \times n$ nonsingular symmetric matrix (where $n \geq 3$), and partition $\mathbf{A}$ as

$$\mathbf{A} = \begin{pmatrix} \mathbf{B} & \mathbf{a} \\ \mathbf{a}' & c \end{pmatrix},$$

where the dimensions of $\mathbf{B}$ are $(n-1) \times (n-1)$. Then, $\mathbf{B}$ is nonsingular or $\mathbf{B}$ contains an $(n-2) \times (n-2)$ nonsingular principal submatrix. Moreover, if $\mathbf{B}$ is singular and $\mathbf{B}_*$ is any $(n-2) \times (n-2)$ nonsingular principal submatrix of $\mathbf{B}$, then $|\mathbf{B}_*||\mathbf{A}| < 0$ (i.e., $|\mathbf{B}_*|$ and $|\mathbf{A}|$ are opposite in sign).

Proof (of Theorem 14.9.8). Suppose that $\mathbf{B}$ is singular, and let $\mathbf{A}_1 = (\mathbf{B}, \mathbf{a})$. Since the rows of $\mathbf{A}$ are linearly independent, the rows of $\mathbf{A}_1$ are also linearly independent, and hence $\text{rank}(\mathbf{A}_1) = n - 1$. Thus, $\mathbf{A}_1$ contains $n - 1$ linearly independent columns, implying that $\mathbf{B}$ contains (at least) $n-2$ linearly independent columns. Since (by supposition) $\mathbf{B}$ is singular, we have that

$$\text{rank}(\mathbf{B}) = n - 2.$$

Moreover, $\mathbf{B}$ is symmetric. We conclude (in light of Corollary 4.4.11) that $\mathbf{B}$ contains an $(n-2) \times (n-2)$ nonsingular principal submatrix.

Let $\mathbf{B}_*$ represent an $(n-2) \times (n-2)$ nonsingular principal submatrix of $\mathbf{B}$, say that obtained by striking out the kth row and column (of $\mathbf{B}$). Let $\mathbf{U} = (\mathbf{U}_*, \mathbf{u})$, where $\mathbf{U}_*$ is the $(n-1) \times (n-2)$ submatrix of $\mathbf{I}_{n-1}$ obtained by striking out the kth column (of $\mathbf{I}_{n-1}$) and $\mathbf{u}$ is the kth column of $\mathbf{I}_{n-1}$. Then, $\mathbf{U}$ is a permutation matrix, and

$$\mathbf{U}'\mathbf{B}\mathbf{U} = \begin{pmatrix} \mathbf{B}_* & \mathbf{U}_*'\mathbf{B}\mathbf{u} \\ \mathbf{u}'\mathbf{B}\mathbf{U}_* & \mathbf{u}'\mathbf{B}\mathbf{u} \end{pmatrix}$$

[so that the $(n-2) \times (n-2)$ leading principal submatrix of $\mathbf{U}'\mathbf{B}\mathbf{U}$ is $\mathbf{B}_*$].

Take $\mathbf{P}$ to be the $n \times n$ matrix

$$\mathbf{P} = \begin{pmatrix} \mathbf{U} & \mathbf{0} \\ \mathbf{0}' & 1 \end{pmatrix}.$$

Then, $\mathbf{P}$ is a permutation matrix, and

$$\mathbf{P}'\mathbf{AP} = \begin{pmatrix} \mathbf{U}'\mathbf{BU} & \mathbf{U}'\mathbf{a} \\ \mathbf{a}'\mathbf{U} & c \end{pmatrix} = \begin{pmatrix} \mathbf{B}_* & \mathbf{U}_*'\mathbf{Bu} & \mathbf{U}_*'\mathbf{a} \\ \mathbf{u}'\mathbf{BU}_* & \mathbf{u}'\mathbf{Bu} & \mathbf{u}'\mathbf{a} \\ \mathbf{a}'\mathbf{U}_* & \mathbf{a}'\mathbf{u} & c \end{pmatrix}.$$

Thus, making use of Theorem 13.3.8, we find that

$$|\mathbf{A}| = |\mathbf{P}|^2|\mathbf{A}| = |\mathbf{P}'\mathbf{AP}| = |\mathbf{B}_*||\mathbf{F}|, \tag{9.7}$$

where $\mathbf{F} = \begin{pmatrix} f_{11} & f_{12} \\ f_{12} & f_{22} \end{pmatrix}$ with

$$f_{11} = \mathbf{u}'\mathbf{Bu} - \mathbf{u}'\mathbf{BU}_*\mathbf{B}_*^{-1}\mathbf{U}_*'\mathbf{Bu},$$
$$f_{12} = \mathbf{u}'\mathbf{a} - \mathbf{u}'\mathbf{BU}_*\mathbf{B}_*^{-1}\mathbf{U}_*'\mathbf{a},$$
$$f_{22} = c - \mathbf{a}'\mathbf{U}_*\mathbf{B}_*^{-1}\mathbf{U}_*'\mathbf{a}.$$

Further,

$$|\mathbf{B}| = |\mathbf{U}|^2|\mathbf{B}| = |\mathbf{U}'\mathbf{BU}| = |\mathbf{B}_*|(\mathbf{u}'\mathbf{Bu} - \mathbf{u}'\mathbf{BU}_*\mathbf{B}_*^{-1}\mathbf{U}_*'\mathbf{Bu}) = |\mathbf{B}_*| f_{11}. \tag{9.8}$$

Since (by supposition) **B** is singular and since (by definition) $\mathbf{B}_*$ is nonsingular, we have that $|\mathbf{B}| = 0$ and $|\mathbf{B}_*| \neq 0$, implying [in light of result (9.8)] that $f_{11} = 0$ and hence that $|\mathbf{F}| = -f_{12}^2$. Thus, making use of result (9.7), we find that

$$|\mathbf{A}| = -f_{12}^2|\mathbf{B}_*|.$$

We conclude that $|\mathbf{B}_*|$ and $|\mathbf{A}|$ are opposite in sign. Q.E.D.

Proof (of Theorem 14.9.7). If **A** is positive definite, then it is clear from Lemma 14.2.8, Corollary 14.2.12, and Lemma 14.9.1 that **A** is nonsingular and that the determinant of every principal submatrix of **A** is positive (and hence nonnegative). Thus, it suffices to prove that if **A** is nonsingular and the determinant of every principal submatrix of **A** is nonnegative, then **A** is positive definite. The proof of this is by mathematical induction.

Consider first the relatively trivial case of a 1×1 matrix $\mathbf{A} = (a_{11})$ that is nonsingular and whose only principal submatrix (**A** itself) has a nonnegative determinant. We have that $a_{11} \neq 0$ and $a_{11} \geq 0$, implying that $a_{11} > 0$ and hence that **A** is positive definite.

Next, consider the case of a 2×2 symmetric matrix $\mathbf{A} = \begin{pmatrix} a_{11} & a_{12} \\ a_{12} & a_{22} \end{pmatrix}$ that is nonsingular and whose three principal submatrices [(a_{11}), (a_{22}), and **A** itself] have nonnegative determinants. We have that $|\mathbf{A}| \neq 0$, implying (since $|\mathbf{A}| \geq 0$) that $|\mathbf{A}| > 0$. And $a_{11} \geq 0$, so that $a_{11} > 0$ (since if $a_{11} = 0$, it would be the case that $|\mathbf{A}| = -a_{12}^2 \leq 0$). Based on Theorem 14.9.5, we conclude that **A** is positive definite.

Now, suppose that every $(n-1) \times (n-1)$ symmetric matrix that is nonsingular and whose principal submatrices all have nonnegative determinants is positive definite, and consider an $n \times n$ symmetric matrix **A** that is nonsingular and whose principal submatrices all have nonnegative determinants. To complete the induction argument, we must show that $\mathbf{A}$ is positive definite.

Observe that $|\mathbf{A}| \neq 0$ (since $\mathbf{A}$ is nonsingular) and that $|\mathbf{A}| \geq 0$ (since $\mathbf{A}$ is a principal submatrix of itself) and hence that $|\mathbf{A}| > 0$. Partition $\mathbf{A}$ as

$$\mathbf{A} = \begin{pmatrix} \mathbf{B} & \mathbf{a} \\ \mathbf{a}' & c \end{pmatrix},$$

where the dimensions of $\mathbf{B}$ are $(n-1) \times (n-1)$. Observe that every principal submatrix of $\mathbf{B}$ is a principal submatrix of $\mathbf{A}$ and hence that every principal submatrix of $\mathbf{B}$ has a nonnegative determinant. Observe also that $\mathbf{B}$ is nonsingular [since if $\mathbf{B}$ were singular, $\mathbf{B}$ would (according to Theorem 14.9.8) contain an $(n-2) \times (n-2)$ principal submatrix whose determinant would be opposite in sign to $|\mathbf{A}|$ and hence negative]. Accordingly, it follows from our supposition that $\mathbf{B}$ is positive definite.

We conclude (on the basis of Lemma 14.9.6) that $\mathbf{A}$ is positive definite. Q.E.D.

For purposes of establishing a necessary and sufficient condition for the nonnegative definiteness of a symmetric matrix, analogous to the necessary and sufficient condition given by Theorem 14.9.7 for the positive definiteness of a symmetric matrix, it is helpful to introduce the following two lemmas.

Lemma 14.9.9. Let $\mathbf{A}$ represent an $n \times n$ symmetric matrix of rank r. Then, corresponding to any $r \times r$ nonsingular principal submatrix $\mathbf{A}_*$ of $\mathbf{A}$, there exists an $r \times n$ matrix $\mathbf{Q}$ (of rank r) such that $\mathbf{A} = \mathbf{Q}'\mathbf{A}_*\mathbf{Q}$.

Proof. Suppose that $\mathbf{A}_*$ is the $r \times r$ nonsingular principal submatrix obtained by striking out all of the rows and columns of $\mathbf{A}$ except the $i_1, i_2, \ldots, i_r$th rows and columns (where $i_1 < i_2 < \cdots < i_r$). Take $\mathbf{P}$ to be any permutation matrix whose first, second, ..., rth columns are the $i_1, i_2, \ldots, i_r$th columns, respectively, of $\mathbf{I}_n$.

Let $\mathbf{B} = \mathbf{P}'\mathbf{A}\mathbf{P}$, and partition $\mathbf{P}$ as $\mathbf{P} = (\mathbf{P}_1, \mathbf{P}_2)$, where $\mathbf{P}_1$ is of dimensions $n \times r$. Then, $\mathbf{A}_* = \mathbf{P}_1'\mathbf{A}\mathbf{P}_1$ and hence

$$\mathbf{B} = \begin{pmatrix} \mathbf{A}_* & \mathbf{B}_{12} \\ \mathbf{B}_{12}' & \mathbf{B}_{22} \end{pmatrix},$$

where $\mathbf{B}_{12} = \mathbf{P}_1'\mathbf{A}\mathbf{P}_2$ and $\mathbf{B}_{22} = \mathbf{P}_2'\mathbf{A}\mathbf{P}_2$. Since $\text{rank}(\mathbf{B}) = \text{rank}(\mathbf{A}) = r$, it follows from Lemma 9.2.2 that $\mathbf{B}_{22} = \mathbf{B}_{12}'\mathbf{A}_*^{-1}\mathbf{B}_{12}$. Thus,

$$\mathbf{B} = (\mathbf{I}_r, \mathbf{A}_*^{-1}\mathbf{B}_{12})'\mathbf{A}_*(\mathbf{I}_r, \mathbf{A}_*^{-1}\mathbf{B}_{12}),$$

so that

$$\mathbf{A} = \mathbf{P}\mathbf{B}\mathbf{P}' = \mathbf{Q}'\mathbf{A}_*\mathbf{Q},$$

where $\mathbf{Q} = (\mathbf{I}_r, \mathbf{A}_*^{-1}\mathbf{B}_{12})\mathbf{P}'$.

Moreover, $\text{rank}(\mathbf{Q}) \geq \text{rank}(\mathbf{A}) = r$, and (since $\mathbf{Q}$ is of dimensions $r \times n$) $\text{rank}(\mathbf{Q}) \leq r$. Thus, $\text{rank}(\mathbf{Q}) = r$. Q.E.D.

Lemma 14.9.10. . Let $\mathbf{A}$ represent an $n \times n$ symmetric matrix of rank r, and let $\mathbf{A}_*$ represent an $r \times r$ nonsingular principal submatrix of $\mathbf{A}$. Then, $\mathbf{A}$ is nonnegative definite if and only if $\mathbf{A}_*$ is positive definite.

Proof. Suppose that $\mathbf{A}$ is nonnegative definite. Then, according to Corollary 14.2.12, $\mathbf{A}_*$ is nonnegative definite. Moreover, since $\mathbf{A}$ is symmetric, $\mathbf{A}_*$ is symmetric. Thus, it follows from Corollary 14.3.12 that $\mathbf{A}_*$ is positive definite.

Conversely, suppose that $\mathbf{A}_*$ is positive definite. Then, according to Lemma 14.9.9, there exists a matrix $\mathbf{Q}$ such that $\mathbf{A} = \mathbf{Q}'\mathbf{A}_*\mathbf{Q}$. Thus, it follows from Theorem 14.2.9 that $\mathbf{A}$ is nonnegative definite. Q.E.D.

The following theorem gives a necessary and sufficient condition for the nonnegative definiteness of a symmetric matrix, analogous to the necessary and sufficient condition given by Theorem 14.9.7 for the positive definiteness of a symmetric matrix.

Theorem 14.9.11. Let $\mathbf{A}$ represent an $n \times n$ symmetric matrix. Then, $\mathbf{A}$ is nonnegative definite if and only if the determinant of every principal submatrix of $\mathbf{A}$ is nonnegative.

Proof. Suppose that the symmetric matrix $\mathbf{A}$ is nonnegative definite. Then (in light of Corollary 14.2.12), every principal submatrix of $\mathbf{A}$ is nonnegative definite (and symmetric). And we conclude (on the basis of Lemma 14.9.1) that the determinant of every principal submatrix of $\mathbf{A}$ is nonnegative.

Conversely, suppose that the determinant of every principal submatrix of $\mathbf{A}$ is nonnegative. Let $r = \text{rank}(\mathbf{A})$. Then, according to Corollary 4.4.11, $\mathbf{A}$ contains an $r \times r$ nonsingular principal submatrix, say $\mathbf{A}_*$. Clearly, every principal submatrix of $\mathbf{A}_*$ is a principal submatrix of $\mathbf{A}$, and hence (by supposition) the determinant of every principal submatrix of $\mathbf{A}_*$ is nonnegative, implying (in light of Theorem 14.9.7) that $\mathbf{A}_*$ is positive definite. We conclude (on the basis of Lemma 14.9.10) that $\mathbf{A}$ is nonnegative definite. Q.E.D.

14.10 Geometrical Considerations

As discussed in Section 6.1, the inner product of two $m \times n$ matrices $\mathbf{A}$ and $\mathbf{B}$ (in a linear space $\mathcal{V}$ of $m \times n$ matrices) is the value $\mathbf{A} \bullet \mathbf{B}$ assigned to $\mathbf{A}$ and $\mathbf{B}$ by a designated function f whose domain consists of all pairs of matrices in $\mathcal{V}$ and that has certain basic properties. The usual inner product of $\mathbf{A}$ and $\mathbf{B}$ is that given by the formula

$$\mathbf{A} \bullet \mathbf{B} = \text{tr}(\mathbf{A}'\mathbf{B}).$$

In the special case of n-dimensional column vectors $\mathbf{x} = \{x_i\}$ and $\mathbf{y} = \{y_i\}$, the formula for the usual inner product simplifies to

$$\mathbf{x} \bullet \mathbf{y} = \mathbf{x}'\mathbf{y} = x_1 y_1 + x_2 y_2 + \cdots + x_n y_n.$$

In the present section, we consider alternative inner products for the linear space $\mathcal{R}^{n \times 1}$ of all n-dimensional column vectors.

a. Expression of inner products as bilinear forms

Consider a bilinear form $\mathbf{x}'\mathbf{A}\mathbf{y}$ (in n-dimensional column vectors $\mathbf{x}$ and $\mathbf{y}$). If the matrix $\mathbf{A}$ is symmetric positive definite, then $\mathbf{x}'\mathbf{A}\mathbf{y}$ qualifies as an inner product for

$\mathcal{R}^{n\times 1}$. To see this, let $\mathbf{x}$, $\mathbf{y}$, and $\mathbf{z}$ represent arbitrary n-dimensional column vectors and k an arbitrary scalar, and observe that if $\mathbf{A}$ is symmetric positive definite, then the bilinear form has the following four properties:

(1) $\mathbf{x}'\mathbf{A}\mathbf{y} = \mathbf{y}'\mathbf{A}\mathbf{x}$;

(2) $\mathbf{x}'\mathbf{A}\mathbf{x} \geq 0$, with equality holding if and only if $\mathbf{x} = \mathbf{0}$;

(3) $(k\mathbf{x})'\mathbf{A}\mathbf{y} = k(\mathbf{x}'\mathbf{A}\mathbf{y})$;

(4) $(\mathbf{x}+\mathbf{y})'\mathbf{A}\mathbf{z} = \mathbf{x}'\mathbf{A}\mathbf{z} + \mathbf{y}'\mathbf{A}\mathbf{z}$.

Are the symmetry and positive definiteness of $\mathbf{A}$ necessary as well as sufficient for the bilinear form $\mathbf{x}'\mathbf{A}\mathbf{y}$ to qualify as an inner product for $\mathcal{R}^{n\times 1}$? The answer is yes. If $\mathbf{A}$ were not symmetric, then (as discussed in Section 14.1) the bilinear form $\mathbf{x}'\mathbf{A}\mathbf{y}$ would not be symmetric; that is, there would exist n-dimensional vectors $\mathbf{x}$ and $\mathbf{y}$ such that $\mathbf{x}'\mathbf{A}\mathbf{y} \neq \mathbf{y}'\mathbf{A}\mathbf{x}$. Moreover, if $\mathbf{A}$ were not positive definite, then $\mathbf{x}'\mathbf{A}\mathbf{x} < 0$ for some $\mathbf{x}$ or $\mathbf{x}'\mathbf{A}\mathbf{x} = 0$ for some nonnull $\mathbf{x}$.

Note (in light of Corollary 14.1.3) that distinct (symmetric positive definite) choices for the matrix $\mathbf{A}$ of the bilinear form $\mathbf{x}'\mathbf{A}\mathbf{y}$ produce distinct inner products.

Is every inner product for $\mathcal{R}^{n\times 1}$ expressible as a bilinear form? The answer is yes. To verify this, let $\mathbf{x}\bullet\mathbf{y}$ represent the value assigned to arbitrary n-dimensional column vectors $\mathbf{x}$ and $\mathbf{y}$ by an inner product for $\mathcal{R}^{n\times 1}$; denote by $\mathbf{e}_1, \ldots, \mathbf{e}_n$ the first, ..., nth columns, respectively, of $\mathbf{I}_n$; and take $\mathbf{A}$ to be the $n \times n$ matrix whose ijth element is $a_{ij} = \mathbf{e}_i \bullet \mathbf{e}_j$. Then,

$$
\begin{aligned}
\mathbf{x}\bullet\mathbf{y} &= \Big(\sum_i x_i\mathbf{e}_i\Big)\bullet\mathbf{y} \\
&= \sum_i x_i(\mathbf{e}_i\bullet\mathbf{y}) \\
&= \sum_i x_i\Big[\mathbf{e}_i\bullet\Big(\sum_j y_j\mathbf{e}_j\Big)\Big] \\
&= \sum_i x_i \sum_j y_j(\mathbf{e}_i\bullet\mathbf{e}_j) = \sum_{i,j} x_i y_j a_{ij} = \mathbf{x}'\mathbf{A}\mathbf{y}.
\end{aligned}
$$

In summary, a bilinear form (in n-dimensional vectors) qualifies as an inner product for $\mathcal{R}^{n\times 1}$ if and only if the matrix of the bilinear form is symmetric and positive definite; distinct (symmetric positive definite) choices for the matrix of the bilinear form produce distinct inner products for $\mathcal{R}^{n\times 1}$; and every inner product for $\mathcal{R}^{n\times 1}$ is expressible as a bilinear form.

b. Schwarz inequality

In light of the discussion of Subsection a, we obtain, as a direct consequence of Theorem 6.3.1, the following version of the Schwarz inequality.

Theorem 14.10.1. Let $\mathbf{W}$ represent an $n \times n$ symmetric positive definite matrix. Then, for any n-dimensional column vectors $\mathbf{x}$ and $\mathbf{y}$,

$$(\mathbf{x}'\mathbf{W}\mathbf{y})^2 \leq (\mathbf{x}'\mathbf{W}\mathbf{x})(\mathbf{y}'\mathbf{W}\mathbf{y}), \tag{10.1}$$

with equality holding if and only if $\mathbf{y} = \mathbf{0}$ or $\mathbf{x} = k\mathbf{y}$ for some scalar k.

The following corollary, which is essentially a special case of Corollary 6.3.2, expresses the result of Theorem 14.10.1 in an alternative form.

Corollary 14.10.2. Let $\mathbf{W}$ represent an $n \times n$ symmetric positive definite matrix. Then, for any n-dimensional column vector $\mathbf{y}$,

$$\max_{\{\mathbf{x}\in\mathcal{R}^n : \mathbf{x}\neq\mathbf{0}\}} \frac{(\mathbf{x}'\mathbf{W}\mathbf{y})^2}{\mathbf{x}'\mathbf{W}\mathbf{x}} = \mathbf{y}'\mathbf{W}\mathbf{y}. \tag{10.2}$$

Moreover, if $\mathbf{y} = \mathbf{0}$, the maximum is attained at every nonnull $\mathbf{x} \in \mathcal{R}^n$, and if $\mathbf{y} \neq \mathbf{0}$, the maximum is attained at every nonnull $\mathbf{x} \in \mathcal{R}^n$ that is proportional to $\mathbf{y}$.

By replacing the vector $\mathbf{y}$ in Theorem 14.10.1 and Corollary 14.10.2 with the vector $\mathbf{W}^{-1}\mathbf{y}$, we obtain the following two corollaries.

Corollary 14.10.3. Let $\mathbf{W}$ represent an $n \times n$ symmetric positive definite matrix. Then, for any n-dimensional column vectors $\mathbf{x}$ and $\mathbf{y}$,

$$(\mathbf{x}'\mathbf{y})^2 \leq (\mathbf{x}'\mathbf{W}\mathbf{x})(\mathbf{y}'\mathbf{W}^{-1}\mathbf{y}), \tag{10.3}$$

with equality holding if and only if $\mathbf{y} = \mathbf{0}$ or $\mathbf{x} = k\mathbf{W}^{-1}\mathbf{y}$ for some scalar k.

Corollary 14.10.4. Let $\mathbf{W}$ represent an $n \times n$ symmetric positive definite matrix. Then, for any n-dimensional column vector $\mathbf{y}$,

$$\max_{\{\mathbf{x}\in\mathcal{R}^n : \mathbf{x}\neq\mathbf{0}\}} \frac{(\mathbf{x}'\mathbf{y})^2}{\mathbf{x}'\mathbf{W}\mathbf{x}} = \mathbf{y}'\mathbf{W}^{-1}\mathbf{y}. \tag{10.4}$$

Moreover, if $\mathbf{y} = \mathbf{0}$, the maximum is attained at every nonnull $\mathbf{x} \in \mathcal{R}^n$, and if $\mathbf{y} \neq \mathbf{0}$, the maximum is attained at every nonnull $\mathbf{x} \in \mathcal{R}^n$ that is proportional to $\mathbf{W}^{-1}\mathbf{y}$.

c. Quasi-inner products

Let $\mathcal{V}$ represent a linear space of $m \times n$ matrices, let f represent a function whose domain consists of all pairs of matrices in $\mathcal{V}$, and, for $\mathbf{A}$ and $\mathbf{B}$ in $\mathcal{V}$, denote by $\mathbf{A} * \mathbf{B}$ the value assigned by f to the pair $\mathbf{A}$ and $\mathbf{B}$. The four basic requirements that would (for all matrices $\mathbf{A}$, $\mathbf{B}$, and $\mathbf{C}$ in $\mathcal{V}$ and for every scalar k) have to be satisfied by f for f to qualify as an inner product are:

(1) $\mathbf{A} * \mathbf{B} = \mathbf{B} * \mathbf{A}$;

(2) $\mathbf{A} * \mathbf{A} \geq 0$, with equality holding if and only if $\mathbf{A} = \mathbf{0}$;

(3) $(k\mathbf{A}) * \mathbf{B} = k(\mathbf{A} * \mathbf{B})$;

(4) $(\mathbf{A} + \mathbf{B}) * \mathbf{C} = (\mathbf{A} * \mathbf{C}) + (\mathbf{B} * \mathbf{C})$.

A less restrictive requirement than requirement (2) is the requirement

(2′) $\mathbf{A} * \mathbf{A} \geq 0$, with equality holding if $\mathbf{A} = \mathbf{0}$;

obtained by allowing $\mathbf{A} * \mathbf{A}$ to equal 0 even if $\mathbf{A} \neq \mathbf{0}$. If f satisfies requirement (2′) [as well as requirements (1), (3), and (4)], but not necessarily requirement (2), let us use the term *quasi-inner product* in referring to f and to the values assigned by f to the various pairs of matrices.

Subsequently, the symbol $\mathbf{A} \bullet \mathbf{B}$ will sometimes be used to represent the quasi-inner product of two matrices $\mathbf{A}$ and $\mathbf{B}$, although this will never be done without calling attention to the (temporary) departure from our standard usage of $\mathbf{A} \bullet \mathbf{B}$ as a symbol for a "true" inner product. In using the dot notation to represent a quasi-inner product, the symbol $\|\mathbf{A}\|$ will continue to be used to represent the quantity $(\mathbf{A} \bullet \mathbf{A})^{1/2}$. In the case of a quasi-inner product, let us refer to this quantity as a *quasi-norm*, rather than a norm.

It is a straightforward exercise to generalize the results of Subsection a to quasi-inner products. We find that a bilinear form (in n-dimensional vectors) qualifies as a quasi-inner product for $\mathcal{R}^{n \times 1}$ if and only if the matrix of the bilinear form is symmetric and nonnegative definite; that distinct (symmetric nonnegative definite) choices for the matrix of the bilinear form produce distinct quasi-inner products for $\mathcal{R}^{n \times 1}$; and that every quasi-inner product for $\mathcal{R}^{n \times 1}$ is expressible as a bilinear form.

The Schwarz inequality (6.3.1) applies to quasi-inner products as well as "true" inner products. To see this, let $\mathcal{V}$ represent a linear space of $m \times n$ matrices, and let $\mathbf{A} \bullet \mathbf{B}$ represent the value assigned by a quasi-inner product to any pair of matrices $\mathbf{A}$ and $\mathbf{B}$ (in $\mathcal{V}$).

That the Schwarz inequality

$$|\mathbf{A} \bullet \mathbf{B}| \leq \|\mathbf{A}\| \|\mathbf{B}\| \tag{10.5}$$

is valid for all $\mathbf{A}$ and $\mathbf{B}$ in $\mathcal{V}$ such that $\|\mathbf{B}\| > 0$ can be established in essentially the same way as in the case of a "true" inner product—refer to the proof of Theorem 6.3.1. And it can be established via an analogous argument that this inequality is valid for all $\mathbf{A}$ and $\mathbf{B}$ in $\mathcal{V}$ such that $\|\mathbf{A}\| > 0$. It remains to establish the validity of inequality (10.5) for those $\mathbf{A}$ and $\mathbf{B}$ in $\mathcal{V}$ such that $\|\mathbf{A}\| = \|\mathbf{B}\| = 0$—this can be accomplished by observing that if $\|\mathbf{A}\| = \|\mathbf{B}\| = 0$, then

$$0 \leq (\mathbf{A} + \mathbf{B}) \bullet (\mathbf{A} + \mathbf{B}) = 2\mathbf{A} \bullet \mathbf{B}$$

and

$$0 \leq (\mathbf{A} - \mathbf{B}) \bullet (\mathbf{A} - \mathbf{B}) = -2\mathbf{A} \bullet \mathbf{B},$$

and, as a consequence, $\mathbf{A} \bullet \mathbf{B} = 0$.

As in the case of a "true" inner product, a sufficient condition for equality to hold in inequality (10.5), that is, for

$$|\mathbf{A} \bullet \mathbf{B}| = \|\mathbf{A}\| \|\mathbf{B}\|, \tag{10.6}$$

is that $\mathbf{B} = \mathbf{0}$ or $\mathbf{A} = k\mathbf{B}$ for some scalar k. However, this condition is no longer necessary. That is, in general, equality (10.6) can be satisfied without $\mathbf{B}$ being null or $\mathbf{A}$ being proportional to $\mathbf{B}$.

In the case of a quasi-inner product, $\|\mathbf{A}\| = 0$ need not imply that $\mathbf{A} = \mathbf{0}$. However, even in the case of a quasi-inner product,

$$\|\mathbf{A}\| = 0 \quad \Rightarrow \quad \mathbf{A} \bullet \mathbf{B} = 0 \tag{10.7}$$

(for every $\mathbf{B}$ in $\mathcal{V}$), as is evident from the Schwarz inequality.

14.11 Some Results on Ranks and Row and Column Spaces and on Linear Systems

The following theorem generalizes Theorem 7.4.1.

Theorem 14.11.1. Let $\mathbf{W}$ represent an $m \times m$ symmetric nonnegative definite matrix. Then, for any $m \times n$ matrix $\mathbf{A}$ and $m \times p$ matrix $\mathbf{B}$, the linear system

$$\mathbf{A}'\mathbf{W}\mathbf{A}\mathbf{X} = \mathbf{A}'\mathbf{W}\mathbf{B} \quad (\text{in } \mathbf{X}) \tag{11.1}$$

is consistent.

Proof. According to Corollary 14.3.8, $\mathbf{W} = \mathbf{P}'\mathbf{P}$ for some matrix $\mathbf{P}$. Linear system (11.1) can be rewritten as

$$\mathbf{A}_*'\mathbf{A}_*\mathbf{X} = \mathbf{A}_*'\mathbf{B}_*,$$

where $\mathbf{A}_* = \mathbf{P}\mathbf{A}$ and $\mathbf{B}_* = \mathbf{P}\mathbf{B}$. Thus, it follows from Theorem 7.4.1 that linear system (11.1) is consistent. Q.E.D.

Some results on the row and column spaces and the rank of the coefficient matrix of linear system (11.1) are given by the following lemma, which generalizes Corollary 7.4.5.

Lemma 14.11.2. For any $m \times n$ matrix $\mathbf{A}$ and any $m \times m$ symmetric nonnegative definite matrix $\mathbf{W}$, $\mathcal{C}(\mathbf{A}'\mathbf{W}\mathbf{A}) = \mathcal{C}(\mathbf{A}'\mathbf{W})$, $\mathcal{R}(\mathbf{A}'\mathbf{W}\mathbf{A}) = \mathcal{R}(\mathbf{W}\mathbf{A})$, and $\text{rank}(\mathbf{A}'\mathbf{W}\mathbf{A}) = \text{rank}(\mathbf{W}\mathbf{A})$.

Proof. According to Corollary 14.3.8, $\mathbf{W} = \mathbf{P}'\mathbf{P}$ for some matrix $\mathbf{P}$. Observing that $\mathbf{A}'\mathbf{W}\mathbf{A} = (\mathbf{P}\mathbf{A})'\mathbf{P}\mathbf{A}$ and making use of Corollary 7.4.5 and Theorem 7.4.3, we find that

$$\mathcal{R}(\mathbf{A}'\mathbf{W}\mathbf{A}) = \mathcal{R}(\mathbf{P}\mathbf{A}) = \mathcal{R}(\mathbf{P}'\mathbf{P}\mathbf{A}) = \mathcal{R}(\mathbf{W}\mathbf{A}).$$

That $\mathcal{C}(\mathbf{A}'\mathbf{W}\mathbf{A}) = \mathcal{C}(\mathbf{A}'\mathbf{W})$ follows from Lemma 4.2.5. Q.E.D.

Since any positive definite matrix is nonsingular, we have (in light of Corollary 8.3.3) the following corollary of Lemma 14.11.2, which (like Lemma 14.11.2 itself) can be regarded as a generalization of Corollary 7.4.5.

Corollary 14.11.3. For any $m \times n$ matrix $\mathbf{A}$ and any $m \times m$ symmetric positive definite matrix $\mathbf{W}$, $\mathcal{C}(\mathbf{A}'\mathbf{W}\mathbf{A}) = \mathcal{C}(\mathbf{A}')$, $\mathcal{R}(\mathbf{A}'\mathbf{W}\mathbf{A}) = \mathcal{R}(\mathbf{A})$, and $\text{rank}(\mathbf{A}'\mathbf{W}\mathbf{A}) = \text{rank}(\mathbf{A})$.

14.12 Projections, Projection Matrices, and Orthogonal Complements

In this section, further consideration is given to the projection of an n-dimensional column vector on a subspace of $\mathcal{R}^{n\times 1}$. By definition, the projection depends on the choice of inner product. A rather extensive discourse on the projection of a column vector was presented in Chapter 12. The results included in that discourse are, for the most part, specific to the case where the inner product is the usual inner product. In this section, these results are generalized to an arbitrary inner product.

a. Some terminology, notation, and basic results

Let $\mathbf{W}$ represent an $n \times n$ symmetric positive definite matrix. As discussed in Section 14.10a, the bilinear form $\mathbf{x}'\mathbf{W}\mathbf{y}$ (in n-dimensional column vectors $\mathbf{x}$ and $\mathbf{y}$) qualifies as an inner product for $\mathcal{R}^{n\times 1}$. (And every inner product is expressible as a bilinear form.) When the bilinear form $\mathbf{x}'\mathbf{W}\mathbf{y}$ is chosen to be the inner product for $\mathcal{R}^{n\times 1}$, $\mathbf{x} \perp \mathbf{y}$ (i.e., two n-dimensional column vectors $\mathbf{x}$ and $\mathbf{y}$ are orthogonal) if and only if $\mathbf{x}'\mathbf{W}\mathbf{y} = 0$, $\mathbf{x} \perp \mathcal{U}$ (i.e., $\mathbf{x}$ is orthogonal to a subspace $\mathcal{U}$ of $\mathcal{R}^{n\times 1}$) if and only if $\mathbf{x}'\mathbf{W}\mathbf{y} = 0$ for every $\mathbf{y}$ in $\mathcal{U}$, and $\mathcal{U} \perp \mathcal{V}$ (i.e., $\mathcal{U}$ is orthogonal to a subspace $\mathcal{V}$) if and only if $\mathbf{x}'\mathbf{W}\mathbf{y} = 0$ for every $\mathbf{x}$ in $\mathcal{U}$ and every $\mathbf{y}$ in $\mathcal{V}$.

To indicate that $\mathbf{x}'\mathbf{W}\mathbf{y} = 0$ for two n-dimensional vectors $\mathbf{x}$ and $\mathbf{y}$, let us (for convenience and/or to emphasize that the orthogonality of two vectors depends on the choice of inner product) write $\mathbf{x} \perp_{\mathbf{W}} \mathbf{y}$ or say that $\mathbf{x}$ and $\mathbf{y}$ are orthogonal with respect to $\mathbf{W}$. Similarly, to indicate that $\mathbf{x}'\mathbf{W}\mathbf{y} = 0$ for every vector $\mathbf{y}$ in a subspace $\mathcal{U}$, let us write $\mathbf{x} \perp_{\mathbf{W}} \mathcal{U}$ or say that $\mathbf{x}$ and $\mathcal{U}$ are orthogonal with respect to $\mathbf{W}$. And, to indicate that $\mathbf{x}'\mathbf{W}\mathbf{y} = 0$ for every $\mathbf{x}$ in a subspace $\mathcal{U}$ and every $\mathbf{y}$ in a subspace $\mathcal{V}$, let us write $\mathcal{U} \perp_{\mathbf{W}} \mathcal{V}$ or say that $\mathcal{U}$ and $\mathcal{V}$ are orthogonal with respect to $\mathbf{W}$. Clearly, saying that $\mathbf{x}$ and $\mathbf{y}$, $\mathbf{x}$ and $\mathcal{U}$, or $\mathcal{U}$ and $\mathcal{V}$ are orthogonal with respect to $\mathbf{I}$ is synonymous with saying that they are orthogonal with respect to the usual inner product.

Furthermore, let us extend our use of this notation and terminology to the case where $\mathbf{W}$ is positive semidefinite rather than positive definite. Thus, for any $n \times n$ symmetric nonnegative definite matrix $\mathbf{W}$, let us write $\mathbf{x} \perp_{\mathbf{W}} \mathbf{y}$ or say that $\mathbf{x}$ and $\mathbf{y}$ are orthogonal with respect to $\mathbf{W}$ to indicate that $\mathbf{x}'\mathbf{W}\mathbf{y} = 0$; let us write $\mathbf{x} \perp_{\mathbf{W}} \mathcal{U}$ or say that $\mathbf{x}$ and $\mathcal{U}$ are orthogonal with respect to $\mathbf{W}$ to indicate that $\mathbf{x}'\mathbf{W}\mathbf{y} = 0$ for every $\mathbf{y}$ in $\mathcal{U}$; and let us write $\mathcal{U} \perp_{\mathbf{W}} \mathcal{V}$ or say that $\mathcal{U}$ and $\mathcal{V}$ are orthogonal with respect to $\mathbf{W}$ to indicate that $\mathbf{x}'\mathbf{W}\mathbf{y} = 0$ for every $\mathbf{x}$ in $\mathcal{U}$ and every $\mathbf{y}$ in $\mathcal{V}$. Note that if $\mathbf{W}$ is only positive semidefinite, then the bilinear form $\mathbf{x}'\mathbf{W}\mathbf{y}$ does not qualify as an inner product (as it does when $\mathbf{W}$ is positive definite)—although it still qualifies as what (in Section 14.10c) is called a quasi-inner product.

By applying Lemma 12.1.1 (in the special case where $\mathcal{V} = \mathcal{R}^{m\times 1}$ and $\mathcal{U}$ and $\mathcal{W}$ are column spaces of matrices), we obtain the following generalization of Corollary 12.1.2.

Lemma 14.12.1. Let $\mathbf{y}$ represent an m-dimensional column vector, $\mathbf{X}$ an $m\times n$ matrix, and $\mathbf{Z}$ an $m \times p$ matrix, and let $\mathbf{W}$ represent an $m \times m$ symmetric positive definite matrix. Then, $\mathbf{y} \perp_{\mathbf{W}} \mathcal{C}(\mathbf{X})$ if and only if $\mathbf{X}'\mathbf{W}\mathbf{y} = \mathbf{0}$ (or equivalently if and only if $\mathbf{y}'\mathbf{W}\mathbf{X} = \mathbf{0}$). Similarly, $\mathcal{C}(\mathbf{X}) \perp_{\mathbf{W}} \mathcal{C}(\mathbf{Z})$ if and only if $\mathbf{X}'\mathbf{W}\mathbf{Z} = \mathbf{0}$ (or equivalently if and only if $\mathbf{Z}'\mathbf{W}\mathbf{X} = \mathbf{0}$).

The restriction in Lemma 14.12.1 that $\mathbf{W}$ be positive definite can be relaxed. The lemma is valid for any $m \times m$ symmetric nonnegative definite matrix $\mathbf{W}$, as can be easily verified.

b. Approach

To extend (to inner products other than the usual inner product) the results presented in Chapter 12 on the projection of a column vector, we could proceed in either of two ways. One approach is to generalize and (when necessary) modify the derivations given in Chapter 12. An alternative approach, which is sometimes employed here, uses the following lemma to take advantage of the results already established (in Chapter 12) in the special case of the usual inner product.

Lemma 14.12.2. Let $\mathbf{L}$ represent an arbitrary matrix of dimensions $m \times n$, and let $\mathcal{U}$ represent an arbitrary subspace of $\mathcal{R}^{n\times 1}$. Further, define

$$\mathcal{V} = \{\mathbf{v} \in \mathcal{R}^m \; : \; \mathbf{v} = \mathbf{L}\mathbf{x} \text{ for some } \mathbf{x} \in \mathcal{U}\},$$

define $\mathbf{W} = \mathbf{L}'\mathbf{L}$, and let $\mathbf{X}$ represent an $n \times p$ matrix whose columns span $\mathcal{U}$. Then,

(1) the set $\mathcal{V}$ is a subspace of $\mathcal{R}^m$;

(2) two vectors $\mathbf{x}$ and $\mathbf{y}$ in $\mathcal{R}^n$ are orthogonal with respect to $\mathbf{W}$ if and only if $\mathbf{L}\mathbf{x}$ and $\mathbf{L}\mathbf{y}$ are orthogonal with respect to the usual inner product (for $\mathcal{R}^m$);

(3) a vector $\mathbf{x}$ (in $\mathcal{R}^n$) and the subspace $\mathcal{U}$ are orthogonal with respect to $\mathbf{W}$ if and only if $\mathbf{L}\mathbf{x}$ and $\mathcal{V}$ are orthogonal with respect to the usual inner product (for $\mathcal{R}^m$);

(4) the subspace $\mathcal{U}$ and a subspace $\mathcal{M}$ (of $\mathcal{R}^{n\times 1}$) are orthogonal with respect to $\mathbf{W}$ if and only if $\mathcal{V}$ and the subspace

$$\mathcal{N} = \{\mathbf{z} \in \mathcal{R}^m \; : \; \mathbf{z} = \mathbf{L}\mathbf{y} \text{ for some } \mathbf{y} \in \mathcal{M}\}$$

are orthogonal with respect to the usual inner product (for $\mathcal{R}^m$);

(5) the columns of $\mathbf{L}\mathbf{X}$ span $\mathcal{V}$.

Proof. (1) By definition, $\mathcal{V}$ is a subset of $\mathcal{R}^m$. Thus, it suffices to show that $\mathcal{V}$ is a linear space. Let $\mathbf{v}$ and $\mathbf{w}$ represent arbitrary vectors in $\mathcal{V}$, and let k be an arbitrary scalar. Then, $\mathbf{v} = \mathbf{L}\mathbf{x}$ and $\mathbf{w} = \mathbf{L}\mathbf{y}$ for some vectors $\mathbf{x}$ and $\mathbf{y}$ in $\mathcal{U}$, and (since $\mathcal{U}$ is a linear space) $\mathbf{x} + \mathbf{y} \in \mathcal{U}$ and $k\mathbf{x} \in \mathcal{U}$. It follows that $\mathbf{v} + \mathbf{w} = \mathbf{L}(\mathbf{x} + \mathbf{y}) \in \mathcal{V}$ and $k\mathbf{v} = \mathbf{L}(k\mathbf{x}) \in \mathcal{V}$ and hence that $\mathcal{V}$ is a linear space.

(2) Observe that $\mathbf{x}'\mathbf{W}\mathbf{y} = (\mathbf{L}\mathbf{x})'\mathbf{L}\mathbf{y}$, and hence that $\mathbf{x}'\mathbf{W}\mathbf{y} = 0$ if and only if $(\mathbf{L}\mathbf{x})'\mathbf{L}\mathbf{y} = 0$.

(3) In light of Part (2), $\mathbf{x}$ and $\mathcal{U}$ are orthogonal with respect to $\mathbf{W}$, or, equivalently, $\mathbf{x}$ and $\mathbf{y}$ are orthogonal (with respect to $\mathbf{W}$) for every $\mathbf{y}$ in $\mathcal{U}$, if and only if $\mathbf{L}\mathbf{x}$ and $\mathbf{L}\mathbf{y}$ are orthogonal (with respect to the usual inner product) for every $\mathbf{y}$

in $\mathcal{U}$. Thus, $\mathbf{x}$ and $\mathcal{U}$ are orthogonal (with respect to $\mathbf{W}$) if and only if $\mathbf{Lx}$ and $\mathbf{v}$ are orthogonal (with respect to the usual inner product) for every $\mathbf{v}$ in $\mathcal{V}$, or, equivalently, if and only if $\mathbf{Lx}$ and $\mathcal{V}$ are orthogonal (with respect to the usual inner product).

(4) In light of Part (2), $\mathcal{U}$ and $\mathcal{M}$ are orthogonal with respect to $\mathbf{W}$, or, equivalently, $\mathbf{x}$ and $\mathbf{y}$ are orthogonal (with respect to $\mathbf{W}$) for every $\mathbf{x}$ in $\mathcal{U}$ and every $\mathbf{y}$ in $\mathcal{M}$, if and only if $\mathbf{Lx}$ and $\mathbf{Ly}$ are orthogonal (with respect to the usual inner product) for every $\mathbf{x}$ in $\mathcal{U}$ and every $\mathbf{y}$ in $\mathcal{M}$. Thus, $\mathcal{U}$ and $\mathcal{M}$ are orthogonal (with respect to $\mathbf{W}$) if and only if $\mathbf{v}$ and $\mathbf{z}$ are orthogonal (with respect to the usual inner product) for every $\mathbf{v}$ in $\mathcal{V}$ and every $\mathbf{z}$ in $\mathcal{N}$, or, equivalently, if and only if $\mathcal{V}$ and $\mathcal{N}$ are orthogonal (with respect to the usual inner product).

(5) Let $\mathbf{v}$ represent an arbitrary vector in $\mathcal{V}$. Then, $\mathbf{v} = \mathbf{Lx}$ for some $\mathbf{x} \in \mathcal{U}$. Moreover, $\mathbf{x} = \mathbf{Xr}$ for some p-dimensional column vector $\mathbf{r}$. Thus, $\mathbf{v} = \mathbf{LXr}$, so that $\mathbf{v}$ is expressible as a linear combination of the columns of $\mathbf{LX}$. We conclude that the columns of $\mathbf{LX}$ span $\mathcal{V}$. Q.E.D.

c. Projections

Let $\mathbf{W}$ represent an $n \times n$ symmetric positive definite matrix, let $\mathbf{y}$ represent an n-dimensional column vector, and let $\mathcal{U}$ represent a subspace of $\mathcal{R}^{n\times 1}$. Define $\mathbf{z}$ to be the projection of $\mathbf{y}$ on $\mathcal{U}$ when the inner product (for $\mathcal{R}^{n\times 1}$) is taken to be the bilinear form $\mathbf{x}'\mathbf{W}\mathbf{y}$. Let us refer to this projection as the *orthogonal projection of* $\mathbf{y}$ *on* $\mathcal{U}$ *with respect to* $\mathbf{W}$, or simply as the *projection of* $\mathbf{y}$ *on* $\mathcal{U}$ *with respect to* $\mathbf{W}$.

Formula (12.2.1) gives the projection of $\mathbf{y}$ on $\mathcal{U}$ with respect to the usual inner product (i.e., with respect to $\mathbf{I}$) in terms of an arbitrary solution to the normal equations (12.2.2). For purposes of generalizing this formula, let $\mathbf{L}$ represent a nonsingular matrix such that $\mathbf{W} = \mathbf{L}'\mathbf{L}$, define

$$\mathcal{V} = \{\mathbf{v} \in \mathcal{R}^n \ : \ \mathbf{v} = \mathbf{Lx} \text{ for some } \mathbf{x} \in \mathcal{U}\},$$

and take $\mathbf{X}$ to be any $n \times p$ matrix whose columns span $\mathcal{U}$.

Let $\mathbf{b}^*$ represent any solution to the linear system

$$\mathbf{X}'\mathbf{W}\mathbf{X}\mathbf{b} = \mathbf{X}'\mathbf{W}\mathbf{y} \quad (\text{in } \mathbf{b}).$$

(The existence of such a solution follows from Theorem 14.11.1.) Note that this linear system can be rewritten as

$$(\mathbf{LX})'\mathbf{LXb} = (\mathbf{LX})'\mathbf{Ly}.$$

By definition, $\mathbf{z}$ is the (unique) vector in $\mathcal{U}$ such that $\mathbf{y} - \mathbf{z} \perp_{\mathbf{W}} \mathcal{U}$. Thus, it follows from Part (3) of Lemma 14.12.2 that $\mathbf{L}(\mathbf{y} - \mathbf{z})$, or, equivalently, $\mathbf{Ly} - \mathbf{Lz}$, is orthogonal to $\mathcal{V}$ with respect to the usual inner product, implying that $\mathbf{Lz}$ is the projection of $\mathbf{Ly}$ on $\mathcal{V}$ with respect to the usual inner product. Since [according to Part (5) of Lemma 14.12.2] the columns of $\mathbf{LX}$ span $\mathcal{V}$, we have, as a consequence of Theorem 12.2.1, that $\mathbf{Lz} = \mathbf{LXb}^*$ and hence (since $\mathbf{L}$ is nonsingular) that $\mathbf{z} = \mathbf{Xb}^*$.

In summary, we have the following generalization of Theorem 12.2.1.

Theorem 14.12.3. Let $\mathbf{z}$ represent the projection of an n-dimensional column vector $\mathbf{y}$ on a subspace $\mathcal{U}$ (of $\mathcal{R}^n$) with respect to a symmetric positive definite matrix $\mathbf{W}$, and let $\mathbf{X}$ represent any $n \times p$ matrix whose columns span $\mathcal{U}$. Then,

$$\mathbf{z} = \mathbf{Xb}^*$$

for any solution $\mathbf{b}^*$ to the (consistent) linear system

$$\mathbf{X}'\mathbf{WXb} = \mathbf{X}'\mathbf{Wy} \quad (\text{in } \mathbf{b}).$$

The following three corollaries (of Theorem 14.12.3) generalize Corollaries 12.2.2–12.2.4, respectively.

Corollary 14.12.4. Let $\mathbf{z}$ represent the projection of an n-dimensional column vector $\mathbf{y}$ on a subspace $\mathcal{U}$ (of $\mathcal{R}^n$) with respect to a symmetric positive definite matrix $\mathbf{W}$, and let $\mathbf{X}$ represent any $n \times p$ matrix whose columns span $\mathcal{U}$. Then,

$$\mathbf{z} = \mathbf{X}(\mathbf{X}'\mathbf{WX})^{-}\mathbf{X}'\mathbf{Wy}.$$

Corollary 14.12.5. Let $\mathbf{y}$ represent an n-dimensional column vector, $\mathbf{X}$ an $n \times p$ matrix, and $\mathbf{U}$ any $n \times q$ matrix such that $\mathcal{C}(\mathbf{U}) = \mathcal{C}(\mathbf{X})$. Then,

$$\mathbf{Ua}^* = \mathbf{Xb}^*$$

for any solution $\mathbf{a}^*$ to the linear system $\mathbf{U}'\mathbf{WUa} = \mathbf{U}'\mathbf{Wy}$ (in $\mathbf{a}$) and any solution $\mathbf{b}^*$ to the linear system $\mathbf{X}'\mathbf{WXb} = \mathbf{X}'\mathbf{Wy}$ (in $\mathbf{b}$).

Corollary 14.12.6. Let $\mathbf{y}$ represent an n-dimensional column vector, and $\mathbf{X}$ an $n \times p$ matrix. Then, $\mathbf{Xb}_1 = \mathbf{Xb}_2$ for any two solutions $\mathbf{b}_1$ and $\mathbf{b}_2$ to the linear system $\mathbf{X}'\mathbf{WXb} = \mathbf{X}'\mathbf{Wy}$ (in $\mathbf{b}$).

The following theorem generalizes Theorem 12.2.5.

Theorem 14.12.7. Let $\mathbf{z}$ represent the projection of an n-dimensional column vector $\mathbf{y}$ on a subspace $\mathcal{U}$ (of $\mathcal{R}^n$) with respect to a symmetric positive definite matrix $\mathbf{W}$, and let $\mathbf{X}$ represent any $n \times p$ matrix whose columns span $\mathcal{U}$. Then, any $p \times 1$ vector $\mathbf{b}^*$ such that $\mathbf{z} = \mathbf{Xb}^*$ is a solution to the linear system $\mathbf{X}'\mathbf{WXb} = \mathbf{X}'\mathbf{Wy}$ (in $\mathbf{b}$).

Proof. As a consequence of Theorem 14.12.3, $\mathbf{X}'\mathbf{WXb} = \mathbf{X}'\mathbf{Wy}$ has a solution, say $\mathbf{a}$, and $\mathbf{z} = \mathbf{Xa}$. Thus,

$$\mathbf{X}'\mathbf{WXb}^* = \mathbf{X}'\mathbf{Wz} = \mathbf{X}'\mathbf{WXa} = \mathbf{X}'\mathbf{Wy}.$$

Q.E.D.

d. Projection matrices

Subsequently, for any $n \times p$ matrix $\mathbf{X}$ and for any $n \times n$ matrix $\mathbf{W}$, the symbol $\mathbf{P}_{\mathbf{X},\mathbf{W}}$ is used to represent the $n \times n$ matrix $\mathbf{X}(\mathbf{X}'\mathbf{W}\mathbf{X})^{-}\mathbf{X}'\mathbf{W}$. Note that $\mathbf{P}_{\mathbf{X},\mathbf{I}} = \mathbf{X}(\mathbf{X}'\mathbf{X})^{-}\mathbf{X}' = \mathbf{P}_{\mathbf{X}}$. Note also that if $\mathbf{W}$ is symmetric and positive definite, then $\mathbf{P}_{\mathbf{X},\mathbf{W}}$ is invariant to the choice of the generalized inverse $(\mathbf{X}'\mathbf{W}\mathbf{X})^{-}$ as (in light of Corollary 14.11.3) is evident from Theorem 9.4.1 and as is implicit in the following theorem.

Theorem 14.12.8. Let $\mathcal{U}$ represent any subspace of the linear space $\mathcal{R}^n$ of all n-dimensional column vectors, and let $\mathbf{W}$ represent any $n \times n$ symmetric positive definite matrix. Then, there exists a unique matrix $\mathbf{A}$ (of dimensions $n \times n$) such that, for every $\mathbf{y}$ in $\mathcal{R}^n$, $\mathbf{A}\mathbf{y}$ is the projection of $\mathbf{y}$ on $\mathcal{U}$ with respect to $\mathbf{W}$. Moreover, $\mathbf{A} = \mathbf{P}_{\mathbf{X},\mathbf{W}}$ for any matrix $\mathbf{X}$ such that $\mathcal{C}(\mathbf{X}) = \mathcal{U}$.

Theorem 14.12.8 generalizes Theorem 12.3.1 and can be proved in essentially the same way as Theorem 12.3.1.

Let us refer to the matrix $\mathbf{A}$ (from Theorem 14.12.8) as the *orthogonal projection matrix for* $\mathcal{U}$ *with respect to* $\mathbf{W}$ or simply as the *projection matrix for* $\mathcal{U}$ *with respect to* $\mathbf{W}$. And let us speak of an $n \times n$ matrix as a *projection matrix with respect to* $\mathbf{W}$ when we wish to indicate that there exists some subspace of $\mathcal{R}^n$ for which that matrix is the projection matrix with respect to $\mathbf{W}$. Moreover, for consistency with our previous terminology, let us refer to the projection matrix for $\mathcal{U}$ with respect to $\mathbf{I}$ simply as the projection matrix for $\mathcal{U}$ and to a projection matrix with respect to $\mathbf{I}$ simply as a projection matrix.

The following two corollaries (of Theorem 14.12.8) generalize Corollaries 12.3.2 and 12.3.3.

Corollary 14.12.9. Let $\mathbf{A}$ represent the projection matrix for a subspace $\mathcal{U}$ (of $\mathcal{R}^{n\times 1}$) with respect to an $n \times n$ symmetric positive definite matrix $\mathbf{W}$. Then, $\mathbf{A} = \mathbf{P}_{\mathbf{X},\mathbf{W}}$ for any matrix $\mathbf{X}$ such that $\mathcal{C}(\mathbf{X}) = \mathcal{U}$.

Corollary 14.12.10. An $n \times n$ matrix $\mathbf{A}$ is a projection matrix with respect to an $n \times n$ symmetric positive definite matrix $\mathbf{W}$ if and only if $\mathbf{A} = \mathbf{P}_{\mathbf{X},\mathbf{W}}$ for some matrix $\mathbf{X}$.

The basic properties of projection matrices encompassed in Theorems 12.3.4 and 12.3.5 are generalized in the following two theorems.

Theorem 14.12.11. Let $\mathbf{X}$ represent any $n \times p$ matrix, and $\mathbf{W}$ any $n \times n$ symmetric positive definite matrix. Then,

(1) $\mathbf{P}_{\mathbf{X},\mathbf{W}}\mathbf{X} = \mathbf{X}$; that is, $\mathbf{X}(\mathbf{X}'\mathbf{W}\mathbf{X})^{-}\mathbf{X}'\mathbf{W}\mathbf{X} = \mathbf{X}$; implying that $\mathbf{W}\mathbf{P}_{\mathbf{X},\mathbf{W}}\mathbf{X} = \mathbf{W}\mathbf{X}$;

(2) $\mathbf{P}_{\mathbf{X},\mathbf{W}} = \mathbf{X}\mathbf{B}^*$ for any solution $\mathbf{B}^*$ to the linear system $\mathbf{X}'\mathbf{W}\mathbf{X}\mathbf{B} = \mathbf{X}'\mathbf{W}$ (in $\mathbf{B}$);

(3) $\mathbf{W}\mathbf{P}_{\mathbf{X},\mathbf{W}}$ is symmetric; that is, $\mathbf{W}\mathbf{X}[(\mathbf{X}'\mathbf{W}\mathbf{X})^{-}]'\mathbf{X}'\mathbf{W} = \mathbf{W}\mathbf{X}(\mathbf{X}'\mathbf{W}\mathbf{X})^{-}\mathbf{X}'\mathbf{W}$;

(3′) $\mathbf{P}_{\mathbf{X},\mathbf{W}}\mathbf{W}^{-1}$ is symmetric; that is, $\mathbf{X}[(\mathbf{X}'\mathbf{W}\mathbf{X})^{-}]'\mathbf{X}' = \mathbf{X}(\mathbf{X}'\mathbf{W}\mathbf{X})^{-}\mathbf{X}'$;

(4) $\mathbf{X}[(\mathbf{X}'\mathbf{W}\mathbf{X})^-]'\mathbf{X}'\mathbf{W}\mathbf{X} = \mathbf{X}$; implying that $\mathbf{P}'_{\mathbf{X},\mathbf{W}}\mathbf{W}\mathbf{X} = \mathbf{W}\mathbf{X}$;

(5) $\mathbf{X}'\mathbf{W}\mathbf{X}(\mathbf{X}'\mathbf{W}\mathbf{X})^-\mathbf{X}' = \mathbf{X}'\mathbf{W}\mathbf{X}[(\mathbf{X}'\mathbf{W}\mathbf{X})^-]'\mathbf{X}' = \mathbf{X}'\mathbf{P}'_{\mathbf{X},\mathbf{W}} = \mathbf{X}'$; implying that $\mathbf{X}'\mathbf{W}\mathbf{P}_{\mathbf{X},\mathbf{W}} = \mathbf{X}'\mathbf{W}$;

(6) $\mathbf{P}^2_{\mathbf{X},\mathbf{W}} = \mathbf{P}_{\mathbf{X},\mathbf{W}}$; that is, $\mathbf{P}_{\mathbf{X},\mathbf{W}}$ is idempotent;

(6′) $\mathbf{P}'_{\mathbf{X},\mathbf{W}}\mathbf{W}\mathbf{P}_{\mathbf{X},\mathbf{W}} = \mathbf{W}\mathbf{P}_{\mathbf{X},\mathbf{W}}$;

(7) $\mathcal{C}(\mathbf{P}_{\mathbf{X},\mathbf{W}}) = \mathcal{C}(\mathbf{X})$, and $\mathcal{R}(\mathbf{P}_{\mathbf{X},\mathbf{W}}) = \mathcal{R}(\mathbf{X}'\mathbf{W})$;

(8) $\text{rank}(\mathbf{P}_{\mathbf{X},\mathbf{W}}) = \text{rank}(\mathbf{X})$;

(9) $(\mathbf{I} - \mathbf{P}_{\mathbf{X},\mathbf{W}})^2 = \mathbf{I} - \mathbf{P}_{\mathbf{X},\mathbf{W}}$; that is, $\mathbf{I} - \mathbf{P}_{\mathbf{X},\mathbf{W}}$ is idempotent;

(9′) $(\mathbf{I} - \mathbf{P}_{\mathbf{X},\mathbf{W}})'\mathbf{W}(\mathbf{I} - \mathbf{P}_{\mathbf{X},\mathbf{W}}) = \mathbf{W}(\mathbf{I} - \mathbf{P}_{\mathbf{X},\mathbf{W}})$;

(10) $\text{rank}(\mathbf{I} - \mathbf{P}_{\mathbf{X},\mathbf{W}}) = n - \text{rank}(\mathbf{X})$.

Proof. According to Corollary 14.3.13, there exists a nonsingular matrix $\mathbf{L}$ such that $\mathbf{W} = \mathbf{L}'\mathbf{L}$.

(1) Making use of Part (1) of Theorem 12.3.4, we find that

$$\mathbf{L}\mathbf{X}(\mathbf{X}'\mathbf{W}\mathbf{X})^-\mathbf{X}'\mathbf{W}\mathbf{X} = \mathbf{L}\mathbf{X}[(\mathbf{L}\mathbf{X})'\mathbf{L}\mathbf{X}]^-(\mathbf{L}\mathbf{X})'\mathbf{L}\mathbf{X} = \mathbf{L}\mathbf{X}.$$

Since $\mathbf{L}$ is nonsingular, we conclude that $\mathbf{X}(\mathbf{X}'\mathbf{W}\mathbf{X})^-\mathbf{X}'\mathbf{W}\mathbf{X} = \mathbf{X}$.

(2) If $\mathbf{B}^*$ is a solution to $\mathbf{X}'\mathbf{W}\mathbf{X}\mathbf{B} = \mathbf{X}'\mathbf{W}$, then (according to Theorem 11.2.4)

$$\mathbf{B}^* = (\mathbf{X}'\mathbf{W}\mathbf{X})^-\mathbf{X}'\mathbf{W} + [\mathbf{I} - (\mathbf{X}'\mathbf{W}\mathbf{X})^-\mathbf{X}'\mathbf{W}\mathbf{X}]\mathbf{Y}$$

for some matrix $\mathbf{Y}$, implying [in light of Part (1)] that

$$\mathbf{X}\mathbf{B}^* = \mathbf{P}_{\mathbf{X},\mathbf{W}} + [\mathbf{X} - \mathbf{X}(\mathbf{X}'\mathbf{W}\mathbf{X})^-\mathbf{X}'\mathbf{W}\mathbf{X}]\mathbf{Y} = \mathbf{P}_{\mathbf{X},\mathbf{W}}.$$

(3) and (3′). It follows from Part (3) of Theorem 12.3.4 that

$$\begin{aligned}\mathbf{L}\mathbf{X}[(\mathbf{X}'\mathbf{W}\mathbf{X})^-]'\mathbf{X}'\mathbf{L}' &= \mathbf{L}\mathbf{X}\{[(\mathbf{L}\mathbf{X})'\mathbf{L}\mathbf{X}]^-\}'(\mathbf{L}\mathbf{X})' \\ &= \mathbf{L}\mathbf{X}[(\mathbf{L}\mathbf{X})'\mathbf{L}\mathbf{X}]^-(\mathbf{L}\mathbf{X})' = \mathbf{L}\mathbf{X}(\mathbf{X}'\mathbf{W}\mathbf{X})^-\mathbf{X}'\mathbf{L}'.\end{aligned}$$

Thus,

$$\begin{aligned}\mathbf{W}\mathbf{X}[(\mathbf{X}'\mathbf{W}\mathbf{X})^-]'\mathbf{X}'\mathbf{W} &= \mathbf{L}'\mathbf{L}\mathbf{X}[(\mathbf{X}'\mathbf{W}\mathbf{X})^-]'\mathbf{X}'\mathbf{L}'\mathbf{L} \\ &= \mathbf{L}'\mathbf{L}\mathbf{X}(\mathbf{X}'\mathbf{W}\mathbf{X})^-\mathbf{X}'\mathbf{L}'\mathbf{L} = \mathbf{W}\mathbf{X}(\mathbf{X}'\mathbf{W}\mathbf{X})^-\mathbf{X}'\mathbf{W},\end{aligned}$$

and

$$\begin{aligned}\mathbf{X}[(\mathbf{X}'\mathbf{W}\mathbf{X})^-]'\mathbf{X}' &= \mathbf{L}^{-1}\mathbf{L}\mathbf{X}[(\mathbf{X}'\mathbf{W}\mathbf{X})^-]'\mathbf{X}'\mathbf{L}'(\mathbf{L}')^{-1} \\ &= \mathbf{L}^{-1}\mathbf{L}\mathbf{X}(\mathbf{X}'\mathbf{W}\mathbf{X})^-\mathbf{X}'\mathbf{L}'(\mathbf{L}')^{-1} = \mathbf{X}(\mathbf{X}'\mathbf{W}\mathbf{X})^-\mathbf{X}'.\end{aligned}$$

(4) Making use of Parts (3) and (1), we find that

$$\begin{aligned}\mathbf{X}[(\mathbf{X}'\mathbf{W}\mathbf{X})^-]'\mathbf{X}'\mathbf{W}\mathbf{X} &= \mathbf{W}^{-1}\{\mathbf{W}\mathbf{X}[(\mathbf{X}'\mathbf{W}\mathbf{X})^-]'\mathbf{X}'\mathbf{W}\}\mathbf{X} \\ &= \mathbf{W}^{-1}[\mathbf{W}\mathbf{X}(\mathbf{X}'\mathbf{W}\mathbf{X})^-\mathbf{X}'\mathbf{W}]\mathbf{X} \\ &= \mathbf{X}(\mathbf{X}'\mathbf{W}\mathbf{X})^-\mathbf{X}'\mathbf{W}\mathbf{X} = \mathbf{X}.\end{aligned}$$

(5) It follows from Part (4) that

$$\mathbf{X}' = \{\mathbf{X}[(\mathbf{X}'\mathbf{W}\mathbf{X})^-]'\mathbf{X}'\mathbf{W}\mathbf{X}\}' = \mathbf{X}'\mathbf{W}\mathbf{X}(\mathbf{X}'\mathbf{W}\mathbf{X})^-\mathbf{X}'$$

and from Part (1) that

$$\mathbf{X}' = (\mathbf{P}_{\mathbf{X},\mathbf{W}}\mathbf{X})' = \mathbf{X}'\mathbf{P}'_{\mathbf{X},\mathbf{W}} = \mathbf{X}'\mathbf{W}\mathbf{X}[(\mathbf{X}'\mathbf{W}\mathbf{X})^-]'\mathbf{X}'.$$

(6) Making use of Part (5), we find that

$$\mathbf{P}^2_{\mathbf{X},\mathbf{W}} = \mathbf{X}(\mathbf{X}'\mathbf{W}\mathbf{X})^-\mathbf{X}'\mathbf{W}\mathbf{P}_{\mathbf{X},\mathbf{W}} = \mathbf{X}(\mathbf{X}'\mathbf{W}\mathbf{X})^-\mathbf{X}'\mathbf{W} = \mathbf{P}_{\mathbf{X},\mathbf{W}}.$$

(6′) Making use of Part (4), we find that

$$\begin{aligned}\mathbf{P}'_{\mathbf{X},\mathbf{W}}\mathbf{W}\mathbf{P}_{\mathbf{X},\mathbf{W}} &= \mathbf{P}'_{\mathbf{X},\mathbf{W}}\mathbf{W}\mathbf{X}(\mathbf{X}'\mathbf{W}\mathbf{X})^-\mathbf{X}'\mathbf{W}\\ &= \mathbf{W}\mathbf{X}(\mathbf{X}'\mathbf{W}\mathbf{X})^-\mathbf{X}'\mathbf{W} = \mathbf{W}\mathbf{P}_{\mathbf{X},\mathbf{W}}.\end{aligned}$$

(7) Recalling Corollary 4.2.3, it is clear from the definition of $\mathbf{P}_{\mathbf{X},\mathbf{W}}$ that $\mathcal{C}(\mathbf{P}_{\mathbf{X},\mathbf{W}}) \subset \mathcal{C}(\mathbf{X})$ and from Part (1) that $\mathcal{C}(\mathbf{X}) \subset \mathcal{C}(\mathbf{P}_{\mathbf{X},\mathbf{W}})$, so that $\mathcal{C}(\mathbf{P}_{\mathbf{X},\mathbf{W}}) = \mathcal{C}(\mathbf{X})$. Similarly, it is clear from the definition of $\mathbf{P}_{\mathbf{X},\mathbf{W}}$ that $\mathcal{R}(\mathbf{P}_{\mathbf{X},\mathbf{W}}) \subset \mathcal{R}(\mathbf{X}'\mathbf{W})$ and from Part (5) that $\mathcal{R}(\mathbf{X}'\mathbf{W}) \subset \mathcal{R}(\mathbf{P}_{\mathbf{X},\mathbf{W}})$, so that $\mathcal{R}(\mathbf{P}_{\mathbf{X},\mathbf{W}}) = \mathcal{R}(\mathbf{X}'\mathbf{W})$.

(8) That $\text{rank}(\mathbf{P}_{\mathbf{X},\mathbf{W}}) = \text{rank}(\mathbf{X})$ is an immediate consequence of Part (7).

(9) Since [according to Part (6)] $\mathbf{P}_{\mathbf{X},\mathbf{W}}$ is idempotent, it follows from Lemma 10.1.2 that $\mathbf{I} - \mathbf{P}_{\mathbf{X},\mathbf{W}}$ is idempotent.

(9′) Making use of Parts (3) and (6′), we find that

$$\begin{aligned}(\mathbf{I} - \mathbf{P}_{\mathbf{X},\mathbf{W}})'\mathbf{W}(\mathbf{I} - \mathbf{P}_{\mathbf{X},\mathbf{W}}) &= \mathbf{W} - (\mathbf{W}\mathbf{P}_{\mathbf{X},\mathbf{W}})' - \mathbf{W}\mathbf{P}_{\mathbf{X},\mathbf{W}} + \mathbf{P}'_{\mathbf{X},\mathbf{W}}\mathbf{W}\mathbf{P}_{\mathbf{X},\mathbf{W}}\\ &= \mathbf{W} - \mathbf{W}\mathbf{P}_{\mathbf{X},\mathbf{W}} - \mathbf{W}\mathbf{P}_{\mathbf{X},\mathbf{W}} + \mathbf{W}\mathbf{P}_{\mathbf{X},\mathbf{W}}\\ &= \mathbf{W}(\mathbf{I} - \mathbf{P}_{\mathbf{X},\mathbf{W}}).\end{aligned}$$

(10) Making use of Lemma 10.2.4 and Parts (6) and (8), we find that

$$\text{rank}(\mathbf{I} - \mathbf{P}_{\mathbf{X},\mathbf{W}}) = n - \text{rank}(\mathbf{P}_{\mathbf{X},\mathbf{W}}) = n - \text{rank}(\mathbf{X}).$$

Q.E.D.

Theorem 14.12.12. Let $\mathbf{X}$ represent an $n \times p$ matrix, $\mathbf{U}$ an $n \times q$ matrix such that $\mathcal{C}(\mathbf{U}) \subset \mathcal{C}(\mathbf{X})$, and $\mathbf{W}$ an $n \times n$ symmetric positive definite matrix. Then,

(1) $\mathbf{P}_{\mathbf{X},\mathbf{W}}\mathbf{U} = \mathbf{U}$, implying that $\mathbf{W}\mathbf{P}_{\mathbf{X},\mathbf{W}}\mathbf{U} = \mathbf{W}\mathbf{U}$,
and $\mathbf{U}'\mathbf{W}\mathbf{X}(\mathbf{X}'\mathbf{W}\mathbf{X})^-\mathbf{X}' = \mathbf{U}'$, implying that $\mathbf{U}'\mathbf{W}\mathbf{P}_{\mathbf{X},\mathbf{W}} = \mathbf{U}'\mathbf{W}$;

(2) $\mathbf{P}_{\mathbf{X},\mathbf{W}}\mathbf{P}_{\mathbf{U},\mathbf{W}} = \mathbf{P}_{\mathbf{U},\mathbf{W}}\mathbf{P}_{\mathbf{X},\mathbf{W}} = \mathbf{P}_{\mathbf{U},\mathbf{W}}$, and $\mathbf{P}'_{\mathbf{X},\mathbf{W}}\mathbf{W}\mathbf{P}_{\mathbf{U},\mathbf{W}} = \mathbf{W}\mathbf{P}_{\mathbf{U},\mathbf{W}}$.

Proof. (1) According to Lemma 4.2.2, there exists a matrix $\mathbf{F}$ such that $\mathbf{U} = \mathbf{X}\mathbf{F}$. Thus, making use of Parts (1) and (5) of Theorem 14.12.11, we find that

$$\mathbf{P}_{\mathbf{X},\mathbf{W}}\mathbf{U} = \mathbf{P}_{\mathbf{X},\mathbf{W}}\mathbf{X}\mathbf{F} = \mathbf{X}\mathbf{F} = \mathbf{U}$$

and

$$\mathbf{U}'\mathbf{W}\mathbf{X}(\mathbf{X}'\mathbf{W}\mathbf{X})^{-}\mathbf{X}' = \mathbf{F}'\mathbf{X}'\mathbf{W}\mathbf{X}(\mathbf{X}'\mathbf{W}\mathbf{X})^{-}\mathbf{X}' = \mathbf{F}'\mathbf{X}' = \mathbf{U}'.$$

(2) Making use of Part (1), we find that

$$\mathbf{P}_{\mathbf{X},\mathbf{W}}\mathbf{P}_{\mathbf{U},\mathbf{W}} = \mathbf{P}_{\mathbf{X},\mathbf{W}}\mathbf{U}(\mathbf{U}'\mathbf{W}\mathbf{U})^{-}\mathbf{U}'\mathbf{W} = \mathbf{U}(\mathbf{U}'\mathbf{W}\mathbf{U})^{-}\mathbf{U}'\mathbf{W} = \mathbf{P}_{\mathbf{U},\mathbf{W}}$$

and similarly that

$$\begin{aligned}\mathbf{P}_{\mathbf{U},\mathbf{W}}\mathbf{P}_{\mathbf{X},\mathbf{W}} &= \mathbf{U}(\mathbf{U}'\mathbf{W}\mathbf{U})^{-}\mathbf{U}'\mathbf{W}\mathbf{X}(\mathbf{X}'\mathbf{W}\mathbf{X})^{-}\mathbf{X}'\mathbf{W} \\ &= \mathbf{U}(\mathbf{U}'\mathbf{W}\mathbf{U})^{-}\mathbf{U}'\mathbf{W} = \mathbf{P}_{\mathbf{U},\mathbf{W}}.\end{aligned}$$

Further, making use of Part (3) of Theorem 14.12.11, we find that

$$\mathbf{P}'_{\mathbf{X},\mathbf{W}}\mathbf{W}\mathbf{P}_{\mathbf{U},\mathbf{W}} = (\mathbf{W}\mathbf{P}_{\mathbf{X},\mathbf{W}})'\mathbf{P}_{\mathbf{U},\mathbf{W}} = \mathbf{W}\mathbf{P}_{\mathbf{X},\mathbf{W}}\mathbf{P}_{\mathbf{U},\mathbf{W}} = \mathbf{W}\mathbf{P}_{\mathbf{U},\mathbf{W}}.$$

Q.E.D.

The following two corollaries generalize Corollaries 12.3.6 and 12.3.7 and can be deduced from Theorem 14.12.12 in essentially the same way that Corollaries 12.3.6 and 12.3.7 were deduced from Theorem 12.3.5.

Corollary 14.12.13. Let $\mathbf{W}$ represent an $n \times n$ symmetric positive definite matrix. Then, for any $n \times p$ matrix $\mathbf{X}$ and any $n \times q$ matrix $\mathbf{U}$ such that $\mathcal{C}(\mathbf{U}) = \mathcal{C}(\mathbf{X})$, $\mathbf{P}_{\mathbf{U},\mathbf{W}} = \mathbf{P}_{\mathbf{X},\mathbf{W}}$.

Corollary 14.12.14. Let $\mathbf{W}$ represent an $n \times n$ symmetric positive definite matrix and $\mathbf{X}$ an $n \times p$ matrix. Then, for any matrix $\mathbf{A}$ that is the projection matrix for some subspace $\mathcal{U}$ of $\mathcal{C}(\mathbf{X})$ with respect to $\mathbf{W}$,

$$\mathbf{P}_{\mathbf{X},\mathbf{W}}\mathbf{A} = \mathbf{A}\mathbf{P}_{\mathbf{X},\mathbf{W}} = \mathbf{A}.$$

The following theorem generalizes Theorem 12.3.8 and can be derived in essentially the same way as Theorem 12.3.8.

Theorem 14.12.15. If an $n \times n$ matrix $\mathbf{A}$ is the projection matrix for some subspace $\mathcal{U}$ of $\mathcal{R}^{n\times 1}$ with respect to some $n\times n$ symmetric positive definite matrix, then $\mathcal{U} = \mathcal{C}(\mathbf{A})$ and $\dim(\mathcal{U}) = \text{rank}(\mathbf{A})$.

The following theorem characterizes matrices that are projection matrices with respect to any particular symmetric positive definite matrix.

Theorem 14.12.16. An $n \times n$ matrix $\mathbf{A}$ is a projection matrix with respect to an $n \times n$ symmetric positive definite matrix $\mathbf{W}$ if and only if $\mathbf{A}'\mathbf{W}\mathbf{A} = \mathbf{W}\mathbf{A}$ or, equivalently, if and only if $(\mathbf{I} - \mathbf{A})'\mathbf{W}\mathbf{A} = \mathbf{0}$.

Proof. Suppose that $\mathbf{A}$ is a projection matrix with respect to $\mathbf{W}$. Then, according to Corollary 14.12.10, $\mathbf{A} = \mathbf{P}_{\mathbf{X},\mathbf{W}}$ for some matrix $\mathbf{X}$. Thus, it follows from Part (6′) of Theorem 14.12.11 that $\mathbf{A}'\mathbf{W}\mathbf{A} = \mathbf{W}\mathbf{A}$.

Conversely, suppose that $\mathbf{A}'\mathbf{W}\mathbf{A} = \mathbf{W}\mathbf{A}$ or, equivalently, that

$$\mathbf{A}'\mathbf{W} = (\mathbf{W}\mathbf{A})' = (\mathbf{A}'\mathbf{W}\mathbf{A})' = \mathbf{A}'\mathbf{W}\mathbf{A}.$$

Then, making use of Part (1) of Theorem 14.12.11, we find that

$$\begin{aligned} \mathbf{A} &= \mathbf{P}_{\mathbf{A},\mathbf{W}} + \mathbf{A} - \mathbf{A}(\mathbf{A}'\mathbf{W}\mathbf{A})^{-}\mathbf{A}'\mathbf{W} \\ &= \mathbf{P}_{\mathbf{A},\mathbf{W}} + \mathbf{A} - \mathbf{A}(\mathbf{A}'\mathbf{W}\mathbf{A})^{-}\mathbf{A}'\mathbf{W}\mathbf{A} \\ &= \mathbf{P}_{\mathbf{A},\mathbf{W}} + \mathbf{A} - \mathbf{A} \\ &= \mathbf{P}_{\mathbf{A},\mathbf{W}}, \end{aligned}$$

which is the projection matrix for $\mathcal{C}(\mathbf{A})$ with respect to $\mathbf{W}$. Q.E.D.

The following corollary, which generalizes Theorem 12.3.9, gives an alternative characterization of matrices that are projection matrices with respect to any particular symmetric positive definite matrix.

Corollary 14.12.17. An $n \times n$ matrix $\mathbf{A}$ is a projection matrix with respect to an $n \times n$ symmetric positive definite matrix $\mathbf{W}$ if and only if $\mathbf{WA}$ is symmetric and $\mathbf{A}$ is idempotent.

Proof. In light of Theorem 14.12.16, it suffices to show that $\mathbf{A}'\mathbf{WA} = \mathbf{WA}$ if and only if $\mathbf{WA}$ is symmetric and $\mathbf{A}$ is idempotent or, equivalently (since $\mathbf{W}$ is nonsingular), if and only if $\mathbf{WA}$ is symmetric and $\mathbf{WA}^2 = \mathbf{WA}$.

If $\mathbf{WA}$ is symmetric and $\mathbf{WA}^2 = \mathbf{WA}$, then

$$\mathbf{A}'\mathbf{WA} = (\mathbf{WA})'\mathbf{A} = \mathbf{WAA} = \mathbf{WA}^2 = \mathbf{WA}.$$

Conversely, if $\mathbf{A}'\mathbf{WA} = \mathbf{WA}$, then

$$(\mathbf{WA})' = (\mathbf{A}'\mathbf{WA})' = \mathbf{A}'\mathbf{WA} = \mathbf{WA}$$

(i.e., $\mathbf{WA}$ is symmetric), and

$$\mathbf{WA}^2 = \mathbf{WAA} = (\mathbf{WA})'\mathbf{A} = \mathbf{A}'\mathbf{WA} = \mathbf{WA}.$$

Q.E.D.

e. When are projections with respect to a symmetric positive definite matrix W projections with respect to another symmetric positive definite matrix V?

Suppose that $\mathcal{U}$ is a subspace of the linear space $\mathcal{R}^n$ of all n-dimensional column vectors. Let $\mathbf{X}$ represent an $n \times p$ matrix whose columns span $\mathcal{U}$, and let $\mathbf{W}$ and $\mathbf{V}$ represent $n \times n$ symmetric positive definite matrices. Under what circumstances is the projection $\mathbf{P}_{\mathbf{X},\mathbf{W}}\mathbf{y}$ of $\mathbf{y}$ on $\mathcal{U}$ with respect to $\mathbf{W}$ the same (for every $\mathbf{y}$ in $\mathcal{R}^n$) as the projection $\mathbf{P}_{\mathbf{X},\mathbf{V}}\mathbf{y}$ of $\mathbf{y}$ on $\mathcal{U}$ with respect to $\mathbf{V}$? Or, equivalently (in light of Lemma 2.3.2), under what circumstances is $\mathbf{P}_{\mathbf{X},\mathbf{W}} = \mathbf{P}_{\mathbf{X},\mathbf{V}}$; that is, under what circumstances is $\mathbf{P}_{\mathbf{X},\mathbf{W}}$ the projection matrix for $\mathcal{U}$ with respect to $\mathbf{V}$ as well as the projection matrix for $\mathcal{U}$ with respect to $\mathbf{W}$? These circumstances can be characterized in terms of the conditions given by the following theorem.

Theorem 14.12.18. Let $\mathcal{U}$ represent a subspace of $\mathcal{R}^{n\times 1}$, let $\mathbf{X}$ represent an $n \times p$ matrix whose columns span $\mathcal{U}$, and let $\mathbf{W}$ and $\mathbf{V}$ represent $n \times n$ symmetric positive definite matrices. Then, each of the following four conditions is necessary and sufficient for the projection $\mathbf{P}_{\mathbf{X},\mathbf{W}}\mathbf{y}$ of $\mathbf{y}$ on $\mathcal{U}$ with respect to $\mathbf{W}$ to be the same (for every $\mathbf{y}$ in $\mathcal{R}^n$) as the projection $\mathbf{P}_{\mathbf{X},\mathbf{V}}\mathbf{y}$ of $\mathbf{y}$ on $\mathcal{U}$ with respect to $\mathbf{V}$ (or, equivalently, for $\mathbf{P}_{\mathbf{X},\mathbf{W}} = \mathbf{P}_{\mathbf{X},\mathbf{V}}$):

(1) $\mathbf{X}'\mathbf{V}\mathbf{P}_{\mathbf{X},\mathbf{W}} = \mathbf{X}'\mathbf{V}$, or, equivalently, $\mathbf{X}'\mathbf{V}(\mathbf{I} - \mathbf{P}_{\mathbf{X},\mathbf{W}}) = \mathbf{0}$;

(2) $\mathbf{P}'_{\mathbf{X},\mathbf{W}}\mathbf{V}\mathbf{P}_{\mathbf{X},\mathbf{W}} = \mathbf{V}\mathbf{P}_{\mathbf{X},\mathbf{W}}$, or, equivalently, $(\mathbf{I} - \mathbf{P}_{\mathbf{X},\mathbf{W}})'\mathbf{V}\mathbf{P}_{\mathbf{X},\mathbf{W}} = \mathbf{0}$;

(3) $\mathbf{V}\mathbf{P}_{\mathbf{X},\mathbf{W}}$ is symmetric;

(4) there exists a $p \times p$ matrix $\mathbf{Q}$ such that $\mathbf{V}\mathbf{X} = \mathbf{W}\mathbf{X}\mathbf{Q}$, or, equivalently, $\mathcal{C}(\mathbf{V}\mathbf{X}) \subset \mathcal{C}(\mathbf{W}\mathbf{X})$.

Proof. (1) It follows from Lemma 14.12.1 that the projection $\mathbf{P}_{\mathbf{X},\mathbf{W}}\mathbf{y}$ of $\mathbf{y}$ on $\mathcal{U}$ with respect to $\mathbf{W}$ is the same (for every $\mathbf{y}$ in $\mathcal{R}^n$) as the projection of $\mathbf{y}$ on $\mathcal{U}$ with respect to $\mathbf{V}$ if and only if, for every $\mathbf{y}$,

$$\mathbf{X}'\mathbf{V}(\mathbf{y} - \mathbf{P}_{\mathbf{X},\mathbf{W}}\mathbf{y}) = \mathbf{0},$$

or, equivalently, if and only if, for every $\mathbf{y}$,

$$\mathbf{X}'\mathbf{V}(\mathbf{I} - \mathbf{P}_{\mathbf{X},\mathbf{W}})\mathbf{y} = \mathbf{0},$$

and hence if and only if

$$\mathbf{X}'\mathbf{V}(\mathbf{I} - \mathbf{P}_{\mathbf{X},\mathbf{W}}) = \mathbf{0}.$$

(2) It suffices to show that Condition (2) is equivalent to Condition (1). If Condition (1) is satisfied, then

$$\begin{aligned}\mathbf{P}'_{\mathbf{X},\mathbf{W}}\mathbf{V}\mathbf{P}_{\mathbf{X},\mathbf{W}} &= (\mathbf{X}'\mathbf{V}\mathbf{P}_{\mathbf{X},\mathbf{W}})'(\mathbf{X}'\mathbf{W}\mathbf{X})^{-}\mathbf{X}'\mathbf{W} \\ &= (\mathbf{X}'\mathbf{V})'(\mathbf{X}'\mathbf{W}\mathbf{X})^{-}\mathbf{X}'\mathbf{W} = \mathbf{V}\mathbf{P}_{\mathbf{X},\mathbf{W}}.\end{aligned}$$

Conversely, if Condition (2) is satisfied, then, making use of Part (5) of Theorem 14.12.11, we find that

$$\begin{aligned}\mathbf{X}'\mathbf{V} &= \mathbf{X}'\mathbf{P}'_{\mathbf{X},\mathbf{W}}\mathbf{V} = \mathbf{X}'(\mathbf{V}\mathbf{P}_{\mathbf{X},\mathbf{W}})' \\ &= \mathbf{X}'(\mathbf{P}'_{\mathbf{X},\mathbf{W}}\mathbf{V}\mathbf{P}_{\mathbf{X},\mathbf{W}})' = \mathbf{X}'\mathbf{P}'_{\mathbf{X},\mathbf{W}}\mathbf{V}\mathbf{P}_{\mathbf{X},\mathbf{W}} = \mathbf{X}'\mathbf{V}\mathbf{P}_{\mathbf{X},\mathbf{W}}.\end{aligned}$$

(3) It suffices to show that Condition (3) is equivalent to Condition (2). If Condition (2) is satisfied, then

$$(\mathbf{V}\mathbf{P}_{\mathbf{X},\mathbf{W}})' = (\mathbf{P}'_{\mathbf{X},\mathbf{W}}\mathbf{V}\mathbf{P}_{\mathbf{X},\mathbf{W}})' = \mathbf{P}'_{\mathbf{X},\mathbf{W}}\mathbf{V}\mathbf{P}_{\mathbf{X},\mathbf{W}} = \mathbf{V}\mathbf{P}_{\mathbf{X},\mathbf{W}},$$

that is, $\mathbf{V}\mathbf{P}_{\mathbf{X},\mathbf{W}}$ is symmetric. Conversely, if Condition (3) is satisfied, then [in light of Part (6) of Theorem 14.12.11]

$$\mathbf{P}'_{\mathbf{X},\mathbf{W}}\mathbf{V}\mathbf{P}_{\mathbf{X},\mathbf{W}} = (\mathbf{V}\mathbf{P}_{\mathbf{X},\mathbf{W}})'\mathbf{P}_{\mathbf{X},\mathbf{W}} = \mathbf{V}\mathbf{P}_{\mathbf{X},\mathbf{W}}\mathbf{P}_{\mathbf{X},\mathbf{W}} = \mathbf{V}\mathbf{P}_{\mathbf{X},\mathbf{W}}.$$

(4) It suffices to show that Condition (4) is equivalent to Condition (1) or, equivalently [since $\mathbf{VX} = (\mathbf{X}'\mathbf{V})'$ and $\mathbf{P}'_{\mathbf{X},\mathbf{W}}\mathbf{VX} = (\mathbf{X}'\mathbf{V}\mathbf{P}_{\mathbf{X},\mathbf{W}})'$], to the condition

$$\mathbf{VX} = \mathbf{P}'_{\mathbf{X},\mathbf{W}}\mathbf{VX}. \tag{12.1}$$

If Condition (12.1) is satisfied, then

$$\mathbf{VX} = \mathbf{WX}[(\mathbf{X}'\mathbf{WX})^{-}]'\mathbf{X}'\mathbf{VX} = \mathbf{WXQ}$$

for $\mathbf{Q} = [(\mathbf{X}'\mathbf{WX})^{-}]'\mathbf{X}'\mathbf{VX}$. Conversely, if $\mathbf{VX} = \mathbf{WXQ}$ for some matrix $\mathbf{Q}$, then, making use of Part (4) of Theorem 14.12.11, we find that

$$\mathbf{P}'_{\mathbf{X},\mathbf{W}}\mathbf{VX} = \mathbf{P}'_{\mathbf{X},\mathbf{W}}\mathbf{WXQ} = \mathbf{WXQ} = \mathbf{VX};$$

that is, condition (12.1) is satisifed. Q.E.D.

Condition (4) of Theorem 14.12.18 has appeal relative to Conditions (1)–(3) in that it does not explicitly involve $\mathbf{P}_{\mathbf{X},\mathbf{W}}$. In the special case where $\mathbf{W} = \mathbf{I}$, $\mathbf{P}_{\mathbf{X},\mathbf{W}}$ reduces to the usual projection matrix $\mathbf{P}_{\mathbf{X}} = \mathbf{X}(\mathbf{X}'\mathbf{X})^{-}\mathbf{X}'$, and (in this special case) the four conditions of Theorem 14.12.18 reduce, respectively, to

(1) $\mathbf{X}'\mathbf{V}\mathbf{P}_{\mathbf{X}} = \mathbf{X}'\mathbf{V}$, or, equivalently, $\mathbf{X}'\mathbf{V}(\mathbf{I} - \mathbf{P}_{\mathbf{X}}) = \mathbf{0}$;

(2) $\mathbf{P}_{\mathbf{X}}\mathbf{V}\mathbf{P}_{\mathbf{X}} = \mathbf{V}\mathbf{P}_{\mathbf{X}}$, or, equivalently, $(\mathbf{I} - \mathbf{P}_{\mathbf{X}})\mathbf{V}\mathbf{P}_{\mathbf{X}} = \mathbf{0}$;

(3) $\mathbf{V}\mathbf{P}_{\mathbf{X}}$ is symmetric;

(4) there exists a $p \times p$ matrix $\mathbf{Q}$ such that $\mathbf{VX} = \mathbf{XQ}$, or, equivalently, $\mathcal{C}(\mathbf{VX}) \subset \mathcal{C}(\mathbf{X})$.

f. Generalized or weighted least squares problem

Suppose that $\mathbf{y} = \{y_i\}$ is an n-dimensional column vector and that $\mathcal{U}$ is a subspace of the linear space $\mathcal{R}^n$ of all n-dimensional column vectors. Let $\mathbf{W} = \{w_{ij}\}$ represent an $n \times n$ symmetric positive definite matrix. Consider the problem of minimizing, for $\mathbf{u} = \{u_i\}$ in $\mathcal{U}$, the quadratic form $(\mathbf{y} - \mathbf{u})'\mathbf{W}(\mathbf{y} - \mathbf{u})$ (in the difference $\mathbf{y} - \mathbf{u}$) or, equivalently, the quantity $[(\mathbf{y} - \mathbf{u})'\mathbf{W}(\mathbf{y} - \mathbf{u})]^{1/2}$, which is the distance between $\mathbf{y}$ and $\mathbf{u}$ when the inner product is taken to be the bilinear form $\mathbf{x}'\mathbf{W}\mathbf{y}$. This minimization problem is known as the *generalized* or *weighted least squares problem.* [Typically, the use of the term weighted least squares is reserved for the special case where $\mathbf{W}$ is a diagonal matrix—in this special case, $(\mathbf{y} - \mathbf{u})'\mathbf{W}(\mathbf{y} - \mathbf{u}) = \sum_{i=1}^{n} w_{ii}(y_i - u_i)^2$, which is a weighted sum of the squared deviations of the elements of $\mathbf{u}$ from the corresponding elements of $\mathbf{y}$ (with weights $w_{11}, w_{22}, \ldots, w_{nn}$, respectively).]

In the special case where $\mathbf{W} = \mathbf{I}$, the generalized or weighted least squares problem reduces to the least squares problem considered in Section 12.4. It is customary (and convenient) to refer to this special case as *simple least squares*, *ordinary least squares*, or *unweighted least squares*.

The solution of the generalized least squares problem, like that of the simple least squares problem, can be obtained as a special case of Theorem 12.4.1. Since

$[(\mathbf{y}-\mathbf{u})'\mathbf{W}(\mathbf{y}-\mathbf{u})]^{1/2}$ is the distance between $\mathbf{y}$ and $\mathbf{u}$ (when the inner product is taken to be the bilinear form $\mathbf{x}'\mathbf{W}\mathbf{y}$), Theorem 12.4.1 implies that $[(\mathbf{y}-\mathbf{u})'\mathbf{W}(\mathbf{y}-\mathbf{u})]^{1/2}$ is minimized uniquely by taking $\mathbf{u}$ to be the projection of $\mathbf{y}$ on $\mathcal{U}$ with respect to $\mathbf{W}$. Thus, based on Theorem 14.12.3, we arrive at the following generalization of Theorem 12.4.2.

Theorem 14.12.19. Let $\mathcal{U}$ represent a subspace of the linear space $\mathcal{R}^n$ of all n-dimensional column vectors, let $\mathbf{X}$ be an $n \times p$ matrix such that $\mathcal{C}(\mathbf{X}) = \mathcal{U}$, let $\mathbf{W}$ be an $n \times n$ symmetric positive definite matrix, and let $\mathbf{y}$ represent a vector in $\mathcal{R}^n$. Then, for $\mathbf{u} \in \mathcal{U}$, the quadratic form $(\mathbf{y}-\mathbf{u})'\mathbf{W}(\mathbf{y}-\mathbf{u})$ (in the difference $\mathbf{y}-\mathbf{u}$) is minimized uniquely by taking $\mathbf{u} = \mathbf{X}\mathbf{b}^*$, where $\mathbf{b}^*$ is any solution to the linear system $\mathbf{X}'\mathbf{W}\mathbf{X}\mathbf{b} = \mathbf{X}'\mathbf{W}\mathbf{y}$ (in $\mathbf{b}$), or, equivalently, by taking $\mathbf{u} = \mathbf{P}_{\mathbf{X},\mathbf{W}}\mathbf{y}$. Furthermore, the minimum value of $(\mathbf{y}-\mathbf{u})'\mathbf{W}(\mathbf{y}-\mathbf{u})$ (for $\mathbf{u} \in \mathcal{U}$) is expressible as

$$(\mathbf{y}-\mathbf{X}\mathbf{b}^*)'\mathbf{W}(\mathbf{y}-\mathbf{X}\mathbf{b}^*) = \mathbf{y}'\mathbf{W}(\mathbf{y}-\mathbf{X}\mathbf{b}^*) = \mathbf{y}'\mathbf{W}(\mathbf{I}-\mathbf{P}_{\mathbf{X},\mathbf{W}})\mathbf{y}.$$

In light of Theorem 14.12.7, we have the following variant of Theorem 14.12.19, which generalizes Theorem 12.4.3.

Theorem 14.12.20. Let $\mathbf{X}$ represent an $n \times p$ matrix, $\mathbf{W}$ an $n \times n$ symmetric positive definite matrix, and $\mathbf{y}$ an n-dimensional column vector. Then, for $\mathbf{b} \in \mathcal{R}^p$, the quadratic form $(\mathbf{y}-\mathbf{X}\mathbf{b})'\mathbf{W}(\mathbf{y}-\mathbf{X}\mathbf{b})$ has a minimum at a point $\mathbf{b}^*$ if and only if $\mathbf{b}^*$ is a solution to the linear system $\mathbf{X}'\mathbf{W}\mathbf{X}\mathbf{b} = \mathbf{X}'\mathbf{W}\mathbf{y}$, in which case $\mathbf{X}\mathbf{b}^* = \mathbf{P}_{\mathbf{X},\mathbf{W}}\mathbf{y}$ and

$$(\mathbf{y}-\mathbf{X}\mathbf{b}^*)'\mathbf{W}(\mathbf{y}-\mathbf{X}\mathbf{b}^*) = \mathbf{y}'\mathbf{W}(\mathbf{y}-\mathbf{X}\mathbf{b}^*) = \mathbf{y}'\mathbf{W}(\mathbf{I}-\mathbf{P}_{\mathbf{X},\mathbf{W}})\mathbf{y}.$$

The results encompassed in Theorems 14.12.19 and 14.12.20 are generally attributed to Aitken (1935). In fact, the p equations that compose the linear system $\mathbf{X}'\mathbf{W}\mathbf{X}\mathbf{b} = \mathbf{X}'\mathbf{W}\mathbf{y}$ have come to be known as the *Aitken equations*.

g. Orthogonal complements of row and column spaces

Let $\mathbf{X}$ represent an $n \times p$ matrix, and take $\mathbf{W}$ to be an $n \times n$ symmetric positive definite matrix. In Section 12.5, consideration was given to $\mathcal{C}^{\perp}(\mathbf{X})$ and $\mathcal{R}^{\perp}(\mathbf{X})$, which are the orthogonal complements of $\mathcal{C}(\mathbf{X})$ and $\mathcal{R}(\mathbf{X})$, respectively, relative to the linear space $\mathcal{R}^n$ (of all n-dimensional column or row vectors) when the inner product is taken to be the usual inner product for $\mathcal{R}^n$. Let us now consider the orthogonal complements of $\mathcal{C}(\mathbf{X})$ and $\mathcal{R}(\mathbf{X})$ relative to $\mathcal{R}^n$ when the inner product is taken to be the bilinear form $\mathbf{x}'\mathbf{W}\mathbf{y}$. Let us denote these orthogonal complements by $\mathcal{C}^{\perp}_{\mathbf{W}}(\mathbf{X})$ and $\mathcal{R}^{\perp}_{\mathbf{W}}(\mathbf{X})$, respectively, and speak of them as the orthogonal complements of $\mathcal{C}(\mathbf{X})$ and $\mathcal{R}(\mathbf{X})$ with respect to $\mathbf{W}$. Note that $\mathcal{C}^{\perp}_{\mathbf{I}}(\mathbf{X}) = \mathcal{C}^{\perp}(\mathbf{X})$ and $\mathcal{R}^{\perp}_{\mathbf{I}}(\mathbf{X}) = \mathcal{R}^{\perp}(\mathbf{X})$.

Since [according to Part (5) of Theorem 14.12.11] $\mathbf{X}(\mathbf{X}'\mathbf{W}\mathbf{X})^-$ is a generalized inverse of $\mathbf{X}'\mathbf{W}$, we have (in light of Lemma 14.12.1 and Corollary 11.2.2) the following generalization of Lemma 12.5.2.

Lemma 14.12.21. For any $n \times p$ matrix $\mathbf{X}$ and any $n \times n$ symmetric positive definite matrix $\mathbf{W}$,

$$\mathcal{C}^{\perp}_{\mathbf{W}}(\mathbf{X}) = \mathcal{N}(\mathbf{X}'\mathbf{W}) = \mathcal{C}(\mathbf{I} - \mathbf{P}_{\mathbf{X},\mathbf{W}}). \tag{12.2}$$

As a consequence of Lemma 11.3.1 or, alternatively, Theorem 14.12.11, we have the following corollary, which is a generalization of Corollary 12.5.3.

Corollary 14.12.22. For any $n \times p$ matrix $\mathbf{X}$ and any $n \times n$ symmetric positive definite matrix $\mathbf{W}$,

$$\dim[\mathcal{C}^{\perp}_{\mathbf{W}}(\mathbf{X})] = n - \text{rank}(\mathbf{X}) = n - \dim[\mathcal{C}(\mathbf{X})]. \tag{12.3}$$

h. Projection on the orthogonal complement of a column space

In light of Theorems 12.5.8 and 14.12.3, we have the following theorem and corollary, which are generalizations of Corollaries 12.5.9 and 12.5.10.

Theorem 14.12.23. Let $\mathbf{X}$ represent an $n \times p$ matrix, $\mathbf{W}$ an $n \times n$ symmetric positive definite matrix, and $\mathbf{y}$ an n-dimensional column vector, and take $\mathbf{z}$ to be the projection of $\mathbf{y}$ on $\mathcal{C}^{\perp}_{\mathbf{W}}(\mathbf{X})$ with respect to $\mathbf{W}$. Then,

$$\mathbf{z} = \mathbf{y} - \mathbf{X}\mathbf{b}^*$$

for any solution $\mathbf{b}^*$ to the linear system $\mathbf{X}'\mathbf{W}\mathbf{X}\mathbf{b} = \mathbf{X}'\mathbf{W}\mathbf{y}$ (in $\mathbf{b}$). In particular,

$$\mathbf{z} = \mathbf{y} - \mathbf{P}_{\mathbf{X},\mathbf{W}}\mathbf{y} = (\mathbf{I} - \mathbf{P}_{\mathbf{X},\mathbf{W}})\mathbf{y}.$$

Corollary 14.12.24. Let $\mathbf{X}$ represent an $n \times p$ matrix and $\mathbf{W}$ an $n \times n$ symmetric positive definite matrix. Then, the projection matrix for $\mathcal{C}^{\perp}_{\mathbf{W}}(\mathbf{X})$ with respect to $\mathbf{W}$ is $\mathbf{I} - \mathbf{P}_{\mathbf{X},\mathbf{W}}$.

i. Extensions

Let $\mathbf{W}$ represent an $n \times n$ symmetric nonnegative definite matrix, and let $\mathcal{U}$ represent a subspace of $\mathcal{R}^{n \times 1}$. In what follows, a vector $\mathbf{z}$ in $\mathcal{U}$ is said to be a *projection of an n-dimensional column vector* $\mathbf{y}$ *on* $\mathcal{U}$ *with respect to* $\mathbf{W}$ if $(\mathbf{y} - \mathbf{z}) \perp_{\mathbf{W}} \mathcal{U}$; and an $n \times n$ matrix $\mathbf{A}$ is said to be a *projection matrix for* $\mathcal{U}$ *with respect to* $\mathbf{W}$ if, for every $\mathbf{y}$ in $\mathcal{R}^{n \times 1}$, $\mathbf{A}\mathbf{y}$ is a projection of $\mathbf{y}$ on $\mathcal{U}$ with respect to $\mathbf{W}$. (To say simply that an $n \times n$ matrix $\mathbf{A}$ is a projection matrix with respect to $\mathbf{W}$ is to indicate that there exists some subspace of $\mathcal{R}^{n \times 1}$ for which $\mathbf{A}$ is a projection matrix with respect to $\mathbf{W}$.) Our previous use of this terminology was restricted to the case where $\mathbf{W}$ is positive definite.

Recall that, for any $n \times p$ matrix $\mathbf{X}$ and any $n \times n$ matrix $\mathbf{W}$, $\mathbf{P}_{\mathbf{X},\mathbf{W}} = \mathbf{X}(\mathbf{X}'\mathbf{W}\mathbf{X})^{-}\mathbf{X}'\mathbf{W}$. Theorem 14.12.11 set forth some basic properties of the matrix $\mathbf{P}_{\mathbf{X},\mathbf{W}}$ in the special case where $\mathbf{W}$ is symmetric and positive definite. The following theorem gives generalizations of those properties for the less restrictive special case where $\mathbf{W}$ is symmetric and nonnegative definite.

Theorem 14.12.25. Let $\mathbf{X}$ represent any $n \times p$ matrix, and $\mathbf{W}$ any $n \times n$ symmetric nonnegative definite matrix. Then,

(1) $\mathbf{W}\mathbf{P}_{\mathbf{X},\mathbf{W}}\mathbf{X} = \mathbf{W}\mathbf{X}$; that is, $\mathbf{W}\mathbf{X}(\mathbf{X}'\mathbf{W}\mathbf{X})^{-}\mathbf{X}'\mathbf{W}\mathbf{X} = \mathbf{W}\mathbf{X}$;

(2) $\mathbf{W}\mathbf{P}_{\mathbf{X},\mathbf{W}} = \mathbf{W}\mathbf{X}\mathbf{B}^*$ for any solution $\mathbf{B}^*$ to the linear system $\mathbf{X}'\mathbf{W}\mathbf{X}\mathbf{B} = \mathbf{X}'\mathbf{W}$ (in $\mathbf{B}$);

(3) $\mathbf{W}\mathbf{P}_{\mathbf{X},\mathbf{W}}$ is symmetric; that is, $\mathbf{W}\mathbf{X}[(\mathbf{X}'\mathbf{W}\mathbf{X})^{-}]'\mathbf{X}'\mathbf{W} = \mathbf{W}\mathbf{X}(\mathbf{X}'\mathbf{W}\mathbf{X})^{-}\mathbf{X}'\mathbf{W}$;

(4) $\mathbf{P}'_{\mathbf{X},\mathbf{W}}\mathbf{W}\mathbf{X} = \mathbf{W}\mathbf{X}$; that is, $\mathbf{W}\mathbf{X}[(\mathbf{X}'\mathbf{W}\mathbf{X})^{-}]'\mathbf{X}'\mathbf{W}\mathbf{X} = \mathbf{W}\mathbf{X}$;

(5) $\mathbf{X}'\mathbf{W}\mathbf{P}_{\mathbf{X},\mathbf{W}} = \mathbf{X}'\mathbf{P}'_{\mathbf{X},\mathbf{W}}\mathbf{W} = \mathbf{X}'\mathbf{W}$; that is, $\mathbf{X}'\mathbf{W}\mathbf{X}(\mathbf{X}'\mathbf{W}\mathbf{X})^{-}\mathbf{X}'\mathbf{W} = \mathbf{X}'\mathbf{W}\mathbf{X}[(\mathbf{X}'\mathbf{W}\mathbf{X})^{-}]'\mathbf{X}'\mathbf{W} = \mathbf{X}'\mathbf{W}$;

(6) $\mathbf{P}^2_{\mathbf{X},\mathbf{W}} = \mathbf{P}_{\mathbf{X},\mathbf{W}}$; that is, $\mathbf{P}_{\mathbf{X},\mathbf{W}}$ is idempotent;

(6′) $\mathbf{P}'_{\mathbf{X},\mathbf{W}}\mathbf{W}\mathbf{P}_{\mathbf{X},\mathbf{W}} = \mathbf{W}\mathbf{P}_{\mathbf{X},\mathbf{W}}$;

(7) $\mathcal{C}(\mathbf{W}\mathbf{P}_{\mathbf{X},\mathbf{W}}) = \mathcal{C}(\mathbf{W}\mathbf{X})$, and $\mathcal{R}(\mathbf{P}_{\mathbf{X},\mathbf{W}}) = \mathcal{R}(\mathbf{X}'\mathbf{W})$;

(8) $\operatorname{rank}(\mathbf{P}_{\mathbf{X},\mathbf{W}}) = \operatorname{rank}(\mathbf{W}\mathbf{P}_{\mathbf{X},\mathbf{W}}) = \operatorname{rank}(\mathbf{W}\mathbf{X})$;

(9) $(\mathbf{I} - \mathbf{P}_{\mathbf{X},\mathbf{W}})^2 = \mathbf{I} - \mathbf{P}_{\mathbf{X},\mathbf{W}}$; that is, $\mathbf{I} - \mathbf{P}_{\mathbf{X},\mathbf{W}}$ is idempotent;

(9′) $(\mathbf{I} - \mathbf{P}_{\mathbf{X},\mathbf{W}})'\mathbf{W}(\mathbf{I} - \mathbf{P}_{\mathbf{X},\mathbf{W}}) = \mathbf{W}(\mathbf{I} - \mathbf{P}_{\mathbf{X},\mathbf{W}})$;

(10) $\operatorname{rank}(\mathbf{I} - \mathbf{P}_{\mathbf{X},\mathbf{W}}) = n - \operatorname{rank}(\mathbf{W}\mathbf{X})$.

The various parts of Theorem 14.12.25 can be established via arguments that are essentially the same as the arguments employed in establishing the corresponding parts of Theorem 14.12.11 or via simple variants of those arguments.

The following theorem extends the result (on projections) given by Theorem 14.12.3.

Theorem 14.12.26. Let $\mathbf{y}$ represent an n-dimensional column vector, let $\mathcal{U}$ represent a subspace of $\mathcal{R}^{n\times 1}$, let $\mathbf{W}$ represent an $n \times n$ symmetric nonnegative definite matrix, and let $\mathbf{X}$ represent any $n \times p$ matrix whose columns span $\mathcal{U}$. Then, (1) there exists a projection of $\mathbf{y}$ on $\mathcal{U}$ with respect to $\mathbf{W}$; (2) $\mathbf{W}\mathbf{z}$ has the same value for every vector $\mathbf{z}$ that is a projection of $\mathbf{y}$ on $\mathcal{U}$ with respect to $\mathbf{W}$; and (3) a vector $\mathbf{z}$ is a projection of $\mathbf{y}$ on $\mathcal{U}$ with respect to $\mathbf{W}$ if and only if $\mathbf{z} = \mathbf{X}\mathbf{b}^*$ for some solution $\mathbf{b}^*$ to the (consistent) linear system

$$\mathbf{X}'\mathbf{W}\mathbf{X}\mathbf{b} = \mathbf{X}'\mathbf{W}\mathbf{y} \quad (\text{in } \mathbf{b}). \tag{12.4}$$

Proof. Let $\mathbf{L}$ represent a matrix such that $\mathbf{W} = \mathbf{L}'\mathbf{L}$, denote by m the number of rows in $\mathbf{L}$, define

$$\mathcal{V} = \{\mathbf{v} \in \mathcal{R}^m : \mathbf{v} = \mathbf{L}\mathbf{x} \text{ for some } \mathbf{x} \in \mathcal{U}\},$$

and let $\mathbf{v}^*$ represent the (unique) projection of $\mathbf{L}\mathbf{y}$ on $\mathcal{V}$ (with respect to $\mathbf{I}$). Then, it follows from Part (3) of Lemma 14.12.2 that a vector $\mathbf{z}$ in $\mathcal{U}$ is a projection of $\mathbf{y}$ on $\mathcal{U}$ with respect to $\mathbf{W}$ if and only if $\mathbf{L}(\mathbf{y} - \mathbf{z})$ (or, equivalently, $\mathbf{L}\mathbf{y} - \mathbf{L}\mathbf{z}$) is orthogonal to $\mathcal{V}$ with respect to the usual inner product and hence if and only if

$$\mathbf{Lz} = \mathbf{v}^*. \tag{12.5}$$

Let $\mathbf{b}^*$ represent any solution to linear system (12.4). (The existence of such a solution follows from Theorem 14.11.1.) According to Part (5) of Lemma 14.12.2, the columns of $\mathbf{LX}$ span $\mathcal{V}$. Moreover, linear system (12.4) can be rewritten as

$$(\mathbf{LX})'\mathbf{LXb} = (\mathbf{LX})'\mathbf{Ly}.$$

Thus, it follows from Theorem 12.2.1 that

$$\mathbf{v}^* = \mathbf{LXb}^*. \tag{12.6}$$

Clearly, $\mathbf{Xb}^*$ is in $\mathcal{U}$. Moreover, it follows from result (12.6) that condition (12.5) can be satisfied by taking $\mathbf{z} = \mathbf{Xb}^*$. Thus, $\mathbf{Xb}^*$ is a projection of $\mathbf{y}$ on $\mathcal{U}$ with respect to $\mathbf{W}$. Furthermore, since a vector $\mathbf{z}$ is a projection of $\mathbf{y}$ on $\mathcal{U}$ with respect to $\mathbf{W}$ only if $\mathbf{z}$ satisfies condition (12.5), $\mathbf{Wz}$ (which equals $\mathbf{L}'\mathbf{Lz}$) has the same value, namely $\mathbf{L}'\mathbf{v}^*$, for every vector $\mathbf{z}$ that is a projection of $\mathbf{y}$ on $\mathcal{U}$ with respect to $\mathbf{W}$.

Suppose now that $\mathbf{z}$ is a projection of $\mathbf{y}$ on $\mathcal{U}$ with respect to $\mathbf{W}$. Then, $\mathbf{z} = \mathbf{Xa}$ for some vector $\mathbf{a}$. Further,

$$\mathbf{WXa} = \mathbf{Wz} = \mathbf{L}'\mathbf{v}^* = \mathbf{L}'\mathbf{LXb}^* = \mathbf{WXb}^*,$$

implying that

$$\mathbf{X}'\mathbf{WXa} = \mathbf{X}'\mathbf{WXb}^* = \mathbf{X}'\mathbf{Wy}.$$

Thus, $\mathbf{a}$ is a solution to linear system (12.4). Q.E.D.

The following three corollaries extend the results of Corollaries 14.12.4–14.12.6.

Corollary 14.12.27. Let $\mathbf{y}$ represent an n-dimensional column vector, let $\mathcal{U}$ represent a subspace of $\mathcal{R}^{n\times 1}$, let $\mathbf{W}$ represent an $n \times n$ symmetric nonnegative definite matrix, and let $\mathbf{X}$ represent any $n \times p$ matrix whose columns span $\mathcal{U}$. Then, a vector $\mathbf{z}$ is a projection of $\mathbf{y}$ on $\mathcal{U}$ with respect to $\mathbf{W}$ if and only if

$$\mathbf{z} = \mathbf{P}_{\mathbf{X},\mathbf{W}}\mathbf{y} + (\mathbf{I} - \mathbf{P}_{\mathbf{X},\mathbf{W}})\mathbf{Xk}$$

for some $p \times 1$ vector $\mathbf{k}$.

Proof. Corollary 14.12.27 follows from Part (3) of Theorem 14.12.26 upon observing that (for any $\mathbf{k}$)

$$\mathbf{P}_{\mathbf{X},\mathbf{W}}\mathbf{y} + (\mathbf{I} - \mathbf{P}_{\mathbf{X},\mathbf{W}})\mathbf{Xk} = \mathbf{X}\{(\mathbf{X}'\mathbf{WX})^-\mathbf{X}'\mathbf{Wy} + [\mathbf{I} - (\mathbf{X}'\mathbf{WX})^-\mathbf{X}'\mathbf{WX}]\mathbf{k}\}$$

and that (in light of Theorem 11.2.4) a vector $\mathbf{b}^*$ is a solution to linear system (12.4) if and only if

$$\mathbf{b}^* = (\mathbf{X}'\mathbf{WX})^-\mathbf{X}'\mathbf{Wy} + [\mathbf{I} - (\mathbf{X}'\mathbf{WX})^-\mathbf{X}'\mathbf{WX}]\mathbf{k}$$

for some $\mathbf{k}$. Q.E.D.

Corollary 14.12.28. Let $\mathbf{y}$ represent an n-dimensional column vector, $\mathbf{W}$ an $n \times n$ symmetric nonnegative definite matrix, $\mathbf{X}$ an $n \times p$ matrix, and $\mathbf{U}$ an $n \times q$ matrix such that $\mathcal{C}(\mathbf{U}) = \mathcal{C}(\mathbf{X})$. Then,

$$\mathbf{WUa}^* = \mathbf{WXb}^*$$

for any solution $\mathbf{a}^*$ to the linear system $\mathbf{U'WUa} = \mathbf{U'Wy}$ (in $\mathbf{a}$) and any solution $\mathbf{b}^*$ to the linear system $\mathbf{X'WXb} = \mathbf{X'Wy}$ (in $\mathbf{b}$).

Corollary 14.12.29. Let $\mathbf{y}$ represent an n-dimensional column vector, $\mathbf{X}$ an $n \times p$ matrix, and $\mathbf{W}$ an $n \times n$ symmetric nonnegative definite matrix. Then, $\mathbf{WXb}_1 = \mathbf{WXb}_2$ for any two solutions $\mathbf{b}_1$ and $\mathbf{b}_2$ to the linear system $\mathbf{X'WXb} = \mathbf{X'Wy}$ (in $\mathbf{b}$).

Note that, for any vector $\mathbf{y}$ in a subspace $\mathcal{U}$ of $\mathcal{R}^{n\times 1}$ and for any symmetric nonnegative definite matrix $\mathbf{W}$, $\mathbf{y}$ itself is a projection of $\mathbf{y}$ on $\mathcal{U}$ with respect to $\mathbf{W}$ (as is evident from the very definition of a projection of $\mathbf{y}$ on $\mathcal{U}$ with respect to $\mathbf{W}$).

The following theorem generalizes Theorem 14.12.7.

Theorem 14.12.30. Let $\mathbf{y}$ represent an n-dimensional column vector, let $\mathcal{U}$ represent a subspace of $\mathcal{R}^{n\times 1}$, let $\mathbf{W}$ represent an $n \times n$ symmetric nonnegative definite matrix, and let $\mathbf{X}$ represent any $n \times p$ matrix whose columns span $\mathcal{U}$. Then, any $p \times 1$ vector $\mathbf{b}^*$ such that $\mathbf{Xb}^*$ is a projection of $\mathbf{y}$ on $\mathcal{U}$ with respect to $\mathbf{W}$ is a solution to the linear system $\mathbf{X'WXb} = \mathbf{X'Wy}$ (in $\mathbf{b}$).

Proof. According to Theorem 14.12.26, $\mathbf{Xb}^* = \mathbf{Xa}$ for some solution $\mathbf{a}$ to $\mathbf{X'WXb} = \mathbf{X'Wy}$. Thus,

$$\mathbf{X'WXb}^* = \mathbf{X'WXa} = \mathbf{X'Wy}.$$

Q.E.D.

Corollary 14.12.27 leads to the following lemma, which extends the result of Corollary 14.12.9.

Lemma 14.12.31. Let $\mathcal{U}$ represent a subspace of $\mathcal{R}^{n\times 1}$, and let $\mathbf{W}$ represent an $n \times n$ symmetric nonnegative definite matrix. Then, for any matrix $\mathbf{X}$ such that $\mathcal{C}(\mathbf{X}) = \mathcal{U}$, $\mathbf{P}_{\mathbf{X},\mathbf{W}}$ is a projection matrix for $\mathcal{U}$ with respect to $\mathbf{W}$.

Thus, $\mathbf{P}_{\mathbf{X},\mathbf{W}}$ is a projection matrix for $\mathcal{U}$ [$= \mathcal{C}(\mathbf{X})$] with respect to $\mathbf{W}$ when $\mathbf{W}$ is positive semidefinite as well as when $\mathbf{W}$ is positive definite. However, when $\mathbf{W}$ is positive semidefinite, there may be more than one projection matrix for $\mathcal{U}$ with respect to $\mathbf{W}$.

The following theorem, whose proof parallels that of Theorem 12.4.1, provides a basis for extending the results of Theorem 14.12.19.

Theorem 14.12.32. Let $\mathcal{U}$ represent a subspace of the linear space $\mathcal{R}^n$ of all n-dimensional column vectors, take $\mathbf{W}$ to be an $n \times n$ symmetric nonnegative definite matrix, and let $\mathbf{y}$ represent a vector in $\mathcal{R}^n$. Then, for $\mathbf{u} \in \mathcal{U}$, the quadratic form $(\mathbf{y} - \mathbf{u})'\mathbf{W}(\mathbf{y} - \mathbf{u})$ (in the difference $\mathbf{y} - \mathbf{u}$) attains its minimum value at $\mathbf{u} = \mathbf{z}$ if and only if $\mathbf{z}$ is a projection of $\mathbf{y}$ on $\mathcal{U}$ with respect to $\mathbf{W}$, in which case

$$(\mathbf{y}-\mathbf{z})'\mathbf{W}(\mathbf{y}-\mathbf{z}) = \mathbf{y}'\mathbf{W}(\mathbf{y}-\mathbf{z}).$$

Proof. Suppose that $\mathbf{z}$ is a projection of $\mathbf{y}$ on $\mathcal{U}$ with respect to $\mathbf{W}$. Then, for any vector $\mathbf{u}$ in $\mathcal{U}$,

$$\begin{aligned}(\mathbf{y}-\mathbf{u})'\mathbf{W}(\mathbf{y}-\mathbf{u}) &= [\mathbf{y}-\mathbf{z}-(\mathbf{u}-\mathbf{z})]'\mathbf{W}[\mathbf{y}-\mathbf{z}-(\mathbf{u}-\mathbf{z})]\\ &= (\mathbf{y}-\mathbf{z})'\mathbf{W}(\mathbf{y}-\mathbf{z}) - 2(\mathbf{y}-\mathbf{z})'\mathbf{W}(\mathbf{u}-\mathbf{z})\\ &\qquad + (\mathbf{u}-\mathbf{z})'\mathbf{W}(\mathbf{u}-\mathbf{z}).\end{aligned}$$

Moreover, $\mathbf{u}-\mathbf{z}$ is in $\mathcal{U}$ and, by definition, $\mathbf{y}-\mathbf{z}$ is orthogonal to every vector in $\mathcal{U}$. Thus, $(\mathbf{y}-\mathbf{z})'\mathbf{W}(\mathbf{u}-\mathbf{z}) = 0$, and hence

$$\begin{aligned}(\mathbf{y}-\mathbf{u})'\mathbf{W}(\mathbf{y}-\mathbf{u}) &= (\mathbf{y}-\mathbf{z})'\mathbf{W}(\mathbf{y}-\mathbf{z}) + (\mathbf{u}-\mathbf{z})'\mathbf{W}(\mathbf{u}-\mathbf{z})\\ &\geq (\mathbf{y}-\mathbf{z})'\mathbf{W}(\mathbf{y}-\mathbf{z}),\end{aligned} \tag{12.7}$$

which implies that $(\mathbf{y}-\mathbf{u})'\mathbf{W}(\mathbf{y}-\mathbf{u})$ attains its minimum value at $\mathbf{u} = \mathbf{z}$.

Furthermore, since $\mathbf{z}$ is in $\mathcal{U}$, we have that $\mathbf{z}'\mathbf{W}(\mathbf{y}-\mathbf{z}) = 0$, with the consequence that

$$(\mathbf{y}-\mathbf{z})'\mathbf{W}(\mathbf{y}-\mathbf{z}) = \mathbf{y}'\mathbf{W}(\mathbf{y}-\mathbf{z}).$$

To complete the proof, it suffices to show that equality is attained in inequality (12.7), or, equivalently, that $(\mathbf{u}-\mathbf{z})'\mathbf{W}(\mathbf{u}-\mathbf{z}) = 0$, only if $\mathbf{u}$ is a projection of $\mathbf{y}$ on $\mathcal{U}$ with respect to $\mathbf{W}$. If $(\mathbf{u}-\mathbf{z})'\mathbf{W}(\mathbf{u}-\mathbf{z}) = 0$, then we find (in light of Corollary 14.3.11) that $\mathbf{W}(\mathbf{u}-\mathbf{z}) = \mathbf{0}$, or, equivalently, that $\mathbf{W}\mathbf{u} = \mathbf{W}\mathbf{z}$, implying that, for every vector $\mathbf{x}$ in $\mathcal{U}$,

$$\mathbf{x}'\mathbf{W}(\mathbf{y}-\mathbf{u}) = \mathbf{x}'\mathbf{W}(\mathbf{y}-\mathbf{z}) = 0$$

and hence that $\mathbf{u}$, like $\mathbf{z}$, is a projection of $\mathbf{y}$ on $\mathcal{U}$ with respect to $\mathbf{W}$. Q.E.D.

The following theorem, which extends the results of Theorem 14.12.19, is obtained by combining the results of Theorem 14.12.32 with those of Theorem 14.12.26 and Corollary 14.12.27.

Theorem 14.12.33. Let $\mathcal{U}$ represent a subspace of the linear space $\mathcal{R}^n$ of all n-dimensional column vectors, take $\mathbf{X}$ to be an $n \times p$ matrix such that $\mathcal{C}(\mathbf{X}) = \mathcal{U}$, take $\mathbf{W}$ to be an $n \times n$ symmetric nonnegative definite matrix, and let $\mathbf{y}$ represent a vector in $\mathcal{R}^n$. Then, for $\mathbf{u} \in \mathcal{U}$, the quadratic form $(\mathbf{y}-\mathbf{u})'\mathbf{W}(\mathbf{y}-\mathbf{u})$ (in the difference $\mathbf{y}-\mathbf{u}$) has a minimum at a point $\mathbf{u}^*$ if and only if $\mathbf{u}^* = \mathbf{X}\mathbf{b}^*$ for some solution $\mathbf{b}^*$ to the linear system $\mathbf{X}'\mathbf{W}\mathbf{X}\mathbf{b} = \mathbf{X}'\mathbf{W}\mathbf{y}$ (in $\mathbf{b}$) or, equivalently, if and only if $\mathbf{u}^* = \mathbf{P}_{\mathbf{X},\mathbf{W}}\mathbf{y} + (\mathbf{I} - \mathbf{P}_{\mathbf{X},\mathbf{W}})\mathbf{X}\mathbf{k}$ for some $p \times 1$ vector $\mathbf{k}$. Furthermore, the minimum value of $(\mathbf{y}-\mathbf{u})'\mathbf{W}(\mathbf{y}-\mathbf{u})$ (for $\mathbf{u} \in \mathcal{U}$) is expressible as

$$(\mathbf{y}-\mathbf{X}\mathbf{b}^*)'\mathbf{W}(\mathbf{y}-\mathbf{X}\mathbf{b}^*) = \mathbf{y}'\mathbf{W}(\mathbf{y}-\mathbf{X}\mathbf{b}^*) = \mathbf{y}'\mathbf{W}(\mathbf{I}-\mathbf{P}_{\mathbf{X},\mathbf{W}})\mathbf{y}$$

(where $\mathbf{b}^*$ is any solution to $\mathbf{X}'\mathbf{W}\mathbf{X}\mathbf{b} = \mathbf{X}'\mathbf{W}\mathbf{y}$).

In light of Theorem 14.12.30, we have the following variant of Theorem 14.12.33, which generalizes Theorem 14.12.20.

Theorem 14.12.34. Let $\mathbf{X}$ represent an $n \times p$ matrix, $\mathbf{W}$ an $n \times n$ symmetric nonnegative definite matrix, and $\mathbf{y}$ an n-dimensional column vector. Then, for $\mathbf{b} \in \mathcal{R}^p$, the quadratic form $(\mathbf{y} - \mathbf{Xb})'\mathbf{W}(\mathbf{y} - \mathbf{Xb})$ has a minimum at a point $\mathbf{b}^*$ if and only if $\mathbf{b}^*$ is a solution to the linear system $\mathbf{X}'\mathbf{WXb} = \mathbf{X}'\mathbf{Wy}$ (in $\mathbf{b}$), in which case $\mathbf{Xb}^* = \mathbf{P}_{\mathbf{X},\mathbf{W}}\mathbf{y} + (\mathbf{I} - \mathbf{P}_{\mathbf{X},\mathbf{W}})\mathbf{Xk}$ for some $p \times 1$ vector $\mathbf{k}$ and

$$(\mathbf{y} - \mathbf{Xb}^*)'\mathbf{W}(\mathbf{y} - \mathbf{Xb}^*) = \mathbf{y}'\mathbf{W}(\mathbf{y} - \mathbf{Xb}^*) = \mathbf{y}'\mathbf{W}(\mathbf{I} - \mathbf{P}_{\mathbf{X},\mathbf{W}})\mathbf{y}.$$

Exercises

Section 14.1

1. Show that a symmetric bilinear form $\mathbf{x}'\mathbf{Ay}$ (in n-dimensional vectors $\mathbf{x}$ and $\mathbf{y}$) can be expressed in terms of the corresponding quadratic form, that is, the quadratic form whose matrix is $\mathbf{A}$. Do so by verifying that

$$\mathbf{x}'\mathbf{Ay} = (1/2)[(\mathbf{x} + \mathbf{y})'\mathbf{A}(\mathbf{x} + \mathbf{y}) - \mathbf{x}'\mathbf{Ax} - \mathbf{y}'\mathbf{Ay}].$$

2. Show that corresponding to any quadratic form $\mathbf{x}'\mathbf{Ax}$ (in the n-dimensional vector $\mathbf{x}$) there exists a unique upper triangular matrix $\mathbf{B}$ such that $\mathbf{x}'\mathbf{Ax}$ and $\mathbf{x}'\mathbf{Bx}$ are identically equal, and express the elements of $\mathbf{B}$ in terms of thc elements of $\mathbf{A}$.

Section 14.2

3. Show, by example, that the sum of two positive semidefinite matrices can be positive definite.

4. Show, via an example, that there exist (nonsymmetric) nonsingular positive semidefinite matrices.

5. Show, by example, that there exist an $n \times n$ positive semidefinite matrix $\mathbf{A}$ and an $n \times m$ matrix $\mathbf{P}$ (where $m < n$) such that $\mathbf{P}'\mathbf{AP}$ is positive definite.

6. Convert results (1)–(3) of Theorem 14.2.9, which are for nonnegative definite (positive definite or positive semidefinite) matrices, into equivalent results for nonpositive definite matrices.

7. Let $\{\mathbf{X}_1, \ldots, \mathbf{X}_r\}$ represent a set of matrices from a linear space $\mathcal{V}$, and let $\mathbf{A} = \{a_{ij}\}$ represent the $r \times r$ matrix whose ijth element is $\mathbf{X}_i \bullet \mathbf{X}_j$—this matrix is referred to as the *Gram matrix* (or the Gramian) of the set $\{\mathbf{X}_1, \ldots, \mathbf{X}_r\}$ and its determinant is referred to as the *Gramian* (or the Gram determinant) of $\{\mathbf{X}_1, \ldots, \mathbf{X}_r\}$.

(a) Show that $\mathbf{A}$ is symmetric and nonnegative definite.

(b) Show that $\mathbf{X}_1, \ldots, \mathbf{X}_r$ are linearly independent if and only if $\mathbf{A}$ is nonsingular.

8. Let $\mathbf{A} = \{a_{ij}\}$ represent an $n \times n$ symmetric positive definite matrix, and let $\mathbf{B} = \{b_{ij}\} = \mathbf{A}^{-1}$. Show that, for $i = 1, \ldots, n$,

$$b_{ii} \geq 1/a_{ii},$$

with equality holding if and only if, for all $j \neq i$, $a_{ij} = 0$.

Section 14.3

9. Let $\mathbf{A}$ represent an $m \times n$ matrix and $\mathbf{D}$ a diagonal matrix such that $\mathbf{A} = \mathbf{PDQ}$ for some matrix $\mathbf{P}$ of full column rank and some matrix $\mathbf{Q}$ of full row rank. Extend the result of Lemma 14.3.1 by showing that $\text{rank}(\mathbf{A})$ equals the number of nonzero diagonal elements in $\mathbf{D}$.

10. Let $\mathbf{A}$ represent an $n \times n$ symmetric idempotent matrix and $\mathbf{V}$ an $n \times n$ symmetric positive definite matrix. Show that $\text{rank}(\mathbf{AVA}) = \text{tr}(\mathbf{A})$.

11. Show that if an $n \times n$ matrix $\mathbf{A}$ is such that $\mathbf{x}'\mathbf{A}\mathbf{x} \neq 0$ for every $n \times 1$ nonnull vector $\mathbf{x}$, then $\mathbf{A}$ is either positive definite or negative definite.

12. (a) Let $\mathbf{A}$ represent an $n \times n$ symmetric matrix of rank r. Take $\mathbf{P}$ to be an $n \times n$ nonsingular matrix and $\mathbf{D}$ an $n \times n$ diagonal matrix such that $\mathbf{A} = \mathbf{P}'\mathbf{DP}$—the existence of such matrices is guaranteed by Corollary 14.3.5. The number, say m, of diagonal elements of $\mathbf{D}$ that are positive is called the *index of inertia* of $\mathbf{A}$ (or of the quadratic form $\mathbf{x}'\mathbf{A}\mathbf{x}$ whose matrix is $\mathbf{A}$). Show that the index of inertia is well-defined in the sense that m does not vary with the choice of $\mathbf{P}$ or $\mathbf{D}$. That is, show that if $\mathbf{P}_1$ and $\mathbf{P}_2$ are nonsingular matrices and $\mathbf{D}_1$ and $\mathbf{D}_2$ are diagonal matrices such that $\mathbf{A} = \mathbf{P}_1'\mathbf{D}_1\mathbf{P}_1 = \mathbf{P}_2'\mathbf{D}_2\mathbf{P}_2$, then $\mathbf{D}_2$ contains the same number of positive diagonal elements as $\mathbf{D}_1$. Show also that the number of diagonal elements of $\mathbf{D}$ that are negative equals $r - m$.

(b) Let $\mathbf{A}$ represent an $n \times n$ symmetric matrix. Show that $\mathbf{A} = \mathbf{P}' \,\text{diag}(\mathbf{I}_m, -\mathbf{I}_{r-m}, \mathbf{0})\mathbf{P}$ for some $n \times n$ nonsingular matrix $\mathbf{P}$ and some nonnegative integers m and r. Show further that m equals the index of inertia of the matrix $\mathbf{A}$ and that $r = \text{rank}(\mathbf{A})$.

(c) An $n \times n$ symmetric matrix $\mathbf{B}$ is said to be *congruent* to an $n \times n$ symmetric matrix $\mathbf{A}$ if there exists an $n \times n$ nonsingular matrix $\mathbf{P}$ such that $\mathbf{B} = \mathbf{P}'\mathbf{AP}$. (If $\mathbf{B}$ is congruent to $\mathbf{A}$, then clearly $\mathbf{A}$ is congruent to $\mathbf{B}$.) Show that $\mathbf{B}$ is congruent to $\mathbf{A}$ if and only if $\mathbf{B}$ has the same rank and the same index of inertia as $\mathbf{A}$. This result is called *Sylvester's law of inertia*, after James Joseph Sylvester (1814–1897).

(d) Let $\mathbf{A}$ represent an $n \times n$ symmetric matrix of rank r with index of inertia m. Show that $\mathbf{A}$ is nonnegative definite if and only if $m = r$, and is positive definite if and only if $m = r = n$.

13. Let $\mathbf{A}$ represent an $n \times n$ symmetric nonnegative definite matrix of rank r. Then, according to Theorem 14.3.7, there exists an $n \times r$ matrix $\mathbf{B}$ (of rank r) such that $\mathbf{A} = \mathbf{BB}'$. Let $\mathbf{X}$ represent any $n \times m$ matrix (where $m \geq r$) such that $\mathbf{A} = \mathbf{XX}'$.

(a) Show that $\mathbf{X} = \mathbf{P_B X}$.

(b) Show that $\mathbf{X} = (\mathbf{B},\ \mathbf{0})\mathbf{Q}$ for some orthogonal matrix $\mathbf{Q}$.

Section 14.4

14. Show that if a symmetric matrix $\mathbf{A}$ has a nonnegative definite generalized inverse, then $\mathbf{A}$ is nonnegative definite.

Section 14.5

15. Suppose that an $n \times n$ matrix $\mathbf{A}$ has an LDU decomposition, say $\mathbf{A} = \mathbf{LDU}$, and let $d_1, d_2, \ldots, d_n$ represent the diagonal elements of the diagonal matrix $\mathbf{D}$. Show that

$$|\mathbf{A}| = d_1 d_2 \cdots d_n.$$

16. (a) Suppose that an $n \times n$ matrix $\mathbf{A}$ (where $n \geq 2$) has a unique LDU decomposition, say $\mathbf{A} = \mathbf{LDU}$, and let $d_1, d_2, \ldots, d_n$ represent the first, second, $\ldots$, nth diagonal elements of $\mathbf{D}$. Show that $d_i \neq 0$ $(i = 1, 2, \ldots, n-1)$ and that $d_n \neq 0$ if and only if $\mathbf{A}$ is nonsingular.

(b) Suppose that an $n \times n$ (symmetric) matrix $\mathbf{A}$ (where $n \geq 2$) has a unique U′DU decomposition, say $\mathbf{A} = \mathbf{U'DU}$, and let $d_1, d_2, \ldots, d_n$ represent the first, second, $\ldots$, nth diagonal elements of $\mathbf{D}$. Show that $d_i \neq 0$ $(i = 1, 2, \ldots, n-1)$ and that $d_n \neq 0$ if and only if $\mathbf{A}$ is nonsingular.

17. Suppose that an $n\times n$ (symmetric) matrix $\mathbf{A}$ has a unique U′DU decomposition, say $\mathbf{A} = \mathbf{U'DU}$. Use the result of Part (b) of Exercise 16 to show that $\mathbf{A}$ has no LDU decompositions other than $\mathbf{A} = \mathbf{U'DU}$.

18. Show that if a nonsingular matrix has an LDU decomposition, then that decomposition is unique.

19. Let $\mathbf{A}$ represent an $n\times n$ matrix (where $n \geq 2$). By for instance using the results of Exercises 16, 17, and 18, show that if $\mathbf{A}$ has a unique LDU decomposition or (in the special case where $\mathbf{A}$ is symmetric) a unique U′DU decomposition, then the leading principal submatrices (of $\mathbf{A}$) of orders $1, 2, \ldots, n-1$ are

nonsingular (thereby establishing the converse of Corollary 14.5.7) and have unique LDU decompositions.

20. (a) Let $\mathbf{A} = \{a_{ij}\}$ represent an $m \times n$ nonnull matrix of rank r. Show that there exist an $m \times m$ permutation matrix $\mathbf{P}$ and an $n \times n$ permutation matrix $\mathbf{Q}$ such that

$$\mathbf{PAQ} = \begin{pmatrix} \mathbf{B}_{11} & \mathbf{B}_{12} \\ \mathbf{B}_{21} & \mathbf{B}_{22} \end{pmatrix},$$

where $\mathbf{B}_{11}$ is an $r \times r$ nonsingular matrix whose leading principal submatrices (of orders $1, 2, \ldots, r-1$) are nonsingular.

(b) Let $\mathbf{B} = \begin{pmatrix} \mathbf{B}_{11} & \mathbf{B}_{12} \\ \mathbf{B}_{21} & \mathbf{B}_{22} \end{pmatrix}$ represent any $m \times n$ nonnull matrix of rank r such that $\mathbf{B}_{11}$ is an $r \times r$ nonsingular matrix whose leading principal submatrices (of orders $1, 2, \ldots, r-1$) are nonsingular. Show that there exists a unique decomposition of $\mathbf{B}$ of the form

$$\mathbf{B} = \begin{pmatrix} \mathbf{L}_1 \\ \mathbf{L}_2 \end{pmatrix} \mathbf{D}(\mathbf{U}_1, \ \mathbf{U}_2),$$

where $\mathbf{L}_1$ is an $r \times r$ unit lower triangular matrix, $\mathbf{U}_1$ is an $r \times r$ unit upper triangular matrix, and $\mathbf{D}$ is an $r \times r$ diagonal matrix. Show further that this decomposition is such that $\mathbf{B}_{11} = \mathbf{L}_1\mathbf{D}\mathbf{U}_1$ is the unique LDU decomposition of $\mathbf{B}_{11}$, $\mathbf{D}$ is nonsingular, $\mathbf{L}_2 = \mathbf{B}_{21}\mathbf{U}_1^{-1}\mathbf{D}^{-1}$ and $\mathbf{U}_2 = \mathbf{D}^{-1}\mathbf{L}_1^{-1}\mathbf{B}_{12}$.

21. Show, by example, that there exist $n \times n$ (nonsymmetric) positive semidefinite matrices that do not have LDU decompositions.

22. Let $\mathbf{A}$ represent an $n \times n$ nonnegative definite (possibly nonsymmetric) matrix that has an LDU decomposition, say $\mathbf{A} = \mathbf{LDU}$. Show that the diagonal elements of the diagonal matrix $\mathbf{D}$ are nonnegative (thereby extending part of Theorem 14.5.9 and part of Corollary 14.5.15).

23 Let $\mathbf{A}$ represent an $m \times k$ matrix of full column rank. And let $\mathbf{A} = \mathbf{QR}$ represent the QR decomposition of $\mathbf{A}$; that is, let $\mathbf{Q}$ represent the unique $m \times k$ matrix whose columns are orthonormal with respect to the usual inner product and let $\mathbf{R}$ represent the unique $k \times k$ upper triangular matrix with positive diagonal elements such that $\mathbf{A} = \mathbf{QR}$. Show that $\mathbf{A}'\mathbf{A} = \mathbf{R}'\mathbf{R}$ (so that $\mathbf{A}'\mathbf{A} = \mathbf{R}'\mathbf{R}$ is the Cholesky decomposition of $\mathbf{A}'\mathbf{A}$).

24. Let $\mathbf{A}$ represent an $m \times k$ matrix of rank r (where r is possibly less than k). Consider the decomposition $\mathbf{A} = \mathbf{QR}_1$, where $\mathbf{Q}$ is an $m \times r$ matrix with orthonormal columns and $\mathbf{R}_1$ is an $r \times k$ submatrix whose rows are the r nonnull rows of a $k \times k$ upper triangular matrix $\mathbf{R}$ having r positive diagonal

elements and $k - r$ null rows. (Such a decomposition can be obtained by using the results of Exercise 6.4—refer to Exercise 6.5.) Generalize the result of Exercise 23 by showing that if the inner product with respect to which the columns of $\mathbf{Q}$ are orthonormal is the usual inner product, then $\mathbf{A}'\mathbf{A} = \mathbf{R}'\mathbf{R}$ (so that $\mathbf{A}'\mathbf{A} = \mathbf{R}'\mathbf{R}$ is the Cholesky decomposition of $\mathbf{A}'\mathbf{A}$).

25. Let $\mathbf{A} = \{a_{ij}\}$ represent an $n \times n$ matrix that has an LDU decomposition, say $\mathbf{A} = \mathbf{LDU}$. And define $\mathbf{G} = \mathbf{U}^{-1}\mathbf{D}^{-}\mathbf{L}^{-1}$ (which, as discussed in Section 14.5f, is a generalized inverse of $\mathbf{A}$).

    (a) Show that

$$\mathbf{G} = \mathbf{D}^{-}\mathbf{L}^{-1} + (\mathbf{I} - \mathbf{U})\mathbf{G} = \mathbf{U}^{-1}\mathbf{D}^{-} + \mathbf{G}(\mathbf{I} - \mathbf{L}).$$

    (b) For $i = 1, \ldots, n$, let d_i represent the ith diagonal element of the diagonal matrix $\mathbf{D}$; and, for $i, j = 1, \ldots, n$, let ℓ_{ij}, u_{ij}, and g_{ij} represent the ijth elements of $\mathbf{L}$, $\mathbf{U}$, and $\mathbf{G}$, respectively. Take $\mathbf{D}^{-} = \operatorname{diag}(d_1^*, \ldots, d_n^*)$, where $d_i^* = 1/d_i$, if $d_i \neq 0$, and d_i^* is an arbitrary scalar, if $d_i = 0$. Show that

$$g_{ii} = d_i^* - \sum_{k=i+1}^{n} u_{ik} g_{ki} = d_i^* - \sum_{k=i+1}^{n} g_{ik} \ell_{ki} \tag{E.1}$$

    and that

$$g_{ij} = \begin{cases} -\sum_{k=j+1}^{n} g_{ik}\ell_{kj}, & \text{for } j < i, \quad \text{(E.2a)} \\ -\sum_{k=i+1}^{n} u_{ik} g_{kj}, & \text{for } j > i \quad \text{(E.2b)} \end{cases}$$

    (where the degenerate sums $\sum_{k=n+1}^{n} g_{ik}\ell_{ki}$ and $\sum_{k=n+1}^{n} u_{ik} g_{ki}$ are to be interpreted as 0).

    (c) Devise a recursive procedure that uses the formulas from Part (b) to generate a generalized inverse of $\mathbf{A}$.

Section 14.6

26. Verify that a principal submatrix of a skew-symmetric matrix is skew-symmetric.

27. (a) Show that the sum of skew-symmetric matrices is skew-symmetric.

    (b) Show that the sum $\mathbf{A}_1 + \mathbf{A}_2 + \cdots + \mathbf{A}_k$ of $n \times n$ nonnegative definite matrices $\mathbf{A}_1, \mathbf{A}_2, \ldots, \mathbf{A}_k$ is skew-symmetric if and only if $\mathbf{A}_1, \mathbf{A}_2, \ldots, \mathbf{A}_k$ are skew-symmetric.

    (c) Show that the sum $\mathbf{A}_1 + \mathbf{A}_2 + \cdots + \mathbf{A}_k$ of $n \times n$ symmetric nonnegative definite matrices $\mathbf{A}_1, \mathbf{A}_2, \ldots, \mathbf{A}_k$ is a null matrix if and only if $\mathbf{A}_1, \mathbf{A}_2, \ldots, \mathbf{A}_k$ are null matrices.

Section 14.7

28. (a) Let $\mathbf{A}_1, \mathbf{A}_2, \ldots, \mathbf{A}_k$ represent $n \times n$ nonnegative definite matrices. Show that $\text{tr}(\sum_{i=1}^k \mathbf{A}_i) \geq 0$, with equality holding if and only if $\sum_{i=1}^k \mathbf{A}_i$ is skew-symmetric or, equivalently, if and only if $\mathbf{A}_1, \mathbf{A}_2, \ldots, \mathbf{A}_k$ are skew-symmetric, thereby generalizing Theorem 14.7.2. [*Note*. That $\sum_{i=1}^k \mathbf{A}_i$ being skew-symmetric is equivalent to $\mathbf{A}_1$, $\mathbf{A}_2$, ..., $\mathbf{A}_k$ being skew-symmetric is the result of Part (b) of Exercise 27.]

    (b) Let $\mathbf{A}_1, \mathbf{A}_2, \ldots, \mathbf{A}_k$ represent $n \times n$ symmetric nonnegative definite matrices. Show that $\text{tr}(\sum_{i=1}^k \mathbf{A}_i) \geq 0$, with equality holding if and only if $\sum_{i=1}^k \mathbf{A}_i = \mathbf{0}$ or, equivalently, if and only if $\mathbf{A}_1, \mathbf{A}_2, \ldots, \mathbf{A}_k$ are null matrices, thereby generalizing Corollary 14.7.3.

29. Show, via an example, that (for $n > 1$) there exist $n \times n$ (nonsymmetric) positive definite matrices $\mathbf{A}$ and $\mathbf{B}$ such that $\text{tr}(\mathbf{AB}) < 0$.

30. (a) Show, via an example, that (for $n > 1$) there exist $n \times n$ symmetric positive definite matrices $\mathbf{A}$ and $\mathbf{B}$ such that the product $\mathbf{AB}$ has one or more negative diagonal elements (and hence such that $\mathbf{AB}$ is not nonnegative definite).

    (b) Show, however, that the product of two $n \times n$ symmetric positive definite matrices cannot be nonpositive definite.

31. Let $\mathbf{A} = \{a_{ij}\}$ and $\mathbf{B} = \{b_{ij}\}$ represent $n \times n$ matrices, and take $\mathbf{C}$ to be the $n \times n$ matrix whose ijth element $c_{ij} = a_{ij}b_{ij}$ is the product of the ijth elements of $\mathbf{A}$ and $\mathbf{B}$. Show that if $\mathbf{A}$ is nonnegative definite and $\mathbf{B}$ is symmetric nonnegative definite, then $\mathbf{C}$ is nonnegative definite. Show further that if $\mathbf{A}$ is positive definite and $\mathbf{B}$ is symmetric positive definite, then $\mathbf{C}$ is positive definite. [*Hint*. Taking $\mathbf{x} = (x_1, \ldots, x_n)'$ to be an arbitrary $n \times 1$ vector and $\mathbf{F} = (\mathbf{f}_1, \ldots, \mathbf{f}_n)$ to be a matrix such that $\mathbf{B} = \mathbf{F}'\mathbf{F}$, begin by showing that $\mathbf{x}'\mathbf{C}\mathbf{x} = \text{tr}(\mathbf{AH})$, where $\mathbf{H} = \mathbf{G}'\mathbf{G}$ with $\mathbf{G} = (x_1\mathbf{f}_1, \ldots, x_n\mathbf{f}_n)$.]

32. Let $\mathbf{A}_1, \mathbf{A}_2, \ldots, \mathbf{A}_k$ and $\mathbf{B}_1, \mathbf{B}_2, \ldots, \mathbf{B}_k$ represent $n \times n$ symmetric nonnegative definite matrices. Show that $\text{tr}(\sum_{i=1}^k \mathbf{A}_i\mathbf{B}_i) \geq 0$, with equality holding if and only if, for $i = 1, 2, \ldots, k$, $\mathbf{A}_i\mathbf{B}_i = \mathbf{0}$, thereby generalizing Corollary 14.7.7 and the results of Part (b) of Exercise 28.

Section 14.8

33. Let $\mathbf{A}$ represent a symmetric nonnegative definite matrix that has been partitioned as

$$\mathbf{A} = \begin{pmatrix} \mathbf{T} & \mathbf{U} \\ \mathbf{V} & \mathbf{W} \end{pmatrix},$$

where $\mathbf{T}$ (and hence $\mathbf{W}$) is square. Show that $\mathbf{V}\mathbf{T}^{-}\mathbf{U}$ and $\mathbf{U}\mathbf{W}^{-}\mathbf{V}$ are symmetric and nonnegative definite.

34. Show, via an example, that there exists an $(m+n) \times (m+n)$ (nonsymmetric) positive semidefinite matrix $\mathbf{A}$ of the form $\mathbf{A} = \begin{pmatrix} \mathbf{T} & \mathbf{U} \\ \mathbf{V} & \mathbf{W} \end{pmatrix}$, where $\mathbf{T}$ is of dimensions $m \times m$, $\mathbf{W}$ of dimensions $n \times n$, $\mathbf{U}$ of dimensions $m \times n$, and $\mathbf{V}$ of dimensions $n \times m$, for which $\mathcal{C}(\mathbf{U}) \not\subset \mathcal{C}(\mathbf{T})$ and/or $\mathcal{R}(\mathbf{V}) \not\subset \mathcal{R}(\mathbf{T})$, expression (9.6.1) does not necessarily equal $\text{rank}(\mathbf{A})$ [i.e., $\text{rank}(\mathbf{T}) + \text{rank}(\mathbf{W} - \mathbf{V}\mathbf{T}^{-}\mathbf{U})$ does not necessarily equal $\text{rank}(\mathbf{A})$], and formula (9.6.2) does not necessarily give a generalized inverse of $\mathbf{A}$.

35. Show, via an example, that there exists an $(m + n) \times (m + n)$ symmetric partitioned matrix $\mathbf{A}$ of the form $\mathbf{A} = \begin{pmatrix} \mathbf{T} & \mathbf{U} \\ \mathbf{U}' & \mathbf{W} \end{pmatrix}$, where $\mathbf{T}$ is of dimensions $m \times m$, $\mathbf{U}$ of dimensions $m \times n$, and $\mathbf{W}$ of dimensions $n \times n$, such that $\mathbf{T}$ is nonnegative definite and (depending on the choice of $\mathbf{T}^{-}$) the Schur complement $\mathbf{W} - \mathbf{U}'\mathbf{T}^{-}\mathbf{U}$ of $\mathbf{T}$ relative to $\mathbf{T}^{-}$ is nonnegative definite, but $\mathbf{A}$ is not nonnegative definite.

36. An $n \times n$ matrix $\mathbf{A} = \{a_{ij}\}$ is said to be *diagonally dominant* if, for $i = 1, 2, \ldots, n$, $|a_{ii}| > \sum_{j=1 (j \neq i)}^{n} |a_{ij}|$. (In the degenerate special case where $n = 1$, $\mathbf{A}$ is said to be diagonally dominant if it is nonnull.)

    (a) Show that a principal submatrix of a diagonally dominant matrix is diagonally dominant.

    (b) Let $\mathbf{A} = \{a_{ij}\}$ represent an $n \times n$ diagonally dominant matrix, partition $\mathbf{A}$ as $\mathbf{A} = \begin{pmatrix} \mathbf{A}_{11} & \mathbf{a} \\ \mathbf{b}' & a_{nn} \end{pmatrix}$ [so that $\mathbf{A}_{11}$ is of dimensions $(n-1) \times (n-1)$], and let $\mathbf{C} = \mathbf{A}_{11} - (1/a_{nn})\mathbf{a}\mathbf{b}'$ represent the Schur complement of a_{nn}. Show that $\mathbf{C}$ is diagonally dominant.

    (c) Show that a diagonally dominant matrix is nonsingular.

    (d) Show that a diagonally dominant matrix has a unique LDU decomposition.

    (e) Let $\mathbf{A} = \{a_{ij}\}$ represent an $n \times n$ symmetric matrix. Show that if $\mathbf{A}$ is diagonally dominant and if the diagonal elements $a_{11}, a_{22}, \ldots, a_{nn}$ of $\mathbf{A}$ are all positive, then $\mathbf{A}$ is positive definite.

Section 14.9

37. Let $\mathbf{A} = \{a_{ij}\}$ represent an $n \times n$ symmetric positive definite matrix. Show that $\det(\mathbf{A}) \leq \prod_{i=1}^{n} a_{ii}$, with equality holding if and only if $\mathbf{A}$ is diagonal.

38. Let $\mathbf{A} = \begin{pmatrix} a & b \\ c & d \end{pmatrix}$, where a, b, c, and d are scalars.

   (a) Show that $\mathbf{A}$ is positive definite if and only if $a > 0$, $d > 0$, and $|b+c|/2 < \sqrt{ad}$.

   (b) Show that, in the special case where $\mathbf{A}$ is symmetric (i.e., where $c = b$), $\mathbf{A}$ is positive definite if and only if $a > 0$, $d > 0$, and $|b| < \sqrt{ad}$.

39. By, for example, making use of the result of Exercise 38, show that if an $n \times n$ matrix $\mathbf{A} = \{a_{ij}\}$ is symmetric positive definite, then, for $j \neq i = 1, \ldots, n$,

$$|a_{ij}| < \sqrt{a_{ii}a_{jj}} \leq \max(a_{ii}, a_{jj}).$$

40. Show, by example, that it is possible for the determinants of both leading principal submatrices of a 2×2 symmetric matrix to be nonnegative without the matrix being nonnegative definite and that, for $n \geq 3$, it is possible for the determinants of all n leading principal submatrices of an $n \times n$ symmetric matrix to be nonnegative *and* for the matrix to be nonsingular without the matrix being nonnegative definite.

Section 14.10

41. Let $\mathcal{V}$ represent a subspace of $\mathcal{R}^{n\times 1}$ of dimension r (where $r \geq 1$). Take $\mathbf{B} = (\mathbf{b}_1, \mathbf{b}_2, \ldots, \mathbf{b}_r)$ to be any $n \times r$ matrix whose columns $\mathbf{b}_1, \mathbf{b}_2, \ldots, \mathbf{b}_r$ form a basis for $\mathcal{V}$, and let $\mathbf{L}$ represent any left inverse of $\mathbf{B}$. Let g represent a function that assigns the value $\mathbf{x} * \mathbf{y}$ to an arbitrary pair of vectors $\mathbf{x}$ and $\mathbf{y}$ in $\mathcal{V}$.

   (a) Let f represent an arbitrary inner product for $\mathcal{R}^{r\times 1}$, and denote by $\mathbf{s} \bullet \mathbf{t}$ the value assigned by f to an arbitrary pair of r-dimensional vectors $\mathbf{s}$ and $\mathbf{t}$. Show that g is an inner product (for $\mathcal{V}$) if and only if there exists an f such that (for all $\mathbf{x}$ and $\mathbf{y}$ in $\mathcal{V}$)

$$\mathbf{x} * \mathbf{y} = (\mathbf{L}\mathbf{x}) \bullet (\mathbf{L}\mathbf{y}).$$

   (b) Show that g is an inner product (for $\mathcal{V}$) if and only if there exists an $r \times r$ symmetric positive definite matrix $\mathbf{W}$ such that (for all $\mathbf{x}$ and $\mathbf{y}$ in $\mathcal{V}$)

$$\mathbf{x} * \mathbf{y} = \mathbf{x}'\mathbf{L}'\mathbf{W}\mathbf{L}\mathbf{y}.$$

   (c) Show that g is an inner product (for $\mathcal{V}$) if and only if there exists an $n \times n$ symmetric positive definite matrix $\mathbf{W}$ such that (for all $\mathbf{x}$ and $\mathbf{y}$ in $\mathcal{V}$)

$$\mathbf{x} * \mathbf{y} = \mathbf{x}'\mathbf{W}\mathbf{y}.$$

42. Let $\mathcal{V}$ represent a linear space of $m \times n$ matrices, and let $\mathbf{A} \bullet \mathbf{B}$ represent the value assigned by a quasi-inner product to any pair of matrices $\mathbf{A}$ and $\mathbf{B}$ (in $\mathcal{V}$). Show that the set

$$\mathcal{U} = \{\mathbf{A} \in \mathcal{V} \,:\, \mathbf{A} \bullet \mathbf{A} = 0\},$$

which comprises every matrix in $\mathcal{V}$ with a zero quasi-norm, is a linear space.

43. Let $\mathbf{W}$ represent an $m \times m$ symmetric positive definite matrix and $\mathbf{V}$ an $n \times n$ symmetric positive definite matrix.

    (a) Show that the function that assigns the value $\mathrm{tr}(\mathbf{A}'\mathbf{W}\mathbf{B}\mathbf{V})$ to an arbitrary pair of $m \times n$ marices $\mathbf{A}$ and $\mathbf{B}$ qualifies as an inner product for the linear space $\mathcal{R}^{m \times n}$.

    (b) Show that the function that assigns the value $\mathrm{tr}(\mathbf{A}'\mathbf{W}\mathbf{B})$ to an arbitrary pair of $m \times n$ matrices $\mathbf{A}$ and $\mathbf{B}$ qualifies as an inner product for $\mathcal{R}^{m \times n}$.

    (c) Show that the function that assigns the value $\mathrm{tr}(\mathbf{A}'\mathbf{W}\mathbf{B}\mathbf{W})$ to an arbitrary pair of $m \times m$ matrices $\mathbf{A}$ and $\mathbf{B}$ qualifies as an inner product for $\mathcal{R}^{m \times m}$.

Section 14.12

44. Let $\mathbf{A}$ represent a $q \times p$ matrix, $\mathbf{B}$ a $p \times n$ matrix, and $\mathbf{C}$ an $m \times q$ matrix. Show that (a) $\mathbf{CAB}(\mathbf{CAB})^{-}\mathbf{C} = \mathbf{C}$ if and only if $\mathrm{rank}(\mathbf{CAB}) = \mathrm{rank}(\mathbf{C})$, and (b) $\mathbf{B}(\mathbf{CAB})^{-}\mathbf{CAB} = \mathbf{B}$ if and only if $\mathrm{rank}(\mathbf{CAB}) = \mathrm{rank}(\mathbf{B})$ [thereby (in light of Corollary 14.11.3) extending the results of Parts (5) and (1) of Theorem 14.12.11].

45. Let $\mathcal{U}$ represent a subspace of $\mathcal{R}^{n \times 1}$, let $\mathbf{X}$ represent an $n \times p$ matrix whose columns span $\mathcal{U}$, and let $\mathbf{W}$ and $\mathbf{V}$ represent $n \times n$ symmetric positive definite matrices. Show that each of the following two conditions is necessary and sufficient for the projection $\mathbf{P}_{\mathbf{X},\mathbf{W}}\mathbf{y}$ of $\mathbf{y}$ on $\mathcal{U}$ with respect to $\mathbf{W}$ to be the same (for every $\mathbf{y}$ in $\mathcal{R}^n$) as the projection $\mathbf{P}_{\mathbf{X},\mathbf{V}}\mathbf{y}$ of $\mathbf{y}$ on $\mathcal{U}$ with respect to $\mathbf{V}$:

    (a) $\mathbf{V} = \mathbf{P}'_{\mathbf{X},\mathbf{W}}\mathbf{V}\mathbf{P}_{\mathbf{X},\mathbf{W}} + (\mathbf{I} - \mathbf{P}_{\mathbf{X},\mathbf{W}})'\mathbf{V}(\mathbf{I} - \mathbf{P}_{\mathbf{X},\mathbf{W}})$;

    (b) there exist a scalar c, a $p \times p$ matrix $\mathbf{K}$, and an $n \times n$ matrix $\mathbf{H}$ such that

$$\mathbf{V} = c\mathbf{W} + \mathbf{WXKX}'\mathbf{W} + (\mathbf{I} - \mathbf{P}_{\mathbf{X},\mathbf{W}})'\mathbf{H}(\mathbf{I} - \mathbf{P}_{\mathbf{X},\mathbf{W}}).$$

46. Let $\mathbf{y}$ represent an n-dimensional column vector, let $\mathcal{U}$ represent a subspace of $\mathcal{R}^{n \times 1}$, and let $\mathbf{X}$ represent an $n \times p$ matrix whose columns span $\mathcal{U}$. It follows from Corollary 12.1.2 that $\mathbf{y} \perp_{\mathbf{I}} \mathcal{U}$ if and only if $\mathbf{X}'\mathbf{y} = \mathbf{0}$ and, more generally, it follows from Lemma 14.12.1 that, for any $n \times n$ symmetric positive definite matrix $\mathbf{W}$, $\mathbf{y} \perp_{\mathbf{W}} \mathcal{U}$ if and only if $\mathbf{X}'\mathbf{W}\mathbf{y} = \mathbf{0}$. Extend this result (in the direction indicated in the discussion following Lemma 14.12.1) by using

Parts (3) and (5) of Lemma 14.12.2 to show that, for any $n \times n$ symmetric nonnegative definite matrix $\mathbf{W}$, $\mathbf{y} \perp_{\mathbf{W}} \mathcal{U}$ if and only if $\mathbf{X}'\mathbf{W}\mathbf{y} = \mathbf{0}$.

47. Let $\mathbf{W}$ represent an $n \times n$ symmetric nonnegative definite matrix.

(a) Generalize Theorem 14.12.12 by showing that, for any $n \times p$ matrix $\mathbf{X}$ and any $n \times q$ matrix $\mathbf{U}$ such that $\mathcal{C}(\mathbf{U}) \subset \mathcal{C}(\mathbf{X})$,
(1) $\mathbf{W}\mathbf{P}_{\mathbf{X},\mathbf{W}}\mathbf{U} = \mathbf{W}\mathbf{U}$, and $\mathbf{U}'\mathbf{W}\mathbf{P}_{\mathbf{X},\mathbf{W}} = \mathbf{U}'\mathbf{W}$;
(2) $\mathbf{P}_{\mathbf{U},\mathbf{W}}\mathbf{P}_{\mathbf{X},\mathbf{W}} = \mathbf{P}_{\mathbf{U},\mathbf{W}}$, and $\mathbf{P}'_{\mathbf{X},\mathbf{W}}\mathbf{W}\mathbf{P}_{\mathbf{U},\mathbf{W}} = \mathbf{W}\mathbf{P}_{\mathbf{X},\mathbf{W}}\mathbf{P}_{\mathbf{U},\mathbf{W}} = \mathbf{W}\mathbf{P}_{\mathbf{U},\mathbf{W}}$.

(b) Generalize Corollary 14.12.13 by showing that, for any $n \times p$ matrix $\mathbf{X}$ and any $n \times q$ matrix $\mathbf{U}$ such that $\mathcal{C}(\mathbf{U}) = \mathcal{C}(\mathbf{X})$,

$$\mathbf{W}\mathbf{P}_{\mathbf{U},\mathbf{W}} = \mathbf{W}\mathbf{P}_{\mathbf{X},\mathbf{W}}.$$

48. Let $\mathbf{U}$ represent a subspace of $\mathcal{R}^{n\times 1}$, let $\mathbf{A}$ represent an $n \times n$ matrix and $\mathbf{W}$ an $n \times n$ symmetric nonnegative definite matrix, and let $\mathbf{X}$ represent any $n \times p$ matrix whose columns span $\mathcal{U}$.

(a) Show that $\mathbf{A}$ is a projection matrix for $\mathcal{U}$ with respect to $\mathbf{W}$ if and only if

$$\mathbf{A} = \mathbf{P}_{\mathbf{X},\mathbf{W}} + (\mathbf{I} - \mathbf{P}_{\mathbf{X},\mathbf{W}})\mathbf{X}\mathbf{K}$$

for some $p \times n$ matrix $\mathbf{K}$.

(b) Show that if $\mathbf{A}$ is a projection matrix for $\mathcal{U}$ with respect to $\mathbf{W}$, then $\mathbf{W}\mathbf{A} = \mathbf{W}\mathbf{P}_{\mathbf{X},\mathbf{W}}$.

49. Let $\mathbf{A}$ represent an $n \times n$ matrix and $\mathbf{W}$ an $n \times n$ symmetric nonnegative definite matrix.

(a) Show (by, e.g., using the results of Exercise 48) that if $\mathbf{A}'\mathbf{W}\mathbf{A} = \mathbf{W}\mathbf{A}$ [or, equivalently, if $(\mathbf{I} - \mathbf{A})'\mathbf{W}\mathbf{A} = \mathbf{0}$], then $\mathbf{A}$ is a projection matrix with respect to $\mathbf{W}$, and in particular $\mathbf{A}$ is a projection matrix for $\mathcal{C}(\mathbf{A})$ with respect to $\mathbf{W}$, and, conversely, show that if $\mathbf{A}$ is a projection matrix with respect to $\mathbf{W}$, then $\mathbf{A}'\mathbf{W}\mathbf{A} = \mathbf{W}\mathbf{A}$ (thereby generalizing Theorem 14.12.16).

(b) Show that if $\mathbf{A}$ is a projection matrix with respect to $\mathbf{W}$, then in particular $\mathbf{A}$ is a projection matrix for $\mathcal{C}(\mathbf{A})$ with respect to $\mathbf{W}$.

(c) Show that $\mathbf{A}$ is a projection matrix with respect to $\mathbf{W}$ if and only if $\mathbf{W}\mathbf{A}$ is symmetric and $\mathbf{W}\mathbf{A}^2 = \mathbf{W}\mathbf{A}$ (thereby generalizing Corollary 14.12.17).

50. Let $\mathcal{U}$ represent a subspace of $\mathcal{R}^{n\times 1}$, let $\mathbf{X}$ represent an $n \times p$ matrix whose columns span $\mathcal{U}$, and let $\mathbf{W}$ and $\mathbf{V}$ represent $n \times n$ symmetric nonnegative definite matrices. Show (by, e.g., making use of the result of Exercise 46) that each of the following two conditions is necessary and sufficient for every

projection of $\mathbf{y}$ on $\mathcal{U}$ with respect to $\mathbf{W}$ to be a projection (for every $\mathbf{y}$ in $\mathcal{R}^n$) of $\mathbf{y}$ on $\mathcal{U}$ with respect to $\mathbf{V}$:

(a) $\mathbf{X}'\mathbf{V}\mathbf{P}_{\mathbf{X},\mathbf{W}} = \mathbf{X}'\mathbf{V}$, or, equivalently, $\mathbf{X}'\mathbf{V}(\mathbf{I} - \mathbf{P}_{\mathbf{X},\mathbf{W}}) = \mathbf{0}$;

(b) there exists a $p \times p$ matrix $\mathbf{Q}$ such that $\mathbf{VX} = \mathbf{WXQ}$, or, equivalently, $\mathcal{C}(\mathbf{VX}) \subset \mathcal{C}(\mathbf{WX})$ [thereby generalizing Parts (1) and (4) of Theorem 14.12.18].

51. Let $\mathbf{X}$ represent an $n \times p$ matrix and $\mathbf{W}$ an $n \times n$ symmetric nonnegative definite matrix. As in the special case where $\mathbf{W}$ is positive definite, let

$$\mathcal{C}_{\mathbf{W}}^{\perp}(\mathbf{X}) = \{\mathbf{y} \in \mathcal{R}^{n\times 1} : \mathbf{y} \perp_{\mathbf{W}} \mathcal{C}(\mathbf{X})\}.$$

(a) By, for example, making use of the result of Exercise 46, show that

$$\mathcal{C}_{\mathbf{W}}^{\perp}(\mathbf{X}) = \mathcal{N}(\mathbf{X}'\mathbf{W}) = \mathcal{C}(\mathbf{I} - \mathbf{P}_{\mathbf{X},\mathbf{W}})$$

(thereby generalizing Lemma 14.12.21).

(b) Show that

$$\dim[\mathcal{C}_{\mathbf{W}}^{\perp}(\mathbf{X})] = n - \operatorname{rank}(\mathbf{WX}) \geq n - \operatorname{rank}(\mathbf{X}) = n - \dim[\mathcal{C}(\mathbf{X})]$$

(thereby generalizing Corollary 14.12.22).

(c) By, for example, making use of the result of Exercise 46, show that, for any solution $\mathbf{b}^*$ to the linear system $\mathbf{X}'\mathbf{WXb} = \mathbf{X}'\mathbf{Wy}$ (in $\mathbf{b}$), the vector $\mathbf{y} - \mathbf{Xb}^*$ is a projection of $\mathbf{y}$ on $\mathcal{C}_{\mathbf{W}}^{\perp}(\mathbf{X})$ with respect to $\mathbf{W}$ (thereby extending the result of Theorem 14.12.23).

15

Matrix Differentiation

It is natural and convenient to use matrix notation and terminology in defining and discussing certain functions of one or more variables. Earlier (in Chapter 14), matrix notation and terminology were used in defining and discussing linear, bilinear, and quadratic forms. In some cases, the use of matrix notation and terminology is essentially unavoidable—consider, for example, a case where the determinant of an $m \times m$ matrix is regarded as a function of its m^2 elements.

Matrix differentiation is the derivation of the first-, second-, or higher-order partial derivatives of a function or functions that have been expressed in terms of matrices. In deriving, presenting, and discussing the partial derivatives of such functions, it is natural, convenient, and in some cases necessary to employ matrix notation and terminology. Not only may the functions be expressed in terms of matrices, but the functions may comprise the elements of a vector or matrix, as would be the case if the elements of the inverse of an $m \times m$ (nonsingular) matrix $\mathbf{A}$ were regarded as functions of the elements of $\mathbf{A}$.

Matrix differentiation is of considerable importance in statistics. It is especially useful in connection with the maximum likelihood estimation of the parameters in a statistical model. The maximum likelihood estimates of the model's parameters satisfy the equations (known as the likelihood equations) obtained by equating to zero the first-order partial derivatives (with respect to the model's parameters) of the logarithm of the so-called likelihood function—in many important cases, the likelihood function involves the determinant and/or inverse of a matrix. Further, an approximation (suitable for large samples) to the variance-covariance matrix of the maximum likelihood estimators can be obtained by inverting the matrix (known as the information matrix) whose ijth element is -1 times the second-order partial derivative (with respect to the ith and jth parameters) of the logarithm of the likelihood function.

15.1 Definitions, Notation, and Other Preliminaries

a. Neighborhoods, interior points, and open sets

Let $\mathbf{c}$ represent an arbitrary m-dimensional column vector. Taking the norm for $\mathcal{R}^{m\times 1}$ to be the usual norm, a *neighborhood* of $\mathbf{c}$ is (by definition) a set of the general form

$$\{\mathbf{x} \in \mathcal{R}^{m\times 1} : \|\mathbf{x} - \mathbf{c}\| < r\},$$

where r is a positive number called the *radius* of the neighborhood. Thus, a neighborhood of $\mathbf{c}$ of radius r is the set of all m-dimensional column vectors whose distance from $\mathbf{c}$ is less than r. Geometrically, a neighborhood of $\mathbf{c}$ is (depending on whether m is $1, 2, 3$, or greater than 3) the interior of an interval, circle, sphere, or "hypersphere" centered at $\mathbf{c}$.

The definition of neighborhood can be extended to row vectors. In fact, it can be extended to matrices of any dimensions. Taking the norm for $\mathcal{R}^{m\times n}$ to be the usual norm, a *neighborhood* of an arbitrary $m \times n$ matrix $\mathbf{C}$ is a set of the general form

$$\{\mathbf{X} \in \mathcal{R}^{m\times n} : \|\mathbf{X} - \mathbf{C}\| < r\}.$$

Let S represent an arbitrary set of m-dimensional column vectors (or more generally of $m \times n$ matrices), that is, let S represent a subset of $\mathcal{R}^{m\times 1}$ (or more generally of $\mathcal{R}^{m\times n}$). Then, a vector $\mathbf{x}$ (or matrix $\mathbf{X}$) in S is said to be an *interior point* of S if there exists a neighborhood of $\mathbf{x}$ (or $\mathbf{X}$), all of whose points belong to S. The set S is said to be *open* if all of its points are interior points. It can be shown that every neighborhood of a point $\mathbf{x}$ (or $\mathbf{X}$) in $\mathcal{R}^m$ (or $\mathcal{R}^{m\times n}$) is an open set.

b. Functions and continuity

In what follows, the word function is to be used restrictively to mean a scalar-valued function whose domain is a set (of m-dimensional column vectors) in $\mathcal{R}^m$ or more generally a set (of $m \times n$ matrices) in $\mathcal{R}^{m\times n}$. The value of a function f corresponding to an arbitrary vector $\mathbf{x}$ or matrix $\mathbf{X}$ is to be denoted by the symbol $f(\mathbf{x})$ or $f(\mathbf{X})$. Further, if $\mathbf{f} = \{f_s\}$ is a $p\times 1$ vector or, more generally, if $\mathbf{F} = \{f_{st}\}$ is a $p\times q$ matrix, each element of which is a function defined on some set in $\mathcal{R}^{m\times n}$, then, for an arbitrary matrix $\mathbf{X}$, the $p \times 1$ vector $[f_1(\mathbf{X}), \ldots, f_p(\mathbf{X})]'$ or the $p \times q$ matrix with stth element $f_{st}(\mathbf{X})$ is to be denoted by $\mathbf{f}(\mathbf{X})$ or $\mathbf{F}(\mathbf{X})$ and referred to as the value of $\mathbf{f}$ or $\mathbf{F}$ at $\mathbf{X}$.

At times, attention will be restricted to functions whose domains are sets of column vectors. This can be done without loss of generality in the sense that the elements of an $m \times n$ matrix can be rearranged in the form of an mn-dimensional column vector.

Let f represent a function whose domain is a set S in $\mathcal{R}^{m\times 1}$. Then, f is said to be *continuous* at an interior point $\mathbf{c}$ of S if

$$\lim_{\mathbf{x}\to\mathbf{c}} f(\mathbf{x}) = f(\mathbf{c})$$

—by definition, $\lim_{\mathbf{x}\to\mathbf{c}} f(\mathbf{x})$ is a scalar b such that, for every positive scalar ϵ, there exists a neighborhood N_ϵ of $\mathbf{c}$ such that $|f(\mathbf{x}) - b| < \epsilon$ for all $\mathbf{x}$ in N_ϵ other than $\mathbf{c}$. A $p \times q$ matrix $\mathbf{F}$ of functions, each of whose domains is the same set S in $\mathcal{R}^{m\times 1}$, will be said to be *continuous* at an interior point $\mathbf{c}$ of S if all pq elements of $\mathbf{F}$ are continuous at $\mathbf{c}$.

c. First-order partial derivatives and continuous differentiability

Let f represent a function, defined on a set S, of a vector $\mathbf{x} = (x_1, \ldots, x_m)'$ of m variables. Suppose that S contains at least some interior points, and let $\mathbf{c} = (c_1, \ldots, c_m)'$ represent an arbitrary one of those points. Further, let $\mathbf{u}_j$ represent the jth column of $\mathbf{I}_m$.

Consider the limit

$$\lim_{t\to 0} \frac{f(\mathbf{c} + t\mathbf{u}_j) - f(\mathbf{c})}{t}.$$

When this limit exists, it is called the *jth (first-order) partial derivative* of f at $\mathbf{c}$ and is denoted by $D_j f(\mathbf{c})$. Note that $\mathbf{c} + t\mathbf{u}_j = (c_1, \ldots, c_{j-1}, c_j + t, c_{j+1}, \ldots, c_m)'$, so that $D_j f(\mathbf{c})$ can be regarded as the ordinary derivative (at c_j) of the function of one variable obtained from $f(\mathbf{x})$ by fixing $x_1, \ldots, x_{j-1}, x_{j+1}, \ldots, x_m$ at $c_1, \ldots, c_{j-1}, c_{j+1}, \ldots, c_m$, respectively.

The scalar $D_j f(\mathbf{c})$ can be regarded as the value assigned to the point $\mathbf{c}$ by a function. This function is denoted by the symbol $D_j f$ (and is referred to as the jth partial derivative of f). Its domain consists of those interior points (of S) at which the jth partial derivative (of f) is defined.

The symbol $\mathbf{D}f$ represents the row vector $(D_1 f, \ldots, D_m f)$ whose elements are the partial derivatives of f, and accordingly the symbol $\mathbf{D}f(\mathbf{c})$ represents the row vector $[D_1 f(\mathbf{c}), \ldots, D_m f(\mathbf{c})]$ whose elements are the values of the functions $D_1 f, \ldots, D_m f$ at $\mathbf{c}$—note that $\mathbf{D}f(\mathbf{c})$ is defined only if $\mathbf{c}$ is such that all m of the partial derivatives of f at $\mathbf{c}$ exist. The column vector $(\mathbf{D}f)'$ is referred to as the *gradient* (or *gradient vector*) of f.

An alternative notation is obtained by writing $\partial f(\mathbf{x})/\partial x_j$ for the jth partial derivative of f at $\mathbf{x}$, writing $\partial f(\mathbf{x})/\partial \mathbf{x}'$ for the row vector $[\partial f(\mathbf{x})/\partial x_1, \ldots, \partial f(\mathbf{x})/\partial x_m]$ of partial derivatives of f at $\mathbf{x}$, and writing $\partial f(\mathbf{x})/\partial \mathbf{x}$ for the column vector $[\partial f(\mathbf{x})/\partial x_1, \ldots, \partial f(\mathbf{x})/\partial x_m]'$ of partial derivatives of f at $\mathbf{x}$. In this context, $\partial f(\mathbf{x})/\partial \mathbf{x}'$ may be called the derivative of $f(\mathbf{x})$ with respect to $\mathbf{x}'$, and $\partial f(\mathbf{x})/\partial \mathbf{x}$ may be called the derivative of $f(\mathbf{x})$ with respect to $\mathbf{x}$. The symbols $\partial f(\mathbf{x})/\partial x_j$, $\partial f(\mathbf{x})/\partial \mathbf{x}'$, and $\partial f(\mathbf{x})/\partial \mathbf{x}$ have the same interpretations as $D_j f(\mathbf{x})$, $\mathbf{D}f(\mathbf{x})$, and $[\mathbf{D}f(\mathbf{x})]'$, respectively, and are sometimes abbreviated to $\partial f/\partial x_j$, $\partial f/\partial \mathbf{x}'$, and $\partial f/\partial \mathbf{x}$. On certain occasions (to be ascertained from the context), these symbols may be used to represent the function $D_j f$ and the vectors of functions $\mathbf{D}f$ and $(\mathbf{D}f)'$ (rather than their values at $\mathbf{x}$).

The function f (with domain S in $\mathcal{R}^{m\times 1}$) will be said to be *continuously differentiable* at the interior point $\mathbf{c}$ (of S) if $D_1 f(\mathbf{x}), \ldots, D_m f(\mathbf{x})$ exist and are continuous at every point $\mathbf{x}$ in some neighborhood of $\mathbf{c}$. It can be shown that if f is continuously differentiable at $\mathbf{c}$, then

$$\lim_{\mathbf{x}\to\mathbf{c}} \frac{f(\mathbf{x}) - [f(\mathbf{c}) + \mathbf{D}f(\mathbf{c})(\mathbf{x}-\mathbf{c})]}{\|\mathbf{x}-\mathbf{c}\|} = 0, \tag{1.1}$$

which indicates that, for $\mathbf{x}$ sufficiently close to $\mathbf{c}$, the first-order Taylor formula $f(\mathbf{c}) + \mathbf{D}f(\mathbf{c})(\mathbf{x}-\mathbf{c})$ approximates $f(\mathbf{x})$ with an error that is of smaller order than $\|\mathbf{x}-\mathbf{c}\|$ (e.g., Magnus and Neudecker, 1988, chap. 5).

If the function f is such that result (1.1) holds, then $f(\mathbf{x}) \to f(\mathbf{c})$, as is evident upon writing

$$f(\mathbf{x}) = f(\mathbf{c}) + \mathbf{D}f(\mathbf{c})(\mathbf{x}-\mathbf{c}) + \|\mathbf{x}-\mathbf{c}\| \frac{f(\mathbf{x}) - [f(\mathbf{c}) + \mathbf{D}f(\mathbf{c})(\mathbf{x}-\mathbf{c})]}{\|\mathbf{x}-\mathbf{c}\|}.$$

Thus, we have the following lemma.

Lemma 15.1.1. If a function f, with domain S in $\mathcal{R}^{m\times 1}$, is continuously differentiable at an interior point $\mathbf{c}$ of S, then f is continuous at $\mathbf{c}$.

A function f for which result (1.1) holds is said to be *differentiable* at $\mathbf{c}$. It is worth mentioning that, while f being continuously differentiable at $\mathbf{c}$ is a sufficient condition for f to be differentiable at $\mathbf{c}$, it is not in general necessary—as demonstrated by, for example, Magnus and Neudecker (1988, secs. 5.9 and 5.10).

d. Second- and higher-order partial derivatives

As in Subsection c, let f represent a function, defined on a set S, of a vector $\mathbf{x} = (x_1, \ldots x_m)'$ of m variables; suppose that S contains at least some interior points; and let $\mathbf{c}$ represent an arbitrary one of those points.

Suppose that $D_1 f(\mathbf{x}), \ldots, D_m f(\mathbf{x})$ exist for all $\mathbf{x}$ in some neighborhood of $\mathbf{c}$ (so that $\mathbf{c}$ is an interior point of the domains of the functions $D_1 f, \ldots, D_m f$). When the ith (first-order) partial derivative of $D_j f$ at $\mathbf{c}$ exists, it is called the *ijth second-order partial derivative* of f at $\mathbf{c}$ and is denoted by $D_{ij}^2 f(\mathbf{c})$. The scalar $D_{ij}^2 f(\mathbf{c})$ can be regarded as the value assigned to the point $\mathbf{c}$ by a function. This function is denoted by the symbol $D_{ij}^2 f$ (and is referred to as the ijth second-order partial derivative of f).

The symbol $\mathbf{H}f$ represents the $m \times m$ matrix whose ijth element is $D_{ij}^2 f$, and accordingly the symbol $\mathbf{H}f(\mathbf{c})$ represents the value of $\mathbf{H}f$ at the point $\mathbf{c}$, that is, the $m \times m$ matrix whose ijth element is $D_{ij}^2 f(\mathbf{c})$. The matrix $\mathbf{H}f$ is called the *Hessian matrix* of f.

An alternative notation is obtained by writing $\partial^2 f(\mathbf{x})/\partial x_i \partial x_j$—or, in the special case where $j = i$, $\partial^2 f(\mathbf{x})/\partial x_i^2$—for the ijth second-order partial derivative $D_{ij}^2 f(\mathbf{x})$ of $\mathbf{f}$ at $\mathbf{x}$ and writing $\partial^2 f(\mathbf{x})/\partial \mathbf{x} \partial \mathbf{x}'$ for the Hessian matrix $\mathbf{H}f(\mathbf{x})$ of f at $\mathbf{x}$. The symbols $\partial^2 f(\mathbf{x})/\partial x_i \partial x_j$, $\partial^2 f(\mathbf{x})/\partial x_i^2$, and $\partial^2 f(\mathbf{x})/\partial \mathbf{x} \partial \mathbf{x}'$ are sometimes abbreviated to $\partial^2 f/\partial x_i \partial x_j$, $\partial^2 f/\partial x_i^2$, and $\partial^2 f/\partial \mathbf{x} \partial \mathbf{x}'$, respectively, and are sometimes used to represent the functions $D_{ij}^2 f$ and $D_{ii}^2 f$ and the matrix of functions $\mathbf{H}f$ (rather than their values at $\mathbf{x}$).

The function f will be said to be *twice* (or *2 times*) *continuously differentiable* at $\mathbf{c}$ if f and all of its first-order partial derivatives are continuously differentiable at $\mathbf{c}$ or, equivalently, if all of the first- and second-order partial derivatives of f exist and are continuous at every point in some neighborhood of $\mathbf{c}$. It can be shown that if f is twice continuously differentiable at $\mathbf{c}$, then the matrix $\mathbf{H}f$ is symmetric at $\mathbf{c}$ [i.e., $\mathbf{H}f(\mathbf{c})$ is symmetric, or, equivalently, $D_{ji}^2 f(\mathbf{c}) = D_{ij}^2 f(\mathbf{c})$ for all i and $j < i$]—refer to, for example, Magnus and Neudecker (1988, sec. 6.7). It can be further shown that if f is twice continuously differentiable at $\mathbf{c}$, then

$$\lim_{\mathbf{x} \to \mathbf{c}} \frac{f(\mathbf{x}) - [f(\mathbf{c}) + \mathbf{D}f(\mathbf{c})(\mathbf{x} - \mathbf{c}) + (1/2)(\mathbf{x} - \mathbf{c})'\mathbf{H}f(\mathbf{c})(\mathbf{x} - \mathbf{c})]}{\|\mathbf{x} - \mathbf{c}\|^2} = 0, \quad (1.2)$$

which indicates that, for $\mathbf{x}$ sufficiently close to $\mathbf{c}$, the second-order Taylor formula

$$f(\mathbf{c}) + \mathbf{D}f(\mathbf{c})(\mathbf{x} - \mathbf{c}) + (1/2)(\mathbf{x} - \mathbf{c})'\mathbf{H}f(\mathbf{c})(\mathbf{x} - \mathbf{c})$$

approximates $f(\mathbf{x})$ with an error that is of smaller order than $\|\mathbf{x} - \mathbf{c}\|^2$ (e.g., Magnus and Neudecker, 1988, sec. 6.9).

Partial derivatives of f of order k (where $k \geq 2$) can be defined recursively. Suppose that the $(k-1)$th-order partial derivatives of f at $\mathbf{x}$ exist for all $\mathbf{x}$ in some neighborhood of $\mathbf{c}$. For $j = 1, \ldots, k$, let i_j represent an arbitrary integer between 1 and m, inclusive. When the i_1th partial derivative of the $i_2 \cdots i_k$th $(k-1)$th-order partial derivative of f at $\mathbf{c}$ exists, it is called the *$i_1 i_2 \cdots i_k$th kth-order partial derivative* of f at $\mathbf{c}$. The function whose value at $\mathbf{c}$ is the $i_1 i_2 \cdots i_k$th kth-order partial derivative of f at $\mathbf{c}$ is referred to as the $i_1 i_2 \cdots i_k$th kth-order partial derivative of f.

The symbol $\partial^k f(\mathbf{x})/\partial x_{i_1} \cdots \partial x_{i_k}$ (or in abbreviated form $\partial^k f/\partial x_{i_1} \cdots \partial x_{i_k}$) can be used to represent the $i_1 \cdots i_k$th kth-order partial derivative of f at $\mathbf{x}$.

The function f will be said to be *k times continuously differentiable* at $\mathbf{c}$ if f and all of its first- through $(k-1)$th-order partial derivatives are continuously differentiable at $\mathbf{c}$ or, equivalently, if all of the first- through kth-order partial derivatives of f exist and are continuous at every point in some neighborhood of $\mathbf{c}$. It can be shown that if f is k times continuously differentiable at $\mathbf{c}$, then, for any permutation $j_1, \ldots, j_k$ of the sequence $i_1, \ldots, i_k$, the $j_1 \cdots j_k$th and $i_1 \cdots i_k$th kth-order partial derivatives of f are identical, and, letting $i_1^*, \ldots, i_s^*$ ($i_1^* < \cdots < i_s^*$) denote the distinct integers represented among $i_1, \ldots, i_k$ and letting k_j represent the number of integers in the sequence $i_1, \ldots, i_k$ that equal i_j^*, we may write $\partial^k f(\mathbf{x})/\partial x_{i_1^*}^{k_1} \cdots \partial x_{i_s^*}^{k_s}$ (or simply $\partial^k f/\partial x_{i_1^*}^{k_1} \cdots \partial x_{i_s^*}^{k_s}$) for $\partial^k f(\mathbf{x})/\partial x_{i_1} \cdots \partial x_{i_k}$. It can be further shown that if f is k times continuously differentiable at $\mathbf{c}$, then, for $\mathbf{x}$ sufficiently close to $\mathbf{c}$, the kth-order Taylor formula approximates $f(\mathbf{x})$ with an error that is of smaller order than $\|\mathbf{x} - \mathbf{c}\|^k$.

Note that if f is k times continuously differentiable at $\mathbf{c}$, then f is continuously differentiable at $\mathbf{c}$ and is also $2, \ldots, k-1$ times continuously differentiable at $\mathbf{c}$.

Moreover, if f is k times continuously differentiable at $\mathbf{c}$, then f is k times continuously differentiable at every point in some neighborhood of $\mathbf{c}$, as can be easily verified.

e. Partial derivatives of a function of an unrestricted or symmetric matrix

Suppose that the domain of the function f to be differentiated is the set $\mathcal{R}^{m \times n}$ of all $m \times n$ matrices or, more generally, is a set S in $\mathcal{R}^{m \times n}$ that contains at least some interior points. Then, f can be regarded as a function of an $m \times n$ matrix $\mathbf{X} = \{x_{ij}\}$ of mn "independent" variables. And, for purposes of differentiating f, the elements of $\mathbf{X}$ can be rearranged in the form of an mn-dimensional column vector $\mathbf{x}$, and f can be reinterpreted as a function of $\mathbf{x}$, in which case the domain of f is the set, say S^*, obtained by rearranging the elements of each $m \times n$ matrix in S in the form of a column vector. (It should be noted that an mn-dimensional column vector is an interior point of S^* if and only if it is a rearrangement of an $m \times n$ matrix that is an interior point of S.)

By definition, the elements $\partial f / \partial x_{ij}$ $(i = 1, \ldots, m; j = 1, \ldots, n)$ of the mn-dimensional column vector $\partial f / \partial \mathbf{x}$ are the first-order partial derivatives of f at $\mathbf{x}$ (and the elements of the $mn \times mn$ matrix $\partial^2 f / \partial \mathbf{x} \partial \mathbf{x}'$ are the second-order partial derivatives of f at $\mathbf{x}$). However, instead of presenting the first-order partial derivatives of f at $\mathbf{x}$ (or, equivalently, $\mathbf{X}$) in the form of the vector $\partial f / \partial \mathbf{x}$, it is natural and in many cases convenient to present them in the form of the $m \times n$ matrix whose ijth element is $\partial f / \partial x_{ij}$. This matrix is to be denoted by the symbol $\partial f(\mathbf{X}) / \partial \mathbf{X}$ (or in abbreviated form $\partial f / \partial \mathbf{X}$) and is to be called the derivative of $f(\mathbf{X})$ with respect to $\mathbf{X}$. Further, let us write $\partial f(\mathbf{X}) / \partial \mathbf{X}'$ (or $\partial f / \partial \mathbf{X}'$) for the $n \times m$ matrix $[\partial f(\mathbf{X}) / \partial \mathbf{X}]'$ and refer to this matrix as the derivative of $f(\mathbf{X})$ with respect to $\mathbf{X}'$.

Suppose now that the function f of interest is one whose domain is restricted to all $m \times m$ symmetric matrices or, more generally, to a subset S of such matrices. Then, f can still be regarded as a function of an $m \times m$ matrix $\mathbf{X} = \{x_{ij}\}$ of m^2 variables. However, because of the restriction to symmetric matrices, $x_{ji} = x_{ij}$ for all i and $j > i$, so that the m^2 elements of $\mathbf{X}$ can no longer be regarded as m^2 independent variables. A closely related point is that S cannot contain any interior points. As a consequence, the previous development—that leading to the introduction of the symbol $\partial f(\mathbf{X}) / \partial \mathbf{X}$ and to the definition of the derivative of $f(\mathbf{X})$ with respect to $\mathbf{X}$—is not applicable. In the present case—that where S contains only symmetric matrices—the symbol $\partial f(\mathbf{X}) / \partial \mathbf{X}$ and the term "derivative of $f(\mathbf{X})$ with respect to $\mathbf{X}$" are to be defined differently. (When not explicitly indicated, the intended interpretation must be ascertained from the context.)

For purposes of differentiating the function f of the symmetric matrix $\mathbf{X}$, f is to be interpreted as a function of an $[m(m+1)/2]$-dimensional column vector $\mathbf{x}$ whose elements are x_{ij} $(j \leq i = 1, \ldots, m)$ or, alternatively, x_{ij} $(j \geq i = 1, \ldots, m)$ [i.e., whose elements are the "independent" elements of $\mathbf{X}$, which are those on and below (or alternatively above) the diagonal]. When f is reinterpreted in this way, the domain of f is the set S^* of $[m(m+1)/2]$-dimensional column vectors obtained by transforming the $m \times m$ matrices in S from $\mathbf{X}$-values into $\mathbf{x}$-values.

Suppose that S^* contains at least some interior points. (The set S^* can contain interior points, even though S cannot.) By definition, the elements $\partial f/\partial x_{ij}$ ($j \leq i = 1, \ldots, m$ or, alternatively, $j \geq i = 1, \ldots, m$) of the $[m(m+1)/2]$-dimensional column vector $\partial f/\partial \mathbf{x}$ are the first-order partial derivatives of f at $\mathbf{x}$ {and the elements of the $[m(m+1)/2] \times [m(m+1)/2]$ matrix $\partial f/\partial \mathbf{x}\partial \mathbf{x}'$ are the second-order partial derivatives of f at $\mathbf{x}$}. However, instead of presenting the first-order partial derivatives of f at $\mathbf{x}$ (or, equivalently, $\mathbf{X}$) in the form of the vector $\partial f/\partial \mathbf{x}$, it is natural, and in many cases convenient, to present them in the form of the $m \times m$ symmetric matrix whose ijth and jith elements are $\partial f/\partial x_{ij}$. In the present case (that where the domain S of f is restricted to symmetric matrices), it is this matrix that is denoted by the symbol $\partial f(\mathbf{X})/\partial \mathbf{X}$ (or $\partial f/\partial \mathbf{X}$) and is called the derivative of f with respect to $\mathbf{X}$.

f. Differentiation of a vector or matrix of functions

Suppose that there is a (column) vector $\mathbf{f} = (f_1, \ldots, f_p)'$ of p functions to be differentiated and that the domain of all of these functions is a set S in $\mathcal{R}^{m \times 1}$. Suppose further that S contains at least some interior points, and let $\mathbf{c}$ represent an arbitrary one of those points.

The symbol $D_j\mathbf{f}$ is used to represent the p-dimensional column vector whose sth element is the jth partial derivative $D_j f_s$ of f_s, and the symbol $\mathbf{Df}$ is used to represent the $p \times m$ matrix whose sjth element is $D_j f_s$ or, equivalently, the $p \times m$ matrix whose jth column is $D_j\mathbf{f}$. Accordingly, $D_j\mathbf{f}(\mathbf{c}) = [D_j f_1(\mathbf{c}), \ldots, D_j f_p(\mathbf{c})]'$, and $\mathbf{Df}(\mathbf{c}) = [D_1\mathbf{f}(\mathbf{c}), \ldots, D_m\mathbf{f}(\mathbf{c})]$ (provided that each of the m partial derivatives of each of the p functions $f_1, \ldots, f_p$ at $\mathbf{c}$ exists). The matrix $\mathbf{Df}$ is called the *Jacobian matrix* of $\mathbf{f}$, and its transpose $(\mathbf{Df})'$ is called the *gradient* (or *gradient matrix*) of $\mathbf{f}$. Note that $(\mathbf{Df})' = [(\mathbf{D}f_1)', \ldots, (\mathbf{D}f_p)']$. In the special case where $p = m$, the determinant of $\mathbf{Df}$ is called the *Jacobian* (or *Jacobian determinant*) of $\mathbf{f}$.

An alternative notation is obtained by letting $\mathbf{x} = (x_1, \ldots, x_m)'$ represent a vector of m variables and writing $\partial \mathbf{f}(\mathbf{x})/\partial \mathbf{x}'$ (or $\partial \mathbf{f}/\partial \mathbf{x}'$) for the $p \times m$ matrix whose sjth element is $\partial f_s(\mathbf{x})/\partial x_j$ and $\partial \mathbf{f}'(\mathbf{x})/\partial \mathbf{x}$ (or $\partial \mathbf{f}'/\partial \mathbf{x}$) for the $m \times p$ matrix $[\partial \mathbf{f}(\mathbf{x})/\partial \mathbf{x}']'$ whose jsth element is $\partial f_s(\mathbf{x})/\partial x_j$. In this context, $\partial \mathbf{f}(\mathbf{x})/\partial \mathbf{x}'$ may be called the derivative of $\mathbf{f}(\mathbf{x})$ with respect to $\mathbf{x}'$, and $\partial \mathbf{f}'(\mathbf{x})/\partial \mathbf{x}$ may be called the derivative of $\mathbf{f}'(\mathbf{x})$ with respect to $\mathbf{x}$. The symbol $\partial \mathbf{f}(\mathbf{x})/\partial \mathbf{x}'$ has the same interpretation as $\mathbf{Df}(\mathbf{x})$ (or alternatively $\mathbf{Df}$).

Suppose now that there is a $p \times q$ matrix $\mathbf{F} = \{f_{st}\}$ of pq functions to be differentiated and that the domain of all of these functions is a set S in $\mathcal{R}^{m \times 1}$ (that contains at least some interior points). As in the special case where $q = 1$, each of the elements of $\mathbf{F}$ can be regarded as a function of a vector $\mathbf{x} = (x_1, \ldots, x_m)'$ of m "independent" variables.

All mpq (first-order) partial derivatives (of the elements of $\mathbf{F}$) can be presented in the form of a single $pq \times m$ (or $m \times pq$) matrix by rearranging the elements of

$\mathbf{F}$ in the form of a column vector $\mathbf{f}$ and by then forming the Jacobian (or gradient) matrix of $\mathbf{f}$. However, with the possible exception of the special case where $p = 1$ or $q = 1$ (i.e., where $\mathbf{F}$ is a row or column vector), it is sometimes preferable to present the jth (first-order) partial derivatives of the elements of $\mathbf{F}$ in a separate $p \times q$ matrix whose stth element is $\partial f_{st}(\mathbf{x})/\partial x_j$ ($j = 1, \ldots, m$). This matrix is to be denoted by the symbol $\partial \mathbf{F}(\mathbf{x})/\partial x_j$ (or, in abbrevated form, $\partial \mathbf{F}/\partial x_j$) and is to be referred to as the jth partial derivative of $\mathbf{F}$—in the special case where $q = 1$ (i.e., where $\mathbf{F}$ is a column vector), $\partial \mathbf{F}(\mathbf{x})/\partial x_j$ has the same interpretation as $D_j\mathbf{F}(\mathbf{x})$. And the $p \times q$ matrix whose stth element is $\partial^k f_{st}(\mathbf{x})/\partial x_{j_1}^{k_1} \cdots \partial x_{j_r}^{k_r}$ is to be denoted by the symbol $\partial^k \mathbf{F}(\mathbf{x})/\partial x_{j_1}^{k_1} \cdots \partial x_{j_r}^{k_r}$ or $\partial^k \mathbf{F}/\partial x_{j_1}^{k_1} \cdots x_{j_r}^{k_r}$ (where $j_1, \ldots, j_r$ is a subsequence of the first m positive integers; $k_1, \ldots, k_r$ are positive integers; and $k = k_1 + \cdots + k_r$).

The matrix $\mathbf{F}$ will be said to be *continuously differentiable* at an interior point $\mathbf{c}$ (of S) if all pq of its elements are continuously differentiable at $\mathbf{c}$, and to be *twice* (or *2 times*) *continuously differentiable* at $\mathbf{c}$ if all pq of its elements are twice continuously differentiable at $\mathbf{c}$. More generally, $\mathbf{F}$ is said to be k *times continuously differentiable* at $\mathbf{c}$ if all pq of its elements are k times continuously differentiable at $\mathbf{c}$.

15.2 Differentiation of (Scalar-Valued) Functions: Some Elementary Results

The techniques employed in matrix differentiation can be regarded as generalizations or extensions of those employed in nonmatrix differentiation. Some elementary results on nonmatrix differentiation are stated (without proof) in the following two lemmas.

Lemma 15.2.1. Let f represent a function, defined on a set S, of a vector $\mathbf{x} = (x_1, \ldots, x_m)'$ of m variables, and suppose that (for $\mathbf{x} \in S$) $f(\mathbf{x})$ is constant or (more generally) does not vary with x_j. Then, for any interior point $\mathbf{c}$ of S, $D_j f(\mathbf{c}) = 0$.

Lemma 15.2.2. Let f and g represent functions, defined on a set S, of a vector $\mathbf{x} = (x_1, \ldots, x_m)'$ of m variables. And let a and b represent constants or (more generally) functions (defined on S) that are continuous at every interior point of S and are such that $a(\mathbf{x})$ and $b(\mathbf{x})$ do not vary with x_j. Define

$$\ell = af + bg, \quad h = fg, \quad \text{and} \quad r = f/g,$$

so that ℓ and h are functions, each of whose domain is S, and r is a function whose domain is $S^* = \{\mathbf{x} \in S : g(\mathbf{x}) \neq 0\}$. If f and g are continuously differentiable at an interior point $\mathbf{c}$ of S, then ℓ and h are also continuously differentiable at $\mathbf{c}$, and

$$D_j\ell(\mathbf{c}) = a(\mathbf{c})D_j f(\mathbf{c}) + b(\mathbf{c})D_j g(\mathbf{c}) \tag{2.1}$$

and

$$D_j h(\mathbf{c}) = f(\mathbf{c}) D_j g(\mathbf{c}) + g(\mathbf{c}) D_j f(\mathbf{c}). \tag{2.2}$$

And if f and g are continuously differentiable at an interior point $\mathbf{c}$ of S^*, then r is also continuously differentiable at $\mathbf{c}$, and

$$D_j r(\mathbf{c}) = [g(\mathbf{c}) D_j f(\mathbf{c}) - f(\mathbf{c}) D_j g(\mathbf{c})]/[g(\mathbf{c})]^2. \tag{2.3}$$

Formulas (2.1)–(2.3) can be expressed less formally and more succinctly as

$$\frac{\partial(af + bg)}{\partial x_j} = a\frac{\partial f}{\partial x_j} + b\frac{\partial g}{\partial x_j}, \tag{2.4}$$

$$\frac{\partial fg}{\partial x_j} = f\frac{\partial g}{\partial x_j} + g\frac{\partial f}{\partial x_j}, \tag{2.5}$$

and

$$\frac{\partial(f/g)}{\partial x_j} = \left[g\frac{\partial f}{\partial x_j} - f\frac{\partial g}{\partial x_j}\right]/g^2. \tag{2.6}$$

Special cases of formula (2.4) include:

$$\frac{\partial(af)}{\partial x_j} = a\frac{\partial f}{\partial x_j}, \quad \frac{\partial(f+g)}{\partial x_j} = \frac{\partial f}{\partial x_j} + \frac{\partial g}{\partial x_j}, \quad \frac{\partial(f-g)}{\partial x_j} = \frac{\partial f}{\partial x_j} - \frac{\partial g}{\partial x_j}. \tag{2.7}$$

And, as a special case of formula (2.6), we have (in light of Lemma 15.2.1) that

$$\frac{\partial(1/g)}{\partial x_j} = -(1/g^2)\frac{\partial g}{\partial x_j}. \tag{2.8}$$

Results (2.1) and (2.2) can be extended (by repeated application) to a linear combination or a product of an arbitrary number of functions. Let $f_1, f_2, \ldots, f_k$ represent k functions, defined on a set S, of a vector $\mathbf{x} = (x_1, \ldots, x_m)'$ of m variables, and let $a_1, a_2, \ldots, a_k$ represent constants or (more generally) functions (defined on S) that are continuous at every interior point of S and are such that $a_1(\mathbf{x}), a_2(\mathbf{x}), \ldots, a_k(\mathbf{x})$ do not vary with x_j. Define

$$\ell = a_1 f_1 + a_2 f_2 + \cdots + a_k f_k \quad \text{and} \quad h = f_1 f_2 \cdots f_k.$$

If $f_1, f_2, \ldots, f_k$ are continuously differentiable at an interior point $\mathbf{c}$ of S, then ℓ and h are also continuously differentiable at $\mathbf{c}$, and

$$D_j \ell(\mathbf{c}) = a_1(\mathbf{c}) D_j f_1(\mathbf{c}) + a_2(\mathbf{c}) D_j f_2(\mathbf{c}) + \cdots + a_k(\mathbf{c}) D_j f_k(\mathbf{c}) \tag{2.9}$$

and

$$D_j h(\mathbf{c}) = \sum_{i=1}^{k} \Big[\prod_{s \neq i} f_s(\mathbf{c})\Big] D_j f_i(\mathbf{c}). \tag{2.10}$$

Formulas (2.9) and (2.10) can be rewritten as

$$\frac{\partial(a_1 f_1 + a_2 f_2 + \cdots a_k f_k)}{\partial x_j} = a_1\frac{\partial f_1}{\partial x_j} + a_2\frac{\partial f_2}{\partial x_j} + \cdots + a_k\frac{\partial f_k}{\partial x_j} \tag{2.11}$$

and

$$\frac{\partial(f_1 f_2 \cdots f_k)}{\partial x_j} = \sum_{i=1}^{k} \Bigl(\prod_{s \neq i} f_s\Bigr) \frac{\partial f_i}{\partial x_j}. \tag{2.12}$$

Note that formulas (2.6) and (2.12) for the partial derivatives of a ratio of two functions or for a product of two or more functions can be recast in vector notation as

$$\frac{\partial(f/g)}{\partial \mathbf{x}} = g^{-2}\Bigl[g\frac{\partial f}{\partial \mathbf{x}} - f\frac{\partial g}{\partial \mathbf{x}}\Bigr] \tag{2.13}$$

and

$$\frac{\partial(f_1 f_2 \cdots f_k)}{\partial \mathbf{x}} = \sum_{i=1}^{k} \Bigl(\prod_{s \neq i} f_s\Bigr) \frac{\partial f_i}{\partial \mathbf{x}}. \tag{2.14}$$

Similarly, formula (2.11) can (in the special case where $a_1, a_2, \ldots, a_k$ are constants) be recast as

$$\frac{\partial(a_1 f_1 + a_2 f_2 + \cdots + a_k f_k)}{\partial \mathbf{x}} = a_1\frac{\partial f_1}{\partial \mathbf{x}} + a_2\frac{\partial f_2}{\partial \mathbf{x}} + \cdots + a_k\frac{\partial f_k}{\partial \mathbf{x}}. \tag{2.15}$$

Note also that by setting each of the k functions $f_1, \ldots, f_k$ in formula (2.12) equal to the same function, say f, we obtain the result that (for any positive integer k)

$$\frac{\partial f^k}{\partial x_j} = k f^{k-1}\frac{\partial f}{\partial x_j}. \tag{2.16}$$

Since clearly

$$\frac{\partial x_j}{\partial x_j} = 1,$$

we have in particular [upon taking $f(\mathbf{x}) = x_j$] that

$$\frac{\partial x_j^k}{\partial x_j} = k x_j^{k-1}. \tag{2.17}$$

15.3 Differentiation of Linear and Quadratic Forms

The results of Section 15.2 can be used to derive formulas for the partial derivatives of a linear form $\mathbf{a}'\mathbf{x}$ or a quadratic form $\mathbf{x}'\mathbf{A}\mathbf{x}$ (in an unconstrained m-dimensional column vector $\mathbf{x}$). Let a_i and x_i represent the ith elements of $\mathbf{a}$ and $\mathbf{x}$, respectively, and let a_{ik} represent the ikth element of $\mathbf{A}$. Then,

$$\mathbf{a}'\mathbf{x} = \sum_i a_i x_i \quad \text{and} \quad \mathbf{x}'\mathbf{A}\mathbf{x} = \sum_{i,k} a_{ik} x_i x_k.$$

In light of Lemma 15.2.1 and results (2.17) and (2.7), we have that

$$\frac{\partial x_i}{\partial x_j} = \begin{cases} 1, & \text{if } i = j, \quad (3.1a) \\ 0, & \text{if } i \neq j, \quad (3.1b) \end{cases}$$

and that

$$\frac{\partial(x_i x_k)}{\partial x_j} = \begin{cases} 2x_j, & \text{if } i = k = j, \quad (3.2a) \\ x_i, & \text{if } k = j \text{ but } i \neq j, \quad (3.2b) \\ x_k, & \text{if } i = j \text{ but } k \neq j, \quad (3.2c) \\ 0, & \text{otherwise (i.e., if } i \neq j \text{ and } k \neq j). \quad (3.2d) \end{cases}$$

Using these results in combination with result (2.11), we find that

$$\frac{\partial(\mathbf{a}'\mathbf{x})}{\partial x_j} = \frac{\partial(\sum_i a_i x_i)}{\partial x_j} = \sum_i a_i \frac{\partial x_i}{\partial x_j} = a_j \tag{3.3}$$

and

$$\begin{aligned} \frac{\partial(\mathbf{x}'\mathbf{A}\mathbf{x})}{\partial x_j} & \\ &= \frac{\partial(\sum_{i,k} a_{ik} x_i x_k)}{\partial x_j} \\ &= \frac{\partial\left(a_{jj}x_j^2 + \sum_{i \neq j} a_{ij} x_i x_j + \sum_{k \neq j} a_{jk} x_j x_k + \sum_{i \neq j, k \neq j} a_{ik} x_i x_k\right)}{\partial x_j} \\ &= a_{jj} \frac{\partial x_j^2}{\partial x_j} + \sum_{i \neq j} a_{ij} \frac{\partial(x_i x_j)}{\partial x_j} + \sum_{k \neq j} a_{jk} \frac{\partial(x_j x_k)}{\partial x_j} \\ &\qquad + \sum_{i \neq j, k \neq j} a_{ik} \frac{\partial(x_i x_k)}{\partial x_j} \\ &= 2a_{jj} x_j + \sum_{i \neq j} a_{ij} x_i + \sum_{k \neq j} a_{jk} x_k + 0 \\ &= \sum_i a_{ij} x_i + \sum_k a_{jk} x_k. \end{aligned} \tag{3.4}$$

Result (3.3) can be recast in vector notation as

$$\frac{\partial(\mathbf{a}'\mathbf{x})}{\partial \mathbf{x}} = \mathbf{a} \tag{3.5}$$

or alternatively as

$$\frac{\partial(\mathbf{a}'\mathbf{x})}{\partial \mathbf{x}'} = \mathbf{a}'. \tag{3.6}$$

And result (3.4) can be recast in matrix notation as

$$\frac{\partial(\mathbf{x}'\mathbf{A}\mathbf{x})}{\partial \mathbf{x}} = (\mathbf{A} + \mathbf{A}')\mathbf{x}, \tag{3.7}$$

as is evident upon observing that $\sum_k a_{jk} x_k$ is the jth element of the column vector $\mathbf{Ax}$ and $\sum_i a_{ij} x_i$ is the jth element of $\mathbf{A}'\mathbf{x}$.

The sjth second-order partial derivative of $\mathbf{x}'\mathbf{A}\mathbf{x}$ can be obtained by finding the sth partial derivative of expression (3.4), giving

$$\frac{\partial^2(\mathbf{x}'\mathbf{A}\mathbf{x})}{\partial x_s \partial x_j} = \sum_i a_{ij}\frac{\partial x_i}{\partial x_s} + \sum_k a_{jk}\frac{\partial x_k}{\partial x_s} = a_{sj} + a_{js}. \tag{3.8}$$

Result (3.8) can be recast in matrix notation as

$$\frac{\partial^2(\mathbf{x}'\mathbf{A}\mathbf{x})}{\partial \mathbf{x}\partial \mathbf{x}'} = \mathbf{A} + \mathbf{A}'. \tag{3.9}$$

Note that all partial derivatives of $\mathbf{a}'\mathbf{x}$ of order higher than one are zero, and that all partial derivatives of $\mathbf{x}'\mathbf{A}\mathbf{x}$ of order higher than two are zero. Note also that, in the special case where $\mathbf{A}$ is symmetric, results (3.7) and (3.9) simplify to

$$\frac{\partial(\mathbf{x}'\mathbf{A}\mathbf{x})}{\partial \mathbf{x}} = 2\mathbf{A}\mathbf{x}, \qquad \frac{\partial^2(\mathbf{x}'\mathbf{A}\mathbf{x})}{\partial \mathbf{x}\partial \mathbf{x}'} = 2\mathbf{A}.$$

Formulas (3.5) and (3.6) for the derivative of the linear form $\mathbf{a}'\mathbf{x}$ with respect to $\mathbf{x}$ or $\mathbf{x}'$ can be extended to a vector of linear forms. Let $\mathbf{B} = \{b_{ij}\}$ represent a $p \times m$ matrix of constants, and consider the column vector $\mathbf{B}\mathbf{x}$. The ith element of $\mathbf{B}\mathbf{x}$ is the linear form $\mathbf{b}_i'\mathbf{x}$, whose coefficient vector is $\mathbf{b}_i' = (b_{i1}, \ldots, b_{im})$. According to result (3.3), the jth partial derivative of this linear form is b_{ij}. Thus, the partial derivative of $\mathbf{B}\mathbf{x}$ with respect to $\mathbf{x}'$ (which is the Jacobian matrix of $\mathbf{B}\mathbf{x}$) is

$$\frac{\partial(\mathbf{B}\mathbf{x})}{\partial \mathbf{x}'} = \mathbf{B}, \tag{3.10}$$

and the partial derivative of $(\mathbf{B}\mathbf{x})'$ with respect to $\mathbf{x}$ (which is the gradient matrix of $\mathbf{B}\mathbf{x}$) is

$$\frac{\partial(\mathbf{B}\mathbf{x})'}{\partial \mathbf{x}} = \mathbf{B}'. \tag{3.11}$$

Note that, in the special case where $\mathbf{B} = \mathbf{I}_m$, results (3.10) and (3.11) reduce to

$$\frac{\partial \mathbf{x}}{\partial \mathbf{x}'} = \frac{\partial \mathbf{x}'}{\partial \mathbf{x}} = \mathbf{I}_m. \tag{3.12}$$

15.4 Differentiation of Matrix Sums, Products, and Transposes (and of Matrices of Constants)

The results presented in Section 15.2 (on the differentiation of functions) can be extended to the differentiation of matrices of functions.

The following lemma generalizes (and follows from) Lemma 15.2.1.

Lemma 15.4.1. Let $\mathbf{F} = \{f_{is}\}$ represent a $p \times q$ matrix of functions, defined on a set S, of a vector $\mathbf{x} = (x_1, \ldots, x_m)'$ of m variables; and suppose that (for $\mathbf{x} \in S$) $\mathbf{F}(\mathbf{x})$ is constant or (more generally) does not vary with x_j. Then, at any interior point of S, $\partial\mathbf{F}/\partial x_j = \mathbf{0}$.

A generalization of formula (2.4) is given by the following lemma.

Lemma 15.4.2. Let $\mathbf{F} = \{f_{is}\}$ and $\mathbf{G} = \{g_{is}\}$ represent $p \times q$ matrices of functions, defined on a set S, of a vector $\mathbf{x} = (x_1, \ldots, x_m)'$ of m variables. And let a and b represent constants or (more generally) functions (defined on S) that are continuous at every interior point of S and are such that $a(\mathbf{x})$ and $b(\mathbf{x})$ do not vary with x_j. Then, at any interior point $\mathbf{c}$ (of S) at which $\mathbf{F}$ and $\mathbf{G}$ are continuously differentiable, $a\mathbf{F} + b\mathbf{G}$ is continuously differentiable and

$$\frac{\partial(a\mathbf{F} + b\mathbf{G})}{\partial x_j} = a\frac{\partial \mathbf{F}}{\partial x_j} + b\frac{\partial \mathbf{G}}{\partial x_j}. \tag{4.1}$$

Proof. Let $\mathbf{L} = a\mathbf{F} + b\mathbf{G}$. The isth element of $\mathbf{L}$ is

$$\ell_{is} = af_{is} + bg_{is}.$$

The functions f_{is} and g_{is} are continuously differentiable at $\mathbf{c}$, implying (in light of Lemma 15.2.2) that ℓ_{is} is continuously differentiable at $\mathbf{c}$ and that (at $\mathbf{x} = \mathbf{c}$)

$$\frac{\partial \ell_{is}}{\partial x_j} = a\frac{\partial f_{is}}{\partial x_j} + b\frac{\partial g_{is}}{\partial x_j}.$$

It follows that $\mathbf{L}$ is continuously differentiable at $\mathbf{c}$, and (since $\partial \ell_{is}/\partial x_j$, $\partial f_{is}/\partial x_j$, and $\partial g_{is}/\partial x_j$ are the isth elements of $\partial \mathbf{L}/\partial x_j$, $\partial \mathbf{F}/\partial x_j$, and $\partial \mathbf{G}/\partial x_j$, respectively) that (at $\mathbf{x} = \mathbf{c}$)

$$\frac{\partial \mathbf{L}}{\partial x_j} = a\frac{\partial \mathbf{F}}{\partial x_j} + b\frac{\partial \mathbf{G}}{\partial x_j}.$$

Q.E.D.

We obtain, as special cases of formula (4.1), the following generalization of result (2.7):

$$\frac{\partial(a\mathbf{F})}{\partial x_j} = a\frac{\partial \mathbf{F}}{\partial x_j}, \quad \frac{\partial(\mathbf{F} + \mathbf{G})}{\partial x_j} = \frac{\partial \mathbf{F}}{\partial x_j} + \frac{\partial \mathbf{G}}{\partial x_j}, \quad \frac{\partial(\mathbf{F} - \mathbf{G})}{\partial x_j} = \frac{\partial \mathbf{F}}{\partial x_j} - \frac{\partial \mathbf{G}}{\partial x_j}. \tag{4.2}$$

Formula (2.5) is generalized in the following lemma.

Lemma 15.4.3. Let $\mathbf{F} = \{f_{is}\}$ and $\mathbf{G} = \{g_{is}\}$ represent $p \times q$ and $q \times r$ matrices of functions, defined on a set S, of a vector $\mathbf{x} = (x_1, \ldots, x_m)'$ of m variables. Then, at any interior point $\mathbf{c}$ (of S) at which $\mathbf{F}$ and $\mathbf{G}$ are continuously differentiable, $\mathbf{FG}$ is continuously differentiable and

$$\frac{\partial \mathbf{FG}}{\partial x_j} = \mathbf{F}\frac{\partial \mathbf{G}}{\partial x_j} + \frac{\partial \mathbf{F}}{\partial x_j}\mathbf{G}. \tag{4.3}$$

Proof. Let $\mathbf{H} = \mathbf{FG}$. The itth element of $\mathbf{H}$ is

$$h_{it} = \sum_{s=1}^{q} f_{is} g_{st}.$$

The functions f_{is} and g_{st} are continuously differentiable at $\mathbf{c}$, implying (in light of Lemma 15.2.2) that $f_{is}g_{st}$ is continuously differentiable at $\mathbf{c}$ and that (at $\mathbf{x} = \mathbf{c}$)

$$\frac{\partial f_{is}g_{st}}{\partial x_j} = f_{is}\frac{\partial g_{st}}{\partial x_j} + \frac{\partial f_{is}}{\partial x_j}g_{st}.$$

Thus, h_{it} is continuously differentiable at $\mathbf{c}$, and (at $\mathbf{x} = \mathbf{c}$)

$$\frac{\partial h_{it}}{\partial x_j} = \sum_{s=1}^{q} \frac{\partial(f_{is}g_{st})}{\partial x_j} = \sum_{s=1}^{q} f_{is}\frac{\partial g_{st}}{\partial x_j} + \sum_{s=1}^{q} \frac{\partial f_{is}}{\partial x_j}g_{st}.$$

We conclude that $\mathbf{H}$ is continuously differentiable at $\mathbf{c}$ and [since $\sum_{s=1}^{q} f_{is}(\partial g_{st}/\partial x_j)$ and $\sum_{s=1}^{q}(\partial f_{is}/\partial x_j)g_{st}$ are the itth elements of $\mathbf{F}(\partial \mathbf{G}/\partial x_j)$ and $(\partial \mathbf{F}/\partial x_j)\mathbf{G}$, respectively] that (at $\mathbf{x} = \mathbf{c}$)

$$\frac{\partial \mathbf{FG}}{\partial x_j} = \mathbf{F}\frac{\partial \mathbf{G}}{\partial x_j} + \frac{\partial \mathbf{F}}{\partial x_j}\mathbf{G}.$$

Q.E.D.

In the special case where (for $\mathbf{x} \in S$) $\mathbf{F}(\mathbf{x})$ is constant or (more generally) does not vary with x_j, formula (4.3) simplifies to

$$\frac{\partial \mathbf{FG}}{\partial x_j} = \mathbf{F}\frac{\partial \mathbf{G}}{\partial x_j}. \tag{4.4}$$

And, in the special case where (for $\mathbf{x} \in S$) $\mathbf{G}(\mathbf{x})$ is constant or (more generally) does not vary with x_j, formula (4.3) simplifies to

$$\frac{\partial (\mathbf{FG})}{\partial x_j} = \frac{\partial \mathbf{F}}{\partial x_j}\mathbf{G}. \tag{4.5}$$

The results of Lemma 15.4.3 can be extended (by repeated application) to the product of three or more matrices. Let $\mathbf{F}$, $\mathbf{G}$, and $\mathbf{H}$ represent $p \times q$, $q \times r$, and $r \times v$ matrices of functions, defined on a set S, of a vector $\mathbf{x} = (x_1, \ldots, x_m)'$ of m variables. Then, at any interior point (of S) at which $\mathbf{F}$, $\mathbf{G}$, and $\mathbf{H}$ are continuously differentiable, $\mathbf{FGH}$ is continuously differentiable and

$$\frac{\partial \mathbf{FGH}}{\partial x_j} = \mathbf{FG}\frac{\partial \mathbf{H}}{\partial x_j} + \mathbf{F}\frac{\partial \mathbf{G}}{\partial x_j}\mathbf{H} + \frac{\partial \mathbf{F}}{\partial x_j}\mathbf{GH}. \tag{4.6}$$

In the special case where (for $\mathbf{x} \in S$) $\mathbf{F}(\mathbf{x})$ and $\mathbf{H}(\mathbf{x})$ are constant or (more generally) do not vary with x_j, formula (4.6) simplifies to

$$\frac{\partial (\mathbf{FGH})}{\partial x_j} = \mathbf{F}\frac{\partial \mathbf{G}}{\partial x_j}\mathbf{H}. \tag{4.7}$$

More generally, let $\mathbf{F}_1, \mathbf{F}_2, \ldots, \mathbf{F}_k$ represent matrices of functions, defined on a set S, of a vector $\mathbf{x} = (x_1, \ldots, x_m)'$ of m variables. Then, at any interior point (of S) at which $\mathbf{F}_1, \mathbf{F}_2, \ldots, \mathbf{F}_k$ are continuously differentiable, $\mathbf{F}_1\mathbf{F}_2\cdots\mathbf{F}_k$ is continuously differentiable and

$$\frac{\partial(\mathbf{F}_1\mathbf{F}_2\cdots\mathbf{F}_k)}{\partial x_j} = \mathbf{F}_1\cdots\mathbf{F}_{k-1}\frac{\partial \mathbf{F}_k}{\partial x_j} + \mathbf{F}_1\cdots\mathbf{F}_{k-2}\frac{\partial \mathbf{F}_{k-1}}{\partial x_j}\mathbf{F}_k + \cdots + \frac{\partial \mathbf{F}_1}{\partial x_j}\mathbf{F}_2\cdots\mathbf{F}_k. \quad (4.8)$$

(assuming that the dimensions of $\mathbf{F}_1, \mathbf{F}_2, \ldots, \mathbf{F}_k$ are such that the product $\mathbf{F}_1\mathbf{F}_2\cdots\mathbf{F}_k$ is defined).

Note that if g is a function, defined on a set S, of a vector $\mathbf{x} = (x_1, \ldots, x_m)'$ of m variables and if $\mathbf{F}$ is a $p \times q$ matrix of functions (defined on S) of $\mathbf{x}$, then, at any interior point (of S) at which g and $\mathbf{F}$ are continuously differentiable, $g\mathbf{F}$ is continuously differentiable and

$$\frac{\partial(g\mathbf{F})}{\partial x_j} = \frac{\partial g}{\partial x_j}\mathbf{F} + g\frac{\partial \mathbf{F}}{\partial x_j}, \quad (4.9)$$

as is evident from Lemma 15.4.3 upon taking $\mathbf{G} = g\mathbf{I}$.

Note also that if $\mathbf{F}$ is a $p \times q$ matrix of functions, defined on a set S, of a vector $\mathbf{x} = (x_1, \ldots, x_m)'$ of m variables, then, at any interior point at which $\mathbf{F}$ is continuously differentiable, $\mathbf{F}'$ is continuously differentiable and

$$\frac{\partial \mathbf{F}'}{\partial x_j} = \left(\frac{\partial \mathbf{F}}{\partial x_j}\right)'. \quad (4.10)$$

15.5 Differentiation of a Vector or (Unrestricted or Symmetric) Matrix With Respect to Its Elements

Let $\mathbf{x} = \{x_s\}$ represent an m-dimensional column vector, and let $\mathbf{u}_j$ represent the jth column of an identity matrix (of unspecified dimensions). The elements of $\mathbf{x}$ can be regarded as functions, defined on $\mathcal{R}^m$, of $\mathbf{x}$. The jth partial derivatives of these functions are given by result (3.1). This result can be reexpressed in matrix notation as

$$\frac{\partial \mathbf{x}}{\partial x_j} = \mathbf{u}_j. \quad (5.1)$$

Result (5.1) can be generalized to an (unrestricted) $m \times n$ matrix $\mathbf{X} = \{x_{st}\}$. The elements of $\mathbf{X}$ can be regarded as functions, defined on $\mathcal{R}^{m\times n}$, of $\mathbf{X}$. For purposes of differentiating x_{st}, the elements of $\mathbf{X}$ can be rearranged in the form of an mn-dimensional column vector $\mathbf{x}$, and x_{st} can be reinterpreted as a function (defined on $\mathcal{R}^{mn}$) of $\mathbf{x}$ (as discussed in Section 15.1e). Then [in light of result (3.1)],

$$\frac{\partial x_{st}}{\partial x_{ij}} = \begin{cases} 1, & \text{if } s = i \text{ and } t = j, \quad (5.2a) \\ 0, & \text{otherwise.} \quad (5.2b) \end{cases}$$

Or, in matrix notation,

$$\frac{\partial \mathbf{X}}{\partial x_{ij}} = \mathbf{u}_i \mathbf{u}_j'. \tag{5.3}$$

Suppose now that $\mathbf{X} = \{x_{st}\}$ is a symmetric (but otherwise unrestricted) matrix (of dimensions $m \times m$). Then, the elements of $\mathbf{X}$ can still be regarded as functions of $\mathbf{X}$; however, the domain of these functions now comprises all $m \times m$ symmetric matrices, which (unless $m = 1$) is a proper subset of $\mathbb{R}^{m \times m}$. For purposes of differentiating x_{st}, x_{st} is (by convention) interpreted as a function (defined on $\mathbb{R}^{m(m+1)/2}$) of an $[m(m+1)/2]$-dimensional column vector $\mathbf{x}$ whose elements are x_{ij} $(j \le i = 1, \ldots, m)$ or alternatively x_{ij} $(j \ge i = 1, \ldots, m)$. This approach gives

$$\frac{\partial x_{st}}{\partial x_{ii}} = \begin{cases} 1, & \text{if } s = t = i, & (5.4a) \\ 0, & \text{otherwise}; & (5.4b) \end{cases}$$

and, for $j < i$ (or alternatively for $j > i$),

$$\frac{\partial x_{st}}{\partial x_{ij}} = \begin{cases} 1, & \text{if } s = i \text{ and } t = j \text{ or if } s = j \text{ and } t = i, & (5.5a) \\ 0, & \text{otherwise}. & (5.5b) \end{cases}$$

Or, in matrix notation,

$$\frac{\partial \mathbf{X}}{\partial x_{ii}} = \mathbf{u}_i \mathbf{u}_i', \tag{5.6}$$

and, for $j < i$ (or alternatively for $j > i$),

$$\frac{\partial \mathbf{X}}{\partial x_{ij}} = \mathbf{u}_i \mathbf{u}_j' + \mathbf{u}_j \mathbf{u}_i'. \tag{5.7}$$

15.6 Differentiation of a Trace of a Matrix

Let $\mathbf{F} = \{f_{is}\}$ represent a $p \times p$ matrix of functions, defined on a set S, of a vector $\mathbf{x} = (x_1, \ldots, x_m)'$ of m variables. Then, at any interior point $\mathbf{c}$ (of S) at which $\mathbf{F}$ is continuously differentiable, $\mathrm{tr}(\mathbf{F})$ is continuously differentiable and

$$\frac{\partial\, \mathrm{tr}(\mathbf{F})}{\partial x_j} = \mathrm{tr}\left(\frac{\partial \mathbf{F}}{\partial x_j}\right), \tag{6.1}$$

as is evident upon observing that $\mathrm{tr}(\mathbf{F}) = f_{11} + f_{22} + \cdots + f_{pp}$ [which establishes that $\mathrm{tr}(\mathbf{F})$ is continuously differentiable at $\mathbf{c}$] and that (at $\mathbf{x} = \mathbf{c}$)

$$\frac{\partial\, \mathrm{tr}(\mathbf{F})}{\partial x_j} = \frac{\partial f_{11}}{\partial x_j} + \frac{\partial f_{22}}{\partial x_j} + \cdots + \frac{\partial f_{pp}}{\partial x_j} = \mathrm{tr}\left(\frac{\partial \mathbf{F}}{\partial x_j}\right).$$

Suppose now that $\mathbf{F}$ is a $p \times q$ matrix of functions, defined on a set S, of a vector $\mathbf{x} = (x_1, \ldots, x_m)'$ of m variables, and suppose that $\mathbf{A}$ is a $q \times p$ matrix of constants or (more generally) a $q \times p$ matrix of functions (defined on S) that is continuous at every interior point of S and such that $\mathbf{A}(\mathbf{x})$ does not vary with x_j. Then, making use of results (6.1) and (4.5) (along with Lemma 5.2.1), we find that, at any interior point (of S) at which $\mathbf{F}$ is continuously differentiable, $\text{tr}(\mathbf{AF})$ or, equivalently, $\text{tr}(\mathbf{FA})$ is continuously differentiable and

$$\frac{\partial\,\text{tr}(\mathbf{AF})}{\partial x_j} = \frac{\partial\,\text{tr}(\mathbf{FA})}{\partial x_j} = \text{tr}\left(\frac{\partial \mathbf{F}}{\partial x_j}\mathbf{A}\right) = \text{tr}\left(\mathbf{A}\frac{\partial \mathbf{F}}{\partial x_j}\right). \tag{6.2}$$

Result (6.2) can be generalized to the trace of the product of "any" two matrices of functions. Let $\mathbf{F}$ and $\mathbf{G}$ represent $p \times q$ and $q \times p$ matrices of functions, defined on a set S, of a vector $\mathbf{x} = (x_1, \ldots, x_m)'$ of m variables. Then, in combination with result (6.1) and Lemma 5.2.1, Lemma 15.4.3 implies that, at any interior point at which $\mathbf{F}$ and $\mathbf{G}$ are continuously differentiable, $\text{tr}(\mathbf{FG})$ or, equivalently, $\text{tr}(\mathbf{GF})$ is continuously differentiable and

$$\frac{\partial\,\text{tr}(\mathbf{FG})}{\partial x_j} = \frac{\partial\,\text{tr}(\mathbf{GF})}{\partial x_j} = \text{tr}\left(\mathbf{F}\frac{\partial \mathbf{G}}{\partial x_j}\right) + \text{tr}\left(\mathbf{G}\frac{\partial \mathbf{F}}{\partial x_j}\right). \tag{6.3}$$

Note that alternative versions of formula (6.3) can be obtained by making either or both of the two substitutions

$$\text{tr}\left(\mathbf{F}\frac{\partial \mathbf{G}}{\partial x_j}\right) = \text{tr}\left(\frac{\partial \mathbf{G}}{\partial x_j}\mathbf{F}\right) \quad \text{and} \quad \text{tr}\left(\mathbf{G}\frac{\partial \mathbf{F}}{\partial x_j}\right) = \text{tr}\left(\frac{\partial \mathbf{F}}{\partial x_j}\mathbf{G}\right).$$

Let us now consider a special case of result (6.2). Take $\mathbf{X} = \{x_{st}\}$ to be an $m \times n$ matrix of "independent" variables, and suppose that the range of $\mathbf{X}$ comprises all of $\mathcal{R}^{m \times n}$. Define $\mathbf{A} = \{a_{ts}\}$ to be an $n \times m$ matrix of constants, and let $\mathbf{u}_j$ represent the jth column of an identity matrix (of unspecified dimensions). Then,

$$\frac{\partial\,\text{tr}(\mathbf{AX})}{\partial x_{ij}} = \frac{\partial\,\text{tr}(\mathbf{XA})}{\partial x_{ij}} = a_{ji}, \tag{6.4}$$

as is evident from result (6.2) upon observing [in light of result (5.3)] that

$$\text{tr}\left(\mathbf{A}\frac{\partial \mathbf{X}}{\partial x_{ij}}\right) = \text{tr}(\mathbf{A}\mathbf{u}_i\mathbf{u}_j') = \text{tr}(\mathbf{u}_j'\mathbf{a}\mathbf{u}_i) = \mathbf{u}_j'\mathbf{a}\mathbf{u}_i = a_{ji}.$$

Result (6.4) can be restated as

$$\frac{\partial\,\text{tr}(\mathbf{AX})}{\partial \mathbf{X}} = \frac{\partial\,\text{tr}(\mathbf{XA})}{\partial \mathbf{X}} = \mathbf{A}'. \tag{6.5}$$

Suppose now that $\mathbf{X}$ is a symmetric (but otherwise unrestricted) matrix (of dimensions $m \times m$). Then, for purposes of differentiating a function of $\mathbf{X}$, the function is reinterpreted as a function of an $[m(m+1)/2]$-dimensional column vector $\mathbf{x}$ whose elements are x_{ij} ($j \le i = 1, \ldots, m$). Consequently, result (6.4) is no longer applicable. Instead, we have that

$$\frac{\partial\,\text{tr}(\mathbf{AX})}{\partial x_{ij}} = \frac{\partial\,\text{tr}(\mathbf{XA})}{\partial x_{ij}} = \begin{cases} a_{ii}, & \text{if } j = i, \quad (6.6\text{a}) \\ a_{ij} + a_{ji}, & \text{if } j < i. \quad (6.6\text{b}) \end{cases}$$

To verify this, apply result (6.2) and observe [in light of results (5.6) and (5.7)] that

$$\operatorname{tr}\left(\mathbf{A}\frac{\partial \mathbf{X}}{\partial x_{ii}}\right) = \operatorname{tr}(\mathbf{A}\mathbf{u}_i\mathbf{u}_i') = \mathbf{u}_i'\mathbf{A}\mathbf{u}_i = a_{ii}$$

and that (for $j < i$)

$$\operatorname{tr}\left(\mathbf{A}\frac{\partial \mathbf{X}}{\partial x_{ij}}\right) = \operatorname{tr}(\mathbf{A}\mathbf{u}_i\mathbf{u}_j') + \operatorname{tr}(\mathbf{A}\mathbf{u}_j\mathbf{u}_i') = \mathbf{u}_j'\mathbf{A}\mathbf{u}_i + \mathbf{u}_i'\mathbf{A}\mathbf{u}_j = a_{ji} + a_{ij}.$$

Result (6.6) can be restated as

$$\frac{\partial \operatorname{tr}(\mathbf{AX})}{\partial \mathbf{X}} = \frac{\partial \operatorname{tr}(\mathbf{XA})}{\partial \mathbf{X}} = \mathbf{A} + \mathbf{A}' - \operatorname{diag}(a_{11}, a_{22}, \ldots, a_{mm}). \tag{6.7}$$

Note that, regardless of whether $\mathbf{X}$ is unrestricted (but square) or symmetric, we have that

$$\frac{\partial \operatorname{tr}(\mathbf{X})}{\partial x_{ij}} = \begin{cases} 1, & \text{if } j = i, \\ 0, & \text{if } j \neq i, \end{cases}$$

or, equivalently, that

$$\frac{\partial \operatorname{tr}(\mathbf{X})}{\partial \mathbf{X}} = \mathbf{I},$$

as is evident from results (6.4)–(6.7) upon taking $\mathbf{A} = \mathbf{I}$.

15.7 The Chain Rule

The chain rule can be very helpful in deriving the partial derivatives of a function f that is the composite of a function g and a vector of functions $\mathbf{h}$, that is, of a function f whose value at an arbitrary point $\mathbf{x}$ is given by the formula $f(\mathbf{x}) = g[\mathbf{h}(\mathbf{x})]$. The chain rule is stated (without proof) in the following theorem.

Theorem 15.7.1 (chain rule). Let $\mathbf{h} = \{h_i\}$ represent an $n \times 1$ vector of functions, defined on a set S, of a vector $\mathbf{x} = (x_1, \ldots, x_m)'$ of m variables. Let g represent a function, defined on a set T, of a vector $\mathbf{y} = (y_1, \ldots, y_n)'$ of n variables. Suppose that $\mathbf{h}(\mathbf{x}) \in T$ for every $\mathbf{x}$ in S, and take f to be the composite function defined (on S) by $f(\mathbf{x}) = g[\mathbf{h}(\mathbf{x})]$. If $\mathbf{h}$ is continuously differentiable at an interior point $\mathbf{c}$ of S and if [assuming that $\mathbf{h}(\mathbf{c})$ is an interior point of T] g is continuously differentiable at $\mathbf{h}(\mathbf{c})$, then f is continuously differentiable at $\mathbf{c}$ and

$$D_j f(\mathbf{c}) = \sum_{i=1}^{n} D_i g[\mathbf{h}(\mathbf{c})] D_j h_i(\mathbf{c}) = \mathbf{D}g[\mathbf{h}(\mathbf{c})] D_j \mathbf{h}(\mathbf{c}). \tag{7.1}$$

Formula (7.1) can be reexpressed less formally as

$$\frac{\partial f}{\partial x_j} = \sum_{i=1}^{n} \frac{\partial g}{\partial y_i}\frac{\partial h_i}{\partial x_j} = \frac{\partial g}{\partial \mathbf{y}'}\frac{\partial \mathbf{h}}{\partial x_j}, \tag{7.2}$$

where $\partial g/\partial y_i$ and $\partial g/\partial \mathbf{y}'$ are to be interpreted as having been evaluated at $\mathbf{y} = \mathbf{h}(\mathbf{x})$. Formula (7.1) or (7.2) can be recast as

$$\mathbf{D}f(\mathbf{c}) = \sum_{i=1}^{n} D_i g[\mathbf{h}(\mathbf{c})]\mathbf{D}h_i(\mathbf{c}) = \mathbf{D}g[\mathbf{h}(\mathbf{c})]\mathbf{D}\mathbf{h}(\mathbf{c}) \tag{7.3}$$

or

$$\frac{\partial f}{\partial \mathbf{x}'} = \sum_{i=1}^{n} \frac{\partial g}{\partial y_i}\frac{\partial h_i}{\partial \mathbf{x}'} = \frac{\partial g}{\partial \mathbf{y}'}\frac{\partial \mathbf{h}}{\partial \mathbf{x}'}. \tag{7.4}$$

The result of Theorem 15.7.1 can be generalized by taking $\mathbf{g} = \{g_s\}$ to be a $p \times 1$ vector of functions (defined on T) of $\mathbf{y}$ and $\mathbf{f} = \{f_s\}$ to be a $p \times 1$ vector of composite functions defined (on S) by $f_s(\mathbf{x}) = g_s[\mathbf{h}(\mathbf{x})]$ $(s = 1, \ldots, p)$ or, equivalently, by $\mathbf{f}(\mathbf{x}) = \mathbf{g}[\mathbf{h}(\mathbf{x})]$. If (in this more general context) $\mathbf{h}$ is continuously differentiable at an interior point $\mathbf{c}$ of S and $\mathbf{g}$ is continuously differentiable at $\mathbf{h}(\mathbf{c})$, then $\mathbf{f}$ is continuously differentiable at $\mathbf{c}$ and

$$D_j\mathbf{f}(\mathbf{c}) = \sum_{i=1}^{n} D_i\mathbf{g}[\mathbf{h}(\mathbf{c})]D_j h_i(\mathbf{c}) = \mathbf{D}\mathbf{g}[\mathbf{h}(\mathbf{c})]D_j\mathbf{h}(\mathbf{c}) \tag{7.5}$$

or, equivalently,

$$\frac{\partial \mathbf{f}}{\partial x_j} = \sum_{i=1}^{n} \frac{\partial \mathbf{g}}{\partial y_i}\frac{\partial h_i}{\partial x_j} = \frac{\partial \mathbf{g}}{\partial \mathbf{y}'}\frac{\partial \mathbf{h}}{\partial x_j} \tag{7.6}$$

[where $\partial \mathbf{g}/\partial y_i$ and $\partial \mathbf{g}/\partial \mathbf{y}'$ are to be interpreted as having been evaluated at $\mathbf{y} = \mathbf{h}(\mathbf{x})$]. Formula (7.5) or (7.6) can be recast as

$$\mathbf{D}\mathbf{f}(\mathbf{c}) = \sum_{i=1}^{n} D_i\mathbf{g}[\mathbf{h}(\mathbf{c})]\mathbf{D}h_i(\mathbf{c}) = \mathbf{D}\mathbf{g}[\mathbf{h}(\mathbf{c})]\mathbf{D}\mathbf{h}(\mathbf{c}) \tag{7.7}$$

or

$$\frac{\partial \mathbf{f}}{\partial \mathbf{x}'} = \sum_{i=1}^{n} \frac{\partial \mathbf{g}}{\partial y_i}\frac{\partial h_i}{\partial \mathbf{x}'} = \frac{\partial \mathbf{g}}{\partial \mathbf{y}'}\frac{\partial \mathbf{h}}{\partial \mathbf{x}'}. \tag{7.8}$$

Let us now consider a different generalization of the result of Theorem 15.7.1—one where $\mathbf{H} = \{h_{is}\}$ is an $n \times r$ matrix of functions, defined on S, of $\mathbf{x}$; where g is a function, defined on a set T, of an $n \times r$ matrix $\mathbf{Y} = \{y_{is}\}$ of nr variables; where $\mathbf{H}(\mathbf{x}) \in T$ for every $\mathbf{x}$ in S; and where f is the composite function defined (on S) by $f(\mathbf{x}) = g[\mathbf{H}(\mathbf{x})]$. Suppose that the elements of $\mathbf{H}$ and $\mathbf{Y}$ are rearranged in the form of column vectors $\mathbf{h}$ and $\mathbf{y}$, respectively, and that, for purposes of differentiation, g is reinterpreted as a function of $\mathbf{y}$. If $\mathbf{h}$ or, equivalently, $\mathbf{H}$ is continuously differentiable at an interior point $\mathbf{c}$ of S and if [assuming that $\mathbf{H}(\mathbf{c})$ is an interior point of T] g is continuously differentiable at $\mathbf{h}(\mathbf{c})$ or, equivalently, $\mathbf{H}(\mathbf{c})$, then f is continuously differentiable at $\mathbf{c}$ and (at $\mathbf{x} = \mathbf{c}$)

$$\frac{\partial f}{\partial x_j} = \sum_{i=1}^{n}\sum_{s=1}^{r} \frac{\partial g}{\partial y_{is}}\frac{\partial h_{is}}{\partial x_j} \tag{7.9}$$

(as is evident from Theorem 15.7.1). Formula (7.9) can be rewritten as

$$\frac{\partial f}{\partial x_j} = \mathrm{tr}\left[\left(\frac{\partial g}{\partial \mathbf{Y}}\right)' \frac{\partial \mathbf{H}}{\partial x_j}\right]. \tag{7.10}$$

[The quantities $\partial g/\partial y_{is}$ and $\partial g/\partial \mathbf{Y}$, which appear in formulas (7.9) and (7.10), are to be interpreted as having been evaluated at $\mathbf{Y} = \mathbf{H}(\mathbf{x})$.]

15.8 First-Order Partial Derivatives of Determinants and Inverse and Adjoint Matrices

Let $\mathbf{X} = \{x_{ij}\}$ represent an $m \times m$ matrix of m^2 "independent" variables (where $m \geq 2$), and suppose that the range of $\mathbf{X}$ comprises all of $\mathcal{R}^{m\times m}$. Denote by ξ_{ij} the cofactor of x_{ij}. Then, the function f of $\mathbf{X}$ defined (on $\mathcal{R}^{m\times m}$) by $f(\mathbf{X}) = \det(\mathbf{X})$ is continuously differentiable at every $\mathbf{X}$ and

$$\frac{\partial \det(\mathbf{X})}{\partial x_{ij}} = \xi_{ij}. \tag{8.1}$$

To see this, rearrange the elements of $\mathbf{X}$ in the form of an m^2-dimensional column vector $\mathbf{x}$, and regard f as a function of $\mathbf{x}$. Since each element of $\mathbf{X}$ is a continuously differentiable function of $\mathbf{x}$, and since (according to Lemma 15.2.2) sums and products of continuously differentiable functions are continuously differentiable, it follows from the very definition of the determinant [given by expression (13.1.2)] that f is continuously differentiable. Moreover, expanding $\det(\mathbf{X})$ as $\det(\mathbf{X}) = \sum_{t=1}^{m} x_{it}\xi_{it}$ [in accordance with result (13.5.1)], observing that $\xi_{i1}, \ldots, \xi_{im}$ do not vary with x_{ij}, and recalling result (5.2), we find that

$$\frac{\partial \det(\mathbf{X})}{\partial x_{ij}} = \sum_{t=1}^{m} \xi_{it} \frac{\partial x_{it}}{\partial x_{ij}} = \xi_{ij}. \tag{8.2}$$

Result (8.1) indicates that the derivative of $\det(\mathbf{X})$ with respect to $\mathbf{X}$ is the matrix of cofactors (of $\mathbf{X}$) or, equivalently, that

$$\frac{\partial \det(\mathbf{X})}{\partial \mathbf{X}} = [\mathrm{adj}(\mathbf{X})]'. \tag{8.3}$$

Let us now extend our results to the differentiation of the determinant of a $p \times p$ matrix $\mathbf{F} = \{f_{is}\}$ of functions, defined on a set S, of a vector $\mathbf{x} = (x_1, \ldots, x_m)'$ of m variables. Thus, the function to be differentiated is the function h of $\mathbf{x}$ defined (on S) by $h(\mathbf{x}) = \det[\mathbf{F}(\mathbf{x})]$. It is convenient (for purposes of differentiating h) to introduce a function g of a $p \times p$ matrix $\mathbf{Y} = \{y_{is}\}$ of p^2 variables, defined (on $\mathcal{R}^{p\times p}$) by $g(\mathbf{Y}) = \det(\mathbf{Y})$, and to express h as the composite of g and $\mathbf{F}$, so that $h(\mathbf{x}) = g[\mathbf{F}(\mathbf{x})]$. Clearly (in light of our earlier results), g is continuously differentiable at every $\mathbf{Y}$ and

$$\frac{\partial g}{\partial \mathbf{Y}} = [\mathrm{adj}(\mathbf{Y})]'.$$

Thus, it follows from the chain rule [and in particular from result (7.10)] that if $\mathbf{F}$ is continuously differentiable at an interior point $\mathbf{c}$ of S, then h is continuously differentiable at $\mathbf{c}$ and (at $\mathbf{x} = \mathbf{c}$)

$$\frac{\partial \det(\mathbf{F})}{\partial x_j} = \mathrm{tr}\left[\mathrm{adj}(\mathbf{F})\frac{\partial \mathbf{F}}{\partial x_j}\right]. \tag{8.4}$$

Moreover, if $\mathbf{F}$ is nonsingular as well as continuously differentiable at $\mathbf{c}$, then [in light of result (13.5.6)]

$$\frac{\partial \det(\mathbf{F})}{\partial x_j} = |\mathbf{F}|\,\mathrm{tr}\left(\mathbf{F}^{-1}\frac{\partial \mathbf{F}}{\partial x_j}\right). \tag{8.5}$$

Suppose now that S is the set of all $\mathbf{x}$-values for which $\det[\mathbf{F}(\mathbf{x})] > 0$ or is a subset of that set, and consider the differentiation of the function ℓ of $\mathbf{x}$ defined (on S) by $\ell(\mathbf{x}) = \log\det[\mathbf{F}(\mathbf{x})]$. To facilitate the differentiation, let g represent a function of a variable y defined (for $y > 0$) by $g(y) = \log y$, and express ℓ as the composite of g and h, so that $\ell(\mathbf{x}) = g[h(\mathbf{x})]$.

Recall (from the calculus of a single variable) that g is continuously differentiable (at all points in its domain) and that (for $y > 0$)

$$\frac{\partial \log y}{\partial y} = \frac{1}{y}.$$

Applying the chain rule, we find [in light of results (8.4) and (8.5)] that if $\mathbf{F}$ is continuously differentiable at an interior point $\mathbf{c}$ of S (in which case h is continuously differentiable at $\mathbf{c}$), then ℓ is continuously differentiable at $\mathbf{c}$ and (at $\mathbf{x} = \mathbf{c}$)

$$\frac{\partial \log\det(\mathbf{F})}{\partial x_j} = \frac{1}{|\mathbf{F}|}\frac{\partial \det(\mathbf{F})}{\partial x_j} = \frac{1}{|\mathbf{F}|}\mathrm{tr}\left[\mathrm{adj}(\mathbf{F})\frac{\partial \mathbf{F}}{\partial x_j}\right] = \mathrm{tr}\left(\mathbf{F}^{-1}\frac{\partial \mathbf{F}}{\partial x_j}\right). \tag{8.6}$$

Let us now consider result (8.6) in the special case where $\mathbf{x}$ is an m^2-dimensional column vector whose elements are the same as those of an $m \times m$ matrix $\mathbf{X} = \{x_{ij}\}$ of m^2 "independent" variables and where $\mathbf{F}(\mathbf{x}) = \mathbf{X}$. Let $\mathbf{c}$ represent an interior point of S [where S consists of some or all x-values for which $\det(\mathbf{X}) > 0$], and let $\mathbf{u}_j$ represent the jth column of $\mathbf{I}_m$.

It follows from result (5.3) that $\mathbf{X}$ is continuously differentiable at $\mathbf{c}$ and that (at $\mathbf{x} = \mathbf{c}$) $\partial\mathbf{X}/\partial x_{ij} = \mathbf{u}_i\mathbf{u}_j'$. We conclude that the function ℓ defined (on S) by $\ell(\mathbf{x}) = \log\det(\mathbf{X})$ is continuously differentiable at $\mathbf{c}$ and that (at $\mathbf{x} = \mathbf{c}$)

$$\frac{\partial \log\det(\mathbf{X})}{\partial x_{ij}} = \mathrm{tr}(\mathbf{X}^{-1}\mathbf{u}_i\mathbf{u}_j') = \mathbf{u}_j'\mathbf{X}^{-1}\mathbf{u}_i = y_{ji}, \tag{8.7}$$

where y_{ji} is the jith element of $\mathbf{X}^{-1}$ or, equivalently, the ijth element of $(\mathbf{X}^{-1})'$. Result (8.7) can be recast as

$$\frac{\partial \log \det(\mathbf{X})}{\partial \mathbf{X}} = (\mathbf{X}^{-1})'. \tag{8.8}$$

Formulas (8.3) and (8.8) [or (8.2) and (8.7)] are applicable when the m^2 elements of $\mathbf{X}$ are independent variables. Suppose now that $\mathbf{X} = \{x_{ij}\}$ is an $m \times m$ symmetric matrix, in which case formulas (8.3) and (8.8) are not applicable. For purposes of differentiation, the function f of $\mathbf{X}$ defined (on a set S comprising some or all $m \times m$ symmetric matrices) by $f(\mathbf{X}) = \det(\mathbf{X})$ and the function ℓ of $\mathbf{X}$ defined (on a set S^* comprising some or all $m \times m$ symmetric matrices with positive determinants) by $\ell(\mathbf{X}) = \log \det(\mathbf{X})$ are by convention to be interpreted as functions of the $[m(m+1)/2]$-dimensional column vector $\mathbf{x}$ whose elements x_{ij} $(j \leq i = 1, \ldots, m)$ are "independent" elements of $\mathbf{X}$.

Let $\mathbf{c}$ represent an interior point of S and $\mathbf{c}^*$ an interior point of S^*. Then, it follows from the results of Section 15.5 that $\mathbf{X}$ is continuously differentiable at $\mathbf{c}$ and $\mathbf{c}^*$, implying that f is continuously differentiable at $\mathbf{c}$ and that ℓ is continuously differentiable at $\mathbf{c}^*$. Moreover, we have [as a special case of result (8.4)] that (at $\mathbf{x} = \mathbf{c}$)

$$\frac{\partial \det(\mathbf{X})}{\partial x_{ij}} = \operatorname{tr}\left[\operatorname{adj}(\mathbf{X}) \frac{\partial \mathbf{X}}{\partial x_{ij}}\right].$$

Making use of results (5.6) and (5.7) and denoting the jth column of $\mathbf{I}_m$ by $\mathbf{u}_j$, we find that

$$\operatorname{tr}\left[\operatorname{adj}(\mathbf{X}) \frac{\partial \mathbf{X}}{\partial x_{ii}}\right] = \operatorname{tr}[\operatorname{adj}(\mathbf{X})\mathbf{u}_i\mathbf{u}_i'] = \mathbf{u}_i' \operatorname{adj}(\mathbf{X})\mathbf{u}_i$$

and (for $j < i$)

$$\begin{aligned}\operatorname{tr}\left[\operatorname{adj}(\mathbf{X}) \frac{\partial \mathbf{X}}{\partial x_{ij}}\right] &= \operatorname{tr}[\operatorname{adj}(\mathbf{X})\mathbf{u}_i\mathbf{u}_j'] + \operatorname{tr}[\operatorname{adj}(\mathbf{X})\mathbf{u}_j\mathbf{u}_i'] \\ &= \mathbf{u}_j' \operatorname{adj}(\mathbf{X})\mathbf{u}_i + \mathbf{u}_i' \operatorname{adj}(\mathbf{X})\mathbf{u}_j.\end{aligned}$$

Thus, denoting the cofactor of x_{ij} by ξ_{ij} (and recalling that the cofactor matrix of a symmetric matrix is symmetric), we have that

$$\frac{\partial \det(\mathbf{X})}{\partial x_{ij}} = \begin{cases} \xi_{ii}, & \text{if } j = i, \quad (8.9\text{a}) \\ 2\xi_{ij}, & \text{if } j < i, \quad (8.9\text{b}) \end{cases}$$

or, equivalently, that

$$\frac{\partial \det(\mathbf{X})}{\partial \mathbf{X}} = 2\operatorname{adj}(\mathbf{X}) - \operatorname{diag}(\xi_{11}, \xi_{22}, \ldots, \xi_{mm}). \tag{8.10}$$

Further, letting y_{ij} represent the ijth element of $\mathbf{X}^{-1}$, we find [in light of results (8.6) and (13.5.7)] that (at $\mathbf{x} = \mathbf{c}^*$)

$$\frac{\partial \log \det(\mathbf{X})}{\partial x_{ij}} = \begin{cases} y_{ii}, & \text{if } j = i, \quad (8.11\text{a}) \\ 2y_{ij}, & \text{if } j < i, \quad (8.11\text{b}) \end{cases}$$

or, equivalently, that

$$\frac{\partial \log \det(\mathbf{X})}{\partial \mathbf{X}} = 2\mathbf{X}^{-1} - \text{diag}(y_{11}, y_{22}, \ldots, y_{mm}). \tag{8.12}$$

Consider now the differentiation of the adjoint matrix of a $p \times p$ matrix $\mathbf{F} = \{f_{is}\}$ of functions, defined on a set S, of a vector $\mathbf{x} = (x_1, \ldots, x_m)'$ of m variables. Let $\mathbf{F}_{si}$ represent the $(p-1) \times (p-1)$ submatrix of $\mathbf{F}$ obtained by striking out the sth row and the ith column (of $\mathbf{F}$), and let $\phi_{si} = (-1)^{s+i} \det(\mathbf{F}_{si})$. Then, by definition, ϕ_{si} is the cofactor of f_{si} and hence the isth element of adj$(\mathbf{F})$. It follows from our discussion of the differentiation of determinants [including result (8.4)] that if $\mathbf{F}$ is continuously differentiable at an interior point $\mathbf{c}$ of S, then ϕ_{si} is continuously differentiable at $\mathbf{c}$ and (at $\mathbf{x} = \mathbf{c}$)

$$\frac{\partial \phi_{si}}{\partial x_j} = (-1)^{s+i} \operatorname{tr}\left[\text{adj}(\mathbf{F}_{si}) \frac{\partial \mathbf{F}_{si}}{\partial x_j}\right]. \tag{8.13}$$

In other words, if $\mathbf{F}$ is continuously differentiable at $\mathbf{c}$, then adj$(\mathbf{F})$ is continuously differentiable at $\mathbf{c}$, and the isth element of the matrix $\partial\, \text{adj}(\mathbf{F})/\partial x_j$ is given by expression (8.13).

Suppose now that S is the set of all $\mathbf{x}$-values for which $\mathbf{F}(\mathbf{x})$ is nonsingular or is a subset of that set, and consider the differentiation of the inverse matrix $\mathbf{F}^{-1}$. Denote by $\mathbf{c}$ any interior point (of S) at which $\mathbf{F}$ is continuously differentiable. Then, ϕ_{si} and $\det(\mathbf{F})$ are continuously differentiable at $\mathbf{c}$, implying (in light of Lemma 15.2.2) that $\phi_{si}/\det(\mathbf{F})$ is continuously differentiable at $\mathbf{c}$. Since (according to Corollary 13.5.4) $\phi_{si}/\det(\mathbf{F})$ equals the isth element of $\mathbf{F}^{-1}$, we conclude that $\mathbf{F}^{-1}$ is continuously differentiable at $\mathbf{c}$. Moreover, $\mathbf{F}\mathbf{F}^{-1} = \mathbf{I}$, implying (in light of Lemmas 15.4.3 and 15.4.1) that (at $\mathbf{x} = \mathbf{c}$)

$$\mathbf{F} \frac{\partial \mathbf{F}^{-1}}{\partial x_j} + \frac{\partial \mathbf{F}}{\partial x_j} \mathbf{F}^{-1} = \frac{\partial \mathbf{I}}{\partial x_j} = \mathbf{0}$$

and hence

$$\mathbf{F} \frac{\partial \mathbf{F}^{-1}}{\partial x_j} = -\frac{\partial \mathbf{F}}{\partial x_j} \mathbf{F}^{-1}. \tag{8.14}$$

Premultiplying both sides of equality (8.14) by $\mathbf{F}^{-1}$, we obtain

$$\frac{\partial \mathbf{F}^{-1}}{\partial x_j} = -\mathbf{F}^{-1} \frac{\partial \mathbf{F}}{\partial x_j} \mathbf{F}^{-1} \tag{8.15}$$

as a formula for $\partial \mathbf{F}^{-1}/\partial x_j$.

An alternative expression for the elements of $\partial\, \text{adj}(\mathbf{F})/\partial x_j$ can [when S is restricted to $\mathbf{x}$-values for which $\mathbf{F}(\mathbf{x})$ is nonsingular] be obtained by making use of formula (4.9). Since (according to Corollary 13.5.4) adj$(\mathbf{F}) = |\mathbf{F}|\mathbf{F}^{-1}$, we have [in light of formula (8.5)] that

$$\begin{aligned}
\frac{\partial\,\mathrm{adj}(\mathbf{F})}{\partial x_j} &= \frac{\partial|\mathbf{F}|}{\partial x_j}\mathbf{F}^{-1} + |\mathbf{F}|\frac{\partial\mathbf{F}^{-1}}{\partial x_j} \\
&= |\mathbf{F}|\,\mathrm{tr}\left(\mathbf{F}^{-1}\frac{\partial\mathbf{F}}{\partial x_j}\right)\mathbf{F}^{-1} + |\mathbf{F}|\left(-\mathbf{F}^{-1}\frac{\partial\mathbf{F}}{\partial x_j}\mathbf{F}^{-1}\right) \\
&= |\mathbf{F}|\left[\mathrm{tr}\left(\mathbf{F}^{-1}\frac{\partial\mathbf{F}}{\partial x_j}\right)\mathbf{F}^{-1} - \mathbf{F}^{-1}\frac{\partial\mathbf{F}}{\partial x_j}\mathbf{F}^{-1}\right].
\end{aligned} \tag{8.16}$$

Formula (8.15) can be generalized by making use of result (4.6). Let **A** and **B** represent $k \times p$ and $p \times r$ matrices of functions (defined on the same set S as the elements of **F**) of **x**. Suppose that **A** and **B** (as well as **F**) are continuously differentiable at **c**. Then, $\mathbf{AF}^{-1}\mathbf{B}$ is continuously differentiable at **c** and (at $\mathbf{x} = \mathbf{c}$)

$$\frac{\partial(\mathbf{AF}^{-1}\mathbf{B})}{\partial x_j} = \mathbf{AF}^{-1}\frac{\partial\mathbf{B}}{\partial x_j} - \mathbf{AF}^{-1}\frac{\partial\mathbf{F}}{\partial x_j}\mathbf{F}^{-1}\mathbf{B} + \frac{\partial\mathbf{A}}{\partial x_j}\mathbf{F}^{-1}\mathbf{B}. \tag{8.17}$$

In the special case where (for $\mathbf{x} \in S$) $\mathbf{A}(\mathbf{x})$ and $\mathbf{B}(\mathbf{x})$ are constant or (more generally) do not vary with x_j, formula (8.17) simplifies to

$$\frac{\partial(\mathbf{AF}^{-1}\mathbf{B})}{\partial x_j} = -\mathbf{AF}^{-1}\frac{\partial\mathbf{F}}{\partial x_j}\mathbf{F}^{-1}\mathbf{B}. \tag{8.18}$$

15.9 Second-Order Partial Derivatives of Determinants and Inverse Matrices

Let $\mathbf{F} = \{f_{is}\}$ represent a $p \times p$ matrix of functions, defined on a set S, of a vector $\mathbf{x} = (x_1, \ldots, x_m)'$ of m variables. Suppose that $\mathbf{F}(\mathbf{x})$ is nonsingular for every **x** in S, and denote by **c** any interior point (of S) at which **F** is twice continuously differentiable. Then, **F** is continuously differentiable at **c**. In fact, **F** is continuously differentiable at every point in some neighborhood N of **c**.

In light of the results of Section 15.8, we have that $\det(\mathbf{F})$ and $\mathbf{F}^{-1}$ are continuously differentiable at every point in N and that (for $\mathbf{x} \in N$)

$$\frac{\partial\det(\mathbf{F})}{\partial x_j} = |\mathbf{F}|\,\mathrm{tr}\left(\mathbf{F}^{-1}\frac{\partial\mathbf{F}}{\partial x_j}\right)$$

and

$$\frac{\partial\mathbf{F}^{-1}}{\partial x_j} = -\mathbf{F}^{-1}\frac{\partial\mathbf{F}}{\partial x_j}\mathbf{F}^{-1}.$$

Moreover, $\partial\mathbf{F}/\partial x_j$ is continuously differentiable at $\mathbf{x} = \mathbf{c}$. Consequently, $\partial\det(\mathbf{F})/\partial x_j$ and $\partial\mathbf{F}^{-1}/\partial x_j$ are continuously differentiable at **c**, and hence $\det(\mathbf{F})$ and $\mathbf{F}^{-1}$ are twice continuously differentiable at **c**.

Further, making use of results (2.5) and (6.3), we find that (at $\mathbf{x} = \mathbf{c}$)

$$\begin{aligned}
\frac{\partial^2 \det(\mathbf{F})}{\partial x_i\,\partial x_j} &= \frac{\partial\{|\mathbf{F}|\operatorname{tr}[\mathbf{F}^{-1}(\partial\mathbf{F}/\partial x_j)]\}}{\partial x_i} \\
&= |\mathbf{F}|\left[\operatorname{tr}\left(\mathbf{F}^{-1}\frac{\partial^2\mathbf{F}}{\partial x_i\,\partial x_j}\right) + \operatorname{tr}\left(-\mathbf{F}^{-1}\frac{\partial\mathbf{F}}{\partial x_i}\mathbf{F}^{-1}\frac{\partial\mathbf{F}}{\partial x_j}\right)\right] \\
&\qquad + |\mathbf{F}|\operatorname{tr}\left(\mathbf{F}^{-1}\frac{\partial\mathbf{F}}{\partial x_i}\right)\operatorname{tr}\left(\mathbf{F}^{-1}\frac{\partial\mathbf{F}}{\partial x_j}\right) \\
&= |\mathbf{F}|\left[\operatorname{tr}\left(\mathbf{F}^{-1}\frac{\partial^2\mathbf{F}}{\partial x_i\,\partial x_j}\right) + \operatorname{tr}\left(\mathbf{F}^{-1}\frac{\partial\mathbf{F}}{\partial x_i}\right)\operatorname{tr}\left(\mathbf{F}^{-1}\frac{\partial\mathbf{F}}{\partial x_j}\right)\right. \\
&\qquad \left. - \operatorname{tr}\left(\mathbf{F}^{-1}\frac{\partial\mathbf{F}}{\partial x_i}\mathbf{F}^{-1}\frac{\partial\mathbf{F}}{\partial x_j}\right)\right]. \qquad (9.1)
\end{aligned}$$

And, making use of result (4.6), we find that (at $\mathbf{x} = \mathbf{c}$)

$$\begin{aligned}
\frac{\partial^2\mathbf{F}^{-1}}{\partial x_i\,\partial x_j} &= \frac{\partial[-\mathbf{F}^{-1}(\partial\mathbf{F}/\partial x_j)\mathbf{F}^{-1}]}{\partial x_i} \\
&= -\mathbf{F}^{-1}\frac{\partial\mathbf{F}}{\partial x_j}\frac{\partial\mathbf{F}^{-1}}{\partial x_i} - \mathbf{F}^{-1}\frac{\partial^2\mathbf{F}}{\partial x_i\,\partial x_j}\mathbf{F}^{-1} - \frac{\partial\mathbf{F}^{-1}}{\partial x_i}\frac{\partial\mathbf{F}}{\partial x_j}\mathbf{F}^{-1} \\
&= -\mathbf{F}^{-1}\frac{\partial\mathbf{F}}{\partial x_j}\left(-\mathbf{F}^{-1}\frac{\partial\mathbf{F}}{\partial x_i}\mathbf{F}^{-1}\right) - \mathbf{F}^{-1}\frac{\partial^2\mathbf{F}}{\partial x_i\,\partial x_j}\mathbf{F}^{-1} \\
&\qquad - \left(-\mathbf{F}^{-1}\frac{\partial\mathbf{F}}{\partial x_i}\mathbf{F}^{-1}\right)\frac{\partial\mathbf{F}}{\partial x_j}\mathbf{F}^{-1} \\
&= -\mathbf{F}^{-1}\frac{\partial^2\mathbf{F}}{\partial x_i\,\partial x_j}\mathbf{F}^{-1} + \mathbf{F}^{-1}\frac{\partial\mathbf{F}}{\partial x_i}\mathbf{F}^{-1}\frac{\partial\mathbf{F}}{\partial x_j}\mathbf{F}^{-1} \\
&\qquad + \mathbf{F}^{-1}\frac{\partial\mathbf{F}}{\partial x_j}\mathbf{F}^{-1}\frac{\partial\mathbf{F}}{\partial x_i}\mathbf{F}^{-1}. \qquad (9.2)
\end{aligned}$$

Suppose now that $\det[\mathbf{F}(\mathbf{x})] > 0$ for every $\mathbf{x}$ in S. Then, in light of our previous results, $\log\det(\mathbf{F})$ is continuously differentiable at every point in N and (for $\mathbf{x} \in N$)

$$\frac{\partial\log\det(\mathbf{F})}{\partial x_j} = \operatorname{tr}\left(\mathbf{F}^{-1}\frac{\partial\mathbf{F}}{\partial x_j}\right).$$

Thus (since $\mathbf{F}^{-1}$ and $\partial\mathbf{F}/\partial x_j$ are continuously differentiable at $\mathbf{c}$), $\partial\log\det(\mathbf{F})/\partial x_j$ is continuously differentiable at $\mathbf{c}$, and hence $\log\det(\mathbf{F})$ is twice continuously differentiable at $\mathbf{c}$. Moreover, making use of result (6.3), we find that (at $\mathbf{x} = \mathbf{c}$)

$$\begin{aligned}
\frac{\partial^2\log\det(\mathbf{F})}{\partial x_i\,\partial x_j} &= \frac{\partial\operatorname{tr}[\mathbf{F}^{-1}(\partial\mathbf{F}/\partial x_j)]}{\partial x_i} \\
&= \operatorname{tr}\left(\mathbf{F}^{-1}\frac{\partial^2\mathbf{F}}{\partial x_i\,\partial x_j}\right) + \operatorname{tr}\left(-\mathbf{F}^{-1}\frac{\partial\mathbf{F}}{\partial x_i}\mathbf{F}^{-1}\frac{\partial\mathbf{F}}{\partial x_j}\right) \\
&= \operatorname{tr}\left(\mathbf{F}^{-1}\frac{\partial^2\mathbf{F}}{\partial x_i\,\partial x_j}\right) - \operatorname{tr}\left(\mathbf{F}^{-1}\frac{\partial\mathbf{F}}{\partial x_i}\mathbf{F}^{-1}\frac{\partial\mathbf{F}}{\partial x_j}\right). \qquad (9.3)
\end{aligned}$$

15.10 Differentiation of Generalized Inverses

Let $\mathbf{F}$ represent a $p \times q$ matrix of functions, defined on a set S, of a vector $\mathbf{x} = (x_1, \ldots, x_m)'$ of m variables. Consider the $q \times p$ matrix $\mathbf{G}$ of functions defined (on S) by taking $\mathbf{G}(\mathbf{x})$ to be any generalized inverse of $\mathbf{F}(\mathbf{x})$.

In the special case where $q = p$ and S is composed exclusively of nonsingular matrices (and hence where $\mathbf{G} = \mathbf{F}^{-1}$), $\mathbf{G}$ is continuously differentiable at every interior point of S at which $\mathbf{F}$ is continuously differentiable, and $\partial\mathbf{G}/\partial x_j = -\mathbf{G}(\partial\mathbf{F}/\partial x_j)\mathbf{G}$ (as discussed in Section 15.8). In this section, we consider the extent to which these results (and various other results on the differentiation of ordinary inverses) can be extended to the case where $\mathbf{F}$ is not necessarily nonsingular (and q not necessarily equal to p). One such extension is given by the following theorem.

Theorem 15.10.1. Let $\mathbf{F}$ represent a $p \times q$ matrix of functions, defined on a set S, of a vector $\mathbf{x} = (x_1, \ldots, x_m)'$ of m variables. Let $\mathbf{c}$ represent any interior point of S at which $\mathbf{F}$ is continuously differentiable. Suppose that $\mathbf{F}$ has constant rank, say r, on some neighborhood of $\mathbf{c}$. Take $i_1, \ldots, i_r$ to be integers chosen from the first p positive integers, $1, \ldots, p$, in such a way that the $i_1, \ldots, i_r$th rows of $\mathbf{F}(\mathbf{c})$ are linearly independent, and take $j_1, \ldots, j_r$ to be integers chosen from the first q positive integers, $1, \ldots, q$, in such a way that the $j_1, \ldots, j_r$th columns of $\mathbf{F}(\mathbf{c})$ are linearly idependent. Further, take $\mathbf{P}$ to be any $p \times p$ permutation matrix whose first r rows are the $i_1, \ldots, i_r$th rows of $\mathbf{I}_p$ and $\mathbf{Q}$ to be any $q \times q$ permutation matrix whose first r columns are the $j_1, \ldots, j_r$th columns of $\mathbf{I}_q$. Let $\mathbf{B} = \mathbf{PFQ}$, and partition $\mathbf{B}$ as

$$\mathbf{B} = \begin{pmatrix} \mathbf{B}_{11} & \mathbf{B}_{12} \\ \mathbf{B}_{21} & \mathbf{B}_{22} \end{pmatrix},$$

where $\mathbf{B}_{11}$ is of dimensions $r \times r$. Then, $\text{rank}(\mathbf{B}_{11}) = \text{rank}(\mathbf{F}) = r$ on some neighborhood N of $\mathbf{c}$. And there exists a generalized inverse $\mathbf{G}$ of $\mathbf{F}$ such that

$$\mathbf{G} = \mathbf{Q}\begin{pmatrix} \mathbf{B}_{11}^{-1} & \mathbf{0} \\ \mathbf{0} & \mathbf{0} \end{pmatrix}\mathbf{P}$$

on N. Moreover, $\mathbf{G}$ is continuously differentiable at $\mathbf{c}$, and (at $\mathbf{x} = \mathbf{c}$)

$$\frac{\partial\mathbf{G}}{\partial x_j} = -\mathbf{Q}\begin{bmatrix} \mathbf{B}_{11}^{-1}(\partial\mathbf{B}_{11}/\partial x_j)\mathbf{B}_{11}^{-1} & \mathbf{0} \\ \mathbf{0} & \mathbf{0} \end{bmatrix}\mathbf{P} = -\mathbf{G}\frac{\partial\mathbf{F}}{\partial x_j}\mathbf{G}.$$

With regard to Theorem 15.10.1, note (in light of the discussion of Section 9.2a) that $\mathbf{F}(\mathbf{c})$ necessarily contains r linearly independent rows and r linearly independent columns, that $\mathbf{B}_{11}$ is the submatrix of $\mathbf{F}$ obtained by striking out all of the rows and columns of $\mathbf{F}$ except the $i_1, \ldots, i_r$th rows and $j_1, \ldots, j_r$th columns (or is an $r \times r$ matrix obtained from that submatrix by permuting its rows and columns), and that $\mathbf{B}_{11}(\mathbf{c})$ is nonsingular. In proving Theorem 15.10.1, it is convenient to make use of the following lemma.

Lemma 15.10.2. Let $\mathbf{F}$ represent a $p \times p$ matrix of functions, defined on a set S, of an m-dimensional column vector $\mathbf{x}$. Suppose that $\mathbf{F}$ is nonsingular at an interior point $\mathbf{c}$ of S and that $\mathbf{F}$ is continuous at $\mathbf{c}$. Then, $\mathbf{F}$ is nonsingular at all points in some neighborhood of $\mathbf{c}$.

Proof (of Lemma 15.10.2). It follows from the very definition of a determinant [given by formula (13.1.2)] and from standard results on the continuity of functions that $\det(\mathbf{F})$ is a continuous function of $\mathbf{x}$ (at $\mathbf{x} = \mathbf{c}$) and hence that

$$\lim_{\mathbf{x}\to\mathbf{c}} \det[\mathbf{F}(\mathbf{x})] = \det[\mathbf{F}(\mathbf{c})].$$

Since $\det[\mathbf{F}(\mathbf{c})] \neq 0$, S contains a neighborhood N of $\mathbf{c}$ such that $|\det[\mathbf{F}(\mathbf{x})] - \det[\mathbf{F}(\mathbf{c})]| < |\det[\mathbf{F}(\mathbf{c})]|$ for $\mathbf{x} \in N$ or, equivalently, such that $|\det[\mathbf{F}(\mathbf{c})]| - |\det[\mathbf{F}(\mathbf{c})] - \det[\mathbf{F}(\mathbf{x})]| > 0$ for $\mathbf{x} \in N$. And, since $|\det[\mathbf{F}(\mathbf{c})]| \leq |\det[\mathbf{F}(\mathbf{x})]| + |\det[\mathbf{F}(\mathbf{c})] - \det[\mathbf{F}(\mathbf{x})]|$ or, equivalently, $|\det[\mathbf{F}(\mathbf{x})]| \geq |\det[\mathbf{F}(\mathbf{c})]| - |\det[\mathbf{F}(\mathbf{c})] - \det[\mathbf{F}(\mathbf{x})]|$, we have that $|\det[\mathbf{F}(\mathbf{x})]| > 0$ for $\mathbf{x} \in N$. We conclude that $\mathbf{F}(\mathbf{x})$ is nonsingular for $\mathbf{x} \in N$. Q.E.D.

Proof (of Theorem 15.10.1). Since (in light of Lemma 15.1.1) $\mathbf{F}$ is continuous at $\mathbf{c}$ (implying that $\mathbf{B}_{11}$ is continuous at $\mathbf{c}$), it follows from Lemma 15.10.2 that $\mathbf{B}_{11}$ is nonsingular at all points in some neighborhood N_1 of $\mathbf{c}$. Let N_2 represent a neighborhood of $\mathbf{c}$ on which $\text{rank}(\mathbf{F}) = r$, and take N to be whichever of the two neighborhoods N_1 and N_2 has the smallest radius. Then, clearly, $\text{rank}(\mathbf{B}_{11}) = \text{rank}(\mathbf{F}) = r$ for $\mathbf{x} \in N$.

The existence of a generalized inverse $\mathbf{G}$ of $\mathbf{F}$ such that

$$\mathbf{G} = \mathbf{Q}\begin{pmatrix} \mathbf{B}_{11}^{-1} & \mathbf{0} \\ \mathbf{0} & \mathbf{0} \end{pmatrix}\mathbf{P}$$

on N follows from Theorem 9.2.3. That $\mathbf{G}$ is continuously differentiable at $\mathbf{c}$ and that (at $\mathbf{x} = \mathbf{c}$)

$$\frac{\partial \mathbf{G}}{\partial x_j} = -\mathbf{Q}\begin{pmatrix} \mathbf{B}_{11}^{-1}(\partial \mathbf{B}_{11}/\partial x_j)\mathbf{B}_{11}^{-1} & \mathbf{0} \\ \mathbf{0} & \mathbf{0} \end{pmatrix}\mathbf{P}$$

follows from the results of Sections 15.4 and 15.8. To complete the proof, observe that $\partial \mathbf{B}/\partial x_j = \mathbf{P}(\partial \mathbf{F}/\partial x_j)\mathbf{Q}$, implying that

$$\frac{\partial \mathbf{F}}{\partial x_j} = \mathbf{P}'\frac{\partial \mathbf{B}}{\partial x_j}\mathbf{Q}' = \mathbf{P}'\begin{pmatrix} \partial \mathbf{B}_{11}/\partial x_j & \partial \mathbf{B}_{12}/\partial x_j \\ \partial \mathbf{B}_{21}/\partial x_j & \partial \mathbf{B}_{22}/\partial x_j \end{pmatrix}\mathbf{Q}'$$

and hence that

$$\begin{aligned} \mathbf{G}\frac{\partial \mathbf{F}}{\partial x_j}\mathbf{G} &= \mathbf{Q}\begin{pmatrix} \mathbf{B}_{11}^{-1} & \mathbf{0} \\ \mathbf{0} & \mathbf{0} \end{pmatrix}\begin{pmatrix} \partial \mathbf{B}_{11}/\partial x_j & \partial \mathbf{B}_{12}/\partial x_j \\ \partial \mathbf{B}_{21}/\partial x_j & \partial \mathbf{B}_{22}/\partial x_j \end{pmatrix}\begin{pmatrix} \mathbf{B}_{11}^{-1} & \mathbf{0} \\ \mathbf{0} & \mathbf{0} \end{pmatrix}\mathbf{P} \\ &= \mathbf{Q}\begin{bmatrix} \mathbf{B}_{11}^{-1}(\partial \mathbf{B}_{11}/\partial x_j)\mathbf{B}_{11}^{-1} & \mathbf{0} \\ \mathbf{0} & \mathbf{0} \end{bmatrix}\mathbf{P}. \end{aligned}$$

Q.E.D.

According to Theorem 15.10.1, a sufficient condition for the existence of a generalized inverse of $\mathbf{F}$ that is continuously differentiable at $\mathbf{c}$ (where $\mathbf{c}$ is an interior point at which $\mathbf{F}$ is continuously differentiable) is that $\mathbf{F}$ have constant rank on some neighborhood of $\mathbf{c}$. Is this condition necessary as well as sufficient? This question is answered (in the affirmative) by the corollary of the following lemma.

Lemma 15.10.3. Let $\mathbf{F}$ represent a $p \times q$ matrix of functions, defined on a set S, of an m-dimensional column vector $\mathbf{x}$. And let $\mathbf{c}$ represent any interior point of S at which $\mathbf{F}$ is continuous. If there exists a generalized inverse of $\mathbf{F}$ that is continuous at $\mathbf{x} = \mathbf{c}$, then $\mathbf{F}$ has constant rank on some neighborhood of $\mathbf{c}$.

Proof. Suppose that there exists a generalized inverse, say $\mathbf{G}$, of $\mathbf{F}$ that is continuous at $\mathbf{x} = \mathbf{c}$. Then, since sums and products of continuous functions are continuous, $\text{tr}(\mathbf{FG})$ is continuous at $\mathbf{x} = \mathbf{c}$. Moreover, according to result (10.2.1), $\text{tr}(\mathbf{FG}) = \text{rank}(\mathbf{F})$. Thus, $\text{rank}(\mathbf{F})$ is continuous at $\mathbf{x} = \mathbf{c}$. Since $\text{rank}(\mathbf{F})$ is integer-valued, we conclude that $\text{rank}(\mathbf{F})$ is constant on some neighborhood of $\mathbf{c}$. Q.E.D.

Corollary 15.10.4. Let $\mathbf{F}$ represent a $p \times q$ matrix of functions, defined on a set S, of an m-dimensional column vector $\mathbf{x}$. Let $\mathbf{c}$ represent any interior point of S at which $\mathbf{F}$ is continuously differentiable. If there exists a generalized inverse of $\mathbf{F}$ that is continuously differentiable at $\mathbf{x} = \mathbf{c}$, then $\mathbf{F}$ has constant rank on some neighborhood of $\mathbf{c}$.

Proof. Suppose that there exists a generalized inverse, say $\mathbf{G}$, of $\mathbf{F}$ that is continuously differentiable at $\mathbf{c}$. Since (according to Lemma 15.1.1) continuously differentiable functions are continuous, $\mathbf{F}$ and $\mathbf{G}$ are both continuous at $\mathbf{x} = \mathbf{c}$. Thus, it follows from Lemma 15.10.3 that $\mathbf{F}$ has constant rank on some neighborhood of $\mathbf{c}$. Q.E.D.

Let $\mathbf{F}$ represent a $p \times q$ matrix of functions, defined on a set S, of a vector $\mathbf{x} = (x_1, \ldots, x_m)'$ of m variables. Further, let $\mathbf{c}$ represent an interior point of S at which $\mathbf{F}$ is continuously differentiable, and suppose that $\mathbf{F}$ has constant rank on some neighborhood of $\mathbf{c}$. Then, according to Theorem 15.10.1, there exists a generalized inverse $\mathbf{G}$ of $\mathbf{F}$ that is continuously differentiable at $\mathbf{c}$ and whose partial derivatives at $\mathbf{x} = \mathbf{c}$ are given by the formula

$$\frac{\partial \mathbf{G}}{\partial x_j} = -\mathbf{G}\frac{\partial \mathbf{F}}{\partial x_j}\mathbf{G}. \tag{10.1}$$

Is formula (10.1) applicable to every generalized inverse $\mathbf{G}$ (of $\mathbf{F}$) that is continuously differentiable at $\mathbf{c}$? Except in special cases, the answer is no. Suppose, for example, that $\mathbf{F} = \mathbf{0}$ for all $\mathbf{x}$ in some neighborhood N of $\mathbf{c}$ and that $\mathbf{G}$ is chosen in such a way that $\mathbf{G}$ is continuously differentiable at $\mathbf{c}$ and $\partial \mathbf{G}/\partial x_j$ is nonnull—since any $q \times p$ matrix is a generalized inverse of a $p \times q$ null matrix, such a choice is clearly possible. Then, the right side of equality (10.1) is null at $\mathbf{x} = \mathbf{c}$, while the left side is nonnull.

The following lemma relates the partial derivatives (at $\mathbf{x} = \mathbf{c}$) of a generalized inverse of $\mathbf{F}$ to the partial derivatives of $\mathbf{F}$ itself—it does so in a less definitive way than formula (10.1), but in a way that is applicable to any generalized inverse (of $\mathbf{F}$) that is continuously differentiable at $\mathbf{c}$.

Lemma 15.10.5. Let $\mathbf{F}$ represent a $p \times q$ matrix of functions, defined on a set S, of a vector $\mathbf{x} = (x_1, \dots, x_m)'$ of m variables. Further, let $\mathbf{c}$ represent an interior point of S at which $\mathbf{F}$ is continuously differentiable, and (assuming that $\mathbf{F}$ has constant rank on some neighborhood of $\mathbf{c}$) let $\mathbf{G}$ represent any generalized inverse of $\mathbf{F}$ that is continuously differentiable at $\mathbf{c}$. Then (at $\mathbf{x} = \mathbf{c}$),

$$\mathbf{F}\frac{\partial \mathbf{G}}{\partial x_j}\mathbf{F} = -\mathbf{F}\mathbf{G}\frac{\partial \mathbf{F}}{\partial x_j}\mathbf{G}\mathbf{F}. \tag{10.2}$$

Proof. Differentiating both sides of the equality $\mathbf{F} = \mathbf{F}\mathbf{G}\mathbf{F}$ [with the help of result (4.6)], we find that (at $\mathbf{x} = \mathbf{c}$)

$$\frac{\partial \mathbf{F}}{\partial x_j} = \mathbf{F}\mathbf{G}\frac{\partial \mathbf{F}}{\partial x_j} + \mathbf{F}\frac{\partial \mathbf{G}}{\partial x_j}\mathbf{F} + \frac{\partial \mathbf{F}}{\partial x_j}\mathbf{G}\mathbf{F}. \tag{10.3}$$

Premultiplying and postmultiplying both sides of equality (10.3) by $\mathbf{F}\mathbf{G}$ and $\mathbf{G}\mathbf{F}$, respectively, gives

$$\begin{aligned}\mathbf{F}\mathbf{G}\frac{\partial \mathbf{F}}{\partial x_j}\mathbf{G}\mathbf{F} &= \mathbf{F}\mathbf{G}\mathbf{F}\mathbf{G}\frac{\partial \mathbf{F}}{\partial x_j}\mathbf{G}\mathbf{F} + \mathbf{F}\mathbf{G}\mathbf{F}\frac{\partial \mathbf{G}}{\partial x_j}\mathbf{F}\mathbf{G}\mathbf{F} + \mathbf{F}\mathbf{G}\frac{\partial \mathbf{F}}{\partial x_j}\mathbf{G}\mathbf{F}\mathbf{G}\mathbf{F} \\ &= \mathbf{F}\mathbf{G}\frac{\partial \mathbf{F}}{\partial x_j}\mathbf{G}\mathbf{F} + \mathbf{F}\frac{\partial \mathbf{G}}{\partial x_j}\mathbf{F} + \mathbf{F}\mathbf{G}\frac{\partial \mathbf{F}}{\partial x_j}\mathbf{G}\mathbf{F}\end{aligned}$$

or, equivalently,

$$\mathbf{F}\frac{\partial \mathbf{G}}{\partial x_j}\mathbf{F} = -\mathbf{F}\mathbf{G}\frac{\partial \mathbf{F}}{\partial x_j}\mathbf{G}\mathbf{F}.$$

Q.E.D.

The following theorem generalizes result (10.1).

Theorem 15.10.6. Let $\mathbf{F}$, $\mathbf{A}$, and $\mathbf{B}$ represent $p \times q$, $k \times q$, and $p \times r$ matrices of functions, defined on a set S, of a vector $\mathbf{x} = (x_1, \dots, x_m)'$ of m variables. Let $\mathbf{c}$ represent an interior point of S at which $\mathbf{F}$, $\mathbf{A}$, and $\mathbf{B}$ are continuously differentiable. Suppose that there exists a neighborhood N of $\mathbf{c}$ such that (1) $\mathbf{F}$ has constant rank on N and (2) $\mathcal{R}(\mathbf{A}) \subset \mathcal{R}(\mathbf{F})$ and $\mathcal{C}(\mathbf{B}) \subset \mathcal{C}(\mathbf{F})$ for all $\mathbf{x}$ in N. Then, for any generalized inverse $\mathbf{G}$ of $\mathbf{F}$, $\mathbf{A}\mathbf{G}\mathbf{B}$ is continuously differentiable at $\mathbf{c}$, and (at $\mathbf{x} = \mathbf{c}$)

$$\frac{\partial \mathbf{A}\mathbf{G}\mathbf{B}}{\partial x_j} = \mathbf{A}\mathbf{G}\frac{\partial \mathbf{B}}{\partial x_j} - \mathbf{A}\mathbf{G}\frac{\partial \mathbf{F}}{\partial x_j}\mathbf{G}\mathbf{B} + \frac{\partial \mathbf{A}}{\partial x_j}\mathbf{G}\mathbf{B}. \tag{10.4}$$

Theorem 15.10.6 indicates that, even if the generalized inverse $\mathbf{G}$ is not continuously differentiable (at $\mathbf{x} = \mathbf{c}$) or the formula

$$\frac{\partial \mathbf{G}}{\partial x_j} = -\mathbf{G}\frac{\partial \mathbf{F}}{\partial x_j}\mathbf{G} \tag{10.5}$$

is not applicable to $\mathbf{G}$, we can for purposes of obtaining the partial derivatives of $\mathbf{AGB}$ (at $\mathbf{x} = \mathbf{c}$) proceed as though $\mathbf{G}$ is continuously differentiable and formula (10.5) is valid. In the special case where (for $\mathbf{x} \in S$) $\mathbf{A}(\mathbf{x})$ and $\mathbf{B}(\mathbf{x})$ are constant or (more generally) do not vary with x_j, formula (10.4) simplifies to

$$\frac{\partial(\mathbf{AGB})}{\partial x_j} = -\mathbf{AG}\frac{\partial \mathbf{F}}{\partial x_j}\mathbf{GB}. \tag{10.6}$$

Preliminary to proving Theorem 15.10.6, it is convenient to establish the following theorem, which is of some interest in its own right.

Theorem 15.10.7. Let $\mathbf{F}$ represent a $p \times q$ matrix of functions, defined on a set S, of an m-dimensional column vector $\mathbf{x}$, and let $\mathbf{G}$ represent a generalized inverse of $\mathbf{F}$. Further, let $\mathbf{c}$ represent any interior point (of S) at which $\mathbf{F}$ is continuously differentiable, and suppose that $\mathbf{F}$ has constant rank on some neighborhood of $\mathbf{c}$. Then, there exists a second generalized inverse $\mathbf{G}_*$ (of $\mathbf{F}$) such that $\mathbf{G}_*$ is continuously differentiable (at $\mathbf{c}$) and $\mathbf{G}_*(\mathbf{c}) = \mathbf{G}(\mathbf{c})$.

Proof (of Theorem 15.10.7). According to Theorem 15.10.1, there exists a generalized inverse, say $\mathbf{H}$, of $\mathbf{F}$ that is continuously differentiable at $\mathbf{c}$. Let

$$\mathbf{G}_* = \mathbf{H} + \mathbf{Z} - \mathbf{HFZFH},$$

where $\mathbf{Z} = \mathbf{G}(\mathbf{c}) - \mathbf{H}(\mathbf{c})$. Then, it follows from Theorem 9.2.7 that $\mathbf{G}_*$ is a generalized inverse of $\mathbf{F}$ and from the results of Section 15.4 that $\mathbf{G}_*$ is continuously differentiable (at $\mathbf{c}$). Moreover,

$$\begin{aligned}
\mathbf{G}_*(\mathbf{c}) &= \mathbf{H}(\mathbf{c}) + \mathbf{Z} - \mathbf{H}(\mathbf{c})\mathbf{F}(\mathbf{c})[\mathbf{G}(\mathbf{c}) - \mathbf{H}(\mathbf{c})]\mathbf{F}(\mathbf{c})\mathbf{H}(\mathbf{c}) \\
&= \mathbf{H}(\mathbf{c}) + \mathbf{Z} - \mathbf{H}(\mathbf{c})\mathbf{F}(\mathbf{c})\mathbf{H}(\mathbf{c}) + \mathbf{H}(\mathbf{c})\mathbf{F}(\mathbf{c})\mathbf{H}(\mathbf{c}) \\
&= \mathbf{H}(\mathbf{c}) + \mathbf{Z} \\
&= \mathbf{G}(\mathbf{c})
\end{aligned}$$

Q.E.D.

Proof (of Theorem 15.10.6). According to Theorem 15.10.7, there exists a generalized inverse $\mathbf{G}_*$ of $\mathbf{F}$ such that $\mathbf{G}_*$ is continuously differentiable (at $\mathbf{c}$) and $\mathbf{G}_*(\mathbf{c}) = \mathbf{G}(\mathbf{c})$. Moreover, according to Theorem 9.4.1, $\mathbf{AGB} = \mathbf{AG}_*\mathbf{B}$ for every $\mathbf{x}$ in N. Thus, it follows from the results of Section 15.4 that $\mathbf{AGB}$ is continuously differentiable at $\mathbf{c}$ and that (at $\mathbf{x} = \mathbf{c}$)

$$\begin{aligned}
\frac{\partial(\mathbf{AGB})}{\partial x_j} = \frac{\partial(\mathbf{AG}_*\mathbf{B})}{\partial x_j} &= \mathbf{AG}_*\frac{\partial \mathbf{B}}{\partial x_j} + \mathbf{A}\frac{\partial \mathbf{G}_*}{\partial x_j}\mathbf{B} + \frac{\partial \mathbf{A}}{\partial x_j}\mathbf{G}_*\mathbf{B} \\
&= \mathbf{AG}\frac{\partial \mathbf{B}}{\partial x_j} + \mathbf{A}\frac{\partial \mathbf{G}_*}{\partial x_j}\mathbf{B} + \frac{\partial \mathbf{A}}{\partial x_j}\mathbf{GB}.
\end{aligned}$$

To complete the proof, observe that $\mathbf{A}(\mathbf{c}) = \mathbf{LF}(\mathbf{c})$ and $\mathbf{B}(\mathbf{c}) = \mathbf{F}(\mathbf{c})\mathbf{R}$ for some matrices $\mathbf{L}$ and $\mathbf{R}$ and hence (in light of Lemma 15.10.5) that (at $\mathbf{x} = \mathbf{c}$)

$$\mathbf{A}\frac{\partial \mathbf{G}_*}{\partial x_j}\mathbf{B} = \mathbf{L}\mathbf{F}\frac{\partial \mathbf{G}_*}{\partial x_j}\mathbf{F}\mathbf{R} = -\mathbf{L}\mathbf{F}\mathbf{G}_*\frac{\partial \mathbf{F}}{\partial x_j}\mathbf{G}_*\mathbf{F}\mathbf{R}$$
$$= -\mathbf{A}\mathbf{G}_*\frac{\partial \mathbf{F}}{\partial x_j}\mathbf{G}_*\mathbf{B} = -\mathbf{A}\mathbf{G}\frac{\partial \mathbf{F}}{\partial x_j}\mathbf{G}\mathbf{B}.$$

Q.E.D.

15.11 Differentiation of Projection Matrices

Recall (from Section 14.12d) that $\mathbf{P}_{\mathbf{X},\mathbf{W}}$ represents the matrix $\mathbf{X}(\mathbf{X}'\mathbf{W}\mathbf{X})^{-}\mathbf{X}'\mathbf{W}$ (where $\mathbf{X}$ is an $n \times p$ matrix and $\mathbf{W}$ an $n \times n$ matrix) and that, for an $n \times n$ symmetric positive definite matrix $\mathbf{W}$, $\mathbf{P}_{\mathbf{X},\mathbf{W}}$ is the projection matrix for $\mathcal{C}(\mathbf{X})$ with respect to $\mathbf{W}$. If the elements of $\mathbf{W}$ and/or $\mathbf{X}$ are functions of a vector, say $\mathbf{z}$, of variables, then the differentiation (with respect to the elements of $\mathbf{z}$) of $\mathbf{P}_{\mathbf{X},\mathbf{W}}$ may be of interest. The following theorem gives some results on the differentiation of $\mathbf{P}_{\mathbf{X},\mathbf{W}}$ and also on the differentiation of $\mathbf{W}\mathbf{P}_{\mathbf{X},\mathbf{W}}$ and $\mathbf{W} - \mathbf{W}\mathbf{P}_{\mathbf{X},\mathbf{W}}$.

Theorem 15.11.1. Let $\mathbf{X}$ represent an $n \times p$ matrix and $\mathbf{W}$ an $n \times n$ symmetric positive definite matrix, and suppose that the elements of $\mathbf{X}$ and $\mathbf{W}$ are functions, defined on a set S, of a vector $\mathbf{z} = (z_1, \ldots, z_m)'$ of m variables. Further, let $\mathbf{c}$ represent any interior point (of S) at which $\mathbf{X}$ and $\mathbf{W}$ are continuously differentiable, and suppose that $\mathbf{X}$ has constant rank on some neighborhood of $\mathbf{c}$. Then, $\mathbf{P}_{\mathbf{X},\mathbf{W}}$, $\mathbf{W}\mathbf{P}_{\mathbf{X},\mathbf{W}}$, and $\mathbf{W} - \mathbf{W}\mathbf{P}_{\mathbf{X},\mathbf{W}}$ are continuously differentiable at $\mathbf{c}$, and (at $\mathbf{z} = \mathbf{c}$)

$$\frac{\partial \mathbf{P}_{\mathbf{X},\mathbf{W}}}{\partial z_j} = (\mathbf{I} - \mathbf{P}_{\mathbf{X},\mathbf{W}})\frac{\partial \mathbf{X}}{\partial z_j}(\mathbf{X}'\mathbf{W}\mathbf{X})^{-}\mathbf{X}'\mathbf{W}$$
$$+ \mathbf{X}(\mathbf{X}'\mathbf{W}\mathbf{X})^{-}\left(\frac{\partial \mathbf{X}}{\partial z_j}\right)'\mathbf{W}(\mathbf{I} - \mathbf{P}_{\mathbf{X},\mathbf{W}})$$
$$+ \mathbf{X}(\mathbf{X}'\mathbf{W}\mathbf{X})^{-}\mathbf{X}'\frac{\partial \mathbf{W}}{\partial z_j}(\mathbf{I} - \mathbf{P}_{\mathbf{X},\mathbf{W}}), \qquad (11.1)$$

$$\frac{\partial (\mathbf{W}\mathbf{P}_{\mathbf{X},\mathbf{W}})}{\partial z_j} = \frac{\partial \mathbf{W}}{\partial z_j} - (\mathbf{I} - \mathbf{P}'_{\mathbf{X},\mathbf{W}})\frac{\partial \mathbf{W}}{\partial z_j}(\mathbf{I} - \mathbf{P}_{\mathbf{X},\mathbf{W}})$$
$$+ \mathbf{W}(\mathbf{I} - \mathbf{P}_{\mathbf{X},\mathbf{W}})\frac{\partial \mathbf{X}}{\partial z_j}(\mathbf{X}'\mathbf{W}\mathbf{X})^{-}\mathbf{X}'\mathbf{W}$$
$$+ \mathbf{W}\mathbf{X}(\mathbf{X}'\mathbf{W}\mathbf{X})^{-}\left(\frac{\partial \mathbf{X}}{\partial z_j}\right)'\mathbf{W}(\mathbf{I} - \mathbf{P}_{\mathbf{X},\mathbf{W}}), \qquad (11.2)$$

$$\frac{\partial (\mathbf{W} - \mathbf{W}\mathbf{P}_{\mathbf{X},\mathbf{W}})}{\partial z_j} = (\mathbf{I} - \mathbf{P}'_{\mathbf{X},\mathbf{W}})\frac{\partial \mathbf{W}}{\partial z_j}(\mathbf{I} - \mathbf{P}_{\mathbf{X},\mathbf{W}})$$
$$- \mathbf{W}(\mathbf{I} - \mathbf{P}_{\mathbf{X},\mathbf{W}})\frac{\partial \mathbf{X}}{\partial z_j}(\mathbf{X}'\mathbf{W}\mathbf{X})^{-}\mathbf{X}'\mathbf{W}$$
$$- \mathbf{W}\mathbf{X}(\mathbf{X}'\mathbf{W}\mathbf{X})^{-}\left(\frac{\partial \mathbf{X}}{\partial z_j}\right)'\mathbf{W}(\mathbf{I} - \mathbf{P}_{\mathbf{X},\mathbf{W}}). \qquad (11.3)$$

The results of Theorem 15.11.1 are special cases of various results encompassed in the following theorem {as is evident upon recalling (from Corollary 14.11.3) that $\text{rank}(\mathbf{X}'\mathbf{W}\mathbf{X}) = \text{rank}(\mathbf{X})$ and (from Theorem 14.12.11) that $\mathbf{X}[(\mathbf{X}'\mathbf{W}\mathbf{X})^-]'\mathbf{X}' = \mathbf{X}(\mathbf{X}'\mathbf{W}\mathbf{X})^-\mathbf{X}'$ and hence that $\mathbf{W}\mathbf{X}(\mathbf{X}'\mathbf{W}\mathbf{X})^-\mathbf{X}' = \mathbf{P}'_{\mathbf{X},\mathbf{W}}$}.

Theorem 15.11.2. Let $\mathbf{W}, \mathbf{X}$, and $\mathbf{Y}$ represent $r \times n$, $n \times p$, and $q \times r$ matrices of functions, defined on a set S, of a vector $\mathbf{z} = (z_1, \ldots, z_m)'$ of m variables. Let $\mathbf{c}$ represent any interior point of S at which $\mathbf{W}$, $\mathbf{X}$, and $\mathbf{Y}$ are continuously differentiable. Suppose that there exists a neighborhood N of $\mathbf{c}$ such that (1) $\mathbf{YWX}$ has constant rank on N and (2) $\text{rank}(\mathbf{X}) = \text{rank}(\mathbf{Y}) = \text{rank}(\mathbf{YWX})$ for all $\mathbf{z}$ in N. Define $\mathbf{K} = \mathbf{X}(\mathbf{YWX})^-\mathbf{Y}$. Then, $\mathbf{K}$, $\mathbf{KW}$, $\mathbf{WKW}$, and $\mathbf{W} - \mathbf{WKW}$ are continuously differentiable at $\mathbf{c}$, and (at $\mathbf{z} = \mathbf{c}$)

$$\frac{\partial \mathbf{K}}{\partial z_j} = (\mathbf{I}-\mathbf{KW})\frac{\partial \mathbf{X}}{\partial z_j}(\mathbf{YWX})^-\mathbf{Y}+\mathbf{X}(\mathbf{YWX})^-\frac{\partial \mathbf{Y}}{\partial z_j}(\mathbf{I}-\mathbf{WK})-\mathbf{K}\frac{\partial \mathbf{W}}{\partial z_j}\mathbf{K}, \quad (11.4)$$

$$\begin{aligned}\frac{\partial (\mathbf{KW})}{\partial z_j} &= (\mathbf{I}-\mathbf{KW})\frac{\partial \mathbf{X}}{\partial z_j}(\mathbf{YWX})^-\mathbf{YW} \\ &\quad + \mathbf{X}(\mathbf{YWX})^-\frac{\partial \mathbf{Y}}{\partial z_j}\mathbf{W}(\mathbf{I}-\mathbf{KW}) + \mathbf{K}\frac{\partial \mathbf{W}}{\partial z_j}(\mathbf{I}-\mathbf{KW}), \quad (11.5)\end{aligned}$$

$$\begin{aligned}\frac{\partial (\mathbf{WKW})}{\partial z_j} &= \frac{\partial \mathbf{W}}{\partial z_j} - (\mathbf{I}-\mathbf{WK})\frac{\partial \mathbf{W}}{\partial z_j}(\mathbf{I}-\mathbf{KW}) \\ &\quad + \mathbf{W}(\mathbf{I}-\mathbf{KW})\frac{\partial \mathbf{X}}{\partial z_j}(\mathbf{YWX})^-\mathbf{YW} \\ &\quad + \mathbf{WX}(\mathbf{YWX})^-\frac{\partial \mathbf{Y}}{\partial z_j}\mathbf{W}(\mathbf{I}-\mathbf{KW}), \quad (11.6)\end{aligned}$$

$$\begin{aligned}\frac{\partial (\mathbf{W}-\mathbf{WKW})}{\partial z_j} &= (\mathbf{I}-\mathbf{WK})\frac{\partial \mathbf{W}}{\partial z_j}(\mathbf{I}-\mathbf{KW}) \\ &\quad - \mathbf{W}(\mathbf{I}-\mathbf{KW})\frac{\partial \mathbf{W}}{\partial z_j}(\mathbf{YWX})^-\mathbf{YW} \\ &\quad - \mathbf{WX}(\mathbf{YWX})^-\frac{\partial \mathbf{Y}}{\partial z_j}\mathbf{W}(\mathbf{I}-\mathbf{KW}). \quad (11.7)\end{aligned}$$

Proof. In light of the results of Section 15.4, we have that $\mathbf{YWX}$ is continuously differentiable at $\mathbf{c}$. And, in light of Corollary 4.4.7, we have that $\mathcal{R}(\mathbf{X}) = \mathcal{R}(\mathbf{YWX})$ and $\mathcal{C}(\mathbf{Y}) = \mathcal{C}(\mathbf{YWX})$ for all $\mathbf{z}$ in N. Applying Theorem 15.10.6 (with $\mathbf{F} = \mathbf{YWX}$, $\mathbf{A} = \mathbf{X}$, and $\mathbf{B} = \mathbf{Y}$), we find that $\mathbf{K}$ is continuously differentiable at $\mathbf{c}$ and that (at $\mathbf{z} = \mathbf{c}$)

$$\begin{aligned}\frac{\partial \mathbf{K}}{\partial z_j} &= \mathbf{X}(\mathbf{YWX})^-\frac{\partial \mathbf{Y}}{\partial z_j} \\ &\quad - \mathbf{X}(\mathbf{YWX})^-\frac{\partial (\mathbf{YWX})}{\partial z_j}(\mathbf{YWX})^-\mathbf{Y} + \frac{\partial \mathbf{X}}{\partial z_j}(\mathbf{YWX})^-\mathbf{Y}. \quad (11.8)\end{aligned}$$

Moreover,

$$\frac{\partial(\mathbf{YWX})}{\partial z_j} = \mathbf{YW}\frac{\partial \mathbf{X}}{\partial z_j} + \mathbf{Y}\frac{\partial \mathbf{W}}{\partial z_j}\mathbf{X} + \frac{\partial \mathbf{Y}}{\partial z_j}\mathbf{WX}. \tag{11.9}$$

Upon substituting expression (11.9) for $\partial(\mathbf{YWX})/\partial z_j$ in expression (11.8), we obtain

$$\begin{aligned}\frac{\partial \mathbf{K}}{\partial z_j} &= \mathbf{X}(\mathbf{YWX})^{-}\frac{\partial \mathbf{Y}}{\partial z_j} - \mathbf{KW}\frac{\partial \mathbf{X}}{\partial z_j}(\mathbf{YWX})^{-}\mathbf{Y} - \mathbf{K}\frac{\partial \mathbf{W}}{\partial z_j}\mathbf{K} \\ &\qquad - \mathbf{X}(\mathbf{YWX})^{-}\frac{\partial \mathbf{Y}}{\partial z_j}\mathbf{WK} + \frac{\partial \mathbf{X}}{\partial z_j}(\mathbf{YWX})^{-}\mathbf{Y} \\ &= (\mathbf{I} - \mathbf{KW})\frac{\partial \mathbf{X}}{\partial z_j}(\mathbf{YWX})^{-}\mathbf{Y} + \mathbf{X}(\mathbf{YWX})^{-}\frac{\partial \mathbf{Y}}{\partial z_j}(\mathbf{I} - \mathbf{WK}) - \mathbf{K}\frac{\partial \mathbf{W}}{\partial z_j}\mathbf{K},\end{aligned}$$

thereby validating formula (11.4).

Further, since $\mathbf{K}$ is continuously differentiable at $\mathbf{c}$, it follows from the results of Section 15.4 that $\mathbf{KW}$, $\mathbf{WKW}$, and $\mathbf{W} - \mathbf{WKW}$ are continuously differentiable at $\mathbf{c}$ and that (at $\mathbf{z} = \mathbf{c}$)

$$\frac{\partial(\mathbf{KW})}{\partial z_j} = \mathbf{K}\frac{\partial \mathbf{W}}{\partial z_j} + \frac{\partial \mathbf{K}}{\partial z_j}\mathbf{W}, \tag{11.10}$$

$$\frac{\partial(\mathbf{WKW})}{\partial z_j} = \mathbf{W}\frac{\partial(\mathbf{KW})}{\partial z_j} + \frac{\partial \mathbf{W}}{\partial z_j}\mathbf{KW}, \tag{11.11}$$

$$\frac{\partial(\mathbf{W} - \mathbf{WKW})}{\partial z_j} = \frac{\partial \mathbf{W}}{\partial z_j} - \frac{\partial(\mathbf{WKW})}{\partial z_j}. \tag{11.12}$$

Upon substituting expression (11.4) for $\partial\mathbf{K}/\partial z_j$ in expression (11.10), we obtain (after a little simplification) expression (11.5). Similarly, upon substituting expression (11.5) for $\partial(\mathbf{KW})/\partial z_j$ in expression (11.11), we obtain expression (11.6), which when substituted for $\partial(\mathbf{WKW})/\partial z_j$ in expression (11.12) gives expression (11.7). Q.E.D.

According to Theorem 15.11.1, a sufficient condition for $\mathbf{P}_{\mathbf{X},\mathbf{W}}$ to be continuously differentiable at $\mathbf{c}$ (where $\mathbf{c}$ is an interior point at which $\mathbf{X}$ and $\mathbf{W}$ are continuously differentiable) is that $\mathbf{X}$ have constant rank on some neighborhood of $\mathbf{c}$. Is this condition necessary as well as sufficient? This question is answered (in the affirmative) by the corollary of the following lemma.

Lemma 15.11.3. Let $\mathbf{X}$ represent an $n \times p$ matrix and $\mathbf{W}$ an $n \times n$ symmetric positive definite matrix, and suppose that the elements of $\mathbf{X}$ and $\mathbf{W}$ are functions, defined on a set S, of an m-dimensional column vector $\mathbf{z}$. And let $\mathbf{c}$ represent any interior point of S at which $\mathbf{X}$ and $\mathbf{W}$ are continuous. If $\mathbf{P}_{\mathbf{X},\mathbf{W}}$ is continuous at $\mathbf{z} = \mathbf{c}$, then $\mathbf{X}$ has constant rank on some neighborhood of $\mathbf{c}$.

Proof. Suppose that $\mathbf{P}_{\mathbf{X},\mathbf{W}}$ is continuous at $\mathbf{z} = \mathbf{c}$. Then, $\text{tr}(\mathbf{P}_{\mathbf{X},\mathbf{W}})$ is continuous at $\mathbf{z} = \mathbf{c}$. Moreover, since (according to Theorem 14.12.11) $\mathbf{P}_{\mathbf{X},\mathbf{W}}$ is idempotent and $\text{rank}(\mathbf{X}) = \text{rank}(\mathbf{P}_{\mathbf{X},\mathbf{W}})$, we have (in light of Corollary 10.2.2) that $\text{rank}(\mathbf{X}) = \text{tr}(\mathbf{P}_{\mathbf{X},\mathbf{W}})$. Thus, $\text{rank}(\mathbf{X})$ is continuous at $\mathbf{z} = \mathbf{c}$. Since $\text{rank}(\mathbf{X})$ is integer-valued, we conclude that $\text{rank}(\mathbf{X})$ is constant on some neighborhood of $\mathbf{c}$. Q.E.D.

Corollary 15.11.4. Let $\mathbf{X}$ represent an $n \times p$ matrix and $\mathbf{W}$ an $n \times n$ symmetric positive definite matrix, and suppose that the elements of $\mathbf{X}$ and $\mathbf{W}$ are functions, defined on a set S, of an m-dimensional column vector $\mathbf{z}$. Let $\mathbf{c}$ represent any interior point of S at which $\mathbf{X}$ and $\mathbf{W}$ are continuously differentiable. If $\mathbf{P}_{\mathbf{X},\mathbf{W}}$ is continuously differentiable at $\mathbf{c}$, then $\mathbf{X}$ has constant rank on some neighborhood of $\mathbf{c}$.

Proof. Suppose that $\mathbf{P}_{\mathbf{X},\mathbf{W}}$ is continuously differentiable at $\mathbf{c}$. Since (according to Lemma 15.1.1) continuously differentiable functions are continuous, $\mathbf{X}$, $\mathbf{W}$, and $\mathbf{P}_{\mathbf{X},\mathbf{W}}$ are all continuous at $\mathbf{z} = \mathbf{c}$. Thus, it follows from Lemma 15.11.3 that $\mathbf{X}$ has constant rank on some neighborhood of $\mathbf{c}$. Q.E.D.

In the special case where $\mathbf{X}$ is a matrix of constants, or, more generally, where (for $\mathbf{z} \in S$) $\mathbf{X}(\mathbf{z})$ does not vary with z_j, formulas (11.1)–(11.3) reduce to

$$\frac{\partial \mathbf{P}_{\mathbf{X},\mathbf{W}}}{\partial z_j} = \mathbf{X}(\mathbf{X}'\mathbf{W}\mathbf{X})^{-}\mathbf{X}'\frac{\partial \mathbf{W}}{\partial z_j}(\mathbf{I} - \mathbf{P}_{\mathbf{X},\mathbf{W}}), \tag{11.13}$$

$$\frac{\partial (\mathbf{W}\mathbf{P}_{\mathbf{X},\mathbf{W}})}{\partial z_j} = \frac{\partial \mathbf{W}}{\partial z_j} - (\mathbf{I} - \mathbf{P}'_{\mathbf{X},\mathbf{W}})\frac{\partial \mathbf{W}}{\partial z_j}(\mathbf{I} - \mathbf{P}_{\mathbf{X},\mathbf{W}}), \tag{11.14}$$

$$\frac{\partial (\mathbf{W} - \mathbf{W}\mathbf{P}_{\mathbf{X},\mathbf{W}})}{\partial z_j} = (\mathbf{I} - \mathbf{P}'_{\mathbf{X},\mathbf{W}})\frac{\partial \mathbf{W}}{\partial z_j}(\mathbf{I} - \mathbf{P}_{\mathbf{X},\mathbf{W}}). \tag{11.15}$$

Formulas (11.1)–(11.3) can be expressed in a more general form by making use of the following lemma.

Lemma 15.11.5. Let $\mathbf{X}$ represent an $n \times p$ matrix and $\mathbf{W}$ an $n \times n$ symmetric positive definite matrix, and suppose that the elements of $\mathbf{X}$ and $\mathbf{W}$ are functions, defined on a set S, of a vector $\mathbf{z} = (z_1, \ldots, z_m)'$ of m variables. Let $\mathbf{c}$ represent any interior point (of S) at which $\mathbf{W}$ and $\mathbf{X}$ are continuously differentiable, and suppose that $\mathbf{X}$ has constant rank on some neighborhood of $\mathbf{c}$. Further, let $\mathbf{B}$ represent any $p \times n$ matrix such that $\mathbf{X}'\mathbf{W}\mathbf{X}\mathbf{B} = \mathbf{X}'\mathbf{W}$. Then, at $\mathbf{z} = \mathbf{c}$,

$$\left(\frac{\partial \mathbf{P}_{\mathbf{X},\mathbf{W}}}{\partial z_j}\right)'\mathbf{W}\mathbf{P}_{\mathbf{X},\mathbf{W}} = (\mathbf{I} - \mathbf{P}'_{\mathbf{X},\mathbf{W}})\frac{\partial \mathbf{W}}{\partial z_j}\mathbf{P}_{\mathbf{X},\mathbf{W}} + \mathbf{W}(\mathbf{I} - \mathbf{P}_{\mathbf{X},\mathbf{W}})\frac{\partial \mathbf{X}}{\partial z_j}\mathbf{B}. \tag{11.16}$$

Proof. Recall (from Theorem 14.12.11) that

$$\mathbf{P}'_{\mathbf{X},\mathbf{W}}\mathbf{W}\mathbf{X} = \mathbf{W}\mathbf{X}$$

(for all $\mathbf{z}$ in S) and (from Theorem 15.11.1) that $\mathbf{P}_{\mathbf{X},\mathbf{W}}$ is continuously differentiable at $\mathbf{c}$. Thus (at $\mathbf{z} = \mathbf{c}$)

$$\frac{\partial (\mathbf{W}\mathbf{X})}{\partial z_j} = \frac{\partial (\mathbf{P}'_{\mathbf{X},\mathbf{W}}\mathbf{W}\mathbf{X})}{\partial z_j} = \mathbf{P}'_{\mathbf{X},\mathbf{W}}\frac{\partial (\mathbf{W}\mathbf{X})}{\partial z_j} + \left(\frac{\partial \mathbf{P}_{\mathbf{X},\mathbf{W}}}{\partial z_j}\right)'\mathbf{W}\mathbf{X},$$

so that

$$\left(\frac{\partial \mathbf{P}_{\mathbf{X},\mathbf{W}}}{\partial z_j}\right)'\mathbf{W}\mathbf{X} = (\mathbf{I} - \mathbf{P}'_{\mathbf{X},\mathbf{W}})\frac{\partial(\mathbf{W}\mathbf{X})}{\partial z_j}$$
$$= (\mathbf{I} - \mathbf{P}'_{\mathbf{X},\mathbf{W}})\frac{\partial \mathbf{W}}{\partial z_j}\mathbf{X} + (\mathbf{I} - \mathbf{P}'_{\mathbf{X},\mathbf{W}})\mathbf{W}\frac{\partial \mathbf{X}}{\partial z_j},$$

implying that

$$\left(\frac{\partial \mathbf{P}_{\mathbf{X},\mathbf{W}}}{\partial z_j}\right)'\mathbf{W}\mathbf{X}\mathbf{B} = (\mathbf{I} - \mathbf{P}'_{\mathbf{X},\mathbf{W}})\frac{\partial \mathbf{W}}{\partial z_j}\mathbf{X}\mathbf{B} + (\mathbf{I} - \mathbf{P}'_{\mathbf{X},\mathbf{W}})\mathbf{W}\frac{\partial \mathbf{X}}{\partial z_j}\mathbf{B}. \tag{11.17}$$

Moreover, by making use of Parts (2) and (3) of Theorem 14.12.11, equality (11.17) can be reexpressed as equality (11.16). Q.E.D.

Since neither $(\partial \mathbf{P}_{\mathbf{X},\mathbf{W}}/\partial z_j)'\mathbf{W}\mathbf{P}_{\mathbf{X},\mathbf{W}}$ nor $(\mathbf{I} - \mathbf{P}'_{\mathbf{X},\mathbf{W}})(\partial \mathbf{W}/\partial z_j)\mathbf{P}_{\mathbf{X},\mathbf{W}}$ involves $\mathbf{B}$, it follows from Lemma 15.11.5 that $\mathbf{W}(\mathbf{I} - \mathbf{P}_{\mathbf{X},\mathbf{W}})(\partial \mathbf{X}/\partial z_j)\mathbf{B}$ is invariant to the choice of $\mathbf{B}$. Thus, since the choices for $\mathbf{B}$ include $(\mathbf{X}'\mathbf{W}\mathbf{X})^-\mathbf{X}'\mathbf{W}$ and $[(\mathbf{X}'\mathbf{W}\mathbf{X})^-]'\mathbf{X}'\mathbf{W}$, we have that

$$\mathbf{W}(\mathbf{I} - \mathbf{P}_{\mathbf{X},\mathbf{W}})\frac{\partial \mathbf{X}}{\partial z_j}(\mathbf{X}'\mathbf{W}\mathbf{X})^-\mathbf{X}'\mathbf{W} = \mathbf{W}(\mathbf{I} - \mathbf{P}_{\mathbf{X},\mathbf{W}})\frac{\partial \mathbf{X}}{\partial z_j}\mathbf{B} \tag{11.18}$$

and

$$\mathbf{W}(\mathbf{I} - \mathbf{P}_{\mathbf{X},\mathbf{W}})\frac{\partial \mathbf{X}}{\partial z_j}[(\mathbf{X}'\mathbf{W}\mathbf{X})^-]'\mathbf{X}'\mathbf{W} = \mathbf{W}(\mathbf{I} - \mathbf{P}_{\mathbf{X},\mathbf{W}})\frac{\partial \mathbf{X}}{\partial z_j}\mathbf{B},$$

which [since $\mathbf{W}(\mathbf{I} - \mathbf{P}_{\mathbf{X},\mathbf{W}})$ is symmetric] is equivalent to

$$\mathbf{W}\mathbf{X}(\mathbf{X}'\mathbf{W}\mathbf{X})^-\left(\frac{\partial \mathbf{X}}{\partial z_j}\right)'\mathbf{W}(\mathbf{I} - \mathbf{P}_{\mathbf{X},\mathbf{W}}) = \left[\mathbf{W}(\mathbf{I} - \mathbf{P}_{\mathbf{X},\mathbf{W}})\frac{\partial \mathbf{X}}{\partial z_j}\mathbf{B}\right]'. \tag{11.19}$$

Moreover, upon premultiplying both sides of equalities (11.18) and (11.19) by $\mathbf{W}^{-1}$, we find that

$$(\mathbf{I} - \mathbf{P}_{\mathbf{X},\mathbf{W}})\frac{\partial \mathbf{X}}{\partial z_j}(\mathbf{X}'\mathbf{W}\mathbf{X})^-\mathbf{X}'\mathbf{W} = (\mathbf{I} - \mathbf{P}_{\mathbf{X},\mathbf{W}})\frac{\partial \mathbf{X}}{\partial z_j}\mathbf{B} \tag{11.20}$$

and

$$\mathbf{X}(\mathbf{X}'\mathbf{W}\mathbf{X})^-\left(\frac{\partial \mathbf{X}}{\partial z_j}\right)'\mathbf{W}(\mathbf{I} - \mathbf{P}_{\mathbf{X},\mathbf{W}}) = \mathbf{W}^{-1}\left[\mathbf{W}(\mathbf{I} - \mathbf{P}_{\mathbf{X},\mathbf{W}})\frac{\partial \mathbf{X}}{\partial z_j}\mathbf{B}\right]'. \tag{11.21}$$

By using results (11.18)–(11.21) [and writing $\mathbf{P}_{\mathbf{X},\mathbf{W}}\mathbf{W}^{-1}$ for $\mathbf{X}(\mathbf{X}'\mathbf{W}\mathbf{X})^-\mathbf{X}'$], formulas (11.1)–(11.3) can be reexpressed as

$$\frac{\partial \mathbf{P}_{\mathbf{X},\mathbf{W}}}{\partial z_j} = (\mathbf{I} - \mathbf{P}_{\mathbf{X},\mathbf{W}})\frac{\partial \mathbf{X}}{\partial z_j}\mathbf{B} + \mathbf{W}^{-1}\left[\mathbf{W}(\mathbf{I} - \mathbf{P}_{\mathbf{X},\mathbf{W}})\frac{\partial \mathbf{X}}{\partial z_j}\mathbf{B}\right]'$$
$$+ \mathbf{P}_{\mathbf{X},\mathbf{W}}\mathbf{W}^{-1}\frac{\partial \mathbf{W}}{\partial z_j}(\mathbf{I} - \mathbf{P}_{\mathbf{X},\mathbf{W}}), \tag{11.22}$$

$$\frac{\partial(\mathbf{W}\mathbf{P}_{\mathbf{X},\mathbf{W}})}{\partial z_j} = \frac{\partial \mathbf{W}}{\partial z_j} - (\mathbf{I} - \mathbf{P}'_{\mathbf{X},\mathbf{W}})\frac{\partial \mathbf{W}}{\partial z_j}(\mathbf{I} - \mathbf{P}_{\mathbf{X},\mathbf{W}}) + \mathbf{W}(\mathbf{I} - \mathbf{P}_{\mathbf{X},\mathbf{W}})\frac{\partial \mathbf{X}}{\partial z_j}\mathbf{B} + \left[\mathbf{W}(\mathbf{I} - \mathbf{P}_{\mathbf{X},\mathbf{W}})\frac{\partial \mathbf{X}}{\partial z_j}\mathbf{B}\right]', \quad (11.23)$$

$$\frac{\partial(\mathbf{W} - \mathbf{W}\mathbf{P}_{\mathbf{X},\mathbf{W}})}{\partial z_j} = (\mathbf{I} - \mathbf{P}'_{\mathbf{X},\mathbf{W}})\frac{\partial \mathbf{W}}{\partial z_j}(\mathbf{I} - \mathbf{P}_{\mathbf{X},\mathbf{W}}) - \mathbf{W}(\mathbf{I} - \mathbf{P}_{\mathbf{X},\mathbf{W}})\frac{\partial \mathbf{X}}{\partial z_j}\mathbf{B} - \left[\mathbf{W}(\mathbf{I} - \mathbf{P}_{\mathbf{X},\mathbf{W}})\frac{\partial \mathbf{X}}{\partial z_j}\mathbf{B}\right]' \quad (11.24)$$

(where $\mathbf{B}$ is any $p \times n$ matrix such that $\mathbf{X}'\mathbf{W}\mathbf{X}\mathbf{B} = \mathbf{X}'\mathbf{W}$).

15.12 Evaluation of Some Multiple Integrals

Let $\mathbf{x} = (x_1, \ldots, x_n)'$ represent a (column) vector of n variables $x_1, \ldots, x_n$. Further, let $\mathbf{W}$ represent an $n \times n$ symmetric positive definite matrix, let $\mathbf{A}$ represent an $n \times n$ matrix, and let $\mathbf{c}$ and $\mathbf{k}$ represent $n \times 1$ vectors. Consider the evaluation of the following three integrals:

$$\int_{\mathbb{R}^n} \exp[-(1/2)(\mathbf{x} - \mathbf{c})'\mathbf{W}(\mathbf{x} - \mathbf{c})]\, d\mathbf{x},$$

$$\int_{\mathbb{R}^n} (\mathbf{k}'\mathbf{x}) \exp[-(1/2)(\mathbf{x} - \mathbf{c})'\mathbf{W}(\mathbf{x} - \mathbf{c})]\, d\mathbf{x},$$

$$\int_{\mathbb{R}^n} (\mathbf{x} - \mathbf{c})'\mathbf{A}(\mathbf{x} - \mathbf{c}) \exp[-(1/2)(\mathbf{x} - \mathbf{c})'\mathbf{W}(\mathbf{x} - \mathbf{c})]\, d\mathbf{x}$$

[where the symbol $\int_{\mathbb{R}^n} f(\mathbf{z})\, d\mathbf{z}$ is used to denote the integral of a function $f(\mathbf{z})$ of a (column) vector $\mathbf{z}$ of n variables over all of $\mathbb{R}^n$].

These three integrals are of considerable importance in probability and statistics, where they arise in connection with the multivariate normal distribution (e.g., Searle 1971, chap. 2). They can be evaluated by introducing a suitable change of variables and by recalling (from the integral calculus of a single variable) that

$$\int_{-\infty}^{\infty} e^{-x^2/2}\, dx = (2\pi)^{1/2},$$

$$\int_{-\infty}^{\infty} x e^{-x^2/2}\, dx = 0,$$

$$\int_{-\infty}^{\infty} x^2 e^{-x^2/2}\, dx = (2\pi)^{1/2}.$$

According to Corollary 14.3.13, there exists an $n \times n$ nonsingular matrix $\mathbf{P}$ such that $\mathbf{W} = \mathbf{P}'\mathbf{P}$. Define

$$\mathbf{z} = \mathbf{P}(\mathbf{x} - \mathbf{c}),$$

so that $\mathbf{x} = \mathbf{P}^{-1}\mathbf{z} + \mathbf{c}$. Further, let J represent the Jacobian of $\mathbf{P}^{-1}\mathbf{z} + \mathbf{c}$ (where $\mathbf{P}^{-1}\mathbf{z} + \mathbf{c}$ is regarded as a vector of functions of $\mathbf{z}$). Then, observing [in light of result (3.10)] that

$$\frac{\partial(\mathbf{P}^{-1}\mathbf{z} + \mathbf{c})}{\partial \mathbf{z}'} = \mathbf{P}^{-1},$$

we find that

$$J = |\mathbf{P}^{-1}| = |\mathbf{P}|^{-1}.$$

And, since $|\mathbf{W}| = |\mathbf{P}'\mathbf{P}| = |\mathbf{P}|^2$, we have that

$$J = \pm|\mathbf{W}|^{-1/2}.$$

Thus, upon making use of standard results on changes of variables (e.g., Bartle 1976, sec. 45), we find that

$$\begin{aligned}
\int_{\mathcal{R}^n} &\exp[-(1/2)(\mathbf{x} - \mathbf{c})'\mathbf{W}(\mathbf{x} - \mathbf{c})]\, d\mathbf{x} \\
&= \int_{\mathcal{R}^n} \exp[-(1/2)\mathbf{z}'\mathbf{z}]\, |J|\, d\mathbf{z} \\
&= |\mathbf{W}|^{-1/2} \int_{-\infty}^{\infty} e^{-z_1^2/2}\, dz_1 \cdots \int_{-\infty}^{\infty} e^{-z_n^2/2} dz_n \\
&= (2\pi)^{n/2}|\mathbf{W}|^{-1/2}. \qquad (12.1)
\end{aligned}$$

Further, letting d_i represent the ith element of the $1 \times n$ vector $\mathbf{k}'\mathbf{P}^{-1}$ [and observing that (for $i = 1, \ldots, n$) $\int_{-\infty}^{\infty} z_i e^{-z_i^2/2}\, dz_i = 0$], we find that

$$\begin{aligned}
\int_{\mathcal{R}^n} &(\mathbf{k}'\mathbf{x}) \exp[-(1/2)(\mathbf{x} - \mathbf{c})'\mathbf{W}(\mathbf{x} - \mathbf{c})]\, d\mathbf{x} \\
&= (\mathbf{k}'\mathbf{c}) \int_{\mathcal{R}^n} \exp[-(1/2)(\mathbf{x} - \mathbf{c})'\mathbf{W}(\mathbf{x} - \mathbf{c})]\, d\mathbf{x} \\
&\qquad + \int_{\mathcal{R}^n} \mathbf{k}'(\mathbf{x} - \mathbf{c}) \exp[-(1/2)(\mathbf{x} - \mathbf{c})'\mathbf{W}(\mathbf{x} - \mathbf{c})]\, d\mathbf{x} \\
&= (\mathbf{k}'\mathbf{c})(2\pi)^{n/2}|\mathbf{W}|^{-1/2} + \int_{\mathcal{R}^n} (\mathbf{k}'\mathbf{P}^{-1}\mathbf{z}) \exp[-(1/2)\mathbf{z}'\mathbf{z}]\, |J|\, d\mathbf{z} \\
&= (\mathbf{k}'\mathbf{c})(2\pi)^{n/2}|\mathbf{W}|^{-1/2} \\
&\qquad + |\mathbf{W}|^{-1/2} \sum_{i=1}^{n} d_i \int_{-\infty}^{\infty} z_i e^{-z_i^2/2}\, dz_i \prod_{s \neq i} \int_{-\infty}^{\infty} e^{-z_s^2/2}\, dz_s \\
&= (\mathbf{k}'\mathbf{c})(2\pi)^{n/2}|\mathbf{W}|^{-1/2}. \qquad (12.2)
\end{aligned}$$

And, letting b_{ij} represent the ijth element of the $n \times n$ matrix $(\mathbf{P}^{-1})'\mathbf{A}\mathbf{P}^{-1}$ and recalling Lemma 5.2.1 [and observing that $(\mathbf{P}^{-1})' = (\mathbf{P}')^{-1}$], we find that

$$
\begin{aligned}
\int_{\mathbb{R}^n} &(\mathbf{x}-\mathbf{c})'\mathbf{A}(\mathbf{x}-\mathbf{c}) \exp[-(1/2)(\mathbf{x}-\mathbf{c})'\mathbf{W}(\mathbf{x}-\mathbf{c})]\,d\mathbf{x} \\
&= \int_{\mathbb{R}^n} [\mathbf{z}'(\mathbf{P}^{-1})'\mathbf{A}\mathbf{P}^{-1}\mathbf{z}] \exp[-(1/2)\mathbf{z}'\mathbf{z}]\,|J|\,d\mathbf{z} \\
&= |\mathbf{W}|^{-1/2} \sum_{i,j} b_{ij} \int_{\mathbb{R}^n} z_i z_j \exp[-(1/2)\mathbf{z}'\mathbf{z}]\,d\mathbf{z} \\
&= |\mathbf{W}|^{-1/2} \Big\{ \sum_i b_{ii} \int_{-\infty}^{\infty} z_i^2 e^{-z_i^2/2}\,dz_i \prod_{s\neq i} \int_{-\infty}^{\infty} e^{-z_s^2/2}\,dz_s \\
&\qquad + \sum_{i,j\neq i} b_{ij} \int_{-\infty}^{\infty} z_i e^{-z_i^2/2}\,dz_i \int_{-\infty}^{\infty} z_j e^{-z_j^2/2}\,dz_j \prod_{s\neq i,j} \int_{-\infty}^{\infty} e^{-z_s^2/2}\,dz_s \Big\} \\
&= |\mathbf{W}|^{-1/2} \sum_i b_{ii} (2\pi)^{1/2} (2\pi)^{(n-1)/2} \\
&= (2\pi)^{n/2} |\mathbf{W}|^{-1/2} \operatorname{tr}[(\mathbf{P}^{-1})'\mathbf{A}\mathbf{P}^{-1}] \\
&= (2\pi)^{n/2} |\mathbf{W}|^{-1/2} \operatorname{tr}[\mathbf{A}\mathbf{P}^{-1}(\mathbf{P}^{-1})'] \\
&= (2\pi)^{n/2} |\mathbf{W}|^{-1/2} \operatorname{tr}(\mathbf{A}\mathbf{W}^{-1}). \qquad (12.3)
\end{aligned}
$$

Formulas (12.1)–(12.3) can be regarded as special cases of the formula given by the following theorem.

Theorem 15.12.1. Let **B** represent an $n \times n$ symmetric positive definite matrix and **A** an $n \times n$ matrix, let **b** represent an $n \times 1$ vector, and let b_0 and a_0 represent scalars, Then,

$$
\begin{aligned}
\int_{\mathbb{R}^n} &(a_0 + \mathbf{a}'\mathbf{x} + \mathbf{x}'\mathbf{A}\mathbf{x}) \exp[-(b_0 + \mathbf{b}'\mathbf{x} + \mathbf{x}'\mathbf{B}\mathbf{x})]\,d\mathbf{x} \\
&= (1/2)\pi^{n/2} |\mathbf{B}|^{-1/2} \exp[(1/4)\mathbf{b}'\mathbf{B}^{-1}\mathbf{b} - b_0] \\
&\qquad \times [\operatorname{tr}(\mathbf{A}\mathbf{B}^{-1}) - \mathbf{a}'\mathbf{B}^{-1}\mathbf{b} + (1/2)\mathbf{b}'\mathbf{B}^{-1}\mathbf{A}\mathbf{B}^{-1}\mathbf{b} + 2a_0]. \qquad (12.4)
\end{aligned}
$$

Proof. It is easy to show that

$$b_0 + \mathbf{b}'\mathbf{x} + \mathbf{x}'\mathbf{B}\mathbf{x} = (1/2)(\mathbf{x}-\mathbf{c})'\mathbf{W}(\mathbf{x}-\mathbf{c}) + f,$$

where $\mathbf{W} = 2\mathbf{B}$, $\mathbf{c} = -(1/2)\mathbf{B}^{-1}\mathbf{b}$, and $f = b_0 - (1/4)\mathbf{b}'\mathbf{B}^{-1}\mathbf{b}$. It is also easy to show that

$$a_0 + \mathbf{a}'\mathbf{x} + \mathbf{x}'\mathbf{A}\mathbf{x} = (\mathbf{x}-\mathbf{c})'\mathbf{A}(\mathbf{x}-\mathbf{c}) + \mathbf{k}'\mathbf{x} + g,$$

where $\mathbf{k} = \mathbf{a} - (1/2)(\mathbf{A} + \mathbf{A}')\mathbf{B}^{-1}\mathbf{b}$ and $g = a_0 - (1/4)\mathbf{b}'\mathbf{B}^{-1}\mathbf{A}\mathbf{B}^{-1}\mathbf{b}$.

Thus, making use of results (12.1)–(12.3) [and observing that $\mathbf{b}'\mathbf{B}^{-1}\mathbf{A}'\mathbf{B}^{-1}\mathbf{b} = (\mathbf{b}'\mathbf{B}^{-1}\mathbf{A}'\mathbf{B}^{-1}\mathbf{b})' = \mathbf{b}'\mathbf{B}^{-1}\mathbf{A}\mathbf{B}^{-1}\mathbf{b}$], we find that

$$\int_{\mathcal{R}^n} (a_0 + \mathbf{a}'\mathbf{x} + \mathbf{x}'\mathbf{A}\mathbf{x}) \exp[-(b_0 + \mathbf{b}'\mathbf{x} + \mathbf{x}'\mathbf{B}\mathbf{x}] \, d\mathbf{x}$$

$$\begin{aligned}
&= e^{-f} \int_{\mathcal{R}^n} [g + \mathbf{k}'\mathbf{x} + (\mathbf{x}-\mathbf{c})'\mathbf{A}(\mathbf{x}-\mathbf{c})] \\
&\qquad\qquad \times \exp[-(1/2)(\mathbf{x}-\mathbf{c})'\mathbf{W}(\mathbf{x}-\mathbf{c})] \, d\mathbf{x} \\
&= e^{-f}[g(2\pi)^{n/2}|\mathbf{W}|^{-1/2} + (\mathbf{k}'\mathbf{c})(2\pi)^{n/2}|\mathbf{W}|^{-1/2} \\
&\qquad\qquad + (2\pi)^{n/2}|\mathbf{W}|^{-1/2}\operatorname{tr}(\mathbf{A}\mathbf{W}^{-1})] \\
&= (2\pi)^{n/2}|\mathbf{W}|^{-1/2}e^{-f}[g + (\mathbf{k}'\mathbf{c}) + \operatorname{tr}(\mathbf{A}\mathbf{W}^{-1})] \\
&= (2\pi)^{n/2}2^{-n/2}|\mathbf{B}|^{-1/2}\exp[(1/4)\mathbf{b}'\mathbf{B}^{-1}\mathbf{b} - b_0] \\
&\qquad \times [a_0 - (1/4)\mathbf{b}'\mathbf{B}^{-1}\mathbf{A}\mathbf{B}^{-1}\mathbf{b} - (1/2)\mathbf{a}'\mathbf{B}^{-1}\mathbf{b} \\
&\qquad\qquad + (1/4)\mathbf{b}'\mathbf{B}^{-1}(\mathbf{A}+\mathbf{A}')\mathbf{B}^{-1}\mathbf{b} + (1/2)\operatorname{tr}(\mathbf{A}\mathbf{B}^{-1})] \\
&= (1/2)\pi^{n/2}|\mathbf{B}|^{-1/2}\exp[(1/4)\mathbf{b}'\mathbf{B}^{-1}\mathbf{b} - b_0] \\
&\qquad \times [\operatorname{tr}(\mathbf{A}\mathbf{B}^{-1}) - \mathbf{a}'\mathbf{B}^{-1}\mathbf{b} + (1/2)\mathbf{b}'\mathbf{B}^{-1}\mathbf{A}\mathbf{B}^{-1}\mathbf{b} + 2a_0].
\end{aligned}$$

Q.E.D.

Exercises

Section 15.1

1. Using the result of Part (c) of Exercise 6.2, show that every neighborhood of a point $\mathbf{x}$ in $\mathcal{R}^{m\times 1}$ is an open set.

2. Let f represent a function, defined on a set S, of a vector $\mathbf{x} = (x_1, \ldots, x_m)'$ of m variables, suppose that S contains at least some interior points, and let $\mathbf{c}$ represent an arbitrary one of those points. Verify that if f is k times continuously differentiable at $\mathbf{c}$, then f is k times continuously differentiable at every point in some neighborhood of $\mathbf{c}$.

3. Let $\mathbf{X} = \{x_{ij}\}$ represent an $m \times n$ matrix of mn variables, and let $\mathbf{x}$ represent an mn-dimensional column vector obtained by rearranging the elements of $\mathbf{X}$ (in the form of a column vector). Further, let S represent a set of $\mathbf{X}$-values, and let S^* represent the corresponding set of $\mathbf{x}$-values (i.e., the set obtained by rearranging the elements of each $m \times n$ matrix in S in the form of a column vector). Verify that an mn-dimensional column vector is an interior point of S^* if and only if it is a rearrangement of an $m \times n$ matrix that is an interior point of S.

4. Let f represent a function whose domain is a set S in $\mathcal{R}^{m\times 1}$ (that contains at least some interior points). Show that the Hessian matrix $\mathbf{H}f$ of f is the gradient matrix of the gradient vector $(\mathbf{D}f)'$ of f.

Section 15.2

5. Let g represent a function, defined on a set S, of a vector $\mathbf{x} = (x_1, \ldots, x_m)'$ of m variables, let $S^* = \{\mathbf{x} \in S : g(\mathbf{x}) \neq 0\}$, and let $\mathbf{c}$ represent any interior point of S^* at which g is continuously differentiable. Use Lemma 15.2.2 and the ensuing discussion [including formulas (2.16) and (2.8)] to show that (for any positive integer k) g^{-k} is continuously differentiable at $\mathbf{c}$ and

$$\frac{\partial g^{-k}}{\partial x_j} = -k g^{-k-1} \frac{\partial g}{\partial x_j},$$

thereby generalizing formula (2.8) and establishing that formula (2.16) is valid for negative (as well as positive) values of k.

Section 15.4

6. Let $\mathbf{F}$ represent a $p \times p$ matrix of functions, defined on a set S, of a vector $\mathbf{x} = (x_1, \ldots, x_m)'$ of m variables. Let $\mathbf{c}$ represent any interior point of S at which $\mathbf{F}$ is continuously differentiable. Show that if $\mathbf{F}$ is idempotent at all points in some neighborhood of $\mathbf{c}$, then (at $\mathbf{x} = \mathbf{c}$)

$$\mathbf{F} \frac{\partial \mathbf{F}}{\partial x_j} \mathbf{F} = \mathbf{0}.$$

7. Let g represent a function, defined on a set S, of a vector $\mathbf{x} = (x_1, \ldots, x_m)'$ of m variables, and let $\mathbf{f}$ represent a $p \times 1$ vector of functions (defined on S) of $\mathbf{x}$. Let $\mathbf{c}$ represent any interior point (of S) at which g and $\mathbf{f}$ are continuously differentiable. Show that $g\mathbf{f}$ is continuously differentiable at $\mathbf{c}$ and that (at $\mathbf{x} = \mathbf{c}$)

$$\frac{\partial (g\mathbf{f})}{\partial \mathbf{x}'} = \mathbf{f} \frac{\partial g}{\partial \mathbf{x}'} + g \frac{\partial \mathbf{f}}{\partial \mathbf{x}'}.$$

Section 15.6

8. (a) Let $\mathbf{X} = \{x_{ij}\}$ represent an $m \times n$ matrix of mn "independent" variables, and suppose that $\mathbf{X}$ is free to range over all of $\mathbb{R}^{m \times n}$.

   (1) Show that, for any $p \times m$ and $n \times p$ matrices of constants $\mathbf{A}$ and $\mathbf{B}$,

$$\frac{\partial \operatorname{tr}(\mathbf{AXB})}{\partial \mathbf{X}} = \mathbf{A}'\mathbf{B}'.$$

   [*Hint*. Observe that $\operatorname{tr}(\mathbf{AXB}) = \operatorname{tr}(\mathbf{BAX})$.]

   (2) Show that, for any m- and n-dimensional column vectors $\mathbf{a}$ and $\mathbf{b}$,

$$\frac{\partial (\mathbf{a}'\mathbf{X}\mathbf{b})}{\partial \mathbf{X}} = \mathbf{a}\mathbf{b}'.$$

   [*Hint*. Observe that $\mathbf{a}'\mathbf{X}\mathbf{b} = \operatorname{tr}(\mathbf{a}'\mathbf{X}\mathbf{b})$.]

(b) Suppose now that $\mathbf{X}$ is a symmetric (but otherwise unrestricted) matrix (of dimensions $m \times m$).

(1) Show that, for any $p \times m$ and $m \times p$ matrices of constants $\mathbf{A}$ and $\mathbf{B}$,

$$\frac{\partial \operatorname{tr}(\mathbf{AXB})}{\partial \mathbf{X}} = \mathbf{C} + \mathbf{C}' - \operatorname{diag}(c_{11}, c_{22}, \ldots, c_{mm}),$$

where $\mathbf{C} = \{c_{ij}\} = \mathbf{BA}$.

(2) Show that, for any m-dimensional column vectors $\mathbf{a} = \{a_i\}$ and $\mathbf{b} = \{b_i\}$,

$$\frac{\partial(\mathbf{a}'\mathbf{X}\mathbf{b})}{\partial \mathbf{X}} = \mathbf{ab}' + \mathbf{ba}' - \operatorname{diag}(a_1 b_1, a_2 b_2, \ldots, a_m b_m).$$

9. (a) Let $\mathbf{X} = \{x_{st}\}$ represent an $m \times n$ matrix of "independent" variables, and suppose that $\mathbf{X}$ is free to range over all of $\mathcal{R}^{m \times n}$. Show that, for any $n \times m$ matrix of constants $\mathbf{A}$,

$$\frac{\partial \operatorname{tr}[(\mathbf{AX})^2]}{\partial \mathbf{X}} = 2(\mathbf{AXA})'.$$

(b) Let $\mathbf{X} = \{x_{st}\}$ represent an $m \times m$ symmetric (but otherwise unrestricted) matrix of variables. Show that, for any $m \times m$ matrix of constants $\mathbf{A}$,

$$\frac{\partial \operatorname{tr}[(\mathbf{AX})^2]}{\partial \mathbf{X}} = 2[\mathbf{B} + \mathbf{B}' - \operatorname{diag}(b_{11}, b_{22}, \ldots, b_{mm})],$$

where $\mathbf{B} = \{b_{st}\} = \mathbf{AXA}$.

10. Let $\mathbf{X} = \{x_{ij}\}$ represent an $m \times m$ matrix of "independent" variables, and suppose that $\mathbf{X}$ is free to range over all of $\mathcal{R}^{m \times m}$. Show that, for $k = 2, 3, \ldots,$

$$\frac{\partial \operatorname{tr}(\mathbf{X}^k)}{\partial \mathbf{X}} = k(\mathbf{X}')^{k-1},$$

thereby generalizing result (2.17).

11. Let $\mathbf{X} = \{x_{st}\}$ represent an $m \times n$ matrix of "independent" variables, and suppose that $\mathbf{X}$ is free to range over all of $\mathcal{R}^{m \times n}$.

(a) Show that, for any $m \times m$ matrix of constants $\mathbf{A}$,

$$\frac{\partial \operatorname{tr}(\mathbf{X}'\mathbf{AX})}{\partial \mathbf{X}} = (\mathbf{A} + \mathbf{A}')\mathbf{X}.$$

(b) Show that, for any $n \times n$ matrix of constants $\mathbf{A}$,

$$\frac{\partial \operatorname{tr}(\mathbf{XAX}')}{\partial \mathbf{X}} = \mathbf{X}(\mathbf{A} + \mathbf{A}').$$

(c) Show (in the special case where $n = m$) that, for any $m \times m$ matrix of constants $\mathbf{A}$,

$$\frac{\partial \operatorname{tr}(\mathbf{XAX})}{\partial \mathbf{X}} = (\mathbf{AX})' + (\mathbf{XA})'.$$

(d) Use the formulas from Parts (a)–(c) to devise simple formulas for $\partial \operatorname{tr}(\mathbf{X}'\mathbf{X})/\partial \mathbf{X}$, $\partial \operatorname{tr}(\mathbf{XX}')/\partial \mathbf{X}$, and (in the special case where $n = m$) $\partial \operatorname{tr}(\mathbf{X}^2)/\partial \mathbf{X}$.

Section 15.7

12. Let $\mathbf{X} = \{x_{ij}\}$ represent an $m \times m$ matrix. Let f represent a function of $\mathbf{X}$ defined on a set S comprising some or all $m \times m$ symmetric matrices. Suppose that, for purposes of differentiation, f is to be interpreted as a function of the $[m(m+1)/2]$-dimensional column vector $\mathbf{x}$ whose elements are x_{ij} ($j \le i = 1, \ldots, m$). Suppose further that there exists a function g, whose domain is a set T of not-necessarily-symmetric matrices that contains S as a proper subset, such that $g(\mathbf{X}) = f(\mathbf{X})$ for $\mathbf{X} \in S$, so that g is a function of $\mathbf{X}$ and f is the function obtained by restricting the domain of g to S. Define $S^* = \{\mathbf{x} : \mathbf{X} \in S\}$. Let $\mathbf{c}$ represent an interior point of S^*, and let $\mathbf{C}$ represent the corresponding value of $\mathbf{X}$. Show that if $\mathbf{C}$ is an interior point of T and if g is continuously differentiable at $\mathbf{C}$, then f is continuously differentiable at $\mathbf{c}$ and that (at $\mathbf{x} = \mathbf{c}$)

$$\frac{\partial f}{\partial \mathbf{X}} = \frac{\partial g}{\partial \mathbf{X}} + \left(\frac{\partial g}{\partial \mathbf{X}}\right)' - \operatorname{diag}\left(\frac{\partial g}{\partial x_{11}}, \frac{\partial g}{\partial x_{22}}, \ldots, \frac{\partial g}{\partial x_{mm}}\right).$$

13. Let $\mathbf{h} = \{h_i\}$ represent an $n \times 1$ vector of functions, defined on a set S, of a vector $\mathbf{x} = (x_1, \ldots, x_m)'$ of m variables. Let g represent a function, defined on a set T, of a vector $\mathbf{y} = (y_1, \ldots, y_n)'$ of n variables. Suppose that $\mathbf{h}(\mathbf{x}) \in T$ for every $\mathbf{x}$ in S, and take f to be the composite function defined (on S) by $f(\mathbf{x}) = g[\mathbf{h}(\mathbf{x})]$. Show that if $\mathbf{h}$ is twice continuously differentiable at an interior point $\mathbf{c}$ of S and if [assuming that $\mathbf{h}(\mathbf{c})$ is an interior point of T] g is twice continuously differentiable at $\mathbf{h}(\mathbf{c})$, then f is twice continuously differentiable at $\mathbf{c}$ and

$$\mathbf{H}f(\mathbf{c}) = [\mathbf{Dh}(\mathbf{c})]'\mathbf{H}g[\mathbf{h}(\mathbf{c})]\mathbf{Dh}(\mathbf{c}) + \sum_{i=1}^{n} D_i g[\mathbf{h}(\mathbf{c})]\mathbf{H}h_i(\mathbf{c}).$$

Section 15.8

14. Let $\mathbf{X} = \{x_{ij}\}$ represent an $m \times m$ matrix of m^2 "independent" variables (where $m \ge 2$), and suppose that the range of $\mathbf{X}$ comprises all of $\mathcal{R}^{m \times m}$. Show that (for any positive integer k) the function f defined (on $\mathcal{R}^{m \times m}$) by $f(\mathbf{X}) = |\mathbf{X}|^k$ is continuously differentiable at every $\mathbf{X}$ and that

$$\frac{\partial |\mathbf{X}|^k}{\partial \mathbf{X}} = k|\mathbf{X}|^{k-1}[\text{adj}(\mathbf{X})]'.$$

15. Let $\mathbf{F} = \{f_{is}\}$ represent a $p \times p$ matrix of functions, defined on a set S, of a vector $\mathbf{x} = (x_1, \ldots, x_m)'$ of m variables. Let $\mathbf{c}$ represent any interior point (of S) at which $\mathbf{F}$ is continuously differentiable. Use the results of Exercise 13.10 to show that (a) if $\text{rank}[\mathbf{F}(\mathbf{c})] = p - 1$, then (at $\mathbf{x} = \mathbf{c}$)

$$\frac{\partial \det(\mathbf{F})}{\partial x_j} = k\mathbf{y}'\frac{\partial \mathbf{F}}{\partial x_j}\mathbf{z},$$

where $\mathbf{z} = \{z_s\}$ and $\mathbf{y} = \{y_i\}$ are any nonnull p-dimensional vectors such that $\mathbf{F}(\mathbf{c})\mathbf{z} = \mathbf{0}$ and $[\mathbf{F}(\mathbf{c})]'\mathbf{y} = \mathbf{0}$ and where [letting ϕ_{is} represent the cofactor of $f_{is}(\mathbf{c})$] k is a scalar that is expressible as $k = \phi_{is}/(y_i z_s)$ for any i and s such that $y_i \neq 0$ and $z_s \neq 0$; and (b) if $\text{rank}[\mathbf{F}(\mathbf{c})] \leq p - 2$, then (at $\mathbf{x} = \mathbf{c}$)

$$\frac{\partial \det(\mathbf{F})}{\partial x_j} = 0.$$

16. Let $\mathbf{X} = \{x_{st}\}$ represent an $m \times n$ matrix of "independent" variables, let $\mathbf{A}$ represent an $m \times m$ matrix of constants, and suppose that the range of $\mathbf{X}$ is a set S comprising some or all $\mathbf{X}$-values for which $\det(\mathbf{X}'\mathbf{A}\mathbf{X}) > 0$. Show that $\log\det(\mathbf{X}'\mathbf{A}\mathbf{X})$ is continuously differentiable at any interior point $\mathbf{C}$ of S and that (at $\mathbf{X} = \mathbf{C}$)

$$\frac{\partial \log\det(\mathbf{X}'\mathbf{A}\mathbf{X})}{\partial \mathbf{X}} = \mathbf{A}\mathbf{X}(\mathbf{X}'\mathbf{A}\mathbf{X})^{-1} + [(\mathbf{X}'\mathbf{A}\mathbf{X})^{-1}\mathbf{X}'\mathbf{A}]'.$$

17. (a) Let $\mathbf{X}$ represent an $m \times n$ matrix of "independent" variables, let $\mathbf{A}$ and $\mathbf{B}$ represent $q \times m$ and $n \times q$ matrices of constants, and suppose that the range of $\mathbf{X}$ is a set S comprising some or all $\mathbf{X}$-values for which $\det(\mathbf{A}\mathbf{X}\mathbf{B}) > 0$. Show that $\log\det(\mathbf{A}\mathbf{X}\mathbf{B})$ is continuously differentiable at any interior point $\mathbf{C}$ of S and that (at $\mathbf{X} = \mathbf{C}$)

$$\frac{\partial \log\det(\mathbf{A}\mathbf{X}\mathbf{B})}{\partial \mathbf{X}} = [\mathbf{B}(\mathbf{A}\mathbf{X}\mathbf{B})^{-1}\mathbf{A}]'.$$

(b) Suppose now that $\mathbf{X} = \{x_{ij}\}$ is an $m \times m$ symmetric matrix; that $\mathbf{A}$ and $\mathbf{B}$ are $q \times m$ and $m \times q$ matrices of constants; that, for purposes of differentiating any function of $\mathbf{X}$, the function is to be interpreted as a function of the column vector $\mathbf{x}$ whose elements are x_{ij} ($j \leq i = 1, \ldots, m$); and that the range of $\mathbf{x}$ is a set S comprising some or all $\mathbf{x}$-values for which $\det(\mathbf{A}\mathbf{X}\mathbf{B}) > 0$. Show that $\log\det(\mathbf{A}\mathbf{X}\mathbf{B})$ is continuously differentiable at any interior point $\mathbf{c}$ (of S) and that (at $\mathbf{x} = \mathbf{c}$)

$$\frac{\partial \log\det(\mathbf{A}\mathbf{X}\mathbf{B})}{\partial \mathbf{X}} = \mathbf{K} + \mathbf{K}' - \text{diag}(k_{11}, k_{22}, \ldots, k_{qq}),$$

where $\mathbf{K} = \{k_{ij}\} = \mathbf{B}(\mathbf{A}\mathbf{X}\mathbf{B})^{-1}\mathbf{A}$.

18. Let $\mathbf{F} = \{f_{is}\}$ represent a $p \times p$ matrix of functions, defined on a set S, of a vector $\mathbf{x} = (x_1, \ldots, x_m)'$ of m variables, and let $\mathbf{A}$ and $\mathbf{B}$ represent $q \times p$ and $p \times q$ matrices of constants. Suppose that S is the set of all $\mathbf{x}$-values for which $\mathbf{F}(\mathbf{x})$ is nonsingular and $\det[\mathbf{A}\mathbf{F}^{-1}(\mathbf{x})\mathbf{B}] > 0$, or is a subset of that set. Show that if $\mathbf{F}$ is continuously differentiable at an interior point $\mathbf{c}$ of S, then $\log\det(\mathbf{A}\mathbf{F}^{-1}\mathbf{B})$ is continuously differentiable at $\mathbf{c}$ and (at $\mathbf{x} = \mathbf{c}$)

$$\frac{\partial \log\det(\mathbf{A}\mathbf{F}^{-1}\mathbf{B})}{\partial x_j} = -\operatorname{tr}\left[\mathbf{F}^{-1}\mathbf{B}(\mathbf{A}\mathbf{F}^{-1}\mathbf{B})^{-1}\mathbf{A}\mathbf{F}^{-1}\frac{\partial \mathbf{F}}{\partial x_j}\right].$$

19. Let $\mathbf{A}$ and $\mathbf{B}$ represent $q \times m$ and $m \times q$ matrices of constants.

    (a) Let $\mathbf{X}$ represent an $m \times m$ matrix of m^2 "independent" variables, and suppose that the range of $\mathbf{X}$ is a set S comprising some or all $\mathbf{X}$-values for which $\mathbf{X}$ is nonsingular and $\det(\mathbf{A}\mathbf{X}^{-1}\mathbf{B}) > 0$. Use the result of Exercise 18 to show that $\log\det(\mathbf{A}\mathbf{X}^{-1}\mathbf{B})$ is continuously differentiable at any interior point $\mathbf{C}$ of S and that (at $\mathbf{X} = \mathbf{C}$)

    $$\frac{\partial \log\det(\mathbf{A}\mathbf{X}^{-1}\mathbf{B})}{\partial \mathbf{X}} = -[\mathbf{X}^{-1}\mathbf{B}(\mathbf{A}\mathbf{X}^{-1}\mathbf{B})^{-1}\mathbf{A}\mathbf{X}^{-1}]'.$$

    (b) Suppose now that $\mathbf{X} = \{x_{ij}\}$ is an $m \times m$ symmetric matrix; that, for purposes of differentiating any function of $\mathbf{X}$, the function is to be interpreted as a function of the column vector $\mathbf{x}$ whose elements are x_{ij} ($j \le i = 1, \ldots, m$); and that the range of $\mathbf{x}$ is a set S comprising some or all $\mathbf{x}$-values for which $\mathbf{X}$ is nonsingular and $\det(\mathbf{A}\mathbf{X}^{-1}\mathbf{B}) > 0$. Use the result of Exercise 18 to show that $\log\det(\mathbf{A}\mathbf{X}^{-1}\mathbf{B})$ is continuously differentiable at any interior point $\mathbf{c}$ of S and that (at $\mathbf{x} = \mathbf{c}$)

    $$\frac{\partial \log\det(\mathbf{A}\mathbf{X}^{-1}\mathbf{B})}{\partial \mathbf{X}} = -\mathbf{K} - \mathbf{K}' + \operatorname{diag}(k_{11}, k_{22}, \ldots, k_{qq}),$$

    where $\mathbf{K} = \{k_{ij}\} = \mathbf{X}^{-1}\mathbf{B}(\mathbf{A}\mathbf{X}^{-1}\mathbf{B})^{-1}\mathbf{A}\mathbf{X}^{-1}$.

20. Let $\mathbf{F} = \{f_{is}\}$ represent a $p \times p$ matrix of functions, defined on a set S, of a vector $\mathbf{x} = (x_1, \ldots, x_m)'$ of m variables. Let $\mathbf{c}$ represent any interior point (of S) at which $\mathbf{F}$ is continuously differentiable. By, for instance, using the result of Part (b) of Exercise 13.10, show that if $\operatorname{rank}[\mathbf{F}(\mathbf{c})] \le p - 3$, then

$$\frac{\partial\, \operatorname{adj}(\mathbf{F})}{\partial x_j} = \mathbf{0}.$$

21. (a) Let $\mathbf{X} = \{x_{ij}\}$ represent an $m \times m$ matrix of m^2 "independent" variables, and suppose that the range of $\mathbf{X}$ is a set S comprising some or all $\mathbf{X}$-values for which $\mathbf{X}$ is nonsingular. Show that (when the elements of $\mathbf{X}^{-1}$

are regarded as functions of $\mathbf{X}$) $\mathbf{X}^{-1}$ is continuously differentiable at any interior point $\mathbf{C}$ of S and that (at $\mathbf{X} = \mathbf{C}$)

$$\frac{\partial \mathbf{X}^{-1}}{\partial x_{ij}} = -\mathbf{y}_i \mathbf{z}_j',$$

where $\mathbf{y}_i$ represents the ith column and $\mathbf{z}_j'$ the jth row of $\mathbf{X}^{-1}$.

(b) Suppose now that $\mathbf{X} = \{x_{ij}\}$ is an $m \times m$ symmetric matrix; that, for purposes of differentiating a function of $\mathbf{X}$, the function is to be interpreted as a function of the column vector $\mathbf{x}$ whose elements are x_{ij} ($j \leq i = 1, \ldots, m$); and that the range of $\mathbf{x}$ is a set S comprising some or all $\mathbf{x}$-values for which $\mathbf{X}$ is nonsingular. Show that $\mathbf{X}^{-1}$ is continuously differentiable at any interior point $\mathbf{c}$ of S and that (at $\mathbf{x} = \mathbf{c}$)

$$\frac{\partial \mathbf{X}^{-1}}{\partial x_{ij}} = \begin{cases} -\mathbf{y}_i \mathbf{y}_i', & \text{if } j = i, \\ -\mathbf{y}_i \mathbf{y}_j' - \mathbf{y}_j \mathbf{y}_i', & \text{if } j < i \end{cases}$$

(where $\mathbf{y}_i$ represents the ith column of $\mathbf{X}^{-1}$).

Section 15.9

22. Let $\mathbf{X} = \{x_{ij}\}$ represent an $m \times m$ matrix of m^2 "independent" variables. Suppose that the range of $\mathbf{X}$ is a set S comprising some or all $\mathbf{X}$-values for which $\mathbf{X}$ is nonsingular, and let $\mathbf{C}$ represent an interior point of S. Denote the ijth element of $\mathbf{X}^{-1}$ by y_{ij}, the jth column of $\mathbf{X}^{-1}$ by $\mathbf{y}_j$, and the ith row of $\mathbf{X}^{-1}$ by $\mathbf{z}_i'$.

(a) Show that $\mathbf{X}^{-1}$ is twice continuously differentiable at $\mathbf{C}$ and that (at $\mathbf{X} = \mathbf{C}$)

$$\frac{\partial^2 \mathbf{X}^{-1}}{\partial x_{ij}\, \partial x_{st}} = y_{js} \mathbf{y}_i \mathbf{z}_t' + y_{ti} \mathbf{y}_s \mathbf{z}_j'.$$

(b) Suppose that $\det(\mathbf{X}) > 0$ for every $\mathbf{X}$ in S. Show that $\log\det(\mathbf{X})$ is twice continuously differentiable at $\mathbf{C}$ and that (at $\mathbf{X} = \mathbf{C}$)

$$\frac{\partial^2 \log\det(\mathbf{X})}{\partial x_{ij}\, \partial x_{st}} = -y_{ti} y_{js}.$$

23. Let $\mathbf{F} = \{f_{is}\}$ represent a $p \times p$ matrix of functions, defined on a set S, of a vector $\mathbf{x} = (x_1, \ldots, x_m)'$ of m variables. For any nonempty set $T = \{t_1, \ldots, t_s\}$, whose members are integers between 1 and m, inclusive, define $\mathbf{D}(T) = \partial^s \mathbf{F} / \partial x_{t_1} \cdots \partial x_{t_s}$. Let k represent a positive integer and, for $i = 1, \ldots, k$, let j_i represent an arbitrary integer between 1 and m, inclusive.

(a) Suppose that $\mathbf{F}$ is nonsingular for every $\mathbf{x}$ in S, and denote by $\mathbf{c}$ any interior point (of S) at which $\mathbf{F}$ is k times continuously differentiable. Show that $\mathbf{F}^{-1}$ is k times continuously differentiable at $\mathbf{c}$ and that (at $\mathbf{x} = \mathbf{c}$)

$$\frac{\partial^k \mathbf{F}^{-1}}{\partial x_{j_1} \cdots \partial x_{j_k}}$$
$$= \sum_{r=1}^{k} \sum_{T_1,\ldots,T_r} (-1)^r \mathbf{F}^{-1}\mathbf{D}(T_1)\mathbf{F}^{-1}\mathbf{D}(T_2)\cdots\mathbf{F}^{-1}\mathbf{D}(T_r)\mathbf{F}^{-1}, \quad \text{(E.1)}$$

where $T_1, \ldots, T_r$ are r nonempty mutually exclusive and exhaustive subsets of $\{j_1, \ldots, j_k\}$ (and where the second summation is over all possible choices for $T_1, \ldots, T_r$).

(b) Suppose that $\det(\mathbf{F}) > 0$ for every $\mathbf{x}$ in S, and denote by $\mathbf{c}$ any interior point (of S) at which $\mathbf{F}$ is k times continuously differentiable. Show that $\log\det(\mathbf{F})$ is k times continuously differentiable at $\mathbf{c}$ and that (at $\mathbf{x} = \mathbf{c}$)

$$\frac{\partial^k \log\det(\mathbf{F})}{\partial x_{j_1} \cdots \partial x_{j_k}}$$
$$= \sum_{r=1}^{k} \sum_{T_1,\ldots,T_r} (-1)^{r+1}\, \mathrm{tr}[\mathbf{F}^{-1}\mathbf{D}(T_1)\mathbf{F}^{-1}\mathbf{D}(T_2)\cdots\mathbf{F}^{-1}\mathbf{D}(T_r)], \quad \text{(E.2)}$$

where $T_1, \ldots, T_r$ are r nonempty mutually exclusive and exhaustive subsets of $\{j_1, \ldots, j_k\}$ with $j_k \in T_r$ (and where the second summation is over all possible choices for $T_1, \ldots, T_r$).

Section 15.10

24. Let $\mathbf{X} = \{x_{ij}\}$ represent an $m \times m$ symmetric matrix, and let $\mathbf{x}$ represent the $[m(m+1)/2]$-dimensional column vector whose elements are x_{ij} $(j \le i = 1, \ldots, m)$. Define S to be the set of all $\mathbf{x}$-values for which $\mathbf{X}$ is nonsingular and S^* to be the set of all $\mathbf{x}$-values for which $\mathbf{X}$ is positive definite. Show that S and S^* are both open sets.

Section 15.11

25. Let $\mathbf{X}$ represent an $n \times p$ matrix of constants, and let $\mathbf{W}$ represent an $n \times n$ symmetric positive definite matrix whose elements are functions, defined on a set S, of a vector $\mathbf{z} = (z_1, \ldots, z_m)'$ of m variables. Further, let $\mathbf{c}$ represent any interior point (of S) at which $\mathbf{W}$ is twice continuously differentiable. Show that $\mathbf{W} - \mathbf{W}\mathbf{P}_{\mathbf{X},\mathbf{W}}$ is twice continuously differentiable at $\mathbf{c}$ and that (at $\mathbf{x} = \mathbf{c}$)

$$
\begin{aligned}
&\frac{\partial^2(\mathbf{W} - \mathbf{W}\mathbf{P}_{\mathbf{X},\mathbf{W}})}{\partial z_i \partial z_j} \\
&\qquad = (\mathbf{I} - \mathbf{P}'_{\mathbf{X},\mathbf{W}})\frac{\partial^2 \mathbf{W}}{\partial z_i \partial z_j}(\mathbf{I} - \mathbf{P}_{\mathbf{X},\mathbf{W}}) \\
&\qquad\quad - (\mathbf{I} - \mathbf{P}'_{\mathbf{X},\mathbf{W}})\frac{\partial \mathbf{W}}{\partial z_i}\mathbf{X}(\mathbf{X}'\mathbf{W}\mathbf{X})^{-}\mathbf{X}'\frac{\partial \mathbf{W}}{\partial z_j}(\mathbf{I} - \mathbf{P}_{\mathbf{X},\mathbf{W}}) \\
&\qquad\quad - [(\mathbf{I} - \mathbf{P}'_{\mathbf{X},\mathbf{W}})\frac{\partial \mathbf{W}}{\partial z_i}\mathbf{X}(\mathbf{X}'\mathbf{W}\mathbf{X})^{-}\mathbf{X}'\frac{\partial \mathbf{W}}{\partial z_j}(\mathbf{I} - \mathbf{P}_{\mathbf{X},\mathbf{W}})]'.
\end{aligned}
$$

26. Provide an alternative derivation of formula (11.23) for $\partial(\mathbf{W}\mathbf{P}_{\mathbf{X},\mathbf{W}})/\partial z_j$ by using Part (6′) of Theorem 14.12.11 to obtain the representation

$$
\frac{\partial(\mathbf{W}\mathbf{P}_{\mathbf{X},\mathbf{W}})}{\partial z_j} = \mathbf{P}'_{\mathbf{X},\mathbf{W}}\mathbf{W}\frac{\partial \mathbf{P}_{\mathbf{X},\mathbf{W}}}{\partial z_j} + \mathbf{P}'_{\mathbf{X},\mathbf{W}}\frac{\partial \mathbf{W}}{\partial z_j}\mathbf{P}_{\mathbf{X},\mathbf{W}} + \left(\frac{\partial \mathbf{P}_{\mathbf{X},\mathbf{W}}}{\partial z_j}\right)'\mathbf{W}\mathbf{P}_{\mathbf{X},\mathbf{W}}
$$

and by then making use of result (11.16).

Bibliographic and Supplementary Notes

Dwyer (1967) played a leading role in the development of matrix differentiation. Book-length treatments of matrix differentiation have been authored by Rogers (1980) and by Magnus and Neudecker (1988). §1. For a more general and complete discussion of the terminology and alternative notations employed in matrix differentiation, refer to Magnus and Neudecker (1988, chaps. 4–10). §2 and §3. Sections 2 and 3 can be read without first reading Section 1e. §10 and §11. Many of the results presented in Sections 10 and 11 are adaptations of results presented by Hearon and Evans (1968), Golub and Pereyra (1973), and Rogers (1980, chap. 16). §12. The presentation in Section 12 is modeled after that of Graybill (1983, secs. 10.3–10.5).

16

Kronecker Products and the Vec and Vech Operators

Some (though by no means all) of the partitioned matrices encountered in statistics are expressible in terms of two (or more) matrices (of relatively small dimensions) in the form of something called a Kronecker product. In Section 16.1, the definition of a Kronecker product of matrices is given, and a number of results on Kronecker products are presented. These results can (when applicable) be exploited for computational (and other) purposes.

In subsequent sections of this chapter, the notion introduced in Chapter 15 (in connection with the differentiation of a function of a matrix) of rearranging the (nonredundant) elements of a matrix in the form of a column vector is formalized by introducing something called the vec or vech operator. A number of results on the vec or vech operator are presented. Kronecker products appear in many of these results.

16.1 The Kronecker Product of Two or More Matrices: Definition and Some Basic Properties

The *Kronecker product* of two matrices, say an $m \times n$ matrix $\mathbf{A} = \{a_{ij}\}$ and a $p \times q$ matrix $\mathbf{B} = \{b_{ij}\}$, is denoted by the symbol $\mathbf{A} \otimes \mathbf{B}$ and is defined to be the $mp \times nq$ matrix

$$\mathbf{A} \otimes \mathbf{B} = \begin{pmatrix} a_{11}\mathbf{B} & a_{12}\mathbf{B} & \dots & a_{1n}\mathbf{B} \\ a_{21}\mathbf{B} & a_{22}\mathbf{B} & \dots & a_{2n}\mathbf{B} \\ \vdots & \vdots & & \vdots \\ a_{m1}\mathbf{B} & a_{m2}\mathbf{B} & \dots & a_{mn}\mathbf{B} \end{pmatrix}$$

obtained by replacing each element a_{ij} of $\mathbf{A}$ with the $p \times q$ matrix $a_{ij}\mathbf{B}$. Thus, the Kronecker product of $\mathbf{A}$ and $\mathbf{B}$ is a partitioned matrix, comprising m rows and n columns of $p \times q$ dimensional blocks, the ijth of which is $a_{ij}\mathbf{B}$.

Clearly, each element of $\mathbf{A} \otimes \mathbf{B}$ is the product of an element of $\mathbf{A}$ and an element of $\mathbf{B}$. Specifically, the element that appears in the $[p(i-1)+r]$th row and $[q(j-1)+s]$th column of $\mathbf{A} \otimes \mathbf{B}$ is the rsth element $a_{ij}b_{rs}$ of $a_{ij}\mathbf{B}$.

Note that, unlike the "ordinary" product **AB** of **A** and **B** (which is defined only in the special case where $p = n$), the Kronecker product $\mathbf{A} \otimes \mathbf{B}$ is defined regardless of the dimensions of **A** and **B**. Note also that, while the Kronecker product $\mathbf{B} \otimes \mathbf{A}$ of **B** and **A** is of the same dimensions ($mp \times nq$) as $\mathbf{A} \otimes \mathbf{B}$, it is only in special cases that $\mathbf{B} \otimes \mathbf{A}$ equals $\mathbf{A} \otimes \mathbf{B}$. For example, if $m = n = q = 2$ and $p = 3$, then

$$\mathbf{A} \otimes \mathbf{B} = \left(\begin{array}{cc|cc} a_{11}b_{11} & a_{11}b_{12} & a_{12}b_{11} & a_{12}b_{12} \\ a_{11}b_{21} & a_{11}b_{22} & a_{12}b_{21} & a_{12}b_{22} \\ a_{11}b_{31} & a_{11}b_{32} & a_{12}b_{31} & a_{12}b_{32} \\ \hline a_{21}b_{11} & a_{21}b_{12} & a_{22}b_{11} & a_{22}b_{12} \\ a_{21}b_{21} & a_{21}b_{22} & a_{22}b_{21} & a_{22}b_{22} \\ a_{21}b_{31} & a_{21}b_{32} & a_{22}b_{31} & a_{22}b_{32} \end{array}\right),$$

$$\mathbf{B} \otimes \mathbf{A} = \left(\begin{array}{cc|cc} b_{11}a_{11} & b_{11}a_{12} & b_{12}a_{11} & b_{12}a_{12} \\ b_{11}a_{21} & b_{11}a_{22} & b_{12}a_{21} & b_{12}a_{22} \\ \hline b_{21}a_{11} & b_{21}a_{12} & b_{22}a_{11} & b_{22}a_{12} \\ b_{21}a_{21} & b_{21}a_{22} & b_{22}a_{21} & b_{22}a_{22} \\ \hline b_{31}a_{11} & b_{31}a_{12} & b_{32}a_{11} & b_{32}a_{12} \\ b_{31}a_{21} & b_{31}a_{22} & b_{32}a_{21} & b_{32}a_{22} \end{array}\right).$$

In some presentations, $\mathbf{A} \otimes \mathbf{B}$ is referred to as the *direct product* or the *tensor product* of **A** and **B** rather than the Kronecker product. And $\mathbf{A} \otimes \mathbf{B}$ and $\mathbf{B} \otimes \mathbf{A}$ are sometimes referred to, respectively, as the *right* and *left Kronecker* (or *direct* or *tensor*) *products* of **A** and **B**.

For any scalar k and any $m \times n$ matrix **A**,

$$k \otimes \mathbf{A} = \mathbf{A} \otimes k = k\mathbf{A}, \tag{1.1}$$

as is evident from the very definition of a Kronecker product. And, letting $\mathbf{a} = \{a_i\}$ and $\mathbf{b} = \{b_j\}$ represent column vectors of dimensions m and n, respectively, it is easy to verify that

$$\mathbf{a} \otimes \mathbf{b} = \begin{pmatrix} a_1\mathbf{b} \\ a_2\mathbf{b} \\ \vdots \\ a_m\mathbf{b} \end{pmatrix} \tag{1.2}$$

(i.e., that $\mathbf{a} \otimes \mathbf{b}$ is an mn-dimensional partitioned column vector comprising m subvectors, the ith of which is $a_i\mathbf{b}$); that

$$\mathbf{a}' \otimes \mathbf{b}' = (a_1\mathbf{b}', a_2\mathbf{b}', \ldots, a_m\mathbf{b}') \tag{1.3}$$

(i.e., that $\mathbf{a}' \otimes \mathbf{b}'$ is an mn-dimensional partitioned row vector comprising m subvectors, the ith of which is $a_i\mathbf{b}'$); and that

$$\mathbf{a} \otimes \mathbf{b}' = \mathbf{b}' \otimes \mathbf{a} = \mathbf{a}\mathbf{b}'. \tag{1.4}$$

In connection with result (1.4), recall that $\mathbf{ab}'$ is an $m \times n$ matrix whose ijth element is $a_i b_j$.

Let $\mathbf{A} = \{a_{ij}\}$ represent a $p \times q$ matrix. Then, clearly,

$$\mathbf{0} \otimes \mathbf{A} = \mathbf{A} \otimes \mathbf{0} = \mathbf{0}. \tag{1.5}$$

Further, for any diagonal matrix $\mathbf{D} = \{d_i\}$ of order m,

$$\mathbf{D} \otimes \mathbf{A} = \text{diag}(d_1\mathbf{A}, d_2\mathbf{A}, \ldots, d_m\mathbf{A}), \tag{1.6}$$

which, in the special case where $\mathbf{A}$ is a diagonal matrix (of order p), is the diagonal matrix of order mp whose $[p(i-1)+r]$th diagonal element is $d_i a_{rr}$ (as is easily verified). In the special case where $\mathbf{D} = \mathbf{I}_m$, result (1.6) simplifies to

$$\mathbf{I} \otimes \mathbf{A} = \text{diag}(\mathbf{A}, \mathbf{A}, \ldots, \mathbf{A}), \tag{1.7}$$

and in the further special case where $\mathbf{A} = \mathbf{I}_p$, it simplifies to

$$\mathbf{I}_m \otimes \mathbf{I}_p = \mathbf{I}_{mp}. \tag{1.8}$$

Note that, aside from certain special cases, the Kronecker product

$$\mathbf{A} \otimes \mathbf{I} = \begin{pmatrix} a_{11}\mathbf{I} & a_{12}\mathbf{I} & \ldots & a_{1q}\mathbf{I} \\ a_{21}\mathbf{I} & a_{22}\mathbf{I} & \ldots & a_{2q}\mathbf{I} \\ \vdots & \vdots & & \vdots \\ a_{p1}\mathbf{I} & a_{p2}\mathbf{I} & \ldots & a_{pq}\mathbf{I} \end{pmatrix} \tag{1.9}$$

of $\mathbf{A}$ and $\mathbf{I}_m$ differs from the Kronecker product $\mathbf{I} \otimes \mathbf{A}$ of $\mathbf{I}_m$ and $\mathbf{A}$.

Letting k represent an arbitrary scalar and $\mathbf{A}$ and $\mathbf{B}$ arbitrary matrices, it is easy to see that the effect on the Kronecker product $\mathbf{A} \otimes \mathbf{B}$ of replacing either $\mathbf{A}$ or $\mathbf{B}$ by the scalar multiple $k\mathbf{A}$ or $k\mathbf{B}$ is as follows:

$$(k\mathbf{A}) \otimes \mathbf{B} = \mathbf{A} \otimes (k\mathbf{B}) = k(\mathbf{A} \otimes \mathbf{B}). \tag{1.10}$$

Another easily verified property of Kronecker products is that, for any $m \times n$ matrices $\mathbf{A}$ and $\mathbf{B}$ and any $p \times q$ matrix $\mathbf{C}$,

$$(\mathbf{A} + \mathbf{B}) \otimes \mathbf{C} = (\mathbf{A} \otimes \mathbf{C}) + (\mathbf{B} \otimes \mathbf{C}), \tag{1.11}$$

$$\mathbf{C} \otimes (\mathbf{A} + \mathbf{B}) = (\mathbf{C} \otimes \mathbf{A}) + (\mathbf{C} \otimes \mathbf{B}). \tag{1.12}$$

Results (1.11) and (1.12) can be used to show that, for any $m \times n$ matrices $\mathbf{A}$ and $\mathbf{B}$ and any $p \times q$ matrices $\mathbf{C}$ and $\mathbf{D}$,

$$(\mathbf{A} + \mathbf{B}) \otimes (\mathbf{C} + \mathbf{D}) = (\mathbf{A} \otimes \mathbf{C}) + (\mathbf{A} \otimes \mathbf{D}) + (\mathbf{B} \otimes \mathbf{C}) + (\mathbf{B} \otimes \mathbf{D}). \tag{1.13}$$

More generally, results (1.11) and (1.12) can be used to show that, for any $m \times n$ matrices $\mathbf{A}_1, \mathbf{A}_2, \ldots, \mathbf{A}_r$ and $p \times q$ matrices $\mathbf{B}_1, \mathbf{B}_2, \ldots, \mathbf{B}_s$,

$$\Big(\sum_{i=1}^{r} \mathbf{A}_i\Big) \otimes \Big(\sum_{j=1}^{s} \mathbf{B}_j\Big) = \sum_{i=1}^{r}\sum_{j=1}^{s} (\mathbf{A}_i \otimes \mathbf{B}_j). \tag{1.14}$$

Note that, in the special case where $r = s = 2$, result (1.14) reduces to what is essentially result (1.13).

Consider now the transpose $(\mathbf{A} \otimes \mathbf{B})'$ of the Kronecker product of an $m \times n$ matrix $\mathbf{A} = \{a_{ij}\}$ and a $p \times q$ matrix $\mathbf{B}$. It follows from result (2.2.3) (together with the very definition of a Kronecker product) that $(\mathbf{A} \otimes \mathbf{B})'$ is a partitioned matrix, comprising n rows and m columns of $q \times p$ dimensional blocks, the ijth of which is $a_{ji}\mathbf{B}'$. Since $a_{ji}\mathbf{B}'$ is also the ijth block of $\mathbf{A}' \otimes \mathbf{B}'$,

$$(\mathbf{A} \otimes \mathbf{B})' = \mathbf{A}' \otimes \mathbf{B}'. \tag{1.15}$$

Note that, in contrast to what might have been conjectured from the formula $[(\mathbf{AB})' = \mathbf{B}'\mathbf{A}']$ for the transpose of the "ordinary" product of two matrices, $(\mathbf{A} \otimes \mathbf{B})'$ is *not* in general equal to $\mathbf{B}' \otimes \mathbf{A}'$. Note also that if $\mathbf{A}$ and $\mathbf{B}$ are both symmetric, then it follows from result (1.15) that

$$(\mathbf{A} \otimes \mathbf{B})' = \mathbf{A} \otimes \mathbf{B};$$

that is, the Kronecker product of symmetric matrices is symmetric.

The following lemma provides a basis for extending the definition of a Kronecker product to three or more matrices.

Lemma 16.1.1. For any $m \times n$ matrix $\mathbf{A} = \{a_{ij}\}$, $p \times q$ matrix $\mathbf{B} = \{b_{ij}\}$, and $u \times v$ matrix $\mathbf{C}$,

$$\mathbf{A} \otimes (\mathbf{B} \otimes \mathbf{C}) = (\mathbf{A} \otimes \mathbf{B}) \otimes \mathbf{C}. \tag{1.16}$$

Proof. Let $\mathbf{G} = \mathbf{A} \otimes (\mathbf{B} \otimes \mathbf{C})$ and $\mathbf{H} = (\mathbf{A} \otimes \mathbf{B}) \otimes \mathbf{C}$. We wish to show that $\mathbf{G} = \mathbf{H}$.

Partition $\mathbf{G}$ into m rows and n columns of $pu \times qv$ dimensional blocks, and denote the ijth of these blocks by $\mathbf{G}^*_{ij}$. By definition, $\mathbf{G}^*_{ij} = a_{ij}(\mathbf{B} \otimes \mathbf{C})$. For each i and j, partition $\mathbf{G}^*_{ij}$ into p rows and q columns of $u \times v$ dimensional blocks, thereby defining a further partitioning of $\mathbf{G}$ into mp rows and nq columns of ($u \times v$ dimensional) blocks. Denote the rsth block of $\mathbf{G}^*_{ij}$ by $\mathbf{G}^{rs}_{ij}$ and the xzth of the ($u \times v$ dimensional) blocks of $\mathbf{G}$ by $\mathbf{G}_{xz}$. Then,

$$\mathbf{G}_{p(i-1)+r,q(j-1)+s} = \mathbf{G}^{rs}_{ij} = a_{ij}b_{rs}\mathbf{C}. \tag{1.17}$$

Now, observe that $\mathbf{H} = \mathbf{F} \otimes \mathbf{C}$, where $\mathbf{F} = \{f_{xz}\} = \mathbf{A} \otimes \mathbf{B}$. Partition $\mathbf{H}$ into mp rows and nq columns of $u \times v$ dimensional blocks, and denote the xzth of these blocks by $\mathbf{H}_{xz}$. Then, $f_{p(i-1)+r,q(j-1)+s} = a_{ij}b_{rs}$, and consequently

$$\mathbf{H}_{p(i-1)+r,q(j-1)+s} = f_{p(i-1)+r,q(j-1)+s}\mathbf{C} = a_{ij}b_{rs}\mathbf{C}. \tag{1.18}$$

Together, results (1.17) and (1.18) imply that (for arbitrary i, j, r, and s)

$$\mathbf{G}_{p(i-1)+r,q(j-1)+s} = \mathbf{H}_{p(i-1)+r,q(j-1)+s}.$$

We conclude that $\mathbf{G} = \mathbf{H}$. Q.E.D.

The symbol $\mathbf{A} \otimes \mathbf{B} \otimes \mathbf{C}$ is used to represent the common value of the left and right sides of equality (1.16), and this value is referred to as the *Kronecker product* of $\mathbf{A}$, $\mathbf{B}$, and $\mathbf{C}$. This notation and terminology extend in an obvious way to any finite number of matrices.

The following lemma establishes a connection between the "ordinary" product of Kronecker products and the Kronecker product of ordinary products.

Lemma 16.1.2. For any $m \times n$ matrix $\mathbf{A} = \{a_{ij}\}$, $p \times q$ matrix $\mathbf{B} = \{b_{ij}\}$, $n \times u$ matrix $\mathbf{C} = \{c_{ij}\}$, and $q \times v$ matrix $\mathbf{D} = \{d_{ij}\}$,

$$(\mathbf{A} \otimes \mathbf{B})(\mathbf{C} \otimes \mathbf{D}) = (\mathbf{AC}) \otimes (\mathbf{BD}). \tag{1.19}$$

Proof. By definition, $\mathbf{A} \otimes \mathbf{B}$ is a partitioned matrix, comprising m rows and n columns of $p \times q$ dimensional blocks, the ijth of which is $a_{ij}\mathbf{B}$; and $\mathbf{C} \otimes \mathbf{D}$ is a partitioned matrix, comprising n rows and u columns of $q \times v$ dimensional blocks, the jrth of which is $c_{jr}\mathbf{D}$. Thus, $(\mathbf{A} \otimes \mathbf{B})(\mathbf{C} \otimes \mathbf{D})$ is a partitioned matrix, comprising m rows and u columns of $p \times v$ dimensional blocks, the irth of which is the matrix

$$\sum_{j=1}^{n} (a_{ij}\mathbf{B})(c_{jr}\mathbf{D}) = \Big(\sum_{j=1}^{n} a_{ij}c_{jr}\Big)\mathbf{BD}.$$

By way of comparison, $(\mathbf{AC}) \otimes (\mathbf{BD})$ is a partitioned matrix, comprising m rows and u columns of $p \times v$ dimensional blocks, the irth of which is the matrix

$$f_{ir}\mathbf{BD},$$

where f_{ir} is the irth element of $\mathbf{AC}$. The proof is complete upon observing that $f_{ir} = \sum_{j=1}^{n} a_{ij}c_{jr}$ and hence that the irth block of $(\mathbf{AC}) \otimes (\mathbf{BD})$ equals the irth block of $(\mathbf{A} \otimes \mathbf{B})(\mathbf{C} \otimes \mathbf{D})$. Q.E.D.

One implication of result (1.19) is that the Kronecker product $\mathbf{A} \otimes \mathbf{B}$ of an $m \times n$ matrix $\mathbf{A}$ and a $p \times q$ matrix $\mathbf{B}$ can be decomposed in either of the following two ways:

$$\mathbf{A} \otimes \mathbf{B} = (\mathbf{A} \otimes \mathbf{I}_p)(\mathbf{I}_n \otimes \mathbf{B}) = (\mathbf{I}_m \otimes \mathbf{B})(\mathbf{A} \otimes \mathbf{I}_q). \tag{1.20}$$

And, in light of result (1.4), a further implication is that, for any $m \times n$ matrix $\mathbf{A}$, $p \times q$ matrix $\mathbf{C}$, $p \times 1$ vector $\mathbf{b}$, and $n \times 1$ vector $\mathbf{d}$,

$$(\mathbf{A} \otimes \mathbf{b}')(\mathbf{d} \otimes \mathbf{C}) = (\mathbf{b}' \otimes \mathbf{A})(\mathbf{C} \otimes \mathbf{d}) = \mathbf{Adb}'\mathbf{C}. \tag{1.21}$$

Result (1.19) can be extended (by repeated application) to the ordinary product of an arbitrary number of Kronecker products. We have that

$$(\mathbf{A}_1 \otimes \mathbf{B}_1)(\mathbf{A}_2 \otimes \mathbf{B}_2)\cdots(\mathbf{A}_k \otimes \mathbf{B}_k) = (\mathbf{A}_1\mathbf{A}_2\cdots\mathbf{A}_k) \otimes (\mathbf{B}_1\mathbf{B}_2\cdots\mathbf{B}_k), \tag{1.22}$$

where (for $i = 1, 2, \ldots, k$) $\mathbf{A}_i$ is an $m_i \times m_{i+1}$ dimensional matrix and $\mathbf{B}_i$ is a $p_i \times p_{i+1}$ dimensional matrix.

The Kronecker product $\mathbf{A} \otimes \mathbf{B}$ of any $m \times m$ nonsingular matrix $\mathbf{A}$ and any $p \times p$ nonsingular matrix $\mathbf{B}$ is invertible, and

$$(\mathbf{A} \otimes \mathbf{B})^{-1} = \mathbf{A}^{-1} \otimes \mathbf{B}^{-1}, \tag{1.23}$$

as is evident upon observing [in light of results (1.19) and (1.8)] that

$$(\mathbf{A} \otimes \mathbf{B})(\mathbf{A}^{-1} \otimes \mathbf{B}^{-1}) = (\mathbf{A}\mathbf{A}^{-1}) \otimes (\mathbf{B}\mathbf{B}^{-1}) = \mathbf{I}_m \otimes \mathbf{I}_p = \mathbf{I}_{mp}.$$

More generally, $\mathbf{A}^- \otimes \mathbf{B}^-$ is a generalized inverse of the Kronecker product $\mathbf{A} \otimes \mathbf{B}$ of any $m \times n$ matrix $\mathbf{A}$ and any $p \times q$ matrix $\mathbf{B}$, as is evident upon observing that

$$(\mathbf{A} \otimes \mathbf{B})(\mathbf{A}^- \otimes \mathbf{B}^-)(\mathbf{A} \otimes \mathbf{B}) = (\mathbf{A}\mathbf{A}^-\mathbf{A}) \otimes (\mathbf{B}\mathbf{B}^-\mathbf{B}) = \mathbf{A} \otimes \mathbf{B}. \tag{1.24}$$

Consider now the trace of the Kronecker product $\mathbf{A} \otimes \mathbf{B}$ of an $m \times m$ matrix $\mathbf{A} = \{a_{ij}\}$ and a $p \times p$ matrix $\mathbf{B}$. Making use of formula (5.1.7), we find that

$$\begin{aligned} \operatorname{tr}(\mathbf{A} \otimes \mathbf{B}) &= \operatorname{tr}(a_{11}\mathbf{B}) + \operatorname{tr}(a_{22}\mathbf{B}) + \cdots + \operatorname{tr}(a_{mm}\mathbf{B}) \\ &= (a_{11} + a_{22} + \cdots + a_{mm}) \operatorname{tr}(\mathbf{B}). \end{aligned}$$

Thus,

$$\operatorname{tr}(\mathbf{A} \otimes \mathbf{B}) = \operatorname{tr}(\mathbf{A}) \operatorname{tr}(\mathbf{B}). \tag{1.25}$$

Next, consider the rank of the Kronecker product of an $m \times n$ matrix $\mathbf{A}$ and a $p \times q$ matrix $\mathbf{B}$. Since $\mathbf{A}^- \otimes \mathbf{B}^-$ is a generalized inverse of $\mathbf{A} \otimes \mathbf{B}$, it follows from results (10.2.1), (1.19), and (1.25) that

$$\operatorname{rank}(\mathbf{A} \otimes \mathbf{B}) = \operatorname{tr}[(\mathbf{A} \otimes \mathbf{B})(\mathbf{A}^- \otimes \mathbf{B}^-)] = \operatorname{tr}[(\mathbf{A}\mathbf{A}^-) \otimes (\mathbf{B}\mathbf{B}^-)] = \operatorname{tr}(\mathbf{A}\mathbf{A}^-) \operatorname{tr}(\mathbf{B}\mathbf{B}^-),$$

and [again making use of result (10.2.1)] we conclude that

$$\operatorname{rank}(\mathbf{A} \otimes \mathbf{B}) = \operatorname{rank}(\mathbf{A}) \operatorname{rank}(\mathbf{B}). \tag{1.26}$$

Note that result (1.26) implies that the $mp \times nq$ dimensional Kronecker product $\mathbf{A} \otimes \mathbf{B}$ of an $m \times n$ matrix $\mathbf{A}$ and a $p \times q$ matrix $\mathbf{B}$ has full row rank if and only if both $\mathbf{A}$ and $\mathbf{B}$ have full row rank, has full column rank if and only if both $\mathbf{A}$ and $\mathbf{B}$ have full column rank, and hence is nonsingular if and only if both $\mathbf{A}$ and $\mathbf{B}$ are nonsingular.

The Kronecker product of an $m \times n$ partitioned matrix

$$\mathbf{A} = \begin{pmatrix} \mathbf{A}_{11} & \mathbf{A}_{12} & \dots & \mathbf{A}_{1c} \\ \mathbf{A}_{21} & \mathbf{A}_{22} & \dots & \mathbf{A}_{2c} \\ \vdots & \vdots & & \vdots \\ \mathbf{A}_{r1} & \mathbf{A}_{r2} & \dots & \mathbf{A}_{rc} \end{pmatrix}$$

and a $p \times q$ matrix $\mathbf{B}$ equals the $mp \times nq$ matrix obtained by replacing each block $\mathbf{A}_{ij}$ of $\mathbf{A}$ with the Kronecker product of $\mathbf{A}_{ij}$ and $\mathbf{B}$; that is,

$$\begin{pmatrix} \mathbf{A}_{11} & \mathbf{A}_{12} & \dots & \mathbf{A}_{1c} \\ \mathbf{A}_{21} & \mathbf{A}_{22} & \dots & \mathbf{A}_{2c} \\ \vdots & \vdots & & \vdots \\ \mathbf{A}_{r1} & \mathbf{A}_{r2} & \dots & \mathbf{A}_{rc} \end{pmatrix} \otimes \mathbf{B} = \begin{pmatrix} \mathbf{A}_{11} \otimes \mathbf{B} & \mathbf{A}_{12} \otimes \mathbf{B} & \dots & \mathbf{A}_{1c} \otimes \mathbf{B} \\ \mathbf{A}_{21} \otimes \mathbf{B} & \mathbf{A}_{22} \otimes \mathbf{B} & \dots & \mathbf{A}_{2c} \otimes \mathbf{B} \\ \vdots & \vdots & & \vdots \\ \mathbf{A}_{r1} \otimes \mathbf{B} & \mathbf{A}_{r2} \otimes \mathbf{B} & \dots & \mathbf{A}_{rc} \otimes \mathbf{B} \end{pmatrix}, \quad (1.27)$$

as is easily verified. And the Kronecker product of an m-dimensional column vector $\mathbf{a} = \{a_i\}$ and a $p \times q$ matrix $\mathbf{B}$ that has been partitioned as $\mathbf{B} = (\mathbf{B}_1, \mathbf{B}_2, \dots, \mathbf{B}_k)$ is the $mp \times q$ matrix obtained by replacing each block $\mathbf{B}_j$ of $\mathbf{B}$ with the Kronecker product of $\mathbf{a}$ and $\mathbf{B}_j$; that is,

$$\mathbf{a} \otimes (\mathbf{B}_1, \mathbf{B}_2, \dots, \mathbf{B}_k) = (\mathbf{a} \otimes \mathbf{B}_1,\ \mathbf{a} \otimes \mathbf{B}_2, \dots,\ \mathbf{a} \otimes \mathbf{B}_k), \quad (1.28)$$

as is also easily verified.

Suppose now that

$$\mathbf{A} = \begin{pmatrix} \mathbf{A}_{11} & \mathbf{A}_{12} & \dots & \mathbf{A}_{1c} \\ \mathbf{A}_{21} & \mathbf{A}_{22} & \dots & \mathbf{A}_{2c} \\ \vdots & \vdots & & \vdots \\ \mathbf{A}_{r1} & \mathbf{A}_{r2} & \dots & \mathbf{A}_{rc} \end{pmatrix}$$

is a partitioned matrix whose rc blocks $\mathbf{A}_{11}, \mathbf{A}_{12}, \dots, \mathbf{A}_{rc}$ are all of the same size, say $m \times n$ (in which case $\mathbf{A}$ is of dimensions $rm \times cn$). Let $\mathbf{U}_{ij}$ represent an $r \times c$ matrix whose ijth element is 1 and whose remaining $rc - 1$ elements are 0. Then,

$$\mathbf{A} = \sum_{i=1}^{r} \sum_{j=1}^{c} \mathbf{U}_{ij} \otimes \mathbf{A}_{ij}, \quad (1.29)$$

as is evident upon observing that $\mathbf{U}_{ij} \otimes \mathbf{A}_{ij}$ is an $rm \times cn$ partitioned matrtix comprising r rows and c columns of $m \times n$ dimensional blocks, the pqth of which equals $\mathbf{A}_{pq}$, if $i = p$ and $j = q$, and equals $\mathbf{0}$, if $i \neq p$ or $j \neq q$.

16.2 The Vec Operator: Definition and Some Basic Properties

It is sometimes convenient (as in the discussion of matrix differentiation in Chapter 15) to rearrange the elements of an $m \times n$ matrix $\mathbf{A} = \{a_{ij}\}$ in the form of an mn-dimensional column vector. The conventional way of doing so is to successively stack the first, second, ..., nth columns $\mathbf{a}_1, \mathbf{a}_2, \dots, \mathbf{a}_n$ of $\mathbf{A}$ one under the other, giving the mn-dimensional column vector

$$\begin{pmatrix} \mathbf{a}_1 \\ \mathbf{a}_2 \\ \vdots \\ \mathbf{a}_n \end{pmatrix}, \quad (2.1)$$

that is, giving a partitioned column vector comprising n subvectors of dimension m, the ith of which is $\mathbf{a}_i$.

The mn-dimensional column vector (2.1) is referred to as the *vec* of $\mathbf{A}$—think of vec as being an abbreviation for vector. It is denoted by the symbol $\text{vec}(\mathbf{A})$ or sometimes (when the parentheses are not needed for clarity) by $\text{vec}\,\mathbf{A}$. $\text{Vec}(\mathbf{A})$ can be regarded as the value assigned to $\mathbf{A}$ by a vector-valued function or operator whose domain is $\mathcal{R}^{m\times n}$—this operator is known as the *vec operator*. By definition, the ijth element of $\mathbf{A}$ is the $[(j-1)m+i]$th element of $\text{vec}(\mathbf{A})$.

Suppose, for example, that $m = 2$ and $n = 3$. Then,

$$\text{vec}(\mathbf{A}) = \begin{pmatrix} a_{11} \\ a_{21} \\ a_{12} \\ a_{22} \\ a_{13} \\ a_{23} \end{pmatrix}.$$

For any column vector $\mathbf{a}$,

$$\text{vec}(\mathbf{a}') = \text{vec}(\mathbf{a}) = \mathbf{a}, \tag{2.2}$$

as is evident from the very definition of the vec operator. And, for any m-dimensional column vector $\mathbf{a} = \{a_i\}$ and n-dimensional column vector $\mathbf{b}$,

$$\text{vec}(\mathbf{ba}') = \mathbf{a} \otimes \mathbf{b}, \tag{2.3}$$

as is evident from result (1.2) upon observing that the ith (of the m columns) of $\mathbf{ba}'$ is $a_i\mathbf{b}$.

Clearly, for any scalar c and any matrix $\mathbf{A}$,

$$\text{vec}(c\mathbf{A}) = c\,\text{vec}(\mathbf{A}), \tag{2.4}$$

and, for any two matrices $\mathbf{A}$ and $\mathbf{B}$ (of the same size),

$$\text{vec}(\mathbf{A}+\mathbf{B}) = \text{vec}(\mathbf{A}) + \text{vec}(\mathbf{B}). \tag{2.5}$$

More generally, for any k scalars $c_1, c_2, \ldots, c_k$ and for any k matrices $\mathbf{A}_1, \mathbf{A}_2, \ldots, \mathbf{A}_k$ (of the same size),

$$\text{vec}\Big(\sum_{i=1}^{k} c_i\mathbf{A}_i\Big) = \sum_{i=1}^{k} c_i\,\text{vec}(\mathbf{A}_i), \tag{2.6}$$

as can be easily established by the repeated application of results (2.5) and (2.4).

Let $\mathbf{A}$ represent an $m \times n$ matrix and $\mathbf{B}$ an $n \times p$ matrix, and denote the first, second, ..., pth columns of $\mathbf{B}$ by $\mathbf{b}_1, \mathbf{b}_2, \ldots, \mathbf{b}_p$, respectively. Then,

$$\text{vec}(\mathbf{AB}) = \begin{pmatrix} \mathbf{Ab}_1 \\ \mathbf{Ab}_2 \\ \vdots \\ \mathbf{Ab}_p \end{pmatrix}, \tag{2.7}$$

as is evident upon observing that the jth column of $\mathbf{AB}$ is $\mathbf{Ab}_j$. Note that result (2.7) implies that

$$\text{vec}(\mathbf{AB}) = \text{diag}(\mathbf{A}, \mathbf{A}, \ldots, \mathbf{A}) \, \text{vec}(\mathbf{B}) \tag{2.8}$$

or equivalently [in light of result (1.7)] that

$$\text{vec}(\mathbf{AB}) = (\mathbf{I}_p \otimes \mathbf{A}) \, \text{vec}(\mathbf{B}). \tag{2.9}$$

Result (2.9) is generalized in the following theorem.

Theorem 16.2.1. For any $m \times n$ matrix $\mathbf{A}$, $n \times p$ matrix $\mathbf{B}$, and $p \times q$ matrix $\mathbf{C}$,

$$\text{vec}(\mathbf{ABC}) = (\mathbf{C}' \otimes \mathbf{A}) \, \text{vec}(\mathbf{B}). \tag{2.10}$$

Proof. According to result (2.4.2), $\mathbf{B}$ is expressible as

$$\mathbf{B} = \sum_{j=1}^{p} \mathbf{b}_j \mathbf{u}_j',$$

where (for $j = 1, \ldots, p$) $\mathbf{b}_j$ is the jth column of $\mathbf{B}$ and $\mathbf{u}_j'$ is the jth row of $\mathbf{I}_p$. Thus, making use of results (2.3) and (1.19), we find that

$$\begin{aligned}
\text{vec}(\mathbf{ABC}) &= \text{vec}\Big[\mathbf{A}\Big(\sum_j \mathbf{b}_j \mathbf{u}_j'\Big)\mathbf{C}\Big] \\
&= \sum_j \text{vec}(\mathbf{Ab}_j \mathbf{u}_j' \mathbf{C}) \\
&= \sum_j [(\mathbf{C}'\mathbf{u}_j) \otimes (\mathbf{Ab}_j)] \\
&= \sum_j (\mathbf{C}' \otimes \mathbf{A})(\mathbf{u}_j \otimes \mathbf{b}_j) \\
&= \sum_j (\mathbf{C}' \otimes \mathbf{A}) \, \text{vec}(\mathbf{b}_j \mathbf{u}_j') \\
&= (\mathbf{C}' \otimes \mathbf{A}) \, \text{vec}\Big(\sum_j \mathbf{b}_j \mathbf{u}_j'\Big) = (\mathbf{C}' \otimes \mathbf{A}) \, \text{vec}\, \mathbf{B}.
\end{aligned}$$

Q.E.D.

Three alternative expressions for the vec of the product $\mathbf{AB}$ of an $m \times n$ matrix $\mathbf{A}$ and an $n \times p$ matrix $\mathbf{B}$ can be obtained as special cases of result (2.10). One of these expressions is given by result (2.9). The other two are

$$\text{vec}(\mathbf{AB}) = (\mathbf{B}' \otimes \mathbf{I}_m) \, \text{vec}(\mathbf{A}), \tag{2.11}$$

$$\text{vec}(\mathbf{AB}) = (\mathbf{B}' \otimes \mathbf{A}) \, \text{vec}(\mathbf{I}_n). \tag{2.12}$$

Consider now the product $\mathbf{ABx}$ of an $m \times n$ matrix $\mathbf{A}$, an $n \times p$ matrix $\mathbf{B}$, and a $p \times 1$ vector $\mathbf{x}$. According to result (2.2),

$$\mathbf{ABx} = \text{vec}(\mathbf{ABx}) = \text{vec}(\mathbf{x}'\mathbf{B}'\mathbf{A}').$$

Thus, it follows from result (2.10) that

$$\mathbf{ABx} = (\mathbf{x}' \otimes \mathbf{A})\,\text{vec}(\mathbf{B}) = (\mathbf{A} \otimes \mathbf{x}')\,\text{vec}(\mathbf{B}'). \tag{2.13}$$

For any $m \times n$ matrices $\mathbf{A}$ and $\mathbf{B}$,

$$\text{tr}(\mathbf{A}'\mathbf{B}) = (\text{vec}\,\mathbf{A})'\,\text{vec}\,\mathbf{B}. \tag{2.14}$$

To see this, let $\mathbf{a}_j$ and $\mathbf{b}_j$ represent the jth columns of $\mathbf{A}$ and $\mathbf{B}$, respectively, and observe that the jth diagonal element of $\mathbf{A}'\mathbf{B}$ equals $\mathbf{a}_j'\mathbf{b}_j$, so that

$$\text{tr}(\mathbf{A}'\mathbf{B}) = \sum_{j=1}^{n} \mathbf{a}_j'\mathbf{b}_j = (\mathbf{a}_1', \mathbf{a}_2', \ldots, \mathbf{a}_n')\begin{pmatrix}\mathbf{b}_1\\ \mathbf{b}_2\\ \vdots\\ \mathbf{b}_n\end{pmatrix} = \begin{pmatrix}\mathbf{a}_1\\ \mathbf{a}_2\\ \vdots\\ \mathbf{a}_n\end{pmatrix}'\begin{pmatrix}\mathbf{b}_1\\ \mathbf{b}_2\\ \vdots\\ \mathbf{b}_n\end{pmatrix}$$
$$= (\text{vec}\,\mathbf{A})'\,\text{vec}\,\mathbf{B}.$$

Note that result (2.14) implies that the (usual) inner product of any two $m \times n$ matrices equals the (usual) inner product of their vecs. Note also that, since [according to result (5.2.8)] $\text{tr}(\mathbf{A}'\mathbf{B}) = \text{tr}(\mathbf{B}'\mathbf{A}) = \text{tr}(\mathbf{BA}') = \text{tr}(\mathbf{AB}')$, alternative versions of formula (2.14) can be obtained by replacing $\text{tr}(\mathbf{A}'\mathbf{B})$ with $\text{tr}(\mathbf{B}'\mathbf{A})$, $\text{tr}(\mathbf{BA}')$, or $\text{tr}(\mathbf{AB}')$.

Result (2.14) is generalized in the following theorem.

Theorem 16.2.2. For any $m \times n$ matrix $\mathbf{A}$, $m \times p$ matrix $\mathbf{B}$, $p \times q$ matrix $\mathbf{C}$, and $n \times q$ matrix $\mathbf{D}$,

$$\text{tr}(\mathbf{A}'\mathbf{BCD}') = (\text{vec}\,\mathbf{A})'(\mathbf{D} \otimes \mathbf{B})\,\text{vec}\,\mathbf{C}. \tag{2.15}$$

Proof. Making use of results (2.14) and (2.10), we find that

$$\text{tr}(\mathbf{A}'\mathbf{BCD}') = (\text{vec}\,\mathbf{A})'\,\text{vec}(\mathbf{BCD}') = (\text{vec}\,\mathbf{A})'(\mathbf{D} \otimes \mathbf{B})\,\text{vec}\,\mathbf{C}.$$

Q.E.D.

Note [in light of results (5.2.3) and (5.1.5)] that $\text{tr}(\mathbf{A}'\mathbf{BCD}') = \text{tr}(\mathbf{D}'\mathbf{A}'\mathbf{BC}) = \text{tr}(\mathbf{CD}'\mathbf{A}'\mathbf{B}) = \text{tr}(\mathbf{BCD}'\mathbf{A}')$ and that $\text{tr}(\mathbf{A}'\mathbf{BCD}') = \text{tr}[(\mathbf{A}'\mathbf{BCD}')'] = \text{tr}(\mathbf{DC}'\mathbf{B}'\mathbf{A}) = \text{tr}(\mathbf{ADC}'\mathbf{B}') = \text{tr}(\mathbf{B}'\mathbf{ADC}') = \text{tr}(\mathbf{C}'\mathbf{B}'\mathbf{AD})$. Thus, alternative versions of formula (2.15) can be obtained by replacing $\text{tr}(\mathbf{A}'\mathbf{BCD}')$ with $\text{tr}(\mathbf{D}'\mathbf{A}'\mathbf{BC})$, $\text{tr}(\mathbf{CD}'\mathbf{A}'\mathbf{B})$, $\text{tr}(\mathbf{BCD}'\mathbf{A}')$, $\text{tr}(\mathbf{DC}'\mathbf{B}'\mathbf{A})$, $\text{tr}(\mathbf{ADC}'\mathbf{B}')$, $\text{tr}(\mathbf{B}'\mathbf{ADC}')$, or $\text{tr}(\mathbf{C}'\mathbf{B}'\mathbf{AD})$. Note also [in light of result (1.7)] that, in the special case where $q = n$ and $\mathbf{D} = \mathbf{I}_n$, result (2.15) reduces in effect to

$$\text{tr}(\mathbf{A}'\mathbf{BC}) = (\text{vec}\,\mathbf{A})'\,\text{diag}(\mathbf{B}, \mathbf{B}, \ldots, \mathbf{B})\,\text{vec}\,\mathbf{C}. \tag{2.16}$$

Let $\mathbf{A} = (\mathbf{A}_1, \mathbf{A}_2, \ldots, \mathbf{A}_k)$ represent an $m \times n$ partitioned matrix comprising a single row of k blocks, and let n_j represent the number of columns in $\mathbf{A}_j$. Then,

$$\text{vec}(\mathbf{A}_1, \mathbf{A}_2, \ldots, \mathbf{A}_k) = \begin{pmatrix} \text{vec}\,\mathbf{A}_1 \\ \text{vec}\,\mathbf{A}_2 \\ \vdots \\ \text{vec}\,\mathbf{A}_k \end{pmatrix}; \tag{2.17}$$

that is, $\text{vec}(\mathbf{A}_1, \mathbf{A}_2, \ldots, \mathbf{A}_k)$ equals an mn-dimensional column vector, comprising k subvectors, the jth of which is the mn_j-dimensional column vector $\text{vec}(\mathbf{A}_j)$. To see this, observe that both $\text{vec}(\mathbf{A}_1, \mathbf{A}_2, \ldots, \mathbf{A}_k)$ and

$$\begin{pmatrix} \text{vec}\,\mathbf{A}_1 \\ \text{vec}\,\mathbf{A}_2 \\ \vdots \\ \text{vec}\,\mathbf{A}_k \end{pmatrix}$$

are equal to an mn-dimensional partitioned column vector, comprising n m-dimensional subvectors, the $(\sum_{s=1}^{j-1} n_s + r)$th of which is the rth column of $\mathbf{A}_j$ (where $\sum_{s=1}^{0} n_s = 0$).

16.3 Vec-Permutation Matrix

a. Definition and some alternative descriptions

Let $\mathbf{A} = \{a_{ij}\}$ represent an $m \times n$ matrix, and denote the first, ..., nth columns of $\mathbf{A}$ by $\mathbf{a}_1, \ldots, \mathbf{a}_n$, respectively, and the first, ..., mth rows of $\mathbf{A}$ by $\mathbf{r}'_1, \ldots, \mathbf{r}'_m$, respectively. Then, $\mathbf{r}_1, \ldots, \mathbf{r}_m$ are the m columns of $\mathbf{A}'$, and, by definition,

$$\text{vec}(\mathbf{A}) = \begin{pmatrix} \mathbf{a}_1 \\ \mathbf{a}_2 \\ \vdots \\ \mathbf{a}_n \end{pmatrix} \quad \text{and} \quad \text{vec}(\mathbf{A}') = \begin{pmatrix} \mathbf{r}_1 \\ \mathbf{r}_2 \\ \vdots \\ \mathbf{r}_m \end{pmatrix}.$$

Both $\text{vec}(\mathbf{A}')$ and $\text{vec}(\mathbf{A})$ are obtained by rearranging the elements of $\mathbf{A}$ in the form of an mn-dimensional column vector; however, they are arranged row by row in $\text{vec}(\mathbf{A}')$ instead of column by column [as in $\text{vec}(\mathbf{A})$]. For example, when $m = 2$ and $n = 3$,

$$\text{vec}(\mathbf{A}) = \begin{pmatrix} a_{11} \\ a_{21} \\ a_{12} \\ a_{22} \\ a_{13} \\ a_{23} \end{pmatrix} \quad \text{and} \quad \text{vec}(\mathbf{A}') = \begin{pmatrix} a_{11} \\ a_{12} \\ a_{13} \\ a_{21} \\ a_{22} \\ a_{23} \end{pmatrix}.$$

Clearly, $\text{vec}(\mathbf{A}')$ can be obtained by permuting the elements of $\text{vec}(\mathbf{A})$. Accordingly, there exists an $mn \times mn$ permutation matrix, to be denoted by the symbol $\mathbf{K}_{mn}$, such that

$$\text{vec}(\mathbf{A}') = \mathbf{K}_{mn}\,\text{vec}(\mathbf{A}). \tag{3.1}$$

(This matrix depends on m and n, but not on the values of $a_{11}, a_{12}, \ldots, a_{mn}$.) For example,

$$\mathbf{K}_{23} = \begin{pmatrix} 1 & 0 & 0 & 0 & 0 & 0 \\ 0 & 0 & 1 & 0 & 0 & 0 \\ 0 & 0 & 0 & 0 & 1 & 0 \\ 0 & 1 & 0 & 0 & 0 & 0 \\ 0 & 0 & 0 & 1 & 0 & 0 \\ 0 & 0 & 0 & 0 & 0 & 1 \end{pmatrix}.$$

The matrix $\mathbf{K}_{mn}$ is referred to as a *vec-permutation matrix* (e.g., Henderson and Searle 1979) or, more commonly and for reasons that will become evident in Subsection c, as a *commutation matrix* (e.g., Magnus and Neudecker 1979).

As discussed in Section 16.2, the ijth element of the $m \times n$ matrix $\mathbf{A}$ is the $[(j-1)m+i]$th element of vec($\mathbf{A}$). And the ijth element of $\mathbf{A}$ is the jith element of the $n \times m$ matrix $\mathbf{A}'$ and accordingly is the $[(i-1)n+j]$th element of vec($\mathbf{A}'$). Thus, the $[(i-1)n+j]$th row of the vec-permutation matrix $\mathbf{K}_{mn}$ is the $[(j-1)m+i]$th row of $\mathbf{I}_{mn}$ ($i = 1, \ldots, m$; $j = 1, \ldots, n$).

Note that, since the transpose $\mathbf{A}'$ of the $m \times n$ matrix $\mathbf{A}$ is of dimensions $n \times m$, it follows from the very definition of a vec-permutation matrix that

$$\text{vec}(\mathbf{A}) = \text{vec}[(\mathbf{A}')'] = \mathbf{K}_{nm}\,\text{vec}(\mathbf{A}'). \tag{3.2}$$

The following theorem gives a useful and informative representation for the vec-permutation matrix $\mathbf{K}_{mn}$.

Theorem 16.3.1. Let m and n represent positive integers, and (for $i = 1, \ldots, m$ and $j = 1, \ldots, n$) let $\mathbf{U}_{ij} = \mathbf{e}_i \mathbf{u}_j'$, where $\mathbf{e}_i$ is the ith column of $\mathbf{I}_m$ and $\mathbf{u}_j$ the jth column of $\mathbf{I}_n$ (in which case $\mathbf{U}_{ij}$ is an $m \times n$ matrix whose ijth element is 1 and whose remaining elements are 0). Then,

$$\mathbf{K}_{mn} = \sum_{i=1}^{m} \sum_{j=1}^{n} (\mathbf{U}_{ij} \otimes \mathbf{U}_{ij}'). \tag{3.3}$$

Proof. Using results (3.1), (2.4.3), (2.4.5), and (2.10), we find that, for any $m \times n$ matrix $\mathbf{A}$,

$$\begin{aligned}
\mathbf{K}_{mn}\,\text{vec}(\mathbf{A}) = \text{vec}(\mathbf{A}') &= \text{vec}\Big(\sum_{i,j} a_{ij}\mathbf{U}_{ij}'\Big) \\
&= \sum_{i,j} \text{vec}(a_{ij}\mathbf{u}_j\mathbf{e}_i') \\
&= \sum_{i,j} \text{vec}(\mathbf{u}_j a_{ij}\mathbf{e}_i') \\
&= \sum_{i,j} \text{vec}[\mathbf{u}_j(\mathbf{e}_i'\mathbf{A}\mathbf{u}_j)\mathbf{e}_i']
\end{aligned}$$

$$= \sum_{i,j} \operatorname{vec}(\mathbf{U}_{ij}'\mathbf{A}\mathbf{U}_{ij}')$$

$$= \sum_{i,j} (\mathbf{U}_{ij} \otimes \mathbf{U}_{ij}') \operatorname{vec}(\mathbf{A})$$

and hence that, for any mn-dimensional column vector $\mathbf{a}$,

$$\mathbf{K}_{mn}\mathbf{a} = \Big(\sum_{i,j} \mathbf{U}_{ij} \otimes \mathbf{U}_{ij}'\Big)\mathbf{a}.$$

Based on Lemma 2.3.2, we conclude that

$$\mathbf{K}_{mn} = \sum_{i,j} (\mathbf{U}_{ij} \otimes \mathbf{U}_{ij}').$$

Q.E.D.

In light of result (1.29), result (3.3) can be reexpressed as

$$\mathbf{K}_{mn} = \begin{pmatrix} \mathbf{U}_{11}' & \mathbf{U}_{12}' & \dots & \mathbf{U}_{1n}' \\ \mathbf{U}_{21}' & \mathbf{U}_{22}' & \dots & \mathbf{U}_{2n}' \\ \vdots & \vdots & & \vdots \\ \mathbf{U}_{m1}' & \mathbf{U}_{m2}' & \dots & \mathbf{U}_{mn}' \end{pmatrix}. \tag{3.4}$$

That is, $\mathbf{K}_{mn}$ can be regarded as an $mn \times mn$ partitioned matrix, comprising m rows and n columns of $n \times m$ dimensional blocks, the ijth of which is the matrix $\mathbf{U}_{ij}' = \mathbf{u}_j\mathbf{e}_i'$ (whose jith element is 1 and whose other $mn - 1$ elements are 0). For example,

$$\mathbf{K}_{23} = \begin{pmatrix} \mathbf{U}_{11}' & \mathbf{U}_{12}' & \mathbf{U}_{13}' \\ \mathbf{U}_{21}' & \mathbf{U}_{22}' & \mathbf{U}_{23}' \end{pmatrix} = \left(\begin{array}{cc|cc|cc} 1 & 0 & 0 & 0 & 0 & 0 \\ 0 & 0 & 1 & 0 & 0 & 0 \\ 0 & 0 & 0 & 0 & 1 & 0 \\ \hline 0 & 1 & 0 & 0 & 0 & 0 \\ 0 & 0 & 0 & 1 & 0 & 0 \\ 0 & 0 & 0 & 0 & 0 & 1 \end{array}\right).$$

Let $\mathbf{a}$ represent an m-dimensional column vector. Then, upon setting $\mathbf{A} = \mathbf{a}$ in equality (3.1) and result (3.2), we find [in light of result (2.2)] that

$$\mathbf{a} = \mathbf{K}_{m1}\mathbf{a}, \qquad \mathbf{a} = \mathbf{K}_{1m}\mathbf{a}.$$

It follows (in light of Lemma 2.3.2) that

$$\mathbf{K}_{m1} = \mathbf{K}_{1m} = \mathbf{I}_m. \tag{3.5}$$

b. Some basic properties

It follows from result (3.2) that, for any $m \times n$ matrix $\mathbf{A}$,

$$\text{vec}(\mathbf{A}) = \mathbf{K}_{nm}\,\text{vec}(\mathbf{A}') = \mathbf{K}_{nm}\mathbf{K}_{mn}\,\text{vec}(\mathbf{A})$$

and hence that, for any mn-dimensional column vector $\mathbf{a}$,

$$\mathbf{a} = \mathbf{K}_{nm}\mathbf{K}_{mn}\mathbf{a},$$

implying (in light of Lemma 2.3.2) that

$$\mathbf{I}_{mn} = \mathbf{K}_{nm}\mathbf{K}_{mn}.$$

Thus (in light of Lemma 8.1.3), $\mathbf{K}_{mn}$ is nonsingular and $\mathbf{K}_{mn}^{-1} = \mathbf{K}_{nm}$. Moreover, since $\mathbf{K}_{mn}$ is a permutation matrix, it is orthogonal, and consequently $\mathbf{K}'_{mn} = \mathbf{K}_{mn}^{-1}$. In summary, we have that $\mathbf{K}_{mn}$ is nonsingular and that

$$\mathbf{K}'_{mn} = \mathbf{K}_{mn}^{-1} = \mathbf{K}_{nm}. \tag{3.6}$$

Note that result (3.6) implies that

$$\mathbf{K}'_{mm} = \mathbf{K}_{mm}; \tag{3.7}$$

that is, $\mathbf{K}_{mm}$ is symmetric (as well as orthogonal).

As a special case of a formula for $\text{tr}(\mathbf{K}_{mn})$ given by, for example, Magnus and Neudecker (1979, p. 383), we have that

$$\text{tr}(\mathbf{K}_{mm}) = m. \tag{3.8}$$

To verify formula (3.8), let $\mathbf{e}_i$ represent the ith column of $\mathbf{I}_m$, let $\mathbf{U}_{ij} = \mathbf{e}_i\mathbf{e}'_j$, and observe that $\mathbf{e}'_j\mathbf{e}_i$ equals 1, if $j = i$, and equals 0, if $j \neq i$. Then, making use of results (3.3), (1.25), and (5.2.6), we find that

$$\begin{aligned}
\text{tr}(\mathbf{K}_{mm}) &= \text{tr}\Big(\sum_{i,j} \mathbf{U}_{ij} \otimes \mathbf{U}'_{ij}\Big) \\
&= \sum_{i,j} \text{tr}(\mathbf{U}_{ij} \otimes \mathbf{U}'_{ij}) \\
&= \sum_{i,j} \text{tr}(\mathbf{U}_{ij})\,\text{tr}(\mathbf{U}'_{ij}) \\
&= \sum_{i,j} [\text{tr}(\mathbf{U}_{ij})]^2 \\
&= \sum_{i,j} [\text{tr}(\mathbf{e}_i\mathbf{e}'_j)]^2 \\
&= \sum_{i,j} (\mathbf{e}'_j\mathbf{e}_i)^2 = \sum_{i} (\mathbf{e}'_i\mathbf{e}_i)^2 = \sum_{i=1}^{m} (1)^2 = m.
\end{aligned}$$

c. Kronecker product of two matrices: the effect of premultiplying or postmultiplying by a vec-permutation matrix

As discussed in Section 16.1, the Kronecker product $\mathbf{B} \otimes \mathbf{A}$ of two matrices $\mathbf{B}$ and $\mathbf{A}$ is not in general equal to the Kronecker product $\mathbf{A} \otimes \mathbf{B}$ of $\mathbf{A}$ and $\mathbf{B}$. However, $\mathbf{B} \otimes \mathbf{A}$ can be "made equal to" $\mathbf{A} \otimes \mathbf{B}$ by permuting the columns of $\mathbf{B} \otimes \mathbf{A}$ and the rows of $\mathbf{A} \otimes \mathbf{B}$, as indicated by the following theorem.

Theorem 16.3.2. For any $m \times n$ matrix $\mathbf{A}$ and $p \times q$ matrix $\mathbf{B}$,

$$(\mathbf{B} \otimes \mathbf{A})\mathbf{K}_{qn} = \mathbf{K}_{pm}(\mathbf{A} \otimes \mathbf{B}), \tag{3.9}$$

$$\mathbf{B} \otimes \mathbf{A} = \mathbf{K}_{pm}(\mathbf{A} \otimes \mathbf{B})\mathbf{K}_{nq}. \tag{3.10}$$

Proof. Making use of Theorem 16.2.1, we find that, for any $q \times n$ matrix $\mathbf{X}$,

$$\begin{aligned}(\mathbf{B} \otimes \mathbf{A})\mathbf{K}_{qn} \operatorname{vec} \mathbf{X} = (\mathbf{B} \otimes \mathbf{A}) \operatorname{vec}(\mathbf{X}') &= \operatorname{vec}(\mathbf{A}\mathbf{X}'\mathbf{B}') \\ &= \operatorname{vec}[(\mathbf{B}\mathbf{X}\mathbf{A}')'] \\ &= \mathbf{K}_{pm} \operatorname{vec}(\mathbf{B}\mathbf{X}\mathbf{A}') \\ &= \mathbf{K}_{pm}(\mathbf{A} \otimes \mathbf{B}) \operatorname{vec} \mathbf{X}.\end{aligned}$$

Thus,

$$(\mathbf{B} \otimes \mathbf{A})\mathbf{K}_{qn}\mathbf{x} = \mathbf{K}_{pm}(\mathbf{A} \otimes \mathbf{B})\mathbf{x}$$

for every $nq \times 1$ vector $\mathbf{x}$, implying (in light of Lemma 2.3.2) that

$$(\mathbf{B} \otimes \mathbf{A})\mathbf{K}_{qn} = \mathbf{K}_{pm}(\mathbf{A} \otimes \mathbf{B}).$$

Further, since $\mathbf{K}_{qn}^{-1} = \mathbf{K}_{nq}$, we have that

$$\mathbf{B} \otimes \mathbf{A} = (\mathbf{B} \otimes \mathbf{A})\mathbf{K}_{qn}\mathbf{K}_{nq} = \mathbf{K}_{pm}(\mathbf{A} \otimes \mathbf{B})\mathbf{K}_{nq}.$$

Q.E.D.

As indicated by result (3.10), the effect on $\mathbf{A} \otimes \mathbf{B}$ of premultiplication and postmultiplication by the vec-permutation matrices $\mathbf{K}_{pm}$ and $\mathbf{K}_{nq}$, respectively, is to interchange or "commute" $\mathbf{A}$ and $\mathbf{B}$. It is because of this effect that vec-permutation matrices are often referred to as commutation matrices.

In the various special cases where $\mathbf{A}$ or $\mathbf{B}$ is a row or column vector, result (3.9) or (3.10) of Theorem 16.3.2 can be stated more simply, as indicated by the following corollary.

Corollary 16.3.3. For any $m \times n$ matrix $\mathbf{A}$ and $p \times 1$ vector $\mathbf{b}$,

$$\mathbf{b} \otimes \mathbf{A} = \mathbf{K}_{pm}(\mathbf{A} \otimes \mathbf{b}), \tag{3.11}$$

$$\mathbf{A} \otimes \mathbf{b} = \mathbf{K}_{mp}(\mathbf{b} \otimes \mathbf{A}), \tag{3.12}$$

$$\mathbf{b}' \otimes \mathbf{A} = (\mathbf{A} \otimes \mathbf{b}')\mathbf{K}_{np}, \tag{3.13}$$

$$\mathbf{A} \otimes \mathbf{b}' = (\mathbf{b}' \otimes \mathbf{A})\mathbf{K}_{pn}. \tag{3.14}$$

Note that in the special case where $\mathbf{B}$ is a row vector, say $\mathbf{b}'$, and $\mathbf{A}$ is a column vector, say $\mathbf{a}$, equality (3.13) or (3.14) reduces to the equality

$$\mathbf{b}' \otimes \mathbf{a} = \mathbf{a} \otimes \mathbf{b}',$$

given previously in result (1.4).

The trace of the matrix obtained by premultiplying or postmultiplying a Kronecker product (of an $n \times m$ matrix and an $m \times n$ matrix) by a vec-permutation matrix is given by the following theorem.

Theorem 16.3.4. For any $m \times n$ matrices $\mathbf{A}$ and $\mathbf{B}$,

$$\operatorname{tr}[(\mathbf{A}' \otimes \mathbf{B})\mathbf{K}_{mn}] = \operatorname{tr}[\mathbf{K}_{mn}(\mathbf{A}' \otimes \mathbf{B})] = \operatorname{tr}(\mathbf{A}'\mathbf{B}) = (\operatorname{vec} \mathbf{A}')'\mathbf{K}_{mn} \operatorname{vec} \mathbf{B}. \tag{3.15}$$

Proof. Let $\mathbf{e}_i$ represent the ith column of $\mathbf{I}_m$ and $\mathbf{u}_j$ the jth column of $\mathbf{I}_n$, and define $\mathbf{U}_{ij} = \mathbf{e}_i\mathbf{u}_j'$. That $\operatorname{tr}[(\mathbf{A}' \otimes \mathbf{B})\mathbf{K}_{mn}] = \operatorname{tr}[\mathbf{K}_{mn}(\mathbf{A}' \otimes \mathbf{B})]$ is an immediate consequence of Lemma 5.2.1. Further, making use of results (3.3), (1.19), (1.25), (5.2.6), (2.4.5), and (5.2.4), we find that

$$\begin{aligned}
\operatorname{tr}[\mathbf{K}_{mn}(\mathbf{A}' \otimes \mathbf{B})] &= \operatorname{tr}\Big[\sum_{i,j} (\mathbf{U}_{ij} \otimes \mathbf{U}_{ij}')(\mathbf{A}' \otimes \mathbf{B})\Big] \\
&= \sum_{i,j} \operatorname{tr}[(\mathbf{U}_{ij} \otimes \mathbf{U}_{ij}')(\mathbf{A}' \otimes \mathbf{B})] \\
&= \sum_{i,j} \operatorname{tr}[(\mathbf{U}_{ij}\mathbf{A}') \otimes (\mathbf{U}_{ij}'\mathbf{B})] \\
&= \sum_{i,j} \operatorname{tr}(\mathbf{U}_{ij}\mathbf{A}') \operatorname{tr}(\mathbf{U}_{ij}'\mathbf{B}) \\
&= \sum_{i,j} \operatorname{tr}(\mathbf{e}_i\mathbf{u}_j'\mathbf{A}') \operatorname{tr}(\mathbf{u}_j\mathbf{e}_i'\mathbf{B}) \\
&= \sum_{i,j} (\mathbf{u}_j'\mathbf{A}'\mathbf{e}_i)\mathbf{e}_i'\mathbf{B}\mathbf{u}_j \\
&= \sum_{i,j} b_{ij}a_{ij} \\
&= \operatorname{tr}(\mathbf{A}'\mathbf{B}).
\end{aligned}$$

And, in light of results (2.14), (3.2), and (3.6), we have that

$$\begin{aligned}
\operatorname{tr}(\mathbf{A}'\mathbf{B}) = (\operatorname{vec} \mathbf{A})' \operatorname{vec} \mathbf{B} &= (\mathbf{K}_{nm} \operatorname{vec} \mathbf{A}')' \operatorname{vec} \mathbf{B} \\
&= (\operatorname{vec} \mathbf{A}')'\mathbf{K}_{nm}' \operatorname{vec} \mathbf{B} \\
&= (\operatorname{vec} \mathbf{A}')'\mathbf{K}_{mn} \operatorname{vec} \mathbf{B}.
\end{aligned}$$

Q.E.D.

d. Relationship between the vec of the Kronecker product of two matrices and the Kronecker product of their vecs

The following theorem relates the vec of the Kronecker product of two matrices **A** and **B** to the Kronecker product of the vecs of **A** and **B**.

Theorem 16.3.5. For any $m \times n$ matrix $\mathbf{A}$ and any $p \times q$ matrix $\mathbf{B}$,

$$\text{vec}(\mathbf{A} \otimes \mathbf{B}) = (\mathbf{I}_n \otimes \mathbf{K}_{qm} \otimes \mathbf{I}_p)[\text{vec}(\mathbf{A}) \otimes \text{vec}(\mathbf{B})]. \tag{3.16}$$

Proof. Let $\mathbf{a}_i$ represent the ith column of $\mathbf{A}$, and $\mathbf{e}_i$ the ith column of $\mathbf{I}_n$ $(i = 1, \ldots, n)$. And let $\mathbf{b}_j$ represent the jth column of $\mathbf{B}$, and $\mathbf{u}_j$ the jth column of $\mathbf{I}_q$ $(j = 1, \ldots, q)$. Then, writing $\mathbf{A}$ and $\mathbf{B}$ as $\mathbf{A} = \sum_i \mathbf{a}_i \mathbf{e}_i'$ and $\mathbf{B} = \sum_j \mathbf{b}_j \mathbf{u}_j'$ and making use of results (1.14), (2.6), (1.15), (1.19), (2.3), (1.16), and (3.11), we find that

$$\begin{aligned}
\text{vec}(\mathbf{A} \otimes \mathbf{B}) &= \text{vec}\Big[\sum_{i,j} (\mathbf{a}_i \mathbf{e}_i') \otimes (\mathbf{b}_j \mathbf{u}_j')\Big] \\
&= \sum_{i,j} \text{vec}[(\mathbf{a}_i \mathbf{e}_i') \otimes (\mathbf{b}_j \mathbf{u}_j')] \\
&= \sum_{i,j} \text{vec}[(\mathbf{a}_i \otimes \mathbf{b}_j)(\mathbf{e}_i \otimes \mathbf{u}_j)'] \\
&= \sum_{i,j} (\mathbf{e}_i \otimes \mathbf{u}_j) \otimes (\mathbf{a}_i \otimes \mathbf{b}_j) \\
&= \sum_{i,j} \mathbf{e}_i \otimes (\mathbf{u}_j \otimes \mathbf{a}_i) \otimes \mathbf{b}_j \\
&= \sum_{i,j} (\mathbf{I}_n \otimes \mathbf{K}_{qm} \otimes \mathbf{I}_p)(\mathbf{e}_i \otimes \mathbf{a}_i \otimes \mathbf{u}_j \otimes \mathbf{b}_j) \\
&= \sum_{i,j} (\mathbf{I}_n \otimes \mathbf{K}_{qm} \otimes \mathbf{I}_p)[\text{vec}(\mathbf{a}_i \mathbf{e}_i') \otimes \text{vec}(\mathbf{b}_j \mathbf{u}_j')] \\
&= (\mathbf{I}_n \otimes \mathbf{K}_{qm} \otimes \mathbf{I}_p)\Big[\sum_i \text{vec}(\mathbf{a}_i \mathbf{e}_i') \otimes \sum_j \text{vec}(\mathbf{b}_j \mathbf{u}_j')\Big] \\
&= (\mathbf{I}_n \otimes \mathbf{K}_{qm} \otimes \mathbf{I}_p)\Big[\text{vec}\Big(\sum_i \mathbf{a}_i \mathbf{e}_i'\Big) \otimes \text{vec}\Big(\sum_j \mathbf{b}_j \mathbf{u}_j'\Big)\Big] \\
&= (\mathbf{I}_n \otimes \mathbf{K}_{qm} \otimes \mathbf{I}_p)[\text{vec}(\mathbf{A}) \otimes \text{vec}(\mathbf{B})].
\end{aligned}$$

Q.E.D.

e. Determinant of a Kronecker product

Preliminary to deriving a formula for the determinant of a Kronecker product of two matrices, observe [in light of results (1.20) and (3.10)] that, for any $m \times n$ matrix $\mathbf{A}$ and any $p \times q$ matrix $\mathbf{B}$,

$$
\begin{aligned}
\mathbf{A} \otimes \mathbf{B} &= (\mathbf{A} \otimes \mathbf{I}_p)(\mathbf{I}_n \otimes \mathbf{B}) \\
&= \mathbf{K}_{mp}(\mathbf{I}_p \otimes \mathbf{A})\mathbf{K}_{pn}(\mathbf{I}_n \otimes \mathbf{B}) \\
&= \mathbf{K}_{mp}\,\text{diag}(\mathbf{A}, \mathbf{A}, \ldots, \mathbf{A})\mathbf{K}_{pn}\,\text{diag}(\mathbf{B}, \mathbf{B}, \ldots, \mathbf{B}). \qquad (3.17)
\end{aligned}
$$

Consider now the determinant of the Kronecker product $\mathbf{A} \otimes \mathbf{B}$ of an $m \times m$ matrix $\mathbf{A}$ and a $p \times p$ matrix $\mathbf{B}$. Applying result (3.17) (in the special case where $n = m$ and $q = p$), we find [in light of results (13.3.9) and (13.3.5)] that

$$|\mathbf{A} \otimes \mathbf{B}| = |\mathbf{K}_{mp}||\mathbf{A}|^p|\mathbf{K}_{pm}||\mathbf{B}|^m.$$

And, in light of result (3.6), we have that

$$|\mathbf{K}_{mp}||\mathbf{K}_{pm}| = |\mathbf{K}_{mp}\mathbf{K}_{pm}| = |\mathbf{I}| = 1.$$

We conclude that (for any $m \times m$ matrix $\mathbf{A}$ and any $p \times p$ matrix $\mathbf{B}$)

$$|\mathbf{A} \otimes \mathbf{B}| = |\mathbf{A}|^p|\mathbf{B}|^m. \qquad (3.18)$$

16.4 The Vech Operator

a. Definition and basic characteristics

The values of all n^2 elements of an $n \times n$ symmetric matrix $\mathbf{A}$ can be determined from the values of those $n(n+1)/2$ elements that are on or below the diagonal [or alternatively from those $n(n+1)/2$ elements that are on or above the diagonal]. In rearranging the elements of $\mathbf{A}$ in the form of a vector, we may wish to exclude the $n(n-1)/2$ "duplicate" elements. This would be the case if, for example, $\mathbf{A}$ were a (symmetric) matrix of variables and our objective were the differentiation of a function of $\mathbf{A}$.

Let $\mathbf{A} = \{a_{ij}\}$ represent an $n \times n$ (symmetric or nonsymmetric) matrix. To obtain the vec of $\mathbf{A}$, we successively stack the first, second, ..., nth columns $\mathbf{a}_1, \mathbf{a}_2, \ldots, \mathbf{a}_n$ of $\mathbf{A}$ one under the other. Consider a modification of this process in which (before or after the stacking) the $n(n-1)/2$ "supradiagonal" elements of $\mathbf{A}$ are eliminated from $\mathbf{a}_1, \mathbf{a}_2, \ldots, \mathbf{a}_n$. The result is the $[n(n+1)/2]$-dimensional vector

$$\begin{pmatrix} \mathbf{a}_1^* \\ \mathbf{a}_2^* \\ \vdots \\ \mathbf{a}_n^* \end{pmatrix}, \qquad (4.1)$$

where (for $i = 1, 2, \ldots, n$) $\mathbf{a}_i^* = (a_{ii}, a_{i+1,i}, \ldots, a_{ni})'$ is the subvector of $\mathbf{a}_i$ obtained by striking out its first $i - 1$ elements. Thus, by definition, the vector (4.1) is a subvector of vec $\mathbf{A}$ obtained by striking out a particular set of what are, in the special case where $\mathbf{A}$ is symmetric, duplicate or redundant elements.

Following Henderson and Searle (1979), let us refer to the vector (4.1) as the *vech* of $\mathbf{A}$— think of vech as being an abbreviation for "vector-half." And denote this vector by the symbol vech(**A**) or possibly (when the parentheses are not needed for clarity) by vech **A**. Like vec **A**, vech **A** can be regarded as the value assigned to **A** by a vector-valued function or operator—the domain of this function is $\mathbb{R}^{n\times n}$.

For $n = 1$, $n = 2$, and $n = 3$,

$$\text{vech}\,\mathbf{A} = (a_{11}),\ \ \text{vech}\,\mathbf{A} = \begin{pmatrix} a_{11} \\ a_{21} \\ a_{22} \end{pmatrix},\ \text{and vech}\,\mathbf{A} = \begin{pmatrix} a_{11} \\ a_{21} \\ a_{31} \\ a_{22} \\ a_{32} \\ a_{33} \end{pmatrix},\ \text{respectively.}$$

By way of comparison,

$$\text{vec}\,\mathbf{A} = (a_{11}),\ \ \text{vec}\,\mathbf{A} = \begin{pmatrix} a_{11} \\ a_{21} \\ a_{12} \\ a_{22} \end{pmatrix},\ \text{and vec}\,\mathbf{A} = \begin{pmatrix} a_{11} \\ a_{21} \\ a_{31} \\ a_{12} \\ a_{22} \\ a_{32} \\ a_{13} \\ a_{23} \\ a_{33} \end{pmatrix},\ \text{respectively.}$$

Observe that the total number of elements in the j vectors $\mathbf{a}_1^*, \mathbf{a}_2^*, \mathbf{a}_3^*, \ldots, \mathbf{a}_j^*$ is

$$\begin{aligned} n + (n-1) + (n-2) + \cdots + n - (j-1) \\ = nj - (0 + 1 + 2 + \cdots + j - 1) = nj - j(j-1)/2 \end{aligned}$$

and that, of the $n-(j-1) = n-j+1$ elements of $\mathbf{a}_j^*$, there are (for $i \geq j$) $n-i$ elements that come after a_{ij}. Since $nj - j(j-1)/2 - (n-i) = (j-1)(n-j/2)+i$, it follows that (for $i \geq j$) the ijth element of **A** is the $[(j-1)(n-j/2)+i]$th element of vech **A**. By way of comparison, the ijth element of **A** is (as discussed in Section 16.2) the $[(j-1)n+i]$th element of vec **A**, so that (for $i \geq j$) the $[(j-1)n+i]$th element of vec **A** is the $[(j-1)(n-j/2)+i]$th element of vech **A**. For example, when $n = 8$, we find [upon setting $i = 7$ and $j = 5$ and observing that $(j-1)n+i = 39$ and that $(j-1)(n-j/2)+i = 29$] that a_{75} is the 39th element of vec **A** and the 29th element of vech **A**.

Let $c_1, \ldots, c_k$ represent any k scalars, and let $\mathbf{A}_1, \ldots, \mathbf{A}_k$ represent any k square matrices (of the same order). Then,

$$\text{vech}\Big(\sum_{i=1}^{k} c_i \mathbf{A}_i\Big) = \sum_{i=1}^{k} c_i\ \text{vech}(\mathbf{A}_i), \tag{4.2}$$

as is evident from result (2.6) upon observing that the vech of any (square) matrix is a subvector of its vec.

b. Duplication matrix

Every element of an $n \times n$ symmetric matrix $\mathbf{A}$, and hence every element of vec $\mathbf{A}$, is either an element of vech $\mathbf{A}$ or a "duplicate" of an element of vech $\mathbf{A}$. Thus, there exists a unique $\{[n^2 \times n(n+1)/2]\text{-dimensional}\}$ matrix, to be denoted by the symbol $\mathbf{G}_n$, such that

$$\text{vec}\,\mathbf{A} = \mathbf{G}_n\,\text{vech}\,\mathbf{A} \tag{4.3}$$

for every $n \times n$ symmetric matrix $\mathbf{A}$. This matrix is called the *duplication matrix*.

Clearly,

$$\mathbf{G}_1 = (1),\ \mathbf{G}_2 = \begin{pmatrix} 1 & 0 & 0 \\ 0 & 1 & 0 \\ 0 & 1 & 0 \\ 0 & 0 & 1 \end{pmatrix},\ \text{and}\ \mathbf{G}_3 = \begin{pmatrix} 1 & 0 & 0 & 0 & 0 & 0 \\ 0 & 1 & 0 & 0 & 0 & 0 \\ 0 & 0 & 1 & 0 & 0 & 0 \\ 0 & 1 & 0 & 0 & 0 & 0 \\ 0 & 0 & 0 & 1 & 0 & 0 \\ 0 & 0 & 0 & 0 & 1 & 0 \\ 0 & 0 & 1 & 0 & 0 & 0 \\ 0 & 0 & 0 & 0 & 1 & 0 \\ 0 & 0 & 0 & 0 & 0 & 1 \end{pmatrix}.$$

More generally (for an arbitrary positive integer n), the matrix $\mathbf{G}_n$ can be described in terms of its rows or columns. For $i \geq j$, the $[(j-1)n+i]$th and $[(i-1)n+j]$th rows of $\mathbf{G}_n$ equal the $[(j-1)(n-j/2)+i]$th row of $\mathbf{I}_{n(n+1)/2}$, that is, they equal the $[n(n+1)/2]$-dimensional row vector whose $[(j-1)(n-j/2)+i]$th element is 1 and whose remaining elements are 0. And (for $i \geq j$) the $[(j-1)(n-j/2)+i]$th column of $\mathbf{G}_n$ is an n^2-dimensional column vector whose $[(j-1)n+i]$th and $[(i-1)n+j]$th elements are 1 and whose remaining n^2-1 (if $i=j$) or n^2-2 (if $i>j$) elements are 0.

The matrix $\mathbf{G}_n$ is of full column rank; that is,

$$\text{rank}(\mathbf{G}_n) = n(n+1)/2. \tag{4.4}$$

To see this, observe that the columns of $\mathbf{G}_n$ are orthogonal (since every row of $\mathbf{G}_n$ contains only one nonzero element) and are nonnull. Or, alternatively, observe that every row of $\mathbf{I}_{n(n+1)/2}$ is a row of $\mathbf{G}_n$ and hence that $\mathbf{G}_n$ contains $n(n+1)/2$ linearly independent rows.

Now, for purposes of deriving a recursive formula for $\mathbf{G}_n$, let $\mathbf{B}$ represent an arbitrary $(n+1) \times (n+1)$ matrix, and partition $\mathbf{B}$ as

$$\mathbf{B} = \begin{pmatrix} c & \mathbf{b}' \\ \mathbf{a} & \mathbf{A} \end{pmatrix}$$

(where $\mathbf{A}$ is $n \times n$). Then, there exists an $(n+1)^2 \times (n+1)^2$ permutation matrix, to be denoted by the symbol $\mathbf{Q}_{n+1}$, such that (for every choice of $\mathbf{B}$)

$$\mathbf{Q}'_{n+1}\,\text{vec}\,\mathbf{B} = \begin{pmatrix} c \\ \mathbf{b} \\ \mathbf{a} \\ \text{vec}\,\mathbf{A} \end{pmatrix}.$$

Further, letting $\mathbf{u}$ represent the first column of $\mathbf{I}_{n+1}$ and letting $\mathbf{U}$ represent the $(n+1) \times n$ submatrix of $\mathbf{I}_{n+1}$ obtained by striking out $\mathbf{u}$, $\mathbf{Q}_{n+1}$ is expressible as $\mathbf{Q}_{n+1} = (\mathbf{Q}_{n+1}^{(1)}, \mathbf{Q}_{n+1}^{(2)})$, where

$$\mathbf{Q}_{n+1}^{(1)} = \operatorname{diag}(\mathbf{u}, \mathbf{u}, \ldots, \mathbf{u}) = \mathbf{I}_{n+1} \otimes \mathbf{u},$$
$$\mathbf{Q}_{n+1}^{(2)} = \operatorname{diag}(\mathbf{U}, \mathbf{U}, \ldots, \mathbf{U}) = \mathbf{I}_{n+1} \otimes \mathbf{U}.$$

Clearly,

$$\operatorname{vech} \mathbf{B} = \begin{pmatrix} c \\ \mathbf{a} \\ \operatorname{vech} \mathbf{A} \end{pmatrix}.$$

Thus, if $\mathbf{B}$ is symmetric (in which case $\mathbf{A}$ is symmetric and $\mathbf{b} = \mathbf{a}$), we have that

$$\begin{pmatrix} c \\ \mathbf{b} \\ \mathbf{a} \\ \operatorname{vec} \mathbf{A} \end{pmatrix} = \begin{pmatrix} 1 & \mathbf{0} & \mathbf{0} \\ \mathbf{0} & \mathbf{I}_n & \mathbf{0} \\ \mathbf{0} & \mathbf{I}_n & \mathbf{0} \\ \mathbf{0} & \mathbf{0} & \mathbf{G}_n \end{pmatrix} \operatorname{vech} \mathbf{B}$$

or, equivalently, that

$$\mathbf{Q}_{n+1}' \operatorname{vec} \mathbf{B} = \begin{pmatrix} 1 & \mathbf{0} & \mathbf{0} \\ \mathbf{0} & \mathbf{I}_n & \mathbf{0} \\ \mathbf{0} & \mathbf{I}_n & \mathbf{0} \\ \mathbf{0} & \mathbf{0} & \mathbf{G}_n \end{pmatrix} \operatorname{vech} \mathbf{B}.$$

We conclude that if $\mathbf{B}$ is symmetric, then

$$\operatorname{vec} \mathbf{B} = \mathbf{Q}_{n+1} \begin{pmatrix} 1 & \mathbf{0} & \mathbf{0} \\ \mathbf{0} & \mathbf{I}_n & \mathbf{0} \\ \mathbf{0} & \mathbf{I}_n & \mathbf{0} \\ \mathbf{0} & \mathbf{0} & \mathbf{G}_n \end{pmatrix} \operatorname{vech} \mathbf{B}.$$

This conclusion gives rise to the recursive formula

$$\mathbf{G}_{n+1} = \mathbf{Q}_{n+1} \begin{pmatrix} 1 & \mathbf{0} & \mathbf{0} \\ \mathbf{0} & \mathbf{I}_n & \mathbf{0} \\ \mathbf{0} & \mathbf{I}_n & \mathbf{0} \\ \mathbf{0} & \mathbf{0} & \mathbf{G}_n \end{pmatrix}. \tag{4.5}$$

Since the duplication matrix $\mathbf{G}_n$ is of full column rank, it has a left inverse. In fact, except in the special case where $n = 1$, $\mathbf{G}_n$ has an infinite number of left inverses. ($\mathbf{G}_1$ is nonsingular and consequently has a unique left inverse.)

In what follows, the symbol $\mathbf{H}_n$ is used to represent an arbitrary left inverse of $\mathbf{G}_n$. Thus, by definition, $\mathbf{H}_n$ is an $n(n+1)/2 \times n^2$ matrix such that

$$\mathbf{H}_n \mathbf{G}_n = \mathbf{I}.$$

One choice for $\mathbf{H}_n$ is

$$\mathbf{H}_n = (\mathbf{G}_n'\mathbf{G}_n)^{-1}\mathbf{G}_n'.$$

(Since $\mathbf{G}_n$ is of full column rank, $\mathbf{G}_n'\mathbf{G}_n$ is nonsingular.)

The premultiplication of the vech of an $n \times n$ symmetric matrix $\mathbf{A}$ by the duplication matrix $\mathbf{G}_n$ transforms vech $\mathbf{A}$ to vec $\mathbf{A}$. Upon premultiplying both sides of equality (4.3) by $\mathbf{H}_n$, we find that

$$\text{vech}\,\mathbf{A} = \mathbf{H}_n \text{vec}\,\mathbf{A} \tag{4.6}$$

for every $n \times n$ symmetric matrix $\mathbf{A}$. Thus, the premultiplication of the vec of an $n \times n$ symmetric matrix $\mathbf{A}$ by the left inverse $\mathbf{H}_n$ transforms vec $\mathbf{A}$ to vech $\mathbf{A}$.

Are there matrices other than left inverses (of $\mathbf{G}_n$) that have this property? That is, does there exist an $n(n+1)/2 \times n^2$ matrix $\mathbf{L}$ such that

$$\text{vech}\,\mathbf{A} = \mathbf{L}\,\text{vec}\,\mathbf{A} \tag{4.7}$$

for every $n \times n$ symmetric matrix $\mathbf{A}$ that does not satisfy the condition $\mathbf{L}\mathbf{G}_n = \mathbf{I}$? The answer is no, as is evident upon observing that, together with equality (4.3), equality (4.7) implies that

$$\text{vech}\,\mathbf{A} = \mathbf{L}\mathbf{G}_n \text{vech}\,\mathbf{A}$$

for every $n \times n$ symmetric matrix $\mathbf{A}$, or, equivalently, that $\mathbf{a} = \mathbf{L}\mathbf{G}_n\mathbf{a}$ for every $[n(n+1)/2]$-dimensional column vector $\mathbf{a}$, and hence that

$$\mathbf{L}\mathbf{G}_n = \mathbf{I}.$$

Thus, as an alternative to characterizing the matrix $\mathbf{H}_n$ as an arbitrary left inverse of $\mathbf{G}_n$, it could be characterized as any $n(n+1)/2 \times n^2$ matrix $\mathbf{L}$ such that equality (4.7) holds for every $n \times n$ symmetric matrix $\mathbf{A}$. In fact, as noted by Henderson and Searle (1979), it is the latter characterization that provides the motivation for denoting this matrix by a capital h—h added to vec gives vech.

We find that

$$\mathbf{H}_1 = (1), \quad \mathbf{H}_2 = \begin{pmatrix} 1 & 0 & 0 & 0 \\ 0 & c & 1-c & 0 \\ 0 & 0 & 0 & 1 \end{pmatrix}, \tag{4.8a}$$

$$\mathbf{H}_3 = \begin{pmatrix} 1 & 0 & 0 & 0 & 0 & 0 & 0 & 0 & 0 \\ 0 & c_1 & 0 & 1-c_1 & 0 & 0 & 0 & 0 & 0 \\ 0 & 0 & c_2 & 0 & 0 & 0 & 1-c_2 & 0 & 0 \\ 0 & 0 & 0 & 0 & 1 & 0 & 0 & 0 & 0 \\ 0 & 0 & 0 & 0 & 0 & c_3 & 0 & 1-c_3 & 0 \\ 0 & 0 & 0 & 0 & 0 & 0 & 0 & 0 & 1 \end{pmatrix}, \tag{4.8b}$$

where c, c_1, c_2, and c_3 are arbitrary scalars. More generally (for an arbitrary positive integer n), the matrix $\mathbf{H}_n$ can be described in terms of its rows. The $[(i-1)(n-i/2)+i]$th row of $\mathbf{H}_n$ is the $[(i-1)n+i]$th row of $\mathbf{I}_{n^2}$, and (for

$i > j$) the $[(j-1)(n-j/2)+i]$th row of $\mathbf{H}_n$ is any n^2-dimensional row vector whose $[(j-1)n+i]$th and $[(i-1)n+j]$th elements sum to 1 and whose remaining n^2-2 elements are 0.

Note that $\mathbf{H}_n$ is of full row rank, that is,

$$\text{rank}(\mathbf{H}_n) = n(n+1)/2 \tag{4.9}$$

(as is evident from Lemma 8.1.1 upon observing that $\mathbf{G}_n$ is a right inverse of $\mathbf{H}_n$). Note also that

$$\text{vec}\,\mathbf{A} = \mathbf{G}_n\mathbf{H}_n\,\text{vec}\,\mathbf{A} \tag{4.10}$$

for every $n \times n$ symmetric matrix $\mathbf{A}$ [as is evident upon substituting expression (4.6) in the right side of equality (4.3)].

Consider now the matrix $(\mathbf{G}_n'\mathbf{G}_n)^{-1}\mathbf{G}_n'$ (which is one choice for $\mathbf{H}_n$). The $n(n+1)/2 \times n(n+1)/2$ matrix $\mathbf{G}_n'\mathbf{G}_n$ is diagonal (since the columns of $\mathbf{G}_n$ are orthogonal). Further, the $[(i-1)(n-i/2)+i]$th diagonal element of $\mathbf{G}_n'\mathbf{G}_n$ equals 1, and (for $i > j$) the $[(j-1)(n-j/2)+i]$th diagonal element of $\mathbf{G}_n'\mathbf{G}_n$ equals 2. Consequently, the $[(i-1)(n-i/2)+i]$th row of $(\mathbf{G}_n'\mathbf{G}_n)^{-1}\mathbf{G}_n'$ is the $[(i-1)n+i]$th row of $\mathbf{I}_{n^2}$, and (for $i > j$) the $[(j-1)(n-j/2)+i]$th row of $(\mathbf{G}_n'\mathbf{G}_n)^{-1}\mathbf{G}_n'$ is an n^2-dimensional row vector whose $[(j-1)n+i]$th and $[(i-1)n+j]$th elements equal 1/2 and whose remaining n^2-2 elements equal 0. For example, $\mathbf{G}_1'\mathbf{G}_1 = (1)$, $\mathbf{G}_2'\mathbf{G}_2 = \text{diag}(1,2,1)$, and $\mathbf{G}_3'\mathbf{G}_3 = \text{diag}(1,2,2,1,2,1)$; and $(\mathbf{G}_1'\mathbf{G}_1)^{-1}\mathbf{G}_1'$, $(\mathbf{G}_2'\mathbf{G}_2)^{-1}\mathbf{G}_2'$, and $(\mathbf{G}_3'\mathbf{G}_3)^{-1}\mathbf{G}_3'$ are the special cases of $\mathbf{H}_1$, $\mathbf{H}_2$, and $\mathbf{H}_3$, respectively, obtained by setting $c = c_1 = c_2 = c_3 = 1/2$ in result (4.8).

Recursive formulas for $\mathbf{G}_n'\mathbf{G}_n$ and for $\mathbf{H}_n$ can be obtained from formula (4.5). Recalling that $\mathbf{Q}_{n+1}$ is orthogonal, we find that

$$\mathbf{G}_{n+1}'\mathbf{G}_{n+1} = \begin{pmatrix} 1 & \mathbf{0} & \mathbf{0} \\ \mathbf{0} & 2\mathbf{I}_n & \mathbf{0} \\ \mathbf{0} & \mathbf{0} & \mathbf{G}_n'\mathbf{G}_n \end{pmatrix} \tag{4.11}$$

and that, for $\mathbf{H}_{n+1} = (\mathbf{G}_{n+1}'\mathbf{G}_{n+1})^{-1}\mathbf{G}_{n+1}'$ and $\mathbf{H}_n = (\mathbf{G}_n'\mathbf{G}_n)^{-1}\mathbf{G}_n'$,

$$\mathbf{H}_{n+1} = \begin{pmatrix} 1 & \mathbf{0} & \mathbf{0} & \mathbf{0} \\ \mathbf{0} & (1/2)\mathbf{I}_n & (1/2)\mathbf{I}_n & \mathbf{0} \\ \mathbf{0} & \mathbf{0} & \mathbf{0} & \mathbf{H}_n \end{pmatrix} \mathbf{Q}_{n+1}'. \tag{4.12}$$

Note that, for $\mathbf{H}_n = (\mathbf{G}_n'\mathbf{G}_n)^{-1}\mathbf{G}_n'$,

$$\mathbf{G}_n\mathbf{H}_n = \mathbf{G}_n(\mathbf{G}_n'\mathbf{G}_n)^{-1}\mathbf{G}_n' = \mathbf{P}_{\mathbf{G}_n}. \tag{4.13}$$

c. Some relationships involving the duplication and vec-permutation matrices

For every $n \times n$ symmetric matrix $\mathbf{A}$,

$$\mathbf{G}_n\,\text{vech}\,\mathbf{A} = \text{vec}\,\mathbf{A} = \text{vec}(\mathbf{A}') = \mathbf{K}_{nn}\,\text{vec}\,\mathbf{A} = \mathbf{K}_{nn}\mathbf{G}_n\,\text{vech}\,\mathbf{A}.$$

Thus, $\mathbf{G}_n\mathbf{a} = \mathbf{K}_{nn}\mathbf{G}_n\mathbf{a}$ for every $[n(n+1)/2]$-dimensional column vector $\mathbf{a}$, leading to the conclusion that

$$\mathbf{K}_{nn}\mathbf{G}_n = \mathbf{G}_n. \tag{4.14}$$

Further, since $\mathbf{K}'_{nn} = \mathbf{K}_{nn}$,

$$(\mathbf{G}'_n\mathbf{G}_n)^{-1}\mathbf{G}'_n\mathbf{K}_{nn} = (\mathbf{G}'_n\mathbf{G}_n)^{-1}(\mathbf{K}_{nn}\mathbf{G}_n)' = (\mathbf{G}'_n\mathbf{G}_n)^{-1}\mathbf{G}'_n,$$

so that, for $\mathbf{H}_n = (\mathbf{G}'_n\mathbf{G}_n)^{-1}\mathbf{G}'_n$,

$$\mathbf{H}_n\mathbf{K}_{nn} = \mathbf{H}_n. \tag{4.15}$$

Note that result (4.14) implies that

$$(1/2)(\mathbf{I}_{n^2} + \mathbf{K}_{nn})\mathbf{G}_n = \mathbf{G}_n. \tag{4.16}$$

Similarly, result (4.15) implies that, for $\mathbf{H}_n = (\mathbf{G}'_n\mathbf{G}_n)^{-1}\mathbf{G}'_n$,

$$\mathbf{H}_n[(1/2)(\mathbf{I}_{n^2} + \mathbf{K}_{nn})] = \mathbf{H}_n. \tag{4.17}$$

Let us now consider (for future reference) some of the properties of the matrix $(1/2)(\mathbf{I}_{n^2} + \mathbf{K}_{nn})$, which enters in results (4.16) and (4.17). Since $\mathbf{K}^2_{nn} = \mathbf{I}$, we have that

$$(1/2)(\mathbf{I}_{n^2} + \mathbf{K}_{nn})\mathbf{K}_{nn} = (1/2)(\mathbf{I}_{n^2} + \mathbf{K}_{nn}) = \mathbf{K}_{nn}[(1/2)(\mathbf{I}_{n^2} + \mathbf{K}_{nn})]. \tag{4.18}$$

Further, recalling that $\mathbf{K}_{nn}$ is symmetric [and using result (4.18)], we find that

$$[(1/2)(\mathbf{I}_{n^2} + \mathbf{K}_{nn})]' = (1/2)(\mathbf{I}_{n^2} + \mathbf{K}_{nn}) = [(1/2)(\mathbf{I}_{n^2} + \mathbf{K}_{nn})]^2, \tag{4.19}$$

that is, $(1/2)(\mathbf{I}_{n^2} + \mathbf{K}_{nn})$ is symmetric and idempotent. And, in light of result (3.8), we have that

$$\text{tr}[(1/2)(\mathbf{I}_{n^2} + \mathbf{K}_{nn})] = n(n+1)/2 \tag{4.20}$$

or, equivalently [since $(1/2)(\mathbf{I}_{n^2} + \mathbf{K}_{nn})$ is idempotent], that

$$\text{rank}[(1/2)(\mathbf{I}_{n^2} + \mathbf{K}_{nn})] = n(n+1)/2. \tag{4.21}$$

Now, observe [in light of results (4.19), (4.16), and (4.17)] that, for $\mathbf{H}_n = (\mathbf{G}'_n\mathbf{G}_n)^{-1}\mathbf{G}'_n$,

$$\begin{aligned}
&[(1/2)(\mathbf{I}_{n^2} + \mathbf{K}_{nn}) - \mathbf{G}_n\mathbf{H}_n]^2 \\
&\quad = [(1/2)(\mathbf{I}_{n^2} + \mathbf{K}_{nn})]^2 - (1/2)(\mathbf{I}_{n^2} + \mathbf{K}_{nn})\mathbf{G}_n\mathbf{H}_n \\
&\qquad\qquad - \mathbf{G}_n\mathbf{H}_n[(1/2)(\mathbf{I}_{n^2} + \mathbf{K}_{nn})] + \mathbf{G}_n\mathbf{H}_n\mathbf{G}_n\mathbf{H}_n \\
&\quad = (1/2)(\mathbf{I}_{n^2} + \mathbf{K}_{nn}) - \mathbf{G}_n\mathbf{H}_n - \mathbf{G}_n\mathbf{H}_n + \mathbf{G}_n\mathbf{H}_n \\
&\quad = (1/2)(\mathbf{I}_{n^2} + \mathbf{K}_{nn}) - \mathbf{G}_n\mathbf{H}_n.
\end{aligned}$$

Moreover, as a consequence of results (5.2.3) and (4.20), we have that

$$\begin{aligned}\operatorname{tr}[(1/2)(\mathbf{I}_{n^2}+\mathbf{K}_{nn})-\mathbf{G}_n\mathbf{H}_n] &= \operatorname{tr}[(1/2)(\mathbf{I}_{n^2}+\mathbf{K}_{nn})]-\operatorname{tr}(\mathbf{G}_n\mathbf{H}_n)\\ &= \operatorname{tr}[(1/2)(\mathbf{I}_{n^2}+\mathbf{K}_{nn})]-\operatorname{tr}(\mathbf{H}_n\mathbf{G}_n)\\ &= \operatorname{tr}[(1/2)(\mathbf{I}_{n^2}+\mathbf{K}_{nn})]-\operatorname{tr}(\mathbf{I}_{n(n+1)/2})\\ &= n(n+1)/2-n(n+1)/2=0.\end{aligned}$$

Thus, for $\mathbf{H}_n = (\mathbf{G}_n'\mathbf{G}_n)^{-1}\mathbf{G}_n'$, the matrix $(1/2)(\mathbf{I}_{n^2}+\mathbf{K}_{nn})-\mathbf{G}_n\mathbf{H}_n$ is idempotent and its trace equals 0. We conclude (on the basis of Corollary 10.2.3) that, for $\mathbf{H}_n = (\mathbf{G}_n'\mathbf{G}_n)^{-1}\mathbf{G}_n'$,

$$(1/2)(\mathbf{I}_{n^2}+\mathbf{K}_{nn})-\mathbf{G}_n\mathbf{H}_n=\mathbf{0}$$

or, equivalently, that [for $\mathbf{H}_n = (\mathbf{G}_n'\mathbf{G}_n)^{-1}\mathbf{G}_n'$]

$$\mathbf{G}_n\mathbf{H}_n=(1/2)(\mathbf{I}_{n^2}+\mathbf{K}_{nn}). \tag{4.22}$$

d. Vech of a product of matrices

Let $\mathbf{A}$ represent an $n \times n$ matrix and $\mathbf{X}$ an $n \times n$ symmetric matrix. Observe (in light of Theorem 16.2.1) that

$$\operatorname{vec}(\mathbf{AXA}')=(\mathbf{A}\otimes\mathbf{A})\operatorname{vec}\mathbf{X}=(\mathbf{A}\otimes\mathbf{A})\mathbf{G}_n\operatorname{vech}\mathbf{X}. \tag{4.23}$$

Observe also that (since $\mathbf{AXA}'$ is symmetric)

$$\operatorname{vech}(\mathbf{AXA}')=\mathbf{H}_n\operatorname{vec}(\mathbf{AXA}'). \tag{4.24}$$

Together, results (4.23) and (4.24) imply that

$$\operatorname{vech}(\mathbf{AXA}')=\mathbf{H}_n(\mathbf{A}\otimes\mathbf{A})\mathbf{G}_n\operatorname{vech}\mathbf{X}. \tag{4.25}$$

Result (4.25) can be regarded as the vech counterpart of formula (2.10) for the vec of a product of matrices. According to this result, the premultiplication of the vech of $\mathbf{X}$ by the $n(n+1)/2 \times n(n+1)/2$ matrix $\mathbf{H}_n(\mathbf{A}\otimes\mathbf{A})\mathbf{G}_n$ transforms $\operatorname{vech}\mathbf{X}$ into $\operatorname{vech}(\mathbf{AXA}')$. Let us now consider various properties of the matrix $\mathbf{H}_n(\mathbf{A}\otimes\mathbf{A})\mathbf{G}_n$.

Observe [in light of results (4.23) and (4.10)] that (for any $n \times n$ symmetric matrix $\mathbf{X}$)

$$\begin{aligned}(\mathbf{A}\otimes\mathbf{A})\mathbf{G}_n\operatorname{vech}\mathbf{X}=\operatorname{vec}(\mathbf{AXA}') &= \mathbf{G}_n\mathbf{H}_n\operatorname{vec}(\mathbf{AXA}')\\ &= \mathbf{G}_n\mathbf{H}_n(\mathbf{A}\otimes\mathbf{A})\mathbf{G}_n\operatorname{vech}\mathbf{X}.\end{aligned}$$

Thus, $(\mathbf{A}\otimes\mathbf{A})\mathbf{G}_n\mathbf{x}=\mathbf{G}_n\mathbf{H}_n(\mathbf{A}\otimes\mathbf{A})\mathbf{G}_n\mathbf{x}$ for every $[n(n+1)/2]$-dimensional column vector $\mathbf{x}$, leading to the conclusion that

$$\mathbf{G}_n\mathbf{H}_n(\mathbf{A}\otimes\mathbf{A})\mathbf{G}_n=(\mathbf{A}\otimes\mathbf{A})\mathbf{G}_n. \tag{4.26}$$

And, for $\mathbf{H}_n = (\mathbf{G}_n'\mathbf{G}_n)^{-1}\mathbf{G}_n'$,

$$\mathbf{G}_n\mathbf{H}_n(\mathbf{A} \otimes \mathbf{A}) = (\mathbf{A} \otimes \mathbf{A})\mathbf{G}_n\mathbf{H}_n, \tag{4.27}$$

as is evident from result (4.22) upon observing [in light of result (3.9)] that

$$\begin{aligned}(1/2)(\mathbf{I}_{n^2} + \mathbf{K}_{nn})(\mathbf{A} \otimes \mathbf{A}) &= (1/2)[(\mathbf{A} \otimes \mathbf{A}) + \mathbf{K}_{nn}(\mathbf{A} \otimes \mathbf{A})] \\ &= (1/2)[(\mathbf{A} \otimes \mathbf{A}) + (\mathbf{A} \otimes \mathbf{A})\mathbf{K}_{nn}] \\ &= (\mathbf{A} \otimes \mathbf{A})[(1/2)(\mathbf{I}_{n^2} + \mathbf{K}_{nn})].\end{aligned}$$

Further, premultiplying both sides of equality (4.27) by $\mathbf{H}_n$ (and recalling that $\mathbf{H}_n\mathbf{G}_n = \mathbf{I}$), we find that [for $\mathbf{H}_n = (\mathbf{G}_n'\mathbf{G}_n)^{-1}\mathbf{G}_n'$]

$$\mathbf{H}_n(\mathbf{A} \otimes \mathbf{A})\mathbf{G}_n\mathbf{H}_n = \mathbf{H}_n(\mathbf{A} \otimes \mathbf{A}), \tag{4.28}$$

analogous to result (4.26).

Several properties of the matrix $\mathbf{H}_n(\mathbf{A} \otimes \mathbf{A})\mathbf{G}_n$ are given in the following theorem.

Theorem 16.4.1. For any $n \times n$ matrix $\mathbf{A}$,

(1) $\mathbf{H}_n(\mathbf{A} \otimes \mathbf{A})\mathbf{G}_n$ is invariant to the choice of $\mathbf{H}_n$;

(2) $\mathbf{H}_n(\mathbf{A}^- \otimes \mathbf{A}^-)\mathbf{G}_n$ is a generalized inverse of $\mathbf{H}_n(\mathbf{A} \otimes \mathbf{A})\mathbf{G}_n$;

(3) $\operatorname{tr}[\mathbf{H}_n(\mathbf{A} \otimes \mathbf{A})\mathbf{G}_n] = (1/2)[\operatorname{tr}(\mathbf{A})]^2 + (1/2)\operatorname{tr}(\mathbf{A}^2)$;

(4) $\operatorname{rank}[\mathbf{H}_n(\mathbf{A} \otimes \mathbf{A})\mathbf{G}_n] = (1/2)[\operatorname{rank}(\mathbf{A})]^2 + (1/2)\operatorname{rank}(\mathbf{A})$;

(5) $\mathbf{H}_n(\mathbf{A} \otimes \mathbf{A})\mathbf{G}_n$ is nonsingular if and only if $\mathbf{A}$ is nonsingular.

Proof. (1) Let $\mathbf{H}_n^{(1)}$ and $\mathbf{H}_n^{(2)}$ represent any two choices for $\mathbf{H}_n$, that is, any two left inverses of $\mathbf{G}_n$. Then, using result (4.26), we find that

$$\mathbf{H}_n^{(2)}(\mathbf{A} \otimes \mathbf{A})\mathbf{G}_n = \mathbf{H}_n^{(2)}\mathbf{G}_n\mathbf{H}_n^{(1)}(\mathbf{A} \otimes \mathbf{A})\mathbf{G}_n = \mathbf{H}_n^{(1)}(\mathbf{A} \otimes \mathbf{A})\mathbf{G}_n.$$

(2) Using result (4.26) and observing (in light of the discussion in Section 16.1) that $\mathbf{A}^- \otimes \mathbf{A}^-$ is a generalized inverse of $\mathbf{A} \otimes \mathbf{A}$, we find that

$$\begin{aligned}&\mathbf{H}_n(\mathbf{A} \otimes \mathbf{A})\mathbf{G}_n\mathbf{H}_n(\mathbf{A}^- \otimes \mathbf{A}^-)\mathbf{G}_n\mathbf{H}_n(\mathbf{A} \otimes \mathbf{A})\mathbf{G}_n \\ &\qquad = \mathbf{H}_n(\mathbf{A} \otimes \mathbf{A})(\mathbf{A}^- \otimes \mathbf{A}^-)\mathbf{G}_n\mathbf{H}_n(\mathbf{A} \otimes \mathbf{A})\mathbf{G}_n \\ &\qquad = \mathbf{H}_n(\mathbf{A} \otimes \mathbf{A})(\mathbf{A}^- \otimes \mathbf{A}^-)(\mathbf{A} \otimes \mathbf{A})\mathbf{G}_n = \mathbf{H}_n(\mathbf{A} \otimes \mathbf{A})\mathbf{G}_n.\end{aligned}$$

(3) Set $\mathbf{H}_n = (\mathbf{G}_n'\mathbf{G}_n)^{-1}\mathbf{G}_n'$—in light of Part (1), it suffices to verify Part (3) for a single choice of $\mathbf{H}_n$. Then, using results (5.2.3), (4.22), (1.25), and (3.15), we find that

$$\begin{aligned}\operatorname{tr}[\mathbf{H}_n(\mathbf{A} \otimes \mathbf{A})\mathbf{G}_n] &= \operatorname{tr}[(\mathbf{A} \otimes \mathbf{A})\mathbf{G}_n\mathbf{H}_n] \\ &= \operatorname{tr}\{(\mathbf{A} \otimes \mathbf{A})[(1/2)(\mathbf{I}_{n^2} + \mathbf{K}_{nn})]\} \\ &= (1/2)\operatorname{tr}(\mathbf{A} \otimes \mathbf{A}) + (1/2)\operatorname{tr}[(\mathbf{A} \otimes \mathbf{A})\mathbf{K}_{nn}] \\ &= (1/2)[\operatorname{tr}(\mathbf{A})]^2 + (1/2)\operatorname{tr}(\mathbf{A}^2).\end{aligned}$$

(4) Since [according to Part (2)] $\mathbf{H}_n(\mathbf{A}^- \otimes \mathbf{A}^-)\mathbf{G}_n$ is a generalized inverse of $\mathbf{H}_n(\mathbf{A} \otimes \mathbf{A})\mathbf{G}_n$, it follows from result (10.2.1) that

$$\text{rank}[\mathbf{H}_n(\mathbf{A} \otimes \mathbf{A})\mathbf{G}_n] = \text{tr}[\mathbf{H}_n(\mathbf{A} \otimes \mathbf{A})\mathbf{G}_n\mathbf{H}_n(\mathbf{A}^- \otimes \mathbf{A}^-)\mathbf{G}_n].$$

Further, using result (4.26), Part (3), and result (10.2.1), we find that

$$\begin{aligned}
\text{tr}[\mathbf{H}_n(\mathbf{A} \otimes \mathbf{A})\mathbf{G}_n\mathbf{H}_n(\mathbf{A}^- \otimes \mathbf{A}^-)\mathbf{G}_n] &= \text{tr}[\mathbf{H}_n(\mathbf{A} \otimes \mathbf{A})(\mathbf{A}^- \otimes \mathbf{A}^-)\mathbf{G}_n] \\
&= \text{tr}\{\mathbf{H}_n[(\mathbf{A}\mathbf{A}^-) \otimes (\mathbf{A}\mathbf{A}^-)]\mathbf{G}_n\} \\
&= (1/2)[\text{tr}(\mathbf{A}\mathbf{A}^-)]^2 + (1/2)\,\text{tr}(\mathbf{A}\mathbf{A}^-\mathbf{A}\mathbf{A}^-) \\
&= (1/2)[\text{tr}(\mathbf{A}\mathbf{A}^-)]^2 + (1/2)\,\text{tr}(\mathbf{A}\mathbf{A}^-) \\
&= (1/2)[\text{rank}(\mathbf{A})]^2 + (1/2)\,\text{rank}(\mathbf{A}).
\end{aligned}$$

(5) Since $\mathbf{H}_n(\mathbf{A} \otimes \mathbf{A})\mathbf{G}_n$ is a square matrix of order $n(n+1)/2$, Part (5) is an immediate consequence of Part (4). Q.E.D.

For purposes of deriving a recursive formula for $\mathbf{H}_n(\mathbf{A} \otimes \mathbf{A})\mathbf{G}_n$, let $\mathbf{B}$ represent an arbitrary $(n+1) \times (n+1)$ matrix, partition $\mathbf{B}$ as

$$\mathbf{B} = \begin{pmatrix} c & \mathbf{b}' \\ \mathbf{a} & \mathbf{A} \end{pmatrix}$$

(where $\mathbf{A}$ is $n \times n$), and define $\mathbf{Q}_{n+1} = (\mathbf{Q}_{n+1}^{(1)}, \mathbf{Q}_{n+1}^{(2)})$, $\mathbf{u}$, and $\mathbf{U}$ as in Subsection b. Then,

$$\mathbf{Q}_{n+1}'(\mathbf{B} \otimes \mathbf{B})\mathbf{Q}_{n+1} = \begin{pmatrix} \mathbf{Q}_{n+1}^{(1)\prime}(\mathbf{B} \otimes \mathbf{B})\mathbf{Q}_{n+1}^{(1)} & \mathbf{Q}_{n+1}^{(1)\prime}(\mathbf{B} \otimes \mathbf{B})\mathbf{Q}_{n+1}^{(2)} \\ \mathbf{Q}_{n+1}^{(2)\prime}(\mathbf{B} \otimes \mathbf{B})\mathbf{Q}_{n+1}^{(1)} & \mathbf{Q}_{n+1}^{(2)\prime}(\mathbf{B} \otimes \mathbf{B})\mathbf{Q}_{n+1}^{(2)} \end{pmatrix}.$$

Further, since

$$\begin{pmatrix} c & \mathbf{b}' \\ \mathbf{a} & \mathbf{A} \end{pmatrix} = \mathbf{I}_{n+1}'\mathbf{B}\mathbf{I}_{n+1} = (\mathbf{u},\ \mathbf{U})'\mathbf{B}(\mathbf{u},\ \mathbf{U}) = \begin{pmatrix} \mathbf{u}'\mathbf{B}\mathbf{u} & \mathbf{u}'\mathbf{B}\mathbf{U} \\ \mathbf{U}'\mathbf{B}\mathbf{u} & \mathbf{U}'\mathbf{B}\mathbf{U} \end{pmatrix},$$

we find that

$$\begin{aligned}
\mathbf{Q}_{n+1}^{(1)\prime}(\mathbf{B} \otimes \mathbf{B})\mathbf{Q}_{n+1}^{(1)} &= (\mathbf{I}_{n+1} \otimes \mathbf{u}')(\mathbf{B} \otimes \mathbf{B})(\mathbf{I}_{n+1} \otimes \mathbf{u}) \\
&= \mathbf{B} \otimes (\mathbf{u}'\mathbf{B}\mathbf{u}) = \mathbf{B} \otimes c = c\mathbf{B}, \\
\mathbf{Q}_{n+1}^{(1)\prime}(\mathbf{B} \otimes \mathbf{B})\mathbf{Q}_{n+1}^{(2)} &= (\mathbf{I}_{n+1} \otimes \mathbf{u}')(\mathbf{B} \otimes \mathbf{B})(\mathbf{I}_{n+1} \otimes \mathbf{U}) \\
&= \mathbf{B} \otimes (\mathbf{u}'\mathbf{B}\mathbf{U}) = \mathbf{B} \otimes \mathbf{b}', \\
\mathbf{Q}_{n+1}^{(2)\prime}(\mathbf{B} \otimes \mathbf{B})\mathbf{Q}_{n+1}^{(1)} &= (\mathbf{I}_{n+1} \otimes \mathbf{U}')(\mathbf{B} \otimes \mathbf{B})(\mathbf{I}_{n+1} \otimes \mathbf{u}) \\
&= \mathbf{B} \otimes (\mathbf{U}'\mathbf{B}\mathbf{u}) = \mathbf{B} \otimes \mathbf{a}, \\
\mathbf{Q}_{n+1}^{(2)\prime}(\mathbf{B} \otimes \mathbf{B})\mathbf{Q}_{n+1}^{(2)} &= (\mathbf{I}_{n+1} \otimes \mathbf{U}')(\mathbf{B} \otimes \mathbf{B})(\mathbf{I}_{n+1} \otimes \mathbf{U}) \\
&= \mathbf{B} \otimes (\mathbf{U}'\mathbf{B}\mathbf{U}) = \mathbf{B} \otimes \mathbf{A}.
\end{aligned}$$

Thus, in light of results (1.27), (1.4), (3.14), and (3.12), we have that

$$\mathbf{Q}'_{n+1}(\mathbf{B}\otimes\mathbf{B})\mathbf{Q}_{n+1} = \begin{pmatrix} c^2 & c\mathbf{b}' & c\mathbf{b}' & \mathbf{b}'\otimes\mathbf{b}' \\ c\mathbf{a} & c\mathbf{A} & \mathbf{ab}' & (\mathbf{b}'\otimes\mathbf{A})\mathbf{K}_{nn} \\ c\mathbf{a} & \mathbf{ab}' & c\mathbf{A} & \mathbf{b}'\otimes\mathbf{A} \\ \mathbf{a}\otimes\mathbf{a} & \mathbf{K}_{nn}(\mathbf{a}\otimes\mathbf{A}) & \mathbf{a}\otimes\mathbf{A} & \mathbf{A}\otimes\mathbf{A} \end{pmatrix}.$$

And, in light of results (4.12), (4.5), (4.15), and (4.16) [as well as Part (1) of Theorem 16.4.1], it follows that, for $\mathbf{H}_n = (\mathbf{G}'_n\mathbf{G}_n)^{-1}\mathbf{G}'_n$,

$$\begin{aligned}
&\mathbf{H}_{n+1}(\mathbf{B}\otimes\mathbf{B})\mathbf{G}_{n+1} \\
&= \begin{pmatrix} 1 & \mathbf{0} & \mathbf{0} & \mathbf{0} \\ \mathbf{0} & (1/2)\mathbf{I}_n & (1/2)\mathbf{I}_n & \mathbf{0} \\ \mathbf{0} & \mathbf{0} & \mathbf{0} & \mathbf{H}_n \end{pmatrix} \mathbf{Q}'_{n+1}(\mathbf{B}\otimes\mathbf{B})\mathbf{Q}_{n+1} \begin{pmatrix} 1 & \mathbf{0} & \mathbf{0} \\ \mathbf{0} & \mathbf{I}_n & \mathbf{0} \\ \mathbf{0} & \mathbf{I}_n & \mathbf{0} \\ \mathbf{0} & \mathbf{0} & \mathbf{G}_n \end{pmatrix} \\
&= \begin{pmatrix} c^2 & c\mathbf{b}' & c\mathbf{b}' & \mathbf{b}'\otimes\mathbf{b}' \\ c\mathbf{a} & (1/2)(c\mathbf{A}+\mathbf{ab}') & (1/2)(c\mathbf{A}+\mathbf{ab}') & \begin{matrix}(\mathbf{b}'\otimes\mathbf{A}) \\ \times[(1/2)(\mathbf{I}_{n^2}+\mathbf{K}_{nn})]\end{matrix} \\ \mathbf{H}_n(\mathbf{a}\otimes\mathbf{a}) & \mathbf{H}_n(\mathbf{a}\otimes\mathbf{A}) & \mathbf{H}_n(\mathbf{a}\otimes\mathbf{A}) & \mathbf{H}_n(\mathbf{A}\otimes\mathbf{A}) \end{pmatrix} \\
&\qquad \times \begin{pmatrix} 1 & \mathbf{0} & \mathbf{0} \\ \mathbf{0} & \mathbf{I}_n & \mathbf{0} \\ \mathbf{0} & \mathbf{I}_n & \mathbf{0} \\ \mathbf{0} & \mathbf{0} & \mathbf{G}_n \end{pmatrix} \\
&= \begin{pmatrix} c^2 & 2c\mathbf{b}' & (\mathbf{b}'\otimes\mathbf{b}')\mathbf{G}_n \\ c\mathbf{a} & c\mathbf{A}+\mathbf{ab}' & (\mathbf{b}'\otimes\mathbf{A})\mathbf{G}_n \\ \mathbf{H}_n(\mathbf{a}\otimes\mathbf{a}) & 2\mathbf{H}_n(\mathbf{a}\otimes\mathbf{A}) & \mathbf{H}_n(\mathbf{A}\otimes\mathbf{A})\mathbf{G}_n \end{pmatrix}. \qquad (4.29)
\end{aligned}$$

Result (4.29) provides what is in effect a recursive formula for $\mathbf{H}_n(\mathbf{A}\otimes\mathbf{A})\mathbf{G}_n$.

Let

$$\mathbf{S} = \begin{pmatrix} c^2 & 2c\mathbf{b}' \\ c\mathbf{a} & c\mathbf{A}+\mathbf{ab}' \end{pmatrix} - \begin{pmatrix} \mathbf{b}'\otimes\mathbf{b}' \\ \mathbf{b}'\otimes\mathbf{A} \end{pmatrix} \mathbf{G}_n\mathbf{H}_n(\mathbf{A}^-\otimes\mathbf{A}^-)\mathbf{G}_n\mathbf{H}_n[\mathbf{a}\otimes\mathbf{a},\ 2(\mathbf{a}\otimes\mathbf{A})]$$

[where $\mathbf{H}_n = (\mathbf{G}'_n\mathbf{G}_n)^{-1}\mathbf{G}'_n$]. And observe [in light of Part (2) of Theorem 16.4.1] that $\mathbf{S}$ is a Schur complement of the $n(n+1)/2 \times n(n+1)/2$ submatrix $\mathbf{H}_n(\mathbf{A}\otimes\mathbf{A})\mathbf{G}_n$ in the lower right corner of partitioned matrix (4.29). Further, using results (4.26), (4.22), and (3.12), we find that

$$\begin{aligned}
&\mathbf{G}_n\mathbf{H}_n(\mathbf{A}^-\otimes\mathbf{A}^-)\mathbf{G}_n\mathbf{H}_n[\mathbf{a}\otimes\mathbf{a},\ 2(\mathbf{a}\otimes\mathbf{A})] \\
&\qquad = (\mathbf{A}^-\otimes\mathbf{A}^-)\mathbf{G}_n\mathbf{H}_n[\mathbf{a}\otimes\mathbf{a},\ 2(\mathbf{a}\otimes\mathbf{A})] \\
&\qquad = (\mathbf{A}^-\otimes\mathbf{A}^-)[(1/2)(\mathbf{I}_{n^2}+\mathbf{K}_{nn})][\mathbf{a}\otimes\mathbf{a},\ 2(\mathbf{a}\otimes\mathbf{A})] \\
&\qquad = (\mathbf{A}^-\otimes\mathbf{A}^-)[\mathbf{a}\otimes\mathbf{a},\ (\mathbf{a}\otimes\mathbf{A})+(\mathbf{A}\otimes\mathbf{a})].
\end{aligned}$$

Thus,

$$\mathbf{S} = \begin{pmatrix} c^2 - (\mathbf{b}'\mathbf{A}^-\mathbf{a}) \otimes (\mathbf{b}'\mathbf{A}^-\mathbf{a}) & \begin{array}{r} 2c\mathbf{b}' - (\mathbf{b}'\mathbf{A}^-\mathbf{a}) \otimes (\mathbf{b}'\mathbf{A}^-\mathbf{A}) \\ - (\mathbf{b}'\mathbf{A}^-\mathbf{A}) \otimes (\mathbf{b}'\mathbf{A}^-\mathbf{a}) \end{array} \\ c\mathbf{a} - (\mathbf{b}'\mathbf{A}^-\mathbf{a}) \otimes (\mathbf{A}\mathbf{A}^-\mathbf{a}) & \begin{array}{r} c\mathbf{A} + \mathbf{a}\mathbf{b}' - (\mathbf{b}'\mathbf{A}^-\mathbf{a}) \otimes \mathbf{A} \\ - (\mathbf{b}'\mathbf{A}^-\mathbf{A}) \otimes (\mathbf{A}\mathbf{A}^-\mathbf{a}) \end{array} \end{pmatrix}$$

$$= \begin{pmatrix} c^2 - (\mathbf{b}'\mathbf{A}^-\mathbf{a})^2 & 2c\mathbf{b}' - 2(\mathbf{b}'\mathbf{A}^-\mathbf{a})\mathbf{b}'\mathbf{A}^-\mathbf{A} \\ c\mathbf{a} - (\mathbf{b}'\mathbf{A}^-\mathbf{a})\mathbf{A}\mathbf{A}^-\mathbf{a} & (c - \mathbf{b}'\mathbf{A}^-\mathbf{a})\mathbf{A} + \mathbf{a}\mathbf{b}' - \mathbf{A}\mathbf{A}^-\mathbf{a}\mathbf{b}'\mathbf{A}^-\mathbf{A} \end{pmatrix},$$

and, in the special case where $\mathbf{A}$ is nonsingular,

$$\mathbf{S} = (c - \mathbf{b}'\mathbf{A}^{-1}\mathbf{a}) \begin{pmatrix} c + \mathbf{b}'\mathbf{A}^{-1}\mathbf{a} & 2\mathbf{b}' \\ \mathbf{a} & \mathbf{A} \end{pmatrix}. \tag{4.30}$$

Consider now the determinant of the matrix $\mathbf{H}_{n+1}(\mathbf{B} \otimes \mathbf{B})\mathbf{G}_{n+1}$. By applying result (13.3.13) to partitioned matrix (4.29), we find that if $\mathbf{A}$ is nonsingular [in which case it follows from Part (5) of Theorem 16.4.1 that $\mathbf{H}_n(\mathbf{A} \otimes \mathbf{A})\mathbf{G}_n$ is also nonsingular], then

$$|\mathbf{H}_{n+1}(\mathbf{B} \otimes \mathbf{B})\mathbf{G}_{n+1}| = |\mathbf{H}_n(\mathbf{A} \otimes \mathbf{A})\mathbf{G}_n||\mathbf{S}|. \tag{4.31}$$

Further, using result (4.30), Corollary 13.2.4, and result (13.3.13), we find that if $\mathbf{A}$ is nonsingular, then

$$\begin{aligned} |\mathbf{S}| &= (c - \mathbf{b}'\mathbf{A}^{-1}\mathbf{a})^{n+1} \begin{vmatrix} c + \mathbf{b}'\mathbf{A}^{-1}\mathbf{a} & 2\mathbf{b}' \\ \mathbf{a} & \mathbf{A} \end{vmatrix} \\ &= (c - \mathbf{b}'\mathbf{A}^{-1}\mathbf{a})^{n+1}|\mathbf{A}|(c + \mathbf{b}'\mathbf{A}^{-1}\mathbf{a} - 2\mathbf{b}'\mathbf{A}^{-1}\mathbf{a}) \\ &= (c - \mathbf{b}'\mathbf{A}^{-1}\mathbf{a})^{n+2}|\mathbf{A}|. \end{aligned} \tag{4.32}$$

Together, results (4.31) and (4.32) imply that (in the special case where $\mathbf{A}$ is nonsingular)

$$|\mathbf{H}_{n+1}(\mathbf{B} \otimes \mathbf{B})\mathbf{G}_{n+1}| = (c - \mathbf{b}'\mathbf{A}^{-1}\mathbf{a})^{n+2}|\mathbf{A}||\mathbf{H}_n(\mathbf{A} \otimes \mathbf{A})\mathbf{G}_n|. \tag{4.33}$$

Result (4.33), which provides what is in effect a recursive formula for the determinant of $\mathbf{H}_n(\mathbf{A} \otimes \mathbf{A})\mathbf{G}_n$, can be used to establish the following theorem.

Theorem 16.4.2. For any $n \times n$ matrix $\mathbf{A}$,

$$|\mathbf{H}_n(\mathbf{A} \otimes \mathbf{A})\mathbf{G}_n| = |\mathbf{A}|^{n+1}. \tag{4.34}$$

Proof. The proof is by mathematical induction. Clearly, equality (4.34) holds for every 1×1 matrix $\mathbf{A}$. Suppose that equality (4.34) holds for every $n \times n$ matrix $\mathbf{A}$. Let $\mathbf{B}$ represent an arbitrary $(n + 1) \times (n + 1)$ matrix, and partition $\mathbf{B}$ as

$$\mathbf{B} = \begin{pmatrix} c & \mathbf{b}' \\ \mathbf{a} & \mathbf{A} \end{pmatrix}$$

(where $\mathbf{A}$ is $n \times n$). To complete the induction argument, it suffices to show that

$$|\mathbf{H}_{n+1}(\mathbf{B} \otimes \mathbf{B})\mathbf{G}_{n+1}| = |\mathbf{B}|^{n+2}. \tag{4.35}$$

In establishing the validity of this equality, it is convenient to consider successively three different cases.

Case (1): **B** *singular.* In this case, both the left and right sides of equality (4.35) equal 0—refer to Part (5) of Theorem 16.4.1.

Case (2): Both **A** *and* **B** *nonsingular.* Since (by supposition) $|\mathbf{H}_n(\mathbf{A}\otimes\mathbf{A})\mathbf{G}_n| = |\mathbf{A}|^{n+1}$, it follows from result (4.33) [along with result (13.3.13)] that

$$|\mathbf{H}_{n+1}(\mathbf{B} \otimes \mathbf{B})\mathbf{G}_{n+1}| = (c - \mathbf{b}'\mathbf{A}^{-1}\mathbf{a})^{n+2}|\mathbf{A}|^{n+2} = |\mathbf{B}|^{n+2}.$$

Case (3): **B** *nonsingular, but* **A** *singular.* Denote the first, ..., nth columns of **A** by $\mathbf{a}_1, \ldots, \mathbf{a}_n$, respectively. Since **B** is nonsingular, the rows of **B** are linearly independent, implying that $\text{rank}(\mathbf{a}, \mathbf{A}) = n$ and also (since **A** is singular) that $\text{rank}(\mathbf{A}) = n - 1$. Consequently, **A** contains $n-1$ linearly independent columns, say $\mathbf{a}_1, \ldots, \mathbf{a}_{i-1}, \mathbf{a}_{i+1}, \ldots, \mathbf{a}_n$, and **a** is not expressible as a linear combination of the columns of **A**.

Let **e** represent the ith column of $\mathbf{I}_n$. Then, the matrix $\mathbf{A} + \mathbf{a}\mathbf{e}'$, whose columns are $\mathbf{a}_1, \ldots, \mathbf{a}_{i-1}, \mathbf{a}_i + \mathbf{a}, \mathbf{a}_{i+1}, \ldots, \mathbf{a}_n$, respectively, is nonsingular (since otherwise there would exist scalars $c_1, \ldots, c_{i-1}, c_{i+1}, \ldots, c_n$ such that $\mathbf{a}_i + \mathbf{a} = \sum_{j \neq i} c_j \mathbf{a}_j$, which would imply that **a** is expressible as a linear combination of the columns of **A**).

Now, let

$$\mathbf{T} = \begin{pmatrix} 1 & \mathbf{e}' \\ \mathbf{0} & \mathbf{I}_n \end{pmatrix}.$$

Then,

$$\mathbf{BT} = \begin{pmatrix} c & \mathbf{b}' + c\mathbf{e}' \\ \mathbf{a} & \mathbf{A} + \mathbf{a}\mathbf{e}' \end{pmatrix}.$$

Further, $|\mathbf{T}| = 1$.

Since $\mathbf{I}_n$ and $\mathbf{A} + \mathbf{a}\mathbf{e}'$ are nonsingular, we have [from the same reasoning as in Case (2)] that

$$|\mathbf{H}_{n+1}(\mathbf{T} \otimes \mathbf{T})\mathbf{G}_{n+1}| = |\mathbf{T}|^{n+2} = 1, \tag{4.36}$$

$$|\mathbf{H}_{n+1}[(\mathbf{BT}) \otimes (\mathbf{BT})]\mathbf{G}_{n+1}| = |\mathbf{BT}|^{n+2} = |\mathbf{B}|^{n+2}|\mathbf{T}|^{n+2} = |\mathbf{B}|^{n+2}. \tag{4.37}$$

And, using result (4.26), we find that

$$\begin{aligned} |\mathbf{H}_{n+1}[(\mathbf{BT}) \otimes (\mathbf{BT})]\mathbf{G}_{n+1}| &= |\mathbf{H}_{n+1}(\mathbf{B} \otimes \mathbf{B})(\mathbf{T} \otimes \mathbf{T})\mathbf{G}_{n+1}| \\ &= |\mathbf{H}_{n+1}(\mathbf{B} \otimes \mathbf{B})\mathbf{G}_{n+1}\mathbf{H}_{n+1}(\mathbf{T} \otimes \mathbf{T})\mathbf{G}_{n+1}| \\ &= |\mathbf{H}_{n+1}(\mathbf{B} \otimes \mathbf{B})\mathbf{G}_{n+1}||\mathbf{H}_{n+1}(\mathbf{T} \otimes \mathbf{T})\mathbf{G}_{n+1}|, \end{aligned}$$

which [in combination with results (4.36) and (4.37)] implies that

$$|\mathbf{H}_{n+1}(\mathbf{B} \otimes \mathbf{B})\mathbf{G}_{n+1}| = |\mathbf{B}|^{n+2}.$$

Q.E.D.

16.5 Reformulation of a Linear System

Consider a linear system

$$\mathbf{AX} = \mathbf{B}, \tag{5.1}$$

with an $m \times n$ coefficient matrix $\mathbf{A}$, $m \times p$ right side $\mathbf{B}$, and $n \times p$ matrix of unknowns $\mathbf{X}$. This linear system can be reformulated as a linear system whose right side is an $mp \times 1$ vector and whose unknowns are arranged in an $np \times 1$ vector. This process is facilitated by the use of the vec operator (and the use of Kronecker products).

Specifically, by applying the vec operator to each side of equation (5.1) and recalling result (2.9), we obtain the equivalent linear system

$$(\mathbf{I}_p \otimes \mathbf{A})\mathbf{x} = \mathbf{b}, \tag{5.2}$$

where $\mathbf{b} = \operatorname{vec}\mathbf{B}$ and where $\mathbf{x} = \operatorname{vec}\mathbf{X}$ is an $np \times 1$ vector of unknowns—the equivalence is in the sense that $\mathbf{X}$ is a solution to linear system (5.1) if and only if $\operatorname{vec}\mathbf{X}$ is a solution to linear system (5.2). Note that linear system (5.2) can be rewritten as

$$\operatorname{diag}(\mathbf{A}, \mathbf{A}, \ldots, \mathbf{A})\mathbf{x} = \mathbf{b}.$$

Let us now generalize from linear system (5.1) to the system

$$\mathbf{AXC} = \mathbf{B}, \tag{5.3}$$

where $\mathbf{A}$ is an $m \times n$ matrix, $\mathbf{C}$ a $p \times q$ matrix, $\mathbf{B}$ an $m \times q$ matrix, and $\mathbf{X}$ an $n \times p$ matrix of unknowns. Upon applying the vec operator to each side of equation (5.3) and recalling result (2.10), we obtain the equivalent (linear) system

$$(\mathbf{C}' \otimes \mathbf{A})\mathbf{x} = \mathbf{b},$$

where $\mathbf{b} = \operatorname{vec}\mathbf{B}$ and where $\mathbf{x} = \operatorname{vec}\mathbf{X}$ is an $np \times 1$ vector of unknowns.

A further generalization is possible. Consider the system

$$\sum_{i=1}^{r} \mathbf{A}_i\mathbf{X}\mathbf{C}_i + \sum_{j=1}^{s} \mathbf{L}_j\mathbf{X}'\mathbf{T}_j = \mathbf{B}, \tag{5.4}$$

where $\mathbf{A}_1, \ldots, \mathbf{A}_r$ are $m \times n$ matrices, $\mathbf{C}_1, \ldots, \mathbf{C}_r$ are $p \times q$ matrices, $\mathbf{L}_1, \ldots, \mathbf{L}_s$ are $m \times p$ matrices, $\mathbf{T}_1, \ldots, \mathbf{T}_s$ are $n \times q$ matrices, $\mathbf{B}$ is an $m \times q$ matrix, and $\mathbf{X}$ is an $n \times p$ matrix of unknowns. Using results (2.6) and (2.10), we find that

$$\begin{aligned}
\operatorname{vec}\Big(\sum_i \mathbf{A}_i\mathbf{X}\mathbf{C}_i + \sum_j \mathbf{L}_j\mathbf{X}'\mathbf{T}_j\Big)& \\
&= \sum_i \operatorname{vec}(\mathbf{A}_i\mathbf{X}\mathbf{C}_i) + \sum_j \operatorname{vec}(\mathbf{L}_j\mathbf{X}'\mathbf{T}_j) \\
&= \sum_i (\mathbf{C}_i' \otimes \mathbf{A}_i)\operatorname{vec}\mathbf{X} + \sum_j (\mathbf{T}_j' \otimes \mathbf{L}_j)\operatorname{vec}\mathbf{X}' \\
&= \Big\{\sum_i (\mathbf{C}_i' \otimes \mathbf{A}_i) + \Big[\sum_j (\mathbf{T}_j' \otimes \mathbf{L}_j)\Big]\mathbf{K}_{np}\Big\}\operatorname{vec}\mathbf{X}. \qquad (5.5)
\end{aligned}$$

Consequently, system (5.4) is equivalent to the (linear) system

$$\Big\{\sum_i (\mathbf{C}_i' \otimes \mathbf{A}_i) + \Big[\sum_j (\mathbf{T}_j' \otimes \mathbf{L}_j)\Big]\mathbf{K}_{np}\Big\}\mathbf{x} = \mathbf{b}, \tag{5.6}$$

where $\mathbf{b} = \operatorname{vec}\mathbf{B}$, and where $\mathbf{x} = \operatorname{vec}\mathbf{X}$ is an $np \times 1$ vector of unknowns. As a special case (that where "$s = 0$"), we have that the system

$$\sum_{i=1}^{r} \mathbf{A}_i \mathbf{X} \mathbf{C}_i = \mathbf{B} \tag{5.7}$$

is equivalent to the (linear) system

$$\Big[\sum_i (\mathbf{C}_i' \otimes \mathbf{A}_i)\Big]\mathbf{x} = \mathbf{b}. \tag{5.8}$$

In our consideration of system (5.4) [and systems (5.1), (5.3), and (5.7)], it has been implicitly assumed that there are no restrictions on $\mathbf{X}$ (other than those imposed by the system)—if there were restrictions on $\mathbf{X}$, system (5.4) would be equivalent to system (5.6) only if equivalent restrictions were imposed on $\mathbf{x}$. Suppose now that $\mathbf{X}$ is restricted to be symmetric [in which case $p = n$ and system (5.4) is no more general than system (5.7)]. Then, since $\operatorname{vec}\mathbf{X} = \mathbf{G}_n \operatorname{vech}\mathbf{X}$, system (5.7) is equivalent to the (linear) system

$$\Big[\sum_i (\mathbf{C}_i' \otimes \mathbf{A}_i)\Big]\mathbf{G}_n\mathbf{x} = \mathbf{b}, \tag{5.9}$$

where now $\mathbf{x} = \operatorname{vech}\mathbf{X}$ is an $n(n+1)/2 \times 1$ (unrestricted) vector of unknowns [and where $\mathbf{b} = \operatorname{vec}\mathbf{B}$]—the equivalence is in the sense that a symmetric matrix $\mathbf{X}$ is a solution to system (5.7) if and only if $\operatorname{vech}\mathbf{X}$ is a solution to system (5.9).

16.6 Some Results on Jacobian Matrices

In this section, attention is given to some Jacobian matrices of a kind encountered in multivariate statistical analysis.

Let $\mathbf{F}$ represent a $p \times q$ matrix of functions, defined on a set S, of a vector $\mathbf{x} = (x_1, \ldots, x_m)'$ of m variables. Then, $\operatorname{vec}\mathbf{F}$ is a pq-dimensional column vector obtained by rearranging the elements of $\mathbf{F}$ (in a particular way). By definition, the Jacobian matrix of $\operatorname{vec}\mathbf{F}$ is the $pq \times m$ matrix $\partial \operatorname{vec}(\mathbf{F})/\partial \mathbf{x}'$, whose jth column is $\partial \operatorname{vec}(\mathbf{F})/\partial x_j$. Note that

$$\frac{\partial \operatorname{vec}\mathbf{F}}{\partial x_j} = \operatorname{vec}\Big(\frac{\partial \mathbf{F}}{\partial x_j}\Big) \tag{6.1}$$

(as is evident from the very definition of the vec operator). Thus, the Jacobian matrix of $\operatorname{vec}\mathbf{F}$ is the $pq \times m$ matrix whose jth column is $\operatorname{vec}(\partial\mathbf{F}/\partial x_j)$.

Consider now the special case where $q = p$. If $\mathbf{F}$ is symmetric (for all $\mathbf{x}$ in S), then $p(p-1)/2$ of the elements of $\text{vec}\,\mathbf{F}$ are redundant. In contrast, $\text{vech}\,\mathbf{F}$ contains no redundancies (that are attributable to symmetry). By definition, the Jacobian matrix of $\text{vech}\,\mathbf{F}$ is the $p(p+1)/2 \times m$ matrix $\partial\,\text{vech}(\mathbf{F})/\partial\mathbf{x}'$, whose jth column is $\partial\,\text{vech}(\mathbf{F})/\partial x_j$. Note that [analogous to result (6.1)]

$$\frac{\partial\,\text{vech}\,\mathbf{F}}{\partial x_j} = \text{vech}\left(\frac{\partial\mathbf{F}}{\partial x_j}\right). \tag{6.2}$$

Thus, the Jacobian matrix of $\text{vech}\,\mathbf{F}$ is the $p(p+1)/2 \times m$ matrix whose jth column is $\text{vech}(\partial\mathbf{F}/\partial x_j)$.

If $\mathbf{F}$ is symmetric, then clearly [in light of result (4.6)]

$$\frac{\partial\,\text{vec}\,\mathbf{F}}{\partial x_j} = \mathbf{G}_p\frac{\partial\,\text{vech}\,\mathbf{F}}{\partial x_j}, \qquad \frac{\partial\,\text{vech}\,\mathbf{F}}{\partial x_j} = \mathbf{H}_p\frac{\partial\,\text{vec}\,\mathbf{F}}{\partial x_j} \tag{6.3}$$

$(j = 1, \ldots, m)$, and consequently

$$\frac{\partial\,\text{vec}\,\mathbf{F}}{\partial\mathbf{x}'} = \mathbf{G}_p\frac{\partial\,\text{vech}\,\mathbf{F}}{\partial\mathbf{x}'}, \qquad \frac{\partial\,\text{vech}\,\mathbf{F}}{\partial\mathbf{x}'} = \mathbf{H}_p\frac{\partial\,\text{vec}\,\mathbf{F}}{\partial\mathbf{x}'}. \tag{6.4}$$

Next, consider the Jacobian matrix of $\text{vec}(\mathbf{AFB})$, where $\mathbf{F}$ is a $p \times q$ matrix of functions, defined on a set S, of a vector $\mathbf{x} = (x_1, \ldots, x_m)'$ of m variables and where $\mathbf{A}$ and $\mathbf{B}$ are $r \times p$ and $q \times s$ matrices of constants. Making use of formula (2.10), we find that

$$\frac{\partial\,\text{vec}(\mathbf{AFB})}{\partial x_j} = (\mathbf{B}' \otimes \mathbf{A})\frac{\partial\,\text{vec}\,\mathbf{F}}{\partial x_j} \tag{6.5}$$

$(j = 1, \ldots, m)$ and hence that

$$\frac{\partial\,\text{vec}(\mathbf{AFB})}{\partial\mathbf{x}'} = (\mathbf{B}' \otimes \mathbf{A})\frac{\partial\,\text{vec}\,\mathbf{F}}{\partial\mathbf{x}'}. \tag{6.6}$$

Formula (6.6) relates the Jacobian matrix of $\text{vec}(\mathbf{AFB})$ to that of $\text{vec}\,\mathbf{F}$.

When $s = q$, $r = p$, and $m = pq$, it follows from results (6.6) and (3.18) that

$$\left|\frac{\partial\,\text{vec}(\mathbf{AFB})}{\partial\mathbf{x}'}\right| = |\mathbf{A}|^q|\mathbf{B}|^p\left|\frac{\partial\,\text{vec}\,\mathbf{F}}{\partial\mathbf{x}'}\right|, \tag{6.7}$$

which relates the Jacobian of $\text{vec}(\mathbf{AFB})$ to that of $\text{vec}\,\mathbf{F}$.

In the special case where $\mathbf{x} = \text{vec}\,\mathbf{X}$ and $\mathbf{F}(\mathbf{x}) = \mathbf{X}$ for some $p \times q$ matrix $\mathbf{X}$ of variables, result (6.6) simplifies [in light of result (15.3.12)] to

$$\frac{\partial\,\text{vec}(\mathbf{AXB})}{\partial(\text{vec}\,\mathbf{X})'} = \mathbf{B}' \otimes \mathbf{A}, \tag{6.8}$$

and (when $s = q$ and $r = p$) result (6.7) simplifies to

$$\left|\frac{\partial\,\text{vec}(\mathbf{AXB})}{\partial(\text{vec}\,\mathbf{X})'}\right| = |\mathbf{A}|^q|\mathbf{B}|^p. \tag{6.9}$$

Suppose now that $\mathbf{F}$ is symmetric and that $\mathbf{A}$ is square (in which case both $\mathbf{F}$ and $\mathbf{A}$ are $p \times p$). Then,

$$\frac{\partial \operatorname{vech}(\mathbf{AFA}')}{\partial \mathbf{x}'} = \mathbf{H}_p(\mathbf{A} \otimes \mathbf{A})\mathbf{G}_p \frac{\partial \operatorname{vech} \mathbf{F}}{\partial \mathbf{x}'}, \tag{6.10}$$

as is evident from result (6.6) upon setting $\mathbf{B} = \mathbf{A}'$ and observing [in light of result (4.6)] that

$$\frac{\partial \operatorname{vech}(\mathbf{AFA}')}{\partial \mathbf{x}'} = \mathbf{H}_p \frac{\partial \operatorname{vec}(\mathbf{AFA}')}{\partial \mathbf{x}'}, \qquad \frac{\partial \operatorname{vec} \mathbf{F}}{\partial \mathbf{x}'} = \mathbf{G}_p \frac{\partial \operatorname{vech} \mathbf{F}}{\partial \mathbf{x}'}.$$

When $m = p(p+1)/2$, it follows from results (6.10) and (4.34) that

$$\left| \frac{\partial \operatorname{vech}(\mathbf{AFA}')}{\partial \mathbf{x}'} \right| = |\mathbf{A}|^{p+1} \left| \frac{\partial \operatorname{vech} \mathbf{F}}{\partial \mathbf{x}'} \right|. \tag{6.11}$$

In the special case where $\mathbf{x} = \operatorname{vech} \mathbf{X}$ and $\mathbf{F}(\mathbf{x}) = \mathbf{X}$ for some $p \times p$ symmetric matrix $\mathbf{X}$, results (6.10) and (6.11) simplify to

$$\frac{\partial \operatorname{vech}(\mathbf{AXA}')}{\partial (\operatorname{vech} \mathbf{X})'} = \mathbf{H}_p(\mathbf{A} \otimes \mathbf{A})\mathbf{G}_p,$$

$$\left| \frac{\partial \operatorname{vech}(\mathbf{AXA}')}{\partial (\operatorname{vech} \mathbf{X})'} \right| = |\mathbf{A}|^{p+1}.$$

Finally, consider the Jacobian matrix of $\operatorname{vec}(\mathbf{F}^{-1})$, where $\mathbf{F}$ is a $p \times p$ matrix of functions, defined on a set S, of a vector $\mathbf{x} = (x_1, \ldots, x_m)'$ of m variables [with $\mathbf{F}(\mathbf{x})$ nonsingular for every $\mathbf{x}$ in S]. Making use of results (15.8.15) and (2.10) [as well as result (6.1)], we find that

$$\begin{aligned} \frac{\partial \operatorname{vec}(\mathbf{F}^{-1})}{\partial x_j} &= \operatorname{vec}\left(\frac{\partial \mathbf{F}^{-1}}{\partial x_j}\right) = \operatorname{vec}\left(-\mathbf{F}^{-1} \frac{\partial \mathbf{F}}{\partial x_j} \mathbf{F}^{-1}\right) \\ &= -[(\mathbf{F}^{-1})' \otimes \mathbf{F}^{-1}] \operatorname{vec}\left(\frac{\partial \mathbf{F}}{\partial x_j}\right) \\ &= -[(\mathbf{F}^{-1})' \otimes \mathbf{F}^{-1}] \frac{\partial \operatorname{vec} \mathbf{F}}{\partial x_j} \end{aligned} \tag{6.12}$$

$(j = 1, \ldots, m)$, which implies that

$$\frac{\partial \operatorname{vec}(\mathbf{F}^{-1})}{\partial \mathbf{x}'} = -[(\mathbf{F}^{-1})' \otimes \mathbf{F}^{-1}] \frac{\partial \operatorname{vec} \mathbf{F}}{\partial \mathbf{x}'}, \tag{6.13}$$

thereby relating the Jacobian matrix of $\operatorname{vec}(\mathbf{F}^{-1})$ to that of $\operatorname{vec} \mathbf{F}$. Moreover, when $m = p^2$, it follows from result (6.13), together with Corollary 13.2.5, result (3.18), and Theorem 13.3.7 [and the fact that $(-1)^{p^2} = (-1)^p$], that

$$\left| \frac{\partial \operatorname{vec}(\mathbf{F}^{-1})}{\partial \mathbf{x}'} \right| = (-1)^p |\mathbf{F}|^{-2p} \left| \frac{\partial \operatorname{vec} \mathbf{F}}{\partial \mathbf{x}'} \right|, \tag{6.14}$$

which relates the Jacobian of $\text{vec}(\mathbf{F}^{-1})$ to that of $\text{vec}\,\mathbf{F}$.

In the special case where $\mathbf{x} = \text{vec}\,\mathbf{X}$ and $\mathbf{F}(\mathbf{x}) = \mathbf{X}$ for some $p \times p$ nonsingular matrix $\mathbf{X}$ of variables, results (6.13) and (6.14) simplify to

$$\frac{\partial\,\text{vec}(\mathbf{X}^{-1})}{\partial(\text{vec}\,\mathbf{X})'} = -(\mathbf{X}^{-1})' \otimes \mathbf{X}^{-1}, \tag{6.15}$$

$$\left|\frac{\partial\,\text{vec}(\mathbf{X}^{-1})}{\partial(\text{vec}\,\mathbf{X})'}\right| = (-1)^p|\mathbf{X}|^{-2p}. \tag{6.16}$$

Suppose now that the $(p \times p)$ matrix $\mathbf{F}$ is symmetric. Then,

$$\frac{\partial\,\text{vech}(\mathbf{F}^{-1})}{\partial \mathbf{x}'} = -\mathbf{H}_p(\mathbf{F}^{-1} \otimes \mathbf{F}^{-1})\mathbf{G}_p \frac{\partial\,\text{vech}\,\mathbf{F}}{\partial \mathbf{x}'}, \tag{6.17}$$

as is evident from result (6.13) upon observing [in light of result (4.6)] that

$$\frac{\partial\,\text{vech}(\mathbf{F}^{-1})}{\partial \mathbf{x}'} = \mathbf{H}_p \frac{\partial\,\text{vec}(\mathbf{F}^{-1})}{\partial \mathbf{x}'}, \qquad \frac{\partial\,\text{vec}\,\mathbf{F}}{\partial \mathbf{x}'} = \mathbf{G}_p \frac{\partial\,\text{vech}\,\mathbf{F}}{\partial \mathbf{x}'}.$$

When $m = p(p+1)/2$, it follows from Corollary 13.2.5, result (4.34), and Theorem 13.3.7 that

$$\left|\frac{\partial\,\text{vech}(\mathbf{F}^{-1})}{\partial \mathbf{x}'}\right| = (-1)^{p(p+1)/2}|\mathbf{F}|^{-(p+1)}\left|\frac{\partial\,\text{vech}\,\mathbf{F}}{\partial \mathbf{x}'}\right|. \tag{6.18}$$

In the special case where $\mathbf{x} = \text{vech}\,\mathbf{X}$ and $\mathbf{F}(\mathbf{x}) = \mathbf{X}$ for some $p \times p$ nonsingular symmetric matrix $\mathbf{X}$, results (6.17) and (6.18) simplify to

$$\frac{\partial\,\text{vech}(\mathbf{X}^{-1})}{\partial(\text{vech}\,\mathbf{X})'} = -\mathbf{H}_p(\mathbf{X}^{-1} \otimes \mathbf{X}^{-1})\mathbf{G}_p, \tag{6.19}$$

$$\left|\frac{\partial\,\text{vech}(\mathbf{X}^{-1})}{\partial(\text{vech}\,\mathbf{X})'}\right| = (-1)^{p(p+1)/2}|\mathbf{X}|^{-(p+1)}. \tag{6.20}$$

Exercises

Section 16.1

1. (a) Verify result (1.13); that is, verify that, for any $m \times n$ matrices $\mathbf{A}$ and $\mathbf{B}$ and any $p \times q$ matrices $\mathbf{C}$ and $\mathbf{D}$,

$$(\mathbf{A} + \mathbf{B}) \otimes (\mathbf{C} + \mathbf{D}) = (\mathbf{A} \otimes \mathbf{C}) + (\mathbf{A} \otimes \mathbf{D}) + (\mathbf{B} \otimes \mathbf{C}) + (\mathbf{B} \otimes \mathbf{D}).$$

   (b) Verify result (1.14); that is, verify that, for any $m \times n$ matrices $\mathbf{A}_1, \mathbf{A}_2, \ldots, \mathbf{A}_r$ and $p \times q$ matrices $\mathbf{B}_1, \mathbf{B}_2, \ldots, \mathbf{B}_s$,

$$\Big(\sum_{i=1}^{r} \mathbf{A}_i\Big) \otimes \Big(\sum_{j=1}^{s} \mathbf{B}_j\Big) = \sum_{i=1}^{r}\sum_{j=1}^{s} (\mathbf{A}_i \otimes \mathbf{B}_j).$$

2. Show that, for any $m \times n$ matrix $\mathbf{A}$ and $p \times q$ matrix $\mathbf{B}$,
$$\mathbf{A} \otimes \mathbf{B} = (\mathbf{A} \otimes \mathbf{I}_p)\,\text{diag}(\mathbf{B}, \mathbf{B}, \ldots, \mathbf{B}).$$

3. Show that, for any $m \times 1$ vector $\mathbf{a}$ and any $p \times 1$ vector $\mathbf{b}$, (1) $\mathbf{a} \otimes \mathbf{b} = (\mathbf{a} \otimes \mathbf{I}_p)\mathbf{b}$ and (2) $\mathbf{a}' \otimes \mathbf{b}' = \mathbf{b}'(\mathbf{a}' \otimes \mathbf{I}_p)$.

4. Let $\mathbf{A}$ and $\mathbf{B}$ represent square matrices.

   (a) Show that if $\mathbf{A}$ and $\mathbf{B}$ are orthogonal, then $\mathbf{A} \otimes \mathbf{B}$ is orthogonal.

   (b) Show that if $\mathbf{A}$ and $\mathbf{B}$ are idempotent, then $\mathbf{A} \otimes \mathbf{B}$ is idempotent.

5. Letting m, n, p, and q represent arbitrary positive integers, show (a) that, for any $p \times q$ matrix $\mathbf{B}$ (having $p > 1$ or $q > 1$), there exists an $m \times n$ matrix $\mathbf{A}$ such that $\mathbf{A} \otimes \mathbf{B}$ has generalized inverses that are not expressible in the form $\mathbf{A}^- \otimes \mathbf{B}^-$; and (b) that, for any $m \times n$ matrix $\mathbf{A}$ (having $m > 1$ or $n > 1$), there exists a $p \times q$ matrix $\mathbf{B}$ such that $\mathbf{A} \otimes \mathbf{B}$ has generalized inverses that are not expressible in the form $\mathbf{A}^- \otimes \mathbf{B}^-$.

6. Let $\mathbf{X} = \mathbf{A} \otimes \mathbf{B}$, where $\mathbf{A}$ is an $m \times n$ matrix and $\mathbf{B}$ a $p \times q$ matrix. Show that $\mathbf{P_X} = \mathbf{P_A} \otimes \mathbf{P_B}$.

7. Show that the Kronecker product $\mathbf{A} \otimes \mathbf{B}$ of an $m \times m$ matrix $\mathbf{A}$ and an $n \times n$ matrix $\mathbf{B}$ is (a) symmetric nonnegative definite if $\mathbf{A}$ and $\mathbf{B}$ are both symmetric nonnegative definite or both symmetric nonpositive definite, and (b) symmetric positive definite if $\mathbf{A}$ and $\mathbf{B}$ are both symmetric positive definite or both symmetric negative definite.

8. Let $\mathbf{A}$ and $\mathbf{B}$ represent $m \times m$ symmetric matrices and $\mathbf{C}$ and $\mathbf{D}$ $n \times n$ symmetric matrices. Using the result of Exercise 7 (or otherwise), show that if $\mathbf{A} - \mathbf{B}$, $\mathbf{C} - \mathbf{D}$, $\mathbf{B}$, and $\mathbf{C}$ are nonnegative definite, then $\mathbf{A} \otimes \mathbf{C} - \mathbf{B} \otimes \mathbf{D}$ is symmetric nonnegative definite.

9. Let $\mathbf{A}$ represent an $m \times n$ matrix and $\mathbf{B}$ a $p \times q$ matrix. Show that, in the case of the usual norm,
$$\|\mathbf{A} \otimes \mathbf{B}\| = \|\mathbf{A}\|\,\|\mathbf{B}\|.$$

10. Verify result (1.27).

11. Show that (a) if $\mathbf{T}$ and $\mathbf{U}$ are both upper triangular matrices, then $\mathbf{T} \otimes \mathbf{U}$ is an upper triangular matrix, and (b) if $\mathbf{T}$ and $\mathbf{L}$ are both lower triangular matrices, then $\mathbf{T} \otimes \mathbf{L}$ is a lower triangular matrix.

12. Let $\mathbf{A}$ represent an $m \times m$ matrix and $\mathbf{B}$ an $n \times n$ matrix. Suppose that $\mathbf{A}$ and $\mathbf{B}$ have LDU decompositions, say $\mathbf{A} = \mathbf{L}_1\mathbf{D}_1\mathbf{U}_1$ and $\mathbf{B} = \mathbf{L}_2\mathbf{D}_2\mathbf{U}_2$. Using the results of Exercise 11, show that $\mathbf{A} \otimes \mathbf{B}$ has the LDU decomposition $\mathbf{A} \otimes \mathbf{B} = \mathbf{LDU}$, where $\mathbf{L} = \mathbf{L}_1 \otimes \mathbf{L}_2$, $\mathbf{D} = \mathbf{D}_1 \otimes \mathbf{D}_2$, and $\mathbf{U} = \mathbf{U}_1 \otimes \mathbf{U}_2$.

Section 16.2

13. Let $\mathbf{A}_1, \mathbf{A}_2, \ldots, \mathbf{A}_k$ represent k matrices (of the same size). Show that $\mathbf{A}_1$, $\mathbf{A}_2, \ldots, \mathbf{A}_k$ are linearly independent if and only if $\text{vec}(\mathbf{A}_1)$, $\text{vec}(\mathbf{A}_2), \ldots, \text{vec}(\mathbf{A}_k)$ are linearly independent.

14. Let m represent a positive integer, let $\mathbf{e}_i$ represent the ith column of $\mathbf{I}_m$ ($i = 1, \ldots, m$), and (for $i, j = 1, \ldots, m$) let $\mathbf{U}_{ij} = \mathbf{e}_i\mathbf{e}_j'$ (in which case $\mathbf{U}_{ij}$ is an $m \times m$ matrix whose ijth element is 1 and whose remaining $m^2 - 1$ elements are 0).

    (a) Show that
    $$\text{vec}(\mathbf{I}_m) = \sum_{i=1}^{m} \mathbf{e}_i \otimes \mathbf{e}_i.$$

    (b) Show that (for $i, j, r, s = 1, \ldots, m$)
    $$\text{vec}(\mathbf{U}_{ri})[\text{vec}(\mathbf{U}_{sj})]' = \mathbf{U}_{ij} \otimes \mathbf{U}_{rs}.$$

    (c) Show that
    $$\sum_{i=1}^{m}\sum_{j=1}^{m} \mathbf{U}_{ij} \otimes \mathbf{U}_{ij} = \text{vec}(\mathbf{I}_m)[\text{vec}(\mathbf{I}_m)]'.$$

15. Let $\mathbf{A}$ represent an $n \times n$ matrix.

    (a) Show that if $\mathbf{A}$ is orthogonal, then $(\text{vec}\,\mathbf{A})'\text{vec}\,\mathbf{A} = n$.

    (b) Show that if $\mathbf{A}$ is idempotent, then $[\text{vec}(\mathbf{A}')]'\text{vec}\,\mathbf{A} = \text{rank}(\mathbf{A})$.

16. Show that for any $m \times n$ matrix $\mathbf{A}$, $p \times n$ matrix $\mathbf{X}$, $p \times p$ matrix $\mathbf{B}$, and $n \times m$ matrix $\mathbf{C}$,
$$\text{tr}(\mathbf{AX}'\mathbf{BXC}) = (\text{vec}\,\mathbf{X})'[(\mathbf{A}'\mathbf{C}') \otimes \mathbf{B}]\,\text{vec}\,\mathbf{X} = (\text{vec}\,\mathbf{X})'[(\mathbf{CA}) \otimes \mathbf{B}']\,\text{vec}\,\mathbf{X}.$$

17. (a) Let $\mathcal{V}$ represent a linear space of $m \times n$ matrices, and let g represent a function that assigns the value $\mathbf{A} * \mathbf{B}$ to each pair of matrices $\mathbf{A}$ and $\mathbf{B}$ in $\mathcal{V}$. Take $\mathcal{U}$ to be the linear space of $mn \times 1$ vectors defined by
    $$\mathcal{U} = \{\mathbf{x} \in \mathcal{R}^{mn\times 1} \; : \; \mathbf{x} = \text{vec}(\mathbf{A}) \text{ for some } \mathbf{A} \in \mathcal{V}\},$$
    and let $\mathbf{x} \bullet \mathbf{y}$ represent the value assigned to each pair of vectors $\mathbf{x}$ and $\mathbf{y}$ in $\mathcal{U}$ by an arbitrary inner product f. Show that g is an inner product (for $\mathcal{V}$) if and only if there exists an f such that (for all $\mathbf{A}$ and $\mathbf{B}$ in $\mathcal{V}$)

$$\mathbf{A} * \mathbf{B} = \text{vec}(\mathbf{A}) \bullet \text{vec}(\mathbf{B}).$$

(b) Let g represent a function that assigns the value $\mathbf{A} * \mathbf{B}$ to an arbitrary pair of matrices $\mathbf{A}$ and $\mathbf{B}$ in $\mathcal{R}^{m \times n}$. Show that g is an inner product (for $\mathcal{R}^{m \times n}$) if and only if there exists an $mn \times mn$ partitioned symmetric positive definite matrix

$$\mathbf{W} = \begin{pmatrix} \mathbf{W}_{11} & \mathbf{W}_{12} & \dots & \mathbf{W}_{1n} \\ \mathbf{W}_{21} & \mathbf{W}_{22} & \dots & \mathbf{W}_{2n} \\ \vdots & \vdots & \ddots & \vdots \\ \mathbf{W}_{n1} & \mathbf{W}_{n2} & \dots & \mathbf{W}_{nn} \end{pmatrix}$$

(where each submatrix is of dimensions $m \times m$) such that (for all $\mathbf{A}$ and $\mathbf{B}$ in $\mathcal{R}^{m \times n}$)

$$\mathbf{A} * \mathbf{B} = \sum_{i,j} \mathbf{a}_i' \mathbf{W}_{ij} \mathbf{b}_j,$$

where $\mathbf{a}_1, \mathbf{a}_2, \dots, \mathbf{a}_n$ and $\mathbf{b}_1, \mathbf{b}_2, \dots, \mathbf{b}_n$ represent the first, second, ..., nth columns of $\mathbf{A}$ and $\mathbf{B}$, respectively.

(c) Let g represent a function that assigns the value $\mathbf{x}' * \mathbf{y}'$ to an arbitrary pair of (row) vectors in $\mathcal{R}^{1 \times n}$. Show that g is an inner product (for $\mathcal{R}^{1 \times n}$) if and only if there exists an $n \times n$ symmetric positive definite matrix $\mathbf{W}$ such that (for every pair of n-dimensional row vectors $\mathbf{x}'$ and $\mathbf{y}'$)

$$\mathbf{x}' * \mathbf{y}' = \mathbf{x}'\mathbf{W}\mathbf{y}.$$

Section 16.3

18. (a) Define (for $m \geq 2$) $\mathbf{P}$ to be the $mn \times mn$ permutation matrix such that, for every $m \times n$ matrix $\mathbf{A}$,

$$\begin{pmatrix} \text{vec}\, \mathbf{A}_* \\ \mathbf{r} \end{pmatrix} = \mathbf{P}\, \text{vec}\, \mathbf{A},$$

where $\mathbf{A}_*$ is the $(m-1) \times n$ matrix whose rows are respectively the first, ..., $(m-1)$th rows of $\mathbf{A}$ and $\mathbf{r}'$ is the mth row of $\mathbf{A}$ [and hence where $\mathbf{A} = \begin{pmatrix} \mathbf{A}_* \\ \mathbf{r}' \end{pmatrix}$ and $\mathbf{A}' = (\mathbf{A}_*', \mathbf{r})$].

(1) Show that $\mathbf{K}_{mn} = \begin{pmatrix} \mathbf{K}_{m-1,n} & \mathbf{0} \\ \mathbf{0} & \mathbf{I}_n \end{pmatrix} \mathbf{P}$.

(2) Show that $|\mathbf{P}| = (-1)^{(m-1)n(n-1)/2}$.

(3) Show that $|\mathbf{K}_{mn}| = (-1)^{(m-1)n(n-1)/2} |\mathbf{K}_{m-1,n}|$.

(b) Show that $|\mathbf{K}_{mn}| = (-1)^{m(m-1)n(n-1)/4}$.

(c) Show that $|\mathbf{K}_{mm}| = (-1)^{m(m-1)/2}$.

19. Show that, for any $m \times n$ matrix $\mathbf{A}$, $p \times 1$ vector $\mathbf{a}$, and $q \times 1$ vector $\mathbf{b}$,

(1) $\mathbf{b}' \otimes \mathbf{A} \otimes \mathbf{a} = \mathbf{K}_{mp}[(\mathbf{ab}') \otimes \mathbf{A}]$;

(2) $\mathbf{a} \otimes \mathbf{A} \otimes \mathbf{b}' = \mathbf{K}_{pm}[\mathbf{A} \otimes (\mathbf{ab}')]$.

20. Let m and n represent positive integers, and let $\mathbf{e}_i$ represent the ith column of $\mathbf{I}_m$ ($i = 1, \ldots, m$) and $\mathbf{u}_j$ represent the jth column of $\mathbf{I}_n$ ($j = 1, \ldots, n$). Show that

$$\mathbf{K}_{mn} = \sum_{j=1}^{n} \mathbf{u}_j' \otimes \mathbf{I}_m \otimes \mathbf{u}_j = \sum_{i=1}^{m} \mathbf{e}_i \otimes \mathbf{I}_n \otimes \mathbf{e}_i'.$$

21. Let m, n, and p represent positive integers. Using the result of Exercise 20, show that

(a) $\mathbf{K}_{mp,n} = \mathbf{K}_{p,mn}\mathbf{K}_{m,np}$;

(b) $\mathbf{K}_{mp,n}\mathbf{K}_{np,m}\mathbf{K}_{mn,p} = \mathbf{I}$;

(c) $\mathbf{K}_{n,mp} = \mathbf{K}_{np,m}\mathbf{K}_{mn,p}$;

(d) $\mathbf{K}_{p,mn}\mathbf{K}_{m,np} = \mathbf{K}_{m,np}\mathbf{K}_{p,mn}$;

(e) $\mathbf{K}_{np,m}\mathbf{K}_{mn,p} = \mathbf{K}_{mn,p}\mathbf{K}_{np,m}$;

(f) $\mathbf{K}_{m,np}\mathbf{K}_{mp,n} = \mathbf{K}_{mp,n}\mathbf{K}_{m,np}$.

[*Hint*. Begin by letting $\mathbf{u}_j$ represent the jth column of $\mathbf{I}_n$ and showing that $\mathbf{K}_{mp,n} = \sum_j (\mathbf{u}_j' \otimes \mathbf{I}_p) \otimes (\mathbf{I}_m \otimes \mathbf{u}_j)$ and then making use of result (3.10).]

22. Let $\mathbf{A}$ represent an $m \times n$ matrix, and define $\mathbf{B} = \mathbf{K}_{mn}(\mathbf{A}' \otimes \mathbf{A})$. Show (a) that $\mathbf{B}$ is symmetric, (b) that $\mathrm{rank}(\mathbf{B}) = [\mathrm{rank}(\mathbf{A})]^2$, (c) that $\mathbf{B}^2 = (\mathbf{AA}') \otimes (\mathbf{A}'\mathbf{A})$, and (d) that $\mathrm{tr}(\mathbf{B}) = \mathrm{tr}(\mathbf{A}'\mathbf{A})$.

23. Show that, for any $m \times n$ matrix $\mathbf{A}$ and any $p \times q$ matrix $\mathbf{B}$,

$$\mathrm{vec}(\mathbf{A} \otimes \mathbf{B}) = (\mathbf{I}_n \otimes \mathbf{G})\,\mathrm{vec}\,\mathbf{A} = (\mathbf{H} \otimes \mathbf{I}_p)\,\mathrm{vec}\,\mathbf{B},$$

where $\mathbf{G} = (\mathbf{K}_{qm} \otimes \mathbf{I}_p)(\mathbf{I}_m \otimes \mathrm{vec}\,\mathbf{B})$ and $\mathbf{H} = (\mathbf{I}_n \otimes \mathbf{K}_{qm})[\mathrm{vec}(\mathbf{A}) \otimes \mathbf{I}_q]$.

Section 16.4

24. Show that, for $\mathbf{H}_n = (\mathbf{G}_n'\mathbf{G}_n)^{-1}\mathbf{G}_n'$,

$$\mathbf{G}_n\mathbf{H}_n\mathbf{H}_n' = \mathbf{H}_n'.$$

25. There exists a unique matrix $\mathbf{L}_n$ such that

$$\mathrm{vech}\,\mathbf{A} = \mathbf{L}_n\,\mathrm{vec}\,\mathbf{A}$$

for *every* $n \times n$ matrix $\mathbf{A}$ (symmetric or not). [The matrix $\mathbf{L}_n$ is one choice for the matrix $\mathbf{H}_n$ discussed in Section 16.4b. It is referred to by Magnus and

Neudecker (1980) as the *elimination matrix*—the effect of premultiplying the vec of an $n \times n$ matrix $\mathbf{A}$ by $\mathbf{L}_n$ is to eliminate (from vec $\mathbf{A}$) the "supradiagonal" elements of $\mathbf{A}$.]

(a) Write out the elements of $\mathbf{L}_1$, $\mathbf{L}_2$, and $\mathbf{L}_3$.

(b) For an arbitrary positive integer n, describe $\mathbf{L}_n$ in terms of its rows.

26. Let $\mathbf{A}$ represent an $n \times n$ matrix and $\mathbf{b}$ an $n \times 1$ vector.

(a) Show that $(1/2)[(\mathbf{A} \otimes \mathbf{b}') + (\mathbf{b}' \otimes \mathbf{A})]\mathbf{G}_n = (\mathbf{A} \otimes \mathbf{b}')\mathbf{G}_n$.

(b) Show that, for $\mathbf{H}_n = (\mathbf{G}_n'\mathbf{G}_n)^{-1}\mathbf{G}_n'$,

(1) $(1/2)\mathbf{H}_n[(\mathbf{b} \otimes \mathbf{A}) + (\mathbf{A} \otimes \mathbf{b})] = \mathbf{H}_n(\mathbf{b} \otimes \mathbf{A})$,

(2) $(\mathbf{A} \otimes \mathbf{b}')\mathbf{G}_n\mathbf{H}_n = (1/2)[(\mathbf{A} \otimes \mathbf{b}') + (\mathbf{b}' \otimes \mathbf{A})]$,

(3) $\mathbf{G}_n\mathbf{H}_n(\mathbf{b} \otimes \mathbf{A}) = (1/2)[(\mathbf{b} \otimes \mathbf{A}) + (\mathbf{A} \otimes \mathbf{b})]$.

27. Let $\mathbf{A} = \{a_{ij}\}$ represent an $n \times n$ (possibly nonsymmetric) matrix.

(a) Show that, for $\mathbf{H}_n = (\mathbf{G}_n'\mathbf{G}_n)^{-1}\mathbf{G}_n'$,

$$\mathbf{H}_n \operatorname{vec} \mathbf{A} = (1/2) \operatorname{vech}(\mathbf{A} + \mathbf{A}').$$

(b) Show that

$$\mathbf{G}_n'\mathbf{G}_n \operatorname{vech} \mathbf{A} = \operatorname{vech}[2\mathbf{A} - \operatorname{diag}(a_{11}, a_{22}, \ldots, a_{nn})].$$

(c) Show that

$$\mathbf{G}_n' \operatorname{vec} \mathbf{A} = \operatorname{vech}[\mathbf{A} + \mathbf{A}' - \operatorname{diag}(a_{11}, a_{22}, \ldots, a_{nn})].$$

28. Let $\mathbf{A}$ represent an $n \times n$ matrix. Show that, for $\mathbf{H}_n = (\mathbf{G}_n'\mathbf{G}_n)^{-1}\mathbf{G}_n'$,

$$\mathbf{G}_n\mathbf{H}_n(\mathbf{A} \otimes \mathbf{A})\mathbf{H}_n' = (\mathbf{A} \otimes \mathbf{A})\mathbf{H}_n'.$$

29. Show that if an $n \times n$ matrix $\mathbf{A} = \{a_{ij}\}$ is upper triangular, lower triangular, or diagonal, then $\mathbf{H}_n(\mathbf{A} \otimes \mathbf{A})\mathbf{G}_n$ is respectively upper triangular, lower triangular, or diagonal with diagonal elements $a_{ii}a_{jj}$ $(i = 1, \ldots, n;\ j = i, \ldots, n)$.

Section 16.5

30. Let $\mathbf{A}_1, \ldots, \mathbf{A}_k$, and $\mathbf{B}$ represent $m \times n$ matrices, and let $\mathbf{b} = \operatorname{vec} \mathbf{B}$.

(a) Show that the matrix equation $\sum_{i=1}^k x_i\mathbf{A}_i = \mathbf{B}$ (in unknowns $x_1, \ldots, x_k$) is equivalent to a linear system of the form $\mathbf{Ax} = \mathbf{b}$, where $\mathbf{x} = (x_1, \ldots, x_k)'$ is a vector of unknowns.

(b) Show that if $\mathbf{A}_1, \ldots, \mathbf{A}_k$, and $\mathbf{B}$ are symmetric, then the matrix equation $\sum_{i=1}^k x_i\mathbf{A}_i = \mathbf{B}$ (in unknowns $x_1, \ldots, x_k$) is equivalent to a linear system of the form $\mathbf{A}^*\mathbf{x} = \mathbf{b}^*$, where $\mathbf{b}^* = \operatorname{vech} \mathbf{B}$ and $\mathbf{x} = (x_1, \ldots, x_k)'$ is a vector of unknowns.

Section 16.6

31 Let $\mathbf{F}$ represent a $p \times p$ matrix of functions, defined on a set S, of a vector $\mathbf{x} = (x_1, \ldots, x_m)'$ of m variables. Show that, for $k = 2, 3, \ldots,$

$$\frac{\partial \operatorname{vec}(\mathbf{F}^k)}{\partial \mathbf{x}'} = \sum_{s=1}^{k} [(\mathbf{F}^{s-1})' \otimes \mathbf{F}^{k-s}] \frac{\partial \operatorname{vec} \mathbf{F}}{\partial \mathbf{x}'}$$

(where $\mathbf{F}^0 = \mathbf{I}_p$).

32. Let $\mathbf{F} = \{f_{is}\}$ and $\mathbf{G}$ represent $p \times q$ and $r \times s$ matrices of functions, defined on a set S, of a vector $\mathbf{x} = (x_1, \ldots, x_m)'$ of m variables.

(a) Show that (for $j = 1, \ldots, m$)

$$\frac{\partial(\mathbf{F} \otimes \mathbf{G})}{\partial x_j} = \left(\mathbf{F} \otimes \frac{\partial \mathbf{G}}{\partial x_j}\right) + \left(\frac{\partial \mathbf{F}}{\partial x_j} \otimes \mathbf{G}\right).$$

(b) Show that (for $j = 1, \ldots, m$)

$$\frac{\partial \operatorname{vec}(\mathbf{F} \otimes \mathbf{G})}{\partial x_j} = (\mathbf{I}_q \otimes \mathbf{K}_{sp} \otimes \mathbf{I}_r)\left[(\operatorname{vec} \mathbf{F}) \otimes \frac{\partial \operatorname{vec} \mathbf{G}}{\partial x_j} + \frac{\partial \operatorname{vec} \mathbf{F}}{\partial x_j} \otimes (\operatorname{vec} \mathbf{G})\right].$$

(c) Show that

$$\frac{\partial \operatorname{vec}(\mathbf{F} \otimes \mathbf{G})}{\partial \mathbf{x}'} = (\mathbf{I}_q \otimes \mathbf{K}_{sp} \otimes \mathbf{I}_r)\left[(\operatorname{vec} \mathbf{F}) \otimes \frac{\partial \operatorname{vec} \mathbf{G}}{\partial \mathbf{x}'} + \frac{\partial \operatorname{vec} \mathbf{F}}{\partial \mathbf{x}'} \otimes (\operatorname{vec} \mathbf{G})\right].$$

(d) Show that, in the special case where $\mathbf{x}' = [(\operatorname{vec} \mathbf{X})', (\operatorname{vec} \mathbf{Y})']$, $\mathbf{F}(\mathbf{x}) = \mathbf{X}$, and $\mathbf{G}(\mathbf{x}) = \mathbf{Y}$ for some $p \times q$ and $r \times s$ matrices $\mathbf{X}$ and $\mathbf{Y}$ of variables, the formula in Part (c) simplifies to

$$\frac{\partial \operatorname{vec}(\mathbf{X} \otimes \mathbf{Y})}{\partial \mathbf{x}'} = (\mathbf{I}_q \otimes \mathbf{K}_{sp} \otimes \mathbf{I}_r)(\mathbf{I}_{pq} \otimes (\operatorname{vec} \mathbf{Y}),\ (\operatorname{vec} \mathbf{X}) \otimes \mathbf{I}_{rs}).$$

Bibliographic and Supplementary Notes

§1. For some historical notes on Kronecker products, refer to Henderson and Searle (1981b, sec. 2.2). *Warning:* There exist presentations (e.g., Graybill 1983) where the left Kronecker product of two matrices **A** and **B**, which is denoted herein by $\mathbf{B} \otimes \mathbf{A}$, is referred to as the Kronecker (or direct) product of **A** and **B** and is denoted by $\mathbf{A} \otimes \mathbf{B}$. §2. The idea of rearranging the elements of a matrix in the form of a vector dates back at least to Sylvester (1884). Refer to Henderson and Searle (1979, sec. 1; 1981b, sec. 2.1) for some historical notes on the development of this idea. In some presentations (e.g., Rogers 1980), the vec of a matrix **A** is called the *pack* of **A**. §3. Refer to Henderson and Searle (1981b) and to Magnus and Neudecker (1986) for some historical notes on vec-permutation matrices and their properties. Their papers include discussions of the alternative names that have been used in referring to vec-permutation matrices and of the various symbols that have been used to denote these matrices. §4. Henderson and Searle (1979) and Magnus and Neudecker (1986) provide references to early work on the vech operator (and on related concepts) and review the properties of the vech operator. They indicate that the idea of including only the distinct elements of a symmetric matrix in its vector representation dates back at least to Aitken (1949). This idea can be extended from symmetric matrices to other matrices that contain "redundant" elements—refer, for example, to Henderson and Searle (1979). §6. The results of Section 6 depend to a very significant extent on the material on matrix differentiation presented in Chapter 15.

17

Intersections and Sums of Subspaces

The orthogonality of a pair of subspaces was defined and discussed in Chapter 12. There are pairs of subspaces that are not orthogonal but that have a weaker property called essential disjointness—orthogonal subspaces are essentially disjoint, but essentially disjoint subspaces are not necessarily orthogonal. Unlike orthogonality, essential disjointness does not depend on the choice of inner product. The essential disjointness of a pair of subspaces is defined and discussed in the present chapter. The concept of essential disjointness arises in a very natural and fundamental way in results (like those of Sections 17.2, 17.3, and 17.5) on the ranks, row and column spaces, and generalized inverses of partitioned matrices and of sums and products of matrices.

It was found in Chapter 12 that a subspace $\mathcal{U}$ of a linear space $\mathcal{V}$ and its orthogonal complement $\mathcal{U}^{\perp}$ (which is a subspace of $\mathcal{V}$ that is by definition orthogonal to $\mathcal{U}$ and for which $\dim \mathcal{U} + \dim \mathcal{U}^{\perp} = \dim \mathcal{V}$) have the property that every matrix $\mathbf{Y}$ in $\mathcal{V}$ can be uniquely expressed as the sum of a matrix in $\mathcal{U}$ (which is the projection of $\mathbf{Y}$ on $\mathcal{U}$) and a matrix in $\mathcal{U}^{\perp}$ (which is the projection of $\mathbf{Y}$ on $\mathcal{U}^{\perp}$)—refer to Theorem 12.5.11. It is shown in Section 17.6 that this property extends to any subspaces $\mathcal{U}$ and $\mathcal{W}$ of $\mathcal{V}$ that are essentially disjoint and that have $\dim \mathcal{U} + \dim \mathcal{W} = \dim \mathcal{V}$. That is, for any such subspaces $\mathcal{U}$ and $\mathcal{W}$, every matrix in $\mathcal{V}$ can be uniquely expressed as the sum of a matrix in $\mathcal{U}$ and a matrix in $\mathcal{W}$.

17.1 Definitions and Some Basic Properties

Let $\mathcal{U}$ and $\mathcal{V}$ represent subsets of the linear space $\mathcal{R}^{m\times n}$ of all $m \times n$ matrices. The *intersection* of $\mathcal{U}$ and $\mathcal{V}$ is the subset (of $\mathcal{R}^{m\times n}$) comprising all $m \times n$ matrices that are common to $\mathcal{U}$ and $\mathcal{V}$ (i.e., that belong to both $\mathcal{U}$ and $\mathcal{V}$) and is denoted by the symbol $\mathcal{U} \cap \mathcal{V}$.

If $\mathcal{U}$ and $\mathcal{V}$ are subspaces (of $\mathcal{R}^{m\times n}$), then their intersection $\mathcal{U} \cap \mathcal{V}$ is also a subspace. To see this, suppose that $\mathcal{U}$ and $\mathcal{V}$ are subspaces. Then, the $m \times n$ null matrix $\mathbf{0}$ belongs to both $\mathcal{U}$ and $\mathcal{V}$ and hence to $\mathcal{U} \cap \mathcal{V}$, so that $\mathcal{U} \cap \mathcal{V}$ is nonempty. Further, for any matrix $\mathbf{A}$ in $\mathcal{U} \cap \mathcal{V}$ and any scalar k, $\mathbf{A}$ is in $\mathcal{U}$ and also in $\mathcal{V}$, implying that $k\mathbf{A}$ is in both $\mathcal{U}$ and $\mathcal{V}$ and consequently in $\mathcal{U} \cap \mathcal{V}$. And, for any matrices $\mathbf{A}$ and $\mathbf{B}$ in $\mathcal{U} \cap \mathcal{V}$, both $\mathbf{A}$ and $\mathbf{B}$ are in $\mathcal{U}$ and both are also in $\mathcal{V}$, implying that $\mathbf{A}+\mathbf{B}$ is in $\mathcal{U}$ and also in $\mathcal{V}$ and hence that $\mathbf{A}+\mathbf{B}$ is in $\mathcal{U} \cap \mathcal{V}$.

More generally, the intersection of k subsets $\mathcal{U}_1, \ldots, \mathcal{U}_k$ of $\mathcal{R}^{m\times n}$ (where $k \geq 2$) is the subset (of $\mathcal{R}^{m\times n}$) comprising all $m \times n$ matrices that are common to $\mathcal{U}_1, \ldots, \mathcal{U}_k$ and is denoted by the symbol $\mathcal{U}_1 \cap \cdots \cap \mathcal{U}_k$ (or alternatively by $\cap_{i=1}^k \mathcal{U}_i$ or simply $\cap_i \mathcal{U}_i$). And, if $\mathcal{U}_1, \ldots, \mathcal{U}_k$ are subspaces (of $\mathcal{R}^{m\times n}$), then $\mathcal{U}_1 \cap \cdots \cap \mathcal{U}_k$ is also a subspace.

Note that if (for subsets $\mathcal{U}$, $\mathcal{V}$, and $\mathcal{W}$ of $\mathcal{R}^{m\times n}$) $\mathcal{W} \subset \mathcal{U}$ and $\mathcal{W} \subset \mathcal{V}$, then $\mathcal{W} \subset \mathcal{U} \cap \mathcal{V}$. Note also that if $\mathcal{U} \subset \mathcal{W}$ and $\mathcal{V} \subset \mathcal{W}$, then $\mathcal{U} \cap \mathcal{V} \subset \mathcal{W}$.

The *sum* of two (nonempty) subsets $\mathcal{U}$ and $\mathcal{V}$ of $\mathcal{R}^{m\times n}$ is defined to be the subset $\{\mathbf{U}+\mathbf{V} : \mathbf{U} \in \mathcal{U}, \mathbf{V} \in \mathcal{V}\}$ comprising every $(m\times n)$ matrix that is expressible as the sum of a matrix in $\mathcal{U}$ and a matrix in $\mathcal{V}$. It is denoted by the symbol $\mathcal{U}+\mathcal{V}$. Note that, except for special cases, $\mathcal{U}+\mathcal{V}$ differs from the union $\mathcal{U} \cup \mathcal{V}$ of $\mathcal{U}$ and $\mathcal{V}$, which by definition is the subset (of $\mathcal{R}^{m\times n}$) comprising all $m \times n$ matrices that belong to $\mathcal{U}$ or $\mathcal{V}$.

The following lemma provides a useful characterization of the sum of two subspaces of $\mathcal{R}^{m\times n}$.

Lemma 17.1.1. Let $\mathcal{U}$ and $\mathcal{V}$ represent subspaces of $\mathcal{R}^{m\times n}$. If $\mathcal{V} = \{\mathbf{0}\}$ (i.e., if there are no nonnull matrices in $\mathcal{V}$), then $\mathcal{U}+\mathcal{V} = \mathcal{U}$, and similarly if $\mathcal{U} = \{\mathbf{0}\}$, then $\mathcal{U}+\mathcal{V} = \mathcal{V}$. Further, if $\mathcal{U}$ is spanned by a (finite nonempty) set of $(m \times n)$ matrices $\mathbf{U}_1, \ldots, \mathbf{U}_s$ and $\mathcal{V}$ is spanned by a set of matrices $\mathbf{V}_1, \ldots, \mathbf{V}_t$, then

$$\mathcal{U}+\mathcal{V} = \text{sp}(\mathbf{U}_1, \ldots, \mathbf{U}_s, \mathbf{V}_1, \ldots, \mathbf{V}_t).$$

Proof. That $\mathcal{U}+\mathcal{V} = \mathcal{U}$ if $\mathcal{V} = \{\mathbf{0}\}$ and that $\mathcal{U}+\mathcal{V} = \mathcal{V}$ if $\mathcal{U} = \{\mathbf{0}\}$ is obvious. Suppose now that $\mathcal{U}$ is spanned by the set $\{\mathbf{U}_1, \ldots, \mathbf{U}_s\}$ and $\mathcal{V}$ by the set $\{\mathbf{V}_1, \ldots, \mathbf{V}_t\}$. Then, for any matrix $\mathbf{U}$ in $\mathcal{U}$ and any matrix $\mathbf{V}$ in $\mathcal{V}$, there exist scalars $c_1, \ldots, c_s$ and $k_1, \ldots, k_t$ such that $\mathbf{U} = \sum_{i=1}^s c_i \mathbf{U}_i$ and $\mathbf{V} = \sum_{j=1}^t k_j \mathbf{V}_j$ and hence such that

$$\mathbf{U}+\mathbf{V} = \sum_{i=1}^{s} c_i \mathbf{U}_i + \sum_{j=1}^{t} k_j \mathbf{V}_j,$$

implying that $\mathbf{U}+\mathbf{V} \in \text{sp}(\mathbf{U}_1, \ldots, \mathbf{U}_s, \mathbf{V}_1, \ldots, \mathbf{V}_t)$. Thus,

$$\mathcal{U}+\mathcal{V} \subset \text{sp}(\mathbf{U}_1, \ldots, \mathbf{U}_s, \mathbf{V}_1, \ldots, \mathbf{V}_t). \tag{1.1}$$

Further, for any matrix $\mathbf{A}$ in $\text{sp}(\mathbf{U}_1, \ldots, \mathbf{U}_s, \mathbf{V}_1, \ldots, \mathbf{V}_t)$, there exist scalars $c_1, \ldots, c_s, k_1, \ldots, k_t$ such that $\mathbf{A} = \sum_{i=1}^s c_i \mathbf{U}_i + \sum_{j=1}^t k_j \mathbf{V}_j$, implying that $\mathbf{A} = \mathbf{U}+\mathbf{V}$ for some matrix $\mathbf{U}$ in $\mathcal{U}$ and some matrix $\mathbf{V}$ in $\mathcal{V}$ (namely, $\mathbf{U} = \sum_{i=1}^s c_i \mathbf{U}_i$ and $\mathbf{V} = \sum_{j=1}^t k_j \mathbf{V}_j$). It follows that $\text{sp}(\mathbf{U}_1, \ldots, \mathbf{U}_s, \mathbf{V}_1, \ldots, \mathbf{V}_t) \subset \mathcal{U}+\mathcal{V}$, leading [in light of result (1.1)] to the conclusion that $\mathcal{U}+\mathcal{V} = \text{sp}(\mathbf{U}_1, \ldots, \mathbf{U}_s, \mathbf{V}_1, \ldots, \mathbf{V}_t)$. Q.E.D.

As an immediate consequence of Lemma 17.1.1, we have the following corollary.

Corollary 17.1.2. The sum $\mathcal{U} + \mathcal{V}$ of two subspaces $\mathcal{U}$ and $\mathcal{V}$ (of $\mathcal{R}^{m \times n}$) is itself a subspace. Further, if both $\mathcal{U}$ and $\mathcal{V}$ are contained in a subspace $\mathcal{W}$ (of $\mathcal{R}^{m \times n}$), then $\mathcal{U} + \mathcal{V}$ is contained in $\mathcal{W}$.

The addition of (nonempty) subsets of $\mathcal{R}^{m \times n}$ (like the addition of matrices in $\mathcal{R}^{m \times n}$) is commutative and associative (as is easily verified). Thus, taking $\mathcal{U}$ and $\mathcal{V}$ to be two (nonempty) subsets of $\mathcal{R}^{m \times n}$,

$$\mathcal{U} + \mathcal{V} = \mathcal{V} + \mathcal{U}, \tag{1.2}$$

and, taking $\mathcal{W}$ to be a third (nonempty) subset,

$$\mathcal{U} + (\mathcal{V} + \mathcal{W}) = (\mathcal{U} + \mathcal{V}) + \mathcal{W}. \tag{1.3}$$

The symbol $\mathcal{U} + \mathcal{V} + \mathcal{W}$ is used to represent the subset that is common to the left and right sides of equality (1.3), and this subset is referred to as the *sum* of $\mathcal{U}$, $\mathcal{V}$, and $\mathcal{W}$. Further,

$$\mathcal{U} + \mathcal{V} + \mathcal{W} = \{\mathbf{U} + \mathbf{V} + \mathbf{W} \; : \; \mathbf{U} \in \mathcal{U}, \; \mathbf{V} \in \mathcal{V}, \; \mathbf{W} \in \mathcal{W}\}.$$

This notation, terminology, and property extend in an obvious way to any finite number of (nonempty) subsets. Moreover, the sum of any (finite) number of subspaces is itself a subspace.

The column space $\mathcal{C}(\mathbf{A})$ and the row space $\mathcal{R}(\mathbf{A})$ of an $m \times n$ matrix $\mathbf{A}$ are subspaces of $\mathcal{R}^m$ and $\mathcal{R}^n$, respectively. Further, $\mathcal{C}(\mathbf{A})$ is spanned by the n columns of $\mathbf{A}$, and $\mathcal{R}(\mathbf{A})$ is spanned by the m rows of $\mathbf{A}$. Thus, it follows from Lemma 17.1.1 that, for any $m \times n$ matrix $\mathbf{A}$ and any $m \times p$ matrix $\mathbf{B}$,

$$\mathcal{C}(\mathbf{A}, \mathbf{B}) = \mathcal{C}(\mathbf{A}) + \mathcal{C}(\mathbf{B}) \tag{1.4}$$

and, similarly, that, for any $m \times n$ matrix $\mathbf{A}$ and any $p \times n$ matrix $\mathbf{C}$,

$$\mathcal{R}\begin{pmatrix} \mathbf{A} \\ \mathbf{C} \end{pmatrix} = \mathcal{R}(\mathbf{A}) + \mathcal{R}(\mathbf{C}). \tag{1.5}$$

Results (1.4) and (1.5) can be extended (by repeated application). For any matrices $\mathbf{A}_1, \mathbf{A}_2, \ldots, \mathbf{A}_k$ having the same number of rows,

$$\mathcal{C}(\mathbf{A}_1, \mathbf{A}_2, \ldots, \mathbf{A}_k) = \mathcal{C}(\mathbf{A}_1) + \mathcal{C}(\mathbf{A}_2) + \cdots + \mathcal{C}(\mathbf{A}_k); \tag{1.6}$$

and similarly, for any matrices $\mathbf{A}_1, \mathbf{A}_2, \ldots, \mathbf{A}_k$ having the same number of columns,

$$\mathcal{R}\begin{pmatrix} \mathbf{A}_1 \\ \mathbf{A}_2 \\ \vdots \\ \mathbf{A}_k \end{pmatrix} = \mathcal{R}(\mathbf{A}_1) + \mathcal{R}(\mathbf{A}_2) + \cdots + \mathcal{R}(\mathbf{A}_k). \tag{1.7}$$

Let $\mathcal{U}$ and $\mathcal{V}$ represent subspaces of the linear space $\mathcal{R}^{m \times n}$. If $\mathcal{U} \cap \mathcal{V} = \{\mathbf{0}\}$, that is, if the only matrix that $\mathcal{U}$ and $\mathcal{V}$ have in common is the $m \times n$ null matrix, then $\mathcal{U}$ and $\mathcal{V}$ are said to be *essentially disjoint*. Note that the intersection of essentially disjoint subspaces of $\mathcal{R}^{m\times n}$ is not the empty set; rather, it is a set with one member (the $m \times n$ null matrix)—this is the reason for referring to essentially disjoint subspaces as essentially disjoint rather than as disjoint. (Since the $m \times n$ null matrix is a member of every subspace of $\mathcal{R}^{m\times n}$, it is not possible for the intersection of two subspaces to be the empty set.)

The following lemma gives an important property of essentially disjoint subspaces.

Lemma 17.1.3. Let $\mathcal{U}$ and $\mathcal{V}$ represent subspaces (of dimension 1 or more) of $\mathcal{R}^{m\times n}$. And, let $\{\mathbf{U}_1, \ldots, \mathbf{U}_r\}$ represent any linearly independent set of matrices in $\mathcal{U}$, and $\{\mathbf{V}_1, \ldots, \mathbf{V}_s\}$ any linearly independent set of matrices in $\mathcal{V}$. If $\mathcal{U}$ and $\mathcal{V}$ are essentially disjoint, then the combined set $\{\mathbf{U}_1, \ldots, \mathbf{U}_r, \mathbf{V}_1, \ldots, \mathbf{V}_s\}$ is linearly independent.

Proof. It suffices to show that if $\{\mathbf{U}_1, \ldots, \mathbf{U}_r, \mathbf{V}_1, \ldots, \mathbf{V}_s\}$ is linearly dependent, then $\mathcal{U}$ and $\mathcal{V}$ are not essentially disjoint.

Suppose that $\{\mathbf{U}_1, \ldots, \mathbf{U}_r, \mathbf{V}_1, \ldots, \mathbf{V}_s\}$ is linearly dependent. Then, for some scalars $c_1, \ldots c_r$ and $k_1, \ldots, k_s$ (not all of which are 0),

$$\sum_{i=1}^{r} c_i \mathbf{U}_i + \sum_{j=1}^{s} k_j \mathbf{V}_j = \mathbf{0}$$

or, equivalently,

$$\sum_{i=1}^{r} c_i \mathbf{U}_i = -\sum_{j=1}^{s} k_j \mathbf{V}_j. \tag{1.8}$$

We have that $\sum_{i=1}^{r} c_i \mathbf{U}_i \neq \mathbf{0}$—since if $\sum_{i=1}^{r} c_i \mathbf{U}_i = \mathbf{0}$ (in which case $\sum_{j=1}^{s} k_j \mathbf{V}_j = \mathbf{0}$), we would be led to the contradictory conclusion that $\{\mathbf{U}_1, \ldots, \mathbf{U}_r\}$ or $\{\mathbf{V}_1, \ldots, \mathbf{V}_s\}$ is linearly dependent. Thus, since $\sum_{i=1}^{r} c_i \mathbf{U}_i \in \mathcal{U}$ and [in light of equality (1.8)] $\sum_{i=1}^{r} c_i \mathbf{U}_i \in \mathcal{V}$, we conclude that $\mathcal{U} \cap \mathcal{V}$ contains a nonnull matrix and hence that $\mathcal{U}$ and $\mathcal{V}$ are not essentially disjoint. Q.E.D.

Note that if $\mathcal{U}$ and $\mathcal{V}$ are essentially disjoint subspaces of $\mathcal{R}^{m\times n}$, then, for any subspace $\mathcal{W}$ of $\mathcal{U}$ and any subspace $\mathcal{X}$ of $\mathcal{V}$, $\mathcal{W}$ and $\mathcal{V}$ are essentially disjoint, $\mathcal{U}$ and $\mathcal{X}$ are essentially disjoint, and more generally $\mathcal{W}$ and $\mathcal{X}$ are essentially disjoint.

The following lemma gives two alternative characterizations of essentially disjoint subspaces.

Lemma 17.1.4. Let $\mathcal{U}$ and $\mathcal{V}$ represent subspaces of $\mathcal{R}^{m\times n}$. Then,

(1) $\mathcal{U}$ and $\mathcal{V}$ are essentially disjoint if and only if, for matrices $\mathbf{U} \in \mathcal{U}$ and $\mathbf{V} \in \mathcal{V}$, the only solution to the matrix equation

$$\mathbf{U} + \mathbf{V} = \mathbf{0} \tag{1.9}$$

is $\mathbf{U} = \mathbf{V} = \mathbf{0}$; and

(2) $\mathcal{U}$ and $\mathcal{V}$ are essentially disjoint if and only if, for every nonnull matrix $\mathbf{U}$ in $\mathcal{U}$ and every nonnull matrix $\mathbf{V}$ in $\mathcal{V}$, $\mathbf{U}$ and $\mathbf{V}$ are linearly independent.

Proof. (1) It suffices to show that $\mathcal{U}$ and $\mathcal{V}$ are not essentially disjoint if and only if, for $\mathbf{U} \in \mathcal{U}$ and $\mathbf{V} \in \mathcal{V}$, equation (1.9) or, equivalently, the equation

$$\mathbf{U} = -\mathbf{V}, \tag{1.10}$$

has a solution other than $\mathbf{U} = \mathbf{V} = \mathbf{0}$.

Suppose that (for $\mathbf{U} \in \mathcal{U}$ and $\mathbf{V} \in \mathcal{V}$) equation (1.10) has a solution other than $\mathbf{U} = \mathbf{V} = \mathbf{0}$, say $\mathbf{U} = \mathbf{U}_*$ and $\mathbf{V} = \mathbf{V}_*$. Then, $\mathbf{U}_*$ is nonnull. Further, since $-\mathbf{V}_* \in \mathcal{V}$, $\mathbf{U}_*$ is a member of $\mathcal{V}$ (as well as of $\mathcal{U}$), and hence $\mathbf{U}_* \in \mathcal{U} \cap \mathcal{V}$. Thus, $\mathcal{U}$ and $\mathcal{V}$ are not essentially disjoint.

Conversely, suppose that $\mathcal{U}$ and $\mathcal{V}$ are not essentially disjoint. Then, $\mathcal{U} \cap \mathcal{V}$ contains a nonnull matrix $\mathbf{U}_*$. And, since $\mathbf{U}_* = -(-\mathbf{U}_*)$, $\mathbf{U}_* \in \mathcal{U}$, and $-\mathbf{U}_* \in \mathcal{V}$, a solution to equation (1.10) (that satisfies the constraints $\mathbf{U} \in \mathcal{U}$ and $\mathbf{V} \in \mathcal{V}$ and differs from $\mathbf{U} = \mathbf{V} = \mathbf{0}$) can be obtained by taking $\mathbf{U} = \mathbf{U}_*$ and $\mathbf{V} = -\mathbf{U}_*$.

(2) Suppose that $\mathcal{U}$ and $\mathcal{V}$ are essentially disjoint. Then, upon applying Lemma 17.1.3 (in the special case where $r = s = 1$), we find that, for every nonnull matrix $\mathbf{U}$ in $\mathcal{U}$ and every nonnull matrix $\mathbf{V}$ in $\mathcal{V}$, $\mathbf{U}$ and $\mathbf{V}$ are linearly independent.

Conversely, suppose that, for every nonnull $\mathbf{U} \in \mathcal{U}$ and every nonnull $\mathbf{V} \in \mathcal{V}$, $\mathbf{U}$ and $\mathbf{V}$ are linearly independent. Then, there does not exist any nonnull matrix belonging to both $\mathcal{U}$ and $\mathcal{V}$, since any such matrix would not be linearly independent of itself. Q.E.D.

When two subspaces $\mathcal{U}$ and $\mathcal{V}$ of $\mathcal{R}^{m\times n}$ are essentially disjoint, their sum $\mathcal{U} + \mathcal{V}$ is sometimes referred to as a *direct sum* and is sometimes denoted by the symbol $\mathcal{U} \oplus \mathcal{V}$. By stating that a subspace $\mathcal{W}$ is the direct sum of two subspaces $\mathcal{U}$ and $\mathcal{V}$ or by writing $\mathcal{W} = \mathcal{U} \oplus \mathcal{V}$, we can simultaneously indicate that $\mathcal{W}$ is the sum of $\mathcal{U}$ and $\mathcal{V}$ and that $\mathcal{U}$ and $\mathcal{V}$ are essentially disjoint. Or, even if it has already been indicated that two subspaces $\mathcal{U}$ and $\mathcal{V}$ are essentially disjoint, we can emphasize their essential disjointness by referring to their sum as a direct sum and/or by using the notation $\mathcal{U} \oplus \mathcal{V}$.

Direct sums have an important property, which is described in the following theorem.

Theorem 17.1.5. Let $\mathcal{U}$ represent a p-dimensional subspace and $\mathcal{V}$ a q-dimensional subspace of $\mathcal{R}^{m\times n}$. Further, let S_1 and S_2 represent bases for $\mathcal{U}$ and $\mathcal{V}$, respectively, and define S to be the set of $p + q$ matrices obtained by combining the p matrices in S_1 with the q matrices in S_2. If $\mathcal{U}$ and $\mathcal{V}$ are essentially disjoint (or equivalently if $\mathcal{U} + \mathcal{V}$ is a direct sum), then S is a basis for $\mathcal{U} + \mathcal{V}$; if $\mathcal{U}$ and $\mathcal{V}$ are not essentially disjoint, then S contains a proper subset that is a basis for $\mathcal{U} + \mathcal{V}$.

Proof. It is clear from Lemma 17.1.1 that S spans $\mathcal{U} + \mathcal{V}$.

Now, suppose that $\mathcal{U}$ and $\mathcal{V}$ are essentially disjoint. Then, it is evident from Lemma 17.1.3 that S is a linearly independent set. Thus, S is a basis for $\mathcal{U} + \mathcal{V}$.

Alternatively, suppose that $\mathcal{U}$ and $\mathcal{V}$ are not essentially disjoint. Then, $\mathcal{U} \cap \mathcal{V}$ contains a nonnull matrix, say $\mathbf{U}$. Further, $\mathbf{U} = \sum_{i=1}^{p} c_i \mathbf{U}_i$, where $c_1, \ldots, c_p$ are scalars (not all of which can be zero) and $\mathbf{U}_1, \ldots, \mathbf{U}_p$ are the matrices in S_1.

Similarly, $\mathbf{U} = \sum_{j=1}^{q} k_j \mathbf{V}_j$, where $k_1, \ldots, k_q$ are scalars and $\mathbf{V}_1, \ldots, \mathbf{V}_q$ are the matrices in S_2. Thus,

$$\sum_{i=1}^{p} c_i \mathbf{U}_i + \sum_{j=1}^{q} (-k_j)\mathbf{V}_j = \mathbf{U} - \mathbf{U} = \mathbf{0},$$

implying that S is a linearly dependent set. We conclude that S itself is not a basis and consequently (in light of Theorem 4.3.11) that S contains a proper subset that is a basis for $\mathcal{U} + \mathcal{V}$. Q.E.D.

The following corollary is an immediate consequence of Theorem 17.1.5.

Corollary 17.1.6. Let $\mathcal{U}$ and $\mathcal{V}$ represent subspaces of $\mathcal{R}^{m \times n}$. If $\mathcal{U}$ and $\mathcal{V}$ are essentially disjoint (or equivalently if $\mathcal{U} + \mathcal{V}$ is a direct sum), then

$$\dim(\mathcal{U} \oplus \mathcal{V}) = \dim(\mathcal{U}) + \dim(\mathcal{V}); \tag{1.11}$$

and if $\mathcal{U}$ and $\mathcal{V}$ are not essentially disjoint, then

$$\dim(\mathcal{U} + \mathcal{V}) < \dim(\mathcal{U}) + \dim(\mathcal{V}).$$

Or, equivalently,

$$\dim(\mathcal{U} + \mathcal{V}) \leq \dim(\mathcal{U}) + \dim(\mathcal{V}), \tag{1.12}$$

with equality holding if and only if $\mathcal{U}$ and $\mathcal{V}$ are essentially disjoint.

Corollary 17.1.6 leads to the following result.

Corollary 17.1.7. Let $\mathcal{U}$ and $\mathcal{V}$ represent subspaces of a linear space $\mathcal{W}$. If $\mathcal{U}$ and $\mathcal{V}$ are essentially disjoint and if $\dim(\mathcal{U}) + \dim(\mathcal{V}) = \dim(\mathcal{W})$, then $\mathcal{U} \oplus \mathcal{V} = \mathcal{W}$.

Proof (of Corollary 17.1.7). Suppose that $\mathcal{U}$ and $\mathcal{V}$ are essentially disjoint and that $\dim(\mathcal{U}) + \dim(\mathcal{V}) = \dim(\mathcal{W})$. Then, it follows from Corollary 17.1.6 that $\dim(\mathcal{U} \oplus \mathcal{V}) = \dim(\mathcal{W})$. Since (according to Corollary 17.1.2) $\mathcal{U} \oplus \mathcal{V} \subset \mathcal{W}$, we conclude (on the basis of Theorem 4.3.10) that $\mathcal{U} \oplus \mathcal{V} = \mathcal{W}$. Q.E.D.

As a special case of Corollary 17.1.7, we have the following corollary.

Corollary 17.1.8. Let $\mathcal{U}$ and $\mathcal{V}$ represent subspaces of $\mathcal{R}^{m \times n}$. If $\mathcal{U}$ and $\mathcal{V}$ are essentially disjoint and if $\dim(\mathcal{U}) + \dim(\mathcal{V}) = mn$, then $\mathcal{U} \oplus \mathcal{V} = \mathcal{R}^{m \times n}$.

The following lemma establishes an important relationship.

Lemma 17.1.9. Let $\mathcal{U}$ and $\mathcal{V}$ represent subspaces of $\mathcal{R}^{m \times n}$. If $\mathcal{U} \perp \mathcal{V}$ (i.e., if $\mathcal{U}$ and $\mathcal{V}$ are orthogonal), then $\mathcal{U} \cap \mathcal{V} = \{\mathbf{0}\}$ (i.e., $\mathcal{U}$ and $\mathcal{V}$ are essentially disjoint).

Proof. Suppose that $\mathcal{U} \perp \mathcal{V}$. Then, for any matrix $\mathbf{U}$ in $\mathcal{U} \cap \mathcal{V}$ (i.e., in both $\mathcal{U}$ and $\mathcal{V}$), $\mathbf{U} \perp \mathbf{U}$, or, equivalently, $\mathbf{U} \bullet \mathbf{U} = 0$, implying that $\mathbf{U} = \mathbf{0}$. We conclude that $\mathcal{U} \cap \mathcal{V} = \{\mathbf{0}\}$. Q.E.D.

The converse of Lemma 17.1.9 is not necessarily true. That is, essentially disjoint subspaces are not necessarily orthogonal. Suppose, for example, that $\mathcal{U}$

is the one-dimensional subspace spanned by the vector $(1, 0)$ and $\mathcal{V}$ is the one-dimensional subspace spanned by $(1, 1)$. Then, $\mathcal{U}$ and $\mathcal{V}$ are essentially disjoint; however, $\mathcal{U}$ and $\mathcal{V}$ are not orthogonal (with respect to the usual inner product).

As an immediate consequence of Lemma 17.1.9, we have the following corollary.

Corollary 17.1.10. Let $\mathcal{U}$ and $\mathcal{V}$ represent subspaces of $\mathcal{R}^{m\times n}$. If $\mathcal{U}$ and $\mathcal{V}$ are orthogonal, then their sum is a direct sum.

17.2 Some Results on Row and Column Spaces and on the Ranks of Partitioned Matrices

The following lemma is an almost immediate consequence of Lemma 4.1.1.

Lemma 17.2.1. For any $m \times n$ matrix $\mathbf{A}$ and $m \times p$ matrix $\mathbf{B}$, $\mathcal{C}(\mathbf{A})$ and $\mathcal{C}(\mathbf{B})$ are essentially disjoint if and only if $\mathcal{R}(\mathbf{A}')$ and $\mathcal{R}(\mathbf{B}')$ are essentially disjoint.

The following lemma can be useful in establishing that a matrix equals a null matrix.

Lemma 17.2.2. (1) Let $\mathbf{A}$ represent an $m \times n$ matrix and $\mathbf{B}$ an $m \times p$ matrix whose column spaces $\mathcal{C}(\mathbf{A})$ and $\mathcal{C}(\mathbf{B})$ are essentially disjoint. And let $\mathbf{X}$ represent an $m \times q$ matrix. If $\mathcal{C}(\mathbf{X}) \subset \mathcal{C}(\mathbf{A})$ and $\mathcal{C}(\mathbf{X}) \subset \mathcal{C}(\mathbf{B})$, or, equivalently, if $\mathbf{X} = \mathbf{AK}$ and $\mathbf{X} = \mathbf{BL}$ for some matrices $\mathbf{K}$ and $\mathbf{L}$, then $\mathbf{X} = \mathbf{0}$. (2) Similarly, let $\mathbf{A}$ represent an $m \times n$ matrix and $\mathbf{B}$ a $p \times n$ matrix whose row spaces $\mathcal{R}(\mathbf{A})$ and $\mathcal{R}(\mathbf{B})$ are essentially disjoint. And let $\mathbf{X}$ represent a $q \times n$ matrix. If $\mathcal{R}(\mathbf{X}) \subset \mathcal{R}(\mathbf{A})$ and $\mathcal{R}(\mathbf{X}) \subset \mathcal{R}(\mathbf{B})$, or, equivalently, if $\mathbf{X} = \mathbf{KA}$ and $\mathbf{X} = \mathbf{LB}$ for some matrices $\mathbf{K}$ and $\mathbf{L}$, then $\mathbf{X} = \mathbf{0}$.

Proof. (1) If $\mathcal{C}(\mathbf{X}) \subset \mathcal{C}(\mathbf{A})$ and $\mathcal{C}(\mathbf{X}) \subset \mathcal{C}(\mathbf{B})$, then

$$\mathcal{C}(\mathbf{X}) \subset \mathcal{C}(\mathbf{A}) \cap \mathcal{C}(\mathbf{B}) = \{\mathbf{0}\},$$

implying that $\text{rank}(\mathbf{X}) = \dim[\mathcal{C}(\mathbf{X})] = 0$ and hence that $\mathbf{X} = \mathbf{0}$.

(2) The proof of Part (2) is analogous to that of Part (1). Q.E.D.

As a variation on Lemma 17.2.2, we have the following corollary.

Corollary 17.2.3. (1) Let $\mathbf{A}$ represent an $m \times n$ matrix and $\mathbf{B}$ an $m \times p$ matrix whose column spaces $\mathcal{C}(\mathbf{A})$ and $\mathcal{C}(\mathbf{B})$ are essentially disjoint. And, let $\mathbf{K}$ represent an $n \times q$ matrix and $\mathbf{L}$ a $p \times q$ matrix. If $\mathbf{AK} = \mathbf{BL}$, then $\mathbf{AK} = \mathbf{0}$ and $\mathbf{BL} = \mathbf{0}$. (2) Similarly, let $\mathbf{A}$ represent an $m \times n$ matrix and $\mathbf{B}$ a $p \times n$ matrix whose row spaces $\mathcal{R}(\mathbf{A})$ and $\mathcal{R}(\mathbf{B})$ are essentially disjoint. And let $\mathbf{K}$ represent a $q \times m$ matrix and $\mathbf{L}$ a $q \times p$ matrix. If $\mathbf{KA} = \mathbf{LB}$, then $\mathbf{KA} = \mathbf{0}$ and $\mathbf{LB} = \mathbf{0}$.

By applying Corollary 17.1.6 in the special case where $\mathcal{U}$ and $\mathcal{V}$ are the column spaces of an $m \times n$ matrix $\mathbf{A}$ and an $m \times p$ matrix $\mathbf{B}$, and in the special case where $\mathcal{U}$ and $\mathcal{V}$ are the row spaces of an $m \times n$ matrix $\mathbf{A}$ and a $q \times n$ matrix $\mathbf{C}$, we obtain [in light of equalities (1.4) and (1.5)] the following theorem, which sharpens the results of Lemma 4.5.7.

Theorem 17.2.4. (1) Let **A** represent an $m \times n$ matrix and **B** an $m \times p$ matrix. If $\mathcal{C}(\mathbf{A})$ and $\mathcal{C}(\mathbf{B})$ are essentially disjoint, then

$$\text{rank}(\mathbf{A},\ \mathbf{B}) = \text{rank}(\mathbf{A}) + \text{rank}(\mathbf{B}),$$

and if $\mathcal{C}(\mathbf{A})$ and $\mathcal{C}(\mathbf{B})$ are not essentially disjoint, then

$$\text{rank}(\mathbf{A},\ \mathbf{B}) < \text{rank}(\mathbf{A}) + \text{rank}(\mathbf{B}).$$

Or, equivalently,

$$\text{rank}(\mathbf{A},\ \mathbf{B}) \leq \text{rank}(\mathbf{A}) + \text{rank}(\mathbf{B}), \tag{2.1}$$

with equality holding if and only if $\mathcal{C}(\mathbf{A})$ and $\mathcal{C}(\mathbf{B})$ are essentially disjoint. (2) Let **A** represent an $m \times n$ matrix and **C** a $q \times n$ matrix. If $\mathcal{R}(\mathbf{A})$ and $\mathcal{R}(\mathbf{C})$ are essentially disjoint, then

$$\text{rank}\begin{pmatrix} \mathbf{A} \\ \mathbf{C} \end{pmatrix} = \text{rank}(\mathbf{A}) + \text{rank}(\mathbf{C}),$$

and if $\mathcal{R}(\mathbf{A})$ and $\mathcal{R}(\mathbf{C})$ are not essentially disjoint, then

$$\text{rank}\begin{pmatrix} \mathbf{A} \\ \mathbf{C} \end{pmatrix} < \text{rank}(\mathbf{A}) + \text{rank}(\mathbf{C}).$$

Or, equivalently,

$$\text{rank}\begin{pmatrix} \mathbf{A} \\ \mathbf{C} \end{pmatrix} \leq \text{rank}(\mathbf{A}) + \text{rank}(\mathbf{C}), \tag{2.2}$$

with equality holding if and only if $\mathcal{R}(\mathbf{A})$ and $\mathcal{R}(\mathbf{C})$ are essentially disjoint.

A further result on row and column spaces is given by the following lemma.

Lemma 17.2.5. (1) Let **A** represent an $m \times n$ matrix and **B** an $m \times p$ matrix such that $\mathcal{C}(\mathbf{A})$ and $\mathcal{C}(\mathbf{B})$ are essentially disjoint and $\text{rank}(\mathbf{A}) + \text{rank}(\mathbf{B}) = m$. Then, corresponding to any $m \times q$ matrix **C**, there exist an $n \times q$ matrix **R** and a $p \times q$ matrix **S** such that

$$\mathbf{C} = \mathbf{AR} + \mathbf{BS}.$$

(2) Similarly, let **A** represent an $m \times n$ matrix and **B** a $p \times n$ matrix such that $\mathcal{R}(\mathbf{A})$ and $\mathcal{R}(\mathbf{B})$ are essentially disjoint and $\text{rank}(\mathbf{A}) + \text{rank}(\mathbf{B}) = n$. Then, corresponding to any $q \times n$ matrix **C**, there exist a $q \times m$ matrix **R** and a $q \times p$ matrix **S** such that

$$\mathbf{C} = \mathbf{RA} + \mathbf{SB}.$$

Proof. (1) It follows from Corollary 17.1.8 that $\mathcal{C}(\mathbf{A}) + \mathcal{C}(\mathbf{B}) = \mathcal{R}^m$ or, equivalently [in light of result (1.4)], that $\mathcal{C}(\mathbf{A},\ \mathbf{B}) = \mathcal{R}^m$. Thus, $\mathcal{C}(\mathbf{C}) \subset \mathcal{C}(\mathbf{A},\ \mathbf{B})$, implying (in light of Lemma 4.2.2) that $\mathbf{C} = (\mathbf{A},\ \mathbf{B})\mathbf{L}$ for some $(n+p) \times q$ matrix **L**. And, upon partitioning **L** as $\mathbf{L} = \begin{pmatrix} \mathbf{R} \\ \mathbf{S} \end{pmatrix}$ (where **R** has n rows), we find that $\mathbf{C} = \mathbf{AR} + \mathbf{BS}$.

(2) The proof of Part (2) is analogous to that of Part (1). Q.E.D.

The following lemma characterizes the idempotency of an $n \times n$ matrix **A** in terms of the essential disjointness of two subspaces of $\mathcal{R}^n$.

Lemma 17.2.6. (1) An $n \times n$ matrix $\mathbf{A}$ is idempotent if and only if $\mathcal{C}(\mathbf{A})$ and $\mathcal{C}(\mathbf{I}-\mathbf{A})$ are essentially disjoint. (2) An $n \times n$ matrix $\mathbf{A}$ is idempotent if and only if $\mathcal{R}(\mathbf{A})$ and $\mathcal{R}(\mathbf{I}-\mathbf{A})$ are essentially disjoint.

Proof. (1) Since $\mathbf{A}-\mathbf{A}^2 = \mathbf{A}(\mathbf{I}-\mathbf{A})$ and $\mathbf{A}-\mathbf{A}^2 = (\mathbf{I}-\mathbf{A})\mathbf{A}$, it follows from Lemma 17.2.2 that if $\mathcal{C}(\mathbf{A})$ and $\mathcal{C}(\mathbf{I}-\mathbf{A})$ are essentially disjoint, then $\mathbf{A}-\mathbf{A}^2 = \mathbf{0}$ or, equivalently, $\mathbf{A}$ is idempotent.

Conversely, suppose that $\mathbf{A}$ is idempotent. Let $\mathbf{x}$ represent an arbitrary vector in $\mathcal{C}(\mathbf{A}) \cap \mathcal{C}(\mathbf{I}-\mathbf{A})$. Then, $\mathbf{x} = \mathbf{A}\mathbf{y}$ and $\mathbf{x} = (\mathbf{I}-\mathbf{A})\mathbf{z}$ for some vectors $\mathbf{y}$ and $\mathbf{z}$, implying that

$$\mathbf{x} = \mathbf{A}\mathbf{A}\mathbf{y} = \mathbf{A}\mathbf{x} = \mathbf{A}(\mathbf{I}-\mathbf{A})\mathbf{z} = (\mathbf{A}-\mathbf{A}^2)\mathbf{z} = \mathbf{0}.$$

Thus, $\mathcal{C}(\mathbf{A})$ and $\mathcal{C}(\mathbf{I}-\mathbf{A})$ are essentially disjoint.

(2) The proof of Part (2) is analogous to that of Part (1). Q.E.D.

The following lemma has some useful consequences.

Lemma 17.2.7. For any $m \times n$ matrix $\mathbf{A}$, $\mathcal{C}(\mathbf{A})$ and $\mathcal{C}(\mathbf{I}-\mathbf{A}\mathbf{A}^-)$ are essentially disjoint, and $\mathcal{R}(\mathbf{A})$ and $\mathcal{R}(\mathbf{I}-\mathbf{A}^-\mathbf{A})$ are essentially disjoint.

Proof. Since (according to Lemma 10.2.5) $\mathbf{A}\mathbf{A}^-$ is idempotent, it follows from Lemma 17.2.6 that $\mathcal{C}(\mathbf{A}\mathbf{A}^-)$ and $\mathcal{C}(\mathbf{I}-\mathbf{A}\mathbf{A}^-)$ are essentially disjoint and hence (in light of Lemma 9.3.7) that $\mathcal{C}(\mathbf{A})$ and $\mathcal{C}(\mathbf{I}-\mathbf{A}\mathbf{A}^-)$ are essentially disjoint. That $\mathcal{R}(\mathbf{A})$ and $\mathcal{R}(\mathbf{I}-\mathbf{A}^-\mathbf{A})$ are essentially disjoint follows from an analogous argument. Q.E.D.

In light of Lemma 4.2.2 and Theorem 17.2.4, we have the following corollary (of Lemma 17.2.7).

Corollary 17.2.8. Let $\mathbf{A}$ represent an $m \times n$ matrix, $\mathbf{B}$ an $m \times q$ matrix, and $\mathbf{C}$ a $q \times n$ matrix. Then, $\mathcal{C}(\mathbf{A})$ and $\mathcal{C}[(\mathbf{I}-\mathbf{A}\mathbf{A}^-)\mathbf{B}]$ are essentially disjoint, and

$$\text{rank}[\mathbf{A},\ (\mathbf{I}-\mathbf{A}\mathbf{A}^-)\mathbf{B}] = \text{rank}(\mathbf{A}) + \text{rank}[(\mathbf{I}-\mathbf{A}\mathbf{A}^-)\mathbf{B}]. \tag{2.3}$$

Similarly, $\mathcal{R}(\mathbf{A})$ and $\mathcal{R}[\mathbf{C}(\mathbf{I}-\mathbf{A}^-\mathbf{A})]$ are essentially disjoint, and

$$\text{rank}\begin{pmatrix} \mathbf{A} \\ \mathbf{C}(\mathbf{I}-\mathbf{A}^-\mathbf{A}) \end{pmatrix} = \text{rank}(\mathbf{A}) + \text{rank}[\mathbf{C}(\mathbf{I}-\mathbf{A}^-\mathbf{A})]. \tag{2.4}$$

More generally, for any $m \times p$ matrix $\mathbf{F}$ such that $\mathcal{C}(\mathbf{F}) \subset \mathcal{C}(\mathbf{A})$ (or, equivalently, for any $m \times p$ matrix $\mathbf{F}$ that is expressible as $\mathbf{F} = \mathbf{A}\mathbf{K}$ for some $n \times p$ matrix $\mathbf{K}$), $\mathcal{C}(\mathbf{F})$ and $\mathcal{C}[(\mathbf{I}-\mathbf{A}\mathbf{A}^-)\mathbf{B}]$ are essentially disjoint, and

$$\text{rank}[\mathbf{F},\ (\mathbf{I}-\mathbf{A}\mathbf{A}^-)\mathbf{B}] = \text{rank}(\mathbf{F}) + \text{rank}[(\mathbf{I}-\mathbf{A}\mathbf{A}^-)\mathbf{B}]. \tag{2.5}$$

And, for any $p \times n$ matrix $\mathbf{H}$ such that $\mathcal{R}(\mathbf{H}) \subset \mathcal{R}(\mathbf{A})$ (or, equivalently, for any $p \times n$ matrix $\mathbf{H}$ that is expressible as $\mathbf{H} = \mathbf{L}\mathbf{A}$ for some $p \times m$ matrix $\mathbf{L}$), $\mathcal{R}(\mathbf{H})$ and $\mathcal{R}[\mathbf{C}(\mathbf{I}-\mathbf{A}^-\mathbf{A})]$ are essentially disjoint, and

$$\text{rank}\begin{pmatrix} \mathbf{H} \\ \mathbf{C}(\mathbf{I}-\mathbf{A}^-\mathbf{A}) \end{pmatrix} = \text{rank}(\mathbf{H}) + \text{rank}[\mathbf{C}(\mathbf{I}-\mathbf{A}^-\mathbf{A})]. \tag{2.6}$$

Upon combining Lemma 4.5.4 with Corollary 17.2.8 [and recalling results (1.4) and (1.5)], we obtain as a second corollary (of Lemma 17.2.7) the following result.

Corollary 17.2.9. For any $m \times n$ matrix $\mathbf{A}$ and $m \times p$ matrix $\mathbf{B}$,

$$\mathcal{C}(\mathbf{A}, \mathbf{B}) = \mathcal{C}[\mathbf{A}, (\mathbf{I} - \mathbf{A}\mathbf{A}^-)\mathbf{B}] = \mathcal{C}(\mathbf{A}) \oplus \mathcal{C}[(\mathbf{I} - \mathbf{A}\mathbf{A}^-)\mathbf{B}], \tag{2.7}$$

$$\text{rank}(\mathbf{A}, \mathbf{B}) = \text{rank}[\mathbf{A}, (\mathbf{I} - \mathbf{A}\mathbf{A}^-)\mathbf{B}] = \text{rank}(\mathbf{A}) + \text{rank}[(\mathbf{I} - \mathbf{A}\mathbf{A}^-)\mathbf{B}]. \tag{2.8}$$

Similarly, for any $m \times n$ matrix $\mathbf{A}$ and $q \times n$ matrix $\mathbf{C}$,

$$\mathcal{R}\begin{pmatrix} \mathbf{A} \\ \mathbf{C} \end{pmatrix} = \mathcal{R}\begin{pmatrix} \mathbf{A} \\ \mathbf{C}(\mathbf{I} - \mathbf{A}^-\mathbf{A}) \end{pmatrix} = \mathcal{R}(\mathbf{A}) \oplus \mathcal{R}[\mathbf{C}(\mathbf{I} - \mathbf{A}^-\mathbf{A})], \tag{2.9}$$

$$\text{rank}\begin{pmatrix} \mathbf{A} \\ \mathbf{C} \end{pmatrix} = \text{rank}\begin{pmatrix} \mathbf{A} \\ \mathbf{C}(\mathbf{I} - \mathbf{A}^-\mathbf{A}) \end{pmatrix} = \text{rank}(\mathbf{A}) + \text{rank}[\mathbf{C}(\mathbf{I} - \mathbf{A}^-\mathbf{A})]. \tag{2.10}$$

Three additional corollaries (of Lemma 17.2.7) are as follows.

Corollary 17.2.10. Let $\mathbf{A}$ represent an $m \times n$ matrix, $\mathbf{B}$ an $m \times p$ matrix, and $\mathbf{C}$ a $q \times n$ matrix. Then, $\text{rank}[(\mathbf{I} - \mathbf{A}\mathbf{A}^-)\mathbf{B}] = \text{rank}(\mathbf{B})$ if and only if $\mathcal{C}(\mathbf{A})$ and $\mathcal{C}(\mathbf{B})$ are essentially disjoint, and, similarly, $\text{rank}[\mathbf{C}(\mathbf{I} - \mathbf{A}^-\mathbf{A})] = \text{rank}(\mathbf{C})$ if and only if $\mathcal{R}(\mathbf{A})$ and $\mathcal{R}(\mathbf{C})$ are essentially disjoint.

Proof. According to Corollary 17.2.9,

$$\text{rank}[(\mathbf{I} - \mathbf{A}\mathbf{A}^-)\mathbf{B}] = \text{rank}(\mathbf{A}, \mathbf{B}) - \text{rank}(\mathbf{A}).$$

Moreover, according to Theorem 17.2.4, $\text{rank}(\mathbf{A}, \mathbf{B}) = \text{rank}(\mathbf{A}) + \text{rank}(\mathbf{B})$, or, equivalently, $\text{rank}(\mathbf{A}, \mathbf{B}) - \text{rank}(\mathbf{A}) = \text{rank}(\mathbf{B})$, if and only if $\mathcal{C}(\mathbf{A})$ and $\mathcal{C}(\mathbf{B})$ are essentially disjoint. We conclude that $\text{rank}[(\mathbf{I} - \mathbf{A}^-\mathbf{A})\mathbf{B}] = \text{rank}(\mathbf{B})$ if and only if $\mathcal{C}(\mathbf{A})$ and $\mathcal{C}(\mathbf{B})$ are essentially disjoint. That $\text{rank}[\mathbf{C}(\mathbf{I} - \mathbf{A}^-\mathbf{A})] = \text{rank}(\mathbf{C})$ if and only if $\mathcal{R}(\mathbf{A})$ and $\mathcal{R}(\mathbf{C})$ are essentially disjoint follows from a similar argument. Q.E.D.

Corollary 17.2.11. Let $\mathbf{A}$ represent an $m \times n$ matrix, $\mathbf{B}$ an $m \times p$ matrix, and $\mathbf{C}$ a $q \times n$ matrix. Then, $\mathcal{R}[(\mathbf{I} - \mathbf{A}\mathbf{A}^-)\mathbf{B}] = \mathcal{R}(\mathbf{B})$ if and only if $\mathcal{C}(\mathbf{A})$ and $\mathcal{C}(\mathbf{B})$ are essentially disjoint, and, similarly, $\mathcal{C}[\mathbf{C}(\mathbf{I} - \mathbf{A}^-\mathbf{A})] = \mathcal{C}(\mathbf{C})$ if and only if $\mathcal{R}(\mathbf{A})$ and $\mathcal{R}(\mathbf{C})$ are essentially disjoint.

Proof. Corollary 17.2.11 follows from Corollary 17.2.10 upon observing (in light of Corollary 4.4.7) that $\mathcal{R}[(\mathbf{I} - \mathbf{A}\mathbf{A}^-)\mathbf{B}] = \mathcal{R}(\mathbf{B})$ if and only if $\text{rank}[(\mathbf{I} - \mathbf{A}\mathbf{A}^-)\mathbf{B}] = \text{rank}(\mathbf{B})$ and that $\mathcal{C}[\mathbf{C}(\mathbf{I} - \mathbf{A}^-\mathbf{A})] = \mathcal{C}(\mathbf{C})$ if and only if $\text{rank}[\mathbf{C}(\mathbf{I} - \mathbf{A}^-\mathbf{A})] = \text{rank}(\mathbf{C})$. Q.E.D.

Corollary 17.2.12. Let $\mathbf{A}$ represent an $m \times n$ matrix, $\mathbf{B}$ an $m \times p$ matrix, and $\mathbf{C}$ a $q \times n$ matrix. If $\mathbf{A}\mathbf{A}^-\mathbf{B} = \mathbf{0}$ or $\mathbf{B}\mathbf{B}^-\mathbf{A} = \mathbf{0}$, then $\mathcal{C}(\mathbf{A})$ and $\mathcal{C}(\mathbf{B})$ are essentially disjoint. Similarly, if $\mathbf{C}\mathbf{A}^-\mathbf{A} = \mathbf{0}$ or $\mathbf{A}\mathbf{C}^-\mathbf{C} = \mathbf{0}$, then $\mathcal{R}(\mathbf{A})$ and $\mathcal{R}(\mathbf{C})$ are essentially disjoint.

Proof. If $\mathbf{A}\mathbf{A}^-\mathbf{B} = \mathbf{0}$ or $\mathbf{B}\mathbf{B}^-\mathbf{A} = \mathbf{0}$, then clearly $\text{rank}[(\mathbf{I} - \mathbf{A}\mathbf{A}^-)\mathbf{B}] = \text{rank}(\mathbf{B})$ or $\text{rank}[(\mathbf{I} - \mathbf{B}\mathbf{B}^-)\mathbf{A}] = \text{rank}(\mathbf{A})$, and it follows from Corollary 17.2.10

that $\mathcal{C}(\mathbf{A})$ and $\mathcal{C}(\mathbf{B})$ are essentially disjoint. Similarly, if $\mathbf{CA}^-\mathbf{A} = \mathbf{0}$ or $\mathbf{AC}^-\mathbf{C} = \mathbf{0}$, then clearly $\text{rank}[\mathbf{C}(\mathbf{I} - \mathbf{A}^-\mathbf{A})] = \text{rank}(\mathbf{C})$ or $\text{rank}[\mathbf{A}(\mathbf{I} - \mathbf{C}^-\mathbf{C})] = \text{rank}(\mathbf{A})$, and it follows from Corollary 17.2.10 that $\mathcal{R}(\mathbf{A})$ and $\mathcal{R}(\mathbf{C})$ are essentially disjoint. Q.E.D.

In connection with Lemma 17.2.7 (and its corollaries), recall [from Part (1) of Theorem 12.3.4] that one choice for $\mathbf{A}^-$ is $(\mathbf{A}'\mathbf{A})^-\mathbf{A}'$, and observe that, for $\mathbf{A}^- = (\mathbf{A}'\mathbf{A})^-\mathbf{A}'$, $\mathbf{AA}^- = \mathbf{P}_\mathbf{A}$. Similarly, recall [from Part (5) of Theorem 12.3.4] that another choice for $\mathbf{A}^-$ is $\mathbf{A}'(\mathbf{AA}')^-$, and observe that, for $\mathbf{A}^- = \mathbf{A}'(\mathbf{AA}')^-$, $\mathbf{A}^-\mathbf{A} = \mathbf{P}_{\mathbf{A}'}$.

As a consequence of Corollary 17.2.9, we have that, for any $m \times n$ matrix $\mathbf{A}$ and $m \times p$ matrix $\mathbf{B}$, there exists a matrix $\mathbf{K}$ (having m rows) such that $\mathcal{C}(\mathbf{A})$ and $\mathcal{C}(\mathbf{K})$ are essentially disjoint and $\mathcal{C}(\mathbf{A}, \mathbf{B}) = \mathcal{C}(\mathbf{A}, \mathbf{K})$ or, equivalently, such that $\mathcal{C}(\mathbf{A}, \mathbf{B})$ equals the direct sum of $\mathcal{C}(\mathbf{A})$ and $\mathcal{C}(\mathbf{K})$. One choice for $\mathbf{K}$ (that given by Corollary 17.2.9) is $\mathbf{K} = (\mathbf{I} - \mathbf{AA}^-)\mathbf{B}$. The following theorem gives, for the special case where $\mathbf{B}$ is an $(m \times m)$ symmetric nonnegative definite matrix, an alternative choice for $\mathbf{K}$.

Theorem 17.2.13. For any $m \times n$ matrix $\mathbf{A}$ and any $m \times m$ symmetric nonnegative definite matrix $\mathbf{V}$, $\mathcal{C}(\mathbf{A})$ and $\mathcal{C}\{\mathbf{V}[\mathbf{I} - (\mathbf{A}')^-\mathbf{A}']\}$ are essentially disjoint and $\mathcal{C}(\mathbf{A}, \mathbf{V}) = \mathcal{C}\{\mathbf{A}, \mathbf{V}[\mathbf{I} - (\mathbf{A}')^-\mathbf{A}']\}$, or, equivalently,

$$\mathcal{C}(\mathbf{A}, \mathbf{V}) = \mathcal{C}(\mathbf{A}) \oplus \mathcal{C}\{\mathbf{V}[\mathbf{I} - (\mathbf{A}')^-\mathbf{A}']\}. \tag{2.11}$$

Proof. Let $\mathbf{Z} = \mathbf{I} - (\mathbf{A}')^-\mathbf{A}'$, and observe that $\mathbf{A}'\mathbf{Z} = \mathbf{0}$ and hence that $\mathbf{Z}'\mathbf{A} = (\mathbf{A}'\mathbf{Z})' = \mathbf{0}$.

Let $\mathbf{x}$ represent an arbitrary $(m \times 1)$ vector in $\mathcal{C}(\mathbf{A}) \cap \mathcal{C}(\mathbf{VZ})$. Then, $\mathbf{x} = \mathbf{Ar}$ for some vector $\mathbf{r}$, and $\mathbf{x} = \mathbf{VZs}$ for some vector $\mathbf{s}$. Moreover,

$$\mathbf{Z}'\mathbf{VZs} = \mathbf{Z}'\mathbf{x} = \mathbf{Z}'\mathbf{Ar} = \mathbf{0},$$

so that

$$(\mathbf{Zs})'\mathbf{VZs} = \mathbf{s}'\mathbf{Z}'\mathbf{VZs} = 0,$$

implying (in light of Corollary 14.3.11) that $\mathbf{VZs} = \mathbf{0}$ or equivalently that $\mathbf{x} = \mathbf{0}$. Thus, $\mathcal{C}(\mathbf{A})$ and $\mathcal{C}(\mathbf{VZ})$ are essentially disjoint.

Now, it is clear from Lemma 4.5.5 that

$$\mathcal{C}(\mathbf{A}, \mathbf{VZ}) \subset \mathcal{C}(\mathbf{A}, \mathbf{V}).$$

And, making use of Theorem 17.2.4 and result (2.8) and observing that

$$\text{rank}(\mathbf{VZ}) = \text{rank}[(\mathbf{VZ})'] = \text{rank}(\{\mathbf{I} - \mathbf{A}[(\mathbf{A}')^-]'\}\mathbf{V})$$

and that $[(\mathbf{A}')^-]'$ is a generalized inverse of $\mathbf{A}$, we find that

$$\begin{aligned}\text{rank}(\mathbf{A}, \mathbf{VZ}) &= \text{rank}(\mathbf{A}) + \text{rank}(\mathbf{VZ}) \\ &= \text{rank}(\mathbf{A}) + \text{rank}(\{\mathbf{I} - \mathbf{A}[(\mathbf{A}')^-]'\}\mathbf{V}) \\ &= \text{rank}(\mathbf{A}, \mathbf{V}).\end{aligned}$$

We conclude, on the basis of Theorem 4.4.6, that $\mathcal{C}(\mathbf{A}, \mathbf{VZ}) = \mathcal{C}(\mathbf{A}, \mathbf{V})$. Q.E.D.

In connection with Theorem 17.2.13, recall (from Corollary 11.2.2) that $\mathcal{C}[\mathbf{I} - (\mathbf{A}')^-\mathbf{A}'] = \mathcal{N}(\mathbf{A}')$, and observe (in light of Corollaries 4.2.4 and 4.5.6) that, for any matrix $\mathbf{Z}$ such that $\mathcal{C}(\mathbf{Z}) = \mathcal{N}(\mathbf{A}')$, $\mathcal{C}\{\mathbf{V}[\mathbf{I} - (\mathbf{A}')^-\mathbf{A}']\} = \mathcal{C}(\mathbf{VZ})$ and $\mathcal{C}\{\mathbf{A}, \mathbf{V}[\mathbf{I} - (\mathbf{A}')^-\mathbf{A}']\} = \mathcal{C}(\mathbf{A}, \mathbf{VZ})$. Thus, we have the following corollary.

Corollary 17.2.14. Let $\mathbf{A}$ represent an $m \times n$ matrix, and let $\mathbf{V}$ represent an $m \times m$ symmetric nonnegative definite matrix. Then, for any matrix $\mathbf{Z}$ such that $\mathcal{C}(\mathbf{Z}) = \mathcal{N}(\mathbf{A}')$ (i.e., any matrix $\mathbf{Z}$ whose columns span the null space of $\mathbf{A}'$), $\mathcal{C}(\mathbf{A})$ and $\mathcal{C}(\mathbf{VZ})$ are essentially disjoint and $\mathcal{C}(\mathbf{A}, \mathbf{V}) = \mathcal{C}(\mathbf{A}, \mathbf{VZ})$, or, equivalently,

$$\mathcal{C}(\mathbf{A}, \mathbf{V}) = \mathcal{C}(\mathbf{A}) \oplus \mathcal{C}(\mathbf{VZ}). \tag{2.12}$$

The following lemma gives a basic result on the essential disjointness of the row or column spaces of partitioned matrices.

Lemma 17.2.15. (1) Let $\mathbf{A} = (\mathbf{A}_1, \mathbf{A}_2)$ represent an $m \times n$ partitioned matrix and $\mathbf{B} = (\mathbf{B}_1, \mathbf{B}_2)$ a $p \times n$ partitioned matrix, where $\mathbf{B}_1$ has the same number of columns, say n_1, as $\mathbf{A}_1$. If $\mathcal{R}(\mathbf{A}_1)$ and $\mathcal{R}(\mathbf{B}_1)$ are essentially disjoint and if $\mathcal{R}(\mathbf{A}_2)$ and $\mathcal{R}(\mathbf{B}_2)$ are essentially disjoint, then $\mathcal{R}(\mathbf{A})$ and $\mathcal{R}(\mathbf{B})$ are essentially disjoint. (2) Similarly, let $\mathbf{A} = \begin{pmatrix} \mathbf{A}_1 \\ \mathbf{A}_2 \end{pmatrix}$ represent an $m \times n$ partitioned matrix, and $\mathbf{B} = \begin{pmatrix} \mathbf{B}_1 \\ \mathbf{B}_2 \end{pmatrix}$ an $m \times p$ partitioned matrix, where $\mathbf{B}_1$ has the same number of rows, say m_1, as $\mathbf{A}_1$. If $\mathcal{C}(\mathbf{A}_1)$ and $\mathcal{C}(\mathbf{B}_1)$ are essentially disjoint and if $\mathcal{C}(\mathbf{A}_2)$ and $\mathcal{C}(\mathbf{B}_2)$ are essentially disjoint, then $\mathcal{C}(\mathbf{A})$ and $\mathcal{C}(\mathbf{B})$ are essentially disjoint.

Proof. (1) Suppose that $\mathcal{R}(\mathbf{A}_1)$ and $\mathcal{R}(\mathbf{B}_1)$ are essentially disjoint and that $\mathcal{R}(\mathbf{A}_2)$ and $\mathcal{R}(\mathbf{B}_2)$ are essentially disjoint. Let $\mathbf{x}'$ represent an arbitrary $(1 \times n)$ vector in $\mathcal{R}(\mathbf{A}) \cap \mathcal{R}(\mathbf{B})$. Then, $\mathbf{x}' = \mathbf{r}'\mathbf{A}$ and $\mathbf{x}' = \mathbf{s}'\mathbf{B}$ for some (row) vectors $\mathbf{r}'$ and $\mathbf{s}'$. Now, partitioning $\mathbf{x}'$ as $\mathbf{x}' = (\mathbf{x}_1', \mathbf{x}_2')$ (where $\mathbf{x}_1'$ is of dimensions $1 \times n_1$), we find that

$$(\mathbf{x}_1', \mathbf{x}_2') = \mathbf{r}'(\mathbf{A}_1, \mathbf{A}_2) = (\mathbf{r}'\mathbf{A}_1, \mathbf{r}'\mathbf{A}_2)$$

and, similarly, that

$$(\mathbf{x}_1', \mathbf{x}_2') = \mathbf{s}'(\mathbf{B}_1, \mathbf{B}_2) = (\mathbf{s}'\mathbf{B}_1, \mathbf{s}'\mathbf{B}_2).$$

Thus, $\mathbf{x}_1' = \mathbf{r}'\mathbf{A}_1$ and $\mathbf{x}_1' = \mathbf{s}'\mathbf{B}_1$, implying (in light of Lemma 17.2.2) that $\mathbf{x}_1' = \mathbf{0}$; and similarly $\mathbf{x}_2' = \mathbf{r}'\mathbf{A}_2$ and $\mathbf{x}_2' = \mathbf{s}'\mathbf{B}_2$, implying that $\mathbf{x}_2' = \mathbf{0}$. It follows that $\mathbf{x}' = (\mathbf{0}, \mathbf{0}) = \mathbf{0}$. We conclude that $\mathcal{R}(\mathbf{A})$ and $\mathcal{R}(\mathbf{B})$ are essentially disjoint.

(2) The proof of Part (2) is analogous to that of Part (1). Q.E.D.

In the special case where one of the submatrices $\mathbf{A}_1$, $\mathbf{A}_2$, $\mathbf{B}_1$, and $\mathbf{B}_2$ is a null matrix, Lemma 17.2.15 can be restated as the following corollary.

Corollary 17.2.16. Let $\mathbf{T}$ represent an $m \times p$ matrix, $\mathbf{U}$ an $m \times q$ matrix, $\mathbf{V}$ an $n \times p$ matrix, and $\mathbf{W}$ an $n \times q$ matrix. (1) If $\mathcal{R}(\mathbf{T})$ and $\mathcal{R}(\mathbf{V})$ are essentially disjoint, then $\mathcal{R}(\mathbf{T}, \mathbf{U})$ and $\mathcal{R}(\mathbf{V}, \mathbf{0})$ are essentially disjoint. Similarly, if $\mathcal{R}(\mathbf{U})$ and $\mathcal{R}(\mathbf{W})$ are essentially disjoint, then $\mathcal{R}(\mathbf{T}, \mathbf{U})$ and $\mathcal{R}(\mathbf{0}, \mathbf{W})$ are essentially disjoint. (2) If $\mathcal{C}(\mathbf{T})$ and $\mathcal{C}(\mathbf{U})$ are essentially disjoint, then $\mathcal{C}\begin{pmatrix}\mathbf{T}\\\mathbf{V}\end{pmatrix}$ and $\mathcal{C}\begin{pmatrix}\mathbf{U}\\\mathbf{0}\end{pmatrix}$ are essentially disjoint. Similarly, if $\mathcal{C}(\mathbf{V})$ and $\mathcal{C}(\mathbf{W})$ are essentially disjoint, then $\mathcal{C}\begin{pmatrix}\mathbf{T}\\\mathbf{V}\end{pmatrix}$ and $\mathcal{C}\begin{pmatrix}\mathbf{0}\\\mathbf{W}\end{pmatrix}$ are essentially disjoint.

Let $\mathbf{A} = (\mathbf{A}_1, \mathbf{A}_2)$ represent an $m \times n$ partitioned matrix, and $\mathbf{B} = (\mathbf{B}_1, \mathbf{B}_2)$ a $p \times n$ partitioned matrix (where $\mathbf{B}_1$ has the same number of columns as $\mathbf{A}_1$). For $\mathcal{R}(\mathbf{A})$ and $\mathcal{R}(\mathbf{B})$ to be essentially disjoint, it is (according to Lemma 17.2.15) sufficient that $\mathcal{R}(\mathbf{A}_1)$ and $\mathcal{R}(\mathbf{B}_1)$ be essentially disjoint and that $\mathcal{R}(\mathbf{A}_2)$ and $\mathcal{R}(\mathbf{B}_2)$ be essentially disjoint. Are these conditions necessary as well as sufficient? The answer is no, as is easily verified.

Results (2.10) and (2.8) are extended in the following theorem.

Theorem 17.2.17. Let $\mathbf{T}$ represent an $m \times p$ matrix, $\mathbf{U}$ an $m \times q$ matrix, and $\mathbf{V}$ an $n \times p$ matrix. Then,

$$\begin{aligned}\operatorname{rank}\begin{pmatrix}\mathbf{T} & \mathbf{U}\\ \mathbf{V} & \mathbf{0}\end{pmatrix} &= \operatorname{rank}\begin{pmatrix}\mathbf{0} & \mathbf{V}\\ \mathbf{U} & \mathbf{T}\end{pmatrix}\\ &= \operatorname{rank}(\mathbf{V}) + \operatorname{rank}[\mathbf{U},\ \mathbf{T}(\mathbf{I} - \mathbf{V}^-\mathbf{V})] \qquad (2.13)\\ &= \operatorname{rank}(\mathbf{U}) + \operatorname{rank}\begin{pmatrix}\mathbf{V}\\ (\mathbf{I} - \mathbf{U}\mathbf{U}^-)\mathbf{T}\end{pmatrix} \qquad (2.14)\\ &= \operatorname{rank}(\mathbf{U}) + \operatorname{rank}(\mathbf{V}) + \operatorname{rank}[(\mathbf{I} - \mathbf{U}\mathbf{U}^-)\mathbf{T}(\mathbf{I} - \mathbf{V}^-\mathbf{V})]. \qquad (2.15)\end{aligned}$$

Proof. Clearly, $\begin{pmatrix}\mathbf{0}\\ \mathbf{V}^-\end{pmatrix}$ is a generalized inverse of $(\mathbf{0},\ \mathbf{V})$, and

$$\mathbf{I} - \begin{pmatrix}\mathbf{0}\\ \mathbf{V}^-\end{pmatrix}(\mathbf{0},\ \mathbf{V}) = \begin{pmatrix}\mathbf{I} & \mathbf{0}\\ \mathbf{0} & \mathbf{I} - \mathbf{V}^-\mathbf{V}\end{pmatrix}.$$

Thus, making use of result (2.10), we find that

$$\begin{aligned}\operatorname{rank}\begin{pmatrix}\mathbf{0} & \mathbf{V}\\ \mathbf{U} & \mathbf{T}\end{pmatrix} &= \operatorname{rank}(\mathbf{0},\ \mathbf{V}) + \operatorname{rank}\left\{(\mathbf{U},\ \mathbf{T})\left[\mathbf{I} - \begin{pmatrix}\mathbf{0}\\ \mathbf{V}^-\end{pmatrix}(\mathbf{0},\ \mathbf{V})\right]\right\}\\ &= \operatorname{rank}(\mathbf{V}) + \operatorname{rank}[\mathbf{U},\ \mathbf{T}(\mathbf{I} - \mathbf{V}^-\mathbf{V})].\end{aligned}$$

It can be established in similar fashion that $\operatorname{rank}\begin{pmatrix}\mathbf{0} & \mathbf{V}\\ \mathbf{U} & \mathbf{T}\end{pmatrix}$ equals expression (2.14).

That $\operatorname{rank}\begin{pmatrix}\mathbf{T} & \mathbf{U}\\ \mathbf{V} & \mathbf{0}\end{pmatrix} = \operatorname{rank}\begin{pmatrix}\mathbf{0} & \mathbf{V}\\ \mathbf{U} & \mathbf{T}\end{pmatrix}$ is evident from Lemma 8.5.1. And it follows from result (2.8) that

$$\operatorname{rank}[\mathbf{U},\ \mathbf{T}(\mathbf{I} - \mathbf{V}^-\mathbf{V})] = \operatorname{rank}(\mathbf{U}) + \operatorname{rank}[(\mathbf{I} - \mathbf{U}\mathbf{U}^-)\mathbf{T}(\mathbf{I} - \mathbf{V}^-\mathbf{V})].$$

Substitution of this expression for $\operatorname{rank}[\mathbf{U},\ \mathbf{T}(\mathbf{I} - \mathbf{V}^-\mathbf{V})]$ into expression (2.13) gives expression (2.15). Q.E.D.

The following theorem can be regarded as a generalization of Corollary 17.2.10.

Theorem 17.2.18. Let $\mathbf{T}$ represent an $m \times p$ matrix, $\mathbf{U}$ an $m \times q$ matrix, and $\mathbf{V}$ an $n \times p$ matrix. Then,

$$\text{rank}[(\mathbf{I} - \mathbf{U}\mathbf{U}^-)\mathbf{T}(\mathbf{I} - \mathbf{V}^-\mathbf{V})] \leq \text{rank}(\mathbf{T}), \tag{2.16}$$

with equality holding if and only if $\mathcal{C}(\mathbf{T})$ and $\mathcal{C}(\mathbf{U})$ are essentially disjoint and $\mathcal{R}(\mathbf{T})$ and $\mathcal{R}(\mathbf{V})$ are essentially disjoint.

Proof. It follows from Corollary 4.4.5 that

$$\text{rank}[(\mathbf{I} - \mathbf{U}\mathbf{U}^-)\mathbf{T}(\mathbf{I} - \mathbf{V}^-\mathbf{V})] \leq \text{rank}[\mathbf{T}(\mathbf{I} - \mathbf{V}^-\mathbf{V})] \leq \text{rank}(\mathbf{T}).$$

Further, in light of Corollary 17.2.10,

$$\begin{aligned}
&\text{rank}[(\mathbf{I} - \mathbf{U}\mathbf{U}^-)\mathbf{T}(\mathbf{I} - \mathbf{V}^-\mathbf{V})] = \text{rank}(\mathbf{T}) \\
&\qquad \Leftrightarrow \text{rank}[(\mathbf{I} - \mathbf{U}\mathbf{U}^-)\mathbf{T}(\mathbf{I} - \mathbf{V}^-\mathbf{V})] = \text{rank}[\mathbf{T}(\mathbf{I} - \mathbf{V}^-\mathbf{V})] = \text{rank}(\mathbf{T}) \\
&\qquad \Leftrightarrow \mathcal{C}(\mathbf{U}) \text{ and } \mathcal{C}[\mathbf{T}(\mathbf{I} - \mathbf{V}^-\mathbf{V})] \text{ are essentially disjoint and} \\
&\qquad\quad\ \mathcal{R}(\mathbf{V}) \text{ and } \mathcal{R}(\mathbf{T}) \text{ are essentially disjoint.}
\end{aligned}$$

And, according to Corollary 17.2.11,

$$\mathcal{R}(\mathbf{V}) \text{ and } \mathcal{R}(\mathbf{T}) \text{ are essentially disjoint } \Leftrightarrow \mathcal{C}[\mathbf{T}(\mathbf{I} - \mathbf{V}^-\mathbf{V})] = \mathcal{C}(\mathbf{T}),$$

and consequently

$$\begin{aligned}
&\mathcal{C}(\mathbf{U}) \text{ and } \mathcal{C}[\mathbf{T}(\mathbf{I} - \mathbf{V}^-\mathbf{V})] \text{ are essentially disjoint and} \\
&\mathcal{R}(\mathbf{V}) \text{ and } \mathcal{R}(\mathbf{T}) \text{ are essentially disjoint} \\
&\qquad\qquad \Leftrightarrow \mathcal{C}(\mathbf{U}) \text{ and } \mathcal{C}(\mathbf{T}) \text{ are essentially disjoint and} \\
&\qquad\qquad\quad\ \mathcal{R}(\mathbf{V}) \text{ and } \mathcal{R}(\mathbf{T}) \text{ are essentially disjoint.}
\end{aligned}$$

Q.E.D.

Upon combining Theorem 17.2.18 with result (2.15), we obtain the following theorem, which extends the results of Theorem 17.2.4.

Theorem 17.2.19. For any $m \times p$ matrix $\mathbf{T}$, $m \times q$ matrix $\mathbf{U}$, and $n \times p$ matrix $\mathbf{V}$,

$$\text{rank}\begin{pmatrix} \mathbf{T} & \mathbf{U} \\ \mathbf{V} & \mathbf{0} \end{pmatrix} \leq \text{rank}(\mathbf{T}) + \text{rank}(\mathbf{U}) + \text{rank}(\mathbf{V}), \tag{2.17}$$

with equality holding if and only if $\mathcal{R}(\mathbf{T})$ and $\mathcal{R}(\mathbf{V})$ are essentially disjoint and $\mathcal{C}(\mathbf{T})$ and $\mathcal{C}(\mathbf{U})$ are essentially disjoint.

17.3 Some Results on Linear Systems and on Generalized Inverses of Partitioned Matrices

The following theorem gives a basic result on a linear system whose left and right sides are the sums of the left and right sides of two other linear systems.

Theorem 17.3.1. Let $\mathbf{A}_1$ represent an $m \times n_1$ matrix, $\mathbf{A}_2$ an $m \times n_2$ matrix, and $\mathbf{B}_1$ and $\mathbf{B}_2$ $m \times p$ matrices. Further, let $n = n_1 + n_2$, $\mathbf{A} = (\mathbf{A}_1, \mathbf{A}_2)$, and $\mathbf{B} = \mathbf{B}_1 + \mathbf{B}_2$. Suppose that $\mathcal{C}(\mathbf{A}_1)$ and $\mathcal{C}(\mathbf{A}_2)$ are essentially disjoint and that $\mathcal{C}(\mathbf{B}_1) \subset \mathcal{C}(\mathbf{A}_1)$ and $\mathcal{C}(\mathbf{B}_2) \subset \mathcal{C}(\mathbf{A}_2)$. Then, for any solution $\mathbf{X}^* = \begin{pmatrix} \mathbf{X}_1^* \\ \mathbf{X}_2^* \end{pmatrix}$ to the linear system $\mathbf{AX} = \mathbf{B}$ (in an $n \times p$ matrix $\mathbf{X}$), or, equivalently, to the linear system

$$\mathbf{A}_1\mathbf{X}_1 + \mathbf{A}_2\mathbf{X}_2 = \mathbf{B}_1 + \mathbf{B}_2$$

(in $n_1 \times p$ and $n_2 \times p$ matrices $\mathbf{X}_1$ and $\mathbf{X}_2$), $\mathbf{X}_1^*$ is a solution to the linear system $\mathbf{A}_1\mathbf{X}_1 = \mathbf{B}_1$, and $\mathbf{X}_2^*$ is a solution to the linear system $\mathbf{A}_2\mathbf{X}_2 = \mathbf{B}_2$.

Proof. According to Lemma 4.2.2, $\mathbf{B}_1 = \mathbf{A}_1\mathbf{F}_1$ and $\mathbf{B}_2 = \mathbf{A}_2\mathbf{F}_2$ for some matrices $\mathbf{F}_1$ and $\mathbf{F}_2$. Thus, we have that

$$\mathbf{A}_1\mathbf{X}_1^* + \mathbf{A}_2\mathbf{X}_2^* = \mathbf{B}_1 + \mathbf{B}_2 = \mathbf{A}_1\mathbf{F}_1 + \mathbf{A}_2\mathbf{F}_2,$$

implying that

$$\mathbf{A}_1(\mathbf{X}_1^* - \mathbf{F}_1) = \mathbf{A}_2(\mathbf{F}_2 - \mathbf{X}_2^*).$$

Since $\mathcal{C}(\mathbf{A}_1)$ and $\mathcal{C}(\mathbf{A}_2)$ are essentially disjoint, it follows (in light of Corollary 17.2.3) that $\mathbf{A}_1(\mathbf{X}_1^* - \mathbf{F}_1) = \mathbf{0}$ and $\mathbf{A}_2(\mathbf{F}_2 - \mathbf{X}_2^*) = \mathbf{0}$ and hence that $\mathbf{A}_1\mathbf{X}_1^* = \mathbf{A}_1\mathbf{F}_1 = \mathbf{B}_1$ and $\mathbf{A}_2\mathbf{X}_2^* = \mathbf{A}_2\mathbf{F}_2 = \mathbf{B}_2$. We conclude that $\mathbf{X}_1^*$ is a solution to $\mathbf{A}_1\mathbf{X}_1 = \mathbf{B}_1$ and that $\mathbf{X}_2^*$ is a solution to $\mathbf{A}_2\mathbf{X}_2 = \mathbf{B}_2$. Q.E.D.

In connection with Theorem 17.3.1, note (in light of Lemmas 4.5.8 and 4.5.5) that the supposition that $\mathcal{C}(\mathbf{B}_1) \subset \mathcal{C}(\mathbf{A}_1)$ and $\mathcal{C}(\mathbf{B}_2) \subset \mathcal{C}(\mathbf{A}_2)$ implies that $\mathcal{C}(\mathbf{B}) \subset \mathcal{C}(\mathbf{A})$ and hence that the linear system $\mathbf{AX} = \mathbf{B}$ is consistent.

Suppose that a linear system $\mathbf{A}_1\mathbf{X} = \mathbf{B}_1$ (in a matrix $\mathbf{X}$ of unknowns) is consistent and that a second linear system $\mathbf{A}_2\mathbf{X} = \mathbf{B}_2$ (in $\mathbf{X}$) is also consistent. Under what circumstances is the combined linear system $\begin{pmatrix} \mathbf{A}_1 \\ \mathbf{A}_2 \end{pmatrix}\mathbf{X} = \begin{pmatrix} \mathbf{B}_1 \\ \mathbf{B}_2 \end{pmatrix}$ consistent, or, equivalently, under what circumstances does $\mathbf{A}_2\mathbf{X} = \mathbf{B}_2$ have solutions in common with $\mathbf{A}_1\mathbf{X} = \mathbf{B}_1$? This question is addressed in the following theorem.

Theorem 17.3.2. Let $\mathbf{A}_1$ represent an $m_1 \times n$ matrix, $\mathbf{A}_2$ an $m_2 \times n$ matrix, $\mathbf{B}_1$ an $m_1 \times p$ matrix, and $\mathbf{B}_2$ an $m_2 \times p$ matrix. And suppose that the linear systems $\mathbf{A}_1\mathbf{X} = \mathbf{B}_1$ and $\mathbf{A}_2\mathbf{X} = \mathbf{B}_2$ (in $\mathbf{X}$) are both consistent [or, equivalently, that $\mathcal{C}(\mathbf{B}_1) \subset \mathcal{C}(\mathbf{A}_1)$ and $\mathcal{C}(\mathbf{B}_2) \subset \mathcal{C}(\mathbf{A}_2)$]. If $\mathcal{R}(\mathbf{A}_1)$ and $\mathcal{R}(\mathbf{A}_2)$ are essentially disjoint, then the combined linear system $\begin{pmatrix} \mathbf{A}_1 \\ \mathbf{A}_2 \end{pmatrix}\mathbf{X} = \begin{pmatrix} \mathbf{B}_1 \\ \mathbf{B}_2 \end{pmatrix}$ (in $\mathbf{X}$) is consistent.

Proof. Suppose that $\mathcal{R}(\mathbf{A}_1)$ and $\mathcal{R}(\mathbf{A}_2)$ are essentially disjoint. Let $\mathbf{k}$ represent any $(m_1 + m_2) \times 1$ vector such that $\mathbf{k}'\begin{pmatrix} \mathbf{A}_1 \\ \mathbf{A}_2 \end{pmatrix} = \mathbf{0}'$. Then, partitioning $\mathbf{k}$ as $\mathbf{k} = \begin{pmatrix} \mathbf{k}_1 \\ \mathbf{k}_2 \end{pmatrix}$ (where $\mathbf{k}_1$ has m_1 elements), $\mathbf{k}_1'\mathbf{A}_1 + \mathbf{k}_2'\mathbf{A}_2 = \mathbf{0}'$, or, equivalently, $\mathbf{k}_1'\mathbf{A}_1 = -\mathbf{k}_2'\mathbf{A}_2$. Thus, $\mathbf{k}_1'\mathbf{A}_1 = \mathbf{0}'$, and $\mathbf{k}_2'\mathbf{A}_2 = \mathbf{0}'$.

Moreover, each of the linear systems $\mathbf{A}_1\mathbf{X} = \mathbf{B}_1$ and $\mathbf{A}_2\mathbf{X} = \mathbf{B}_2$ is (in light of Theorem 7.3.1) compatible, so that $\mathbf{k}_1'\mathbf{B}_1 = \mathbf{0}'$ and $\mathbf{k}_2'\mathbf{B}_2 = \mathbf{0}'$ and hence $\mathbf{k}'\begin{pmatrix}\mathbf{B}_1\\ \mathbf{B}_2\end{pmatrix} = \mathbf{0}'$. We conclude that the combined linear system $\begin{pmatrix}\mathbf{A}_1\\ \mathbf{A}_2\end{pmatrix}\mathbf{X} = \begin{pmatrix}\mathbf{B}_1\\ \mathbf{B}_2\end{pmatrix}$ is compatible and hence (in light of Theorem 7.3.1) consistent. Q.E.D.

Under what condition(s) are the blocks $\mathbf{G}_1$ and $\mathbf{G}_2$ of a generalized inverse $\begin{pmatrix}\mathbf{G}_1\\ \mathbf{G}_2\end{pmatrix}$ of a partitioned matrix $(\mathbf{A}, \mathbf{B})$ generalized inverses of the blocks ($\mathbf{A}$ and $\mathbf{B}$) of $(\mathbf{A}, \mathbf{B})$? And, similarly, under what condition(s) are the blocks $\mathbf{H}_1$ and $\mathbf{H}_2$ of a generalized inverse $(\mathbf{H}_1, \mathbf{H}_2)$ of a partitioned matrix $\begin{pmatrix}\mathbf{A}\\ \mathbf{C}\end{pmatrix}$ generalized inverses of the blocks of $\begin{pmatrix}\mathbf{A}\\ \mathbf{C}\end{pmatrix}$? These questions are answered in the following theorem.

Theorem 17.3.3. Let $\mathbf{A}$ represent an $m \times n$ matrix, $\mathbf{B}$ an $m \times p$ matrix, and $\mathbf{C}$ a $q \times n$ matrix. (1) For any generalized inverse $\mathbf{G} = \begin{pmatrix}\mathbf{G}_1\\ \mathbf{G}_2\end{pmatrix}$ (where $\mathbf{G}_1$ has n rows) of the $m \times (n + p)$ partitioned matrix $(\mathbf{A}, \mathbf{B})$, $\mathbf{G}_1$ is a generalized inverse of $\mathbf{A}$, and $\mathbf{G}_2$ a generalized inverse of $\mathbf{B}$, if and only if $\mathcal{C}(\mathbf{A})$ and $\mathcal{C}(\mathbf{B})$ are essentially disjoint. (2) Similarly, for any generalized inverse $\mathbf{H} = (\mathbf{H}_1, \mathbf{H}_2)$ (where $\mathbf{H}_1$ has m columns) of the $(m + q) \times n$ partitioned matrix $\begin{pmatrix}\mathbf{A}\\ \mathbf{C}\end{pmatrix}$, $\mathbf{H}_1$ is a generalized inverse of $\mathbf{A}$, and $\mathbf{H}_2$ a generalized inverse of $\mathbf{C}$, if and only if $\mathcal{R}(\mathbf{A})$ and $\mathcal{R}(\mathbf{C})$ are essentially disjoint.

Proof. (1) Clearly,

$$(\mathbf{AG}_1\mathbf{A} + \mathbf{BG}_2\mathbf{A},\ \mathbf{AG}_1\mathbf{B} + \mathbf{BG}_2\mathbf{B}) = (\mathbf{A}, \mathbf{B})\mathbf{G}(\mathbf{A}, \mathbf{B}) = (\mathbf{A}, \mathbf{B}),$$

so that

$$\mathbf{BG}_2\mathbf{A} = \mathbf{A} - \mathbf{AG}_1\mathbf{A}, \qquad \mathbf{AG}_1\mathbf{B} = \mathbf{B} - \mathbf{BG}_2\mathbf{B}. \tag{3.1}$$

Now, suppose that $\mathcal{C}(\mathbf{A})$ and $\mathcal{C}(\mathbf{B})$ are essentially disjoint. Then, rewriting equalities (3.1) as

$$\mathbf{BG}_2\mathbf{A} = \mathbf{A}(\mathbf{I} - \mathbf{G}_1\mathbf{A}), \qquad \mathbf{AG}_1\mathbf{B} = \mathbf{B}(\mathbf{I} - \mathbf{G}_2\mathbf{B})$$

and applying Corollary 17.2.3, we find that

$$\mathbf{A}(\mathbf{I} - \mathbf{G}_1\mathbf{A}) = \mathbf{0}, \qquad \mathbf{B}(\mathbf{I} - \mathbf{G}_2\mathbf{B}) = \mathbf{0},$$

or, equivalently, that $\mathbf{AG}_1\mathbf{A} = \mathbf{A}$ and $\mathbf{BG}_2\mathbf{B} = \mathbf{B}$. Thus, $\mathbf{G}_1$ is a generalized inverse of $\mathbf{A}$ and $\mathbf{G}_2$ a generalized inverse of $\mathbf{B}$.

Conversely, suppose that $\mathbf{G}_1$ is a generalized inverse of $\mathbf{A}$ and $\mathbf{G}_2$ a generalized inverse of $\mathbf{B}$. Then, it follows from result (3.1) that $\mathbf{BG}_2\mathbf{A} = \mathbf{0}$, and we conclude (on the basis of Corollary 17.2.12) that $\mathcal{C}(\mathbf{A})$ and $\mathcal{C}(\mathbf{B})$ are essentially disjoint.

(2) The proof of Part (2) is analogous to that of Part (1). Q.E.D.

The following theorem gives sufficient conditions for the blocks $\mathbf{G}_{11}$, $\mathbf{G}_{21}$, $\mathbf{G}_{12}$, and $\mathbf{G}_{22}$ of a generalized inverse $\begin{pmatrix} \mathbf{G}_{11} & \mathbf{G}_{12} \\ \mathbf{G}_{21} & \mathbf{G}_{22} \end{pmatrix}$ of a partitioned matrix $\mathbf{A} = \begin{pmatrix} \mathbf{T} & \mathbf{U} \\ \mathbf{V} & \mathbf{W} \end{pmatrix}$ to be generalized inverses of the blocks ($\mathbf{T}$, $\mathbf{U}$, $\mathbf{V}$, and $\mathbf{W}$) of $\mathbf{A}$.

Theorem 17.3.4. Let $\mathbf{T}$ represent an $m \times p$ matrix, $\mathbf{U}$ an $m \times q$ matrix, $\mathbf{V}$ an $n \times p$ matrix, and $\mathbf{W}$ an $n \times q$ matrix. And let $\begin{pmatrix} \mathbf{G}_{11} & \mathbf{G}_{12} \\ \mathbf{G}_{21} & \mathbf{G}_{22} \end{pmatrix}$ (where $\mathbf{G}_{11}$ is of dimensions $p \times m$) represent a generalized inverse of the partitioned matrix $\begin{pmatrix} \mathbf{T} & \mathbf{U} \\ \mathbf{V} & \mathbf{W} \end{pmatrix}$.

(1) If $\mathcal{C}(\mathbf{T})$ and $\mathcal{C}(\mathbf{U})$ are essentially disjoint and $\mathcal{R}(\mathbf{T})$ and $\mathcal{R}(\mathbf{V})$ are essentially disjoint, then $\mathbf{G}_{11}$ is a generalized inverse of $\mathbf{T}$;

(2) if $\mathcal{C}(\mathbf{U})$ and $\mathcal{C}(\mathbf{T})$ are essentially disjoint and $\mathcal{R}(\mathbf{U})$ and $\mathcal{R}(\mathbf{W})$ are essentially disjoint, then $\mathbf{G}_{21}$ is a generalized inverse of $\mathbf{U}$;

(3) if $\mathcal{C}(\mathbf{V})$ and $\mathcal{C}(\mathbf{W})$ are essentially disjoint and $\mathcal{R}(\mathbf{V})$ and $\mathcal{R}(\mathbf{T})$ are essentially disjoint, then $\mathbf{G}_{12}$ is a generalized inverse of $\mathbf{V}$; and

(4) if $\mathcal{C}(\mathbf{W})$ and $\mathcal{C}(\mathbf{V})$ are essentially disjoint and $\mathcal{R}(\mathbf{W})$ and $\mathcal{R}(\mathbf{U})$ are essentially disjoint, then $\mathbf{G}_{22}$ is a generalized inverse of $\mathbf{W}$.

Proof. Clearly,

$$\begin{pmatrix} \mathbf{T}\mathbf{G}_{11}\mathbf{T} + \mathbf{U}\mathbf{G}_{21}\mathbf{T} & \mathbf{T}\mathbf{G}_{11}\mathbf{U} + \mathbf{U}\mathbf{G}_{21}\mathbf{U} \\ \quad + \mathbf{T}\mathbf{G}_{12}\mathbf{V} + \mathbf{U}\mathbf{G}_{22}\mathbf{V} & \quad + \mathbf{T}\mathbf{G}_{12}\mathbf{W} + \mathbf{U}\mathbf{G}_{22}\mathbf{W} \\ \mathbf{V}\mathbf{G}_{11}\mathbf{T} + \mathbf{W}\mathbf{G}_{21}\mathbf{T} & \mathbf{V}\mathbf{G}_{11}\mathbf{U} + \mathbf{W}\mathbf{G}_{21}\mathbf{U} \\ \quad + \mathbf{V}\mathbf{G}_{12}\mathbf{V} + \mathbf{W}\mathbf{G}_{22}\mathbf{V} & \quad + \mathbf{V}\mathbf{G}_{12}\mathbf{W} + \mathbf{W}\mathbf{G}_{22}\mathbf{W} \end{pmatrix}$$
$$= \begin{pmatrix} \mathbf{T} & \mathbf{U} \\ \mathbf{V} & \mathbf{W} \end{pmatrix} \begin{pmatrix} \mathbf{G}_{11} & \mathbf{G}_{12} \\ \mathbf{G}_{21} & \mathbf{G}_{22} \end{pmatrix} \begin{pmatrix} \mathbf{T} & \mathbf{U} \\ \mathbf{V} & \mathbf{W} \end{pmatrix} = \begin{pmatrix} \mathbf{T} & \mathbf{U} \\ \mathbf{V} & \mathbf{W} \end{pmatrix}. \quad (3.2)$$

(1) It follows from result (3.2) that

$$\mathbf{T}\mathbf{G}_{11}\mathbf{T} + \mathbf{U}\mathbf{G}_{21}\mathbf{T} + \mathbf{T}\mathbf{G}_{12}\mathbf{V} + \mathbf{U}\mathbf{G}_{22}\mathbf{V} = \mathbf{T}. \quad (3.3)$$

Now, suppose that $\mathcal{C}(\mathbf{T})$ and $\mathcal{C}(\mathbf{U})$ are essentially disjoint and $\mathcal{R}(\mathbf{T})$ and $\mathcal{R}(\mathbf{V})$ are essentially disjoint. Then, rewriting equality (3.3) as

$$\mathbf{U}(\mathbf{G}_{21}\mathbf{T} + \mathbf{G}_{22}\mathbf{V}) = \mathbf{T}(\mathbf{I} - \mathbf{G}_{11}\mathbf{T} - \mathbf{G}_{12}\mathbf{V})$$

and applying Corollary 17.2.3, we find that

$$\mathbf{T}(\mathbf{I} - \mathbf{G}_{11}\mathbf{T} - \mathbf{G}_{12}\mathbf{V}) = \mathbf{0}. \quad (3.4)$$

Further, rewriting equality (3.4) as

$$(\mathbf{I} - \mathbf{T}\mathbf{G}_{11})\mathbf{T} = \mathbf{T}\mathbf{G}_{12}\mathbf{V}$$

and again applying Corollary 17.2.3, we find that $(\mathbf{I} - \mathbf{T}\mathbf{G}_{11})\mathbf{T} = \mathbf{0}$ or, equivalently, that $\mathbf{T} = \mathbf{T}\mathbf{G}_{11}\mathbf{T}$. Thus, $\mathbf{G}_{11}$ is a generalized inverse of $\mathbf{T}$.

(2), (3), and (4). The proofs of Parts (2), (3), and (4) are analogous to the proof of Part (1). Q.E.D.

Note that the conditions of Parts (1) and (4) of Theorem 17.3.4 [as well as those of Parts (2) and (3)] are satisfied in particular if $\mathbf{U} = \mathbf{0}$ and $\mathbf{V} = \mathbf{0}$—if $\mathbf{U} = \mathbf{0}$ and $\mathbf{V} = \mathbf{0}$, $\begin{pmatrix} \mathbf{T} & \mathbf{U} \\ \mathbf{V} & \mathbf{W} \end{pmatrix}$ reduces to the block-diagonal matrix $\text{diag}(\mathbf{T}, \mathbf{W})$.

17.4 Subspaces: Sum of Their Dimensions Versus Dimension of Their Sum

The following theorem quantifies the difference between the right and left sides of inequality (1.12).

Theorem 17.4.1. For any subspaces $\mathcal{U}$ and $\mathcal{V}$ of $\mathcal{R}^{m\times n}$,

$$\dim(\mathcal{U} + \mathcal{V}) = \dim(\mathcal{U}) + \dim(\mathcal{V}) - \dim(\mathcal{U} \cap \mathcal{V}), \tag{4.1}$$

or equivalently

$$\dim(\mathcal{U} \cap \mathcal{V}) = \dim(\mathcal{U}) + \dim(\mathcal{V}) - \dim(\mathcal{U} + \mathcal{V}). \tag{4.2}$$

Proof. Let $r = \dim(\mathcal{U})$, $s = \dim(\mathcal{V})$, and $t = \dim(\mathcal{U} \cap \mathcal{V})$. If $t = 0$, $\mathcal{U}$ and $\mathcal{V}$ are essentially disjoint, and result (4.1) follows from result (1.11). Consider now the case where $t > 0$.

Let $\{\mathbf{F}_1, \ldots, \mathbf{F}_t\}$ represent a basis for the subspace $\mathcal{U} \cap \mathcal{V}$. Since $\mathcal{U} \cap \mathcal{V} \subset \mathcal{U}$, $\mathbf{F}_1, \ldots, \mathbf{F}_t$ are contained in $\mathcal{U}$, and it follows from Theorem 4.3.12 that there exist $r-t$ additional matrices $\mathbf{A}_1, \ldots, \mathbf{A}_{r-t}$ such that the set $\{\mathbf{F}_1, \ldots, \mathbf{F}_t, \mathbf{A}_1, \ldots, \mathbf{A}_{r-t}\}$ is a basis for $\mathcal{U}$—if $r = t$, the set $\{\mathbf{F}_1, \ldots, \mathbf{F}_t, \mathbf{A}_1, \ldots, \mathbf{A}_{r-t}\}$ reduces to the set $\{\mathbf{F}_1, \ldots, \mathbf{F}_t\}$. Similarly, $\mathbf{F}_1, \ldots, \mathbf{F}_t$ are contained in $\mathcal{V}$, and there exist $s - t$ additional matrices $\mathbf{B}_1, \ldots, \mathbf{B}_{s-t}$ such that the set $\{\mathbf{F}_1, \ldots, \mathbf{F}_t, \mathbf{B}_1, \ldots, \mathbf{B}_{s-t}\}$ is a basis for $\mathcal{V}$.

To validate result (4.1), it suffices to establish that the set $\{\mathbf{F}_1, \ldots, \mathbf{F}_t, \mathbf{A}_1, \ldots, \mathbf{A}_{r-t}, \mathbf{B}_1, \ldots, \mathbf{B}_{s-t}\}$ spans $\mathcal{U}+\mathcal{V}$ and is linearly independent [thereby establishing that this $(r + s - t)$-dimensional set is a basis for $\mathcal{U} + \mathcal{V}$]. Clearly, if $\mathbf{X} \in \mathcal{U}$ and $\mathbf{Y} \in \mathcal{V}$, then each of the matrices $\mathbf{X}$ and $\mathbf{Y}$ is expressible as a linear combination of the matrices $\mathbf{F}_1, \ldots, \mathbf{F}_t, \mathbf{A}_1, \ldots, \mathbf{A}_{r-t}, \mathbf{B}_1, \ldots, \mathbf{B}_{s-t}$ and consequently $\mathbf{X} + \mathbf{Y}$ is expressible as a linear combination of these matrices. Thus, the set $\{\mathbf{F}_1, \ldots, \mathbf{F}_t, \mathbf{A}_1, \ldots, \mathbf{A}_{r-t}, \mathbf{B}_1, \ldots, \mathbf{B}_{s-t}\}$ spans $\mathcal{U} + \mathcal{V}$.

To complete the proof, we must show that this set is linearly independent. For this purpose, suppose that $s > t$—if $s = t$, the set $\{\mathbf{F}_1, \ldots, \mathbf{F}_t, \mathbf{A}_1, \ldots, \mathbf{A}_{r-t}, \mathbf{B}_1, \ldots, \mathbf{B}_{s-t}\}$ reduces to the set $\{\mathbf{F}_1, \ldots, \mathbf{F}_t, \mathbf{A}_1, \ldots, \mathbf{A}_{r-t}\}$, which is a basis for $\mathcal{U}$ and is therefore linearly independent.

Let $k_1, \ldots, k_t, c_1, \ldots, c_{r-t}, d_1, \ldots, d_{s-t}$ represent any scalars such that

$$k_1\mathbf{F}_1+\cdots+k_t\mathbf{F}_t+c_1\mathbf{A}_1+\cdots+c_{r-t}\mathbf{A}_{r-t}+d_1\mathbf{B}_1+\cdots+d_{s-t}\mathbf{B}_{s-t}=\mathbf{0}. \quad (4.3)$$

And define

$$\mathbf{H}=k_1\mathbf{F}_1+\cdots+k_t\mathbf{F}_t+c_1\mathbf{A}_1+\cdots+c_{r-t}\mathbf{A}_{r-t}. \quad (4.4)$$

Then,

$$\mathbf{H}=-d_1\mathbf{B}_1-\cdots-d_{s-t}\mathbf{B}_{s-t}. \quad (4.5)$$

Equality (4.4) implies that $\mathbf{H}\in\mathcal{U}$, and equality (4.5) implies that $\mathbf{H}\in\mathcal{V}$. Thus, $\mathbf{H}\in\mathcal{U}\cap\mathcal{V}$, and, consequently, there exist scalars $z_1,\ldots,z_t$ such that

$$\mathbf{H}=z_1\mathbf{F}_1+\cdots+z_t\mathbf{F}_t. \quad (4.6)$$

Together, equalities (4.5) and (4.6) imply that

$$z_1\mathbf{F}_1+\cdots+z_t\mathbf{F}_t+d_1\mathbf{B}_1+\cdots+d_{s-t}\mathbf{B}_{s-t}=\mathbf{H}-\mathbf{H}=\mathbf{0}.$$

Since $\{\mathbf{F}_1,\ldots,\mathbf{F}_t,\mathbf{B}_1,\ldots,\mathbf{B}_{s-t}\}$ is a basis for $\mathcal{V}$ and hence is linearly independent, it follows that

$$d_1=\cdots=d_{s-t}=0 \quad (4.7)$$

(and that $z_1=\cdots=z_t=0$) and [in light of equality (4.3)] that

$$k_1\mathbf{F}_1+\cdots+k_t\mathbf{F}_t+c_1\mathbf{A}_1+\cdots+c_{r-t}\mathbf{A}_{r-t}=\mathbf{0}. \quad (4.8)$$

And since $\{\mathbf{F}_1,\ldots,\mathbf{F}_t,\mathbf{A}_1,\ldots,\mathbf{A}_{r-t}\}$ is a basis for $\mathcal{U}$ and hence is linearly independent, result (4.8) implies that

$$k_1=\cdots=k_t=c_1=\cdots=c_{r-t}=0. \quad (4.9)$$

We conclude [on the basis of results (4.7) and (4.9)] that the set $\{\mathbf{F}_1,\ldots,\mathbf{F}_t, \mathbf{A}_1,\ldots,\mathbf{A}_{r-t},\mathbf{B}_1,\ldots,\mathbf{B}_{s-t}\}$ is linearly independent. Q.E.D.

By applying Theorem 17.4.1 in the special case where $\mathcal{U}$ and $\mathcal{V}$ are the column spaces of an $m\times n$ matrix $\mathbf{A}$ and an $m\times p$ matrix $\mathbf{B}$ and in the special case where $\mathcal{U}$ and $\mathcal{V}$ are the row spaces of an $m\times n$ matrix $\mathbf{A}$ and a $q\times n$ matrix $\mathbf{C}$, we obtain [in light of equalities (1.4) and (1.5)] the following theorem, which quantifies the difference between the right and left sides of inequality (2.1) and the difference between the right and left sides of inequality (2.2).

Theorem 17.4.2. Let $\mathbf{A}$ represent an $m\times n$ matrix, $\mathbf{B}$ an $m\times p$ matrix, and $\mathbf{C}$ a $q\times n$ matrix. Then,

$$\text{rank}(\mathbf{A},\ \mathbf{B})=\text{rank}(\mathbf{A})+\text{rank}(\mathbf{B})-\dim[\mathcal{C}(\mathbf{A})\cap\mathcal{C}(\mathbf{B})], \quad (4.10)$$

$$\text{rank}\begin{pmatrix}\mathbf{A}\\ \mathbf{C}\end{pmatrix}=\text{rank}(\mathbf{A})+\text{rank}(\mathbf{C})-\dim[\mathcal{R}(\mathbf{A})\cap\mathcal{R}(\mathbf{C})]. \quad (4.11)$$

Or, equivalently,

$$\dim[\mathcal{C}(\mathbf{A})\cap\mathcal{C}(\mathbf{B})]=\text{rank}(\mathbf{A})+\text{rank}(\mathbf{B})-\text{rank}(\mathbf{A},\ \mathbf{B}), \quad (4.12)$$

$$\dim[\mathcal{R}(\mathbf{A})\cap\mathcal{R}(\mathbf{C})]=\text{rank}(\mathbf{A})+\text{rank}(\mathbf{C})-\text{rank}\begin{pmatrix}\mathbf{A}\\ \mathbf{C}\end{pmatrix}. \quad (4.13)$$

Upon combining result (4.10) with result (2.8) and result (4.11) with result (2.10), we obtain the following corollary of Theorem 17.4.2.

Corollary 17.4.3. Let $\mathbf{A}$ represent an $m \times n$ matrix, $\mathbf{B}$ an $m \times p$ matrix, and $\mathbf{C}$ a $q \times n$ matrix. Then,

$$\text{rank}[(\mathbf{I} - \mathbf{AA}^-)\mathbf{B}] = \text{rank}(\mathbf{B}) - \dim[\mathcal{C}(\mathbf{A}) \cap \mathcal{C}(\mathbf{B})], \tag{4.14}$$

$$\text{rank}[\mathbf{C}(\mathbf{I} - \mathbf{A}^-\mathbf{A})] = \text{rank}(\mathbf{C}) - \dim[\mathcal{R}(\mathbf{A}) \cap \mathcal{R}(\mathbf{C})]. \tag{4.15}$$

Result (4.15) can be restated in terms of the following set:

$$\mathcal{U} = \{\mathbf{x} \in \mathcal{R}^{q\times 1} \;:\; \mathbf{x} = \mathbf{Cz} \text{ for some } n \times 1 \text{ vector } \mathbf{z} \text{ in } \mathcal{N}(\mathbf{A})\}.$$

We have that

$$\mathcal{U} = \mathcal{C}[\mathbf{C}(\mathbf{I} - \mathbf{A}^-\mathbf{A})], \tag{4.16}$$

as is evident upon observing (in light of Corollary 11.2.2) that

$$\begin{aligned} \mathbf{x} \in \mathcal{C}[\mathbf{C}(\mathbf{I} - \mathbf{A}^-\mathbf{A})] &\Leftrightarrow \mathbf{x} = \mathbf{C}(\mathbf{I} - \mathbf{A}^-\mathbf{A})\mathbf{y} \text{ for some } n \times 1 \text{ vector } \mathbf{y} \\ &\Leftrightarrow \mathbf{x} = \mathbf{Cz} \text{ for some } \mathbf{z} \text{ in } \mathcal{C}(\mathbf{I} - \mathbf{A}^-\mathbf{A}) \\ &\Leftrightarrow \mathbf{x} = \mathbf{Cz} \text{ for some } \mathbf{z} \text{ in } \mathcal{N}(\mathbf{A}). \end{aligned}$$

Thus, $\mathcal{U}$ is a linear space, and, as a consequence of result (4.15),

$$\dim(\mathcal{U}) = \text{rank}[\mathbf{C}(\mathbf{I} - \mathbf{A}^-\mathbf{A})] = \text{rank}(\mathbf{C}) - \dim[\mathcal{R}(\mathbf{A}) \cap \mathcal{R}(\mathbf{C})]. \tag{4.17}$$

17.5 Some Results on the Rank of a Product of Matrices

a. Initial presentation of results

Upper bounds for the product of an $m \times n$ matrix $\mathbf{A}$ and an $n \times p$ matrix $\mathbf{B}$ are, as previously indicated (in Corollary 4.4.5):

$$\text{rank}(\mathbf{AB}) \le \text{rank}(\mathbf{A}), \qquad \text{rank}(\mathbf{AB}) \le \text{rank}(\mathbf{B}). \tag{5.1}$$

Note that inequalities (5.1) can be combined and restated as

$$\text{rank}(\mathbf{AB}) \le \min[\text{rank}(\mathbf{A}),\ \text{rank}(\mathbf{B})]. \tag{5.2}$$

For purposes of obtaining a lower bound for the rank of a product of matrices, let $\mathbf{A}$ represent an $m \times n$ matrix, $\mathbf{C}$ an $n \times q$ matrix, and $\mathbf{B}$ a $q \times p$ matrix. According to Theorem 9.6.4,

$$\text{rank}\begin{pmatrix} \mathbf{C} & \mathbf{CB} \\ \mathbf{AC} & \mathbf{0} \end{pmatrix} = \text{rank}(\mathbf{C}) + \text{rank}(-\mathbf{ACB}) = \text{rank}(\mathbf{C}) + \text{rank}(\mathbf{ACB}), \tag{5.3}$$

and, according to Theorem 17.2.17,

$$\text{rank}\begin{pmatrix} \mathbf{C} & \mathbf{CB} \\ \mathbf{AC} & \mathbf{0} \end{pmatrix} = \text{rank}(\mathbf{AC}) + \text{rank}(\mathbf{CB}) + \text{rank}\{[\mathbf{I} - \mathbf{CB}(\mathbf{CB})^-]\mathbf{C}[\mathbf{I} - (\mathbf{AC})^-\mathbf{AC}]\}. \quad (5.4)$$

Equating expression (5.4) to expression (5.3), we obtain the following result on the rank of a product of three matrices.

Theorem 17.5.1. Let $\mathbf{A}$ represent an $m \times n$ matrix, $\mathbf{C}$ an $n \times q$ matrix, and $\mathbf{B}$ a $q \times p$ matrix. Then,

$$\text{rank}(\mathbf{ACB}) = \text{rank}(\mathbf{AC}) + \text{rank}(\mathbf{CB}) - \text{rank}(\mathbf{C}) + \text{rank}\{[\mathbf{I} - \mathbf{CB}(\mathbf{CB})^-]\mathbf{C}[\mathbf{I} - (\mathbf{AC})^-\mathbf{AC}]\}. \quad (5.5)$$

Further,

$$\text{rank}(\mathbf{ACB}) \geq \text{rank}(\mathbf{AC}) + \text{rank}(\mathbf{CB}) - \text{rank}(\mathbf{C}), \quad (5.6)$$

with equality holding if and only if

$$[\mathbf{I} - \mathbf{CB}(\mathbf{CB})^-]\mathbf{C}[\mathbf{I} - (\mathbf{AC})^-\mathbf{AC}] = \mathbf{0}. \quad (5.7)$$

In the special case where $\mathbf{C} = \mathbf{I}_n$, Theorem 17.5.1 reduces to the following corollary on the rank of a product of two matrices.

Corollary 17.5.2. Let $\mathbf{A}$ represent an $m \times n$ matrix and $\mathbf{B}$ an $n \times p$ matrix. Then,

$$\text{rank}(\mathbf{AB}) = \text{rank}(\mathbf{A}) + \text{rank}(\mathbf{B}) - n + \text{rank}[(\mathbf{I} - \mathbf{BB}^-)(\mathbf{I} - \mathbf{A}^-\mathbf{A})]. \quad (5.8)$$

Further,

$$\text{rank}(\mathbf{AB}) \geq \text{rank}(\mathbf{A}) + \text{rank}(\mathbf{B}) - n, \quad (5.9)$$

with equality holding if and only if

$$(\mathbf{I} - \mathbf{BB}^-)(\mathbf{I} - \mathbf{A}^-\mathbf{A}) = \mathbf{0}. \quad (5.10)$$

Inequality (5.9) is called *Sylvester's law of nullity*, after James Joseph Sylvester (1814–1897), and inequality (5.6) is called the *Frobenius inequality*, after Ferdinand Georg Frobenius (1849–1917).

According to Corollary 17.5.2, the lower bound (5.9) (on the rank of a product of two matrices) is attained if and only if equality (5.10) holds. Under what condition(s) are the upper bounds [given by result (5.1)] attained? To answer this question, observe (in light of Corollary 17.2.9) that

$$\text{rank}(\mathbf{B},\ \mathbf{I} - \mathbf{A}^-\mathbf{A}) = \text{rank}(\mathbf{B}) + \text{rank}[(\mathbf{I} - \mathbf{BB}^-)(\mathbf{I} - \mathbf{A}^-\mathbf{A})], \quad (5.11)$$

$$\text{rank}\begin{pmatrix} \mathbf{A} \\ \mathbf{I} - \mathbf{BB}^- \end{pmatrix} = \text{rank}(\mathbf{A}) + \text{rank}[(\mathbf{I} - \mathbf{BB}^-)(\mathbf{I} - \mathbf{A}^-\mathbf{A})]. \quad (5.12)$$

A comparison of expressions (5.11) and (5.12) with expression (5.8) leads to the following theorem.

Theorem 17.5.3. Let **A** represent an $m \times n$ matrix and **B** an $n \times p$ matrix. Then,

$$\text{rank}(\mathbf{AB}) = \text{rank}(\mathbf{A}) - n + \text{rank}(\mathbf{B},\ \mathbf{I} - \mathbf{A}^{-}\mathbf{A}) \quad (5.13)$$

$$= \text{rank}(\mathbf{B}) - n + \text{rank}\begin{pmatrix} \mathbf{A} \\ \mathbf{I} - \mathbf{BB}^{-} \end{pmatrix}. \quad (5.14)$$

Further, equality is attained in the inequality $\text{rank}(\mathbf{AB}) \le \text{rank}(\mathbf{A})$ if and only if $\text{rank}(\mathbf{B},\ \mathbf{I} - \mathbf{A}^{-}\mathbf{A}) = n$ or, equivalently, if and only if $(\mathbf{B},\ \mathbf{I} - \mathbf{A}^{-}\mathbf{A})$ is of full row rank. And equality is attained in the inequality $\text{rank}(\mathbf{AB}) \le \text{rank}(\mathbf{B})$ if and only if $\text{rank}\begin{pmatrix} \mathbf{A} \\ \mathbf{I} - \mathbf{BB}^{-} \end{pmatrix} = n$ or, equivalently, if and only if $\begin{pmatrix} \mathbf{A} \\ \mathbf{I} - \mathbf{BB}^{-} \end{pmatrix}$ is of full column rank.

b. Reexpression of results in more "geometrically meaningful" terms

Let **A** represent an $m \times n$ matrix and **B** an $n \times p$ matrix. Then, making use of Theorem 17.4.2, Lemma 10.2.6, and Corollary 11.2.2, we find that

$$\begin{aligned} \text{rank}(\mathbf{B},\ \mathbf{I} - \mathbf{A}^{-}\mathbf{A}) &= \text{rank}(\mathbf{B}) + \text{rank}(\mathbf{I} - \mathbf{A}^{-}\mathbf{A}) \\ &\quad - \dim[\mathcal{C}(\mathbf{B}) \cap \mathcal{C}(\mathbf{I} - \mathbf{A}^{-}\mathbf{A})] \\ &= \text{rank}(\mathbf{B}) + n - \text{rank}(\mathbf{A}) - \dim[\mathcal{C}(\mathbf{B}) \cap \mathcal{N}(\mathbf{A})]. \quad (5.15) \end{aligned}$$

Further, recalling that $(\mathbf{B}^{-})'$ is a generalized inverse of $\mathbf{B}'$, we find in similar fashion that

$$\begin{aligned} \text{rank}\begin{pmatrix} \mathbf{A} \\ \mathbf{I} - \mathbf{BB}^{-} \end{pmatrix} &= \text{rank}\left[\begin{pmatrix} \mathbf{A} \\ \mathbf{I} - \mathbf{BB}^{-} \end{pmatrix}'\right] \\ &= \text{rank}[\mathbf{A}',\ \mathbf{I} - (\mathbf{B}^{-})'\mathbf{B}'] \\ &= \text{rank}(\mathbf{A}) + \text{rank}[\mathbf{I} - (\mathbf{B}^{-})'\mathbf{B}'] \\ &\quad - \dim\{\mathcal{C}(\mathbf{A}') \cap \mathcal{C}[\mathbf{I} - (\mathbf{B}^{-})'\mathbf{B}']\} \\ &= \text{rank}(\mathbf{A}) + n - \text{rank}(\mathbf{B}) \\ &\quad - \dim[\mathcal{C}(\mathbf{A}') \cap \mathcal{N}(\mathbf{B}')]. \quad (5.16) \end{aligned}$$

Substituting expressions (5.15) and (5.16) into expressions (5.13) and (5.14), respectively, we obtain the following alternative to Theorem 17.5.3.

Theorem 17.5.4. Let **A** represent an $m \times n$ matrix and **B** an $n \times p$ matrix. Then,

$$\text{rank}(\mathbf{AB}) = \text{rank}(\mathbf{B}) - \dim[\mathcal{C}(\mathbf{B}) \cap \mathcal{N}(\mathbf{A})] \quad (5.17)$$

$$= \text{rank}(\mathbf{A}) - \dim[\mathcal{C}(\mathbf{A}') \cap \mathcal{N}(\mathbf{B}')]. \quad (5.18)$$

Further, equality is attained in the inequality $\text{rank}(\mathbf{AB}) \le \text{rank}(\mathbf{B})$ if and only if $\mathcal{C}(\mathbf{B})$ and $\mathcal{N}(\mathbf{A})$ are essentially disjoint. And equality is attained in the inequality $\text{rank}(\mathbf{AB}) \le \text{rank}(\mathbf{A})$ if and only if $\mathcal{C}(\mathbf{A}')$ and $\mathcal{N}(\mathbf{B}')$ are essentially disjoint.

By equating expression (5.8) for the rank of a product of two matrices to each of the alternative expressions (5.17) and (5.18), we find (in light of Lemma 11.3.1) that (for any $m \times n$ matrix $\mathbf{A}$ and any $n \times p$ matrix $\mathbf{B}$)

$$\begin{aligned}\text{rank}[(\mathbf{I} - \mathbf{B}\mathbf{B}^-)(\mathbf{I} - \mathbf{A}^-\mathbf{A})] &= n - \text{rank}(\mathbf{A}) - \dim[\mathcal{C}(\mathbf{B}) \cap \mathcal{N}(\mathbf{A})] \\ &= \dim[\mathcal{N}(\mathbf{A})] - \dim[\mathcal{C}(\mathbf{B}) \cap \mathcal{N}(\mathbf{A})] \qquad (5.19)\end{aligned}$$

and similarly that

$$\begin{aligned}\text{rank}[(\mathbf{I} - \mathbf{B}\mathbf{B}^-)(\mathbf{I} - \mathbf{A}^-\mathbf{A})] &= n - \text{rank}(\mathbf{B}') - \dim[\mathcal{C}(\mathbf{A}') \cap \mathcal{N}(\mathbf{B}')] \\ &= \dim[\mathcal{N}(\mathbf{B}')] - \dim[\mathcal{C}(\mathbf{A}') \cap \mathcal{N}(\mathbf{B}')]. \qquad (5.20)\end{aligned}$$

Based on equalities (5.19) and (5.20), we obtain the following theorem, which allows us to reexpress condition (5.10) [under which equality holds in inequality (5.9)] in more "geometrically meaningful" terms.

Theorem 17.5.5. Let $\mathbf{A}$ represent an $m \times n$ matrix and $\mathbf{B}$ an $n \times p$ matrix. Then,

$$(\mathbf{I} - \mathbf{B}\mathbf{B}^-)(\mathbf{I} - \mathbf{A}^-\mathbf{A}) = \mathbf{0} \;\Leftrightarrow\; \mathcal{N}(\mathbf{A}) \subset \mathcal{C}(\mathbf{B}) \;\Leftrightarrow\; \mathcal{N}(\mathbf{B}') \subset \mathcal{C}(\mathbf{A}').$$

Let $\mathbf{A}$ represent an $m \times n$ matrix, $\mathbf{C}$ an $n \times q$ matrix, and $\mathbf{B}$ a $q \times p$ matrix. Then, making use of result (5.17), we find that

$$\text{rank}(\mathbf{ACB}) = \text{rank}[\mathbf{A}(\mathbf{CB})] = \text{rank}(\mathbf{CB}) - \dim[\mathcal{C}(\mathbf{CB}) \cap \mathcal{N}(\mathbf{A})] \qquad (5.21)$$

and, similarly, making use of result (5.18), that

$$\text{rank}(\mathbf{ACB}) = \text{rank}[(\mathbf{AC})\mathbf{B}] = \text{rank}(\mathbf{AC}) - \dim\{\mathcal{C}[(\mathbf{AC})'] \cap \mathcal{N}(\mathbf{B}')\}. \qquad (5.22)$$

Further, equating expression (5.5) [for rank($\mathbf{ACB}$)] to expression (5.21) and making use of result (5.17), we find that

$$\begin{aligned}&\text{rank}\{[\mathbf{I} - \mathbf{CB}(\mathbf{CB})^-]\mathbf{C}[\mathbf{I} - (\mathbf{AC})^-\mathbf{AC}]\} \\ &\qquad = \text{rank}(\mathbf{C}) - \text{rank}(\mathbf{AC}) - \dim[\mathcal{C}(\mathbf{CB}) \cap \mathcal{N}(\mathbf{A})] \\ &\qquad = \dim[\mathcal{C}(\mathbf{C}) \cap \mathcal{N}(\mathbf{A})] - \dim[\mathcal{C}(\mathbf{CB}) \cap \mathcal{N}(\mathbf{A})] \qquad (5.23)\end{aligned}$$

and, similarly, equating expression (5.5) to expression (5.22) and making use of result (5.18), that

$$\begin{aligned}&\text{rank}\{[\mathbf{I} - \mathbf{CB}(\mathbf{CB})^-]\mathbf{C}[\mathbf{I} - (\mathbf{AC})^-\mathbf{AC}]\} \\ &\qquad = \text{rank}(\mathbf{C}) - \text{rank}(\mathbf{CB}) - \dim[\mathcal{C}(\mathbf{C}'\mathbf{A}') \cap \mathcal{N}(\mathbf{B}')] \\ &\qquad = \dim[\mathcal{C}(\mathbf{C}') \cap \mathcal{N}(\mathbf{B}')] - \dim[\mathcal{C}(\mathbf{C}'\mathbf{A}') \cap \mathcal{N}(\mathbf{B}')]. \qquad (5.24)\end{aligned}$$

Based on equalities (5.23) and (5.24), we obtain the following theorem, which allows us to reexpress condition (5.7) [under which equality holds in the Frobenius inequality] in more geometrically meaningful terms.

Theorem 17.5.6. Let $\mathbf{A}$ represent an $m \times n$ matrix, $\mathbf{C}$ an $n \times q$ matrix, and $\mathbf{B}$ a $q \times p$ matrix. Then,

$$\begin{aligned}[\mathbf{I} - \mathbf{CB}(\mathbf{CB})^{-}]\mathbf{C}[\mathbf{I} - (\mathbf{AC})^{-}\mathbf{AC}] = \mathbf{0} \\ \Leftrightarrow \mathcal{C}(\mathbf{CB}) \cap \mathcal{N}(\mathbf{A}) = \mathcal{C}(\mathbf{C}) \cap \mathcal{N}(\mathbf{A}) \\ \Leftrightarrow \mathcal{C}(\mathbf{C}'\mathbf{A}') \cap \mathcal{N}(\mathbf{B}') = \mathcal{C}(\mathbf{C}') \cap \mathcal{N}(\mathbf{B}').\end{aligned}$$

17.6 Projections Along a Subspace

a. Some general results and terminology

Let $\mathcal{V}$ represent a linear space of $m \times n$ matrices, and let $\mathcal{U}$ represent a subspace of $\mathcal{V}$. The subspace $\mathcal{U}$ and its orthogonal complement $\mathcal{U}^{\perp}$ (which is also a subspace of $\mathcal{V}$) are orthogonal (as is evident from the very definition of $\mathcal{U}^{\perp}$) and hence are (as a consequence of Lemma 17.1.9) essentially disjoint. Moreover, it follows from Corollary 17.1.2 that $\mathcal{U}+\mathcal{U}^{\perp} \subset \mathcal{V}$ and from Theorem 12.5.11 that $\mathcal{V} \subset \mathcal{U}+\mathcal{U}^{\perp}$. Thus,

$$\mathcal{V} = \mathcal{U} \oplus \mathcal{U}^{\perp} \tag{6.1}$$

(i.e., $\mathcal{V}$ equals the direct sum of $\mathcal{U}$ and $\mathcal{U}^{\perp}$).

Let $\mathbf{X}$ represent an $m \times p$ matrix, and take $\mathbf{W}$ to be an $m \times m$ symmetric positive definite matrix. Then, as a special case of result (6.1) [that where $\mathcal{V} = \mathcal{R}^m$ and $\mathcal{U} = \mathcal{C}(\mathbf{X})$ and where the inner product for $\mathcal{R}^m$ is taken to be the bilinear form whose matrix is $\mathbf{W}$], we have that

$$\mathcal{R}^m = \mathcal{C}(\mathbf{X}) \oplus \mathcal{C}^{\perp}_{\mathbf{W}}(\mathbf{X}), \tag{6.2}$$

which, in the further special case where $\mathbf{W} = \mathbf{I}$, reduces to

$$\mathcal{R}^m = \mathcal{C}(\mathbf{X}) \oplus \mathcal{C}^{\perp}(\mathbf{X}). \tag{6.3}$$

According to Theorem 12.1.3, a subspace $\mathcal{U}$ (of a linear space $\mathcal{V}$ of $m \times n$ matrices) and its orthogonal complement $\mathcal{U}^{\perp}$ have the following property: corresponding to any $m \times n$ matrix $\mathbf{Y}$ in $\mathcal{V}$, there exists a unique matrix $\mathbf{Z}$ in $\mathcal{U}$ such that $\mathbf{Y} - \mathbf{Z} \in \mathcal{U}^{\perp}$. The following theorem indicates that any two essentially disjoint subspaces whose sum is $\mathcal{V}$ have this property.

Theorem 17.6.1. Let $\mathbf{Y}$ represent a matrix in a linear space $\mathcal{V}$ of $m \times n$ matrices, and let $\mathcal{U}$ and $\mathcal{W}$ represent essentially disjoint subspaces whose sum is $\mathcal{V}$. Then, $\mathcal{U}$ contains a unique matrix $\mathbf{Z}$ such that $\mathbf{Y} - \mathbf{Z} \in \mathcal{W}$.

Proof. Since $\mathcal{V} = \mathcal{U} + \mathcal{W}$, it follows from the very definition of a sum (of subspaces) that $\mathcal{U}$ contains a matrix $\mathbf{Z}$ and $\mathcal{W}$ a matrix $\mathbf{W}$ such that $\mathbf{Y} = \mathbf{Z} + \mathbf{W}$. Clearly, $\mathbf{Y} - \mathbf{Z} = \mathbf{W}$, and hence $\mathbf{Y} - \mathbf{Z} \in \mathcal{W}$.

For purposes of establishing the uniqueness of $\mathbf{Z}$, let $\mathbf{Z}_*$ represent a matrix (potentially different from $\mathbf{Z}$) in $\mathcal{U}$ such that $\mathbf{Y} - \mathbf{Z}_* \in \mathcal{W}$. Then, $\mathbf{Z}_* - \mathbf{Z} \in \mathcal{U}$ and $\mathbf{Y} - \mathbf{Z}_* - (\mathbf{Y} - \mathbf{Z}) \in \mathcal{W}$. And

$$\mathbf{Z}_* - \mathbf{Z} + [\mathbf{Y} - \mathbf{Z}_* - (\mathbf{Y} - \mathbf{Z})] = \mathbf{0}.$$

We conclude, on the basis of Part (1) of Lemma 17.1.4, that $\mathbf{Z}_* - \mathbf{Z} = \mathbf{0}$ [and that $\mathbf{Y} - \mathbf{Z}_* - (\mathbf{Y} - \mathbf{Z}) = \mathbf{0}$] and hence that $\mathbf{Z}_* = \mathbf{Z}$, thereby establishing the uniqueness of $\mathbf{Z}$. Q.E.D.

Theorem 17.6.1 can be restated in the form of the following corollary.

Corollary 17.6.2. Let $\mathbf{Y}$ represent a matrix in a linear space $\mathcal{V}$ of $m \times n$ matrices, and let $\mathcal{U}$ and $\mathcal{W}$ represent essentially disjoint subspaces whose sum is $\mathcal{V}$. Then, $\mathcal{U}$ contains a unique matrix $\mathbf{Z}$ and $\mathcal{W}$ a unique matrix $\mathbf{W}$ such that $\mathbf{Y} = \mathbf{Z} + \mathbf{W}$.

Suppose that a linear space $\mathcal{V}$ is the direct sum $\mathcal{U} \oplus \mathcal{W}$ of two subspaces $\mathcal{U}$ and $\mathcal{W}$ (as in Theorem 17.6.1). Then, as indicated by the theorem, there exists a unique matrix $\mathbf{Z}$ in $\mathcal{U}$ such that the difference between $\mathbf{Y}$ and $\mathbf{Z}$ is in $\mathcal{W}$. This matrix (the matrix $\mathbf{Z}$) is referred to as the *projection of* $\mathbf{Y}$ *on* $\mathcal{U}$ *along* $\mathcal{W}$.

The following theorem, which is easy to verify, relates the projection of $\mathbf{Y}$ on $\mathcal{W}$ along $\mathcal{U}$ to the projection of $\mathbf{Y}$ on $\mathcal{U}$ along $\mathcal{W}$.

Theorem 17.6.3. Let $\mathbf{Y}$ represent a matrix in a linear space $\mathcal{V}$, and let $\mathcal{U}$ and $\mathcal{W}$ represent essentially disjoint subspaces whose sum is $\mathcal{V}$. Then, the projection of $\mathbf{Y}$ on $\mathcal{W}$ along $\mathcal{U}$ equals $\mathbf{Y} - \mathbf{Z}$, where $\mathbf{Z}$ is the projection of $\mathbf{Y}$ on $\mathcal{U}$ along $\mathcal{W}$.

The following theorem gives an explicit expression for the projection of a matrix $\mathbf{Y}$ (in a linear space $\mathcal{V}$) on a subspace $\mathcal{U}$ along a subspace $\mathcal{W}$ (where $\mathcal{V} = \mathcal{U} \oplus \mathcal{W}$) and also for the projection of $\mathbf{Y}$ on $\mathcal{W}$ along $\mathcal{U}$.

Theorem 17.6.4. Let $\mathbf{Y}$ represent a matrix in a linear space $\mathcal{V}$ of $m \times n$ matrices, and let $\mathcal{U}$ and $\mathcal{W}$ represent essentially disjoint subspaces whose sum is $\mathcal{V}$. Further, let $\{\mathbf{U}_1, \ldots, \mathbf{U}_s\}$ represent any set of matrices that spans $\mathcal{U}$, let $\{\mathbf{W}_1, \ldots, \mathbf{W}_t\}$ represent any set of matrices that spans $\mathcal{W}$, and define $c_1, \ldots, c_s$ and $k_1, \ldots, k_t$ to be any scalars such that

$$c_1\mathbf{U}_1 + \cdots + c_s\mathbf{U}_s + k_1\mathbf{W}_1 + \cdots + k_t\mathbf{W}_t = \mathbf{Y}. \tag{6.4}$$

Then, the projection of $\mathbf{Y}$ on $\mathcal{U}$ along $\mathcal{W}$ equals $c_1\mathbf{U}_1 + \cdots + c_s\mathbf{U}_s$, and the projection of $\mathbf{Y}$ on $\mathcal{W}$ along $\mathcal{U}$ equals $k_1\mathbf{W}_1 + \cdots + k_t\mathbf{W}_t$.

In connection with Theorem 17.6.4, note that the existence of scalars $c_1, \ldots, c_s$ and $k_1, \ldots, k_t$ that satisfy condition (6.4) is guaranteed by Lemma 17.1.1.

Proof (of Theorem 17.6.4). Clearly, $\sum_{i=1}^{s} c_i\mathbf{U}_i \in \mathcal{U}$, and

$$\mathbf{Y} - \sum_{i=1}^{s} c_i\mathbf{U}_i = \sum_{j=1}^{t} k_j\mathbf{W}_j \in \mathcal{W}.$$

Thus, $\sum_{i=1}^{s} c_i\mathbf{U}_i$ is (by definition) the projection of $\mathbf{Y}$ on $\mathcal{U}$ along $\mathcal{W}$, and $\sum_{j=1}^{t} k_j\mathbf{W}_j$ is (in light of Theorem 17.6.3) the projection of $\mathbf{Y}$ on $\mathcal{W}$ along $\mathcal{U}$. Q.E.D.

In the special case where $n = 1$ (i.e., where $\mathcal{V}$ is a linear space of column vectors), Theorem 17.6.4 can be restated as the following corollary.

Corollary 17.6.5. Let $\mathbf{y}$ represent a vector in a linear space $\mathcal{V}$ of m-dimensional column vectors, and let $\mathcal{U}$ and $\mathcal{W}$ represent essentially disjoint subspaces whose sum is $\mathcal{V}$. Further, let $\mathbf{U} = (\mathbf{u}_1, \ldots, \mathbf{u}_s)$ and $\mathbf{W} = (\mathbf{w}_1, \ldots, \mathbf{w}_t)$, where $\mathbf{u}_1, \ldots, \mathbf{u}_s$ are any vectors that span $\mathcal{U}$ and $\mathbf{w}_1, \ldots, \mathbf{w}_t$ any vectors that span $\mathcal{W}$, or, equivalently, let $\mathcal{U}$ represent any $m \times s$ matrix such that $\mathcal{C}(\mathbf{U}) = \mathcal{U}$ and $\mathbf{W}$ any $m \times t$ matrix such that $\mathcal{C}(\mathbf{W}) = \mathcal{W}$. And define $\mathbf{x}^*$ to be any solution to the linear system $(\mathbf{U}, \mathbf{W})\mathbf{x} = \mathbf{y}$ [in an $(s + t)$-dimensional column vector $\mathbf{x}$], and partition $\mathbf{x}^*$ as $\mathbf{x}^* = \begin{pmatrix} \mathbf{x}_1^* \\ \mathbf{x}_2^* \end{pmatrix}$ (where $\mathbf{x}_1^*$ has s elements). Then, the projection of $\mathbf{y}$ on $\mathcal{U}$ along $\mathcal{W}$ equals $\mathbf{U}\mathbf{x}_1^*$, and the projection of $\mathbf{y}$ on $\mathcal{W}$ along $\mathcal{U}$ equals $\mathbf{W}\mathbf{x}_2^*$.

The following theorem, whose validity is apparent from result (6.1) and Theorems 12.5.8 and 17.6.3, establishes a connection to the results of Chapter 12 (on orthogonal projections).

Theorem 17.6.6. Let $\mathbf{Y}$ represent a matrix in a linear space $\mathcal{V}$, and let $\mathcal{U}$ represent a subspace of $\mathcal{V}$. Then, the projection of $\mathbf{Y}$ on $\mathcal{U}$ along $\mathcal{U}^\perp$ equals the orthogonal projection of $\mathbf{Y}$ on $\mathcal{U}$. And the projection of $\mathbf{Y}$ on $\mathcal{U}^\perp$ along $\mathcal{U}$ equals the orthogonal projection of $\mathbf{Y}$ on $\mathcal{U}^\perp$.

The orthogonal projection of an $m \times n$ matrix $\mathbf{Y}$ on a subspace $\mathcal{U}$ (of a linear space $\mathcal{V}$ of $m \times n$ matrices) equals $\mathbf{Y}$ if and only if $\mathbf{Y} \in \mathcal{U}$—refer to the last part of Theorem 12.1.3. An extension of this result is given in the following lemma.

Lemma 17.6.7. Let $\mathbf{Y}$ represent a matrix in a linear space $\mathcal{V}$, let $\mathcal{U}$ and $\mathcal{W}$ represent essentially disjoint subspaces whose sum is $\mathcal{V}$, and let $\mathbf{Z}$ represent the projection of $\mathbf{Y}$ on $\mathcal{U}$ along $\mathcal{W}$. Then, $\mathbf{Z} = \mathbf{Y}$ if and only if $\mathbf{Y} \in \mathcal{U}$. (And $\mathbf{Z} = \mathbf{0}$ if and only if $\mathbf{Y} \in \mathcal{W}$.)

Proof. If $\mathbf{Z} = \mathbf{Y}$, then (since, by definition, $\mathbf{Z} \in \mathcal{U}$), $\mathbf{Y} \in \mathcal{U}$. Conversely, if $\mathbf{Y} \in \mathcal{U}$, then, since $\mathbf{Y} - \mathbf{Y} = \mathbf{0} \in \mathcal{W}$, $\mathbf{Z} = \mathbf{Y}$. (If $\mathbf{Z} = \mathbf{0}$, then $\mathbf{Y} = \mathbf{Y} - \mathbf{Z} \in \mathcal{W}$; conversely, if $\mathbf{Y} \in \mathcal{W}$, then clearly $\mathbf{Z} = \mathbf{0}$.) Q.E.D.

The following theorem can be regarded as an extension of Theorem 12.1.4.

Theorem 17.6.8. Let $\mathbf{Y}_1, \ldots, \mathbf{Y}_p$ represent matrices in a linear space $\mathcal{V}$, let $\mathcal{U}$ and $\mathcal{W}$ represent essentially disjoint subspaces whose sum is $\mathcal{V}$, and let $\mathbf{Z}_1, \ldots, \mathbf{Z}_p$ represent the projections of $\mathbf{Y}_1, \ldots, \mathbf{Y}_p$, respectively, on $\mathcal{U}$ along $\mathcal{W}$. Then, for any scalars $k_1, \ldots, k_p$, the projection of the linear combination $k_1\mathbf{Y}_1 + \cdots + k_p\mathbf{Y}_p$ on $\mathcal{U}$ along $\mathcal{W}$ is the corresponding linear combination $k_1\mathbf{Z}_1 + \cdots + k_p\mathbf{Z}_p$ of $\mathbf{Z}_1, \ldots, \mathbf{Z}_p$.

Proof. By definition, $\mathbf{Z}_i \in \mathcal{U}$ and $\mathbf{Y}_i - \mathbf{Z}_i \in \mathcal{W}$ $(i = 1, \ldots, p)$. Thus, $k_1\mathbf{Z}_1 + \cdots + k_p\mathbf{Z}_p \in \mathcal{U}$, and

$$\begin{aligned} k_1\mathbf{Y}_1 + \cdots + k_p\mathbf{Y}_p - (k_1\mathbf{Z}_1 + \cdots + k_p\mathbf{Z}_p) \\ = k_1(\mathbf{Y}_1 - \mathbf{Z}_1) + \cdots + k_p(\mathbf{Y}_p - \mathbf{Z}_p) \in \mathcal{W}. \end{aligned}$$

We conclude that $k_1\mathbf{Z}_1 + \cdots + k_p\mathbf{Z}_p$ is the projection of $k_1\mathbf{Y}_1 + \cdots + k_p\mathbf{Y}_p$ on $\mathcal{U}$ along $\mathcal{W}$. Q.E.D.

b. Two-dimensional example

Let us find the projection of an m-dimensional column vector $\mathbf{y}$ on a subspace $\mathcal{U}$ of $\mathcal{R}^m$ along a subspace $\mathcal{W}$ of $\mathcal{R}^m$ in the special case where $m = 2$, $\mathbf{y} = (4,\ 8)'$, $\mathcal{U} = \text{sp}\{\mathbf{u}\}$, and $\mathcal{W} = \text{sp}\{\mathbf{w}\}$, with $\mathbf{u} = (3,\ 1)'$ and $\mathbf{w} = (1,\ 4)'$. (Clearly, $\mathcal{R}^2 = \mathcal{U} \oplus \mathcal{W}$.)

It follows from Corollary 17.6.5 that the projection of $\mathbf{y}$ on $\mathcal{U}$ along $\mathcal{W}$ equals $x_1^*\mathbf{u}$ and the projection of $\mathbf{y}$ on $\mathcal{W}$ along $\mathcal{U}$ equals $x_2^*\mathbf{w}$, where $\mathbf{x}^* = (x_1^*,\ x_2^*)'$ is any solution to the linear system

$$(\mathbf{u},\ \mathbf{w})\mathbf{x} = \mathbf{y} \quad \Leftrightarrow \quad \begin{pmatrix} 3 & 1 \\ 1 & 4 \end{pmatrix}\begin{pmatrix} x_1 \\ x_2 \end{pmatrix} = \begin{pmatrix} 4 \\ 8 \end{pmatrix} \tag{6.5}$$

[in the 2×1 vector $\mathbf{x} = (x_1,\ x_2)'$]. The (unique) solution to linear system (6.5) is $\mathbf{x} = (8/11,\ 20/11)'$, as is easily verified. Thus, the projection of $\mathbf{y}$ on $\mathcal{U}$ along $\mathcal{W}$ is the vector

$$(8/11)\begin{pmatrix} 3 \\ 1 \end{pmatrix} = \begin{pmatrix} 24/11 \\ 8/11 \end{pmatrix},$$

and the projection of $\mathbf{y}$ on $\mathcal{W}$ along $\mathcal{U}$ is the vector

$$(20/11)\begin{pmatrix} 1 \\ 4 \end{pmatrix} = \begin{pmatrix} 20/11 \\ 80/11 \end{pmatrix}.$$

Both of these vectors are displayed in Figure 17.1.

It is informative to compare Figure 17.1 with Figures 12.1 and 12.3; the latter figures depict the orthogonal projection of $\mathbf{y}$ on $\mathcal{U}$ (which is identical to the projection of $\mathbf{y}$ on $\mathcal{U}$ along $\mathcal{U}^\perp$) and the orthogonal projection of $\mathbf{y}$ on $\mathcal{U}^\perp$ (which is identical to the projection of $\mathbf{y}$ on $\mathcal{U}^\perp$ along $\mathcal{U}$).

c. Three-dimensional example

Let us find the projection of an m-dimensional column vector $\mathbf{y}$ on a subspace $\mathcal{U}$ of $\mathcal{R}^m$ along a subspace $\mathcal{W}$ of $\mathcal{R}^m$ in the special case where $m = 3$, $\mathbf{y} = (3,\ -38/5,\ 74/5)'$, $\mathcal{U} = \text{sp}\{\mathbf{u}_1, \mathbf{u}_2, \mathbf{u}_3\}$, and $\mathcal{W} = \text{sp}\{\mathbf{w}\}$, with

$$\mathbf{u}_1 = \begin{pmatrix} 0 \\ 3 \\ 6 \end{pmatrix}, \quad \mathbf{u}_2 = \begin{pmatrix} -2 \\ 2 \\ 4 \end{pmatrix}, \quad \mathbf{u}_3 = \begin{pmatrix} -2 \\ 1 \\ 2 \end{pmatrix}, \quad \mathbf{w} = \begin{pmatrix} -1 \\ -3 \\ 6 \end{pmatrix}.$$

{It is easy to show that $\mathbf{u}_3 = \mathbf{u}_2 - (1/3)\mathbf{u}_1$ and that $\mathbf{u}_1$, $\mathbf{u}_2$, and $\mathbf{w}$ are linearly independent, implying (in light of Lemma 17.1.1) that

$$\dim(\mathcal{U}) + \dim(\mathcal{W}) = 2 + 1 = 3 = \dim[\text{sp}(\mathbf{u}_1, \mathbf{u}_2, \mathbf{u}_3, \mathbf{w})] = \dim(\mathcal{U} + \mathcal{W})$$

and hence (in light of Corollaries 17.1.6 and 17.1.8) that $\mathcal{R}^3 = \mathcal{U} \oplus \mathcal{W}$.}

Let $\mathbf{U} = (\mathbf{u}_1, \mathbf{u}_2, \mathbf{u}_3)$, and take $\mathbf{x}^*$ to be any solution to the linear system

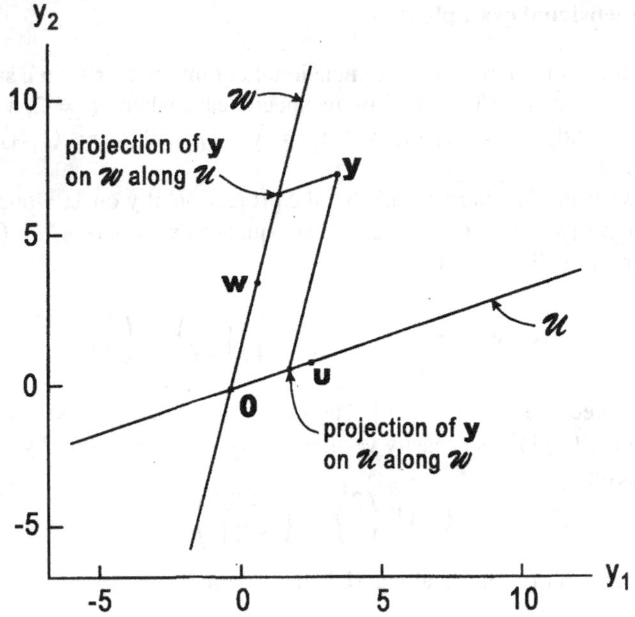

FIGURE 17.1: The projection of a two-dimensional column vector **y** on a one-dimensional subspace $\mathcal{U}$ of $\mathcal{R}^2$ along a one-dimensional subspace $\mathcal{W}$ of $\mathcal{R}^2$.

$$(\mathbf{U},\, \mathbf{w})\mathbf{x} = \mathbf{y} \;\Leftrightarrow\; \begin{pmatrix} 0 & -2 & -2 & -1 \\ 3 & 2 & 1 & -3 \\ 6 & 4 & 2 & 6 \end{pmatrix} \mathbf{x} = \begin{pmatrix} 3 \\ -38/5 \\ 74/5 \end{pmatrix} \tag{6.6}$$

(in the 4×1 vector **x**). Then, partitioning $\mathbf{x}^{*\prime}$ as $\mathbf{x}^{*\prime} = (\mathbf{x}_1^{*\prime},\, x_2^*)$ (where $\mathbf{x}_1^*$ has 3 elements), it follows from Corollary 17.6.5 that the projection of **y** on $\mathcal{U}$ along $\mathcal{W}$ equals $\mathbf{U}\mathbf{x}_1^*$ and that the projection of **y** on $\mathcal{W}$ along $\mathcal{U}$ equals $x_2^*\mathbf{w}$.

It is easy to verify that one solution to linear system (6.6) is $\mathbf{x} = (9/5,\, -11/4,\, 0,\, 5/2)'$. Thus, the projection of **y** on $\mathcal{U}$ along $\mathcal{W}$ is the vector

$$\begin{pmatrix} 0 & -2 & -2 \\ 3 & 2 & 1 \\ 6 & 4 & 2 \end{pmatrix} \begin{pmatrix} 9/5 \\ -11/4 \\ 0 \end{pmatrix} = \begin{pmatrix} 11/2 \\ -1/10 \\ -1/5 \end{pmatrix},$$

and the projection of **y** on $\mathcal{W}$ along $\mathcal{U}$ is the vector

$$(5/2) \begin{pmatrix} -1 \\ -3 \\ 6 \end{pmatrix} = \begin{pmatrix} -5/2 \\ -15/2 \\ 15 \end{pmatrix}.$$

Both of these vectors are displayed in Figure 17.2.

It is informative to compare Figure 17.2 with Figures 12.2 and 12.4; the latter figures depict the orthogonal projection of **y** on $\mathcal{U}$ (which is identical to the pro-

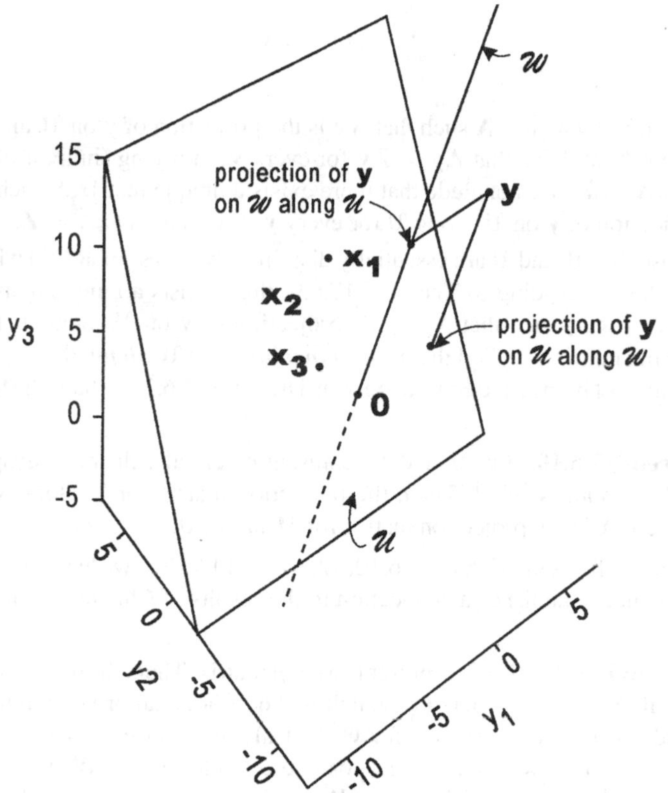

FIGURE 17.2: The projection of a three-dimensional column vector **y** on a two-dimensional subspace $\mathcal{U}$ of $\mathcal{R}^3$ along a one-dimensional subspace $\mathcal{W}$ of $\mathcal{R}^3$.

jection of **y** on $\mathcal{U}$ along $\mathcal{U}^\perp$) and the orthogonal projection of **y** on $\mathcal{U}^\perp$ (which is identical to the projection of **y** on $\mathcal{U}^\perp$ along $\mathcal{U}$).

d. Projection matrices

The concept of a projection matrix was introduced in Section 12.3 and was extended in Section 14.12d. The following theorem provides a basis for a further extension of this concept.

Theorem 17.6.9. Let $\mathcal{U}$ and $\mathcal{V}$ represent essentially disjoint subspaces (of $\mathcal{R}^{n\times 1}$) whose sum is $\mathcal{R}^{n\times 1}$. Then, there exists a unique matrix $\mathbf{A}$ (of dimensions $n \times n$) such that $\mathbf{Ay}$ is the projection of $\mathbf{y}$ on $\mathcal{U}$ along $\mathcal{V}$ for every $\mathbf{y}$ in $\mathcal{R}^n$. Moreover, $\mathbf{A} = (\mathbf{z}_1, \ldots, \mathbf{z}_n)$, where, letting $\mathbf{e}_j$ represent the jth column of $\mathbf{I}_n$, $\mathbf{z}_j$ is the projection of $\mathbf{e}_j$ on $\mathcal{U}$ along $\mathcal{V}$ $(j = 1, \ldots, n)$.

Proof. Define $\mathbf{Z} = (\mathbf{z}_1, \ldots, \mathbf{z}_n)$. Then, for every n-dimensional column vector $\mathbf{y} = (y_1, \ldots, y_n)'$, we have that $\mathbf{y} = \sum_{j=1}^n y_j \mathbf{e}_j$, implying (in light of Theorem 17.6.8) that the projection of $\mathbf{y}$ on $\mathcal{U}$ along $\mathcal{V}$ equals

$$\sum_{j=1}^{n} y_j \mathbf{z}_j = \mathbf{Z}\mathbf{y}.$$

Moreover, for any matrix $\mathbf{A}$ such that $\mathbf{Ay}$ is the projection of $\mathbf{y}$ on $\mathcal{U}$ along $\mathcal{V}$ for every $\mathbf{y}$ in $\mathcal{R}^n$, we have that $\mathbf{Ay} = \mathbf{Zy}$ for every $\mathbf{y}$, implying (in light of Lemma 2.3.2) that $\mathbf{A} = \mathbf{Z}$. We conclude that there exists a unique matrix $\mathbf{A}$ such that $\mathbf{Ay}$ is the projection of $\mathbf{y}$ on $\mathcal{U}$ along $\mathcal{V}$ for every $\mathbf{y} \in \mathcal{R}^n$ and that $\mathbf{A} = \mathbf{Z}$. Q.E.D.

Suppose that $\mathcal{U}$ and $\mathcal{V}$ are essentially disjoint subspaces (of $\mathcal{R}^{n\times 1}$) whose sum is $\mathcal{R}^{n\times 1}$. Then, according to Theorem 17.6.9, there exists a unique matrix $\mathbf{A}$ (of dimensions $n \times n$) such that $\mathbf{Ay}$ is the projection of $\mathbf{y}$ on $\mathcal{U}$ along $\mathcal{V}$ for every $\mathbf{y} \in \mathcal{R}^n$. This matrix is called the *projection matrix for* $\mathcal{U}$ *along* $\mathcal{V}$.

As an almost immediate consequence of Theorem 17.6.3, we have the following result.

Theorem 17.6.10. Let $\mathcal{U}$ and $\mathcal{V}$ represent essentially disjoint subspaces (of $\mathcal{R}^{n\times 1}$) whose sum is $\mathcal{R}^{n\times 1}$. Then, the projection matrix for $\mathcal{V}$ along $\mathcal{U}$ equals $\mathbf{I} - \mathbf{A}$, where $\mathbf{A}$ is the projection matrix for $\mathcal{U}$ along $\mathcal{V}$.

In light of Theorems 17.6.6, 17.6.10, 12.3.1, and 14.12.8, we have the following theorem, which establishes a connection to the results of Chapter 12 and Section 14.12.

Theorem 17.6.11. Let $\mathbf{X}$ represent an $n \times p$ matrix. Then, the projection matrix for $\mathcal{C}(\mathbf{X})$ along $\mathcal{C}^{\perp}(\mathbf{X})$ equals $\mathbf{P_X}$, which is the orthogonal projection matrix for $\mathcal{C}(\mathbf{X})$, and the projection matrix for $\mathcal{C}^{\perp}(\mathbf{X})$ along $\mathcal{C}(\mathbf{X})$ equals $\mathbf{I} - \mathbf{P_X}$. More generally, for any $n \times n$ symmetric positive definite matrix $\mathbf{W}$, the projection matrix for $\mathcal{C}(\mathbf{X})$ along $\mathcal{C}^{\perp}_{\mathbf{W}}(\mathbf{X})$ equals $\mathbf{P}_{\mathbf{X},\mathbf{W}}$, which is the orthogonal projection matrix for $\mathcal{C}(\mathbf{X})$ with respect to $\mathbf{W}$, and the projection matrix for $\mathcal{C}^{\perp}_{\mathbf{W}}(\mathbf{X})$ along $\mathcal{C}(\mathbf{X})$ equals $\mathbf{I} - \mathbf{P}_{\mathbf{X},\mathbf{W}}$.

The following theorem extends the results of Theorems 12.3.8 and 14.12.15.

Theorem 17.6.12. If an $n \times n$ matrix $\mathbf{A}$ is the projection matrix for a subspace $\mathcal{U}$ of $\mathcal{R}^{n\times 1}$ along a subspace $\mathcal{V}$ of $\mathcal{R}^{n\times 1}$ (where $\mathcal{U} \oplus \mathcal{V} = \mathcal{R}^{n\times 1}$), then $\mathcal{U} = \mathcal{C}(\mathbf{A})$ and $\mathcal{V} = \mathcal{C}(\mathbf{I} - \mathbf{A})$.

Proof. Suppose that $\mathbf{A}$ is the projection matrix for $\mathcal{U}$ along $\mathcal{V}$. Then, for any $\mathbf{y} \in \mathcal{R}^n$, $\mathbf{Ay}$ is the projection of $\mathbf{y}$ on $\mathcal{U}$ along $\mathcal{V}$ and hence is in $\mathcal{U}$. Thus, $\mathcal{C}(\mathbf{A}) \subset \mathcal{U}$. Moreover, as a consequence of Lemma 17.6.7, we have that, for any $\mathbf{y} \in \mathcal{U}$, $\mathbf{y} = \mathbf{Ay} \in \mathcal{C}(\mathbf{A})$. It follows that $\mathcal{U} \subset \mathcal{C}(\mathbf{A})$ and hence [since $\mathcal{C}(\mathbf{A}) \subset \mathcal{U}$] that $\mathcal{U} = \mathcal{C}(\mathbf{A})$.

Further, since (in light of Theorem 17.6.10) $\mathbf{I} - \mathbf{A}$ is the projection matrix for $\mathcal{V}$ along $\mathcal{U}$, it follows from a similar argument that $\mathcal{V} = \mathcal{C}(\mathbf{I} - \mathbf{A})$. Q.E.D.

The following theorem expands on the results of Theorem 12.3.9 and Corollary 14.12.17.

Theorem 17.6.13. An $n \times n$ matrix $\mathbf{A}$ is the projection matrix for some subspace $\mathcal{U}$ of $\mathcal{R}^{n\times 1}$ along some subspace $\mathcal{V}$ of $\mathcal{R}^{n\times 1}$ (where $\mathcal{U} \oplus \mathcal{V} = \mathcal{R}^{n\times 1}$) if and only if $\mathbf{A}$ is idempotent.

Proof. Suppose that $\mathbf{A}$ is idempotent. Then, according to Lemma 17.2.6, $\mathcal{C}(\mathbf{A})$ and $\mathcal{C}(\mathbf{I}-\mathbf{A})$ are essentially disjoint. And every $\mathbf{y}$ in $\mathcal{R}^n$ is expressible as $\mathbf{y} = \mathbf{A}\mathbf{y} + (\mathbf{I}-\mathbf{A})\mathbf{y}$. Thus, $\mathcal{C}(\mathbf{A}) \oplus \mathcal{C}(\mathbf{I}-\mathbf{A}) = \mathcal{R}^n$. Further, for every $\mathbf{y}$ in $\mathcal{R}^n$, $\mathbf{y} - \mathbf{A}\mathbf{y} = (\mathbf{I}-\mathbf{A})\mathbf{y}$, implying [since $\mathbf{A}\mathbf{y} \in \mathcal{C}(\mathbf{A})$ and $(\mathbf{I}-\mathbf{A})\mathbf{y} \in \mathcal{C}(\mathbf{I}-\mathbf{A})$] that $\mathbf{A}\mathbf{y}$ is the projection of $\mathbf{y}$ on $\mathcal{C}(\mathbf{A})$ along $\mathcal{C}(\mathbf{I}-\mathbf{A})$. It follows that $\mathbf{A}$ is the projection matrix for $\mathcal{C}(\mathbf{A})$ along $\mathcal{C}(\mathbf{I}-\mathbf{A})$.

Conversely, suppose that $\mathbf{A}$ is the projection matrix for some subspace $\mathcal{U}$ along some subspace $\mathcal{V}$ (where $\mathcal{U} \oplus \mathcal{V} = \mathcal{R}^{n\times 1}$). Then, it follows from Theorem 17.6.12 that $\mathcal{C}(\mathbf{A}) = \mathcal{U}$ and $\mathcal{C}(\mathbf{I}-\mathbf{A}) = \mathcal{V}$, implying that $\mathcal{C}(\mathbf{A})$ and $\mathcal{C}(\mathbf{I}-\mathbf{A})$ are essentially disjoint and hence (in light of Lemma 17.2.6) that $\mathbf{A}$ is idempotent. Q.E.D.

The results of Theorems 17.6.12 and 17.6.13 can be combined and (in light of Theorem 11.7.1) restated as the following theorem.

Theorem 17.6.14. If an $n \times n$ matrix $\mathbf{A}$ is the projection matrix for a subspace $\mathcal{U}$ of $\mathcal{R}^{n\times 1}$ along a subspace $\mathcal{V}$ of $\mathcal{R}^{n\times 1}$ (where $\mathcal{U} \oplus \mathcal{V} = \mathcal{R}^{n\times 1}$), then $\mathbf{A}$ is idempotent, $\mathcal{U} = \mathcal{C}(\mathbf{A})$, and $\mathcal{V} = \mathcal{C}(\mathbf{I}-\mathbf{A}) = \mathcal{N}(\mathbf{A})$. And, if an $n \times n$ matrix $\mathbf{A}$ is idempotent, then $\mathcal{C}(\mathbf{A}) \oplus \mathcal{C}(\mathbf{I}-\mathbf{A}) = \mathcal{R}^n$ [or, equivalently, $\mathcal{C}(\mathbf{A}) \oplus \mathcal{N}(\mathbf{A}) = \mathcal{R}^n$], and $\mathbf{A}$ is the projection matrix for $\mathcal{C}(\mathbf{A})$ along $\mathcal{C}(\mathbf{I}-\mathbf{A})$ [or, equivalently, along $\mathcal{N}(\mathbf{A})$].

17.7 Some Further Results on the Essential Disjointness and Orthogonality of Subspaces and on Projections and Projection Matrices

Let $\mathcal{U}$ and $\mathcal{W}$ represent two subspaces of $\mathcal{R}^{n\times 1}$. Recall (from Section 17.1) that if $\mathcal{U}$ and $\mathcal{W}$ are orthogonal, then they are essentially disjoint. Recall also that $\mathcal{U}$ and $\mathcal{W}$ can be essentially disjoint without being orthogonal with respect to the usual inner product (or with respect to whatever other inner product may have been adopted). Are essentially disjoint subspaces of $\mathcal{R}^{n\times 1}$ necessarily orthogonal with respect to some inner product? This question is answered (in the affirmative) by the following theorem.

Theorem 17.7.1. Let $\mathcal{U}$ and $\mathcal{W}$ represent essentially disjoint subspaces of $\mathcal{R}^{n\times 1}$. Further, letting $s = \dim(\mathcal{U})$ and $t = \dim(\mathcal{W})$ (and supposing that $s > 0$ and $t > 0$), take $\mathbf{U}$ to be any $n \times s$ matrix such that $\mathcal{C}(\mathbf{U}) = \mathcal{U}$, $\mathbf{W}$ to be any $n \times t$ matrix such that $\mathcal{C}(\mathbf{W}) = \mathcal{W}$, and $\mathbf{Z}$ to be any $n \times (n-s-t)$ matrix such that the $n \times n$ partitioned matrix $(\mathbf{U}, \mathbf{W}, \mathbf{Z})$ is nonsingular. Define $\mathbf{B} = (\mathbf{U}, \mathbf{W}, \mathbf{Z})^{-1}$, partition $\mathbf{B}$ as $\mathbf{B} = \begin{pmatrix} \mathbf{B}_1 \\ \mathbf{B}_2 \\ \mathbf{B}_3 \end{pmatrix}$ (where $\mathbf{B}_1$ has s rows and $\mathbf{B}_2$ has t rows), and let $\mathbf{H} = \mathbf{B}'\mathbf{B}$ or (more generally) let $\mathbf{H}$ represent any matrix of the form

$$\mathbf{H} = \mathbf{B}_1'\mathbf{A}_1\mathbf{B}_1 + \mathbf{B}_2'\mathbf{A}_2\mathbf{B}_2 + \mathbf{B}_3'\mathbf{A}_3\mathbf{B}_3,$$

where $\mathbf{A}_1, \mathbf{A}_2$, and $\mathbf{A}_3$ are symmetric positive definite matrices. Then, $\mathcal{U}$ and $\mathcal{W}$ are orthogonal with respect to $\mathbf{H}$.

In connection with Theorem 17.7.1, observe (in light of Theorem 17.2.4) that $\operatorname{rank}(\mathbf{U}, \mathbf{W}) = s + t$ [or, equivalently, that the columns of $(\mathbf{U}, \mathbf{W})$ are linearly independent], insuring (in light of Theorem 4.3.12) the existence of $n - s - t$ n-dimensional column vectors (the columns of $\mathbf{Z}$) that, together with the columns of $(\mathbf{U}, \mathbf{W})$, form a basis for $\mathcal{R}^n$ (and hence form a set of n linearly independent vectors). Observe also that $\mathbf{H} = \mathbf{B}'\operatorname{diag}(\mathbf{A}_1, \mathbf{A}_2, \mathbf{A}_3)\mathbf{B}$ and that (according to Lemma 14.8.3) $\operatorname{diag}(\mathbf{A}_1, \mathbf{A}_2, \mathbf{A}_3)$ is positive definite, insuring (in light of Corollary 14.2.10) that $\mathbf{H}$ is positive definite. Further, in the degenerate special case of Theorem 17.7.1 where $s + t = n$ (or, equivalently, where $\mathcal{U} \oplus \mathcal{W} = \mathcal{R}^{n\times 1}$), interpret $(\mathbf{U}, \mathbf{W}, \mathbf{Z})$ as $(\mathbf{U}, \mathbf{W})$, $\begin{pmatrix} \mathbf{B}_1 \\ \mathbf{B}_2 \\ \mathbf{B}_3 \end{pmatrix}$ as $\begin{pmatrix} \mathbf{B}_1 \\ \mathbf{B}_2 \end{pmatrix}$, and $\mathbf{B}_1'\mathbf{A}_1\mathbf{B}_1 + \mathbf{B}_2'\mathbf{A}_2\mathbf{B}_2 + \mathbf{B}_3'\mathbf{A}_3\mathbf{B}_3$ as $\mathbf{B}_1'\mathbf{A}_1\mathbf{B}_1 + \mathbf{B}_2'\mathbf{A}_2\mathbf{B}_2$.

Proof (of Theorem 17.7.1). In light of Lemma 14.12.1, it suffices to show that $\mathbf{U}'\mathbf{H}\mathbf{W} = \mathbf{0}$. By definition,

$$\begin{pmatrix} \mathbf{B}_1\mathbf{U} & \mathbf{B}_1\mathbf{W} & \mathbf{B}_1\mathbf{Z} \\ \mathbf{B}_2\mathbf{U} & \mathbf{B}_2\mathbf{W} & \mathbf{B}_2\mathbf{Z} \\ \mathbf{B}_3\mathbf{U} & \mathbf{B}_3\mathbf{W} & \mathbf{B}_3\mathbf{Z} \end{pmatrix} = \mathbf{B}(\mathbf{U}, \mathbf{W}, \mathbf{Z}) = \mathbf{I}_n = \begin{pmatrix} \mathbf{I}_s & \mathbf{0} & \mathbf{0} \\ \mathbf{0} & \mathbf{I}_t & \mathbf{0} \\ \mathbf{0} & \mathbf{0} & \mathbf{I}_{n-s-t} \end{pmatrix},$$

implying in particular that $\mathbf{B}_1\mathbf{W} = \mathbf{0}$, $\mathbf{B}_2\mathbf{U} = \mathbf{0}$, and $\mathbf{B}_3\mathbf{W} = \mathbf{0}$. Thus,

$$\mathbf{U}'\mathbf{H}\mathbf{W} = \mathbf{U}'\mathbf{B}_1'\mathbf{A}_1\mathbf{B}_1\mathbf{W} + (\mathbf{B}_2\mathbf{U})'\mathbf{A}_2\mathbf{B}_2\mathbf{W} + \mathbf{U}'\mathbf{B}_3'\mathbf{A}_3\mathbf{B}_3\mathbf{W} = \mathbf{0} + \mathbf{0} + \mathbf{0} = \mathbf{0}.$$

Q.E.D.

In discussing the implications of Theorem 17.7.1, it will be convenient to have at our disposal the following lemma.

Lemma 17.7.2. Let $\mathcal{U}$ and $\mathcal{W}$ represent subspaces of a linear space $\mathcal{V}$ of $m \times n$ matrices. If $\mathcal{U} \perp \mathcal{W}$ and if $\mathcal{U} + \mathcal{W} = \mathcal{V}$, then $\mathcal{W} = \mathcal{U}^\perp$ and $\mathcal{U} = \mathcal{W}^\perp$.

Proof. Suppose that $\mathcal{U} \perp \mathcal{W}$ and that $\mathcal{U} + \mathcal{W} = \mathcal{V}$. Then, clearly, $\mathcal{W} \subset \mathcal{U}^\perp$, and $\mathcal{U} \subset \mathcal{W}^\perp$. Moreover, $\mathcal{V}$ and $\mathcal{W}$ are (in light of Lemma 17.1.9) essentially disjoint, implying (in light of Corollary 17.1.6) that

$$\dim(\mathcal{V}) = \dim(\mathcal{U}) + \dim(\mathcal{W}).$$

Thus, making use of Theorem 12.5.12, we find that

$$\dim(\mathcal{W}) = \dim(\mathcal{V}) - \dim(\mathcal{U}) = \dim(\mathcal{U}^\perp)$$

and, similarly, that

$$\dim(\mathcal{U}) = \dim(\mathcal{V}) - \dim(\mathcal{W}) = \dim(\mathcal{W}^\perp).$$

We conclude (on the basis of Theorem 4.3.10) that $\mathcal{W} = \mathcal{U}^\perp$ and that $\mathcal{U} = \mathcal{W}^\perp$.

Q.E.D.

Let $\mathcal{V}$ represent a linear space of $n \times 1$ vectors, and let $\mathcal{U}$ and $\mathcal{W}$ represent essentially disjoint subspaces whose sum is $\mathcal{V}$. Then, according to Theorem 17.7.1,

$\mathcal{U}$ and $\mathcal{V}$ are orthogonal with respect to some symmetric positive definite matrix $\mathbf{H}$, implying (in light of Lemma 17.7.2) that $\mathcal{W}$ equals the orthogonal complement $\mathcal{U}^{\perp}$ of $\mathcal{U}$ (where the orthogonality in the orthogonal complement is with respect to the bilinear form $\mathbf{x}'\mathbf{H}\mathbf{y}$). Thus, it follows from Theorem 17.6.6 that, for any vector $\mathbf{y}$ in $\mathcal{V}$, the projection of $\mathbf{y}$ on $\mathcal{U}$ along $\mathcal{W}$ equals the orthogonal projection of $\mathbf{y}$ on $\mathcal{U}$ with respect to $\mathbf{H}$. Further, letting $\mathbf{X}$ represent any matrix such that $\mathcal{C}(\mathbf{X}) = \mathcal{U}$, it follows from Theorem 17.6.11 that, in the special case where $\mathcal{V} = \mathcal{R}^{n\times 1}$ (and hence where $\mathcal{U} \oplus \mathcal{W} = \mathcal{R}^{n\times 1}$), the projection matrix for $\mathcal{U}$ along $\mathcal{W}$ equals the orthogonal projection matrix $\mathbf{P}_{\mathbf{X},\mathbf{H}}$ for $\mathcal{U}$ with respect to $\mathbf{H}$.

Exercises

Section 17.1

1. Let $\mathcal{U}$ and $\mathcal{V}$ represent subspaces of $\mathcal{R}^{m\times n}$.

   (a) Show that $\mathcal{U} \cup \mathcal{V} \subset \mathcal{U} + \mathcal{V}$.

   (b) Show that $\mathcal{U} + \mathcal{V}$ is the smallest subspace (of $\mathcal{R}^{m\times n}$) that contains $\mathcal{U} \cup \mathcal{V}$, or, equivalently [in light of Part (a)], show that, for any subspace $\mathcal{W}$ such that $\mathcal{U} \cup \mathcal{V} \subset \mathcal{W}$, $\mathcal{U} + \mathcal{V} \subset \mathcal{W}$.

2. Let $\mathbf{A} = \begin{pmatrix} 1 & 0 \\ 0 & 1 \\ 0 & 0 \end{pmatrix}$ and $\mathbf{B} = \begin{pmatrix} 0 & 2 \\ 1 & 1 \\ 2 & 3 \end{pmatrix}$. Find (a) a basis for $\mathcal{C}(\mathbf{A}) + \mathcal{C}(\mathbf{B})$, (b) a basis for $\mathcal{C}(\mathbf{A}) \cap \mathcal{C}(\mathbf{B})$, and (c) a vector in $\mathcal{C}(\mathbf{A}) + \mathcal{C}(\mathbf{B})$ that is not in $\mathcal{C}(\mathbf{A}) \cup \mathcal{C}(\mathbf{B})$.

3. Let $\mathcal{U}$, $\mathcal{W}$, and $\mathcal{X}$ represent subspaces of a linear space $\mathcal{V}$ of matrices, and let $\mathbf{Y}$ represent an arbitrary matrix in $\mathcal{V}$.

   (a) Show (1) that if $\mathbf{Y} \perp \mathcal{W}$ and $\mathbf{Y} \perp \mathcal{X}$, then $\mathbf{Y} \perp (\mathcal{W} + \mathcal{X})$, and (2) that if $\mathcal{U} \perp \mathcal{W}$ and $\mathcal{U} \perp \mathcal{X}$, then $\mathcal{U} \perp (\mathcal{W} + \mathcal{X})$.

   (b) Show (1) that $(\mathcal{U} + \mathcal{W})^{\perp} = \mathcal{U}^{\perp} \cap \mathcal{W}^{\perp}$ and (2) that $(\mathcal{U} \cap \mathcal{W})^{\perp} = \mathcal{U}^{\perp} + \mathcal{W}^{\perp}$.

4. Let $\mathcal{U}$, $\mathcal{W}$, and $\mathcal{X}$ represent subspaces of a linear space $\mathcal{V}$ of matrices.

   (a) Show that $(\mathcal{U} \cap \mathcal{W}) + (\mathcal{U} \cap \mathcal{X}) \subset \mathcal{U} \cap (\mathcal{W} + \mathcal{X})$.

   (b) Show (via an example) that $\mathcal{U} \cap \mathcal{W} = \{\mathbf{0}\}$ and $\mathcal{U} \cap \mathcal{X} = \{\mathbf{0}\}$ does not necessarily imply that $\mathcal{U} \cap (\mathcal{W} + \mathcal{X}) = \{\mathbf{0}\}$.

   (c) Show that if $\mathcal{W} \subset \mathcal{U}$, then (1) $\mathcal{U} + \mathcal{W} = \mathcal{U}$ and (2) $\mathcal{U} \cap (\mathcal{W} + \mathcal{X}) = \mathcal{W} + (\mathcal{U} \cap \mathcal{X})$.

5. Let $\mathcal{U}_1, \mathcal{U}_2, \ldots, \mathcal{U}_k$ represent subspaces of $\mathcal{R}^{m\times n}$. Show that if, for $j = 1, 2, \ldots, k$, $\mathcal{U}_j$ is spanned by a (finite nonempty) set of $(m \times n)$ matrices $\mathbf{U}_1^{(j)}, \ldots, \mathbf{U}_{r_j}^{(j)}$, then

$$\mathcal{U}_1 + \mathcal{U}_2 + \cdots + \mathcal{U}_k = \operatorname{sp}(\mathbf{U}_1^{(1)}, \ldots, \mathbf{U}_{r_1}^{(1)}, \mathbf{U}_1^{(2)}, \ldots, \mathbf{U}_{r_2}^{(2)}, \ldots, \mathbf{U}_1^{(k)}, \ldots, \mathbf{U}_{r_k}^{(k)}).$$

6. Let $\mathcal{U}_1, \ldots, \mathcal{U}_k$ represent subspaces of $\mathcal{R}^{m\times n}$. The k subspaces $\mathcal{U}_1, \ldots, \mathcal{U}_k$ are said to be *independent* if, for matrices $\mathbf{U}_1 \in \mathcal{U}_1, \ldots, \mathbf{U}_k \in \mathcal{U}_k$, the only solution to the matrix equation

$$\mathbf{U}_1 + \cdots + \mathbf{U}_k = \mathbf{0} \tag{E.1}$$

is $\mathbf{U}_1 = \cdots = \mathbf{U}_k = \mathbf{0}$.

(a) Show that $\mathcal{U}_1, \ldots, \mathcal{U}_k$ are independent if and only if, for $i = 2, \ldots, k$, $\mathcal{U}_i$ and $\mathcal{U}_1 + \cdots + \mathcal{U}_{i-1}$ are essentially disjoint.

(b) Show that $\mathcal{U}_1, \ldots, \mathcal{U}_k$ are independent if and only if, for $i = 1, \ldots, k$, $\mathcal{U}_i$ and $\mathcal{U}_1 + \cdots + \mathcal{U}_{i-1} + \mathcal{U}_{i+1} + \cdots + \mathcal{U}_k$ are essentially disjoint.

(c) Use the results of Exercise 3 [along with Part (a) or (b)] to show that if $\mathcal{U}_1, \ldots, \mathcal{U}_k$ are (pairwise) orthogonal, then they are independent.

(d) Assuming that $\mathcal{U}_1, \mathcal{U}_2, \ldots, \mathcal{U}_k$ are of dimension one or more and letting $\{\mathbf{U}_1^{(j)}, \ldots, \mathbf{U}_{r_j}^{(j)}\}$ represent any linearly independent set of matrices in $\mathcal{U}_j$ $(j = 1, 2, \ldots, k)$, show that if $\mathcal{U}_1, \mathcal{U}_2, \ldots, \mathcal{U}_k$ are independent, then the combined set $\{\mathbf{U}_1^{(1)}, \ldots, \mathbf{U}_{r_1}^{(1)}, \mathbf{U}_1^{(2)}, \ldots, \mathbf{U}_{r_2}^{(2)}, \ldots, \mathbf{U}_1^{(k)}, \ldots, \mathbf{U}_{r_k}^{(k)}\}$ is linearly independent.

(e) Assuming that $\mathcal{U}_1, \mathcal{U}_2, \ldots, \mathcal{U}_k$ are of dimension one or more, show that $\mathcal{U}_1, \mathcal{U}_2, \ldots, \mathcal{U}_k$ are independent if and only if, for every nonnull matrix $\mathbf{U}_1$ in $\mathcal{U}_1$, every nonnull matrix $\mathbf{U}_2$ in $\mathcal{U}_2$, ..., and every nonnull matrix $\mathbf{U}_k$ in $\mathcal{U}_k$, $\mathbf{U}_1, \mathbf{U}_2, \ldots, \mathbf{U}_k$ are linearly independent.

(f) For $j = 1, \ldots, k$, let $p_j = \dim(\mathcal{U}_j)$, and let S_j represent a basis for $\mathcal{U}_j$ $(j = 1, \ldots, k)$. Define S to be the set of $\sum_{j=1}^{k} p_j$ matrices obtained by combining all of the matrices in $S_1, \ldots, S_k$ into a single set. Use the result of Exercise 5 [along with Part (d)] to show that (1) if $\mathcal{U}_1, \ldots, \mathcal{U}_k$ are independent, then S is a basis for $\mathcal{U}_1 + \cdots + \mathcal{U}_k$; and (2) if $\mathcal{U}_1, \ldots, \mathcal{U}_k$ are not independent, then S contains a proper subset that is a basis for $\mathcal{U}_1 + \cdots + \mathcal{U}_k$.

(g) Show that (1) if $\mathcal{U}_1, \ldots, \mathcal{U}_k$ are independent, then

$$\dim(\mathcal{U}_1 + \cdots + \mathcal{U}_k) = \dim(\mathcal{U}_1) + \cdots + \dim(\mathcal{U}_k);$$

and (2) if $\mathcal{U}_1, \ldots, \mathcal{U}_k$ are not independent, then

$$\dim(\mathcal{U}_1 + \cdots + \mathcal{U}_k) < \dim(\mathcal{U}_1) + \cdots + \dim(\mathcal{U}_k).$$

Section 17.2

7. Let $\mathbf{A}_1, \ldots, \mathbf{A}_k$ represent matrices having the same number of rows, and let $\mathbf{B}_1, \ldots, \mathbf{B}_k$ represent matrices having the same number of columns. Adopting the terminology of Exercise 6, use Part (g) of that exercise to show (a) that if $\mathcal{C}(\mathbf{A}_1), \ldots, \mathcal{C}(\mathbf{A}_k)$ are independent, then

$$\text{rank}(\mathbf{A}_1, \ldots, \mathbf{A}_k) = \text{rank}(\mathbf{A}_1) + \cdots + \text{rank}(\mathbf{A}_k),$$

and if $\mathcal{C}(\mathbf{A}_1), \ldots, \mathcal{C}(\mathbf{A}_k)$ are not independent, then

$$\text{rank}(\mathbf{A}_1, \ldots, \mathbf{A}_k) < \text{rank}(\mathbf{A}_1) + \cdots + \text{rank}(\mathbf{A}_k),$$

and (b) that if $\mathcal{R}(\mathbf{B}_1), \ldots, \mathcal{R}(\mathbf{B}_k)$ are independent, then

$$\text{rank}\begin{pmatrix} \mathbf{B}_1 \\ \vdots \\ \mathbf{B}_k \end{pmatrix} = \text{rank}(\mathbf{B}_1) + \cdots + \text{rank}(\mathbf{B}_k),$$

and if $\mathcal{R}(\mathbf{B}_1), \ldots, \mathcal{R}(\mathbf{B}_k)$ are not independent, then

$$\text{rank}\begin{pmatrix} \mathbf{B}_1 \\ \vdots \\ \mathbf{B}_k \end{pmatrix} < \text{rank}(\mathbf{B}_1) + \cdots + \text{rank}(\mathbf{B}_k).$$

8. Show [by, e.g., using the result of Part (c)-(2) of Exercise 4 in combination with Corollary 17.2.9] that, for any $m \times n$ matrix $\mathbf{A}$ and any $m \times p$ matrix $\mathbf{B}$,

   (a) $\mathcal{C}[(\mathbf{I} - \mathbf{A}\mathbf{A}^-)\mathbf{B}] = \mathcal{C}(\mathbf{I} - \mathbf{A}\mathbf{A}^-) \cap \mathcal{C}(\mathbf{A}, \mathbf{B})$ and

   (b) $\mathcal{C}[(\mathbf{I} - \mathbf{P}_{\mathbf{A}})\mathbf{B}] = \mathcal{N}(\mathbf{A}') \cap \mathcal{C}(\mathbf{A}, \mathbf{B})$.

9. Let $\mathbf{A} = (\mathbf{T}, \mathbf{U})$ and $\mathbf{B} = (\mathbf{V}, \mathbf{0})$, where $\mathbf{T}$ is an $m \times p$ matrix, $\mathbf{U}$ an $m \times q$ matrix, and $\mathbf{V}$ an $n \times p$ matrix, and suppose that $\mathbf{U}$ is of full row rank. Show that $\mathcal{R}(\mathbf{A})$ and $\mathcal{R}(\mathbf{B})$ are essentially disjoint [even if $\mathcal{R}(\mathbf{T})$ and $\mathcal{R}(\mathbf{V})$ are not essentially disjoint].

10. To what extent does formula (2.15) simplify in (a) the special case where $\mathcal{C}(\mathbf{T})$ and $\mathcal{C}(\mathbf{U})$ are essentially disjoint [but $\mathcal{R}(\mathbf{T})$ and $\mathcal{R}(\mathbf{V})$ are not necessarily essentially disjoint] and (b) the special case where $\mathcal{R}(\mathbf{T})$ and $\mathcal{R}(\mathbf{V})$ are essentially disjoint.

Section 17.3

11. Let $\mathbf{T}$ represent an $m \times p$ matrix, $\mathbf{U}$ an $m \times q$ matrix, and $\mathbf{V}$ an $n \times p$ matrix. Further, define $\mathbf{E}_T = \mathbf{I} - \mathbf{T}\mathbf{T}^-$, $\mathbf{F}_T = \mathbf{I} - \mathbf{T}^-\mathbf{T}$, $\mathbf{X} = \mathbf{E}_T\mathbf{U}$, and $\mathbf{Y} = \mathbf{V}\mathbf{F}_T$. Show (a) that the partitioned matrix $\begin{pmatrix} \mathbf{T}^- - \mathbf{T}^-\mathbf{U}\mathbf{X}^-\mathbf{E}_T \\ \mathbf{X}^-\mathbf{E}_T \end{pmatrix}$ is a generalized

inverse of the partitioned matrix $(\mathbf{T}, \mathbf{U})$ and (b) that the partitioned matrix $(\mathbf{T}^- - \mathbf{F}_T\mathbf{Y}^-\mathbf{V}\mathbf{T}^-, \mathbf{F}_T\mathbf{Y}^-)$ is a generalized inverse of the partitioned matrix $\begin{pmatrix}\mathbf{T}\\ \mathbf{V}\end{pmatrix}$. Do so by applying formula (E.1) from Part (a) of Exercise 10.10 to the partitioned matrices $\begin{pmatrix}\mathbf{T} & \mathbf{U}\\ \mathbf{0} & \mathbf{0}\end{pmatrix}$ and $\begin{pmatrix}\mathbf{T} & \mathbf{0}\\ \mathbf{V} & \mathbf{0}\end{pmatrix}$ and by making use of Theorem 17.3.3.

12. Let $\mathbf{T}$ represent an $m \times p$ matrix, $\mathcal{U}$ an $m \times q$ matrix, and $\mathbf{V}$ an $n \times p$ matrix. And let $\begin{pmatrix}\mathbf{G}_{11} & \mathbf{G}_{12}\\ \mathbf{G}_{21} & \mathbf{G}_{22}\end{pmatrix}$ (where $\mathbf{G}_{11}$ is of dimensions $p \times m$) represent a generalized inverse of the partitioned matrix $\begin{pmatrix}\mathbf{T} & \mathbf{U}\\ \mathbf{V} & \mathbf{0}\end{pmatrix}$. Show that (a) if $\mathbf{G}_{11}$ is a generalized inverse of $\mathbf{T}$ and $\mathbf{G}_{12}$ a generalized inverse of $\mathbf{V}$, then $\mathcal{R}(\mathbf{T})$ and $\mathcal{R}(\mathbf{V})$ are essentially disjoint, and (b) if $\mathbf{G}_{11}$ is a generalized inverse of $\mathbf{T}$ and $\mathbf{G}_{21}$ a generalized inverse of $\mathbf{U}$, then $\mathcal{C}(\mathbf{T})$ and $\mathcal{C}(\mathbf{U})$ are essentially disjoint.

Section 17.4

13. (a) Generalize Theorem 17.4.1 by showing that, for any subspaces $\mathcal{U}_1, \ldots, \mathcal{U}_k$ of $\mathcal{R}^{m\times n}$,

$$\begin{aligned}\dim(\mathcal{U}_1 + \cdots + \mathcal{U}_k) &\\ = \dim(\mathcal{U}_1) + \cdots + \dim(\mathcal{U}_k) &\\ - \sum_{i=2}^{k} \dim[(\mathcal{U}_1 + \cdots + \mathcal{U}_{i-1}) \cap \mathcal{U}_i]. &\quad \text{(E.2)}\end{aligned}$$

(b) Generalize Theorem 17.4.2 by showing that, for any matrices $\mathbf{A}_1, \ldots, \mathbf{A}_k$ having the same number of rows,

$$\operatorname{rank}(\mathbf{A}_1, \ldots, \mathbf{A}_k) = \operatorname{rank}(\mathbf{A}_1) + \cdots + \operatorname{rank}(\mathbf{A}_k) - \sum_{i=2}^{k} \dim[\mathcal{C}(\mathbf{A}_1, \ldots \mathbf{A}_{i-1}) \cap \mathcal{C}(\mathbf{A}_i)]$$

and, for any matrices $\mathbf{B}_1, \ldots, \mathbf{B}_k$ having the same number of columns,

$$\operatorname{rank}\begin{pmatrix}\mathbf{B}_1\\ \vdots\\ \mathbf{B}_k\end{pmatrix} = \operatorname{rank}(\mathbf{B}_1) + \cdots + \operatorname{rank}(\mathbf{B}_k) - \sum_{i=2}^{k} \dim\left[\mathcal{R}\begin{pmatrix}\mathbf{B}_1\\ \vdots\\ \mathbf{B}_{i-1}\end{pmatrix} \cap \mathcal{R}(\mathbf{B}_i)\right].$$

Section 17.5

14. Show that, for any $m \times n$ matrix $\mathbf{A}$, $n \times q$ matrix $\mathbf{C}$, and $q \times p$ matrix $\mathbf{B}$,

$$\begin{aligned}\text{rank}\{[\mathbf{I} - \mathbf{CB}(\mathbf{CB})^{-}]\mathbf{C}[\mathbf{I} - (\mathbf{AC})^{-}\mathbf{AC}]\} &= \text{rank}(\mathbf{A}) + \text{rank}(\mathbf{C}) - \text{rank}(\mathbf{AC}) - n \\ &\quad + \text{rank}\{[\mathbf{I} - \mathbf{CB}(\mathbf{CB})^{-}](\mathbf{I} - \mathbf{A}^{-}\mathbf{A})\}.\end{aligned}$$

[*Hint*. Apply equality (5.8) to the product $\mathbf{A}(\mathbf{CB})$, and make use of equality (5.5).]

Section 17.6

15. Show that if an $n \times n$ matrix $\mathbf{A}$ is the projection matrix for a subspace $\mathcal{U}$ of $\mathcal{R}^{n\times 1}$ along a subspace $\mathcal{V}$ of $\mathcal{R}^{n\times 1}$ (where $\mathcal{U} \oplus \mathcal{V} = \mathcal{R}^{n\times 1}$), then $\mathbf{A}'$ is the projection matrix for $\mathcal{V}^{\perp}$ along $\mathcal{U}^{\perp}$ [where $\mathcal{U}^{\perp}$ and $\mathcal{V}^{\perp}$ are the orthogonal complements (with respect to the usual inner product and relative to $\mathcal{R}^{n\times 1}$) of $\mathcal{U}$ and $\mathcal{V}$, respectively].

16. Show that, for any $n \times p$ matrix $\mathbf{X}$, $\mathbf{XX}^{-}$ is the projection matrix for $\mathcal{C}(\mathbf{X})$ along $\mathcal{N}(\mathbf{XX}^{-})$.

17. Let $\mathbf{Y}$ represent a matrix in a linear space $\mathcal{V}$ of $m \times n$ matrices, and let $\mathcal{U}_1, \ldots, \mathcal{U}_k$ represent subspaces of $\mathcal{V}$. Adopting the terminology and using the results of Exercise 6, show that if $\mathcal{U}_1, \ldots, \mathcal{U}_k$ are independent and if $\mathcal{U}_1 + \cdots + \mathcal{U}_k = \mathcal{V}$, then (a) there exist unique matrices $\mathbf{Z}_1, \ldots, \mathbf{Z}_k$ in $\mathcal{U}_1, \ldots, \mathcal{U}_k$, respectively, such that $\mathbf{Y} = \mathbf{Z}_1 + \cdots + \mathbf{Z}_k$, and (b) for $i = 1, \ldots, k$, $\mathbf{Z}_i$ equals the projection of $\mathbf{Y}$ on $\mathcal{U}_i$ along $\mathcal{U}_1 + \cdots + \mathcal{U}_{i-1} + \mathcal{U}_{i+1} + \cdots + \mathcal{U}_k$.

18. Let $\mathcal{U}$ and $\mathcal{W}$ represent essentially disjoint subspaces (of $\mathcal{R}^{n\times 1}$) whose sum is $\mathcal{R}^{n\times 1}$, and let $\mathbf{U}$ represent any $n \times s$ matrix such that $\mathcal{C}(\mathbf{U}) = \mathcal{U}$ and $\mathbf{W}$ any $n \times t$ matrix such that $\mathcal{C}(\mathbf{W}) = \mathcal{W}$.

    (a) Show that the $n \times (s + t)$ partitioned matrix $(\mathbf{U}, \mathbf{W})$ has a right inverse.

    (b) Taking $\mathbf{R}$ to be an arbitrary right inverse of $(\mathbf{U}, \mathbf{W})$ and partitioning $\mathbf{R}$ as $\mathbf{R} = \begin{pmatrix}\mathbf{R}_1 \\ \mathbf{R}_2\end{pmatrix}$ (where $\mathbf{R}_1$ has s rows), show that the projection matrix for $\mathcal{U}$ along $\mathcal{W}$ equals $\mathbf{UR}_1$ and that the projection matrix for $\mathcal{W}$ along $\mathcal{U}$ equals $\mathbf{WR}_2$.

19. Let $\mathbf{A}$ represent the $(n \times n)$ projection matrix for a subspace $\mathcal{U}$ of $\mathcal{R}^{n\times 1}$ along a subspace $\mathcal{V}$ of $\mathcal{R}^{n\times 1}$ (where $\mathcal{U} \oplus \mathcal{V} = \mathcal{R}^{n\times 1}$), let $\mathbf{B}$ represent the $(n \times n)$ projection matrix for a subspace $\mathcal{W}$ of $\mathcal{R}^{n\times 1}$ along a subspace $\mathcal{X}$ of $\mathcal{R}^{n\times 1}$ (where $\mathcal{W} \oplus \mathcal{X} = \mathcal{R}^{n\times 1}$), and suppose that $\mathbf{A}$ and $\mathbf{B}$ commute (i.e., that $\mathbf{BA} = \mathbf{AB}$).

(a) Show that $\mathbf{AB}$ is the projection matrix for $\mathcal{U} \cap \mathcal{W}$ along $\mathcal{V} + \mathcal{X}$.

(b) Show that $\mathbf{A} + \mathbf{B} - \mathbf{AB}$ is the projection matrix for $\mathcal{U} + \mathcal{W}$ along $\mathcal{V} \cap \mathcal{X}$. [*Hint for Part (b)*. Observe that $\mathbf{I} - (\mathbf{A} + \mathbf{B} - \mathbf{AB}) = (\mathbf{I} - \mathbf{A})(\mathbf{I} - \mathbf{B})$, and make use of Part (a).]

20. Let $\mathcal{V}$ represent a linear space of n-dimensional column vectors, and let $\mathcal{U}$ and $\mathcal{W}$ represent essentially disjoint subspaces whose sum is $\mathcal{V}$. Then, an $n \times n$ matrix $\mathbf{A}$ is said to be a *projection matrix for* $\mathcal{U}$ *along* $\mathcal{W}$ if $\mathbf{Ay}$ is the projection of $\mathbf{y}$ on $\mathcal{U}$ along $\mathcal{W}$ for every $\mathbf{y} \in \mathcal{V}$—this represents an extension of the definition of a projection matrix for $\mathcal{U}$ along $\mathcal{W}$ given in Section 17.6d (in the special case where $\mathcal{V} = \mathcal{R}^n$). Further, let $\mathbf{U}$ represent an $n \times s$ matrix such that $\mathcal{C}(\mathbf{U}) = \mathcal{U}$, and let $\mathbf{W}$ represent an $n \times t$ matrix such that $\mathcal{C}(\mathbf{W}) = \mathcal{W}$.

(a) Show that an $n \times n$ matrix $\mathbf{A}$ is a projection matrix for $\mathcal{U}$ along $\mathcal{W}$ if and only if $\mathbf{AU} = \mathbf{U}$ and $\mathbf{AW} = \mathbf{0}$ or, equivalently, if and only if $\mathbf{A}'$ is a solution to the linear system $\begin{pmatrix} \mathbf{U}' \\ \mathbf{W}' \end{pmatrix} \mathbf{B} = \begin{pmatrix} \mathbf{U}' \\ \mathbf{0} \end{pmatrix}$ (in an $n \times n$ matrix $\mathbf{B}$).

(b) Establish the existence of a projection matrix for $\mathcal{U}$ along $\mathcal{W}$.

(c) Show that if $\mathbf{A}$ is a projection matrix for $\mathcal{U}$ along $\mathcal{W}$, then $\mathbf{I} - \mathbf{A}$ is a projection matrix for $\mathcal{W}$ along $\mathcal{U}$ (thereby generalizing Theorem 17.6.10).

(d) Let $\mathbf{X}$ represent any $n \times p$ matrix whose columns span $\mathcal{N}(\mathbf{W}')$ or, equivalently, $\mathcal{W}^{\perp}$. Show that an $n \times n$ matrix $\mathbf{A}$ is a projection matrix for $\mathcal{U}$ along $\mathcal{W}$ if and only if $\mathbf{A}' = \mathbf{XR}_*$ for some solution $\mathbf{R}_*$ to the linear system $\mathbf{U}'\mathbf{XR} = \mathbf{U}'$ (in a $p \times n$ matrix $\mathbf{R}$).

Section 17.7

21. Let $\mathcal{U}_1, \ldots, \mathcal{U}_k$ represent independent subspaces of $\mathcal{R}^{n \times 1}$ such that $\mathcal{U}_1 + \cdots + \mathcal{U}_k = \mathcal{R}^{n \times 1}$ (where the independence of subspaces is as defined in Exercise 6). Further, letting $s_i = \dim(\mathcal{U}_i)$ (and supposing that $s_i > 0$), take $\mathbf{U}_i$ to be any $n \times s_i$ matrix such that $\mathcal{C}(\mathbf{U}_i) = \mathcal{U}_i$ $(i = 1, \ldots, k)$. And define $\mathbf{B} = (\mathbf{U}_1, \ldots, \mathbf{U}_k)^{-1}$, partition $\mathbf{B}$ as $\mathbf{B} = \begin{pmatrix} \mathbf{B}_1 \\ \vdots \\ \mathbf{B}_k \end{pmatrix}$ (where, for $i = 1, \ldots, k$, $\mathbf{B}_i$ has s_i rows), and let $\mathbf{H} = \mathbf{B}'\mathbf{B}$ or (more generally) let $\mathbf{H}$ represent any matrix of the form

$$\mathbf{H} = \mathbf{B}_1'\mathbf{A}_1\mathbf{B}_1 + \mathbf{B}_2'\mathbf{A}_2\mathbf{B}_2 + \cdots + \mathbf{B}_k'\mathbf{A}_k\mathbf{B}_k, \qquad \text{(E.3)}$$

where $\mathbf{A}_1, \mathbf{A}_2, \ldots, \mathbf{A}_k$ are symmetric positive definite matrices.

(a) Using the result of Part (g)-(1) of Exercise 6 (or otherwise), verify that the partitioned matrix $(\mathbf{U}_1, \ldots, \mathbf{U}_k)$ is nonsingular (i.e., is square and of rank n).

(b) Show that $\mathbf{H}$ is positive definite.

(c) Show that (for $j \neq i = 1, \ldots, k$) $\mathcal{U}_i$ and $\mathcal{U}_j$ are orthogonal with respect to $\mathbf{H}$.

(d) Using the result of Part(a)-(2) of Exercise 3 (or otherwise), show that, for $i = 1, \ldots, k$, (1) $\mathcal{U}_1 + \cdots + \mathcal{U}_{i-1} + \mathcal{U}_{i+1} + \cdots + \mathcal{U}_k$ equals the orthogonal complement $\mathcal{U}_i^{\perp}$ of $\mathcal{U}_i$ (where the orthogonality in the orthogonal complement is with respect to the bilinear form $\mathbf{x}'\mathbf{H}\mathbf{y}$) and (2) the projection of any $n \times 1$ vector $\mathbf{y}$ on $\mathcal{U}_i$ along $\mathcal{U}_1 + \cdots + \mathcal{U}_{i-1} + \mathcal{U}_{i+1} + \cdots + \mathcal{U}_k$ equals the orthogonal projection of $\mathbf{y}$ on $\mathcal{U}_i$ with respect to $\mathbf{H}$.

(e) Show that if, for $j \neq i = 1, \ldots, k$, $\mathcal{U}_i$ and $\mathcal{U}_j$ are orthogonal with respect to some symmetric positive definite matrix $\mathbf{H}_*$, then $\mathbf{H}_*$ is expressible in the form (E.3).

Bibliographic and Supplementary Notes

§2. Theorem 17.2.13 and its corollary (Corollary 17.2.14) are based on results given by C. R. Rao—refer, for example, to Section 1.3 of Rao's (1989) paper. §5. The approach taken in Section 5 is essentially that taken by Marsaglia and Styan (1974) in their Section 3. Section 4 is a prerequisite for Subsection b. §7. Section 7 (and Exercise 21) are based on results given by Rao and Mitra (1971, sec. 5.3).

18

Sums (and Differences) of Matrices

It is common in such areas of statistics as linear statistical models and Bayesian statistics to encounter matrices of the form $\mathbf{R} + \mathbf{STU}$, where the dimensions (number of rows and/or number of columns) of $\mathbf{T}$ are small relative to those of $\mathbf{R}$ and where $\mathbf{R}$ and possibly $\mathbf{T}$ are diagonal matrices or are of some other form that, for example, makes them easy to "invert." Typically, there is a need to find the determinant, ordinary or generalized inverse, or rank of a matrix of this form and/or to solve a linear system having a coefficient matrix of this form. This chapter includes (in Sections 18.1a, 18.2d-e, and 18.5a) formulas that can be very useful in such a situation. It also includes (in the remainder of Sections 18.1, 18.2, and 18.5 and in Section 18.3) a wide variety of relatively basic results (having a myriad of applications in statistics) on the determinants, ordinary or generalized inverses, and ranks of sums, differences, or linear combinations of matrices and on the solution of a linear system whose coefficient matrix is a sum of matrices.

Section 18.4 gives necessary and sufficient conditions for (square) matrices whose sum is idempotent to be individually idempotent. These conditions can be used to establish a classical result on the statistical distribution of quadratic forms that is due to Cochran (1934) and is known as Cochran's theorem. This result can in turn be used to establish the conditions under which the sums of squares in a statistical analysis of variance are distributed independently as scalar multiples of central or noncentral chi-square random variables.

18.1 Some Results on Determinants

a. Determinants of matrices of the form R + STU

The following theorem gives a very useful formula for the determinant of the sum $\mathbf{R} + \mathbf{STU}$ of an $n \times n$ nonsingular matrix $\mathbf{R}$ and the product $\mathbf{STU}$ of an $n \times m$ matrix $\mathbf{S}$, $m \times m$ nonsingular matrix $\mathbf{T}$, and $m \times n$ matrix $\mathbf{U}$.

Theorem 18.1.1. Let $\mathbf{R}$ represent an $n \times n$ matrix, $\mathbf{S}$ an $n \times m$ matrix, $\mathbf{T}$ an $m \times m$ matrix, and $\mathbf{U}$ an $m \times n$ matrix. If $\mathbf{R}$ and $\mathbf{T}$ are nonsingular, then

$$|\mathbf{R} + \mathbf{S}\mathbf{T}\mathbf{U}| = |\mathbf{R}||\mathbf{T}||\mathbf{T}^{-1} + \mathbf{U}\mathbf{R}^{-1}\mathbf{S}|. \tag{1.1}$$

Proof. Suppose that $\mathbf{R}$ and $\mathbf{T}$ are nonsingular. Then, making use of Theorem 13.3.8, we find that

$$\begin{vmatrix} \mathbf{R} & -\mathbf{S} \\ \mathbf{U} & \mathbf{T}^{-1} \end{vmatrix} = |\mathbf{T}^{-1}||\mathbf{R} - (-\mathbf{S})(\mathbf{T}^{-1})^{-1}\mathbf{U}| = |\mathbf{T}^{-1}||\mathbf{R} + \mathbf{S}\mathbf{T}\mathbf{U}|$$

and also that

$$\begin{vmatrix} \mathbf{R} & -\mathbf{S} \\ \mathbf{U} & \mathbf{T}^{-1} \end{vmatrix} = |\mathbf{R}||\mathbf{T}^{-1} - \mathbf{U}\mathbf{R}^{-1}(-\mathbf{S})| = |\mathbf{R}||\mathbf{T}^{-1} + \mathbf{U}\mathbf{R}^{-1}\mathbf{S}|.$$

Thus,

$$|\mathbf{T}^{-1}||\mathbf{R} + \mathbf{S}\mathbf{T}\mathbf{U}| = |\mathbf{R}||\mathbf{T}^{-1} + \mathbf{U}\mathbf{R}^{-1}\mathbf{S}|,$$

or, equivalently (since $|\mathbf{T}^{-1}| = 1/|\mathbf{T}|$),

$$|\mathbf{R} + \mathbf{S}\mathbf{T}\mathbf{U}| = |\mathbf{R}||\mathbf{T}||\mathbf{T}^{-1} + \mathbf{U}\mathbf{R}^{-1}\mathbf{S}|.$$

Q.E.D.

In the special case where $\mathbf{R} = \mathbf{I}_n$ and $\mathbf{T} = \mathbf{I}_m$, Theorem 18.1.1 simplifies to the following result.

Corollary 18.1.2. For any $n \times m$ matrix $\mathbf{S}$ and any $m \times n$ matrix $\mathbf{U}$,

$$|\mathbf{I}_n + \mathbf{S}\mathbf{U}| = |\mathbf{I}_m + \mathbf{U}\mathbf{S}|. \tag{1.2}$$

In the special case where $m = 1$, Corollary 18.1.2 can be restated as the following corollary.

Corollary 18.1.3. For any n-dimensional column vectors $\mathbf{s} = \{s_i\}$ and $\mathbf{u} = \{u_i\}$,

$$|\mathbf{I}_n + \mathbf{s}\mathbf{u}'| = 1 + \mathbf{u}'\mathbf{s} = 1 + \mathbf{s}'\mathbf{u} = 1 + \sum_{i=1}^{n} s_i u_i. \tag{1.3}$$

Corollary 18.1.3 reduces the problem of evaluating the determinant of an $n \times n$ matrix of the form $\mathbf{I}_n + \mathbf{s}\mathbf{u}'$ to one of evaluating a scalar-valued expression. More generally, Corollary 18.1.2 converts the problem of evaluating the determinant of an $n \times n$ matrix of the form $\mathbf{I}_n + \mathbf{S}\mathbf{U}$ into one of evaluating the determinant of an $m \times m$ matrix (where m equals the number of columns in $\mathbf{S}$ and the number of rows in $\mathbf{U}$), which can be advantageous if $m < n$. Still more generally, Theorem 18.1.1 converts the problem of evaluating the determinant of an $n \times n$ matrix of the form $\mathbf{R} + \mathbf{S}\mathbf{T}\mathbf{U}$ into one of evaluating the determinant and inverse of the $n \times n$ matrix $\mathbf{R}$, the determinant and inverse of the $m \times m$ matrix $\mathbf{T}$, and the determinant of a further $m \times m$ matrix—this can be advantageous if $m < n$ and if $|\mathbf{R}|$, $|\mathbf{T}|$, $\mathbf{R}^{-1}$, and $\mathbf{T}^{-1}$ are easy to obtain (as would be the case if, e.g., $\mathbf{R}$ and $\mathbf{T}$ were diagonal).

In connection with Theorem 18.1.1, note that

$$\mathbf{R} + \mathbf{STU} = \mathbf{R} + (\mathbf{ST})\mathbf{T}^{-1}(\mathbf{TU}). \tag{1.4}$$

Upon applying formula (1.1) to the right side of equality (1.4) [i.e., applying formula (1.1) with $\mathbf{ST}$, $\mathbf{T}^{-1}$, and $\mathbf{TU}$ in place of $\mathbf{S}$, $\mathbf{T}$, and $\mathbf{U}$, respectively] and observing that $(\mathbf{T}^{-1})^{-1} = \mathbf{T}$ and $|\mathbf{T}^{-1}| = 1/|\mathbf{T}|$, we obtain (as an additional corollary of Theorem 18.1.1) the following, alternative formula.

Corollary 18.1.4. Let $\mathbf{R}$ represent an $n \times n$ matrix, $\mathbf{S}$ an $n \times m$ matrix, $\mathbf{T}$ an $m \times m$ matrix, and $\mathbf{U}$ an $m \times n$ matrix. If $\mathbf{R}$ and $\mathbf{T}$ are nonsingular, then

$$|\mathbf{R} + \mathbf{STU}| = |\mathbf{R}||\mathbf{T} + \mathbf{TUR}^{-1}\mathbf{ST}|/|\mathbf{T}|. \tag{1.5}$$

Finally, as a consequence of Theorem 18.1.1 and Corollary 18.1.4, we have the following corollary.

Corollary 18.1.5. Let $\mathbf{R}$ represent an $n \times n$ matrix, $\mathbf{S}$ an $n \times m$ matrix, $\mathbf{T}$ an $m \times m$ matrix, and $\mathbf{U}$ an $m \times n$ matrix. Suppose that $\mathbf{R}$ and $\mathbf{T}$ are nonsingular. Then, $\mathbf{R} + \mathbf{STU}$ is nonsingular if and only if $\mathbf{T}^{-1} + \mathbf{UR}^{-1}\mathbf{S}$ is nonsingular or, equivalently, if and only if $\mathbf{T} + \mathbf{TUR}^{-1}\mathbf{ST}$ is nonsingular.

b. Some inequalities

The following theorem gives a basic result on the determinant $|\mathbf{A} + \mathbf{B}|$ of the sum of a symmetric positive definite matrix $\mathbf{A}$ and a symmetric nonnegative definite matrix $\mathbf{B}$.

Theorem 18.1.6. For any $n \times n$ symmetric positive definite matrix $\mathbf{A}$ and any $n \times n$ symmetric nonnegative definite matrix $\mathbf{B}$,

$$|\mathbf{A} + \mathbf{B}| \geq |\mathbf{A}|,$$

with equality holding if and only if $\mathbf{B} = \mathbf{0}$.

Proof. According to Corollary 14.3.8, there exists a matrix $\mathbf{K}$ such that $\mathbf{B} = \mathbf{K}'\mathbf{K}$. Denote by r the number of rows in $\mathbf{K}$; and (for $i = 1, \ldots, r$) let $\mathbf{k}_i'$ represent the ith row of $\mathbf{K}$, and let $\mathbf{K}_i$ represent the $i \times n$ submatrix (of $\mathbf{K}$) obtained by striking out the last $r - i$ rows of $\mathbf{K}$. Further, define $\mathbf{A}_0 = \mathbf{A}$ and (for $i = 1, \ldots, r$) $\mathbf{A}_i = \mathbf{A} + \mathbf{K}_i'\mathbf{K}_i$. Observe (in light of Lemma 14.2.4 and Corollary 14.2.14) that $\mathbf{A}_i$ is symmetric and positive definite.

It is clear that (for $i = 2, \ldots, r$) $\mathbf{K}_i = \begin{bmatrix} \mathbf{K}_{i-1} \\ \mathbf{k}_i' \end{bmatrix}$ and hence that $\mathbf{K}_i'\mathbf{K}_i = \mathbf{K}_{i-1}'\mathbf{K}_{i-1} + \mathbf{k}_i\mathbf{k}_i'$. Thus, for $i = 1, \ldots, r$,

$$\mathbf{A}_i = \mathbf{A}_{i-1} + \mathbf{k}_i\mathbf{k}_i' = \mathbf{A}_{i-1} + \mathbf{k}_i(1)\mathbf{k}_i',$$

and, applying formula (1.1), we find that

$$|\mathbf{A}_i| = |\mathbf{A}_{i-1}||(1)||(1 + \mathbf{k}_i'\mathbf{A}_{i-1}^{-1}\mathbf{k}_i)| = |\mathbf{A}_{i-1}|(1 + \mathbf{k}_i'\mathbf{A}_{i-1}^{-1}\mathbf{k}_i).$$

Since (according to Corollary 14.2.11) $\mathbf{A}_{i-1}^{-1}$ is positive definite, we have (for $i = 1, \ldots, r$) that

$$|\mathbf{A}_i| \geq |\mathbf{A}_{i-1}|,$$

with equality holding if and only if $\mathbf{k}_i'\mathbf{A}_{i-1}^{-1}\mathbf{k}_i = 0$ or, equivalently, if and only if $\mathbf{k}_i = \mathbf{0}$.

It is now clear that

$$|\mathbf{A}_r| \geq |\mathbf{A}_{r-1}| \geq \cdots \geq |\mathbf{A}_1| \geq |\mathbf{A}_0|, \tag{1.6}$$

implying that

$$|\mathbf{A}_r| \geq |\mathbf{A}_0|. \tag{1.7}$$

Further, equality holds in inequality (1.7) if and only if equality holds in all r of the inequalities (1.6). Thus, equality holds in inequality (1.7) if and only if $\mathbf{k}_1, \ldots, \mathbf{k}_r$ are null, or, equivalently, if and only if $\mathbf{K} = \mathbf{0}$, and hence (in light of Corollary 5.3.2) if and only if $\mathbf{B} = \mathbf{0}$. Since clearly $\mathbf{A}_r = \mathbf{A}+\mathbf{B}$ (and $\mathbf{A}_0 = \mathbf{A}$), we conclude that

$$|\mathbf{A}+\mathbf{B}| \geq |\mathbf{A}|,$$

with equality holding if and only if $\mathbf{B} = \mathbf{0}$. Q.E.D.

Theorem 18.1.6 can be restated in the form of the following corollary.

Corollary 18.1.7. For any $n \times n$ symmetric positive definite matrix $\mathbf{A}$ and for any $n \times n$ symmetric matrix $\mathbf{C}$ such that $\mathbf{C} - \mathbf{A}$ is nonnegative definite,

$$|\mathbf{C}| \geq |\mathbf{A}|,$$

with equality holding if and only if $\mathbf{C} = \mathbf{A}$.

Proof. Upon setting $\mathbf{B} = \mathbf{C} - \mathbf{A}$ in Theorem 18.1.6 (and observing that $\mathbf{C} - \mathbf{A} = \mathbf{0} \Leftrightarrow \mathbf{C} = \mathbf{A}$), we obtain Corollary 18.1.7. Q.E.D.

A variation on Corollary 18.1.7 (in which the requirement that $\mathbf{A}$ be symmetric positive definite is replaced by the weaker requirement that $\mathbf{A}$ be symmetric nonnegative definite) is given by the following corollary.

Corollary 18.1.8. For any $n \times n$ symmetric nonnegative definite matrix $\mathbf{A}$ and for any $n \times n$ symmetric matrix $\mathbf{C}$ such that $\mathbf{C} - \mathbf{A}$ is nonnegative definite,

$$|\mathbf{C}| \geq |\mathbf{A}|, \tag{1.8}$$

with equality holding if and only if $\mathbf{C}$ is singular or $\mathbf{C} = \mathbf{A}$.

Proof. Let us begin by observing (in light of Lemma 14.9.1) that

$$|\mathbf{A}| \geq 0 \tag{1.9}$$

and that, since $\mathbf{C} = \mathbf{A} + (\mathbf{C} - \mathbf{A})$ (implying that $\mathbf{C}$ is nonnegative definite),

$$|\mathbf{C}| \geq 0. \tag{1.10}$$

Now, consider the inequality $|\mathbf{C}| \geq |\mathbf{A}|$. Either $\mathbf{A}$ is positive definite, or $\mathbf{A}$ is positive semidefinite. If $\mathbf{A}$ is positive definite, then it follows from Corollary 18.1.7 that $|\mathbf{C}| \geq |\mathbf{A}|$. Alternatively, if $\mathbf{A}$ is positive semidefinite, then $|\mathbf{A}| = 0$, and it follows from inequality (1.10) that $|\mathbf{C}| \geq |\mathbf{A}|$.

It remains to show that $|\mathbf{C}| = |\mathbf{A}|$ if and only if $\mathbf{C}$ is singular or $\mathbf{C} = \mathbf{A}$. If $\mathbf{C} = \mathbf{A}$, then obviously $|\mathbf{C}| = |\mathbf{A}|$. If $\mathbf{C}$ is singular, then $|\mathbf{C}| = 0$, which together with inequalities (1.8) and (1.9) implies that $|\mathbf{C}| = |\mathbf{A}|$.

Suppose now that $|\mathbf{C}| = |\mathbf{A}|$. If $\mathbf{A}$ is positive definite, then it follows from Corollary 18.1.7 that $\mathbf{C} = \mathbf{A}$. Alternatively, if $\mathbf{A}$ is positive semidefinite, then (in light of Lemma 14.9.1) $|\mathbf{A}| = 0$ and hence $|\mathbf{C}| = 0$, implying that $\mathbf{C}$ is singular. Q.E.D.

18.2 Some Results on Inverses and Generalized Inverses and on Linear Systems

a. Some basic results on (ordinary) inverses

The following lemma gives some very basic (but useful) results on the inverse of a (nonsingular) sum of two (square) matrices.

Lemma 18.2.1. Let $\mathbf{A}$ and $\mathbf{B}$ represent $n \times n$ matrices. If $\mathbf{A}+\mathbf{B}$ is nonsingular, then

$$(\mathbf{A}+\mathbf{B})^{-1}\mathbf{A} = \mathbf{I} - (\mathbf{A}+\mathbf{B})^{-1}\mathbf{B}, \tag{2.1}$$

$$\mathbf{A}(\mathbf{A}+\mathbf{B})^{-1} = \mathbf{I} - \mathbf{B}(\mathbf{A}+\mathbf{B})^{-1}, \tag{2.2}$$

$$\mathbf{B}(\mathbf{A}+\mathbf{B})^{-1}\mathbf{A} = \mathbf{A}(\mathbf{A}+\mathbf{B})^{-1}\mathbf{B}.$$

Proof. Suppose that $\mathbf{A}+\mathbf{B}$ is nonsingular. Then, clearly,

$$(\mathbf{A}+\mathbf{B})^{-1}\mathbf{A} + (\mathbf{A}+\mathbf{B})^{-1}\mathbf{B} = (\mathbf{A}+\mathbf{B})^{-1}(\mathbf{A}+\mathbf{B}) = \mathbf{I},$$

and hence $(\mathbf{A}+\mathbf{B})^{-1}\mathbf{A} = \mathbf{I} - (\mathbf{A}+\mathbf{B})^{-1}\mathbf{B}$. Similarly,

$$\mathbf{A}(\mathbf{A}+\mathbf{B})^{-1} + \mathbf{B}(\mathbf{A}+\mathbf{B})^{-1} = (\mathbf{A}+\mathbf{B})(\mathbf{A}+\mathbf{B})^{-1} = \mathbf{I},$$

and hence $\mathbf{A}(\mathbf{A}+\mathbf{B})^{-1} = \mathbf{I} - \mathbf{B}(\mathbf{A}+\mathbf{B})^{-1}$. That $\mathbf{B}(\mathbf{A}+\mathbf{B})^{-1}\mathbf{A} = \mathbf{A}(\mathbf{A}+\mathbf{B})^{-1}\mathbf{B}$ is evident upon premultiplying both sides of equality (2.1) and postmultiplying both sides of equality (2.2) by $\mathbf{B}$. Q.E.D.

Note that Lemma 18.2.1 implies in particular that, for $n \times n$ symmetric matrices $\mathbf{A}$ and $\mathbf{B}$ (such that $\mathbf{A}+\mathbf{B}$ is nonsingular), $\mathbf{B}(\mathbf{A}+\mathbf{B})^{-1}\mathbf{A}$ is symmetric.

In the special case where $\mathbf{A} = \mathbf{I}$, Lemma 18.2.1 reduces to the following result.

Corollary 18.2.2. For any (square) matrix $\mathbf{B}$ such that $\mathbf{I}+\mathbf{B}$ is nonsingular,

$$(\mathbf{I}+\mathbf{B})^{-1} = \mathbf{I} - (\mathbf{I}+\mathbf{B})^{-1}\mathbf{B}, \tag{2.3}$$

$$(\mathbf{I}+\mathbf{B})^{-1} = \mathbf{I} - \mathbf{B}(\mathbf{I}+\mathbf{B})^{-1}, \tag{2.4}$$

$$\mathbf{B}(\mathbf{I}+\mathbf{B})^{-1} = (\mathbf{I}+\mathbf{B})^{-1}\mathbf{B}.$$

A further basic result on inverse matrices is given by the following lemma.

Lemma 18.2.3. Let $\mathbf{A}$ represent an $m \times n$ matrix and $\mathbf{B}$ an $n \times m$ matrix. Then, $\mathbf{I}_n + \mathbf{BA}$ is nonsingular if and only if $\mathbf{I}_m + \mathbf{AB}$ is nonsingular, in which case

$$(\mathbf{I}_n + \mathbf{BA})^{-1}\mathbf{B} = \mathbf{B}(\mathbf{I}_m + \mathbf{AB})^{-1}. \tag{2.5}$$

Proof. That $\mathbf{I} + \mathbf{BA}$ is nonsingular if and only if $\mathbf{I} + \mathbf{AB}$ is nonsingular is evident from Corollary 18.1.5 (upon setting $\mathbf{R} = \mathbf{I}_n$, $\mathbf{S} = \mathbf{B}$, $\mathbf{T} = \mathbf{I}_m$, and $\mathbf{U} = \mathbf{A}$).

Suppose now that $\mathbf{I} + \mathbf{AB}$ is nonsingular (in which case $\mathbf{I} + \mathbf{BA}$ is also nonsingular), and observe that

$$\mathbf{B}(\mathbf{I} + \mathbf{AB}) = (\mathbf{I} + \mathbf{BA})\mathbf{B}. \tag{2.6}$$

Then, upon premultiplying both sides of equality (2.6) by $(\mathbf{I} + \mathbf{BA})^{-1}$ and postmultiplying both sides by $(\mathbf{I} + \mathbf{AB})^{-1}$, we obtain the equality

$$(\mathbf{I} + \mathbf{BA})^{-1}\mathbf{B} = \mathbf{B}(\mathbf{I} + \mathbf{AB})^{-1}.$$

Q.E.D.

The following theorem establishes a relationship between the inverse of a (nonsingular) difference $\mathbf{A} - \mathbf{B}$ between two nonsingular matrices $\mathbf{A}$ and $\mathbf{B}$ and the inverse of the difference $\mathbf{A}^{-1} - \mathbf{B}^{-1}$ between $\mathbf{A}^{-1}$ and $\mathbf{B}^{-1}$.

Theorem 18.2.4. Let $\mathbf{A}$ and $\mathbf{B}$ represent $n \times n$ nonsingular matrices. Then,

$$\text{rank}(\mathbf{A}^{-1} - \mathbf{B}^{-1}) = \text{rank}(\mathbf{B} - \mathbf{A}) = \text{rank}(\mathbf{A} - \mathbf{B}). \tag{2.7}$$

Further, $\mathbf{A}^{-1} - \mathbf{B}^{-1}$ is nonsingular if and only if $\mathbf{B} - \mathbf{A}$ is nonsingular (or, equivalently, if and only if $\mathbf{A} - \mathbf{B}$ is nonsingular), in which case

$$(\mathbf{A}^{-1} - \mathbf{B}^{-1})^{-1} = \mathbf{A} + \mathbf{A}(\mathbf{B} - \mathbf{A})^{-1}\mathbf{A} \tag{2.8}$$

$$= \mathbf{A} - \mathbf{A}(\mathbf{A} - \mathbf{B})^{-1}\mathbf{A}. \tag{2.9}$$

Proof. Clearly,

$$\mathbf{A}^{-1} - \mathbf{B}^{-1} = (\mathbf{I} - \mathbf{B}^{-1}\mathbf{A})\mathbf{A}^{-1}, \tag{2.10}$$

$$\mathbf{B} - \mathbf{A} = \mathbf{B}(\mathbf{I} - \mathbf{B}^{-1}\mathbf{A}). \tag{2.11}$$

Thus, recalling (from Corollary 8.3.3) that the premultiplication or postmultiplication of a matrix by a nonsingular matrix does not affect its rank [and observing that $\mathbf{A} - \mathbf{B} = -(\mathbf{B} - \mathbf{A})$], we find that

$$\text{rank}(\mathbf{A}^{-1} - \mathbf{B}^{-1}) = \text{rank}(\mathbf{I} - \mathbf{B}^{-1}\mathbf{A}) = \text{rank}(\mathbf{B} - \mathbf{A}) = \text{rank}(\mathbf{A} - \mathbf{B}).$$

It follows in particular that $\mathbf{A}^{-1} - \mathbf{B}^{-1}$ is nonsingular if and only if $\mathbf{B} - \mathbf{A}$ is nonsingular (or equivalently if and only if $\mathbf{A} - \mathbf{B}$ is nonsingular).

Suppose now that $\mathbf{B}-\mathbf{A}$ is nonsingular (in which case $\mathbf{I}-\mathbf{B}^{-1}\mathbf{A}$ is also nonsingular). Then, observing [in light of equality (2.11)] that

$$(\mathbf{B}-\mathbf{A})^{-1} = [\mathbf{B}(\mathbf{I}-\mathbf{B}^{-1}\mathbf{A})]^{-1} = (\mathbf{I}-\mathbf{B}^{-1}\mathbf{A})^{-1}\mathbf{B}^{-1},$$

we find [in light of equality (2.10)] that

$$\begin{aligned}(\mathbf{A}^{-1}-\mathbf{B}^{-1})&[\mathbf{A}+\mathbf{A}(\mathbf{B}-\mathbf{A})^{-1}\mathbf{A}]\\ &= (\mathbf{I}-\mathbf{B}^{-1}\mathbf{A})\mathbf{A}^{-1}[\mathbf{A}+\mathbf{A}(\mathbf{I}-\mathbf{B}^{-1}\mathbf{A})^{-1}\mathbf{B}^{-1}\mathbf{A}]\\ &= \mathbf{I}-\mathbf{B}^{-1}\mathbf{A}+\mathbf{B}^{-1}\mathbf{A}\\ &= \mathbf{I},\end{aligned}$$

thereby validating result (2.8) and also {since $(\mathbf{A}-\mathbf{B})^{-1} = [-(\mathbf{B}-\mathbf{A})]^{-1} = -(\mathbf{B}-\mathbf{A})^{-1}$} result (2.9). Q.E.D.

b. Some basic results on generalized inverses

Under what conditions is a generalized inverse $(\mathbf{A}+\mathbf{B})^{-}$ of the sum of two $m \times n$ matrices $\mathbf{A}$ and $\mathbf{B}$ a generalized inverse of $\mathbf{A}$ and a generalized inverse of $\mathbf{B}$? This question is addressed in the following theorem.

Theorem 18.2.5. Let $\mathbf{A}$ and $\mathbf{B}$ represent $m \times n$ matrices. If $\mathcal{R}(\mathbf{A})$ and $\mathcal{R}(\mathbf{B})$ are essentially disjoint and if $\mathcal{C}(\mathbf{A})$ and $\mathcal{C}(\mathbf{B})$ are essentially disjoint, then any generalized inverse of $\mathbf{A}+\mathbf{B}$ is a generalized inverse of $\mathbf{A}$ and a generalized inverse of $\mathbf{B}$.

Theorem 18.2.5 is an immediate consequence of the following result.

Theorem 18.2.6. For any $m \times n$ matrices $\mathbf{A}$ and $\mathbf{B}$,

$$\begin{pmatrix}\mathbf{A}\\ \mathbf{B}\end{pmatrix}(\mathbf{A}+\mathbf{B})^{-}(\mathbf{A},\,\mathbf{B}) = \begin{pmatrix}\mathbf{A} & \mathbf{0}\\ \mathbf{0} & \mathbf{B}\end{pmatrix} \tag{2.12}$$

[or, equivalently, $\mathbf{A}(\mathbf{A}+\mathbf{B})^{-}\mathbf{A} = \mathbf{A}$, $\mathbf{B}(\mathbf{A}+\mathbf{B})^{-}\mathbf{B} = \mathbf{B}$, $\mathbf{A}(\mathbf{A}+\mathbf{B})^{-}\mathbf{B} = \mathbf{0}$, and $\mathbf{B}(\mathbf{A}+\mathbf{B})^{-}\mathbf{A} = \mathbf{0}$] if and only if $\mathcal{R}(\mathbf{A})$ and $\mathcal{R}(\mathbf{B})$ are essentially disjoint and $\mathcal{C}(\mathbf{A})$ and $\mathcal{C}(\mathbf{B})$ are essentially disjoint.

Proof (of Theorem 18.2.6). Suppose that $\mathcal{R}(\mathbf{A})$ and $\mathcal{R}(\mathbf{B})$ are essentially disjoint and $\mathcal{C}(\mathbf{A})$ and $\mathcal{C}(\mathbf{B})$ are essentially disjoint. And observe that

$$\begin{aligned}\mathbf{A}(\mathbf{A}+\mathbf{B})^{-}\mathbf{A}+\mathbf{A}(\mathbf{A}+\mathbf{B})^{-}\mathbf{B}&+\mathbf{B}(\mathbf{A}+\mathbf{B})^{-}\mathbf{A}+\mathbf{B}(\mathbf{A}+\mathbf{B})^{-}\mathbf{B}\\ &= (\mathbf{A}+\mathbf{B})(\mathbf{A}+\mathbf{B})^{-}(\mathbf{A}+\mathbf{B}) = \mathbf{A}+\mathbf{B}\end{aligned}$$

and hence that

$$\mathbf{A}(\mathbf{A}+\mathbf{B})^{-}\mathbf{A}+\mathbf{B}(\mathbf{A}+\mathbf{B})^{-}\mathbf{A}-\mathbf{A} = \mathbf{B}-\mathbf{A}(\mathbf{A}+\mathbf{B})^{-}\mathbf{B}-\mathbf{B}(\mathbf{A}+\mathbf{B})^{-}\mathbf{B}. \tag{2.13}$$

Since the left side of equality (2.13) is expressible in the form $\mathbf{KA}$ for some matrix $\mathbf{K}$ and the right side is expressible in the form $\mathbf{LB}$ for some matrix $\mathbf{L}$, it follows from Part (2) of Corollary 17.2.3 that

$$\mathbf{A}(\mathbf{A}+\mathbf{B})^{-}\mathbf{A}+\mathbf{B}(\mathbf{A}+\mathbf{B})^{-}\mathbf{A}-\mathbf{A}=\mathbf{0},$$
$$\mathbf{B}-\mathbf{A}(\mathbf{A}+\mathbf{B})^{-}\mathbf{B}-\mathbf{B}(\mathbf{A}+\mathbf{B})^{-}\mathbf{B}=\mathbf{0},$$

or, equivalently, that

$$\mathbf{B}(\mathbf{A}+\mathbf{B})^{-}\mathbf{A}=\mathbf{A}-\mathbf{A}(\mathbf{A}+\mathbf{B})^{-}\mathbf{A}, \tag{2.14}$$
$$\mathbf{A}(\mathbf{A}+\mathbf{B})^{-}\mathbf{B}=\mathbf{B}-\mathbf{B}(\mathbf{A}+\mathbf{B})^{-}\mathbf{B}. \tag{2.15}$$

And, since the right side of equality (2.14) is expressible in the form $\mathbf{AK}$ for some matrix $\mathbf{K}$ and the right side of equality (2.15) is expressible in the form $\mathbf{BL}$ for some matrix $\mathbf{L}$, it follows from Part (1) of Corollary 17.2.3 that

$$\mathbf{B}(\mathbf{A}+\mathbf{B})^{-}\mathbf{A}=\mathbf{0}, \quad \mathbf{A}-\mathbf{A}(\mathbf{A}+\mathbf{B})^{-}\mathbf{A}=\mathbf{0},$$
$$\mathbf{A}(\mathbf{A}+\mathbf{B})^{-}\mathbf{B}=\mathbf{0}, \quad \mathbf{B}-\mathbf{B}(\mathbf{A}+\mathbf{B})^{-}\mathbf{B}=\mathbf{0},$$

or, equivalently, that

$$\begin{pmatrix}\mathbf{A}\\ \mathbf{B}\end{pmatrix}(\mathbf{A}+\mathbf{B})^{-}(\mathbf{A},\ \mathbf{B})=\begin{pmatrix}\mathbf{A} & \mathbf{0}\\ \mathbf{0} & \mathbf{B}\end{pmatrix}.$$

Conversely, suppose that

$$\begin{pmatrix}\mathbf{A}\\ \mathbf{B}\end{pmatrix}(\mathbf{A}+\mathbf{B})^{-}(\mathbf{A},\ \mathbf{B})=\begin{pmatrix}\mathbf{A} & \mathbf{0}\\ \mathbf{0} & \mathbf{B}\end{pmatrix}.$$

Then, making use of Corollary 4.4.5 and inequality (4.5.1), we find that

$$\begin{aligned}\operatorname{rank}(\mathbf{A})+\operatorname{rank}(\mathbf{B})=\operatorname{rank}\begin{pmatrix}\mathbf{A} & \mathbf{0}\\ \mathbf{0} & \mathbf{B}\end{pmatrix}&=\operatorname{rank}\left[\begin{pmatrix}\mathbf{A}\\ \mathbf{B}\end{pmatrix}(\mathbf{A}+\mathbf{B})^{-}(\mathbf{A},\ \mathbf{B})\right]\\ &\le \operatorname{rank}(\mathbf{A},\ \mathbf{B})\le \operatorname{rank}(\mathbf{A})+\operatorname{rank}(\mathbf{B})\end{aligned}$$

and hence that

$$\operatorname{rank}(\mathbf{A},\ \mathbf{B})=\operatorname{rank}(\mathbf{A})+\operatorname{rank}(\mathbf{B}).$$

It can be established in similar fashion that

$$\operatorname{rank}\begin{pmatrix}\mathbf{A}\\ \mathbf{B}\end{pmatrix}=\operatorname{rank}(\mathbf{A})+\operatorname{rank}(\mathbf{B}).$$

We conclude, on the basis of Theorem 17.2.4, that $\mathcal{C}(\mathbf{A})$ and $\mathcal{C}(\mathbf{B})$ are essentially disjoint and $\mathcal{R}(\mathbf{A})$ and $\mathcal{R}(\mathbf{B})$ are essentially disjoint. Q.E.D.

c. A result on linear systems of the form $(\mathbf{A}+\mathbf{B})\mathbf{X}=\mathbf{C}+\mathbf{D}$ (in $\mathbf{X}$)

Under what conditions is a solution to a linear system of the form $(\mathbf{A}+\mathbf{B})\mathbf{X}=\mathbf{C}+\mathbf{D}$ (in $\mathbf{X}$) a solution to the linear system $\mathbf{AX}=\mathbf{C}$ and to the linear system $\mathbf{BX}=\mathbf{D}$? This question is addressed in the following theorem.

Theorem 18.2.7. Let $\mathbf{A}$ and $\mathbf{B}$ represent $m \times n$ matrices, and let $\mathbf{C}$ and $\mathbf{D}$ represent $m \times p$ matrices such that $\mathcal{C}(\mathbf{C}) \subset \mathcal{C}(\mathbf{A})$ and $\mathcal{C}(\mathbf{D}) \subset \mathcal{C}(\mathbf{B})$. (1) If $\mathcal{R}(\mathbf{A})$ and $\mathcal{R}(\mathbf{B})$ are essentially disjoint, then the linear system $(\mathbf{A}+\mathbf{B})\mathbf{X} = \mathbf{C}+\mathbf{D}$ (in $\mathbf{X}$) is consistent. (2) If $\mathcal{C}(\mathbf{A})$ and $\mathcal{C}(\mathbf{B})$ are essentially disjoint, then any solution to the linear system $(\mathbf{A}+\mathbf{B})\mathbf{X} = \mathbf{C}+\mathbf{D}$ is a solution to the linear system $\mathbf{AX} = \mathbf{C}$ and also to the linear system $\mathbf{BX} = \mathbf{D}$.

Proof. (1) Suppose that $\mathcal{R}(\mathbf{A})$ and $\mathcal{R}(\mathbf{B})$ are essentially disjoint. Then, it follows from Theorem 17.3.2 that the linear system $\begin{pmatrix}\mathbf{A}\\ \mathbf{B}\end{pmatrix}\mathbf{X} = \begin{pmatrix}\mathbf{C}\\ \mathbf{D}\end{pmatrix}$ (in $\mathbf{X}$) is consistent and hence has a solution, say $\mathbf{X}_*$. Thus, $\mathbf{AX}_* = \mathbf{C}$ and $\mathbf{BX}_* = \mathbf{D}$, implying that $(\mathbf{A}+\mathbf{B})\mathbf{X}_* = \mathbf{C}+\mathbf{D}$, so that $\mathbf{X}_*$ is a solution to the linear system $(\mathbf{A}+\mathbf{B})\mathbf{X} = \mathbf{C}+\mathbf{D}$. And we conclude that the linear system $(\mathbf{A}+\mathbf{B})\mathbf{X} = \mathbf{C}+\mathbf{D}$ is consistent.

(2) Suppose that $\mathcal{C}(\mathbf{A})$ and $\mathcal{C}(\mathbf{B})$ are essentially disjoint, and let $\mathbf{X}_*$ represent any solution to the linear system $(\mathbf{A}+\mathbf{B})\mathbf{X} = \mathbf{C}+\mathbf{D}$. Then, $\mathbf{AX}_* + \mathbf{BX}_* = \mathbf{C}+\mathbf{D}$, and a solution to the linear system $\mathbf{AX}_1 + \mathbf{BX}_2 = \mathbf{C}+\mathbf{D}$ (in $n \times p$ matrices $\mathbf{X}_1$ and $\mathbf{X}_2$) is obtained by setting $\mathbf{X}_1 = \mathbf{X}_2 = \mathbf{X}_*$. We conclude, on the basis of Theorem 17.3.1, that $\mathbf{X}_*$ is a solution to the linear system $\mathbf{AX} = \mathbf{C}$ and also to the linear system $\mathbf{BX} = \mathbf{D}$. Q.E.D.

d. (Ordinary) inverses of matrices of the form R + STU

Let $\mathbf{R}$ represent an $n \times n$ matrix, $\mathbf{S}$ an $n \times m$ matrix, $\mathbf{T}$ an $m \times m$ matrix, and $\mathbf{U}$ an $m \times n$ matrix. And suppose that $\mathbf{R}$ and $\mathbf{T}$ are nonsingular. Then, result (1.1) [or (1.5)] can be used to express the determinant of $\mathbf{R}+\mathbf{STU}$ in terms of the determinants of $\mathbf{R}$, $\mathbf{T}$, and $\mathbf{T}^{-1} + \mathbf{UR}^{-1}\mathbf{S}$ [or $\mathbf{T} + \mathbf{TUR}^{-1}\mathbf{ST}$]. Can the inverse of $\mathbf{R}+\mathbf{STU}$ (assuming that $\mathbf{R}+\mathbf{STU}$ is nonsingular) be expressed in a comparable way? In what follows, the answer to this question is shown to be yes.

Suppose now that $\mathbf{R}+\mathbf{STU}$ is nonsingular, and let $\mathbf{A} = \mathbf{TUR}^{-1}$. Then, since $\mathbf{R}+\mathbf{STU} = (\mathbf{I}+\mathbf{SA})\mathbf{R}$ (and since, e.g., the premultiplication or postmultiplication of a matrix by a nonsingular matrix does not affect its rank), $\mathbf{I}+\mathbf{SA}$ is nonsingular, and

$$(\mathbf{R}+\mathbf{STU})^{-1} = \mathbf{R}^{-1}(\mathbf{I}+\mathbf{SA})^{-1}. \tag{2.16}$$

Further, it follows from Corollary 18.2.2 that

$$(\mathbf{I}+\mathbf{SA})^{-1} = \mathbf{I} - (\mathbf{I}+\mathbf{SA})^{-1}\mathbf{SA}. \tag{2.17}$$

And it follows from Lemma 18.2.3 that $\mathbf{I}+\mathbf{AS}$ $(= \mathbf{I}+\mathbf{TUR}^{-1}\mathbf{S})$ is nonsingular and that

$$(\mathbf{I}+\mathbf{SA})^{-1}\mathbf{S} = \mathbf{S}(\mathbf{I}+\mathbf{AS})^{-1}. \tag{2.18}$$

Substituting expression (2.18) into expression (2.17) and then substituting the resultant expression into expression (2.16), we find that

$$\begin{aligned}(\mathbf{R}+\mathbf{STU})^{-1} &= \mathbf{R}^{-1}[\mathbf{I} - \mathbf{S}(\mathbf{I}+\mathbf{AS})^{-1}\mathbf{A}]\\ &= \mathbf{R}^{-1} - \mathbf{R}^{-1}\mathbf{S}(\mathbf{I}+\mathbf{AS})^{-1}\mathbf{TUR}^{-1}.\end{aligned} \tag{2.19}$$

And, since $\mathbf{I} + \mathbf{AS} = \mathbf{T}(\mathbf{T}^{-1} + \mathbf{UR}^{-1}\mathbf{S})$ and since also $\mathbf{I} + \mathbf{AS} = (\mathbf{T} + \mathbf{TUR}^{-1}\mathbf{ST})\mathbf{T}^{-1}$, $\mathbf{T}^{-1} + \mathbf{UR}^{-1}\mathbf{S}$ and $\mathbf{T} + \mathbf{TUR}^{-1}\mathbf{ST}$ are nonsingular, and

$$(\mathbf{I} + \mathbf{AS})^{-1} = (\mathbf{T}^{-1} + \mathbf{UR}^{-1}\mathbf{S})^{-1}\mathbf{T}^{-1} \tag{2.20}$$

$$= \mathbf{T}(\mathbf{T} + \mathbf{TUR}^{-1}\mathbf{ST})^{-1}. \tag{2.21}$$

Substituting expression (2.20) into expression (2.19), we obtain

$$(\mathbf{R} + \mathbf{STU})^{-1} = \mathbf{R}^{-1} - \mathbf{R}^{-1}\mathbf{S}(\mathbf{T}^{-1} + \mathbf{UR}^{-1}\mathbf{S})^{-1}\mathbf{UR}^{-1}.$$

Alternatively, substituting expression (2.21) into expression (2.19), we obtain

$$(\mathbf{R} + \mathbf{STU})^{-1} = \mathbf{R}^{-1} - \mathbf{R}^{-1}\mathbf{ST}(\mathbf{T} + \mathbf{TUR}^{-1}\mathbf{ST})^{-1}\mathbf{TUR}^{-1}.$$

In summary, we have the following theorem (which incorporates the content of Corollary 18.1.5).

Theorem 18.2.8. Let $\mathbf{R}$ represent an $n \times n$ matrix, $\mathbf{S}$ an $n \times m$ matrix, $\mathbf{T}$ an $m \times m$ matrix, and $\mathbf{U}$ an $m \times n$ matrix. Suppose that $\mathbf{R}$ and $\mathbf{T}$ are nonsingular. Then, $\mathbf{R} + \mathbf{STU}$ is nonsingular if and only if $\mathbf{T}^{-1} + \mathbf{UR}^{-1}\mathbf{S}$ is nonsingular, or, equivalently, if and only if $\mathbf{T} + \mathbf{TUR}^{-1}\mathbf{ST}$ is nonsingular, in which case

$$(\mathbf{R} + \mathbf{STU})^{-1} = \mathbf{R}^{-1} - \mathbf{R}^{-1}\mathbf{S}(\mathbf{T}^{-1} + \mathbf{UR}^{-1}\mathbf{S})^{-1}\mathbf{UR}^{-1} \tag{2.22}$$

$$= \mathbf{R}^{-1} - \mathbf{R}^{-1}\mathbf{ST}(\mathbf{T} + \mathbf{TUR}^{-1}\mathbf{ST})^{-1}\mathbf{TUR}^{-1}. \tag{2.23}$$

Some special cases of Theorem 18.2.8 are given in the following three corollaries.

Corollary 18.2.9. Let $\mathbf{S}$ represent an $n \times m$ matrix and $\mathbf{U}$ an $m \times n$ matrix. Then, $\mathbf{I}_n + \mathbf{SU}$ is nonsingular if and only if $\mathbf{I}_m + \mathbf{US}$ is nonsingular, in which case

$$(\mathbf{I}_n + \mathbf{SU})^{-1} = \mathbf{I}_n - \mathbf{S}(\mathbf{I}_m + \mathbf{US})^{-1}\mathbf{U}. \tag{2.24}$$

Corollary 18.2.10. Let $\mathbf{R}$ represent an $n \times n$ nonsingular matrix, and let $\mathbf{s}$ and $\mathbf{u}$ represent n-dimensional column vectors. Then, $\mathbf{R} + \mathbf{su}'$ is nonsingular if and only if $\mathbf{u}'\mathbf{R}^{-1}\mathbf{s} \neq -1$, in which case

$$(\mathbf{R} + \mathbf{su}')^{-1} = \mathbf{R}^{-1} - (1 + \mathbf{u}'\mathbf{R}^{-1}\mathbf{s})^{-1}\mathbf{R}^{-1}\mathbf{su}'\mathbf{R}^{-1}. \tag{2.25}$$

Corollary 18.2.11. Let $\mathbf{s} = \{s_i\}$ and $\mathbf{u} = \{u_i\}$ represent n-dimensional column vectors. Then, $\mathbf{I} + \mathbf{su}'$ is nonsingular if and only if $\mathbf{u}'\mathbf{s} \neq -1$ (or, equivalently, if and only if $\sum_i s_i u_i \neq -1$), in which case

$$(\mathbf{I} + \mathbf{su}')^{-1} = \mathbf{I} - (1 + \mathbf{u}'\mathbf{s})^{-1}\mathbf{su}' = \mathbf{I} - \Big(1 + \sum_i s_i u_i\Big)^{-1}\mathbf{su}'. \tag{2.26}$$

Formula (2.25) is generally attributed to Sherman and Morrison (1949 and 1950) and/or to Bartlett (1951). The more general formulas (2.22) and (2.23) are often attributed to Woodbury (1950), and one or the other of them is sometimes referred to as Woodbury's formula [although a formula that is essentially the same as formula (2.22) was given earlier by Duncan (1944) and by Guttman (1946)]. Refer to Henderson and Searle (1981a) for additional information about the history of formulas (2.22) and (2.23).

Depending upon the circumstances, formula (2.22) or (2.23) can be used to great advantage in the inversion of a (nonsingular) matrix of the form $\mathbf{R}+\mathbf{STU}$. The circumstances under which the use of formula (2.22) can be advantageous are similar to the circumstances (discussed in Section 18.1a) under which the use of formula (1.1) (for the determinant of $\mathbf{R}+\mathbf{STU}$) can be advantageous. Specifically, the use of formula (2.22) can be advantageous if the dimensions of $\mathbf{T}^{-1}+\mathbf{UR}^{-1}\mathbf{S}$ (or, equivalently, of $\mathbf{T}$) are small relative to those of $\mathbf{R}+\mathbf{STU}$ (or, equivalently, of $\mathbf{R}$) and if $\mathbf{R}^{-1}$ and $\mathbf{T}^{-1}$ are easy to obtain. Formula (2.23) calls for more matrix multiplications than formula (2.22), but does not call for the inversion of $\mathbf{T}$.

Formulas (2.22) and (2.23) were derived by making use of Corollary 18.2.2 and Lemma 18.2.3. An alternative derivation of these formulas is obtained by applying Theorem 8.5.11 (on the inverse of a partitioned matrix) to the partitioned matrices $\begin{pmatrix} \mathbf{R} & -\mathbf{S} \\ \mathbf{U} & \mathbf{T}^{-1} \end{pmatrix}$ and $\begin{pmatrix} \mathbf{R} & -\mathbf{ST} \\ \mathbf{TU} & \mathbf{T} \end{pmatrix}$.

If $\mathbf{T}^{-1}+\mathbf{UR}^{-1}\mathbf{S}$ is nonsingular, then (according to Theorem 8.5.11) $\begin{pmatrix} \mathbf{R} & -\mathbf{S} \\ \mathbf{U} & \mathbf{T}^{-1} \end{pmatrix}$ is nonsingular, $\mathbf{R}+\mathbf{STU}$ is also nonsingular, and

$$\begin{pmatrix} (\mathbf{R}+\mathbf{STU})^{-1} & (\mathbf{R}+\mathbf{STU})^{-1}\mathbf{ST} \\ -\mathbf{TU}(\mathbf{R}+\mathbf{STU})^{-1} & \mathbf{T}-\mathbf{TU}(\mathbf{R}+\mathbf{STU})^{-1}\mathbf{ST} \end{pmatrix}$$
$$= \begin{pmatrix} \mathbf{R} & -\mathbf{S} \\ \mathbf{U} & \mathbf{T}^{-1} \end{pmatrix}^{-1}$$
$$= \begin{pmatrix} \mathbf{R}^{-1}-\mathbf{R}^{-1}\mathbf{S}(\mathbf{T}^{-1}+\mathbf{UR}^{-1}\mathbf{S})^{-1}\mathbf{UR}^{-1} & \mathbf{R}^{-1}\mathbf{S}(\mathbf{T}^{-1}+\mathbf{UR}^{-1}\mathbf{S})^{-1} \\ -(\mathbf{T}^{-1}+\mathbf{UR}^{-1}\mathbf{S})^{-1}\mathbf{UR}^{-1} & (\mathbf{T}^{-1}+\mathbf{UR}^{-1}\mathbf{S})^{-1} \end{pmatrix},$$

implying in particular that

$$(\mathbf{R}+\mathbf{STU})^{-1} = \mathbf{R}^{-1}-\mathbf{R}^{-1}\mathbf{S}(\mathbf{T}^{-1}+\mathbf{UR}^{-1}\mathbf{S})^{-1}\mathbf{UR}^{-1}.$$

Similarly, if $\mathbf{T}+\mathbf{TUR}^{-1}\mathbf{ST}$ is nonsingular, then $\begin{pmatrix} \mathbf{R} & -\mathbf{ST} \\ \mathbf{TU} & \mathbf{T} \end{pmatrix}$ is nonsingular, $\mathbf{R}+\mathbf{STU}$ is also nonsingular, and

$$\begin{pmatrix} (\mathbf{R}+\mathbf{STU})^{-1} & (\mathbf{R}+\mathbf{STU})^{-1}\mathbf{S} \\ -\mathbf{U}(\mathbf{R}+\mathbf{STU})^{-1} & \mathbf{T}^{-1}-\mathbf{U}(\mathbf{R}+\mathbf{STU})^{-1}\mathbf{S} \end{pmatrix}$$
$$= \begin{pmatrix} \mathbf{R} & -\mathbf{ST} \\ \mathbf{TU} & \mathbf{T} \end{pmatrix}^{-1}$$
$$= \begin{pmatrix} \begin{array}{l} \mathbf{R}^{-1}-\mathbf{R}^{-1}\mathbf{ST}(\mathbf{T} \\ \qquad +\mathbf{TUR}^{-1}\mathbf{ST})^{-1}\mathbf{TUR}^{-1} \end{array} & \mathbf{R}^{-1}\mathbf{ST}(\mathbf{T}+\mathbf{TUR}^{-1}\mathbf{ST})^{-1} \\ -(\mathbf{T}+\mathbf{TUR}^{-1}\mathbf{ST})^{-1}\mathbf{TUR}^{-1} & (\mathbf{T}+\mathbf{TUR}^{-1}\mathbf{ST})^{-1} \end{pmatrix},$$

implying in particular that

$$(\mathbf{R} + \mathbf{STU})^{-1} = \mathbf{R}^{-1} - \mathbf{R}^{-1}\mathbf{ST}(\mathbf{T} + \mathbf{TUR}^{-1}\mathbf{ST})^{-1}\mathbf{TUR}^{-1}.$$

Either of the two formulas (2.22) and (2.23) can be obtained as a "special case" of the other. To see this, reexpress $\mathbf{R} + \mathbf{STU}$ as

$$\mathbf{R} + \mathbf{STU} = \mathbf{R} + (\mathbf{ST})\mathbf{T}^{-1}(\mathbf{TU}).$$

Then, applying formula (2.22) with $\mathbf{ST}$, $\mathbf{T}^{-1}$, and $\mathbf{TU}$ in place of $\mathbf{S}$, $\mathbf{T}$, and $\mathbf{U}$, respectively, we obtain

$$\begin{aligned}(\mathbf{R} + \mathbf{STU})^{-1} &= \mathbf{R}^{-1} - \mathbf{R}^{-1}\mathbf{ST}[(\mathbf{T}^{-1})^{-1} + \mathbf{TUR}^{-1}\mathbf{ST}]^{-1}\mathbf{TUR}^{-1} \\ &= \mathbf{R}^{-1} - \mathbf{R}^{-1}\mathbf{ST}(\mathbf{T} + \mathbf{TUR}^{-1}\mathbf{ST})^{-1}\mathbf{TUR}^{-1}.\end{aligned}$$

Similarly, applying formula (2.23) with $\mathbf{ST}$, $\mathbf{T}^{-1}$, and $\mathbf{TU}$ in place of $\mathbf{S}$, $\mathbf{T}$, and $\mathbf{U}$, respectively, we obtain

$$\begin{aligned}(\mathbf{R} + \mathbf{STU})^{-1} &= \mathbf{R}^{-1} - \mathbf{R}^{-1}\mathbf{STT}^{-1}(\mathbf{T}^{-1} \\ &\qquad + \mathbf{T}^{-1}\mathbf{TUR}^{-1}\mathbf{STT}^{-1})^{-1}\mathbf{T}^{-1}\mathbf{TUR}^{-1} \\ &= \mathbf{R}^{-1} - \mathbf{R}^{-1}\mathbf{S}(\mathbf{T}^{-1} + \mathbf{UR}^{-1}\mathbf{S})^{-1}\mathbf{UR}^{-1}.\end{aligned}$$

e. Generalized inverses of matrices of the form $\mathbf{R} + \mathbf{STU}$

Let $\mathbf{R}$ represent an $n \times q$ matrix, $\mathbf{S}$ an $n \times m$ matrix, $\mathbf{T}$ an $m \times p$ matrix, and $\mathbf{U}$ a $p \times q$ matrix. In the special case where $\mathbf{R}$, $\mathbf{T}$, and $\mathbf{T} + \mathbf{TUR}^{-1}\mathbf{ST}$ are nonsingular, $\mathbf{R} + \mathbf{STU}$ is nonsingular, and formula (2.23) gives the inverse of $\mathbf{R} + \mathbf{STU}$ in terms of the inverses of $\mathbf{R}$ and $\mathbf{T} + \mathbf{TUR}^{-1}\mathbf{ST}$.

Consider now the general case where $\mathbf{R}$ and $\mathbf{T}$ may not be nonsingular or even square. Does the expression obtained from formula (2.23) by replacing $\mathbf{R}^{-1}$ and $(\mathbf{T} + \mathbf{TUR}^{-1}\mathbf{ST})^{-1}$ with generalized inverses $\mathbf{R}^{-}$ and $(\mathbf{T} + \mathbf{TUR}^{-}\mathbf{ST})^{-}$ give a generalized inverse of $\mathbf{R} + \mathbf{STU}$? That is, is the matrix

$$\mathbf{R}^{-} - \mathbf{R}^{-}\mathbf{ST}(\mathbf{T} + \mathbf{TUR}^{-}\mathbf{ST})^{-}\mathbf{TUR}^{-}$$

necessarily a generalized inverse of $\mathbf{R} + \mathbf{STU}$?

To answer this question, let $\mathbf{D} = \mathbf{STU}$ and $\mathbf{Q} = \mathbf{T} + \mathbf{TUR}^{-}\mathbf{ST}$, and observe that

$$\begin{aligned}(\mathbf{R} + \mathbf{D})\mathbf{R}^{-}\mathbf{ST} &= \mathbf{RR}^{-}\mathbf{ST} + \mathbf{S}(\mathbf{Q} - \mathbf{T}) = \mathbf{SQ} - (\mathbf{I} - \mathbf{RR}^{-})\mathbf{ST}, \\ \mathbf{TUR}^{-}(\mathbf{R} + \mathbf{D}) &= \mathbf{TUR}^{-}\mathbf{R} + (\mathbf{Q} - \mathbf{T})\mathbf{U} = \mathbf{QU} - \mathbf{TU}(\mathbf{I} - \mathbf{R}^{-}\mathbf{R}), \\ \mathbf{SQU} &= \mathbf{D} + \mathbf{DR}^{-}\mathbf{D},\end{aligned}$$

so that

$$
\begin{aligned}
&(\mathbf{R}+\mathbf{D})(\mathbf{R}^- - \mathbf{R}^-\mathbf{STQ}^-\mathbf{TUR}^-)(\mathbf{R}+\mathbf{D}) \\
&\quad = \mathbf{R} + \mathbf{DR}^-\mathbf{R} + \mathbf{RR}^-\mathbf{D} + \mathbf{DR}^-\mathbf{D} - (\mathbf{R}+\mathbf{D})\mathbf{R}^-\mathbf{STQ}^-\mathbf{TUR}^-(\mathbf{R}+\mathbf{D}) \\
&\quad = \mathbf{R} + \mathbf{DR}^-\mathbf{R} + \mathbf{RR}^-\mathbf{D} + \mathbf{DR}^-\mathbf{D} \\
&\qquad\qquad - [\mathbf{SQ} - (\mathbf{I}-\mathbf{RR}^-)\mathbf{ST}]\mathbf{Q}^-[\mathbf{QU} - \mathbf{TU}(\mathbf{I}-\mathbf{R}^-\mathbf{R})] \\
&\quad = \mathbf{R} + \mathbf{DR}^-\mathbf{R} + \mathbf{RR}^-\mathbf{D} + \mathbf{DR}^-\mathbf{D} - \mathbf{SQU} + \mathbf{SQQ}^-\mathbf{TU}(\mathbf{I}-\mathbf{R}^-\mathbf{R}) \\
&\qquad\qquad + (\mathbf{I}-\mathbf{RR}^-)\mathbf{STQ}^-\mathbf{QU} - (\mathbf{I}-\mathbf{RR}^-)\mathbf{STQ}^-\mathbf{TU}(\mathbf{I}-\mathbf{R}^-\mathbf{R}) \\
&\quad = \mathbf{R} + \mathbf{DR}^-\mathbf{R} + \mathbf{RR}^-\mathbf{D} + \mathbf{DR}^-\mathbf{D} - \mathbf{D} - \mathbf{DR}^-\mathbf{D} + \mathbf{SQQ}^-\mathbf{TU}(\mathbf{I}-\mathbf{R}^-\mathbf{R}) \\
&\qquad\qquad + (\mathbf{I}-\mathbf{RR}^-)\mathbf{STQ}^-\mathbf{QU} - (\mathbf{I}-\mathbf{RR}^-)\mathbf{STQ}^-\mathbf{TU}(\mathbf{I}-\mathbf{R}^-\mathbf{R}) \\
&\quad = \mathbf{R} + \mathbf{D} - \mathbf{STU}(\mathbf{I}-\mathbf{R}^-\mathbf{R}) - (\mathbf{I}-\mathbf{RR}^-)\mathbf{STU} + \mathbf{SQQ}^-\mathbf{TU}(\mathbf{I}-\mathbf{R}^-\mathbf{R}) \\
&\qquad\qquad + (\mathbf{I}-\mathbf{RR}^-)\mathbf{STQ}^-\mathbf{QU} - (\mathbf{I}-\mathbf{RR}^-)\mathbf{STQ}^-\mathbf{TU}(\mathbf{I}-\mathbf{R}^-\mathbf{R}) \\
&\quad = \mathbf{R} + \mathbf{D} - \mathbf{S}(\mathbf{I}-\mathbf{QQ}^-)\mathbf{TU}(\mathbf{I}-\mathbf{R}^-\mathbf{R}) - (\mathbf{I}-\mathbf{RR}^-)\mathbf{ST}(\mathbf{I}-\mathbf{Q}^-\mathbf{Q})\mathbf{U} \\
&\qquad\qquad - (\mathbf{I}-\mathbf{RR}^-)\mathbf{STQ}^-\mathbf{TU}(\mathbf{I}-\mathbf{R}^-\mathbf{R}).
\end{aligned}
$$

Thus, we have the following theorem.

Theorem 18.2.12. Let $\mathbf{R}$ represent an $n \times q$ matrix, $\mathbf{S}$ an $n \times m$ matrix, $\mathbf{T}$ an $m \times p$ matrix, and $\mathbf{U}$ a $p \times q$ matrix, and define $\mathbf{Q} = \mathbf{T} + \mathbf{TUR}^-\mathbf{ST}$. Then, the matrix

$$\mathbf{R}^- - \mathbf{R}^-\mathbf{STQ}^-\mathbf{TUR}^- \tag{2.27}$$

is a generalized inverse of the matrix $\mathbf{R} + \mathbf{STU}$ if and only if

$$
\begin{aligned}
&\mathbf{S}(\mathbf{I}-\mathbf{QQ}^-)\mathbf{TU}(\mathbf{I}-\mathbf{R}^-\mathbf{R}) + (\mathbf{I}-\mathbf{RR}^-)\mathbf{ST}(\mathbf{I}-\mathbf{Q}^-\mathbf{Q})\mathbf{U} \\
&\qquad + (\mathbf{I}-\mathbf{RR}^-)\mathbf{STQ}^-\mathbf{TU}(\mathbf{I}-\mathbf{R}^-\mathbf{R}) = \mathbf{0}.
\end{aligned} \tag{2.28}
$$

It is clear (in light of Lemma 9.3.5) that condition (2.28) is satisfied if $\mathcal{R}(\mathbf{TU}) \subset \mathcal{R}(\mathbf{R})$ and $\mathcal{C}(\mathbf{ST}) \subset \mathcal{C}(\mathbf{R})$. In fact, condition (2.28) is satisfied if $\mathcal{R}(\mathbf{STU}) \subset \mathcal{R}(\mathbf{R})$ and $\mathcal{C}(\mathbf{STU}) \subset \mathcal{C}(\mathbf{R})$, as is evident from the following lemma.

Lemma 18.2.13. Let $\mathbf{R}$ represent an $n \times q$ matrix, $\mathbf{S}$ an $n \times m$ matrix, $\mathbf{T}$ an $m \times p$ matrix, and $\mathbf{U}$ a $p \times q$ matrix, and define $\mathbf{Q} = \mathbf{T} + \mathbf{TUR}^-\mathbf{ST}$. If $\mathcal{R}(\mathbf{STU}) \subset \mathcal{R}(\mathbf{R})$ and $\mathcal{C}(\mathbf{STU}) \subset \mathcal{C}(\mathbf{R})$, then

$$(\mathbf{I}-\mathbf{QQ}^-)\mathbf{TU}(\mathbf{I}-\mathbf{R}^-\mathbf{R}) = \mathbf{0}, \tag{2.29}$$

$$(\mathbf{I}-\mathbf{RR}^-)\mathbf{ST}(\mathbf{I}-\mathbf{Q}^-\mathbf{Q}) = \mathbf{0}, \tag{2.30}$$

$$(\mathbf{I}-\mathbf{RR}^-)\mathbf{STQ}^-\mathbf{TU}(\mathbf{I}-\mathbf{R}^-\mathbf{R}) = \mathbf{0}. \tag{2.31}$$

Proof. Let $\mathbf{D} = \mathbf{STU}$, and observe that $\mathbf{QQ}^-(\mathbf{T} + \mathbf{TUR}^-\mathbf{ST}) = \mathbf{T} + \mathbf{TUR}^-\mathbf{ST}$, or, equivalently, that $(\mathbf{I}-\mathbf{QQ}^-)\mathbf{T} = -(\mathbf{I}-\mathbf{QQ}^-)\mathbf{TUR}^-\mathbf{ST}$, and hence that

$$(\mathbf{I}-\mathbf{QQ}^-)\mathbf{TU} = -(\mathbf{I}-\mathbf{QQ}^-)\mathbf{TUR}^-\mathbf{D}. \tag{2.32}$$

It can be shown in similar fashion that

$$\mathbf{ST}(\mathbf{I}-\mathbf{Q}^-\mathbf{Q}) = -\mathbf{DR}^-\mathbf{ST}(\mathbf{I}-\mathbf{Q}^-\mathbf{Q}). \tag{2.33}$$

Further,

$$(\mathbf{T} + \mathbf{TUR^-ST})\mathbf{Q}^-(\mathbf{T} + \mathbf{TUR^-ST}) = \mathbf{T} + \mathbf{TUR^-ST},$$

or, equivalently,

$$\mathbf{TQ^-T} = \mathbf{T} + \mathbf{TUR^-ST} - \mathbf{TUR^-STQ^-T} \\ - \mathbf{TQ^-TUR^-ST} - \mathbf{TUR^-STQ^-TUR^-ST},$$

implying that

$$\mathbf{STQ^-TU} = \mathbf{D} + \mathbf{DR^-D} - \mathbf{DR^-STQ^-TU} \\ - \mathbf{STQ^-TUR^-D} - \mathbf{DR^-STQ^-TUR^-D}. \quad (2.34)$$

And, if $\mathcal{R}(\mathbf{D}) \subset \mathcal{R}(\mathbf{R})$ and $\mathcal{C}(\mathbf{D}) \subset \mathcal{C}(\mathbf{R})$ [in which case $\mathbf{D}(\mathbf{I} - \mathbf{R^-R}) = \mathbf{0}$ and $(\mathbf{I} - \mathbf{RR^-})\mathbf{D} = \mathbf{0}$], results (2.29)–(2.31) follow from results (2.32)–(2.34). Q.E.D.

In light of Lemma 18.2.13, we have (as a consequence of Theorem 18.2.12) the following result.

Theorem 18.2.14. Let $\mathbf{R}$ represent an $n \times q$ matrix, $\mathbf{S}$ an $n \times m$ matrix, $\mathbf{T}$ an $m \times p$ matrix, and $\mathbf{U}$ a $p \times q$ matrix. If $\mathcal{R}(\mathbf{STU}) \subset \mathcal{R}(\mathbf{R})$ and $\mathcal{C}(\mathbf{STU}) \subset \mathcal{C}(\mathbf{R})$, then the matrix

$$\mathbf{R}^- - \mathbf{R^-ST}(\mathbf{T} + \mathbf{TUR^-ST})^-\mathbf{TUR}^-$$

is a generalized inverse of the matrix $\mathbf{R} + \mathbf{STU}$.

Observe (in connection with Theorem 18.2.14) that $\mathbf{R} + \mathbf{STU}$ can be reexpressed as

$$\mathbf{R} + \mathbf{STU} = \mathbf{R} + (\mathbf{ST})\mathbf{T}^-(\mathbf{TU}).$$

Thus, applying Theorem 18.2.14 with $\mathbf{ST}$, $\mathbf{T}^-$, and $\mathbf{TU}$ in place of $\mathbf{S}$, $\mathbf{T}$, and $\mathbf{U}$, respectively, we obtain the following corollary, which generalizes result (2.22).

Corollary 18.2.15. Let $\mathbf{R}$ represent an $n \times q$ matrix, $\mathbf{S}$ an $n \times m$ matrix, $\mathbf{T}$ an $m \times p$ matrix, and $\mathbf{U}$ a $p \times q$ matrix. If $\mathcal{R}(\mathbf{STU}) \subset \mathcal{R}(\mathbf{R})$ and $\mathcal{C}(\mathbf{STU}) \subset \mathcal{C}(\mathbf{R})$, then the matrix

$$\mathbf{R}^- - \mathbf{R^-STT^-}(\mathbf{T}^- + \mathbf{T^-TUR^-STT^-})^-\mathbf{T^-TUR}^-$$

is a generalized inverse of the matrix $\mathbf{R} + \mathbf{STU}$.

f. An infinite-series representation of the inverse of a matrix of the form $\mathbf{I} - \mathbf{A}$

Let $\mathbf{A}_1, \mathbf{A}_2, \ldots$ represent a sequence of $m \times n$ matrices, and (for $i = 1, \ldots, m$, $j = 1, \ldots, n$, and $k = 1, 2, \ldots$) let $a_{ij}^{(k)}$ represent the ijth element of $\mathbf{A}_k$. If for every i and j (i.e., if for $i = 1, \ldots, m$ and $j = 1, \ldots, n$) there exists a scalar a_{ij} such that a_{ij} is the limit of the sequence $a_{ij}^{(1)}, a_{ij}^{(2)}, \ldots$, we say that the $m \times n$

matrix $\mathbf{A}$, whose ijth element is a_{ij}, is the *limit* of the sequence $\mathbf{A}_1, \mathbf{A}_2, \ldots$ (or that the sequence $\mathbf{A}_1, \mathbf{A}_2, \ldots$ *converges* to $\mathbf{A}$), and write $\lim_{k\to\infty} \mathbf{A}_k = \mathbf{A}$ (or $\mathbf{A}_k \to \mathbf{A}$).

If the sequence $\mathbf{A}_1, \mathbf{A}_2, \ldots$ has a limit, it is said to be *convergent*. If the sequence $\mathbf{A}_1, \mathbf{A}_2, \ldots$ does not have a limit (i.e., if for some i and j the sequence $a_{ij}^{(1)}, a_{ij}^{(2)}, \ldots$ does not have a limit), it is said to be *divergent*.

Four basic properties of sequences of matrices (which are almost immediate consequences of basic properties of sequences of scalars) are as follows:

(1) For any $m \times n$ matrix $\mathbf{A}$, the sequence $\mathbf{A}, \mathbf{A}, \ldots$ (each of whose members equals $\mathbf{A}$) converges to $\mathbf{A}$.

(2) For any scalar c (including $c = -1$) and for any sequence $\mathbf{A}_1, \mathbf{A}_2, \ldots$ of $m \times n$ matrices that converge to an $(m \times n)$ matrix $\mathbf{A}$, $\lim_{k\to\infty} (c\mathbf{A}_k) = c\mathbf{A}$.

(3) If $\mathbf{A}_1, \mathbf{A}_2, \ldots$ and $\mathbf{B}_1, \mathbf{B}_2, \ldots$ are sequences of $m \times n$ matrices that converge to $(m \times n)$ matrices $\mathbf{A}$ and $\mathbf{B}$, respectively, then

$$\lim_{k\to\infty} (\mathbf{A}_k + \mathbf{B}_k) = \mathbf{A} + \mathbf{B}. \tag{2.35}$$

(4) If $\mathbf{A}_1, \mathbf{A}_2, \ldots$ is a sequence of $m \times n$ matrices that converges to an $(m \times n)$ matrix $\mathbf{A}$ and $\mathbf{B}_1, \mathbf{B}_2, \ldots$ a sequence of $n \times p$ matrices that converges to an $(n \times p)$ matrix $\mathbf{B}$, then

$$\lim_{k\to\infty} \mathbf{A}_k\mathbf{B}_k = \mathbf{AB}. \tag{2.36}$$

Results (2.35) and (2.36) extend in a straightforward way to an arbitrary (finite) number of sequences of matrices.

Corresponding to any sequence $\mathbf{A}_1, \mathbf{A}_2, \ldots$ of $m \times n$ matrices is another sequence $\mathbf{S}_1, \mathbf{S}_2, \ldots,$ where $\mathbf{S}_1 = \mathbf{A}_1$, $\mathbf{S}_2 = \mathbf{A}_1 + \mathbf{A}_2$, and more generally (for $k = 1, 2, \ldots$) $\mathbf{S}_k = \sum_{i=1}^{k} \mathbf{A}_i$. The sequence $\mathbf{S}_1, \mathbf{S}_2, \ldots$ is referred to as the *infinite series* (or simply as the *series*) generated by the sequence $\mathbf{A}_1, \mathbf{A}_2, \ldots$. Further, the matrix $\mathbf{A}_k$ is referred to as the kth *term* of the infinite series, and the matrix $\mathbf{S}_k$ is referred to as the kth *partial sum*. And, if the infinite series $\mathbf{S}_1, \mathbf{S}_2, \ldots$ converges, its limit is referred to as the *sum* of the infinite series and is denoted by the symbol $\sum_{k=1}^{\infty} \mathbf{A}_k$ or by the expression $\mathbf{A}_1 + \mathbf{A}_2 + \cdots$.

It is convenient to use the symbol $\sum_{k=1}^{\infty} \mathbf{A}_k$, or the expression $\mathbf{A}_1 + \mathbf{A}_2 + \cdots$, not only to represent the sum of the infinite series generated by the sequence $\mathbf{A}_1, \mathbf{A}_2, \ldots$ (when the infinite series converges) but also to represent the infinite series itself—the intended usage must be determined from the context. Further, the symbol $\sum_{k=p}^{\infty} \mathbf{A}_k$ (or the expression $\mathbf{A}_p + \mathbf{A}_{p+1} + \cdots$) is used to represent an infinite series $(\mathbf{A}_p, \mathbf{A}_p + \mathbf{A}_{p+1}, \cdots)$ generated by a sequence $\mathbf{A}_p, \mathbf{A}_{p+1}, \ldots$ of $m \times n$ matrices and also to represent the sum of that infinite series (where p is an integer, not necessarily equal to 1).

The following theorem gives a useful result on the inversion of a matrix of the form $\mathbf{I} - \mathbf{A}$.

Theorem 18.2.16. Let $\mathbf{A}$ represent an $n \times n$ matrix. Then, the infinite series $\mathbf{I} + \mathbf{A} + \mathbf{A}^2 + \mathbf{A}^3 + \cdots$ converges if and only if $\lim_{k\to\infty} \mathbf{A}^k = \mathbf{0}$, in which case

$\mathbf{I} - \mathbf{A}$ is nonsingular and

$$(\mathbf{I} - \mathbf{A})^{-1} = \sum_{k=0}^{\infty} \mathbf{A}^k = \mathbf{I} + \mathbf{A} + \mathbf{A}^2 + \mathbf{A}^3 + \cdots \tag{2.37}$$

(where $\mathbf{A}^0 = \mathbf{I}$).

Preliminary to proving Theorem 18.2.16, it is convenient to establish the following lemma.

Lemma 18.2.17. Let $\mathbf{A}_0, \mathbf{A}_1, \mathbf{A}_2, \ldots$ represent a sequence of $m \times n$ matrices. If the infinite series $\sum_{k=0}^{\infty} \mathbf{A}_k$ converges, then $\lim_{k\to\infty} \mathbf{A}_k = \mathbf{0}$.

Proof (of Lemma 18.2.17). Suppose that $\sum_{k=0}^{\infty} \mathbf{A}_k$ converges. And, for $k = 0, 1, 2, \ldots$, let $\mathbf{S}_k = \sum_{i=0}^{k} \mathbf{A}_i$. Then (in light of our supposition), the sequence $\mathbf{S}_0, \mathbf{S}_1, \mathbf{S}_2, \ldots$ has a limit, say $\mathbf{S}$. Thus, observing that (for $k = 1, 2, \ldots$) $\mathbf{A}_k = \mathbf{S}_k - \mathbf{S}_{k-1}$ and that $\lim_{k\to\infty} \mathbf{S}_{k-1} = \mathbf{S}$, it follows that

$$\lim_{k\to\infty} \mathbf{A}_k = \lim_{k\to\infty} (\mathbf{S}_k - \mathbf{S}_{k-1}) = \lim_{k\to\infty} \mathbf{S}_k - \lim_{k\to\infty} \mathbf{S}_{k-1} = \mathbf{S} - \mathbf{S} = \mathbf{0}.$$

Q.E.D.

Proof (of Theorem 18.2.16). If the infinite series $\mathbf{I} + \mathbf{A} + \mathbf{A}^2 + \mathbf{A}^3 + \cdots$ converges, then it is clear from Lemma 18.2.17 that $\lim_{k\to\infty} \mathbf{A}^k = \mathbf{0}$.

Conversely, suppose that $\lim_{k\to\infty} \mathbf{A}^k = \mathbf{0}$. And observe that

$$(\mathbf{I} + \mathbf{A} + \mathbf{A}^2 + \cdots + \mathbf{A}^k)(\mathbf{I} - \mathbf{A}) = \mathbf{I} - \mathbf{A}^{k+1} \tag{2.38}$$

and that

$$\lim_{k\to\infty} (\mathbf{I} - \mathbf{A}^{k+1}) = \mathbf{I} - \lim_{k\to\infty} \mathbf{A}^{k+1} = \mathbf{I} - \mathbf{0} = \mathbf{I}. \tag{2.39}$$

Then, for any $n \times 1$ vector $\mathbf{x}$ such that $(\mathbf{I} - \mathbf{A})\mathbf{x} = \mathbf{0}$, we find that

$$(\mathbf{I} - \mathbf{A}^{k+1})\mathbf{x} = (\mathbf{I} + \mathbf{A} + \mathbf{A}^2 + \cdots + \mathbf{A}^k)(\mathbf{I} - \mathbf{A})\mathbf{x} = \mathbf{0}$$

($k = 0, 1, 2, \ldots$) and consequently that

$$\mathbf{x} = \mathbf{I}\mathbf{x} = [\lim_{k\to\infty} (\mathbf{I} - \mathbf{A}^{k+1})]\mathbf{x} = \lim_{k\to\infty} (\mathbf{I} - \mathbf{A}^{k+1})\mathbf{x} = \lim_{k\to\infty} \mathbf{0} = \mathbf{0}.$$

Thus, $\mathbf{I} - \mathbf{A}$ is nonsingular.

Further, postmultiplying both sides of equality (2.38) by $(\mathbf{I} - \mathbf{A})^{-1}$, we obtain the equality

$$\mathbf{I} + \mathbf{A} + \mathbf{A}^2 + \cdots + \mathbf{A}^k = (\mathbf{I} - \mathbf{A}^{k+1})(\mathbf{I} - \mathbf{A})^{-1}.$$

And, in light of result (2.39), we conclude that the infinite series $\sum_{k=0}^{\infty} \mathbf{A}^k$ converges and that

$$\begin{aligned}\sum_{k=0}^{\infty} \mathbf{A}^k &= \lim_{k\to\infty} (\mathbf{I} - \mathbf{A}^{k+1})(\mathbf{I} - \mathbf{A})^{-1} \\ &= [\lim_{k\to\infty} (\mathbf{I} - \mathbf{A}^{k+1})](\mathbf{I} - \mathbf{A})^{-1} = \mathbf{I}(\mathbf{I} - \mathbf{A})^{-1} = (\mathbf{I} - \mathbf{A})^{-1}.\end{aligned}$$

Q.E.D.

Theorem 18.2.16 can be generalized as follows.

Theorem 18.2.18. Let $\mathbf{A}$ and $\mathbf{B}$ represent $n \times n$ matrices. Suppose that $\mathbf{B}$ is nonsingular, and define $\mathbf{F} = \mathbf{B}^{-1}\mathbf{A}$. Then, the infinite series $\mathbf{B}^{-1} + \mathbf{F}\mathbf{B}^{-1} + \mathbf{F}^2\mathbf{B}^{-1} + \mathbf{F}^3\mathbf{B}^{-1} + \cdots$ converges if and only if $\lim_{k\to\infty} \mathbf{F}^k = \mathbf{0}$, in which case $\mathbf{B} - \mathbf{A}$ is nonsingular and

$$(\mathbf{B} - \mathbf{A})^{-1} = \sum_{k=0}^{\infty} \mathbf{F}^k\mathbf{B}^{-1} = \mathbf{B}^{-1} + \mathbf{F}\mathbf{B}^{-1} + \mathbf{F}^2\mathbf{B}^{-1} + \mathbf{F}^3\mathbf{B}^{-1} + \cdots \quad (2.40)$$

(where $\mathbf{F}^0 = \mathbf{I}$).

Proof. If the infinite series $\sum_{k=0}^{\infty} \mathbf{F}^k\mathbf{B}^{-1}$ converges, then, as a consequence of Lemma 18.2.17, $\lim_{k\to\infty} \mathbf{F}^k\mathbf{B}^{-1} = \mathbf{0}$, and hence

$$\lim_{k\to\infty} \mathbf{F}^k = \lim_{k\to\infty} \mathbf{F}^k\mathbf{B}^{-1}\mathbf{B} = (\lim_{k\to\infty} \mathbf{F}^k\mathbf{B}^{-1})\mathbf{B} = \mathbf{0}\mathbf{B} = \mathbf{0}.$$

Conversely, suppose that $\lim_{k\to\infty} \mathbf{F}^k = \mathbf{0}$. Then, it follows from Theorem 18.2.16 that $\mathbf{I} - \mathbf{F}$ is nonsingular and that

$$(\mathbf{I} - \mathbf{F})^{-1} = \sum_{k=0}^{\infty} \mathbf{F}^k.$$

Further, $\mathbf{B} - \mathbf{A} = \mathbf{B}(\mathbf{I} - \mathbf{F})$, implying (since $\mathbf{B}$ is nonsingular) that $\mathbf{B} - \mathbf{A}$ is nonsingular and [in light of result (2.36)] that

$$\begin{aligned}(\mathbf{B} - \mathbf{A})^{-1} = (\mathbf{I} - \mathbf{F})^{-1}\mathbf{B}^{-1} &= \Big(\sum_{k=0}^{\infty} \mathbf{F}^k\Big)\mathbf{B}^{-1} \\ &= \Big(\lim_{p\to\infty} \sum_{k=0}^{p} \mathbf{F}^k\Big)\mathbf{B}^{-1} \\ &= \lim_{p\to\infty} \sum_{k=0}^{p} \mathbf{F}^k\mathbf{B}^{-1} = \sum_{k=0}^{\infty} \mathbf{F}^k\mathbf{B}^{-1}.\end{aligned}$$

Q.E.D.

Formula (2.37) is applicable if (and only if) $\lim_{k\to\infty} \mathbf{A}^k = \mathbf{0}$, and, more generally, formula (2.40) is applicable if (and only if) $\lim_{k\to\infty} \mathbf{F}^k = \mathbf{0}$. Depending on the nature of $\mathbf{A}$ (or $\mathbf{F}$), it may be difficult to determine whether the condition $\lim_{k\to\infty} \mathbf{A}^k = \mathbf{0}$ (or the condition $\lim_{k\to\infty} \mathbf{F}^k = \mathbf{0}$) is satisfied. A condition that is more stringent, but that is typically easier to check, can be obtained from the following theorem.

Theorem 18.2.19. Let $\mathbf{A}$ represent an $n \times n$ matrix. If $\|\mathbf{A}\| < 1$ (where the norm is the usual norm), then $\lim_{k\to\infty} \mathbf{A}^k = \mathbf{0}$.

Preliminary to proving Theorem 18.2.19, it is convenient to establish the following two lemmas, which are of some interest in their own right.

Lemma 18.2.20. Let $\mathbf{A}_1, \mathbf{A}_2, \ldots$ represent a sequence of $m \times n$ matrices. Then, $\mathbf{A}_1, \mathbf{A}_2, \ldots$ converges to an $(m \times n)$ matrix $\mathbf{A}$ if and only if $\|\mathbf{A}_k - \mathbf{A}\| \to 0$ (where the norm is the usual norm).

Proof (of Lemma 18.2.20). Let a_{ij} represent the ijth element of $\mathbf{A}$, and (for $k = 1, 2, \ldots$) let $a_{ij}^{(k)}$ represent the ijth element of $\mathbf{A}_k$. Then, it is clear from the very definition of the usual norm that (for $k = 1, 2, \ldots$)

$$\max_{i,j} |a_{ij}^{(k)} - a_{ij}| \leq \|\mathbf{A}_k - \mathbf{A}\| \leq (mn)^{1/2} \max_{i,j} |a_{ij}^{(k)} - a_{ij}|. \tag{2.41}$$

Suppose now that $\|\mathbf{A}_k - \mathbf{A}\| \to 0$. Then, corresponding to each positive scalar ϵ, there exists a positive integer p such that, for $k > p$, $\|\mathbf{A}_k - \mathbf{A}\| < \epsilon$ and hence [in light of result (2.41)] such that, for $k > p$ (and for $i = 1, \ldots, m$ and $j = 1, \ldots, n$), $\|a_{ij}^{(k)} - a_{ij}\| < \epsilon$. Thus, $\lim_{k\to\infty} a_{ij}^{(k)} = a_{ij}$ (for $i = 1, \ldots, m$ and $j = 1, \ldots, n$), and consequently $\mathbf{A}_k \to \mathbf{A}$.

Conversely, suppose that $\mathbf{A}_k \to \mathbf{A}$. Then, by definition, $a_{ij}^{(k)} \to a_{ij}$ (for $i = 1, \ldots, m$ and $j = 1, \ldots, n$). Thus, corresponding to any positive scalar ϵ, there exists a positive integer p such that, for $k > p$ (and for $i = 1, \ldots, m$ and $j = 1, \ldots, n$), $|a_{ij}^{(k)} - a_{ij}| < (mn)^{-1/2}\epsilon$ and hence [in light of result (2.41)] such that, for $k > p$, $\|\mathbf{A}_k - \mathbf{A}\| < \epsilon$. Thus, $\|\mathbf{A}_k - \mathbf{A}\| \to 0$. Q.E.D.

Lemma 18.2.21. For any $m \times n$ matrix $\mathbf{A}$ and $n \times p$ matrix $\mathbf{B}$,

$$\|\mathbf{AB}\| \leq \|\mathbf{A}\|\|\mathbf{B}\|$$

(where the norms are the usual norms).

Proof (of Lemma 18.2.21). Denote the first, ..., pth columns of $\mathbf{B}$ by $\mathbf{b}_1$, ..., $\mathbf{b}_p$, respectively. Then, the jth column of $\mathbf{AB}$ is $\mathbf{Ab}_j$, and [in light of result (5.2.5)] we have that

$$\|\mathbf{AB}\|^2 = \text{tr}[(\mathbf{AB})'\mathbf{AB}] = \sum_{j=1}^{p} (\mathbf{Ab}_j)'\mathbf{Ab}_j. \tag{2.42}$$

Further, letting $\mathbf{a}_i'$ represent the ith row of $\mathbf{A}$ (for $i = 1, \ldots, m$) and making use of inequality (6.3.2) (the Schwarz inequality), we find that (for $j = 1, \ldots, p$)

$$(\mathbf{Ab}_j)'\mathbf{Ab}_j = \sum_{i=1}^{m} (\mathbf{a}_i'\mathbf{b}_j)^2 \leq \sum_{i=1}^{m} (\mathbf{a}_i'\mathbf{a}_i)(\mathbf{b}_j'\mathbf{b}_j) = \|\mathbf{A}\|^2\mathbf{b}_j'\mathbf{b}_j. \tag{2.43}$$

And, upon combining result (2.43) with result (2.42), we obtain

$$\|\mathbf{AB}\|^2 \leq \|\mathbf{A}\|^2 \sum_{j=1}^{p} \mathbf{b}_j'\mathbf{b}_j = \|\mathbf{A}\|^2\|\mathbf{B}\|^2,$$

or, equivalently,

$$\|\mathbf{AB}\| \leq \|\mathbf{A}\|\|\mathbf{B}\|.$$

Q.E.D.

Proof (of Theorem 18.2.19). By making repeated use of Lemma 18.2.21, we find that (for any positive integer k)

$$\|\mathbf{A}^k\| \leq \|\mathbf{A}\|^k. \tag{2.44}$$

Suppose now that $\|\mathbf{A}\| < 1$. Then, it follows from a basic result on limits [which is example 14.8(c) in Bartle's (1976) book] that $\|\mathbf{A}\|^k \to 0$, implying [in light of inequality (2.44)] that $\|\mathbf{A}^k\| \to 0$. We conclude, on the basis of Lemma 18.2.20, that $\lim_{k\to\infty} \mathbf{A}^k = \mathbf{0}$. Q.E.D.

18.3 Some Results on Positive (and Nonnegative) Definiteness

a. Matrices of the form $\mathbf{V} + \sum_{i=1}^k x_i \mathbf{A}_i$

The following theorem gives a basic result on matrices of the form $\mathbf{V}+\sum_{i=1}^k x_i \mathbf{A}_i$ (where $\mathbf{V}$ is symmetric and positive definite and $\mathbf{A}_1, \ldots, \mathbf{A}_k$ are symmetric).

Theorem 18.3.1. Let $\mathbf{V}$ represent an $n\times n$ symmetric positive definite matrix, and let $\mathbf{A}_1, \ldots, \mathbf{A}_k$ represent k symmetric matrices of order n. And let $\mathbf{x} = \{x_i\}$ represent a $k \times 1$ vector. Then, there exists a neighborhood N of the $k \times 1$ null vector $\mathbf{0}$ such that, for $\mathbf{x} \in N$, $\mathbf{V} + \sum_{i=1}^k x_i \mathbf{A}_i$ is (symmetric) positive definite.

Proof. For $s = 1, \ldots, n$, let $\mathbf{A}_1^{(s)}, \ldots, \mathbf{A}_k^{(s)}$, and $\mathbf{V}^{(s)}$ represent the $s\times s$ leading principal submatrices of $\mathbf{A}_1, \ldots, \mathbf{A}_k$ and $\mathbf{V}$, respectively. Then, when regarded as a function of $\mathbf{x}$, $|\mathbf{V}^{(s)} + \sum_{i=1}^k x_i \mathbf{A}_i^{(s)}|$ is continuous at $\mathbf{0}$ (and at every other point in $\mathcal{R}^k$), as is evident from the very definition of a determinant [given by formula (13.1.2) or (13.1.6)]. Thus,

$$\lim_{\mathbf{x}\to\mathbf{0}} |\mathbf{V}^{(s)} + \sum_{i=1}^k x_i \mathbf{A}_i^{(s)}| = |\mathbf{V}^{(s)} + \sum_{i=1}^k 0\mathbf{A}_i^{(s)}| = |\mathbf{V}^{(s)}|.$$

And, since (according to Theorem 14.9.5) $|\mathbf{V}^{(s)}| > 0$, there exists a neighborhood N_s of $\mathbf{0}$ such that, for $\mathbf{x} \in N_s$, $|\mathbf{V}^{(s)} + \sum_{i=1}^k x_i \mathbf{A}_i^{(s)}| > 0$.

Now, take N to be the smallest of the n neighborhoods $N_1, \ldots, N_n$. Then, for $\mathbf{x}\in N$, $|\mathbf{V}^{(1)} + \sum_{i=1}^k x_i \mathbf{A}_i^{(1)}| > 0, \ldots, |\mathbf{V}^{(n)} + \sum_{i=1}^k x_i \mathbf{A}_i^{(n)}| > 0$. And, observing that $\mathbf{V}^{(1)} + \sum_{i=1}^k x_i \mathbf{A}_i^{(1)}, \ldots, \mathbf{V}^{(n)} + \sum_{i=1}^k x_i \mathbf{A}_i^{(n)}$ are the leading principal submatrices of $\mathbf{V} + \sum_{i=1}^k x_i \mathbf{A}_i$, we conclude (on the basis of Theorem 14.9.5) that, for $\mathbf{x} \in N$, $\mathbf{V} + \sum_{i=1}^k x_i \mathbf{A}_i$ is positive definite. Q.E.D.

In the special case where $k = 1$, Theorem 18.3.1 can be restated as the following corollary.

Corollary 18.3.2. Let $\mathbf{A}$ represent an $n \times n$ symmetric matrix, and let $\mathbf{V}$ represent an $n \times n$ symmetric positive definite matrix. Then, there exists a (strictly) positive scalar c such that $\mathbf{V} + x\mathbf{A}$ is (symmetric) positive definite for every scalar x in the interval $-c < x < c$.

The following corollary gives a variation on Corollary 18.3.2.

Corollary 18.3.3. Let $\mathbf{A}$ represent an $n \times n$ symmetric matrix, and let $\mathbf{V}$ represent an $n \times n$ symmetric positive definite matrix. Then, there exists a scalar c such that $\mathbf{A} + x\mathbf{V}$ is (symmetric) positive definite for every (scalar) $x > c$.

Proof. Let w represent an arbitrary scalar. Then, it follows from Corollary 18.3.2 that there exists a (strictly) positive scalar d such that $\mathbf{V} + w\mathbf{A}$ is positive definite for $-d < w < d$. And, since $x > 1/d \Rightarrow 0 < 1/x < d$ and since (for $x \neq 0$) $\mathbf{A} + x\mathbf{V} = x[\mathbf{V} + (1/x)\mathbf{A}]$, it follows from Lemma 14.2.3 that $\mathbf{A} + x\mathbf{V}$ is positive definite for $x > 1/d$. Q.E.D.

b. Matrices of the form $\mathbf{A}^{-1} - \mathbf{B}^{-1}$

The following theorem gives sufficient conditions for the difference $\mathbf{A}^{-1} - \mathbf{B}^{-1}$ between the inverses of a symmetric positive definite matrix $\mathbf{A}$ and a (nonsingular) matrix $\mathbf{B}$ to be positive definite or positive semidefinite.

Theorem 18.3.4. Let $\mathbf{A}$ represent an $n \times n$ symmetric positive definite matrix, and let $\mathbf{B}$ represent an $n \times n$ matrix.

(1) If $\mathbf{B} - \mathbf{A}$ is positive definite or, more generally, if $\mathbf{B} - \mathbf{A}$ is nonnegative definite and nonsingular (in which case $\mathbf{B}$ is positive definite), then $\mathbf{A}^{-1} - \mathbf{B}^{-1}$ is positive definite.

(2) If $\mathbf{B} - \mathbf{A}$ is symmetric and positive semidefinite (in which case $\mathbf{B}$ is symmetric and positive definite), then $\mathbf{A}^{-1} - \mathbf{B}^{-1}$ is positive semidefinite.

Proof. (1) Suppose that $\mathbf{B} - \mathbf{A}$ is nonnegative definite and nonsingular [which, since $\mathbf{B} = \mathbf{A} + (\mathbf{B} - \mathbf{A})$, implies (in light of Lemma 14.2.4) that $\mathbf{B}$ is positive definite (and hence nonsingular)]. Then, it follows from Theorem 18.2.4 that $\mathbf{A}^{-1} - \mathbf{B}^{-1}$ is nonsingular and that

$$(\mathbf{A}^{-1} - \mathbf{B}^{-1})^{-1} = \mathbf{A} + \mathbf{A}(\mathbf{B} - \mathbf{A})^{-1}\mathbf{A} = \mathbf{A} + \mathbf{A}'(\mathbf{B} - \mathbf{A})^{-1}\mathbf{A}.$$

Moreover, it follows from Corollary 14.2.11 that $(\mathbf{B} - \mathbf{A})^{-1}$ is nonnegative definite, implying (in light of Theorem 14.2.9) that $\mathbf{A}'(\mathbf{B} - \mathbf{A})^{-1}\mathbf{A}$ is nonnegative definite and hence (in light of Lemma 14.2.4) that $\mathbf{A} + \mathbf{A}'(\mathbf{B} - \mathbf{A})^{-1}\mathbf{A}$ is positive definite. Thus, $(\mathbf{A}^{-1} - \mathbf{B}^{-1})^{-1}$ is positive definite. Since $\mathbf{A}^{-1} - \mathbf{B}^{-1} = [(\mathbf{A}^{-1} - \mathbf{B}^{-1})^{-1}]^{-1}$, we conclude (on the basis of Corollary 14.2.11) that $\mathbf{A}^{-1} - \mathbf{B}^{-1}$ is positive definite.

(2) Suppose that $\mathbf{B} - \mathbf{A}$ is symmetric and positive semidefinite [which, since $\mathbf{B} = \mathbf{A} + (\mathbf{B} - \mathbf{A})$, implies that $\mathbf{B}$ is symmetric and (in light of Lemma 14.2.4) that $\mathbf{B}$ is positive definite (and hence nonsingular)]. And let $r = \text{rank}(\mathbf{B} - \mathbf{A})$ [and observe (in light of Corollary 14.3.12) that $r < n$]. Then, according to Theorem 14.3.7, there exists an $r \times n$ matrix $\mathbf{P}$ (of rank r) such that $\mathbf{B} - \mathbf{A} = \mathbf{P}'\mathbf{P}$. Thus,

$$\mathbf{B} = \mathbf{A} + (\mathbf{B} - \mathbf{A}) = \mathbf{A} + \mathbf{P}'\mathbf{P} = \mathbf{A} + \mathbf{P}'\mathbf{I}_r\mathbf{P};$$

and it follows from Theorem 18.2.8 that $\mathbf{I}_r + \mathbf{P}\mathbf{A}^{-1}\mathbf{P}'$ is nonsingular and that

$$\mathbf{B}^{-1} = \mathbf{A}^{-1} - \mathbf{A}^{-1}\mathbf{P}'(\mathbf{I}_r + \mathbf{P}\mathbf{A}^{-1}\mathbf{P}')^{-1}\mathbf{P}\mathbf{A}^{-1},$$

or, equivalently, that

$$\mathbf{A}^{-1} - \mathbf{B}^{-1} = (\mathbf{P}\mathbf{A}^{-1})'(\mathbf{I}_r + \mathbf{P}\mathbf{A}^{-1}\mathbf{P}')^{-1}\mathbf{P}\mathbf{A}^{-1}.$$

Moreover, since $\mathbf{A}^{-1}$ is positive definite, $\mathbf{I}_r + \mathbf{P}\mathbf{A}^{-1}\mathbf{P}'$ is positive definite and hence $(\mathbf{I}_r + \mathbf{P}\mathbf{A}^{-1}\mathbf{P}')^{-1}$ is positive definite. And, since $\operatorname{rank}(\mathbf{P}\mathbf{A}^{-1}) = r \ (< n)$, it follows from Theorem 14.2.9 that $(\mathbf{P}\mathbf{A}^{-1})'(\mathbf{I}_r + \mathbf{P}\mathbf{A}^{-1}\mathbf{P}')^{-1}\mathbf{P}\mathbf{A}^{-1}$ is positive semidefinite. Q.E.D.

18.4 Some Results on Idempotency

a. Basic results

Suppose that the sum of k (square) matrices $\mathbf{A}_1, \ldots, \mathbf{A}_k$ is idempotent. Under what condition(s) are $\mathbf{A}_1, \ldots, \mathbf{A}_k$ idempotent? This question is answered in the following theorem.

Theorem 18.4.1. Let $\mathbf{A}_1, \ldots, \mathbf{A}_k$ represent $n \times n$ matrices, and define $\mathbf{A} = \mathbf{A}_1 + \cdots + \mathbf{A}_k$. Suppose that $\mathbf{A}$ is idempotent. Then, each of the following three conditions implies the other two:

(1) $\mathbf{A}_i\mathbf{A}_j = \mathbf{0}$ (for $j \neq i = 1, \ldots, k$) and $\operatorname{rank}(\mathbf{A}_i^2) = \operatorname{rank}(\mathbf{A}_i)$ (for $i = 1, \ldots, k$);

(2) $\mathbf{A}_i^2 = \mathbf{A}_i$ (for $i = 1, \ldots, k$);

(3) $\operatorname{rank}(\mathbf{A}_1) + \cdots + \operatorname{rank}(\mathbf{A}_k) = \operatorname{rank}(\mathbf{A})$.

Preliminary to proving Theorem 18.4.1, it is convenient to establish the following two lemmas.

Lemma 18.4.2. An $n \times n$ matrix $\mathbf{A}$ is idempotent if and only if $\operatorname{rank}(\mathbf{A}) + \operatorname{rank}(\mathbf{I} - \mathbf{A}) = n$.

Proof (of Lemma 18.4.2). According to Theorem 11.7.1, $\mathbf{A}$ is idempotent if and only if $\mathcal{N}(\mathbf{A}) = \mathcal{C}(\mathbf{I} - \mathbf{A})$. Moreover, since [according to result (11.1.1)] $\mathcal{N}(\mathbf{A}) \subset \mathcal{C}(\mathbf{I} - \mathbf{A})$, it follows from Theorem 4.3.10 that $\mathcal{N}(\mathbf{A}) = \mathcal{C}(\mathbf{I} - \mathbf{A})$ if and only if $\dim[\mathcal{N}(\mathbf{A})] = \dim[\mathcal{C}(\mathbf{I} - \mathbf{A})]$. Since (according to Lemma 11.3.1) $\dim[\mathcal{N}(\mathbf{A})] = n - \operatorname{rank}(\mathbf{A})$ and since (by definition) $\dim[\mathcal{C}(\mathbf{I} - \mathbf{A})] = \operatorname{rank}(\mathbf{I} - \mathbf{A})$, we conclude that $\mathbf{A}$ is idempotent if and only if $n - \operatorname{rank}(\mathbf{A}) = \operatorname{rank}(\mathbf{I} - \mathbf{A})$ or, equivalently, if and only if $\operatorname{rank}(\mathbf{A}) + \operatorname{rank}(\mathbf{I} - \mathbf{A}) = n$. Q.E.D.

Lemma 18.4.3. For any idempotent matrices $\mathbf{A}$ and $\mathbf{B}$ (of the same size), $\mathbf{A} + \mathbf{B}$ is idempotent if and only if $\mathbf{B}\mathbf{A} = \mathbf{A}\mathbf{B} = \mathbf{0}$.

Proof (of Lemma 18.4.3). Clearly,

$$(\mathbf{A}+\mathbf{B})^2 = \mathbf{A}^2 + \mathbf{B}^2 + \mathbf{AB} + \mathbf{BA} = \mathbf{A} + \mathbf{B} + \mathbf{AB} + \mathbf{BA}.$$

Thus, $\mathbf{A}+\mathbf{B}$ is idempotent if and only if $\mathbf{AB}+\mathbf{BA} = \mathbf{0}$.

Suppose now that $\mathbf{BA} = \mathbf{AB} = \mathbf{0}$. Then, obviously, $\mathbf{AB}+\mathbf{BA} = \mathbf{0}$, and consequently $\mathbf{A}+\mathbf{B}$ is idempotent.

Conversely, suppose that $\mathbf{AB}+\mathbf{BA} = \mathbf{0}$. Then,

$$\mathbf{AB} + \mathbf{ABA} = \mathbf{A}^2\mathbf{B} + \mathbf{ABA} = \mathbf{A}(\mathbf{AB}+\mathbf{BA}) = \mathbf{0}$$

and

$$\mathbf{ABA} + \mathbf{BA} = \mathbf{ABA} + \mathbf{BA}^2 = (\mathbf{AB}+\mathbf{BA})\mathbf{A} = \mathbf{0},$$

implying that

$$\mathbf{AB} - \mathbf{BA} = \mathbf{AB} + \mathbf{ABA} - (\mathbf{ABA}+\mathbf{BA}) = \mathbf{0} - \mathbf{0} = \mathbf{0}$$

and hence that

$$\mathbf{AB} = \mathbf{BA}.$$

Further,

$$\mathbf{AB} = (1/2)(\mathbf{AB}+\mathbf{AB}) = (1/2)(\mathbf{AB}+\mathbf{BA}) = \mathbf{0}.$$

Q.E.D.

Proof (of Theorem 18.4.1). The proof consists of successively showing that Condition (1) implies Condition (2), Condition (2) implies Condition (3), and Condition (3) implies Condition (1).

(1) ⇒ (2). Suppose that Condition (1) is satisfied. Then (for $i = 1, \ldots, k$)

$$\mathbf{A}_i^2 = \mathbf{A}_i\mathbf{A} = \mathbf{A}_i\mathbf{A}\mathbf{A} = \mathbf{A}_i^2\mathbf{A} = \mathbf{A}_i^3.$$

Moreover, since $\text{rank}(\mathbf{A}_i^2) = \text{rank}(\mathbf{A}_i)$, it follows from Corollary 4.4.7 that $\mathcal{C}(\mathbf{A}_i^2) = \mathcal{C}(\mathbf{A}_i)$, so that $\mathbf{A}_i = \mathbf{A}_i^2\mathbf{L}_i$ for some matrix $\mathbf{L}_i$. Thus,

$$\mathbf{A}_i = \mathbf{A}_i^2\mathbf{L}_i = \mathbf{A}_i^3\mathbf{L}_i = \mathbf{A}_i(\mathbf{A}_i^2\mathbf{L}_i) = \mathbf{A}_i^2.$$

(2) ⇒ (3). Suppose that Condition (2) is satisfied. Then, making use of Corollary 10.2.2, we find that

$$\sum_{i=1}^{k} \text{rank}(\mathbf{A}_i) = \sum_{i=1}^{k} \text{tr}(\mathbf{A}_i) = \text{tr}\Big(\sum_{i=1}^{k} \mathbf{A}_i\Big) = \text{tr}(\mathbf{A}) = \text{rank}(\mathbf{A}).$$

(3) ⇒ (1). Suppose that Condition (3) is satisfied. And define $\mathbf{A}_0 = \mathbf{I}_n - \mathbf{A}$. Then, $\sum_{i=0}^{k} \mathbf{A}_i = \mathbf{I}$. Moreover, it follows from Lemma 10.2.4 that $\text{rank}(\mathbf{A}_0) = n - \text{rank}(\mathbf{A})$, so that $\sum_{i=0}^{k} \text{rank}(\mathbf{A}_i) = n$.

Thus, making use of result (4.5.10), we find (for $i = 1, \ldots, k$) that

$$\text{rank}(\mathbf{I}-\mathbf{A}_i) = \text{rank}\Big(\sum_{m=0\,(m\neq i)}^{k} \mathbf{A}_m\Big) \leq \sum_{m=0\,(m\neq i)}^{k} \text{rank}(\mathbf{A}_m) = n - \text{rank}(\mathbf{A}_i),$$

so that $\text{rank}(\mathbf{A}_i)+\text{rank}(\mathbf{I}-\mathbf{A}_i) \leq n$, implying [since $\text{rank}(\mathbf{A}_i)+\text{rank}(\mathbf{I}-\mathbf{A}_i) \geq \text{rank}(\mathbf{A}_i+\mathbf{I}-\mathbf{A}_i) = \text{rank}(\mathbf{I}_n) = n$] that

$$\text{rank}(\mathbf{A}_i) + \text{rank}(\mathbf{I}-\mathbf{A}_i) = n$$

and hence implying (in light of Lemma 18.4.2) that $\mathbf{A}_i$ is idempotent [and that $\text{rank}(\mathbf{A}_i^2) = \text{rank}(\mathbf{A}_i)$]. Similarly, again making use of result (4.5.10), we find (for $j \neq i = 1,\ldots,k$) that

$$\begin{aligned}\text{rank}(\mathbf{I}-\mathbf{A}_i-\mathbf{A}_j) &= \text{rank}\Big(\sum_{m=0\,(m\neq i,j)}^{k} \mathbf{A}_m\Big)\\ &\leq \sum_{m=0\,(m\neq i,j)}^{k} \text{rank}(\mathbf{A}_m) = n - \text{rank}(\mathbf{A}_i) - \text{rank}(\mathbf{A}_j)\\ &\leq n - \text{rank}(\mathbf{A}_i+\mathbf{A}_j),\end{aligned}$$

so that $\text{rank}(\mathbf{A}_i+\mathbf{A}_j)+\text{rank}(\mathbf{I}-\mathbf{A}_i-\mathbf{A}_j) \leq n$, implying [since $\text{rank}(\mathbf{A}_i+\mathbf{A}_j)+\text{rank}(\mathbf{I}-\mathbf{A}_i-\mathbf{A}_j) \geq \text{rank}(\mathbf{A}_i+\mathbf{A}_j+\mathbf{I}-\mathbf{A}_i-\mathbf{A}_j) = \text{rank}(\mathbf{I}_n) = n$] that

$$\text{rank}(\mathbf{A}_i+\mathbf{A}_j) + \text{rank}[\mathbf{I}-(\mathbf{A}_i+\mathbf{A}_j)] = n$$

and hence (in light of Lemma 18.4.2) that $\mathbf{A}_i+\mathbf{A}_j$ is idempotent. We conclude, on the basis of Lemma 18.4.3, that $\mathbf{A}_i\mathbf{A}_j = \mathbf{0}$ for $j \neq i = 1,\ldots,k$. Q.E.D.

The following theorem complements Theorem 18.4.1.

Theorem 18.4.4. Let $\mathbf{A}_1,\ldots,\mathbf{A}_k$ represent $n \times n$ matrices, and define $\mathbf{A} = \mathbf{A}_1+\cdots+\mathbf{A}_k$. If $\mathbf{A}_1,\ldots,\mathbf{A}_k$ are idempotent and if $\mathbf{A}_i\mathbf{A}_j = \mathbf{0}$ for all i and $j \neq i$, then $\mathbf{A}$ is idempotent and $\text{rank}(\mathbf{A}_1)+\cdots+\text{rank}(\mathbf{A}_k) = \text{rank}(\mathbf{A})$.

Proof. Suppose that $\mathbf{A}_1,\ldots,\mathbf{A}_k$ are idempotent and that $\mathbf{A}_i\mathbf{A}_j = \mathbf{0}$ for all i and $j \neq i$. Then, $\mathbf{A}$ is idempotent, as is evident upon observing that

$$\mathbf{A}^2 = \sum_i \mathbf{A}_i^2 + \sum_{i,j\neq i} \mathbf{A}_i\mathbf{A}_j = \sum_i \mathbf{A}_i = \mathbf{A}.$$

Further, it follows from Theorem 18.4.1 that $\text{rank}(\mathbf{A}_1)+\cdots+\text{rank}(\mathbf{A}_k) = \text{rank}(\mathbf{A})$. Q.E.D.

In connection with Theorem 18.4.1 (and Theorem 18.4.4), observe that, in the special case where $\mathbf{A} = \mathbf{I}_n$, $\text{rank}(\mathbf{A}) = n$. Observe also that if $\mathbf{A}_i$ is a symmetric matrix, then $\mathbf{A}_i^2 = \mathbf{A}_i'\mathbf{A}_i$, implying (in light of Corollary 7.4.5) that $\text{rank}(\mathbf{A}_i^2) = \text{rank}(\mathbf{A}_i)$. Thus, in the special case where $\mathbf{A}_1,\ldots,\mathbf{A}_k$ are symmetric matrices, Condition (1) of Theorem 18.4.1 reduces to the condition

$$\mathbf{A}_i\mathbf{A}_j = \mathbf{0} \quad (\text{for } j \neq i = 1,\ldots,k).$$

In light of these observations, we obtain, as a special case of Theorem 18.4.1, the following theorem.

Theorem 18.4.5. Let $\mathbf{A}_1, \ldots, \mathbf{A}_k$ represent $n \times n$ symmetric matrices such that $\mathbf{A}_1 + \cdots + \mathbf{A}_k = \mathbf{I}$. Then, each of the following three conditions implies the other two:

(1) $\mathbf{A}_i\mathbf{A}_j = \mathbf{0}$ (for $j \neq i = 1, \ldots, k$);

(2) $\mathbf{A}_i^2 = \mathbf{A}_i$ (for $i = 1, \ldots, k$);

(3) $\text{rank}(\mathbf{A}_1) + \cdots + \text{rank}(\mathbf{A}_k) = n$.

Theorem 18.4.5 provides the underpinnings for Cochran's (1934) theorem (on the statistical distribution of quadratic forms) and is itself sometimes referred to as Cochran's theorem.

b. Extensions

Theorems 18.4.1 and 18.4.4 are generalized in the following two theorems.

Theorem 18.4.6. Let $\mathbf{A}_1, \ldots, \mathbf{A}_k$ represent $n \times n$ matrices, and let $\mathbf{V}$ represent an $n \times n$ symmetric nonnegative definite matrix. Define $\mathbf{A} = \mathbf{A}_1 + \cdots + \mathbf{A}_k$, and take $\mathbf{R}$ to be any matrix such that $\mathbf{V} = \mathbf{R}'\mathbf{R}$. Suppose that $\mathbf{RAR}'$ is idempotent. Then, each of the following three conditions implies the other two:

(1) $\mathbf{RA}_i\mathbf{VA}_j\mathbf{R}' = \mathbf{0}$ (for $j \neq i = 1, \ldots, k$) and $\text{rank}(\mathbf{RA}_i\mathbf{VA}_i\mathbf{R}') = \text{rank}(\mathbf{RA}_i\mathbf{R}')$ (for $i = 1, \ldots, k$);

(2) $\mathbf{RA}_i\mathbf{VA}_i\mathbf{R}' = \mathbf{RA}_i\mathbf{R}'$ (for $i = 1, \ldots, k$);

(3) $\text{rank}(\mathbf{RA}_1\mathbf{R}') + \cdots + \text{rank}(\mathbf{RA}_k\mathbf{R}') = \text{rank}(\mathbf{RAR}')$.

Theorem 18.4.7. Let $\mathbf{A}_1, \ldots, \mathbf{A}_k$ represent $n \times n$ matrices, and let $\mathbf{V}$ represent an $n \times n$ symmetric nonnegative definite matrix. Define $\mathbf{A} = \mathbf{A}_1 + \cdots + \mathbf{A}_k$, and take $\mathbf{R}$ to be any matrix such that $\mathbf{V} = \mathbf{R}'\mathbf{R}$. If $\mathbf{RA}_i\mathbf{VA}_i\mathbf{R}' = \mathbf{RA}_i\mathbf{R}'$ for all i and if $\mathbf{RA}_i\mathbf{VA}_j\mathbf{R}' = \mathbf{0}$ for all i and $j \neq i$, then $\mathbf{RAR}'$ is idempotent and $\text{rank}(\mathbf{RA}_1\mathbf{R}') + \cdots + \text{rank}(\mathbf{RA}_k\mathbf{R}') = \text{rank}(\mathbf{RAR}')$.

In connection with Theorems 18.4.6 and 18.4.7, note that the existence of a matrix $\mathbf{R}$ such that $\mathbf{V} = \mathbf{R}'\mathbf{R}$ is guaranteed by Corollary 14.3.8. Note also that, in the special case where $\mathbf{A}_1, \ldots, \mathbf{A}_k$ are symmetric, Condition (1) of Theorem 18.4.6 reduces to the condition

$$\mathbf{RA}_i\mathbf{VA}_j\mathbf{R}' = \mathbf{0} \quad (\text{for } j \neq i = 1, \ldots, k)$$

{since, in that special case, $\text{rank}(\mathbf{RA}_i\mathbf{VA}_i\mathbf{R}') = \text{rank}[(\mathbf{RA}_i\mathbf{R}')'\mathbf{RA}_i\mathbf{R}'] = \text{rank}(\mathbf{RA}_i\mathbf{R}')$}. And note that, in the special case where $\mathbf{V} = \mathbf{I}$ (and where $\mathbf{R}$ is chosen to be $\mathbf{I}$), Theorems 18.4.6 and 18.4.7 reduce to Theorems 18.4.1 and 18.4.4.

Theorems 18.4.6 and 18.4.7 can be derived from Theorems 18.4.1 and 18.4.4—upon applying Theorems 18.4.1 and 18.4.4 with $\mathbf{RA}_1\mathbf{R}', \ldots, \mathbf{RA}_k\mathbf{R}'$ in place of $\mathbf{A}_1, \ldots, \mathbf{A}_k$, respectively, we obtain Theorems 18.4.6 and 18.4.7.

Theorems 18.4.6 and 18.4.7 can be restated in terms that do not involve the matrix $\mathbf{R}$. For this purpose, observe that

$$\begin{aligned}(\mathbf{RAR}')^2 = \mathbf{RAR}' &\Rightarrow \mathbf{R}'\mathbf{RAVAR}'\mathbf{R} = \mathbf{R}'\mathbf{RAR}'\mathbf{R},\\ \mathbf{RA}_i\mathbf{VA}_i\mathbf{R}' = \mathbf{RA}_i\mathbf{R}' &\Rightarrow \mathbf{R}'\mathbf{RA}_i\mathbf{VA}_i\mathbf{R}'\mathbf{R} = \mathbf{R}'\mathbf{RA}_i\mathbf{R}'\mathbf{R},\\ \mathbf{RA}_i\mathbf{VA}_j\mathbf{R}' = \mathbf{0} &\Rightarrow \mathbf{R}'\mathbf{RA}_i\mathbf{VA}_j\mathbf{R}'\mathbf{R} = \mathbf{0},\end{aligned}$$

and conversely (in light of Corollary 5.3.3) that

$$\begin{aligned}\mathbf{R}'\mathbf{RAVAR}'\mathbf{R} = \mathbf{R}'\mathbf{RAR}'\mathbf{R} &\Rightarrow \mathbf{RAVAR}'\mathbf{R} = \mathbf{RAR}'\mathbf{R}\\ &\Rightarrow (\mathbf{RAR}')^2 = \mathbf{RAR}',\end{aligned}$$

$$\begin{aligned}\mathbf{R}'\mathbf{RA}_i\mathbf{VA}_i\mathbf{R}'\mathbf{R} = \mathbf{R}'\mathbf{RA}_i\mathbf{R}'\mathbf{R} &\Rightarrow \mathbf{RA}_i\mathbf{VA}_i\mathbf{R}'\mathbf{R} = \mathbf{RA}_i\mathbf{R}'\mathbf{R}\\ &\Rightarrow \mathbf{RA}_i\mathbf{VA}_i\mathbf{R}' = \mathbf{RA}_i\mathbf{R}',\end{aligned}$$

$$\mathbf{R}'\mathbf{RA}_i\mathbf{VA}_j\mathbf{R}'\mathbf{R} = \mathbf{0} \Rightarrow \mathbf{RA}_i\mathbf{VA}_j\mathbf{R}'\mathbf{R} = \mathbf{0} \Rightarrow \mathbf{RA}_i\mathbf{VA}_j\mathbf{R}' = \mathbf{0}.$$

Thus,

$$(\mathbf{RAR}')^2 = \mathbf{RAR}' \Leftrightarrow \mathbf{VAVAV} = \mathbf{VAV}, \tag{4.1}$$

$$\mathbf{RA}_i\mathbf{VA}_i\mathbf{R}' = \mathbf{RA}_i\mathbf{R}' \Leftrightarrow \mathbf{VA}_i\mathbf{VA}_i\mathbf{V} = \mathbf{VA}_i\mathbf{V}, \tag{4.2}$$

$$\mathbf{RA}_i\mathbf{VA}_j\mathbf{R}' = \mathbf{0} \Leftrightarrow \mathbf{VA}_i\mathbf{VA}_j\mathbf{V} = \mathbf{0}. \tag{4.3}$$

Further, in light of Corollary 7.4.4, we have that

$$\text{rank}(\mathbf{RAR}') = \text{rank}(\mathbf{R}'\mathbf{RAR}') = \text{rank}(\mathbf{R}'\mathbf{RAR}'\mathbf{R}) = \text{rank}(\mathbf{VAV}), \tag{4.4}$$

$$\text{rank}(\mathbf{RA}_i\mathbf{R}') = \text{rank}(\mathbf{R}'\mathbf{RA}_i\mathbf{R}') = \text{rank}(\mathbf{R}'\mathbf{RA}_i\mathbf{R}'\mathbf{R}) = \text{rank}(\mathbf{VA}_i\mathbf{V}), \tag{4.5}$$

$$\begin{aligned}\text{rank}(\mathbf{RA}_i\mathbf{VA}_i\mathbf{R}') &= \text{rank}(\mathbf{R}'\mathbf{RA}_i\mathbf{VA}_i\mathbf{R}')\\ &= \text{rank}(\mathbf{R}'\mathbf{RA}_i\mathbf{VA}_i\mathbf{R}'\mathbf{R}) = \text{rank}(\mathbf{VA}_i\mathbf{VA}_i\mathbf{V}).\end{aligned} \tag{4.6}$$

Based on results (4.1)–(4.6), Theorems 18.4.6 and 18.4.7 can be restated as the following two theorems.

Theorem 18.4.8. Let $\mathbf{A}_1, \ldots, \mathbf{A}_k$ represent $n \times n$ matrices, let $\mathbf{V}$ represent an $n \times n$ symmetric nonnegative definite matrix, and define $\mathbf{A} = \mathbf{A}_1 + \cdots + \mathbf{A}_k$. Suppose that $\mathbf{VAVAV} = \mathbf{VAV}$. Then, each of the following three conditions implies the other two:

(1) $\mathbf{VA}_i\mathbf{VA}_j\mathbf{V} = \mathbf{0}$ (for $j \neq i = 1, \ldots, k$) and
$\text{rank}(\mathbf{VA}_i\mathbf{VA}_i\mathbf{V}) = \text{rank}(\mathbf{VA}_i\mathbf{V})$ (for $i = 1, \ldots, k$);

(2) $\mathbf{VA}_i\mathbf{VA}_i\mathbf{V} = \mathbf{VA}_i\mathbf{V}$ (for $i = 1, \ldots, k$);

(3) $\text{rank}(\mathbf{VA}_1\mathbf{V}) + \cdots + \text{rank}(\mathbf{VA}_k\mathbf{V}) = \text{rank}(\mathbf{VAV})$.

Theorem 18.4.9. Let $\mathbf{A}_1, \ldots, \mathbf{A}_k$ represent $n \times n$ matrices, let $\mathbf{V}$ represent an $n \times n$ symmetric nonnegative definite matrix, and define $\mathbf{A} = \mathbf{A}_1 + \cdots + \mathbf{A}_k$. If $\mathbf{VA}_i\mathbf{VA}_i\mathbf{V} = \mathbf{VA}_i\mathbf{V}$ for all i and if $\mathbf{VA}_i\mathbf{VA}_j\mathbf{V} = \mathbf{0}$ for all i and $j \neq i$, then $\mathbf{VAVAV} = \mathbf{VAV}$ and $\text{rank}(\mathbf{VA}_1\mathbf{V}) + \cdots + \text{rank}(\mathbf{VA}_k\mathbf{V}) = \text{rank}(\mathbf{VAV})$.

Note that, in the special case where $\mathbf{A}_1, \ldots, \mathbf{A}_k$ are symmetric, Condition (1) of Theorem 18.4.8 reduces to the condition

$$\mathbf{VA}_i\mathbf{VA}_j\mathbf{V} = \mathbf{0} \quad (\text{for } j \neq i = 1, \ldots, k)$$

{since, in that special case, $\text{rank}(\mathbf{VA}_i\mathbf{VA}_i\mathbf{V}) = \text{rank}(\mathbf{RA}_i\mathbf{VA}_i\mathbf{R}') = \text{rank}[(\mathbf{RA}_i\mathbf{R}')'\mathbf{RA}_i\mathbf{R}'] = \text{rank}(\mathbf{RA}_i\mathbf{R}') = \text{rank}(\mathbf{VA}_i\mathbf{V})$}. Note also that, in the special case where $\mathbf{V}$ is positive definite (and hence nonsingular), simplifications can be effected in Theorem 18.4.8 (and in Theorem 18.4.9) by observing that

$$\mathbf{VAVAV} = \mathbf{VAV} \Leftrightarrow (\mathbf{AV})^2 = \mathbf{AV} \Leftrightarrow \mathbf{AVA} = \mathbf{A},$$
$$\mathbf{VA}_i\mathbf{VA}_i\mathbf{V} = \mathbf{VA}_i\mathbf{V} \Leftrightarrow (\mathbf{A}_i\mathbf{V})^2 = \mathbf{A}_i\mathbf{V} \Leftrightarrow \mathbf{A}_i\mathbf{VA}_i = \mathbf{A}_i,$$
$$\mathbf{VA}_i\mathbf{VA}_j\mathbf{V} = \mathbf{0} \Leftrightarrow \mathbf{A}_i\mathbf{VA}_j\mathbf{V} = \mathbf{0} \Leftrightarrow \mathbf{A}_i\mathbf{VA}_j = \mathbf{0},$$

and that

$$\text{rank}(\mathbf{VAV}) = \text{rank}(\mathbf{AV}) = \text{rank}(\mathbf{A}),$$
$$\text{rank}(\mathbf{VA}_i\mathbf{V}) = \text{rank}(\mathbf{A}_i\mathbf{V}) = \text{rank}(\mathbf{A}_i),$$
$$\text{rank}(\mathbf{VA}_i\mathbf{VA}_i\mathbf{V}) = \text{rank}[(\mathbf{A}_i\mathbf{V})^2] = \text{rank}(\mathbf{A}_i\mathbf{VA}_i).$$

18.5 Some Results on Ranks

a. Ranks of matrices of the form $\mathbf{R} + \mathbf{STU}$

In Sections 18.1a and 18.2d–e, expressions were obtained for the determinant and for the ordinary or generalized inverse of a matrix of the form $\mathbf{R} + \mathbf{STU}$. These expressions (when applicable) give the determinant of $\mathbf{R} + \mathbf{STU}$ in terms of the determinants of $\mathbf{R}$, $\mathbf{T}$, and $\mathbf{T} + \mathbf{TUR}^{-1}\mathbf{ST}$ or the determinants of $\mathbf{R}$, $\mathbf{T}$, and $\mathbf{T}^{-1} + \mathbf{UR}^{-1}\mathbf{S}$ and give the ordinary or generalized inverse of $\mathbf{R} + \mathbf{STU}$ in terms of the ordinary or generalized inverses of $\mathbf{R}$ and $\mathbf{T} + \mathbf{TUR}^{-1}\mathbf{ST}$ (or $\mathbf{T} + \mathbf{TUR}^{-}\mathbf{ST}$) or the ordinary or generalized inverses of $\mathbf{R}$, $\mathbf{T}$, and $\mathbf{T}^{-1} + \mathbf{UR}^{-1}\mathbf{S}$ (or $\mathbf{T}^{-} + \mathbf{T}^{-}\mathbf{TUR}^{-}\mathbf{STT}^{-}$). The corollary of the following theorem gives a comparable expression for the rank of $\mathbf{R} + \mathbf{STU}$.

Theorem 18.5.1. For any $n \times q$ matrix $\mathbf{R}$, $n \times m$ matrix $\mathbf{S}$, $m \times p$ matrix $\mathbf{T}$, and $p \times q$ matrix $\mathbf{U}$,

$$\text{rank}(\mathbf{T}) + \text{rank}(\mathbf{R} + \mathbf{RR}^{-}\mathbf{STUR}^{-}\mathbf{R}) = \text{rank}(\mathbf{R}) + \text{rank}(\mathbf{T} + \mathbf{TUR}^{-}\mathbf{RR}^{-}\mathbf{ST}). \quad (5.1)$$

Proof. Let $\mathbf{A} = \begin{pmatrix} \mathbf{R} & -\mathbf{RR}^{-}\mathbf{ST} \\ \mathbf{TUR}^{-}\mathbf{R} & \mathbf{T} \end{pmatrix}$. Then, observing that

$$\mathbf{A}\begin{pmatrix} \mathbf{I} & \mathbf{0} \\ -\mathbf{UR}^{-}\mathbf{R} & \mathbf{I} \end{pmatrix} = \begin{pmatrix} \mathbf{R} + \mathbf{RR}^{-}\mathbf{STUR}^{-}\mathbf{R} & -\mathbf{RR}^{-}\mathbf{ST} \\ \mathbf{0} & \mathbf{T} \end{pmatrix},$$
$$\begin{pmatrix} \mathbf{I} & \mathbf{0} \\ -\mathbf{TUR}^{-} & \mathbf{I} \end{pmatrix}\mathbf{A} = \begin{pmatrix} \mathbf{R} & -\mathbf{RR}^{-}\mathbf{ST} \\ \mathbf{0} & \mathbf{T} + \mathbf{TUR}^{-}\mathbf{RR}^{-}\mathbf{ST} \end{pmatrix}$$

and making use of Lemma 8.5.2 and Corollary 9.6.2, we find that

$$\begin{aligned}\operatorname{rank}(\mathbf{A}) &= \operatorname{rank}\begin{pmatrix}\mathbf{R}+\mathbf{RR}^-\mathbf{STUR}^-\mathbf{R} & -\mathbf{RR}^-\mathbf{ST}\\ \mathbf{0} & \mathbf{T}\end{pmatrix}\\ &= \operatorname{rank}(\mathbf{T}) + \operatorname{rank}(\mathbf{R}+\mathbf{RR}^-\mathbf{STUR}^-\mathbf{R}),\end{aligned}$$

$$\begin{aligned}\operatorname{rank}(\mathbf{A}) &= \operatorname{rank}\begin{pmatrix}\mathbf{R} & -\mathbf{RR}^-\mathbf{ST}\\ \mathbf{0} & \mathbf{T}+\mathbf{TUR}^-\mathbf{RR}^-\mathbf{ST}\end{pmatrix}\\ &= \operatorname{rank}(\mathbf{R}) + \operatorname{rank}(\mathbf{T}+\mathbf{TUR}^-\mathbf{RR}^-\mathbf{ST}),\end{aligned}$$

implying that

$$\operatorname{rank}(\mathbf{T})+\operatorname{rank}(\mathbf{R}+\mathbf{RR}^-\mathbf{STUR}^-\mathbf{R}) = \operatorname{rank}(\mathbf{R})+\operatorname{rank}(\mathbf{T}+\mathbf{TUR}^-\mathbf{RR}^-\mathbf{ST}).$$

Or, alternatively, Theorem 18.5.1 can be proved by applying result (9.6.1)—observing that $\mathcal{C}(\mathbf{TUR}^-\mathbf{R}) \subset \mathcal{C}(\mathbf{T})$ and $\mathcal{R}(-\mathbf{RR}^-\mathbf{ST}) \subset \mathcal{R}(\mathbf{T})$ and that $\mathcal{C}(-\mathbf{RR}^-\mathbf{ST}) \subset \mathcal{C}(\mathbf{R})$ and $\mathcal{R}(\mathbf{TUR}^-\mathbf{R}) \subset \mathcal{R}(\mathbf{R})$, we find that

$$\begin{aligned}&\operatorname{rank}(\mathbf{T}) + \operatorname{rank}(\mathbf{R}+\mathbf{RR}^-\mathbf{STUR}^-\mathbf{R})\\ &\qquad = \operatorname{rank}(\mathbf{A}) = \operatorname{rank}(\mathbf{R}) + \operatorname{rank}(\mathbf{T}+\mathbf{TUR}^-\mathbf{RR}^-\mathbf{ST}).\end{aligned}$$

Q.E.D.

Corollary 18.5.2. Let $\mathbf{R}$ represent an $n \times q$ matrix, $\mathbf{S}$ an $n \times m$ matrix, $\mathbf{T}$ an $m \times p$ matrix, and $\mathbf{U}$ a $p \times q$ matrix. If $\mathcal{R}(\mathbf{STU}) \subset \mathcal{R}(\mathbf{R})$ and $\mathcal{C}(\mathbf{STU}) \subset \mathcal{C}(\mathbf{R})$, then

$$\operatorname{rank}(\mathbf{R}+\mathbf{STU}) = \operatorname{rank}(\mathbf{R}) + \operatorname{rank}(\mathbf{T}+\mathbf{TUR}^-\mathbf{ST}) - \operatorname{rank}(\mathbf{T}). \tag{5.2}$$

Proof. Suppose that $\mathcal{R}(\mathbf{STU}) \subset \mathcal{R}(\mathbf{R})$ and $\mathcal{C}(\mathbf{STU}) \subset \mathcal{C}(\mathbf{R})$. Then, in light of Lemma 9.3.5,

$$\mathbf{R}+\mathbf{RR}^-\mathbf{STUR}^-\mathbf{R} = \mathbf{R}+\mathbf{RR}^-\mathbf{STU} = \mathbf{R}+\mathbf{STU}.$$

Now, let $\mathbf{Q} = \mathbf{T}+\mathbf{TUR}^-\mathbf{ST}$ and $\mathbf{H} = \mathbf{T}+\mathbf{TUR}^-\mathbf{RR}^-\mathbf{ST}$, and observe that

$$\mathbf{Q}-\mathbf{H} = \mathbf{TU}(\mathbf{I}-\mathbf{R}^-\mathbf{R})\mathbf{R}^-\mathbf{ST}.$$

Clearly, to complete the proof, it suffices [in light of result (5.1)] to show that $\operatorname{rank}(\mathbf{H}) = \operatorname{rank}(\mathbf{Q})$.

In light of result (2.29),

$$(\mathbf{I}-\mathbf{QQ}^-)(\mathbf{Q}-\mathbf{H}) = (\mathbf{I}-\mathbf{QQ}^-)\mathbf{TU}(\mathbf{I}-\mathbf{R}^-\mathbf{R})\mathbf{R}^-\mathbf{ST} = \mathbf{0},$$

implying (in light of Lemma 9.3.5) that $\mathcal{C}(\mathbf{Q}-\mathbf{H}) \subset \mathcal{C}(\mathbf{Q})$, so that

$$\mathbf{Q}-\mathbf{H} = \mathbf{QK}$$

for some matrix $\mathbf{K}$. Moreover, upon substituting $\mathbf{R}^-\mathbf{RR}^-$ (which, like $\mathbf{R}^-$ itself, is a generalized inverse of $\mathbf{R}$) for $\mathbf{R}^-$ in result (2.29), we find that

$$(\mathbf{I} - \mathbf{HH}^-)\mathbf{TU}(\mathbf{I} - \mathbf{R}^-\mathbf{R}) = (\mathbf{I} - \mathbf{HH}^-)\mathbf{TU}(\mathbf{I} - \mathbf{R}^-\mathbf{RR}^-\mathbf{R}) = \mathbf{0},$$

implying that

$$(\mathbf{I} - \mathbf{HH}^-)(\mathbf{Q} - \mathbf{H}) = (\mathbf{I} - \mathbf{HH}^-)\mathbf{TU}(\mathbf{I} - \mathbf{R}^-\mathbf{R})\mathbf{R}^-\mathbf{ST} = \mathbf{0}$$

and hence (in light of Lemma 9.3.5) that $\mathcal{C}(\mathbf{Q} - \mathbf{H}) \subset \mathcal{C}(\mathbf{H})$, so that

$$\mathbf{Q} - \mathbf{H} = \mathbf{HL}$$

for some matrix $\mathbf{L}$. Thus,

$$\mathbf{H} = \mathbf{Q} - (\mathbf{Q} - \mathbf{H}) = \mathbf{Q}(\mathbf{I} - \mathbf{K}), \qquad \mathbf{Q} = \mathbf{H} + (\mathbf{Q} - \mathbf{H}) = \mathbf{H}(\mathbf{I} + \mathbf{L}),$$

implying that $\mathcal{C}(\mathbf{H}) \subset \mathcal{C}(\mathbf{Q})$ and $\mathcal{C}(\mathbf{Q}) \subset \mathcal{C}(\mathbf{H})$ and hence that $\mathcal{C}(\mathbf{H}) = \mathcal{C}(\mathbf{Q})$, and it follows that

$$\text{rank}(\mathbf{H}) = \text{rank}(\mathbf{Q}).$$

Q.E.D.

In connection with Corollary 18.5.2, note that equality (5.2) [which is applicable if $\mathcal{R}(\mathbf{STU}) \subset \mathcal{R}(\mathbf{R})$ and $\mathcal{C}(\mathbf{STU}) \subset \mathcal{C}(\mathbf{R})$] is equivalent to the equality

$$\text{rank}(\mathbf{R}) - \text{rank}(\mathbf{R} + \mathbf{STU}) = \text{rank}(\mathbf{T}) - \text{rank}(\mathbf{T} + \mathbf{TUR}^-\mathbf{ST}). \tag{5.3}$$

Thus, if $\mathcal{R}(\mathbf{STU}) \subset \mathcal{R}(\mathbf{R})$ and $\mathcal{C}(\mathbf{STU}) \subset \mathcal{C}(\mathbf{R})$, then the difference in rank between $\mathbf{R}$ and $\mathbf{R}+\mathbf{STU}$ is the same as that between $\mathbf{T}$ and $\mathbf{T}+\mathbf{TUR}^-\mathbf{ST}$. Note also that this difference in rank is nonnegative [as is evident, e.g., upon observing that $\mathbf{T} + \mathbf{TUR}^-\mathbf{ST} = \mathbf{T}(\mathbf{I} + \mathbf{UR}^-\mathbf{ST})$].

b. Lower (and upper) bounds on the rank of a sum of two matrices

Let $\mathbf{A}$ and $\mathbf{B}$ represent $m \times n$ matrices. In Section 4.5, it was established that

$$\text{rank}(\mathbf{A} + \mathbf{B}) \le \text{rank}(\mathbf{A}, \mathbf{B}) \le \text{rank}(\mathbf{A}) + \text{rank}(\mathbf{B}), \tag{5.4}$$

$$\text{rank}(\mathbf{A} + \mathbf{B}) \le \text{rank}\begin{pmatrix}\mathbf{A}\\ \mathbf{B}\end{pmatrix} \le \text{rank}(\mathbf{A}) + \text{rank}(\mathbf{B}). \tag{5.5}$$

Thus, $\text{rank}(\mathbf{A} + \mathbf{B})$ is bounded from above by $\text{rank}(\mathbf{A}, \mathbf{B})$, by $\text{rank}\begin{pmatrix}\mathbf{A}\\ \mathbf{B}\end{pmatrix}$, and by $\text{rank}(\mathbf{A}) + \text{rank}(\mathbf{B})$. Further, in Section 17.2, it was established that equality holds in the inequality $\text{rank}(\mathbf{A}, \mathbf{B}) \le \text{rank}(\mathbf{A}) + \text{rank}(\mathbf{B})$ if and only if $\mathcal{C}(\mathbf{A})$ and $\mathcal{C}(\mathbf{B})$ are essentially disjoint, and that equality holds in the inequality $\text{rank}\begin{pmatrix}\mathbf{A}\\ \mathbf{B}\end{pmatrix} \le \text{rank}(\mathbf{A}) + \text{rank}(\mathbf{B})$ if and only if $\mathcal{R}(\mathbf{A})$ and $\mathcal{R}(\mathbf{B})$ are essentially disjoint.

In this section, these results are augmented with results on lower bounds for $\text{rank}(\mathbf{A} + \mathbf{B})$ and with results on the conditions under which these lower bounds are attained [and also with results on the conditions under which the upper bounds $\text{rank}(\mathbf{A}, \mathbf{B})$ and $\text{rank}\begin{pmatrix}\mathbf{A}\\ \mathbf{B}\end{pmatrix}$ are attained]. These additional results are derived from the following theorem.

Theorem 18.5.3. Let $\mathbf{A}$ and $\mathbf{B}$ represent $m \times n$ matrices. And let $c = \dim[\mathcal{C}(\mathbf{A}) \cap \mathcal{C}(\mathbf{B})]$, $d = \dim[\mathcal{R}(\mathbf{A}) \cap \mathcal{R}(\mathbf{B})]$, and

$$\mathbf{H} = \left[\mathbf{I} - \begin{pmatrix}\mathbf{A}\\ \mathbf{B}\end{pmatrix}\begin{pmatrix}\mathbf{A}\\ \mathbf{B}\end{pmatrix}^{-}\right]\begin{pmatrix}\mathbf{A} & \mathbf{0}\\ \mathbf{0} & \mathbf{B}\end{pmatrix}[\mathbf{I} - (\mathbf{A}, \mathbf{B})^{-}(\mathbf{A}, \mathbf{B})]. \tag{5.6}$$

Then,

$$\begin{aligned}
&\operatorname{rank}(\mathbf{A} + \mathbf{B}) \\
&\quad = \operatorname{rank}(\mathbf{A}, \mathbf{B}) + \operatorname{rank}\begin{pmatrix}\mathbf{A}\\ \mathbf{B}\end{pmatrix} - \operatorname{rank}(\mathbf{A}) - \operatorname{rank}(\mathbf{B}) + \operatorname{rank}(\mathbf{H}) &(5.7)\\
&\quad = \operatorname{rank}(\mathbf{A}, \mathbf{B}) - d + \operatorname{rank}(\mathbf{H}) &(5.8)\\
&\quad = \operatorname{rank}\begin{pmatrix}\mathbf{A}\\ \mathbf{B}\end{pmatrix} - c + \operatorname{rank}(\mathbf{H}) &(5.9)\\
&\quad = \operatorname{rank}(\mathbf{A}) + \operatorname{rank}(\mathbf{B}) - c - d + \operatorname{rank}(\mathbf{H}). &(5.10)
\end{aligned}$$

Proof. Equality (5.7) is the special case of equality (17.5.5) obtained by putting $(\mathbf{I}_m, \mathbf{I}_m)$ in place of $\mathbf{A}$, $\begin{pmatrix}\mathbf{I}_n\\ \mathbf{I}_n\end{pmatrix}$ in place of $\mathbf{B}$, and $\begin{pmatrix}\mathbf{A} & \mathbf{0}\\ \mathbf{0} & \mathbf{B}\end{pmatrix}$ in place of $\mathbf{C}$ and by observing that $(\mathbf{I}_m, \mathbf{I}_m)\begin{pmatrix}\mathbf{A} & \mathbf{0}\\ \mathbf{0} & \mathbf{B}\end{pmatrix} = (\mathbf{A}, \mathbf{B})$, $\begin{pmatrix}\mathbf{A} & \mathbf{0}\\ \mathbf{0} & \mathbf{B}\end{pmatrix}\begin{pmatrix}\mathbf{I}_n\\ \mathbf{I}_n\end{pmatrix} = \begin{pmatrix}\mathbf{A}\\ \mathbf{B}\end{pmatrix}$, $(\mathbf{I}_m, \mathbf{I}_m)\begin{pmatrix}\mathbf{A} & \mathbf{0}\\ \mathbf{0} & \mathbf{B}\end{pmatrix}\begin{pmatrix}\mathbf{I}_n\\ \mathbf{I}_n\end{pmatrix} = \mathbf{A} + \mathbf{B}$, and $\operatorname{rank}\begin{pmatrix}\mathbf{A} & \mathbf{0}\\ \mathbf{0} & \mathbf{B}\end{pmatrix} = \operatorname{rank}(\mathbf{A}) + \operatorname{rank}(\mathbf{B})$. Further, equalities (5.8) and (5.9) are obtained from equality (5.7) by substituting for $\operatorname{rank}\begin{pmatrix}\mathbf{A}\\ \mathbf{B}\end{pmatrix}$ or $\operatorname{rank}(\mathbf{A}, \mathbf{B})$, respectively, on the basis of formula (17.4.11) or (17.4.10). Similarly, equality (5.10) is obtained from equality (5.8) or (5.9) by substituing for $\operatorname{rank}(\mathbf{A}, \mathbf{B})$ or $\operatorname{rank}\begin{pmatrix}\mathbf{A}\\ \mathbf{B}\end{pmatrix}$, respectively, on the basis of formula (17.4.10) or (17.4.11). Q.E.D.

Lower and upper bounds for the rank of the sum of two matrices $\mathbf{A}$ and $\mathbf{B}$ can be obtained from the following theorem, which gives various inequalities relating $\operatorname{rank}(\mathbf{A} + \mathbf{B})$ to $\operatorname{rank}(\mathbf{A}) + \operatorname{rank}(\mathbf{B})$ and to $\operatorname{rank}(\mathbf{A}, \mathbf{B})$ or $\operatorname{rank}\begin{pmatrix}\mathbf{A}\\ \mathbf{B}\end{pmatrix}$ and which also gives the conditions under which these inequalities hold as equalities.

Theorem 18.5.4. Let $\mathbf{A}$ and $\mathbf{B}$ represent $m \times n$ matrices. And let $c = \dim[\mathcal{C}(\mathbf{A}) \cap \mathcal{C}(\mathbf{B})]$, $d = \dim[\mathcal{R}(\mathbf{A}) \cap \mathcal{R}(\mathbf{B})]$, and

$$\mathbf{H} = \left[\mathbf{I} - \begin{pmatrix}\mathbf{A}\\ \mathbf{B}\end{pmatrix}\begin{pmatrix}\mathbf{A}\\ \mathbf{B}\end{pmatrix}^{-}\right]\begin{pmatrix}\mathbf{A} & \mathbf{0}\\ \mathbf{0} & \mathbf{B}\end{pmatrix}[\mathbf{I} - (\mathbf{A}, \mathbf{B})^{-}(\mathbf{A}, \mathbf{B})]. \tag{5.11}$$

(1) Then,

$$\begin{aligned}
\text{rank}(\mathbf{A}) + \text{rank}(\mathbf{B}) - c - d \qquad & \\
= \text{rank}(\mathbf{A},\ \mathbf{B}) - d & \qquad (5.12a)\\
\le \text{rank}(\mathbf{A}+\mathbf{B}) \le \text{rank}(\mathbf{A},\ \mathbf{B}) = \text{rank}(\mathbf{A}) + \text{rank}(\mathbf{B}) - c & \qquad (5.12b)\\
\le \text{rank}(\mathbf{A}) + \text{rank}(\mathbf{B}), & \qquad (5.12c)
\end{aligned}$$

with equality holding under the following conditions:

$$\begin{aligned}
&\text{rank}(\mathbf{A}+\mathbf{B}) = \text{rank}(\mathbf{A},\ \mathbf{B}) - d\\
&\qquad [\text{or, equivalently, } \text{rank}(\mathbf{A}+\mathbf{B}) = \text{rank}(\mathbf{A}) + \text{rank}(\mathbf{B}) - c - d]\\
&\qquad\qquad \Leftrightarrow \ \text{rank}(\mathbf{H}) = 0; \qquad (5.13)
\end{aligned}$$

$$\text{rank}(\mathbf{A}+\mathbf{B}) = \text{rank}(\mathbf{A},\ \mathbf{B}) \ \Leftrightarrow \ \text{rank}(\mathbf{H}) = d; \qquad (5.14)$$

$$\begin{aligned}
&\text{rank}(\mathbf{A},\ \mathbf{B}) = \text{rank}(\mathbf{A}) + \text{rank}(\mathbf{B})\\
&\qquad [\text{or, equivalently, } \text{rank}(\mathbf{A}) + \text{rank}(\mathbf{B}) - c = \text{rank}(\mathbf{A}) + \text{rank}(\mathbf{B})]\\
&\qquad\qquad \Leftrightarrow \ c = 0; \qquad (5.15)
\end{aligned}$$

(2) Similarly,

$$\begin{aligned}
\text{rank}(\mathbf{A}) + \text{rank}(\mathbf{B}) - c - d \qquad & \\
= \text{rank}\begin{pmatrix}\mathbf{A}\\ \mathbf{B}\end{pmatrix} - c & \qquad (5.16a)\\
\le \text{rank}(\mathbf{A}+\mathbf{B}) \le \text{rank}\begin{pmatrix}\mathbf{A}\\ \mathbf{B}\end{pmatrix} = \text{rank}(\mathbf{A}) + \text{rank}(\mathbf{B}) - d & \qquad (5.16b)\\
\le \text{rank}(\mathbf{A}) + \text{rank}(\mathbf{B}), & \qquad (5.16c)
\end{aligned}$$

with equality holding under the following conditions:

$$\begin{aligned}
&\text{rank}(\mathbf{A}+\mathbf{B}) = \text{rank}\begin{pmatrix}\mathbf{A}\\ \mathbf{B}\end{pmatrix} - c\\
&\qquad [\text{or, equivalently, } \text{rank}(\mathbf{A}+\mathbf{B}) = \text{rank}(\mathbf{A}) + \text{rank}(\mathbf{B}) - c - d]\\
&\qquad\qquad \Leftrightarrow \ \text{rank}(\mathbf{H}) = 0; \qquad (5.17)
\end{aligned}$$

$$\text{rank}(\mathbf{A}+\mathbf{B}) = \text{rank}\begin{pmatrix}\mathbf{A}\\ \mathbf{B}\end{pmatrix} \ \Leftrightarrow \ \text{rank}(\mathbf{H}) = c; \qquad (5.18)$$

$$\begin{aligned}
&\text{rank}\begin{pmatrix}\mathbf{A}\\ \mathbf{B}\end{pmatrix} = \text{rank}(\mathbf{A}) + \text{rank}(\mathbf{B})\\
&\qquad [\text{or, equivalently, } \text{rank}(\mathbf{A}) + \text{rank}(\mathbf{B}) - d = \text{rank}(\mathbf{A}) + \text{rank}(\mathbf{B})]\\
&\qquad\qquad \Leftrightarrow \ d = 0; \qquad (5.19)
\end{aligned}$$

Proof. (1) That $\text{rank}(\mathbf{A}) + \text{rank}(\mathbf{B}) - c - d = \text{rank}(\mathbf{A},\ \mathbf{B}) - d$ is an immediate consequence of result (17.4.10). That $\text{rank}(\mathbf{A},\ \mathbf{B}) - d \le \text{rank}(\mathbf{A}+\mathbf{B})$, with equality holding if and only if $\text{rank}(\mathbf{H}) = 0$, is evident from result (5.8).

That $\text{rank}(\mathbf{A}+\mathbf{B}) \le \text{rank}(\mathbf{A}, \mathbf{B})$ was established earlier (in Lemma 4.5.8). That $\text{rank}(\mathbf{A}+\mathbf{B}) = \text{rank}(\mathbf{A}, \mathbf{B})$ if and only if $\text{rank}(\mathbf{H}) = d$ is clear from result (5.8). That $\text{rank}(\mathbf{A}, \mathbf{B}) = \text{rank}(\mathbf{A}) + \text{rank}(\mathbf{B}) - c$ was established earlier [as result (17.4.10)]. And that $\text{rank}(\mathbf{A}) + \text{rank}(\mathbf{B}) - c \le \text{rank}(\mathbf{A}) + \text{rank}(\mathbf{B})$, with equality holding if and only if $c = 0$, is obvious.

(2) The proof of Part (2) of Theorem 18.5.4 is analogous to that of Part (1). Q.E.D.

Together, results (5.12) and (5.16) lead to the following corollary.

Corollary 18.5.5. For any matrices $\mathbf{A}$ and $\mathbf{B}$ (of the same size),

$$\begin{aligned}&\text{rank}(\mathbf{A}) + \text{rank}(\mathbf{B}) - c - d \\ &\qquad \le \text{rank}(\mathbf{A}+\mathbf{B}) \le \text{rank}(\mathbf{A}) + \text{rank}(\mathbf{B}) - \max(c, d), \qquad (5.20)\end{aligned}$$

where $c = \dim[\mathcal{C}(\mathbf{A}) \cap \mathcal{C}(\mathbf{B})]$ and $d = \dim[\mathcal{R}(\mathbf{A}) \cap \mathcal{R}(\mathbf{B})]$.

The matrix $\mathbf{H}$, defined by expression (5.11) [or by expression (5.6)], plays a prominent role in Theorems 18.5.4 and 18.5.3. The following theorem gives some results on the rank of this matrix.

Theorem 18.5.6. Let $\mathbf{A}$ and $\mathbf{B}$ represent $m \times n$ matrices. And let $c = \dim[\mathcal{C}(\mathbf{A}) \cap \mathcal{C}(\mathbf{B})]$, $d = \dim[\mathcal{R}(\mathbf{A}) \cap \mathcal{R}(\mathbf{B})]$, and

$$\mathbf{H} = \left[\mathbf{I} - \begin{pmatrix}\mathbf{A}\\ \mathbf{B}\end{pmatrix}\begin{pmatrix}\mathbf{A}\\ \mathbf{B}\end{pmatrix}^-\right]\begin{pmatrix}\mathbf{A} & \mathbf{0}\\ \mathbf{0} & \mathbf{B}\end{pmatrix}[\mathbf{I} - (\mathbf{A}, \mathbf{B})^-(\mathbf{A}, \mathbf{B})].$$

Then,

$$\begin{aligned}0 \le \text{rank}(\mathbf{H}) &= c - \left[\text{rank}\begin{pmatrix}\mathbf{A}\\ \mathbf{B}\end{pmatrix} - \text{rank}(\mathbf{A}+\mathbf{B})\right] \le c, \\ 0 \le \text{rank}(\mathbf{H}) &= d - [\text{rank}(\mathbf{A}, \mathbf{B}) - \text{rank}(\mathbf{A}+\mathbf{B})] \le d.\end{aligned}$$

Proof. The equalities $\text{rank}(\mathbf{H}) = c - \left[\text{rank}\begin{pmatrix}\mathbf{A}\\ \mathbf{B}\end{pmatrix} - \text{rank}(\mathbf{A}+\mathbf{B})\right]$ and $\text{rank}(\mathbf{H}) = d - [\text{rank}(\mathbf{A}, \mathbf{B}) - \text{rank}(\mathbf{A}+\mathbf{B})]$ are equivalent to equalities (5.9) and (5.8). Further, since [according to results (5.5) and (5.4)] $\text{rank}\begin{pmatrix}\mathbf{A}\\ \mathbf{B}\end{pmatrix} \ge \text{rank}(\mathbf{A}+\mathbf{B})$ and $\text{rank}(\mathbf{A}, \mathbf{B}) \ge \text{rank}(\mathbf{A}+\mathbf{B})$, it is clear that $c - \left[\text{rank}\begin{pmatrix}\mathbf{A}\\ \mathbf{B}\end{pmatrix} - \text{rank}(\mathbf{A}+\mathbf{B})\right] \le c$ and that $d - [\text{rank}(\mathbf{A}, \mathbf{B}) - \text{rank}(\mathbf{A}+\mathbf{B})] \le d$. Q.E.D.

In connection with Theorem 18.5.6, note that the condition $\text{rank}(\mathbf{H}) = d$, which appears in Part (1) of Theorem 18.5.4 [as a necessary and sufficient condition for $\text{rank}(\mathbf{A}+\mathbf{B}) = \text{rank}(\mathbf{A}, \mathbf{B})$], is satisfied if $d = 0$—if $d = 0$, then $\text{rank}(\mathbf{H}) = 0 = d$. And, similarly, the condition $\text{rank}(\mathbf{H}) = c$, which appears in Part (2) of Theorem 18.5.4 [as a necessary and sufficient condition for $\text{rank}(\mathbf{A}+\mathbf{B}) = \text{rank}\begin{pmatrix}\mathbf{A}\\ \mathbf{B}\end{pmatrix}$], is satisfied if $c = 0$—if $c = 0$, then $\text{rank}(\mathbf{H}) = 0 = c$.

c. Rank additivity

The following theorem gives necessary and sufficient conditions for rank additivity, that is, for the rank of the sum of two matrices to equal the sum of their ranks.

Theorem 18.5.7. For any two matrices $\mathbf{A}$ and $\mathbf{B}$ (of the same size), $\text{rank}(\mathbf{A} + \mathbf{B}) = \text{rank}(\mathbf{A}) + \text{rank}(\mathbf{B})$ if and only if $\mathcal{R}(\mathbf{A})$ and $\mathcal{R}(\mathbf{B})$ are essentially disjoint and $\mathcal{C}(\mathbf{A})$ and $\mathcal{C}(\mathbf{B})$ are essentially disjoint.

Proof. Let $c = \dim[\mathcal{C}(\mathbf{A}) \cap \mathcal{C}(\mathbf{B})]$ and $d = \dim[\mathcal{R}(\mathbf{A}) \cap \mathcal{R}(\mathbf{B})]$. Suppose that $\mathcal{R}(\mathbf{A})$ and $\mathcal{R}(\mathbf{B})$ are essentially disjoint and $\mathcal{C}(\mathbf{A})$ and $\mathcal{C}(\mathbf{B})$ are essentially disjoint (or, equivalently, that $c = 0$ and $d = 0$). Then, it follows from Corollary 18.5.5 that

$$\text{rank}(\mathbf{A}) + \text{rank}(\mathbf{B}) \leq \text{rank}(\mathbf{A} + \mathbf{B}) \leq \text{rank}(\mathbf{A}) + \text{rank}(\mathbf{B})$$

and hence that $\text{rank}(\mathbf{A} + \mathbf{B}) = \text{rank}(\mathbf{A}) + \text{rank}(\mathbf{B})$.

Conversely, suppose that $\text{rank}(\mathbf{A} + \mathbf{B}) = \text{rank}(\mathbf{A}) + \text{rank}(\mathbf{B})$. Then, in light of inequalities (5.4) and (5.5), we have that $\text{rank}(\mathbf{A}, \mathbf{B}) = \text{rank}(\mathbf{A}) + \text{rank}(\mathbf{B})$ and $\text{rank}\begin{pmatrix} \mathbf{A} \\ \mathbf{B} \end{pmatrix} = \text{rank}(\mathbf{A}) + \text{rank}(\mathbf{B})$ and hence [according to results (17.4.12) and (17.4.13)] that $c = 0$ and $d = 0$. Q.E.D.

Exercises

Section 18.1

1. Provide an alternative derivation of formula (1.5) by applying the results of Theorem 13.3.8 to $\begin{vmatrix} \mathbf{R} & -\mathbf{ST} \\ \mathbf{TU} & \mathbf{T} \end{vmatrix}$ (and by proceeding as in the proof of Theorem 18.1.1).

2. Let $\mathbf{R}$ represent an $n \times n$ matrix, $\mathbf{S}$ an $n \times m$ matrix, $\mathbf{T}$ an $m \times m$ matrix, and $\mathbf{U}$ an $m \times n$ matrix. Use Theorem 18.1.1 (or Corollary 18.1.2 or 18.1.4) to show that if $\mathbf{R}$ is nonsingular, then

$$|\mathbf{R} + \mathbf{STU}| = |\mathbf{R}||\mathbf{I}_m + \mathbf{UR}^{-1}\mathbf{ST}| = |\mathbf{R}||\mathbf{I}_m + \mathbf{TUR}^{-1}\mathbf{S}|.$$

3. Let $\mathbf{A}$ represent an $n \times n$ symmetric nonnegative definite matrix. Show that if $\mathbf{I} - \mathbf{A}$ is nonnegative definite and if $|\mathbf{A}| = 1$, then $\mathbf{A} = \mathbf{I}$.

4. Show that, for any $n \times n$ symmetric nonnegative definite matrix $\mathbf{B}$ and for any $n \times n$ symmetric matrix $\mathbf{C}$ such that $\mathbf{C} - \mathbf{B}$ is nonnegative definite,

$$|\mathbf{C}| \geq |\mathbf{C} - \mathbf{B}|,$$

with equality holding if and only if $\mathbf{C}$ is singular or $\mathbf{B} = \mathbf{0}$.

5. Let $\mathbf{A}$ represent a symmetric nonnegative definite matrix that has been partitioned as

$$\mathbf{A} = \begin{pmatrix} \mathbf{T} & \mathbf{U} \\ \mathbf{U}' & \mathbf{W} \end{pmatrix},$$

where $\mathbf{T}$ is of dimensions $m \times m$ and $\mathbf{W}$ of dimensions $n \times n$ (and where $\mathbf{U}$ is of dimensions $m \times n$). Define $\mathbf{Q} = \mathbf{W} - \mathbf{U}'\mathbf{T}^{-}\mathbf{U}$ (which is the Schur complement of $\mathbf{T}$).

(a) Using Theorem 14.8.4 and the result of Exercise 14.33 (or otherwise), show that

$$|\mathbf{W}| \geq |\mathbf{U}'\mathbf{T}^{-}\mathbf{U}|,$$

with equality holding if and only if $\mathbf{W}$ is singular or $\mathbf{Q} = \mathbf{0}$.

(b) Suppose that $n = m$ and that $\mathbf{T}$ is nonsingular. Show that

$$|\mathbf{W}||\mathbf{T}| \geq |\mathbf{U}|^2,$$

with equality holding if and only if $\mathbf{W}$ is singular or $\text{rank}(\mathbf{A}) = m$.

(c) Suppose that $n = m$ and that $\mathbf{A}$ is positive definite. Show that

$$|\mathbf{W}||\mathbf{T}| > |\mathbf{U}|^2.$$

6. Show that, for any $n \times p$ matrix $\mathbf{X}$ and any symmetric positive definite matrix $\mathbf{W}$,

$$|\mathbf{X}'\mathbf{W}\mathbf{X}||\mathbf{X}'\mathbf{W}^{-1}\mathbf{X}| \geq |\mathbf{X}'\mathbf{X}|^2. \tag{E.1}$$

[*Hint*. Begin by showing that the matrices $\mathbf{X}'\mathbf{X}(\mathbf{X}'\mathbf{W}\mathbf{X})^{-}\mathbf{X}'\mathbf{X}$ and $\mathbf{X}'\mathbf{W}^{-1}\mathbf{X} - \mathbf{X}'\mathbf{X}(\mathbf{X}'\mathbf{W}\mathbf{X})^{-}\mathbf{X}'\mathbf{X}$ are symmetric and nonnegative definite.]

7. (a) Show that, for any $n \times n$ skew-symmetric matrix $\mathbf{C}$,

$$|\mathbf{I}_n + \mathbf{C}| \geq 1,$$

with equality holding if and only if $\mathbf{C} = \mathbf{0}$.

(b) Generalize the result of Part (a) by showing that, for any $n \times n$ symmetric positive definite matrix $\mathbf{A}$ and any $n \times n$ skew-symmetric matrix $\mathbf{B}$,

$$|\mathbf{A} + \mathbf{B}| \geq |\mathbf{A}|,$$

with equality holding if and only if $\mathbf{B} = \mathbf{0}$.

Section 18.2

8. (a) Let $\mathbf{R}$ represent an $n \times n$ nonsingular matrix, and let $\mathbf{B}$ represent an $n \times n$ matrix of rank one. Show that $\mathbf{R} + \mathbf{B}$ is nonsingular if and only if $\text{tr}(\mathbf{R}^{-1}\mathbf{B}) \neq -1$, in which case

$$(\mathbf{R} + \mathbf{B})^{-1} = \mathbf{R}^{-1} - [1 + \text{tr}(\mathbf{R}^{-1}\mathbf{B})]^{-1}\mathbf{R}^{-1}\mathbf{B}\mathbf{R}^{-1}.$$

(b) To what does the result of Part (a) simplify in the special case where $\mathbf{R} = \mathbf{I}_n$?

9. Let $\mathbf{R}$ represent an $n \times n$ matrix, $\mathbf{S}$ an $n \times m$ matrix, $\mathbf{T}$ an $m \times m$ matrix, and $\mathbf{U}$ an $m \times n$ matrix. Suppose that $\mathbf{R}$ is nonsingular. Use Theorem 18.2.8 to show (a) that $\mathbf{R} + \mathbf{STU}$ is nonsingular if and only if $\mathbf{I}_m + \mathbf{UR}^{-1}\mathbf{ST}$ is nonsingular, in which case

$$(\mathbf{R} + \mathbf{STU})^{-1} = \mathbf{R}^{-1} - \mathbf{R}^{-1}\mathbf{ST}(\mathbf{I}_m + \mathbf{UR}^{-1}\mathbf{ST})^{-1}\mathbf{UR}^{-1},$$

and (b) that $\mathbf{R} + \mathbf{STU}$ is nonsingular if and only if $\mathbf{I}_m + \mathbf{TUR}^{-1}\mathbf{S}$ is nonsingular, in which case

$$(\mathbf{R} + \mathbf{STU})^{-1} = \mathbf{R}^{-1} - \mathbf{R}^{-1}\mathbf{S}(\mathbf{I}_m + \mathbf{TUR}^{-1}\mathbf{S})^{-1}\mathbf{TUR}^{-1}.$$

[*Hint*. Reexpress $\mathbf{R} + \mathbf{STU}$ as $\mathbf{R} + \mathbf{STU} = \mathbf{R} + (\mathbf{ST})\mathbf{I}_m\mathbf{U}$ and as $\mathbf{R} + \mathbf{STU} = \mathbf{R} + \mathbf{SI}_m\mathbf{TU}$.]

10. Let $\mathbf{R}$ represent an $n \times q$ matrix, $\mathbf{S}$ an $n \times m$ matrix, $\mathbf{T}$ an $m \times p$ matrix, and $\mathbf{U}$ a $p \times q$ matrix. Extend the results of Exercise 9 by showing that if $\mathcal{R}(\mathbf{STU}) \subset \mathcal{R}(\mathbf{R})$ and $\mathcal{C}(\mathbf{STU}) \subset \mathcal{C}(\mathbf{R})$, then the matrix

$$\mathbf{R}^- - \mathbf{R}^-\mathbf{ST}(\mathbf{I}_p + \mathbf{UR}^-\mathbf{ST})^-\mathbf{UR}^-$$

and the matrix

$$\mathbf{R}^- - \mathbf{R}^-\mathbf{S}(\mathbf{I}_m + \mathbf{TUR}^-\mathbf{S})^-\mathbf{TUR}^-$$

are both generalized inverses of the matrix $\mathbf{R} + \mathbf{STU}$.

11. Let $\mathbf{R}$ represent an $n \times q$ matrix, $\mathbf{S}$ an $n \times m$ matrix, $\mathbf{T}$ an $m \times p$ matrix, and $\mathbf{U}$ a $p \times q$ matrix.

(a) Let $\mathbf{G}$ represent a generalized inverse of the partitioned matrix $\begin{pmatrix} \mathbf{R} & -\mathbf{ST} \\ \mathbf{TU} & \mathbf{T} \end{pmatrix}$, and partition $\mathbf{G}$ as $\mathbf{G} = \begin{pmatrix} \mathbf{G}_{11} & \mathbf{G}_{12} \\ \mathbf{G}_{21} & \mathbf{G}_{22} \end{pmatrix}$ (where $\mathbf{G}_{11}$ is of dimensions $q \times n$). Use Theorem 9.6.5 to show that $\mathbf{G}_{11}$ is a generalized inverse of the matrix $\mathbf{R} + \mathbf{STU}$.

(b) Let $\mathbf{E}_R = \mathbf{I} - \mathbf{RR}^-$, $\mathbf{F}_R = \mathbf{I} - \mathbf{R}^-\mathbf{R}$, $\mathbf{X} = \mathbf{E}_R\mathbf{ST}$, $\mathbf{Y} = \mathbf{TUF}_R$, $\mathbf{E}_Y = \mathbf{I} - \mathbf{YY}^-$, $\mathbf{F}_X = \mathbf{I} - \mathbf{X}^-\mathbf{X}$, $\mathbf{Q} = \mathbf{T} + \mathbf{TUR}^-\mathbf{ST}$, $\mathbf{Z} = \mathbf{E}_Y\mathbf{QF}_X$, and $\mathbf{Q}^* = \mathbf{F}_X\mathbf{Z}^-\mathbf{E}_Y$. Use the result of Part (a) of Exercise 10.10 to show that the matrix

$$\begin{aligned}\mathbf{R}^- - \mathbf{R}^-\mathbf{STQ}^*\mathbf{TUR}^- - \mathbf{R}^-\mathbf{ST}(\mathbf{I} - \mathbf{Q}^*\mathbf{Q})\mathbf{X}^-\mathbf{E}_R \\ - \mathbf{F}_R\mathbf{Y}^-(\mathbf{I} - \mathbf{QQ}^*)\mathbf{TUR}^- + \mathbf{F}_R\mathbf{Y}^-(\mathbf{I} - \mathbf{QQ}^*)\mathbf{QX}^-\mathbf{E}_R \quad \text{(E.2)}\end{aligned}$$

is a generalized inverse of the matrix $\mathbf{R} + \mathbf{STU}$.

(c) Show that if $\mathcal{R}(\mathbf{TU}) \subset \mathcal{R}(\mathbf{R})$ and $\mathcal{C}(\mathbf{ST}) \subset \mathcal{C}(\mathbf{R})$, then formula (2.27) for a generalized inverse of $\mathbf{R} + \mathbf{STU}$ can be obtained as a special case of formula (E.2).

12. Let $\mathbf{A}_1, \mathbf{A}_2, \ldots$ represent a sequence of $m \times n$ matrices, and let $\mathbf{A}$ represent another $m \times n$ matrix.

(a) Using the result of Exercise 6.1 (i.e., the triangle inequality), show that if $\|\mathbf{A}_k - \mathbf{A}\| \to 0$, then $\|\mathbf{A}_k\| \to \|\mathbf{A}\|$.

(b) Show that if $\mathbf{A}_k \to \mathbf{A}$, then $\|\mathbf{A}_k\| \to \|\mathbf{A}\|$ (where the norms are the usual norms).

13. Let $\mathbf{A}$ represent an $n \times n$ matrix. Using the results of Exercise 6.1 and of Part (b) of Exercise 12, show that if $\|\mathbf{A}\| < 1$, then (for $k = 0, 1, 2, \ldots$)

$$\|(\mathbf{I} - \mathbf{A})^{-1} - (\mathbf{I} + \mathbf{A} + \mathbf{A}^2 + \cdots + \mathbf{A}^k)\| \le \|\mathbf{A}\|^{k+1}/(1 - \|\mathbf{A}\|)$$

(where the norms are the usual norms). (*Note.* It follows from Theorems 18.2.16 and 18.2.19 that if $\|\mathbf{A}\| < 1$, then $\mathbf{I} - \mathbf{A}$ is nonsingular.)

14. Let $\mathbf{A}$ and $\mathbf{B}$ represent $n \times n$ matrices. Suppose that $\mathbf{B}$ is nonsingular, and define $\mathbf{F} = \mathbf{B}^{-1}\mathbf{A}$. Using the result of Exercise 13, show that if $\|\mathbf{F}\| < 1$, then (for $k = 0, 1, 2, \ldots$)

$$\begin{aligned}\|(\mathbf{B} - \mathbf{A})^{-1} - (\mathbf{B}^{-1} + \mathbf{F}\mathbf{B}^{-1} + \mathbf{F}^2\mathbf{B}^{-1} + \cdots + \mathbf{F}^k\mathbf{B}^{-1})\| \\ \le \|\mathbf{B}^{-1}\|\|\mathbf{F}\|^{k+1}/(1 - \|\mathbf{F}\|)\end{aligned}$$

(where the norms are the usual norms). (*Note.* It follows from Theorems 18.2.18 and 18.2.19 that if $\|\mathbf{F}\| < 1$, then $\mathbf{B} - \mathbf{A}$ is nonsingular.)

Section 18.3

15. Let $\mathbf{A}$ represent an $n \times n$ symmetric nonnegative definite matrix, and let $\mathbf{B}$ represent an $n \times n$ matrix. Show that if $\mathbf{B} - \mathbf{A}$ is nonnegative definite (in which case $\mathbf{B}$ is also nonnegative definite), then $\mathcal{R}(\mathbf{A}) \subset \mathcal{R}(\mathbf{B})$ and $\mathcal{C}(\mathbf{A}) \subset \mathcal{C}(\mathbf{B})$.

Section 18.4

16. Let $\mathbf{A}$ represent an $n \times n$ symmetric idempotent matrix, and let $\mathbf{B}$ represent an $n \times n$ symmetric nonnegative definite matrix. Show that if $\mathbf{I} - \mathbf{A} - \mathbf{B}$ is nonnegative definite, then $\mathbf{BA} = \mathbf{AB} = \mathbf{0}$. (*Hint.* Show that $\mathbf{A}'(\mathbf{I} - \mathbf{A} - \mathbf{B})\mathbf{A} = -\mathbf{A}'\mathbf{BA}$, and then consider the implications of this equality.)

17. Let $\mathbf{A}_1, \ldots, \mathbf{A}_k$ represent $n \times n$ symmetric matrices, and let $\mathbf{A} = \mathbf{A}_1 + \cdots + \mathbf{A}_k$. Suppose that $\mathbf{A}$ is idempotent. Suppose further that $\mathbf{A}_1, \ldots, \mathbf{A}_{k-1}$ are idempotent and that $\mathbf{A}_k$ is nonnegative definite. Using the result of Exercise

16 (or otherwise), show that $\mathbf{A}_i\mathbf{A}_j = \mathbf{0}$ (for $j \neq i = 1, \ldots, k$), that $\mathbf{A}_k$ is idempotent, and that $\text{rank}(\mathbf{A}_k) = \text{rank}(\mathbf{A}) - \sum_{i=1}^{k-1} \text{rank}(\mathbf{A}_i)$.

18. Let $\mathbf{A}_1, \ldots, \mathbf{A}_k$ represent $n \times n$ symmetric matrices, and define $\mathbf{A} = \mathbf{A}_1 + \cdots + \mathbf{A}_k$. Suppose that $\mathbf{A}$ is idempotent. Show that if $\mathbf{A}_1, \ldots, \mathbf{A}_k$ are nonnegative definite and if $\text{tr}(\mathbf{A}) \leq \sum_{i=1}^{k} \text{tr}(\mathbf{A}_i^2)$, then $\mathbf{A}_i\mathbf{A}_j = \mathbf{0}$ (for $j \neq i = 1, \ldots, k$) and $\mathbf{A}_1, \ldots, \mathbf{A}_k$ are idempotent. (*Hint*. Show that $\sum_{i,j \neq i} \text{tr}(\mathbf{A}_i\mathbf{A}_j) \leq 0$ and then make use of Corollary 14.7.7.)

19. Let $\mathbf{A}_1, \ldots, \mathbf{A}_k$ represent $n \times n$ symmetric matrices such that $\mathbf{A}_1 + \cdots + \mathbf{A}_k = \mathbf{I}$. Show that if $\text{rank}(\mathbf{A}_1) + \cdots + \text{rank}(\mathbf{A}_k) = n$, then, for any (strictly) positive scalars $c_1, \ldots, c_k$, the matrix $c_1\mathbf{A}_1 + \cdots + c_k\mathbf{A}_k$ is positive definite.

20. Let $\mathbf{A}_1, \ldots, \mathbf{A}_k$ represent $n \times n$ symmetric idempotent matrices such that $\mathbf{A}_i\mathbf{A}_j = \mathbf{0}$ for $j \neq i = 1, 2, \ldots, k$. Show that, for any (strictly) positive scalar c_0 and any nonnegative scalars $c_1, \ldots, c_k$, the matrix $c_0\mathbf{I} + \sum_{i=1}^{k} c_i\mathbf{A}_i$ is positive definite (and hence nonsingular), and

$$\left(c_0\mathbf{I} + \sum_{i=1}^{k} c_i\mathbf{A}_i\right)^{-1} = d_0\mathbf{I} + \sum_{i=1}^{k} d_i\mathbf{A}_i,$$

where $d_0 = 1/c_0$ and (for $i = 1, \ldots, k$) $d_i = -c_i/[c_0(c_0 + c_i)]$.

21. Let $\mathbf{A}_1, \ldots, \mathbf{A}_k$ represent $n \times n$ symmetric idempotent matrices such that (for $j \neq i = 1, \ldots, k$) $\mathbf{A}_i\mathbf{A}_j = \mathbf{0}$, and let $\mathbf{A}$ represent an $n \times n$ symmetric idempotent matrix such that (for $i = 1, \ldots, k$) $\mathcal{C}(\mathbf{A}_i) \subset \mathcal{C}(\mathbf{A})$. Show that if $\text{rank}(\mathbf{A}_1) + \cdots + \text{rank}(\mathbf{A}_k) = \text{rank}(\mathbf{A})$, then $\mathbf{A}_1 + \cdots + \mathbf{A}_k = \mathbf{A}$.

22. Let $\mathbf{A}$ represent an $m \times n$ matrix and $\mathbf{B}$ an $n \times m$ matrix. If $\mathbf{B}$ is a generalized inverse of $\mathbf{A}$, then it follows from Lemma 10.2.6 that $\text{rank}(\mathbf{I} - \mathbf{BA}) = n - \text{rank}(\mathbf{A})$. Show that the converse is also true; that is, show that if $\text{rank}(\mathbf{I} - \mathbf{BA}) = n - \text{rank}(\mathbf{A})$, then $\mathbf{B}$ is a generalized inverse of $\mathbf{A}$.

23. Let $\mathbf{A}$ represent the $(n \times n)$ projection matrix for a subspace $\mathcal{U}$ of $\mathcal{R}^{n\times 1}$ along a subspace $\mathcal{V}$ of $\mathcal{R}^{n\times 1}$ (where $\mathcal{U} \oplus \mathcal{V} = \mathcal{R}^{n\times 1}$), and let $\mathbf{B}$ represent the $(n \times n)$ projection matrix for a subspace $\mathcal{W}$ of $\mathcal{R}^{n\times 1}$ along a subspace $\mathcal{X}$ of $\mathcal{R}^{n\times 1}$ (where $\mathcal{W} \oplus \mathcal{X} = \mathcal{R}^{n\times 1}$).

(a) Show that $\mathbf{A} + \mathbf{B}$ is the projection matrix for some subspace $\mathcal{L}$ of $\mathcal{R}^{n\times 1}$ along some subspace $\mathcal{M}$ of $\mathcal{R}^{n\times 1}$ (where $\mathcal{L} \oplus \mathcal{M} = \mathcal{R}^{n\times 1}$) if and only if $\mathbf{BA} = \mathbf{AB} = \mathbf{0}$, in which case $\mathcal{L} = \mathcal{U} \oplus \mathcal{W}$ and $\mathcal{M} = \mathcal{V} \cap \mathcal{X}$.

(b) Show that $\mathbf{A} - \mathbf{B}$ is the projection matrix for some subspace $\mathcal{L}$ of $\mathcal{R}^{n\times 1}$ along some subspace $\mathcal{M}$ of $\mathcal{R}^{n\times 1}$ (where $\mathcal{L} \oplus \mathcal{M} = \mathcal{R}^{n\times 1}$) if and only if $\mathbf{BA} = \mathbf{AB} = \mathbf{B}$, in which case $\mathcal{L} = \mathcal{U} \cap \mathcal{X}$ and $\mathcal{M} = \mathcal{V} \oplus \mathcal{W}$. [*Hint*.

Observe (in light of Theorem 17.6.13 and Lemma 10.1.2) that $\mathbf{A} - \mathbf{B}$ is the projection matrix for some subspace $\mathcal{L}$ along some subspace $\mathfrak{M}$ if and only if $\mathbf{I} - (\mathbf{A} - \mathbf{B}) = (\mathbf{I} - \mathbf{A}) + \mathbf{B}$ is the projection matrix for some subspace $\mathcal{L}^*$ along some subspace $\mathfrak{M}^*$, and then make use of Part (a) and Theorem 17.6.10.]

24. (a) Let $\mathbf{B}$ represent an $n \times n$ symmetric matrix, and let $\mathbf{W}$ represent an $n \times n$ symmetric nonnegative definite matrix. Show that

$$\mathbf{WBWBW} = \mathbf{WBW} \Leftrightarrow (\mathbf{BW})^3 = (\mathbf{BW})^2$$
$$\Leftrightarrow \operatorname{tr}[(\mathbf{BW})^2] = \operatorname{tr}[(\mathbf{BW})^3] = \operatorname{tr}[(\mathbf{BW})^4].$$

(b) Indicate how, in the special case of Theorems 18.4.8 and 18.4.9 where $\mathbf{A}_1, \ldots, \mathbf{A}_k$ are symmetric, the conditions $\mathbf{VAVAV} = \mathbf{VAV}$ and $\mathbf{VA}_i\mathbf{VA}_i\mathbf{V} = \mathbf{VA}_i\mathbf{V}$ (which appear in both theorems) can be reexpressed by applying the results of Part (a) (of the current exercise).

Section 18.5

25. Let $\mathbf{R}$ represent an $n \times q$ matrix, $\mathbf{S}$ an $n \times m$ matrix, $\mathbf{T}$ an $m \times p$ matrix, and $\mathbf{U}$ a $p \times q$ matrix.

(a) Show that

$$\operatorname{rank}(\mathbf{R} + \mathbf{STU}) = \operatorname{rank}\begin{pmatrix} \mathbf{R} & -\mathbf{ST} \\ \mathbf{TU} & \mathbf{T} \end{pmatrix} - \operatorname{rank}(\mathbf{T}). \tag{E.3}$$

(b) Let $\mathbf{E}_R = \mathbf{I} - \mathbf{RR}^-$, $\mathbf{F}_R = \mathbf{I} - \mathbf{R}^-\mathbf{R}$, $\mathbf{X} = \mathbf{E}_R\mathbf{ST}$, $\mathbf{Y} = \mathbf{TUF}_R$, $\mathbf{E}_Y = \mathbf{I} - \mathbf{YY}^-$, $\mathbf{F}_X = \mathbf{I} - \mathbf{X}^-\mathbf{X}$, $\mathbf{Q} = \mathbf{T} + \mathbf{TUR}^-\mathbf{ST}$, and $\mathbf{Z} = \mathbf{E}_Y\mathbf{QF}_X$. Use the result of Part (b) of Exercise 10.10 to show that

$$\operatorname{rank}(\mathbf{R} + \mathbf{STU}) = \operatorname{rank}(\mathbf{R}) + \operatorname{rank}(\mathbf{X}) + \operatorname{rank}(\mathbf{Y}) + \operatorname{rank}(\mathbf{Z}) - \operatorname{rank}(\mathbf{T}).$$

26. Show that, for any $m \times n$ matrices $\mathbf{A}$ and $\mathbf{B}$,

$$\operatorname{rank}(\mathbf{A} + \mathbf{B}) \geq |\operatorname{rank}(\mathbf{A}) - \operatorname{rank}(\mathbf{B})|.$$

27. Show that, for any $n \times n$ symmetric nonnegative definite matrices $\mathbf{A}$ and $\mathbf{B}$,

$$\mathcal{C}(\mathbf{A} + \mathbf{B}) = \mathcal{C}(\mathbf{A}, \mathbf{B}), \qquad \mathcal{R}(\mathbf{A} + \mathbf{B}) = \mathcal{R}\begin{pmatrix} \mathbf{A} \\ \mathbf{B} \end{pmatrix},$$
$$\operatorname{rank}(\mathbf{A} + \mathbf{B}) = \operatorname{rank}(\mathbf{A}, \mathbf{B}) = \operatorname{rank}\begin{pmatrix} \mathbf{A} \\ \mathbf{B} \end{pmatrix}.$$

28. Let $\mathbf{A}$ and $\mathbf{B}$ represent $m \times n$ matrices.

(a) Show that (1) $\mathcal{C}(\mathbf{A}) \subset \mathcal{C}(\mathbf{A}+\mathbf{B})$ if and only if $\text{rank}(\mathbf{A}, \mathbf{B}) = \text{rank}(\mathbf{A}+\mathbf{B})$ and (2) $\mathcal{R}(\mathbf{A}) \subset \mathcal{R}(\mathbf{A}+\mathbf{B})$ if and only if $\text{rank}\begin{pmatrix}\mathbf{A}\\ \mathbf{B}\end{pmatrix} = \text{rank}(\mathbf{A}+\mathbf{B})$.

(b) Show that (1) if $\mathcal{R}(\mathbf{A})$ and $\mathcal{R}(\mathbf{B})$ are essentially disjoint, then $\mathcal{C}(\mathbf{A}) \subset \mathcal{C}(\mathbf{A}+\mathbf{B})$ and (2) if $\mathcal{C}(\mathbf{A})$ and $\mathcal{C}(\mathbf{B})$ are essentially disjoint, then $\mathcal{R}(\mathbf{A}) \subset \mathcal{R}(\mathbf{A}+\mathbf{B})$.

29. Let $\mathbf{A}$ and $\mathbf{B}$ represent $m \times n$ matrices. Show that *each* of the following five conditions is necessary and sufficient for rank additivity [i.e., for $\text{rank}(\mathbf{A}+\mathbf{B}) = \text{rank}(\mathbf{A}) + \text{rank}(\mathbf{B})$]:

(a) $\text{rank}(\mathbf{A}, \mathbf{B}) = \text{rank}\begin{pmatrix}\mathbf{A}\\ \mathbf{B}\end{pmatrix} = \text{rank}(\mathbf{A}) + \text{rank}(\mathbf{B})$;

(b) $\text{rank}(\mathbf{A}) = \text{rank}[\mathbf{A}(\mathbf{I}-\mathbf{B}^-\mathbf{B})] = \text{rank}[(\mathbf{I}-\mathbf{B}\mathbf{B}^-)\mathbf{A}]$;

(c) $\text{rank}(\mathbf{B}) = \text{rank}[\mathbf{B}(\mathbf{I}-\mathbf{A}^-\mathbf{A})] = \text{rank}[(\mathbf{I}-\mathbf{A}\mathbf{A}^-)\mathbf{B}]$;

(d) $\text{rank}(\mathbf{A}) = \text{rank}[\mathbf{A}(\mathbf{I}-\mathbf{B}^-\mathbf{B})]$ and $\text{rank}(\mathbf{B}) = \text{rank}[(\mathbf{I}-\mathbf{A}\mathbf{A}^-)\mathbf{B}]$;

(e) $\text{rank}(\mathbf{A}) = \text{rank}[(\mathbf{I}-\mathbf{B}\mathbf{B}^-)\mathbf{A}]$ and $\text{rank}(\mathbf{B}) = \text{rank}[\mathbf{B}(\mathbf{I}-\mathbf{A}^-\mathbf{A})]$.

30. Let $\mathbf{A}$ and $\mathbf{B}$ represent $m \times n$ matrices. And let

$$\mathbf{H} = \left[\mathbf{I} - \begin{pmatrix}\mathbf{A}\\ \mathbf{B}\end{pmatrix}\begin{pmatrix}\mathbf{A}\\ \mathbf{B}\end{pmatrix}^-\right]\begin{pmatrix}\mathbf{A} & \mathbf{0}\\ \mathbf{0} & -\mathbf{B}\end{pmatrix}[\mathbf{I} - (\mathbf{A}, \mathbf{B})^-(\mathbf{A}, \mathbf{B})].$$

(a) Using result (5.7) (with $-\mathbf{B}$ in place of $\mathbf{B}$), show that

$$\text{rank}(\mathbf{A}-\mathbf{B}) = \text{rank}(\mathbf{A}) - \text{rank}(\mathbf{B}) + [\text{rank}(\mathbf{A}, \mathbf{B}) - \text{rank}(\mathbf{A})] + \left[\text{rank}\begin{pmatrix}\mathbf{A}\\ \mathbf{B}\end{pmatrix} - \text{rank}(\mathbf{A})\right] + \text{rank}(\mathbf{H}).$$

(b) Show that $\mathbf{A}$ and $\mathbf{B}$ are rank subtractive [in the sense that $\text{rank}(\mathbf{A}-\mathbf{B}) = \text{rank}(\mathbf{A}) - \text{rank}(\mathbf{B})$] if and only if $\text{rank}(\mathbf{A}, \mathbf{B}) = \text{rank}\begin{pmatrix}\mathbf{A}\\ \mathbf{B}\end{pmatrix} = \text{rank}(\mathbf{A})$ and $\mathbf{H} = \mathbf{0}$.

(c) Show that if $\text{rank}(\mathbf{A}, \mathbf{B}) = \text{rank}\begin{pmatrix}\mathbf{A}\\ \mathbf{B}\end{pmatrix} = \text{rank}(\mathbf{A})$, then (1) $(\mathbf{A}^-, \mathbf{0})$ and $\begin{pmatrix}\mathbf{A}^-\\ \mathbf{0}\end{pmatrix}$ are generalized inverses of $\begin{pmatrix}\mathbf{A}\\ \mathbf{B}\end{pmatrix}$ and $(\mathbf{A}, \mathbf{B})$, respectively, and (2) for $\begin{pmatrix}\mathbf{A}\\ \mathbf{B}\end{pmatrix}^- = (\mathbf{A}^-, \mathbf{0})$ and $(\mathbf{A}, \mathbf{B})^- = \begin{pmatrix}\mathbf{A}^-\\ \mathbf{0}\end{pmatrix}$, $\mathbf{H} = \begin{pmatrix}\mathbf{0} & \mathbf{0}\\ \mathbf{0} & \mathbf{B}\mathbf{A}^-\mathbf{B}-\mathbf{B}\end{pmatrix}$.

(d) Show that *each* of the following three conditions is necessary and sufficient for rank subtractivity [i.e., for $\text{rank}(\mathbf{A}-\mathbf{B}) = \text{rank}(\mathbf{A}) - \text{rank}(\mathbf{B})$]:

(1) $\text{rank}(\mathbf{A}, \mathbf{B}) = \text{rank}\begin{pmatrix}\mathbf{A}\\ \mathbf{B}\end{pmatrix} = \text{rank}(\mathbf{A})$ and $\mathbf{B}\mathbf{A}^-\mathbf{B} = \mathbf{B}$;

(2) $\mathcal{C}(\mathbf{B}) \subset \mathcal{C}(\mathbf{A})$, $\mathcal{R}(\mathbf{B}) \subset \mathcal{R}(\mathbf{A})$, and $\mathbf{BA^-B} = \mathbf{B}$;

(3) $\mathbf{AA^-B} = \mathbf{BA^-A} = \mathbf{BA^-B} = \mathbf{B}$.

(e) Using the result of Exercise 29 (or otherwise), show that $\text{rank}(\mathbf{A} - \mathbf{B}) = \text{rank}(\mathbf{A}) - \text{rank}(\mathbf{B})$ if and only if $\text{rank}(\mathbf{A} - \mathbf{B}) = \text{rank}[\mathbf{A}(\mathbf{I} - \mathbf{B^-B})] = \text{rank}[(\mathbf{I} - \mathbf{BB^-})\mathbf{A}]$.

31. Let $\mathbf{A}_1, \ldots, \mathbf{A}_k$ represent $m \times n$ matrices. Adopting the terminology of Exercise 17.6, use Part (a) of that exercise to show that if $\mathcal{R}(\mathbf{A}_1), \ldots, \mathcal{R}(\mathbf{A}_k)$ are independent and $\mathcal{C}(\mathbf{A}_1), \ldots, \mathcal{C}(\mathbf{A}_k)$ are independent, then $\text{rank}(\mathbf{A}_1 + \cdots + \mathbf{A}_k) = \text{rank}(\mathbf{A}_1) + \cdots + \text{rank}(\mathbf{A}_k)$.

32. Let $\mathbf{T}$ represent an $m \times p$ matrix, $\mathbf{U}$ an $m \times q$ matrix, $\mathbf{V}$ an $n \times p$ matrix, and $\mathbf{W}$ an $n \times q$ matrix, and define $\mathbf{Q} = \mathbf{W} - \mathbf{VT^-U}$. Further, let $\mathbf{E}_T = \mathbf{I} - \mathbf{TT^-}$, $\mathbf{F}_T = \mathbf{I} - \mathbf{T^-T}$, $\mathbf{X} = \mathbf{E}_T\mathbf{U}$, and $\mathbf{Y} = \mathbf{VF}_T$.

(a) Show that

$$\text{rank}\begin{pmatrix} \mathbf{T} & \mathbf{U} \\ \mathbf{V} & \mathbf{W} \end{pmatrix} = \text{rank}(\mathbf{T}) + \text{rank}\begin{pmatrix} \mathbf{0} & \mathbf{X} \\ \mathbf{Y} & \mathbf{Q} \end{pmatrix}. \tag{E.4}$$

(b) Show that

$$\text{rank}\begin{pmatrix} \mathbf{0} & \mathbf{V} \\ \mathbf{U} & \mathbf{T} \end{pmatrix} = \text{rank}(\mathbf{T}) + \text{rank}\begin{pmatrix} -\mathbf{VT^-U} & \mathbf{Y} \\ \mathbf{X} & \mathbf{0} \end{pmatrix}.$$

[*Hint*. Observe (in light of Lemma 8.5.1) that $\text{rank}\begin{pmatrix} \mathbf{0} & \mathbf{V} \\ \mathbf{U} & \mathbf{T} \end{pmatrix} = \text{rank}\begin{pmatrix} \mathbf{T} & \mathbf{U} \\ \mathbf{V} & \mathbf{0} \end{pmatrix}$, and make use of Part (a).]

(c) Show that

$$\text{rank}\begin{pmatrix} \mathbf{0} & \mathbf{V} \\ \mathbf{U} & \mathbf{T} \end{pmatrix} = \text{rank}(\mathbf{T}) + \text{rank}(\mathbf{X}) + \text{rank}(\mathbf{Y}) + \text{rank}(\mathbf{E}_Y\mathbf{VT^-UF}_X),$$

where $\mathbf{E}_Y = \mathbf{I} - \mathbf{YY^-}$ and $\mathbf{F}_X = \mathbf{I} - \mathbf{X^-X}$. [*Hint*. Use Theorem 17.2.17 in combination with Part (b).]

(d) Show that

$$\text{rank}\begin{pmatrix} \mathbf{T} & \mathbf{U} \\ \mathbf{V} & \mathbf{W} \end{pmatrix} = \text{rank}(\mathbf{T}) + \text{rank}(\mathbf{Q}) + \text{rank}(\mathbf{A}) + \text{rank}(\mathbf{B})$$
$$+ \text{rank}[(\mathbf{I} - \mathbf{AA^-})\mathbf{XQ^-Y}(\mathbf{I} - \mathbf{B^-B})],$$

where $\mathbf{A} = \mathbf{X}(\mathbf{I} - \mathbf{Q^-Q})$ and $\mathbf{B} = (\mathbf{I} - \mathbf{QQ^-})\mathbf{Y}$. [*Hint*. Use Part (c) in combination with Part (a).]

33. Let $\mathbf{R}$ represent an $n \times q$ matrix, $\mathbf{S}$ an $n \times m$ matrix, $\mathbf{T}$ an $m \times p$ matrix, and $\mathbf{U}$ a $p \times q$ matrix, and define $\mathbf{Q} = \mathbf{T} + \mathbf{TUR}^{-}\mathbf{ST}$. Further, let $\mathbf{E}_R = \mathbf{I} - \mathbf{RR}^{-}$, $\mathbf{F}_R = \mathbf{I} - \mathbf{R}^{-}\mathbf{R}$, $\mathbf{X} = \mathbf{E}_R\mathbf{ST}$, $\mathbf{Y} = \mathbf{TUF}_R$, $\mathbf{A} = \mathbf{X}(\mathbf{I} - \mathbf{Q}^{-}\mathbf{Q})$, and $\mathbf{B} = (\mathbf{I} - \mathbf{QQ}^{-})\mathbf{Y}$. Use the result of Part (d) of Exercise 32 in combination with the result of Part (a) of Exercise 25 to show that

$$\begin{aligned}\text{rank}(\mathbf{R} + \mathbf{STU}) = \text{rank}(\mathbf{R}) + \text{rank}(\mathbf{Q}) - \text{rank}(\mathbf{T}) + \text{rank}(\mathbf{A}) + \text{rank}(\mathbf{B}) \\ + \text{rank}[(\mathbf{I} - \mathbf{AA}^{-})\mathbf{XQ}^{-}\mathbf{Y}(\mathbf{I} - \mathbf{B}^{-}\mathbf{B})],\end{aligned}$$

thereby generalizing formula (5.2).

Bibliographic and Supplementary Notes

§2. The presentation in Subsection a (and in the first part of Subsection d) is similar to that of Miller (1987, pp. 10–11). The material in Subsection e (and in Exercises 10 and 11) includes a number of the results of Harville (1997). The presentation in Subsection f requires of the reader some familiarity with basic results on sequences of scalars and on the infinite series generated by such sequences. For a more general and more complete treatment of the subject of Subsection f, refer, e.g., to Faddeeva (1959, sec. 5) or to Golub and Van Loan (1989, sec. 2.3). §3. Exercise 15 is based on Lemma 2.4 of Milliken and Akdeniz (1977). §4. The result of Exercise 16 is given by Loynes (1966) and is referred to by Searle (1971, p. 61) as Loynes' lemma. §5. The results of Corollary 18.5.2 and of Exercises 25 and 33 were given by Harville (1997). The results covered in Subsections b and c (and in Exercises 28, 29, 30, and 32) are adaptations of results presented by Marsaglia and Styan (1974) in their Sections 4, 5, 7, and 8.

19

Minimization of a Second-Degree Polynomial (in *n* Variables) Subject to Linear Constraints

The subject of this chapter is the minimization of a quadratic form, or, more generally, a second-degree polynomial, in some number, say n, of variables that may be subject to linear constraints. Special cases of this minimization problem are encountered in various areas of statistics and in many related disciplines. In particular, they are encountered in estimating the parameters of a linear statistical model. One approach to the estimation goes by the acronym BLUE (for best linear unbiased estimation); in this approach, consideration is restricted to estimators that are linear (i.e., that are expressible as linear combinations of the data) and that are unbiased (i.e., whose "expected values" equal the parameters), and the estimator of each parameter is chosen to have minimum variance among all estimators that are linear and unbiased. The minimization problem encompassed in this approach can be formulated as one of minimizing a quadratic form (in the coefficients of the linear combination) subject to linear constraints—the constraints arise from the restriction to unbiased estimators.

Another approach is to regard the estimation of the parameters as a least squares problem—the least squares problem was considered (from a geometrical perspective) in Chapter 12. This approach consists of minimizing the sum of the squared deviations between the data points and the linear combinations of the parameters (in the model) that correspond to those points. The sum of the squared deviations is a second-degree polynomial (in the parameters). In some cases, the parameters of the model may be subject to linear constraints—knowledge of the process that gave rise to the data may suggest such constraints (along with the other aspects of the model). In the presence of such constraints, the minimization problem encompassed in the least squares appraoch is one of minimizing a second-degree polynomial (in the parameters) subject to linear constraints (on the parameters).

19.1 Unconstrained Minimization

As a preliminary to considering the minimization of a second–degree polynomial (in n variables) subject to linear constraints, it is instructive to consider the unconstrained minimization of a second–degree polynomial.

a. Basic results

Let $\mathbf{V} = \{v_{ij}\}$ represent an $n \times n$ symmetric nonnegative definite matrix, and let $\mathbf{b} = \{b_i\}$ represent an $n \times 1$ vector such that $\mathbf{b} \in \mathcal{C}(\mathbf{V})$. Further, let $\mathbf{a} = (a_1, \ldots, a_n)'$ represent a (column) vector of n variables, and consider the minimization of the function f of $\mathbf{a}$ defined (for $\mathbf{a} \in \mathcal{R}^n$) by

$$f(\mathbf{a}) = \mathbf{a}'\mathbf{V}\mathbf{a} - 2\mathbf{b}'\mathbf{a}.$$

Note that $f(\mathbf{a})$ can be reexpressed as

$$f(\mathbf{a}) = \mathbf{a}'\mathbf{V}\mathbf{a} - 2\mathbf{a}'\mathbf{b} = \mathbf{a}'\mathbf{V}\mathbf{a} - \mathbf{b}'\mathbf{a} - \mathbf{a}'\mathbf{b}$$

or in nonmatrix form as the second-degree polynomial

$$f(\mathbf{a}) = \sum_{i,j} v_{ij} a_i a_j - 2 \sum_i b_i a_i.$$

For f to have a minimum at a point $\mathbf{a}_*$, it is necessary that $\mathbf{a}_*$ be a stationary point of f, that is, it is necessary that $\partial f/\partial \mathbf{a} = \mathbf{0}$ at $\mathbf{a} = \mathbf{a}_*$ (e.g., Magnus and Neudecker 1988, pp. 119–120). As a consequence of formulas (15.3.5) and (15.3.7), we have that

$$\frac{\partial f}{\partial \mathbf{a}} = 2\mathbf{V}\mathbf{a} - 2\mathbf{b} = 2(\mathbf{V}\mathbf{a} - \mathbf{b}). \tag{1.1}$$

Thus, for f to have a minimum at a point $\mathbf{a}_*$, it is necessary that $\mathbf{V}\mathbf{a}_* = \mathbf{b}$ or, equivalently, that $\mathbf{a}_*$ be a solution to the linear system $\mathbf{V}\mathbf{a} = \mathbf{b}$ (in $\mathbf{a}$).

The following theorem makes a stronger claim—for f to have a minimum at a point $\mathbf{a}_*$, it is sufficient, as well as necessary, that $\mathbf{V}\mathbf{a}_* = \mathbf{b}$.

Theorem 19.1.1. Let $\mathbf{a}$ represent an $n \times 1$ vector of (unconstrained) variables, and define $f(\mathbf{a}) = \mathbf{a}'\mathbf{V}\mathbf{a} - 2\mathbf{b}'\mathbf{a}$, where $\mathbf{V}$ is an $n \times n$ symmetric nonnegative definite matrix and $\mathbf{b}$ is an $n \times 1$ vector such that $\mathbf{b} \in \mathcal{C}(\mathbf{V})$. Then, the linear system $\mathbf{V}\mathbf{a} = \mathbf{b}$ (in $\mathbf{a}$) is consistent. Further, $f(\mathbf{a})$ attains its minimum value at a point $\mathbf{a}_*$ if and only if $\mathbf{a}_*$ is a solution to $\mathbf{V}\mathbf{a} = \mathbf{b}$, in which case $f(\mathbf{a}_*) = -\mathbf{b}'\mathbf{a}_* = -\mathbf{a}_*'\mathbf{b}$.

Theorem 19.1.1 is a special case of the following result (that where $s = 1$).

Theorem 19.1.2. Let $\mathbf{A}$ represent an $n \times s$ matrix of (unconstrained) variables, and define $\mathbf{F}(\mathbf{A}) = \mathbf{A}'\mathbf{V}\mathbf{A} - \mathbf{B}'\mathbf{A} - \mathbf{A}'\mathbf{B}$, where $\mathbf{V}$ is an $n \times n$ symmetric nonnegative definite matrix and $\mathbf{B}$ is an $n \times s$ matrix such that $\mathcal{C}(\mathbf{B}) \subset \mathcal{C}(\mathbf{V})$. Then, the linear system $\mathbf{V}\mathbf{A} = \mathbf{B}$ (in $\mathbf{A}$) is consistent. Further, $\mathbf{F}(\mathbf{A})$ attains its minimum value at a matrix $\mathbf{A}_*$ (in the sense that $\mathbf{F}(\mathbf{A}) - \mathbf{F}(\mathbf{A}_*)$ is nonnegative definite for every $\mathbf{A} \in \mathcal{R}^{n \times s}$) if and only if $\mathbf{A}_*$ is a solution to $\mathbf{V}\mathbf{A} = \mathbf{B}$, in which case $\mathbf{F}(\mathbf{A}_*) = -\mathbf{B}'\mathbf{A}_* = -\mathbf{A}_*'\mathbf{B}$.

Proof (of Theorem 19.1.2). Since $\mathcal{C}(\mathbf{B}) \subset \mathcal{C}(\mathbf{V})$, the consistency of the linear system $\mathbf{VA} = \mathbf{B}$ follows from Theorem 7.2.1.

Now, let $\mathbf{A}_*$ represent any solution to $\mathbf{VA} = \mathbf{B}$, and let $\mathbf{A}$ represent an arbitrary $n \times s$ matrix. Then,

$$\begin{aligned}\mathbf{A}'\mathbf{VA} &= [\mathbf{A}_* + (\mathbf{A} - \mathbf{A}_*)]'\mathbf{V}[\mathbf{A}_* + (\mathbf{A} - \mathbf{A}_*)] \\ &= \mathbf{A}_*'\mathbf{VA}_* + \mathbf{A}_*'\mathbf{V}(\mathbf{A} - \mathbf{A}_*) + (\mathbf{A} - \mathbf{A}_*)'\mathbf{VA}_* + (\mathbf{A} - \mathbf{A}_*)'\mathbf{V}(\mathbf{A} - \mathbf{A}_*) \\ &= \mathbf{A}_*'\mathbf{VA}_* + \mathbf{B}'(\mathbf{A} - \mathbf{A}_*) + (\mathbf{A} - \mathbf{A}_*)'\mathbf{B} + (\mathbf{A} - \mathbf{A}_*)'\mathbf{V}(\mathbf{A} - \mathbf{A}_*).\end{aligned}$$

Thus,

$$\begin{aligned}\mathbf{F}(\mathbf{A}) - \mathbf{F}(\mathbf{A}_*) &= \mathbf{A}'\mathbf{VA} - \mathbf{A}_*'\mathbf{VA}_* - \mathbf{B}'(\mathbf{A} - \mathbf{A}_*) - (\mathbf{A} - \mathbf{A}_*)'\mathbf{B} \\ &= (\mathbf{A} - \mathbf{A}_*)'\mathbf{V}(\mathbf{A} - \mathbf{A}_*), \qquad (1.2)\end{aligned}$$

implying (in light of Theorem 14.2.9) that $\mathbf{F}(\mathbf{A}) - \mathbf{F}(\mathbf{A}_*)$ is nonnegative definite.

Further, let $\mathbf{A}_0$ represent any $n \times s$ matrix such that $\mathbf{F}(\mathbf{A}) - \mathbf{F}(\mathbf{A}_0)$ is nonnegative definite for every $\mathbf{A} \in \mathcal{R}^{n \times s}$. Then, $\mathbf{F}(\mathbf{A}_*) - \mathbf{F}(\mathbf{A}_0)$ is nonnegative definite, or, equivalently, $\mathbf{F}(\mathbf{A}_0) - \mathbf{F}(\mathbf{A}_*)$ {which $= -[\mathbf{F}(\mathbf{A}_*) - \mathbf{F}(\mathbf{A}_0)]$} is nonpositive definite, and, since $\mathbf{F}(\mathbf{A}_0) - \mathbf{F}(\mathbf{A}_*)$ is nonnegative definite, it follows from Lemma 14.2.2 that $\mathbf{F}(\mathbf{A}_0) - \mathbf{F}(\mathbf{A}_*) = \mathbf{0}$ and hence [upon applying equality (1.2) with $\mathbf{A} = \mathbf{A}_0$] that $(\mathbf{A}_0 - \mathbf{A}_*)'\mathbf{V}(\mathbf{A}_0 - \mathbf{A}_*) = \mathbf{0}$. Thus, as a consequence of Corollary 14.3.11, $\mathbf{V}(\mathbf{A}_0 - \mathbf{A}_*) = \mathbf{0}$, implying that $\mathbf{VA}_0 = \mathbf{VA}_* = \mathbf{B}$ and hence that $\mathbf{A}_0$ is a solution to $\mathbf{VA} = \mathbf{B}$.

Finally, observe that

$$\mathbf{F}(\mathbf{A}_*) = \mathbf{A}_*'\mathbf{VA}_* - \mathbf{B}'\mathbf{A}_* - \mathbf{A}_*'\mathbf{B} = \mathbf{A}_*'\mathbf{B} - \mathbf{B}'\mathbf{A}_* - \mathbf{A}_*'\mathbf{B} = -\mathbf{B}'\mathbf{A}_*$$

and similarly that

$$\mathbf{F}(\mathbf{A}_*) = (\mathbf{VA}_*)'\mathbf{A}_* - \mathbf{B}'\mathbf{A}_* - \mathbf{A}_*'\mathbf{B} = \mathbf{B}'\mathbf{A}_* - \mathbf{B}'\mathbf{A}_* - \mathbf{A}_*'\mathbf{B} = -\mathbf{A}_*'\mathbf{B}.$$

Q.E.D.

Note, in connection with Theorem 19.1.1, that in the special case where $\mathbf{V}$ is positive definite (and hence nonsingular), the linear system $\mathbf{Va} = \mathbf{b}$ has a unique solution, namely, $\mathbf{a} = \mathbf{V}^{-1}\mathbf{b}$, and that $\mathbf{V}^{-1}\mathbf{b}$ is the only value of $\mathbf{a}$ at which $f(\mathbf{a})$ attains its minimum value. Similarly (and more generally), note, in connection with Theorem 19.1.2, that, in the special case where $\mathbf{V}$ is positive definite, the linear system $\mathbf{VA} = \mathbf{B}$ has a unique solution, namely, $\mathbf{A} = \mathbf{V}^{-1}\mathbf{B}$, and that there is only one $n \times s$ matrix $\mathbf{A}_*$ such that $\mathbf{F}(\mathbf{A}) - \mathbf{F}(\mathbf{A}_*)$ is nonnegative definite for every $\mathbf{A} \in \mathcal{R}^{n \times s}$, namely, $\mathbf{A}_* = \mathbf{V}^{-1}\mathbf{B}$.

b. Least squares approach to derivation of main result

Let $\mathbf{a}$ represent an $n \times 1$ vector of (unconstrained) variables, and define $f(\mathbf{a}) = \mathbf{a}'\mathbf{Va} - 2\mathbf{b}'\mathbf{a}$, where $\mathbf{V}$ is an $n \times n$ symmetric nonnegative definite matrix and $\mathbf{b}$ is an $n \times 1$ vector such that $\mathbf{b} \in \mathcal{C}(\mathbf{V})$. Further, let $\mathbf{a}_*$ represent any solution to the (consistent) linear system $\mathbf{Va} = \mathbf{b}$. It was established in Subsection a (specifically in Theorem 19.1.1) that $f(\mathbf{a})$ attains its minimum value at a point $\mathbf{a}_*$ if and only if $\mathbf{a}_*$ is a solution to the (consistent) linear system $\mathbf{Va} = \mathbf{b}$, in which case $f(\mathbf{a}_*) = -\mathbf{b}'\mathbf{a}_*$. This result can be derived from the results of Section 12.4 (on the least squares problem), thereby providing an alternative to the proof given in Subsection a—actually, the proof given in Subsection a is for the more general result given by Theorem 19.1.2.

Let $\mathbf{R}$ represent a matrix such that $\mathbf{V} = \mathbf{R}'\mathbf{R}$—the existence of such a matrix is guaranteed by Corollary 14.3.8—let $\mathbf{t}$ represent an $n \times 1$ vector such that $\mathbf{b} = \mathbf{Vt}$, and let $\mathbf{y} = \mathbf{Rt}$. Then,

$$f(\mathbf{a}) = \mathbf{a}'\mathbf{Va} - 2\mathbf{t}'\mathbf{Va} = \mathbf{a}'\mathbf{R}'\mathbf{Ra} - 2\mathbf{y}'\mathbf{Ra} = (\mathbf{y} - \mathbf{Ra})'(\mathbf{y} - \mathbf{Ra}) - \mathbf{y}'\mathbf{y}.$$

Since $\mathbf{y}'\mathbf{y}$ does not depend on $\mathbf{a}$, the minimization of $f(\mathbf{a})$ is equivalent to the minimization of $(\mathbf{y} - \mathbf{Ra})'(\mathbf{y} - \mathbf{Ra})$. Moreover, since $\mathbf{R}'\mathbf{y} = \mathbf{R}'\mathbf{Rt} = \mathbf{Vt} = \mathbf{b}$, the linear system $\mathbf{Va} = \mathbf{b}$ is equivalent to the linear system $\mathbf{R}'\mathbf{Ra} = \mathbf{R}'\mathbf{y}$ of normal equations. Thus, it follows from Theorem 12.4.3 that $(\mathbf{y} - \mathbf{Ra})'(\mathbf{y} - \mathbf{Ra})$ attains its minimum value at a point $\mathbf{a}_*$ if and only if $\mathbf{a}_*$ is a solution to $\mathbf{Va} = \mathbf{b}$, in which case

$$(\mathbf{y} - \mathbf{Ra}_*)'(\mathbf{y} - \mathbf{Ra}_*) = \mathbf{y}'\mathbf{y} - \mathbf{y}'\mathbf{Ra}_* = \mathbf{y}'\mathbf{y} - \mathbf{b}'\mathbf{a}_*.$$

We conclude that $f(\mathbf{a})$ attains its minimum value at a point $\mathbf{a}_*$ if and only if $\mathbf{a}_*$ is a solution to $\mathbf{Va} = \mathbf{b}$, in which case

$$f(\mathbf{a}_*) = (\mathbf{y} - \mathbf{Ra}_*)'(\mathbf{y} - \mathbf{Ra}_*) - \mathbf{y}'\mathbf{y} = -\mathbf{b}'\mathbf{a}_*.$$

c. Use of convexity to establish main result

Let $\mathbf{a}$ represent an $n \times 1$ vector of (unconstrained) variables, and define $f(\mathbf{a}) = \mathbf{a}'\mathbf{Va} - 2\mathbf{b}'\mathbf{a}$, where $\mathbf{V}$ is an $n \times n$ symmetric nonnegative definite matrix and $\mathbf{b}$ is an $n \times 1$ vector such that $\mathbf{b} \in \mathcal{C}(\mathbf{V})$. Further, let $\mathbf{a}_*$ represent any solution to the (consistent) linear system $\mathbf{Va} = \mathbf{b}$. Then, $f(\mathbf{a})$ attains its minimum value at $\mathbf{a} = \mathbf{a}_*$, as was established in Subsection a and as was reconfirmed (via a second approach) in Subsection b. There is yet another way of verifying that $f(\mathbf{a})$ attains its minimum value at $\mathbf{a} = \mathbf{a}_*$.

Using the results of Section 15.3, we find that

$$\frac{\partial^2 f}{\partial \mathbf{a} \partial \mathbf{a}'} = 2\mathbf{V}.$$

Thus, the Hessian matrix of f is nonnegative definite (for all $\mathbf{a}$), and consequently f is a convex function—refer, for example, to Theorem 7.7 of Magnus and Neudecker (1988). And, recalling (from Subsection a) that $\mathbf{a}_*$ is a stationary point of f, it follows from a well-known result on convex functions [which is Theorem 7.8 of Magnus and Neudecker (1988)] that $f(\mathbf{a})$ attains its minimum value at $\mathbf{a}_*$.

19.2 Constrained Minimization

a. Basic results

Let $\mathbf{a} = (a_1, \ldots, a_n)'$ represent an $n \times 1$ vector of variables, and impose on $\mathbf{a}$ the constraint $\mathbf{X}'\mathbf{a} = \mathbf{d}$, where $\mathbf{X} = \{x_{ij}\}$ is an $n \times p$ matrix and $\mathbf{d} = \{d_j\}$ is a $p \times 1$ vector such that $\mathbf{d} \in \mathcal{C}(\mathbf{X}')$. Further, define (for $\mathbf{a} \in \mathcal{R}^n$) $f(\mathbf{a}) = \mathbf{a}'\mathbf{V}\mathbf{a} - 2\mathbf{b}'\mathbf{a}$, where $\mathbf{V}$ is an $n \times n$ symmetric nonnegative definite matrix and $\mathbf{b}$ is an $n \times 1$ vector such that $\mathbf{b} \in \mathcal{C}(\mathbf{V}, \mathbf{X})$. And consider the constrained (by $\mathbf{X}'\mathbf{a} = \mathbf{d}$) minimization of $f(\mathbf{a})$.

Note that the (vector-valued) constraint $\mathbf{X}'\mathbf{a} = \mathbf{d}$ can be rewritten as p scalar-valued constraints

$$\mathbf{x}_j'\mathbf{a} = d_j \qquad (j = 1, \ldots, p),$$

where $\mathbf{x}_j = (x_{1j}, \ldots, x_{nj})'$ is the jth column of $\mathbf{X}$, or in completely nonmatrix notation as

$$\sum_{i=1}^{n} x_{ij} a_i = d_j \qquad (j = 1, \ldots, p).$$

Letting $\mathbf{r} = \{r_j\}$ represent an arbitrary $p \times 1$ vector, the Lagrangian function for our constrained minimization problem can be expressed as

$$g(\mathbf{a}) = f(\mathbf{a}) - 2\mathbf{r}'(\mathbf{d} - \mathbf{X}'\mathbf{a})$$

or, equivalently, as

$$g(\mathbf{a}) = f(\mathbf{a}) + \sum_{j=1}^{p} (-2r_j)(d_j - \mathbf{x}_j'\mathbf{a}).$$

The p scalars $-2r_1, \ldots, -2r_p$ are the Lagrange multipliers. (Expressing the Lagrangian function so that the vector of Lagrange multipliers is represented by $-2\mathbf{r}$ rather than by $\mathbf{r}$ will be convenient in what follows.)

Suppose now that $\text{rank}(\mathbf{X}) = p$. Then, upon applying Lagrange's theorem [which is Theorem 7.10 of Magnus and Neudecker (1988)], we find that, for $f(\mathbf{a})$ to have a minimum at a point $\mathbf{a}_*$ under the constraint $\mathbf{X}'\mathbf{a} = \mathbf{d}$, it is necessary that there exist a $p \times 1$ vector $\mathbf{r}_*$ such that, for $\mathbf{r} = \mathbf{r}_*$, $\mathbf{a}_*$ is a stationary point of g (i.e., such that, for $\mathbf{r} = \mathbf{r}_*$, $\partial g / \partial \mathbf{a} = \mathbf{0}$ at $\mathbf{a} = \mathbf{a}_*$). Of course, if is also necessary that $\mathbf{a}_*$ satisfy the constraint (i.e., that $\mathbf{X}'\mathbf{a}_* = \mathbf{d}$).

Making use of result (1.1) and of formula (15.3.5), we find that

$$\frac{\partial g}{\partial \mathbf{a}} = \frac{\partial f}{\partial \mathbf{a}} + 2\frac{\partial(\mathbf{r}'\mathbf{X}'\mathbf{a})}{\partial \mathbf{a}} = 2(\mathbf{V}\mathbf{a} - \mathbf{b} + \mathbf{X}\mathbf{r}).$$

Thus, for f to have a minimum at a point $\mathbf{a}_*$ under the constraint $\mathbf{X}'\mathbf{a} = \mathbf{d}$, it is necessary that there exist a $p \times 1$ vector $\mathbf{r}_*$ such that $\mathbf{V}\mathbf{a}_* + \mathbf{X}\mathbf{r}_* = \mathbf{b}$ (and that $\mathbf{X}'\mathbf{a}_* = \mathbf{d}$) or, equivalently, it is necessary that there exist a $p \times 1$ vector $\mathbf{r}_*$ such that $\mathbf{a}_*$ and $\mathbf{r}_*$ are, respectively, the first and second parts of a solution to the linear system comprising the two equations

$$\mathbf{V}\mathbf{a} + \mathbf{X}\mathbf{r} = \mathbf{b}, \tag{2.1}$$

$$\mathbf{X}'\mathbf{a} = \mathbf{d} \tag{2.2}$$

(in $\mathbf{a}$ and $\mathbf{r}$). Note that equations (2.1) and (2.2) can be reformulated as

$$\begin{pmatrix} \mathbf{V} & \mathbf{X} \\ \mathbf{X}' & \mathbf{0} \end{pmatrix} \begin{pmatrix} \mathbf{a} \\ \mathbf{r} \end{pmatrix} = \begin{pmatrix} \mathbf{b} \\ \mathbf{d} \end{pmatrix}.$$

Are the conditions obtained by applying Lagrange's theorem sufficient as well as necessary for $f(\mathbf{a})$ to have a minimum at a point $\mathbf{a}_*$ under the constraint $\mathbf{X}'\mathbf{a} = \mathbf{d}$? And does their necessity extend to the degenerate case where $\text{rank}(\mathbf{X}) < p$? The following theorem answers these questions (in the affirmative).

Theorem 19.2.1. Let $\mathbf{a}$ represent an $n \times 1$ vector of variables; let $\mathbf{X}$ represent an $n \times p$ matrix and $\mathbf{d}$ a $p \times 1$ vector such that $\mathbf{d} \in \mathcal{C}(\mathbf{X}')$; and define $f(\mathbf{a}) = \mathbf{a}'\mathbf{V}\mathbf{a} - 2\mathbf{b}'\mathbf{a}$, where $\mathbf{V}$ is an $n \times n$ symmetric nonnegative definite matrix and $\mathbf{b}$ is an $n \times 1$ vector such that $\mathbf{b} \in \mathcal{C}(\mathbf{V}, \mathbf{X})$. Then, the linear system

$$\begin{pmatrix} \mathbf{V} & \mathbf{X} \\ \mathbf{X}' & \mathbf{0} \end{pmatrix} \begin{pmatrix} \mathbf{a} \\ \mathbf{r} \end{pmatrix} = \begin{pmatrix} \mathbf{b} \\ \mathbf{d} \end{pmatrix} \tag{2.3}$$

(in $\mathbf{a}$ and a $p \times 1$ vector $\mathbf{r}$) is consistent, and the values of $\mathbf{V}\mathbf{a}$ and $\mathbf{X}\mathbf{r}$ are invariant to the choice of solution to this linear system. Further, $f(\mathbf{a})$ attains its minimum value at a point $\mathbf{a}_*$ under the constraint $\mathbf{X}'\mathbf{a} = \mathbf{d}$ if and only if $\mathbf{a}_*$ is the first part of some solution to linear system (2.3). And, for the first and second parts $\mathbf{a}_*$ and $\mathbf{r}_*$ of any solution to linear system (2.3),

$$f(\mathbf{a}_*) = -\mathbf{b}'\mathbf{a}_* - \mathbf{d}'\mathbf{r}_* = -\mathbf{a}_*'\mathbf{b} - \mathbf{r}_*'\mathbf{d}. \tag{2.4}$$

Theorem 19.2.1 is a special case of the following result (that where $s = 1$).

Theorem 19.2.2. Let $\mathbf{A}$ represent an $n \times s$ matrix of variables; let $\mathbf{X}$ represent an $n \times p$ matrix and $\mathbf{D}$ a $p \times s$ matrix such that $\mathcal{C}(\mathbf{D}) \subset \mathcal{C}(\mathbf{X}')$; and define $\mathbf{F}(\mathbf{A}) = \mathbf{A}'\mathbf{V}\mathbf{A} - \mathbf{B}'\mathbf{A} - \mathbf{A}'\mathbf{B}$, where $\mathbf{V}$ is an $n \times n$ symmetric nonnegative definite matrix and $\mathbf{B}$ is an $n \times s$ matrix such that $\mathcal{C}(\mathbf{B}) \subset \mathcal{C}(\mathbf{V}, \mathbf{X})$. Then, the linear system

$$\begin{pmatrix} \mathbf{V} & \mathbf{X} \\ \mathbf{X}' & \mathbf{0} \end{pmatrix} \begin{pmatrix} \mathbf{A} \\ \mathbf{R} \end{pmatrix} = \begin{pmatrix} \mathbf{B} \\ \mathbf{D} \end{pmatrix} \tag{2.5}$$

(in $\mathbf{A}$ and a $p \times s$ matrix $\mathbf{R}$) is consistent, and the values of $\mathbf{V}\mathbf{A}$ and $\mathbf{X}\mathbf{R}$ are invariant to the choice of solution to this linear system. Further, an $n \times s$ matrix $\mathbf{A}_*$ satisfies the constraint $\mathbf{X}'\mathbf{A} = \mathbf{D}$ and is such that $\mathbf{F}(\mathbf{A}) - \mathbf{F}(\mathbf{A}_*)$ is nonnegative definite for all $\mathbf{A}$ satisfying $\mathbf{X}'\mathbf{A} = \mathbf{D}$ if and only if $\mathbf{A}_*$ is the first part of some solution to linear system (2.5). And, for the first and second parts $\mathbf{A}_*$ and $\mathbf{R}_*$ of any solution to linear system (2.5),

$$\mathbf{F}(\mathbf{A}_*) = -\mathbf{B}'\mathbf{A}_* - \mathbf{D}'\mathbf{R}_* = -\mathbf{A}_*'\mathbf{B} - \mathbf{R}_*'\mathbf{D}. \tag{2.6}$$

Proof (of Theorem 19.2.2). There exists a matrix $\mathbf{L}$ such that $\mathbf{V} = \mathbf{L}'\mathbf{L}$. And there exist matrices $\mathbf{T}$ and $\mathbf{U}$ such that $\mathbf{B} = \mathbf{VT} + \mathbf{XU}$ or, equivalently, such that $\mathbf{B} = \mathbf{L}'\mathbf{LT} + \mathbf{XU}$. Thus, the consistency of linear system (2.5) follows from Corollary 7.4.9.

Now, let $\mathbf{A}_*$ and $\mathbf{R}_*$ represent the first and second parts of any solution to linear system (2.5), and let $\mathbf{A}$ represent an arbitrary $n \times s$ matrix satisfying $\mathbf{X}'\mathbf{A} = \mathbf{D}$. Then, clearly, $\mathbf{X}'\mathbf{A}_* = \mathbf{D}$. And

$$\begin{aligned}(\mathbf{A} - \mathbf{A}_*)'\mathbf{VA}_* &= (\mathbf{A} - \mathbf{A}_*)'(\mathbf{B} - \mathbf{XR}_*)\\ &= (\mathbf{A} - \mathbf{A}_*)'\mathbf{B} - (\mathbf{X}'\mathbf{A})'\mathbf{R}_* + (\mathbf{X}'\mathbf{A}_*)'\mathbf{R}_*\\ &= (\mathbf{A} - \mathbf{A}_*)'\mathbf{B} - \mathbf{D}'\mathbf{R}_* + \mathbf{D}'\mathbf{R}_*\\ &= (\mathbf{A} - \mathbf{A}_*)'\mathbf{B},\end{aligned}$$

and consequently

$$\begin{aligned}\mathbf{A}'\mathbf{VA} &= [\mathbf{A}_* + (\mathbf{A} - \mathbf{A}_*)]'\mathbf{V}[\mathbf{A}_* + (\mathbf{A} - \mathbf{A}_*)]\\ &= \mathbf{A}_*'\mathbf{VA}_* + (\mathbf{A} - \mathbf{A}_*)'\mathbf{VA}_*\\ &\qquad + [(\mathbf{A} - \mathbf{A}_*)'\mathbf{VA}_*]' + (\mathbf{A} - \mathbf{A}_*)'\mathbf{V}(\mathbf{A} - \mathbf{A}_*)\\ &= \mathbf{A}_*'\mathbf{VA}_* + (\mathbf{A} - \mathbf{A}_*)'\mathbf{B} + \mathbf{B}'(\mathbf{A} - \mathbf{A}_*) + (\mathbf{A} - \mathbf{A}_*)'\mathbf{V}(\mathbf{A} - \mathbf{A}_*).\end{aligned}$$

Thus,

$$\begin{aligned}\mathbf{F}(\mathbf{A}) - \mathbf{F}(\mathbf{A}_*) &= \mathbf{A}'\mathbf{VA} - \mathbf{A}_*'\mathbf{VA}_* - \mathbf{B}'(\mathbf{A} - \mathbf{A}_*) - (\mathbf{A} - \mathbf{A}_*)'\mathbf{B}\\ &= (\mathbf{A} - \mathbf{A}_*)'\mathbf{V}(\mathbf{A} - \mathbf{A}_*),\end{aligned} \tag{2.7}$$

implying (in light of Theorem 14.2.9) that $\mathbf{F}(\mathbf{A}) - \mathbf{F}(\mathbf{A}_*)$ is nonnegative definite.

Further, let $\mathbf{A}_0$ represent any $n \times s$ matrix that satisfies the constraint $\mathbf{X}'\mathbf{A} = \mathbf{D}$ and is such that $\mathbf{F}(\mathbf{A}) - \mathbf{F}(\mathbf{A}_0)$ is nonnegative definite for all $\mathbf{A}$ satisfying $\mathbf{X}'\mathbf{A} = \mathbf{D}$. Then, $\mathbf{F}(\mathbf{A}_*) - \mathbf{F}(\mathbf{A}_0)$ is nonnegative definite, or, equivalently, $\mathbf{F}(\mathbf{A}_0) - \mathbf{F}(\mathbf{A}_*)$ {which equals $-[\mathbf{F}(\mathbf{A}_*) - \mathbf{F}(\mathbf{A}_0)]$} is nonpositive definite, and, since $\mathbf{F}(\mathbf{A}_0) - \mathbf{F}(\mathbf{A}_*)$ is nonnegative definite, it follows from Lemma 14.2.2 that $\mathbf{F}(\mathbf{A}_0) - \mathbf{F}(\mathbf{A}_*) = \mathbf{0}$ and hence [upon applying equality (2.7) with $\mathbf{A} = \mathbf{A}_0$] that $(\mathbf{A}_0 - \mathbf{A}_*)'\mathbf{V}(\mathbf{A}_0 - \mathbf{A}_*) = \mathbf{0}$. Thus, as a consequence of Corollary 14.3.11, $\mathbf{V}(\mathbf{A}_0 - \mathbf{A}_*) = \mathbf{0}$, or, equivalently,

$$\mathbf{VA}_0 = \mathbf{VA}_*, \tag{2.8}$$

implying that

$$\mathbf{VA}_0 + \mathbf{XR}_* = \mathbf{VA}_* + \mathbf{XR}_* = \mathbf{B}$$

and hence (since $\mathbf{X}'\mathbf{A}_0 = \mathbf{D}$) that $\mathbf{A}_0$ and $\mathbf{R}_*$ are the first and second parts of a solution to linear system (2.5).

Next, let $\mathbf{A}_1$ and $\mathbf{R}_1$ represent the first and second parts of a possibly different (from $\mathbf{A}_*$ and $\mathbf{R}_*$) solution to linear system (2.5). Then, $\mathbf{X}'\mathbf{A}_1 = \mathbf{D}$, and (in light of what has already been established) $\mathbf{F}(\mathbf{A}) - \mathbf{F}(\mathbf{A}_1)$ is nonnegative definite for every $\mathbf{A}$ satisfying $\mathbf{X}'\mathbf{A} = \mathbf{D}$. Thus, as a consequence of result (2.8), we have that

$$\mathbf{VA}_1 = \mathbf{VA}_*$$

and that

$$\mathbf{XR}_1 = \mathbf{B} - \mathbf{VA}_1 = \mathbf{B} - \mathbf{VA}_* = \mathbf{XR}_*,$$

implying that $\mathbf{VA}$ and $\mathbf{XR}$ are invariant to the choice of solution to linear system (2.5).

Finally, observe that

$$\begin{aligned} F(\mathbf{A}_*) = \mathbf{A}_*'\mathbf{VA}_* - \mathbf{B}'\mathbf{A}_* - \mathbf{A}_*'\mathbf{B} &= \mathbf{A}_*'(\mathbf{B} - \mathbf{XR}_*) - \mathbf{B}'\mathbf{A}_* - \mathbf{A}_*'\mathbf{B} \\ &= -(\mathbf{X}'\mathbf{A}_*)'\mathbf{R}_* - \mathbf{B}'\mathbf{A}_* \\ &= -\mathbf{D}'\mathbf{R}_* - \mathbf{B}'\mathbf{A}_* \end{aligned}$$

and, similarly, that

$$\begin{aligned} F(\mathbf{A}_*) = (\mathbf{VA}_*)'\mathbf{A}_* - \mathbf{B}'\mathbf{A}_* - \mathbf{A}_*'\mathbf{B} &= (\mathbf{B} - \mathbf{XR}_*)'\mathbf{A}_* - \mathbf{B}'\mathbf{A}_* - \mathbf{A}_*'\mathbf{B} \\ &= -\mathbf{R}_*'\mathbf{X}'\mathbf{A}_* - \mathbf{A}_*'\mathbf{B} \\ &= -\mathbf{R}_*'\mathbf{D} - \mathbf{A}_*'\mathbf{B}. \end{aligned}$$

Q.E.D.

b. Interpretation of the Lagrange multipliers (in the special case where X is of full column rank)

Let $\mathbf{a}$ represent an $n \times 1$ vector of variables; impose on $\mathbf{a}$ the constraint $\mathbf{X}'\mathbf{a} = \mathbf{d}$, where $\mathbf{X}$ is an $n \times p$ matrix and $\mathbf{d}$ is a $p \times 1$ vector such that $\mathbf{d} \in \mathcal{C}(\mathbf{X}')$; define $f(\mathbf{a}) = \mathbf{a}'\mathbf{Va} - 2\mathbf{b}'\mathbf{a}$, where $\mathbf{V}$ is an $n \times n$ symmetric nonnegative definite matrix and $\mathbf{b}$ is an $n \times 1$ vector such that $\mathbf{b} \in \mathcal{C}(\mathbf{V}, \mathbf{X})$; and consider the constrained (by $\mathbf{X}'\mathbf{a} = \mathbf{d}$) minimization of $f(\mathbf{a})$.

According to Theorem 19.2.1, $f(\mathbf{a})$ attains its minimum value at a point $\mathbf{a}_*$ (under the constraint $\mathbf{X}'\mathbf{a} = \mathbf{d}$) if and only if $\mathbf{a}_*$ is the first $(n \times 1)$ part of a solution to the linear system

$$\begin{pmatrix} \mathbf{V} & \mathbf{X} \\ \mathbf{X}' & \mathbf{0} \end{pmatrix} \begin{pmatrix} \mathbf{a} \\ \mathbf{r} \end{pmatrix} = \begin{pmatrix} \mathbf{b} \\ \mathbf{d} \end{pmatrix}. \tag{2.9}$$

What (if any) significance can be attributed to the second $(p \times 1)$ part of a solution to linear system (2.9)?

Suppose that $\text{rank}(\mathbf{X}) = p$ (i.e., that $\mathbf{X}$ is of full column rank). Then, the second part of the solution to linear system (2.9) is unique—if $\mathbf{r}_1$ and $\mathbf{r}_2$ are the second parts of two solutions to linear system (2.9), then (according to Theorem 19.2.1) $\mathbf{Xr}_1 = \mathbf{Xr}_2$, or, equivalently, $\mathbf{X}(\mathbf{r}_1 - \mathbf{r}_2) = \mathbf{0}$, implying (since, by supposition, the columns of $\mathbf{X}$ are linearly independent) that $\mathbf{r}_1 - \mathbf{r}_2 = \mathbf{0}$, or, equivalently, that $\mathbf{r}_1 = \mathbf{r}_2$. Further, defining (for $\mathbf{d} \in \mathcal{R}^{p \times 1}$)

$$h(\mathbf{d}) = \min_{\{\mathbf{a} \in \mathcal{R}^n : \mathbf{X}'\mathbf{a} = \mathbf{d}\}} f(\mathbf{a}),$$

it follows from well-known results on Lagrange multipliers [described, for example, by Magnus and Neudecker (1988, sec. 7.16)] that

$$\frac{\partial h}{\partial \mathbf{d}} = -2\mathbf{r}_*, \tag{2.10}$$

where $\mathbf{r}_*$ is the (unique) second part of the solution to linear system (2.9). Thus, the elements of the vector $-2\mathbf{r}_*$ indicate how h is affected by small changes in the elements of $\mathbf{d}$.

An alternative way of arriving at formula (2.10) is by direct differentiation. Let $\mathbf{G}$ represent a generalized inverse of $\begin{pmatrix} \mathbf{V} & \mathbf{X} \\ \mathbf{X}' & \mathbf{0} \end{pmatrix}$, and partition $\mathbf{G}$ as $\mathbf{G} = \begin{pmatrix} \mathbf{G}_{11} & \mathbf{G}_{12} \\ \mathbf{G}_{21} & \mathbf{G}_{22} \end{pmatrix}$ (where $\mathbf{G}_{11}$ is of dimensions $n \times n$). Then, $\mathbf{G}\begin{pmatrix} \mathbf{b} \\ \mathbf{d} \end{pmatrix} = \begin{pmatrix} \mathbf{G}_{11}\mathbf{b} + \mathbf{G}_{12}\mathbf{d} \\ \mathbf{G}_{21}\mathbf{b} + \mathbf{G}_{22}\mathbf{d} \end{pmatrix}$ is a solution to linear system (2.9). Moreover, since [in light of the symmetry of $\begin{pmatrix} \mathbf{V} & \mathbf{X} \\ \mathbf{X}' & \mathbf{0} \end{pmatrix}$] $\mathbf{G}'$ is a generalized inverse of $\begin{pmatrix} \mathbf{V} & \mathbf{X} \\ \mathbf{X}' & \mathbf{0} \end{pmatrix}$, $\mathbf{G}'\begin{pmatrix} \mathbf{b} \\ \mathbf{d} \end{pmatrix} = \begin{pmatrix} \mathbf{G}'_{11}\mathbf{b} + \mathbf{G}'_{21}\mathbf{d} \\ \mathbf{G}'_{12}\mathbf{b} + \mathbf{G}'_{22}\mathbf{d} \end{pmatrix}$ is also a solution to linear system (2.9). Further, as a consequence of result (2.4), we have that

$$\begin{aligned} h(\mathbf{d}) &= -\mathbf{b}'\mathbf{G}_{11}\mathbf{b} - \mathbf{d}'\mathbf{G}_{21}\mathbf{b} - \mathbf{b}'\mathbf{G}_{12}\mathbf{d} - \mathbf{d}'\mathbf{G}_{22}\mathbf{d} \\ &= -\mathbf{b}'\mathbf{G}_{11}\mathbf{b} - (\mathbf{G}_{21}\mathbf{b})'\mathbf{d} - (\mathbf{G}'_{12}\mathbf{b})'\mathbf{d} - \mathbf{d}'\mathbf{G}_{22}\mathbf{d}. \end{aligned} \tag{2.11}$$

And, using formulas (15.3.5) and (15.3.7) to differentiate expression (2.11), we find that

$$\begin{aligned} \frac{\partial h}{\partial \mathbf{d}} &= -\mathbf{G}_{21}\mathbf{b} - \mathbf{G}'_{12}\mathbf{b} - (\mathbf{G}_{22} + \mathbf{G}'_{22})\mathbf{d} \\ &= -(\mathbf{G}_{21}\mathbf{b} + \mathbf{G}_{22}\mathbf{d}) - (\mathbf{G}'_{12}\mathbf{b} + \mathbf{G}'_{22}\mathbf{d}) = -\mathbf{r}_* - \mathbf{r}_* = -2\mathbf{r}_*. \end{aligned}$$

c. Use of convexity to establish main result

Let $\mathbf{a}$ represent an $n \times 1$ vector of variables; impose on $\mathbf{a}$ the constraint $\mathbf{X}'\mathbf{a} = \mathbf{d}$, where $\mathbf{X}$ is an $n \times p$ matrix and $\mathbf{d}$ is a $p \times 1$ vector such that $\mathbf{d} \in \mathcal{C}(\mathbf{X}')$; and define $f(\mathbf{a}) = \mathbf{a}'\mathbf{V}\mathbf{a} - 2\mathbf{b}'\mathbf{a}$, where $\mathbf{V}$ is an $n \times n$ symmetric nonnegative definite matrix and $\mathbf{b}$ is an $n \times 1$ vector such that $\mathbf{b} \in \mathcal{C}(\mathbf{V}, \mathbf{X})$. Further, let $\mathbf{a}_*$ and $\mathbf{r}_*$ represent the first and second parts of any solution to the (consistent) linear system $\begin{pmatrix} \mathbf{V} & \mathbf{X} \\ \mathbf{X}' & \mathbf{0} \end{pmatrix}\begin{pmatrix} \mathbf{a} \\ \mathbf{r} \end{pmatrix} = \begin{pmatrix} \mathbf{b} \\ \mathbf{d} \end{pmatrix}$.

As discussed in Subsection a, the Lagrangian function for the problem of minimizing $f(\mathbf{a})$ under the constraint $\mathbf{X}'\mathbf{a} = \mathbf{d}$ is

$$g(\mathbf{a}) = f(\mathbf{a}) - 2\mathbf{r}'(\mathbf{d} - \mathbf{X}'\mathbf{a})$$

(where $-2\mathbf{r}$ is the vector of Lagrange multipliers). Using the results of Section 15.3, we find that

$$\frac{\partial^2 g}{\partial \mathbf{a} \partial \mathbf{a}'} = 2\mathbf{V}. \tag{2.12}$$

Thus, the Hessian matrix of g is nonnegative definite (for all $\mathbf{a}$), and, consequently, g is a convex function (regardless of the value of $\mathbf{r}$)—refer, for example, to Theorem 7.7 of Magnus and Neudecker (1988). And, recalling (from Subsection a) that (for $\mathbf{r} = \mathbf{r}_*$) $\partial g/\partial \mathbf{a} = \mathbf{0}$ at $\mathbf{a} = \mathbf{a}_*$ and that $\mathbf{X}'\mathbf{a}_* - \mathbf{d} = \mathbf{0}$, it follows from a well-known result on convex functions [which is Theorem 7.13 of Magnus and Neudecker (1988)] that $f(\mathbf{a})$ attains its minimum value (under the constraint $\mathbf{X}'\mathbf{a} = \mathbf{d}$) at $\mathbf{a} = \mathbf{a}_*$.

19.3 Explicit Expressions for Solutions to the Constrained Minimization Problem

In Section 19.2a (specifically, in Theorem 19.2.2), we considered a constrained "minimization" problem that includes as a special case the problem of minimizing a second-degree polynomial (in n variables) subject to linear constraints. And we found that this problem "reduces" to the problem of finding the first ($n \times s$) part of a solution to a (consistent) linear system of the form

$$\begin{pmatrix} \mathbf{V} & \mathbf{X} \\ \mathbf{X}' & \mathbf{0} \end{pmatrix} \begin{pmatrix} \mathbf{A} \\ \mathbf{R} \end{pmatrix} = \begin{pmatrix} \mathbf{B} \\ \mathbf{D} \end{pmatrix} \tag{3.1}$$

(in an $n \times s$ matrix $\mathbf{A}$ and a $p \times s$ matrix $\mathbf{R}$), where $\mathbf{V}$ is an $n \times n$ symmetric nonnegative definite matrix, $\mathbf{X}$ is an $n \times p$ matrix, $\mathbf{B}$ is an $n \times s$ matrix such that $\mathcal{C}(\mathbf{B}) \subset \mathcal{C}(\mathbf{V}, \mathbf{X})$, and $\mathbf{D}$ is a $p \times s$ matrix such that $\mathcal{C}(\mathbf{D}) \subset \mathcal{C}(\mathbf{X}')$. This linear system comprises the two equations

$$\mathbf{VA} + \mathbf{XR} = \mathbf{B}, \tag{3.2}$$

$$\mathbf{X}'\mathbf{A} = \mathbf{D}. \tag{3.3}$$

The objective in the present section is to obtain "explicit" expressions for solutions to linear system (3.1). Let us begin by attacking the important special case where $\mathbf{V}$ is positive definite and then the less restrictive special case where $\mathcal{C}(\mathbf{X}) \subset \mathcal{C}(\mathbf{V})$, before proceeding to the general case.

a. Special case: V positive definite

Suppose that the matrix $\mathbf{V}$ is positive definite and hence nonsingular. Then, linear system (3.1) is equivalent to the system comprising the two equations

$$\mathbf{A} = \mathbf{V}^{-1}\mathbf{B} - \mathbf{V}^{-1}\mathbf{XR}, \tag{3.4}$$

$$\mathbf{X}'\mathbf{V}^{-1}\mathbf{XR} = \mathbf{X}'\mathbf{V}^{-1}\mathbf{B} - \mathbf{D}. \tag{3.5}$$

To verify the equivalence of these two systems, suppose that $\mathbf{A}$ and $\mathbf{R}$ satisfy equation (3.1), or, equivalently, equations (3.2) and (3.3). Then, clearly, $\mathbf{A}$ and $\mathbf{R}$ satisfy equation (3.4) and, upon replacing $\mathbf{A}$ in equation (3.3) with the right side of equation (3.4), we find that

$$\mathbf{X}'\mathbf{V}^{-1}\mathbf{B} - \mathbf{X}'\mathbf{V}^{-1}\mathbf{X}\mathbf{R} = \mathbf{D}$$

or, equivalently, that $\mathbf{R}$ satisfies equation (3.5).

Conversely, suppose that $\mathbf{A}$ and $\mathbf{R}$ satisfy equations (3.4) and (3.5). Then, clearly, $\mathbf{A}$ and $\mathbf{R}$ satisfy equation (3.2). Further,

$$\begin{aligned}\mathbf{X}'\mathbf{A} = \mathbf{X}'(\mathbf{V}^{-1}\mathbf{B} - \mathbf{V}^{-1}\mathbf{X}\mathbf{R}) &= \mathbf{X}'\mathbf{V}^{-1}\mathbf{B} - \mathbf{X}'\mathbf{V}^{-1}\mathbf{X}\mathbf{R} \\ &= \mathbf{X}'\mathbf{V}^{-1}\mathbf{B} - (\mathbf{X}'\mathbf{V}^{-1}\mathbf{B} - \mathbf{D}) = \mathbf{D},\end{aligned}$$

so that $\mathbf{A}$ satisfies equation (3.3).

The equivalence of linear system (3.1) to the system comprising equations (3.4) and (3.5) suggests a "computational" scheme for solving this system. First, solve equation (3.5) for $\mathbf{R}$. Then, substitute this solution in expression (3.4) to obtain the value of $\mathbf{A}$. If the inverse of $\mathbf{V}$ is easy to obtain (as would be the case if, e.g., $\mathbf{V}$ were diagonal), then in effect this scheme reduces the problem of solving linear system (3.1) (which comprises $n + p$ equations in $n + p$ rows of unknowns) to one of solving a linear system comprising p equations in p rows of unknowns.

According to Corollary 14.11.3, $\text{rank}(\mathbf{X}'\mathbf{V}^{-1}\mathbf{X}) = \text{rank}(\mathbf{X})$. Thus, aside from the special case where $\text{rank}(\mathbf{X}) = p$ (i.e., where $\mathbf{X}$ is of full column rank), equation (3.5) has an infinite number of solutions (for $\mathbf{R}$). However, in light of Theorem 19.2.2 [and of the equivalence of linear system (3.1) and the system comprising equations (3.4) and (3.5)], the matrix $\mathbf{X}\mathbf{R}$ is invariant to the choice of solution. Further, the right side of equation (3.4) is invariant to the choice of solution (for $\mathbf{R}$), which implies that the first ($n \times s$) part of the solution to linear system (3.1) is unique. [The uniqueness of the first part of the solution to linear system (3.1) can also be seen to follow from the invariance of $\mathbf{V}\mathbf{A}$ to the choice of solution to linear system (3.1)—if $\mathbf{A}_1$ and $\mathbf{A}_2$ are the first parts of two solutions to linear system (3.1), then $\mathbf{V}\mathbf{A}_1 = \mathbf{V}\mathbf{A}_2$, implying that $\mathbf{A}_1 = \mathbf{V}^{-1}\mathbf{V}\mathbf{A}_1 = \mathbf{V}^{-1}\mathbf{V}\mathbf{A}_2 = \mathbf{A}_2$.]

b. Special case: $\mathcal{C}(\mathbf{X}) \subset \mathcal{C}(\mathbf{V})$

The results obtained in Subsection a [on the solution of linear system (3.1)], which are for the special case where $\mathbf{V}$ is positive definite, can be extended in a relatively straightforward way to the less restrictive special case where $\mathcal{C}(\mathbf{X}) \subset \mathcal{C}(\mathbf{V})$.

Suppose that $\mathcal{C}(\mathbf{X}) \subset \mathcal{C}(\mathbf{V})$. Then, the matrices $\mathbf{A}$ and $\mathbf{R}$ satisfy the two equations (3.2) and (3.3), which compose linear system (3.1), if and only if $\mathbf{R}$ satisfies the equation

$$\mathbf{X}'\mathbf{V}^{-}\mathbf{X}\mathbf{R} = \mathbf{X}'\mathbf{V}^{-}\mathbf{B} - \mathbf{D} \tag{3.6}$$

and

$$\mathbf{A} = \mathbf{V}^{-}\mathbf{B} - \mathbf{V}^{-}\mathbf{X}\mathbf{R} + (\mathbf{I} - \mathbf{V}^{-}\mathbf{V})\mathbf{L} \tag{3.7}$$

for some $(n \times s)$ matrix $\mathbf{L}$.

For purposes of verifying this, observe (in light of Lemma 9.3.5 and Lemma 4.2.5) that

$$\mathbf{V}\mathbf{V}^{-}\mathbf{X} = \mathbf{X} \tag{3.8}$$

$$\mathbf{X}'\mathbf{V}^{-}\mathbf{V} = \mathbf{X}'. \tag{3.9}$$

Observe also that $\mathbf{B} = \mathbf{V}\mathbf{E} + \mathbf{X}\mathbf{F}$ for some matrices $\mathbf{E}$ and $\mathbf{F}$ and hence that

$$\mathbf{V}\mathbf{V}^{-}\mathbf{B} = \mathbf{V}\mathbf{V}^{-}\mathbf{V}\mathbf{E} + \mathbf{V}\mathbf{V}^{-}\mathbf{X}\mathbf{F} = \mathbf{V}\mathbf{E} + \mathbf{X}\mathbf{F} = \mathbf{B}. \tag{3.10}$$

Now, suppose that $\mathbf{A}$ and $\mathbf{R}$ satisfy equations (3.2) and (3.3). Then, $\mathbf{V}\mathbf{A} = \mathbf{B} - \mathbf{X}\mathbf{R}$, and it follows from Theorem 11.2.4 that equality (3.7) holds for some $\mathbf{L}$. Moreover, upon replacing $\mathbf{A}$ in equation (3.3) with the right side of equality (3.7), we find [in light of result (3.9)] that

$$\mathbf{X}'\mathbf{V}^{-}\mathbf{B} - \mathbf{X}'\mathbf{V}^{-}\mathbf{X}\mathbf{R} = \mathbf{D}$$

or, equivalently, that $\mathbf{R}$ satisfies equation (3.6).

Conversely, suppose that $\mathbf{R}$ satisfies equation (3.6) and that equality (3.7) holds for some $\mathbf{L}$. Then, making use of results (3.8)–(3.10), we find that

$$\begin{aligned} \mathbf{V}\mathbf{A} + \mathbf{X}\mathbf{R} &= \mathbf{V}\mathbf{V}^{-}\mathbf{B} - \mathbf{V}\mathbf{V}^{-}\mathbf{X}\mathbf{R} + \mathbf{V}(\mathbf{I} - \mathbf{V}^{-}\mathbf{V})\mathbf{L} + \mathbf{X}\mathbf{R} \\ &= \mathbf{B} - \mathbf{X}\mathbf{R} + \mathbf{0} + \mathbf{X}\mathbf{R} = \mathbf{B} \end{aligned}$$

and

$$\begin{aligned} \mathbf{X}'\mathbf{A} &= \mathbf{X}'\mathbf{V}^{-}\mathbf{B} - \mathbf{X}'\mathbf{V}^{-}\mathbf{X}\mathbf{R} + \mathbf{X}'(\mathbf{I} - \mathbf{V}^{-}\mathbf{V})\mathbf{L} \\ &= \mathbf{X}'\mathbf{V}^{-}\mathbf{B} - \mathbf{X}'\mathbf{V}^{-}\mathbf{X}\mathbf{R} = \mathbf{X}'\mathbf{V}^{-}\mathbf{B} - (\mathbf{X}'\mathbf{V}^{-}\mathbf{B} - \mathbf{D}) = \mathbf{D}, \end{aligned}$$

so that $\mathbf{A}$ and $\mathbf{R}$ satisfy equations (3.2) and (3.3).

Our results suggest a "computational" scheme for solving linear system (3.1). First, solve equation (3.6) for $\mathbf{R}$. Then, substitute this solution in expression (3.7) to obtain a value for $\mathbf{A}$—the matrix $\mathbf{L}$ in expression (3.7) can be chosen arbitrarily; in particular, $\mathbf{L}$ can be set equal to $\mathbf{0}$.

We have that

$$\operatorname{rank}(\mathbf{X}'\mathbf{V}^{-}\mathbf{X}) = \operatorname{rank}(\mathbf{X}). \tag{3.11}$$

To see this, observe that $\mathbf{X} = \mathbf{V}\mathbf{K}$ for some matrix $\mathbf{K}$ and hence that $\mathbf{X}'\mathbf{V}^{-}\mathbf{X} = \mathbf{K}'\mathbf{V}\mathbf{V}^{-}\mathbf{V}\mathbf{K} = \mathbf{K}'\mathbf{V}\mathbf{K}$, implying [in light of Part (1) of Theorem 14.12.25] that

$$\mathbf{X}(\mathbf{X}'\mathbf{V}^{-}\mathbf{X})^{-}\mathbf{X}'\mathbf{V}^{-}\mathbf{X} = \mathbf{V}\mathbf{K}(\mathbf{K}'\mathbf{V}\mathbf{K})^{-}\mathbf{K}'\mathbf{V}\mathbf{K} = \mathbf{V}\mathbf{K} = \mathbf{X}. \tag{3.12}$$

And result (3.12) implies result (3.11), as is evident from the following lemma.

Lemma 19.3.1. Let $\mathbf{W}$ represent an $n \times n$ matrix and $\mathbf{X}$ an $n \times p$ matrix. Then, the following three conditions are equivalent:

(1) $\operatorname{rank}(\mathbf{X}'\mathbf{W}\mathbf{X}) = \operatorname{rank}(\mathbf{X})$;

(2) $\mathbf{X}(\mathbf{X}'\mathbf{W}\mathbf{X})^{-}\mathbf{X}'\mathbf{W}\mathbf{X} = \mathbf{X}$;

(3) $\mathbf{X}'\mathbf{W}\mathbf{X}(\mathbf{X}'\mathbf{W}\mathbf{X})^{-}\mathbf{X}' = \mathbf{X}'$.

Proof. If Condition (1) is satisfied, then (according to Corollary 4.4.7) $\mathcal{R}(\mathbf{X}'\mathbf{W}\mathbf{X}) = \mathcal{R}(\mathbf{X})$, implying (in light of Lemma 9.3.5) that Condition (2) is satisfied. Conversely, if Condition (2) is satisfied, then (according to Corollary 4.4.5) $\operatorname{rank}(\mathbf{X}'\mathbf{W}\mathbf{X}) \geq \operatorname{rank}(\mathbf{X})$, implying [since clearly $\operatorname{rank}(\mathbf{X}'\mathbf{W}\mathbf{X}) \leq \operatorname{rank}(\mathbf{X})$] that Condition (1) is satisfied. Thus, Conditions (1) and (2) are equivalent.

It can be shown in similar fashion that Conditions (1) and (3) are equivalent. Q.E.D.

It is clear from result (3.11) that, aside from the special case where $\operatorname{rank}(\mathbf{X}) = p$, equation (3.6) has an infinite number of solutions (for $\mathbf{R}$). However, since any solution to equation (3.6) is the second part of a solution to linear system (3.1), it follows from Theorem 19.2.2 that the value of $\mathbf{XR}$ is invariant to the choice of a solution.

Since a solution (for $\mathbf{R}$) to equation (3.6) can be obtained by premultiplying the right side of the equation by $(\mathbf{X}'\mathbf{V}^{-}\mathbf{X})^{-}$, we find [upon substituting this solution in expression (3.7)] that $\mathbf{A}$ is the first part of some solution to linear system (3.1) if and only if

$$\mathbf{A} = [\mathbf{V}^{-} - \mathbf{V}^{-}\mathbf{X}(\mathbf{X}'\mathbf{V}^{-}\mathbf{X})^{-}\mathbf{X}'\mathbf{V}^{-}]\mathbf{B} + \mathbf{V}^{-}\mathbf{X}(\mathbf{X}'\mathbf{V}^{-}\mathbf{X})^{-}\mathbf{D} + (\mathbf{I} - \mathbf{V}^{-}\mathbf{V})\mathbf{L}$$

for some matrix $\mathbf{L}$, or, equivalently, if and only if

$$\mathbf{A} = \mathbf{V}^{-}(\mathbf{I} - \mathbf{P}_{\mathbf{X},\mathbf{V}^{-}})\mathbf{B} + \mathbf{V}^{-}\mathbf{X}(\mathbf{X}'\mathbf{V}^{-}\mathbf{X})^{-}\mathbf{D} + (\mathbf{I} - \mathbf{V}^{-}\mathbf{V})\mathbf{L} \tag{3.13}$$

for some $\mathbf{L}$. Clearly, the first part of the solution to linear system (3.1) is unique if and only if $\mathbf{V}$ is nonsingular (which is the special case considered in Subsection a).

c. General case

The following, easily verifiable lemma provides a basis for extending the results of Subsection b on the solution of linear system (3.1) [which are for the special case where $\mathcal{C}(\mathbf{X}) \subset \mathcal{C}(\mathbf{V})$] to the general case.

Lemma 19.3.2. Let $\mathbf{V}$ represent an $m \times n$ matrix, $\mathbf{X}$ an $m \times p$ matrix, $\mathbf{Y}$ a $q \times n$ matrix, $\mathbf{B}$ an $m \times s$ matrix, $\mathbf{D}$ a $q \times s$ matrix, and $\mathbf{U}$ a $p \times q$ matrix. Then, $\mathbf{A}_*$ and $\mathbf{R}_*$ are, respectively, the first ($n \times s$) and second ($p \times s$) parts of a solution to the linear system

$$\begin{pmatrix} \mathbf{V} & \mathbf{X} \\ \mathbf{Y} & \mathbf{0} \end{pmatrix} \begin{pmatrix} \mathbf{A} \\ \mathbf{R} \end{pmatrix} = \begin{pmatrix} \mathbf{B} \\ \mathbf{D} \end{pmatrix}$$

(in an $n \times s$ matrix $\mathbf{A}$ and a $p \times s$ matrix $\mathbf{R}$) if and only if $\mathbf{A}_*$ and $\mathbf{R}_* - \mathbf{UD}$ are, respectively, the first ($n \times s$) and second ($p \times s$) parts of a solution to the linear system

$$\begin{pmatrix} \mathbf{V} + \mathbf{XUY} & \mathbf{X} \\ \mathbf{Y} & \mathbf{0} \end{pmatrix} \begin{pmatrix} \mathbf{A} \\ \mathbf{T} \end{pmatrix} = \begin{pmatrix} \mathbf{B} \\ \mathbf{D} \end{pmatrix}$$

(in an $n \times s$ matrix $\mathbf{A}$ and a $p \times s$ matrix $\mathbf{T}$).

The following theorem gives "explicit" expressions for the first and second parts of solutions to linear system (3.1).

Theorem 19.3.3. Let $\mathbf{V}$ represent an $n \times n$ symmetric nonnegative definite matrix, $\mathbf{X}$ an $n \times p$ matrix, $\mathbf{B}$ an $n \times s$ matrix such that $\mathcal{C}(\mathbf{B}) \subset \mathcal{C}(\mathbf{V}, \mathbf{X})$, and $\mathbf{D}$ a $p \times s$ matrix such that $\mathcal{C}(\mathbf{D}) \subset \mathcal{C}(\mathbf{X}')$. Further, let $\mathbf{U}$ represent any $p \times p$ matrix such that $\mathcal{C}(\mathbf{X}) \subset \mathcal{C}(\mathbf{V} + \mathbf{XUX}')$, and let $\mathbf{W}$ represent an arbitrary generalized inverse of $\mathbf{V} + \mathbf{XUX}'$. Then, $\mathbf{A}_*$ and $\mathbf{R}_*$ are, respectively, the first ($n \times s$) and second ($p \times s$) parts of a solution to the (consistent) linear system

$$\begin{pmatrix} \mathbf{V} & \mathbf{X} \\ \mathbf{X}' & \mathbf{0} \end{pmatrix} \begin{pmatrix} \mathbf{A} \\ \mathbf{R} \end{pmatrix} = \begin{pmatrix} \mathbf{B} \\ \mathbf{D} \end{pmatrix} \tag{3.14}$$

(in an $n \times s$ matrix $\mathbf{A}$ and a $p \times s$ matrix $\mathbf{R}$) if and only if

$$\mathbf{R}_* = \mathbf{T}_* + \mathbf{UD} \tag{3.15}$$

and

$$\mathbf{A}_* = \mathbf{WB} - \mathbf{WXT}_* + [\mathbf{I} - \mathbf{W}(\mathbf{V} + \mathbf{XUX}')]\mathbf{L} \tag{3.16}$$

for some solution $\mathbf{T}_*$ to the (consistent) linear system

$$\mathbf{X}'\mathbf{WXT} = \mathbf{X}'\mathbf{WB} - \mathbf{D} \tag{3.17}$$

(in a $p \times s$ matrix $\mathbf{T}$) and for some $n \times s$ matrix $\mathbf{L}$.

A natural question (in connection with Theorem 19.3.3) is whether a $p \times p$ matrix $\mathbf{U}$ such that $\mathcal{C}(\mathbf{X}) \subset \mathcal{C}(\mathbf{V} + \mathbf{XUX}')$ necessarily exists. The answer is yes. To see this, note that (according to Corollary 14.3.8) $\mathbf{V} = \mathbf{S}'\mathbf{S}$ for some matrix $\mathbf{S}$, implying (in light of Corollary 7.4.5) that, for any nonzero scalar k,

$$\mathcal{C}(\mathbf{X}) = \mathcal{C}(k\mathbf{X}) \subset \mathcal{C}(\mathbf{S}', k\mathbf{X}) = \mathcal{C}[(\mathbf{S}', k\mathbf{X})(\mathbf{S}', k\mathbf{X})'] = \mathcal{C}(\mathbf{V} + k^2\mathbf{XX}').$$

Thus, $\mathcal{C}(\mathbf{X}) \subset \mathcal{C}(\mathbf{V} + \mathbf{XUX}')$ for $\mathbf{U} = k^2\mathbf{I}$ and in particular for $\mathbf{U} = \mathbf{I}$.

In the special case where $\mathcal{C}(\mathbf{X}) \subset \mathcal{C}(\mathbf{V})$, the condition $\mathcal{C}(\mathbf{X}) \subset \mathcal{C}(\mathbf{V} + \mathbf{XUX}')$ can be satisfied by setting $\mathbf{U} = \mathbf{0}$. When $\mathbf{U} = \mathbf{0}$, formula (3.16) reduces to formula (3.7).

Preliminary to proving Theorem 19.3.3, it is convenient to establish the following lemma.

Lemma 19.3.4. Let $\mathbf{V}$ represent an $n \times n$ symmetric matrix, $\mathbf{X}$ an $n \times p$ matrix, and $\mathbf{U}$ a $p \times p$ matrix. Then, the following ten conditions are equivalent:

(1) $\mathcal{C}(\mathbf{X}) \subset \mathcal{C}(\mathbf{V} + \mathbf{XUX}')$;

(2) $\mathcal{C}(\mathbf{V} + \mathbf{XUX}') = \mathcal{C}(\mathbf{V}, \mathbf{X})$;

(3) $\text{rank}(\mathbf{V} + \mathbf{XUX}') = \text{rank}(\mathbf{V}, \mathbf{X})$;

(4) $\text{rank}(\mathbf{V} + \mathbf{XU}'\mathbf{X}') = \text{rank}(\mathbf{V}, \mathbf{X})$;

(5) $\mathcal{C}(\mathbf{V} + \mathbf{XU}'\mathbf{X}') = \mathcal{C}(\mathbf{V}, \mathbf{X})$;

(6) $\mathcal{C}(\mathbf{X}) \subset \mathcal{C}(\mathbf{V} + \mathbf{X}\mathbf{U}'\mathbf{X}')$;

(7) $\mathcal{R}(\mathbf{X}') \subset \mathcal{R}(\mathbf{V} + \mathbf{X}\mathbf{U}\mathbf{X}')$;

(8) $\mathcal{R}(\mathbf{X}') \subset \mathcal{R}(\mathbf{V} + \mathbf{X}\mathbf{U}'\mathbf{X}')$;

(9) $\mathbf{X} = (\mathbf{V} + \mathbf{X}\mathbf{U}\mathbf{X}')(\mathbf{V} + \mathbf{X}\mathbf{U}\mathbf{X}')^{-}\mathbf{X}$;

(10) $\mathbf{X}' = \mathbf{X}'(\mathbf{V} + \mathbf{X}\mathbf{U}\mathbf{X}')^{-}(\mathbf{V} + \mathbf{X}\mathbf{U}\mathbf{X}')$.

Proof (of Lemma 19.3.4). Clearly,

$$\mathbf{V} + \mathbf{X}\mathbf{U}\mathbf{X}' = (\mathbf{V}, \mathbf{X})\begin{pmatrix}\mathbf{I} \\ \mathbf{U}\mathbf{X}'\end{pmatrix},$$

so that

$$\mathcal{C}(\mathbf{V} + \mathbf{X}\mathbf{U}\mathbf{X}') \subset \mathcal{C}(\mathbf{V}, \mathbf{X}). \tag{3.18}$$

Now, if $\mathcal{C}(\mathbf{X}) \subset \mathcal{C}(\mathbf{V} + \mathbf{X}\mathbf{U}\mathbf{X}')$, then $\mathbf{X} = (\mathbf{V} + \mathbf{X}\mathbf{U}\mathbf{X}')\mathbf{L}$ for some matrix $\mathbf{L}$, in which case

$$\begin{aligned}\mathbf{V} = \mathbf{V} + \mathbf{X}\mathbf{U}\mathbf{X}' - \mathbf{X}\mathbf{U}\mathbf{X}' &= \mathbf{V} + \mathbf{X}\mathbf{U}\mathbf{X}' - (\mathbf{V} + \mathbf{X}\mathbf{U}\mathbf{X}')\mathbf{L}\mathbf{U}\mathbf{X}' \\ &= (\mathbf{V} + \mathbf{X}\mathbf{U}\mathbf{X}')(\mathbf{I} - \mathbf{L}\mathbf{U}\mathbf{X}')\end{aligned}$$

and consequently

$$(\mathbf{V}, \mathbf{X}) = (\mathbf{V} + \mathbf{X}\mathbf{U}\mathbf{X}')(\mathbf{I} - \mathbf{L}\mathbf{U}\mathbf{X}', \mathbf{L}),$$

implying that $\mathcal{C}(\mathbf{V}, \mathbf{X}) \subset \mathcal{C}(\mathbf{V} + \mathbf{X}\mathbf{U}\mathbf{X}')$ and hence [in light of result (3.18)] that $\mathcal{C}(\mathbf{V} + \mathbf{X}\mathbf{U}\mathbf{X}') = \mathcal{C}(\mathbf{V}, \mathbf{X})$. Thus, (1) $\Rightarrow$ (2), and, since clearly (2) $\Rightarrow$ (1), (1) $\Leftrightarrow$ (2).

Moreover, in light of Theorem 4.4.6, it follows from result (3.18) that (3) $\Rightarrow$ (2), and, since clearly (2) $\Rightarrow$ (3), (2) $\Leftrightarrow$ (3). We conclude that Conditions (1), (2), and (3) are equivalent. And, based on essentially the same line of reasoning (with $\mathbf{U}'$ in place of $\mathbf{U}$), we further conclude that Conditions (4), (5), and (6) are equivalent.

Further, observe that $\mathbf{V} + \mathbf{X}\mathbf{U}'\mathbf{X}' = (\mathbf{V} + \mathbf{X}\mathbf{U}\mathbf{X}')'$. Thus, $\text{rank}(\mathbf{V} + \mathbf{X}\mathbf{U}'\mathbf{X}') = \text{rank}(\mathbf{V} + \mathbf{X}\mathbf{U}\mathbf{X}')$, and consequently (4) $\Leftrightarrow$ (3). And, in light of Lemma 4.2.5, (7) $\Leftrightarrow$ (6), and (8) $\Leftrightarrow$ (1). Finally, in light of Lemma 9.3.5, (9) $\Leftrightarrow$ (1), and (10) $\Leftrightarrow$ (7). Q.E.D.

Proof (of Theorem 19.3.3). Observe (in light of Lemma 19.3.4) that

$$(\mathbf{V} + \mathbf{X}\mathbf{U}\mathbf{X}')\mathbf{W}\mathbf{X} = \mathbf{X} \tag{3.19}$$

and

$$\mathbf{X}'\mathbf{W}(\mathbf{V} + \mathbf{X}\mathbf{U}\mathbf{X}') = \mathbf{X}'. \tag{3.20}$$

Observe also that $\mathbf{B} = \mathbf{V}\mathbf{E} + \mathbf{X}\mathbf{F}$ for some matrices $\mathbf{E}$ and $\mathbf{F}$ [so that $\mathbf{B} = (\mathbf{V} + \mathbf{X}\mathbf{U}\mathbf{X}')\mathbf{E} + \mathbf{X}(\mathbf{F} - \mathbf{U}\mathbf{X}'\mathbf{E})$] and hence that

$$\begin{aligned}(\mathbf{V} + \mathbf{X}\mathbf{U}\mathbf{X}')\mathbf{W}\mathbf{B} &= (\mathbf{V} + \mathbf{X}\mathbf{U}\mathbf{X}')\mathbf{W}(\mathbf{V} + \mathbf{X}\mathbf{U}\mathbf{X}')\mathbf{E} \\ &\qquad + (\mathbf{V} + \mathbf{X}\mathbf{U}\mathbf{X}')\mathbf{W}\mathbf{X}(\mathbf{F} - \mathbf{U}\mathbf{X}'\mathbf{E}) \\ &= (\mathbf{V} + \mathbf{X}\mathbf{U}\mathbf{X}')\mathbf{E} + \mathbf{X}(\mathbf{F} - \mathbf{U}\mathbf{X}'\mathbf{E}) = \mathbf{B}.\end{aligned} \tag{3.21}$$

Now, suppose that $\mathbf{A}_*$ and $\mathbf{R}_*$ are the first and second parts of a solution to linear system (3.14), and let $\mathbf{T}_0 = \mathbf{R}_* - \mathbf{UD}$. Then, $\mathbf{R}_* = \mathbf{T}_0 + \mathbf{UD}$. And we have, as a consequence of Lemma 19.3.2, that

$$\begin{pmatrix} \mathbf{V} + \mathbf{XUX}' & \mathbf{X} \\ \mathbf{X}' & \mathbf{0} \end{pmatrix} \begin{pmatrix} \mathbf{A}_* \\ \mathbf{T}_0 \end{pmatrix} = \begin{pmatrix} \mathbf{B} \\ \mathbf{D} \end{pmatrix}$$

or, equivalently, that

$$\begin{aligned} (\mathbf{V} + \mathbf{XUX}')\mathbf{A}_* &= \mathbf{B} - \mathbf{XT}_0 \\ \mathbf{X}'\mathbf{A}_* &= \mathbf{D}. \end{aligned} \tag{3.22}$$

Thus, it follows from Theorem 11.2.4 that

$$\mathbf{A}_* = \mathbf{WB} - \mathbf{WXT}_0 + [\mathbf{I} - \mathbf{W}(\mathbf{V} + \mathbf{XUX}')]\mathbf{L} \tag{3.23}$$

for some matrix $\mathbf{L}$. Moreover, upon replacing $\mathbf{A}_*$ in equality (3.22) with expression (3.23), we find [in light of result (3.20)] that

$$\mathbf{X}'\mathbf{WB} - \mathbf{X}'\mathbf{WXT}_0 = \mathbf{D},$$

or, equivalently, that $\mathbf{X}'\mathbf{WXT}_0 = \mathbf{X}'\mathbf{WB} - \mathbf{D}$, so that $\mathbf{T}_0$ is a solution to linear system (3.17).

Conversely, suppose that $\mathbf{R}_* = \mathbf{T}_* + \mathbf{UD}$ and

$$\mathbf{A}_* = \mathbf{WB} - \mathbf{WXT}_* + [\mathbf{I} - \mathbf{W}(\mathbf{V} + \mathbf{XUX}')]\mathbf{L}$$

for some solution $\mathbf{T}_*$ to linear system (3.17) and for some matrix $\mathbf{L}$. Then, making use of results (3.19)–(3.21), we find that

$$\begin{aligned} (\mathbf{V} + \mathbf{XUX}')\mathbf{A}_* + \mathbf{XT}_* &= (\mathbf{V} + \mathbf{XUX}')\mathbf{WB} - (\mathbf{V} + \mathbf{XUX}')\mathbf{WXT}_* \\ &\qquad + (\mathbf{V} + \mathbf{XUX}')[\mathbf{I} - \mathbf{W}(\mathbf{V} + \mathbf{XUX}')]\mathbf{L} + \mathbf{XT}_* \\ &= \mathbf{B} - \mathbf{XT}_* + \mathbf{0} + \mathbf{XT}_* = \mathbf{B} \end{aligned}$$

and

$$\begin{aligned} \mathbf{X}'\mathbf{A}_* &= \mathbf{X}'\mathbf{WB} - \mathbf{X}'\mathbf{WXT}_* + \mathbf{X}'[\mathbf{I} - \mathbf{W}(\mathbf{V} + \mathbf{XUX}')]\mathbf{L} \\ &= \mathbf{X}'\mathbf{WB} - \mathbf{X}'\mathbf{WXT}_* = \mathbf{X}'\mathbf{WB} - (\mathbf{X}'\mathbf{WB} - \mathbf{D}) = \mathbf{D}, \end{aligned}$$

so that

$$\begin{pmatrix} \mathbf{V} + \mathbf{XUX}' & \mathbf{X} \\ \mathbf{X}' & \mathbf{0} \end{pmatrix} \begin{pmatrix} \mathbf{A}_* \\ \mathbf{T}_* \end{pmatrix} = \begin{pmatrix} \mathbf{B} \\ \mathbf{D} \end{pmatrix}.$$

Since $\mathbf{T}_* = \mathbf{R}_* - \mathbf{UD}$, it follows from Lemma 19.3.2 that $\mathbf{A}_*$ and $\mathbf{R}_*$ are the first and second parts of a solution to linear system (3.14).

Finally, note that the consistency of linear system (3.14) was established in Theorem 19.2.2 and that its consistency implies the consistency of linear system (3.17). Q.E.D.

Theorem 19.3.3 suggests a "computational" scheme for solving linear system (3.1) or, equivalently, linear system (3.14). First, obtain a solution $\mathbf{T}_*$ to linear system (3.17). Then, use expressions (3.16) and (3.15) to obtain the first and second parts of a solution to linear system (3.14)—the matrix $\mathbf{L}$ in expression (3.16) can be chosen arbitrarily; in particular, $\mathbf{L}$ can be set equal to $\mathbf{0}$.

Result (3.11), which is for the special case where $\mathcal{C}(\mathbf{X}) \subset \mathcal{C}(\mathbf{V})$, is generalized in the following theorem.

Theorem 19.3.5. Let $\mathbf{V}$ represent an $n \times n$ symmetric nonnegative definite matrix and $\mathbf{X}$ an $n \times p$ matrix. Further, let $\mathbf{U}$ represent any $p \times p$ matrix such that $\mathcal{C}(\mathbf{X}) \subset \mathcal{C}(\mathbf{V} + \mathbf{XUX}')$, and let $\mathbf{W}$ represent an arbitrary generalized inverse of $\mathbf{V} + \mathbf{XUX}'$. Then,

$$\text{rank}(\mathbf{X}'\mathbf{WX}) = \text{rank}(\mathbf{X}). \tag{3.24}$$

Proof. Observe (in light of Lemma 19.3.4) that $\mathcal{R}(\mathbf{X}') \subset \mathcal{R}(\mathbf{V} + \mathbf{XUX}')$ and hence (in light of Lemma 9.3.5) that

$$\mathbf{X}'\mathbf{W}(\mathbf{V} + \mathbf{XUX}') = \mathbf{X}'.$$

Now, let $\mathbf{Z} = \mathbf{I} - (\mathbf{X}')^-\mathbf{X}'$. Then,

$$\mathbf{X}'\mathbf{WVZ} = \mathbf{X}'\mathbf{WVZ} + \mathbf{X}'\mathbf{WXUX}'\mathbf{Z} = \mathbf{X}'\mathbf{W}(\mathbf{V} + \mathbf{XUX}')\mathbf{Z} = \mathbf{X}'\mathbf{Z} = \mathbf{0}.$$

Thus, since (as a consequence of Theorem 17.2.13 and Lemma 19.3.4)

$$\mathcal{C}(\mathbf{X},\ \mathbf{VZ}) = \mathcal{C}(\mathbf{X},\ \mathbf{V}) = \mathcal{C}(\mathbf{V} + \mathbf{XUX}'),$$

we find (in light of Corollary 4.2.4) that

$$\begin{aligned}\mathcal{C}(\mathbf{X}'\mathbf{WX}) &= \mathcal{C}(\mathbf{X}'\mathbf{WX},\ \mathbf{0})\\ &= \mathcal{C}(\mathbf{X}'\mathbf{WX},\ \mathbf{X}'\mathbf{WVZ})\\ &= \mathcal{C}[\mathbf{X}'\mathbf{W}(\mathbf{X},\ \mathbf{VZ})] = \mathcal{C}[\mathbf{X}'\mathbf{W}(\mathbf{V} + \mathbf{XUX}')] = \mathcal{C}(\mathbf{X}'),\end{aligned}$$

implying that $\text{rank}(\mathbf{X}'\mathbf{WX}) = \text{rank}(\mathbf{X})$. Q.E.D.

In light of Lemma 19.3.1, we have the following corollary of Theorem 19.3.5.

Corollary 19.3.6. Let $\mathbf{V}$ represent an $n \times n$ symmetric nonnegative definite matrix and $\mathbf{X}$ an $n \times p$ matrix. Further, let $\mathbf{U}$ represent any $p \times p$ matrix such that $\mathcal{C}(\mathbf{X}) \subset \mathcal{C}(\mathbf{V} + \mathbf{XUX}')$, and let $\mathbf{W}$ represent an arbitrary generalized inverse of $\mathbf{V} + \mathbf{XUX}'$. Then,

$$\mathbf{X}(\mathbf{X}'\mathbf{WX})^-\mathbf{X}'\mathbf{WX} = \mathbf{X}, \tag{3.25}$$

$$\mathbf{X}'\mathbf{WX}(\mathbf{X}'\mathbf{WX})^-\mathbf{X}' = \mathbf{X}'. \tag{3.26}$$

In connection with Theorem 19.3.3, it is clear from result (3.24) that, aside from the special case where $\text{rank}(\mathbf{X}) = p$, linear system (3.17) has an infinite number of solutions (for $\mathbf{T}$). However, since, for any solution $\mathbf{T}_*$ to linear system (3.17), $\mathbf{T}_* + \mathbf{UD}$ is the second part of a solution to linear system (3.14), it follows from

Theorem 19.2.2 that the value of $\mathbf{X}(\mathbf{T}+\mathbf{U}\mathbf{D})$ is invariant to the choice of solution and hence that $\mathbf{X}\mathbf{T}$ is invariant to the choice of solution—since $\mathcal{R}(\mathbf{X}'\mathbf{W}\mathbf{X}) = \mathcal{R}(\mathbf{X})$ (as is evident from Theorem 19.3.5 and Corollary 4.4.7), the invariance of $\mathbf{X}\mathbf{T}$ is also apparent from Theorem 11.10.1.

Since the matrix $(\mathbf{X}'\mathbf{W}\mathbf{X})^-(\mathbf{X}'\mathbf{W}\mathbf{B}-\mathbf{D})$ is one solution to linear system (3.17), we find [upon substituting this solution in expression (3.16)] that the $n \times s$ matrix $\mathbf{A}$ is the first part of some solution to linear system (3.14) if and only if

$$\mathbf{A} = [\mathbf{W} - \mathbf{W}\mathbf{X}(\mathbf{X}'\mathbf{W}\mathbf{X})^-\mathbf{X}'\mathbf{W}]\mathbf{B} + \mathbf{W}\mathbf{X}(\mathbf{X}'\mathbf{W}\mathbf{X})^-\mathbf{D} + [\mathbf{I} - \mathbf{W}(\mathbf{V}+\mathbf{X}\mathbf{U}\mathbf{X}')]\mathbf{L}$$

for some matrix $\mathbf{L}$, or, equivalently, if and only if

$$\mathbf{A} = \mathbf{W}(\mathbf{I} - \mathbf{P}_{\mathbf{X},\mathbf{W}})\mathbf{B} + \mathbf{W}\mathbf{X}(\mathbf{X}'\mathbf{W}\mathbf{X})^-\mathbf{D} + [\mathbf{I} - \mathbf{W}(\mathbf{V}+\mathbf{X}\mathbf{U}\mathbf{X}')]\mathbf{L} \tag{3.27}$$

for some $\mathbf{L}$. Clearly, the first part of the solution to linear system (3.14) is unique if and only if $\mathbf{V}+\mathbf{X}\mathbf{U}\mathbf{X}'$ is nonsingular or, equivalently (in light of Lemma 19.3.4), if and only if $\text{rank}(\mathbf{V}, \mathbf{X}) = n$.

The matrix $\mathbf{W}$ in Theorem 19.3.3 (and in Theorem 19.3.5 and Corollary 19.3.6) is by definition a generalized inverse of the matrix $\mathbf{V}+\mathbf{X}\mathbf{U}\mathbf{X}'$—recall that $\mathbf{V}$ is a symmetric nonnegative definite matrix and that $\mathbf{U}$ is any matrix such that $\mathcal{C}(\mathbf{X}) \subset \mathcal{C}(\mathbf{V}+\mathbf{X}\mathbf{U}\mathbf{X}')$. There is no requirement that $\mathbf{W}$ be a generalized inverse of $\mathbf{V}$. Nevertheless, $\mathbf{U}$ can be chosen in such a way that $\mathbf{W}$ will be a generalized inverse of $\mathbf{V}$ as well as of $\mathbf{V}+\mathbf{X}\mathbf{U}\mathbf{X}'$.

To verify this, it suffices (in light of Theorem 18.2.5) to show that $\mathbf{U}$ can be chosen in such a way that $\mathcal{C}(\mathbf{V})$ and $\mathcal{C}(\mathbf{X}\mathbf{U}\mathbf{X}')$ are essentially disjoint and $\mathcal{R}(\mathbf{V})$ and $\mathcal{R}(\mathbf{X}\mathbf{U}\mathbf{X}')$ are essentially disjoint [and also, of course, in such a way that $\mathcal{C}(\mathbf{X}) \subset \mathcal{C}(\mathbf{V}+\mathbf{X}\mathbf{U}\mathbf{X}')$]. For purposes of constructing such a $\mathbf{U}$, let $\mathbf{R}$ represent any matrix such that $\mathbf{V} = \mathbf{R}\mathbf{R}'$—the existence of which follows from Corollary 14.3.8—let $\mathbf{T}$ represent any matrix whose columns span the null space of $\mathbf{V}$, and observe (in light of Corollaries 7.4.5, 4.5.6, and 17.2.14) that

$$\begin{aligned}\mathcal{C}(\mathbf{V}, \mathbf{X}) = \mathcal{C}(\mathbf{V}, \mathbf{X}\mathbf{X}') &= \mathcal{C}(\mathbf{V}, \mathbf{X}\mathbf{X}'\mathbf{T}) \\ &= \mathcal{C}(\mathbf{R}, \mathbf{X}\mathbf{X}'\mathbf{T}) \\ &= \mathcal{C}[(\mathbf{R}, \mathbf{X}\mathbf{X}'\mathbf{T})(\mathbf{R}, \mathbf{X}\mathbf{X}'\mathbf{T})'] \\ &= \mathcal{C}(\mathbf{V}+\mathbf{X}\mathbf{X}'\mathbf{T}\mathbf{T}'\mathbf{X}\mathbf{X}').\end{aligned}$$

Now, set $\mathbf{U} = \mathbf{X}'\mathbf{T}\mathbf{T}'\mathbf{X}$. Then, $\mathcal{C}(\mathbf{V}+\mathbf{X}\mathbf{U}\mathbf{X}') = \mathcal{C}(\mathbf{V}, \mathbf{X})$, or, equivalently (in light of Lemma 19.3.4), $\mathcal{C}(\mathbf{X}) \subset \mathcal{C}(\mathbf{V}+\mathbf{X}\mathbf{U}\mathbf{X}')$. Moreover, $\mathcal{C}(\mathbf{X}\mathbf{U}\mathbf{X}') = \mathcal{C}[\mathbf{X}\mathbf{X}'\mathbf{T}(\mathbf{X}\mathbf{X}'\mathbf{T})'] = \mathcal{C}(\mathbf{X}\mathbf{X}'\mathbf{T})$, implying (in light of Corollary 17.2.14) that $\mathcal{C}(\mathbf{V})$ and $\mathcal{C}(\mathbf{X}\mathbf{U}\mathbf{X}')$ are essentially disjoint. And, since clearly $\mathbf{X}\mathbf{U}\mathbf{X}'$ is symmetric, it follows from Lemma 17.2.1 that $\mathcal{R}(\mathbf{V})$ and $\mathcal{R}(\mathbf{X}\mathbf{U}\mathbf{X}')$ are also essentially disjoint.

19.4 Some Results on Generalized Inverses of Partitioned Matrices

In Subsection a, some expressions are obtained for generalized inverses of partitioned matrices of the form $\begin{pmatrix} \mathbf{V} & \mathbf{X} \\ \mathbf{X}' & \mathbf{0} \end{pmatrix}$ (where $\mathbf{V}$ is symmetric and nonnegative definite), and some basic properties of such generalized inverses are described. Actually, some of these results are applicable to generalized inverses of partitioned matrices of the more general form $\begin{pmatrix} \mathbf{V} & \mathbf{X} \\ \mathbf{Y} & \mathbf{0} \end{pmatrix}$ (where $\mathbf{Y}$ is not necessarily equal to $\mathbf{X}'$) and do not require that $\mathbf{V}$ be symmetric or nonnegative definite (or even square). Then, in Subsection b, an alternative derivation of the results of Section 19.3c [on the solution of linear system (3.1)] is devised by applying the results of Subsection a.

a. Some expressions and basic properties

It is easy to verify the following lemma.

Lemma 19.4.1. Let $\mathbf{V}$ represent an $m \times n$ matrix, $\mathbf{X}$ an $m \times p$ matrix, and $\mathbf{Y}$ a $q \times n$ matrix. And let $\mathbf{G}_{11}$ represent an $n \times m$ matrix, $\mathbf{G}_{12}$ an $n \times q$ matrix, $\mathbf{G}_{21}$ a $p \times m$ matrix, and $\mathbf{G}_{22}$ a $p \times q$ matrix. Then,

$$\begin{pmatrix} \mathbf{G}_{11} & \mathbf{G}_{12} \\ \mathbf{G}_{21} & \mathbf{G}_{22} \end{pmatrix} \begin{pmatrix} \mathbf{V} & \mathbf{X} \\ \mathbf{Y} & \mathbf{0} \end{pmatrix} = \begin{pmatrix} \mathbf{G}_{11}\mathbf{V}+\mathbf{G}_{12}\mathbf{Y} & \mathbf{G}_{11}\mathbf{X} \\ \mathbf{G}_{21}\mathbf{V}+\mathbf{G}_{22}\mathbf{Y} & \mathbf{G}_{21}\mathbf{X} \end{pmatrix}, \tag{4.1}$$

$$\begin{pmatrix} \mathbf{V} & \mathbf{X} \\ \mathbf{Y} & \mathbf{0} \end{pmatrix} \begin{pmatrix} \mathbf{G}_{11} & \mathbf{G}_{12} \\ \mathbf{G}_{21} & \mathbf{G}_{22} \end{pmatrix} = \begin{pmatrix} \mathbf{V}\mathbf{G}_{11}+\mathbf{X}\mathbf{G}_{21} & \mathbf{V}\mathbf{G}_{12}+\mathbf{X}\mathbf{G}_{22} \\ \mathbf{Y}\mathbf{G}_{11} & \mathbf{Y}\mathbf{G}_{12} \end{pmatrix}. \tag{4.2}$$

Further, $\begin{pmatrix} \mathbf{G}_{11} & \mathbf{G}_{12} \\ \mathbf{G}_{21} & \mathbf{G}_{22} \end{pmatrix}$ is a generalized inverse of the partitioned matrix $\begin{pmatrix} \mathbf{V} & \mathbf{X} \\ \mathbf{Y} & \mathbf{0} \end{pmatrix}$ if and only if

$$\mathbf{V}(\mathbf{G}_{11}\mathbf{V} + \mathbf{G}_{12}\mathbf{Y}) + \mathbf{X}(\mathbf{G}_{21}\mathbf{V} + \mathbf{G}_{22}\mathbf{Y}) = \mathbf{V}$$
$$[\text{or, equivalently, } (\mathbf{V}\mathbf{G}_{11} + \mathbf{X}\mathbf{G}_{21})\mathbf{V} + (\mathbf{V}\mathbf{G}_{12} + \mathbf{X}\mathbf{G}_{22})\mathbf{Y} = \mathbf{V}], \tag{4.3a}$$

$$(\mathbf{V}\mathbf{G}_{11} + \mathbf{X}\mathbf{G}_{21})\mathbf{X} = \mathbf{X}, \tag{4.3b}$$

$$\mathbf{Y}(\mathbf{G}_{11}\mathbf{V} + \mathbf{G}_{12}\mathbf{Y}) = \mathbf{Y}, \quad \text{and} \tag{4.3c}$$

$$\mathbf{Y}\mathbf{G}_{11}\mathbf{X} = \mathbf{0}. \tag{4.3d}$$

Some basic properties of generalized inverses of matrices of the form $\begin{pmatrix} \mathbf{V} & \mathbf{X} \\ \mathbf{X}' & \mathbf{0} \end{pmatrix}$ (where $\mathbf{V}$ is symmetric and nonnegative definite) are established in the following theorem.

Theorem 19.4.2. Let $\mathbf{V}$ represent an $n \times n$ symmetric nonnegative definite matrix and $\mathbf{X}$ an $n \times p$ matrix. And let $\mathbf{G} = \begin{pmatrix} \mathbf{G}_{11} & \mathbf{G}_{12} \\ \mathbf{G}_{21} & \mathbf{G}_{22} \end{pmatrix}$ (where $\mathbf{G}_{11}$ is of dimensions $n \times n$) represent an arbitrary generalized inverse of the partitioned matrix $\begin{pmatrix} \mathbf{V} & \mathbf{X} \\ \mathbf{X}' & \mathbf{0} \end{pmatrix}$. Then,

(1) $\mathbf{X}'\mathbf{G}_{12}\mathbf{X}' = \mathbf{X}'$ and $\mathbf{X}\mathbf{G}_{21}\mathbf{X} = \mathbf{X}$ (i.e., $\mathbf{G}_{12}$ is a generalized inverse of $\mathbf{X}'$ and $\mathbf{G}_{21}$ a generalized inverse of $\mathbf{X}$);

(2) $\mathbf{V}\mathbf{G}_{12}\mathbf{X}' = \mathbf{X}\mathbf{G}_{21}\mathbf{V} = -\mathbf{X}\mathbf{G}_{22}\mathbf{X}'$;

(3) $\mathbf{V}\mathbf{G}_{11}\mathbf{X} = \mathbf{0}$, $\mathbf{X}'\mathbf{G}_{11}\mathbf{V} = \mathbf{0}$, and $\mathbf{X}'\mathbf{G}_{11}\mathbf{X} = \mathbf{0}$;

(4) $\mathbf{V} = \mathbf{V}\mathbf{G}_{11}\mathbf{V} - \mathbf{X}\mathbf{G}_{22}\mathbf{X}'$; and

(5) $\mathbf{V}\mathbf{G}_{11}\mathbf{V}$, $\mathbf{V}\mathbf{G}_{12}\mathbf{X}'$, $\mathbf{X}\mathbf{G}_{21}\mathbf{V}$, and $\mathbf{X}\mathbf{G}_{22}\mathbf{X}'$ are symmetric and are invariant to the choice of the generalized inverse $\mathbf{G}$.

Proof. (1)–(2) Consider the linear system

$$\begin{pmatrix} \mathbf{V} & \mathbf{X} \\ \mathbf{X}' & \mathbf{0} \end{pmatrix} \begin{pmatrix} \mathbf{A} \\ \mathbf{R} \end{pmatrix} = \begin{pmatrix} \mathbf{0} \\ \mathbf{X}' \end{pmatrix}$$

(in an $n \times n$ matrix $\mathbf{A}$ and a $p \times n$ matrix $\mathbf{R}$). Since (according to Corollary 14.3.8) $\mathbf{V} = \mathbf{K}'\mathbf{K}$ for some matrix $\mathbf{K}$, it follows from Theorem 7.4.8 that this linear system is consistent. Thus,

$$\begin{pmatrix} \mathbf{V} & \mathbf{X} \\ \mathbf{X}' & \mathbf{0} \end{pmatrix} \begin{pmatrix} \mathbf{G}_{11} & \mathbf{G}_{12} \\ \mathbf{G}_{21} & \mathbf{G}_{22} \end{pmatrix} \begin{pmatrix} \mathbf{0} \\ \mathbf{X}' \end{pmatrix} = \begin{pmatrix} \mathbf{0} \\ \mathbf{X}' \end{pmatrix}, \tag{4.4}$$

and also [since, in light of the symmetry of $\begin{pmatrix} \mathbf{V} & \mathbf{X} \\ \mathbf{X}' & \mathbf{0} \end{pmatrix}$, $\mathbf{G}'$ is a generalized inverse of $\begin{pmatrix} \mathbf{V} & \mathbf{X} \\ \mathbf{X}' & \mathbf{0} \end{pmatrix}$]

$$\begin{pmatrix} \mathbf{V} & \mathbf{X} \\ \mathbf{X}' & \mathbf{0} \end{pmatrix} \begin{pmatrix} \mathbf{G}_{11}' & \mathbf{G}_{21}' \\ \mathbf{G}_{12}' & \mathbf{G}_{22}' \end{pmatrix} \begin{pmatrix} \mathbf{0} \\ \mathbf{X}' \end{pmatrix} = \begin{pmatrix} \mathbf{0} \\ \mathbf{X}' \end{pmatrix}. \tag{4.5}$$

Moreover,

$$\begin{pmatrix} \mathbf{V} & \mathbf{X} \\ \mathbf{X}' & \mathbf{0} \end{pmatrix} \begin{pmatrix} \mathbf{G}_{11} & \mathbf{G}_{12} \\ \mathbf{G}_{21} & \mathbf{G}_{22} \end{pmatrix} \begin{pmatrix} \mathbf{0} \\ \mathbf{X}' \end{pmatrix} = \begin{pmatrix} \mathbf{V}\mathbf{G}_{12}\mathbf{X}' + \mathbf{X}\mathbf{G}_{22}\mathbf{X}' \\ \mathbf{X}'\mathbf{G}_{12}\mathbf{X}' \end{pmatrix}, \tag{4.6}$$

$$\begin{pmatrix} \mathbf{V} & \mathbf{X} \\ \mathbf{X}' & \mathbf{0} \end{pmatrix} \begin{pmatrix} \mathbf{G}_{11}' & \mathbf{G}_{21}' \\ \mathbf{G}_{12}' & \mathbf{G}_{22}' \end{pmatrix} \begin{pmatrix} \mathbf{0} \\ \mathbf{X}' \end{pmatrix} = \begin{pmatrix} \mathbf{V}\mathbf{G}_{21}'\mathbf{X}' + \mathbf{X}\mathbf{G}_{22}'\mathbf{X}' \\ \mathbf{X}'\mathbf{G}_{21}'\mathbf{X}' \end{pmatrix} \tag{4.7}$$

[as can be easily verified by taking advantage of result (4.2)]. Clearly, results (4.6) and (4.4) imply that $\mathbf{V}\mathbf{G}_{12}\mathbf{X}' + \mathbf{X}\mathbf{G}_{22}\mathbf{X}' = \mathbf{0}$, or, equivalently, that $\mathbf{V}\mathbf{G}_{12}\mathbf{X}' = -\mathbf{X}\mathbf{G}_{22}\mathbf{X}'$, and that $\mathbf{X}'\mathbf{G}_{12}\mathbf{X}' = \mathbf{X}'$. And results (4.7) and (4.5) imply that $\mathbf{V}\mathbf{G}_{21}'\mathbf{X}' + \mathbf{X}\mathbf{G}_{22}'\mathbf{X}' = \mathbf{0}$ and $\mathbf{X}'\mathbf{G}_{21}'\mathbf{X}' = \mathbf{X}'$, and hence that $\mathbf{X}\mathbf{G}_{21}\mathbf{V} = (\mathbf{V}\mathbf{G}_{21}'\mathbf{X}')' = (-\mathbf{X}\mathbf{G}_{22}'\mathbf{X}')' = -\mathbf{X}\mathbf{G}_{22}\mathbf{X}'$ and $\mathbf{X}\mathbf{G}_{21}\mathbf{X} = (\mathbf{X}'\mathbf{G}_{21}'\mathbf{X}')' = (\mathbf{X}')' = \mathbf{X}$.

(3) That $\mathbf{X}'\mathbf{G}_{11}\mathbf{X} = \mathbf{0}$ is evident from Lemma 19.4.1. And, as a further consequence of Lemma 19.4.1, we have that $(\mathbf{V}\mathbf{G}_{11} + \mathbf{X}\mathbf{G}_{21})\mathbf{X} = \mathbf{X}$ and $\mathbf{X}'(\mathbf{G}_{11}\mathbf{V} + \mathbf{G}_{12}\mathbf{X}') = \mathbf{X}'$, implying [in light of Part (1)] that $\mathbf{V}\mathbf{G}_{11}\mathbf{X} = \mathbf{X} - \mathbf{X}\mathbf{G}_{21}\mathbf{X} = \mathbf{0}$ and $\mathbf{X}'\mathbf{G}_{11}\mathbf{V} = \mathbf{X}' - \mathbf{X}'\mathbf{G}_{12}\mathbf{X}' = \mathbf{0}$.

(4) It follows from Lemma 19.4.1 that

$$\mathbf{V} = \mathbf{V}\mathbf{G}_{11}\mathbf{V} + \mathbf{V}\mathbf{G}_{12}\mathbf{X}' + \mathbf{X}\mathbf{G}_{21}\mathbf{V} + \mathbf{X}\mathbf{G}_{22}\mathbf{X}'.$$

And, upon applying Part (2), we find that

$$\mathbf{V} = \mathbf{V}\mathbf{G}_{11}\mathbf{V} - \mathbf{X}\mathbf{G}_{22}\mathbf{X}' - \mathbf{X}\mathbf{G}_{22}\mathbf{X}' + \mathbf{X}\mathbf{G}_{22}\mathbf{X}' = \mathbf{V}\mathbf{G}_{11}\mathbf{V} - \mathbf{X}\mathbf{G}_{22}\mathbf{X}'.$$

(5) Let $\mathbf{G}^* = \begin{pmatrix} \mathbf{G}_{11}^* & \mathbf{G}_{12}^* \\ \mathbf{G}_{21}^* & \mathbf{G}_{22}^* \end{pmatrix}$ represent an arbitrary generalized inverse of $\begin{pmatrix} \mathbf{V} & \mathbf{X} \\ \mathbf{X}' & \mathbf{0} \end{pmatrix}$ (where $\mathbf{G}_{11}^*$ is of dimensions $n \times n$ and where $\mathbf{G}^*$ may differ from $\mathbf{G}$). Then, making use of Parts (3) and (4), we find that

$$\begin{aligned} \mathbf{V}\mathbf{G}_{11}^*\mathbf{V} &= \mathbf{V}\mathbf{G}_{11}^*(\mathbf{V}\mathbf{G}_{11}\mathbf{V} - \mathbf{X}\mathbf{G}_{22}\mathbf{X}') \\ &= \mathbf{V}\mathbf{G}_{11}^*\mathbf{V}\mathbf{G}_{11}\mathbf{V} - \mathbf{V}\mathbf{G}_{11}^*\mathbf{X}\mathbf{G}_{22}\mathbf{X}' \\ &= \mathbf{V}\mathbf{G}_{11}^*\mathbf{V}\mathbf{G}_{11}\mathbf{V} \quad (\text{since } \mathbf{V}\mathbf{G}_{11}^*\mathbf{X} = \mathbf{0}) \\ &= \mathbf{V}\mathbf{G}_{11}^*\mathbf{V}\mathbf{G}_{11}\mathbf{V} - \mathbf{X}\mathbf{G}_{22}^*\mathbf{X}'\mathbf{G}_{11}\mathbf{V} \quad (\text{since } \mathbf{X}'\mathbf{G}_{11}\mathbf{V} = \mathbf{0}) \\ &= (\mathbf{V}\mathbf{G}_{11}^*\mathbf{V} - \mathbf{X}\mathbf{G}_{22}^*\mathbf{X}')\mathbf{G}_{11}\mathbf{V} \\ &= \mathbf{V}\mathbf{G}_{11}\mathbf{V}. \end{aligned}$$

And, upon setting $\mathbf{G}^* = \mathbf{G}'$, we find, in particular, that $\mathbf{V}\mathbf{G}_{11}'\mathbf{V} = \mathbf{V}\mathbf{G}_{11}\mathbf{V}$ or, equivalently, that $(\mathbf{V}\mathbf{G}_{11}\mathbf{V})' = \mathbf{V}\mathbf{G}_{11}\mathbf{V}$. Thus, $\mathbf{V}\mathbf{G}_{11}\mathbf{V}$ is symmetric and is invariant to the choice of the generalized inverse $\mathbf{G}$.

Further, since [as a consequence of Part (4)] $\mathbf{X}\mathbf{G}_{22}\mathbf{X}' = \mathbf{V}\mathbf{G}_{11}\mathbf{V} - \mathbf{V}$, $\mathbf{X}\mathbf{G}_{22}\mathbf{X}'$ is symmetric and is invariant to the choice of $\mathbf{G}$, and, in light of Part (2), $\mathbf{V}\mathbf{G}_{12}\mathbf{X}'$ and $\mathbf{X}\mathbf{G}_{21}\mathbf{V}$ are symmetric and are invariant to the choice of $\mathbf{G}$. Q.E.D.

Let $\mathbf{V}$ represent an $n \times n$ symmetric nonnegative definite matrix and $\mathbf{X}$ an $n \times p$ matrix. In the special case where $\mathcal{C}(\mathbf{X}) \subset \mathcal{C}(\mathbf{V})$, Theorem 9.6.1 can be used to obtain a generalized inverse for the partitioned matrix $\begin{pmatrix} \mathbf{V} & \mathbf{X} \\ \mathbf{X}' & \mathbf{0} \end{pmatrix}$ in terms of $\mathbf{V}^-$ and $(\mathbf{X}'\mathbf{V}^-\mathbf{X})^-$. Upon applying formula (9.6.2) [and observing that $-(\mathbf{X}'\mathbf{V}^-\mathbf{X})^-$ is a generalized inverse of $-\mathbf{X}'\mathbf{V}^-\mathbf{X}$], we find that [in the special case where $\mathcal{C}(\mathbf{X}) \subset \mathcal{C}(\mathbf{V})$] the partitioned matrix

$$\begin{pmatrix} \mathbf{V}^- - \mathbf{V}^-\mathbf{X}(\mathbf{X}'\mathbf{V}^-\mathbf{X})^-\mathbf{X}'\mathbf{V}^- & \mathbf{V}^-\mathbf{X}(\mathbf{X}'\mathbf{V}^-\mathbf{X})^- \\ (\mathbf{X}'\mathbf{V}^-\mathbf{X})^-\mathbf{X}'\mathbf{V}^- & -(\mathbf{X}'\mathbf{V}^-\mathbf{X})^- \end{pmatrix} \tag{4.8}$$

is a generalized inverse of $\begin{pmatrix} \mathbf{V} & \mathbf{X} \\ \mathbf{X}' & \mathbf{0} \end{pmatrix}$.

A generalization of formula (4.8) that applies without restriction [i.e., that applies even if $\mathcal{C}(\mathbf{X})$ is not contained in $\mathcal{C}(\mathbf{V})$] is given by the following theorem.

Theorem 19.4.3. Let $\mathbf{V}$ represent an $n \times n$ symmetric nonnegative definite matrix and $\mathbf{X}$ an $n \times p$ matrix. Further, let $\mathbf{U}$ represent any $p \times p$ matrix such that $\mathcal{C}(\mathbf{X}) \subset \mathcal{C}(\mathbf{V} + \mathbf{X}\mathbf{U}\mathbf{X}')$, and let $\mathbf{W}$ represent an arbitrary generalized inverse of $\mathbf{V} + \mathbf{X}\mathbf{U}\mathbf{X}'$. Then, the partitioned matrix

$$\begin{pmatrix} \mathbf{W} - \mathbf{W}\mathbf{X}(\mathbf{X}'\mathbf{W}\mathbf{X})^-\mathbf{X}'\mathbf{W} & \mathbf{W}\mathbf{X}(\mathbf{X}'\mathbf{W}\mathbf{X})^- \\ (\mathbf{X}'\mathbf{W}\mathbf{X})^-\mathbf{X}'\mathbf{W} & -(\mathbf{X}'\mathbf{W}\mathbf{X})^- + \mathbf{U} \end{pmatrix}$$

is a generalized inverse of the partitioned matrix $\begin{pmatrix} \mathbf{V} & \mathbf{X} \\ \mathbf{X}' & \mathbf{0} \end{pmatrix}$.

Preliminary to proving Theorem 19.4.3, it is convenient to establish the following theorem, which is of some interest in its own right.

Theorem 19.4.4. Let $\mathbf{V}$ represent an $m \times n$ matrix, $\mathbf{X}$ an $m \times p$ matrix, $\mathbf{Y}$ a $q \times n$ matrix, and $\mathbf{U}$ a $p \times q$ matrix. (1) Then, for any generalized inverse $\begin{pmatrix} \mathbf{G}_{11} & \mathbf{G}_{12} \\ \mathbf{G}_{21} & \mathbf{G}_{22} \end{pmatrix}$ of the partitioned matrix $\begin{pmatrix} \mathbf{V} & \mathbf{X} \\ \mathbf{Y} & \mathbf{0} \end{pmatrix}$ (where $\mathbf{G}_{11}$ is of dimensions $n \times m$), the partitioned matrix $\begin{pmatrix} \mathbf{G}_{11} & \mathbf{G}_{12} \\ \mathbf{G}_{21} & \mathbf{G}_{22} - \mathbf{U} \end{pmatrix}$ is a generalized inverse of the partitioned matrix $\begin{pmatrix} \mathbf{V} + \mathbf{XUY} & \mathbf{X} \\ \mathbf{Y} & \mathbf{0} \end{pmatrix}$. (2) Conversely, for any generalized inverse $\begin{pmatrix} \mathbf{H}_{11} & \mathbf{H}_{12} \\ \mathbf{H}_{21} & \mathbf{H}_{22} \end{pmatrix}$ of the partitioned matrix $\begin{pmatrix} \mathbf{V} + \mathbf{XUY} & \mathbf{X} \\ \mathbf{Y} & \mathbf{0} \end{pmatrix}$ (where $\mathbf{H}_{11}$ is of dimensions $n \times m$), the partitioned matrix $\begin{pmatrix} \mathbf{H}_{11} & \mathbf{H}_{12} \\ \mathbf{H}_{21} & \mathbf{H}_{22} + \mathbf{U} \end{pmatrix}$ is a generalized inverse of the partitioned matrix $\begin{pmatrix} \mathbf{V} & \mathbf{X} \\ \mathbf{Y} & \mathbf{0} \end{pmatrix}$.

Proof (of Theorem 19.4.4). (1) Let

$$\mathbf{C} = \begin{pmatrix} \mathbf{V} + \mathbf{XUY} & \mathbf{X} \\ \mathbf{Y} & \mathbf{0} \end{pmatrix} \begin{pmatrix} \mathbf{G}_{11} & \mathbf{G}_{12} \\ \mathbf{G}_{21} & \mathbf{G}_{22} - \mathbf{U} \end{pmatrix} \begin{pmatrix} \mathbf{V} + \mathbf{XUY} & \mathbf{X} \\ \mathbf{Y} & \mathbf{0} \end{pmatrix}.$$

and partition $\mathbf{C}$ as $\mathbf{C} = \begin{pmatrix} \mathbf{C}_{11} & \mathbf{C}_{12} \\ \mathbf{C}_{21} & \mathbf{C}_{22} \end{pmatrix}$ (where $\mathbf{C}_{11}$ is of dimensions $m \times n$).

Then, making use of the results of Lemma 19.4.1, we find that

$$\mathbf{C}_{22} = \mathbf{YG}_{11}\mathbf{X} = \mathbf{0},$$

$$\begin{aligned} \mathbf{C}_{21} &= \mathbf{Y}(\mathbf{G}_{11}\mathbf{V} + \mathbf{G}_{11}\mathbf{XUY} + \mathbf{G}_{12}\mathbf{Y}) \\ &= \mathbf{Y}(\mathbf{G}_{11}\mathbf{V} + \mathbf{G}_{12}\mathbf{Y}) + \mathbf{YG}_{11}\mathbf{XUY} \\ &= \mathbf{Y}(\mathbf{G}_{11}\mathbf{V} + \mathbf{G}_{12}\mathbf{Y}) \quad (\text{since } \mathbf{YG}_{11}\mathbf{X} = \mathbf{0}) \\ &= \mathbf{Y}, \end{aligned}$$

$$\begin{aligned} \mathbf{C}_{12} &= (\mathbf{VG}_{11} + \mathbf{XUYG}_{11} + \mathbf{XG}_{21})\mathbf{X} \\ &= (\mathbf{VG}_{11} + \mathbf{XG}_{21})\mathbf{X} + \mathbf{XUYG}_{11}\mathbf{X} \\ &= (\mathbf{VG}_{11} + \mathbf{XG}_{21})\mathbf{X} \\ &= \mathbf{X}, \end{aligned}$$

$$\begin{aligned}
\mathbf{C}_{11} &= (\mathbf{V}\mathbf{G}_{11} + \mathbf{X}\mathbf{U}\mathbf{Y}\mathbf{G}_{11} + \mathbf{X}\mathbf{G}_{21})(\mathbf{V} + \mathbf{X}\mathbf{U}\mathbf{Y}) \\
&\qquad\qquad + (\mathbf{V}\mathbf{G}_{12} + \mathbf{X}\mathbf{U}\mathbf{Y}\mathbf{G}_{12} + \mathbf{X}\mathbf{G}_{22} - \mathbf{X}\mathbf{U})\mathbf{Y} \\
&= (\mathbf{V}\mathbf{G}_{11} + \mathbf{X}\mathbf{G}_{21})\mathbf{V} + \mathbf{X}\mathbf{U}\mathbf{Y}\mathbf{G}_{11}\mathbf{V} \\
&\qquad\qquad + (\mathbf{V}\mathbf{G}_{11} + \mathbf{X}\mathbf{G}_{21})\mathbf{X}\mathbf{U}\mathbf{Y} + \mathbf{X}\mathbf{U}\mathbf{Y}\mathbf{G}_{11}\mathbf{X}\mathbf{U}\mathbf{Y} \\
&\qquad\qquad + (\mathbf{V}\mathbf{G}_{12} + \mathbf{X}\mathbf{G}_{22})\mathbf{Y} + \mathbf{X}\mathbf{U}\mathbf{Y}\mathbf{G}_{12}\mathbf{Y} - \mathbf{X}\mathbf{U}\mathbf{Y} \\
&= (\mathbf{V}\mathbf{G}_{11} + \mathbf{X}\mathbf{G}_{21})\mathbf{V} + \mathbf{X}\mathbf{U}\mathbf{Y}\mathbf{G}_{11}\mathbf{V} + \mathbf{X}\mathbf{U}\mathbf{Y} + \mathbf{0} \\
&\qquad + (\mathbf{V}\mathbf{G}_{12} + \mathbf{X}\mathbf{G}_{22})\mathbf{Y} + \mathbf{X}\mathbf{U}\mathbf{Y}\mathbf{G}_{12}\mathbf{Y} - \mathbf{X}\mathbf{U}\mathbf{Y} \quad (\text{since } \mathbf{Y}\mathbf{G}_{11}\mathbf{X} = \mathbf{0}) \\
&= (\mathbf{V}\mathbf{G}_{11} + \mathbf{X}\mathbf{G}_{21})\mathbf{V} + (\mathbf{V}\mathbf{G}_{12} + \mathbf{X}\mathbf{G}_{22})\mathbf{Y} + \mathbf{X}\mathbf{U}\mathbf{Y}\mathbf{G}_{11}\mathbf{V} + \mathbf{X}\mathbf{U}\mathbf{Y}\mathbf{G}_{12}\mathbf{Y} \\
&= \mathbf{V} + \mathbf{X}\mathbf{U}\mathbf{Y}\mathbf{G}_{11}\mathbf{V} + \mathbf{X}\mathbf{U}\mathbf{Y}\mathbf{G}_{12}\mathbf{Y} \\
&= \mathbf{V} + \mathbf{X}\mathbf{U}\mathbf{Y}(\mathbf{G}_{11}\mathbf{V} + \mathbf{G}_{12}\mathbf{Y}) \\
&= \mathbf{V} + \mathbf{X}\mathbf{U}\mathbf{Y}.
\end{aligned}$$

Thus, $\begin{pmatrix} \mathbf{G}_{11} & \mathbf{G}_{12} \\ \mathbf{G}_{21} & \mathbf{G}_{22} - \mathbf{U} \end{pmatrix}$ is a generalized inverse of $\begin{pmatrix} \mathbf{V} + \mathbf{X}\mathbf{U}\mathbf{Y} & \mathbf{X} \\ \mathbf{Y} & \mathbf{0} \end{pmatrix}$.

(2) Rewriting $\begin{pmatrix} \mathbf{V} & \mathbf{X} \\ \mathbf{Y} & \mathbf{0} \end{pmatrix}$ as $\begin{pmatrix} (\mathbf{V} + \mathbf{X}\mathbf{U}\mathbf{Y}) + \mathbf{X}(-\mathbf{U})\mathbf{Y} & \mathbf{X} \\ \mathbf{Y} & \mathbf{0} \end{pmatrix}$ and applying Part (1) (with $\mathbf{V} + \mathbf{X}\mathbf{U}\mathbf{Y}$ in place of $\mathbf{V}$ and $-\mathbf{U}$ in place of $\mathbf{U}$), we find that the partitioned matrix $\begin{pmatrix} \mathbf{H}_{11} & \mathbf{H}_{12} \\ \mathbf{H}_{21} & \mathbf{H}_{22} - (-\mathbf{U}) \end{pmatrix}$, which equals $\begin{pmatrix} \mathbf{H}_{11} & \mathbf{H}_{12} \\ \mathbf{H}_{21} & \mathbf{H}_{22} + \mathbf{U} \end{pmatrix}$, is a generalized inverse of $\begin{pmatrix} \mathbf{V} & \mathbf{X} \\ \mathbf{Y} & \mathbf{0} \end{pmatrix}$. Q.E.D.

Proof (of Theorem 19.4.3). Since (in light of Lemma 19.3.4) $\mathcal{R}(\mathbf{X}') \subset \mathcal{R}(\mathbf{V} + \mathbf{X}\mathbf{U}\mathbf{X}')$, formula (9.6.2) (for the generalized inverse of a partitioned matrix) is applicable to the partitioned matrix $\begin{pmatrix} \mathbf{V} + \mathbf{X}\mathbf{U}\mathbf{X}' & \mathbf{X} \\ \mathbf{X}' & \mathbf{0} \end{pmatrix}$. Upon applying that formula [and observing that $-(\mathbf{X}'\mathbf{W}\mathbf{X})^-$ is a generalized inverse of $-\mathbf{X}'\mathbf{W}\mathbf{X}$], we find that the partitioned matrix

$$\begin{pmatrix} \mathbf{W} - \mathbf{W}\mathbf{X}(\mathbf{X}'\mathbf{W}\mathbf{X})^-\mathbf{X}'\mathbf{W} & \mathbf{W}\mathbf{X}(\mathbf{X}'\mathbf{W}\mathbf{X})^- \\ (\mathbf{X}'\mathbf{W}\mathbf{X})^-\mathbf{X}'\mathbf{W} & -(\mathbf{X}'\mathbf{W}\mathbf{X})^- \end{pmatrix}$$

is a generalized inverse of $\begin{pmatrix} \mathbf{V} + \mathbf{X}\mathbf{U}\mathbf{X}' & \mathbf{X} \\ \mathbf{X}' & \mathbf{0} \end{pmatrix}$. Thus, it follows fom Part (2) of Theorem 19.4.4 that the partitioned matrix

$$\begin{pmatrix} \mathbf{W} - \mathbf{W}\mathbf{X}(\mathbf{X}'\mathbf{W}\mathbf{X})^-\mathbf{X}'\mathbf{W} & \mathbf{W}\mathbf{X}(\mathbf{X}'\mathbf{W}\mathbf{X})^- \\ (\mathbf{X}'\mathbf{W}\mathbf{X})^-\mathbf{X}'\mathbf{W} & -(\mathbf{X}'\mathbf{W}\mathbf{X})^- + \mathbf{U} \end{pmatrix}$$

is a generalized inverse of the partitioned matrix $\begin{pmatrix} \mathbf{V} & \mathbf{X} \\ \mathbf{X}' & \mathbf{0} \end{pmatrix}$. Q.E.D.

b. Alternative derivation of the results of Section 19.3c

For purposes of obtaining an alternative derivation of the results of Section 19.3c, let $\mathbf{V}$ represent an $n \times n$ symmetric nonnegative definite matrix and $\mathbf{X}$ an $n \times p$

matrix. Further, let $\mathbf{U}$ represent any $p \times p$ matrix such that $\mathcal{C}(\mathbf{X}) \subset \mathcal{C}(\mathbf{V}+\mathbf{XUX}')$, and let $\mathbf{W}$ represent an arbitrary generalized inverse of $\mathbf{V}+\mathbf{XUX}'$.

Then, upon taking the generalized inverse $\mathbf{G}$ [of the partitioned matrix $\begin{pmatrix} \mathbf{V} & \mathbf{X} \\ \mathbf{X}' & \mathbf{0} \end{pmatrix}$] in Theorem 19.4.2 to be the generalized inverse given by Theorem 19.4.3, it follows from Part (1) of Theorem 19.4.2 that

$$\mathbf{X}'\mathbf{WX}(\mathbf{X}'\mathbf{WX})^-\mathbf{X}' = \mathbf{X}', \tag{4.9}$$
$$\mathbf{X}(\mathbf{X}'\mathbf{WX})^-\mathbf{X}'\mathbf{WX} = \mathbf{X}. \tag{4.10}$$

And, in light of Lemma 19.3.1, it is evident from equality (4.9) or (4.10) that

$$\operatorname{rank}(\mathbf{X}'\mathbf{WX}) = \operatorname{rank}(\mathbf{X}). \tag{4.11}$$

Now, let $\mathbf{B}$ represent an $n \times s$ matrix such that $\mathcal{C}(\mathbf{B}) \subset \mathcal{C}(\mathbf{V}, \mathbf{X})$ and $\mathbf{D}$ a $p \times s$ matrix such that $\mathcal{C}(\mathbf{D}) \subset \mathcal{C}(\mathbf{X}')$. And let

$$\mathbf{G}_{11} = \mathbf{W} - \mathbf{WX}(\mathbf{X}'\mathbf{WX})^-\mathbf{X}'\mathbf{W},$$
$$\mathbf{G}_{12} = \mathbf{WX}(\mathbf{X}'\mathbf{WX})^-,$$
$$\mathbf{G}_{21} = (\mathbf{X}'\mathbf{WX})^-\mathbf{X}'\mathbf{W},$$
$$\mathbf{G}_{22} = -(\mathbf{X}'\mathbf{WX})^- + \mathbf{U}.$$

Then, according to Theorem 19.4.3, the partitioned matrix $\begin{pmatrix} \mathbf{G}_{11} & \mathbf{G}_{12} \\ \mathbf{G}_{21} & \mathbf{G}_{22} \end{pmatrix}$ is a generalized inverse of the partitioned matrix $\begin{pmatrix} \mathbf{V} & \mathbf{X} \\ \mathbf{X}' & \mathbf{0} \end{pmatrix}$, and it follows from Theorem 11.2.4 that $\mathbf{A}_*$ and $\mathbf{R}_*$ are, respectively, the first ($n \times s$) and second ($p \times s$) parts of a solution to the (consistent) linear system

$$\begin{pmatrix} \mathbf{V} & \mathbf{X} \\ \mathbf{X}' & \mathbf{0} \end{pmatrix} \begin{pmatrix} \mathbf{A} \\ \mathbf{R} \end{pmatrix} = \begin{pmatrix} \mathbf{B} \\ \mathbf{D} \end{pmatrix} \tag{4.12}$$

(in an $n \times s$ matrix $\mathbf{A}$ and a $p \times s$ matrix $\mathbf{R}$) if and only if

$$\begin{pmatrix} \mathbf{A}_* \\ \mathbf{R}_* \end{pmatrix} = \begin{pmatrix} \mathbf{G}_{11} & \mathbf{G}_{12} \\ \mathbf{G}_{21} & \mathbf{G}_{22} \end{pmatrix} \begin{pmatrix} \mathbf{B} \\ \mathbf{D} \end{pmatrix} + \left[\begin{pmatrix} \mathbf{I}_n & \mathbf{0} \\ \mathbf{0} & \mathbf{I}_p \end{pmatrix} - \begin{pmatrix} \mathbf{G}_{11} & \mathbf{G}_{12} \\ \mathbf{G}_{21} & \mathbf{G}_{22} \end{pmatrix} \begin{pmatrix} \mathbf{V} & \mathbf{X} \\ \mathbf{X}' & \mathbf{0} \end{pmatrix} \right] \begin{pmatrix} \mathbf{L} \\ \mathbf{M} \end{pmatrix}$$

for some $n \times s$ matrix $\mathbf{L}$ and some $p \times s$ matrix $\mathbf{M}$.

Note that, as a consequence of Lemma 19.3.4, $\mathbf{X}' = \mathbf{X}'\mathbf{W}(\mathbf{V}+\mathbf{XUX}')$, or, equivalently, that $\mathbf{X}'(\mathbf{I}-\mathbf{WV}) = \mathbf{X}'\mathbf{WXUX}'$. Thus, making use of results (4.1) and (4.10), we find that

$$\begin{aligned} &\begin{pmatrix} \mathbf{G}_{11} & \mathbf{G}_{12} \\ \mathbf{G}_{21} & \mathbf{G}_{22} \end{pmatrix} \begin{pmatrix} \mathbf{V} & \mathbf{X} \\ \mathbf{X}' & \mathbf{0} \end{pmatrix} \\ &\quad = \begin{pmatrix} \mathbf{WV} + \mathbf{WX}(\mathbf{X}'\mathbf{WX})^-\mathbf{X}'(\mathbf{I}-\mathbf{WV}) & \mathbf{0} \\ -(\mathbf{X}'\mathbf{WX})^-\mathbf{X}'(\mathbf{I}-\mathbf{WV}) + \mathbf{UX}' & (\mathbf{X}'\mathbf{WX})^-\mathbf{X}'\mathbf{WX} \end{pmatrix} \\ &\quad = \begin{pmatrix} \mathbf{W}(\mathbf{V}+\mathbf{XUX}') & \mathbf{0} \\ [\mathbf{I}-(\mathbf{X}'\mathbf{WX})^-\mathbf{X}'\mathbf{WX}]\mathbf{UX}' & (\mathbf{X}'\mathbf{WX})^-\mathbf{X}'\mathbf{WX} \end{pmatrix}. \end{aligned}$$

We conclude that $\mathbf{A}_*$ and $\mathbf{R}_*$ are, respectively, the first and second parts of a solution to linear system (4.12) if and only if

$$\mathbf{A}_* = [\mathbf{W} - \mathbf{WX}(\mathbf{X}'\mathbf{WX})^-\mathbf{X}'\mathbf{W}]\mathbf{B} + \mathbf{WX}(\mathbf{X}'\mathbf{WX})^-\mathbf{D} + [\mathbf{I} - \mathbf{W}(\mathbf{V} + \mathbf{XUX}')]\mathbf{L},$$

$$\mathbf{R}_* = (\mathbf{X}'\mathbf{WX})^-(\mathbf{X}'\mathbf{WB} - \mathbf{D}) + \mathbf{UD} + [\mathbf{I} - (\mathbf{X}'\mathbf{WX})^-\mathbf{X}'\mathbf{WX}](\mathbf{M} - \mathbf{UX}'\mathbf{L})$$

for some matrices $\mathbf{L}$ and $\mathbf{M}$, or, equivalently, if and only if

$$\mathbf{A}_* = [\mathbf{W} - \mathbf{WX}(\mathbf{X}'\mathbf{WX})^-\mathbf{X}'\mathbf{W}]\mathbf{B} + \mathbf{WX}(\mathbf{X}'\mathbf{WX})^-\mathbf{D} + [\mathbf{I} - \mathbf{W}(\mathbf{V} + \mathbf{XUX}')]\mathbf{L}, \quad (4.13)$$

$$\mathbf{R}_* = (\mathbf{X}'\mathbf{WX})^-(\mathbf{X}'\mathbf{WB} - \mathbf{D}) + \mathbf{UD} + [\mathbf{I} - (\mathbf{X}'\mathbf{WX})^-\mathbf{X}'\mathbf{WX}]\mathbf{Z} \quad (4.14)$$

for some matrices $\mathbf{L}$ and $\mathbf{Z}$.

This result can be reexpressed in terms of the linear system

$$\mathbf{X}'\mathbf{WXT} = \mathbf{X}'\mathbf{WB} - \mathbf{D} \quad (4.15)$$

(in a $p \times s$ matrix $\mathbf{T}$). Since clearly $\mathbf{D} = \mathbf{X}'\mathbf{K}$ for some matrix $\mathbf{K}$, it follows from result (4.9) that

$$\mathbf{X}'\mathbf{WX}(\mathbf{X}'\mathbf{WX})^-(\mathbf{X}'\mathbf{WB} - \mathbf{D}) = \mathbf{X}'\mathbf{WB} - \mathbf{D}$$

and hence (in light of Lemma 9.5.1) that linear system (4.15) is consistent. Further, in light of Corollary 4.4.7, it follows from result (4.11) that $\mathcal{R}(\mathbf{X}'\mathbf{WX}) = \mathcal{R}(\mathbf{X})$, implying (in light of Theorem 11.10.1) that $\mathbf{XT}$ is invariant to the choice of solution to linear system (4.15). Then, upon recalling Theorem 11.2.4 and observing that the matrix $(\mathbf{X}'\mathbf{WX})^-(\mathbf{X}'\mathbf{WB} - \mathbf{D})$ is a solution to linear system (4.15), we find that equalities (4.13) and (4.14) hold for some matrices $\mathbf{L}$ and $\mathbf{Z}$ if and only if $\mathbf{R}_* = \mathbf{T}_* + \mathbf{UD}$ and

$$\mathbf{A}_* = \mathbf{WB} - \mathbf{WXT}_* + [\mathbf{I} - \mathbf{W}(\mathbf{V} + \mathbf{XUX}')]\mathbf{L}$$

for some solution $\mathbf{T}_*$ to linear system (4.15) and for some matrix $\mathbf{L}$.

19.5 Some Additional Results on the Form of Solutions to the Constrained Minimization Problem

Let $\mathbf{a}$ represent an $n \times 1$ vector of variables. And consider the minimization of the quadratic form $\mathbf{a}'\mathbf{Va}$ subject to the constraint $\mathbf{X}'\mathbf{a} = \mathbf{d}$, where $\mathbf{V}$ is an $n \times n$ symmetric nonnegative definite matrix, $\mathbf{X}$ is an $n \times p$ matrix, and $\mathbf{d}$ is a $p \times 1$ vector such that $\mathbf{d} \in \mathcal{C}(\mathbf{X}')$. This minimization problem is a special case of the minimization problem considered in Sections 19.2–19.4.

It follows from the results of Sections 19.2–19.4 that the problem of minimizing $\mathbf{a}'\mathbf{Va}$ subject to $\mathbf{X}'\mathbf{a} = \mathbf{d}$ has a solution or solutions of the form

$$\mathbf{a} = \mathbf{WX}(\mathbf{X'WX})^{-}\mathbf{d}, \tag{5.1}$$

where $\mathbf{W}$ is an $n \times n$ matrix. For the vector (5.1) to be a solution to this minimization problem, it is sufficient (in light of the results of Section 19.3c) that $\mathbf{W}$ equal any generalized inverse of $\mathbf{V} + \mathbf{XUX'}$, where $\mathbf{U}$ is any $p \times p$ matrix such that $\mathcal{C}(\mathbf{X}) \subset \mathcal{C}(\mathbf{V} + \mathbf{XUX'})$.

In this section, conditions are given that are necessary (as well as sufficient) for the vector (5.1) to solve, for every $\mathbf{d} \in \mathcal{C}(\mathbf{X'})$, the problem of minimizing $\mathbf{a'Va}$ subject to $\mathbf{X'a} = \mathbf{d}$ (when $\mathbf{W}$ does not vary with $\mathbf{d}$). For fixed $\mathbf{V}$, these conditions characterize those vectors of the form (5.1) that are solutions [for every $\mathbf{d} \in \mathcal{C}(\mathbf{X'})$] to this minimization problem. For fixed $\mathbf{W}$, these conditions characterize those nonnegative definite quadratic forms (in $\mathbf{a}$) for which the vector (5.1) is a solution to the constrained minimization problem.

The following theorem gives what are perhaps the most basic of the necessary and sufficient conditions.

Theorem 19.5.1. Let $\mathbf{V}$ represent an $n \times n$ symmetric nonnegative definite matrix, $\mathbf{W}$ an $n \times n$ matrix, $\mathbf{X}$ an $n \times p$ matrix, and $\mathbf{d}$ a $p \times 1$ vector. Then, for the vector $\mathbf{WX}(\mathbf{X'WX})^{-}\mathbf{d}$ to be a solution, for every $\mathbf{d} \in \mathcal{C}(\mathbf{X'})$, to the problem of minimizing the quadratic form $\mathbf{a'Va}$ (in $\mathbf{a}$) subject to $\mathbf{X'a} = \mathbf{d}$, it is necessary and sufficient (assuming that the elements of $\mathbf{W}$ are not functionally dependent on $\mathbf{d}$) that

$$\mathcal{C}(\mathbf{VWX}) \subset \mathcal{C}(\mathbf{X}) \tag{5.2}$$

and

$$\text{rank}(\mathbf{X'WX}) = \text{rank}(\mathbf{X}). \tag{5.3}$$

Proof. It follows from Theorem 19.2.1 that $\mathbf{a'Va}$ has a minimum at $\mathbf{WX}(\mathbf{X'WX})^{-}\mathbf{d}$ under the constraint $\mathbf{X'a} = \mathbf{d}$ [where $\mathbf{d} \in \mathcal{C}(\mathbf{X'})$] if and only if $\mathbf{VWX}(\mathbf{X'WX})^{-}\mathbf{d} + \mathbf{Xr} = \mathbf{0}$ for some vector $\mathbf{r}$ and $\mathbf{X'WX}(\mathbf{X'WX})^{-}\mathbf{d} = \mathbf{d}$, or, equivalently, if and only if $\mathbf{VWX}(\mathbf{X'WX})^{-}\mathbf{d} \in \mathcal{C}(\mathbf{X})$ and $\mathbf{X'WX}(\mathbf{X'WX})^{-}\mathbf{d} = \mathbf{d}$. Thus, for $\mathbf{WX}(\mathbf{X'WX})^{-}\mathbf{d}$ to be a solution, for every $\mathbf{d} \in \mathcal{C}(\mathbf{X'})$, to the problem of minimizing $\mathbf{a'Va}$ subject to $\mathbf{X'a} = \mathbf{d}$, it is necessary and sufficient that, for every $n \times 1$ vector $\mathbf{u}$, $\mathbf{VWX}(\mathbf{X'WX})^{-}\mathbf{X'u} \in \mathcal{C}(\mathbf{X})$ and $\mathbf{X'WX}(\mathbf{X'WX})^{-}\mathbf{X'u} = \mathbf{X'u}$ or, equivalently, that

$$\mathcal{C}[\mathbf{VWX}(\mathbf{X'WX})^{-}\mathbf{X'}] \subset \mathcal{C}(\mathbf{X}) \tag{5.4}$$

and

$$\mathbf{X'WX}(\mathbf{X'WX})^{-}\mathbf{X'} = \mathbf{X'}. \tag{5.5}$$

It remains to show that conditions (5.4) and (5.5) are equivalent to conditions (5.2) and (5.3). If condition (5.2) is satisfied, then, since $\mathcal{C}[\mathbf{VWX}(\mathbf{X'WX})^{-}\mathbf{X'}] \subset \mathcal{C}(\mathbf{VWX})$, condition (5.4) is satisfied; and if condition (5.3) is satisfied, then, as a consequence of Lemma 19.3.1, condition (5.5) is satisfied.

Conversely, suppose that conditions (5.4) and (5.5) are satisfied. Then, it follows from Lemma 19.3.1 that condition (5.3) is satisfied. And, as a further consequence of Lemma 19.3.1, we have that

$$\mathbf{VWX(X'WX)^{-}X'WX = VWX}.$$

Thus,

$$\mathcal{C}(\mathbf{VWX}) \subset \mathcal{C}[\mathbf{VWX(X'WX)^{-}X'}],$$

and hence condition (5.2) is satisfied. Q.E.D.

In connection with Theorem 19.5.1, note that the condition $\mathcal{C}(\mathbf{VWX}) \subset \mathcal{C}(\mathbf{X})$ is equivalent to the condition

$$\mathbf{VWX = XQ} \text{ for some matrix } \mathbf{Q}. \tag{5.6}$$

And, letting $\mathbf{Z}$ represent any matrix whose columns span $\mathcal{C}^{\perp}(\mathbf{X})$ or, equivalently, span $\mathcal{N}(\mathbf{X}')$, it is also equivalent to the condition

$$\mathbf{Z'VWX = 0} \tag{5.7}$$

(as is evident from Corollary 12.5.7).

Moreover, the condition $\text{rank}(\mathbf{X'WX}) = \text{rank}(\mathbf{X})$ is equivalent to each of the following conditions:

$$\mathbf{X(X'WX)^{-}X'WX = X}, \tag{5.8}$$

$$\mathbf{X'WX(X'WX)^{-}X' = X'}, \tag{5.9}$$

$$\mathcal{R}(\mathbf{X'WX}) = \mathcal{R}(\mathbf{X}), \tag{5.10}$$

$$\mathcal{C}(\mathbf{X'WX}) = \mathcal{C}(\mathbf{X}') \tag{5.11}$$

(as is evident from Lemma 19.3.1 and Corollary 4.4.7). Thus, variations on Theorem 19.5.1 can be obtained by replacing condition (5.2) with condition (5.6) or (5.7) and/or by replacing condition (5.3) with any one of conditions (5.8)–(5.11).

Note (in light of Corollary 14.11.3) that one circumstance under which the condition $\text{rank}(\mathbf{X'WX}) = \text{rank}(\mathbf{X})$ is satisfied is that where $\mathbf{W}$ is symmetric and positive definite.

The following theorem provides alternatives to the necessary and sufficient conditions given by Theorem 19.5.1 [conditions (5.2) and (5.3)] and to the variations on those conditions [conditions (5.6)–(5.7) and (5.8)–(5.11)].

Theorem 19.5.2. Let $\mathbf{V}$ represent an $n \times n$ symmetric nonnegative definite matrix, $\mathbf{W}$ an $n \times n$ matrix, $\mathbf{X}$ an $n \times p$ matrix, and $\mathbf{d}$ a $p \times 1$ vector. Then, for the vector $\mathbf{WX(X'WX)^{-}d}$ to be a solution, for every $\mathbf{d} \in \mathcal{C}(\mathbf{X}')$, to the problem of minimizing the quadratic form $\mathbf{a'Va}$ (in $\mathbf{a}$) subject to $\mathbf{X'a = d}$, it is necessary and sufficient that there exist a $p \times p$ matrix $\mathbf{U}$ such that

$$\mathbf{(V + XUX')WX = X}. \tag{5.12}$$

It is also necessary and sufficient that $\mathbf{W}$ be expressible in the form

$$\mathbf{W = G + K_{*}}, \tag{5.13}$$

where, for some $p \times p$ matrix $\mathbf{U}$ such that $\mathcal{C}(\mathbf{X}) \subset \mathcal{C}(\mathbf{V + XUX'})$, $\mathbf{G}$ is an arbitrary generalized inverse of $\mathbf{V + XUX'}$, and where $\mathbf{K}_{*}$ is some solution to the system $\begin{pmatrix}\mathbf{V} \\ \mathbf{X'}\end{pmatrix}\mathbf{KX = 0}$ (in an $n \times n$ matrix $\mathbf{K}$).

Preliminary to proving Theorem 19.5.2, it is convenient to prove the following lemma.

Lemma 19.5.3. For any $n \times n$ symmetric nonnegative definite matrix $\mathbf{V}$, $n \times n$ matrix $\mathbf{W}$, and $n \times p$ matrix $\mathbf{X}$,

$$\mathcal{C}(\mathbf{VWX}) \subset \mathcal{C}(\mathbf{X}) \Rightarrow \mathcal{R}(\mathbf{VWX}) \subset \mathcal{R}(\mathbf{X}'\mathbf{WX}).$$

Proof (of Lemma 19.5.3). Suppose that $\mathcal{C}(\mathbf{VWX}) \subset \mathcal{C}(\mathbf{X})$. Then, as a consequence of Corollary 4.2.4,

$$\mathcal{C}(\mathbf{X}'\mathbf{W}'\mathbf{VWX}) \subset \mathcal{C}(\mathbf{X}'\mathbf{W}'\mathbf{X}),$$

and (in light of Lemma 4.2.5) we have that

$$\mathcal{R}[(\mathbf{WX})'\mathbf{VWX}] = \mathcal{R}[(\mathbf{X}'\mathbf{W}'\mathbf{VWX})'] \subset \mathcal{R}[(\mathbf{X}'\mathbf{W}'\mathbf{X})'] = \mathcal{R}(\mathbf{X}'\mathbf{WX}).$$

Since (in light of Lemma 14.11.2) $\mathcal{R}[(\mathbf{WX})'\mathbf{VWX}] = \mathcal{R}(\mathbf{VWX})$, we conclude that $\mathcal{R}(\mathbf{VWX}) \subset \mathcal{R}(\mathbf{X}'\mathbf{WX})$. Q.E.D.

Proof (of Theorem 19.5.2). To prove the first part of the theorem, it suffices (in light of Theorem 19.5.1) to show that there exists a matrix $\mathbf{U}$ that satisfies condition (5.12) if and only if $\mathcal{C}(\mathbf{VWX}) \subset \mathcal{C}(\mathbf{X})$ and $\text{rank}(\mathbf{X}'\mathbf{WX}) = \text{rank}(\mathbf{X})$ or, equivalently (in light of Corollary 4.4.7), if and only if $\mathcal{C}(\mathbf{VWX}) \subset \mathcal{C}(\mathbf{X})$ and $\mathcal{R}(\mathbf{X}'\mathbf{WX}) = \mathcal{R}(\mathbf{X})$.

Upon observing that condition (5.12) is equivalent to the condition

$$\mathbf{XUX}'\mathbf{WX} = \mathbf{X} - \mathbf{VWX},$$

it is clear from Corollary 9.7.2 that there exists a matrix $\mathbf{U}$ that satisfies condition (5.12) if and only if $\mathcal{C}(\mathbf{X} - \mathbf{VWX}) \subset \mathcal{C}(\mathbf{X})$ and $\mathcal{R}(\mathbf{X} - \mathbf{VWX}) \subset \mathcal{R}(\mathbf{X}'\mathbf{WX})$ or, equivalently (in light of Lemma 4.5.10), if and only if $\mathcal{C}(\mathbf{VWX}) \subset \mathcal{C}(\mathbf{X})$ and $\mathcal{R}(\mathbf{X} - \mathbf{VWX}) \subset \mathcal{R}(\mathbf{X}'\mathbf{WX})$.

Now, suppose that $\mathcal{C}(\mathbf{VWX}) \subset \mathcal{C}(\mathbf{X})$, implying (according to Lemma 19.5.3) that $\mathcal{R}(\mathbf{VWX}) \subset \mathcal{R}(\mathbf{X}'\mathbf{WX})$. Then, if $\mathcal{R}(\mathbf{X}'\mathbf{WX}) = \mathcal{R}(\mathbf{X})$, it follows from Lemma 4.5.10 that $\mathcal{R}(\mathbf{X} - \mathbf{VWX}) \subset \mathcal{R}(\mathbf{X}'\mathbf{WX})$. Conversely, if $\mathcal{R}(\mathbf{X} - \mathbf{VWX}) \subset \mathcal{R}(\mathbf{X}'\mathbf{WX})$, it follows from Lemma 4.5.10 that $\mathcal{R}(\mathbf{X}) \subset \mathcal{R}(\mathbf{X}'\mathbf{WX})$ or, equivalently [since clearly $\mathcal{R}(\mathbf{X}'\mathbf{WX}) \subset \mathcal{R}(\mathbf{X})$], that $\mathcal{R}(\mathbf{X}'\mathbf{WX}) = \mathcal{R}(\mathbf{X})$. Thus, $\mathcal{C}(\mathbf{VWX}) \subset \mathcal{C}(\mathbf{X})$ and $\mathcal{R}(\mathbf{X} - \mathbf{VWX}) \subset \mathcal{R}(\mathbf{X}'\mathbf{WX})$ if and only if $\mathcal{C}(\mathbf{VWX}) \subset \mathcal{C}(\mathbf{X})$ and $\mathcal{R}(\mathbf{X}'\mathbf{WX}) = \mathcal{R}(\mathbf{X})$. And we conclude that there exists a matrix $\mathbf{U}$ that satisfies condition (5.12) if and only if $\mathcal{C}(\mathbf{VWX}) \subset \mathcal{C}(\mathbf{X})$ and $\mathcal{R}(\mathbf{X}'\mathbf{WX}) = \mathcal{R}(\mathbf{X})$.

To complete the proof of Theorem 19.5.2, it suffices to show that $\mathbf{W}$ is expressible in the form (5.13) if and only if there exists a matrix $\mathbf{U}$ that satisfies condition (5.12).

Suppose that $\mathbf{W}$ is expressible in the form (5.13). Then, making use of Lemma 19.3.4, we find that

$$\begin{aligned}(\mathbf{V}+\mathbf{XUX}')\mathbf{WX} &= (\mathbf{V}+\mathbf{XUX}')\mathbf{GX} + \mathbf{VK}_*\mathbf{X} + \mathbf{XUX}'\mathbf{K}_*\mathbf{X} \\ &= \mathbf{X} + \mathbf{0} + \mathbf{XU0} = \mathbf{X}.\end{aligned}$$

Thus, there exists a matrix $\mathbf{U}$ that satisfies condition (5.12).

Conversely, suppose that there exists a matrix $\mathbf{U}$ that satisfies condition (5.12). Then, clearly, $\mathcal{C}(\mathbf{X}) \subset \mathcal{C}(\mathbf{V}+\mathbf{XUX}')$. And it follows from Theorem 11.12.1 that

$$\mathbf{W} = (\mathbf{V}+\mathbf{XUX}')^-\mathbf{XX}^- + \mathbf{Z}_*$$

for some solution $\mathbf{Z}_*$ to the system $(\mathbf{V}+\mathbf{XUX}')\mathbf{ZX} = \mathbf{0}$ (in $\mathbf{Z}$).

Now, let $\mathbf{K}_* = \mathbf{Z}_* - (\mathbf{V}+\mathbf{XUX}')^-(\mathbf{I}-\mathbf{XX}^-)$. Then,

$$\begin{aligned}\mathbf{W} &= (\mathbf{V}+\mathbf{XUX}')^-\mathbf{XX}^- + (\mathbf{V}+\mathbf{XUX}')^-(\mathbf{I}-\mathbf{XX}^-) + \mathbf{K}_* \\ &= (\mathbf{V}+\mathbf{XUX}')^- + \mathbf{K}_*.\end{aligned}$$

Moreover, it follows from Lemma 19.3.4 that $\mathcal{C}(\mathbf{V}+\mathbf{XU}'\mathbf{X}') = \mathcal{C}(\mathbf{V}, \mathbf{X})$, implying (in light of Lemma 4.2.2) that $(\mathbf{V}, \mathbf{X}) = (\mathbf{V}+\mathbf{XU}'\mathbf{X}')\mathbf{L}$ for some matrix $\mathbf{L}$ and hence that

$$\begin{pmatrix}\mathbf{V}\\ \mathbf{X}'\end{pmatrix} = (\mathbf{V}, \mathbf{X})' = \mathbf{L}'(\mathbf{V}+\mathbf{XU}'\mathbf{X}')' = \mathbf{L}'(\mathbf{V}+\mathbf{XUX}').$$

Thus,

$$\begin{aligned}\begin{pmatrix}\mathbf{V}\\ \mathbf{X}'\end{pmatrix}\mathbf{K}_*\mathbf{X} &= \begin{pmatrix}\mathbf{V}\\ \mathbf{X}'\end{pmatrix}\mathbf{Z}_*\mathbf{X} - \begin{pmatrix}\mathbf{V}\\ \mathbf{X}'\end{pmatrix}(\mathbf{V}+\mathbf{XUX}')^-(\mathbf{I}-\mathbf{XX}^-)\mathbf{X} \\ &= \mathbf{L}'(\mathbf{V}+\mathbf{XUX}')\mathbf{Z}_*\mathbf{X} - \begin{pmatrix}\mathbf{V}\\ \mathbf{X}'\end{pmatrix}(\mathbf{V}+\mathbf{XUX}')^-(\mathbf{X}-\mathbf{XX}^-\mathbf{X}) \\ &= \mathbf{L}'\mathbf{0} - \begin{pmatrix}\mathbf{V}\\ \mathbf{X}'\end{pmatrix}(\mathbf{V}+\mathbf{XUX}')^-\mathbf{0} = \mathbf{0},\end{aligned}$$

so that $\mathbf{K}_*$ is a solution to the system $\begin{pmatrix}\mathbf{V}\\ \mathbf{X}'\end{pmatrix}\mathbf{KX} = \mathbf{0}$. Q.E.D.

As discussed earlier, the first of the necessary and sufficient conditions in Theorem 19.5.1 [condition (5.2)] is equivalent to condition (5.6) and also to condition (5.7), and, consequently, can be replaced by either of those two conditions. It can also be replaced by any condition that in combination with condition (5.3) gives a pair of conditions that is equivalent to conditions (5.2) and (5.3). The following theorem gives some variations on Theorem 19.5.1 obtained by replacing condition (5.2) by a condition of the latter kind.

Theorem 19.5.4. Let $\mathbf{V}$ represent an $n \times n$ symmetric nonnegative definite matrix, $\mathbf{W}$ an $n \times n$ matrix, $\mathbf{X}$ an $n \times p$ matrix, and $\mathbf{d}$ a $p \times 1$ vector. Then, for the vector $\mathbf{WX}(\mathbf{X}'\mathbf{WX})^-\mathbf{d}$ to be a solution, for every $\mathbf{d} \in \mathcal{C}(\mathbf{X}')$, to the problem of minimizing the quadratic form $\mathbf{a}'\mathbf{Va}$ (in $\mathbf{a}$) subject to $\mathbf{X}'\mathbf{a} = \mathbf{d}$, it is necessary and sufficient that

$$\mathbf{P}_{\mathbf{X},\mathbf{W}'}\mathbf{V}(\mathbf{I}-\mathbf{P}'_{\mathbf{X},\mathbf{W}'}) = \mathbf{0} \quad \text{and} \quad \text{rank}(\mathbf{X}'\mathbf{WX}) = \text{rank}(\mathbf{X})$$

or, equivalently, that

$$\mathbf{P}_{\mathbf{X},\mathbf{W}'}\mathbf{V} = \mathbf{P}_{\mathbf{X},\mathbf{W}'}\mathbf{V}\mathbf{P}'_{\mathbf{X},\mathbf{W}'} \text{ and } \operatorname{rank}(\mathbf{X}'\mathbf{W}\mathbf{X}) = \operatorname{rank}(\mathbf{X}).$$

It is also necessary and sufficient that

$$\mathbf{P}_{\mathbf{X},\mathbf{W}'}\mathbf{V} = \mathbf{V}\mathbf{P}'_{\mathbf{X},\mathbf{W}'} \text{ and } \operatorname{rank}(\mathbf{X}'\mathbf{W}\mathbf{X}) = \operatorname{rank}(\mathbf{X})$$

or, equivalently, that $\mathbf{P}_{\mathbf{X},\mathbf{W}'}\mathbf{V}$ be symmetric and $\operatorname{rank}(\mathbf{X}'\mathbf{W}\mathbf{X}) = \operatorname{rank}(\mathbf{X})$.

Preliminary to proving Theorem 19.5.4, it is convenient to establish the following lemma.

Lemma 19.5.5. Let $\mathbf{W}$ represent an $n \times n$ matrix and $\mathbf{X}$ an $n \times p$ matrix. If $\operatorname{rank}(\mathbf{X}'\mathbf{W}\mathbf{X}) = \operatorname{rank}(\mathbf{X})$, then (1) $\mathbf{P}_{\mathbf{X},\mathbf{W}}\mathbf{X} = \mathbf{X}$, (2) $\mathbf{X}'\mathbf{W}\mathbf{P}_{\mathbf{X},\mathbf{W}} = \mathbf{X}'\mathbf{W}$, (3) $\mathbf{P}^2_{\mathbf{X},\mathbf{W}} = \mathbf{P}_{\mathbf{X},\mathbf{W}}$ (i.e., $\mathbf{P}_{\mathbf{X},\mathbf{W}}$ is idempotent), (4) $\operatorname{rank}(\mathbf{P}_{\mathbf{X},\mathbf{W}}) = \operatorname{rank}(\mathbf{X})$, and (5) $\operatorname{rank}(\mathbf{I} - \mathbf{P}_{\mathbf{X},\mathbf{W}}) = n - \operatorname{rank}(\mathbf{X})$.

Proof (of Lemma 19.5.5). Suppose that $\operatorname{rank}(\mathbf{X}'\mathbf{W}\mathbf{X}) = \operatorname{rank}(\mathbf{X})$ [and recall that $\mathbf{P}_{\mathbf{X},\mathbf{W}} = \mathbf{X}(\mathbf{X}'\mathbf{W}\mathbf{X})^-\mathbf{X}'\mathbf{W}$].

(1)–(2) That $\mathbf{P}_{\mathbf{X},\mathbf{W}}\mathbf{X} = \mathbf{X}$ and $\mathbf{X}'\mathbf{W}\mathbf{P}_{\mathbf{X},\mathbf{W}} = \mathbf{X}'\mathbf{W}$ is evident from Lemma 19.3.1.

(3) Using Part (1), we find that

$$\mathbf{P}^2_{\mathbf{X},\mathbf{W}} = \mathbf{P}_{\mathbf{X},\mathbf{W}}\mathbf{X}(\mathbf{X}'\mathbf{W}\mathbf{X})^-\mathbf{X}'\mathbf{W} = \mathbf{X}(\mathbf{X}'\mathbf{W}\mathbf{X})^-\mathbf{X}'\mathbf{W} = \mathbf{P}_{\mathbf{X},\mathbf{W}}.$$

(4)–(5) By definition, $\mathbf{P}_{\mathbf{X},\mathbf{W}} = \mathbf{X}\mathbf{G}$, where $\mathbf{G} = (\mathbf{X}'\mathbf{W}\mathbf{X})^-\mathbf{X}'\mathbf{W}$. Since [according to Part (1)] $\mathbf{X}\mathbf{G}\mathbf{X} = \mathbf{X}$ (i.e., since $\mathbf{G}$ is a generalized inverse of $\mathbf{X}$), it follows from result (10.2.1) that $\operatorname{rank}(\mathbf{P}_{\mathbf{X},\mathbf{W}}) = \operatorname{rank}(\mathbf{X})$ and from Lemma 10.2.6 that $\operatorname{rank}(\mathbf{I} - \mathbf{P}_{\mathbf{X},\mathbf{W}}) = n - \operatorname{rank}(\mathbf{X})$. Q.E.D.

Proof (of Theorem 19.5.4). To establish the necessity and sufficiency of the conditions $\mathbf{P}_{\mathbf{X},\mathbf{W}'}\mathbf{V}(\mathbf{I} - \mathbf{P}'_{\mathbf{X},\mathbf{W}'}) = \mathbf{0}$ and $\operatorname{rank}(\mathbf{X}'\mathbf{W}\mathbf{X}) = \operatorname{rank}(\mathbf{X})$, it suffices (in light of Theorem 19.5.1) to show that these conditions are equivalent to the conditions $\mathcal{C}(\mathbf{V}\mathbf{W}\mathbf{X}) \subset \mathcal{C}(\mathbf{X})$ and $\operatorname{rank}(\mathbf{X}'\mathbf{W}\mathbf{X}) = \operatorname{rank}(\mathbf{X})$.

Suppose that $\operatorname{rank}(\mathbf{X}'\mathbf{W}\mathbf{X}) = \operatorname{rank}(\mathbf{X})$. Then, since $\mathbf{X}'\mathbf{W}'\mathbf{X} = (\mathbf{X}'\mathbf{W}\mathbf{X})'$, $\operatorname{rank}(\mathbf{X}'\mathbf{W}'\mathbf{X}) = \operatorname{rank}(\mathbf{X})$. Thus, applying Parts (1) and (2) of Lemma 19.5.5 (with $\mathbf{W}'$ in place of $\mathbf{W}$), we find that $\mathbf{P}_{\mathbf{X},\mathbf{W}'}\mathbf{X} = \mathbf{X}$ and $\mathbf{X}'\mathbf{W}'\mathbf{P}_{\mathbf{X},\mathbf{W}'} = \mathbf{X}'\mathbf{W}'$. Letting $\mathbf{Z} = \mathbf{I} - \mathbf{P}'_{\mathbf{X},\mathbf{W}'}$ and applying Part (5) of Lemma 19.5.5, we find also that $\mathbf{X}'\mathbf{Z} = (\mathbf{Z}'\mathbf{X})' = [(\mathbf{I} - \mathbf{P}_{\mathbf{X},\mathbf{W}'})\mathbf{X}]' = \mathbf{0}' = \mathbf{0}$ and $\operatorname{rank}(\mathbf{Z}) = \operatorname{rank}(\mathbf{Z}') = n - \operatorname{rank}(\mathbf{X})$, implying (in light of Lemma 11.4.1) that $\mathcal{C}(\mathbf{Z}) = \mathcal{N}(\mathbf{X}')$.

Now, if $\mathcal{C}(\mathbf{V}\mathbf{W}\mathbf{X}) \subset \mathcal{C}(\mathbf{X})$ or, equivalently, if $\mathbf{V}\mathbf{W}\mathbf{X} = \mathbf{X}\mathbf{Q}$ for some matrix $\mathbf{Q}$, then

$$\begin{aligned}
\mathbf{P}_{\mathbf{X},\mathbf{W}'}\mathbf{V}(\mathbf{I} - \mathbf{P}'_{\mathbf{X},\mathbf{W}'}) &= [(\mathbf{I} - \mathbf{P}_{\mathbf{X},\mathbf{W}'})\mathbf{V}\mathbf{P}'_{\mathbf{X},\mathbf{W}'}]' \\
&= \{(\mathbf{I} - \mathbf{P}_{\mathbf{X},\mathbf{W}'})\mathbf{V}\mathbf{W}\mathbf{X}[(\mathbf{X}'\mathbf{W}'\mathbf{X})^-]'\mathbf{X}'\}' \\
&= \{(\mathbf{I} - \mathbf{P}_{\mathbf{X},\mathbf{W}'})\mathbf{X}\mathbf{Q}[(\mathbf{X}'\mathbf{W}'\mathbf{X})^-]'\mathbf{X}'\}' \\
&= \{(\mathbf{X} - \mathbf{X})\mathbf{Q}[(\mathbf{X}'\mathbf{W}'\mathbf{X})^-]'\mathbf{X}'\}' = \mathbf{0}.
\end{aligned}$$

Conversely, if $\mathbf{P}_{\mathbf{X},\mathbf{W}'}\mathbf{V}(\mathbf{I}-\mathbf{P}'_{\mathbf{X},\mathbf{W}'}) = \mathbf{0}$, then

$$(\mathbf{VWX})'\mathbf{Z} = \mathbf{X}'\mathbf{W}'\mathbf{V}(\mathbf{I}-\mathbf{P}'_{\mathbf{X},\mathbf{W}'}) = \mathbf{X}'\mathbf{W}'\mathbf{P}_{\mathbf{X},\mathbf{W}'}\mathbf{V}(\mathbf{I}-\mathbf{P}'_{\mathbf{X},\mathbf{W}'}) = \mathbf{0},$$

implying (in light of Corollary 12.5.7) that $\mathcal{C}(\mathbf{VWX}) \subset \mathcal{C}(\mathbf{X})$.

We conclude that the conditions $\mathbf{P}_{\mathbf{X},\mathbf{W}'}\mathbf{V}(\mathbf{I}-\mathbf{P}'_{\mathbf{X},\mathbf{W}'}) = \mathbf{0}$ and $\text{rank}(\mathbf{X}'\mathbf{WX}) = \text{rank}(\mathbf{X})$ are equivalent to the conditions $\mathcal{C}(\mathbf{VWX}) \subset \mathcal{C}(\mathbf{X})$ and $\text{rank}(\mathbf{X}'\mathbf{WX}) = \text{rank}(\mathbf{X})$.

To complete the proof of Theorem 19.5.4, it suffices to show that the conditions $\mathbf{P}_{\mathbf{X},\mathbf{W}'}\mathbf{V} = \mathbf{VP}'_{\mathbf{X},\mathbf{W}'}$ and $\text{rank}(\mathbf{X}'\mathbf{WX}) = \text{rank}(\mathbf{X})$ are equivalent to the conditions $\mathbf{P}_{\mathbf{X},\mathbf{W}'}\mathbf{V} = \mathbf{P}_{\mathbf{X},\mathbf{W}'}\mathbf{VP}'_{\mathbf{X},\mathbf{W}'}$ and $\text{rank}(\mathbf{X}'\mathbf{WX}) = \text{rank}(\mathbf{X})$.

If $\mathbf{P}_{\mathbf{X},\mathbf{W}'}\mathbf{V} = \mathbf{VP}'_{\mathbf{X},\mathbf{W}'}$ and $\text{rank}(\mathbf{X}'\mathbf{WX}) = \text{rank}(\mathbf{X})$, then [since $\mathbf{X}'\mathbf{W}'\mathbf{X} = (\mathbf{X}'\mathbf{WX})'$] $\text{rank}(\mathbf{X}'\mathbf{W}'\mathbf{X}) = \text{rank}(\mathbf{X})$, implying [in light of Part (3) of Lemma 19.5.5] that $\mathbf{P}^2_{\mathbf{X},\mathbf{W}'} = \mathbf{P}_{\mathbf{X},\mathbf{W}'}$, so that

$$\mathbf{P}_{\mathbf{X},\mathbf{W}'}\mathbf{V} = \mathbf{P}_{\mathbf{X},\mathbf{W}'}(\mathbf{P}_{\mathbf{X},\mathbf{W}'}\mathbf{V}) = \mathbf{P}_{\mathbf{X},\mathbf{W}'}\mathbf{VP}'_{\mathbf{X},\mathbf{W}'}.$$

Conversely, if $\mathbf{P}_{\mathbf{X},\mathbf{W}'}\mathbf{V} = \mathbf{P}_{\mathbf{X},\mathbf{W}'}\mathbf{VP}'_{\mathbf{X},\mathbf{W}'}$, then $\mathbf{P}_{\mathbf{X},\mathbf{W}'}\mathbf{V}$ is symmetric, so that

$$\mathbf{P}_{\mathbf{X},\mathbf{W}'}\mathbf{V} = (\mathbf{P}_{\mathbf{X},\mathbf{W}'}\mathbf{V})' = \mathbf{VP}'_{\mathbf{X},\mathbf{W}'}.$$

Thus, the conditions $\mathbf{P}_{\mathbf{X},\mathbf{W}'}\mathbf{V} = \mathbf{VP}'_{\mathbf{X},\mathbf{W}'}$ and $\text{rank}(\mathbf{X}'\mathbf{WX}) = \text{rank}(\mathbf{X})$ are equivalent to the conditions $\mathbf{P}_{\mathbf{X},\mathbf{W}'}\mathbf{V} = \mathbf{P}_{\mathbf{X},\mathbf{W}'}\mathbf{VP}'_{\mathbf{X},\mathbf{W}'}$ and $\text{rank}(\mathbf{X}'\mathbf{WX}) = \text{rank}(\mathbf{X})$. Q.E.D.

19.6 Transformation of the Constrained Minimization Problem to an Unconstrained Minimization Problem

Let $\mathbf{a}$ represent an $n \times 1$ vector of variables, and impose on $\mathbf{a}$ the constraint $\mathbf{X}'\mathbf{a} = \mathbf{d}$, where $\mathbf{X}$ is an $n \times p$ matrix and $\mathbf{d}$ is a $p \times 1$ vector such that $\mathbf{d} \in \mathcal{C}(\mathbf{X}')$. Define $f(\mathbf{a}) = \mathbf{a}'\mathbf{Va} - 2\mathbf{b}'\mathbf{a}$, where $\mathbf{V}$ is an $n \times n$ symmetric nonnegative definite matrix and $\mathbf{b}$ is an $n \times 1$ vector such that $\mathbf{b} \in \mathcal{C}(\mathbf{V}, \mathbf{X})$.

In Section 19.3, an expression was obtained for the general form of a solution to the problem of minimizing $f(\mathbf{a})$ (under the constraint $\mathbf{X}'\mathbf{a} = \mathbf{d}$). This expression was derived by "solving" the linear system

$$\begin{pmatrix} \mathbf{V} & \mathbf{X} \\ \mathbf{X}' & \mathbf{0} \end{pmatrix} \begin{pmatrix} \mathbf{a} \\ \mathbf{r} \end{pmatrix} = \begin{pmatrix} \mathbf{b} \\ \mathbf{d} \end{pmatrix}$$

(in $\mathbf{a}$ and $\mathbf{r}$), which is the linear system obtained by equating the derivative of the Lagrangian function (with respect to $\mathbf{a}$) to the null vector and by adjoining the constraint.

An alternative expression for the general form of a solution to the constrained minimization problem can be derived by transforming this problem to an unconstrained minimization problem (and by then making use of the results of Section 19.1).

For purposes of deriving such an expression, let $\mathbf{a}_0 = (\mathbf{X}')^- \mathbf{d}$, and let $\mathbf{Z} = \mathbf{I} - (\mathbf{X}')^- \mathbf{X}'$. Or, more generally, let $\mathbf{a}_0$ represent any $n \times 1$ vector such that $\mathbf{X}'\mathbf{a}_0 = \mathbf{d}$, and let $\mathbf{Z}$ represent any $n \times k$ matrix whose columns span the null space of $\mathbf{X}'$. Then, in light of Theorem 11.2.3, $\mathbf{X}'\mathbf{a} = \mathbf{d}$ if and only if

$$\mathbf{a} = \mathbf{a}_0 + \mathbf{Z}\mathbf{u}$$

for some $k \times 1$ vector $\mathbf{u}$.

Now, for $\mathbf{u} \in \mathcal{R}^{k \times 1}$, define $h(\mathbf{u}) = f(\mathbf{a}_0 + \mathbf{Z}\mathbf{u})$. Then,

$$\begin{aligned} h(\mathbf{u}) &= \mathbf{u}'\mathbf{Z}'\mathbf{V}\mathbf{Z}\mathbf{u} + 2\mathbf{a}_0'\mathbf{V}\mathbf{Z}\mathbf{u} + \mathbf{a}_0'\mathbf{V}\mathbf{a}_0 - 2\mathbf{b}'\mathbf{a}_0 - 2\mathbf{b}'\mathbf{Z}\mathbf{u} \\ &= \mathbf{u}'\mathbf{Z}'\mathbf{V}\mathbf{Z}\mathbf{u} - 2[\mathbf{Z}'(\mathbf{b} - \mathbf{V}\mathbf{a}_0)]'\mathbf{u} + c, \end{aligned} \tag{6.1}$$

where $c = (\mathbf{V}\mathbf{a}_0 - 2\mathbf{b})'\mathbf{a}_0$ (which is a constant). And it is clear that the minimization of $f(\mathbf{a})$ (with respect to $\mathbf{a}$), subject to the constraint $\mathbf{X}'\mathbf{a} = \mathbf{d}$, is equivalent to the unconstrained minimization of $h(\mathbf{u})$ (with respect to $\mathbf{u}$). The equivalence is in the sense that $f(\mathbf{a})$ attains its minimum value (under the constraint $\mathbf{X}'\mathbf{a} = \mathbf{d}$) at a point $\mathbf{a}_*$ if and only if $\mathbf{a}_* = \mathbf{a}_0 + \mathbf{Z}\mathbf{u}_*$ for some $k \times 1$ vector $\mathbf{u}_*$ at which $h(\mathbf{u})$ attains its minimum value.

Since $\mathbf{b} \in \mathcal{C}(\mathbf{V}, \mathbf{X})$, $\mathbf{b} = \mathbf{V}\mathbf{r} + \mathbf{X}\mathbf{s}$ for some vectors $\mathbf{r}$ and $\mathbf{s}$, so that

$$\begin{aligned} \mathbf{Z}'(\mathbf{b} - \mathbf{V}\mathbf{a}_0) = \mathbf{Z}'(\mathbf{V}\mathbf{r} + \mathbf{X}\mathbf{s} - \mathbf{V}\mathbf{a}_0) &= \mathbf{Z}'\mathbf{V}(\mathbf{r} - \mathbf{a}_0) + (\mathbf{X}'\mathbf{Z})'\mathbf{s} \\ &= \mathbf{Z}'\mathbf{V}(\mathbf{r} - \mathbf{a}_0) + \mathbf{0} = \mathbf{Z}'\mathbf{V}(\mathbf{r} - \mathbf{a}_0), \end{aligned}$$

implying that $\mathbf{Z}'(\mathbf{b} - \mathbf{V}\mathbf{a}_0) \in \mathcal{C}(\mathbf{Z}'\mathbf{V})$ and hence (in light of Lemma 14.11.2) that $\mathbf{Z}'(\mathbf{b} - \mathbf{V}\mathbf{a}_0) \in \mathcal{C}(\mathbf{Z}'\mathbf{V}\mathbf{Z})$. Thus, it follows from Theorem 19.1.1 that the linear system $\mathbf{Z}'\mathbf{V}\mathbf{Z}\mathbf{u} = \mathbf{Z}'(\mathbf{b} - \mathbf{V}\mathbf{a}_0)$ (in $\mathbf{u}$) is consistent and that $h(\mathbf{u})$ has a minimum value at a point $\mathbf{u}_*$ if and only if $\mathbf{u}_*$ is a solution to $\mathbf{Z}'\mathbf{V}\mathbf{Z}\mathbf{u} = \mathbf{Z}'(\mathbf{b} - \mathbf{V}\mathbf{a}_0)$, in which case

$$h(\mathbf{u}_*) = -(\mathbf{b} - \mathbf{V}\mathbf{a}_0)'\mathbf{Z}\mathbf{u}_* + c = -\mathbf{b}'\mathbf{a}_0 - (\mathbf{b} - \mathbf{V}\mathbf{a}_0)'(\mathbf{a}_0 + \mathbf{Z}\mathbf{u}_*). \tag{6.2}$$

We conclude that $f(\mathbf{a})$ attains its minimum value (under the constraint $\mathbf{X}'\mathbf{a} = \mathbf{d}$) at a point $\mathbf{a}_*$ if and only if $\mathbf{a}_* = \mathbf{a}_0 + \mathbf{Z}\mathbf{u}_*$ for some solution $\mathbf{u}_*$ to $\mathbf{Z}'\mathbf{V}\mathbf{Z}\mathbf{u} = \mathbf{Z}'(\mathbf{b} - \mathbf{V}\mathbf{a}_0)$, or, equivalently (in light of Theorem 11.2.4), if and only if

$$\mathbf{a}_* = \mathbf{a}_0 + \mathbf{Z}(\mathbf{Z}'\mathbf{V}\mathbf{Z})^-\mathbf{Z}'(\mathbf{b} - \mathbf{V}\mathbf{a}_0) + \mathbf{Z}[\mathbf{I} - (\mathbf{Z}'\mathbf{V}\mathbf{Z})^-\mathbf{Z}'\mathbf{V}\mathbf{Z}]\mathbf{w} \tag{6.3}$$

for some $k \times 1$ vector $\mathbf{w}$, in which case

$$f(\mathbf{a}_*) = -\mathbf{b}'\mathbf{a}_0 - (\mathbf{b} - \mathbf{V}\mathbf{a}_0)'\mathbf{a}_*. \tag{6.4}$$

19.7 The Effect of Constraints on the Generalized Least Squares Problem

Let $\mathbf{X}$ represent an $n \times p$ matrix, $\mathbf{W}$ an $n \times n$ symmetric nonnegative definite matrix, and $\mathbf{y}$ an n-dimensional column vector. Further, let $\mathbf{b}$ represent a $p \times 1$ vector of variables, and consider the minimization of the quadratic form $(\mathbf{y}-\mathbf{Xb})'\mathbf{W}(\mathbf{y}-\mathbf{Xb})$ (in the difference between $\mathbf{y}$ and $\mathbf{Xb}$), subject to the constraint

$$\mathbf{A}'\mathbf{b} = \mathbf{d},$$

where $\mathbf{A}$ is a $p \times q$ matrix and $\mathbf{d}$ is a $q \times 1$ vector in $\mathcal{C}(\mathbf{A}')$.

The unconstrained minimization of $(\mathbf{y} - \mathbf{Xb})'\mathbf{W}(\mathbf{y} - \mathbf{Xb})$ was considered in Section 12.4 (in the special case where $\mathbf{W} = \mathbf{I}$), in Section 14.12f (in the special case where $\mathbf{W}$ is positive definite), and in Section 14.12i (in the general case). It was found that (in the absence of constraints) $(\mathbf{y} - \mathbf{Xb})'\mathbf{W}(\mathbf{y} - \mathbf{Xb})$ attains its minimum value at a point $\mathbf{b}_*$ if and only if $\mathbf{b}_*$ is a solution to the linear system

$$\mathbf{X}'\mathbf{W}\mathbf{X}\mathbf{b} = \mathbf{X}'\mathbf{W}\mathbf{y}$$

(in $\mathbf{b}$), in which case

$$(\mathbf{y} - \mathbf{Xb}_*)'\mathbf{W}(\mathbf{y} - \mathbf{Xb}_*) = \mathbf{y}'\mathbf{W}(\mathbf{y} - \mathbf{Xb}_*).$$

The results of Section 19.3 can be used to generalize our results on the unconstrained minimization of $(\mathbf{y} - \mathbf{Xb})'\mathbf{W}(\mathbf{y} - \mathbf{Xb})$ to the case where $\mathbf{b}$ is subject to the constraint $\mathbf{A}'\mathbf{b} = \mathbf{d}$. For this purpose, reexpress $(\mathbf{y} - \mathbf{Xb})'\mathbf{W}(\mathbf{y} - \mathbf{Xb})$ as

$$(\mathbf{y} - \mathbf{Xb})'\mathbf{W}(\mathbf{y} - \mathbf{Xb}) = \mathbf{b}'\mathbf{X}'\mathbf{W}\mathbf{X}\mathbf{b} - 2(\mathbf{X}'\mathbf{W}\mathbf{y})'\mathbf{b} + \mathbf{y}'\mathbf{W}\mathbf{y},$$

and observe (in light of Theorem 14.2.9) that $\mathbf{X}'\mathbf{W}\mathbf{X}$ is nonnegative definite (and symmetric) and (in light of Lemma 14.11.2) that $\mathbf{X}'\mathbf{W}\mathbf{y} \in \mathcal{C}(\mathbf{X}'\mathbf{W}\mathbf{X})$.

Now, applying Theorem 19.2.1 (with $\mathbf{b}$, $\mathbf{A}$, $\mathbf{X}'\mathbf{W}\mathbf{X}$, and $\mathbf{X}'\mathbf{W}\mathbf{y}$ in place of $\mathbf{a}$, $\mathbf{X}$, $\mathbf{V}$, and $\mathbf{b}$, respectively), we find that the linear system

$$\begin{pmatrix} \mathbf{X}'\mathbf{W}\mathbf{X} & \mathbf{A} \\ \mathbf{A}' & \mathbf{0} \end{pmatrix} \begin{pmatrix} \mathbf{b} \\ \mathbf{r} \end{pmatrix} = \begin{pmatrix} \mathbf{X}'\mathbf{W}\mathbf{y} \\ \mathbf{d} \end{pmatrix} \tag{7.1}$$

(in $\mathbf{b}$ and the $q \times 1$ vector $\mathbf{r}$) is consistent (and that the values of $\mathbf{X}'\mathbf{W}\mathbf{X}\mathbf{b}$ and $\mathbf{Ar}$ are invariant to the choice of solution to this linear system). Further, $(\mathbf{y} - \mathbf{Xb})'\mathbf{W}(\mathbf{y} - \mathbf{Xb})$ has a minimum value (under the constraint $\mathbf{A}'\mathbf{b} = \mathbf{d}$) at a point $\mathbf{b}_*$ if and only if $\mathbf{b}_*$ is the first part of some solution to linear system (7.1). And, for the first and second parts $\mathbf{b}_*$ and $\mathbf{r}_*$ of any solution to linear system (7.1),

$$\begin{aligned} (\mathbf{y} - \mathbf{Xb}_*)'\mathbf{W}(\mathbf{y} - \mathbf{Xb}_*) &= \mathbf{y}'\mathbf{W}\mathbf{y} - (\mathbf{X}'\mathbf{W}\mathbf{y})'\mathbf{b}_* - \mathbf{d}'\mathbf{r}_* && (7.2) \\ &= \mathbf{y}'\mathbf{W}(\mathbf{y} - \mathbf{Xb}_*) - \mathbf{d}'\mathbf{r}_*. && (7.3) \end{aligned}$$

Linear system (7.1) comprises the two equations

$$\mathbf{X'WXb} + \mathbf{Ar} = \mathbf{X'Wy}, \tag{7.4}$$
$$\mathbf{A'b} = \mathbf{d}. \tag{7.5}$$

If $\mathcal{C}(\mathbf{A}) \subset \mathcal{C}(\mathbf{X'WX})$, then (according to the results of Section 19.3b) $\mathbf{b}$ and $\mathbf{r}$ satisfy equations (7.4) and (7.5) if and only if $\mathbf{r}$ satisfies the equation

$$\mathbf{A'(X'WX)^{-}Ar} = \mathbf{A'(X'WX)^{-}X'Wy} - \mathbf{d} \tag{7.6}$$

and

$$\mathbf{b} = \mathbf{(X'WX)^{-}X'Wy} - \mathbf{(X'WX)^{-}Ar} + [\mathbf{I} - \mathbf{(X'WX)^{-}X'WX}]\mathbf{k} \tag{7.7}$$

for some $p \times 1$ vector $\mathbf{k}$. Note (in light of Lemma 14.11.2 and Corollary 14.11.3) that the condition $\mathcal{C}(\mathbf{A}) \subset \mathcal{C}(\mathbf{X'WX})$ is equivalent to the condition $\mathcal{C}(\mathbf{A}) \subset \mathcal{C}(\mathbf{X'W})$ and that, in the special case where $\mathbf{W}$ is positive definite, it is equivalent to the condition $\mathcal{C}(\mathbf{A}) \subset \mathcal{C}(\mathbf{X'})$.

Turning now to the "solution" of linear system (7.1) in the general case [where $\mathcal{C}(\mathbf{A})$ is not necessarily contained in $\mathcal{C}(\mathbf{X'WX})$], let $\mathbf{U}$ represent any $q \times q$ matrix such that $\mathcal{C}(\mathbf{A}) \subset \mathcal{C}(\mathbf{X'WX} + \mathbf{AUA'})$, and let $\mathbf{G}$ represent any generalized inverse of $\mathbf{X'WX} + \mathbf{AUA'}$. Then, according to Theorem 19.3.3, $\mathbf{b}_*$ and $\mathbf{r}_*$ are, respectively, the first and second parts of a solution to linear system (7.1) if and only if

$$\mathbf{r}_* = \mathbf{t}_* + \mathbf{Ud} \tag{7.8}$$

and

$$\mathbf{b}_* = \mathbf{GX'Wy} + \mathbf{GAt}_* + [\mathbf{I} - \mathbf{G(X'WX} + \mathbf{AUA')}]\mathbf{k} \tag{7.9}$$

for some solution $\mathbf{t}_*$ to the (consistent) linear system

$$\mathbf{A'GAt} = \mathbf{A'GX'Wy} - \mathbf{d} \tag{7.10}$$

(in a $q \times 1$ vector $\mathbf{t}$) and for some $p \times 1$ vector $\mathbf{k}$. In particular, $\mathbf{b}_*$ is the first part of some solution to linear system (7.1) if and only if

$$\mathbf{b}_* = \mathbf{G(I} - \mathbf{P_{A,G})X'Wy} + \mathbf{GA(A'GA)^{-}d} \\ + [\mathbf{I} - \mathbf{G(X'WX} + \mathbf{AUA')}]\mathbf{k} \tag{7.11}$$

for some vector $\mathbf{k}$.

Exercises

Section 19.1

1. Let $\mathbf{a}$ represent an $n \times 1$ vector of (unconstrained) variables, and define $f(\mathbf{a}) = \mathbf{a'Va} - 2\mathbf{b'a}$, where $\mathbf{V}$ is an $n \times n$ matrix and $\mathbf{b}$ an $n \times 1$ vector. Show that if $\mathbf{V}$ is not nonnegative definite or if $\mathbf{b} \notin \mathcal{C}(\mathbf{V})$, then $f(\mathbf{a})$ is unbounded from below; that is, corresponding to any scalar c, there exists a vector $\mathbf{a}_*$ such that $f(\mathbf{a}_*) < c$.

Section 19.3

2. Let $\mathbf{V}$ represent an $n \times n$ symmetric matrix and $\mathbf{X}$ an $n \times p$ matrix. Show that, for any $p \times p$ matrix $\mathbf{U}$ such that $\mathcal{C}(\mathbf{X}) \subset \mathcal{C}(\mathbf{V} + \mathbf{XUX}')$,

   (1) $(\mathbf{V} + \mathbf{XUX}')(\mathbf{V} + \mathbf{XUX}')^{-}\mathbf{V} = \mathbf{V}$;

   (2) $\mathbf{V}(\mathbf{V} + \mathbf{XUX}')^{-}(\mathbf{V} + \mathbf{XUX}') = \mathbf{V}$.

3. In Section 19.3c, Theorem 19.3.3 was proved "from scratch." Devise a shorter proof of Theorem 19.3.3 by taking advantage of Theorem 11.11.1.

4. Let $\mathbf{V}$ represent an $n \times n$ symmetric matrix and $\mathbf{X}$ an $n \times p$ matrix. Further, let $\mathbf{U} = \mathbf{X}'\mathbf{TT}'\mathbf{X}$, where $\mathbf{T}$ is any matrix whose columns span the null space of $\mathbf{V}$. Show that $\mathcal{C}(\mathbf{X}) \subset \mathcal{C}(\mathbf{V} + \mathbf{XUX}')$ and that $\mathcal{C}(\mathbf{V})$ and $\mathcal{C}(\mathbf{XUX}')$ are essentially disjoint and $\mathcal{R}(\mathbf{V})$ and $\mathcal{R}(\mathbf{XUX}')$ are essentially disjoint, as has already been established (in Section 19.3c) in the special case where $\mathbf{V}$ is nonnegative definite.

5. Let $\mathbf{V}$ represent an $n \times n$ symmetric nonnegative definite matrix and $\mathbf{X}$ an $n \times p$ matrix. Further, let $\mathbf{Z}$ represent any matrix whose columns span $\mathcal{N}(\mathbf{X}')$ or, equivalently, $\mathcal{C}^{\perp}(\mathbf{X})$. And adopt the same terminology as in Exercise 17.20.

   (a) Using the result of Part (a) of Exercise 17.20 (or otherwise) show that an $n \times n$ matrix $\mathbf{H}$ is a projection matrix for $\mathcal{C}(\mathbf{X})$ along $\mathcal{C}(\mathbf{VZ})$ if and only if $\mathbf{H}'$ is the first $(n \times n)$ part of a solution to the consistent linear system

$$\begin{pmatrix} \mathbf{V} & \mathbf{X} \\ \mathbf{X}' & \mathbf{0} \end{pmatrix} \begin{pmatrix} \mathbf{A} \\ \mathbf{R} \end{pmatrix} = \begin{pmatrix} \mathbf{0} \\ \mathbf{X}' \end{pmatrix} \tag{E.1}$$

   (in an $n \times n$ matrix $\mathbf{A}$ and a $p \times n$ matrix $\mathbf{R}$).

   (b) Letting $\mathbf{U}$ represent any $p \times p$ matrix such that $\mathcal{C}(\mathbf{X}) \subset \mathcal{C}(\mathbf{V} + \mathbf{XUX}')$ and letting $\mathbf{W}$ represent an arbitrary generalized inverse of $\mathbf{V} + \mathbf{XUX}'$, show that an $n \times n$ matrix $\mathbf{H}$ is a projection matrix for $\mathcal{C}(\mathbf{X})$ along $\mathcal{C}(\mathbf{VZ})$ if and only if

$$\mathbf{H} = \mathbf{P}_{\mathbf{X},\mathbf{W}} + \mathbf{K}[\mathbf{I} - (\mathbf{V} + \mathbf{XUX}')\mathbf{W}]$$

   for some $n \times n$ matrix $\mathbf{K}$.

Section 19.5

6. Let $\mathbf{V}$ represent an $n \times n$ symmetric nonnegative definite matrix, $\mathbf{W}$ an $n \times n$ matrix, and $\mathbf{X}$ an $n \times p$ matrix. Show that, for the matrix $\mathbf{WX}(\mathbf{X}'\mathbf{WX})^{-}\mathbf{X}'$ to be the first $(n \times n)$ part of some solution to the (consistent) linear system

$$\begin{pmatrix} \mathbf{V} & \mathbf{X} \\ \mathbf{X}' & \mathbf{0} \end{pmatrix} \begin{pmatrix} \mathbf{A} \\ \mathbf{R} \end{pmatrix} = \begin{pmatrix} \mathbf{0} \\ \mathbf{X}' \end{pmatrix} \tag{E.2}$$

(in an $n \times n$ matrix $\mathbf{A}$ and a $p \times n$ matrix $\mathbf{R}$), it is necessary and sufficient that $\mathcal{C}(\mathbf{VWX}) \subset \mathcal{C}(\mathbf{X})$ and $\text{rank}(\mathbf{X}'\mathbf{WX}) = \text{rank}(\mathbf{X})$.

7. Let $\mathbf{a}$ represent an $n \times 1$ vector of variables, and impose on $\mathbf{a}$ the constraint $\mathbf{X}'\mathbf{a} = \mathbf{d}$, where $\mathbf{X}$ is an $n \times p$ matrix and $\mathbf{d}$ is a $p \times 1$ vector such that $\mathbf{d} \in \mathcal{C}(\mathbf{X}')$. Define $f(\mathbf{a}) = \mathbf{a}'\mathbf{Va} - 2\mathbf{b}'\mathbf{a}$, where $\mathbf{V}$ is an $n \times n$ symmetric nonnegative definite matrix and $\mathbf{b}$ is an $n \times 1$ vector such that $\mathbf{b} \in \mathcal{C}(\mathbf{V}, \mathbf{X})$. Further, define $g(\mathbf{a}) = \mathbf{a}'(\mathbf{V} + \mathbf{W})\mathbf{a} - 2(\mathbf{b} + \mathbf{c})'\mathbf{a}$, where $\mathbf{W}$ is any $n \times n$ matrix such that $\mathcal{C}(\mathbf{W}) \subset \mathcal{C}(\mathbf{X})$ and $\mathcal{R}(\mathbf{W}) \subset \mathcal{R}(\mathbf{X}')$ and where $\mathbf{c}$ is any $n \times 1$ vector in $\mathcal{C}(\mathbf{X})$. Show that the constrained (by $\mathbf{X}'\mathbf{a} = \mathbf{d}$) minimization of $g(\mathbf{a})$ is equivalent to the constrained minimization of $f(\mathbf{a})$ [in the sense that $g(\mathbf{a})$ and $f(\mathbf{a})$ attain their minimum values at the same points].

8. Let $\mathbf{V}$ represent an $n \times n$ symmetric nonnegative definite matrix, $\mathbf{W}$ an $n \times n$ matrix, $\mathbf{X}$ an $n \times p$ matrix, $\mathbf{f}$ an $n \times 1$ vector, and $\mathbf{d}$ a $p \times 1$ vector. Further, let $\mathbf{b}$ represent an $n \times 1$ vector such that $\mathbf{b} \in \mathcal{C}(\mathbf{V}, \mathbf{X})$. Generalize Theorem 19.5.1 by showing that, for the vector $\mathbf{W}(\mathbf{I} - \mathbf{P}_{\mathbf{X},\mathbf{W}})\mathbf{f} + \mathbf{WX}(\mathbf{X}'\mathbf{WX})^{-}\mathbf{d}$ to be a solution, for every $\mathbf{d} \in \mathcal{C}(\mathbf{X}')$, to the problem of minimizing the second-degree polynomial $\mathbf{a}'\mathbf{Va} - 2\mathbf{b}'\mathbf{a}$ (in $\mathbf{a}$) subject to $\mathbf{X}'\mathbf{a} = \mathbf{d}$, it is necessary and sufficient that

$$\mathbf{VWf} - \mathbf{b} \in \mathcal{C}(\mathbf{X}), \tag{E.3}$$

$$\mathcal{C}(\mathbf{VWX}) \subset \mathcal{C}(\mathbf{X}), \tag{E.4}$$

and

$$\text{rank}(\mathbf{X}'\mathbf{WX}) = \text{rank}(\mathbf{X}). \tag{E.5}$$

9. Let $\mathbf{V}$ and $\mathbf{W}$ represent $n \times n$ matrices, and let $\mathbf{X}$ represent an $n \times p$ matrix. Show that if $\mathbf{V}$ and $\mathbf{W}$ are nonsingular, then the condition $\mathcal{C}(\mathbf{VWX}) \subset \mathcal{C}(\mathbf{X})$ is equivalent to the condition $\mathcal{C}(\mathbf{V}^{-1}\mathbf{X}) \subset \mathcal{C}(\mathbf{WX})$ and is also equivalent to the condition $\mathcal{C}(\mathbf{W}^{-1}\mathbf{V}^{-1}\mathbf{X}) \subset \mathcal{C}(\mathbf{X})$.

10. Let $\mathbf{V}$ represent an $n \times n$ symmetric positive definite matrix, $\mathbf{W}$ an $n \times n$ matrix, $\mathbf{X}$ an $n \times p$ matrix, and $\mathbf{d}$ a $p \times 1$ vector. Show that, for the vector $\mathbf{WX}(\mathbf{X}'\mathbf{WX})^{-}\mathbf{d}$ to be a solution, for every $\mathbf{d} \in \mathcal{C}(\mathbf{X}')$, to the problem of minimizing the quadratic form $\mathbf{a}'\mathbf{Va}$ (in $\mathbf{a}$) subject to $\mathbf{X}'\mathbf{a} = \mathbf{d}$, it is necessary and sufficient that $\mathbf{V}^{-1}\mathbf{P}_{\mathbf{X},\mathbf{W}'}$ be symmetric, and $\text{rank}(\mathbf{X}'\mathbf{WX}) = \text{rank}(\mathbf{X})$. Show that it is also necessary and sufficient that $(\mathbf{I} - \mathbf{P}'_{\mathbf{X},\mathbf{W}'})\mathbf{V}^{-1}\mathbf{P}_{\mathbf{X},\mathbf{W}'} = \mathbf{0}$ and $\text{rank}(\mathbf{X}'\mathbf{WX}) = \text{rank}(\mathbf{X})$.

11. Let $\mathbf{V}$ represent an $n \times n$ symmetric nonnegative definite matrix, $\mathbf{X}$ an $n \times p$ matrix, and $\mathbf{d}$ a $p \times 1$ vector. Show that each of the following six conditions is necessary and sufficient for the vector $\mathbf{X}(\mathbf{X}'\mathbf{X})^{-}\mathbf{d}$ to be a solution for every $\mathbf{d} \in \mathcal{C}(\mathbf{X}')$ to the problem of minimizing the quadratic form $\mathbf{a}'\mathbf{Va}$ (in $\mathbf{a}$) subject to $\mathbf{X}'\mathbf{a} = \mathbf{d}$:

(a) $\mathcal{C}(\mathbf{VX}) \subset \mathcal{C}(\mathbf{X})$ (or, equivalently, $\mathbf{VX} = \mathbf{XQ}$ for some matrix $\mathbf{Q}$);

(b) $\mathbf{P_X V}(\mathbf{I} - \mathbf{P_X}) = \mathbf{0}$ (or, equivalently, $\mathbf{P_X V} = \mathbf{P_X V P_X}$);

(c) $\mathbf{P_X V} = \mathbf{V P_X}$ (or, equivalently, $\mathbf{P_X V}$ is symmetric);

(d) $\mathcal{C}(\mathbf{V P_X}) \subset \mathcal{C}(\mathbf{P_X})$;

(e) $\mathcal{C}(\mathbf{V P_X}) = \mathcal{C}(\mathbf{V}) \cap \mathcal{C}(\mathbf{P_X})$;

(f) $\mathcal{C}(\mathbf{VX}) = \mathcal{C}(\mathbf{V}) \cap \mathcal{C}(\mathbf{X})$.

12. Let $\mathbf{V}$ represent an $n \times n$ symmetric nonnegative definite matrix, $\mathbf{W}$ an $n \times n$ matrix, $\mathbf{X}$ an $n \times p$ matrix, and $\mathbf{d}$ a $p \times 1$ vector. Further, let $\mathbf{K}$ represent any $n \times q$ matrix such that $\mathcal{C}(\mathbf{K}) = \mathcal{C}(\mathbf{I} - \mathbf{P}_{\mathbf{X},\mathbf{W}'})$. Show that if $\text{rank}(\mathbf{X}'\mathbf{W}\mathbf{X}) = \text{rank}(\mathbf{X})$, then each of the following two conditions is necessary and sufficient for the vector $\mathbf{WX}(\mathbf{X}'\mathbf{WX})^{-}\mathbf{d}$ to be a solution, for every $\mathbf{d} \in \mathcal{C}(\mathbf{X}')$, to the problem of minimizing the quadratic form $\mathbf{a}'\mathbf{V}\mathbf{a}$ (in $\mathbf{a}$) subject to $\mathbf{X}'\mathbf{a} = \mathbf{d}$:

(a) $\mathbf{V} = \mathbf{X}\mathbf{R}_1\mathbf{X}' + (\mathbf{I} - \mathbf{P}_{\mathbf{X},\mathbf{W}'})\mathbf{R}_2(\mathbf{I} - \mathbf{P}_{\mathbf{X},\mathbf{W}'})'$
for some $p \times p$ matrix $\mathbf{R}_1$ and some $n \times n$ matrix $\mathbf{R}_2$;

(b) $\mathbf{V} = \mathbf{X}\mathbf{S}_1\mathbf{X}' + \mathbf{K}\mathbf{S}_2\mathbf{K}'$
for some $p \times p$ matrix $\mathbf{S}_1$ and some $q \times q$ matrix $\mathbf{S}_2$.

And show that if $\text{rank}(\mathbf{X}'\mathbf{W}\mathbf{X}) = \text{rank}(\mathbf{X})$ and $\mathbf{W}$ is nonsingular, then another necessary and sufficient condition is:

(c) $\mathbf{V} = t\mathbf{W}^{-1} + \mathbf{X}\mathbf{T}_1\mathbf{X}' + \mathbf{K}\mathbf{T}_2\mathbf{K}'$
for some scalar t, some $p \times p$ matrix $\mathbf{T}_1$, and some $q \times q$ matrix $\mathbf{T}_2$.

[*Hint*. To establish the necessity of Condition (a), begin by observing that $\mathbf{V} = \mathbf{CC}'$ for some matrix $\mathbf{C}$ and by expressing $\mathbf{C}$ as $\mathbf{C} = \mathbf{P}_{\mathbf{X},\mathbf{W}'}\mathbf{C} + (\mathbf{I} - \mathbf{P}_{\mathbf{X},\mathbf{W}'})\mathbf{C}$.]

Bibliographic and Supplementary Notes

§2–§5. Many of the results presented in Sections 2–5 are adaptations of results reported by Rao (1989), Baksalary and Puntanen (1989), and Puntanen and Styan (1989). §6. The approach taken in Section 6 is essentially the same as that taken by Albert (1972, sec. 6.1).

20

The Moore-Penrose Inverse

By definition, a generalized inverse of an $m \times n$ matrix $\mathbf{A}$ is any $n \times m$ matrix $\mathbf{G}$ such that $\mathbf{AGA} = \mathbf{A}$. Except for the special case where $\mathbf{A}$ is a (square) nonsingular matrix, $\mathbf{A}$ has an infinite number of generalized inverses (as discussed in Section 9.2a). While for many purposes one generalized inverse is as good as another, there is a unique one of the generalized inverses, known as the Moore-Penrose inverse, that is sometimes singled out for special attention and that is the primary subject of the present chapter.

20.1 Definition, Existence, and Uniqueness (of the Moore-Penrose Inverse)

The Moore-Penrose inverse of an $m \times n$ matrix $\mathbf{A}$ can be described in terms of the conditions given by the following theorem.

Theorem 20.1.1. Corresponding to any $m \times n$ matrix $\mathbf{A}$, there is a unique $n \times m$ matrix $\mathbf{G}$ such that

(1) $\mathbf{AGA} = \mathbf{A}$ (i.e., $\mathbf{G}$ is a generalized inverse of $\mathbf{A}$);

(2) $\mathbf{GAG} = \mathbf{G}$ (i.e., $\mathbf{A}$ is a generalized inverse of $\mathbf{G}$);

(3) $(\mathbf{AG})' = \mathbf{AG}$ (i.e., $\mathbf{AG}$ is symmetric); and

(4) $(\mathbf{GA})' = \mathbf{GA}$ (i.e., $\mathbf{GA}$ is symmetric).

Moreover, if $\mathbf{A} = \mathbf{0}$, then $\mathbf{G} = \mathbf{0}$; and if $\mathbf{A} \neq \mathbf{0}$, then

$$\mathbf{G} = \mathbf{T}'(\mathbf{B}'\mathbf{A}\mathbf{T}')^{-1}\mathbf{B}', \tag{1.1}$$

where $\mathbf{B}$ is any matrix of full column rank and $\mathbf{T}$ any matrix of full row rank such that $\mathbf{A} = \mathbf{BT}$.

In connection with Theorem 20.1.1, note that the existence of a matrix $\mathbf{B}$ of full column rank and a matrix $\mathbf{T}$ of full row rank such that $\mathbf{A} = \mathbf{BT}$ follows from Theorem 4.4.8. Note also that $\mathbf{B}'\mathbf{A}\mathbf{T}' = \mathbf{B}'\mathbf{B}\mathbf{T}\mathbf{T}'$ and that [since $\text{rank}(\mathbf{B}'\mathbf{B}) =$

rank($\mathbf{B}$) and rank($\mathbf{T}\mathbf{T}'$) = rank($\mathbf{T}$)] $\mathbf{B}'\mathbf{B}$ and $\mathbf{T}\mathbf{T}'$ are nonsingular, implying in particular that $\mathbf{B}'\mathbf{A}\mathbf{T}'$ is nonsingular. Further, $(\mathbf{B}'\mathbf{A}\mathbf{T}')^{-1} = (\mathbf{T}\mathbf{T}')^{-1}(\mathbf{B}'\mathbf{B})^{-1}$, so that, as an alternative to formula (1.1), we have the formula

$$\mathbf{G} = \mathbf{T}'(\mathbf{T}\mathbf{T}')^{-1}(\mathbf{B}'\mathbf{B})^{-1}\mathbf{B}'. \tag{1.2}$$

Proof (of Theorem 20.1.1). If $\mathbf{A} = \mathbf{0}$, then it is obvious that Conditions (1)–(4) are satisfied by taking $\mathbf{G} = \mathbf{0}$.

Suppose now that $\mathbf{A} \neq \mathbf{0}$, and take $\mathbf{G} = \mathbf{T}'(\mathbf{B}'\mathbf{A}\mathbf{T}')^{-1}\mathbf{B}'$ or, equivalently, take $\mathbf{G} = \mathbf{T}'(\mathbf{T}\mathbf{T}')^{-1}(\mathbf{B}'\mathbf{B})^{-1}\mathbf{B}'$. Then,

$$\mathbf{AGA} = \mathbf{B}\mathbf{T}\mathbf{T}'(\mathbf{T}\mathbf{T}')^{-1}(\mathbf{B}'\mathbf{B})^{-1}\mathbf{B}'\mathbf{B}\mathbf{T} = \mathbf{B}\mathbf{T} = \mathbf{A},$$

and

$$\begin{aligned}\mathbf{GAG} &= \mathbf{T}'(\mathbf{T}\mathbf{T}')^{-1}(\mathbf{B}'\mathbf{B})^{-1}\mathbf{B}'\mathbf{B}\mathbf{T}\mathbf{T}'(\mathbf{T}\mathbf{T}')^{-1}(\mathbf{B}'\mathbf{B})^{-1}\mathbf{B}'\\ &= \mathbf{T}'(\mathbf{T}\mathbf{T}')^{-1}(\mathbf{B}'\mathbf{B})^{-1}\mathbf{B}' = \mathbf{G},\end{aligned}$$

so that Conditions (1) and (2) are satisfied. Further,

$$\mathbf{AG} = \mathbf{B}\mathbf{T}\mathbf{T}'(\mathbf{T}\mathbf{T}')^{-1}(\mathbf{B}'\mathbf{B})^{-1}\mathbf{B}' = \mathbf{B}(\mathbf{B}'\mathbf{B})^{-1}\mathbf{B}' = \mathbf{P}_{\mathbf{B}},$$

and

$$\mathbf{GA} = \mathbf{T}'(\mathbf{T}\mathbf{T}')^{-1}(\mathbf{B}'\mathbf{B})^{-1}\mathbf{B}'\mathbf{B}\mathbf{T} = \mathbf{P}_{\mathbf{T}'},$$

implying [in light of Part (3) of Theorem 12.3.4] that $\mathbf{AG}$ and $\mathbf{GA}$ are symmetric.

We have established that Conditions (1)–(4) are satisfied by taking $\mathbf{G} = \mathbf{G}_*$, where $\mathbf{G}_*$ is the $n \times m$ matrix defined as follows: $\mathbf{G}_* = \mathbf{0}$, if $\mathbf{A} = \mathbf{0}$; $\mathbf{G}_* = \mathbf{T}'(\mathbf{B}'\mathbf{A}\mathbf{T}')^{-1}\mathbf{B}'$, if $\mathbf{A} \neq \mathbf{0}$. Moreover, for any $n \times m$ matrix $\mathbf{G}$ that satisfies Conditions (1)–(4),

$$\begin{aligned}\mathbf{G} &= \mathbf{GAG} = \mathbf{G}(\mathbf{AG})' = \mathbf{GG}'\mathbf{A}' = \mathbf{GG}'(\mathbf{AG}_*\mathbf{A})' = \mathbf{GG}'\mathbf{A}'(\mathbf{AG}_*)'\\ &= \mathbf{GG}'\mathbf{A}'\mathbf{AG}_* = \mathbf{G}(\mathbf{AG})'\mathbf{AG}_* = \mathbf{GAGAG}_* = \mathbf{GAG}_*\\ &= \mathbf{GAG}_*\mathbf{AG}_* = (\mathbf{GA})'(\mathbf{G}_*\mathbf{A})'\mathbf{G}_* = \mathbf{A}'\mathbf{G}'\mathbf{A}'\mathbf{G}_*'\mathbf{G}_* = (\mathbf{AGA})'\mathbf{G}_*'\mathbf{G}_*\\ &= \mathbf{A}'\mathbf{G}_*'\mathbf{G}_* = (\mathbf{G}_*\mathbf{A})'\mathbf{G}_* = \mathbf{G}_*\mathbf{AG}_* = \mathbf{G}_*.\end{aligned}$$

Thus, there is no choice for $\mathbf{G}$ other than $\mathbf{G} = \mathbf{G}_*$ that satisfies Conditions (1)–(4). Q.E.D.

The $n \times m$ matrix $\mathbf{G}$ defined (uniquely) by Conditions (1)–(4) of Theorem 20.1.1 is called the *Moore-Penrose inverse* (or *Moore-Penrose pseudoinverse*) of $\mathbf{A}$. (The practice of referring to this matrix as the Moore-Penrose inverse is widespread, although not universal—some other authors refer to this matrix as the generalized inverse or the pseudoinverse.)

The existence and uniqueness of the $n \times m$ matrix $\mathbf{G}$ that satisfies Conditions (1)–(4) was established by Penrose (1955). An earlier effort by Moore (1920) to extend the concept of an inverse matrix to (square) singular matrices and to rectangular matrices had led to a different characterization of the same matrix.

Conditions (1)–(4) are referred to as the *Penrose conditions* or (less commonly and perhaps less appropriately) as the *Moore-Penrose conditions*.

The symbol $\mathbf{A}^+$ is used to denote the Moore-Penrose inverse of a matrix $\mathbf{A}$. Clearly, $\mathbf{A}^+$ is a particular one of what may be an infinite number of generalized inverses of $\mathbf{A}$.

20.2 Some Special Cases

For any $n \times n$ nonsingular matrix $\mathbf{A}$,

$$\mathbf{A}^+ = \mathbf{A}^{-1},$$

as is evident from Lemma 9.1.1 or upon recalling that $\mathbf{A}\mathbf{A}^{-1} = \mathbf{A}^{-1}\mathbf{A} = \mathbf{I}_n$ and verifying that $\mathbf{A}^{-1}$ satisfies the Penrose conditions. More generally, for any $m \times n$ matrix $\mathbf{A}$ of rank m (i.e., of full row rank),

$$\mathbf{A}^+ = \mathbf{A}'(\mathbf{A}\mathbf{A}')^{-1} \tag{2.1}$$

—this result is a special case of result (1.1) (that where $\mathbf{B} = \mathbf{I}_m$ and $\mathbf{T} = \mathbf{A}$). Similarly, for any $m \times n$ matrix $\mathbf{A}$ of rank n (i.e., of full column rank)

$$\mathbf{A}^+ = (\mathbf{A}'\mathbf{A})^{-1}\mathbf{A}'. \tag{2.2}$$

For an $n \times n$ diagonal matrix $\mathbf{D} = \{d_i\}$,

$$\mathbf{D}^+ = \text{diag}(d_1^+, d_2^+, \ldots, d_n^+),$$

where (for $i = 1, \ldots, n$)

$$d_i^+ = \begin{cases} 1/d_i, & \text{if } d_i \neq 0, \\ 0, & \text{if } d_i = 0. \end{cases}$$

More generally, for any matrices $\mathbf{A}_1, \mathbf{A}_2, \ldots, \mathbf{A}_k$,

$$[\text{diag}(\mathbf{A}_1, \mathbf{A}_2, \ldots, \mathbf{A}_k)]^+ = \text{diag}(\mathbf{A}_1^+, \mathbf{A}_2^+, \ldots, \mathbf{A}_k^+),$$

as is easily verified.

The following lemmas are easy to verify.

Lemma 20.2.1. For any symmetric idempotent matrix $\mathbf{A}$, $\mathbf{A}^+ = \mathbf{A}$.

Lemma 20.2.2. For any $m \times n$ matrix $\mathbf{P}$ whose n columns form an orthonormal (with respect to the usual inner product) set (of m-dimensional vectors) or, equivalently, for any $m \times n$ matrix $\mathbf{P}$ such that $\mathbf{P}'\mathbf{P} = \mathbf{I}_n$, $\mathbf{P}^+ = \mathbf{P}'$.

20.3 Special Types of Generalized Inverses

If in choosing a generalized inverse $\mathbf{G}$ of a matrix $\mathbf{A}$ [or, equivalently, a matrix $\mathbf{G}$ that satisfies Penrose Condition (1)] we require that $\mathbf{G}$ satisfy Penrose Conditions (2)–(4), we obtain the Moore-Penrose inverse $\mathbf{A}^{+}$ of $\mathbf{A}$. A less restrictive approach is to require only that $\mathbf{G}$ satisfy a particular one or a particular two of Penrose Conditions (2)–(4). Generalized inverses that satisfy Penrose Condition (2), (4), or (3) are considered in Subsections a, b, and c, respectively.

a. Reflexive generalized inverses

A generalized inverse $\mathbf{G}$ of a matrix $\mathbf{A}$ is said to be *reflexive* if $\mathbf{GAG} = \mathbf{G}$. Thus, an $n \times m$ matrix $\mathbf{G}$ is a reflexive generalized inverse of an $m \times n$ matrix $\mathbf{A}$ if it satisfies Penrose Conditions (1) and (2).

The following theorem gives an expression for the general form of a reflexive generalized inverse.

Theorem 20.3.1. Let $\mathbf{A}$ represent an $m \times n$ nonnull matrix, and let $\mathbf{B}$ represent any matrix of full column rank and $\mathbf{T}$ any matrix of full row rank such that $\mathbf{A} = \mathbf{BT}$. Then, an $n \times m$ matrix $\mathbf{G}$ is a reflexive generalized inverse of $\mathbf{A}$ if and only if $\mathbf{G} = \mathbf{RL}$ for some right inverse $\mathbf{R}$ of $\mathbf{T}$ and some left inverse $\mathbf{L}$ of $\mathbf{B}$.

Proof. Suppose that $\mathbf{G} = \mathbf{RL}$ for some right inverse $\mathbf{R}$ of $\mathbf{T}$ and some left inverse $\mathbf{L}$ of $\mathbf{B}$. Then, according to Theorem 9.1.3, $\mathbf{G}$ is a generalized inverse of $\mathbf{A}$. Further,

$$\mathbf{GAG} = \mathbf{RLBTRL} = \mathbf{RIIL} = \mathbf{RL} = \mathbf{G},$$

so that $\mathbf{G}$ is reflexive.

Conversely, suppose that $\mathbf{G}$ is a reflexive generalized inverse of $\mathbf{A}$. Then,

$$\mathbf{BTGBT} = \mathbf{AGA} = \mathbf{A} = \mathbf{BT} = \mathbf{BIT},$$

and it follows from Part (3) of Lemma 8.3.1 that $\mathbf{TGB} = \mathbf{I}$. Thus, $\mathbf{GB}$ is a right inverse of $\mathbf{T}$, and $\mathbf{TG}$ is a left inverse of $\mathbf{B}$. Further,

$$\mathbf{G} = \mathbf{GAG} = (\mathbf{GB})\mathbf{TG}.$$

Q.E.D.

Any nonreflexive generalized inverse of a matrix $\mathbf{A}$ can be converted into a reflexive generalized inverse by making use of the corollary of the following lemma.

Lemma 20.3.2. Let $\mathbf{G}_1$ and $\mathbf{G}_2$ represent any (possibly different) generalized inverses of a matrix $\mathbf{A}$. Then, the matrix $\mathbf{G}_1\mathbf{A}\mathbf{G}_2$ is a reflexive generalized inverse of $\mathbf{A}$.

Proof. That $\mathbf{G}_1\mathbf{A}\mathbf{G}_2$ is a reflexive generalized inverse of $\mathbf{A}$ is evident upon observing that

$$\mathbf{A}(\mathbf{G}_1\mathbf{A}\mathbf{G}_2)\mathbf{A} = (\mathbf{A}\mathbf{G}_1\mathbf{A})\mathbf{G}_2\mathbf{A} = \mathbf{A}\mathbf{G}_2\mathbf{A} = \mathbf{A},$$

and

$$(\mathbf{G}_1\mathbf{A}\mathbf{G}_2)\mathbf{A}(\mathbf{G}_1\mathbf{A}\mathbf{G}_2) = \mathbf{G}_1(\mathbf{A}\mathbf{G}_2\mathbf{A})\mathbf{G}_1\mathbf{A}\mathbf{G}_2 = \mathbf{G}_1\mathbf{A}\mathbf{G}_1\mathbf{A}\mathbf{G}_2 = \mathbf{G}_1\mathbf{A}\mathbf{G}_2.$$

Q.E.D.

Corollary 20.3.3. Let $\mathbf{G}$ represent any generalized inverse of a matrix $\mathbf{A}$. Then, the matrix $\mathbf{GAG}$ is a reflexive generalized inverse of $\mathbf{A}$.

Proof. The corollary is essentially a special case of Lemma 20.3.2, namely, the special case where $\mathbf{G}_2 = \mathbf{G}_1$. Q.E.D.

Note (as a rather obvious converse of Corollary 20.3.3) that any reflexive generalized inverse $\mathbf{B}$ of a matrix $\mathbf{A}$ is expressible in the form $\mathbf{B} = \mathbf{GAG}$ for some generalized inverse $\mathbf{G}$ of $\mathbf{A}$—in particular, $\mathbf{B} = \mathbf{GAG}$ for $\mathbf{G} = \mathbf{B}$.

A very basic characteristic of reflexive generalized inverses is described in the following theorem.

Theorem 20.3.4. A generalized inverse $\mathbf{G}$ of a matrix $\mathbf{A}$ is reflexive if and only if $\text{rank}(\mathbf{G}) = \text{rank}(\mathbf{A})$.

Proof. Suppose that $\mathbf{G}$ is reflexive. Then, as a consequence of Lemma 9.3.9, $\text{rank}(\mathbf{G}) \geq \text{rank}(\mathbf{A})$ and also (since $\mathbf{A}$ is a generalized inverse of $\mathbf{G}$) $\text{rank}(\mathbf{A}) \geq \text{rank}(\mathbf{G})$. Thus, $\text{rank}(\mathbf{G}) = \text{rank}(\mathbf{A})$.

Conversely, suppose that $\text{rank}(\mathbf{G}) = \text{rank}(\mathbf{A})$. Then, in light of Lemma 10.2.5 and of Corollary 9.3.8 (or in light of Theorem 10.2.7), $\mathbf{GA}$ is idempotent, and

$$\text{rank}(\mathbf{GA}) = \text{rank}(\mathbf{A}) = \text{rank}(\mathbf{G}).$$

Thus, it follows from Theorem 10.2.7 that $\mathbf{A}$ is a generalized inverse of $\mathbf{G}$ (i.e., that $\mathbf{GAG} = \mathbf{G}$) and hence that $\mathbf{G}$ is a reflexive generalized inverse of $\mathbf{A}$. Q.E.D.

If an $n \times m$ matrix $\mathbf{G}$ is a generalized inverse of an $m \times n$ matrix $\mathbf{A}$, then (according to Lemma 9.3.3) $\mathbf{G}'$ is a generalized inverse of $\mathbf{A}'$. Similarly, if $\mathbf{A}$ is a generalized inverse of $\mathbf{G}$, then $\mathbf{A}'$ is a generalized inverse of $\mathbf{G}'$. Thus, we have the following lemma.

Lemma 20.3.5. If an $n \times m$ matrix $\mathbf{G}$ is a reflexive generalized inverse of an $m \times n$ matrix $\mathbf{A}$, then $\mathbf{G}'$ is a reflexive generalized inverse of $\mathbf{A}'$.

b. Minimum norm generalized inverses

A generalized inverse $\mathbf{G}$ of a matrix $\mathbf{A}$ is said to be a *minimum norm generalized inverse* (of $\mathbf{A}$) if $(\mathbf{GA})' = \mathbf{GA}$. Thus, an $n \times m$ matrix $\mathbf{G}$ is a minimum norm generalized inverse of an $m \times n$ matrix $\mathbf{A}$ if it satisfies Penrose Conditions (1) and (4).

The rationale for referring to such a matrix as a minimum norm generalized inverse comes from the following theorem.

Theorem 20.3.6. Let $\mathbf{A}$ represent an $m \times n$ matrix and $\mathbf{G}$ an $n \times m$ matrix. If $\mathbf{G}$ is a minimum norm generalized inverse of $\mathbf{A}$, then, for every $\mathbf{b} \in \mathcal{C}(\mathbf{A})$, $\|\mathbf{x}\|$ attains its minimum value over the set $\{\mathbf{x} : \mathbf{Ax} = \mathbf{b}\}$ [comprising all solutions to the linear system $\mathbf{Ax} = \mathbf{b}$ (in $\mathbf{x}$)] uniquely at $\mathbf{x} = \mathbf{Gb}$ (where the norm is the usual norm). Conversely, if, for every $\mathbf{b} \in \mathcal{C}(\mathbf{A})$, $\|\mathbf{x}\|$ attains its minimum value over the set $\{\mathbf{x} : \mathbf{Ax} = \mathbf{b}\}$ at $\mathbf{x} = \mathbf{Gb}$, then $\mathbf{G}$ is a minimum norm generalized inverse of $\mathbf{A}$.

Proof. Let $\mathbf{H}$ represent a minimum norm generalized inverse of $\mathbf{A}$. And let $\mathbf{x}$ represent an arbitrary solution to $\mathbf{Ax} = \mathbf{b}$ [where $\mathbf{b} \in \mathcal{C}(\mathbf{A})$]. Then, according to Theorem 11.2.4,

$$\mathbf{x} = \mathbf{Hb} + (\mathbf{I} - \mathbf{HA})\mathbf{y}$$

for some vector $\mathbf{y}$.

Further,

$$\begin{aligned}\|\mathbf{x}\|^2 &= [\mathbf{Hb} + (\mathbf{I} - \mathbf{HA})\mathbf{y}] \bullet [\mathbf{Hb} + (\mathbf{I} - \mathbf{HA})\mathbf{y}] \\ &= \|\mathbf{Hb}\|^2 + \|(\mathbf{I} - \mathbf{HA})\mathbf{y}\|^2 + 2(\mathbf{Hb}) \bullet (\mathbf{I} - \mathbf{HA})\mathbf{y}\end{aligned}$$

(where the inner product is the usual inner product). And, since $\mathbf{b} = \mathbf{Ar}$ for some vector $\mathbf{r}$ and since (according to Lemma 10.2.5) $\mathbf{HA}$ is idempotent,

$$\begin{aligned}(\mathbf{Hb}) \bullet (\mathbf{I} - \mathbf{HA})\mathbf{y} = (\mathbf{HAr}) \bullet (\mathbf{I} - \mathbf{HA})\mathbf{y} &= (\mathbf{HAr})'(\mathbf{I} - \mathbf{HA})\mathbf{y} \\ &= \mathbf{r}'(\mathbf{HA})'(\mathbf{I} - \mathbf{HA})\mathbf{y} \\ &= \mathbf{r}'\mathbf{HA}(\mathbf{I} - \mathbf{HA})\mathbf{y} \\ &= \mathbf{r}'[\mathbf{HA} - (\mathbf{HA})^2]\mathbf{y} \\ &= \mathbf{r}'(\mathbf{HA} - \mathbf{HA})\mathbf{y} = 0.\end{aligned}$$

Thus,

$$\|\mathbf{x}\|^2 = \|\mathbf{Hb}\|^2 + \|(\mathbf{I} - \mathbf{HA})\mathbf{y}\|^2,$$

so that $\|\mathbf{x}\|^2 \geq \|\mathbf{Hb}\|^2$, or, equivalently, $\|\mathbf{x}\| \geq \|\mathbf{Hb}\|$, with equality holding if and only if $\|(\mathbf{I} - \mathbf{HA})\mathbf{y}\| = 0$. Since $\|(\mathbf{I} - \mathbf{HA})\mathbf{y}\| = 0$ if and only if $(\mathbf{I} - \mathbf{HA})\mathbf{y} = \mathbf{0}$, and hence if and only if $\mathbf{x} = \mathbf{Hb}$ (and since $\mathbf{Hb}$ is a solution to $\mathbf{Ax} = \mathbf{b}$), we conclude that $\|\mathbf{x}\|$ attains its minimum value over the set $\{\mathbf{x} : \mathbf{Ax} = \mathbf{b}\}$ uniquely at $\mathbf{x} = \mathbf{Hb}$.

To complete the proof, suppose that, for every $\mathbf{b} \in \mathcal{C}(\mathbf{A})$, $\|\mathbf{x}\|$ attains its minimum value over the set $\{\mathbf{x} : \mathbf{Ax} = \mathbf{b}\}$ at $\mathbf{x} = \mathbf{Gb}$. Then, clearly, $\mathbf{Gb} = \mathbf{Hb}$ for every $\mathbf{b} \in \mathcal{C}(\mathbf{A})$, or, equivalently, $\mathbf{GAr} = \mathbf{HAr}$ for every $(n \times 1)$ vector $\mathbf{r}$, implying (in light of Lemma 2.3.2) that $\mathbf{GA} = \mathbf{HA}$. Thus, $(\mathbf{GA})' = (\mathbf{HA})' = \mathbf{HA} = \mathbf{GA}$. And, since (in light of Theorem 9.1.2) $\mathbf{G}$ is a generalized inverse of $\mathbf{A}$, we conclude that $\mathbf{G}$ is a minimum norm generalized inverse of $\mathbf{A}$. Q.E.D.

Two alternative characterizations of a minimum norm generalized inverse are given by the following theorem and corollary.

Theorem 20.3.7. An $n \times m$ matrix $\mathbf{G}$ is a minimum norm generalized inverse of an $m \times n$ matrix $\mathbf{A}$ if and only if $\mathbf{GAA}' = \mathbf{A}'$, or, equivalently, if and only if $\mathbf{AA}'\mathbf{G}' = \mathbf{A}$.

Proof. Suppose that $\mathbf{GAA}' = \mathbf{A}'$ or, equivalently, that $\mathbf{AA}'\mathbf{G}' = \mathbf{A}$. Then, $\mathbf{AGAA}' = \mathbf{AA}'$, implying (in light of Corollary 5.3.3) that $\mathbf{AGA} = \mathbf{A}$. Moreover,

$\mathbf{GA} = \mathbf{GAA'G'} = \mathbf{GA}(\mathbf{GA})'$, so that $\mathbf{GA}$ is symmetric. Thus, $\mathbf{G}$ is a minimum norm generalized inverse of $\mathbf{A}$.

Conversely, if $\mathbf{G}$ is a minimum norm generalized inverse of $\mathbf{A}$, then

$$\mathbf{GAA'} = (\mathbf{GA})'\mathbf{A'} = \mathbf{A'G'A'} = (\mathbf{AGA})' = \mathbf{A'}.$$

Q.E.D.

Corollary 20.3.8. An $n \times m$ matrix $\mathbf{G}$ is a minimum norm generalized inverse of an $m \times n$ matrix $\mathbf{A}$ if and only if $\mathbf{GA} = \mathbf{P_{A'}}$, or, equivalently, if and only if $\mathbf{A'G'} = \mathbf{P_{A'}}$.

Proof. In light of Theorem 20.3.7, it suffices to show that $\mathbf{GA} = \mathbf{P_{A'}} \Leftrightarrow \mathbf{GAA'} = \mathbf{A'}$.

If $\mathbf{GA} = \mathbf{P_{A'}}$, then [in light of Part (1) of Theorem 12.3.4] $\mathbf{GAA'} = \mathbf{P_{A'}A'} = \mathbf{A'}$. Conversely, if $\mathbf{GAA'} = \mathbf{A'}$, then [in light of Part (5) of Theorem 12.3.4]

$$\mathbf{GA} = \mathbf{GAA'}(\mathbf{AA'})^{-}\mathbf{A} = \mathbf{A'}(\mathbf{AA'})^{-}\mathbf{A} = \mathbf{P_{A'}}.$$

Q.E.D.

Let $\mathbf{A}$ represent an $m \times n$ matrix. Then, for any generalized inverse $\mathbf{G}$ of $\mathbf{A}$, $\mathbf{GA}$ is idempotent, $\mathcal{R}(\mathbf{GA}) = \mathcal{R}(\mathbf{A})$, and $\text{rank}(\mathbf{GA}) = \text{rank}(\mathbf{A})$, as was established earlier (in Lemmas 10.2.5 and 9.3.7 and Corollary 9.3.8). Further, for any minimum norm generalized inverse $\mathbf{G}$ of $\mathbf{A}$, $\mathbf{GA}$ is (according to the very definition of a minimum norm generalized inverse) symmetric. Two other properties of the matrix product $\mathbf{GA}$ obtained by premultiplying $\mathbf{A}$ by a minimum norm generalized inverse $\mathbf{G}$ (of $\mathbf{A}$) are described in the following, additional corollary of Theorem 20.3.7.

Corollary 20.3.9. Let $\mathbf{G}$ represent a minimum norm generalized inverse of a matrix $\mathbf{A}$. Then,
(1) $\mathbf{GA}$ is invariant to the choice of $\mathbf{G}$; and
(2) $\mathcal{C}(\mathbf{GA}) = \mathcal{C}(\mathbf{A'})$.

Proof. According to Corollary 20.3.8, $\mathbf{GA} = \mathbf{P_{A'}}$. Thus, $\mathbf{GA}$ is invariant to the choice of $\mathbf{G}$, and [in light of Part (7) of Theorem 12.3.4] $\mathcal{C}(\mathbf{GA}) = \mathcal{C}(\mathbf{P_{A'}}) = \mathcal{C}(\mathbf{A'})$. Q.E.D.

A generalized inverse $\mathbf{G}$ of a matrix $\mathbf{A}$ is said to be a *minimum norm reflexive generalized inverse* (of $\mathbf{A}$) if it is a minimum norm generalized inverse (of $\mathbf{A}$) and it is reflexive. Thus, an $n \times m$ matrix $\mathbf{G}$ is a minimum norm reflexive generalized inverse of an $m \times n$ matrix $\mathbf{A}$ if it satisfies Penrose Conditions (1), (2), and (4).

The following lemma gives a basic property of a minimum norm reflexive generalized inverse.

Lemma 20.3.10. For any minimum norm reflexive generalized inverse $\mathbf{G}$ of a matrix $\mathbf{A}$, $\mathcal{C}(\mathbf{G}) = \mathcal{C}(\mathbf{A'})$.

Proof. Recalling Corollary 4.2.3, it follows from Part (2) of Corollary 20.3.9 that $\mathcal{C}(\mathbf{A'}) \subset \mathcal{C}(\mathbf{G})$. Moreover, in light of Theorem 20.3.4, $\text{rank}(\mathbf{G}) = \text{rank}(\mathbf{A}) = \text{rank}(\mathbf{A'})$. Thus, as a consequence of Theorem 4.4.6, we have that $\mathcal{C}(\mathbf{G}) = \mathcal{C}(\mathbf{A'})$. Q.E.D.

Since [according to Part (1) of Theorem 12.3.4] $\mathbf{A}'(\mathbf{A}\mathbf{A}')^{-}\mathbf{A}\mathbf{A}' = \mathbf{A}'$, it follows from Theorem 20.3.7 that the matrix $\mathbf{A}'(\mathbf{A}\mathbf{A}')^{-}$ is a minimum norm generalized inverse of $\mathbf{A}$. The following theorem makes a stronger statement.

Theorem 20.3.11. An $n \times m$ matrix $\mathbf{G}$ is a minimum norm reflexive generalized inverse of an $m \times n$ matrix $\mathbf{A}$ if and only if $\mathbf{G} = \mathbf{A}'\mathbf{H}$ for some generalized inverse $\mathbf{H}$ of $\mathbf{A}\mathbf{A}'$.

Proof. According to Part (1) of Theorem 12.3.4, $\mathbf{A}'(\mathbf{A}\mathbf{A}')^{-}\mathbf{A}\mathbf{A}' = \mathbf{A}'$. Thus, as previously indicated, it follows from Theorem 20.3.7 that the matrix $\mathbf{A}'(\mathbf{A}\mathbf{A}')^{-}$ is a minimum norm generalized inverse of $\mathbf{A}$. Further,

$$[\mathbf{A}'(\mathbf{A}\mathbf{A}')^{-}]\mathbf{A}[\mathbf{A}'(\mathbf{A}\mathbf{A}')^{-}] = [\mathbf{A}'(\mathbf{A}\mathbf{A}')^{-}\mathbf{A}\mathbf{A}'](\mathbf{A}\mathbf{A}')^{-} = \mathbf{A}'(\mathbf{A}\mathbf{A}')^{-},$$

so that $\mathbf{A}'(\mathbf{A}\mathbf{A}')^{-}$ is a minimum norm reflexive generalized inverse of $\mathbf{A}$.

Now, suppose that $\mathbf{G}$ is a minimum norm reflexive generalized inverse of $\mathbf{A}$. Then, it follows from Lemma 20.3.10 that $\mathcal{C}(\mathbf{G}) = \mathcal{C}(\mathbf{A}')$ and hence that there exists a matrix $\mathbf{H}$ such that $\mathbf{G} = \mathbf{A}'\mathbf{H}$. And $\mathbf{A} = \mathbf{A}\mathbf{G}\mathbf{A} = \mathbf{A}\mathbf{A}'\mathbf{H}\mathbf{A}$, implying that $\mathbf{A}\mathbf{A}' = \mathbf{A}\mathbf{A}'\mathbf{H}\mathbf{A}\mathbf{A}'$, so that $\mathbf{H}$ is a generalized inverse of $\mathbf{A}\mathbf{A}'$. Q.E.D.

c. Least squares generalized inverses

A generalized inverse $\mathbf{G}$ of a matrix $\mathbf{A}$ is said to be a *least squares generalized inverse* (of $\mathbf{A}$) if $(\mathbf{A}\mathbf{G})' = \mathbf{A}\mathbf{G}$. Thus, an $n \times m$ matrix $\mathbf{G}$ is a least squares generalized inverse of an $m \times n$ matrix $\mathbf{A}$ if it satisfies Penrose Conditions (1) and (3).

Two alternative characterizations of a least squares generalized inverse are given by the following theorem and corollary—these characterizations are analogous to the characterizations of a minimum norm generalized inverse given, respectively, by Theorem 20.3.7 and Corollary 20.3.8.

Theorem 20.3.12. An $n \times m$ matrix $\mathbf{G}$ is a least squares generalized inverse of an $m \times n$ matrix $\mathbf{A}$ if and only if $\mathbf{A}'\mathbf{A}\mathbf{G} = \mathbf{A}'$, or, equivalently, if and only if $\mathbf{G}'\mathbf{A}'\mathbf{A} = \mathbf{A}$.

Proof. Suppose that $\mathbf{A}'\mathbf{A}\mathbf{G} = \mathbf{A}'$ or, equivalently, that $\mathbf{G}'\mathbf{A}'\mathbf{A} = \mathbf{A}$. Then, $\mathbf{A}'\mathbf{A}\mathbf{G}\mathbf{A} = \mathbf{A}'\mathbf{A}$, implying (in light of Corollary 5.3.3) that $\mathbf{A}\mathbf{G}\mathbf{A} = \mathbf{A}$. Moreover, $\mathbf{A}\mathbf{G} = \mathbf{G}'\mathbf{A}'\mathbf{A}\mathbf{G} = (\mathbf{A}\mathbf{G})'\mathbf{A}\mathbf{G}$, so that $\mathbf{A}\mathbf{G}$ is symmetric. Thus, $\mathbf{G}$ is a least squares generalized inverse of $\mathbf{A}$.

Conversely, if $\mathbf{G}$ is a least squares generalized inverse of $\mathbf{A}$, then

$$\mathbf{A}'\mathbf{A}\mathbf{G} = \mathbf{A}'(\mathbf{A}\mathbf{G})' = (\mathbf{A}\mathbf{G}\mathbf{A})' = \mathbf{A}'.$$

Q.E.D.

Corollary 20.3.13. An $n \times m$ matrix $\mathbf{G}$ is a least squares generalized inverse of an $m \times n$ matrix $\mathbf{A}$ if and only if $\mathbf{A}\mathbf{G} = \mathbf{P}_{\mathbf{A}}$.

Proof. In light of Theorem 20.3.12, it suffices to show that $\mathbf{A}\mathbf{G} = \mathbf{P}_{\mathbf{A}} \Leftrightarrow \mathbf{A}'\mathbf{A}\mathbf{G} = \mathbf{A}'$.

If $\mathbf{AG} = \mathbf{P_A}$, then [in light of Part (5) of Theorem 12.3.4] $\mathbf{A'AG} = \mathbf{A'P_A} = \mathbf{A'}$. Conversely, if $\mathbf{A'AG} = \mathbf{A'}$, then [in light of Part (1) of Theorem 12.3.4]

$$\mathbf{AG} = \mathbf{A}(\mathbf{A'A})^{-}\mathbf{A'AG} = \mathbf{A}(\mathbf{A'A})^{-}\mathbf{A'} = \mathbf{P_A}.$$

Q.E.D.

The aptness of our use of the term least squares generalized inverse is evident from the following corollary (of Theorem 20.3.12).

Corollary 20.3.14. A $p \times n$ matrix $\mathbf{G}$ is a least squares generalized inverse of an $n \times p$ matrix $\mathbf{X}$ if and only if, for every $\mathbf{y} \in \mathcal{R}^{n\times 1}$, the norm $\|\mathbf{y} - \mathbf{Xb}\|$ (of the difference between $\mathbf{y}$ and $\mathbf{Xb}$) attains its minimum value (with respect to the $p \times 1$ vector $\mathbf{b}$) at $\mathbf{b} = \mathbf{Gy}$ (where the norm is the usual norm), or, equivalently, if and only if, for every $\mathbf{y} \in \mathcal{R}^{n\times 1}$, the sum of squares $(\mathbf{y} - \mathbf{Xb})'(\mathbf{y} - \mathbf{Xb})$ (of the elements of the difference between $\mathbf{y}$ and $\mathbf{Xb}$) attains its minimum value at $\mathbf{b} = \mathbf{Gy}$.

Proof. In light of Theorem 20.3.12, it suffices to show that $\mathbf{X'XG} = \mathbf{X'}$ if and only if, for every $\mathbf{y} \in \mathcal{R}^{n\times 1}$, $(\mathbf{y} - \mathbf{Xb})'(\mathbf{y} - \mathbf{Xb})$ attains its minimum value at $\mathbf{b} = \mathbf{Gy}$. Or, equivalently (in light of Theorem 12.4.3), it suffices to show that $\mathbf{X'XG} = \mathbf{X'}$ if and only if, for every $\mathbf{y} \in \mathcal{R}^{n\times 1}$, the vector $\mathbf{Gy}$ is a solution to the normal equations $\mathbf{X'Xb} = \mathbf{X'y}$ (i.e., if and only if, for every $\mathbf{y} \in \mathcal{R}^{n\times 1}$, $\mathbf{X'XGy} = \mathbf{X'y}$).

If, for every $\mathbf{y} \in \mathcal{R}^{n\times 1}$, $\mathbf{X'XGy} = \mathbf{X'y}$, then it is evident from Lemma 2.3.2 that $\mathbf{X'XG} = \mathbf{X'}$. Conversely, if $\mathbf{X'XG} = \mathbf{X'}$, then it is obvious that, for every $\mathbf{y} \in \mathcal{R}^{n\times 1}$, $\mathbf{X'XGy} = \mathbf{X'y}$. Q.E.D.

Let $\mathbf{A}$ represent an $m\times n$ matrix. Then, for any generalized inverse $\mathbf{G}$ of $\mathbf{A}$, $\mathbf{AG}$ is idempotent, $\mathcal{C}(\mathbf{AG}) = \mathcal{C}(\mathbf{A})$, and $\text{rank}(\mathbf{AG}) = \text{rank}(\mathbf{A})$, as was established earlier (in Lemmas 10.2.5 and 9.3.7 and Corollary 9.3.8). Further, for any least squares generalized inverse $\mathbf{G}$ of $\mathbf{A}$, $\mathbf{AG}$ is (according to the very definition of a least squares generalized inverse) symmetric. Two other properties of the matrix product $\mathbf{AG}$ obtained by postmultiplying $\mathbf{A}$ by a least squares generalized inverse $\mathbf{G}$ (of $\mathbf{A}$) are described in the following, additional corollary of Theorem 20.3.12.

Corollary 20.3.15. Let $\mathbf{G}$ represent a least squares generalized inverse of a matrix $\mathbf{A}$. Then,

(1) $\mathbf{AG}$ is invariant to the choice of $\mathbf{G}$; and

(2) $\mathcal{R}(\mathbf{AG}) = \mathcal{R}(\mathbf{A'})$.

Proof. According to Corollary 20.3.13, $\mathbf{AG} = \mathbf{P_A}$. Thus, $\mathbf{AG}$ is invariant to the choice of $\mathbf{G}$, and [in light of Part (7) of Theorem 12.3.4] $\mathcal{R}(\mathbf{AG}) = \mathcal{R}(\mathbf{P_A}) = \mathcal{R}(\mathbf{A'})$. Q.E.D.

A generalized inverse $\mathbf{G}$ of a matrix $\mathbf{A}$ is said to be a *least squares reflexive generalized inverse* (of $\mathbf{A}$) if it is a least squares generalized inverse (of $\mathbf{A}$) and it is reflexive. Thus, an $n \times m$ matrix $\mathbf{G}$ is a least squares reflexive generalized inverse of an $m \times n$ matrix $\mathbf{A}$ if it satisfies Penrose Conditions (1), (2), and (3).

The following lemma gives some basic properties of a least squares reflexive generalized inverse.

Lemma 20.3.16. For any least squares reflexive generalized inverse $\mathbf{G}$ of a matrix $\mathbf{A}$, $\mathcal{R}(\mathbf{G}) = \mathcal{R}(\mathbf{A}')$ and $\mathcal{N}(\mathbf{G}) = \mathcal{C}^{\perp}(\mathbf{A}) = \mathcal{N}(\mathbf{A}')$.

Proof. Recalling Corollary 4.2.3, it follows from Part (2) of Corollary 20.3.15 that $\mathcal{R}(\mathbf{A}') \subset \mathcal{R}(\mathbf{G})$. Moreover, in light of Theorem 20.3.4, $\text{rank}(\mathbf{G}) = \text{rank}(\mathbf{A}) = \text{rank}(\mathbf{A}')$. Thus, as a consequence of Theorem 4.4.6, we have that $\mathcal{R}(\mathbf{G}) = \mathcal{R}(\mathbf{A}')$.

Further, it follows from Part (2) of Lemma 4.2.5 that $\mathcal{C}(\mathbf{G}') = \mathcal{C}(\mathbf{A})$. And, making use of Lemma 12.5.2, we find that

$$\mathcal{N}(\mathbf{G}) = \mathcal{C}^{\perp}(\mathbf{G}') = \mathcal{C}^{\perp}(\mathbf{A}) = \mathcal{N}(\mathbf{A}').$$

Q.E.D.

Since [according to Part (5) of Theorem 12.3.4] $\mathbf{A}'\mathbf{A}(\mathbf{A}'\mathbf{A})^{-}\mathbf{A}' = \mathbf{A}'$, it follows from Theorem 20.3.12 that the matrix $(\mathbf{A}'\mathbf{A})^{-}\mathbf{A}'$ is a least squares generalized inverse of $\mathbf{A}$. The following theorem makes a stronger statement.

Theorem 20.3.17. An $n \times m$ matrix $\mathbf{G}$ is a least squares reflexive generalized inverse of an $m \times n$ matrix $\mathbf{A}$ if and only if $\mathbf{G} = \mathbf{H}\mathbf{A}'$ for some generalized inverse $\mathbf{H}$ of $\mathbf{A}'\mathbf{A}$.

Proof. According to Part (5) of Theorem 12.3.4, $\mathbf{A}'\mathbf{A}(\mathbf{A}'\mathbf{A})^{-}\mathbf{A}' = \mathbf{A}'$. Thus, as previously indicated, it follows from Theorem 20.3.12 that the matrix $(\mathbf{A}'\mathbf{A})^{-}\mathbf{A}'$ is a least squares generalized inverse of $\mathbf{A}$. Further,

$$[(\mathbf{A}'\mathbf{A})^{-}\mathbf{A}']\mathbf{A}[(\mathbf{A}'\mathbf{A})^{-}\mathbf{A}'] = (\mathbf{A}'\mathbf{A})^{-}[\mathbf{A}'\mathbf{A}(\mathbf{A}'\mathbf{A})^{-}\mathbf{A}'] = (\mathbf{A}'\mathbf{A})^{-}\mathbf{A}',$$

so that $(\mathbf{A}'\mathbf{A})^{-}\mathbf{A}'$ is a least squares reflexive generalized inverse of $\mathbf{A}$.

Now, suppose that $\mathbf{G}$ is a least squares reflexive generalized inverse of $\mathbf{A}$. Then, it follows from Lemma 20.3.16 that $\mathcal{R}(\mathbf{G}) = \mathcal{R}(\mathbf{A}')$ and hence that there exists a matrix $\mathbf{H}$ such that $\mathbf{G} = \mathbf{H}\mathbf{A}'$. And $\mathbf{A} = \mathbf{AGA} = \mathbf{AHA}'\mathbf{A}$, implying that $\mathbf{A}'\mathbf{A} = \mathbf{A}'\mathbf{AHA}'\mathbf{A}$, so that $\mathbf{H}$ is a generalized inverse of $\mathbf{A}'\mathbf{A}$. Q.E.D.

d. Some equivalences

The four Penrose conditions can be reexpressed in accordance with the equivalences given by the following, easily verifiable lemma.

Lemma 20.3.18. Let $\mathbf{A}$ represent an $m \times n$ matrix and $\mathbf{G}$ an $n \times m$ matrix. Then,

(1) $\mathbf{AGA} = \mathbf{A} \Leftrightarrow \mathbf{A}'\mathbf{G}'\mathbf{A}' = \mathbf{A}'$;

(2) $\mathbf{GAG} = \mathbf{G} \Leftrightarrow \mathbf{G}'\mathbf{A}'\mathbf{G}' = \mathbf{G}'$;

(3) $(\mathbf{AG})' = \mathbf{AG} \Leftrightarrow (\mathbf{G}'\mathbf{A}')' = \mathbf{G}'\mathbf{A}'$; and

(4) $(\mathbf{GA})' = \mathbf{GA} \Leftrightarrow (\mathbf{A}'\mathbf{G}')' = \mathbf{A}'\mathbf{G}'$.

As an immediate consequence of Lemma 20.3.18, we have the following corollary (which expands on Lemma 20.3.5).

Corollary 20.3.19. An $n \times m$ matrix $\mathbf{G}$ is a reflexive generalized inverse of an $m \times n$ matrix $\mathbf{A}$ if and only if $\mathbf{G}'$ is a reflexive generalized inverse of $\mathbf{A}'$. Further, $\mathbf{G}$ is a least squares generalized inverse of $\mathbf{A}$ if and only if $\mathbf{G}'$ is a minimum norm generalized inverse of $\mathbf{A}'$ and is a least squares reflexive generalized inverse of $\mathbf{A}$ if and only if $\mathbf{G}'$ is a minimum norm reflexive generalized inverse of $\mathbf{A}'$.

20.4 Some Alternative Representations and Characterizations

The Moore-Penrose inverse can be characterized in the ways described in the following theorem and two corollaries.

Theorem 20.4.1. An $n \times m$ matrix $\mathbf{G}$ is the Moore-Penrose inverse of an $m \times n$ matrix $\mathbf{A}$ if and only if $\mathbf{G}$ is a least squares generalized inverse of $\mathbf{A}$ and $\mathbf{A}$ is a least squares generalized inverse of $\mathbf{G}$.

Proof. By definition, $\mathbf{G}$ is a least squares generalized inverse of $\mathbf{A}$ if and only if $\mathbf{AGA} = \mathbf{A}$ and $(\mathbf{AG})' = \mathbf{AG}$ [which are Penrose Conditions (1) and (3)], and $\mathbf{A}$ is a least squares generalized inverse of $\mathbf{G}$ if and only if $\mathbf{GAG} = \mathbf{G}$ and $(\mathbf{GA})' = \mathbf{GA}$ [which, in the relevant context, are Penrose Conditions (2) and (4)]. Thus, $\mathbf{G}$ is the Moore-Penrose inverse of $\mathbf{A}$ if and only if $\mathbf{G}$ is a least squares generalized inverse of $\mathbf{A}$ and $\mathbf{A}$ is a least squares generalized inverse of $\mathbf{G}$. Q.E.D.

Corollary 20.4.2. An $n \times m$ matrix $\mathbf{G}$ is the Moore-Penrose inverse of an $m \times n$ matrix $\mathbf{A}$ if and only if $\mathbf{A}'\mathbf{AG} = \mathbf{A}'$ and $\mathbf{G}'\mathbf{GA} = \mathbf{G}'$.

Proof. Corollary 20.4.2 follows from Theorem 20.4.1 upon observing (in light of Theorem 20.3.12) that $\mathbf{G}$ is a least squares generalized inverse of $\mathbf{A}$ if and only if $\mathbf{A}'\mathbf{AG} = \mathbf{A}'$ and that $\mathbf{A}$ is a least squares generalized inverse of $\mathbf{G}$ if and only if $\mathbf{G}'\mathbf{GA} = \mathbf{G}'$. Q.E.D.

Corollary 20.4.3. An $n \times m$ matrix $\mathbf{G}$ is the Moore-Penrose inverse of an $m \times n$ matrix $\mathbf{A}$ if and only if $\mathbf{AG} = \mathbf{P}_{\mathbf{A}}$ and $\mathbf{GA} = \mathbf{P}_{\mathbf{G}}$.

Proof. Corollary 20.4.3 follows from Theorem 20.4.1 upon observing (in light of Corollary 20.3.13) that $\mathbf{G}$ is a least squares generalized inverse of $\mathbf{A}$ if and only if $\mathbf{AG} = \mathbf{P}_{\mathbf{A}}$ and that $\mathbf{A}$ is a least squares generalized inverse of $\mathbf{G}$ if and only if $\mathbf{GA} = \mathbf{P}_{\mathbf{G}}$. Q.E.D.

The following theorem expresses the Moore-Penrose inverse of a matrix $\mathbf{A}$ in terms of solutions to the two linear systems $\mathbf{A}'\mathbf{AX} = \mathbf{A}'$ and $\mathbf{AA}'\mathbf{Y} = \mathbf{A}$.

Theorem 20.4.4. For any $m \times n$ matrix $\mathbf{A}$,

$$\mathbf{A}^{+} = \mathbf{Y}_{*}'\mathbf{AX}_{*}, \tag{4.1}$$

where $\mathbf{X}_{*}$ is any solution to the (consistent) linear system $\mathbf{A}'\mathbf{AX} = \mathbf{A}'$ (in an $n \times m$ matrix $\mathbf{X}$) and $\mathbf{Y}_{*}$ is any solution to the (consistent) linear system $\mathbf{AA}'\mathbf{Y} = \mathbf{A}$ (in an $m \times n$ matrix $\mathbf{Y}$).

Proof. Note (in light of Corollary 7.4.2) that the two linear systems $\mathbf{A}'\mathbf{AX} = \mathbf{A}'$ and $\mathbf{AA}'\mathbf{Y} = \mathbf{A}$ are consistent. And observe (in light of Theorems 20.3.12

and 20.3.7) that $\mathbf{AX_*A} = \mathbf{A}$ and $(\mathbf{AX_*})' = \mathbf{AX_*}$ and that $\mathbf{AY_*'A} = \mathbf{A}$ and $(\mathbf{Y_*'A})' = \mathbf{Y_*'A}$. Thus,

$$\mathbf{A(Y_*'AX_*)A} = \mathbf{AY_*'(AX_*A)} = \mathbf{AY_*'A} = \mathbf{A};$$
$$\mathbf{(Y_*'AX_*)A(Y_*'AX_*)} = \mathbf{Y_*'(AX_*A)Y_*'AX_*} = \mathbf{Y_*'AY_*'AX_*} = \mathbf{Y_*'AX_*};$$
$$[\mathbf{A(Y_*'AX_*)}]' = [\mathbf{(AY_*'A)X_*}]' = (\mathbf{AX_*})' = \mathbf{AX_*} = \mathbf{(AY_*'A)X_*} = \mathbf{A(Y_*'AX_*)};$$
$$[\mathbf{(Y_*'AX_*)A}]' = [\mathbf{Y_*'(AX_*A)}]' = (\mathbf{Y_*'A})' = \mathbf{Y_*'A} = \mathbf{Y_*'(AX_*A)} = \mathbf{(Y_*'AX_*)A}.$$

We conclude that $\mathbf{Y_*'AX_*}$ satisfies all four of the Penrose conditions and hence that $\mathbf{A}^+ = \mathbf{Y_*'AX_*}$. Q.E.D.

The matrix $(\mathbf{A'A})^-\mathbf{A'}$ is one solution to the linear system $\mathbf{A'AX} = \mathbf{A'}$ and {since $\mathbf{AA'}$ is symmetric and hence since $[(\mathbf{AA'})^-]'$ is a generalized inverse of $\mathbf{AA'}$} the matrix $[(\mathbf{AA'})^-]'\mathbf{A}$ is one solution to the linear system $\mathbf{AA'Y} = \mathbf{A}$. Thus, as a special case of formula (4.1) {that obtained by setting $\mathbf{X_*} = (\mathbf{A'A})^-\mathbf{A'}$ and $\mathbf{Y_*} = [(\mathbf{AA'})^-]'\mathbf{A}$}, we have that

$$\mathbf{A}^+ = \mathbf{A'(AA')^-A(A'A)^-A'}. \tag{4.2}$$

Note that the right side of equality (4.2) can be reexpressed as $\mathbf{A'(AA')^-AA^-A(A'A)^-A'}$, giving rise to the alternative representation

$$\mathbf{A}^+ = \mathbf{P_{A'}A^-P_A}. \tag{4.3}$$

The following theorem gives three expressions for the Moore-Penrose inverse of a (nonnull) symmetric nonnegative definite matrix.

Theorem 20.4.5. Let $\mathbf{A}$ represent an $n \times n$ (nonnull) symmetric nonnegative definite matrix. Then, for any matrix $\mathbf{T}$ of full row rank such that $\mathbf{A} = \mathbf{T'T}$,

$$\mathbf{A}^+ = \mathbf{T'(TAT')^{-1}T} \tag{4.4}$$
$$= \mathbf{T'(TT')^{-2}T} \tag{4.5}$$
$$= \mathbf{T^+(T^+)'} \tag{4.6}$$

{where $(\mathbf{TT'})^{-2} = [(\mathbf{TT'})^{-1}]^2$}.

In connection with Theorem 20.4.5, note that Theorem 14.3.7 guarantees the existence of a matrix $\mathbf{T}$ of full row rank such that $\mathbf{A} = \mathbf{T'T}$.

Proof. Results (4.4) and (4.5) are special cases of results (1.1) and (1.2), respectively. Specifically, they are the special cases obtained by setting $\mathbf{B} = \mathbf{T'}$. And, in light of result (2.1) [and the symmetry of $(\mathbf{TT'})^{-1}$], result (4.6) follows from result (4.5). Q.E.D.

20.5 Some Basic Properties and Relationships

The Moore-Penrose inverse of a matrix $\mathbf{A}$ is, by definition, a reflexive generalized inverse, a minimum norm generalized inverse, a least squares generalized inverse, a minimum norm reflexive generalized inverse, and a least squares reflexive generalized inverse (of $\mathbf{A}$). Accordingly, certain properties of Moore-Penrose inverses can be deduced from the properties (discussed in Section 3) of reflexive, minimum norm, least squares, minimum norm reflexive, and least squares reflexive generalized inverses. In particular, the following theorem describes some properties of Moore-Penrose inverses that are immediate consequences of Theorem 20.3.4 and of Corollaries 20.3.8 and 20.3.13 and Lemmas 20.3.10 and 20.3.16.

Theorem 20.5.1. For any matrix $\mathbf{A}$,

(1) $\text{rank}(\mathbf{A}^+) = \text{rank}(\mathbf{A})$;

(2) $\mathbf{A}^+\mathbf{A} = \mathbf{P}_{\mathbf{A}'}$ and $\mathbf{A}\mathbf{A}^+ = \mathbf{P}_{\mathbf{A}}$;

(3) $\mathcal{C}(\mathbf{A}^+) = \mathcal{C}(\mathbf{A}')$ and $\mathcal{R}(\mathbf{A}^+) = \mathcal{R}(\mathbf{A}')$; and

(4) $\mathcal{N}(\mathbf{A}^+) = \mathcal{C}^\perp(\mathbf{A}) = \mathcal{N}(\mathbf{A}')$.

In light of Parts (3) and (6) of Theorem 12.3.4, we have the following corollary of Theorem 20.5.1.

Corollary 20.5.2. For any matrix $\mathbf{A}$, $\mathbf{A}^+\mathbf{A}$ and $\mathbf{A}\mathbf{A}^+$ are symmetric and idempotent.

Some additional properties of Moore-Penrose inverses are described in the following theorem.

Theorem 20.5.3. Let $\mathbf{A}$ represent an $m \times n$ matrix. Then,

(1) $(\mathbf{A}')^+ = (\mathbf{A}^+)'$;

(2) if $\mathbf{A}$ is symmetric, then $\mathbf{A}^+$ is symmetric;

(3) if $\mathbf{A}$ is symmetric and positive semidefinite, then $\mathbf{A}^+$ is symmetric and positive semidefinite; if $\mathbf{A}$ is symmetric and positive definite, then $\mathbf{A}^+$ is symmetric and positive definite (and if $\mathbf{A}$ is symmetric and nonnegative definite, then $\mathbf{A}^+$ is symmetric and nonnegative definite);

(4) $(\mathbf{A}^+)^+ = \mathbf{A}$; and

(5) for any nonzero scalar c, $(c\mathbf{A})^+ = (1/c)\mathbf{A}^+$.

Proof. (1) It is clear from Lemma 20.3.18 that $(\mathbf{A}^+)'$ satisfies all four of the Penrose conditions (as applied to the matrix $\mathbf{A}'$) and hence that $(\mathbf{A}^+)'$ is the Moore-Penrose inverse of $\mathbf{A}'$.

(2) Part (2) is an immediate consequence of Part (1).

(3) Suppose that $\mathbf{A}$ is symmetric and nonnegative definite. Then, according to Part (2), $\mathbf{A}^+$ is symmetric. And, since (according to Theorem 20.5.1) $\text{rank}(\mathbf{A}^+) = \text{rank}(\mathbf{A})$, $\mathbf{A}^+$ is (in light of Corollary 14.3.12) singular if $\mathbf{A}$ is positive semidefinite and nonsingular if $\mathbf{A}$ is positive definite. Further, $\mathbf{A}^+ = \mathbf{A}^+\mathbf{A}\mathbf{A}^+ = (\mathbf{A}^+)'\mathbf{A}\mathbf{A}^+$. Thus, it follows from Parts (2) and (3) of Theorem 14.2.9 that $\mathbf{A}^+$ is positive semidefinite if $\mathbf{A}$ is positive semidefinite and positive definite if $\mathbf{A}$ is positive definite.

(4) That $(\mathbf{A}^+)^+ = \mathbf{A}$ is evident from the definition of the Moore-Penrose inverse (as a matrix that satisfies the four Penrose conditions).

(5) Clearly,

$$c\mathbf{A}[(1/c)\mathbf{A}^+]c\mathbf{A} = c\mathbf{A}\mathbf{A}^+\mathbf{A} = c\mathbf{A},$$
$$[(1/c)\mathbf{A}^+]c\mathbf{A}[(1/c)\mathbf{A}^+] = (1/c)\mathbf{A}^+\mathbf{A}\mathbf{A}^+ = (1/c)\mathbf{A}^+,$$
$$\{c\mathbf{A}[(1/c)\mathbf{A}^+]\}' = (\mathbf{A}\mathbf{A}^+)' = \mathbf{A}\mathbf{A}^+ = c\mathbf{A}[(1/c)\mathbf{A}^+], \quad \text{and}$$
$$\{[(1/c)\mathbf{A}^+]c\mathbf{A}\}' = (\mathbf{A}^+\mathbf{A})' = \mathbf{A}^+\mathbf{A} = [(1/c)\mathbf{A}^+]c\mathbf{A},$$

so that $(1/c)\mathbf{A}^+$ satisfies Penrose Conditions (1)–(4) (as applied to the matrix $c\mathbf{A}$). Q.E.D.

Note that, as a special case of Part (5) of Theorem 20.5.3 (that where $c = -1$), we have that, for any matrix $\mathbf{A}$,

$$(-\mathbf{A})^+ = -\mathbf{A}^+. \tag{5.1}$$

The Moore-Penrose inverse of a matrix $\mathbf{A}$ is related to minimum norm generalized inverses of $\mathbf{A}'\mathbf{A}$ and to least squares generalized inverses of $\mathbf{A}\mathbf{A}'$, as described in the following theorem.

Theorem 20.5.4. Let $\mathbf{A}$ represent an $m \times n$ matrix. Then, for any least squares generalized inverse $\mathbf{F}$ of $\mathbf{A}\mathbf{A}'$, $\mathbf{A}^+ = \mathbf{A}'\mathbf{F}$, and, for any minimum norm generalized inverse $\mathbf{H}$ of $\mathbf{A}'\mathbf{A}$, $\mathbf{A}^+ = \mathbf{H}\mathbf{A}'$.

Proof. According to Theorem 20.3.11, $\mathbf{A}'\mathbf{F}$ is a minimum norm reflexive generalized inverse of $\mathbf{A}$, and, according to Theorem 20.3.17, $\mathbf{H}\mathbf{A}'$ is a least squares reflexive generalized inverse of $\mathbf{A}$. Moreover, $(\mathbf{A}\mathbf{A}'\mathbf{F})' = \mathbf{A}\mathbf{A}'\mathbf{F}$ and $(\mathbf{H}\mathbf{A}'\mathbf{A})' = \mathbf{H}\mathbf{A}'\mathbf{A}$, so that $\mathbf{A}'\mathbf{F}$ satisfies Penrose Condition (3) and $\mathbf{H}\mathbf{A}'$ satisfies Penrose Condition (4). Thus, both $\mathbf{A}'\mathbf{F}$ and $\mathbf{H}\mathbf{A}'$ satisfy all four Penrose conditions, and hence both equal $\mathbf{A}^+$. Q.E.D.

As an immediate consequence of Theorem 20.5.4, we have the following corollary, which relates the Moore-Penrose inverse of a matrix $\mathbf{A}$ to the Moore-Penrose inverse of $\mathbf{A}\mathbf{A}'$ and to the Moore-Penrose inverse of $\mathbf{A}'\mathbf{A}$.

Corollary 20.5.5. For any matrix $\mathbf{A}$,

$$\mathbf{A}^+ = \mathbf{A}'(\mathbf{A}\mathbf{A}')^+ = (\mathbf{A}'\mathbf{A})^+\mathbf{A}'. \tag{5.2}$$

The following theorem relates the Moore-Penrose inverse of a matrix of the form $\mathbf{P}\mathbf{A}\mathbf{Q}'$, where $\mathbf{P}$ and $\mathbf{Q}$ are orthogonal matrices (or more generally where $\mathbf{P}$ and $\mathbf{Q}$ are such that $\mathbf{P}'\mathbf{P} = \mathbf{I}$ and $\mathbf{Q}'\mathbf{Q} = \mathbf{I}$), to the Moore-Penrose inverse of $\mathbf{A}$ itself.

Theorem 20.5.6. Let $\mathbf{A}$ represent an $m \times n$ matrix. Then, for any $r \times m$ matrix $\mathbf{P}$ whose columns form an orthonormal (with respect to the usual inner product) set (of r-dimensional vectors) and any $s \times n$ matrix $\mathbf{Q}$ whose columns form an orthonormal (with respect to the usual inner product) set (of s-dimensional vectors) or, equivalently, for any $r \times m$ matrix $\mathbf{P}$ such that $\mathbf{P}'\mathbf{P} = \mathbf{I}_m$ and any $s \times n$ matrix $\mathbf{Q}$ such that $\mathbf{Q}'\mathbf{Q} = \mathbf{I}_n$,

$$(\mathbf{P}\mathbf{A}\mathbf{Q}')^+ = \mathbf{Q}\mathbf{A}^+\mathbf{P}'.$$

Proof. It suffices to show that $\mathbf{QA}^+\mathbf{P}'$ satisfies the four Penrose conditions. Clearly,

(1) $\mathbf{PAQ}'(\mathbf{QA}^+\mathbf{P}')\mathbf{PAQ}' = \mathbf{PAA}^+\mathbf{AQ}' = \mathbf{PAQ}'$;

(2) $\mathbf{QA}^+\mathbf{P}'(\mathbf{PAQ}')\mathbf{QA}^+\mathbf{P}' = \mathbf{QA}^+\mathbf{AA}^+\mathbf{P}' = \mathbf{QA}^+\mathbf{P}'$;

(3) $$[\mathbf{PAQ}'(\mathbf{QA}^+\mathbf{P}')]' = (\mathbf{PAA}^+\mathbf{P}')' = \mathbf{P}(\mathbf{AA}^+)'\mathbf{P}' = \mathbf{PAA}^+\mathbf{P}' \\ = \mathbf{PAQ}'(\mathbf{QA}^+\mathbf{P}');$$

(4) $$[\mathbf{QA}^+\mathbf{P}'(\mathbf{PAQ}')]' = (\mathbf{QA}^+\mathbf{AQ}')' = \mathbf{Q}(\mathbf{A}^+\mathbf{A})'\mathbf{Q}' = \mathbf{QA}^+\mathbf{AQ}' \\ = \mathbf{QA}^+\mathbf{P}'(\mathbf{PAQ}').$$

Q.E.D.

For any $m \times n$ matrix $\mathbf{A}$ and any $p \times q$ matrix $\mathbf{B}$, $\mathbf{A}^- \otimes \mathbf{B}^-$ is a generalized inverse of the Kronecker product $\mathbf{A} \otimes \mathbf{B}$ (as discussed in Section 16.1). This result is generalized in the following theorem.

Theorem 20.5.7. Let $\mathbf{A}$ represent an $m \times n$ matrix, $\mathbf{B}$ a $p \times q$ matrix, $\mathbf{G}$ an $n \times m$ matrix, and $\mathbf{H}$ a $q \times p$ matrix. Then, $\mathbf{G} \otimes \mathbf{H}$ satisfies any of the Penrose Conditions (1)–(4) (when the Penrose conditions are applied to the matrix $\mathbf{A} \otimes \mathbf{B}$) that are satisfied by both $\mathbf{G}$ and $\mathbf{H}$ (when the Penrose conditions are applied to $\mathbf{A}$ and $\mathbf{B}$, respectively).

Proof. (1) If $\mathbf{AGA} = \mathbf{A}$ and $\mathbf{BHB} = \mathbf{B}$, then (as indicated in Section 16.1)

$$(\mathbf{A} \otimes \mathbf{B})(\mathbf{G} \otimes \mathbf{H})(\mathbf{A} \otimes \mathbf{B}) = (\mathbf{AGA}) \otimes (\mathbf{BHB}) = \mathbf{A} \otimes \mathbf{B}.$$

(2) Similarly, if $\mathbf{GAG} = \mathbf{G}$ and $\mathbf{HBH} = \mathbf{H}$, then

$$(\mathbf{G} \otimes \mathbf{H})(\mathbf{A} \otimes \mathbf{B})(\mathbf{G} \otimes \mathbf{H}) = (\mathbf{GAG}) \otimes (\mathbf{HBH}) = \mathbf{G} \otimes \mathbf{H}.$$

(3) If $(\mathbf{AG})' = \mathbf{AG}$ and $(\mathbf{BH})' = \mathbf{BH}$, then

$$[(\mathbf{A} \otimes \mathbf{B})(\mathbf{G} \otimes \mathbf{H})]' = [(\mathbf{AG}) \otimes (\mathbf{BH})]' = (\mathbf{AG})' \otimes (\mathbf{BH})' \\ = (\mathbf{AG}) \otimes (\mathbf{BH}) \\ = (\mathbf{A} \otimes \mathbf{B})(\mathbf{G} \otimes \mathbf{H}).$$

(4) Similarly, if $(\mathbf{GA})' = \mathbf{GA}$ and $(\mathbf{HB})' = \mathbf{HB}$, then

$$[(\mathbf{G} \otimes \mathbf{H})(\mathbf{A} \otimes \mathbf{B})]' = [(\mathbf{GA}) \otimes (\mathbf{HB})]' = (\mathbf{GA})' \otimes (\mathbf{HB})' \\ = (\mathbf{GA}) \otimes (\mathbf{HB}) \\ = (\mathbf{G} \otimes \mathbf{H})(\mathbf{A} \otimes \mathbf{B}).$$

Q.E.D.

As an obvious consequence of Theorem 20.5.7, we have the following corollary.

Corollary 20.5.8. For any matrices $\mathbf{A}$ and $\mathbf{B}$,

$$(\mathbf{A} \otimes \mathbf{B})^+ = \mathbf{A}^+ \otimes \mathbf{B}^+.$$

20.6 Minimum Norm Solution to the Least Squares Problem

For any $n \times p$ matrix $\mathbf{X}$ and any $n \times 1$ vector $\mathbf{y}$, the (least squares) problem of minimizing the (usual) norm $\|\mathbf{y} - \mathbf{Xb}\|$ (of the difference between $\mathbf{y}$ and $\mathbf{Xb}$) with respect to the $p \times 1$ vector $\mathbf{b}$ can be solved by taking $\mathbf{b} = \mathbf{Gy}$, where $\mathbf{G}$ is any least squares generalized inverse of $\mathbf{X}$ (as indicated in Corollary 20.3.14). One choice for $\mathbf{G}$ is $\mathbf{G} = \mathbf{X}^+$. The solution ($\mathbf{b} = \mathbf{X}^+\mathbf{y}$) obtained by choosing $\mathbf{G} = \mathbf{X}^+$ has the property described in the following theorem.

Theorem 20.6.1. Let $\mathbf{X}$ represent an $n \times p$ matrix and $\mathbf{y}$ an n-dimensional column vector. Taking the norm of a vector to be the usual norm, define S to be the set comprising those values of a $p \times 1$ vector $\mathbf{b}$ at which $\|\mathbf{y} - \mathbf{Xb}\|$ attains its minimum value or, equivalently, to be the solution set $\{\mathbf{b} : \mathbf{X}'\mathbf{Xb} = \mathbf{X}'\mathbf{y}\}$ of the normal equations. Then, $\|\mathbf{b}\|$ attains its minimum value over the set S uniquely at $\mathbf{b} = \mathbf{X}^+\mathbf{y}$. Moreover, if $\mathbf{G}$ is a $p \times n$ matrix such that, for every $\mathbf{y}$, $\|\mathbf{b}\|$ attains its minimum value over the set S at $\mathbf{b} = \mathbf{Gy}$, then $\mathbf{G} = \mathbf{X}^+$.

In connection with Theorem 20.6.1, note that in general the set S varies with $\mathbf{y}$ (although the dependence of S on $\mathbf{y}$ is suppressed in the notation).

Proof. That the set comprising those values of $\mathbf{b}$ at which $\|\mathbf{y} - \mathbf{Xb}\|$ attains its minimum value is the same as the solution set of the normal equations is evident from Theorem 12.4.3.

Now, let $\mathbf{H}$ represent a minimum norm generalized inverse of $\mathbf{X}'\mathbf{X}$. Then, since $S = \{\mathbf{b} : \mathbf{X}'\mathbf{Xb} = \mathbf{X}'\mathbf{y}\}$, it follows from Theorem 20.3.6 that $\|\mathbf{b}\|$ attains its minimum value over the set S uniquely at $\mathbf{b} = \mathbf{HX}'\mathbf{y}$. And, according to Theorem 20.5.4, $\mathbf{HX}' = \mathbf{X}^+$. Thus, $\|\mathbf{b}\|$ attains its minimum value over the set S uniquely at $\mathbf{b} = \mathbf{X}^+\mathbf{y}$.

Further, if $\mathbf{G}$ is a $p \times n$ matrix such that, for every $\mathbf{y}$, $\|\mathbf{b}\|$ attains its minimum value over the set S at $\mathbf{b} = \mathbf{Gy}$, then, for every $\mathbf{y}$, $\mathbf{Gy} = \mathbf{X}^+\mathbf{y}$, implying (in light of Lemma 2.3.2) that $\mathbf{G} = \mathbf{X}^+$. Q.E.D.

20.7 Expression of the Moore-Penrose Inverse as a Limit

Let $\mathbf{F} = \{f_{ij}\}$ represent an $m \times n$ matrix whose elements are functions defined on some set S of scalars. And let $f_{ij}(x)$ represent the value of f_{ij} at an arbitrary point x in S. Then, the $m \times n$ matrix whose ijth element is $f_{ij}(x)$ will be referred to as the *value of* $\mathbf{F}$ *at* x and will be denoted by the symbol $\mathbf{F}(x)$. Further, if every element of $\mathbf{F}$ has a limit at a point c (in $\mathcal{R}$), then the $m \times n$ matrix whose ijth element is $\lim_{x \to c} f_{ij}(x)$ (the limit of f_{ij} at c) will be referred to as the *limit of* $\mathbf{F}$ *at* c and will be denoted by the symbol $\lim_{x \to c} \mathbf{F}(x)$.

Clearly, if $\mathbf{F}$ is an $m \times n$ matrix of functions defined on some set S of scalars and if all mn of these functions are continuous at an interior point c of S, then

$$\lim_{x \to c} \mathbf{F}(x) = \mathbf{F}(c).$$

Various properties of the limits of matrices of functions can be readily deduced from well-known properties of scalar-valued functions. In particular, for any $p \times q$ matrix $\mathbf{G}$ of functions defined on some set S (of scalars) that have limits at a point c and for any $m \times p$ matrix of constants $\mathbf{A}$ and any $q \times n$ matrix of constants $\mathbf{B}$, the matrix of functions $\mathbf{F}$, defined by $\mathbf{F} = \mathbf{AGB}$, has a limit at c, and

$$\lim_{x \to c} \mathbf{F}(x) = \mathbf{A}[\lim_{x \to c} \mathbf{G}(x)]\mathbf{B}.$$

The following theorem expresses the Moore-Penrose inverse of a matrix as a limit.

Theorem 20.7.1. For any $m \times n$ matrix $\mathbf{A}$,

$$\mathbf{A}^+ = \lim_{\delta \to 0} (\mathbf{A}'\mathbf{A} + \delta^2 \mathbf{I}_n)^{-1}\mathbf{A}' = \lim_{\delta \to 0} \mathbf{A}'(\mathbf{A}\mathbf{A}' + \delta^2 \mathbf{I}_m)^{-1}. \tag{7.1}$$

In connection with Theorem 20.7.1, note (in light of Lemmas 14.2.1 and 14.2.4 and Corollary 14.2.14) that, for $\delta \neq 0$, the matrices $\mathbf{A}'\mathbf{A} + \delta^2\mathbf{I}_n$ and $\mathbf{A}\mathbf{A}' + \delta^2\mathbf{I}_m$ are positive definite and hence (in light of Lemma 14.2.8) nonsingular.

Proof. Suppose that $\mathbf{A} \neq \mathbf{0}$—that the theorem is valid for $\mathbf{A} = \mathbf{0}$ is easy to verify—and let $\mathbf{B}$ represent an $m \times r$ matrix of full column rank r and $\mathbf{T}$ an $r \times n$ matrix of full row rank r such that $\mathbf{A} = \mathbf{BT}$—the existence of such matrices was established in Theorem 4.4.8. And let $\mathbf{L} = \mathbf{B}'\mathbf{B}\mathbf{T}$, so that $\mathbf{A}'\mathbf{A} = \mathbf{T}'\mathbf{L}$. Further, let δ represent an arbitrary nonzero scalar [and write δ^{-2} for $(\delta^{-1})^2$].

Then, since $\mathbf{A}'\mathbf{A} + \delta^2\mathbf{I}_n = \delta^2(\delta^{-2}\mathbf{T}'\mathbf{L} + \mathbf{I}_n)$, $\delta^{-2}\mathbf{T}'\mathbf{L} + \mathbf{I}_n$ (like $\mathbf{A}'\mathbf{A} + \delta^2\mathbf{I}_n$) is nonsingular, and

$$(\mathbf{A}'\mathbf{A} + \delta^2\mathbf{I}_n)^{-1} = \delta^{-2}(\delta^{-2}\mathbf{T}'\mathbf{L} + \mathbf{I}_n)^{-1}.$$

And, in light of Lemma 18.2.3, $\delta^{-2}\mathbf{L}\mathbf{T}' + \mathbf{I}_r$ is nonsingular, and

$$(\mathbf{A}'\mathbf{A} + \delta^2\mathbf{I}_n)^{-1}\mathbf{T}' = \delta^{-2}(\delta^{-2}\mathbf{T}'\mathbf{L} + \mathbf{I}_n)^{-1}\mathbf{T}' = \delta^{-2}\mathbf{T}'(\delta^{-2}\mathbf{L}\mathbf{T}' + \mathbf{I}_r)^{-1}.$$

Moreover, since $\mathbf{L}\mathbf{T}' + \delta^2\mathbf{I}_r = \delta^2(\delta^{-2}\mathbf{L}\mathbf{T}' + \mathbf{I}_r)$, $\mathbf{L}\mathbf{T}' + \delta^2\mathbf{I}_r$ is nonsingular, and

$$(\mathbf{L}\mathbf{T}' + \delta^2\mathbf{I}_r)^{-1} = \delta^{-2}(\delta^{-2}\mathbf{L}\mathbf{T}' + \mathbf{I}_r)^{-1}.$$

Thus,

$$(\mathbf{A}'\mathbf{A} + \delta^2\mathbf{I}_n)^{-1}\mathbf{A}' = (\mathbf{A}'\mathbf{A} + \delta^2\mathbf{I}_n)^{-1}\mathbf{T}'\mathbf{B}' = \mathbf{T}'(\mathbf{L}\mathbf{T}' + \delta^2\mathbf{I}_r)^{-1}\mathbf{B}'. \tag{7.2}$$

Now, let δ represent a completely arbitrary (not necessarily nonzero) scalar. Recalling (from Section 20.1) that $\mathbf{L}\mathbf{T}'$ ($= \mathbf{B}'\mathbf{A}\mathbf{T}'$) is nonsingular, observe that $\mathbf{L}\mathbf{T}' + \delta^2\mathbf{I}_r$ is nonsingular for $\delta = 0$ as well as for $\delta \neq 0$. Observe also (in light of Corollary 13.5.4) that (for $i, j = 1, \ldots, r$) the ijth element of $(\mathbf{L}\mathbf{T}' + \delta^2\mathbf{I}_r)^{-1}$ equals $\alpha_{ji}/|\mathbf{L}\mathbf{T}' + \delta^2\mathbf{I}_r|$, where α_{ji} is the jith element of the cofactor matrix of $\mathbf{L}\mathbf{T}' + \delta^2\mathbf{I}_r$.

It is clear from Corollary 13.7.4 (or, more basically, from the very definition of a determinant) that $|\mathbf{L}\mathbf{T}' + \delta^2\mathbf{I}_r|$ and the elements of the cofactor matrix of

$\mathbf{LT}' + \delta^2\mathbf{I}_r$ are polynomials in δ. Thus, every element of $(\mathbf{LT}' + \delta^2\mathbf{I}_r)^{-1}$ is a continuous function of δ (at 0 or at any other point)—refer, for example, to Bartle (1976, sec. 20). We conclude that

$$\lim_{\delta\to 0} (\mathbf{LT}' + \delta^2\mathbf{I}_r)^{-1} = (\mathbf{LT}' + 0^2\mathbf{I}_r)^{-1} = (\mathbf{LT}')^{-1}$$

and hence [in light of results (7.2) and (1.1)] that

$$\begin{aligned}\lim_{\delta\to 0} (\mathbf{A}'\mathbf{A} + \delta^2\mathbf{I}_n)^{-1}\mathbf{A}' &= \mathbf{T}'[\lim_{\delta\to 0} (\mathbf{LT}' + \delta^2\mathbf{I}_r)^{-1}]\mathbf{B}' \\ &= \mathbf{T}'(\mathbf{LT}')^{-1}\mathbf{B}' = \mathbf{T}'(\mathbf{B}'\mathbf{AT}')^{-1}\mathbf{B}' = \mathbf{A}^+.\end{aligned}$$

To complete the proof, observe that (for every δ)

$$(\mathbf{A}'\mathbf{A} + \delta^2\mathbf{I}_n)\mathbf{A}' = \mathbf{A}'(\mathbf{AA}' + \delta^2\mathbf{I}_m), \tag{7.3}$$

so that, for $\delta \neq 0$,

$$\mathbf{A}'(\mathbf{AA}' + \delta^2\mathbf{I}_m)^{-1} = (\mathbf{A}'\mathbf{A} + \delta^2\mathbf{I}_n)^{-1}\mathbf{A}',$$

as is evident upon premultiplying both sides of equality (7.3) by $(\mathbf{A}'\mathbf{A} + \delta^2\mathbf{I}_n)^{-1}$ and postmultiplying both sides by $(\mathbf{AA}' + \delta^2\mathbf{I}_m)^{-1}$. It follows that

$$\lim_{\delta\to 0} \mathbf{A}'(\mathbf{AA}' + \delta^2\mathbf{I}_m)^{-1} = \lim_{\delta\to 0} (\mathbf{A}'\mathbf{A} + \delta^2\mathbf{I}_n)^{-1}\mathbf{A}'.$$

Q.E.D.

Result (7.1) is of some relevance in statistics. It can be used in particular to relate the estimates of the parameters in a linear statistical model obtained via a Bayesian approach to those obtained by approaching the estimation of the parameters as a least squares problem. More specifically, it can be used to show that, in the limit (as the "amount of prior information decreases"), the vector of Bayesian estimates equals the minimum norm solution to the least squares problem.

20.8 Differentiation of the Moore-Penrose Inverse

Earlier (in Section 15.10), some results were given on the differentiation of generalized inverses of a matrix of functions. These results were (for the most part) applicable to "any" generalized inverse. In the present section, some results are given that are specific to the Moore-Penrose inverse.

The Moore-Penrose inverse of a matrix of functions is continuously differentiable under the conditions set forth in the following lemma.

Lemma 20.8.1. Let $\mathbf{F}$ represent a $p \times q$ matrix of functions, defined on a set S, of an m-dimensional column vector $\mathbf{x}$. And let $\mathbf{c}$ represent any interior point of S at which $\mathbf{F}$ is continuously differentiable. If $\mathbf{F}$ has constant rank on some neighborhood of $\mathbf{c}$, then $\mathbf{F}^+$ is continuously differentiable at $\mathbf{c}$.

Proof. Suppose that $\mathbf{F}$ has constant rank on some neighborhood of $\mathbf{c}$. Then, since $\text{rank}(\mathbf{FF}') = \text{rank}(\mathbf{F}'\mathbf{F}) = \text{rank}(\mathbf{F})$, each of the matrices $\mathbf{FF}'$ and $\mathbf{F}'\mathbf{F}$ has constant rank on some neighborhood of $\mathbf{c}$. And, in light of the results of Section 15.4, each of the matrices $\mathbf{FF}'$ and $\mathbf{F}'\mathbf{F}$ is continuously differentiable at $\mathbf{c}$.

Thus, it follows from Theorem 15.10.1 that there exist generalized inverses $\mathbf{G}$ and $\mathbf{H}$ of $\mathbf{FF}'$ and $\mathbf{F}'\mathbf{F}$, respectively, such that $\mathbf{G}$ and $\mathbf{H}$ are continuously differentiable at $\mathbf{c}$. Since [according to result (4.2)] $\mathbf{F}^+ = \mathbf{F}'\mathbf{GFHF}'$, we conclude (in light of the results of Section 15.4) that $\mathbf{F}^+$ is continuously differentiable at $\mathbf{c}$. Q.E.D.

In connection with Lemma 20.8.1, note (in light of Corollary 15.10.4) that, for $\mathbf{F}^+$ to be continuously differentiable at $\mathbf{c}$, it is necessary, as well as sufficient, that $\mathbf{F}$ have constant rank on some neighborhood of $\mathbf{c}$. In fact, unless $\mathbf{F}$ has constant rank on some neighborhood of $\mathbf{c}$, there is (according to Corollary 15.10.4) no generalized inverse of $\mathbf{F}$ that is continuously differentiable at $\mathbf{c}$.

The following theorem gives a formula for the partial derivatives of the Moore-Penrose inverse of a matrix of functions.

Theorem 20.8.2. Let $\mathbf{F}$ represent a $p \times q$ matrix of functions, defined on a set S, of a vector $\mathbf{x} = (x_1, \ldots, x_m)'$ of m variables. Further, let $\mathbf{c}$ represent an interior point of S at which $\mathbf{F}$ is continuously differentiable, and suppose that $\mathbf{F}$ has constant rank on some neighborhood of $\mathbf{c}$. Then, at $\mathbf{x} = \mathbf{c}$,

$$\frac{\partial \mathbf{F}^+}{\partial x_j} = -\mathbf{F}^+ \frac{\partial \mathbf{F}}{\partial x_j}\mathbf{F}^+ + \mathbf{F}^+(\mathbf{F}^+)'\left(\frac{\partial \mathbf{F}}{\partial x_j}\right)'(\mathbf{I} - \mathbf{FF}^+) + (\mathbf{I} - \mathbf{F}^+\mathbf{F})\left(\frac{\partial \mathbf{F}}{\partial x_j}\right)'(\mathbf{F}^+)'\mathbf{F}^+. \tag{8.1}$$

Preliminary to proving Theorem 20.8.2, it is convenient to establish the following theorem.

Theorem 20.8.3. Let $\mathbf{F}$ represent a $p \times q$ matrix of functions, defined on a set S, of a vector $\mathbf{x} = (x_1, \ldots, x_m)'$ of m variables. Further, let $\mathbf{c}$ represent an interior point of S at which $\mathbf{F}$ is continuously differentiable, and suppose that $\mathbf{F}$ has constant rank on some neighborhood of $\mathbf{c}$. Then, $\mathbf{F}^+\mathbf{F}$ and $\mathbf{FF}^+$ are continuously differentiable at $\mathbf{c}$, and (at $\mathbf{x} = \mathbf{c}$)

$$\frac{\partial \mathbf{F}^+\mathbf{F}}{\partial x_j} = \mathbf{F}^+ \frac{\partial \mathbf{F}}{\partial x_j}(\mathbf{I} - \mathbf{F}^+\mathbf{F}) + \left[\mathbf{F}^+ \frac{\partial \mathbf{F}}{\partial x_j}(\mathbf{I} - \mathbf{F}^+\mathbf{F})\right]', \tag{8.2}$$

$$\frac{\partial \mathbf{FF}^+}{\partial x_j} = (\mathbf{I} - \mathbf{FF}^+)\frac{\partial \mathbf{F}}{\partial x_j}\mathbf{F}^+ + \left[(\mathbf{I} - \mathbf{FF}^+)\frac{\partial \mathbf{F}}{\partial x_j}\mathbf{F}^+\right]'. \tag{8.3}$$

Proof (of Theorem 20.8.3). That $\mathbf{F}^+\mathbf{F}$ and $\mathbf{FF}^+$ are continuously differentiable at $\mathbf{c}$ follows from Lemmas 15.4.3 and 20.8.1.

Now, recall (from Corollary 20.5.2) that $\mathbf{F}^+\mathbf{F}$ and $\mathbf{FF}^+$ are symmetric and idempotent, and observe (in light of Lemma 15.4.3) that (at $\mathbf{x} = \mathbf{c}$)

$$\frac{\partial \mathbf{F}^+\mathbf{F}}{\partial x_j} = \left(\frac{\partial \mathbf{F}^+\mathbf{F}}{\partial x_j}\right)' = \left(\frac{\partial \mathbf{F}^+\mathbf{F}\mathbf{F}^+\mathbf{F}}{\partial x_j}\right)'$$
$$= \left(\frac{\partial \mathbf{F}^+\mathbf{F}}{\partial x_j}\mathbf{F}^+\mathbf{F} + \mathbf{F}^+\mathbf{F}\frac{\partial \mathbf{F}^+\mathbf{F}}{\partial x_j}\right)'$$
$$= \mathbf{F}^+\mathbf{F}\frac{\partial \mathbf{F}^+\mathbf{F}}{\partial x_j} + \left(\mathbf{F}^+\mathbf{F}\frac{\partial \mathbf{F}^+\mathbf{F}}{\partial x_j}\right)' \tag{8.4}$$

and similarly

$$\frac{\partial \mathbf{F}\mathbf{F}^+}{\partial x_j} = \left(\frac{\partial \mathbf{F}\mathbf{F}^+}{\partial x_j}\right)' = \left(\frac{\partial \mathbf{F}\mathbf{F}^+\mathbf{F}\mathbf{F}^+}{\partial x_j}\right)'$$
$$= \left(\mathbf{F}\mathbf{F}^+\frac{\partial \mathbf{F}\mathbf{F}^+}{\partial x_j} + \frac{\partial \mathbf{F}\mathbf{F}^+}{\partial x_j}\mathbf{F}\mathbf{F}^+\right)'$$
$$= \frac{\partial \mathbf{F}\mathbf{F}^+}{\partial x_j}\mathbf{F}\mathbf{F}' + \left(\frac{\partial \mathbf{F}\mathbf{F}^+}{\partial x_j}\mathbf{F}\mathbf{F}^+\right)'. \tag{8.5}$$

Moreover, at $\mathbf{x} = \mathbf{c}$,

$$\frac{\partial \mathbf{F}}{\partial x_j} = \frac{\partial \mathbf{F}\mathbf{F}^+\mathbf{F}}{\partial x_j} = \frac{\partial \mathbf{F}}{\partial x_j}\mathbf{F}^+\mathbf{F} + \mathbf{F}\frac{\partial \mathbf{F}^+\mathbf{F}}{\partial x_j},$$

and similarly

$$\frac{\partial \mathbf{F}}{\partial x_j} = \frac{\partial \mathbf{F}\mathbf{F}^+\mathbf{F}}{\partial x_j} = \frac{\partial \mathbf{F}\mathbf{F}^+}{\partial x_j}\mathbf{F} + \mathbf{F}\mathbf{F}^+\frac{\partial \mathbf{F}}{\partial x_j},$$

implying that (at $\mathbf{x} = \mathbf{c}$)

$$\mathbf{F}\frac{\partial \mathbf{F}^+\mathbf{F}}{\partial x_j} = \frac{\partial \mathbf{F}}{\partial x_j}(\mathbf{I} - \mathbf{F}^+\mathbf{F}) \tag{8.6}$$

and

$$\frac{\partial \mathbf{F}\mathbf{F}^+}{\partial x_j}\mathbf{F} = (\mathbf{I} - \mathbf{F}\mathbf{F}^+)\frac{\partial \mathbf{F}}{\partial x_j}. \tag{8.7}$$

Upon substituting expressions (8.6) and (8.7) into expressions (8.4) and (8.5), respectively, we obtain expressions (8.2) and (8.3). Q.E.D.

Proof (of Theorem 20.8.2). Observe (in light of Lemmas 20.8.1 and 15.4.3) that $\mathbf{F}^+$, $\mathbf{F}^+\mathbf{F}$, and $\mathbf{F}\mathbf{F}^+$ are continuously differentiable at $\mathbf{c}$.

Now, using Lemma 15.4.3, we find that (at $\mathbf{x} = \mathbf{c}$)

$$\frac{\partial \mathbf{F}^+}{\partial x_j} = \frac{\partial \mathbf{F}^+\mathbf{F}\mathbf{F}^+}{\partial x_j} = \frac{\partial \mathbf{F}^+\mathbf{F}}{\partial x_j}\mathbf{F}^+ + \mathbf{F}^+\mathbf{F}\frac{\partial \mathbf{F}^+}{\partial x_j}. \tag{8.8}$$

Moreover, at $\mathbf{x} = \mathbf{c}$,

$$\frac{\partial \mathbf{F}\mathbf{F}^+}{\partial x_j} = \frac{\partial \mathbf{F}}{\partial x_j}\mathbf{F}^+ + \mathbf{F}\frac{\partial \mathbf{F}^+}{\partial x_j},$$

implying that (at $\mathbf{x} = \mathbf{c}$)

$$\mathbf{F}\frac{\partial \mathbf{F}^+}{\partial x_j} = \frac{\partial \mathbf{F}\mathbf{F}^+}{\partial x_j} - \frac{\partial \mathbf{F}}{\partial x_j}\mathbf{F}^+. \tag{8.9}$$

And, upon substituting expression (8.9) into the second term of expression (8.8) and then making use of results (8.3) and (8.2), we find that (at $\mathbf{x} = \mathbf{c}$)

$$\begin{aligned}\frac{\partial \mathbf{F}^+}{\partial x_j} &= -\mathbf{F}^+\frac{\partial \mathbf{F}}{\partial x_j}\mathbf{F}^+ + \mathbf{F}^+\frac{\partial \mathbf{F}\mathbf{F}^+}{\partial x_j} + \frac{\partial \mathbf{F}^+\mathbf{F}}{\partial x_j}\mathbf{F}^+ \\ &= -\mathbf{F}^+\frac{\partial \mathbf{F}}{\partial x_j}\mathbf{F}^+ + \mathbf{F}^+(\mathbf{I}-\mathbf{F}\mathbf{F}^+)\frac{\partial \mathbf{F}}{\partial x_j}\mathbf{F}^+ + \mathbf{F}^+\left[(\mathbf{I}-\mathbf{F}\mathbf{F}^+)\frac{\partial \mathbf{F}}{\partial x_j}\mathbf{F}^+\right]' \\ &\qquad + \mathbf{F}^+\frac{\partial \mathbf{F}}{\partial x_j}(\mathbf{I}-\mathbf{F}^+\mathbf{F})\mathbf{F}^+ + \left[\mathbf{F}^+\frac{\partial \mathbf{F}}{\partial x_j}(\mathbf{I}-\mathbf{F}^+\mathbf{F})\right]'\mathbf{F}^+.\end{aligned}$$

Thus, since $\mathbf{F}^+(\mathbf{I}-\mathbf{F}\mathbf{F}^+) = \mathbf{0}$ and $(\mathbf{I}-\mathbf{F}^+\mathbf{F})\mathbf{F}^+ = \mathbf{0}$ and since (according to Corollary 20.5.2) $\mathbf{F}^+\mathbf{F}$ and $\mathbf{F}\mathbf{F}^+$ are symmetric,

$$\begin{aligned}\frac{\partial \mathbf{F}^+}{\partial x_j} &= -\mathbf{F}^+\frac{\partial \mathbf{F}}{\partial x_j}\mathbf{F}^+ + \mathbf{F}^+(\mathbf{F}^+)'\left(\frac{\partial \mathbf{F}}{\partial x_j}\right)'(\mathbf{I}-\mathbf{F}\mathbf{F}^+) \\ &\qquad + (\mathbf{I}-\mathbf{F}^+\mathbf{F})\left(\frac{\partial \mathbf{F}}{\partial x_j}\right)'(\mathbf{F}^+)'\mathbf{F}^+.\end{aligned}$$

Q.E.D.

Exercises

Section 20.2

1. Show that, for any $m \times n$ matrix $\mathbf{B}$ of full column rank and for any $n \times p$ matrix $\mathbf{C}$ of full row rank,

$$(\mathbf{B}\mathbf{C})^+ = \mathbf{C}^+\mathbf{B}^+.$$

2. Show that, for any $m \times n$ matrix $\mathbf{A}$, $\mathbf{A}^+ = \mathbf{A}'$ if and only if $\mathbf{A}'\mathbf{A}$ is idempotent.

Section 20.3

3. In connection with Theorem 9.6.1, show that if $\mathcal{C}(\mathbf{U}) \subset \mathcal{C}(\mathbf{T})$ and $\mathcal{R}(\mathbf{V}) \subset \mathcal{R}(\mathbf{T})$ and if the generalized inverses $\mathbf{T}^-$ and $\mathbf{Q}^-$ (of $\mathbf{T}$ and $\mathbf{Q}$, respectively) are both reflexive, then partitioned matrices (9.6.2) and (9.6.3) are reflexive generalized inverses of $\begin{pmatrix}\mathbf{T} & \mathbf{U}\\ \mathbf{V} & \mathbf{W}\end{pmatrix}$ and $\begin{pmatrix}\mathbf{W} & \mathbf{V}\\ \mathbf{U} & \mathbf{T}\end{pmatrix}$, respectively.

4. Determine which of Penrose Conditions (1)–(4) are necessarily satisfied by a left inverse of an $m \times n$ matrix $\mathbf{A}$ (when a left inverse exists). Which of the Penrose conditions are necessarily satisfied by a right inverse of an $m \times n$ matrix $\mathbf{A}$ (when a right inverse exists)?

Section 20.4

5. Let $\mathbf{A}$ represent an $m \times n$ matrix and $\mathbf{G}$ an $n \times m$ matrix.

   (a) Show that $\mathbf{G}$ is the Moore-Penrose inverse of $\mathbf{A}$ if and only if $\mathbf{G}$ is a minimum norm generalized inverse of $\mathbf{A}$ and $\mathbf{A}$ is a minimum norm generalized inverse of $\mathbf{G}$.

   (b) Show that $\mathbf{G}$ is the Moore-Penrose inverse of $\mathbf{A}$ if and only if $\mathbf{GAA}' = \mathbf{A}'$ and $\mathbf{AGG}' = \mathbf{G}'$.

   (c) Show that $\mathbf{G}$ is the Moore-Penrose inverse of $\mathbf{A}$ if and only if $\mathbf{GA} = \mathbf{P}_{\mathbf{A}'}$ and $\mathbf{AG} = \mathbf{P}_{\mathbf{G}'}$.

6. (a) Show that, for any $m \times n$ matrices $\mathbf{A}$ and $\mathbf{B}$ such that $\mathbf{A}'\mathbf{B} = \mathbf{0}$ and $\mathbf{BA}' = \mathbf{0}$, $(\mathbf{A} + \mathbf{B})^+ = \mathbf{A}^+ + \mathbf{B}^+$.

   (b) Let $\mathbf{A}_1, \mathbf{A}_2, \ldots, \mathbf{A}_k$ represent $m \times n$ matrices such that, for $j > i = 1, \ldots, k-1$, $\mathbf{A}_i'\mathbf{A}_j = \mathbf{0}$ and $\mathbf{A}_j\mathbf{A}_i' = \mathbf{0}$. Generalize the result of Part (a) by showing that $(\mathbf{A}_1 + \mathbf{A}_2 + \cdots + \mathbf{A}_k)^+ = \mathbf{A}_1^+ + \mathbf{A}_2^+ + \cdots + \mathbf{A}_k^+$.

Section 20.5

7. Show that, for any $m \times n$ matrix $\mathbf{A}$, $(\mathbf{A}^+\mathbf{A})^+ = \mathbf{A}^+\mathbf{A}$, and $(\mathbf{AA}^+)^+ = \mathbf{AA}^+$.

8. Show that, for any $n \times n$ symmetric matrix $\mathbf{A}$, $\mathbf{AA}^+ = \mathbf{A}^+\mathbf{A}$.

9. Let $\mathbf{V}$ represent an $n \times n$ symmetric nonnegative definite matrix, $\mathbf{X}$ an $n \times p$ matrix, and $\mathbf{d}$ a $p \times 1$ vector. Using the results of Exercises 8 and 19.11 (or otherwise), show that, for the vector $\mathbf{X}(\mathbf{X}'\mathbf{X})^-\mathbf{d}$ to be a solution, for every $\mathbf{d} \in \mathcal{C}(\mathbf{X}')$, to the problem of minimizing the quadratic form $\mathbf{a}'\mathbf{Va}$ (in $\mathbf{a}$) subject to $\mathbf{X}'\mathbf{a} = \mathbf{d}$, it is necessary and sufficient that $\mathcal{C}(\mathbf{V}^+\mathbf{X}) \subset \mathcal{C}(\mathbf{X})$.

10. Use Theorem 20.4.5 to devise an alternative proof of Part (3) of Theorem 20.5.3.

11. Let $\mathbf{C}$ represent an $m \times n$ matrix. Show that, for any $m \times m$ idempotent matrix $\mathbf{A}$, $(\mathbf{AC})^+\mathbf{A}' = (\mathbf{AC})^+$ and that, for any $n \times n$ idempotent matrix $\mathbf{B}$, $\mathbf{B}'(\mathbf{CB})^+ = (\mathbf{CB})^+$.

12. Let $\mathbf{a}$ represent an $n \times 1$ vector of variables, and impose on $\mathbf{a}$ the constraint $\mathbf{X}'\mathbf{a} = \mathbf{d}$, where $\mathbf{X}$ is an $n \times p$ matrix and $\mathbf{d}$ a $p \times 1$ vector such that $\mathbf{d} \in \mathcal{C}(\mathbf{X}')$.

Define $f(\mathbf{a}) = \mathbf{a}'\mathbf{V}\mathbf{a} - 2\mathbf{b}'\mathbf{a}$, where $\mathbf{V}$ is an $n \times n$ symmetric nonnegative definite matrix and $\mathbf{b}$ is an $n \times 1$ vector such that $\mathbf{b} \in \mathcal{C}(\mathbf{V}, \mathbf{X})$. Further, let $\mathbf{R}$ represent any matrix such that $\mathbf{V} = \mathbf{R}'\mathbf{R}$, let $\mathbf{a}_0$ represent any $n \times 1$ vector such that $\mathbf{X}'\mathbf{a}_0 = \mathbf{d}$, and take $\mathbf{s}$ to be any $n \times 1$ vector such that $\mathbf{b} = \mathbf{V}\mathbf{s} + \mathbf{X}\mathbf{t}$ for some $p \times 1$ vector $\mathbf{t}$. Using the results of Section 19.6 and of Exercise 11 (or otherwise), show that $f(\mathbf{a})$ attains its minimum value (under the constraint $\mathbf{X}'\mathbf{a} = \mathbf{d}$) at a point $\mathbf{a}_*$ if and only if

$$\mathbf{a}_* = \mathbf{a}_0 + [\mathbf{R}(\mathbf{I} - \mathbf{P}_\mathbf{X})]^+\mathbf{R}(\mathbf{s} - \mathbf{a}_0) + \{\mathbf{I} - [\mathbf{R}(\mathbf{I} - \mathbf{P}_\mathbf{X})]^+\mathbf{R}\}(\mathbf{I} - \mathbf{P}_\mathbf{X})\mathbf{w}$$

for some $n \times 1$ vector $\mathbf{w}$.

13. Let $\mathbf{A}$ represent an $n \times n$ symmetric nonnegative definite matrix, and let $\mathbf{B}$ represent an $n \times n$ matrix. Suppose that $\mathbf{B} - \mathbf{A}$ is symmetric and nonnegative definite (in which case $\mathbf{B}$ is symmetric and nonnegative definite). Using Theorems 18.3.4 and 20.4.5 and the results of Exercises 1 and 18.15 (or otherwise), show that $\mathbf{A}^+ - \mathbf{B}^+$ is nonnegative definite if and only if $\text{rank}(\mathbf{A}) = \text{rank}(\mathbf{B})$.

Bibliographic and Supplementary Notes

§3(c). The proof of Corollary 20.3.14 makes use of the results of Section 12.4 (on the least squares problem). §6. The result of Exercise 13 is Theorem 3.1 of Milliken and Akdeniz (1977). §8. Formula (8.1) (for the partial derivatives of the Moore-Penrose inverse of a matrix of functions) was given by Golub and Pereyra (1973, thm. 4.3). The derivation of this formula given in Section 8 is patterned after that of Magnus and Neudecker (1988, sec. 8.5).

21

Eigenvalues and Eigenvectors

The decomposition of a matrix $\mathbf{A}$ into a product of two or three matrices can (depending on the characteristics of those matrices) be a very useful first step in computing such things as the rank, the determinant, or an (ordinary or generalized) inverse (of $\mathbf{A}$) as well as a solution to a linear system having $\mathbf{A}$ as its coefficient matrix. In particular, decompositions like the QR, LDU, U$'$DU, and Cholesky decompositions—refer to Sections 6.4 and 14.5—in which the component matrices are diagonal or triangular, or have orthonormal rows or columns, can be very useful for computational purposes. Moreover, establishing that a matrix has a decomposition of a certain type can be instructive about the nature of the matrix. In particular, if it can be shown that a matrix $\mathbf{A}$ is expressible as $\mathbf{A} = \mathbf{R}\mathbf{R}'$ for some matrix $\mathbf{R}$, then it can be concluded—refer to Corollary 14.2.14—that $\mathbf{A}$ is nonnegative definite (and symmetric).

In the present chapter, consideration is given to a decomposition known as the spectral decomposition, which exists for symmetric matrices, and to a related decomposition known as the singular-value decomposition, which exists for all matrices. In the spectral or singular-value decomposition of a matrix $\mathbf{A}$, $\mathbf{A}$ is decomposed into $\mathbf{A} = \mathbf{QDP}$, where $\mathbf{P}$ and $\mathbf{Q}$ are orthogonal matrices and where $\mathbf{D}$ is a diagonal matrix or, more generally, where $\mathbf{D} = \mathrm{diag}(\mathbf{D}_1, \mathbf{0})$ for some diagonal matrix $\mathbf{D}_1$—in the case of the spectral decomposition, $\mathbf{P} = \mathbf{Q}'$.

The first part of the present chapter is devoted to scalars and (column) vectors that satisfy the definitions of what are, respectively, known as an eigenvalue and an eigenvector of a (square) matrix. And, in subsequent parts of the chapter, it is shown that the existence and construction of the spectral or singular-value decomposition of a matrix $\mathbf{A}$ is equivalent to the existence and computation of linearly independent eigenvectors of $\mathbf{A}$ or of $\mathbf{A}'\mathbf{A}$ or $\mathbf{A}\mathbf{A}'$.

The existence of the spectral or singular-value decomposition can (once it has been established) be very useful in establishing various other results on matrices, including a number of the results that were established earlier in the book without resort to the existence of this decomposition. And, once the spectral or singular-value decomposition of a matrix has been constructed, it can be very useful for computational purposes. However, there are drawbacks in using the existence of the spectral or singular-value decomposition in "theoretical" arguments (when there are other options) and in constructing this decomposition for computational purposes. Any proof of existence implicitly or explicitly involves some relatively deep mathematics, and in general the construction of the spectral or singular-value decomposition requires iterative methods and is much more computationally intensive than the construction of various other decompositions.

Most (if not all) of the more basic theoretical results in linear statistical models and in related areas of statistics can be established without implicit or explicit use of the existence of the spectral or singular-value decomposition. And, in addressing the computational problems encountered in this area of statistics, it is typically not efficient to adopt an approach that requires the construction of the spectral or singular-vlaue decomposition. However, there are exceptions—refer, for example, to Harville and Fenech (1985).

21.1 Definitions, Terminology, and Some Basic Results

A scalar (real number) λ is said to be an *eigenvalue* of an $n \times n$ matrix $\mathbf{A} = \{a_{ij}\}$ if there exists an $n \times 1$ nonnull vector $\mathbf{x}$ such that

$$\mathbf{Ax} = \lambda \mathbf{x}$$

or, equivalently, such that

$$(\mathbf{A} - \lambda \mathbf{I}_n)\mathbf{x} = \mathbf{0}.$$

Clearly, λ is an eigenvalue of $\mathbf{A}$ if and only if the matrix $\mathbf{A} - \lambda\mathbf{I}$ is singular. The set comprising the different scalars that are eigenvalues of $\mathbf{A}$ is called the *spectrum* of $\mathbf{A}$.

An $n \times 1$ nonnull vector $\mathbf{x}$ is said to be an *eigenvector* of the matrix $\mathbf{A}$ if there exists a scalar (real number) λ such that $\mathbf{Ax} = \lambda\mathbf{x}$, in which case λ is (by definition) an eigenvalue of $\mathbf{A}$. For any particular eigenvector $\mathbf{x}$ (of $\mathbf{A}$), there is only one eigenvalue λ such that $\mathbf{Ax} = \lambda\mathbf{x}$, which (since $\mathbf{Ax} = \lambda\mathbf{x} \Rightarrow \mathbf{x}'\mathbf{Ax} = \lambda\mathbf{x}'\mathbf{x}$) is

$$\lambda = \frac{\mathbf{x}'\mathbf{Ax}}{\mathbf{x}'\mathbf{x}}\,.$$

The eigenvector $\mathbf{x}$ is said to correspond to (or belong to) this eigenvalue. Note that if $\mathbf{x}$ is an eigenvector of $\mathbf{A}$, then, for any nonzero scalar c, the scalar multiple $c\mathbf{x}$ is also an eigenvector of $\mathbf{A}$ and $c\mathbf{x}$ corresponds to the same eigenvalue as $\mathbf{x}$.

Clearly, a nonnull vector $\mathbf{x}$ is an eigenvector (of the $n \times n$ matrix $\mathbf{A}$) corresponding to any particular eigenvalue λ if and only if it is a solution to the homogeneous linear system $(\mathbf{A} - \lambda\mathbf{I})\mathbf{z} = \mathbf{0}$ (in $\mathbf{z}$) or, equivalently, if and only if $\mathbf{x} \in \mathcal{N}(\mathbf{A} - \lambda\mathbf{I})$. Thus, the set comprising all of the eigenvectors (of $\mathbf{A}$) corresponding to the eigenvalue λ is the set obtained from $\mathcal{N}(\mathbf{A} - \lambda\mathbf{I})$ by deleting the $(n \times 1)$ null vector. The linear space $\mathcal{N}(\mathbf{A} - \lambda\mathbf{I})$ is sometimes referred to as the *eigenspace* of the eigenvalue λ, and the dimension of this space is referred to as the *geometric multiplicity* of λ. As a consequence of Lemma 11.3.1,

$$\dim[\mathcal{N}(\mathbf{A} - \lambda\mathbf{I})] = n - \text{rank}(\mathbf{A} - \lambda\mathbf{I}). \tag{1.1}$$

A scalar (real number) λ is an eigenvalue of the $n \times n$ matrix $\mathbf{A}$ if and only if

$$|\mathbf{A} - \lambda\mathbf{I}| = 0 \tag{1.2}$$

(since $\mathbf{A} - \lambda\mathbf{I}$ is singular if and only if $|\mathbf{A} - \lambda\mathbf{I}| = 0$). Let δ_{ij} represent the ijth element of $\mathbf{I}_n$, so that $\delta_{ij} = 1$, if $j = i$, and $\delta_{ij} = 0$, if $j \neq i$ $(i, j = 1, \ldots, n)$. Then, it follows from the very definition of a determinant [given by formula (13.1.2)] that

$$|\mathbf{A} - \lambda\mathbf{I}| = \sum (-1)^{\phi_n(j_1,\ldots,j_n)}(a_{1j_1} - \delta_{1j_1}\lambda)\cdots(a_{nj_n} - \delta_{nj_n}\lambda), \tag{1.3}$$

where $j_1, \ldots, j_n$ is a permutation of the first n positive integers and the summation is over all such permutations.

Upon inspecting the sum (1.3), we find that the term

$$(a_{11} - \lambda)(a_{22} - \lambda)\cdots(a_{nn} - \lambda), \tag{1.4}$$

corresponding to the permutation $j_1 = 1, j_2 = 2, \ldots, j_n = n$, is a polynomial (in λ) of degree n and that each of the remaining $(n! - 1)$ terms is a polynomial (in λ) of degree $n - 2$ or less. It follows that

$$|\mathbf{A} - \lambda\mathbf{I}| = p(\lambda), \tag{1.5}$$

where

$$p(\lambda) = c_0 + c_1\lambda + \cdots + c_{n-1}\lambda^{n-1} + c_n\lambda^n \tag{1.6}$$

is a polynomial (in λ) whose coefficients $c_0, c_1, \ldots, c_{n-1}, c_n$ depend on the elements of $\mathbf{A}$. And the term (1.4) is the only term of the sum (1.3) that contributes to c_n and c_{n-1}, so that

$$c_n = (-1)^n, \qquad c_{n-1} = (-1)^{n-1}\sum_{i=1}^{n} a_{ii} = (-1)^{n-1}\,\text{tr}(\mathbf{A}). \tag{1.7}$$

Further,

$$c_0 = p(0) = |\mathbf{A}|. \tag{1.8}$$

Alternatively, results (1.7) and (1.8) could have been deduced from Corollary 13.7.4. In fact, Corollary 13.7.4 provides expressions for $c_1, c_2, \ldots, c_{n-2}$ as well as for c_0, c_{n-1}, and c_n.

Condition (1.2) can be reexpressed as

$$p(\lambda) = 0 \tag{1.9}$$

and can be regarded as an equation (in λ). This equation is called the *characteristic equation* of $\mathbf{A}$, and the polynomial p is called the *characteristic polynomial* of $\mathbf{A}$. [Equation (1.9) and the polynomial p differ by a factor of $(-1)^n$ from what some authors call the characteristic equation and characteristic polynomial—those authors refer to $|\lambda\mathbf{I} - \mathbf{A}|$, which equals $(-1)^n|\mathbf{A} - \lambda\mathbf{I}|$, as the characteristic polynomial and/or to $|\lambda\mathbf{I} - \mathbf{A}| = 0$ as the characteristic equation.] Clearly, a scalar (real number) is an eigenvalue of $\mathbf{A}$ if and only if it is a root of the characteristic polynomial or, equivalently, if and only if it is a solution to the characteristic equation.

Let $\{\lambda_1, \ldots, \lambda_k\}$ represent the spectrum of the $n \times n$ matrix $\mathbf{A}$, that is, the set of (distinct) scalars that are eigenvalues of $\mathbf{A}$. It follows from standard results on polynomials—refer, for example, to Beaumont and Pierce (1963, secs. 9-5 and 9-7)—that the characteristic polynomial p of $\mathbf{A}$ has a unique (aside from the order of the factors) representation of the form

$$p(\lambda) = (-1)^n(\lambda - \lambda_1)^{\gamma_1} \cdots (\lambda - \lambda_k)^{\gamma_k} q(\lambda), \tag{1.10}$$

where $\gamma_1, \ldots, \gamma_k$ are positive integers and q is a polynomial (of degree $n - \sum_{i=1}^k \gamma_i$) that has no real roots.

The exponent γ_i of the factor $(\lambda - \lambda_i)^{\gamma_i}$ in expression (1.10) is referred to as the *algebraic multiplicity* of the eigenvalue λ_i $(i = 1, \ldots, k)$. Since the characteristic polynomial p is of degree n, it is clear that the sum $\sum_{i=1}^k \gamma_i$ of the algebraic multiplicities of the k eigenvalues of the $n \times n$ matrix $\mathbf{A}$ cannot exceed n (and also that the number k of distinct eigenvalues of $\mathbf{A}$ cannot exceed n). The sum $\sum_{i=1}^k \gamma_i$ equals n if and only if (for all λ) $q(\lambda) = 1$—compare the coefficient of λ^n in expression (1.10) with that in expression (1.6)—in which case expression (1.10) reduces to

$$p(\lambda) = (-1)^n(\lambda - \lambda_1)^{\gamma_1} \cdots (\lambda - \lambda_{k-1})^{\gamma_{k-1}}(\lambda - \lambda_k)^{\gamma_k}. \tag{1.11}$$

In general, the algebraic multiplicity of an eigenvalue of an $n \times n$ matrix is not the same as its geometric multiplicity. In fact, it will be shown in Section 3 that the algebraic multiplicity of any eigenvalue is greater than or equal to its geometric multiplicity.

Suppose, for example, that $n = 2$ and that $\mathbf{A} = \begin{pmatrix} 1 & 1 \\ 0 & 1 \end{pmatrix}$. Then, the characteristic polynomial of $\mathbf{A}$ is

$$p(\lambda) = \begin{vmatrix} 1-\lambda & 1 \\ 0 & 1-\lambda \end{vmatrix} = (1-\lambda)^2 = (\lambda - 1)^2.$$

Thus, the spectrum of $\mathbf{A}$ comprises a single eigenvalue 1, and the algebraic multiplicity of this eigenvalue is 2. However,

$$\mathbf{A} - (1)\mathbf{I} = \begin{pmatrix} 0 & 1 \\ 0 & 0 \end{pmatrix},$$

so that $\dim\{\mathcal{N}[\mathbf{A} - (1)\mathbf{I}]\} = n - \text{rank}[\mathbf{A} - (1)\mathbf{I}] = 2 - 1 = 1$. Thus, the geometric multiplicity of $\mathbf{A}$ is only 1.

Does every (square) matrix have at least one eigenvalue?

The fundamental theorem of algebra guarantees that every polynomial of degree one or more has a possibly complex root. Refer, for example, to Beaumont and Pierce (1963, sec. 9-8 and app. 3) for a formal statement of the fundamental theorem of algebra and for a proof—the proof of this theorem is well beyond the scope and mathematical level of our presentation.

Thus, the characteristic polynomial of every (square) matrix has at least one possibly complex root. However, there exist matrices whose characteristic polynomials have no real roots. These matrices have no eigenvalues.

Consider, for example, the matrix $\begin{pmatrix} 0 & 1 \\ -1 & 0 \end{pmatrix}$. The characteristic polynomial of this matrix is

$$p(\lambda) = \begin{vmatrix} -\lambda & 1 \\ -1 & -\lambda \end{vmatrix} = \lambda^2 + 1,$$

which has no real roots (although it has two imaginary roots, $\lambda = i$ and $\lambda = -i$). Thus, the matrix $\begin{pmatrix} 0 & 1 \\ -1 & 0 \end{pmatrix}$ has no eigenvalues.

In the degenerate special case where the $n \times n$ matrix $\mathbf{A}$ has no eigenvalues, expression (1.10) for the characteristic polynomial is to be read as $p(\lambda) = (-1)^n q(\lambda)$.

Every eigenvalue has an algebraic and geometric multiplicity of at least one. An eigenvalue is referred to as a *simple eigenvalue* or a *multiple eigenvalue*, depending on whether its algebraic multiplicity equals one or exceeds one. Eigenvectors that correspond to the same simple eigenvalue are scalar multiples of each other—it will be shown in Section 3 that the geometric multiplicity of a simple eigenvalue is necessarily equal to one—and consequently can be distinguished from each other by, for example, specifying the value of any nonzero element. In what follows, scalars that are not eigenvalues will sometimes (for convenience in stating various results) be regarded as eigenvalues whose algebraic and geometric multiplicities equal zero.

If the algebraic multiplicities of the eigenvalues of an $n \times n$ matrix $\mathbf{A}$ sum to n, then every root of the characteristic polynomial of $\mathbf{A}$ is a real number [as is evident from expression (1.11)]. The fundamental theorem of algebra can be used to establish the converse of this result. If the polynomial q in expression (1.10) is of degree one or more, then (according to the fundamental theorem) q has a root (which by definition is not real). Thus, if every root of the characteristic polynomial of the $n \times n$ matrix $\mathbf{A}$ is real, then the algebraic multiplicities of the eigenvalues of $\mathbf{A}$ sum to n.

Our use of the terms eigenvalue and eigenvector is more restrictive than that of many authors. In reference to an $n \times n$ matrix $\mathbf{A}$, they apply the term eigenvalue to any possibly complex number λ for which there exists a vector $\mathbf{x}$ (of possibly complex numbers) such that $\mathbf{Ax} = \lambda\mathbf{x}$ or, equivalently, to any possibly complex number that is a root of the characteristic polynomial. Similarly, they apply the term eigenvector to any vector $\mathbf{x}$ (of possibly complex numbers) for which there exists a possibly complex number λ such that $\mathbf{Ax} = \lambda\mathbf{x}$. In contrast, we confine our use of these terms to what those other authors call real eigenvalues and real eigenvectors.

It should also be noted that some authors use other terms in place of eigenvalue and eigenvector. In the case of eigenvalues, the alternative terms include characteristic value or root, proper value, and latent value. Similarly, the alternatives to eigenvector include characteristic vector, proper vector, and latent vector.

In referring to an $n \times n$ matrix, it is convenient to adopt a terminology that allows us to distinguish between the elements of its spectrum and the elements of a certain closely related set. Letting $m = \sum_{i=1}^{k} \gamma_i$, expression (1.10) for the characteristic polynomial of an $n \times n$ matrix $\mathbf{A}$ can be reexpressed as

$$p(\lambda) = (-1)^n (\lambda - d_1)(\lambda - d_2) \cdots (\lambda - d_m) q(\lambda),$$

where $d_1, d_2, \ldots, d_m$ are m scalars, γ_i of which equal λ_i (i.e., equal the ith element of the spectrum of $\mathbf{A}$). Let us refer to the elements of the set $\{d_1, d_2, \ldots, d_m\}$ as the eigenvalues of the matrix $\mathbf{A}$ or (for emphasis) as the not-necessarily-distinct eigenvalues of $\mathbf{A}$. And, to avoid confusion, let us subsequently refer to the elements of the spectrum of $\mathbf{A}$ as the distinct eigenvalues of $\mathbf{A}$.

Consider, for example, the 7×7 diagonal matrix $\mathbf{D} = \text{diag}(3, 1, -5, -5, 1, 1, 0)$. Its characteristic polynomial is

$$p(\lambda) = (\lambda-3)(\lambda-1)(\lambda+5)(\lambda+5)(\lambda-1)(\lambda-1)\lambda = (\lambda-3)(\lambda-1)^3(\lambda+5)^2\lambda.$$

The (not necessarily distinct) eigenvalues of $\mathbf{D}$ are $3, 1, -5, -5, 1, 1$, and 0; and the distinct eigenvalues of $\mathbf{D}$ are $3, 1, -5$, and 0.

Let $\mathbf{A}$ represent an $n \times n$ matrix. Further, let $\mathbf{x}_1, \ldots, \mathbf{x}_m$ represent $n \times 1$ vectors, and let $\mathbf{X} = (\mathbf{x}_1, \ldots, \mathbf{x}_m)$. Then, all m of the vectors $\mathbf{x}_1, \ldots, \mathbf{x}_m$ are eigenvectors of $\mathbf{A}$ if and only if there exists a diagonal matrix $\mathbf{D} = \{d_j\}$ such that

$$\mathbf{AX} = \mathbf{XD},$$

in which case $d_1, \ldots, d_m$ are the eigenvalues to which $\mathbf{x}_1, \ldots, \mathbf{x}_m$, respectively, correspond. (To see this, observe that the jth columns of $\mathbf{AX}$ and $\mathbf{XD}$ are, respectively, $\mathbf{Ax}_j$ and $d_j\mathbf{x}_j$.)

The eigenvectors of an $n \times n$ matrix $\mathbf{A}$ have a noteworthy geometrical property. Let λ represent any nonzero eigenvalue of $\mathbf{A}$, let $\mathbf{x}$ represent any eigenvector that corresponds to λ, and take the inner product for $\mathcal{R}^{n\times 1}$ to be the usual inner product. Then, the direction of the vector $\mathbf{Ax}$ (which, like $\mathbf{x}$, is an $n \times 1$ vector) is either the same as that of $\mathbf{x}$ or opposite to that of $\mathbf{x}$, depending on whether $\lambda > 0$ or $\lambda < 0$. That is, the angle between $\mathbf{Ax}$ and $\mathbf{x}$ is either 0 or π (180°), depending on whether $\lambda > 0$ or $\lambda < 0$. And the length of $\mathbf{Ax}$ is $|\lambda|$ times that of $\mathbf{x}$.

A subspace $\mathcal{U}$ of the linear space $\mathcal{R}^{n\times 1}$ is said to be *invariant* relative to an $n \times n$ matrix $\mathbf{A}$ if, for every vector $\mathbf{x}$ in $\mathcal{U}$, the vector $\mathbf{Ax}$ is also in $\mathcal{U}$.

Let $\mathbf{x}_1, \ldots, \mathbf{x}_m$ represent any eigenvectors of $\mathbf{A}$, and let $\mathbf{X} = (\mathbf{x}_1, \ldots, \mathbf{x}_m)$. Then, there exists a (diagonal) matrix $\mathbf{D}$ such that $\mathbf{AX} = \mathbf{XD}$. And, corresponding to any vector $\mathbf{u}$ in $\mathcal{C}(\mathbf{X})$, there exists an $m \times 1$ vector $\mathbf{r}$ such that $\mathbf{u} = \mathbf{Xr}$, so that

$$\mathbf{Au} = \mathbf{AXr} = \mathbf{XDr} \in \mathcal{C}(\mathbf{X}).$$

Thus, the subspace $\mathcal{C}(\mathbf{X})$ (of $\mathcal{R}^{n\times 1}$) spanned by the eigenvectors $\mathbf{x}_1, \dots, \mathbf{x}_m$ is invariant (relative to $\mathbf{A}$).

With the help of result (1.1), it is easy to verify the following lemma.

Lemma 21.1.1. The scalar 0 is an eigenvalue of an $n \times n$ matrix $\mathbf{A}$ if and only if $\mathbf{A}$ is singular, in which case the geometric multiplicity of the eigenvalue 0 equals $n - \text{rank}(\mathbf{A})$ or, equivalently, $\text{rank}(\mathbf{A})$ equals n minus the geometric multiplicity of the eigenvalue 0.

The following two lemmas relate the eigenvalues of the transpose of an $n \times n$ matrix $\mathbf{A}$ to those of $\mathbf{A}$ itself and relate the eigenvalues and eigenvectors of $\mathbf{A}^k$ (where k is a positive integer) and of $\mathbf{A}^{-1}$ to those of $\mathbf{A}$.

Lemma 21.1.2. Let $\mathbf{A}$ represent an $n \times n$ matrix. Then, $\mathbf{A}'$ has the same characteristic polynomial and the same spectrum as $\mathbf{A}$, and every scalar in the spectrum (of $\mathbf{A}$) has the same algebraic multiplicity when regarded as an eigenvalue of $\mathbf{A}'$ as when regarded as an eigenvalue of $\mathbf{A}$.

Proof. It suffices to show that $\mathbf{A}'$ and $\mathbf{A}$ have the same characteristic polynomial. That their characteristic polynomials are the same is evident upon observing (in light of Lemma 13.2.1) that, for any scalar λ,

$$|\mathbf{A}' - \lambda\mathbf{I}| = |(\mathbf{A} - \lambda\mathbf{I})'| = |\mathbf{A} - \lambda\mathbf{I}|.$$

Q.E.D.

Lemma 21.1.3. Let λ represent an eigenvalue of an $n \times n$ matrix $\mathbf{A}$, and let $\mathbf{x}$ represent any eigenvector (of $\mathbf{A}$) corresponding to λ. Then, (1) for any positive integer k, λ^k is an eigenvalue of $\mathbf{A}^k$, and $\mathbf{x}$ is an eigenvector of $\mathbf{A}^k$ corresponding to λ^k; and (2) if $\mathbf{A}$ is nonsingular (in which case $\lambda \neq 0$), $1/\lambda$ is an eigenvalue of $\mathbf{A}^{-1}$, and $\mathbf{x}$ is an eigenvector of $\mathbf{A}^{-1}$ corresponding to $1/\lambda$.

Proof. (1) The proof is by mathematical induction. By definition, λ^1 is an eigenvalue of $\mathbf{A}^1$, and $\mathbf{x}$ is an eigenvector of $\mathbf{A}^1$ corresponding to λ^1. Now, suppose that λ^{k-1} is an eigenvalue of $\mathbf{A}^{k-1}$ and that $\mathbf{x}$ is an eigenvector of $\mathbf{A}^{k-1}$ corresponding to λ^{k-1} (where $k \geq 2$). Then,

$$\mathbf{A}^k\mathbf{x} = \mathbf{A}^{k-1}\mathbf{A}\mathbf{x} = \mathbf{A}^{k-1}(\lambda\mathbf{x}) = \lambda\mathbf{A}^{k-1}\mathbf{x} = \lambda\lambda^{k-1}\mathbf{x} = \lambda^k\mathbf{x}.$$

(2) Suppose that $\mathbf{A}$ is nonsingular (in which case λ is clearly nonzero). Then,

$$\mathbf{x} = \mathbf{A}^{-1}\mathbf{A}\mathbf{x} = \mathbf{A}^{-1}(\lambda\mathbf{x}) = \lambda\mathbf{A}^{-1}\mathbf{x},$$

implying that $\mathbf{A}^{-1}\mathbf{x} = (1/\lambda)\mathbf{x}$. Q.E.D.

In connection with Lemma 21.1.2, it should be noted that (unless $\mathbf{A}$ is symmetric) an eigenvector of $\mathbf{A}$ is not in general an eigenvector of $\mathbf{A}'$.

21.2 Eigenvalues of Triangular or Block-Triangular Matrices and of Diagonal or Block-Diagonal Matrices

The eigenvalues of a matrix depend on the elements of the matrix in a very complicated way, and in general their computation requires the use of rather sophisticated and time-consuming numerical procedures. However, for certain types of matrices such as block-triangular or block-diagonal matrices, the eigenvalues can be reexpressed in terms of the eigenvalues of matrices of smaller dimensions or in other relatively simple ways, and the computational requirements can be substantially reduced.

The following lemma gives some results on the eigenvalues of block-triangular matrices.

Lemma 21.2.1. Let $\mathbf{A}$ represent an $n \times n$ matrix that is partitioned as

$$\mathbf{A} = \begin{pmatrix} \mathbf{A}_{11} & \mathbf{A}_{12} & \dots & \mathbf{A}_{1r} \\ \mathbf{A}_{21} & \mathbf{A}_{22} & \dots & \mathbf{A}_{2r} \\ \vdots & \vdots & \ddots & \vdots \\ \mathbf{A}_{r1} & \mathbf{A}_{r2} & \dots & \mathbf{A}_{rr} \end{pmatrix},$$

where (for $i, j = 1, \dots, r$) $\mathbf{A}_{ij}$ is of dimensions $n_i \times n_j$. Suppose that $\mathbf{A}$ is (upper or lower) block-triangular. Then,

(1) the characteristic polynomial $p(\lambda)$ of $\mathbf{A}$ is such that (for $\lambda \in \mathcal{R}$) $p(\lambda) = \prod_{i=1}^{r} p_i(\lambda)$, where (for $i = 1, \dots, r$) $p_i(\lambda)$ is the characteristic polynomial of $\mathbf{A}_{ii}$;

(2) a scalar λ is an eigenvalue of $\mathbf{A}$ if and only if λ is an eigenvalue of at least one of the diagonal blocks $\mathbf{A}_{11}, \dots, \mathbf{A}_{rr}$, or, equivalently, the spectrum of $\mathbf{A}$ is the union of the spectra of $\mathbf{A}_{11}, \dots, \mathbf{A}_{rr}$;

(3) the algebraic multiplicity of an eigenvalue λ of $\mathbf{A}$ equals $\sum_{i=1}^{r} \gamma^{(i)}$, where (for $i = 1, \dots, r$) $\gamma^{(i)}$ is the algebraic multiplicity of λ when λ is regarded as an eigenvalue of $\mathbf{A}_{ii}$—if λ is not an eigenvalue of $\mathbf{A}_{ii}$, then $\gamma^{(i)} = 0$; and

(4) the (not necessarily distinct) eigenvalues of $\mathbf{A}$ are $d_1^{(1)}, \dots, d_{m_1}^{(1)}, d_1^{(2)}, \dots, d_{m_2}^{(2)}, \dots, d_1^{(r)}, \dots, d_{m_r}^{(r)}$, where (for $i = 1, \dots, r$) $d_1^{(i)}, \dots, d_{m_i}^{(i)}$ are the (not necessarily distinct) eigenvalues of $\mathbf{A}_{ii}$.

Proof. (1) For $\lambda \in \mathcal{R}$, the matrix $\mathbf{A} - \lambda\mathbf{I}_n$ is block triangular with diagonal blocks $\mathbf{A}_{11} - \lambda\mathbf{I}_{n_1}, \dots, \mathbf{A}_{rr} - \lambda\mathbf{I}_{n_r}$, respectively. Thus, making use of results (13.3.3) and (13.3.4), we find that (for $\lambda \in \mathcal{R}$)

$$p(\lambda) = |\mathbf{A} - \lambda\mathbf{I}| = \prod_{i=1}^{r} |\mathbf{A}_{ii} - \lambda\mathbf{I}| = \prod_{i=1}^{r} p_i(\lambda).$$

(2)–(4). Parts (2)–(4) are almost immediate consequences of Part (1). Q.E.D.

Observe that the characteristic polynomial of a 1×1 matrix (c) (whose only element is the scalar c) is $p(\lambda) = (-1)(\lambda - c)$, that the spectrum of (c) is the set

$\{c\}$ whose only member is c, and that the algebraic (and geometric) multiplicities of the eigenvalue c of (c) equal 1. Then, by regarding a triangular matrix as a block-triangular matrix with 1×1 blocks, we obtain the following corollary of Lemma 21.2.1.

Corollary 21.2.2. For any $n \times n$ (upper or lower) triangular matrix $\mathbf{A} = \{a_{ij}\}$,

(1) the characteristic polynomial of $\mathbf{A}$ is $p(\lambda) = (-1)^n \prod_{i=1}^{n} (\lambda - a_{ii})$;

(2) a scalar λ is an eigenvalue of $\mathbf{A}$ if and only if it equals one or more of the diagonal elements $a_{11}, \ldots, a_{nn}$, or, equivalently, the spectrum of $\mathbf{A}$ comprises the distinct scalars represented among the diagonal elements of $\mathbf{A}$;

(3) the algebraic multiplicity of an eigenvalue λ of $\mathbf{A}$ equals the number of diagonal elements of $\mathbf{A}$ that equal λ; and

(4) the (not necessarily distinct) eigenvalues of $\mathbf{A}$ are the diagonal elements of $\mathbf{A}$.

In light of Corollary 21.2.2, it is a trivial matter to determine the eigenvalues of a triangular matrix. One need only inspect its diagonal elements.

Lemma 21.2.1 gives various results on the characteristic polynomial, the spectrum, and the algebraic multiplicities of the distinct eigenvalues of a block-triangular matrix (with square diagonal blocks). Since block-diagonal matrices are block-triangular, these results are applicable to block-diagonal matrices (or, more precisely, to block-diagonal matrices with square diagonal blocks). The following lemma gives some results on the geometric multiplicities of the distinct eigenvalues of a block-diagonal matrix and on the eigenvectors of a block-diagonal matrix—these results are specific to block-diagonal matrices, that is, do not (in general) apply to block-triangular matrices.

Lemma 21.2.3. For any $n \times n$ block-diagonal matrix $\mathbf{A}$ with square diagonal blocks $\mathbf{A}_{11}, \ldots, \mathbf{A}_{rr}$ of order $n_1, \ldots, n_r$, respectively,

(1) the geometric multiplicity of an eigenvalue λ of $\mathbf{A}$ equals $\sum_{i=1}^{r} \nu^{(i)}$, where (for $i = 1, \ldots, r$) $\nu^{(i)}$ is the geometric multiplicity of λ when λ is regarded as an eigenvalue of $\mathbf{A}_{ii}$—if λ is not an eigenvalue of $\mathbf{A}_{ii}$, then $\nu^{(i)} = 0$; and

(2) if an $n_i \times 1$ vector $\mathbf{x}_*$ is an eigenvector of $\mathbf{A}_{ii}$ corresponding to an eigenvalue λ, then the $n \times 1$ vector $\mathbf{x} = (\mathbf{0}', \ldots \mathbf{0}', \mathbf{x}_*', \mathbf{0}', \ldots, \mathbf{0}')'$, whose $(1 + \sum_{s=1}^{i-1} n_s)$th, $\ldots$, $(\sum_{s=1}^{i} n_s)$th elements are respectively the first, $\ldots$, n_ith elements of $\mathbf{x}_*$, is an eigenvector of $\mathbf{A}$ corresponding to λ—λ is an eigenvalue of $\mathbf{A}$ as well as of $\mathbf{A}_{ii}$ $(1 \leq i \leq r)$.

Proof. (1) Recalling results (1.1) and (4.5.14) and observing that $\mathbf{A} - \lambda\mathbf{I} = \text{diag}(\mathbf{A}_{11} - \lambda\mathbf{I}, \ldots, \mathbf{A}_{rr} - \lambda\mathbf{I})$, we find that the geometric multiplicity of λ equals

$$n - \text{rank}(\mathbf{A} - \lambda\mathbf{I}) = n - \sum_{i=1}^{r} \text{rank}(\mathbf{A}_{ii} - \lambda\mathbf{I}) = \sum_{i=1}^{r} [n_i - \text{rank}(\mathbf{A}_{ii} - \lambda\mathbf{I})]$$

$$= \sum_{i=1}^{r} \nu^{(i)}.$$

(2) Suppose that $\mathbf{x}_*$ is an eigenvector of $\mathbf{A}_{ii}$ corresponding to an eigenvalue λ. Then,

$$\mathbf{A}\mathbf{x} = \begin{pmatrix} \mathbf{0} \\ \vdots \\ \mathbf{0} \\ \mathbf{A}_{ii}\mathbf{x}_* \\ \mathbf{0} \\ \vdots \\ \mathbf{0} \end{pmatrix} = \begin{pmatrix} \mathbf{0} \\ \vdots \\ \mathbf{0} \\ \lambda\mathbf{x}_* \\ \mathbf{0} \\ \vdots \\ \mathbf{0} \end{pmatrix} = \lambda\mathbf{x}.$$

Q.E.D.

By regarding a diagonal matrix as a block-diagonal matrix with 1×1 blocks, we obtain the following corollary of Lemma 21.2.3.

Corollary 21.2.4. For any $n \times n$ diagonal matrix $\mathbf{D} = \{d_i\}$,

(1) the geometric multiplicity of an eigenvalue λ of $\mathbf{D}$ equals the number of diagonal elements of $\mathbf{D}$ that equal λ; and

(2) the $n \times 1$ vector $c\mathbf{u}_i$, where $\mathbf{u}_i$ is the ith column of $\mathbf{I}_n$ and c is an arbitrary nonzero scalar, is an eigenvector of $\mathbf{D}$ corresponding to d_i—d_i is an eigenvalue of $\mathbf{D}$ $(1 \leq i \leq n)$.

Corollary 21.2.4 gives results on diagonal matrices that augment those given by Corollary 21.2.2—since diagonal matrices are triangular, the results of Corollary 21.2.2 are applicable to diagonal matrices. Note that the geometric multiplicity of any eigenvalue of a diagonal matrix is the same as its algebraic multiplicity, as evidenced by Part (1) of Corollary 21.2.4 and Part (3) of Corollary 21.2.2.

21.3 Similar Matrices

An $n \times n$ matrix $\mathbf{B}$ is said to be *similar* to an $n \times n$ matrix $\mathbf{A}$ if there exists an $n \times n$ nonsingular matrix $\mathbf{C}$ such that $\mathbf{B} = \mathbf{C}^{-1}\mathbf{A}\mathbf{C}$ or, equivalently, such that $\mathbf{C}\mathbf{B} = \mathbf{A}\mathbf{C}$. Upon observing that (for any $n \times n$ nonsingular matrix $\mathbf{C}$)

$$\mathbf{B} = \mathbf{C}^{-1}\mathbf{A}\mathbf{C} \Rightarrow \mathbf{A} = \mathbf{C}\mathbf{B}\mathbf{C}^{-1} = (\mathbf{C}^{-1})^{-1}\mathbf{B}\mathbf{C}^{-1},$$

it is clear that if $\mathbf{B}$ is similar to $\mathbf{A}$, then $\mathbf{A}$ is similar to $\mathbf{B}$.

Some of the ways in which similar matrices are similar are described in the following theorem.

Theorem 21.3.1. For any $n \times n$ matrix $\mathbf{A}$ and any $n \times n$ nonsingular matrix $\mathbf{C}$,

(1) $\operatorname{rank}(\mathbf{C}^{-1}\mathbf{A}\mathbf{C}) = \operatorname{rank}(\mathbf{A})$;

(2) $\det(\mathbf{C}^{-1}\mathbf{A}\mathbf{C}) = \det(\mathbf{A})$;

(3) $\operatorname{tr}(\mathbf{C}^{-1}\mathbf{A}\mathbf{C}) = \operatorname{tr}(\mathbf{A})$;

(4) $\mathbf{C}^{-1}\mathbf{A}\mathbf{C}$ and $\mathbf{A}$ have the same characteristic polynomial and the same spectrum;

(5) an eigenvalue has the same algebraic and geometric multiplicities when regarded as an eigenvalue of $\mathbf{C}^{-1}\mathbf{A}\mathbf{C}$ as when regarded as an eigenvalue of $\mathbf{A}$;

(6) if $\mathbf{x}$ is an eigenvector of $\mathbf{A}$ corresponding to an eigenvalue λ, then $\mathbf{C}^{-1}\mathbf{x}$ is an eigenvector of $\mathbf{C}^{-1}\mathbf{A}\mathbf{C}$ corresponding to λ, and, similarly, if $\mathbf{y}$ is an eigenvector of $\mathbf{C}^{-1}\mathbf{A}\mathbf{C}$ corresponding to λ, then $\mathbf{C}\mathbf{y}$ is an eigenvector of $\mathbf{A}$ corresponding to λ; and

(7) the (not necessarily distinct) eigenvalues of $\mathbf{C}^{-1}\mathbf{A}\mathbf{C}$ are the same as those of $\mathbf{A}$.

Proof. (1) That $\text{rank}(\mathbf{C}^{-1}\mathbf{A}\mathbf{C}) = \text{rank}(\mathbf{A})$ is evident from Corollary 8.3.3.
(2) Using results (13.3.9) and (13.3.12), we find that

$$|\mathbf{C}^{-1}\mathbf{A}\mathbf{C}| = |\mathbf{C}^{-1}||\mathbf{A}||\mathbf{C}| = (1/|\mathbf{C}|)|\mathbf{A}||\mathbf{C}| = |\mathbf{A}|.$$

(3) As a consequence of Lemma 5.2.1,

$$\text{tr}(\mathbf{C}^{-1}\mathbf{A}\mathbf{C}) = \text{tr}(\mathbf{A}\mathbf{C}\mathbf{C}^{-1}) = \text{tr}(\mathbf{A}).$$

(4) Applying Part (2) (with $\mathbf{A} - \lambda\mathbf{I}$ in place of $\mathbf{A}$), we find that, for any scalar λ,

$$|\mathbf{C}^{-1}\mathbf{A}\mathbf{C} - \lambda\mathbf{I}| = |\mathbf{C}^{-1}(\mathbf{A} - \lambda\mathbf{I})\mathbf{C}| = |\mathbf{A} - \lambda\mathbf{I}|.$$

Thus, $\mathbf{C}^{-1}\mathbf{A}\mathbf{C}$ and $\mathbf{A}$ have the same characteristic polynomial (which implies that they have the same spectrum).

(5) Let λ represent an eigenvalue of $\mathbf{A}$ or, equivalently [in light of Part (4)], of $\mathbf{C}^{-1}\mathbf{A}\mathbf{C}$. Since [according to Part (4)] $\mathbf{C}^{-1}\mathbf{A}\mathbf{C}$ and $\mathbf{A}$ have the same characteristic polynomial, λ has the same algebraic multiplicity when regarded as an eigenvalue of $\mathbf{C}^{-1}\mathbf{A}\mathbf{C}$ as when regarded as an eigenvalue of $\mathbf{A}$. Further, the geometric multiplicities of λ when regarded as an eigenvalue of $\mathbf{C}^{-1}\mathbf{A}\mathbf{C}$ and when regarded as an eigenvalue of $\mathbf{A}$ are, respectively [in light of result (1.1)], $n - \text{rank}(\mathbf{C}^{-1}\mathbf{A}\mathbf{C} - \lambda\mathbf{I})$ and $n - \text{rank}(\mathbf{A} - \lambda\mathbf{I})$ and hence $\{$since $\text{rank}(\mathbf{C}^{-1}\mathbf{A}\mathbf{C} - \lambda\mathbf{I}) = \text{rank}[\mathbf{C}^{-1}(\mathbf{A} - \lambda\mathbf{I})\mathbf{C}] = \text{rank}(\mathbf{A} - \lambda\mathbf{I})\}$ are the same.

(6) If $\mathbf{A}\mathbf{x} = \lambda\mathbf{x}$, then

$$(\mathbf{C}^{-1}\mathbf{A}\mathbf{C})\mathbf{C}^{-1}\mathbf{x} = \mathbf{C}^{-1}\mathbf{A}\mathbf{x} = \mathbf{C}^{-1}(\lambda\mathbf{x}) = \lambda\mathbf{C}^{-1}\mathbf{x},$$

and if $\mathbf{C}^{-1}\mathbf{A}\mathbf{C}\mathbf{y} = \lambda\mathbf{y}$, then

$$\mathbf{A}\mathbf{C}\mathbf{y} = \mathbf{C}\mathbf{C}^{-1}\mathbf{A}\mathbf{C}\mathbf{y} = \mathbf{C}(\lambda\mathbf{y}) = \lambda\mathbf{C}\mathbf{y}.$$

(7) Since [according to Part (4)] $\mathbf{C}^{-1}\mathbf{A}\mathbf{C}$ and $\mathbf{A}$ have the same characteristic polynomial, the (not necessarily distinct) eigenvalues of $\mathbf{C}^{-1}\mathbf{A}\mathbf{C}$ are the same as those of $\mathbf{A}$. Q.E.D.

Theorem 21.3.1 can be very useful. If, for example, it can be demonstrated that a matrix $\mathbf{A}$ is similar to a diagonal or triangular matrix or to some other matrix whose eigenvalues are relatively easy to obtain, then the determination of the eigenvalues of $\mathbf{A}$ may be greatly facilitated.

Subsequently, we shall have occasion to make use of the following theorem.

Theorem 21.3.2. Let $\mathbf{A}$ represent an $n \times n$ matrix, $\mathbf{B}$ a $k \times k$ matrix, and $\mathbf{U}$ an $n \times k$ matrix with orthonormal columns (i.e., with $\mathbf{U}'\mathbf{U} = \mathbf{I}$) such that $\mathbf{AU} = \mathbf{UB}$. Then, there exists an $n \times (n-k)$ matrix $\mathbf{V}$ such that the $n \times n$ matrix $(\mathbf{U}, \mathbf{V})$ is orthogonal. And, for any such matrix $\mathbf{V}$,

$$(\mathbf{U}, \mathbf{V})'\mathbf{A}(\mathbf{U}, \mathbf{V}) = \begin{pmatrix} \mathbf{B} & \mathbf{U}'\mathbf{A}\mathbf{V} \\ \mathbf{0} & \mathbf{V}'\mathbf{A}\mathbf{V} \end{pmatrix}$$

[so that $\mathbf{A}$ is similar to the block-triangular matrix $\begin{pmatrix} \mathbf{B} & \mathbf{U}'\mathbf{A}\mathbf{V} \\ \mathbf{0} & \mathbf{V}'\mathbf{A}\mathbf{V} \end{pmatrix}$], and, in the special case where $\mathbf{A}$ is symmetric,

$$(\mathbf{U}, \mathbf{V})'\mathbf{A}(\mathbf{U}, \mathbf{V}) = \begin{pmatrix} \mathbf{B} & \mathbf{0} \\ \mathbf{0} & \mathbf{V}'\mathbf{A}\mathbf{V} \end{pmatrix}$$

[so that $\mathbf{A}$ is similar to $\text{diag}(\mathbf{B}, \mathbf{V}'\mathbf{A}\mathbf{V})$].

Proof. As a consequence of Theorem 6.4.5, there exists an $n \times (n-k)$ matrix $\mathbf{V}$ such that the columns of the $n \times n$ matrix $(\mathbf{U}, \mathbf{V})$ form an orthonormal (with respect to the usual inner product) basis for $\mathcal{R}^n$ or, equivalently, such that the matrix $(\mathbf{U}, \mathbf{V})$ is orthogonal.

Now, let $\mathbf{V}$ represent any such matrix. Then, observing that

$$\mathbf{U}'\mathbf{A}\mathbf{U} = \mathbf{U}'\mathbf{U}\mathbf{B} = \mathbf{I}\mathbf{B} = \mathbf{B}$$

and that

$$\mathbf{V}'\mathbf{A}\mathbf{U} = \mathbf{V}'\mathbf{U}\mathbf{B} = \mathbf{0}\mathbf{B} = \mathbf{0},$$

we find that

$$(\mathbf{U}, \mathbf{V})'\mathbf{A}(\mathbf{U}, \mathbf{V}) = \begin{pmatrix} \mathbf{B} & \mathbf{U}'\mathbf{A}\mathbf{V} \\ \mathbf{0} & \mathbf{V}'\mathbf{A}\mathbf{V} \end{pmatrix}.$$

And, in the special case where $\mathbf{A}$ is symmetric,

$$\mathbf{U}'\mathbf{A}\mathbf{V} = \mathbf{U}'\mathbf{A}'\mathbf{V} = (\mathbf{V}'\mathbf{A}\mathbf{U})' = \mathbf{0}' = \mathbf{0},$$

so that

$$(\mathbf{U}, \mathbf{V})'\mathbf{A}(\mathbf{U}, \mathbf{V}) = \text{diag}(\mathbf{B}, \mathbf{V}'\mathbf{A}\mathbf{V}).$$

Q.E.D.

In the special case where $k = 1$, Theorem 21.3.2 can be restated as the following corollary.

Corollary 21.3.3. Let $\mathbf{A}$ represent an $n \times n$ matrix. And let λ represent any eigenvalue of $\mathbf{A}$ (assuming that one exists), and $\mathbf{u}$ represent an eigenvector of (usual) norm 1 that corresponds to λ. Then, there exists an $n \times (n-1)$ matrix $\mathbf{V}$ such that the $n \times n$ matrix $(\mathbf{u}, \mathbf{V})$ is orthogonal. Moreover, for any such matrix $\mathbf{V}$,

$$(\mathbf{u}, \mathbf{V})'\mathbf{A}(\mathbf{u}, \mathbf{V}) = \begin{pmatrix} \lambda & \mathbf{u}'\mathbf{A}\mathbf{V} \\ \mathbf{0} & \mathbf{V}'\mathbf{A}\mathbf{V} \end{pmatrix}$$

[so that $\mathbf{A}$ is similar to the block-triangular matrix $\begin{pmatrix} \lambda & \mathbf{u}'\mathbf{A}\mathbf{V} \\ \mathbf{0} & \mathbf{V}'\mathbf{A}\mathbf{V} \end{pmatrix}$], and, in the special case where $\mathbf{A}$ is symmetric,

$$(\mathbf{u}, \mathbf{V})'\mathbf{A}(\mathbf{u}, \mathbf{V}) = \begin{pmatrix} \lambda & \mathbf{0} \\ \mathbf{0} & \mathbf{V}'\mathbf{A}\mathbf{V} \end{pmatrix}$$

[so that $\mathbf{A}$ is similar to $\mathrm{diag}(\lambda, \mathbf{V}'\mathbf{A}\mathbf{V})$].

The following theorem gives an inequality that was mentioned (but not verified) in Section 1.

Theorem 21.3.4. The geometric multiplicity of an eigenvalue λ of an $n \times n$ matrix $\mathbf{A}$ is less than or equal to the algebraic multiplicity of λ.

Proof. Let $\nu = \dim[\mathcal{N}(\mathbf{A} - \lambda\mathbf{I})]$, which by definition is the geometric multiplicity of λ. Then, as a consequence of Theorem 6.4.3, there exists an $n \times \nu$ matrix $\mathbf{U}$ with orthonormal (with respect to the usual inner product) columns that form a basis for $\mathcal{N}(\mathbf{A} - \lambda\mathbf{I})$. And, since $(\mathbf{A} - \lambda\mathbf{I})\mathbf{U} = \mathbf{0}$ and hence since $\mathbf{A}\mathbf{U} = \lambda\mathbf{U} = \mathbf{U}(\lambda\mathbf{I}_\nu)$, it follows from Theorem 21.3.2 that $\mathbf{A}$ is similar to a matrix that is expressible as

$$\begin{pmatrix} \lambda\mathbf{I}_\nu & \mathbf{R} \\ \mathbf{0} & \mathbf{S} \end{pmatrix},$$

where $\mathbf{R}$ is a $\nu \times (n-\nu)$ matrix and $\mathbf{S}$ is an $(n-\nu) \times (n-\nu)$ matrix.

Now, it follows from Theorem 21.3.1 that λ is an eigenvalue of $\begin{pmatrix} \lambda\mathbf{I}_\nu & \mathbf{R} \\ \mathbf{0} & \mathbf{S} \end{pmatrix}$—that λ is an eigenvalue of $\begin{pmatrix} \lambda\mathbf{I}_\nu & \mathbf{R} \\ \mathbf{0} & \mathbf{S} \end{pmatrix}$ is also evident from Lemma 21.2.1—and that λ has the same algebraic multiplicity when regarded as an eigenvalue of $\begin{pmatrix} \lambda\mathbf{I}_\nu & \mathbf{R} \\ \mathbf{0} & \mathbf{S} \end{pmatrix}$ as when regarded as an eigenvalue of $\mathbf{A}$. Moreover, it is clear from Lemma 21.2.1 and from Corollary 21.2.2 that the algebraic multiplicity of λ when regarded as an eigenvalue of $\begin{pmatrix} \lambda\mathbf{I}_\nu & \mathbf{R} \\ \mathbf{0} & \mathbf{S} \end{pmatrix}$ is at least ν. We conclude that the algebraic multiplicity of λ when regarded as an eigenvalue of $\mathbf{A}$ is at least ν. Q.E.D.

The following four corollaries give various implications of Theorem 21.3.4.

Corollary 21.3.5. If the algebraic multiplicity of an eigenvalue λ of an $n \times n$ matrix equals 1, then the geometric multiplicity of λ also equals 1.

Proof. The validity of Corollary 21.3.5 is evident from Theorem 21.3.4 upon observing that the geometric multiplicity of λ is at least one. Q.E.D.

Corollary 21.3.6. If an $n \times n$ matrix $\mathbf{A}$ has n distinct eigenvalues, then each of them has a geometric (and algebraic) multiplicity of 1.

Proof. The algebraic multiplicities of the distinct eigenvalues of $\mathbf{A}$ are greater than or equal to 1, and the sum of the algebraic multiplicities cannot exceed n. Thus, if $\mathbf{A}$ has n distinct eigenvalues, then each of them has an algebraic multiplicity of 1 and hence (in light of Corollary 21.3.5) a geometric multiplicity of 1. Q.E.D.

Corollary 21.3.7. Let $\mathbf{A}$ represent an $n \times n$ matrix that has k distinct eigenvalues $\lambda_1, \ldots, \lambda_k$ with algebraic multiplicities $\gamma_1, \ldots, \gamma_k$, respectively, and geometric multiplicities $\nu_1, \ldots, \nu_k$, respectively. Then, $\sum_{i=1}^k \nu_i \leq \sum_{i=1}^k \gamma_i \leq n$. And, if $\sum_{i=1}^k \nu_i = n$, then $\sum_{i=1}^k \gamma_i = n$. Moreover, if $\sum_{i=1}^k \nu_i = \sum_{i=1}^k \gamma_i$, then (for $i = 1, \ldots, k$) $\nu_i = \gamma_i$.

Proof. That $\sum_{i=1}^k \nu_i \leq \sum_{i=1}^k \gamma_i$ is an immediate consequence of Theorem 21.3.4, and that $\sum_{i=1}^k \gamma_i \leq n$ was established earlier (in Section 1). And, if $\sum_{i=1}^k \nu_i = n$, then together these inequalities imply that $\sum_{i=1}^k \gamma_i = n$.

Now, suppose that $\sum_{i=1}^k \nu_i = \sum_{i=1}^k \gamma_i$. Then, $\sum_{i=1}^k (\gamma_i - \nu_i) = 0$. And, according to Theorem 21.3.4, $\gamma_i \geq \nu_i$ or, equivalently, $\gamma_i - \nu_i \geq 0$ $(i = 1, \ldots, k)$. Thus, $\gamma_i - \nu_i = 0$, or, equivalently, $\nu_i = \gamma_i$ $(i = 1, \ldots, k)$. Q.E.D.

Corollary 21.3.8. Let $\mathbf{A}$ represent an $n \times n$ matrix that has k distinct eigenvalues $\lambda_1, \ldots, \lambda_k$ with geometric multiplicities $\nu_1, \ldots, \nu_k$, respectively. Then, the sum of the geometric multiplicities of the (k or $k-1$) nonzero distinct eigenvalues (of $\mathbf{A}$) is less than or equal to $\operatorname{rank}(\mathbf{A})$, with equality holding if and only if $\sum_{i=1}^k \nu_i = n$.

Proof. It is convenient to consider separately two cases.

Case 1: $\operatorname{rank}(\mathbf{A}) = n$. In this case, all k of the distinct eigenvalues $\lambda_1, \ldots, \lambda_k$ are nonzero (as is evident from Lemma 11.3.1), and it is apparent from Corollary 21.3.7 that the sum $\sum_{i=1}^k \nu_i$ of the geometric multiplicities is less than or equal to $\operatorname{rank}(\mathbf{A})$. (And it is obvious that this sum equals $\operatorname{rank}(\mathbf{A})$ if and only if $\sum_{i=1}^k \nu_i = n$.)

Case 2: $\operatorname{rank}(\mathbf{A}) < n$. In this case, it follows from Lemma 11.3.1 that one of the distinct eigenvalues, say λ_s, equals 0 and that $\nu_s = n - \operatorname{rank}(\mathbf{A})$. Thus, making use of Corollary 21.3.7, we find that

$$\sum_{i \neq s}^{k} \nu_i = \sum_{i=1}^{k} \nu_i - [n - \operatorname{rank}(\mathbf{A})] \leq n - [n - \operatorname{rank}(\mathbf{A})] = \operatorname{rank}(\mathbf{A}).$$

And, clearly, $\sum_{i \neq s}^k \nu_i = \operatorname{rank}(\mathbf{A})$ if and only if $n = \sum_{i=1}^k \nu_i$. Q.E.D.

21.4 Linear Independence of Eigenvectors

A basic property of eigenvectors is described in the following theorem.

Theorem 21.4.1. Let $\mathbf{A}$ represent an $n \times n$ matrix, let $\mathbf{x}_1, \ldots, \mathbf{x}_k$ represent k eigenvectors of $\mathbf{A}$, and let $\lambda_1, \ldots, \lambda_k$ represent the eigenvalues (of $\mathbf{A}$) to which $\mathbf{x}_1, \ldots, \mathbf{x}_k$ correspond. If $\lambda_1, \ldots, \lambda_k$ are distinct, then $\mathbf{x}_1, \ldots, \mathbf{x}_k$ are linearly independent.

Theorem 21.4.1 is a special case of the following result.

Theorem 21.4.2. Let $\mathbf{A}$ represent an $n \times n$ matrix. And let $\mathbf{x}_i^{(1)}, \ldots, \mathbf{x}_i^{(m_i)}$ represent linearly independent eigenvectors of $\mathbf{A}$ that correspond to some eigenvalue λ_i $(i = 1, \ldots, k)$. If $\lambda_1, \ldots, \lambda_k$ are distinct, then the combined set $\mathbf{x}_1^{(1)}, \ldots, \mathbf{x}_1^{(m_1)}, \ldots, \mathbf{x}_k^{(1)}, \ldots, \mathbf{x}_k^{(m_k)}$ of $\sum_{i=1}^k m_i$ eigenvectors is linearly independent.

Proof (of Theorem 21.4.2). Assume that $\lambda_1, \ldots, \lambda_k$ are distinct. To show that the combined set is linearly independent, let us proceed by mathematical induction. The set of m_1 eigenvectors $\mathbf{x}_1^{(1)}, \ldots \mathbf{x}_1^{(m_1)}$ is by definition linearly independent. Suppose now that the set comprising the first $\sum_{i=1}^s m_i$ eigenvectors $\mathbf{x}_1^{(1)}, \ldots, \mathbf{x}_1^{(m_1)}, \ldots, \mathbf{x}_s^{(1)}, \ldots, \mathbf{x}_s^{(m_s)}$ is linearly independent (where $1 \le s \le k-1$). Then, to complete the induction argument, it suffices to show that the set comprising the first $\sum_{i=1}^{s+1} m_i$ eigenvectors $\mathbf{x}_1^{(1)}, \ldots, \mathbf{x}_1^{(m_1)}, \ldots, \mathbf{x}_{s+1}^{(1)}, \ldots, \mathbf{x}_{s+1}^{(m_{s+1})}$ is linearly independent.

Let $c_1^{(1)}, \ldots, c_1^{(m_1)}, \ldots, c_{s+1}^{(1)}, \ldots, c_{s+1}^{(m_{s+1})}$ represent any scalars such that

$$c_1^{(1)}\mathbf{x}_1^{(1)} + \cdots + c_1^{(m_1)}\mathbf{x}_1^{(m_1)} + \cdots + c_{s+1}^{(1)}\mathbf{x}_{s+1}^{(1)} + \cdots + c_{s+1}^{(m_{s+1})}\mathbf{x}_{s+1}^{(m_{s+1})} = \mathbf{0}. \quad (4.1)$$

And observe that, for $j = 1, \ldots, m_{s+1}$,

$$(\mathbf{A} - \lambda_{s+1}\mathbf{I})\mathbf{x}_{s+1}^{(j)} = \mathbf{0}$$

and that, for $i = 1, \ldots, s$ and $j = 1, \ldots, m_i$,

$$(\mathbf{A} - \lambda_{s+1}\mathbf{I})\mathbf{x}_i^{(j)} = [\mathbf{A} - \lambda_i\mathbf{I} + (\lambda_i - \lambda_{s+1})\mathbf{I}]\mathbf{x}_i^{(j)} = (\lambda_i - \lambda_{s+1})\mathbf{x}_i^{(j)}.$$

Then, premultiplying both sides of equality (4.1) by $\mathbf{A} - \lambda_{s+1}\mathbf{I}$, we find that

$$(\lambda_1 - \lambda_{s+1})(c_1^{(1)}\mathbf{x}_1^{(1)} + \cdots + c_1^{(m_1)}\mathbf{x}_1^{(m_1)}) + \cdots + (\lambda_s - \lambda_{s+1})(c_s^{(1)}\mathbf{x}_s^{(1)} + \cdots + c_s^{(m_s)}\mathbf{x}_s^{(m_s)}) = \mathbf{0}.$$

And, since $\lambda_i - \lambda_{s+1} \neq 0$ for $i \neq s+1$ and (by supposition) the $\sum_{i=1}^s m_i$ vectors $\mathbf{x}_1^{(1)}, \ldots, \mathbf{x}_1^{(m_1)}, \ldots, \mathbf{x}_s^{(1)}, \ldots, \mathbf{x}_s^{(m_s)}$ are linearly independent, it follows that

$$c_1^{(1)} = \cdots = c_1^{(m_1)} = \cdots = c_s^{(1)} = \cdots = c_s^{(m_s)} = 0. \quad (4.2)$$

Moreover, substituting from result (4.2) into equality (4.1), we find that

$$c_{s+1}^{(1)}\mathbf{x}_{s+1}^{(1)} + \cdots + c_{s+1}^{(m_{s+1})}\mathbf{x}_{s+1}^{(m_{s+1})} = \mathbf{0}$$

and hence, since the m_{s+1} vectors $\mathbf{x}_{s+1}^{(1)}, \ldots, \mathbf{x}_{s+1}^{(m_{s+1})}$ are by definition linearly independent, that

$$c_{s+1}^{(1)} = \cdots = c_{s+1}^{(m_{s+1})} = 0. \tag{4.3}$$

Together, results (4.2) and (4.3) imply that the $\sum_{i=1}^{s+1} m_i$ vectors $\mathbf{x}_1^{(1)}, \ldots, \mathbf{x}_1^{(m_1)}$, $\ldots, \mathbf{x}_{s+1}^{(1)}, \ldots, \mathbf{x}_{s+1}^{(m_{s+1})}$ are linearly independent. Q.E.D.

How many linearly independent eigenvectors are there? This question is answered (in terms of the geometric multiplicities of the distinct eigenvalues) by the following theorem.

Theorem 21.4.3. Let $\lambda_1, \ldots, \lambda_k$ represent the distinct eigenvalues of an $n \times n$ matrix $\mathbf{A}$ (i.e., the members of the spectrum of $\mathbf{A}$), and let $\nu_1, \ldots, \nu_k$ represent their respective geometric multiplicities. Then, for $i = 1, \ldots, k$, there exists a set of ν_i linearly independent eigenvectors $\mathbf{x}_i^{(1)}, \ldots, \mathbf{x}_i^{(\nu_i)}$ of $\mathbf{A}$ corresponding to λ_i. The combined set $S = \{\mathbf{x}_1^{(1)}, \ldots, \mathbf{x}_1^{(\nu_1)}, \ldots, \mathbf{x}_k^{(1)}, \ldots, \mathbf{x}_k^{(\nu_k)}\}$, comprising $\sum_{i=1}^{k} \nu_i$ eigenvectors, is linearly independent; and any set of more than $\sum_{i=1}^{k} \nu_i$ eigenvectors is linearly dependent. Moreover, if $\sum_{i=1}^{k} \nu_i = n$, then S is a basis for $\mathcal{R}^n$, and the set S_* obtained from S by deleting any eigenvectors that correspond to 0 (in the event that 0 is an eigenvalue of $\mathbf{A}$) is a basis for $\mathcal{C}(\mathbf{A})$.

Proof. For $i = 1, \ldots, k$, any basis for the eigenspace $\mathcal{N}(\mathbf{A} - \lambda_i \mathbf{I})$ is a set of ν_i linearly independent eigenvectors corresponding to λ_i, so that such a set clearly exists.

That the combined set S is linearly independent is an immediate consequence of Theorem 21.4.2. Suppose now (for purposes of establishing a contradiction) that there existed a linearly independent set of more than $\sum_{i=1}^{k} \nu_i$ eigenvectors. Then, for some integer t ($1 \le t \le k$), the number of eigenvectors in that set that corresponded to λ_t would exceed ν_t. And, as a consequence, $\mathcal{N}(\mathbf{A} - \lambda_t \mathbf{I})$ would contain more than ν_t linearly independent vectors, which {since by definition $\nu_t = \dim[\mathcal{N}(\mathbf{A} - \lambda_t \mathbf{I})]$} establishes the desired contradiction. Thus, any set of more than $\sum_{i=1}^{k} \nu_i$ eigenvectors is linearly dependent.

To complete the proof, suppose that $\sum_{i=1}^{k} \nu_i = n$. Then, S comprises n (linearly independent, n-dimensional) vectors and hence is a basis for $\mathcal{R}^n$. Further, let r represent the number of vectors in S_*, let $\mathbf{X}$ represent an $n \times r$ matrix whose columns are the vectors in S_*, and let $\mathbf{\Delta}$ represent an $r \times r$ diagonal matrix whose diagonal elements are the eigenvalues to which the columns of $\mathbf{X}$ correspond, so that $\mathbf{AX} = \mathbf{X\Delta}$, implying (since the diagonal elements of $\mathbf{\Delta}$ are nonzero) that $\mathbf{X} = \mathbf{AX\Delta}^{-1}$. And, upon observing that $\operatorname{rank}(\mathbf{X}) = r$ and (in light of Corollary 21.3.8) that $r = \operatorname{rank}(\mathbf{A})$, it follows from Corollary 4.4.7 that $\mathcal{C}(\mathbf{X}) = \mathcal{C}(\mathbf{A})$. Thus, S_* is a basis for $\mathcal{C}(\mathbf{A})$. Q.E.D.

As an immediate consequence of Theorem 21.4.3, we have the following corollary.

Corollary 21.4.4. Let $\nu_1, \ldots, \nu_k$ represent the geometric multiplicities of the distinct eigenvalues of an $n \times n$ matrix $\mathbf{A}$. Then, there exists a linearly independent set of n eigenvectors (of $\mathbf{A}$) if and only if $\sum_{i=1}^{k} \nu_i = n$.

Eigenvectors that correspond to distinct eigenvalues are linearly independent (as indicated by Theorem 21.4.1). If the eigenvectors are eigenvectors of a symmetric matrix, then we can make the following, stronger statement.

Theorem 21.4.5. If two eigenvectors $\mathbf{x}_1$ and $\mathbf{x}_2$ of an $n \times n$ symmetric matrix $\mathbf{A}$ correspond to different eigenvalues, then $\mathbf{x}_1$ and $\mathbf{x}_2$ are orthogonal (with respect to the usual inner product).

Proof. Let λ_1 and λ_2 represent the eigenvalues to which $\mathbf{x}_1$ and $\mathbf{x}_2$ correspond. Then, by definition, $\mathbf{A}\mathbf{x}_1 = \lambda_1\mathbf{x}_1$ and $\mathbf{A}\mathbf{x}_2 = \lambda_2\mathbf{x}_2$. Premultiplying both sides of the first of these two equalities by $\mathbf{x}_2'$ and both sides of the second by $\mathbf{x}_1'$, we find that $\mathbf{x}_2'\mathbf{A}\mathbf{x}_1 = \lambda_1\mathbf{x}_2'\mathbf{x}_1$ and $\mathbf{x}_1'\mathbf{A}\mathbf{x}_2 = \lambda_2\mathbf{x}_1'\mathbf{x}_2$. And, since $\mathbf{A}$ is symmetric, we have that

$$\lambda_1\mathbf{x}_2'\mathbf{x}_1 = \mathbf{x}_2'\mathbf{A}'\mathbf{x}_1 = (\mathbf{x}_1'\mathbf{A}\mathbf{x}_2)' = (\lambda_2\mathbf{x}_1'\mathbf{x}_2)' = \lambda_2\mathbf{x}_2'\mathbf{x}_1,$$

implying that $(\lambda_1 - \lambda_2)\mathbf{x}_2'\mathbf{x}_1 = 0$ and hence if $\lambda_1 \neq \lambda_2$ that $\mathbf{x}_2'\mathbf{x}_1 = 0$. Thus, if $\lambda_1 \neq \lambda_2$, then $\mathbf{x}_1$ and $\mathbf{x}_2$ are orthogonal. Q.E.D.

21.5 Diagonalization

a. Definitions and some basic properties

An $n \times n$ matrix $\mathbf{A}$ is said to be *diagonalizable* (or *diagonable*) if there exists an $n \times n$ nonsingular matrix $\mathbf{Q}$ such that $\mathbf{Q}^{-1}\mathbf{A}\mathbf{Q} = \mathbf{D}$ for some diagonal matrix $\mathbf{D}$, in which case $\mathbf{Q}$ is said to *diagonalize* $\mathbf{A}$ (or $\mathbf{A}$ is said to be *diagonalized* by $\mathbf{Q}$). Note (in connection with this definition) that

$$\mathbf{Q}^{-1}\mathbf{A}\mathbf{Q} = \mathbf{D} \Leftrightarrow \mathbf{A}\mathbf{Q} = \mathbf{Q}\mathbf{D} \Leftrightarrow \mathbf{A} = \mathbf{Q}\mathbf{D}\mathbf{Q}^{-1}. \tag{5.1}$$

Clearly, an $n \times n$ matrix is diagonalizable if and only if it is similar to a diagonal matrix.

An $n \times n$ matrix $\mathbf{A}$ is said to be *orthogonally diagonalizable* if it is diagonalizable by an orthogonal matrix; that is, if there exists an $n \times n$ orthogonal matrix $\mathbf{Q}$ such that $\mathbf{Q}'\mathbf{A}\mathbf{Q}$ is diagonal.

The process of constructing an $n \times n$ nonsingular matrix that diagonalizes an $n \times n$ matrix $\mathbf{A}$ is referred to as the *diagonalization* of $\mathbf{A}$. This process is intimately related to the process of finding eigenvalues and eigenvectors of $\mathbf{A}$, as is evident from the following theorem.

Theorem 21.5.1. Let $\mathbf{A}$ represent an $n \times n$ matrix. And suppose that there exists an $n \times n$ nonsingular matrix $\mathbf{Q}$ such that $\mathbf{Q}^{-1}\mathbf{A}\mathbf{Q} = \mathbf{D}$ for some diagonal matrix $\mathbf{D}$; denote the first, ..., nth columns of $\mathbf{Q}$ by $\mathbf{q}_1, \ldots, \mathbf{q}_n$, respectively, and the first, ..., nth diagonal elements of $\mathbf{D}$ by $d_1, \ldots, d_n$, respectively. Then,

(1) rank($\mathbf{A}$) equals the number of nonzero diagonal elements of $\mathbf{D}$;

(2) $\det(\mathbf{A}) = d_1 d_2 \cdots d_n$;

(3) $\text{tr}(\mathbf{A}) = d_1 + d_2 + \cdots + d_n$;

(4) the characteristic polynomial of **A** is

$$p(\lambda) = (-1)^n(\lambda - d_1)(\lambda - d_2)\cdots(\lambda - d_n);$$

(5) the spectrum of **A** comprises the distinct scalars represented among the diagonal elements of **D**;

(6) the algebraic and geometric multiplicities of an eigenvalue λ of **A** equal the number of diagonal elements of **D** that equal λ;

(7) the (not necessarily distinct) eigenvalues of **A** are the diagonal elements of **D**; and

(8) the columns of **Q** are (linearly independent) eigenvectors of **A** (with the ith column $\mathbf{q}_i$ corresponding to the eigenvalue d_i).

Proof. Upon recalling various results on diagonal matrices [specifically, result (4.5.14) and the results given by Corollaries 13.1.2, 21.2.2, and 21.2.4], Parts (1)–(7) of Theorem 21.5.1 follow from Theorem 21.3.1. That $\mathbf{q}_1, \ldots, \mathbf{q}_n$ are eigenvectors of **A** [as claimed in Part (8)] is evident upon observing [in light of result (5.1)] that $\mathbf{AQ} = \mathbf{QD}$ (and that, for $i = 1, \ldots, k$, the ith columns of **AQ** and **QD** are, respectively, $\mathbf{Aq}_i$ and $d_i\mathbf{q}_i$). Q.E.D.

b. Necessary and sufficient conditions for a matrix to be diagonalizable

If an $n \times n$ matrix **A** is diagonalizable by an $n \times n$ nonsingular matrix **Q**, then, according to Part (8) of Theorem 21.5.1, the columns of **Q** are (linearly independent) eigenvectors of **A**. The following theorem makes a stronger statement.

Theorem 21.5.2. An $n \times n$ matrix **A** is diagonalizable by an $n \times n$ nonsingular matrix **Q** if and only if the columns of **Q** are (linearly independent) eigenvectors of **A**.

Proof. In light of Part (8) of Theorem 21.5.1, it suffices to prove the "if part" of Theorem 21.5.2.

Suppose that the columns $\mathbf{q}_1, \ldots, \mathbf{q}_n$ of **Q** are eigenvectors of **A**. Then, there exists a diagonal matrix $\mathbf{D} = \{d_i\}$ such that $\mathbf{AQ} = \mathbf{QD}$ (as discussed in Section 1 and as is evident upon observing that, for $i = 1, \ldots, n$, the ith columns of **AQ** and **QD** are, respectively, $\mathbf{Aq}_i$ and $d_i\mathbf{q}_i$). We conclude [in light of result (5.1)] that **Q** diagonalizes **A**. Q.E.D.

As an immediate consequence of Theorem 21.5.2, we have the following corollary.

Corollary 21.5.3. An $n \times n$ matrix **A** is diagonalizable if and only if there exists a linearly independent set of n eigenvectors (of **A**).

In light of Corollary 21.4.4 (and of Corollary 21.3.8), we have the following, additional corollary.

Corollary 21.5.4. An $n \times n$ matrix **A**, whose spectrum comprises k eigenvalues with geometric multiplicities $\nu_1, \ldots, \nu_k$, respectively, is diagonalizable if and only if $\sum_{i=1}^{k} \nu_i = n$ or, equivalently, if and only if the geometric multiplicities of the k or $k-1$ nonzero distinct eigenvalues (of $\mathbf{A}$) sum to $\text{rank}(\mathbf{A})$.

c. A basic result on symmetric matrices

In Subsection d, the discussion of the diagonalization of matrices is specialized to the diagonalization of symmetric matrices. The results of Subsection d depend in a fundamental way on the following property of symmetric matrices.

Theorem 21.5.5. Every symmetric matrix has an eigenvalue.

The conventional way of proving this theorem (e.g., Searle 1982, chap. 11) is to implicitly or explicitly use the fundamental theorem of algebra to establish that the characteristic polynomial of any (square) matrix has a possibly complex root and to then show that if the matrix is symmetric, this root is necessarily real. The resultant proof is accessible only to those who are at least somewhat knowledgeable about complex numbers. Moreover, the fundamental theorem of algebra is a very deep result—its proof is well beyond the scope and mathematical level of our presentation.

Here, we take a different approach. This approach consists of establishing the following theorem, of which Theorem 21.5.5 is an immediate consequence.

Theorem 21.5.6. Let $\mathbf{A}$ represent an $n \times n$ matrix. Then, there exist nonnull vectors $\mathbf{x}_1$ and $\mathbf{x}_2$ such that

$$\frac{\mathbf{x}_1'\mathbf{A}\mathbf{x}_1}{\mathbf{x}_1'\mathbf{x}_1} \leq \frac{\mathbf{x}'\mathbf{A}\mathbf{x}}{\mathbf{x}'\mathbf{x}} \leq \frac{\mathbf{x}_2'\mathbf{A}\mathbf{x}_2}{\mathbf{x}_2'\mathbf{x}_2}$$

for every nonnull vector $\mathbf{x}$ in $\mathcal{R}^n$ (or, equivalently, such that $\mathbf{x}_1'\mathbf{A}\mathbf{x}_1/\mathbf{x}_1'\mathbf{x}_1 = \min_{\mathbf{x}\neq\mathbf{0}} \mathbf{x}'\mathbf{A}\mathbf{x}/\mathbf{x}'\mathbf{x}$ and $\mathbf{x}_2'\mathbf{A}\mathbf{x}_2/\mathbf{x}_2'\mathbf{x}_2 = \max_{\mathbf{x}\neq\mathbf{0}} \mathbf{x}'\mathbf{A}\mathbf{x}/\mathbf{x}'\mathbf{x}$). Moreover, if $\mathbf{A}$ is symmetric, then $\mathbf{x}_1'\mathbf{A}\mathbf{x}_1/\mathbf{x}_1'\mathbf{x}_1$ and $\mathbf{x}_2'\mathbf{A}\mathbf{x}_2/\mathbf{x}_2'\mathbf{x}_2$ are eigenvalues of $\mathbf{A}$—in fact, they are, respectively, the smallest and largest eigenvalues of $\mathbf{A}$—and $\mathbf{x}_1$ and $\mathbf{x}_2$ are eigenvectors corresponding to $\mathbf{x}_1'\mathbf{A}\mathbf{x}_1/\mathbf{x}_1'\mathbf{x}_1$ and $\mathbf{x}_2'\mathbf{A}\mathbf{x}_2/\mathbf{x}_2'\mathbf{x}_2$, respectively.

Proof. Let $\mathbf{x}$ represent an $n \times 1$ vector of variables, and define $S = \{\mathbf{x} : \mathbf{x}'\mathbf{x} = 1\}$. The quadratic form $\mathbf{x}'\mathbf{A}\mathbf{x}$ is a continuous function of $\mathbf{x}$, and the set S is closed and bounded. Since any continuous function attains a maximum value and a minimum value over any closed and bounded subset of its domain (e.g., Bartle 1976, thm. 22.6; Magnus and Neudecker 1988, thm. 7.1), S contains vectors $\mathbf{x}_1$ and $\mathbf{x}_2$ such that

$$\mathbf{x}_1'\mathbf{A}\mathbf{x}_1 \leq \mathbf{x}'\mathbf{A}\mathbf{x} \leq \mathbf{x}_2'\mathbf{A}\mathbf{x}_2$$

for every $\mathbf{x}$ in S. Moreover, for any nonnull $\mathbf{x}$ (in $\mathcal{R}^n$), $(\mathbf{x}'\mathbf{x})^{-1/2}\mathbf{x} \in S$ and

$$[(\mathbf{x}'\mathbf{x})^{-1/2}\mathbf{x}]'\mathbf{A}[(\mathbf{x}'\mathbf{x})^{-1/2}\mathbf{x}] = \mathbf{x}'\mathbf{A}\mathbf{x}/\mathbf{x}'\mathbf{x}.$$

Thus,

$$\frac{\mathbf{x}_1'\mathbf{A}\mathbf{x}_1}{\mathbf{x}_1'\mathbf{x}_1} = \mathbf{x}_1'\mathbf{A}\mathbf{x}_1 \leq \frac{\mathbf{x}'\mathbf{A}\mathbf{x}}{\mathbf{x}'\mathbf{x}} \leq \mathbf{x}_2'\mathbf{A}\mathbf{x}_2 = \frac{\mathbf{x}_2'\mathbf{A}\mathbf{x}_2}{\mathbf{x}_2'\mathbf{x}_2} \tag{5.2}$$

for every nonnull $\mathbf{x}$.

Now, suppose that $\mathbf{A}$ is symmetric. Then, in light of results (15.2.13) and (15.3.7), we find that (for nonnull $\mathbf{x}$)

$$\begin{aligned}\frac{\partial(\mathbf{x}'\mathbf{A}\mathbf{x}/\mathbf{x}'\mathbf{x})}{\partial\mathbf{x}} &= (\mathbf{x}'\mathbf{x})^{-2}\left[(\mathbf{x}'\mathbf{x})\frac{\partial\mathbf{x}'\mathbf{A}\mathbf{x}}{\partial\mathbf{x}} - (\mathbf{x}'\mathbf{A}\mathbf{x})\frac{\partial\mathbf{x}'\mathbf{x}}{\partial\mathbf{x}}\right] \\ &= (\mathbf{x}'\mathbf{x})^{-2}[(\mathbf{x}'\mathbf{x})2\mathbf{A}\mathbf{x} - (\mathbf{x}'\mathbf{A}\mathbf{x})2\mathbf{x}].\end{aligned}$$

Moreover, $\mathbf{x}'\mathbf{A}\mathbf{x}/\mathbf{x}'\mathbf{x}$ attains (for $\mathbf{x} \neq \mathbf{0}$) minimum and maximum values at $\mathbf{x}_1$ and $\mathbf{x}_2$, respectively [as is evident from result (5.2)], so that $\partial(\mathbf{x}'\mathbf{A}\mathbf{x}/\mathbf{x}'\mathbf{x})/\partial\mathbf{x} = \mathbf{0}$ for $\mathbf{x} = \mathbf{x}_1$ and $\mathbf{x} = \mathbf{x}_2$. Thus, for $\mathbf{x} = \mathbf{x}_1$ and $\mathbf{x} = \mathbf{x}_2$,

$$(\mathbf{x}'\mathbf{x})2\mathbf{A}\mathbf{x} - (\mathbf{x}'\mathbf{A}\mathbf{x})2\mathbf{x} = \mathbf{0},$$

or, equivalently,

$$\mathbf{A}\mathbf{x} = (\mathbf{x}'\mathbf{A}\mathbf{x}/\mathbf{x}'\mathbf{x})\mathbf{x}.$$

We conclude that $\mathbf{x}_1'\mathbf{A}\mathbf{x}_1/\mathbf{x}_1'\mathbf{x}_1$ and $\mathbf{x}_2'\mathbf{A}\mathbf{x}_2/\mathbf{x}_2'\mathbf{x}_2$ are eigenvalues of $\mathbf{A}$ and that $\mathbf{x}_1$ and $\mathbf{x}_2$ are eigenvectors that correspond to $\mathbf{x}_1'\mathbf{A}\mathbf{x}_1/\mathbf{x}_1'\mathbf{x}_1$ and $\mathbf{x}_2'\mathbf{A}\mathbf{x}_2/\mathbf{x}_2'\mathbf{x}_2$, respectively. Q.E.D.

d. Orthogonal diagonalization

Let $\mathbf{A}$ represent an $n \times n$ matrix, and suppose that there exists an orthogonal matrix $\mathbf{Q}$ such that $\mathbf{Q}'\mathbf{A}\mathbf{Q} = \mathbf{D}$ for some diagonal matrix $\mathbf{D}$. Then, since [in light of result (5.1)] $\mathbf{A} = \mathbf{Q}\mathbf{D}\mathbf{Q}'$,

$$\mathbf{A}' = (\mathbf{Q}\mathbf{D}\mathbf{Q}')' = \mathbf{Q}\mathbf{D}'\mathbf{Q}' = \mathbf{Q}\mathbf{D}\mathbf{Q}' = \mathbf{A}.$$

Thus, a necessary condition for $\mathbf{A}$ to be orthogonally diagonalizable is that $\mathbf{A}$ be symmetric (so that while certain nonsymmetric matrices are diagonalizable, they are not orthogonally diagonalizable). This condition is also sufficient, as indicated by the following theorem.

Theorem 21.5.7. Every symmetric matrix is orthogonally diagonalizable.

Proof. The proof is by mathematical induction. Clearly, every 1×1 matrix is orthogonally diagonalizable. Suppose now that every $(n-1) \times (n-1)$ symmetric matrix is orthogonally diagonalizable (where $n \geq 2$). Then, it suffices to show that every $n \times n$ symmetric matrix is orthogonally diagonalizable.

Let $\mathbf{A}$ represent an $n \times n$ symmetric matrix. And let λ represent an eigenvalue of $\mathbf{A}$ (the existence of which is guaranteed by Theorem 21.5.5), and $\mathbf{u}$ represent an eigenvector of (usual) norm 1 that corresponds to λ. Then, according to Corollary 21.3.3,

$$(\mathbf{u},\ \mathbf{V})'\mathbf{A}(\mathbf{u},\ \mathbf{V}) = \text{diag}(\lambda,\ \mathbf{V}'\mathbf{A}\mathbf{V})$$

for some $n \times (n-1)$ matrix $\mathbf{V}$ such that the $n \times n$ matrix $(\mathbf{u},\ \mathbf{V})$ is orthogonal.

Clearly, $\mathbf{V}'\mathbf{A}\mathbf{V}$ is a symmetric matrix of order $n-1$, so that (by supposition) there exists an orthogonal matrix $\mathbf{R}$ such that $\mathbf{R}'(\mathbf{V}'\mathbf{A}\mathbf{V})\mathbf{R} = \mathbf{F}$ for some diagonal matrix $\mathbf{F}$. Define $\mathbf{S} = \text{diag}(1, \mathbf{R})$ and $\mathbf{P} = (\mathbf{u}, \mathbf{V})\mathbf{S}$. Then,

$$\mathbf{S}'\mathbf{S} = \text{diag}(1, \mathbf{R}'\mathbf{R}) = \text{diag}(1, \mathbf{I}_{n-1}) = \mathbf{I}_n,$$

so that $\mathbf{S}$ is orthogonal and hence (according to Lemma 8.4.1) $\mathbf{P}$ is orthogonal. Further,

$$\begin{aligned}\mathbf{P}'\mathbf{A}\mathbf{P} &= \mathbf{S}'(\mathbf{u}, \mathbf{V})'\mathbf{A}(\mathbf{u}, \mathbf{V})\mathbf{S} = \mathbf{S}'\,\text{diag}(\lambda, \mathbf{V}'\mathbf{A}\mathbf{V})\mathbf{S}\\ &= \text{diag}(\lambda, \mathbf{R}'\mathbf{V}'\mathbf{A}\mathbf{V}\mathbf{R}) = \text{diag}(\lambda, \mathbf{F}),\end{aligned}$$

so that $\mathbf{P}'\mathbf{A}\mathbf{P}$ equals a diagonal matrix. Thus, $\mathbf{A}$ is orthogonally diagonalizable. Q.E.D.

The following corollary (of Theorem 21.5.7) can be deduced from Corollaries 21.5.4 and 21.3.7 (or from Theorem 21.5.1).

Corollary 21.5.8. Let $\mathbf{A}$ represent an $n \times n$ symmetric matrix that has k distinct eigenvalues $\lambda_1, \ldots, \lambda_k$ with algebraic multiplicities $\gamma_1, \ldots, \gamma_k$, respectively, and geometric multiplicities $\nu_1, \ldots, \nu_k$, respectively. Then, $\sum_{i=1}^k \gamma_i = \sum_{i=1}^k \nu_i = n$, and (for $i = 1, \ldots, k$) $\nu_i = \gamma_i$.

As a further corollary of Theorem 21.5.7, we have the following result.

Corollary 21.5.9. Let $\mathbf{A}$ represent an $n \times n$ symmetric matrix, and let $d_1, \ldots, d_n$ represent the (not necessarily distinct) eigenvalues of $\mathbf{A}$ (in arbitrary order). Then, there exists an $n \times n$ orthogonal matrix $\mathbf{Q}$ such that

$$\mathbf{Q}'\mathbf{A}\mathbf{Q} = \text{diag}(d_1, \ldots, d_n).$$

Proof. According to Theorem 21.5.7, there exists an $n \times n$ orthogonal matrix $\mathbf{R}$ such that $\mathbf{R}'\mathbf{A}\mathbf{R} = \mathbf{F}$ for some diagonal matrix $\mathbf{F} = \{f_i\}$. It follows from Part (7) of Theorem 21.5.1 that there exists some permutation $k_1, \ldots, k_n$ of the first n positive integers $1, \ldots, n$ such that (for $i = 1, \ldots, n$) $d_i = f_{k_i}$. Now, let $\mathbf{Q} = \mathbf{R}\mathbf{P}$, where $\mathbf{P}$ is the $n \times n$ permutation matrix whose first, ..., nth columns are, respectively, the $k_1, \ldots, k_n$th columns of $\mathbf{I}_n$. Then, since a permutation matrix is orthogonal, $\mathbf{Q}$ is (in light of Lemma 8.4.1) orthogonal. Further,

$$\mathbf{Q}'\mathbf{A}\mathbf{Q} = \mathbf{P}'\mathbf{R}'\mathbf{A}\mathbf{R}\mathbf{P} = \mathbf{P}'\mathbf{F}\mathbf{P} = \text{diag}(d_1, \ldots, d_n).$$

Q.E.D.

Let $\mathbf{A}$ represent an $n \times n$ symmetric matrix. Then, according to Theorem 21.5.7, there exists an orthogonal matrix that diagonalizes $\mathbf{A}$. There also exist nonorthogonal matrices that diagonalize $\mathbf{A}$. If $\mathbf{A}$ has n distinct eigenvalues, then it follows from Theorem 21.4.5 that the columns of any diagonalizing matrix are orthogonal [since, according to Part (8) of Theorem 21.5.1, they are necessarily eigenvectors]. However, the norms of these columns need not equal one. More generally, if λ is an eigenvalue of $\mathbf{A}$ of multiplicity ν, then we can include as columns of the diagonalizing matrix any ν linearly independent eigenvectors that correspond to λ. We need not choose these ν linearly independent eigenvectors to be orthonormal or even (if $\nu > 1$) orthogonal.

The (usual) norm of a symmetric matrix can be expressed in terms of its eigenvalues, as described in the following theorem.

Theorem 21.5.10. For any $n \times n$ symmetric matrix $\mathbf{A}$,

$$\|\mathbf{A}\| = \Big(\sum_{i=1}^{n} d_i^2\Big)^{1/2},$$

where $d_1, \ldots, d_n$ are the (not necessarily distinct) eigenvalues of $\mathbf{A}$ (and where the norm is the usual norm).

Proof. According to Corollary 21.5.9, there exists an $n \times n$ orthogonal matrix $\mathbf{Q}$ such that $\mathbf{Q}'\mathbf{A}\mathbf{Q} = \mathbf{D}$, where $\mathbf{D} = \text{diag}(d_1, \ldots, d_n)$. Thus, making use of Lemma 5.2.1, we find that

$$\begin{aligned}\|\mathbf{A}\|^2 = \text{tr}(\mathbf{A}'\mathbf{A}) &= \text{tr}(\mathbf{A}\mathbf{A}) \\ &= \text{tr}(\mathbf{A}\mathbf{I}_n\mathbf{A}\mathbf{I}_n) \\ &= \text{tr}(\mathbf{A}\mathbf{Q}\mathbf{Q}'\mathbf{A}\mathbf{Q}\mathbf{Q}') \\ &= \text{tr}(\mathbf{Q}'\mathbf{A}\mathbf{Q}\mathbf{Q}'\mathbf{A}\mathbf{Q}) \\ &= \text{tr}(\mathbf{D}^2) = \text{tr}[\text{diag}(d_1^2, \ldots, d_n^2)] = \sum_{i=1}^{n} d_i^2.\end{aligned}$$

Q.E.D.

e. Orthogonal "triangularization"

Let $\mathbf{A}$ represent an $n \times n$ matrix. If $\mathbf{A}$ is nonsymmetric, then there does not exist any $n \times n$ orthogonal matrix that diagonalizes $\mathbf{A}$. However, there may exist an $n \times n$ orthogonal matrix that "triangularizes" $\mathbf{A}$, as indicated by the following theorem.

Theorem 21.5.11. Let $\mathbf{A}$ represent an $n \times n$ matrix whose distinct eigenvalues have algebraic multiplicities that sum to n, and let $d_1, \ldots, d_n$ represent the not-necessarily-distinct eigenvalues of $\mathbf{A}$ (in arbitrary order). Then, there exists an $n \times n$ orthogonal matrix $\mathbf{Q}$ such that

$$\mathbf{Q}'\mathbf{A}\mathbf{Q} = \mathbf{T},$$

where $\mathbf{T}$ is an $(n \times n)$ upper triangular matrix whose first, ..., nth diagonal elements are $d_1, \ldots, d_n$, respectively.

Proof. The proof is by mathematical induction. Clearly, the theorem is valid for any 1×1 matrix. Suppose now that it is valid for any $(n-1) \times (n-1)$ matrix whose distinct eigenvalues have algebraic multiplicities that sum to $n-1$ (where $n \geq 2$), and consider the validity of the theorem for an $n \times n$ matrix $\mathbf{A}$ whose algebraic multiplicities sum to n and whose (not necessarily distinct) eigenvalues are $d_1, \ldots, d_n$.

Let $\mathbf{u}$ represent an eigenvector (of $\mathbf{A}$) of (usual) norm 1 that corresponds to d_1. Then, according to Corollary 21.3.3,

$$(\mathbf{u}, \mathbf{V})'\mathbf{A}(\mathbf{u}, \mathbf{V}) = \begin{pmatrix} d_1 & \mathbf{u}'\mathbf{A}\mathbf{V} \\ \mathbf{0} & \mathbf{V}'\mathbf{A}\mathbf{V} \end{pmatrix}$$

for some $n \times (n-1)$ matrix $\mathbf{V}$ such that the $n \times n$ matrix $(\mathbf{u}, \mathbf{V})$ is orthogonal. Moreover, it follows from Theorem 21.3.1 and Lemma 21.2.1 that the algebraic multiplicities of the distinct eigenvalues of the $(n-1) \times (n-1)$ matrix $\mathbf{V}'\mathbf{A}\mathbf{V}$ sum to $n-1$ and that the (not necessarily distinct) eigenvalues of $\mathbf{V}'\mathbf{A}\mathbf{V}$ are $d_2, \ldots, d_n$. Thus, by supposition, there exists an orthogonal matrix $\mathbf{R}$ such that $\mathbf{R}'(\mathbf{V}'\mathbf{A}\mathbf{V})\mathbf{R} = \mathbf{T}_*$, where $\mathbf{T}_*$ is an $(n-1) \times (n-1)$ upper triangular matrix whose first, ..., $(n-1)$th diagonal elements are $d_2, \ldots, d_n$, respectively.

Now, define $\mathbf{S} = \text{diag}(1, \mathbf{R})$ and $\mathbf{Q} = (\mathbf{u}, \mathbf{V})\mathbf{S}$. Then,

$$\mathbf{S}'\mathbf{S} = \text{diag}(1, \mathbf{R}'\mathbf{R}) = \text{diag}(1, \mathbf{I}_{n-1}) = \mathbf{I}_n,$$

so that $\mathbf{S}$ is orthogonal and hence (according to Lemma 8.4.1) $\mathbf{Q}$ is orthogonal. Further,

$$\begin{aligned}\mathbf{Q}'\mathbf{A}\mathbf{Q} = \mathbf{S}'(\mathbf{u}, \mathbf{V})'\mathbf{A}(\mathbf{u}, \mathbf{V})\mathbf{S} &= \begin{pmatrix} 1 & \mathbf{0} \\ \mathbf{0} & \mathbf{R}' \end{pmatrix} \begin{pmatrix} d_1 & \mathbf{u}'\mathbf{A}\mathbf{V} \\ \mathbf{0} & \mathbf{V}'\mathbf{A}\mathbf{V} \end{pmatrix} \begin{pmatrix} 1 & \mathbf{0} \\ \mathbf{0} & \mathbf{R} \end{pmatrix} \\ &= \begin{pmatrix} d_1 & \mathbf{u}'\mathbf{A}\mathbf{V}\mathbf{R} \\ \mathbf{0} & \mathbf{R}'\mathbf{V}'\mathbf{A}\mathbf{V}\mathbf{R} \end{pmatrix} = \begin{pmatrix} d_1 & \mathbf{u}'\mathbf{A}\mathbf{V}\mathbf{R} \\ \mathbf{0} & \mathbf{T}_* \end{pmatrix}.\end{aligned}$$

And, upon observing that $\begin{pmatrix} d_1 & \mathbf{u}'\mathbf{A}\mathbf{V}\mathbf{R} \\ \mathbf{0} & \mathbf{T}_* \end{pmatrix}$ is an $(n \times n)$ upper triangular matrix whose first, ..., nth diagonal elements are $d_1, \ldots, d_n$, respectively, the induction argument is complete. Q.E.D.

Let $\mathbf{A}$ represent an $n \times n$ matrix. If there exists an orthogonal matrix $\mathbf{Q}$ such that $\mathbf{Q}'\mathbf{A}\mathbf{Q} = \mathbf{T}$ for some upper triangular matrix $\mathbf{T}$, then the diagonal elements of $\mathbf{T}$ are the not-necessarily-distinct eigenvalues of $\mathbf{A}$, as is evident from Part (4) of Corollary 21.2.2 and Part (7) of Theorem 21.3.1. Theorem 21.5.11 indicates that a sufficient (as well as necessary) condition for the existence of such an orthogonal matrix $\mathbf{Q}$ is that the algebraic multiplicities of the not-necessarily-distinct eigenvalues of $\mathbf{A}$ sum to n and also indicates that (when this condition is satisfied) $\mathbf{Q}$ can be chosen so that the order in which the eigenvalues (of $\mathbf{A}$) appear as diagonal elements of the triangular matrix $\mathbf{T}$ is arbitrary.

Note (in light of Corollary 21.5.8) that Theorem 21.5.11 is applicable to any symmetric matrix. In the special case where the matrix $\mathbf{A}$ in Theorem 21.5.11 is symmetric, the triangular matrix $\mathbf{T}$ is a diagonal matrix (since if $\mathbf{A}$ is symmetric, $\mathbf{Q}'\mathbf{A}\mathbf{Q}$ is symmetric and hence $\mathbf{T}$ is symmetric). Thus, in this special case, Theorem 21.5.11 simplifies to Corollary 21.5.9.

f. Spectral decomposition (of a symmetric matrix)

Let $\mathbf{A}$ represent an $n \times n$ symmetric matrix. And let $\mathbf{Q}$ represent an $n \times n$ orthogonal matrix and $\mathbf{D} = \{d_i\}$ an $n \times n$ diagonal matrix such that $\mathbf{Q}'\mathbf{A}\mathbf{Q} = \mathbf{D}$—the existence of such matrices is guaranteed by Theorem 21.5.7. Or, equivalently, let $\mathbf{Q}$ represent an $n \times n$ matrix whose columns are orthonormal (with respect to the usual inner product) eigenvectors of $\mathbf{A}$, and take $\mathbf{D} = \{d_i\}$ to be the diagonal matrix whose first, ..., nth diagonal elements $d_1, \ldots, d_n$ are the eigenvalues (of $\mathbf{A}$) to which the first, ..., nth columns of $\mathbf{Q}$ correspond—the equivalence is evident from Theorems 21.5.1 and 21.5.2. Further, denote the first, ..., nth columns of $\mathbf{Q}$ by $\mathbf{q}_1, \ldots, \mathbf{q}_n$, respectively.

In light of result (5.1), $\mathbf{A}$ can be expressed as

$$\mathbf{A} = \mathbf{QDQ}' \tag{5.3}$$

or as

$$\mathbf{A} = \sum_{i=1}^{n} d_i \mathbf{q}_i \mathbf{q}_i'. \tag{5.4}$$

And, letting $\lambda_1, \ldots, \lambda_k$ represent the distinct eigenvalues of $\mathbf{A}$, letting $\nu_1, \ldots, \nu_k$ represent the (algebraic or geometric) multiplicities of $\lambda_1, \ldots, \lambda_k$, respectively, and (for $j = 1, \ldots, k$) letting $S_j = \{i : d_i = \lambda_j\}$ represent the set comprising the ν_j values of i such that $d_i = \lambda_j$, $\mathbf{A}$ can also be expressed as

$$\mathbf{A} = \sum_{j=1}^{k} \lambda_j \mathbf{E}_j, \tag{5.5}$$

where (for $j = 1, \ldots, k$) $\mathbf{E}_j = \sum_{i \in S_j} \mathbf{q}_i \mathbf{q}_i'$.

Expression (5.5) is called the *spectral decomposition* or *spectral representation* of the $n \times n$ symmetric matrix $\mathbf{A}$. [Sometimes, the term spectral decomposition is used also in reference to expression (5.3) or (5.4).]

The decomposition $\mathbf{A} = \sum_{j=1}^{k} \lambda_j \mathbf{E}_j$ is unique (aside from the ordering of the terms), whereas (in general) the decomposition $\mathbf{A} = \sum_{i=1}^{n} d_i \mathbf{q}_i \mathbf{q}_i'$ is not. To see this (i.e., the uniqueness of the decomposition $\mathbf{A} = \sum_{j=1}^{k} \lambda_j \mathbf{E}_j$), suppose that $\mathbf{P}$ is an $n \times n$ orthogonal matrix and $\mathbf{D}^* = \{d_i^*\}$ an $n \times n$ diagonal matrix such that $\mathbf{P}'\mathbf{AP} = \mathbf{D}^*$ (where $\mathbf{P}$ and $\mathbf{D}^*$ are possibly different from $\mathbf{Q}$ and $\mathbf{D}$). Further, denote the first, ..., nth columns of $\mathbf{P}$ by $\mathbf{p}_1, \ldots, \mathbf{p}_n$, respectively, and (for $j = 1, \ldots, k$) let $S_j^* = \{i : d_i^* = \lambda_j\}$. Then, analogous to the decomposition $\mathbf{A} = \sum_{j=1}^{k} \lambda_j \mathbf{E}_j$, we have the decomposition

$$\mathbf{A} = \sum_{j=1}^{k} \lambda_j \mathbf{F}_j,$$

where (for $j = 1, \ldots, k$) $\mathbf{F}_j = \sum_{i \in S_j^*} \mathbf{p}_i \mathbf{p}_i'$.

Now, for $j = 1, \ldots, k$, let $\mathbf{Q}_j = (\mathbf{q}_{i_1}, \ldots, \mathbf{q}_{i_{\nu_j}})$ and $\mathbf{P}_j = (\mathbf{p}_{i_1^*}, \ldots, \mathbf{p}_{i_{\nu_j}^*})$, where $i_1, \ldots, i_{\nu_j}$ and $i_1^*, \ldots, i_{\nu_j}^*$ are the elements of S_j and S_j^*, respectively. Then, $\mathcal{C}(\mathbf{P}_j) = \mathcal{N}(\mathbf{A} - \lambda_j \mathbf{I}) = \mathcal{C}(\mathbf{Q}_j)$, so that $\mathbf{P}_j = \mathbf{Q}_j \mathbf{L}_j$ for some $\nu_j \times \nu_j$ matrix $\mathbf{L}_j$. Moreover, since clearly $\mathbf{Q}_j'\mathbf{Q}_j = \mathbf{I}_{\nu_j}$ and $\mathbf{P}_j'\mathbf{P}_j = \mathbf{I}_{\nu_j}$,

$$\mathbf{L}_j'\mathbf{L}_j = \mathbf{L}_j'\mathbf{Q}_j'\mathbf{Q}_j\mathbf{L}_j = \mathbf{P}_j'\mathbf{P}_j = \mathbf{I},$$

implying that $\mathbf{L}_j$ is an orthogonal matrix. Thus,

$$\mathbf{F}_j = \mathbf{P}_j\mathbf{P}_j' = \mathbf{Q}_j\mathbf{L}_j\mathbf{L}_j'\mathbf{Q}_j' = \mathbf{Q}_j\mathbf{I}\mathbf{Q}_j' = \mathbf{Q}_j\mathbf{Q}_j' = \mathbf{E}_j.$$

We conclude that the decomposition $\mathbf{A} = \sum_{j=1}^{k} \lambda_j \mathbf{F}_j$ is identical to the decomposition $\mathbf{A} = \sum_{j=1}^{k} \lambda_j \mathbf{E}_j$, and hence that the decomposition $\mathbf{A} = \sum_{j=1}^{k} \lambda_j \mathbf{E}_j$ is unique (aside from the ordering of terms).

The spectral decomposition of the mth power $\mathbf{A}^m$ of the $n \times n$ symmetric matrix $\mathbf{A}$ (where m is a positive integer) and (if $\mathbf{A}$ is nonsingular) the spectral decomposition of $\mathbf{A}^{-1}$ are closely related to the spectral decomposition of $\mathbf{A}$ itself. It follows from Lemma 21.1.3 that $\mathbf{q}_1, \ldots, \mathbf{q}_n$ are eigenvectors of $\mathbf{A}^m$ and (if $\mathbf{A}$ is nonsingular) of $\mathbf{A}^{-1}$ (as well as of $\mathbf{A}$) and that the eigenvalues to which they correspond are $d_1^m, \ldots, d_n^m$ (in the case of $\mathbf{A}^m$) and $d_1^{-1}, \ldots, d_n^{-1}$ (in the case of $\mathbf{A}^{-1}$). Thus, the spectral decompositions of $\mathbf{A}^m$ and $\mathbf{A}^{-1}$ can be obtained from that of $\mathbf{A}$ by replacing each eigenvalue of $\mathbf{A}$ by its mth power or by its reciprocal. More explicitly, observing that $\mathbf{D}^m = \text{diag}(d_1^m, \ldots, d_n^m)$ and $\mathbf{D}^{-1} = \text{diag}(d_1^{-1}, \ldots, d_n^{-1})$, the spectral decompositions of $\mathbf{A}^m$ and $\mathbf{A}^{-1}$ are

$$\mathbf{A}^m = \mathbf{Q}\mathbf{D}^m\mathbf{Q}' = \sum_{i=1}^{n} d_i^m \mathbf{q}_i\mathbf{q}_i' = \sum_{j=1}^{k} \lambda_j^m \mathbf{E}_j, \tag{5.6}$$

$$\mathbf{A}^{-1} = \mathbf{Q}\mathbf{D}^{-1}\mathbf{Q}' = \sum_{i=1}^{n} d_i^{-1} \mathbf{q}_i\mathbf{q}_i' = \sum_{j=1}^{k} \lambda_j^{-1} \mathbf{E}_j. \tag{5.7}$$

21.6 Expressions for the Trace and Determinant of a Matrix

Theorem 21.5.1 gives various results on $n \times n$ diagonalizable matrices or, equivalently (in light of Corollary 21.5.4), on $n \times n$ matrices whose distinct eigenvalues have geometric multiplicities that sum to n. Included among these results are expressions for the determinant and the trace of such a matrix. These expressions [which are Parts (2) and (3) of Theorem 21.5.1] are in terms of the eigenvalues. They can be extended to a broader class of matrices, as indicated by the following theorem.

Theorem 21.6.1. Let $\mathbf{A}$ represent an $n \times n$ matrix having distinct eigenvalues $\lambda_1, \ldots, \lambda_k$ with algebraic multiplicities $\gamma_1, \ldots, \gamma_k$, respectively, that sum to n or, equivalently, having n not-necessarily-distinct eigenvalues $d_1, \ldots, d_n$. Then,

$$\det(\mathbf{A}) = \prod_{i=1}^{n} d_i = \prod_{j=1}^{k} \lambda_j^{\gamma_j}, \tag{6.1}$$

$$\text{tr}(\mathbf{A}) = \sum_{i=1}^{n} d_i = \sum_{j=1}^{k} \gamma_j \lambda_j. \tag{6.2}$$

Proof. As a consequence of Theorem 21.5.11, there exists an $n \times n$ orthogonal matrix $\mathbf{Q}$ such that $\mathbf{Q}'\mathbf{A}\mathbf{Q} = \mathbf{T}$, where $\mathbf{T}$ is an $(n \times n)$ upper triangular matrix whose diagonal elements are $d_1, \ldots, d_n$, respectively. Then, recalling Corollary 13.3.6, result (13.3.9), and Lemmas 13.2.1 and 13.1.1, we find that

$$|\mathbf{A}| = |\mathbf{Q}^2||\mathbf{A}| = |\mathbf{Q}'\mathbf{A}\mathbf{Q}| = |\mathbf{T}| = \prod_{i=1}^{n} d_i .$$

And, recalling Lemma 5.2.1, we find that

$$\mathrm{tr}(\mathbf{A}) = \mathrm{tr}(\mathbf{A}\mathbf{I}_n) = \mathrm{tr}(\mathbf{A}\mathbf{Q}\mathbf{Q}') = \mathrm{tr}(\mathbf{Q}'\mathbf{A}\mathbf{Q}) = \mathrm{tr}(\mathbf{T}) = \sum_{i=1}^{n} d_i .$$

Q.E.D.

There is an alternative way of proving Theorem 21.6.1. Again take $\mathbf{Q}$ to be an $n \times n$ orthogonal matrix such that $\mathbf{Q}'\mathbf{A}\mathbf{Q} = \mathbf{T}$, where $\mathbf{T}$ is a triangular matrix whose diagonal elements are $d_1, \ldots, d_n$. And recall (from Section 1) that the characteristic polynomial of $\mathbf{A}$ is

$$p(\lambda) = c_0 + c_1\lambda + \cdots + c_{n-1}\lambda^{n-1} + c_n\lambda^n,$$

with first and next-to-last coefficients that are expressible as

$$c_0 = |\mathbf{A}|, \qquad c_{n-1} = (-1)^{n-1}\mathrm{tr}(\mathbf{A}). \tag{6.3}$$

Clearly, $\mathbf{T}$ is similar to $\mathbf{A}$ and hence (according to Theorem 21.3.1) has the same characteristic polynomial as $\mathbf{A}$. Thus, recalling Lemma 13.1.1, we find that

$$c_0 = |\mathbf{T}| = \prod_{i=1}^{n} d_i, \qquad c_{n-1} = (-1)^{n-1}\mathrm{tr}(\mathbf{T}) = (-1)^{n-1}\sum_{i=1}^{n} d_i .$$

And, upon comparing these expressions for c_0 and c_{n-1} with those in result (6.3), we conclude that $|\mathbf{A}| = \prod_{i=1}^{n} d_i$ and that $\mathrm{tr}(\mathbf{A}) = \sum_{i=1}^{n} d_i$.

21.7 Some Results on the Moore-Penrose Inverse of a Symmetric Matrix

The following lemma relates the eigenvalues and eigenvectors of the Moore-Penrose inverse of an $n \times n$ symmetric matrix $\mathbf{A}$ to those of $\mathbf{A}$ itself.

Lemma 21.7.1. Let λ represent an eigenvalue of an $n \times n$ symmetric matrix $\mathbf{A}$, and let $\mathbf{x}$ represent any eigenvector (of $\mathbf{A}$) corresponding to λ. Further, define

$$\lambda^+ = \begin{cases} 1/\lambda, & \text{if } \lambda \neq 0, \\ 0, & \text{if } \lambda = 0. \end{cases}$$

Then, λ^+ is an eigenvalue of $\mathbf{A}^+$, and $\mathbf{x}$ is an eigenvector of $\mathbf{A}^+$ corresponding to λ^+.

Proof. By definition, $\mathbf{Ax} = \lambda\mathbf{x}$. If $\lambda \neq 0$, then $\mathbf{x} = \lambda^{-1}\mathbf{Ax}$, and, upon observing (in light of Theorem 20.5.3) that $\mathbf{A}^+$ is symmetric and hence that $\mathbf{A}^+\mathbf{A} = (\mathbf{A}^+\mathbf{A})' = \mathbf{A}'(\mathbf{A}^+)' = \mathbf{AA}^+$, we find that

$$\begin{aligned}\mathbf{A}^+\mathbf{x} = \mathbf{A}^+(\lambda^{-1}\mathbf{Ax}) = \lambda^{-1}\mathbf{A}^+\mathbf{Ax} &= \lambda^{-1}\mathbf{AA}^+\mathbf{x} \\ &= \lambda^{-1}\mathbf{AA}^+(\lambda^{-1}\mathbf{Ax}) \\ &= (\lambda^{-1})^2\mathbf{AA}^+\mathbf{Ax} \\ &= (\lambda^{-1})^2\mathbf{Ax} = \lambda^{-1}(\lambda^{-1}\mathbf{Ax}) = \lambda^{-1}\mathbf{x}.\end{aligned}$$

Alternatively, if $\lambda = 0$, then $\mathbf{Ax} = \mathbf{0}$, and hence [since, according to Theorem 20.5.1, $\mathcal{N}(\mathbf{A}^+) = \mathcal{N}(\mathbf{A})$] $\mathbf{A}^+\mathbf{x} = \mathbf{0} = 0\mathbf{x}$. Q.E.D.

Let $\mathbf{A}$ represent an $n \times n$ symmetric matrix. And, as in Section 5f, let $\mathbf{Q}$ represent an $n \times n$ orthogonal matrix and $\mathbf{D} = \{d_i\}$ an $n \times n$ diagonal matrix such that $\mathbf{Q}'\mathbf{AQ} = \mathbf{D}$; or, equivalently, let $\mathbf{Q}$ represent an $n \times n$ matrix whose columns are orthonormal eigenvectors of $\mathbf{A}$, and take $\mathbf{D} = \{d_i\}$ to be the diagonal matrix whose first, ..., nth diagonal elements $d_1, \ldots, d_n$ are the eigenvalues (of $\mathbf{A}$) to which the first, ..., nth columns of $\mathbf{Q}$ correspond. Further, denote the first, ..., nth columns of $\mathbf{Q}$ by $\mathbf{q}_1, \ldots, \mathbf{q}_n$, respectively; let $\lambda_1, \ldots, \lambda_k$ represent the distinct eigenvalues of $\mathbf{A}$; and (for $j = 1, \ldots, k$) let $\mathbf{E}_j = \sum_{i \in S_j} \mathbf{q}_i\mathbf{q}_i'$, where $S_j = \{i : d_i = \lambda_j\}$. Then, by definition, the spectral decomposition of $\mathbf{A}$ is

$$\mathbf{A} = \mathbf{QDQ}' = \sum_{i=1}^{n} d_i\mathbf{q}_i\mathbf{q}_i' = \sum_{j=1}^{k} \lambda_j\mathbf{E}_j.$$

The spectral decomposition of the Moore-Penrose inverse $\mathbf{A}^+$ of the $n \times n$ symmetric matrix $\mathbf{A}$ is closely related to that of $\mathbf{A}$ itself. Letting

$$d_i^+ = \begin{cases} 1/d_i, & \text{if } d_i \neq 0, \\ 0, & \text{if } d_i = 0 \end{cases}$$

(for $i = 1, \ldots, n$), it follows from Lemma 21.7.1 that $\mathbf{q}_1, \ldots, \mathbf{q}_n$ are eigenvectors of $\mathbf{A}^+$ and that the eigenvalues to which they correspond are $d_1^+, \ldots, d_n^+$. Thus, the spectral decomposition of $\mathbf{A}^+$ can be obtained from that of $\mathbf{A}$ by replacing each nonzero eigenvalue of $\mathbf{A}$ by its reciprocal. More explicitly, observing (in light of the results of Section 20.2) that $\mathbf{D}^+ = \text{diag}(d_1^+, \ldots, d_n^+)$ and letting (for $j = 1, \ldots, k$)

$$\lambda_j^+ = \begin{cases} 1/\lambda_j, & \text{if } \lambda_j \neq 0, \\ 0, & \text{if } \lambda_j = 0, \end{cases}$$

the spectral decomposition of $\mathbf{A}^+$ is

$$\mathbf{A}^+ = \mathbf{QD}^+\mathbf{Q}' = \sum_{i=1}^{n} d_i^+\mathbf{q}_i\mathbf{q}_i' = \sum_{j=1}^{k} \lambda_j^+\mathbf{E}_j. \tag{7.1}$$

21.8 Eigenvalues of Orthogonal, Idempotent, and Nonnegative Definite Matrices

The following two theorems characterize the eigenvalues of orthogonal matrices and idempotent matrices.

Theorem 21.8.1. If a scalar (real number) λ is an eigenvalue of an $n \times n$ orthogonal matrix $\mathbf{P}$, then $\lambda = \pm 1$.

Proof. Suppose that λ is an eigenvalue of $\mathbf{P}$. Then, by definition, there exists a nonnull vector $\mathbf{x}$ such that $\mathbf{Px} = \lambda\mathbf{x}$, and, consequently,

$$\mathbf{x}'\mathbf{x} = \mathbf{x}'\mathbf{I}_n\mathbf{x} = \mathbf{x}'\mathbf{P}'\mathbf{Px} = (\mathbf{Px})'\mathbf{Px} = (\lambda\mathbf{x})'(\lambda\mathbf{x}) = \lambda^2\mathbf{x}'\mathbf{x}. \qquad (8.1)$$

Since $\mathbf{x} \neq \mathbf{0}$, we have that $\mathbf{x}'\mathbf{x} \neq 0$, and, upon dividing both sides of equality (8.1) by $\mathbf{x}'\mathbf{x}$, we find that $\lambda^2 = 1$ or, equivalently, that $\lambda = \pm 1$. Q.E.D.

Theorem 21.8.2. An $n \times n$ idempotent matrix $\mathbf{A}$ has no eigenvalues other than 0 or 1. Moreover, 0 is an eigenvalue of $\mathbf{A}$ if and only if $\mathbf{A} \neq \mathbf{I}_n$, in which case it has a geometric and algebraic multiplicity of $n - \text{rank}(\mathbf{A})$. Similarly, 1 is an eigenvalue of $\mathbf{A}$ if and only if $\mathbf{A} \neq \mathbf{0}$, in which case it has a geometric and algebraic multiplicity of $\text{rank}(\mathbf{A})$. And the geometric and algebraic multiplicities of the (one or two) distinct eigenvalues of $\mathbf{A}$ sum to n.

Proof. Suppose that a scalar λ is an eigenvalue of $\mathbf{A}$. Then, by definition, there exists a nonnull vector $\mathbf{x}$ such that $\mathbf{Ax} = \lambda\mathbf{x}$, and, consequently,

$$\lambda\mathbf{x} = \mathbf{Ax} = \mathbf{A}^2\mathbf{x} = \mathbf{A}(\mathbf{Ax}) = \mathbf{A}(\lambda\mathbf{x}) = \lambda\mathbf{Ax} = \lambda^2\mathbf{x}.$$

Since $\mathbf{x} \neq \mathbf{0}$, it follows that $\lambda^2 = \lambda$, or, equivalently, that $\lambda(\lambda - 1) = 0$, and hence that $\lambda = 0$ or $\lambda = 1$.

Moreover, if $\mathbf{A} = \mathbf{I}_n$, then (in light of Lemma 21.1.1) 0 is not an eigenvalue of $\mathbf{A}$. Alternatively, if $\mathbf{A} \neq \mathbf{I}_n$, then (according to Lemma 10.1.1) $\mathbf{A}$ is singular, and it follows from Lemma 21.1.1 that 0 is an eigenvalue of $\mathbf{A}$ of geometric multiplicity $n - \text{rank}(\mathbf{A})$.

Similarly, if $\mathbf{A} = \mathbf{0}$, then there exists no nonnull vector $\mathbf{x}$ such that $\mathbf{Ax} = \mathbf{x}$, and, consequently, 1 is not an eigenvalue of $\mathbf{A}$. Alternatively, if $\mathbf{A} \neq \mathbf{0}$, then (in light of Lemma 10.2.4)

$$\text{rank}(\mathbf{A} - 1\mathbf{I}) = \text{rank}[-(\mathbf{I} - \mathbf{A})] = \text{rank}(\mathbf{I} - \mathbf{A}) = n - \text{rank}(\mathbf{A}) < n,$$

implying that 1 is an eigenvalue of $\mathbf{A}$ and that its geometric multiplicity is $n - \text{rank}(\mathbf{A} - 1\mathbf{I}) = \text{rank}(\mathbf{A})$.

And if $\mathbf{A} = \mathbf{I}_n$ or $\mathbf{A} = \mathbf{0}$, then $\mathbf{A}$ has one distinct eigenvalue (1 or 0) of geometric multiplicity n; otherwise, $\mathbf{A}$ has two distinct eigenvalues (0 and 1) with geometric multiplicities $[n - \text{rank}(\mathbf{A})$ and $\text{rank}(\mathbf{A})]$ that sum to n. It remains only to observe (on the basis of Corollary 21.3.7) that the algebraic multiplicities of the distinct eigenvalues of $\mathbf{A}$ equal the geometric multiplicities. Q.E.D.

Idempotent matrices of order n have a maximum of two distinct eigenvalues, and the geometric multiplicities of their distinct eigenvalues sum to n. Orthogonal matrices of order n also have a maximum of two distinct eigenvalues, however the geometric (or algebraic) multiplicities of their distinct eigenvalues do not necessarily sum to n. In fact, for any positive integer n that is divisible by 2, there exist $n \times n$ orthogonal matrices that have no eigenvalues (i.e., whose characteristic polynomials have no real roots).

In the special case of an $n \times n$ symmetric matrix or other $n \times n$ diagonalizable matrix, the first part of Theorem 21.8.2 can be equipped with a converse and restated as the following theorem.

Theorem 21.8.3. An $n \times n$ symmetric matrix $\mathbf{A}$ or, more generally, an $n \times n$ diagonalizable matrix $\mathbf{A}$ is idempotent if and only if it has no eigenvalues other than 0 or 1.

Proof. It suffices to prove the "if" part of the theorem—the "only if" part is a direct consequence of the first part of Theorem 21.8.2.

Suppose that $\mathbf{A}$ has no eigenvalues other than 0 or 1. By definition, there exists an $n \times n$ nonsingular matrix $\mathbf{Q}$ such that $\mathbf{Q}^{-1}\mathbf{A}\mathbf{Q} = \mathbf{D}$ for some $n \times n$ diagonal matrix $\mathbf{D}$. And, according to Part (7) of Theorem 21.5.1, the diagonal elements of $\mathbf{D}$ are the (not necessarily distinct) eigenvalues of $\mathbf{A}$. Thus, every diagonal element of $\mathbf{D}$ equals 0 or 1, implying that $\mathbf{D}^2 = \mathbf{D}$. Further, $\mathbf{A} = \mathbf{Q}\mathbf{Q}^{-1}\mathbf{A}\mathbf{Q}\mathbf{Q}^{-1} = \mathbf{Q}\mathbf{D}\mathbf{Q}^{-1}$, and it follows that

$$\mathbf{A}^2 = \mathbf{Q}\mathbf{D}\mathbf{Q}^{-1}\mathbf{Q}\mathbf{D}\mathbf{Q}^{-1} = \mathbf{Q}\mathbf{D}^2\mathbf{Q}^{-1} = \mathbf{Q}\mathbf{D}\mathbf{Q}^{-1} = \mathbf{A}.$$

Q.E.D.

The following theorem characterizes the eigenvalues of nonnegative definite matrices and positive definite matrices.

Theorem 21.8.4. Every eigenvalue of a nonnegative definite matrix is nonnegative; every eigenvalue of a positive definite matrix is positive.

Proof. Let λ represent an eigenvalue of a (square) matrix $\mathbf{A}$, and let $\mathbf{x}$ represent an eigenvector of $\mathbf{A}$ corresponding to λ. Then, by definition, $\mathbf{x} \neq \mathbf{0}$, and, consequently, $\mathbf{x}'\mathbf{x} > 0$. Further,

$$\lambda = \frac{\mathbf{x}'\mathbf{A}\mathbf{x}}{\mathbf{x}'\mathbf{x}}$$

(as discussed in Section 1 and as is evident upon observing that $\mathbf{A}\mathbf{x} = \lambda\mathbf{x}$ and hence that $\mathbf{x}'\mathbf{A}\mathbf{x} = \lambda\mathbf{x}'\mathbf{x}$).

Now, if $\mathbf{A}$ is nonnegative definite, then $\mathbf{x}'\mathbf{A}\mathbf{x} \geq 0$, and consequently $\lambda \geq 0$. And if $\mathbf{A}$ is positive definite, then $\mathbf{x}'\mathbf{A}\mathbf{x} > 0$, and consequently $\lambda > 0$. Q.E.D.

The following theorem gives conditions that can be used to determine whether a symmetric matrix is nonnegative definite and, if so, whether the matrix is positive definite or positive semidefinite.

Theorem 21.8.5. An $n \times n$ symmetric matrix $\mathbf{A}$ is nonnegative definite if and only if every eigenvalue of $\mathbf{A}$ is nonnegative, is positive definite if and only if every eigenvalue of $\mathbf{A}$ is (strictly) positive, and is positive semidefinite if and only if every eigenvalue of $\mathbf{A}$ is nonnegative and at least one eigenvalue is zero.

Proof. The theorem follows immediately from Corollary 14.2.15 upon recalling (from Section 5) that $\mathbf{A} = \mathbf{QDQ}'$, where $\mathbf{Q}$ is an $n \times n$ orthogonal matrix and $\mathbf{D}$ is an $n \times n$ diagonal matrix whose diagonal elements are the (not necessarily distinct) eigenvalues of $\mathbf{A}$. Q.E.D.

21.9 Square Root of a Symmetric Nonnegative Definite Matrix

A symmetric nonnegative definite matrix can be decomposed in the way described in the following theorem.

Theorem 21.9.1. Corresponding to any $n \times n$ symmetric nonnegative definite matrix $\mathbf{A}$, there exists an $n \times n$ symmetric nonnegative definite matrix $\mathbf{R}$ such that $\mathbf{A} = \mathbf{R}^2$. Moreover, $\mathbf{R}$ is unique and is expressible as

$$\mathbf{R} = \mathbf{Q}\,\mathrm{diag}\big(\sqrt{d_1}, \ldots, \sqrt{d_n}\big)\mathbf{Q}', \tag{9.1}$$

where $d_1, \ldots, d_n$ are the (not necessarily distinct) eigenvalues of $\mathbf{A}$ and, letting $\mathbf{D} = \mathrm{diag}(d_1, \ldots, d_n)$, $\mathbf{Q}$ is any $n \times n$ orthogonal matrix such that $\mathbf{Q}'\mathbf{AQ} = \mathbf{D}$ or, equivalently, such that $\mathbf{A} = \mathbf{QDQ}'$.

Note that in Theorem 21.9.1 it is implicitly assumed that the eigenvalues of $\mathbf{A}$ are nonnegative and that there exists an $n \times n$ orthogonal matrix $\mathbf{Q}$ such that $\mathbf{Q}'\mathbf{AQ} = \mathbf{D}$. The justification for these assumptions comes from Theorem 21.8.5 and Corollary 21.5.9.

Proof. Let $\mathbf{S} = \mathrm{diag}(\sqrt{d_1}, \ldots, \sqrt{d_n})$, so that $\mathbf{Q}\,\mathrm{diag}(\sqrt{d_1}, \ldots, \sqrt{d_n})\mathbf{Q}' = \mathbf{QSQ}'$. Then,

$$(\mathbf{QSQ}')^2 = \mathbf{QSI}_n\mathbf{SQ}' = \mathbf{QS}^2\mathbf{Q}' = \mathbf{QDQ}' = \mathbf{A}.$$

Moreover, $\mathbf{QSQ}'$ is symmetric and nonnegative definite (as is evident from Corollary 14.2.15). It remains to establish the uniqueness of $\mathbf{QSQ}'$; that is, to show that there is no $n \times n$ symmetric nonnegative definite matrix other than $\mathbf{QSQ}'$ with a "squared value" equal to $\mathbf{A}$.

Let $\mathbf{R}$ represent any $n \times n$ symmetric nonnegative definite matrix such that $\mathbf{A} = \mathbf{R}^2$. Then,

$$\begin{aligned}(\mathbf{R} - \mathbf{QSQ}')'(\mathbf{R} - \mathbf{QSQ}') &= (\mathbf{R} - \mathbf{QSQ}')(\mathbf{R} - \mathbf{QSQ}') \\ &= \mathbf{R}^2 - \mathbf{QSQ}'\mathbf{R} - \mathbf{RQSQ}' + (\mathbf{QSQ}')^2 \\ &= 2\mathbf{A} - \mathbf{QSQ}'\mathbf{R} - (\mathbf{QSQ}'\mathbf{R})'.\end{aligned}$$

Thus, to establish the uniqueness of $\mathbf{QSQ}'$, it suffices to show that $\mathbf{QSQ}'\mathbf{R} = \mathbf{A}$ [since then $(\mathbf{R} - \mathbf{QSQ}')'(\mathbf{R} - \mathbf{QSQ}') = \mathbf{0}$, implying that $\mathbf{R} - \mathbf{QSQ}' = \mathbf{0}$ or, equivalently, that $\mathbf{R} = \mathbf{QSQ}'$].

Since $\mathbf{R}$ is symmetric, there exists an $n \times n$ orthogonal matrix $\mathbf{P}$ and an $n \times n$ diagonal matrix $\mathbf{T} = \{t_i\}$ such that $\mathbf{R} = \mathbf{PTP}'$. And we find that

$$\begin{aligned}\mathbf{Q'PT^2} &= \mathbf{Q'PTI}_n\mathbf{TI}_n \\ &= \mathbf{Q'PTP'PTP'P} = \mathbf{Q'R^2P} = \mathbf{Q'AP} = \mathbf{Q'(QSQ')^2P} \\ &= \mathbf{I}_n\mathbf{SI}_n\mathbf{SQ'P} = \mathbf{S^2Q'P}.\end{aligned}$$

Now, for $i, j = 1, \ldots, n$, let b_{ij} represent the ijth element of $\mathbf{Q'P}$, and observe that the ijth elements of $\mathbf{Q'PT^2}$ and $\mathbf{S^2Q'P}$ are, respectively, $t_j^2 b_{ij}$ and $d_i b_{ij}$. So, we have that $t_j^2 b_{ij} = d_i b_{ij}$. And, since (in light of Corollary 14.2.15) $t_j \geq 0$, it follows that $t_j b_{ij} = \sqrt{d_i} b_{ij}$. (If $b_{ij} = 0$, then $t_j b_{ij} = 0 = \sqrt{d_i} b_{ij}$; alternatively, if $b_{ij} \neq 0$, then $t_j^2 b_{ij} = d_i b_{ij} \Rightarrow t_j^2 = d_i \Rightarrow t_j = \sqrt{d_i} \Rightarrow t_j b_{ij} = \sqrt{d_i} b_{ij}$.) Moreover, $t_j b_{ij}$ and $\sqrt{d_i} b_{ij}$ are the ijth elements of $\mathbf{Q'PT}$ and $\mathbf{SQ'P}$, respectively. Thus, $\mathbf{Q'PT} = \mathbf{SQ'P}$, and, consequently,

$$\mathbf{QSQ'R} = \mathbf{QSQ'PTP'} = \mathbf{QSSQ'PP'} = \mathbf{QDQ'I}_n = \mathbf{A}.$$

Q.E.D.

The $n \times n$ symmetric nonnegative definite matrix $\mathbf{R}$ given by expression (9.1) is called the *square root* of the $n \times n$ symmetric nonnegative definite matrix $\mathbf{A}$ and is denoted by the symbol $\mathbf{A}^{1/2}$. Note that if the matrix $\mathbf{A}$ is positive definite, then (as a consequence of Theorem 21.8.5) its eigenvalues $d_1, \ldots, d_n$ are (strictly) positive, implying (in light of Corollary 14.2.15) that the matrix $\mathbf{R}$ is positive definite. Thus, the square root of a symmetric positive definite matrix is positive definite.

21.10 Some Relationships

Let $\mathbf{A}$ represent an $n \times n$ matrix, and let k represent an arbitrary scalar. The eigenvalues and eigenvectors of the difference $\mathbf{A} - k\mathbf{I}_n$ and of the scalar multiple $k\mathbf{A}$ are related to those of $\mathbf{A}$ itself in a relatively simple way.

Clearly, the condition $\mathbf{Ax} = \lambda\mathbf{x}$ (where λ is a scalar and $\mathbf{x}$ a vector) is equivalent to the condition $(\mathbf{A} - k\mathbf{I})\mathbf{x} = (\lambda - k)\mathbf{x}$. Thus, a scalar λ is an eigenvalue of $\mathbf{A}$ if and only if $\lambda - k$ is an eigenvalue of $\mathbf{A} - k\mathbf{I}$. And an $n \times 1$ vector $\mathbf{x}$ is an eigenvector of $\mathbf{A}$ corresponding to the eigenvalue λ (of $\mathbf{A}$) if and only if $\mathbf{x}$ is an eigenvector of $\mathbf{A} - k\mathbf{I}$ corresponding to the eigenvalue $\lambda - k$ (of $\mathbf{A} - k\mathbf{I}$).

Moreover, for any scalar λ,

$$|\mathbf{A} - \lambda\mathbf{I}| = |\mathbf{A} - k\mathbf{I} - (\lambda - k)\mathbf{I}|,$$

so that the characteristic polynomials, say $p(\cdot)$ and $q(\cdot)$, of $\mathbf{A}$ and $\mathbf{A} - k\mathbf{I}$, respectively, are such that (for all λ) $p(\lambda) = q(\lambda - k)$. And the algebraic and geometric multiplicities of an eigenvalue λ of $\mathbf{A}$ are the same as those of the corresponding eigenvalue $\lambda - k$ of $\mathbf{A} - k\mathbf{I}$.

If $k = 0$, then $k\mathbf{A}$ has only one distinct eigenvalue (namely, 0), any nonnull $n \times 1$ vector is an eigenvector of $k\mathbf{A}$, the characteristic polynomial of $k\mathbf{A}$ is $r(\lambda) = (-\lambda)^n$, and the algebraic and geometric multiplicities of the eigenvalue 0 (of $k\mathbf{A}$) equal n.

Suppose now that $k \neq 0$. Then, the condition $\mathbf{Ax} = \lambda\mathbf{x}$ (where λ is a scalar and $\mathbf{x}$ a vector) is equivalent to the condition $k\mathbf{Ax} = k\lambda\mathbf{x}$. Thus, a scalar λ is an eigenvalue of $\mathbf{A}$ if and only if $k\lambda$ is an eigenvalue of $k\mathbf{A}$. And an $n \times 1$ vector $\mathbf{x}$ is an eigenvector of $\mathbf{A}$ corresponding to the eigenvalue λ (of $\mathbf{A}$) if and only if $\mathbf{x}$ is an eigenvector of $k\mathbf{A}$ corresponding to the eigenvalue $k\lambda$ (of $k\mathbf{A}$).

Moreover, for any scalar λ,

$$|\mathbf{A} - \lambda\mathbf{I}| = k^{-n}|k\mathbf{A} - k\lambda\mathbf{I}|,$$

so that the characteristic polynomials, say $p(\cdot)$ and $r(\cdot)$ of $\mathbf{A}$ and $k\mathbf{A}$, respectively, are such that (for all λ) $p(\lambda) = k^{-n}r(k\lambda)$. And the algebraic and geometric multiplicities of an eigenvalue λ of $\mathbf{A}$ are the same as those of the corresponding eigenvalue $k\lambda$ of $k\mathbf{A}$.

The following theorem relates the eigenvalues and eigenvectors of the $n \times n$ product $\mathbf{BA}$ of an $n \times m$ matrix $\mathbf{B}$ and an $m \times n$ matrix $\mathbf{A}$ to those of the $m \times m$ product $\mathbf{AB}$.

Theorem 21.10.1. Let $\mathbf{A}$ represent an $m \times n$ matrix and $\mathbf{B}$ an $n \times m$ matrix. Then,

(1) the characteristic polynomials, say $q(\lambda)$ and $p(\lambda)$ of $\mathbf{BA}$ and $\mathbf{AB}$, respectively, are such that (for all λ) $q(\lambda) = (-\lambda)^{n-m}p(\lambda)$, $(-\lambda)^{m-n}q(\lambda) = p(\lambda)$, or $q(\lambda) = p(\lambda)$, depending on whether $m < n$, $m > n$, or $m = n$;

(2) the nonzero distinct eigenvalues and nonzero not-necessarily-distinct eigenvalues of $\mathbf{BA}$ are the same as those of $\mathbf{AB}$;

(3) for any matrix $\mathbf{X}$ whose columns are linearly independent eigenvectors of $\mathbf{AB}$ corresponding to a nonzero eigenvalue λ, the columns of $\mathbf{BX}$ are linearly independent eigenvectors of $\mathbf{BA}$ corresponding to λ; and, similarly, for any matrix $\mathbf{Y}$ whose columns are linearly independent eigenvectors of $\mathbf{BA}$ corresponding to a nonzero eigenvalue λ, the columns of $\mathbf{AY}$ are linearly independent eigenvectors of $\mathbf{AB}$ corresponding to λ;

(4) each nonzero eigenvalue of $\mathbf{AB}$ (or, equivalently, of $\mathbf{BA}$) has the same algebraic and geometric multiplicities when regarded as an eigenvalue of $\mathbf{BA}$ as when regarded as an eigenvalue of $\mathbf{AB}$; and

(5) the algebraic multiplicity of 0 when 0 is regarded as an eigenvalue (or potential eigenvalue) of $\mathbf{BA}$ equals $n - m$ plus the algebraic multiplicity of 0 when 0 is regarded as an eigenvalue (or potential eigenvalue) of $\mathbf{AB}$.

Proof. (1) Making use of Corollaries 13.2.4 and 18.1.2, we find that, for $\lambda \neq 0$,

$$p(\lambda) = |\mathbf{AB} - \lambda\mathbf{I}_m| = (-\lambda)^m|\mathbf{I}_m - \lambda^{-1}\mathbf{AB}|$$

and

$$q(\lambda) = |\mathbf{BA} - \lambda\mathbf{I}_n| = (-\lambda)^n|\mathbf{I}_n - \lambda^{-1}\mathbf{BA}| = (-\lambda)^n|\mathbf{I}_m - \lambda^{-1}\mathbf{AB}|.$$

Thus, for $\lambda \neq 0$, $q(\lambda) = (-\lambda)^{n-m}p(\lambda)$, $(-\lambda)^{m-n}q(\lambda) = p(\lambda)$, and, in the special case where $m = n$, $q(\lambda) = p(\lambda)$. And the proof of Part (1) is complete upon observing that if two polynomials in λ are equal for all λ in some nondegenerate interval, then they are equal for all λ—refer to Theorem 21.A.5 (in the Appendix)–or, alternatively, upon observing that if $m < n$, then $\mathbf{BA}$ is singular and hence $q(0) = 0$; that if $m > n$, then $\mathbf{AB}$ is singular and hence $p(0) = 0$; and that if $m = n$, then $q(0) = |\mathbf{BA}| = |\mathbf{A}||\mathbf{B}| = |\mathbf{AB}| = p(0)$.

(2) Part (2) is an almost immediate consequence of Part (1).

(3) By definition,

$$\mathbf{ABX} = \lambda\mathbf{X}.$$

Thus,

$$\text{rank}(\mathbf{X}) \geq \text{rank}(\mathbf{BX}) \geq \text{rank}(\mathbf{ABX}) = \text{rank}(\lambda\mathbf{X}) = \text{rank}(\mathbf{X}),$$

so that $\text{rank}(\mathbf{BX}) = \text{rank}(\mathbf{X})$ and consequently (since $\mathbf{BX}$ has the same number of columns as $\mathbf{X}$) the columns of $\mathbf{BX}$ are linearly independent. Moreover,

$$\mathbf{BA}(\mathbf{BX}) = \mathbf{B}(\mathbf{ABX}) = \mathbf{B}(\lambda\mathbf{X}) = \lambda\mathbf{BX},$$

implying (since the columns of $\mathbf{BX}$ are linearly independent and hence nonnull) that the columns of $\mathbf{BX}$ are eigenvectors of $\mathbf{BA}$ corresponding to λ.

That the columns of $\mathbf{AY}$ are linearly independent eigenvectors of $\mathbf{AB}$ corresponding to λ follows from an analogous argument.

(4) It follows from Part (3) that, for each nonzero eigenvalue λ of $\mathbf{AB}$ (or, equivalently, of $\mathbf{BA}$), $\dim[\mathcal{N}(\mathbf{BA} - \lambda\mathbf{I})] \geq \dim[\mathcal{N}(\mathbf{AB} - \lambda\mathbf{I})]$ and $\dim[\mathcal{N}(\mathbf{AB} - \lambda\mathbf{I})] \geq \dim[\mathcal{N}(\mathbf{BA} - \lambda\mathbf{I})]$ and hence $\dim[\mathcal{N}(\mathbf{BA} - \lambda\mathbf{I})] = \dim[\mathcal{N}(\mathbf{AB} - \lambda\mathbf{I})]$. Thus, λ has the same geometric multiplicity when regarded as an eigenvalue of $\mathbf{BA}$ as when regarded as an eigenvalue of $\mathbf{AB}$. That λ has the same algebraic multiplicity when regarded as an eigenvalue of $\mathbf{BA}$ as when regarded as an eigenvalue of $\mathbf{AB}$ is evident from Part (1).

(5) Part (5) is an almost immediate consequence of Part (1). Q.E.D.

In the special case where $\mathbf{B} = \mathbf{A}'$, the results of Theorem 21.10.1 can be augmented by the following result.

Lemma 21.10.2. For any $m \times n$ matrix $\mathbf{A}$, the geometric multiplicity of 0 when 0 is regarded as an eigenvalue (or potential eigenvalue) of $\mathbf{A}'\mathbf{A}$ equals $n - m$ plus the geometric multiplicity of 0 when 0 is regarded as an eigenvalue (or potential eigenvalue) of $\mathbf{AA}'$.

Proof. Making use of result (1.1) (or Lemma 11.3.1) and of Corollary 7.4.5, we find that

$$\dim[\mathcal{N}(\mathbf{A}'\mathbf{A} - 0\mathbf{I}_n)] = n - \text{rank}(\mathbf{A}'\mathbf{A}) = n - \text{rank}(\mathbf{A})$$

and that

$$\dim[\mathcal{N}(\mathbf{AA}' - 0\mathbf{I}_m)] = m - \text{rank}(\mathbf{AA}') = m - \text{rank}(\mathbf{A}).$$

Thus,

$$\dim[\mathcal{N}(\mathbf{A}'\mathbf{A} - 0\mathbf{I}_n)] = n - m + \dim[\mathcal{N}(\mathbf{AA}' - 0\mathbf{I}_m)].$$

Q.E.D.

21.11 Eigenvalues and Eigenvectors of Kronecker Products of (Square) Matrices

Eigenvalues and eigenvectors of a Kronecker product $\mathbf{A} \otimes \mathbf{B}$ of two square matrices $\mathbf{A}$ and $\mathbf{B}$ can be obtained by combining eigenvalues and eigenvectors of $\mathbf{A}$ with eigenvalues and eigenvectors of $\mathbf{B}$ in the way described in the following theorem.

Theorem 21.11.1. Let λ represent an eigenvalue of an $m \times m$ matrix $\mathbf{A}$, and let $\mathbf{x}$ represent an eigenvector of $\mathbf{A}$ corresponding to λ. Similarly, let τ represent an eigenvalue of a $p \times p$ matrix $\mathbf{B}$, and let $\mathbf{y}$ represent an eigenvector of $\mathbf{B}$ corresponding to τ. Then, $\lambda\tau$ is an eigenvalue of $\mathbf{A} \otimes \mathbf{B}$, and $\mathbf{x} \otimes \mathbf{y}$ is an eigenvector of $\mathbf{A} \otimes \mathbf{B}$ corresponding to $\lambda\tau$.

Proof. By definition, $\mathbf{Ax} = \lambda\mathbf{x}$, and $\mathbf{By} = \tau\mathbf{y}$. Thus,

$$(\mathbf{A} \otimes \mathbf{B})(\mathbf{x} \otimes \mathbf{y}) = (\mathbf{Ax}) \otimes (\mathbf{By}) = (\lambda\mathbf{x}) \otimes (\tau\mathbf{y}) = \lambda\tau(\mathbf{x} \otimes \mathbf{y}).$$

Moreover, $\mathbf{x} \otimes \mathbf{y}$ is clearly nonnull. Q.E.D.

In connection with Theorem 21.11.1, it should be noted that $\mathbf{A} \otimes \mathbf{B}$ may have eigenvectors that are not expressible as the Kronecker product $\mathbf{x} \otimes \mathbf{y}$ of an eigenvector $\mathbf{x}$ of $\mathbf{A}$ and an eigenvector $\mathbf{y}$ of $\mathbf{B}$. Following Magnus and Neudecker (1988, p. 29), suppose, for example, that $\mathbf{B} = \mathbf{A} = \begin{pmatrix} 0 & 1 \\ 0 & 0 \end{pmatrix}$, and let $\mathbf{e}_1 = \begin{pmatrix} 1 \\ 0 \end{pmatrix}$ and $\mathbf{e}_2 = \begin{pmatrix} 0 \\ 1 \end{pmatrix}$. Then, $\mathbf{A}$ (or $\mathbf{B}$) has only one distinct eigenvalue, namely 0; and every eigenvector of $\mathbf{A}$ (or of $\mathbf{B}$) is a scalar multiple of $\mathbf{e}_1$. And, in accordance with Theorem 21.11.1, 0 is an eigenvalue of $\mathbf{A} \otimes \mathbf{B}$, and $\mathbf{e}_1 \otimes \mathbf{e}_1$ is an eigenvector of $\mathbf{A} \otimes \mathbf{B}$ corresponding to 0. However, $\mathbf{e}_1 \otimes \mathbf{e}_2$ and $\mathbf{e}_2 \otimes \mathbf{e}_1$ are also eigenvectors of $\mathbf{A} \otimes \mathbf{B}$ corresponding to 0 (as is easily verified).

The following theorem gives some basic results on the similarity of two Kronecker products.

Theorem 21.11.2. Let $\mathbf{R}$ represent an $m \times m$ matrix that is similar to an $m \times m$ matrix $\mathbf{A}$. Similarly, let $\mathbf{S}$ represent a $p \times p$ matrix that is similar to a $p \times p$ matrix $\mathbf{B}$. Then, $\mathbf{R} \otimes \mathbf{S}$ is similar to $\mathbf{A} \otimes \mathbf{B}$. And, for any $m \times m$ nonsingular matrix $\mathbf{C}$ such that $\mathbf{R} = \mathbf{C}^{-1}\mathbf{AC}$ and for any $p \times p$ nonsingular matrix $\mathbf{T}$ such that $\mathbf{S} = \mathbf{T}^{-1}\mathbf{BT}$,

$$\mathbf{R} \otimes \mathbf{S} = (\mathbf{C} \otimes \mathbf{T})^{-1}(\mathbf{A} \otimes \mathbf{B})(\mathbf{C} \otimes \mathbf{T}).$$

Proof. Since $\mathbf{R}$ is similar to $\mathbf{A}$ and $\mathbf{S}$ similar to $\mathbf{B}$, there exist nonsingular matrices $\mathbf{C}$ and $\mathbf{T}$ such that $\mathbf{R} = \mathbf{C}^{-1}\mathbf{AC}$ and $\mathbf{S} = \mathbf{T}^{-1}\mathbf{BT}$. Moreover, recalling formula (16.1.23) (and that the Kronecker product of nonsingular matrices is nonsingular), we find that, for any such nonsingular matrices $\mathbf{C}$ and $\mathbf{T}$, $\mathbf{C} \otimes \mathbf{T}$ is nonsingular, and

$$\mathbf{R} \otimes \mathbf{S} = (\mathbf{C}^{-1} \otimes \mathbf{T}^{-1})(\mathbf{A} \otimes \mathbf{B})(\mathbf{C} \otimes \mathbf{T}) = (\mathbf{C} \otimes \mathbf{T})^{-1}(\mathbf{A} \otimes \mathbf{B})(\mathbf{C} \otimes \mathbf{T}).$$

Q.E.D.

Upon recalling that (by definition) a (square) matrix $\mathbf{A}$ is diagonalizable if and only if there exists a diagonal matrix that is similar to $\mathbf{A}$ and that the Kronecker product of diagonal matrices is diagonal and observing [in light of results (16.1.15) and (16.1.23)] that a Kronecker product of orthogonal matrices is orthogonal, we obtain the following corollary.

Corollary 21.11.3. Let $\mathbf{A}$ represent an $m \times m$ matrix and $\mathbf{B}$ a $p \times p$ matrix. If both $\mathbf{A}$ and $\mathbf{B}$ are diagonalizable, then $\mathbf{A} \otimes \mathbf{B}$ is diagonalizable; if both $\mathbf{A}$ and $\mathbf{B}$ are orthogonally diagonalizable, then $\mathbf{A} \otimes \mathbf{B}$ is orthogonally diagonalizable. And if $\mathbf{A}$ is diagonalized by an $m \times m$ nonsingular matrix $\mathbf{P}$ and $\mathbf{B}$ by a $p \times p$ nonsingular matrix $\mathbf{Q}$, then $\mathbf{A} \otimes \mathbf{B}$ is diagonalized by the (nonsingular) matrix $\mathbf{P} \otimes \mathbf{Q}$. Moreover, if $\mathbf{P}$ and $\mathbf{Q}$ are orthogonal, then $\mathbf{P} \otimes \mathbf{Q}$ is also orthogonal.

The following theorem gives a result on the not-necessarily-distinct eigenvalues of a Kronecker product of two (square) matrices.

Theorem 21.11.4. Let $\mathbf{A}$ represent an $m \times m$ matrix and $\mathbf{B}$ a $p \times p$ matrix. Suppose that $\mathbf{A}$ has m (not necessarily distinct) eigenvalues, say $d_1, \ldots, d_m$, the last s of which equal 0. And let $f_1, \ldots, f_k$ represent the (not necessarily distinct) eigenvalues of $\mathbf{B}$ (where possibly $k < p$). Then, $\mathbf{A} \otimes \mathbf{B}$ has $(m-s)k + sp$ (not necessarily distinct) eigenvalues: $d_1 f_1, \ldots, d_1 f_k, \ldots, d_{m-s} f_1, \ldots, d_{m-s} f_k, 0, \ldots, 0$.

Proof. As a consequence of Theorem 21.5.11, there exists an $m \times m$ orthogonal matrix $\mathbf{Q}$ such that $\mathbf{Q}'\mathbf{A}\mathbf{Q} = \mathbf{T}$, where $\mathbf{T}$ is an upper triangular matrix with diagonal elements $d_1, \ldots, d_m$. Further, $\mathbf{Q} \otimes \mathbf{I}_p$ is an orthogonal matrix (as is easily verified), and

$$(\mathbf{Q} \otimes \mathbf{I}_p)'(\mathbf{A} \otimes \mathbf{B})(\mathbf{Q} \otimes \mathbf{I}_p) = (\mathbf{Q}'\mathbf{A}\mathbf{Q}) \otimes \mathbf{B} = \mathbf{T} \otimes \mathbf{B}.$$

Thus, $\mathbf{A} \otimes \mathbf{B}$ is similar to $\mathbf{T} \otimes \mathbf{B}$, and it follows from Part (7) of Theorem 21.3.1 that the (not necessarily distinct) eigenvalues of $\mathbf{A} \otimes \mathbf{B}$ are the same as those of $\mathbf{T} \otimes \mathbf{B}$. And $\mathbf{T} \otimes \mathbf{B}$ is an (upper) block-triangular matrix with diagonal blocks $d_1\mathbf{B}, \ldots, d_m\mathbf{B}$, so that [in light of Part (4) of Lemma 21.2.1 and the results of Section 10] $\mathbf{T} \otimes \mathbf{B}$ has $(m-s)k + sp$ (not necessarily distinct) eigenvalues: $d_1 f_1, \ldots, d_1 f_k, \ldots, d_{m-s} f_1, \ldots, d_{m-s} f_k, 0, \ldots, 0$. Q.E.D.

In the special case where $k = p$, Theorem 21.11.4 can be restated as the following corollary.

Corollary 21.11.5. Let $\mathbf{A}$ represent an $m \times m$ matrix and $\mathbf{B}$ a $p \times p$ matrix. Suppose that $\mathbf{A}$ has m (not necessarily distinct) eigenvalues, say $d_1, \ldots, d_m$, and that $\mathbf{B}$ has p (not necessarily distinct) eigenvalues, say $f_1, \ldots, f_p$. Then, $\mathbf{A} \otimes \mathbf{B}$ has mp (not necessarily distinct) eigenvalues: $d_i f_j$ ($i = 1, \ldots, m$; $j = 1, \ldots, p$).

Let $\mathbf{A}$ represent an $m \times m$ matrix and $\mathbf{B}$ a $p \times p$ matrix. According to Theorem 21.11.4 (and Corollary 21.11.5), a sufficient condition for $\mathbf{A} \otimes \mathbf{B}$ to have mp (not necessarily distinct) eigenvalues is that $\mathbf{A}$ have m (not necessarily distinct) eigenvalues and $\mathbf{B}$ have p (not necessarily distinct) eigenvalues or that $\mathbf{A} = \mathbf{0}$. However, this condition is not necessary.

Suppose, for example, that

$$\mathbf{B} = \mathbf{A} = \left(\begin{array}{cc|c} 0 & 1 & 0 \\ -1 & 0 & 0 \\ \hline 0 & 0 & 0 \end{array}\right).$$

Then, the characteristic polynomial of $\mathbf{A}$ (or $\mathbf{B}$) is

$$p(\lambda) = |\mathbf{A} - \lambda\mathbf{I}| = -\lambda(\lambda^2 + 1)$$

(as can be easily seen by, e.g., making use of Theorem 13.3.1). Consequently, 0 is an eigenvalue of $\mathbf{A}$ and has an algebraic multiplicity of 1. Moreover, $\mathbf{A}$ has no eigenvalues other than 0—the characteristic polynomial of $\mathbf{A}$ has two imaginary roots, namely, i and $-i$.

By way of comparison, the characteristic polynomial of $\mathbf{B} \otimes \mathbf{A}$ is (in light of Theorem 13.3.1)

$$q(\lambda) = |\mathbf{B} \otimes \mathbf{A} - \lambda\mathbf{I}| = \left|\begin{array}{cc|c} -\lambda\mathbf{I} & \mathbf{A} & \mathbf{0} \\ -\mathbf{A} & -\lambda\mathbf{I} & \mathbf{0} \\ \hline \mathbf{0} & \mathbf{0} & -\lambda\mathbf{I} \end{array}\right| = (-\lambda)^3 \begin{vmatrix} -\lambda\mathbf{I}_3 & \mathbf{A} \\ -\mathbf{A} & -\lambda\mathbf{I}_3 \end{vmatrix}.$$

And, making use of results (13.3.13) and (13.2.2), we find that (for $\lambda \neq 0$)

$$\begin{aligned} \begin{vmatrix} -\lambda\mathbf{I}_3 & \mathbf{A} \\ -\mathbf{A} & -\lambda\mathbf{I}_3 \end{vmatrix} &= |-\lambda\mathbf{I}_3||-\lambda\mathbf{I} - (-\mathbf{A})(-\lambda\mathbf{I})^{-1}\mathbf{A}| \\ &= (-\lambda)^3|-\lambda\mathbf{I}_3 - \lambda^{-1}\mathbf{A}^2| \\ &= |(-\lambda)(-\lambda\mathbf{I} - \lambda^{-1}\mathbf{A}^2)| \\ &= |\lambda^2\mathbf{I} + \mathbf{A}^2| \\ &= \begin{vmatrix} \lambda^2 - 1 & 0 & 0 \\ 0 & \lambda^2 - 1 & 0 \\ 0 & 0 & \lambda^2 \end{vmatrix} = (\lambda^2 - 1)^2\lambda^2 = \lambda^2[(\lambda + 1)(\lambda - 1)]^2 \end{aligned}$$

and hence that (for all λ)

$$q(\lambda) = (-\lambda)^3\lambda^2[(\lambda + 1)(\lambda - 1)]^2 = (-1)^9\lambda^5(\lambda + 1)^2(\lambda - 1)^2.$$

So, $\mathbf{B} \otimes \mathbf{A}$ has three distinct eigenvalues: $0, -1$, and 1, with algebraic multiplicities of $5, 2$, and 2, respectively. Thus, $\mathbf{B} \otimes \mathbf{A}$ has 9 (not necessarily distinct) eigenvalues, whereas each of the matrices $\mathbf{A}$ and $\mathbf{B}$ (which are of order 3) has only one.

21.12 Singular Value Decomposition

a. Definition, existence, and some basic properties and relationships

Let $\mathbf{A}$ represent an $n \times n$ symmetric nonnegative definite matrix. Then, it follows from the results of Section 5 (and from the nonnegativity of the eigenvalues of a nonnegative definite matrix) that $\mathbf{A}$ is expressible in the form

$$\mathbf{A} = \mathbf{Q}\begin{pmatrix} \mathbf{D}_1 & \mathbf{0} \\ \mathbf{0} & \mathbf{0} \end{pmatrix}\mathbf{Q}', \tag{12.1}$$

where $\mathbf{Q}$ is an $n \times n$ orthogonal matrix and where $\mathbf{D}_1$ is a diagonal matrix with (strictly) positive diagonal elements. In fact, decomposition (12.1) is the spectral decomposition of $\mathbf{A}$. The following theorem can be used to establish a generalization of this decomposition that is applicable to any matrix.

Theorem 21.12.1. Let $\mathbf{A}$ represent an $m \times n$ matrix of rank r. And take $\mathbf{Q}$ to be any $n \times n$ orthogonal matrix and $\mathbf{D}_1$ to be any $r \times r$ nonsingular diagonal matrix such that

$$\mathbf{Q}'\mathbf{A}'\mathbf{A}\mathbf{Q} = \begin{pmatrix} \mathbf{D}_1^2 & \mathbf{0} \\ \mathbf{0} & \mathbf{0} \end{pmatrix} \tag{12.2}$$

[where, when $r = 0$ or $r = n$, $\begin{pmatrix} \mathbf{D}_1^2 & \mathbf{0} \\ \mathbf{0} & \mathbf{0} \end{pmatrix}$ equals $\mathbf{0}$ or $\mathbf{D}_1^2$, respectively]. Further, partition $\mathbf{Q}$ as $\mathbf{Q} = (\mathbf{Q}_1, \mathbf{Q}_2)$, where $\mathbf{Q}_1$ has r columns, and let $\mathbf{P} = (\mathbf{P}_1, \mathbf{P}_2)$, where $\mathbf{P}_1 = \mathbf{A}\mathbf{Q}_1\mathbf{D}_1^{-1}$ and where $\mathbf{P}_2$ is any $m \times (m - r)$ matrix such that

$$\mathbf{P}_1'\mathbf{P}_2 = \mathbf{0}. \tag{12.3}$$

(When $r = 0$, $\mathbf{Q} = \mathbf{Q}_2$, $\mathbf{P} = \mathbf{P}_2$, and $\mathbf{P}_2$ is arbitrary; when $r = n$, $\mathbf{Q} = \mathbf{Q}_1$; and when $r = m$, $\mathbf{P} = \mathbf{P}_1$.) Then,

$$\mathbf{P}'\mathbf{A}\mathbf{Q} = \begin{pmatrix} \mathbf{D}_1 & \mathbf{0} \\ \mathbf{0} & \mathbf{0} \end{pmatrix}$$

[where, when $r = 0$, $r = m$, $r = n$, or $r = m = n$, $\begin{pmatrix} \mathbf{D}_1 & \mathbf{0} \\ \mathbf{0} & \mathbf{0} \end{pmatrix}$ equals $\mathbf{0}$, $(\mathbf{D}_1, \mathbf{0})$, $\begin{pmatrix} \mathbf{D}_1 \\ \mathbf{0} \end{pmatrix}$, or $\mathbf{D}_1$, respectively].

Proof. Since

$$\mathbf{Q}'\mathbf{A}'\mathbf{A}\mathbf{Q} = \begin{pmatrix} \mathbf{Q}_1'\mathbf{A}'\mathbf{A}\mathbf{Q}_1 & \mathbf{Q}_1'\mathbf{A}'\mathbf{A}\mathbf{Q}_2 \\ \mathbf{Q}_2'\mathbf{A}'\mathbf{A}\mathbf{Q}_1 & \mathbf{Q}_2'\mathbf{A}'\mathbf{A}\mathbf{Q}_2 \end{pmatrix},$$

we have that $\mathbf{Q}_1'\mathbf{A}'\mathbf{A}\mathbf{Q}_1 = \mathbf{D}_1^2$. Further, $(\mathbf{A}\mathbf{Q}_2)'\mathbf{A}\mathbf{Q}_2 = \mathbf{Q}_2'\mathbf{A}'\mathbf{A}\mathbf{Q}_2 = \mathbf{0}$, implying (in light of Corollary 5.3.2) that $\mathbf{A}\mathbf{Q}_2 = \mathbf{0}$. Thus, upon observing that $\mathbf{P}_1' = \mathbf{D}_1^{-1}\mathbf{Q}_1'\mathbf{A}'$ and that $\mathbf{A}\mathbf{Q}_1 = \mathbf{P}_1\mathbf{D}_1$, we find that

$$\begin{aligned} \mathbf{P}'\mathbf{A}\mathbf{Q} = \begin{pmatrix} \mathbf{P}_1'\mathbf{A}\mathbf{Q}_1 & \mathbf{P}_1'\mathbf{A}\mathbf{Q}_2 \\ \mathbf{P}_2'\mathbf{A}\mathbf{Q}_1 & \mathbf{P}_2'\mathbf{A}\mathbf{Q}_2 \end{pmatrix} &= \begin{pmatrix} \mathbf{D}_1^{-1}\mathbf{Q}_1'\mathbf{A}'\mathbf{A}\mathbf{Q}_1 & \mathbf{P}_1'\mathbf{0} \\ \mathbf{P}_2'\mathbf{P}_1\mathbf{D}_1 & \mathbf{P}_2'\mathbf{0} \end{pmatrix} \\ &= \begin{pmatrix} \mathbf{D}_1^{-1}\mathbf{D}_1^2 & \mathbf{0} \\ (\mathbf{P}_1'\mathbf{P}_2)'\mathbf{D}_1 & \mathbf{0} \end{pmatrix} = \begin{pmatrix} \mathbf{D}_1 & \mathbf{0} \\ \mathbf{0} & \mathbf{0} \end{pmatrix}. \end{aligned}$$

Q.E.D.

There exist an $r \times r$ diagonal matrix $\mathbf{D}_1$ with (strictly) positive diagonal elements and an $n \times n$ orthogonal matrix $\mathbf{Q}$ that satisfy condition (12.2) of Theorem

21.12.1, and there exists an $m \times (m-r)$ matrix $\mathbf{P}_2$ that not only satisfies condition (12.3) but is such that $\mathbf{P}$ is orthogonal. To see this, take $\mathbf{D}_1 = \text{diag}(s_1, \ldots, s_r)$, where $s_1, \ldots, s_r$ are the positive square roots of the r nonzero (and hence positive) not-necessarily-distinct eigenvalues of $\mathbf{A}'\mathbf{A}$. [Since $\mathbf{A}'\mathbf{A}$ is symmetric and (in light of Corollary 14.2.14) nonnegative definite and since $\text{rank}(\mathbf{A}'\mathbf{A}) = \text{rank}(\mathbf{A}) = r$, it follows from Corollaries 21.3.8 and 21.5.8 that $\mathbf{A}'\mathbf{A}$ has r nonzero not-necessarily-distinct eigenvalues and from Theorem 21.8.5 that those eigenvalues are positive.] Now, choose the $n \times n$ orthogonal matrix $\mathbf{Q}$ to satisfy condition (12.2)—that this is possible is evident from Corollary 21.5.9.

Further, observe (as in the proof of Theorem 21.12.1) that $\mathbf{Q}_1'\mathbf{A}'\mathbf{A}\mathbf{Q}_1 = \mathbf{D}_1^2$ and hence that

$$\mathbf{P}_1'\mathbf{P}_1 = \mathbf{D}_1^{-1}\mathbf{Q}_1'\mathbf{A}'\mathbf{A}\mathbf{Q}_1\mathbf{D}_1^{-1} = \mathbf{D}_1^{-1}\mathbf{D}_1^2\mathbf{D}_1^{-1} = \mathbf{I}_r.$$

And, observing (in light of Lemma 11.3.1) that

$$\dim[\mathcal{N}(\mathbf{P}_1')] = m - \text{rank}(\mathbf{P}_1') = m - \text{rank}(\mathbf{P}_1'\mathbf{P}_1) = m - r,$$

take $\mathbf{P}_2$ to be any $m \times (m-r)$ matrix whose columns form an orthonormal basis for $\mathcal{N}(\mathbf{P}_1')$. Then,

$$\mathbf{P}'\mathbf{P} = \begin{pmatrix} \mathbf{P}_1'\mathbf{P}_1 & \mathbf{P}_1'\mathbf{P}_2 \\ \mathbf{P}_2'\mathbf{P}_1 & \mathbf{P}_2'\mathbf{P}_2 \end{pmatrix} = \begin{pmatrix} \mathbf{I}_r & \mathbf{0} \\ \mathbf{0} & \mathbf{I}_{m-r} \end{pmatrix} = \mathbf{I}_m.$$

In light of this discussion, we have the following corollary of Theorem 21.12.1.

Corollary 21.12.2. Corresponding to any $m \times n$ matrix $\mathbf{A}$ of rank r, there exist an $m \times m$ orthogonal matrix $\mathbf{P}$ and an $n \times n$ orthogonal matrix $\mathbf{Q}$ such that

$$\mathbf{P}'\mathbf{A}\mathbf{Q} = \begin{pmatrix} \mathbf{D}_1 & \mathbf{0} \\ \mathbf{0} & \mathbf{0} \end{pmatrix},$$

where $\mathbf{D}_1$ is an $r \times r$ diagonal matrix with diagonal elements that are (strictly) positive.

Let $\mathbf{A}$ represent an $m \times n$ matrix. And let $\mathbf{P}$ represent an $m \times m$ orthogonal matrix, $\mathbf{Q}$ an $n \times n$ orthogonal matrix, and $\mathbf{D}_1 = \{s_i\}$ an $r \times r$ diagonal matrix with (strictly) positive diagonal elements such that

$$\mathbf{P}'\mathbf{A}\mathbf{Q} = \begin{pmatrix} \mathbf{D}_1 & \mathbf{0} \\ \mathbf{0} & \mathbf{0} \end{pmatrix}$$

—the existence of such matrices is guaranteed by Corollary 21.12.2. Further, partition $\mathbf{P}$ and $\mathbf{Q}$ as $\mathbf{P} = (\mathbf{P}_1, \mathbf{P}_2)$ and $\mathbf{Q} = (\mathbf{Q}_1, \mathbf{Q}_2)$, where $\mathbf{P}_1$ has r columns, say $\mathbf{p}_1, \ldots, \mathbf{p}_r$, respectively, and $\mathbf{Q}_1$ has r columns, say $\mathbf{q}_1, \ldots, \mathbf{q}_r$, respectively. Then, $\mathbf{A}$ can be expressed as

$$\mathbf{A} = \mathbf{P}\begin{pmatrix} \mathbf{D}_1 & \mathbf{0} \\ \mathbf{0} & \mathbf{0} \end{pmatrix}\mathbf{Q}' \tag{12.4}$$

or as

$$\mathbf{A} = \mathbf{P}_1\mathbf{D}_1\mathbf{Q}_1' \tag{12.5}$$

or

$$\mathbf{A} = \sum_{i=1}^{r} s_i \mathbf{p}_i \mathbf{q}_i'. \tag{12.6}$$

Letting $\alpha_1, \ldots, \alpha_k$ represent the distinct values represented among $s_1, \ldots, s_r$ and (for $j = 1, \ldots, k$) letting $L_j = \{i : s_i = \alpha_j\}$, $\mathbf{A}$ can also be expressed as

$$\mathbf{A} = \sum_{j=1}^{k} \alpha_j \mathbf{U}_j, \tag{12.7}$$

where (for $j = 1, \ldots, k$) $\mathbf{U}_j = \sum_{i \in L_j} \mathbf{p}_i \mathbf{q}_i'$.

Expression (12.4) is called the *singular-value decomposition* of the $m \times n$ matrix $\mathbf{A}$. Sometimes, the term singular-value decomposition is used also in referring to expression (12.5), (12.6), or (12.7).

The following theorem provides some insight into the nature of the various components of the singular-valuc decomposition and the extent to which those components are unique.

Theorem 21.12.3. Let $\mathbf{A}$ represent an $m \times n$ matrix. And take $\mathbf{P}$ to be an $m \times m$ orthogonal matrix, $\mathbf{Q}$ an $n \times n$ orthogonal matrix, and $\mathbf{D}_1$ an $r \times r$ nonsingular diagonal matrix such that

$$\mathbf{P}'\mathbf{A}\mathbf{Q} = \begin{pmatrix} \mathbf{D}_1 & \mathbf{0} \\ \mathbf{0} & \mathbf{0} \end{pmatrix}. \tag{12.8}$$

Further, partition $\mathbf{P}$ and $\mathbf{Q}$ as $\mathbf{P} = (\mathbf{P}_1, \mathbf{P}_2)$ and $\mathbf{Q} = (\mathbf{Q}_1, \mathbf{Q}_2)$, where each of the matrices $\mathbf{P}_1$ and $\mathbf{Q}_1$ has r columns. Then,

$$r = \operatorname{rank}(\mathbf{A}), \tag{12.9}$$

$$\mathbf{Q}'\mathbf{A}'\mathbf{A}\mathbf{Q} = \begin{pmatrix} \mathbf{D}_1^2 & \mathbf{0} \\ \mathbf{0} & \mathbf{0} \end{pmatrix}, \tag{12.10}$$

$$\mathbf{P}'\mathbf{A}\mathbf{A}'\mathbf{P} = \begin{pmatrix} \mathbf{D}_1^2 & \mathbf{0} \\ \mathbf{0} & \mathbf{0} \end{pmatrix}, \tag{12.11}$$

$$\mathbf{P}_1 = \mathbf{A}\mathbf{Q}_1\mathbf{D}_1^{-1}, \tag{12.12}$$

$$\mathbf{Q}_1 = \mathbf{A}'\mathbf{P}_1\mathbf{D}_1^{-1}. \tag{12.13}$$

Proof. The veracity of result (12.9) is evident from equality (12.8) upon observing that $\operatorname{rank}(\mathbf{P}'\mathbf{A}\mathbf{Q}) = \operatorname{rank}(\mathbf{A})$. Further,

$$\mathbf{Q}'\mathbf{A}'\mathbf{A}\mathbf{Q} = \mathbf{Q}'\mathbf{A}'\mathbf{P}\mathbf{P}'\mathbf{A}\mathbf{Q} = (\mathbf{P}'\mathbf{A}\mathbf{Q})'\mathbf{P}'\mathbf{A}\mathbf{Q} = \begin{pmatrix} \mathbf{D}_1 & \mathbf{0} \\ \mathbf{0} & \mathbf{0} \end{pmatrix}' \begin{pmatrix} \mathbf{D}_1 & \mathbf{0} \\ \mathbf{0} & \mathbf{0} \end{pmatrix} = \begin{pmatrix} \mathbf{D}_1^2 & \mathbf{0} \\ \mathbf{0} & \mathbf{0} \end{pmatrix},$$

which verifies equality (12.10). Equality (12.11) can be verified in similar fashion.

And, observing that $\mathbf{P}'\mathbf{A}\mathbf{Q} = \begin{pmatrix} \mathbf{P}_1'\mathbf{A}\mathbf{Q}_1 & \mathbf{P}_1'\mathbf{A}\mathbf{Q}_2 \\ \mathbf{P}_2'\mathbf{A}\mathbf{Q}_1 & \mathbf{P}_2'\mathbf{A}\mathbf{Q}_2 \end{pmatrix}$ [and making use of equality (12.8)], we find that

$$\begin{aligned} \mathbf{P}_1 = \mathbf{P}_1\mathbf{D}_1\mathbf{D}_1^{-1} &= \mathbf{P}_1(\mathbf{P}_1'\mathbf{A}\mathbf{Q}_1)\mathbf{D}_1^{-1} \\ &= (\mathbf{P}_1\mathbf{P}_1')\mathbf{A}\mathbf{Q}_1\mathbf{D}_1^{-1} \\ &= (\mathbf{I}_m - \mathbf{P}_2\mathbf{P}_2')\mathbf{A}\mathbf{Q}_1\mathbf{D}_1^{-1} \\ &= \mathbf{A}\mathbf{Q}_1\mathbf{D}_1^{-1} - \mathbf{P}_2(\mathbf{P}_2'\mathbf{A}\mathbf{Q}_1)\mathbf{D}_1^{-1} \\ &= \mathbf{A}\mathbf{Q}_1\mathbf{D}_1^{-1} - \mathbf{P}_2\mathbf{0}\mathbf{D}_1^{-1} = \mathbf{A}\mathbf{Q}_1\mathbf{D}_1^{-1}. \end{aligned}$$

That $\mathbf{Q}_1 = \mathbf{A}'\mathbf{P}_1\mathbf{D}_1^{-1}$ can be established via an analogous argument. Q.E.D.

In light of Theorem 21.5.1, it is clear from Theorem 21.12.3 that the scalars $s_1, \ldots, s_r$, which appear in the singular-value decomposition (12.4) (as the diagonal elements of the diagonal matrix $\mathbf{D}_1$) are the positive square roots of the nonzero (not necessarily distinct) eigenvalues of $\mathbf{A}'\mathbf{A}$ (or, equivalently, of the nonzero not-necessarily-distinct eigenvalues of $\mathbf{A}\mathbf{A}'$). Moreover, they are unique (i.e., they do not vary with the choice of the orthogonal matrices $\mathbf{P}$ and $\mathbf{Q}$ that appear in the singular-value decomposition), and they are equal in number to rank($\mathbf{A}$). The scalars $s_1, \ldots, s_r$ are referred to as the *singular values* of the matrix $\mathbf{A}$. (In some presentations, the positive square roots of all n or m eigenvalues of $\mathbf{A}'\mathbf{A}$ or $\mathbf{A}\mathbf{A}'$, including those that equal 0, are regarded as singular values of $\mathbf{A}$.) The scalars $\alpha_1, \ldots, \alpha_k$, which appear in decomposition (12.7) and which (by definition) are the distinct singular values of $\mathbf{A}$, are the positive square roots of the distinct nonzero eigenvalues of $\mathbf{A}'\mathbf{A}$ (or, equivalently, of the distinct nonzero eigenvalues of $\mathbf{A}\mathbf{A}'$).

Theorem 21.12.3 is also informative about the nature of the orthogonal matrices $\mathbf{P}$ and $\mathbf{Q}$, which appear in the singular-value decomposition (12.4). The m columns of $\mathbf{P}$ are eigenvectors of $\mathbf{A}\mathbf{A}'$, with the first r columns corresponding to the nonzero eigenvalues $s_1^2, \ldots, s_r^2$ and the remaining $m - r$ columns corresponding to the 0 eigenvalues. Similarly, the n columns of $\mathbf{Q}$ are eigenvectors of $\mathbf{A}'\mathbf{A}$, with the first r columns corresponding to the nonzero eigenvalues $s_1^2, \ldots, s_r^2$ and the remaining $n - r$ columns corresponding to the 0 eigenvalues. Moreover, once the first r columns of $\mathbf{Q}$ are specified, the first r columns of $\mathbf{P}$ are uniquely determined [by result (12.12)]. Similarly, once the first r columns of $\mathbf{P}$ are specified, the first r columns of $\mathbf{Q}$ are uniquely determined [by result (12.13)].

For any fixed ordering of the distinct singular values $\alpha_1, \ldots, \alpha_k$, decomposition (12.7) is unique. To see this, observe [in light of result (12.12)] that (for $i = 1, \ldots, r$) $\mathbf{p}_i = s_i^{-1}\mathbf{A}\mathbf{q}_i$. Thus, for $j = 1, \ldots, k$,

$$\mathbf{U}_j = \sum_{i \in L_j} \mathbf{p}_i\mathbf{q}_i' = \sum_{i \in L_j} s_i^{-1}\mathbf{A}\mathbf{q}_i\mathbf{q}_i' = \alpha_j^{-1}\mathbf{A}\mathbf{E}_j,$$

where $\mathbf{E}_j = \sum_{i \in L_j} \mathbf{q}_i\mathbf{q}_i'$. Further, since (for $i \in L_j$) $\mathbf{q}_i$ is an eigenvector of $\mathbf{A}'\mathbf{A}$ corresponding to the eigenvalue α_j^2, it follows from the results of Section 5f (on the uniqueness of the spectral decomposition) that $\mathbf{E}_j$ does not vary with the choice of $\mathbf{P}$, $\mathbf{Q}$, and $\mathbf{D}_1$ and hence that $\mathbf{U}_j$ does not vary with this choice. We conclude that decomposition (12.7) is unique (aside from the ordering of the terms).

The (usual) norm of the $m \times n$ matrix $\mathbf{A}$ with singular-value decomposition (12.4) can be expressed in terms of the singular values $s_1, \ldots, s_r$. Making use of Lemma 5.2.1, we find that

$$\begin{aligned}
\|\mathbf{A}\|^2 &= \operatorname{tr}(\mathbf{A}'\mathbf{A}) \\
&= \operatorname{tr}\left[\mathbf{Q}\begin{pmatrix}\mathbf{D}_1 & \mathbf{0}\\ \mathbf{0} & \mathbf{0}\end{pmatrix}'\mathbf{P}'\mathbf{P}\begin{pmatrix}\mathbf{D}_1 & \mathbf{0}\\ \mathbf{0} & \mathbf{0}\end{pmatrix}\mathbf{Q}'\right] \\
&= \operatorname{tr}\left[\begin{pmatrix}\mathbf{D}_1 & \mathbf{0}\\ \mathbf{0} & \mathbf{0}\end{pmatrix}'\mathbf{P}'\mathbf{P}\begin{pmatrix}\mathbf{D}_1 & \mathbf{0}\\ \mathbf{0} & \mathbf{0}\end{pmatrix}\mathbf{Q}'\mathbf{Q}\right] \\
&= \operatorname{tr}\left[\begin{pmatrix}\mathbf{D}_1 & \mathbf{0}\\ \mathbf{0} & \mathbf{0}\end{pmatrix}'\mathbf{I}_m\begin{pmatrix}\mathbf{D}_1 & \mathbf{0}\\ \mathbf{0} & \mathbf{0}\end{pmatrix}\mathbf{I}_n\right] = \operatorname{tr}\left[\begin{pmatrix}\mathbf{D}_1^2 & \mathbf{0}\\ \mathbf{0} & \mathbf{0}\end{pmatrix}\right] = s_1^2 + \cdots + s_r^2.
\end{aligned}$$

Thus,

$$\|\mathbf{A}\| = (s_1^2 + \cdots + s_r^2)^{1/2}. \tag{12.14}$$

Let us now consider the singular-value decomposition of an $n \times n$ symmetric matrix $\mathbf{A}$. Let $d_1, \ldots, d_n$ represent the (not necessarily distinct) eigenvalues of $\mathbf{A}$, ordered in such a way that $d_1, \ldots, d_r$ are nonzero and $d_{r+1} = \cdots = d_n = 0$. And let $\mathbf{q}_1, \ldots, \mathbf{q}_n$ represent orthonormal eigenvectors of $\mathbf{A}$ corresponding to $d_1, \ldots, d_n$, respectively, and (for $i = 1, \ldots, n$) let

$$\Delta_i = \begin{cases} +1, & \text{if } d_i \geq 0, \\ -1, & \text{if } d_i < 0. \end{cases}$$

Further, define $\mathbf{D} = \operatorname{diag}(d_1, \ldots, d_n)$, $\mathbf{Q} = (\mathbf{q}_1, \ldots, \mathbf{q}_n)$, and $\boldsymbol{\Delta} = \operatorname{diag}(\Delta_1, \ldots, \Delta_n)$; and take $\mathbf{P} = \mathbf{Q}\boldsymbol{\Delta}$ to be the $n \times n$ matrix whose ith column $\mathbf{p}_i$ is either $\mathbf{q}_i$ or $-\mathbf{q}_i$ depending on whether $d_i \geq 0$ or $d_i < 0$.

Then, $\mathbf{P}$ and $\mathbf{Q}$ are orthogonal. And $\mathbf{Q}'\mathbf{A}\mathbf{Q} = \mathbf{D}$, so that

$$\begin{aligned}
\mathbf{P}'\mathbf{A}\mathbf{Q} &= \boldsymbol{\Delta}\mathbf{Q}'\mathbf{A}\mathbf{Q} = \boldsymbol{\Delta}\mathbf{D} \\
&= \operatorname{diag}(\Delta_1 d_1, \ldots, \Delta_n d_n) = \operatorname{diag}(|d_1|, \ldots, |d_n|) = \begin{pmatrix}\mathbf{D}_1 & \mathbf{0}\\ \mathbf{0} & \mathbf{0}\end{pmatrix},
\end{aligned}$$

where $\mathbf{D}_1 = \operatorname{diag}(|d_1|, \ldots, |d_n|)$. Thus, the singular values of $\mathbf{A}$ are the absolute values $|d_1|, \ldots, |d_r|$ of its nonzero eigenvalues. And the singular-value decomposition of $\mathbf{A}$ [decomposition (12.6)] is

$$\mathbf{A} = \sum_{i=1}^{r} |d_i|\mathbf{p}_i\mathbf{q}_i' = \sum_{i=1}^{r} |d_i|(\Delta_i\mathbf{q}_i\mathbf{q}_i').$$

By way of comparison, the spectral decomposition of $\mathbf{A}$ [decomposition (5.4)] is

$$\mathbf{A} = \sum_{i=1}^{n} d_i\mathbf{q}_i\mathbf{q}_i' = \sum_{i=1}^{r} d_i\mathbf{q}_i\mathbf{q}_i'.$$

Note that, in the special case where the symmetric matrix $\mathbf{A}$ is nonnegative definite, $d_1, \ldots, d_r$ are positive, so that (for $i = 1, \ldots, n$) $\Delta_i = 1$, $|d_i| = d_i$, and $\mathbf{p}_i = \mathbf{q}_i$. Thus, in this special case, the singular values of $\mathbf{A}$ are its nonzero eigenvalues, and the singular-value decomposition of $\mathbf{A}$ is essentially the same as the spectral decomposition of $\mathbf{A}$.

b. Singular-value decomposition of the Moore-Penrose inverse

Let $\mathbf{A}$ represent an $m \times n$ matrix. And let $\mathbf{P}$ represent an $m \times m$ orthogonal matrix, $\mathbf{Q}$ an $n \times n$ orthogonal matrix, and $\mathbf{D}_1 = \{s_i\}$ an $r \times r$ diagonal matrix with (strictly) positive diagonal elements such that

$$\mathbf{P}'\mathbf{A}\mathbf{Q} = \begin{pmatrix} \mathbf{D}_1 & \mathbf{0} \\ \mathbf{0} & \mathbf{0} \end{pmatrix}.$$

Then, by definition, the singular-value decomposition of $\mathbf{A}$ is

$$\mathbf{A} = \mathbf{P}\begin{pmatrix} \mathbf{D}_1 & \mathbf{0} \\ \mathbf{0} & \mathbf{0} \end{pmatrix}\mathbf{Q}'.$$

According to Theorem 20.5.6,

$$\mathbf{A}^+ = \mathbf{Q}\begin{pmatrix} \mathbf{D}_1 & \mathbf{0} \\ \mathbf{0} & \mathbf{0} \end{pmatrix}^+\mathbf{P}'.$$

And, in light of the results of Section 20.2,

$$\begin{pmatrix} \mathbf{D}_1 & \mathbf{0} \\ \mathbf{0} & \mathbf{0} \end{pmatrix}^+ = \begin{pmatrix} \mathbf{D}_1^+ & \mathbf{0} \\ \mathbf{0} & \mathbf{0} \end{pmatrix} = \begin{pmatrix} \mathbf{E}_1 & \mathbf{0} \\ \mathbf{0} & \mathbf{0} \end{pmatrix},$$

where $\mathbf{E}_1 = \text{diag}(1/s_1, \ldots, 1/s_r)$. Thus,

$$\mathbf{A}^+ = \mathbf{Q}\begin{pmatrix} \mathbf{E}_1 & \mathbf{0} \\ \mathbf{0} & \mathbf{0} \end{pmatrix}\mathbf{P}', \tag{12.15}$$

or, equivalently,

$$\mathbf{Q}'\mathbf{A}^+\mathbf{P} = \begin{pmatrix} \mathbf{E}_1 & \mathbf{0} \\ \mathbf{0} & \mathbf{0} \end{pmatrix}.$$

Clearly, expression (12.15) is the singular-value decomposition of $\mathbf{A}^+$. And the singular values of $\mathbf{A}^+$ are the reciprocals of the singular values of $\mathbf{A}$.

c. Approximation of a matrix by a matrix of smaller rank

The following theorem suggests that if some of the singular values of an $m \times n$ matrix $\mathbf{A}$ are relatively small, the $m \times n$ matrix obtained from the singular-value decomposition (of $\mathbf{A}$) by setting those singular values equal to zero may provide a "good" approximation to $\mathbf{A}$.

Theorem 21.12.4. Let **A** represent an $m \times n$ matrix (of rank r) with singular values $s_1, s_2, \ldots, s_r$ ordered so that $s_1 \geq s_2 \geq \cdots \geq s_r$. Let $\mathbf{D}_1 = \text{diag}(s_1, s_2, \ldots, s_r)$, and let **P** represent an $m \times m$ orthogonal matrix and **Q** an $n \times n$ orthogonal matrix such that $\mathbf{P}'\mathbf{A}\mathbf{Q} = \text{diag}(\mathbf{D}_1, \mathbf{0})$, so that

$$\mathbf{A} = \mathbf{P}\begin{pmatrix} \mathbf{D}_1 & \mathbf{0} \\ \mathbf{0} & \mathbf{0} \end{pmatrix}\mathbf{Q}'$$

is the singular-value decomposition of **A**. Then, for any $m \times n$ matrix **B** of rank k or less (where $k < r$),

$$\|\mathbf{B} - \mathbf{A}\|^2 \geq s_{k+1}^2 + \cdots + s_r^2 \tag{12.16}$$

(where the norm is the usual norm). Moreover, equality is attained in inequality (12.16) by taking

$$\mathbf{B} = \mathbf{P}\begin{pmatrix} \mathbf{D}_1^* & \mathbf{0} \\ \mathbf{0} & \mathbf{0} \end{pmatrix}\mathbf{Q}',$$

where $\mathbf{D}_1^* = \text{diag}(s_1, \ldots, s_k)$.

As a preliminary to proving Theorem 21.12.4, it is convenient to establish the following theorem, which is of some interest in its own right.

Theorem 21.12.5. Let **A** represent an $n \times n$ symmetric matrix with (not necessarily distinct) eigenvalues $d_1, d_2, \ldots, d_n$ ordered so that $d_1 \geq d_2 \geq \cdots \geq d_n$. Then, for any $n \times k$ matrix **X** such that $\mathbf{X}'\mathbf{X} = \mathbf{I}_k$ or, equivalently, for any $n \times k$ matrix **X** with orthonormal columns (where $k \leq n$),

$$\text{tr}(\mathbf{X}'\mathbf{A}\mathbf{X}) \leq \sum_{i=1}^{k} d_i,$$

with equality holding if the columns of **X** are orthonormal eigenvectors of **A** corresponding to $d_1, d_2, \ldots, d_k$, respectively.

Proof (of Theorem 21.12.5). Let **U** represent an $n \times n$ matrix whose first, $\ldots, n$th columns are orthonormal (with respect to the usual inner product) eigenvectors of **A** corresponding to $d_1, \ldots, d_n$, respectively. Then, there exists an $n \times k$ matrix $\mathbf{R} = \{r_{ij}\}$ such that $\mathbf{X} = \mathbf{U}\mathbf{R}$. Further, $\mathbf{U}'\mathbf{A}\mathbf{U} = \mathbf{D}$, where $\mathbf{D} = \text{diag}(d_1, \ldots, d_n)$. Thus,

$$\begin{aligned} \text{tr}(\mathbf{X}'\mathbf{A}\mathbf{X}) = \text{tr}(\mathbf{R}'\mathbf{U}'\mathbf{A}\mathbf{U}\mathbf{R}) &= \text{tr}(\mathbf{R}'\mathbf{D}\mathbf{R}) \\ &= \sum_{j=1}^{k}\sum_{i=1}^{n} d_i r_{ij}^2 = \sum_{i=1}^{n} w_i d_i, \end{aligned}$$

where (for $i = 1, \ldots, n$) $w_i = \sum_{j=1}^{k} r_{ij}^2$.

The scalars $w_1, \ldots, w_n$ are such that

$$0 \leq w_i \leq 1 \tag{12.17}$$

(for $i = 1, \ldots, n$) and

$$\sum_{i=1}^{n} w_i = k. \tag{12.18}$$

To see this, observe that

$$\mathbf{R}'\mathbf{R} = \mathbf{R}'\mathbf{I}_n\mathbf{R} = \mathbf{R}'\mathbf{U}'\mathbf{U}\mathbf{R} = \mathbf{X}'\mathbf{X} = \mathbf{I}_k$$

(which indicates that the columns of $\mathbf{R}$ are orthonormal). Then, as a consequence of Theorem 6.4.5, there exists an $n \times (n-k)$ matrix $\mathbf{S}$ such that the columns of the matrix $(\mathbf{R}, \mathbf{S})$ form an orthonormal basis for $\mathcal{R}^n$ or, equivalently, such that $(\mathbf{R}, \mathbf{S})$ is an orthogonal matrix. And $\mathbf{R}\mathbf{R}' + \mathbf{S}\mathbf{S}' = (\mathbf{R}, \mathbf{S})(\mathbf{R}, \mathbf{S})' = \mathbf{I}_n$, so that

$$\mathbf{I}_n - \mathbf{R}\mathbf{R}' = \mathbf{S}\mathbf{S}',$$

implying that $\mathbf{I}_n - \mathbf{R}\mathbf{R}'$ is nonnegative definite. Thus, since clearly the ith diagonal element of $\mathbf{I}_n - \mathbf{R}\mathbf{R}'$ equals $1 - w_i$, we have that $1 - w_i \geq 0$, which (together with the obvious inequality $w_i \geq 0$) establishes result (12.17). Moreover,

$$\sum_{i=1}^{n} w_i = \sum_{j=1}^{k}\sum_{i=1}^{n} r_{ij}^2 = \operatorname{tr}(\mathbf{R}'\mathbf{R}) = \operatorname{tr}(\mathbf{I}_k) = k,$$

which establishes result (12.18).

Results (12.17) and (12.18) imply that $\sum_{i=1}^{n} w_i d_i \leq \sum_{i=1}^{k} d_i$ (as is easily verified) and hence that

$$\operatorname{tr}(\mathbf{X}'\mathbf{A}\mathbf{X}) \leq \sum_{i=1}^{k} d_i.$$

Moreover, in the special case where the columns of $\mathbf{X}$ are orthonormal eigenvectors of $\mathbf{A}$ corresponding to $d_1, \ldots, d_k$, respectively, $\mathbf{A}\mathbf{X} = \mathbf{X}\operatorname{diag}(d_1, \ldots, d_k)$ and hence

$$\mathbf{X}'\mathbf{A}\mathbf{X} = \mathbf{X}'\mathbf{X}\operatorname{diag}(d_1, \ldots, d_k) = \operatorname{diag}(d_1, \ldots, d_k).$$

Thus, in that special case, $\operatorname{tr}(\mathbf{X}'\mathbf{A}\mathbf{X}) = \sum_{i=1}^{k} d_i$. Q.E.D.

Proof (of Theorem 21.12.4). Corresponding to any $m \times n$ matrix $\mathbf{B}$ of rank k or less, there exists an $m \times k$ matrix $\mathbf{U}$ with orthonormal columns such that $\mathcal{C}(\mathbf{B}) \subset \mathcal{C}(\mathbf{U})$ or, equivalently, such that $\mathbf{B} = \mathbf{U}\mathbf{L}$ for some $k \times n$ matrix $\mathbf{L}$.

Now, let $\mathbf{U}$ represent an arbitrary $m \times k$ matrix with orthonormal columns or, equivalently, an arbitrary $m \times k$ matrix such that $\mathbf{U}'\mathbf{U} = \mathbf{I}_k$. And let $\mathbf{L}$ represent an arbitrary $k \times n$ matrix. Then, to verify inequality (12.16), it suffices to show that

$$\|\mathbf{U}\mathbf{L} - \mathbf{A}\|^2 \geq s_{k+1}^2 + \cdots + s_r^2.$$

We have that

$$\|\mathbf{U}\mathbf{L} - \mathbf{A}\|^2 = \operatorname{tr}[(\mathbf{U}\mathbf{L} - \mathbf{A})'(\mathbf{U}\mathbf{L} - \mathbf{A})] = \operatorname{tr}(\mathbf{L}'\mathbf{L} - \mathbf{A}'\mathbf{U}\mathbf{L} - \mathbf{L}'\mathbf{U}'\mathbf{A}) + \operatorname{tr}(\mathbf{A}'\mathbf{A}).$$

And, since

$$\mathbf{L}'\mathbf{L} - \mathbf{A}'\mathbf{U}\mathbf{L} - \mathbf{L}'\mathbf{U}'\mathbf{A} + \mathbf{A}'\mathbf{U}\mathbf{U}'\mathbf{A} = (\mathbf{L} - \mathbf{U}'\mathbf{A})'(\mathbf{L} - \mathbf{U}'\mathbf{A})$$

and since $(\mathbf{L} - \mathbf{U}'\mathbf{A})'(\mathbf{L} - \mathbf{U}'\mathbf{A})$ is a nonnegative definite matrix, we find (in light of Theorem 14.7.2) that

$$\text{tr}(\mathbf{L}'\mathbf{L} - \mathbf{A}'\mathbf{U}\mathbf{L} - \mathbf{L}'\mathbf{U}'\mathbf{A} + \mathbf{A}'\mathbf{U}\mathbf{U}'\mathbf{A}) \geq 0$$

and hence that

$$\text{tr}(\mathbf{L}'\mathbf{L} - \mathbf{A}'\mathbf{U}\mathbf{L} - \mathbf{L}'\mathbf{U}'\mathbf{A}) \geq -\text{tr}(\mathbf{A}'\mathbf{U}\mathbf{U}'\mathbf{A}) = -\text{tr}(\mathbf{U}'\mathbf{A}\mathbf{A}'\mathbf{U}).$$

Thus,

$$\|\mathbf{U}\mathbf{L} - \mathbf{A}\|^2 \geq \text{tr}(\mathbf{A}'\mathbf{A}) - \text{tr}(\mathbf{U}'\mathbf{A}\mathbf{A}'\mathbf{U}).$$

Moreover, since (in light of the results of Subsection a) $s_1, \ldots, s_r$ are the positive square roots of the nonzero eigenvalues of $\mathbf{A}'\mathbf{A}$ (or, equivalently, of the nonzero eigenvalues of $\mathbf{A}\mathbf{A}'$), it follows from Theorem 21.6.1 that

$$\text{tr}(\mathbf{A}'\mathbf{A}) = \sum_{i=1}^{r} s_i^2$$

and from Theorem 21.12.5 that

$$\text{tr}(\mathbf{U}'\mathbf{A}\mathbf{A}'\mathbf{U}) \leq \sum_{i=1}^{k} s_i^2.$$

We conclude that

$$\|\mathbf{U}\mathbf{L} - \mathbf{A}\|^2 \geq \sum_{i=1}^{r} s_i^2 - \sum_{i=1}^{k} s_i^2 = s_{k+1}^2 + \cdots + s_r^2.$$

And the proof is complete upon observing (in light of Lemma 8.4.2) that

$$\left\|\mathbf{P}\begin{pmatrix}\mathbf{D}_1^* & \mathbf{0}\\ \mathbf{0} & \mathbf{0}\end{pmatrix}\mathbf{Q}' - \mathbf{A}\right\|^2 = \left\|\mathbf{P}\begin{pmatrix}\mathbf{D}_1^* & \mathbf{0}\\ \mathbf{0} & \mathbf{0}\end{pmatrix}\mathbf{Q}' - \mathbf{P}\begin{pmatrix}\mathbf{D}_1 & \mathbf{0}\\ \mathbf{0} & \mathbf{0}\end{pmatrix}\mathbf{Q}'\right\|^2$$
$$= \left\|\begin{pmatrix}\mathbf{D}_1^* & \mathbf{0}\\ \mathbf{0} & \mathbf{0}\end{pmatrix} - \begin{pmatrix}\mathbf{D}_1 & \mathbf{0}\\ \mathbf{0} & \mathbf{0}\end{pmatrix}\right\|^2 = s_{k+1}^2 + \cdots + s_r^2.$$

Q.E.D.

In the special case where the matrix $\mathbf{A}$ is symmetric, Theorem 21.12.4 can (in light of the results of Subsection a) be restated in terms related to the spectral decomposition of $\mathbf{A}$, as described in the following corollary.

Corollary 21.12.6. Let $\mathbf{A}$ represent an $n \times n$ symmetric matrix (of rank r) with nonzero (not necessarily distinct) eigenvalues $d_1, \ldots, d_r$ ordered so that $|d_1| \geq |d_2| \geq \cdots \geq |d_r|$. Let $\mathbf{D}_1 = \text{diag}(d_1, \ldots, d_r)$, and let $\mathbf{Q}$ represent an $n \times n$ orthogonal matrix such that $\mathbf{Q}'\mathbf{A}\mathbf{Q} = \text{diag}(\mathbf{D}_1, \mathbf{0})$, so that

$$\mathbf{A} = \mathbf{Q}\begin{pmatrix}\mathbf{D}_1 & \mathbf{0}\\ \mathbf{0} & \mathbf{0}\end{pmatrix}\mathbf{Q}'$$

is the spectral decomposition of $\mathbf{A}$. Then, for any $n \times n$ matrix $\mathbf{B}$ of rank k or less (where $k < r$)

$$\|\mathbf{B} - \mathbf{A}\|^2 \geq d_{k+1}^2 + \cdots + d_r^2. \tag{12.19}$$

Moreover, equality is attained in inequality (12.19) by taking

$$\mathbf{B} = \mathbf{Q}\begin{pmatrix} \mathbf{D}_1^* & \mathbf{0} \\ \mathbf{0} & \mathbf{0} \end{pmatrix}\mathbf{Q}',$$

where $\mathbf{D}_1^* = \text{diag}(d_1, \ldots, d_k)$.

21.13 Simultaneous Diagonalization

In Section 21.5, we considered the diagonalization of an $n \times n$ matrix. Now, let $\mathbf{A}_1, \ldots, \mathbf{A}_k$ represent k matrices of dimensions $n \times n$, and consider the conditions under which there exists a single $n \times n$ nonsingular matrix $\mathbf{Q}$ that diagonalizes all k of these matrices; that is, the conditions under which there exists an $n \times n$ nonsingular matrix $\mathbf{Q}$ such that $\mathbf{Q}^{-1}\mathbf{A}_1\mathbf{Q} = \mathbf{D}_1, \ldots, \mathbf{Q}^{-1}\mathbf{A}_k\mathbf{Q} = \mathbf{D}_k$ for some diagonal matrices $\mathbf{D}_1, \ldots, \mathbf{D}_k$. When such a nonsingular matrix $\mathbf{Q}$ exists, $\mathbf{Q}$ is said to *simultaneously diagonalize* $\mathbf{A}_1, \ldots, \mathbf{A}_k$ (or $\mathbf{A}_1, \ldots, \mathbf{A}_k$ are said to be *simultaneously diagonalized* by $\mathbf{Q}$), and $\mathbf{A}_1, \ldots, \mathbf{A}_k$ are referred to as being *simultaneously diagonalizable.*

Suppose that there exists an $n \times n$ nonsingular matrix $\mathbf{Q}$ such that $\mathbf{Q}^{-1}\mathbf{A}_1\mathbf{Q} = \mathbf{D}_1, \ldots, \mathbf{Q}^{-1}\mathbf{A}_k\mathbf{Q} = \mathbf{D}_k$ for some diagonal matrices $\mathbf{D}_1, \ldots, \mathbf{D}_k$. Then, for $s \neq i = 1, \ldots, k$,

$$\begin{aligned} \mathbf{Q}^{-1}\mathbf{A}_s\mathbf{A}_i\mathbf{Q} &= \mathbf{Q}^{-1}\mathbf{A}_s\mathbf{Q}\mathbf{Q}^{-1}\mathbf{A}_i\mathbf{Q} \\ &= \mathbf{D}_s\mathbf{D}_i = \mathbf{D}_i\mathbf{D}_s \\ &= \mathbf{Q}^{-1}\mathbf{A}_i\mathbf{Q}\mathbf{Q}^{-1}\mathbf{A}_s\mathbf{Q} = \mathbf{Q}^{-1}\mathbf{A}_i\mathbf{A}_s\mathbf{Q}, \end{aligned}$$

implying that

$$\mathbf{A}_s\mathbf{A}_i = \mathbf{Q}(\mathbf{Q}^{-1}\mathbf{A}_s\mathbf{A}_i\mathbf{Q})\mathbf{Q}^{-1} = \mathbf{Q}(\mathbf{Q}^{-1}\mathbf{A}_i\mathbf{A}_s\mathbf{Q})\mathbf{Q}^{-1} = \mathbf{A}_i\mathbf{A}_s.$$

Thus, a necessary condition for $\mathbf{A}_1, \ldots, \mathbf{A}_k$ to be simultaneously diagonalizable is that

$$\mathbf{A}_s\mathbf{A}_i = \mathbf{A}_i\mathbf{A}_s \qquad (s > i = 1, \ldots, k) \tag{13.1}$$

(i.e., that $\mathbf{A}_1, \ldots, \mathbf{A}_k$ commute in pairs).

For symmetric matrices $\mathbf{A}_1, \ldots, \mathbf{A}_k$, condition (13.1) is sufficient, as well as necessary, for $\mathbf{A}_1, \ldots, \mathbf{A}_k$ to be simultaneously diagonalizable. In fact, symmetric matrices $\mathbf{A}_1, \ldots, \mathbf{A}_k$ that satisfy condition (13.1) can be simultaneously diagonalized by an orthogonal matrix. To verify this, let us proceed by mathematical induction.

In the case of one symmetric matrix $\mathbf{A}_1$, we know (from Theorem 21.5.7) that there exists an orthogonal matrix that "simultaneously" diagonalizes $\mathbf{A}_1$—when

$k = 1$, condition (13.1) is vacuous. Suppose now that any $k-1$ symmetric matrices (of the same order) that commute in pairs can be simultaneously diagonalized by an orthogonal matrix (where $k \geq 2$). And let $\mathbf{A}_1, \ldots, \mathbf{A}_k$ represent k symmetric matrices of arbitrary order n that commute in pairs. Then, to complete the induction argument, it suffices to show that $\mathbf{A}_1, \ldots, \mathbf{A}_k$ can be simultaneously diagonalized by an orthogonal matrix.

Let $\lambda_1, \ldots, \lambda_r$ represent the distinct eigenvalues of $\mathbf{A}_k$, and let $\nu_1, \ldots, \nu_r$ represent their respective multiplicities. Take $\mathbf{Q}_j$ to be an $n \times \nu_j$ matrix whose columns are orthonormal (with respect to the usual inner product) eigenvectors of $\mathbf{A}_k$ corresponding to the eigenvalue λ_j or, equivalently, whose columns form an orthonormal basis for the ν_j-dimensional linear space $\mathcal{N}(\mathbf{A}_k - \lambda_j \mathbf{I})$ $(j = 1, \ldots, r)$. And define $\mathbf{Q} = (\mathbf{Q}_1, \ldots, \mathbf{Q}_r)$.

Since (according to Theorem 21.5.7) $\mathbf{A}_k$ is diagonalizable, it follows from Corollary 21.5.4 that $\sum_{j=1}^{r} \nu_j = n$, so that $\mathbf{Q}$ has n (orthonormal) columns and hence is an orthogonal matrix. Further,

$$\mathbf{Q}'\mathbf{A}_k\mathbf{Q} = \text{diag}(\lambda_1 \mathbf{I}_{\nu_1}, \ldots, \lambda_r \mathbf{I}_{\nu_r}) \tag{13.2}$$

(as is evident from Theorems 21.5.2 and 21.5.1).

Since $\mathbf{A}_k$ commutes with each of the matrices $\mathbf{A}_1, \ldots, \mathbf{A}_{k-1}$, we find that, for $i = 1, \ldots, k-1$ and $j = 1, \ldots, r$,

$$\mathbf{A}_k(\mathbf{A}_i\mathbf{Q}_j) = \mathbf{A}_i(\mathbf{A}_k\mathbf{Q}_j) = \mathbf{A}_i(\lambda_j\mathbf{Q}_j) = \lambda_j\mathbf{A}_i\mathbf{Q}_j.$$

Thus, $(\mathbf{A}_k - \lambda_j\mathbf{I})\mathbf{A}_i\mathbf{Q}_j = \mathbf{0}$, implying (in light of Lemma 11.4.1) that $\mathcal{C}(\mathbf{A}_i\mathbf{Q}_j) \subset \mathcal{N}(\mathbf{A}_k - \lambda_j\mathbf{I})$. And, since the columns of $\mathbf{Q}_j$ form a basis for $\mathcal{N}(\mathbf{A}_k - \lambda_j\mathbf{I})$, there exists a $\nu_j \times \nu_j$ matrix $\mathbf{B}_{ij}$ such that

$$\mathbf{A}_i\mathbf{Q}_j = \mathbf{Q}_j\mathbf{B}_{ij} \tag{13.3}$$

$(i = 1, \ldots, k-1;\ j = 1, \ldots, r)$, and we find that (for $i = 1, \ldots, k-1$)

$$\begin{aligned} \mathbf{Q}'\mathbf{A}_i\mathbf{Q} = \mathbf{Q}'(\mathbf{A}_i\mathbf{Q}_1, \ldots, \mathbf{A}_i\mathbf{Q}_r) &= \mathbf{Q}'(\mathbf{Q}_1\mathbf{B}_{i1}, \ldots, \mathbf{Q}_r\mathbf{B}_{ir}) \\ &= \mathbf{Q}'\mathbf{Q}\,\text{diag}(\mathbf{B}_{i1}, \ldots, \mathbf{B}_{ir}) \\ &= \text{diag}(\mathbf{B}_{i1}, \ldots, \mathbf{B}_{ir}). \end{aligned} \tag{13.4}$$

Now, using result (13.3), we find that, for $j = 1, \ldots, r$,

$$\mathbf{B}_{ij} = \mathbf{I}_{\nu_j}\mathbf{B}_{ij} = \mathbf{Q}_j'\mathbf{Q}_j\mathbf{B}_{ij} = \mathbf{Q}_j'\mathbf{A}_i\mathbf{Q}_j$$

$(i = 1, \ldots, k-1)$ and that

$$\mathbf{B}_{sj}\mathbf{B}_{ij} = \mathbf{Q}_j'\mathbf{A}_s\mathbf{Q}_j\mathbf{B}_{ij} = \mathbf{Q}_j'\mathbf{A}_s\mathbf{A}_i\mathbf{Q}_j = \mathbf{Q}_j'\mathbf{A}_i\mathbf{A}_s\mathbf{Q}_j = \mathbf{Q}_j'\mathbf{A}_i\mathbf{Q}_j\mathbf{B}_{sj} = \mathbf{B}_{ij}\mathbf{B}_{sj}$$

$(s > i = 1, \ldots, k-1)$, so that the $k-1$ matrices $\mathbf{B}_{1j}, \ldots, \mathbf{B}_{k-1,j}$ (which are of order ν_j) are symmetric and they commute in pairs. Thus, by supposition, $\mathbf{B}_{1j}, \ldots, \mathbf{B}_{k-1,j}$ can be simultaneously diagonalized by an orthogonal matrix;

that is, there exists a $\nu_j \times \nu_j$ orthogonal matrix $\mathbf{S}_j$ and $\nu_j \times \nu_j$ diagonal matrices $\mathbf{D}_{1j}, \ldots, \mathbf{D}_{k-1,j}$ such that (for $i = 1, \ldots, k-1$)

$$\mathbf{S}_j'\mathbf{B}_{ij}\mathbf{S}_j = \mathbf{D}_{ij}. \tag{13.5}$$

Define $\mathbf{S} = \text{diag}(\mathbf{S}_1, \ldots, \mathbf{S}_r)$ and $\mathbf{P} = \mathbf{QS}$. Then, clearly, $\mathbf{S}$ is orthogonal and, hence, according to Lemma 8.4.1, $\mathbf{P}$ is also orthogonal. Further, using results (13.4), (13.5), and (13.2), we find that, for $i = 1, \ldots, k-1$,

$$\begin{aligned}\mathbf{P}'\mathbf{A}_i\mathbf{P} &= \mathbf{S}'\mathbf{Q}'\mathbf{A}_i\mathbf{QS} \\ &= \text{diag}(\mathbf{S}_1', \ldots, \mathbf{S}_r')\,\text{diag}(\mathbf{B}_{i1}, \ldots, \mathbf{B}_{ir})\,\text{diag}(\mathbf{S}_1, \ldots, \mathbf{S}_r) \\ &= \text{diag}(\mathbf{S}_1'\mathbf{B}_{i1}\mathbf{S}_1, \ldots, \mathbf{S}_r'\mathbf{B}_{ir}\mathbf{S}_r) \\ &= \text{diag}(\mathbf{D}_{i1}, \ldots, \mathbf{D}_{ir})\end{aligned}$$

and that

$$\begin{aligned}\mathbf{P}'\mathbf{A}_k\mathbf{P} &= \mathbf{S}'\mathbf{Q}'\mathbf{A}_k\mathbf{QS} \\ &= \text{diag}(\mathbf{S}_1', \ldots, \mathbf{S}_r')\,\text{diag}(\lambda_1\mathbf{I}_{\nu_1}, \ldots, \lambda_r\mathbf{I}_{\nu_r})\,\text{diag}(\mathbf{S}_1, \ldots, \mathbf{S}_r) \\ &= \text{diag}(\lambda_1\mathbf{S}_1'\mathbf{S}_1, \ldots, \lambda_r\mathbf{S}_r'\mathbf{S}_r) \\ &= \text{diag}(\lambda_1\mathbf{I}_{\nu_1}, \ldots, \lambda_r\mathbf{I}_{\nu_r}),\end{aligned}$$

so that all k of the matrices $\mathbf{A}_1, \ldots, \mathbf{A}_k$ are simultaneously diagonalized by the orthogonal matrix $\mathbf{P}$.

Summarizing, we have the following theorem.

Theorem 21.13.1. If $n \times n$ matrices $\mathbf{A}_1, \ldots, \mathbf{A}_k$ are simultaneously diagonalizable, then they commute in pairs, that is, for $s > i = 1, \ldots, k$, $\mathbf{A}_s\mathbf{A}_i = \mathbf{A}_i\mathbf{A}_s$. If $n \times n$ symmetric matrices $\mathbf{A}_1, \ldots, \mathbf{A}_k$ commute in pairs, then they can be simultaneously diagonalized by an orthogonal matrix; that is, there exist an $n \times n$ orthogonal matrix $\mathbf{P}$ and diagonal matrices $\mathbf{D}_1, \ldots, \mathbf{D}_k$ such that, for $i = 1, \ldots, k$,

$$\mathbf{P}'\mathbf{A}_i\mathbf{P} = \mathbf{D}_i.$$

In connection with Theorem 21.13.1, note that symmetric matrices $\mathbf{A}_1, \ldots, \mathbf{A}_k$ commute in pairs if and only if each of the $k(k-1)/2$ matrix products $\mathbf{A}_1\mathbf{A}_2$, $\mathbf{A}_1\mathbf{A}_3, \ldots, \mathbf{A}_{k-1}\mathbf{A}_k$ is symmetric.

In the special case where $k = 2$, Theorem 21.13.1 can be restated as the following corollary.

Corollary 21.13.2. If two $n \times n$ matrices $\mathbf{A}$ and $\mathbf{B}$ are simultaneously diagonalizable, then they commute, that is, $\mathbf{BA} = \mathbf{AB}$. If two $n \times n$ symmetric matrices $\mathbf{A}$ and $\mathbf{B}$ commute (or, equivalently, if their product $\mathbf{AB}$ is symmetric), then they can be simultaneously diagonalized by an orthogonal matrix; that is, there exists an $n \times n$ orthogonal matrix $\mathbf{P}$ such that $\mathbf{P}'\mathbf{AP} = \mathbf{D}_1$ and $\mathbf{P}'\mathbf{BP} = \mathbf{D}_2$ for some diagonal matrices $\mathbf{D}_1$ and $\mathbf{D}_2$.

21.14 Generalized Eigenvalue Problem

Let $\mathbf{A}$ represent an $n \times n$ symmetric matrix, and let $\mathbf{B}$ represent an $n \times n$ symmetric positive definite matrix. Then, according to Corollary 14.3.14, there exists an $n \times n$ nonsingular matrix $\mathbf{Q}$ such that the quadratic form $\mathbf{x}'\mathbf{B}\mathbf{x}$ (in an $n \times 1$ vector $\mathbf{x}$) is expressible as the sum of squares $\sum_{i=1}^{n} y_i^2$ of the elements $y_1, \ldots, y_n$ of the transformed vector $\mathbf{y} = \mathbf{Q}'\mathbf{x}$. And, according to Corollary 14.3.6, there exists an $n \times n$ nonsingular matrix $\mathbf{Q}$ and n scalars $d_1, \ldots, d_n$ such that the quadratic form $\mathbf{x}'\mathbf{A}\mathbf{x}$ is expressible as a linear combination $\sum_{i=1}^{n} d_i y_i^2$ of the squares of the elements $y_1, \ldots, y_n$ of the transformed vector $\mathbf{y} = \mathbf{Q}'\mathbf{x}$.

Can the same $n \times n$ nonsingular matrix $\mathbf{Q}$ be used to so express both $\mathbf{x}'\mathbf{B}\mathbf{x}$ and $\mathbf{x}'\mathbf{A}\mathbf{x}$? Or, equivalently, does there exist an $n \times n$ nonsingular matrix $\mathbf{Q}$ such that $\mathbf{B} = \mathbf{Q}\mathbf{Q}'$ and $\mathbf{A} = \mathbf{Q}\mathbf{D}\mathbf{Q}'$ for some diagonal matrix $\mathbf{D}$? The answer is yes, as is shown in the following development.

Let $\mathbf{R}$ represent any $n \times n$ nonsingular matrix such that $\mathbf{B}^{-1} = \mathbf{R}'\mathbf{R}$. And let $\mathbf{C} = \mathbf{R}\mathbf{A}\mathbf{R}'$, and define $\mathbf{P}$ to be an $n \times n$ orthogonal matrix such that $\mathbf{P}'\mathbf{C}\mathbf{P} = \mathbf{D}$ for some diagonal matrix $\mathbf{D} = \{d_i\}$. Now, take $\mathbf{Q} = \mathbf{R}^{-1}\mathbf{P}$. Then,

$$\mathbf{Q}\mathbf{Q}' = \mathbf{R}^{-1}\mathbf{P}\mathbf{P}'(\mathbf{R}')^{-1} = \mathbf{R}^{-1}(\mathbf{R}')^{-1} = \mathbf{B}, \tag{14.1}$$

and

$$\mathbf{Q}\mathbf{D}\mathbf{Q}' = \mathbf{R}^{-1}\mathbf{P}\mathbf{P}'\mathbf{C}\mathbf{P}\mathbf{P}'(\mathbf{R}')^{-1} = \mathbf{R}^{-1}\mathbf{C}(\mathbf{R}')^{-1} = \mathbf{R}^{-1}\mathbf{R}\mathbf{A}\mathbf{R}'(\mathbf{R}')^{-1} = \mathbf{A}. \tag{14.2}$$

In addition to establishing the existence of an $n \times n$ nonsingular matrix $\mathbf{Q}$ such that $\mathbf{B} = \mathbf{Q}\mathbf{Q}'$ and $\mathbf{A} = \mathbf{Q}\mathbf{D}\mathbf{Q}'$ (for some diagonal matrix $\mathbf{D}$), the preceding development suggests an indirect "strategy" for constructing such a matrix. This strategy comprises four steps: (1) construction of the $n \times n$ nonsingular matrix $\mathbf{R}$ (such that $\mathbf{B}^{-1} = \mathbf{R}'\mathbf{R}$); (2) formation of the matrix $\mathbf{C} = \mathbf{R}\mathbf{A}\mathbf{R}'$; (3) determination of the $n \times n$ orthogonal matrix $\mathbf{P}$ that diagonalizes $\mathbf{C}$; and (4) the formation of the matrix $\mathbf{Q} = \mathbf{R}^{-1}\mathbf{P}$. In connection with Step (3), note that the columns of the $n \times n$ orthogonal matrix $\mathbf{P}$ can be any n orthonormal eigenvectors of $\mathbf{C}$.

Let us now consider an alternative approach to the construction of an $n \times n$ nonsingular matrix $\mathbf{Q}$ such that $\mathbf{B} = \mathbf{Q}\mathbf{Q}'$ and $\mathbf{A} = \mathbf{Q}\mathbf{D}\mathbf{Q}'$ (for some diagonal matrix $\mathbf{D}$). In the alternative approach, the columns of the matrix $(\mathbf{Q}')^{-1}$ are taken to be orthonormal "generalized eigenvectors" of $\mathbf{A}$, where the orthonormality of these vectors is with respect to an inner product other than the usual inner product.

Let $\mathbf{X} = (\mathbf{Q}')^{-1}$ (where $\mathbf{Q}$ represents an arbitrary $n \times n$ nonsingular matrix), and let $\mathbf{D} = \{d_i\}$ represent an arbitrary $n \times n$ diagonal matrix. Then,

$$\begin{aligned} \mathbf{B} = \mathbf{Q}\mathbf{Q}' \text{ and } \mathbf{A} = \mathbf{Q}\mathbf{D}\mathbf{Q}' \quad &\Leftrightarrow \quad \mathbf{B}\mathbf{X} = \mathbf{Q} \text{ and } \mathbf{A}\mathbf{X} = \mathbf{Q}\mathbf{D} \\ &\Leftrightarrow \quad \mathbf{B}\mathbf{X} = \mathbf{Q} \text{ and } \mathbf{A}\mathbf{X} = \mathbf{B}\mathbf{X}\mathbf{D} \\ &\Leftrightarrow \quad \mathbf{X}'\mathbf{B}\mathbf{X} = \mathbf{I} \text{ and } \mathbf{A}\mathbf{X} = \mathbf{B}\mathbf{X}\mathbf{D}. \end{aligned}$$

And, letting $\mathbf{x}_i$ represent the ith column of $\mathbf{X}$ $(i = 1, \ldots, n)$,

$$\mathbf{X}'\mathbf{B}\mathbf{X} = \mathbf{I} \quad \Leftrightarrow \quad \mathbf{x}_j'\mathbf{B}\mathbf{x}_i = \begin{cases} 1, & \text{for } j = i = 1, \ldots, n, \\ 0, & \text{for } j \neq i = 1, \ldots, n, \end{cases}$$

and

$$\mathbf{A}\mathbf{X} = \mathbf{B}\mathbf{X}\mathbf{D} \quad \Leftrightarrow \quad \mathbf{A}\mathbf{x}_i = d_i\mathbf{B}\mathbf{x}_i, \text{ for } i = 1, \ldots, n.$$

Accordingly, the alternative approach to the construction of an $n \times n$ nonsingular matrix $\mathbf{Q}$ such that $\mathbf{B} = \mathbf{Q}\mathbf{Q}'$ and $\mathbf{A} = \mathbf{Q}\mathbf{D}\mathbf{Q}'$ (for some diagonal matrix $\mathbf{D}$) consists of finding n solutions $(d_1, \mathbf{x}_1); \ldots; (d_n, \mathbf{x}_n)$ (for a scalar λ and an $n \times 1$ vector $\mathbf{x}$) to the equation

$$\mathbf{A}\mathbf{x} = \lambda\mathbf{B}\mathbf{x} \tag{14.3}$$

such that $\mathbf{x}_1, \ldots, \mathbf{x}_n$ are orthogonal with respect to the inner product $\mathbf{x}'\mathbf{B}\mathbf{y}$ and of then taking $(\mathbf{Q}')^{-1}$ to be the matrix whose columns are $\mathbf{x}_1, \ldots, \mathbf{x}_n$.

Various properties of the "$\mathbf{x}$- and λ-parts" of solutions to equation (14.3) can be deduced by observing that

$$\mathbf{A}\mathbf{x} = \lambda\mathbf{B}\mathbf{x} \quad \Leftrightarrow \quad \mathbf{C}(\mathbf{R}^{-1})'\mathbf{x} = \lambda(\mathbf{R}^{-1})'\mathbf{x} \tag{14.4}$$

and that (for any $n \times 1$ vectors $\mathbf{x}$ and $\mathbf{y}$)

$$\mathbf{x}'\mathbf{B}\mathbf{y} = [(\mathbf{R}^{-1})'\mathbf{x}]'(\mathbf{R}^{-1})'\mathbf{y}. \tag{14.5}$$

Thus, a scalar λ and an n-dimensional vector $\mathbf{x}$ form a solution to equation (14.3) if and only if λ is an eigenvalue of $\mathbf{C}$ and $(\mathbf{R}^{-1})'\mathbf{x}$ is an eigenvector of $\mathbf{C}$ corresponding to λ. Further, an n-dimensional vector $\mathbf{x}$ has unit norm with respect to the inner product $\mathbf{x}'\mathbf{B}\mathbf{y}$ if and only if $(\mathbf{R}^{-1})'\mathbf{x}$ has unit norm with respect to the usual inner product. And two n-dimensional vectors $\mathbf{x}$ and $\mathbf{y}$ are orthogonal with respect to the inner product $\mathbf{x}'\mathbf{B}\mathbf{y}$ if and only if $(\mathbf{R}^{-1})'\mathbf{x}$ and $(\mathbf{R}^{-1})'\mathbf{y}$ are orthogonal with respect to the usual inner product. Recalling Theorem 21.4.5, we conclude in particular that $\mathbf{x}_1'\mathbf{B}\mathbf{x}_2 = 0$ for any two solutions $(\lambda_1, \mathbf{x}_1)$ and $(\lambda_2, \mathbf{x}_2)$ to equation (14.3) such that $\lambda_2 \neq \lambda_1$.

Since

$$\mathbf{A}\mathbf{x} = \lambda\mathbf{B}\mathbf{x} \quad \Leftrightarrow \quad (\mathbf{A} - \lambda\mathbf{B})\mathbf{x} = \mathbf{0}, \tag{14.6}$$

$\mathbf{A}\mathbf{x} = \lambda\mathbf{B}\mathbf{x}$ for some n-dimensional vector $\mathbf{x}$ if and only if $|\mathbf{A} - \lambda\mathbf{B}| = 0$. The problem of finding the roots of $|\mathbf{A} - \lambda\mathbf{B}|$ (i.e., the solutions for λ to the equation $|\mathbf{A} - \lambda\mathbf{B}| = 0$) is called the *generalized eigenvalue problem.*

Note that

$$|\mathbf{A} - \lambda\mathbf{B}| = |\mathbf{R}|^{-2}|\mathbf{C} - \lambda\mathbf{I}| = |\mathbf{B}||\mathbf{A}\mathbf{B}^{-1} - \lambda\mathbf{I}| = |\mathbf{B}||\mathbf{B}^{-1}\mathbf{A} - \lambda\mathbf{I}|, \tag{14.7}$$

so that the generalized eigenvalue problem is equivalent to the problem of finding the eigenvalues of the symmetric matrix $\mathbf{C}$ or of the not-necessarily-symmetric matrix $\mathbf{A}\mathbf{B}^{-1}$ or $\mathbf{B}^{-1}\mathbf{A}$. Note also that, in the special case where $\mathbf{B} = \mathbf{I}$, the generalized eigenvalue problem reduces to the problem of finding the eigenvalues of $\mathbf{A}$.

21.15 Differentiation of Eigenvalues and Eigenvectors

The following lemma provides an appropriate context for our discussion of the differentiation of eigenvalues and eigenvectors.

Lemma 21.15.1. Let $\mathbf{F}$ represent a $p \times p$ symmetric matrix of functions, defined on a set S, of a vector $\mathbf{x} = (x_1, \ldots, x_m)'$ of m variables. Further, let $\mathbf{c}$ represent any interior point of S at which $\mathbf{F}$ is continuously differentiable. And let λ^* represent a simple eigenvalue of $\mathbf{F}(\mathbf{c})$ (i.e., an eigenvalue of multiplicity 1), and let $\mathbf{u}^* = (u_1^*, \ldots, u_p^*)'$ represent an eigenvector [of $\mathbf{F}(\mathbf{c})$] with (usual) norm 1 corresponding to λ^*. Then, for some neighborhood N of $\mathbf{c}$, there exists a function λ (of $\mathbf{x}$), defined on N, and a vector $\mathbf{u} = (u_1, \ldots, u_p)'$ of functions (of $\mathbf{x}$), defined on N, with the following properties: (1) $\lambda(\mathbf{c}) = \lambda^*$ and $\mathbf{u}(\mathbf{c}) = \mathbf{u}^*$; (2) for $\mathbf{x} \in N$, $\lambda(\mathbf{x})$ is an eigenvalue of $\mathbf{F}(\mathbf{x})$, and $\mathbf{u}(\mathbf{x})$ is an eigenvector [of $\mathbf{F}(\mathbf{x})$] with (usual) norm 1 corresponding to $\lambda(\mathbf{x})$ [and the signs of the nonzero elements of $\mathbf{u}(\mathbf{x})$ are the same as those of the corresponding elements of $\mathbf{u}^*$]; and (3) λ and $\mathbf{u}$ are continuously differentiable at $\mathbf{c}$.

Proof. Regarding (for the moment) λ and $\mathbf{u}$ as a variable and a vector of variables, define (for $\mathbf{x} \in S$, $\lambda \in \mathcal{R}$, and $\mathbf{u} \in \mathcal{R}^p$)

$$\mathbf{h}(\mathbf{u}, \lambda; \mathbf{x}) = \begin{bmatrix} (\mathbf{F}(\mathbf{x}) - \lambda \mathbf{I})\mathbf{u} \\ \mathbf{u}'\mathbf{u} - 1 \end{bmatrix}.$$

Further, let

$$\mathbf{J}(\mathbf{u}, \lambda; \mathbf{x}) = \left[\frac{\partial \mathbf{h}(\mathbf{u}, \lambda; \mathbf{x})}{\partial \mathbf{u}'}, \ \frac{\partial \mathbf{h}(\mathbf{u}, \lambda; \mathbf{x})}{\partial \lambda}\right].$$

And observe [in light of results (15.3.10) and (15.3.7)] that

$$\mathbf{J}(\mathbf{u}, \lambda; \mathbf{x}) = \begin{pmatrix} \mathbf{F}(\mathbf{x}) - \lambda \mathbf{I} & -\mathbf{u} \\ 2\mathbf{u}' & 0 \end{pmatrix}.$$

Now, since λ^* is a simple eigenvalue of $\mathbf{F}(\mathbf{c})$,

$$\text{rank}[\mathbf{F}(\mathbf{c}) - \lambda^* \mathbf{I}] = p - 1.$$

And, since $[\mathbf{F}(\mathbf{c}) - \lambda^* \mathbf{I}]'\mathbf{u}^* = [\mathbf{F}(\mathbf{c}) - \lambda^* \mathbf{I}]\mathbf{u}^* = \mathbf{0}$, $\mathcal{C}(\mathbf{u}^*)$ is (in light of Corollary 12.1.2) orthogonal to $\mathcal{C}[\mathbf{F}(\mathbf{c}) - \lambda^* \mathbf{I}]$, implying (in light of Lemma 17.1.9) that $\mathcal{C}(\mathbf{u}^*)$ and $\mathcal{C}[\mathbf{F}(\mathbf{c}) - \lambda^* \mathbf{I}]$ are essentially disjoint and also (in light of Lemma 17.2.1) that $\mathcal{R}(\mathbf{u}^{*\prime})$ and $\mathcal{R}[\mathbf{F}(\mathbf{c}) - \lambda^* \mathbf{I}]$ are essentially disjoint. Thus, making use of Theorem 17.2.19, we find that

$$\begin{aligned} \text{rank}[\mathbf{J}(\mathbf{u}^*, \lambda^*; \mathbf{c})] &= \text{rank}[\mathbf{F}(\mathbf{c}) - \lambda^* \mathbf{I}] + \text{rank}(-\mathbf{u}^*) + \text{rank}(2\mathbf{u}^{*\prime}) \\ &= p - 1 + 1 + 1 = p + 1. \end{aligned}$$

Moreover,

$$\mathbf{h}(\mathbf{u}^*, \lambda^*; \mathbf{c}) = \mathbf{0},$$

and $\mathbf{h}$ is continuously differentiable at the point $(\mathbf{u}^*, \lambda^*; \mathbf{c})$.

The upshot of these results is that the conditions of the implicit function theorem (e.g., Magnus and Neudecker 1988, ch. 7, thm. A.2; Bartle 1976, p. 384) are satisfied. And it follows from this theorem that, for some neighborhood N^* of $\mathbf{c}$, there exist a function λ (of $\mathbf{x}$), defined on N^*, and a vector $\mathbf{u} = (u_1, \ldots, u_p)'$ of functions (of $\mathbf{x}$), defined on N^*, with the following properties: (1) $\lambda(\mathbf{c}) = \lambda^*$ and $\mathbf{u}(\mathbf{c}) = \mathbf{u}^*$; (2) for $\mathbf{x} \in N^*$, $\lambda(\mathbf{x})$ is an eigenvalue of $\mathbf{F}(\mathbf{x})$, and $\mathbf{u}(\mathbf{x})$ is an eigenvector [of $\mathbf{F}(\mathbf{x})$] with (usual) norm 1 corresponding to $\lambda(\mathbf{x})$; and (3) λ and $\mathbf{u}$ are continuously differentiable at $\mathbf{c}$. Moreover, in light of Lemma 15.1.1, the elements of $\mathbf{u}$ are continuous at $\mathbf{c}$, so that there exists some neighborhood N of $\mathbf{c}$ (where $N \subset N^*$) such that the signs of the nonzero elements of $\mathbf{u}(\mathbf{x})$ are the same as those of the corresponding elements of $\mathbf{u}^*$. Q.E.D.

Let us now consider the differentiation of the functions $\lambda, u_1, \ldots, u_n$ described in Lemma 21.15.1.

For $\mathbf{x} \in N$, we have that

$$\mathbf{Fu} = \lambda\mathbf{u}.$$

Upon differentiating both sides of this equality with respect to x_j, we find that (at $\mathbf{x} = \mathbf{c}$)

$$\mathbf{F}\frac{\partial \mathbf{u}}{\partial x_j} + \frac{\partial \mathbf{F}}{\partial x_j}\mathbf{u} = \lambda\frac{\partial \mathbf{u}}{\partial x_j} + \frac{\partial \lambda}{\partial x_j}\mathbf{u} \tag{15.1}$$

($j = 1, \ldots, m$). And, upon premultiplying both sides of equality (15.1) by $\mathbf{u}'$, we obtain

$$\mathbf{u}'\mathbf{F}\frac{\partial \mathbf{u}}{\partial x_j} + \mathbf{u}'\frac{\partial \mathbf{F}}{\partial x_j}\mathbf{u} = \lambda\mathbf{u}'\frac{\partial \mathbf{u}}{\partial x_j} + \frac{\partial \lambda}{\partial x_j}\mathbf{u}'\mathbf{u},$$

or, equivalently [since $\mathbf{u}'\mathbf{F} = (\mathbf{Fu})' = \lambda\mathbf{u}'$ and $\mathbf{u}'\mathbf{u} = 1$],

$$\lambda\mathbf{u}'\frac{\partial \mathbf{u}}{\partial x_j} + \mathbf{u}'\frac{\partial \mathbf{F}}{\partial x_j}\mathbf{u} = \lambda\mathbf{u}'\frac{\partial \mathbf{u}}{\partial x_j} + \frac{\partial \lambda}{\partial x_j}.$$

And it follows that (at $\mathbf{x} = \mathbf{c}$)

$$\frac{\partial \lambda}{\partial x_j} = \mathbf{u}'\frac{\partial \mathbf{F}}{\partial x_j}\mathbf{u} \tag{15.2}$$

($j = 1, \ldots, m$).

As a formula for the jth partial derivative of $\mathbf{u}$, we have that (at $\mathbf{x} = \mathbf{c}$)

$$\frac{\partial \mathbf{u}}{\partial x_j} = -\mathbf{H}^{+}\frac{\partial \mathbf{F}}{\partial x_j}\mathbf{u}, \tag{15.3}$$

where $\mathbf{H} = \mathbf{F} - \lambda\mathbf{I}$.

To verify formula (15.3), rewrite equality (15.1) as

$$\mathbf{H}\frac{\partial \mathbf{u}}{\partial x_j} = \frac{\partial \lambda}{\partial x_j}\mathbf{u} - \frac{\partial \mathbf{F}}{\partial x_j}\mathbf{u}. \tag{15.4}$$

Premultiplying both sides of equality (15.4) by $\mathbf{H}^{+}$, we obtain

$$\mathbf{H}^+\mathbf{H}\frac{\partial \mathbf{u}}{\partial x_j} = \frac{\partial \lambda}{\partial x_j}\mathbf{H}^+\mathbf{u} - \mathbf{H}^+\frac{\partial \mathbf{F}}{\partial x_j}\mathbf{u}. \tag{15.5}$$

Since $\mathbf{H}^+\mathbf{H} = (\mathbf{H}^+\mathbf{H})' = \mathbf{H}(\mathbf{H}^+)' = \mathbf{H}\mathbf{H}^+$ and since [observing that $\mathbf{H}\mathbf{u} = \mathbf{0}$ and recalling (from Theorem 20.5.1) that $\mathcal{N}(\mathbf{H}^+) = \mathcal{N}(\mathbf{H})$] $\mathbf{H}^+\mathbf{u} = \mathbf{0}$, equality (15.5) can be rewritten as

$$\mathbf{H}\mathbf{H}^+\frac{\partial \mathbf{u}}{\partial x_j} = -\mathbf{H}^+\frac{\partial \mathbf{F}}{\partial x_j}\mathbf{u}.$$

To complete the verification of formula (15.3), we need to show that (at $\mathbf{x} = \mathbf{c}$)

$$\mathbf{H}\mathbf{H}^+\frac{\partial \mathbf{u}}{\partial x_j} = \frac{\partial \mathbf{u}}{\partial x_j}. \tag{15.6}$$

For this purpose, observe that

$$\left(\frac{\partial \mathbf{u}}{\partial x_j}\right)'\mathbf{u} + \mathbf{u}'\frac{\partial \mathbf{u}}{\partial x_j} = 0$$

(as is clear upon differentiating both sides of the equality $\mathbf{u}'\mathbf{u} = 1$) and hence $\{$since $(\partial \mathbf{u}/\partial x_j)'\mathbf{u} = [(\partial \mathbf{u}/\partial x_j)'\mathbf{u}]' = \mathbf{u}'(\partial \mathbf{u}/\partial x_j)\}$ that

$$\mathbf{u}'\frac{\partial \mathbf{u}}{\partial x_j} = 0.$$

Thus,

$$\mathbf{u}'\left(\mathbf{H},\ \frac{\partial \mathbf{u}}{\partial x_j}\right) = \mathbf{0}',$$

implying that the p rows of the matrix $(\mathbf{H},\ \partial \mathbf{u}/\partial x_j)$ are linearly dependent and hence that

$$\text{rank}\left(\mathbf{H},\ \frac{\partial \mathbf{u}}{\partial x_j}\right) \le p - 1. \tag{15.7}$$

Moreover, at $\mathbf{x} = \mathbf{c}$,

$$p - 1 = \text{rank}(\mathbf{H}) \le \text{rank}\left(\mathbf{H},\ \frac{\partial \mathbf{u}}{\partial x_j}\right). \tag{15.8}$$

Together, results (15.7) and (15.8) imply that (at $\mathbf{x} = \mathbf{c}$)

$$\text{rank}\left(\mathbf{H},\ \frac{\partial \mathbf{u}}{\partial x_j}\right) = \text{rank}(\mathbf{H}).$$

It follows (in light of Corollary 4.5.2) that $\partial \mathbf{u}/\partial x_j \in \mathcal{C}(\mathbf{H})$, which (upon applying Corollary 9.3.6) gives result (15.6).

Let us now consider formulas (15.2) and (15.3) in the special case where $\mathbf{x}$ is an $[m(m+1)/2]$-dimensional column vector whose elements consist of those elements of an $m \times m$ symmetric matrix $\mathbf{X} = \{x_{ij}\}$ that lie on or below the diagonal and where $\mathbf{F}(\mathbf{x}) = \mathbf{X}$. Assume that $S = \mathcal{R}^{m(m+1)/2}$ (i.e., that there are no restrictions on $\mathbf{X}$ other than symmetry), observe that (in this special case) the interior points of S at which $\mathbf{F}$ is continuously differentiable consist of all of S, and let $\mathbf{e}_j$ represent the jth column of $\mathbf{I}_m$.

Then, upon substituting from result (15.5.6) or (15.5.7) into equality (15.2), we find that

$$\frac{\partial \lambda}{\partial x_{ii}} = \mathbf{u}'\mathbf{e}_i\mathbf{e}_i'\mathbf{u} = u_i^2$$

and, for $j < i$,

$$\frac{\partial \lambda}{\partial x_{ij}} = \mathbf{u}'(\mathbf{e}_i\mathbf{e}_j' + \mathbf{e}_j\mathbf{e}_i')\mathbf{u} = u_i u_j + u_j u_i = 2u_i u_j.$$

Thus,

$$\frac{\partial \lambda}{\partial x_{ij}} = \begin{cases} u_i^2, & \text{if } j = i, \quad (15.9a) \\ 2u_i u_j, & \text{if } j < i. \quad (15.9b) \end{cases}$$

Or, equivalently,

$$\frac{\partial \lambda}{\partial \mathbf{X}} = 2\mathbf{u}\mathbf{u}' - \text{diag}(u_1^2, \ldots, u_m^2). \tag{15.10}$$

Further, let $\mathbf{g}_j$ represent the jth column of $\mathbf{H}^+$ and g_{ij} the ijth element of $\mathbf{H}^+$. Then, upon substituting from result (15.5.6) or (15.5.7) into equality (15.3), we find that

$$\frac{\partial \mathbf{u}}{\partial x_{ij}} = \begin{cases} -u_i \mathbf{g}_i, & \text{if } j = i, \quad (15.11a) \\ -(u_j \mathbf{g}_i + u_i \mathbf{g}_j), & \text{if } j < i. \quad (15.11b) \end{cases}$$

And the sth element of the (m-dimensional) vector $\partial \mathbf{u}/\partial x_{ij}$ is expressible as

$$\frac{\partial u_s}{\partial x_{ij}} = \begin{cases} -u_i g_{si}, & \text{if } j = i, \quad (15.12a) \\ -(u_j g_{si} + u_i g_{sj}), & \text{if } j < i. \quad (15.12b) \end{cases}$$

Further, letting $\mathbf{x}_j = (x_{j1}, \ldots, x_{j,j-1}, x_{jj}, x_{j+1,j}, \ldots, x_{mj})'$,

$$\frac{\partial \mathbf{u}}{\partial \mathbf{x}_j'} = -u_j\mathbf{H}^+ - \mathbf{g}_j\mathbf{u}' + u_j\mathbf{g}_j\mathbf{e}_j'. \tag{15.13}$$

21.16 An Equivalence (Involving Determinants and Polynomials)

The following theorem is of considerable importance in statistics (where it is used in establishing a fundamental result on the statistical independence of quadratic forms)—refer, for example, to Driscoll and Gundberg (1986) and Reid and Driscoll (1988).

Theorem 21.16.1. Let $\mathbf{A}$ and $\mathbf{B}$ represent $n \times n$ symmetric matrices, and let c and d represent (strictly) positive scalars. Then, a necessary and sufficient condition for

$$|\mathbf{I} - t\mathbf{A} - u\mathbf{B}| = |\mathbf{I} - t\mathbf{A}||\mathbf{I} - u\mathbf{B}| \tag{16.1}$$

for all (scalars) t and u satisfying $|t| < c$ and $|u| < d$ (or, equivalently, for all t and u) is that $\mathbf{AB} = \mathbf{0}$.

Proof. The sufficiency of the condition $\mathbf{AB} = \mathbf{0}$ is evident upon observing that

$$|\mathbf{I} - t\mathbf{A}||\mathbf{I} - u\mathbf{B}| = |(\mathbf{I} - t\mathbf{A})(\mathbf{I} - u\mathbf{B})| = |\mathbf{I} - t\mathbf{A} - u\mathbf{B} + tu\mathbf{AB}|.$$

Now, for purposes of establishing the necessity of this condition, suppose that equality (16.1) holds for all t and u satisfying $|t| < c$ and $|u| < d$. And let c^* ($\leq c$) represent a positive scalar such that $\mathbf{I} - t\mathbf{A}$ is positive definite whenever $|t| < c^*$—the existence of such a scalar is guaranteed by Corollary 18.3.2—and (for t satisfying $|t| < c^*$) let

$$\mathbf{H} = (\mathbf{I} - t\mathbf{A})^{-1}.$$

If $|t| < c^*$, then $|\mathbf{I} - u\mathbf{B}| = |\mathbf{I} - u\mathbf{BH}|$ and $|\mathbf{I} - (-u)\mathbf{B}| = |\mathbf{I} - (-u)\mathbf{BH}|$, implying that

$$|\mathbf{I} - u^2\mathbf{B}^2| = |\mathbf{I} - u^2(\mathbf{BH})^2|. \tag{16.2}$$

Since each side of equality (16.2) is a polynomial in u^2, we have that

$$|\mathbf{I} - r\mathbf{B}^2| = |\mathbf{I} - r(\mathbf{BH})^2|$$

for all r (and for t such that $|t| < c^*$)—refer to Theorem 21.A.5 (in the Appendix).

Applying result (15.8.5), we find that (if $|t| < c^*$)

$$\operatorname{tr}(\mathbf{B}^2) = \left.\frac{-\partial|\mathbf{I} - r\mathbf{B}^2|}{\partial r}\right|_{r=0} = \left.\frac{-\partial|\mathbf{I} - r(\mathbf{BH})^2|}{\partial r}\right|_{r=0} = \operatorname{tr}[(\mathbf{BH})^2].$$

Thus,

$$\frac{\partial^2 \operatorname{tr}[(\mathbf{BH})^2]}{\partial t^2} = \frac{\partial^2 \operatorname{tr}(\mathbf{B}^2)}{\partial t^2} = 0 \tag{16.3}$$

(for t such that $|t| < c^*$). And, using results (15.8.15), (15.9.2), (15.6.1), (15.4.8), and (5.2.3), we find that

$$\frac{\partial \mathbf{H}}{\partial t} = \mathbf{HAH}, \qquad \frac{\partial^2 \mathbf{H}}{\partial t^2} = 2\mathbf{HAHAH},$$

$$\frac{\partial \operatorname{tr}[(\mathbf{BH})^2]}{\partial t} = \operatorname{tr}\left[\frac{\partial (\mathbf{BH})^2}{\partial t}\right] = \operatorname{tr}\left[\mathbf{B}\frac{\partial \mathbf{H}}{\partial t}\mathbf{BH} + \mathbf{BHB}\frac{\partial \mathbf{H}}{\partial t}\right] = 2\operatorname{tr}\left[\mathbf{BHB}\frac{\partial \mathbf{H}}{\partial t}\right],$$

and

$$\frac{\partial^2 \operatorname{tr}[(\mathbf{BH})^2]}{\partial t^2} = 2\operatorname{tr}\left[\mathbf{B}\frac{\partial \mathbf{H}}{\partial t}\mathbf{B}\frac{\partial \mathbf{H}}{\partial t} + \mathbf{BHB}\frac{\partial^2 \mathbf{H}}{\partial t^2}\right]$$
$$= 2\operatorname{tr}[\mathbf{BHAHBHAH} + 2\mathbf{BHBHAHAH}]. \tag{16.4}$$

Combining results (16.4) and (16.3) and setting $t = 0$ gives

$$\begin{aligned}
0 &= \operatorname{tr}[(\mathbf{BA})^2] + 2\operatorname{tr}(\mathbf{B}^2\mathbf{A}^2)\\
&= \operatorname{tr}[(\mathbf{BA})^2 + \mathbf{B}^2\mathbf{A}^2] + \operatorname{tr}(\mathbf{B}^2\mathbf{A}^2)\\
&= \operatorname{tr}[\mathbf{B}(\mathbf{AB}+\mathbf{BA})\mathbf{A}] + \operatorname{tr}(\mathbf{B}^2\mathbf{A}^2)\\
&= (1/2)\operatorname{tr}[\mathbf{B}(\mathbf{AB}+\mathbf{BA})\mathbf{A}] + (1/2)\operatorname{tr}\{[\mathbf{B}(\mathbf{AB}+\mathbf{BA})\mathbf{A}]'\} + \operatorname{tr}(\mathbf{BA}^2\mathbf{B})\\
&= (1/2)\operatorname{tr}[(\mathbf{AB}+\mathbf{BA})\mathbf{AB}] + (1/2)\operatorname{tr}[\mathbf{A}(\mathbf{AB}+\mathbf{BA})'\mathbf{B}] + \operatorname{tr}(\mathbf{BA}^2\mathbf{B})\\
&= (1/2)\operatorname{tr}[(\mathbf{AB}+\mathbf{BA})'\mathbf{AB}] + (1/2)\operatorname{tr}[(\mathbf{AB}+\mathbf{BA})'\mathbf{BA}] + \operatorname{tr}(\mathbf{BA}^2\mathbf{B})\\
&= (1/2)\operatorname{tr}[(\mathbf{AB}+\mathbf{BA})'(\mathbf{AB}+\mathbf{BA})] + \operatorname{tr}[(\mathbf{AB})'\mathbf{AB}].
\end{aligned} \tag{16.5}$$

Both terms of expression (16.5) are nonnegative and hence equal to 0. Moreover, $\operatorname{tr}[(\mathbf{AB})'\mathbf{AB}] = 0$ implies that $\mathbf{AB} = \mathbf{0}$—refer to Lemma 5.3.1. Q.E.D.

In light of Theorem 18.3.1, we have the following variation on Theorem 21.16.1.

Corollary 21.16.2. Let $\mathbf{A}$ and $\mathbf{B}$ represent $n \times n$ symmetric matrices. Then, there exist (strictly) positive scalars c and d such that $\mathbf{I} - t\mathbf{A}$, $\mathbf{I} - u\mathbf{B}$, and $\mathbf{I} - t\mathbf{A} - u\mathbf{B}$ are positive definite for all (scalars) t and u satisfying $|t| < c$ and $|u| < d$. And a necessary and sufficient condition for

$$\log\left[\frac{|\mathbf{I} - t\mathbf{A} - u\mathbf{B}|}{|\mathbf{I} - t\mathbf{A}||\mathbf{I} - u\mathbf{B}|}\right] = 0$$

for all t and u satisfying $|t| < c$ and $|u| < d$ is that $\mathbf{AB} = \mathbf{0}$.

The following theorem (which, like Theorem 21.16.1, can be useful in establishing results on the statistical independence of quadratic forms) generalizes Corollary 21.16.2.

Theorem 21.16.3. Let $\mathbf{A}$ and $\mathbf{B}$ represent $n \times n$ symmetric matrices. Then, there exist (strictly) positive scalars c and d such that $\mathbf{I} - t\mathbf{A}$, $\mathbf{I} - u\mathbf{B}$, and $\mathbf{I} - t\mathbf{A} - u\mathbf{B}$ are positive definite for all (scalars) t and u satisfying $|t| < c$ and $|u| < d$. And, letting $h(t, u)$ represent a polynomial (in t and u), necessary and sufficient conditions for

$$\log\left[\frac{|\mathbf{I} - t\mathbf{A} - u\mathbf{B}|}{|\mathbf{I} - t\mathbf{A}||\mathbf{I} - u\mathbf{B}|}\right] = \frac{h(t, u)}{|\mathbf{I} - t\mathbf{A}||\mathbf{I} - u\mathbf{B}||\mathbf{I} - t\mathbf{A} - u\mathbf{B}|} \tag{16.6}$$

for all t and u satisfying $|t| < c$ and $|u| < d$ are that $\mathbf{AB} = \mathbf{0}$ and that, for all t and u satisfying $|t| < c$ and $|u| < d$ (or, equivalently, for all t and u), $h(t, u) = 0$.

As a preliminary to proving Theorem 21.16.3, it is convenient to establish the following result on polynomials.

Theorem 21.16.4. Let $r_1(x)$, $s_1(x)$, and $s_2(x)$ represent polynomials in a real variable x. And let

$$r_2(x) = \gamma(x - \lambda_1)^{m_1}(x - \lambda_2)^{m_2} \cdots (x - \lambda_k)^{m_k},$$

where k is a nonnegative integer; $m_1, m_2, \ldots, m_k$ are (strictly) positive integers; γ is a nonzero real number; and $\lambda_1, \lambda_2, \ldots, \lambda_k$ are real numbers. [When $k = 0$, $r_2(x) = \gamma$.] Suppose that

$$\log\left[\frac{s_1(x)}{s_2(x)}\right] = \frac{r_1(x)}{r_2(x)} \tag{16.7}$$

for all x in some nondegenerate interval I [that does not include $\lambda_1, \lambda_2, \ldots, \lambda_k$ or any roots of $s_2(x)$]. Then, there exists a real number α such that $r_1(x) = \alpha r_2(x)$ for all x. Further, $s_1(x) = e^{\alpha} s_2(x)$ for all x.

Proof (of Theorem 21.16.4). The proof makes considerable use of various basic properties of polynomials—refer to the Appendix for a review of the relevant properties.

Assume (without loss of generality) that $\lambda_1, \lambda_2, \ldots, \lambda_k$ are distinct and that $\lambda_1, \lambda_2, \ldots, \lambda_k$ are not roots of $r_1(x)$. [To see that the latter assumption can be made without loss of generality, suppose that λ_1 were a root of $r_1(x)$. Then, in light of Theorem 21.A.3, there would exist a polynomial $p_1(x)$ such that $r_1(x) = (x - \lambda_1)p_1(x)$, so that, for $x \in I$, $r_1(x)/r_2(x) = p_1(x)/p_2(x)$, with

$$p_2(x) = \gamma(x - \lambda_1)^{m_1 - 1}(x - \lambda_2)^{m_2} \cdots (x - \lambda_k)^{m_k}.$$

Moreover, to show that $r_1(x) = \alpha r_2(x)$ for some α, it would suffice to show that $p_1(x) = \alpha p_2(x)$ for some α. This process of canceling common factors could be continued until the right side of equality (16.7) were expressed as a ratio of two polynomials that have no common roots.] Assume also (again without loss of generality) that λ_i is not a root of both $s_1(x)$ and $s_2(x)$ $(i = 1, 2, \ldots, k)$.

Making frequent use of an abbreviated notation for polynomials (discussed in the Appendix) and letting $r_1^*(x)$, $r_2^*(x)$, $s_1^*(x)$, and $s_2^*(x)$ represent the derivatives of $r_1(x)$, $r_2(x)$, $s_1(x)$, and $s_2(x)$, respectively, we find [upon differentiating both sides of equality (16.7)] that

$$\frac{r_1^* r_2 - r_1 r_2^*}{r_2^2} = \frac{s_1^* s_2 - s_1 s_2^*}{s_1 s_2}$$

for $x \in I$. And, in light of Theorem 21.A.5, it follows that

$$(r_1^* r_2 - r_1 r_2^*) s_1 s_2 = r_2^2 (s_1^* s_2 - s_1 s_2^*) \tag{16.8}$$

for all x.

We now come to the heart of the proof, which consists of showing that equality (16.8) implies that $k = 0$ (i.e., that $r_2 \equiv \gamma$). Letting i represent an arbitrary one of the first m positive integers, we can rewrite $r_2(x)$ as

$$r_2(x) = (x - \lambda_i)^{m_i} t(x),$$

where $t(x) = \gamma \prod_{j \neq i} (x - \lambda_j)^{m_j}$. Then, letting $t^*(x)$ represent the derivative of $t(x)$, we can rewrite equality (16.8) as

$$[(x - \lambda_i)(r_1^* t - r_1 t^*) - m_i r_1 t](x - \lambda_i)^{m_i - 1} s_1 s_2 = (x - \lambda_i)^{2m_i} t^2 (s_1^* s_2 - s_1 s_2^*),$$

so that (for all x)

$$[(x - \lambda_i)(r_1^* t - r_1 t^*) - m_i r_1 t] s_1 s_2 = (x - \lambda_i)^{m_i + 1} t^2 (s_1^* s_2 - s_1 s_2^*). \quad (16.9)$$

Neither r_1 nor t has λ_i as a root, implying that $r_1 t$ does not have λ_i as a root and hence that

$$(x - \lambda_i)(r_1^* t - r_1 t^*) - m_i r_1 t$$

does not have λ_i as a root. Thus, in light of Theorem 21.A.4,

$$s_1 s_2 = (x - \lambda_i)^{m_i + 1} b$$

for some polynomial $b(x)$.

By assumption, λ_i is not a root of both s_1 and s_2, so that $(x - \lambda_i)^{m_i + 1}$ is a factor of either s_1 or s_2. Suppose that it is s_1 that has λ_i as a root, so that

$$s_1(x) = (x - \lambda_i)^{m_i + 1} d(x)$$

for some polynomial $d(x)$. Then, letting $d^*(x)$ represent the derivative of $d(x)$, the right side of equality (16.9) can be rewritten as the polynomial

$$(x - \lambda_i)^{m_i + 1} t^2 \{[(m_i + 1)(x - \lambda_i)^{m_i} d + (x - \lambda_i)^{m_i + 1} d^*] s_2 - (x - \lambda_i)^{m_i + 1} d s_2^*\},$$

which has $(x - \lambda_i)^{2m_i + 1}$ as a factor. It follows that $(x - \lambda_i)^{2m_i + 1}$ is a factor of $s_1 s_2$ and hence of s_1.

This argument can be repeated to establish that $(x - \lambda_i)^{3m_i + 1}$ is a factor of s_1. Further repetition reveals that, for an arbitrarily large integer n, $(x - \lambda_i)^{nm_i + 1}$ is a factor of s_1. Alternatively, if it is s_2 (rather than s_1) that has λ_i as a root, then we find that, for arbitrarily large n, $(x - \lambda_i)^{nm_i + 1}$ is a factor of s_2.

Thus, $k = 0$, that is, $r_2 \equiv \gamma$ (since otherwise we arrive at a contradiction of the fact that s_1 and s_2 are polynomials of fixed degree). And, upon substituting γ for r_2 in equality (16.8), we find that (for all x)

$$\gamma r_1^* s_1 s_2 = \gamma^2 (s_1^* s_2 - s_1 s_2^*). \quad (16.10)$$

Now, if r_1 were not a constant and if (for $x \in I$) s_1/s_2 were of constant value, then (for $x \in I$) the left side of equality (16.7) would be of constant value but the right side would not be of constant value. If r_1 were not a constant and if (for $x \in I$) s_1/s_2 were not of constant value (in which case s_1 or s_2 is not a constant), then the degree of the polynomial that forms the left side of equality (16.10) would exceed the degree of the polynomial that forms the right side. In either case, we arrive at a contradiction. Thus, r_1 is a constant; that is, $r_1(x) = \tau$ for some real number τ, and hence $r_1 \equiv (\tau/\gamma)r_2$. Further, $s_1 \equiv \exp(\tau/\gamma)s_2$ [since, for $x \in I$, $s_1/s_2 = \exp(r_1/r_2) = \exp(\tau/\gamma)$]. Q.E.D.

Proof (of Theorem 21.16.3). The sufficiency of these conditions is an immediate consequence of Corollary 21.16.2 [as is the existence of positive scalars c and d such that $\mathbf{I} - t\mathbf{A}$, $\mathbf{I} - u\mathbf{B}$, and $\mathbf{I} - t\mathbf{A} - u\mathbf{B}$ are positive definite for all t and u satisfying $|t| < c$ and $|u| < d$].

For purposes of establishing their necessity, take u to be an arbitrary scalar satisfying $|u| < d$, and observe (in light of Corollaries 14.2.11 and 14.3.13) that there exists an $n \times n$ nonsingular matrix $\mathbf{S}$ such that $(\mathbf{I} - u\mathbf{B})^{-1} = \mathbf{S}'\mathbf{S}$. And let $d_1, \ldots, d_r$ represent the nonzero (not necessarily distinct) eigenvalues of $\mathbf{A}$ and $f_1, \ldots, f_r$ the nonzero (not necessarily distinct) eigenvalues of $\mathbf{SAS}'$. Then, according to Corollary 21.5.9, there exist $n \times n$ orthogonal matrices $\mathbf{P}$ and $\mathbf{Q}$ such that $\mathbf{P}'\mathbf{AP} = \text{diag}(d_1, \ldots, d_r, 0, \ldots, 0)$ and $\mathbf{Q}'\mathbf{SAS}'\mathbf{Q} = \text{diag}(f_1, \ldots, f_r, 0, \ldots, 0)$. Thus, in light of result (13.3.9), we have that

$$\begin{aligned}|\mathbf{I} - t\mathbf{A}| = |\mathbf{P}(\mathbf{I} - t\mathbf{P}'\mathbf{AP})\mathbf{P}'| &= |\mathbf{P}||\text{diag}(1 - td_1, \ldots, 1 - td_r, 1, \ldots, 1)||\mathbf{P}'| \\ &= \prod_{i=1}^{r}(1 - td_i) \\ &= \prod_{i=1}^{r}(-d_i)(t - d_i^{-1})\end{aligned}$$

and that

$$\begin{aligned}|\mathbf{I} - t\mathbf{A} - u\mathbf{B}| &= |\mathbf{S}^{-1}(\mathbf{S}')^{-1} - t\mathbf{A}| \\ &= |\mathbf{S}^{-1}(\mathbf{I} - t\mathbf{SAS}')(\mathbf{S}')^{-1}| \\ &= |\mathbf{S}|^{-2}|\mathbf{I} - t\mathbf{SAS}'| \\ &= |\mathbf{S}|^{-2}|\mathbf{Q}(\mathbf{I} - t\mathbf{Q}'\mathbf{SAS}'\mathbf{Q})\mathbf{Q}'| \\ &= |\mathbf{S}|^{-2}|\mathbf{Q}||\text{diag}(1 - tf_1, \ldots, 1 - tf_r, 1, \ldots, 1)||\mathbf{Q}'| \\ &= |\mathbf{S}|^{-2}\prod_{i=1}^{r}(1 - tf_i) = |\mathbf{S}|^{-2}\prod_{i=1}^{r}(-f_i)(t - f_i^{-1}),\end{aligned}$$

so that (for fixed u) $|\mathbf{I} - t\mathbf{A} - u\mathbf{B}|$ and $|\mathbf{I} - t\mathbf{A}||\mathbf{I} - u\mathbf{B}|$ are polynomials in t. Further,

$$\begin{aligned}&|\mathbf{I} - t\mathbf{A}||\mathbf{I} - u\mathbf{B}||\mathbf{I} - t\mathbf{A} - u\mathbf{B}| \\ &\qquad = |\mathbf{I} - u\mathbf{B}||\mathbf{S}|^{-2}\prod_{i=1}^{r} d_i f_i (t - d_i^{-1})(t - f_i^{-1}), \qquad (16.11)\end{aligned}$$

which (for fixed u) is a polynomial in t of degree $2r$ with roots $d_1^{-1}, \ldots, d_r^{-1}$, $f_1^{-1}, \ldots, f_r^{-1}$.

Now, regarding u as fixed, suppose that equality (16.6) holds for all t satisfying $|t| < c$. Then, in light of equality (16.11), it follows from Theorem 21.16.4 that there exists a real number α such that, for all t,

$$h(t,u) = \alpha|\mathbf{I} - t\mathbf{A}||\mathbf{I} - u\mathbf{B}||\mathbf{I} - t\mathbf{A} - u\mathbf{B}|$$

and

$$|\mathbf{I} - t\mathbf{A} - u\mathbf{B}| = e^{\alpha}|\mathbf{I} - t\mathbf{A}||\mathbf{I} - u\mathbf{B}|. \tag{16.12}$$

[In applying Theorem 21.16.4, take $x = t$, $s_1(t) = |\mathbf{I} - t\mathbf{A} - u\mathbf{B}|$, $s_2(t) = |\mathbf{I} - t\mathbf{A}||\mathbf{I} - u\mathbf{B}|$, $r_1(t) = h(t,u)$, and $r_2(t) = |\mathbf{I} - t\mathbf{A}||\mathbf{I} - u\mathbf{B}||\mathbf{I} - t\mathbf{A} - u\mathbf{B}|$.] Moreover, upon setting $t = 0$ in equality (16.12), we find that

$$|\mathbf{I} - u\mathbf{B}| = e^{\alpha}|\mathbf{I} - u\mathbf{B}|,$$

implying that $e^{\alpha} = 1$ or, equivalently, that $\alpha = 0$. Thus, for all t, $h(t,u) = 0$ and $|\mathbf{I} - t\mathbf{A} - u\mathbf{B}| = |\mathbf{I} - t\mathbf{A}||\mathbf{I} - u\mathbf{B}|$.

We conclude that if equality (16.6) holds for all t and u satisfying $|t| < c$ and $|u| < d$, then $h(t,u) = 0$ and $|\mathbf{I} - t\mathbf{A} - u\mathbf{B}| = |\mathbf{I} - t\mathbf{A}||\mathbf{I} - u\mathbf{B}|$ for all t and u satisfying $|u| < d$, implying (in light of Theorem 21.A.5) that $h(t,u) = 0$ for all t and u and (in light of Theorem 21.16.1) that $\mathbf{AB} = \mathbf{0}$. Q.E.D.

Appendix: Some Properties of Polynomials (in a Single Variable)

An expression of the form

$$p(x) = a_0 + a_1 x + a_2 x^2 + \cdots + a_n x^n,$$

where x is a real variable, n is a nonnegative integer, and the coefficients a_0, a_1, $a_2, \ldots, a_n$ are real numbers, is referred to as a *polynomial* in x. The polynomial $p(x)$ is said to be *nonzero* if one or more of the coefficients $a_0, a_1, a_2, \ldots, a_n$ is nonzero, in which case the largest nonnegative integer k such that $a_k \neq 0$ is referred to as the *degree* of $p(x)$ and is denoted by the symbol $\deg[p(x)]$. When it causes no confusion, $p(x)$ may be abbreviated to p {and $\deg[p(x)]$ to $\deg(p)$}.

A polynomial q is said to be a *factor* of a polynomial p if there exists a polynomial r such that $p \equiv qr$. And a real number c is said to be a *root* (or a *zero*) of a polynomial p if $p(c) = 0$.

The following three theorems give some basic properties of polynomials—refer, for example, to Beaumont and Pierce (1963) for proofs of these theorems, which are equivalent to their Theorems 9-3.3, 9-3.5, and 9-7.5.

Theorem 21.A.1 (the division algorithm). Let p and q represent polynomials. Suppose that q is nonzero. Then, there exist unique polynomials b and r such that

$$p \equiv bq + r,$$

where either $r \equiv 0$ or else $\deg(r) < \deg(q)$.

Theorem 21.A.2. Let $p(x)$ represent a nonzero polynomial (in x) of degree n. Then, for any real number c, $p(x)$ has a unique representation of the form

$$p(x) = b_0 + b_1(x - c) + b_2(x - c)^2 + \cdots + b_n(x - c)^n,$$

where $b_0, b_1, b_2, \ldots, b_n$ are real numbers.

Theorem 21.A.3. A real number c is a root of a polynomial $p(x)$ (in x) if and only if the polynomial $x - c$ is a factor of $p(x)$.

The following two theorems give some additional properties of polynomials.

Theorem 21.A.4. Let $p(x)$, $q(x)$, and $r(x)$ represent polynomials (in x). And suppose that

$$p(x)q(x) = (x - c)^m r(x), \tag{A.1}$$

where m is a positive integer and where c is a real number that is not a root of $p(x)$. Then, $(x - c)^m$ is a factor of $q(x)$; that is, there exists a polynomial $s(x)$ such that

$$q(x) = (x - c)^m s(x).$$

Proof. The proof consists of using mathematical induction to show that, for $k = 1, \ldots, m$, there exists a polynomial $s_k(x)$ such that $q(x) = (x - c)^k s_k(x)$. Clearly, c is a root of $q(x)$. Thus, it follows from Theorem 21.A.3 that there exists a polynomial $s_1(x)$ such that $q(x) = (x - c)^1 s_1(x)$.

Suppose now that there exists a polynomial $s_{k-1}(x)$ (where $2 \le k \le m$) such that

$$q(x) = (x - c)^{k-1} s_{k-1}(x). \tag{A.2}$$

Then, substituting expression (A.2) into expression (A.1), we find that

$$(x - c)^{k-1} p(x) s_{k-1}(x) = (x - c)^m r(x), \tag{A.3}$$

implying that

$$p(x) s_{k-1}(x) = (x - c)^{m-k+1} r(x)$$

for $x \neq c$ and hence that

$$\begin{aligned} p(c) s_{k-1}(c) = \lim_{x \to c} p(x) s_{k-1}(x) &= \lim_{x \to c} (x - c)^{m-k+1} r(x) \\ &= (c - c)^{m-k+1} r(c) = 0. \end{aligned} \tag{A.4}$$

Equality (A.4) implies that c is a root of $s_{k-1}(x)$. Thus, there exists a polynomial $s_k(x)$ such that

$$s_{k-1}(x) = (x - c) s_k(x). \tag{A.5}$$

And, substituting expression (A.5) into expression (A.2), we find that

$$q(x) = (x - c)^k s_k(x),$$

thereby completing the induction argument. Q.E.D.

Theorem 21.A.5. Let $p(x)$ and $q(x)$ represent polynomials (in x). And suppose that $p(x) = q(x)$ for all x in some nondegenerate interval I. Then, $p(x) = q(x)$ for all x.

Proof. Let c represent an interior point of I, and observe (in light of Theorem 21.A.2) that $p(x)$ and $q(x)$ are expressible as

$$p(x) = a_0 + a_1(x-c) + a_2(x-c)^2 + \cdots + a_n(x-c)^n,$$
$$q(x) = b_0 + b_1(x-c) + b_2(x-c)^2 + \cdots + b_n(x-c)^n,$$

respectively, where n is a nonnegative integer [with $n \geq \deg(p)$ and $n \geq \deg(q)$] and $a_0, a_1, a_2, \ldots, a_n$ and $b_0, b_1, b_2, \ldots, b_n$ are real numbers. And we find that, for $k = 1, \ldots, n$,

$$\frac{\partial^k p(x)}{\partial x^k} = \sum_{m=k}^{n} m(m-1)\cdots(m-k+1)\, a_m(x-c)^{m-k},$$
$$\frac{\partial^k q(x)}{\partial x^k} = \sum_{m=k}^{n} m(m-1)\cdots(m-k+1)\, b_m(x-c)^{m-k}.$$

Thus,

$$b_0 = q(c) = p(c) = a_0,$$

and, for $k = 1, \ldots, n$,

$$k!\, b_k = \left.\frac{\partial^k q(x)}{\partial x^k}\right|_{x=c} = \left.\frac{\partial^k p(x)}{\partial x^k}\right|_{x=c} = k!\, a_k.$$

We conclude that $b_k = a_k$ for $k = 0, 1, \ldots, n$ and that $p(x) = q(x)$ for all x. Q.E.D.

Exercises

Section 21.1

1. Show that an $n \times n$ skew-symmetric matrix $\mathbf{A}$ has no nonzero eigenvalues.

2. Let $\mathbf{A}$ represent an $n \times n$ matrix, $\mathbf{B}$ a $k \times k$ matrix, and $\mathbf{X}$ an $n \times k$ matrix such that $\mathbf{AX} = \mathbf{XB}$.

   (a) Show that $\mathcal{C}(\mathbf{X})$ is an invariant subspace (of $\mathcal{R}^{n\times 1}$) relative to $\mathbf{A}$.

   (b) Show that if $\mathbf{X}$ is of full column rank, then every eigenvalue of $\mathbf{B}$ is an eigenvalue of $\mathbf{A}$.

3. Let $p(\lambda)$ represent the characteristic polynomial of an $n \times n$ matrix $\mathbf{A}$, and let $c_0, c_1, c_2, \ldots, c_n$ represent the respective coefficients of the characteristic polynomial, so that

$$p(\lambda) = c_0\lambda^0 + c_1\lambda + c_2\lambda^2 + \cdots + c_n\lambda^n = \sum_{s=0}^{n} c_s\lambda^s$$

(for $\lambda \in \mathcal{R}$). Further, let $\mathbf{P}$ represent the $n \times n$ matrix obtained from $p(\lambda)$ by formally replacing the scalar λ with the $n \times n$ matrix $\mathbf{A}$ (and by setting $\mathbf{A}^0 = \mathbf{I}_n$). That is, let

$$\mathbf{P} = c_0\mathbf{I} + c_1\mathbf{A} + c_2\mathbf{A}^2 + \cdots + c_n\mathbf{A}^n = \sum_{s=0}^{n} c_s\mathbf{A}^s.$$

Show that $\mathbf{P} = \mathbf{0}$ (a result that is known as the Cayley-Hamilton theorem) by carrying out the following four steps.

(a) Letting $\mathbf{B}(\lambda) = \mathbf{A} - \lambda\mathbf{I}$ and letting $\mathbf{H}(\lambda)$ represent the adjoint matrix of $\mathbf{B}(\lambda)$, show that (for $\lambda \in \mathcal{R}$)

$$\mathbf{H}(\lambda) = \mathbf{K}_0 + \lambda\mathbf{K}_1 + \lambda^2\mathbf{K}_2 + \cdots + \lambda^{n-1}\mathbf{K}_{n-1},$$

where $\mathbf{K}_0, \mathbf{K}_1, \mathbf{K}_2, \ldots, \mathbf{K}_{n-1}$ are $n \times n$ matrices (that do not vary with λ).

(b) Letting $\mathbf{T}_0 = \mathbf{A}\mathbf{K}_0$, $\mathbf{T}_n = -\mathbf{K}_{n-1}$, and (for $s = 1, \ldots, n-1$) $\mathbf{T}_s = \mathbf{A}\mathbf{K}_s - \mathbf{K}_{s-1}$, show that (for $\lambda \in \mathcal{R}$)

$$\mathbf{T}_0 + \lambda\mathbf{T}_1 + \lambda^2\mathbf{T}_2 + \cdots + \lambda^n\mathbf{T}_n = p(\lambda)\mathbf{I}_n.$$

[*Hint.* It follows from Theorem 13.5.3 that (for $\lambda \in \mathcal{R}$) $\mathbf{B}(\lambda)\mathbf{H}(\lambda) = |\mathbf{B}(\lambda)|\mathbf{I}_n = p(\lambda)\mathbf{I}_n$.]

(c) Show that, for $s = 0, 1, \ldots, n$, $\mathbf{T}_s = c_s\mathbf{I}$.

(d) Show that

$$\mathbf{P} = \mathbf{T}_0 + \mathbf{A}\mathbf{T}_1 + \mathbf{A}^2\mathbf{T}_2 + \cdots + \mathbf{A}^n\mathbf{T}_n = \mathbf{0}.$$

4. Let $c_0, c_1, \ldots, c_{n-1}, c_n$ represent the respective coefficients of the characteristic polynomial $p(\lambda)$ of an $n \times n$ matrix $\mathbf{A}$ [so that $p(\lambda) = c_0 + c_1\lambda + \cdots + c_{n-1}\lambda^{n-1} + c_n\lambda^n$ (for $\lambda \in \mathcal{R}$)]. Using the result of Exercise 3 (the Cayley-Hamilton theorem), show that if $\mathbf{A}$ is nonsingular, then $c_0 \neq 0$, and

$$\mathbf{A}^{-1} = -(1/c_0)(c_1\mathbf{I} + c_2\mathbf{A} + \cdots + c_n\mathbf{A}^{n-1}).$$

Section 21.3

5. Show that if an $n \times n$ matrix $\mathbf{B}$ is similar to an $n \times n$ matrix $\mathbf{A}$, then (1) $\mathbf{B}^k$ is similar to $\mathbf{A}^k$ ($k = 2, 3, \ldots$) and (2) $\mathbf{B}'$ is similar to $\mathbf{A}'$.

6. Show that if an $n \times n$ matrix $\mathbf{B}$ is similar to an ($n \times n$) idempotent matrix, then $\mathbf{B}$ is idempotent.

7. Let $\mathbf{A} = \begin{pmatrix} 1 & 0 \\ 0 & 1 \end{pmatrix}$ and $\mathbf{B} = \begin{pmatrix} 1 & 1 \\ 0 & 1 \end{pmatrix}$. Show that $\mathbf{B}$ has the same rank, determinant, trace, and characteristic polynomial as $\mathbf{A}$, but that, nevertheless, $\mathbf{B}$ is not similar to $\mathbf{A}$.

8. Expand on the result of Exercise 7 by showing (for an arbitrary positive integer n) that, for an $n \times n$ matrix $\mathbf{B}$ to be similar to an $n \times n$ matrix $\mathbf{A}$, it is not sufficient for $\mathbf{B}$ to have the same rank, determinant, trace, and characteristic polynomial as $\mathbf{A}$.

9. Let $\mathbf{A}$ represent an $n \times n$ matrix, $\mathbf{B}$ a $k \times k$ matrix, and $\mathbf{X}$ an $n \times k$ matrix such that $\mathbf{AX} = \mathbf{XB}$. Show that if $\mathbf{X}$ is of full column rank, then there exists an orthogonal matrix $\mathbf{Q}$ such that $\mathbf{Q}'\mathbf{AQ} = \begin{pmatrix} \mathbf{T}_{11} & \mathbf{T}_{12} \\ \mathbf{0} & \mathbf{T}_{22} \end{pmatrix}$, where $\mathbf{T}_{11}$ is a $k \times k$ matrix that is similar to $\mathbf{B}$.

10. Show that if 0 is an eigenvalue of an $n \times n$ matrix $\mathbf{A}$, then its algebraic multiplicity is greater than or equal to $n - \text{rank}(\mathbf{A})$.

11. Let $\mathbf{A}$ represent an $n \times n$ matrix. Show that if a scalar λ is an eigenvalue of $\mathbf{A}$ of algebraic multiplicity γ, then $\text{rank}(\mathbf{A} - \lambda\mathbf{I}) \geq n - \gamma$.

12. Let γ_1 represent the algebraic multiplicity and ν_1 the geometric multiplicity of 0 when 0 is regarded as an eigenvalue of an $n \times n$ (singular) matrix $\mathbf{A}$. And let γ_2 represent the algebraic multiplicity and ν_2 the geometric multiplicity of 0 when 0 is regarded as an eigenvalue of $\mathbf{A}^2$. Show that if $\nu_1 = \gamma_1$, then $\nu_2 = \gamma_2 = \nu_1$.

Section 21.4

13. Let $\mathbf{x}_1$ and $\mathbf{x}_2$ represent eigenvectors of an $n \times n$ matrix $\mathbf{A}$, and let c_1 and c_2 represent nonzero scalars. Under what circumstances is the vector $\mathbf{x} = c_1\mathbf{x}_1 + c_2\mathbf{x}_2$ an eigenvector of $\mathbf{A}$?

Section 21.5

14. Let $\mathbf{A}$ represent an $n \times n$ matrix, and suppose that there exists an $n \times n$ nonsingular matrix $\mathbf{Q}$ such that $\mathbf{Q}^{-1}\mathbf{AQ} = \mathbf{D}$ for some diagonal matrix $\mathbf{D} = \{d_i\}$. Further, for $i = 1, \ldots, n$, let $\mathbf{r}_i'$ represent the ith row of $\mathbf{Q}^{-1}$. Show (a) that $\mathbf{A}'$ is diagonalized by $(\mathbf{Q}^{-1})'$, (b) that the diagonal elements of $\mathbf{D}$ are the (not necessarily distinct) eigenvalues of $\mathbf{A}'$, and (c) that $\mathbf{r}_1, \ldots, \mathbf{r}_n$ are eigenvectors of $\mathbf{A}'$ (with $\mathbf{r}_i$ corresponding to the eigenvalue d_i).

15. Show that if an $n \times n$ nonsingular matrix $\mathbf{A}$ is diagonalized by an $n \times n$ nonsingular matrix $\mathbf{Q}$, then $\mathbf{A}^{-1}$ is also diagonalized by $\mathbf{Q}$.

16. Let $\mathbf{A}$ represent an $n \times n$ matrix whose spectrum comprises k eigenvalues $\lambda_1, \ldots, \lambda_k$ with algebraic multiplicities $\gamma_1, \ldots, \gamma_k$, respectively, that sum to n. Show that $\mathbf{A}$ is diagonalizable if and only if, for $i = 1, \ldots, k$, $\text{rank}(\mathbf{A} - \lambda_i \mathbf{I}) = n - \gamma_i$.

17. Let $\mathbf{A}$ represent an $n \times n$ symmetric matrix with not-necessarily-distinct eigenvalues $d_1, \ldots, d_n$ that have been ordered so that $d_1 \le d_2 \le \cdots \le d_n$. And let $\mathbf{Q}$ represent an $n \times n$ orthogonal matrix such that $\mathbf{Q}'\mathbf{A}\mathbf{Q} = \text{diag}(d_1, \ldots, d_n)$—the existence of which is guaranteed by Corollary 21.5.9. Further, for $m = 2, \ldots, n-1$, define $S_m = \{\mathbf{x} \in \mathcal{R}^{n \times 1} : \mathbf{x} \neq \mathbf{0},\ \mathbf{Q}_m'\mathbf{x} = \mathbf{0}\}$ and $T_m = \{\mathbf{x} \in \mathcal{R}^{n \times 1} : \mathbf{x} \neq \mathbf{0},\ \mathbf{P}_m'\mathbf{x} = \mathbf{0}\}$, where $\mathbf{Q}_m = (\mathbf{q}_1, \ldots, \mathbf{q}_{m-1})$ and $\mathbf{P}_m = (\mathbf{q}_{m+1}, \ldots, \mathbf{q}_n)$. Show that, for $m = 2, \ldots, n-1$,

$$d_m = \min_{\mathbf{x} \in S_m} \frac{\mathbf{x}'\mathbf{A}\mathbf{x}}{\mathbf{x}'\mathbf{x}} = \max_{\mathbf{x} \in T_m} \frac{\mathbf{x}'\mathbf{A}\mathbf{x}}{\mathbf{x}'\mathbf{x}}.$$

18. Let $\mathbf{A}$ represent an $n \times n$ symmetric matrix, and adopt the same notation as in Section 21.5f.

 (a) Show that the matrices $\mathbf{E}_1, \ldots, \mathbf{E}_k$, which appear in the spectral decomposition (5.5) of $\mathbf{A}$, have the following properties:

 (1) $\mathbf{E}_1 + \cdots + \mathbf{E}_k = \mathbf{I}$;

 (2) $\mathbf{E}_1, \ldots, \mathbf{E}_k$ are nonnull, symmetric, and idempotent;

 (3) for $t \neq j = 1, \ldots, k$, $\mathbf{E}_t\mathbf{E}_j = \mathbf{0}$; and

 (4) $\text{rank}(\mathbf{E}_1) + \cdots + \text{rank}(\mathbf{E}_k) = n$.

 (b) Take $\mathbf{F}_1, \ldots, \mathbf{F}_r$ to be $n \times n$ nonnull idempotent matrices such that $\mathbf{F}_1 + \cdots + \mathbf{F}_r = \mathbf{I}$. And suppose that, for some distinct scalars $\tau_1, \ldots, \tau_r$,

$$\mathbf{A} = \tau_1\mathbf{F}_1 + \cdots + \tau_r\mathbf{F}_r.$$

 Show that $r = k$ and that there exists a permutation $t_1, \ldots, t_r$ of the first r positive integers such that (for $j = 1, \ldots, r$) $\tau_j = \lambda_{t_j}$ and $\mathbf{F}_j = \mathbf{E}_{t_j}$.

19. Let $\mathbf{A}$ represent an $n \times n$ symmetric matrix, and let $d_1, \ldots, d_n$ represent the (not necessarily distinct) eigenvalues of $\mathbf{A}$. Show that $\lim_{k \to \infty} \mathbf{A}^k = \mathbf{0}$ if and only if, for $i = 1, \ldots, n$, $|d_i| < 1$.

Section 21.7

20. (a) Show that if 0 is an eigenvalue of an $n \times n$ not-necessarily-symmetric matrix $\mathbf{A}$, then it is also an eigenvalue of $\mathbf{A}^+$ and that the geometric multiplicity of 0 is the same when it is regarded as an eigenvalue of $\mathbf{A}^+$ as when it is regarded as an eigenvalue of $\mathbf{A}$.

 (b) Show (via an example) that the reciprocals of the nonzero eigenvalues of a square nonsymmetric matrix $\mathbf{A}$ are not necessarily eigenvalues of $\mathbf{A}^+$.

Section 21.8

21. Show that, for any positive integer n that is divisible by 2, there exists an $n \times n$ orthogonal matrix that has no eigenvalues. [*Hint*. Find a 2×2 orthogonal matrix $\mathbf{Q}$ that has no eigenvalues, and then consider the block-diagonal matrix $\text{diag}(\mathbf{Q}, \mathbf{Q}, \ldots, \mathbf{Q})$.]

22. Let $\mathbf{Q}$ represent an $n \times n$ orthogonal matrix, and let $p(\lambda)$ represent the characteristic polynomial of $\mathbf{Q}$. Show that (for $\lambda \neq 0$)

$$p(\lambda) = \pm\lambda^n p(1/\lambda).$$

23. Let $\mathbf{A}$ represent an $n\times n$ matrix, and suppose that the scalar 1 is an eigenvalue of $\mathbf{A}$ of geometric multiplicity ν. Show that $\nu \leq \text{rank}(\mathbf{A})$ and that if $\nu = \text{rank}(\mathbf{A})$, then $\mathbf{A}$ is idempotent.

Section 21.10

24. Let $\mathbf{A}$ represent an $n\times n$ nonsingular matrix. And let λ represent an eigenvalue of $\mathbf{A}$, and $\mathbf{x}$ represent an eigenvector of $\mathbf{A}$ corresponding to λ. Show that $|\mathbf{A}|/\lambda$ is an eigenvalue of $\text{adj}(\mathbf{A})$ and that $\mathbf{x}$ is an eigenvector of $\text{adj}(\mathbf{A})$ corresponding to $|\mathbf{A}|/\lambda$.

25. Let $\mathbf{A}$ represent an $n \times n$ matrix, and let $p(\lambda)$ represent the characteristic polynomial of $\mathbf{A}$. And let $\lambda_1, \ldots, \lambda_k$ represent the distinct eigenvalues of $\mathbf{A}$, and $\gamma_1, \ldots, \gamma_k$ represent their respective algebraic multiplicities, so that (for all λ)

$$p(\lambda) = (-1)^n q(\lambda) \prod_{j=1}^{k} (\lambda - \lambda_j)^{\gamma_j}$$

for some polynomial $q(\lambda)$ (of degree $n - \sum_{j=1}^{k} \gamma_j$) that has no real roots. Further, define $\mathbf{B} = \mathbf{A} - \lambda_1 \mathbf{U}\mathbf{V}'$, where $\mathbf{U} = (\mathbf{u}_1, \ldots, \mathbf{u}_{\gamma_1})$ is an $n \times \gamma_1$ matrix whose columns $\mathbf{u}_1, \ldots, \mathbf{u}_{\gamma_1}$ are (not necessarily linearly independent) eigenvectors of $\mathbf{A}$ corresponding to λ_1 and where $\mathbf{V} = (\mathbf{v}_1, \ldots, \mathbf{v}_{\gamma_1})$ is an $n \times \gamma_1$ matrix such that $\mathbf{V}'\mathbf{U}$ is diagonal.

(a) Show that the characteristic polynomial, say $r(\lambda)$, of $\mathbf{B}$ is such that (for all λ)

$$r(\lambda) = (-1)^n q(\lambda) \prod_{i=1}^{\gamma_1} [\lambda - (1 - \mathbf{v}_i'\mathbf{u}_i)\lambda_1] \prod_{j=2}^{k} (\lambda - \lambda_j)^{\gamma_j}. \qquad \text{(E.1)}$$

[*Hint*. Since the left and right sides of equality (E.1) are polynomials (in λ), it suffices to show that they are equal for all λ other than $\lambda_1, \ldots, \lambda_k$.]

(b) Show that in the special case where $\mathbf{U}'\mathbf{V} = c\mathbf{I}$ for some nonzero scalar c, the distinct eigenvalues of $\mathbf{B}$ are either $\lambda_2, \ldots, \lambda_{s-1}, \lambda_s, \lambda_{s+1}, \ldots, \lambda_k$ with algebraic multiplicities $\gamma_2, \ldots, \gamma_{s-1}, \gamma_s + \gamma_1, \gamma_{s+1}, \ldots, \gamma_k$, respectively, or $(1-c)\lambda_1, \lambda_2, \ldots, \lambda_k$ with algebraic multiplicities $\gamma_1, \gamma_2, \ldots, \gamma_k$, respectively [depending on whether or not $(1-c)\lambda_1 = \lambda_s$ for some s $(2 \le s \le k)$].

(c) Show that, in the special case where $\gamma_1 = 1$, (1) $\mathbf{u}_1$ is an eigenvector of $\mathbf{B}$ corresponding to the eigenvalue $(1 - \mathbf{v}_1'\mathbf{u}_1)\lambda_1$ and (2) for any eigenvector $\mathbf{x}$ of $\mathbf{B}$ corresponding to an eigenvalue λ [other than $(1 - \mathbf{v}_1'\mathbf{u}_1)\lambda_1$], the vector

$$\mathbf{x} - \lambda_1(\lambda_1 - \lambda)^{-1}(\mathbf{v}_1'\mathbf{x})\mathbf{u}_1$$

is an eigenvector of $\mathbf{A}$ corresponding to λ.

Section 21.12

26. Let $\mathbf{A}$ represent an $m \times n$ matrix of rank r. And take $\mathbf{P}$ to be any $m \times m$ orthogonal matrix and $\mathbf{D}_1$ to be any $r \times r$ nonsingular diagonal matrix such that

$$\mathbf{P}'\mathbf{A}\mathbf{A}'\mathbf{P} = \begin{pmatrix} \mathbf{D}_1^2 & \mathbf{0} \\ \mathbf{0} & \mathbf{0} \end{pmatrix}.$$

Further, partition $\mathbf{P}$ as $\mathbf{P} = (\mathbf{P}_1, \mathbf{P}_2)$, where $\mathbf{P}_1$ has r columns, and let $\mathbf{Q} = (\mathbf{Q}_1, \mathbf{Q}_2)$, where $\mathbf{Q}_1 = \mathbf{A}'\mathbf{P}_1\mathbf{D}_1^{-1}$ and where $\mathbf{Q}_2$ is any $n \times (n-r)$ matrix such that $\mathbf{Q}_1'\mathbf{Q}_2 = \mathbf{0}$. Show that

$$\mathbf{P}'\mathbf{A}\mathbf{Q} = \begin{pmatrix} \mathbf{D}_1 & \mathbf{0} \\ \mathbf{0} & \mathbf{0} \end{pmatrix}.$$

27. Show that the matrices $\mathbf{U}_1, \ldots, \mathbf{U}_k$, which appear in decomposition (12.7), are such that $\mathbf{U}_j\mathbf{U}_j'\mathbf{U}_j = \mathbf{U}_j$ (for $j = 1, \ldots, k$) and $\mathbf{U}_t'\mathbf{U}_j = \mathbf{0}$ and $\mathbf{U}_t\mathbf{U}_j' = \mathbf{0}$ (for $t \neq j = 1, \ldots, k$).

28. Let $\mathbf{A}$ represent an $m \times n$ matrix. And, as in Theorem 21.12.3, take $\mathbf{P}$ to be an $m \times m$ orthogonal matrix, $\mathbf{Q}$ an $n \times n$ orthogonal matrix, and $\mathbf{D}_1$ an $r \times r$ nonsingular diagonal matrix such that

$$\mathbf{P}'\mathbf{A}\mathbf{Q} = \begin{pmatrix} \mathbf{D}_1 & \mathbf{0} \\ \mathbf{0} & \mathbf{0} \end{pmatrix}.$$

Further, partition $\mathbf{P}$ and $\mathbf{Q}$ as $\mathbf{P} = (\mathbf{P}_1, \mathbf{P}_2)$ and $\mathbf{Q} = (\mathbf{Q}_1, \mathbf{Q}_2)$, where each of the matrices $\mathbf{P}_1$ and $\mathbf{Q}_1$ has r columns. Show that $\mathcal{C}(\mathbf{A}) = \mathcal{C}(\mathbf{P}_1)$ and that $\mathcal{N}(\mathbf{A}) = \mathcal{C}(\mathbf{Q}_2)$.

Section 21.13

29. Let $\mathbf{A}_1, \dots, \mathbf{A}_k$ represent $n \times n$ not-necessarily-symmetric matrices, each of which is diagonalizable. Show that if $\mathbf{A}_1, \dots, \mathbf{A}_k$ commute in pairs, then $\mathbf{A}_1, \dots, \mathbf{A}_k$ are simultaneously diagonalizable.

30. Let $\mathbf{V}$ represent an $n \times n$ symmetric nonnegative definite matrix, $\mathbf{X}$ an $n \times p$ matrix of rank r, and $\mathbf{d}$ a $p \times 1$ vector. Using the results of Exercise 19.11 (or otherwise), show that each of the following three conditions is necessary and sufficient for the vector $\mathbf{X}(\mathbf{X}'\mathbf{X})^{-}\mathbf{d}$ to be a solution, for every $\mathbf{d} \in \mathcal{C}(\mathbf{X}')$, to the problem of minimizing the quadratic form $\mathbf{a}'\mathbf{V}\mathbf{a}$ (in $\mathbf{a}$) subject to $\mathbf{X}'\mathbf{a} = \mathbf{d}$:

    (a) there exists an orthogonal matrix that simultaneously diagonalizes $\mathbf{V}$ and $\mathbf{P}_{\mathbf{X}}$;

    (b) there exists a subset of r orthonormal eigenvectors of $\mathbf{V}$ that is a basis for $\mathcal{C}(\mathbf{X})$;

    (c) there exists a subset of r eigenvectors of $\mathbf{V}$ that is a basis for $\mathcal{C}(\mathbf{X})$.

Section 21.14

31. Let $\mathbf{A}$ represent an $n \times n$ symmetric matrix, and let $\mathbf{B}$ represent an $n \times n$ symmetric positive definite matrix. And let $\lambda_{\max}$ and $\lambda_{\min}$ represent, respectively, the largest and smallest roots of $|\mathbf{A} - \lambda\mathbf{B}|$. Show that

$$\lambda_{\min} \leq \frac{\mathbf{x}'\mathbf{A}\mathbf{x}}{\mathbf{x}'\mathbf{B}\mathbf{x}} \leq \lambda_{\max}$$

    for every nonnull vector $\mathbf{x}$ in $\mathcal{R}^n$.

32. Let $\mathbf{A}$ represent an $n \times n$ symmetric matrix, and let $\mathbf{B}$ represent an $n \times n$ symmetric positive definite matrix. Show that $\mathbf{A} - \mathbf{B}$ is nonnegative definite if and only if all n (not necessarily distinct) roots of $|\mathbf{A} - \lambda\mathbf{B}|$ are greater than or equal to 1 and is positive definite if and only if all n roots are (strictly) greater than 1.

Bibliographic and Supplementary Notes

§15. The presentation in Section 15 was heavily influenced by that of Magnus and Neudecker (1988, sec. 8.8). For a more extensive discussion of the differentiation of eigenvalues and eigenvectors, refer, for example, to Sections 8.7–8.12 of Magnus and Neudecker (and to the sources cited by them). Their coverage includes some results on the second-order differentiation of eigenvalues, on the differentiation of multiple eigenvalues, and on the differentiation of eigenvalues and eigenvectors of nonsymmetric matrices. §16. The proof of the result on polynomials given by Theorem 21.16.4 is that of Harville and Kempthorne (1997).

22

Linear Transformations

In more advanced presentations of the topics of Chapters 1–21, matrices typically play a subordinate role. In such presentations, the main results are in terms of something called a linear transformation, which is regarded as a more fundamental concept than that of a matrix. The emphasis on linear transformations results in an approach that is more abstract, more "elegant," and more conducive to geometrical interpretation. It results in more generality in one sense, although not in another sense. It tends to appeal to the "mathematically more sophisticated," who typically prefer "geometrical reasoning" to "algebraic arguments."

In the first part of the present (and final) chapter of the book, the concept of a linear transformation is formally introduced, and a number of results on linear transformations are derived from first principles (i.e., "independently" of matrices). And, in the latter parts of the chapter, it is shown that linear transformations can be put into 1–1 correspondence with matrices, and then, based on this correspondence, various results and terminology pertaining to matrices are extended to linear transformations.

22.1 Some Definitions, Terminology, and Basic Results

A *transformation* (also known as a function, operator, map, or mapping) T from a set $\mathcal{V}$ (consisting of scalars, row or column vectors, matrices, or other "objects") into a set $\mathcal{W}$ is a correspondence that assigns to each member X of $\mathcal{V}$ a unique member of $\mathcal{W}$. This unique member of $\mathcal{W}$ is denoted by the symbol $T(X)$ and is referred to as the *image* of X. The use of the term "into" (in the definition of T) alludes to the possibility that some members of $\mathcal{W}$ may not appear as the image of any member of $\mathcal{V}$.

The set $\mathcal{V}$ is called the *domain* of T. And the set $\{Y \in \mathcal{W} : Y = T(X)$ for some $X \in \mathcal{V}\}$, consisting of every member of $\mathcal{W}$ that appears as the image of one or more members of $\mathcal{V}$, is called the *range* of T. In the event that the range of T consists of all of $\mathcal{W}$, T is said to be *onto* and may be referred to as a transformation from $\mathcal{V}$ onto $\mathcal{W}$. If each member of the range of T appears as the image of only one member of $\mathcal{V}$, T is referred to as a *1–1 (one to one) transformation.*

Corresponding to any $m \times n$ matrix $\mathbf{A}$ is the transformation T from $\mathcal{R}^{n\times 1}$ into $\mathcal{R}^{m\times 1}$ that assigns to each $n \times 1$ vector $\mathbf{x}$ the $m \times 1$ vector $\mathbf{Ax}$. Clearly, the range of T equals $\mathcal{C}(\mathbf{A})$. Further, T is onto if and only if $\mathbf{A}$ is of full row rank, that is, if and only if $\text{rank}(\mathbf{A}) = m$. And T is 1–1 if and only if the columns of $\mathbf{A}$ are linearly independent, that is, if and only if $\text{rank}(\mathbf{A}) = n$.

In what follows, the emphasis is on linear transformations. Let T represent a transformation from a linear space $\mathcal{V}$ (of $m \times n$ matrices) into a linear space $\mathcal{W}$ (of $p \times q$ matrices). Then, T is said to be *linear* if it satisfies the following two conditions:

(1) for all $\mathbf{X}$ and $\mathbf{Z}$ in $\mathcal{V}$, $T(\mathbf{X}+\mathbf{Z}) = T(\mathbf{X}) + T(\mathbf{Z})$; and

(2) for every scalar c and for all $\mathbf{X}$ in $\mathcal{V}$, $T(c\mathbf{X}) = cT(\mathbf{X})$.

It is easy to verify that T is linear if and only if, for all scalars c and k and for all $\mathbf{X}$ and $\mathbf{Z}$ in $\mathcal{V}$,

$$T(c\mathbf{X} + k\mathbf{Z}) = cT(\mathbf{X}) + kT(\mathbf{Z}).$$

More generally, letting r represent a positive integer (greater than 1), T is linear if and only if, for all scalars $c_1, \ldots, c_r$ and for all $\mathbf{X}_1, \ldots, \mathbf{X}_r$ in $\mathcal{V}$,

$$T\Big(\sum_{i=1}^{r} c_i \mathbf{X}_i\Big) = \sum_{i=1}^{r} c_i T(\mathbf{X}_i). \tag{1.1}$$

Clearly, for any $m \times n$ matrix $\mathbf{A}$, the transformation from the linear space $\mathcal{R}^{n\times 1}$ into the linear space $\mathcal{R}^{m\times 1}$ that assigns to each $n \times 1$ vector $\mathbf{x}$ the $m \times 1$ vector $\mathbf{Ax}$ is a linear transformation. More generally, for any $r \times p$ matrix $\mathbf{A}$ and any $q \times s$ matrix $\mathbf{B}$, the transformation from the linear space $\mathcal{R}^{p\times q}$ into the linear space $\mathcal{R}^{r\times s}$ that assigns to each $p \times q$ matrix $\mathbf{X}$ the $r \times s$ matrix $\mathbf{AXB}$ is a linear transformation (as is easily verified).

Let T represent a transformation from a set $\mathcal{V}$ into a set $\mathcal{W}$. Then, corresponding to any subset $\mathcal{U}$ of $\mathcal{V}$, is the subset $\{Y \in \mathcal{W} : Y = T(X) \text{ for some } X \in \mathcal{U}\}$, consisting of every member of $\mathcal{W}$ that appears as the image of one or more members of $\mathcal{U}$. This subset is denoted by the symbol $T(\mathcal{U})$ and is referred to as the *image* of the set $\mathcal{U}$. Note that, by definition, the image $T(\mathcal{V})$ of the domain $\mathcal{V}$ of T is the range of T.

Now, suppose that T is a linear transformation from a linear space $\mathcal{V}$ into a linear space $\mathcal{W}$ and that $\mathcal{U}$ is a subspace of $\mathcal{V}$. Then, for any matrices $\mathbf{X}_1, \ldots, \mathbf{X}_r$ that span $\mathcal{U}$,

$$T(\mathcal{U}) = \text{sp}[T(\mathbf{X}_1), \ldots, T(\mathbf{X}_r)],$$

as is evident from result (1.1). Thus, $T(\mathcal{U})$ is a linear space; it is a subspace of $\mathcal{W}$. In particular, the range $T(\mathcal{V})$ of T is a subspace of $\mathcal{W}$ and, accordingly, may be referred to as the *range space* of T.

A linear transformation T from a linear space $\mathcal{V}$ of $m \times n$ matrices into a linear space $\mathcal{W}$ of $p \times q$ matrices has the property that $T(\mathbf{0}) = T(0\mathbf{0}) = 0T(\mathbf{0}) = \mathbf{0}$.

Thus, T assigns the $(p \times q)$ null matrix in $\mathcal{W}$ to the $(m \times n)$ null matrix in $\mathcal{V}$. The linear transformation T may also assign the null matrix in $\mathcal{W}$ to various nonnull matrices in $\mathcal{V}$.

The set

$$\{\mathbf{X} \in \mathcal{V} : T(\mathbf{X}) = \mathbf{0}\}, \tag{1.2}$$

comprising all matrices in $\mathcal{V}$ to which T assigns the null matrix, is nonempty. Moreover, for any matrices $\mathbf{X}$ and $\mathbf{Z}$ in the set (1.2) and for any scalar c, $T(\mathbf{X} + \mathbf{Z}) = T(\mathbf{X}) + T(\mathbf{Z}) = \mathbf{0} + \mathbf{0} = \mathbf{0}$ and $T(c\mathbf{X}) = cT(\mathbf{X}) = c\mathbf{0} = \mathbf{0}$, so that $\mathbf{X} + \mathbf{Z}$ and $c\mathbf{X}$ are in the set (1.2). Thus, the set (1.2) is a linear space; it is a subspace of $\mathcal{V}$. This subspace is called the *null space* (or the *kernel*) of the linear transformation T and is denoted by the symbol $\mathfrak{N}(T)$.

There is a simple and fundamental relationship between the range and null spaces of a linear transformation. This relationship is described in the following theorem.

Theorem 22.1.1. Let T represent a linear transformation from a linear space $\mathcal{V}$ into a linear space $\mathcal{W}$. Then,

$$\dim(\mathcal{V}) = \dim(\mathfrak{M}) + \dim(\mathfrak{N}), \tag{1.3}$$

where $\mathfrak{M}$ and $\mathfrak{N}$ are, respectively, the range and null spaces of T.

Proof. Let $k = \dim(\mathcal{V})$ and $s = \dim(\mathfrak{N})$. Further, let $r = k - s$. And assume that $r > 0$—if $r = 0$, then (in light of Theorem 4.3.10) $\mathcal{V} = \mathfrak{N}$, in which case $\mathfrak{M} = \{\mathbf{0}\}$ and equality (1.3) is clearly valid—and take $\{\mathbf{Z}_1, \ldots, \mathbf{Z}_s\}$ to be a basis for $\mathfrak{N}$—if $s = 0$, interpret $\{\mathbf{Z}_1, \ldots, \mathbf{Z}_s\}$ as the empty set. Then, it follows from Theorem 4.3.12 that there exist matrices $\mathbf{X}_1, \ldots, \mathbf{X}_r$ in $\mathcal{V}$ such that $\{\mathbf{X}_1, \ldots, \mathbf{X}_r, \mathbf{Z}_1, \ldots, \mathbf{Z}_s\}$ is a basis for $\mathcal{V}$—if $s = 0$, interpret $\{\mathbf{X}_1, \ldots, \mathbf{X}_r, \mathbf{Z}_1, \ldots, \mathbf{Z}_s\}$ as $\{\mathbf{X}_1, \ldots, \mathbf{X}_r\}$.

Now, for $i = 1, \ldots, r$, let $\mathbf{Y}_i = T(\mathbf{X}_i)$ (and observe that $\mathbf{Y}_i \in \mathfrak{M}$). To prove Theorem 22.1.1, it suffices to show that $\{\mathbf{Y}_1, \ldots, \mathbf{Y}_r\}$ is a basis for $\mathfrak{M}$. Let us begin by showing that this set spans $\mathfrak{M}$.

Let $\mathbf{Y}$ represent an arbitrary matrix in $\mathfrak{M}$. Then, there exists a matrix $\mathbf{X}$ in $\mathcal{V}$ such that $\mathbf{Y} = T(\mathbf{X})$. Further, if $s > 0$, there exist scalars $a_1, \ldots, a_r, b_1, \ldots, b_s$ such that

$$\mathbf{X} = \sum_{i=1}^{r} a_i \mathbf{X}_i + \sum_{j=1}^{s} b_j \mathbf{Z}_j,$$

in which case

$$\mathbf{Y} = T(\mathbf{X}) = \sum_{i=1}^{r} a_i T(\mathbf{X}_i) + \sum_{j=1}^{s} b_j T(\mathbf{Z}_j) = \sum_{i=1}^{r} a_i T(\mathbf{X}_i) = \sum_{i=1}^{r} a_i \mathbf{Y}_i.$$

Alternatively, if $s = 0$, there exist scalars $a_1, \ldots, a_r$ such that $\mathbf{X} = \sum_{i=1}^{r} a_i \mathbf{X}_i$, in which case

$$\mathbf{Y} = T(\mathbf{X}) = \sum_{i=1}^{r} a_i T(\mathbf{X}_i) = \sum_{i=1}^{r} a_i \mathbf{Y}_i.$$

Thus, whether $s > 0$ or $s = 0$, $\mathcal{M} \subset \text{sp}(\mathbf{Y}_1, \dots, \mathbf{Y}_r)$, and, consequently, $\{\mathbf{Y}_1, \dots, \mathbf{Y}_r\}$ spans $\mathcal{M}$.

It remains to show that $\{\mathbf{Y}_1, \dots, \mathbf{Y}_r\}$ is linearly independent. Let $c_1, \dots, c_r$ represent any scalars such that $\sum_{i=1}^r c_i \mathbf{Y}_i = \mathbf{0}$. And define $\mathbf{Z} = \sum_{i=1}^r c_i \mathbf{X}_i$. Then, we have that

$$T(\mathbf{Z}) = T\Big(\sum_{i=1}^r c_i \mathbf{X}_i\Big) = \sum_{i=1}^r c_i T(\mathbf{X}_i) = \sum_{i=1}^r c_i \mathbf{Y}_i = \mathbf{0},$$

implying that $\mathbf{Z}$ belongs to $\mathcal{N}$. Thus, if $s > 0$, there exist scalars $d_1, \dots, d_s$ such that $\mathbf{Z} = \sum_{j=1}^s d_j \mathbf{Z}_j$, in which case

$$c_1\mathbf{X}_1 + \cdots + c_r\mathbf{X}_r - d_1\mathbf{Z}_1 - \cdots - d_s\mathbf{Z}_s = \mathbf{Z} - \mathbf{Z} = \mathbf{0},$$

and (since $\{\mathbf{X}_1, \dots, \mathbf{X}_r, \mathbf{Z}_1, \dots, \mathbf{Z}_s\}$ is a basis for $\mathcal{V}$ and hence is linearly independent) it follows that $c_1 = \cdots = c_r = 0$ (and that $d_1 = \cdots = d_s = 0$). Alternatively, if $s = 0$, then $\mathbf{Z} = \mathbf{0}$, and (since $\{\mathbf{X}_1, \dots, \mathbf{X}_r\}$ is linearly independent) it follows that $c_1 = \cdots = c_r = 0$. In either case, we conclude that $\{\mathbf{Y}_1, \dots, \mathbf{Y}_r\}$ is linearly independent (and hence is a basis for $\mathcal{M}$). Q.E.D.

In connection with Theorem 22.1.1, equality (1.3) can be rewritten to give the following expressions for $\dim(\mathcal{M})$ and $\dim(\mathcal{N})$:

$$\dim(\mathcal{M}) = \dim(\mathcal{V}) - \dim(\mathcal{N}), \tag{1.4}$$

$$\dim(\mathcal{N}) = \dim(\mathcal{V}) - \dim(\mathcal{M}). \tag{1.5}$$

Moreover, as an immediate consequence of Theorem 22.1.1, we have the following corollary.

Corollary 22.1.2. For any linear transformation T from a linear space $\mathcal{V}$ into a linear space $\mathcal{W}$, the dimension of the range space of T is less than or equal to $\dim(\mathcal{V})$, that is, less than or equal to the dimension of the domain of T.

One very basic transformation is the transformation from a linear space $\mathcal{V}$ into a linear space $\mathcal{W}$ that assigns to every matrix in $\mathcal{V}$ the null matrix (in $\mathcal{W}$). This transformation is called the *zero transformation* and is denoted by the symbol 0. Clearly, the zero transformation is linear. Further, its range space is the set whose only member is the null matrix, and its null space equals its domain $\mathcal{V}$.

Another very basic transformation is the identity transformation. The *identity transformation* is the transformation from a linear space $\mathcal{V}$ into (the same linear space) $\mathcal{V}$ defined by $T(\mathbf{X}) = \mathbf{X}$; it is denoted by the symbol I. The identity transformation is linear (as is readily apparent). Further, it is onto; that is, its range space equals $\mathcal{V}$. And, clearly, it is 1–1, implying in particular that its null space is the set whose only member is the null matrix.

Let T represent a linear transformation from a linear space $\mathcal{V}$ into a linear space $\mathcal{W}$, and let $\mathcal{U}$ represent any subspace of $\mathcal{V}$. Then, the transformation, say R, from $\mathcal{U}$ into $\mathcal{W}$ defined by $R(\mathbf{X}) = T(\mathbf{X})$ (which assigns to each matrix in $\mathcal{U}$ the same matrix in $\mathcal{W}$ assigned by T) is called the *restriction* of T to $\mathcal{U}$. Clearly, the restriction of T to $\mathcal{U}$ is a linear transformation. Further, its range space equals $T(\mathcal{U})$, and its null space equals $\mathcal{U} \cap \mathcal{N}(T)$. And if T is 1–1, then so is the restriction of T to $\mathcal{U}$.

The following four lemmas give some basic results on linear transformations.

Lemma 22.1.3. Let T represent a linear transformation from a linear space $\mathcal{V}$ into a linear space $\mathcal{W}$. Then, T is 1–1 if and only if $\mathcal{N}(T) = \{\mathbf{0}\}$ [i.e., if and only if $\mathcal{N}(T)$ contains no nonnull matrices].

Proof. Suppose that T is 1–1. Then, since the image of the null matrix (in $\mathcal{V}$) is the null matrix (in $\mathcal{W}$), $\mathcal{V}$ cannot contain any other matrix (i.e., any nonnull matrix) whose image is the null matrix. Thus, $\mathcal{N}(T)$ contains no nonnull matrices.

Conversely, suppose that $\mathcal{N}(T)$ contains no nonnull matrices. Then, for any matrices $\mathbf{Z}$ and $\mathbf{X}$ in $\mathcal{V}$ for which $T(\mathbf{Z}) = T(\mathbf{X})$, we have that $T(\mathbf{Z} - \mathbf{X}) = T(\mathbf{Z}) - T(\mathbf{X}) = \mathbf{0}$, implying that $\mathbf{Z} - \mathbf{X} \in \mathcal{N}(T)$ and hence that $\mathbf{Z} - \mathbf{X} = \mathbf{0}$ or, equivalently, $\mathbf{Z} = \mathbf{X}$. Thus, T is 1–1. Q.E.D.

Lemma 22.1.4. Let T represent a linear transformation from a linear space $\mathcal{V}$ into a linear space $\mathcal{W}$. And let $\mathbf{X}_1, \ldots, \mathbf{X}_k$ represent matrices in $\mathcal{V}$. If $\mathbf{X}_1, \ldots, \mathbf{X}_k$ are linearly dependent, then their images $T(\mathbf{X}_1), \ldots, T(\mathbf{X}_k)$ are also linearly dependent.

Proof. Suppose that $\mathbf{X}_1, \ldots, \mathbf{X}_k$ are linearly dependent. Then, there exist scalars $c_1, \ldots, c_k$, one or more of which is nonzero, such that $\sum_{i=1}^k c_i \mathbf{X}_i = \mathbf{0}$. And

$$\sum_{i=1}^{k} c_i T(\mathbf{X}_i) = T\Big(\sum_{i=1}^{k} c_i \mathbf{X}_i\Big) = T(\mathbf{0}) = \mathbf{0}.$$

Thus, $T(\mathbf{X}_1), \ldots, T(\mathbf{X}_k)$ are linearly dependent. Q.E.D.

Lemma 22.1.5. Let T represent a 1–1 linear transformation from a linear space $\mathcal{V}$ into a linear space $\mathcal{W}$. And let $\mathbf{X}_1, \ldots, \mathbf{X}_k$ represent matrices in $\mathcal{V}$. Then, $T(\mathbf{X}_1), \ldots, T(\mathbf{X}_k)$ are linearly independent if and only if $\mathbf{X}_1, \ldots, \mathbf{X}_k$ are linearly independent.

Proof. It suffices to show that $T(\mathbf{X}_1), \ldots, T(\mathbf{X}_k)$ are linearly dependent if and only if $\mathbf{X}_1, \ldots, \mathbf{X}_k$ are linearly dependent. That $T(\mathbf{X}_1), \ldots, T(\mathbf{X}_k)$ are linearly dependent if $\mathbf{X}_1, \ldots, \mathbf{X}_k$ are linearly dependent is an immediate consequence of Lemma 22.1.4.

Now, suppose that $T(\mathbf{X}_1), \ldots, T(\mathbf{X}_k)$ are linearly dependent. Then, there exist scalars $c_1, \ldots, c_k$, one or more of which is nonzero, such that $\sum_{i=1}^k c_i T(\mathbf{X}_i) = \mathbf{0}$. Further, $T(\sum_{i=1}^k c_i \mathbf{X}_i) = \sum_{i=1}^k c_i T(\mathbf{X}_i) = \mathbf{0}$. And it follows from Lemma 22.1.3 that $\sum_{i=1}^k c_i \mathbf{X}_i = \mathbf{0}$. Thus, $\mathbf{X}_1, \ldots, \mathbf{X}_k$ are linearly dependent. Q.E.D.

Lemma 22.1.6. Let T represent a linear transformation from $\mathcal{V}$ into $\mathcal{W}$, where $\mathcal{V}$ and $\mathcal{W}$ are linear spaces of the same dimension. Then, T is 1–1 if and only if $T(\mathcal{V}) = \mathcal{W}$ (i.e., if and only if T is onto).

Proof. According to Lemma 22.1.3, T is 1–1 if and only if $\mathcal{N}(T) = \{\mathbf{0}\}$ or, equivalently, if and only if $\dim[\mathcal{N}(T)] = 0$. Moreover, according to Theorem 22.1.1,

$$\dim[\mathcal{N}(T)] = \dim(\mathcal{V}) - \dim[T(\mathcal{V})].$$

Thus, T is 1–1 if and only if $\dim(\mathcal{V}) - \dim[T(\mathcal{V})] = 0$ or, equivalently, if and only if $\dim[T(\mathcal{V})] = \dim(\mathcal{V})$. And, since $\dim(\mathcal{V}) = \dim(\mathcal{W})$, T is 1–1 if and only if $\dim[T(\mathcal{V})] = \dim(\mathcal{W})$.

It is now clear that if $T(\mathcal{V}) = \mathcal{W}$, then T is 1–1. And, conversely, if T is 1–1, then [since $T(\mathcal{V}) \subset \mathcal{W}$] we have [in light of Theorem 4.3.10] that $T(\mathcal{V}) = \mathcal{W}$. Q.E.D.

As a special case of Lemma 22.1.6, we have the following corollary.

Corollary 22.1.7. Let T represent a linear transformation from $\mathcal{V}$ into $\mathcal{V}$. Then, T is 1–1 if and only if $T(\mathcal{V}) = \mathcal{V}$ (i.e., if and only if T is onto).

The following lemma is sometimes useful.

Lemma 22.1.8. Let $\mathcal{V}$ and $\mathcal{W}$ represent linear spaces. And, denoting $\dim(\mathcal{V})$ by r (and supposing that $r > 0$), let $\{\mathbf{X}_1, \ldots, \mathbf{X}_r\}$ represent a basis for $\mathcal{V}$, and let $\mathbf{Y}_1, \ldots, \mathbf{Y}_r$ represent any r (not necessarily distinct) matrices in $\mathcal{W}$. Then, the transformation T (from $\mathcal{V}$ into $\mathcal{W}$) defined by

$$T(\mathbf{X}) = \sum_{i=1}^{r} c_i \mathbf{Y}_i,$$

where $c_1, \ldots, c_r$ are the (unique) scalars that satisfy $\mathbf{X} = \sum_{i=1}^{r} c_i \mathbf{X}_i$, is linear. And $T(\mathcal{V}) = \text{sp}(\mathbf{Y}_1, \ldots, \mathbf{Y}_r)$. Moreover, T is 1–1 if and only if $\mathbf{Y}_1, \ldots, \mathbf{Y}_r$ are linearly independent.

Proof. Letting k represent an arbitrary scalar, letting $\mathbf{X}$ and $\mathbf{Z}$ represent arbitrary matrices in $\mathcal{V}$, and taking $c_1, \ldots, c_r$ and $d_1, \ldots, d_r$ to be the (unique) scalars that satisfy $\mathbf{X} = \sum_{i=1}^{r} c_i \mathbf{X}_i$ and $\mathbf{Z} = \sum_{i=1}^{r} d_i \mathbf{X}_i$, we find that $k\mathbf{X} = \sum_{i=1}^{r} (kc_i)\mathbf{X}_i$ and $\mathbf{X} + \mathbf{Z} = \sum_{i=1}^{r} (c_i + d_i)\mathbf{X}_i$ and hence that

$$T(k\mathbf{X}) = \sum_{i=1}^{r} (kc_i)\mathbf{Y}_i = k \sum_{i=1}^{r} c_i \mathbf{Y}_i = kT(\mathbf{X})$$

and

$$T(\mathbf{X} + \mathbf{Z}) = \sum_{i=1}^{r} (c_i + d_i)\mathbf{Y}_i = \sum_{i=1}^{r} c_i \mathbf{Y}_i + \sum_{i=1}^{r} d_i \mathbf{Y}_i = T(\mathbf{X}) + T(\mathbf{Z}).$$

Thus, T is linear.

To verify that $T(\mathcal{V}) = \text{sp}(\mathbf{Y}_1, \ldots, \mathbf{Y}_r)$, let $\mathbf{Y}$ represent any matrix in $\text{sp}(\mathbf{Y}_1, \ldots, \mathbf{Y}_r)$. Then, there exist scalars $c_1, \ldots, c_r$ such that $\mathbf{Y} = \sum_{i=1}^{r} c_i \mathbf{Y}_i$. Clearly, $\sum_{i=1}^{r} c_i \mathbf{X}_i \in \mathcal{V}$ and $T(\sum_{i=1}^{r} c_i \mathbf{X}_i) = \mathbf{Y}$, implying that $\mathbf{Y} \in T(\mathcal{V})$. Thus, $\text{sp}(\mathbf{Y}_1, \ldots, \mathbf{Y}_r) \subset T(\mathcal{V})$, and [since obviously $T(\mathcal{V}) \subset \text{sp}(\mathbf{Y}_1, \ldots, \mathbf{Y}_r)$] it follows that $T(\mathcal{V}) = \text{sp}(\mathbf{Y}_1, \ldots, \mathbf{Y}_r)$.

To prove the final part of the lemma, observe (in light of Lemma 22.1.3) that T is 1–1 if and only if $\mathcal{N}(T) = \{\mathbf{0}\}$ and (since $\sum_{i=1}^{r} c_i \mathbf{X}_i = \mathbf{0} \Leftrightarrow c_1 = \cdots = c_r = 0$) that $\mathcal{N}(T) = \{\mathbf{0}\}$ if and only if the only scalars $c_1, \ldots, c_r$ for which

$\sum_{i=1}^{r} c_i \mathbf{Y}_i = \mathbf{0}$ are $c_1 = \cdots = c_r = 0$. Thus, T is 1–1 if and only if $\mathbf{Y}_1, \ldots, \mathbf{Y}_r$ are linearly independent. Q.E.D.

Let S and T represent transformations from a set $\mathcal{V}$ into a set $\mathcal{W}$. If $S(\mathbf{X}) = T(\mathbf{X})$ for every member $\mathbf{X}$ of $\mathcal{V}$, S and T are said to be equal (or identical), and we may write $S = T$.

The following lemma gives a necessary and sufficient condition for two linear transformations to be equal.

Lemma 22.1.9. Let S and T represent linear transformations from a linear space $\mathcal{V}$ into a linear space $\mathcal{W}$. And let $\mathbf{X}_1, \ldots, \mathbf{X}_r$ represent any matrices (in $\mathcal{V}$) that form a basis for $\mathcal{V}$ or, more generally, any matrices such that $\text{sp}(\mathbf{X}_1, \ldots, \mathbf{X}_r) = \mathcal{V}$. Then (supposing that $r > 0$), $S = T$ if and only if, for $i = 1, \ldots, r$, $S(\mathbf{X}_i) = T(\mathbf{X}_i)$.

Proof. That $S(\mathbf{X}_i) = T(\mathbf{X}_i)$ (for $i = 1, \ldots, r$) if $S = T$ is obvious.

Now, suppose that, for $i = 1, \ldots, r$, $S(\mathbf{X}_i) = T(\mathbf{X}_i)$. And let $\mathbf{X}$ represent an arbitrary matrix in $\mathcal{V}$. Then, there exist scalars $c_1, \ldots, c_r$ such that $\mathbf{X} = \sum_{i=1}^{r} c_i \mathbf{X}_i$. Further,

$$S(\mathbf{X}) = S\Big(\sum_{i=1}^{r} c_i \mathbf{X}_i\Big) = \sum_{i=1}^{r} c_i S(\mathbf{X}_i) = \sum_{i=1}^{r} c_i T(\mathbf{X}_i) = T\Big(\sum_{i=1}^{r} c_i \mathbf{X}_i\Big) = T(\mathbf{X}).$$

Thus, $S = T$. Q.E.D.

22.2 Scalar Multiples, Sums, and Products of Linear Transformations

a. Scalar multiples

Let T represent a transformation from a linear space $\mathcal{V}$ into a linear space $\mathcal{W}$, and let k represent an arbitrary scalar. Then, the transformation from $\mathcal{V}$ into $\mathcal{W}$ that assigns to each matrix $\mathbf{X}$ in $\mathcal{V}$ the matrix $kT(\mathbf{X})$ (in $\mathcal{W}$) is referred to as a *scalar multiple* of T and is denoted by the symbol (kT) or (in the absence of ambiguity) kT. Thus, $(kT)(\mathbf{X}) = kT(\mathbf{X})$. If the transformation T is linear, then the scalar multiple kT is also linear, as is easily verified.

Clearly,

$$1T = T, \tag{2.1}$$

and letting c (like k) represent a scalar, we find that

$$c(kT) = (ck)T = (kc)T = k(cT). \tag{2.2}$$

In the special case where $k = 0$ or $T = 0$, the scalar multiple kT reduces to the zero transformation (from $\mathcal{V}$ into $\mathcal{W}$). That is, $0T = 0$, and $k0 = 0$. Further, it is customary to refer to the scalar multiple $(-1)T$ as the *negative* of T and to abbreviate $(-1)T$ to $(-T)$ or $-T$.

b. Sums

Let T and S represent transformations from a linear space $\mathcal{V}$ into a linear space $\mathcal{W}$. Then, the transformation from $\mathcal{V}$ into $\mathcal{W}$ that assigns to each matrix $\mathbf{X}$ in $\mathcal{V}$ the matrix $T(\mathbf{X}) + S(\mathbf{X})$ (in $\mathcal{W}$) is referred to as the *sum* of T and S and is denoted by the symbol $(T + S)$ or $T + S$. Thus, $(T + S)(\mathbf{X}) = T(\mathbf{X}) + S(\mathbf{X})$. If the transformations T and S are both linear, then their sum $T + S$ is also linear, as is easily verified.

The addition of transformations is commutative, that is,

$$T + S = S + T. \tag{2.3}$$

The addition of transformations is also associative, that is, taking R to be a third transformation from $\mathcal{V}$ into $\mathcal{W}$,

$$T + (S + R) = (T + S) + R. \tag{2.4}$$

The symbol $(T + S + R)$ or $T + S + R$ is used to represent the (common) transformation given by the left and right sides of equality (2.4), and this transformation is referred to as the *sum* of T, S, and R. This notation and terminology extend in an obvious way to any finite number of transformations from $\mathcal{V}$ into $\mathcal{W}$.

We find (via a straightforward exercise) that, for any scalar c,

$$c(T + S) = (cT) + (cS) \tag{2.5}$$

and that, for any scalars c and k,

$$(c + k)T = (cT) + (kT). \tag{2.6}$$

Let us write $(T - S)$ or $T - S$ for the sum $T + (-S)$ or, equivalently, for the transformation (from $\mathcal{V}$ into $\mathcal{W}$) that assigns to each matrix $\mathbf{X}$ in $\mathcal{V}$ the matrix $T(\mathbf{X}) - S(\mathbf{X})$ (in $\mathcal{W}$). And let us refer to this transformation as the *difference* between T and S.

Clearly,

$$T - T = 0. \tag{2.7}$$

Further,

$$T + 0 = 0 + T = T. \tag{2.8}$$

c. Products

Let S represent a transformation from a linear space $\mathcal{U}$ into a linear space $\mathcal{V}$, and let T represent a transformation from $\mathcal{V}$ into a linear space $\mathcal{W}$. Consider the transformation from $\mathcal{U}$ into $\mathcal{W}$ that assigns to each matrix $\mathbf{X}$ in $\mathcal{U}$ the matrix $T[S(\mathbf{X})]$ (in $\mathcal{W}$) obtained by applying S to $\mathbf{X}$ and by then applying T to $S(\mathbf{X})$. This transformation is called the *product* (or the *composition*) of T and S and is denoted by the symbol (TS) or TS. Thus, $(TS)(\mathbf{X}) = T[S(\mathbf{X})]$. If the transformations T and S are both linear, then their product TS is also linear, as is easily verified.

Let R represent a transformation from a linear space $\mathcal{M}$ into a linear space $\mathcal{U}$, S a transformation from $\mathcal{U}$ into a linear space $\mathcal{V}$, and T a transformation from $\mathcal{V}$ into a linear space $\mathcal{W}$. Then, for any matrix $\mathbf{X}$ in $\mathcal{M}$,

$$(T(SR))(\mathbf{X}) = T[(SR)(\mathbf{X})] = T\{S[R(\mathbf{X})]\} = (TS)[R(\mathbf{X})] = ((TS)R)(\mathbf{X}).$$

Thus,

$$T(SR) = (TS)R. \tag{2.9}$$

The symbol (TSR) or TSR is used to represent the (common) transformation given by the left and right sides of equality (2.9), and this transformation is referred to as the *product* of T, S, and R—if the transformations R, S, and T are all linear, then clearly the transformation TSR is also linear. This terminology and notation extend in an obvious way to an arbitrary number of transformations.

Now, let R and S represent linear transformations from a linear space $\mathcal{U}$ into a linear space $\mathcal{V}$, and let T represent a linear transformation from $\mathcal{V}$ into a linear space $\mathcal{W}$. Then, for any matrix $\mathbf{X}$ in $\mathcal{U}$,

$$\begin{aligned}(T(S+R))(\mathbf{X}) = T[(S+R)(\mathbf{X})] &= T[S(\mathbf{X}) + R(\mathbf{X})] \\ &= T[S(\mathbf{X})] + T[R(\mathbf{X})] \\ &= (TS)(\mathbf{X}) + (TR)(\mathbf{X}).\end{aligned}$$

Thus,

$$T(S+R) = (TS) + (TR). \tag{2.10}$$

[And the transformation $T(S+R)$ (like R, S, and T) is linear.]

Alternatively, suppose that R is a linear transformation from a linear space $\mathcal{U}$ into a linear space $\mathcal{V}$ and that S and T are linear transformations from $\mathcal{V}$ into a linear space $\mathcal{W}$. Then, employing an argument similar to that used to establish equality (2.10), we find that

$$(T+S)R = (TR) + (SR). \tag{2.11}$$

[And the transformation $(T+S)R$ is linear.]

Result (2.10) extends in an obvious way to the product of the linear transformation T (from $\mathcal{V}$ into $\mathcal{W}$) and the sum of an arbitrary number of linear transformations (from $\mathcal{U}$ into $\mathcal{V}$). Similarly, result (2.11) extends in an obvious way to the product of the sum of an arbitrary number of linear transformations (from $\mathcal{V}$ into $\mathcal{W}$) and the linear transformation R (from $\mathcal{U}$ into $\mathcal{V}$).

For any transformation S from a linear space $\mathcal{U}$ into a linear space $\mathcal{V}$ and any transformation T from $\mathcal{V}$ into a linear space $\mathcal{W}$, and for any scalar k,

$$k(TS) = (kT)S = T(kS), \tag{2.12}$$

as is easily verified. The symbol kTS may be used to represent transformation (2.12). (Further, if the transformations S and T are linear, then so is the transformation kTS.)

For any transformation T from a linear space $\mathcal{V}$ into a linear space $\mathcal{W}$,

$$TI = IT = T, \tag{2.13}$$

$$T0 = 0, \qquad 0T = 0 \tag{2.14}$$

(where, depending on the context, I is the identity transformation from $\mathcal{V}$ onto $\mathcal{V}$ or from $\mathcal{W}$ onto $\mathcal{W}$, and 0 is the zero transformation from a linear space $\mathcal{U}$ into $\mathcal{V}$ or $\mathcal{W}$ or from $\mathcal{W}$ or $\mathcal{V}$ into $\mathcal{U}$).

The following theorem gives a basic result on the product of two linear transformations.

Theorem 22.2.1. Let S represent a linear transformation from a linear space $\mathcal{U}$ into a linear space $\mathcal{V}$, and let T represent a linear transformation from $\mathcal{V}$ into a linear space $\mathcal{W}$. Then,

$$\dim(\mathcal{M}_{TS}) = \dim(\mathcal{M}_S) - \dim(\mathcal{M}_S \cap \mathcal{N}_T), \tag{2.15}$$

where $\mathcal{M}_S$ and $\mathcal{M}_{TS}$ are the range spaces of S and TS, respectively, and $\mathcal{N}_T$ is the null space of T.

Proof. Let R represent the restriction of T to $\mathcal{M}_S$. And denote the range space and the null space of R by $\mathcal{M}_R$ and $\mathcal{N}_R$, respectively.

According to Theorem 22.1.1,

$$\dim(\mathcal{M}_R) = \dim(\mathcal{M}_S) - \dim(\mathcal{N}_R).$$

Moreover, it follows from the discussion (of restrictions) in Section 22.1 that

$$\mathcal{M}_R = T(\mathcal{M}_S) = T[S(\mathcal{U})] = TS(\mathcal{U}) = \mathcal{M}_{TS}$$

and $\mathcal{N}_R = \mathcal{M}_S \cap \mathcal{N}_T$. Thus,

$$\dim(\mathcal{M}_{TS}) = \dim(\mathcal{M}_S) - \dim(\mathcal{M}_S \cap \mathcal{N}_T).$$

Q.E.D.

As an immediate consequence of Theorem 22.2.1, we have the following corollary.

Corollary 22.2.2. Let S represent a linear transformation from a linear space $\mathcal{U}$ into a linear space $\mathcal{V}$, and let T represent a linear transformation from $\mathcal{V}$ into a linear space $\mathcal{W}$. Then,

$$\dim(\mathcal{M}_{TS}) \leq \dim(\mathcal{M}_S), \tag{2.16}$$

where $\mathcal{M}_{TS}$ and $\mathcal{M}_S$ are the range spaces of TS and S, respectively.

22.3 Inverse Transformations and Isomorphic Linear Spaces

Let T represent a transformation from a linear space $\mathcal{V}$ into a linear space $\mathcal{W}$. Suppose that T is 1–1 and onto. Then, T is said to be *invertible*. And there exists a (unique) transformation from $\mathcal{W}$ into $\mathcal{V}$ that assigns to each matrix $\mathbf{Y}$ in $\mathcal{W}$ the (unique) matrix $\mathbf{X}$ (in $\mathcal{V}$) such that $T(\mathbf{X}) = \mathbf{Y}$. This transformation is called the *inverse* of T and is denoted by the symbol T^{-1}. Thus,

$$T^{-1}(\mathbf{Y}) = \mathbf{X}$$

[where $\mathbf{X}$ is such that $T(\mathbf{X}) = \mathbf{Y}$].

The inverse transformation T^{-1} is 1–1 and onto, as can be easily verified. Moreover, if T is linear, then so is T^{-1}. To see this, suppose that T is linear. Then, letting c represent an arbitrary scalar, letting $\mathbf{Y}$ and $\mathbf{Z}$ represent arbitrary matrices in $\mathcal{W}$, and letting $\mathbf{X}$ and $\mathbf{U}$ represent the (unique) matrices (in $\mathcal{V}$) such that $T(\mathbf{X}) = \mathbf{Y}$ and $T(\mathbf{U}) = \mathbf{Z}$, we find that $T(c\mathbf{X}) = cT(\mathbf{X}) = c\mathbf{Y}$ and $T(\mathbf{X}+\mathbf{U}) = T(\mathbf{X}) + T(\mathbf{U}) = \mathbf{Y}+\mathbf{Z}$, so that

$$T^{-1}(c\mathbf{Y}) = c\mathbf{X} = cT^{-1}(\mathbf{Y})$$

and

$$T^{-1}(\mathbf{Y}+\mathbf{Z}) = \mathbf{X}+\mathbf{U} = T^{-1}(\mathbf{Y}) + T^{-1}(\mathbf{Z}).$$

And it follows that T^{-1} is linear.

For any matrix $\mathbf{X}$ in $\mathcal{V}$, we have that

$$(T^{-1}T)(\mathbf{X}) = T^{-1}[T(\mathbf{X})] = \mathbf{X} = I(\mathbf{X})$$

(where the identity transformation I is that from $\mathcal{V}$ onto $\mathcal{V}$). Thus,

$$T^{-1}T = I. \tag{3.1}$$

The inverse transformation T^{-1} is itself invertible (since it is 1–1 and onto). Moreover, upon applying result (3.1) (with T^{-1} in place of T), we find that

$$(T^{-1})^{-1}T^{-1} = I \tag{3.2}$$

(where the identity transformation I is that from $\mathcal{W}$ onto $\mathcal{W}$). It follows that

$$(T^{-1})^{-1} = (T^{-1})^{-1}I = (T^{-1})^{-1}(T^{-1}T) = ((T^{-1})^{-1}T^{-1})T = IT = T.$$

Thus, T is the inverse of T^{-1}. Result (3.2) can be restated as

$$TT^{-1} = I. \tag{3.3}$$

Let $\mathcal{V}$ and $\mathcal{W}$ represent linear spaces. If there exists a 1–1 linear transformation, say T, from $\mathcal{V}$ onto $\mathcal{W}$, then $\mathcal{V}$ and $\mathcal{W}$ are said to be *isomorphic*, and T is said to be an *isomorphism* of $\mathcal{V}$ onto $\mathcal{W}$.

Note that the identity transformation I from a linear space $\mathcal{V}$ onto $\mathcal{V}$ is an isomorphism (from $\mathcal{V}$ onto $\mathcal{V}$) and hence that any linear space is isomorphic to itself. Note also that I is invertible and that $I^{-1} = I$.

The following theorem gives a necessary and sufficient condition for two linear spaces to be isomorphic.

Theorem 22.3.1. Two linear spaces $\mathcal{V}$ and $\mathcal{W}$ are isomorphic if and only if $\dim(\mathcal{V}) = \dim(\mathcal{W})$.

Proof. Suppose that $\dim(\mathcal{V}) = \dim(\mathcal{W}) = r$ for some integer r. And suppose that $r > 0$—if $r = 0$, we have that $\mathcal{V} = \{\mathbf{0}\}$ and $\mathcal{W} = \{\mathbf{0}\}$, and these two linear spaces are clearly isomorphic. Then, there exist r linearly independent matrices $\mathbf{Y}_1, \ldots, \mathbf{Y}_r$ in $\mathcal{W}$ (consisting of any r matrices that form a basis for $\mathcal{W}$), and it follows from Lemma 22.1.8 that there exists a 1–1 linear transformation from $\mathcal{V}$ onto $\mathcal{W}$. Thus, $\mathcal{V}$ and $\mathcal{W}$ are isomorphic.

Conversely, suppose that $\mathcal{V}$ and $\mathcal{W}$ are isomorphic. Then, by definition, there exists an isomorphism, say T, of $\mathcal{V}$ onto $\mathcal{W}$. And, since (according to Lemma 22.1.3) $\mathcal{N}(T) = \{\mathbf{0}\}$, it follows from Theorem 22.1.1 that

$$\dim(\mathcal{V}) = \dim[T(\mathcal{V})] + \dim[\mathcal{N}(T)] = \dim(\mathcal{W}) + 0 = \dim(\mathcal{W}).$$

Q.E.D.

As an immediate consequence of Theorem 22.3.1, we have the following corollary.

Corollary 22.3.2. Every n-dimensional linear space is isomorphic to $\mathcal{R}^n$.

Let $\mathcal{W}$ represent an n-dimensional linear space, and let B represent a set of matrices $\mathbf{Y}_1, \ldots, \mathbf{Y}_n$ (in $\mathcal{W}$) that form a basis for $\mathcal{W}$. Subsequently, the symbol L_B is used to represent the transformation from $\mathcal{R}^{n\times 1}$ into $\mathcal{W}$ that assigns to each vector $\mathbf{x} = (x_1, \ldots, x_n)'$ in $\mathcal{R}^{n\times 1}$ the matrix $x_1\mathbf{Y}_1 + \cdots + x_n\mathbf{Y}_n$ in $\mathcal{W}$. Clearly, L_B is 1–1 and onto (and hence invertible) and is linear, so that L_B is an isomorphism of $\mathcal{R}^n$ onto $\mathcal{W}$.

Note that the inverse L_B^{-1} of L_B is the (linear) transformation that assigns to each matrix $\mathbf{Y}$ in $\mathcal{W}$ the column vector $\mathbf{x}$ in $\mathcal{R}^n$ whose elements ($x_1, \ldots, x_n$, respectively) are (uniquely) determined by the condition $\mathbf{Y} = x_1\mathbf{Y}_1 + \cdots + x_n\mathbf{Y}_n$, that is, whose elements are the coefficients of the basis matrices $\mathbf{Y}_1, \ldots, \mathbf{Y}_n$ when $\mathbf{Y}$ is expressed as a linear combination of those matrices.

For $i = 1, \ldots, m$ and $j = 1, \ldots, n$, let $\mathbf{U}_{ij}$ represent that $m \times n$ matrix whose ijth element equals 1 and whose remaining $mn - 1$ elements equal 0. Then, the set, say B, whose members are, respectively, the mn matrices $\mathbf{U}_{11}, \mathbf{U}_{21}, \ldots, \mathbf{U}_{m1}, \ldots, \mathbf{U}_{1n}, \mathbf{U}_{2n}, \ldots, \mathbf{U}_{mn}$, is a basis for $\mathcal{R}^{m\times n}$. It is easy to see that, for any $m \times n$ matrix $\mathbf{A}$,

$$L_B^{-1}(\mathbf{A}) = \text{vec}(\mathbf{A}). \tag{3.4}$$

Similarly, for $i = 1, \ldots, n$, let $\mathbf{U}_{ii}^*$ represent the $n \times n$ matrix whose ith diagonal element equals 1 and whose remaining $n^2 - 1$ elements equal 0. And, for $j < i = 1, \ldots, n$, let $\mathbf{U}_{ij}^*$ represent the $n \times n$ matrix whose ijth and jith elements equal 1 and whose remaining $n^2 - 2$ elements equal 0. Then, the set, say B^*, whose members are, respectively, the $n(n+1)/2$ matrices $\mathbf{U}_{11}^*, \mathbf{U}_{21}^*, \ldots, \mathbf{U}_{n1}^*, \ldots, \mathbf{U}_{ii}^*, \mathbf{U}_{i+1,i}^*, \ldots, \mathbf{U}_{ni}^*, \ldots, \mathbf{U}_{nn}^*$ is a basis for the linear space of all $n \times n$ symmetric matrices. Clearly, for any $n \times n$ symmetric matrix $\mathbf{A}$,

$$L_{B^*}^{-1}(\mathbf{A}) = \text{vech}(\mathbf{A}). \tag{3.5}$$

22.4 Matrix Representation of a Linear Transformation

a. Background, definition, and some basic properties

Any linear transformation T from an n-dimensional linear space $\mathcal{V}$ into an m-dimensional linear space $\mathcal{W}$ can be replaced by a linear transformation from $\mathcal{R}^{n\times 1}$ into $\mathcal{R}^{m\times 1}$. Let B represent a set of matrices $\mathbf{V}_1, \ldots, \mathbf{V}_n$ (in $\mathcal{V}$) that form a basis for $\mathcal{V}$, and let C represent a set of matrices $\mathbf{W}_1, \ldots, \mathbf{W}_m$ (in $\mathcal{W}$) that form a basis for $\mathcal{W}$. And, adopting the notation introduced in Section 22.3, consider the product $L_C^{-1}TL_B$ of the three linear transformations L_C^{-1}, T, and L_B.

For any $n \times 1$ vector $\mathbf{x}$ and any $m \times 1$ vector $\mathbf{y}$,

$$(L_C^{-1}TL_B)(\mathbf{x}) = \mathbf{y} \quad \Leftrightarrow \quad T[L_B(\mathbf{x})] = L_C(\mathbf{y}), \tag{4.1}$$

as is evident upon observing that

$$(L_C^{-1}TL_B)(\mathbf{x}) = (L_C^{-1}(TL_B))(\mathbf{x}) = L_C^{-1}[(TL_B)(\mathbf{x})]$$

and $T[L_B(\mathbf{x})] = (TL_B)(\mathbf{x})$ and hence that result (4.1) can be reexpressed as

$$L_C^{-1}[(TL_B)(\mathbf{x})] = \mathbf{y} \quad \Leftrightarrow \quad (TL_B)(\mathbf{x}) = L_C(\mathbf{y}).$$

And, for any matrix $\mathbf{Z}$ in $\mathcal{V}$ and any matrix $\mathbf{Y}$ in $\mathcal{W}$,

$$T(\mathbf{Z}) = \mathbf{Y} \quad \Leftrightarrow \quad (L_C^{-1}TL_B)[L_B^{-1}(\mathbf{Z})] = L_C^{-1}(\mathbf{Y}), \tag{4.2}$$

as is evident from result (4.1) upon setting $\mathbf{x} = L_B^{-1}(\mathbf{Z})$ and $\mathbf{y} = L_C^{-1}(\mathbf{Y})$.

It is now clear that the linear transformation T from $\mathcal{V}$ into $\mathcal{W}$ can be replaced by the transformation $L_C^{-1}TL_B$, which is a linear transformation from $\mathcal{R}^{n\times 1}$ into $\mathcal{R}^{m\times 1}$. While T transforms each matrix $\mathbf{Z}$ in $\mathcal{V}$ into a matrix $T(\mathbf{Z})$ in $\mathcal{W}$, $L_C^{-1}TL_B$ transforms the n-dimensional column vector $\mathbf{x} = (x_1, \ldots, x_n)'$, whose elements are the coefficients of the basis matrices $\mathbf{V}_1, \ldots, \mathbf{V}_n$ in the representation $\mathbf{Z} = x_1\mathbf{V}_1 + \cdots + x_n\mathbf{V}_n$, into the m-dimensional column vector $\mathbf{y} = (y_1, \ldots, y_m)'$, whose elements are the coefficients of the basis matrices $\mathbf{W}_1, \ldots, \mathbf{W}_m$ in the representation $T(\mathbf{Z}) = y_1\mathbf{W}_1 + \cdots + y_m\mathbf{W}_m$.

The image $(L_C^{-1}TL_B)(\mathbf{x})$ of an arbitrary $n \times 1$ vector $\mathbf{x} = (x_1, \ldots, x_n)'$ under the transformation $L_C^{-1}TL_B$ can be obtained by premultiplying $\mathbf{x}$ by a certain matrix. To see this, observe that (for $j = 1, \ldots, n$) the image $T(\mathbf{V}_j)$ of the jth basis matrix $\mathbf{V}_j$ for $\mathcal{V}$ (under the transformation T) can be uniquely expressed as a linear combination of the basis matrices for $\mathcal{W}$; that is, there exist unique scalars $a_{1j}, a_{2j}, \ldots, a_{mj}$ such that

$$T(\mathbf{V}_j) = a_{1j}\mathbf{W}_1 + a_{2j}\mathbf{W}_2 + \cdots + a_{mj}\mathbf{W}_m. \tag{4.3}$$

And let $\mathbf{A}$ represent the $m \times n$ matrix whose jth column is $(a_{1j}, a_{2j}, \ldots, a_{mj})'$ or, equivalently, whose ijth element is a_{ij}. This matrix, which is dependent on the choice of the bases B and C, is referred to as the *matrix representation* of the transformation T (or simply as the matrix of T).

Now, let $\mathbf{Z} = \sum_{j=1}^{n} x_j \mathbf{V}_j$ [where $\mathbf{x} = (x_1, \ldots, x_n)'$ is an arbitrary $n \times 1$ vector]. Then, $\mathbf{Z}$ is in $\mathcal{V}$, and there exists a unique vector $\mathbf{y} = (y_1, \ldots, y_m)'$ in $\mathcal{R}^{m\times 1}$ such that

$$T(\mathbf{Z}) = \sum_{i=1}^{m} y_i \mathbf{W}_i. \tag{4.4}$$

Moreover, it follows from the definition of a linear transformation that

$$T(\mathbf{Z}) = T\Big(\sum_{j=1}^{n} x_j \mathbf{V}_j\Big) = \sum_{j=1}^{n} x_j T(\mathbf{V}_j) = \sum_{j=1}^{n} x_j \sum_{i=1}^{m} a_{ij} \mathbf{W}_i = \sum_{i=1}^{m} \Big(\sum_{j=1}^{n} x_j a_{ij}\Big) \mathbf{W}_i.$$

And, in light of the uniqueness of the coefficients $y_1, \ldots, y_m$ in representation (4.4), we have that

$$y_i = \sum_{j=1}^{n} x_j a_{ij} \quad (i = 1, \ldots, m). \tag{4.5}$$

The m equalities (4.5) can be rewritten as the matrix equality

$$\mathbf{y} = \mathbf{A}\mathbf{x}. \tag{4.6}$$

Moreover, since clearly $T[L_B(\mathbf{x})] = T(\mathbf{Z}) = L_C(\mathbf{y})$, it follows from result (4.1) that $\mathbf{y} = (L_C^{-1} T L_B)(\mathbf{x})$. Thus,

$$(L_C^{-1} T L_B)(\mathbf{x}) = \mathbf{A}\mathbf{x}; \tag{4.7}$$

that is, the image of $\mathbf{x}$ under the transformation $L_C^{-1} T L_B$ equals $\mathbf{A}\mathbf{x}$.

Let S represent a linear transformation from $\mathcal{V}$ into $\mathcal{W}$ such that $S \neq T$. Further, define $\mathbf{F} = \{f_{ij}\}$ to be the matrix representation of S (with respect to the bases B and C), and observe that, as a consequence of Lemma 22.1.9, there exists an integer j (between 1 and n, inclusive) such that $S(\mathbf{V}_j) \neq T(\mathbf{V}_j)$. Then, since (by definition) $S(\mathbf{V}_j) = \sum_{i=1}^{m} f_{ij} \mathbf{W}_i$, it follows that $f_{ij} \neq a_{ij}$ for some integer i and hence that $\mathbf{F} \neq \mathbf{A}$. Thus, distinct transformations from $\mathcal{V}$ into $\mathcal{W}$ have distinct matrix representations (with respect to the same bases).

b. Some special cases

It is instructive to consider the matrix representation of a linear transformation in the following special cases.

1. Let T represent a linear transformation from an n-dimensional linear space $\mathcal{V}$ into $\mathcal{R}^{m\times 1}$. Further, let B represent a set of matrices $\mathbf{V}_1, \ldots, \mathbf{V}_n$ that form a basis for $\mathcal{V}$, and let C represent a set of (m-dimensional) column vectors $\mathbf{w}_1, \ldots, \mathbf{w}_m$ that form a basis for $\mathcal{R}^{m\times 1}$. Then, letting $\mathbf{W}$ represent the $m \times m$ (nonsingular) matrix whose first, ..., mth columns are $\mathbf{w}_1, \ldots, \mathbf{w}_m$, respectively, the matrix representation of T with respect to B and C is (by definition) the unique $m \times n$ matrix $\mathbf{A}$ such that

$$\mathbf{W}\mathbf{A} = [T(\mathbf{V}_1), \ldots, T(\mathbf{V}_n)], \tag{4.8}$$

that is, the matrix

$$\mathbf{W}^{-1}[T(\mathbf{V}_1), \ldots, T(\mathbf{V}_n)]. \tag{4.9}$$

Note that when C is taken to be the natural basis for $\mathcal{R}^{m\times 1}$, $\mathbf{W} = \mathbf{I}$, and the matrix (4.9) simplifies to $[T(\mathbf{V}_1), \ldots, T(\mathbf{V}_n)]$.

2. Let $\mathbf{A}$ represent an $m \times n$ matrix. Then, as discussed in Section 22.1, the transformation T from $\mathcal{R}^{n\times 1}$ into $\mathcal{R}^{m\times 1}$ defined by $T(\mathbf{x}) = \mathbf{A}\mathbf{x}$ is a linear transformation. The matrix representation of T with respect to the natural bases for $\mathcal{R}^{n\times 1}$ and $\mathcal{R}^{m\times 1}$ (which consist of the columns of $\mathbf{I}_n$ and $\mathbf{I}_m$, respectively) is the matrix $\mathbf{A}$. However, except for special cases, the matrix representation of T with respect to other bases differs from $\mathbf{A}$.

3. Let $\mathcal{V}$ represent an n-dimensional linear space. And let $B = \{\mathbf{V}_1, \ldots, \mathbf{V}_n\}$ and $C = \{\mathbf{W}_1, \ldots, \mathbf{W}_n\}$ represent arbitrary bases for $\mathcal{V}$. Then, the matrix representation (with respect to B and C) of the identity transformation I from $\mathcal{V}$ onto $\mathcal{V}$ is the unique $n \times n$ matrix $\mathbf{A} = \{a_{ij}\}$ such that (for $j = 1, \ldots, n$)

$$\mathbf{V}_j = a_{1j}\mathbf{W}_1 + a_{2j}\mathbf{W}_2 + \cdots + a_{nj}\mathbf{W}_n. \tag{4.10}$$

Moreover, the columns of $\mathbf{A}$ are linearly independent and hence $\mathbf{A}$ is nonsingular (since if the columns of $\mathbf{A}$ were linearly dependent, then, in light of the linear independence of $\mathbf{V}_1, \ldots, \mathbf{V}_n$, we would arrive at a contradiction of Lemma 3.2.4).

Note that if $\mathbf{x} = (x_1, \ldots, x_n)'$ is the n-dimensional column vector whose elements are the coefficients of the basis matrices $\mathbf{V}_1, \ldots, \mathbf{V}_n$ when a matrix $\mathbf{Z}$ (in $\mathcal{V}$) is expressed as $\mathbf{Z} = x_1\mathbf{V}_1 + \cdots + x_n\mathbf{V}_n$, then the first, ..., nth elements $y_1, \ldots, y_n$ of the n-dimensional column vector $\mathbf{y} = \mathbf{A}\mathbf{x}$ are the coefficients of the basis matrices $\mathbf{W}_1, \ldots, \mathbf{W}_n$ when $\mathbf{Z}$ is expressed as $\mathbf{Z} = y_1\mathbf{W}_1 + \cdots + y_n\mathbf{W}_n$. Note also that $\mathbf{A} = \mathbf{I}_n$ if (and only if) $B = C$.

Further, for any $n \times n$ nonsingular matrix $\mathbf{A} = \{a_{ij}\}$ and for any choice of the basis C, the basis B can be chosen so that $\mathbf{A}$ is the matrix representation of the identity transformation I (from $\mathcal{V}$ onto $\mathcal{V}$) with respect to B and C. This can be done by choosing (for $j = 1, \ldots, n$) $\mathbf{V}_j = a_{1j}\mathbf{W}_1 + a_{2j}\mathbf{W}_2 + \cdots + a_{nj}\mathbf{W}_n$; that $\mathbf{V}_1, \ldots, \mathbf{V}_n$ are linearly independent (and hence form a basis for $\mathcal{V}$) is evident from Lemma 3.2.4.

Now, suppose that $\mathcal{V} = \mathcal{R}^{n\times 1}$. Then, B comprises n column vectors $\mathbf{v}_1, \ldots, \mathbf{v}_n$, and similarly C comprises n column vectors $\mathbf{w}_1, \ldots, \mathbf{w}_n$. Defining $\mathbf{V} = (\mathbf{v}_1, \ldots, \mathbf{v}_n)$ and $\mathbf{W} = (\mathbf{w}_1, \ldots, \mathbf{w}_n)$ (both of which are $n \times n$ nonsingular matrices), the matrix representation of the identity transformation I from $\mathcal{R}^{n\times 1}$ onto $\mathcal{R}^{n\times 1}$ with respect to B and C is the (unique) $n \times n$ (nonsingular) matrix $\mathbf{A}$ such that $\mathbf{W}\mathbf{A} = \mathbf{V}$ [as is evident from equality (4.10)], that is, the matrix $\mathbf{W}^{-1}\mathbf{V}$.

4. Let B represent a set of matrices $\mathbf{Y}_1, \ldots, \mathbf{Y}_n$ that form a basis for an n-dimensional linear space $\mathcal{W}$, and let C represent the set comprising the first, ..., nth columns, say $\mathbf{e}_1, \ldots, \mathbf{e}_n$, of $\mathbf{I}_n$ (which is the natural basis for $\mathcal{R}^{n\times 1}$). Further, let $\mathbf{A} = \{a_{ij}\}$ denote the matrix representation of the linear transformation L_B (with respect to C and B). For $j = 1, \ldots, n$, we have that

$$\mathbf{Y}_j = L_B(\mathbf{e}_j) = \sum_{i=1}^{n} a_{ij}\mathbf{Y}_i,$$

implying that $a_{jj} = 1$ and that, for $i \neq j$, $a_{ij} = 0$. Thus, $\mathbf{A} = \mathbf{I}_n$.

5. Let T represent a linear transformation from $\mathcal{R}^{m\times n}$ into $\mathcal{R}^{p\times q}$. Further, let C represent the natural basis for $\mathcal{R}^{m\times n}$, comprising the mn matrices $\mathbf{U}_{11}, \mathbf{U}_{21}, \ldots, \mathbf{U}_{m1}, \ldots, \mathbf{U}_{1n}, \mathbf{U}_{2n}, \ldots, \mathbf{U}_{mn}$, where (for $i = 1, \ldots, m$ and $j = 1, \ldots, n$) $\mathbf{U}_{ij}$ is the $m \times n$ matrix whose ijth element equals 1 and whose remaining $mn - 1$ elements equal 0; and, similarly, let D represent the natural basis for $\mathcal{R}^{p\times q}$. Then, making use of result (3.4), we find that, for any $mn \times 1$ vector $\mathbf{x}$,

$$\begin{aligned}(L_D^{-1}TL_C)(\mathbf{x}) = (L_D^{-1}(TL_C))(\mathbf{x}) &= L_D^{-1}[(TL_C)(\mathbf{x})] \\ &= \text{vec}[(TL_C)(\mathbf{x})] = \text{vec}\{T[L_C(\mathbf{x})]\}.\end{aligned}$$

Further, for any $m \times n$ matrix $\mathbf{X} = \{x_{ij}\}$,

$$\begin{aligned}(L_D^{-1}TL_C)(\text{vec}\,\mathbf{X}) &= \text{vec}\{T[L_C(\text{vec}\,\mathbf{X})]\} \\ &= \text{vec}(T\{L_C[L_C^{-1}(\mathbf{X})]\}) = \text{vec}[T(\mathbf{X})],\end{aligned} \tag{4.11}$$

implying in particular that

$$\begin{aligned}(L_D^{-1}TL_C)(\text{vec}\,\mathbf{X}) &= \text{vec}\Big[T\Big(\sum_{i,j} x_{ij}\mathbf{U}_{ij}\Big)\Big] \\ &= \text{vec}\Big[\sum_{i,j} x_{ij}T(\mathbf{U}_{ij})\Big] \\ &= \sum_{i,j} x_{ij}\,\text{vec}[T(\mathbf{U}_{ij})] \\ &= [\text{vec}\,T(\mathbf{U}_{11}), \ldots, \text{vec}\,T(\mathbf{U}_{m1}), \\ &\qquad \ldots, \text{vec}\,T(\mathbf{U}_{1n}), \ldots, \text{vec}\,T(\mathbf{U}_{mn})]\,\text{vec}(\mathbf{X}).\end{aligned}$$

And, in light of result (4.7), it follows that the matrix representation of T (with respect to the natural bases C and D) equals the $pq \times mn$ matrix

$$[\text{vec}\,T(\mathbf{U}_{11}), \ldots, \text{vec}\,T(\mathbf{U}_{m1}), \ldots, \text{vec}\,T(\mathbf{U}_{1n}), \ldots, \text{vec}\,T(\mathbf{U}_{mn})]. \tag{4.12}$$

Suppose, for instance, that (for every $m \times n$ matrix $\mathbf{X}$)

$$T(\mathbf{X}) = \mathbf{AXB},$$

where $\mathbf{A}$ is a $p \times m$ matrix and $\mathbf{B}$ an $n \times q$ matrix. Then, the matrix representation of T (with respect to the natural bases C and D) is $\mathbf{B}' \otimes \mathbf{A}$.

To see this, observe [in light of result (4.11) and Theorem 16.2.1] that

$$(L_D^{-1}TL_C)(\text{vec}\,\mathbf{X}) = \text{vec}[T(\mathbf{X})] = (\mathbf{B}' \otimes \mathbf{A})\,\text{vec}\,\mathbf{X},$$

and apply result (4.7). Or, alternatively, letting $\mathbf{a}_i$ represent the ith column of $\mathbf{A}$ and $\mathbf{b}_j'$ the jth row of $\mathbf{B}$, observe [in light of result (16.2.3)] that

$$\text{vec}\, T(\mathbf{U}_{ij}) = \text{vec}(\mathbf{a}_i \mathbf{b}_j') = \mathbf{b}_j \otimes \mathbf{a}_i$$

($i = 1, \ldots, m$; $j = 1, \ldots, n$) and hence [in light of results (16.1.28) and (16.1.27)] that expression (4.12) simplifies to

$$\begin{aligned}(\mathbf{b}_1 \otimes \mathbf{a}_1, \ldots, \mathbf{b}_1 \otimes \mathbf{a}_m, \ldots, \mathbf{b}_n \otimes \mathbf{a}_1, \ldots, \mathbf{b}_n \otimes \mathbf{a}_m)& \\ = (\mathbf{b}_1 \otimes \mathbf{A}, \ldots, \mathbf{b}_n \otimes \mathbf{A}) = (\mathbf{b}_1, \ldots, \mathbf{b}_n) \otimes \mathbf{A} &= \mathbf{B}' \otimes \mathbf{A}.\end{aligned}$$

6. Let $\mathcal{V}$ represent the linear space of all $n \times n$ symmetric matrices and $\mathcal{W}$ the linear space of all $p \times p$ symmetric matrices, and let T represent a linear transformation from $\mathcal{V}$ into $\mathcal{W}$. Further, let B represent the usual basis for $\mathcal{V}$, comprising the $n(n+1)/2$ matrices $\mathbf{U}_{11}^*, \mathbf{U}_{21}^*, \ldots, \mathbf{U}_{n1}^*, \ldots, \mathbf{U}_{ii}^*, \mathbf{U}_{i+1,i}^*, \ldots, \mathbf{U}_{ni}^*, \ldots, \mathbf{U}_{nn}^*$, where (for $i = 1, \ldots, n$) $\mathbf{U}_{ii}^*$ is the $n \times n$ matrix whose ith diagonal element equals 1 and whose remaining $n^2 - 1$ elements equal 0 and where (for $j < i = 1, \ldots, n$) $\mathbf{U}_{ij}^*$ is the $n \times n$ matrix whose ijth and jith elements equal 1 and whose remaining $n^2 - 2$ elements equal 0. And let C represent the usual basis for $\mathcal{W}$.

Then, making use of result (3.5), we find that, for any $n(n+1)/2 \times 1$ vector $\mathbf{x}$,

$$\begin{aligned}(L_C^{-1} T L_B)(\mathbf{x}) = (L_C^{-1}(T L_B))(\mathbf{x}) &= L_C^{-1}[(T L_B)(\mathbf{x})] \\ &= \text{vech}[(T L_B)(\mathbf{x})] = \text{vech}\{T[L_B(\mathbf{x})]\}.\end{aligned}$$

Further, for any $n \times n$ symmetric matrix $\mathbf{X} = \{x_{ij}\}$,

$$\begin{aligned}(L_C^{-1} T L_B)(\text{vech}\, \mathbf{X}) &= \text{vech}\{T[L_B(\text{vech}\, \mathbf{X})]\} \\ &= \text{vech}(T\{L_B[L_B^{-1}(\mathbf{X})]\}) = \text{vech}[T(\mathbf{X})], \qquad (4.13)\end{aligned}$$

implying in particular that

$$\begin{aligned}(L_C^{-1} T L_B)(\text{vech}\, \mathbf{X}) &= \text{vech}\Big[T\Big(\sum_{i,j<i} x_{ij} \mathbf{U}_{ij}^*\Big)\Big] \\ &= \text{vech}\Big[\sum_{i,j<i} x_{ij} T(\mathbf{U}_{ij}^*)\Big] \\ &= \sum_{i,j<i} x_{ij}\, \text{vech}[T(\mathbf{U}_{ij}^*)] \\ &= [\text{vech}\, T(\mathbf{U}_{11}^*), \ldots, \text{vech}\, T(\mathbf{U}_{n1}^*), \ldots, \text{vech}\, T(\mathbf{U}_{ii}^*), \\ &\qquad \ldots, \text{vech}\, T(\mathbf{U}_{ni}^*), \ldots, \text{vech}\, T(\mathbf{U}_{nn}^*)]\, \text{vech}(\mathbf{X}).\end{aligned}$$

And, in light or result (4.7), it follows that the matrix representation of T (with respect to the usual bases B and C) equals the $p(p+1)/2 \times n(n+1)/2$ matrix

$$[\text{vech}\, T(\mathbf{U}_{11}^*), \ldots, \text{vech}\, T(\mathbf{U}_{n1}^*), \ldots, \text{vech}\, T(\mathbf{U}_{ii}^*), \\ \ldots, \text{vech}\, T(\mathbf{U}_{ni}^*), \ldots, \text{vech}\, T(\mathbf{U}_{nn}^*)]. \quad (4.14)$$

Suppose, for instance, that $p = n$ (implying that $\mathcal{W} = \mathcal{V}$ and $C = B$) and that (for every $n \times n$ symmetric matrix $\mathbf{X}$)

$$T(\mathbf{X}) = \mathbf{AXA}',$$

where $\mathbf{A}$ is an $n \times n$ matrix. Then, the matrix representation of T (with respect to the usual bases) is $\mathbf{H}_n(\mathbf{A} \otimes \mathbf{A})\mathbf{G}_n$ (where $\mathbf{G}_n$ and $\mathbf{H}_n$ are as defined in Section 16.4b), as can be verified by, for example, observing [in light of results (4.13) and (16.4.25)] that

$$(L_C^{-1} T L_B)(\text{vech}\, \mathbf{X}) = \text{vech}[T(\mathbf{X})] = [\mathbf{H}_n(\mathbf{A} \otimes \mathbf{A})\mathbf{G}_n]\, \text{vech}\, \mathbf{X}$$

and applying result (4.7).

c. Matrix representations of scalar multiples, sums, and products of linear transformations

The following two theorems give some basic results on the matrix representations of scalar multiples, sums, and products of linear transformations.

Theorem 22.4.1. Let k represent a scalar, and let T and S represent linear transformations from an n-dimensional linear space $\mathcal{V}$ into an m-dimensional linear space $\mathcal{W}$. And let C and D represent bases for $\mathcal{V}$ and $\mathcal{W}$, respectively. Further, let $\mathbf{A} = \{a_{ij}\}$ represent the matrix representation of T with respect to C and D, and $\mathbf{B} = \{b_{ij}\}$ the matrix representation of S with respect to C and D. Then, $k\mathbf{A}$ is the matrix representation of kT (with respect to C and D) and $\mathbf{A}+\mathbf{B}$ is the matrix representation of $T + S$ (with respect to C and D).

Proof. Let $\mathbf{V}_1, \ldots, \mathbf{V}_n$ represent the matrices that form the basis C, and $\mathbf{W}_1, \ldots, \mathbf{W}_m$ the matrices that form the basis D. Then, for $j = 1, \ldots, n$,

$$(kT)(\mathbf{V}_j) = kT(\mathbf{V}_j) = k \sum_{i=1}^{m} a_{ij} \mathbf{W}_i = \sum_{i=1}^{m} (ka_{ij}) \mathbf{W}_i,$$

and

$$(T+S)(\mathbf{V}_j) = T(\mathbf{V}_j) + S(\mathbf{V}_j) = \sum_{i=1}^{m} a_{ij} \mathbf{W}_i + \sum_{i=1}^{m} b_{ij} \mathbf{W}_i = \sum_{i=1}^{m} (a_{ij} + b_{ij}) \mathbf{W}_i.$$

Thus, the matrix representation of kT (with respect to C and D) is the $m \times n$ matrix whose ijth element is ka_{ij}, that is, the matrix $k\mathbf{A}$; and the matrix representation of $T + S$ (with respect to C and D) is the $m \times n$ matrix whose ijth element is $a_{ij} + b_{ij}$, that is, the matrix $\mathbf{A} + \mathbf{B}$. Q.E.D.

Theorem 22.4.2. Let S represent a linear transformation from a p-dimensional linear space $\mathcal{U}$ into an n-dimensional linear space $\mathcal{V}$, and let T represent a linear transformation from $\mathcal{V}$ into an m-dimensional linear space $\mathcal{W}$. And let C, D, and E represent bases for $\mathcal{U}$, $\mathcal{V}$, and $\mathcal{W}$, respectively. Further, let $\mathbf{A} = \{a_{ij}\}$ represent the matrix representation of S with respect to C and D, and $\mathbf{B} = \{b_{ij}\}$ the matrix representation of T with respect to D and E. Then, $\mathbf{BA}$ is the matrix representation of TS (with respect to C and E).

Proof. Let $\mathbf{U}_1, \ldots, \mathbf{U}_p$ represent the matrices that form the basis C, $\mathbf{V}_1, \ldots, \mathbf{V}_n$ the matrices that form the basis D, and $\mathbf{W}_1, \ldots, \mathbf{W}_m$ the matrices that form the basis E. Then, for $j = 1, \ldots, p$,

$$\begin{aligned}(TS)(\mathbf{U}_j) = T[S(\mathbf{U}_j)] &= T\Big(\sum_{k=1}^{n} a_{kj}\mathbf{V}_k\Big)\\ &= \sum_{k=1}^{n} a_{kj}\,T(\mathbf{V}_k)\\ &= \sum_{k=1}^{n} a_{kj}\sum_{i=1}^{m} b_{ik}\mathbf{W}_i = \sum_{i=1}^{m}\Big(\sum_{k=1}^{n} b_{ik}a_{kj}\Big)\mathbf{W}_i.\end{aligned}$$

Thus, the matrix representation of TS (with respect to C and E) is the $m \times p$ matrix whose ijth element is $\sum_{k=1}^{n} b_{ik}a_{kj}$, that is, the matrix $\mathbf{BA}$. Q.E.D.

The results given by Theorems 22.4.1 and 22.4.2 on the matrix representations of a sum and a product of two linear transformations can be extended (by repeated application) to a sum and a product of any finite number of linear transformations. If $T_1, T_2, \ldots, T_k$ are linear transformations from a linear space $\mathcal{V}$ into a linear space $\mathcal{W}$, and if $\mathbf{A}_1, \mathbf{A}_2, \ldots, \mathbf{A}_k$ are their respective matrix representations (with respect to the same bases C and D), then $\mathbf{A}_1 + \mathbf{A}_2 + \cdots + \mathbf{A}_k$ is the matrix representation of $T_1 + T_2 + \cdots + T_k$ (with respect to C and D). Similarly, if $B_0, B_1, \ldots, B_k$ are bases for linear spaces $\mathcal{V}_0, \mathcal{V}_1, \ldots, \mathcal{V}_k$, respectively, and if (for $i = 1, \ldots, k$) T_i is a linear transformation from $\mathcal{V}_{i-1}$ into $\mathcal{V}_i$, and $\mathbf{A}_i$ is the matrix representation of T_i with respect to B_{i-1} and B_i, then $\mathbf{A}_k \cdots \mathbf{A}_2\mathbf{A}_1$ is the matrix representation of $T_k \cdots T_2T_1$ with respect to B_0 and B_k.

As a corollary of Theorem 22.4.2, we have the following result.

Corollary 22.4.3. Let T represent an invertible linear transformation from a linear space $\mathcal{V}$ onto a linear space $\mathcal{W}$ (in which case $\mathcal{V}$ and $\mathcal{W}$ are of the same dimension). And let B and C represent bases for $\mathcal{V}$ and $\mathcal{W}$, respectively. Further, let $\mathbf{A}$ represent the matrix representation of T with respect to B and C. Then, $\mathbf{A}$ is nonsingular, and $\mathbf{A}^{-1}$ is the matrix representation of T^{-1} (which is a linear transformation from $\mathcal{W}$ onto $\mathcal{V}$) with respect to C and B.

Proof. Let $\mathbf{H}$ represent the matrix representation of T^{-1} (with respect to C and B). Then, since [according to result (3.1)] $T^{-1}T = I$ and since (as discussed in Part 3 of Subsection b) the matrix representation of the identity transformation from $\mathcal{V}$ onto $\mathcal{V}$ (with respect to B and B) is $\mathbf{I}$, it follows from Theorem 22.4.2

that $\mathbf{HA} = \mathbf{I}$. And, based on Lemma 8.1.3, we conclude that $\mathbf{A}$ is nonsingular and that $\mathbf{H} = \mathbf{A}^{-1}$. Q.E.D.

d. Change of bases

The effect of a change of bases on the matrix representation of a linear transformation is described in the following theorem.

Theorem 22.4.4. Let T represent a linear transformation from a linear space $\mathcal{V}$ into a linear space $\mathcal{W}$, and let $\mathbf{A}$ represent the matrix representation of T with respect to bases B and C (for $\mathcal{V}$ and $\mathcal{W}$, respectively). Then, the matrix representation of T with respect to alternative bases E and F (for $\mathcal{V}$ and $\mathcal{W}$, respectively) equals $\mathbf{S}^{-1}\mathbf{AR}$, where $\mathbf{R}$ is the matrix representation of the identity transformation from $\mathcal{V}$ onto $\mathcal{V}$ with respect to E and B and $\mathbf{S}$ is the matrix representation of the identity transformation from $\mathcal{W}$ onto $\mathcal{W}$ with respect to F and C.

Proof. To avoid confusion, let us write I_1 for the identity transformation from $\mathcal{V}$ onto $\mathcal{V}$ and I_2 for the identity transformation from $\mathcal{W}$ onto $\mathcal{W}$. Then, since $I^{-1} = I$, we have that $T = I_2^{-1} T I_1$. And, upon observing (in light of Corollary 22.4.3) that $\mathbf{S}^{-1}$ is the matrix representation of I_2^{-1} with respect to C and F, and upon applying Theorem 22.4.2, we find that the matrix representation of T with respect to E and F equals $\mathbf{S}^{-1}\mathbf{AR}$. Q.E.D.

As special cases of Theorem 22.4.4, we have the following two corollaries.

Corollary 22.4.5. Let T represent a linear transformation from a linear space $\mathcal{V}$ into a linear space $\mathcal{W}$, and let $\mathbf{A}$ represent the matrix representation of T with respect to bases B and C (for $\mathcal{V}$ and $\mathcal{W}$, respectively). Then, the matrix representation of T with respect to an alternative basis E (for $\mathcal{V}$) and the basis C equals $\mathbf{AR}$, where $\mathbf{R}$ is the matrix representation of the identity transformation from $\mathcal{V}$ onto $\mathcal{V}$ with respect to E and B. And the matrix representation of T with respect to B and an alternative basis F (for $\mathcal{W}$) equals $\mathbf{S}^{-1}\mathbf{A}$, where $\mathbf{S}$ is the matrix representation of the identity transformation from $\mathcal{W}$ onto $\mathcal{W}$ with respect to F and C.

Corollary 22.4.6. Let $\mathcal{V}$ represent a linear space, and let C and F represent bases for $\mathcal{V}$. Further, let T represent a linear transformation from $\mathcal{V}$ into $\mathcal{V}$, and let $\mathbf{A}$ represent the matrix representation of T with respect to C and C. Then, the matrix representation of T with respect to F and F equals $\mathbf{S}^{-1}\mathbf{AS}$, where $\mathbf{S}$ is the matrix representation of the identity transformation (from $\mathcal{V}$ onto $\mathcal{V}$) with respect to F and C.

Let $\mathcal{V}$ represent an n-dimensional linear space and $\mathcal{W}$ an m-dimensional linear space. Further, let B represent a basis for $\mathcal{V}$ and C a basis for $\mathcal{W}$. And let $\mathbf{R}$ represent an $n \times n$ nonsingular matrix and $\mathbf{S}$ an $m \times m$ nonsingular matrix. Then, in light of the discussion in Part 3 of Subsection b, there exists a basis E for $\mathcal{V}$ such that $\mathbf{R}$ is the matrix representation of the identity transformation (from $\mathcal{V}$ onto $\mathcal{V}$) with respect to E and B, and similarly there exists a basis F for $\mathcal{W}$ such that $\mathbf{S}$ is the matrix representation of the identity transformation (from $\mathcal{W}$ onto $\mathcal{W}$) with respect to F and C. Thus, as a consequence of Theorem 22.4.4 and Corollary 22.4.6, we have the following two results.

Theorem 22.4.7. Let $\mathcal{V}$ represent an n-dimensional linear space and $\mathcal{W}$ an m-dimensional linear space. Further, let B represent a basis for $\mathcal{V}$ and C a basis for $\mathcal{W}$. And let T represent a linear transformation from $\mathcal{V}$ into $\mathcal{W}$, and $\mathbf{A}$ represent the matrix representation of T with respect to B and C. Then, an $m \times n$ matrix $\mathbf{H}$ is a matrix representation of T with respect to some bases E and F if and only if $\mathbf{H} = \mathbf{S}^{-1}\mathbf{AR}$ for some $n \times n$ (nonsingular) matrix $\mathbf{R}$ and some $m \times m$ (nonsingular) matrix $\mathbf{S}$.

Theorem 22.4.8. Let $\mathcal{V}$ represent an n-dimensional linear space, and let C represent any basis for $\mathcal{V}$. Further, let T represent a linear transformation from $\mathcal{V}$ into $\mathcal{V}$, and let $\mathbf{A}$ represent the matrix representation of T with respect to C and C. And let $\mathbf{H}$ represent an $n \times n$ matrix. Then, there exists a basis F for $\mathcal{V}$ such that $\mathbf{H}$ is the matrix representation of T with respect to F and F if and only if $\mathbf{H} = \mathbf{S}^{-1}\mathbf{AS}$ for some $n \times n$ (nonsingular) matrix $\mathbf{S}$, that is, if and only if $\mathbf{H}$ is similar to $\mathbf{A}$.

22.5 Terminology and Properties Shared by a Linear Transformation and Its Matrix Representation

Let $\mathcal{V}$ represent an n-dimensional linear space and $\mathcal{W}$ an m-dimensional linear space. And let B represent a set of matrices $\mathbf{V}_1, \ldots, \mathbf{V}_n$ (in $\mathcal{V}$) that form a basis for $\mathcal{V}$, and C represent a set of matrices $\mathbf{W}_1, \ldots, \mathbf{W}_m$ (in $\mathcal{W}$) that form a basis for $\mathcal{W}$. Then, corresponding to each linear transformation T from $\mathcal{V}$ into $\mathcal{W}$ is its matrix representation $\mathbf{A} = \{a_{ij}\}$ (with respect to B and C), which is the $m \times n$ matrix whose jth column is uniquely defined (for $j = 1, \ldots, n$) by equation (4.3).

Conversely, corresponding to any $m \times n$ matrix $\mathbf{A}$, there is a unique linear transformation from $\mathcal{V}$ into $\mathcal{W}$ that has $\mathbf{A}$ as its matrix representation (with respect to B and C), namely, the transformation $L_C S L_B^{-1}$, where S is the (linear) transformation from $\mathcal{R}^n$ into $\mathcal{R}^m$ defined by $S(\mathbf{x}) = \mathbf{Ax}$. To see this, observe (in light of Part 4 of Section 22.4b) that the matrix representation of L_B (with respect to the natural basis for $\mathcal{R}^{n\times 1}$ and the basis B) is $\mathbf{I}_n$, and that the matrix representation of L_C (with respect to the natural basis for $\mathcal{R}^{m\times 1}$ and the basis C) is $\mathbf{I}_m$. Then, making use of Theorem 22.4.2 and Corollary 22.4.3, we find (in light of Part 2 of Section 22.4b) that the matrix representation of $L_C S L_B^{-1}$ (with respect to B and C) is $\mathbf{I}_m \mathbf{A} \mathbf{I}_n^{-1} = \mathbf{A}$. [Moreover, it follows from the discussion in Section 22.4a that $L_C S L_B^{-1}$ is the only linear transformation from $\mathcal{V}$ into $\mathcal{W}$ that has $\mathbf{A}$ as its matrix representation (with respect to B and C).]

Thus, by associating every linear transformation (from $\mathcal{V}$ into $\mathcal{W}$) with its matrix representation (with respect to B and C), we obtain a 1–1 correspondence between transformations (from $\mathcal{V}$ into $\mathcal{W}$) and $m \times n$ matrices. It is customary to use the same or similar terminology in speaking of various kinds of linear transformations as in speaking of their matrix counterparts (and vice versa). For example, in the special case where $\mathcal{W} = \mathcal{V}$ and $C = B$, the *identity matrix* is the matrix representation of the *identity transformation* from $\mathcal{V}$ onto $\mathcal{V}$. Results on linear transformations can be translated into results on matrices (and conversely).

As discussed in Section 22.4a, a linear transformation T from $\mathcal{V}$ into $\mathcal{W}$ can be replaced by $L_C^{-1}TL_B$, which is a linear transformation from $\mathcal{R}^{n\times 1}$ into $\mathcal{R}^{m\times 1}$. $L_C^{-1}TL_B$ transforms an $n \times 1$ vector $\mathbf{x} = (x_1, \ldots, x_n)'$, whose elements can be regarded as the coefficients or "coordinates" of a matrix $\mathbf{Z}$ in $\mathcal{V}$ (when $\mathbf{Z}$ is expressed as a linear combination $\mathbf{Z} = x_1\mathbf{V}_1 + \cdots + x_n\mathbf{V}_n$ of the basis matrices $\mathbf{V}_1, \ldots, \mathbf{V}_n$), into the $m \times 1$ vector $\mathbf{y} = (y_1, \ldots, y_m)'$, whose elements are the coefficients or "coordinates" of the matrix $T(\mathbf{Z})$ in $\mathcal{W}$ [when $T(\mathbf{Z})$ is expressed as a linear combination $T(\mathbf{Z}) = y_1\mathbf{W}_1 + \cdots + y_m\mathbf{W}_m$ of the basis matrices $\mathbf{W}_1, \ldots, \mathbf{W}_m$]. Further, the image $(L_C^{-1}TL_B)(\mathbf{x})$ of $\mathbf{x}$ can be obtained by matrix multiplication, using the formula $(L_C^{-1}TL_B)(\mathbf{x}) = \mathbf{Ax}$, where $\mathbf{A}$ is the matrix representation of T (with respect to B and C).

Accordingly, one approach to the subject of linear transformations is to restrict attention to linear transformations defined, in terms of an $m \times n$ matrix $\mathbf{A}$, by specifying that $\mathbf{Ax}$ is the image of an arbitrary $n \times 1$ vector $\mathbf{x}$. In a certain sense, there is no loss of generality in this approach. Its appeal is that it allows results on linear transformations to be developed as results on matrices and to be expressed in the language of matrices.

An alternative approach is to derive results on linear transformations from first principles (starting with the definition of a linear transformation), with matrices playing a subordinate role. This approach, which was followed to a considerable extent in Sections 22.1–22.3, is sometimes referred to as the *coordinate-free approach*.

Let us now consider a few of the more important relationships between linear transformations and their matrix representations, beginning with the following result on null spaces and on range or column spaces.

Theorem 22.5.1. Let T represent a linear transformation from an n-dimensional linear space $\mathcal{V}$ into an m-dimensional linear space $\mathcal{W}$, and let $\mathbf{A}$ represent the matrix representation of T with respect to bases B and C (for $\mathcal{V}$ and $\mathcal{W}$, respectively).

(1) An $m \times 1$ vector $\mathbf{y}$ is in $\mathcal{C}(\mathbf{A})$ if and only if the corresponding matrix $L_C(\mathbf{y})$ is in the range space $T(\mathcal{V})$ of T, or, equivalently, a matrix $\mathbf{Y}$ (in $\mathcal{W}$) is in $T(\mathcal{V})$ if and only if the corresponding vector $L_C^{-1}(\mathbf{Y})$ is in $\mathcal{C}(\mathbf{A})$.

(2) An $n \times 1$ vector $\mathbf{x}$ is in $\mathcal{N}(\mathbf{A})$ if and only if the corresponding matrix $L_B(\mathbf{x})$ is in $\mathcal{N}(T)$, or, equivalently, a matrix $\mathbf{X}$ (in $\mathcal{V}$) is in $\mathcal{N}(T)$ if and only if the corresponding vector $L_B^{-1}(\mathbf{X})$ is in $\mathcal{N}(\mathbf{A})$.

Proof. (1) Making use of results (4.7) and (4.1), we find that

$$\begin{aligned}
\mathbf{y} \in \mathcal{C}(\mathbf{A}) \quad &\Leftrightarrow \quad \mathbf{y} = \mathbf{Ax} \text{ for some vector } \mathbf{x} \\
&\Leftrightarrow \quad \mathbf{y} = (L_C^{-1}TL_B)(\mathbf{x}) \text{ for some vector } \mathbf{x} \\
&\Leftrightarrow \quad L_C(\mathbf{y}) = T[L_B(\mathbf{x})] \text{ for some vector } \mathbf{x} \\
&\Leftrightarrow \quad L_C(\mathbf{y}) \in T(\mathcal{V}).
\end{aligned}$$

(2) Again making use of results (4.7) and (4.1), we find that

$$\mathbf{x} \in \mathcal{N}(\mathbf{A}) \;\Leftrightarrow\; \mathbf{A}\mathbf{x} = \mathbf{0} \;\Leftrightarrow\; (L_C^{-1} T L_B)(\mathbf{x}) = \mathbf{0}$$
$$\Leftrightarrow\; T[L_B(\mathbf{x})] = L_C(\mathbf{0})$$
$$\Leftrightarrow\; T[L_B(\mathbf{x})] = \mathbf{0} \;\Leftrightarrow\; L_B(\mathbf{x}) \in \mathcal{N}(T).$$

Q.E.D.

The dimension $\dim[T(\mathcal{V})]$ of the range space of a linear transformation T from a linear space $\mathcal{V}$ into a linear space $\mathcal{W}$ is referred to as the *rank* of T and is denoted by the symbol rank T or rank(T). This use of the term rank is consistent with its use in connection with matrices, as described in the following theorem.

Theorem 22.5.2. Let T represent a linear transformation from an n-dimensional linear space $\mathcal{V}$ into an m-dimensional linear space $\mathcal{W}$, and let $\mathbf{A}$ represent the matrix representation of T with respect to bases B and C (for $\mathcal{V}$ and $\mathcal{W}$, respectively). Then,

$$\operatorname{rank} T = \operatorname{rank} \mathbf{A}. \tag{5.1}$$

Proof. Part (1) of Theorem 22.5.1 implies that $L_C[\mathcal{C}(\mathbf{A})] = T(\mathcal{V})$ (as can be easily verified). Thus, there exists a 1–1 linear transformation from $\mathcal{C}(\mathbf{A})$ onto $T(\mathcal{V})$, namely, the linear transformation R defined [for $\mathbf{y} \in \mathcal{C}(\mathbf{A})$] by $R(\mathbf{y}) = L_C(\mathbf{y})$. And it follows that $\mathcal{C}(\mathbf{A})$ and $T(\mathcal{V})$ are isomorphic. Based on Theorem 22.3.1, we conclude that $\dim[T(\mathcal{V})] = \dim[\mathcal{C}(\mathbf{A})]$ or, equivalently, that rank T = rank $\mathbf{A}$.

Q.E.D.

In connection with Theorem 22.5.2, note that result (5.1) does not depend on the choice of the bases B and C. Note also that, as a consequence of Theorem 22.1.1 and Lemma 11.3.1, we have the following corollary.

Corollary 22.5.3. Let T represent a linear transformation from an n-dimensional linear space $\mathcal{V}$ into an m-dimensional linear space $\mathcal{W}$, and let $\mathbf{A}$ represent the matrix representation of T with respect to bases B and C (for $\mathcal{V}$ and $\mathcal{W}$, respectively). Then,

$$\dim[\mathcal{N}(T)] = n - \operatorname{rank}(T) = n - \operatorname{rank}(\mathbf{A}) = \dim[\mathcal{N}(\mathbf{A})]. \tag{5.2}$$

Now, let T represent a linear transformation from a linear space $\mathcal{V}$ into $\mathcal{V}$. Then, a scalar (real number) λ is said to be an *eigenvalue* of T if there exists a nonnull matrix $\mathbf{X}$ (in $\mathcal{V}$) such that

$$T(\mathbf{X}) = \lambda\mathbf{X},$$

in which case $\mathbf{X}$ may be referred to as an *eigenmatrix* (or, in the special case where $\mathcal{V}$ is a linear space of vectors, as an *eigenvector*) corresponding to λ.

As might be suspected, the eigenvalues and eigenmatrices of a linear transformation T are related to the eigenvalues and eigenvectors of the matrix representation of T. This relationship is described in the following theorem.

Theorem 22.5.4. Let T represent a linear transformation from an n-dimensional linear space $\mathcal{V}$ into $\mathcal{V}$, let B represent any basis for $\mathcal{V}$, and let $\mathbf{A}$ represent the matrix representation of T (with respect to B and B). Then, a scalar λ is an eigenvalue of $\mathbf{A}$ and an $n \times 1$ vector $\mathbf{x}$ is an eigenvector of $\mathbf{A}$ corresponding to λ if and only if λ is an eigenvalue of T and $L_B(\mathbf{x})$ is an eigenmatrix of T corresponding to λ.

Proof. In light of results (4.7) and (4.1),

$$\begin{aligned}\mathbf{Ax} = \lambda\mathbf{x} \quad &\Leftrightarrow \quad (L_B^{-1}TL_B)(\mathbf{x}) = \lambda\mathbf{x} \\ &\Leftrightarrow \quad T[L_B(\mathbf{x})] = L_B(\lambda\mathbf{x}) \quad \Leftrightarrow \quad T[L_B(\mathbf{x})] = \lambda L_B(\mathbf{x}).\end{aligned}$$

Moreover, $\mathbf{x}$ is nonnull if and only if $L_B(\mathbf{x})$ is nonnull. Q.E.D.

A subspace $\mathcal{U}$ of an n-dimensional linear space $\mathcal{V}$ is said to be *invariant* relative to a linear transformation T from $\mathcal{V}$ into $\mathcal{V}$ if $T(\mathcal{U}) \subset \mathcal{U}$, that is, if for every matrix $\mathbf{Z}$ in $\mathcal{U}$, the image $T(\mathbf{Z})$ of $\mathbf{Z}$ is also in $\mathcal{U}$. As described in the following theorem, the invariance of a subspace of an n-dimensional linear space $\mathcal{V}$ relative to a linear transformation T (from $\mathcal{V}$ into $\mathcal{V}$) is related to the invariance of a subspace of $\mathcal{R}^{n\times 1}$ relative to the matrix representation of T.

Theorem 22.5.5. Let T represent a linear transformation from an n-dimensional linear space $\mathcal{V}$ into $\mathcal{V}$, let B represent any basis for $\mathcal{V}$, and let $\mathbf{A}$ represent the matrix representation of T (with respect to B and B). Then, a subspace $\mathcal{U}$ of $\mathcal{R}^{n\times 1}$ is invariant relative to $\mathbf{A}$ if and only if the corresponding subspace $L_B(\mathcal{U})$ of matrices in $\mathcal{V}$ is invariant relative to T.

Proof. In light of result (4.7), $\mathcal{U}$ is invariant relative to $\mathbf{A}$ if and only if $\mathcal{U}$ is invariant relative to $L_B^{-1}TL_B$ (which is a linear transformation from $\mathcal{R}^{n\times 1}$ into $\mathcal{R}^{n\times 1}$). Thus, it suffices to show that $\mathcal{U}$ is invariant relative to $L_B^{-1}TL_B$ if and only if $L_B(\mathcal{U})$ is invariant relative to T.

Suppose that $L_B(\mathcal{U})$ is invariant relative to T. And let $\mathbf{x}$ represent an arbitrary vector in $\mathcal{U}$ [so that $L_B(\mathbf{x}) \in L_B(\mathcal{U})$]. Then, $T[L_B(\mathbf{x})] = \mathbf{Y}$ for some matrix $\mathbf{Y}$ in $L_B(\mathcal{U})$, or, equivalently, $T[L_B(\mathbf{x})] = L_B(\mathbf{y})$ for some vector $\mathbf{y}$ in $\mathcal{U}$. And it follows from result (4.1) that $(L_B^{-1}TL_B)(\mathbf{x}) = \mathbf{y}$ for some $\mathbf{y}$ in $\mathcal{U}$ and hence that $(L_B^{-1}TL_B)(\mathbf{x}) \in \mathcal{U}$. Thus, $\mathcal{U}$ is invariant relative to $L_B^{-1}TL_B$.

Conversely, suppose that $\mathcal{U}$ is invariant relative to $L_B^{-1}TL_B$. And let $\mathbf{Z}$ represent an arbitrary matrix in $L_B(\mathcal{U})$. Then, there exists a vector $\mathbf{x}$ in $\mathcal{U}$ such that $\mathbf{Z} = L_B(\mathbf{x})$. Moreover, $(L_B^{-1}TL_B)(\mathbf{x}) = \mathbf{y}$ for some vector $\mathbf{y}$ in $\mathcal{U}$. Thus, making use of result (4.1), we find that

$$T(\mathbf{Z}) = T[L_B(\mathbf{x})] = L_B(\mathbf{y}) \in L_B(\mathcal{U}).$$

And it follows that $L_B(\mathcal{U})$ is invariant relative to T. Q.E.D.

22.6 Linear Functionals and Dual Transformations

In the special case where $\mathcal{W} = \mathcal{R}^1$, it is customary to refer to a linear transformation from a linear space $\mathcal{V}$ into a linear space $\mathcal{W}$ as a *linear functional* on $\mathcal{V}$. Thus, a linear functional f on a linear space $\mathcal{V}$ transforms each matrix $\mathbf{X}$ in $\mathcal{V}$ into a scalar $f(\mathbf{X})$ and does so in such a way that $f(\sum_{i=1}^r c_i\mathbf{X}_i) = \sum_{i=1}^r c_i f(\mathbf{X}_i)$ for any r

matrices $\mathbf{X}_1, \ldots, \mathbf{X}_r$ (in $\mathcal{V}$) and any r scalars $c_1, \ldots, c_r$ (where r is an arbitrary positive integer). Note that if the linear functional f is the zero functional [i.e., if $f(\mathbf{X}) = 0$ for every matrix $\mathbf{X}$ in the domain $\mathcal{V}$ of f], then the range of f is $\{0\}$ and that otherwise the range of f is $\mathcal{R}^1$.

It follows from the discussion in Section 14.1 that, corresponding to any linear functional f on $\mathcal{R}^{n\times 1}$, there exists a unique vector $\mathbf{a}$ in $\mathcal{R}^{n\times 1}$ such that

$$f(\mathbf{x}) = \mathbf{a}'\mathbf{x}$$

for every vector $\mathbf{x}$ in $\mathcal{R}^{n\times 1}$, that is, such that $f(\mathbf{x})$ equals the usual inner product of $\mathbf{a}$ and $\mathbf{x}$. And, for any vector $\mathbf{a}$ in $\mathcal{R}^{n\times 1}$, the function with domain $\mathcal{R}^{n\times 1}$ that transforms each vector $\mathbf{x}$ in $\mathcal{R}^{n\times 1}$ into the usual inner product $\mathbf{a}'\mathbf{x}$ of $\mathbf{a}$ and $\mathbf{x}$ is a linear functional on $\mathcal{R}^{n\times 1}$. These results are generalized (to inner products other than the usual inner product and to linear spaces other than $\mathcal{R}^{n\times 1}$) in the following lemma.

Lemma 22.6.1. (1) Corresponding to any linear functional f on a linear space $\mathcal{V}$ (and to any inner product for $\mathcal{V}$), there exists a matrix $\mathbf{A}$ in $\mathcal{V}$ such that

$$f(\mathbf{X}) = \mathbf{A} \bullet \mathbf{X}$$

for every matrix $\mathbf{X}$ in $\mathcal{V}$. Moreover, this matrix is unique and is expressible as

$$\mathbf{A} = f(\mathbf{X}_1)\mathbf{X}_1 + \cdots + f(\mathbf{X}_r)\mathbf{X}_r,$$

where $\mathbf{X}_1, \ldots, \mathbf{X}_r$ are any matrices that form an orthonormal basis for $\mathcal{V}$. (2) For any matrix $\mathbf{A}$ in $\mathcal{V}$, the function f (with domain $\mathcal{V}$) that transforms each matrix $\mathbf{X}$ in $\mathcal{V}$ into the inner product $\mathbf{A} \bullet \mathbf{X}$ of $\mathbf{A}$ and $\mathbf{X}$ is a linear functional on $\mathcal{V}$.

Proof. (1) Let $\mathbf{X}$ represent an arbitrary matrix in $\mathcal{V}$. Then, according to Theorem 6.4.4, $\mathbf{X} = \sum_{i=1}^{r} (\mathbf{X} \bullet \mathbf{X}_i)\mathbf{X}_i$. And we find that

$$f(\mathbf{X}) = \sum_{i=1}^{r} (\mathbf{X} \bullet \mathbf{X}_i) f(\mathbf{X}_i) = \sum_{i=1}^{r} f(\mathbf{X}_i)(\mathbf{X}_i \bullet \mathbf{X}) = \Big[\sum_{i=1}^{r} f(\mathbf{X}_i)\mathbf{X}_i\Big] \bullet \mathbf{X} = \mathbf{A} \bullet \mathbf{X},$$

where $\mathbf{A} = \sum_{i=1}^{r} f(\mathbf{X}_i)\mathbf{X}_i$.

Now, let $\mathbf{B}$ represent any matrix in $\mathcal{V}$ such that $f(\mathbf{X}) = \mathbf{B} \bullet \mathbf{X}$ for every $\mathbf{X}$ in $\mathcal{V}$. Then, to complete the proof of Part (1), it suffices to show that $\mathbf{B} = \mathbf{A}$. For every $\mathbf{X}$ in $\mathcal{V}$, we have that $\mathbf{B} \bullet \mathbf{X} = f(\mathbf{X}) = \mathbf{A} \bullet \mathbf{X}$ and hence that

$$(\mathbf{B} - \mathbf{A}) \bullet \mathbf{X} = (\mathbf{B} \bullet \mathbf{X}) - (\mathbf{A} \bullet \mathbf{X}) = 0.$$

In particular, $(\mathbf{B} - \mathbf{A}) \bullet \mathbf{X} = 0$ for $\mathbf{X} = \mathbf{B} - \mathbf{A}$; that is, $(\mathbf{B} - \mathbf{A}) \bullet (\mathbf{B} - \mathbf{A}) = 0$, implying that $\mathbf{B} - \mathbf{A} = \mathbf{0}$ or, equivalently, that $\mathbf{B} = \mathbf{A}$.

(2) That the function f is a linear functional on $\mathcal{V}$ is an immediate consequence of the properties of inner products. Q.E.D.

Let $\mathbf{A}$ represent an $m \times n$ matrix. And take $\mathbf{x} \bullet \mathbf{z}$ to be the usual inner product $\mathbf{x}'\mathbf{z}$ of arbitrary vectors $\mathbf{x}$ and $\mathbf{z}$ in $\mathcal{R}^{n\times 1}$, and write $\mathbf{u} * \mathbf{y}$ for the usual inner product

$\mathbf{u}'\mathbf{y}$ of arbitrary vectors $\mathbf{u}$ and $\mathbf{y}$ in $\mathcal{R}^{m\times 1}$. Then, there exists a unique $n \times m$ matrix $\mathbf{B}$ such that

$$\mathbf{x} \bullet (\mathbf{B}\mathbf{y}) = (\mathbf{A}\mathbf{x}) * \mathbf{y} \tag{6.1}$$

for every $\mathbf{x}$ in $\mathcal{R}^{n\times 1}$ and every $\mathbf{y}$ in $\mathcal{R}^{m\times 1}$, and $\mathbf{B} = \mathbf{A}'$ (as can be easily verified). The following theorem extends this result to transformations.

Theorem 22.6.2. Let T represent a linear transformation from an n-dimensional linear space $\mathcal{V}$ into an m-dimensional linear space $\mathcal{W}$. And write $\mathbf{X} \bullet \mathbf{Z}$ for the inner product of arbitrary matrices $\mathbf{X}$ and $\mathbf{Z}$ in $\mathcal{V}$, and $\mathbf{U} * \mathbf{Y}$ for the inner product of arbitrary matrices $\mathbf{U}$ and $\mathbf{Y}$ in $\mathcal{W}$. Then, there exists a linear transformation S from $\mathcal{W}$ into $\mathcal{V}$ such that

$$\mathbf{X} \bullet S(\mathbf{Y}) = T(\mathbf{X}) * \mathbf{Y}$$

for every matrix $\mathbf{X}$ in $\mathcal{V}$ and every matrix $\mathbf{Y}$ in $\mathcal{W}$. Moreover, this linear transformation is unique, and (for every $\mathbf{Y}$ in $\mathcal{W}$)

$$S(\mathbf{Y}) = [\mathbf{Y} * T(\mathbf{X}_1)]\mathbf{X}_1 + [\mathbf{Y} * T(\mathbf{X}_2)]\mathbf{X}_2 + \cdots + [\mathbf{Y} * T(\mathbf{X}_n)]\mathbf{X}_n,$$

where $\mathbf{X}_1, \mathbf{X}_2, \ldots, \mathbf{X}_n$ are any matrices that form an orthonormal basis for $\mathcal{V}$.

Proof. Let $\mathbf{X}$ represent an arbitrary matrix in $\mathcal{V}$, and $\mathbf{Y}$ an arbitrary matrix in $\mathcal{W}$. And observe (in light of Theorem 6.4.4) that $\mathbf{X} = \sum_{j=1}^{n} (\mathbf{X} \bullet \mathbf{X}_j)\mathbf{X}_j$ and that (for $j = 1, \ldots, n$) $\mathbf{X}_j = \sum_{i=1}^{n} (\mathbf{X}_j \bullet \mathbf{X}_i)\mathbf{X}_i$. Then,

$$\begin{aligned}
T(\mathbf{X}) * \mathbf{Y} &= T\Big[\sum_{j=1}^{n} (\mathbf{X} \bullet \mathbf{X}_j)\mathbf{X}_j\Big] * \mathbf{Y} \\
&= \Big[\sum_{j=1}^{n} (\mathbf{X} \bullet \mathbf{X}_j)T(\mathbf{X}_j)\Big] * \mathbf{Y} \\
&= \sum_{j=1}^{n} (\mathbf{X} \bullet \mathbf{X}_j)[T(\mathbf{X}_j) * \mathbf{Y}] \\
&= \sum_{j=1}^{n} (\mathbf{X} \bullet \mathbf{X}_j)\Big\{T\Big[\sum_{i=1}^{n} (\mathbf{X}_i \bullet \mathbf{X}_j)\mathbf{X}_i\Big] * \mathbf{Y}\Big\} \\
&= \sum_{j=1}^{n} (\mathbf{X} \bullet \mathbf{X}_j)\Big\{\Big[\sum_{i=1}^{n} (\mathbf{X}_i \bullet \mathbf{X}_j)T(\mathbf{X}_i)\Big] * \mathbf{Y}\Big\} \\
&= \sum_{j=1}^{n} (\mathbf{X} \bullet \mathbf{X}_j) \sum_{i=1}^{n} (\mathbf{X}_i \bullet \mathbf{X}_j)[T(\mathbf{X}_i) * \mathbf{Y}] \\
&= \sum_{j=1}^{n} (\mathbf{X} \bullet \mathbf{X}_j) \sum_{i=1}^{n} [\mathbf{Y} * T(\mathbf{X}_i)](\mathbf{X}_i \bullet \mathbf{X}_j) \\
&= \sum_{j=1}^{n} (\mathbf{X} \bullet \mathbf{X}_j)\Big\{\Big(\sum_{i=1}^{n} [\mathbf{Y} * T(\mathbf{X}_i)]\mathbf{X}_i\Big) \bullet \mathbf{X}_j\Big\}
\end{aligned}$$

$$= \sum_{j=1}^{n} (\mathbf{X}\bullet\mathbf{X}_j)\Big\{\mathbf{X}_j \bullet \Big(\sum_{i=1}^{n} [\mathbf{Y} * T(\mathbf{X}_i)]\mathbf{X}_i\Big)\Big\}$$

$$= \Big[\sum_{j=1}^{n} (\mathbf{X}\bullet\mathbf{X}_j)\mathbf{X}_j\Big] \bullet \Big(\sum_{i=1}^{n} [\mathbf{Y} * T(\mathbf{X}_i)]\mathbf{X}_i\Big)$$

$$= \mathbf{X}\bullet S(\mathbf{Y}),$$

where S is the transformation from $\mathcal{W}$ into $\mathcal{V}$ that transforms each matrix $\mathbf{Y}$ in $\mathcal{W}$ into the matrix $\sum_{i=1}^{n} [\mathbf{Y} * T(\mathbf{X}_i)]\mathbf{X}_i$. Moreover, S is linear, as is evident upon observing that (for any scalar k and for any matrices $\mathbf{Y}$ and $\mathbf{U}$ in $\mathcal{W}$)

$$S(k\mathbf{Y}) = \sum_{i=1}^{n} [(k\mathbf{Y}) * T(\mathbf{X}_i)]\mathbf{X}_i = k\sum_{i=1}^{n} [\mathbf{Y} * T(\mathbf{X}_i)]\mathbf{X}_i = kS(\mathbf{Y})$$

and

$$S(\mathbf{U}+\mathbf{Y}) = \sum_{i=1}^{n} [(\mathbf{U}+\mathbf{Y}) * T(\mathbf{X}_i)]\mathbf{X}_i$$

$$= \sum_{i=1}^{n} [\mathbf{U} * T(\mathbf{X}_i)]\mathbf{X}_i + \sum_{i=1}^{n} [\mathbf{Y} * T(\mathbf{X}_i)]\mathbf{X}_i = S(\mathbf{U}) + S(\mathbf{Y}).$$

Now, let R represent any linear transformation from $\mathcal{W}$ into $\mathcal{V}$ such that $\mathbf{X}\bullet R(\mathbf{Y}) = T(\mathbf{X}) * \mathbf{Y}$ for every $\mathbf{X} \in \mathcal{V}$ and every $\mathbf{Y} \in \mathcal{W}$. Then, to complete the proof, it suffices to show that $R = S$. For every $\mathbf{X} \in \mathcal{V}$ and every $\mathbf{Y} \in \mathcal{W}$, we have that $\mathbf{X}\bullet R(\mathbf{Y}) = \mathbf{X}\bullet S(\mathbf{Y})$ and hence that

$$\mathbf{X}\bullet[R(\mathbf{Y}) - S(\mathbf{Y})] = \mathbf{X}\bullet R(\mathbf{Y}) - \mathbf{X}\bullet S(\mathbf{Y}) = 0.$$

And, upon setting $\mathbf{X} = R(\mathbf{Y}) - S(\mathbf{Y})$, we find that $[R(\mathbf{Y}) - S(\mathbf{Y})]\bullet[R(\mathbf{Y}) - S(\mathbf{Y})] = 0$ for every $\mathbf{Y}$ and hence that $R(\mathbf{Y}) - S(\mathbf{Y}) = \mathbf{0}$ or, equivalently, $R(\mathbf{Y}) = S(\mathbf{Y})$ (for every $\mathbf{Y}$). Thus, $R = S$. Q.E.D.

The linear transformation S in Theorem 22.6.2 is referred to as the *dual transformation* of T or the *adjoint transformation* of T (or simply as the *adjoint* of T)—referring to this transformation as the adjoint transformation risks confusion with the adjoint matrix (which was introduced in Section 13.5 and is unrelated to the dual or adjoint transformation). Note that the dual transformation of T depends on the choice of inner products for $\mathcal{V}$ and $\mathcal{W}$.

Now, let $\mathbf{A}$ represent an $m \times n$ matrix. And consider the dual transformation S of the linear transformation T from $\mathcal{R}^{n\times 1}$ into $\mathcal{R}^{m\times 1}$ defined by $T(\mathbf{x}) = \mathbf{A}\mathbf{x}$ (where $\mathbf{x}$ is an arbitrary $n \times 1$ vector), taking the inner products for $\mathcal{R}^{n\times 1}$ and $\mathcal{R}^{m\times 1}$ to be the usual inner products. Then, for every $\mathbf{x}$ in $\mathcal{R}^{n\times 1}$ and every $\mathbf{y}$ in $\mathcal{R}^{m\times 1}$,

$$\mathbf{x}'S(\mathbf{y}) = (\mathbf{A}\mathbf{x})'\mathbf{y} = \mathbf{x}'\mathbf{A}'\mathbf{y}.$$

And, in light of the uniqueness of S, it follows that (for all $\mathbf{y}$ in $\mathcal{R}^{m\times 1}$)

$$S(\mathbf{y}) = \mathbf{A}'\mathbf{y} \tag{6.2}$$

[which is consistent with the earlier observation that the unique matrix $\mathbf{B}$ that satisfies condition (6.1) (for all $\mathbf{x} \in \mathcal{R}^{n\times 1}$ and all $\mathbf{y} \in \mathcal{R}^{m\times 1}$) equals $\mathbf{A}'$].

Exercises

Section 22.1

1. Let $\mathcal{U}$ represent a subspace of a linear space $\mathcal{V}$, and let S represent a linear transformation from $\mathcal{U}$ into a linear space $\mathcal{W}$. Show that there exists a linear transformation T from $\mathcal{V}$ into $\mathcal{W}$ such that S is the restriction of T to $\mathcal{U}$.

2. Let T represent a 1–1 linear transformation from a linear space $\mathcal{V}$ into a linear space $\mathcal{W}$. And write $\mathbf{U} \bullet \mathbf{Y}$ for the inner product of arbitrary matrices $\mathbf{U}$ and $\mathbf{Y}$ in $\mathcal{W}$. Further, define $\mathbf{X} * \mathbf{Z} = T(\mathbf{X}) \bullet T(\mathbf{Z})$ for all matrices $\mathbf{X}$ and $\mathbf{Z}$ in $\mathcal{V}$. Show that the "$*$-operation" satisfies the four properties required of an inner product for $\mathcal{V}$.

3. Let T represent a linear transformation from a linear space $\mathcal{V}$ into a linear space $\mathcal{W}$, and let $\mathcal{U}$ represent any subspace of $\mathcal{V}$ such that $\mathcal{U}$ and $\mathcal{N}(T)$ are essentially disjoint. Further, let $\{\mathbf{X}_1, \ldots, \mathbf{X}_r\}$ represent a linearly independent set of r matrices in $\mathcal{U}$.

   (a) Show that $T(\mathbf{X}_1), \ldots, T(\mathbf{X}_r)$ are linearly independent.

   (b) Show that if $r = \dim(\mathcal{U})$ (or, equivalently, if $\mathbf{X}_1, \ldots, \mathbf{X}_r$ form a basis for $\mathcal{U}$) and if $\mathcal{U} \oplus \mathcal{N}(T) = \mathcal{V}$, then $T(\mathbf{X}_1), \ldots, T(\mathbf{X}_r)$ form a basis for $T(\mathcal{V})$.

Section 22.2

4. Let T and S represent linear transformations from a linear space $\mathcal{V}$ into a linear space $\mathcal{W}$, and let k represent an arbitrary scalar.

   (a) Verify that the transformation kT is linear.

   (b) Verify that the transformation $T + S$ is linear.

5. Let S represent a linear transformation from a linear space $\mathcal{U}$ into a linear space $\mathcal{V}$, and let T represent a linear transformation from $\mathcal{V}$ into a linear space $\mathcal{W}$. Show that the transformation TS is linear.

6. Let T represent a linear transformation from a linear space $\mathcal{V}$ into a linear space $\mathcal{W}$, and let R represent a linear transformation from a linear space $\mathcal{U}$ into $\mathcal{W}$. Show that if $T(\mathcal{V}) \subset R(\mathcal{U})$, then there exists a linear transformation S from $\mathcal{V}$ into $\mathcal{U}$ such that $T = RS$.

Section 22.3

7. Let T represent a transformation from a linear space $\mathcal{V}$ into a linear space $\mathcal{W}$, and let S and R represent transformations from $\mathcal{W}$ into $\mathcal{V}$. And suppose that $RT = I$ (where the identity transformation I is from $\mathcal{V}$ onto $\mathcal{V}$) and that $TS = I$ (where the identity transformation I is from $\mathcal{W}$ onto $\mathcal{W}$).

   (a) Show that T is invertible.

   (b) Show that $R = S = T^{-1}$.

8. Let T represent an invertible transformation from a linear space $\mathcal{V}$ into a linear space $\mathcal{W}$, let S represent an invertible transformation from a linear space $\mathcal{U}$ into $\mathcal{V}$, and let k represent an arbitrary nonzero scalar. Using the results of Exercise 7 (or otherwise), show that

   (a) kT is invertible and $(kT)^{-1} = (1/k)T^{-1}$, and that

   (b) TS is invertible and $(TS)^{-1} = S^{-1}T^{-1}$.

Section 22.4

9. Let T represent a linear transformation from an n-dimensional linear space $\mathcal{V}$ into an m-dimensional linear space $\mathcal{W}$. And write $\mathbf{U} \bullet \mathbf{Y}$ for the inner product of arbitrary matrices $\mathbf{U}$ and $\mathbf{Y}$ in $\mathcal{W}$. Further, let B represent a set of matrices $\mathbf{V}_1, \ldots, \mathbf{V}_n$ (in $\mathcal{V}$) that form a basis for $\mathcal{V}$, and let C represent a set of matrices $\mathbf{W}_1, \ldots, \mathbf{W}_m$ (in $\mathcal{W}$) that form an orthonormal basis for $\mathcal{W}$. Show that the matrix representation for T with respect to B and C is the $m \times n$ matrix whose ijth element is $T(\mathbf{V}_j) \bullet \mathbf{W}_i$.

10. Let T represent the linear transformation from $\mathcal{R}^{m \times n}$ into $\mathcal{R}^{n \times m}$ defined by $T(\mathbf{X}) = \mathbf{X}'$. And let C represent the natural basis for $\mathcal{R}^{m \times n}$, comprising the mn matrices $\mathbf{U}_{11}, \mathbf{U}_{21}, \ldots, \mathbf{U}_{m1}, \ldots, \mathbf{U}_{1n}, \mathbf{U}_{2n}, \ldots, \mathbf{U}_{mn}$, where (for $i = 1, \ldots, m$ and $j = 1, \ldots, n$) $\mathbf{U}_{ij}$ is the $m \times n$ matrix whose ijth element equals 1 and whose remaining $mn - 1$ elements equal 0; and, similarly, let D represent the natural basis for $\mathcal{R}^{n \times m}$. Show that the matrix representation for T with respect to the bases C and D is the vec-permutation matrix $\mathbf{K}_{mn}$.

11. Let $\mathcal{W}$ represent the linear space of all $p \times p$ symmetric matrices, and let T represent a linear transformation from $\mathcal{R}^{m \times n}$ into $\mathcal{W}$. Further, let B represent the natural basis for $\mathcal{R}^{m \times n}$, comprising the mn matrices $\mathbf{U}_{11}, \mathbf{U}_{21}, \ldots, \mathbf{U}_{m1}, \ldots, \mathbf{U}_{1n}, \mathbf{U}_{2n}, \ldots, \mathbf{U}_{mn}$, where (for $i = 1, \ldots, m$ and $j = 1, \ldots, n$) $\mathbf{U}_{ij}$ is the $m \times n$ matrix whose ijth element equals 1 and whose remaining $mn - 1$ elements equal 0. And let C represent the usual basis for $\mathcal{W}$.

    (a) Show that, for any $m \times n$ matrix $\mathbf{X}$,

$$(L_C^{-1} T L_B)(\text{vec}\,\mathbf{X}) = \text{vech}[T(\mathbf{X})].$$

(b) Show that the matrix representation for T (with respect to B and C) equals the $p(p+1)/2 \times mn$ matrix

$$[\text{vech}\, T(\mathbf{U}_{11}), \ldots, \text{vech}\, T(\mathbf{U}_{m1}), \ldots, \text{vech}\, T(\mathbf{U}_{1n}), \ldots, \text{vech}\, T(\mathbf{U}_{mn})].$$

(c) Suppose that $p = m = n$ and that (for every $n \times n$ matrix $\mathbf{X}$)

$$T(\mathbf{X}) = (1/2)(\mathbf{X} + \mathbf{X}').$$

Show that the matrix representation of T (with respect to B and C) equals $(\mathbf{G}_n'\mathbf{G}_n)^{-1}\mathbf{G}_n'$ (where $\mathbf{G}_n$ is the duplication matrix).

12. Let $\mathcal{V}$ represent an n-dimensional linear space. Further, let $B = \{\mathbf{V}_1, \mathbf{V}_2, \ldots, \mathbf{V}_n\}$ represent a basis for $\mathcal{V}$, and let $\mathbf{A}$ represent an $n \times n$ nonsingular matrix. And, for $j = 1, \ldots, n$, let

$$\mathbf{W}_j = f_{1j}\mathbf{V}_1 + f_{2j}\mathbf{V}_2 + \cdots + f_{nj}\mathbf{V}_n,$$

where (for $i = 1, \ldots, n$) f_{ij} is the ijth element of $\mathbf{A}^{-1}$. Show that the set C comprising the matrices $\mathbf{W}_1, \mathbf{W}_2, \ldots, \mathbf{W}_n$ is a basis for $\mathcal{V}$ and that $\mathbf{A}$ is the matrix representation of the identity transformation I (from $\mathcal{V}$ onto $\mathcal{V}$) with respect to B and C.

Section 22.5

13. Let T represent the linear transformation from $\mathcal{R}^{4\times 1}$ into $\mathcal{R}^{3\times 1}$ defined by

$$T(\mathbf{x}) = (x_1 + x_2,\; x_2 + x_3 - x_4,\; x_1 - x_3 + x_4)',$$

where $\mathbf{x} = (x_1, x_2, x_3, x_4)'$. Further, let B represent the natural basis for $\mathcal{R}^{4\times 1}$ (comprising the columns of $\mathbf{I}_4$), and let E represent the basis (for $\mathcal{R}^{4\times 1}$) comprising the four vectors $(1, -1, 0, -1)'$, $(0, 0, 1, 1)'$, $(0, 0, 0, 1)'$, and $(1, 1, 0, 0)'$. And let C represent the natural basis for $\mathcal{R}^{3\times 1}$ (comprising the columns of $\mathbf{I}_3$), and F represent the basis (for $\mathcal{R}^{3\times 1}$) comprising the three vectors $(1, 0, 1)'$, $(1, 1, 0)'$, and $(-1, 0, 0)'$.

(a) Find the matrix representation of T with respect to B and C.

(b) Find (1) the matrix representation of the identity transformation from $\mathcal{R}^{4\times 1}$ onto $\mathcal{R}^{4\times 1}$ with respect to E and B and (2) the matrix representation of the identity transformation from $\mathcal{R}^{3\times 1}$ onto $\mathcal{R}^{3\times 1}$ with respect to C and F.

(c) Find the matrix representation of T with respect to E and F via each of two approaches: (1) a direct approach, using equality (4.8); and (2) an indirect approach, using Theorem 22.4.4 together with the results of Parts (a) and (b).

(d) Using Theorem 22.5.2 and Corollary 22.5.3 (or otherwise), find rank T and $\dim[\mathcal{N}(T)]$.

14. Let T represent a linear transformation of rank k (where $k > 0$) from an n-dimensional linear space $\mathcal{V}$ into an m-dimensional linear space $\mathcal{W}$. Show that there exist a basis E for $\mathcal{V}$ and a basis F for $\mathcal{W}$ such that the matrix representation of T with respect to E and F is of the form $\begin{pmatrix} \mathbf{I}_k & \mathbf{0} \\ \mathbf{0} & \mathbf{0} \end{pmatrix}$.

15. Let T represent a linear transformation from an n-dimensional linear space $\mathcal{V}$ into an m-dimensional linear space $\mathcal{W}$, and let $\mathbf{A}$ represent the matrix representation of T with respect to bases B and C (for $\mathcal{V}$ and $\mathcal{W}$, respectively). Use Part (2) of Theorem 22.5.1 to devise a "direct" proof that $\dim[\mathcal{N}(T)] = \dim[\mathcal{N}(\mathbf{A})]$ (as opposed to deriving this equality as a corollary of Theorem 22.5.2).

16. Let T represent a linear transformation from an n-dimensional linear space $\mathcal{V}$ into $\mathcal{V}$.

(a) Let $\mathcal{U}$ represent an r-dimensional subspace of $\mathcal{V}$, and suppose that $\mathcal{U}$ is invariant relative to T. Show that there exists a basis B for $\mathcal{V}$ such that the matrix representation of T with respect to B and B is of the (upper block-triangular) form $\begin{pmatrix} \mathbf{E} & \mathbf{F} \\ \mathbf{0} & \mathbf{H} \end{pmatrix}$ (where $\mathbf{E}$ is of dimensions $r \times r$).

(b) Let $\mathcal{U}$ and $\mathcal{W}$ represent subspaces of $\mathcal{V}$ such that $\mathcal{U} \oplus \mathcal{W} = \mathcal{V}$ (i.e., essentially disjoint subspaces of $\mathcal{V}$ whose sum is $\mathcal{V}$). Suppose that both $\mathcal{U}$ and $\mathcal{W}$ are invariant relative to T. Show that there exists a basis B for $\mathcal{V}$ such that the matrix representation of T with respect to B and B is of the (block-diagonal) form $\text{diag}(\mathbf{E}, \mathbf{H})$ [where the dimensions of $\mathbf{E}$ equal $\dim(\mathcal{U})$].

Section 22.6

17. Let $\mathcal{V}$, $\mathcal{W}$, and $\mathcal{U}$ represent linear spaces.

(a) Show that the dual transformation of the identity transformation I from $\mathcal{V}$ onto $\mathcal{V}$ is I.

(b) Show that the dual transformation of the zero transformation 0_1 from $\mathcal{V}$ into $\mathcal{W}$ is the zero transformation 0_2 from $\mathcal{W}$ into $\mathcal{V}$.

(c) Let S represent the dual transformation of a linear transformation T from $\mathcal{V}$ into $\mathcal{W}$. Show that T is the dual transformation of S.

(d) Let k represent a scalar, and let S represent the dual transformation of a linear transformation T from $\mathcal{V}$ into $\mathcal{W}$. Show that kS is the dual transformation of kT.

(e) Let T_1 and T_2 represent linear transformations from $\mathcal{V}$ into $\mathcal{W}$, and let S_1 and S_2 represent the dual transformations of T_1 and T_2, respectively. Show that $S_1 + S_2$ is the dual transformation of $T_1 + T_2$.

(f) Let P represent the dual transformation of a linear transformation S from $\mathcal{U}$ into $\mathcal{V}$, and let Q represent the dual transformation of a linear transformation T from $\mathcal{V}$ into $\mathcal{W}$. Show that PQ is the dual transformation of TS.

18. Let S represent the dual transformation of a linear transformation T from an n-dimensional linear space $\mathcal{V}$ into an m-dimensional linear space $\mathcal{W}$. And let $\mathbf{A} = \{a_{ij}\}$ represent the matrix representation of T with respect to orthonormal bases C and D, and $\mathbf{B} = \{b_{ij}\}$ represent the matrix representation of S with respect to D and C. Using the result of Exercise 9 (or otherwise), show that $\mathbf{B} = \mathbf{A}'$.

19. Let $\mathbf{A}$ represent an $m \times n$ matrix, let $\mathbf{V}$ represent an $n \times n$ symmetric positive definite matrix, and $\mathbf{W}$ represent an $m \times m$ symmetric positive definite matrix. And let S represent the dual transformation of the linear transformation T from $\mathcal{R}^{n\times 1}$ into $\mathcal{R}^{m\times 1}$ defined by $T(\mathbf{x}) = \mathbf{Ax}$ (where $\mathbf{x}$ is an arbitrary $n \times 1$ vector). Taking the inner product of arbitrary vectors $\mathbf{x}$ and $\mathbf{z}$ in $\mathcal{R}^{n\times 1}$ to be $\mathbf{x}'\mathbf{Vz}$ and taking the inner product of arbitrary vectors $\mathbf{u}$ and $\mathbf{y}$ in $\mathcal{R}^{m\times 1}$ to be $\mathbf{u}'\mathbf{Wy}$, generalize result (6.2) by obtaining a formula for $S(\mathbf{y})$.

20. Let S represent the dual transformation of a linear transformation T from a linear space $\mathcal{V}$ into a linear space $\mathcal{W}$.

(a) Show that $[S(\mathcal{W})]^{\perp} = \mathcal{N}(T)$ (i.e., that the orthogonal complement of the range space of S equals the null space of T).

(b) Using the result of Part (c) of Exercise 17 (or otherwise), show that $[\mathcal{N}(S)]^{\perp} = T(\mathcal{V})$ (i.e., that the orthogonal complement of the null space of S equals the range space of T).

(c) Show that rank S = rank T.

Errata

Matrix Algebra From a Statistician's Perspective

David A. Harville

IBM T. J. Watson Research Center, Mathematical Sciences Department, Yorktown Heights, NY 10598-0218, USA

D.A. Harville, *Matrix Algebra From a Statistician's Perspective*,
DOI: 10.1007/978-0-387-22677-4, © Springer Science+Business Media, LLC 2008

DOI 978-0-387-22677-4_24

The paperback and online versions of the book contain some errors, and the corrections to these versions are given on the following pages.

Preface

Page vi (Preface), line 16. Delete the comma after the word And.

8
Inverse Matrices

Page 87, line 2. Delete the parentheses enclosing $\mathbf{a}_{k_1}, \mathbf{a}_{k_2}, \ldots, \mathbf{a}_{k_n}$.

Page 87, line 13. Delete the parentheses enclosing $\mathbf{a}'_{k_1}, \mathbf{a}'_{k_2}, \ldots, \mathbf{a}'_{k_n}$.

Page 87, line 14. Replace k (in the condition $k = 3$) with n.

The online version of the original chapter can be found at
http://dx.doi.org/10.1007/978-0-387-22677-4_8

13
Determinants

Page 194, line 9. Replace a_{ij} with $a_{i'j}$.

Page 194, line 16. Replace the phrase *adjoint matrix* with the phrase *adjoint* (or *adjoint matrix*).

Page 194, line 8 from the bottom. Insert the word as after the word and.

The online version of the original chapter can be found at
http://dx.doi.org/10.1007/978-0-387-22677-4_13

21
Eigenvalues and Eigenvectors

Theorem 21.5.6 and its proof (pages 539–540). Make the following changes:

(1) Take the equation number (5.2) to be the equation number for the inequalities that appear on line 3 of the theorem rather than for the result that appears on line 11 of the proof.
(2) Replace the first sentence of paragraph 2 of the proof with the following sentence: Now, suppose that $\mathbf{A}$ is symmetric, and take $\mathbf{x}_1$ and $\mathbf{x}_2$ to be any nonnull vectors for which the inequalities (5.2) hold for every nonnull vector $\mathbf{x}$ in $\mathcal{R}^n$.

(3) Modify line 6 of paragraph 2 of the proof by deleting the phrase [as is evident from result (5.2)].

(4) (Optional) Dispensing with (3) above, replace the first sentence of paragraph 2 of the proof as in (2) above and replace the remainder of paragraph 2 with the following development: And observe that, for $\mathbf{x} \neq \mathbf{0}$,

$$\frac{1}{\mathbf{x}'\mathbf{x}}\mathbf{x}'[\mathbf{A} - (\mathbf{x}_1'\mathbf{A}\mathbf{x}_1/\mathbf{x}_1'\mathbf{x}_1)\mathbf{I}_n]\mathbf{x} \geq 0$$

or, equivalently, $\mathbf{x}'[\mathbf{A}-(\mathbf{x}_1'\mathbf{A}\mathbf{x}_1/\mathbf{x}_1'\mathbf{x}_1)\mathbf{I}_n]\mathbf{x} \geq 0$. Thus, $\mathbf{A}-(\mathbf{x}_1'\mathbf{A}\mathbf{x}_1/\mathbf{x}_1'\mathbf{x}_1)\mathbf{I}_n$ is a symmetric nonnegative definite matrix, and upon observing that

$$\mathbf{x}_1'[\mathbf{A} - (\mathbf{x}_1'\mathbf{A}\mathbf{x}_1/\mathbf{x}_1'\mathbf{x}_1)\mathbf{I}_n]\mathbf{x}_1 = 0,$$

it follows from Corollary 14.3.11 that

$$[\mathbf{A} - (\mathbf{x}_1'\mathbf{A}\mathbf{x}_1/\mathbf{x}_1'\mathbf{x}_1)\mathbf{I}_n]\mathbf{x}_1 = \mathbf{0}.$$

It is now clear that $\mathbf{x}_1'\mathbf{A}\mathbf{x}_1/\mathbf{x}_1'\mathbf{x}_1$ is an eigenvalue of $\mathbf{A}$, that $\mathbf{x}_1$ is an eigenvector of $\mathbf{A}$ corresponding to $\mathbf{x}_1'\mathbf{A}\mathbf{x}_1/\mathbf{x}_1'\mathbf{x}_1$, and (since if λ is an eigenvalue of $\mathbf{A}$, $\lambda = \mathbf{x}'\mathbf{A}\mathbf{x}/\mathbf{x}'\mathbf{x}$ for some nonnull vector $\mathbf{x}$) that $\mathbf{x}_1'\mathbf{A}\mathbf{x}_1/\mathbf{x}_1'\mathbf{x}_1$ is the smallest eigenvalue of $\mathbf{A}$. It follows from a similar argument that $\mathbf{x}_2'\mathbf{A}\mathbf{x}_2/\mathbf{x}_2'\mathbf{x}_2$ is an eigenvalue of $\mathbf{A}$, that $\mathbf{x}_2$ is an eigenvector of $\mathbf{A}$ corresponding to $\mathbf{x}_2'\mathbf{A}\mathbf{x}_2/\mathbf{x}_2'\mathbf{x}_2$, and that $\mathbf{x}_2'\mathbf{A}\mathbf{x}_2/\mathbf{x}_2'\mathbf{x}_2$ is the largest eigenvalue of $\mathbf{A}$.

The online version of the original chapter can be found at
http://dx.doi.org/10.1007/978-0-387-22677-4_21

References

Aitken, A. C. (1935), "On Least Squares and Linear Combination of Observations," *Proceedings of the Royal Society of Edinburgh*, 55, 42–48.

Aitken, A. C. (1949), "On the Wishart Distribution in Statistics," *Biometrika*, 36, 59–62.

Albert, A. (1972), *Regression and the Moore-Penrose Pseudoinverse*, New York: Academic Press.

Baksalary, J. K., and Puntanen, S. (1989), "Weighted-Least-Squares Estimation in the General Gauss-Markov Model," in *Statistical Data Analysis and Inference*, ed. Y. Dodge, Amsterdam: North-Holland, pp. 355–368.

Bartle, R. G. (1976), *The Elements of Real Analysis* (2nd ed.), New York: John Wiley.

Bartlett, M. S. (1951), "An Inverse Matrix Adjustment Arising in Discriminant Analysis," *Annals of Mathematical Statistics*, 22, 107–111.

Basilevsky, A. (1983), *Applied Matrix Algebra in the Statistical Sciences*, New York: North-Holland.

Beaumont, R. A., and Pierce, R. S. (1963), *The Algebraic Foundations of Mathematics*, Reading, MA: Addison-Wesley.

Christensen, R. (1996), *Plane Answers to Complex Questions: The Theory of Linear Models* (2nd ed.), New York: Springer-Verlag.

Cochran, W. G. (1934), "The Distribution of Quadratic Forms in a Normal System, With Applications to the Analysis of Covariance," *Proceedings of the Cambridge Philosophical Society*, 30, 178–191.

Driscoll, M. F., and Gundberg, W. R., Jr. (1986), "A History of the Development of Craig's Theorem," *The American Statistician*, 40, 65–70.

Duncan, W. J. (1944), "Some Devices for the Solution of Large Sets of Simultaneous Linear Equations (With an Appendix on the Reciprocation of Patterned Matrices)," *The London, Edinburgh, and Dublin Philosophical Magazine and Journal of Science, Seventh Series*, 35, 660–670.

Dwyer, P. S. (1967), "Some Applications of Matrix Derivatives in Multivariate Analysis," *Journal of the American Statistical Association*, 62, 607–625.

Faddeeva, V. N. (1959), *Computational Methods of Linear Algebra*, New York: Dover.

Fedorov, V. V. (1972), *Theory of Optimal Experiments*, New York: Academic Press.

Golub, G. H., and Pereyra, V. (1973), "The Differentiation of Pseudo-Inverses and Nonlinear Least Squares Problems Whose Variables Separate," *SIAM Journal on Numerical Analysis*, 10, 413–432.

Golub, G. H., and Van Loan, C. F. (1989), *Matrix Computations* (2nd ed.), Baltimore: Johns Hopkins University Press.

Goodnight, J. H. (1979), "A Tutorial on the SWEEP Operator," *The American Statistician*, 33, 149–158.

Graybill, F. A. (1983), *Matrices With Applications in Statistics* (2nd ed.), Belmont, CA: Wadsworth.

Guttman, L. (1946), "Enlargement Methods for Computing the Inverse Matrix," *Annals of Mathematical Statistics*, 17, 336–343.

Halmos, P. R. (1958), *Finite-Dimensional Vector Spaces* (2nd ed.), Princeton, NJ: Van Nostrand.

Harville, D. A. (1996–97), "Generalized Inverses and Ranks of Modified Matrices," *Journal of the Indian Society of Agricultural Statistics*, 49 (Golden Jubilee Number), 67–78.

Harville, D. A., and Fenech, A. P. (1985), "Confidence Intervals for a Variance Ratio, or for Heritability, in an Unbalanced Mixed Linear Model," *Biometrics*, 41, 137–152.

Harville, D. A., and Kempthorne, O. (1997), "An Alternative Way to Establish the Necessity Part of the Classical Result on the Statistical Independence of Quadratic Forms," *Linear Algebra and Its Applications*, 264 (Sixth Special Issue on Linear Algebra and Statistics), 205–215.

Hearon, J. Z., and Evans, J. W. (1968), "Differentiable Generalized Inverses," *Journal of Research of the National Bureau of Standards*, 72B, 109–113.

Henderson, H. V., and Searle, S. R. (1979), "Vec and Vech Operators for Matrices, With Some Uses in Jacobians and Multivariate Statistics," *The Canadian Journal of Statistics*, 7, 65–81.

Henderson, H. V., and Searle, S. R. (1981a), "On Deriving the Inverse of a Sum of Matrices," *SIAM Review*, 23, 53–60.

Henderson, H. V., and Searle, S. R. (1981b), "The Vec-Permutation Matrix, the Vec Operator and Kronecker Products: A Review," *Linear and Multilinear Algebra*, 9, 271–288.

Lancaster, P., and Tismenetsky, M. (1985), *The Theory of Matrices* (2nd ed.), New York: Academic Press.

Loynes, R. M. (1966), "On Idempotent Matrices," *Annals of Mathematical Statistics*, 37, 295–296.

Magnus, J. R., and Neudecker, H. (1979), "The Commutation Matrix: Some Properties and Applications," *The Annals of Statistics*, 7, 381–394.

Magnus, J. R., and Neudecker, H. (1980), "The Elimination Matrix: Some Lemmas and Applications," *SIAM Journal on Algebra and Discrete Mathematics*, 1, 422–449.

Magnus, J. R., and Neudecker, H. (1986), "Symmetry, 0-1 Matrices and Jacobians: A Review," *Econometric Theory*, 2, 157–190.

Magnus, J. R., and Neudecker, H. (1988), *Matrix Differential Calculus With Applications in Statistics and Econometrics*, New York: John Wiley.

Marsaglia, G., and Styan, G. P. H. (1974), "Equalities and Inequalities for Ranks of Matrices," *Linear and Multilinear Algebra*, 2, 269–292.

Meyer, C. D. (1973), "Generalized Inverses and Ranks of Block Matrices," *SIAM Journal on Applied Mathematics*, 25, 597–602.

Miller, K. S. (1987), *Some Eclectic Matrix Theory*, Malabar, FL: Robert E. Krieger Publishing Company.

Milliken, G. A., and Akdeniz, F. (1977), "A Theorem on the Difference of the Generalized Inverses of Two Nonnegative Matrices," *Communications in Statistics*, Part A—*Theory and Methods*, 6, 73–79.

Moore, E. H. (1920), "On the Reciprocal of the General Algebraic Matrix," *Bulletin of the American Mathematical Society*, 26, 394–395.

Penrose, R. (1955), "A Generalized Inverse for Matrices," *Proceedings of the Cambridge Philosophical Society*, 51, 406–413.

Puntanen, S., and Styan, G. P. H. (1989), "The Equality of the Ordinary Least Squares Estimator and the Best Linear Unbiased Estimator" (with discussion), *The American Statistician*, 43, 153–164.

Rao, C. R. (1989), "A Lemma on Optimization of a Matrix Function and a Review of the Unified Theory of Linear Estimation," in *Statistical Data Analysis and Inference*, ed. Y. Dodge, Amsterdam: North-Holland, pp. 397–417.

Rao, C. R., and Mitra, S. K. (1971), *Generalized Inverse of Matrices and Its Applications*, New York: John Wiley.

Reid, J. G., and Driscoll, M. F. (1988), "An Accessible Proof of Craig's Theorem in the Noncentral Case," *The American Statistician*, 42, 139–142.

Rogers, G. S. (1980), *Matrix Derivatives*, New York: Marcel Dekker.

Scheffé, H. (1959), *The Analysis of Variance*, New York: John Wiley.

Searle, S. R. (1971), *Linear Models*, New York: John Wiley.

Searle, S. R. (1982), *Matrix Algebra Useful for Statistics*, New York: John Wiley.

Sherman, J., and Morrison, W. J. (1949), "Adjustment of an Inverse Matrix Corresponding to Changes in the Elements of a Given Column or a Given Row of the Original Matrix" (abstract), *Annals of Mathematical Statistics*, 20, 621.

Sherman, J., and Morrison, W. J. (1950), "Adjustment of an Inverse Matrix Corresponding to a Change in One Element of a Given Matrix," *Annals of Mathematical Statistics*, 21, 124–127.

Stewart, G. W. (1973), *Introduction to Matrix Computations*, New York: Academic Press.

Sylvester, J. (1884), "Sur la Solution du Cas le Plus Général des Équations Linéares en Quantités Binares, C'est-á-dire en Quaternions ou en Matrices du Second Ordre," *Comptes Rendus des Séances de l'Académie des Sciences*, 99, 117–118.

Woodbury, M. A. (1950), "Inverting Modified Matrices," Memorandum Report 42, Princeton University, Statistical Research Group.

Index